SELECTED CHAPTERS FROM BROWN: INTRODUCTION TO ORGANIC CHEMISTRY, 3RD EDITION AND SPERLING: INTRODUCTION TO PHYSICAL POLYMER SCIENCE, 4TH EDITION

A Wiley Canada Custom Publication
for the

University of Toronto

MSE245

Cover Photo Credit: Brown: Ray Coleman/PhotoResearchers, Inc.

Marketing Manager: Patty Maher
Editorial Assistant: Rachel Coffey

Printed and bound in the United States of America
10 9 8 7 6 5 4 3 2 1

WILEY

John Wiley & Sons Canada, Ltd
6045 Freemont Blvd.
Mississauga, Ontario
L5R 4J3
Visit our website at: www.wiley.ca

3rd Edition

Introduction to ORGANIC CHEMISTRY

William H. Brown
Beloit College

Thomas Poon
Claremont McKenna College
Scripps College
Pitzer College

John Wiley & Sons, Inc.

Project Editor: *Jennifer Yee*
Acquistions Editor: *Kevin Molloy*
Senior Media Editor: *Martin Batey*
Marketing Manager: *Amanda Wygal*
Production Editor: *Sandra Dumas*
Senior Designer: *Kevin Murphy*
Cover and Interior Design: *Nancy Field*
Cover Photo: *Ray Coleman/Photo Researchers, Inc.*
Senior Photo Editor: *Lisa Gee*
Illustration Editor: *Sandra Rigby*
Editorial Program Assistant: *Catherine Donovan*
Production Management Services: *Preparé*

This book was typeset in 10/12 New Baskerville Roman by Preparé and printed and bound by Von Hoffmann Corporation. The cover was printed by Von Hoffmann Corporation.

The paper in this book was manufactured by a mill whose forest management programs include sustained yield harvesting of its timberlands. Sustained yield harvesting principles ensure that the number of trees cut each year does not exceed the amount of new growth.

This book is printed on acid-free paper. ∞

Brown, William, H., Poon, Thomas
Introduction To Organic Chemistry—Third Edition.

ISBN 0-471-44451-0
Wiley International Edition 0-471-45161-4

Printed in the United States of America.

10 9 8 7 6 5 4 3 2 1

3 Alkanes and Cycloalkanes

Bunsen burners burn natural gas, which is primarily methane with small amounts of ethane, propane, butane, and 2-methylbutane. Inset: A model of methane. *(Charles D. Winters)*

3.1 INTRODUCTION

Hydrocarbon A compound that contains only carbon atoms and hydrogen atoms.

Alkane A saturated hydrocarbon whose carbon atoms are arranged in an open chain.

Saturated hydrocarbon A hydrocarbon containing only carbon–carbon single bonds.

Aliphatic hydrocarbon An alternative word to describe an alkane.

In this chapter, we begin our study of organic compounds with the physical and chemical properties of alkanes, the simplest types of organic compounds. Actually, alkanes are members of a larger class of organic compounds called **hydrocarbons**. A hydrocarbon is a compound composed of only carbon and hydrogen. Figure 3.1 shows the four classes of hydrocarbons, along with the characteristic type of bonding between carbon atoms in each.

Alkanes are **saturated hydrocarbons**; that is, they contain only carbon–carbon single bonds. In this context, "saturated" means that each carbon has the maximum number of hydrogens bonded to it. We often refer to alkanes as **aliphatic hydro-**

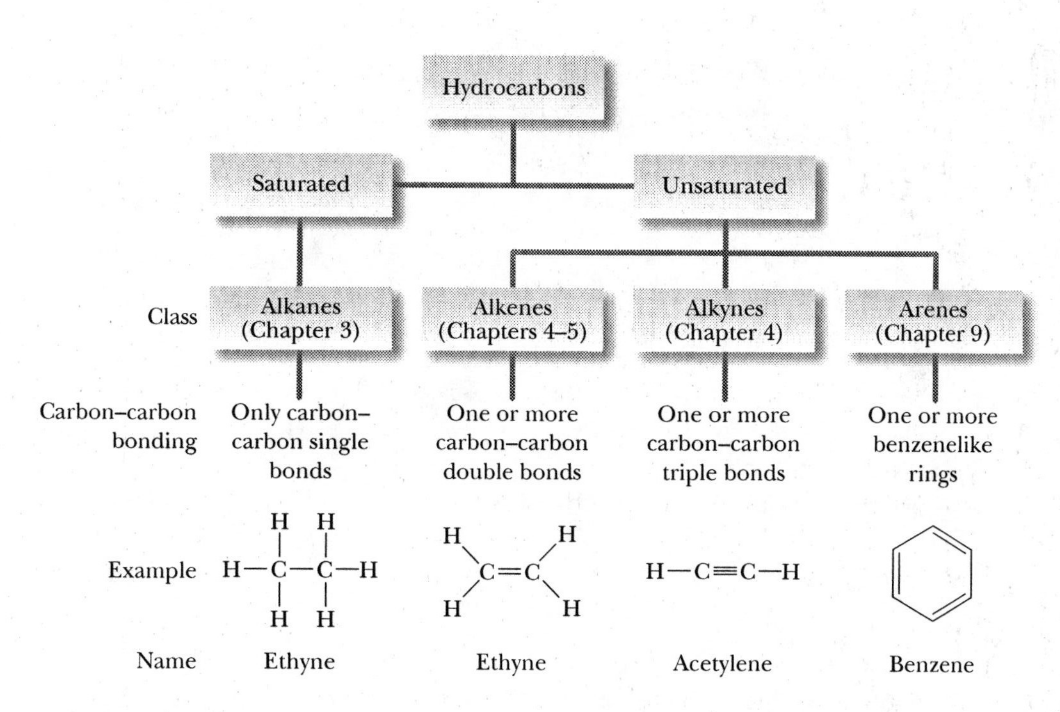

Figure 3.1
The four classes of hydrocarbons.

carbons, because the physical properties of the higher members of this class resemble those of the long carbon-chain molecules we find in animal fats and plant oils (Greek: *aleiphar*, fat or oil).

A hydrocarbon that contains one or more carbon–carbon double bonds, triple bonds, or benzene rings is classified as an **unsaturated hydrocarbon**. We study alkanes (saturated hydrocarbons) in this chapter. We study alkenes and alkynes (both unsaturated hydrocarbons) in Chapters 4 and 5, and arenes (also unsaturated hydrocarbons) in Chapter 9.

Butane is the fuel in this lighter. Butane molecules are present in the liquid and gaseous states in the lighter. *(Charles D. Winters)*

3.2 STRUCTURE OF ALKANES

Methane (CH_4) and ethane (C_2H_6) are the first two members of the alkane family. Figure 3.2 shows Lewis structures and ball-and-stick models for these molecules. The shape of methane is tetrahedral, and all H—C—H bond angles are 109.5°. Each carbon atom in ethane is also tetrahedral, and all bond angles are approximately 109.5°.

Although the three-dimensional shapes of larger alkanes are more complex than those of methane and ethane, the four bonds about each carbon atom are still arranged in a tetrahedral manner, and all bond angles are still approximately 109.5°.

The next members of the alkane family are propane, butane, and pentane. In the representations that follow, these hydrocarbons are drawn first as condensed structural formulas that show all carbons and hydrogens. They are also drawn in an even more abbreviated form called a **line-angle formula**. In this type of representation, a line represents a carbon—carbon bond and an angle represents a carbon

Line-angle formula An abbreviated way to draw structural formulas in which each line ending represents a carbon atom and a line represents a bond.

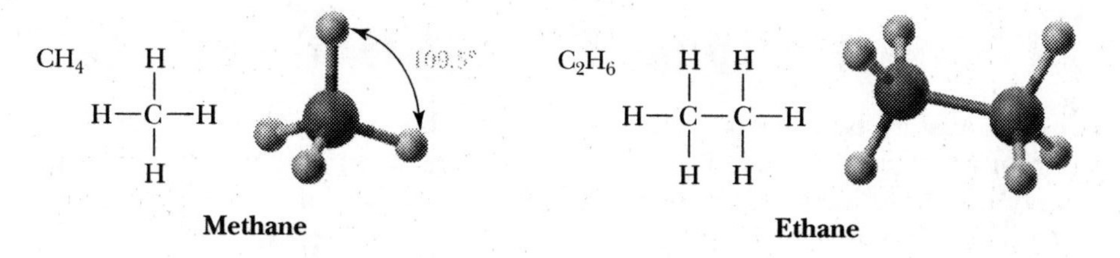

Figure 3.2
Methane and ethane.

A tank for propane fuel. *(Charles D. Winters)*

TABLE 3.1 Names, Molecular Formulas, and Condensed Structural Formulas for the First 20 Alkanes with Unbranched Chains

Name	Molecular Formula	Condensed Structural Formula	Name	Molecular Formula	Condensed Structural Formula
methane	CH_4	CH_4	undecane	$C_{11}H_{24}$	$CH_3(CH_2)_9CH_3$
ethane	C_2H_6	CH_3CH_3	dodecane	$C_{12}H_{26}$	$CH_3(CH_2)_{10}CH_3$
propane	C_3H_8	$CH_3CH_2CH_3$	tridecane	$C_{13}H_{28}$	$CH_3(CH_2)_{11}CH_3$
butane	C_4H_{10}	$CH_3(CH_2)_2CH_3$	tetradecane	$C_{14}H_{30}$	$CH_3(CH_2)_{12}CH_3$
pentane	C_5H_{12}	$CH_3(CH_2)_3CH_3$	pentadecane	$C_{15}H_{32}$	$CH_3(CH_2)_{13}CH_3$
hexane	C_6H_{14}	$CH_3(CH_2)_4CH_3$	hexadecane	$C_{16}H_{34}$	$CH_3(CH_2)_{14}CH_3$
heptane	C_7H_{16}	$CH_3(CH_2)_5CH_3$	heptadecane	$C_{17}H_{36}$	$CH_3(CH_2)_{15}CH_3$
octane	C_8H_{18}	$CH_3(CH_2)_6CH_3$	octadecane	$C_{18}H_{38}$	$CH_3(CH_2)_{16}CH_3$
nonane	C_9H_{20}	$CH_3(CH_2)_7CH_3$	nonadecane	$C_{19}H_{40}$	$CH_3(CH_2)_{17}CH_3$
decane	$C_{10}H_{22}$	$CH_3(CH_2)_8CH_3$	eicosane	$C_{20}H_{42}$	$CH_3(CH_2)_{18}CH_3$

atom. A line ending represents a $—CH_3$ group. Although hydrogen atoms are not shown in line-angle formulas, they are assumed to be there in sufficient numbers to give each carbon four bonds.

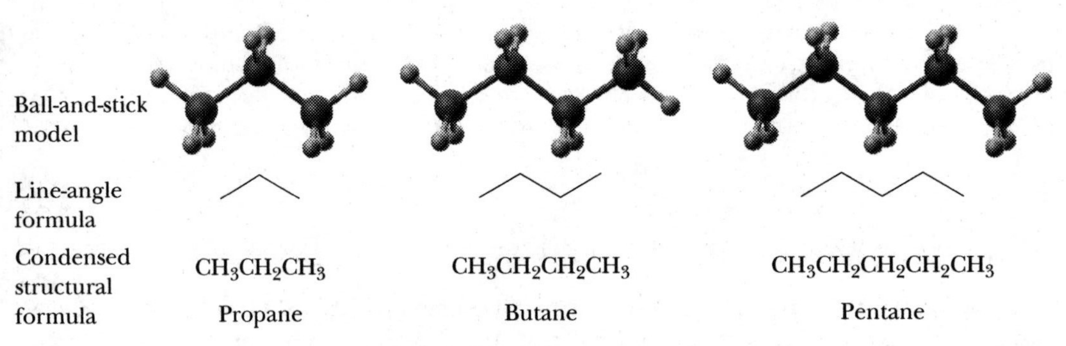

We can write structural formulas for alkanes in still another abbreviated form. The structural formula of pentane, for example, contains three CH_2 (methylene) groups in the middle of the chain. We can collect these groups together and write the structural formula as $CH_3(CH_2)_3CH_3$. Table 3.1 gives the names and molecular formulas of the first 20 alkanes. Note that the names of all these alkanes end in -ane. We will have more to say about naming alkanes in Section 3.4.

Alkanes have the general molecular formula C_nH_{2n+2}. Thus, given the number of carbon atoms in an alkane, it is easy to determine the number of hydrogens in the molecule and also its molecular formula. For example, decane, with 10 carbon atoms, must have $(2 \times 10) + 2 = 22$ hydrogens and a molecular formula of $C_{10}H_{22}$.

3.3 CONSTITUTIONAL ISOMERISM IN ALKANES

Constitutional isomers Compounds with the same molecular formula, but a different order of attachment of their atoms.

Constitutional isomers are compounds that have the same molecular formula, but different structural formulas. By "different structural formulas," we mean that these compounds differ in the kinds of bonds they have (single, double, or triple) or in their connectivity (the order of attachment among their atoms).

For the molecular formulas CH_4, C_2H_6, and C_3H_8, only one order of attachment of atoms is possible. For the molecular formula C_4H_{10}, two orders of attachment of atoms are possible. In one of these, named butane, the four carbons are bonded in a chain; in the other, named 2-methylpropane, three carbons are bonded in a chain, with the fourth carbon as a branch on the chain.

$CH_3CH_2CH_2CH_3$

Butane
(boiling point
= −0.5°C)

CH_3
|
CH_3CHCH_3

2-Methylpropane
(boiling point
= −11.6°C)

Butane and 2-methylpropane are constitutional isomers; they are different compounds and have different physical and chemical properties. Their boiling points, for example, differ by approximately 11°C.

In Section 1.8, we encountered several examples of constitutional isomers, although we did not call them that at the time. We saw that there are two alcohols with molecular formula C_3H_8O, two aldehydes with molecular formula C_4H_8O, and two carboxylic acids with molecular formula $C_4H_8O_2$.

To find out whether two or more structural formulas represent constitutional isomers, write the molecular formula of each and then compare them. All compounds that have the same molecular formula, but different structural formulas, are constitutional isomers.

Example 3.1

Do the structural formulas in each pair represent the same compound or constitutional isomers?

(a) $CH_3CH_2CH_2CH_2CH_2CH_3$ and $CH_3CH_2CH_2$–$CH_2CH_2CH_3$ (the second CH_2 bonded down to $CH_2CH_2CH_3$) (each is C_6H_{14})

(b) $CH_3\overset{CH_3}{C}HCH_2\underset{CH_3}{\overset{CH_3}{C}}H$ and $CH_3CH_2\overset{CH_3}{C}H\underset{CH_3}{C}HCH_3$ (each is C_7H_{16})

Solution

To determine whether these structural formulas represent the same compound or constitutional isomers, first find the longest chain of carbon atoms in each. Note that it makes no difference whether the chain is drawn straight or bent. Second, number the longest chain from the end nearest the first branch. Third, compare the lengths of each chain and the sizes and locations of any branches. Structural formulas that have the same order of attachment of atoms represent the same compound; those which have different orders of attachment of atoms represent constitutional isomers.

(a) Each structural formula has an unbranched chain of six carbons. The two structures are identical and represent the same compound:

$CH_3CH_2CH_2CH_2CH_2CH_3$ and $CH_3CH_2CH_2$–$CH_2CH_2CH_3$

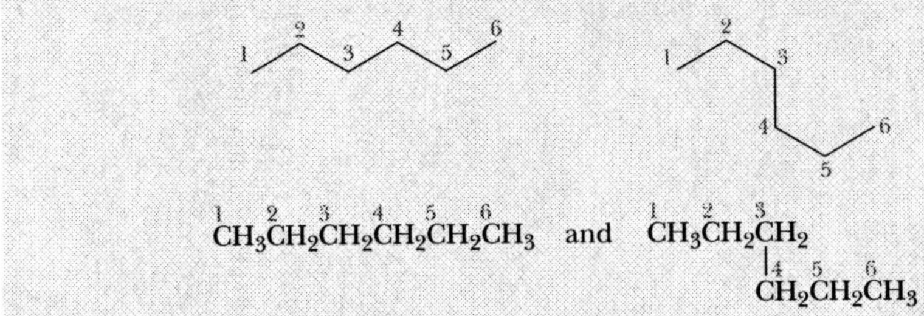

(b) Each structural formula has a chain of five carbons with two CH_3 branches. Although the branches are identical, they are at different locations on the chains. Therefore, these structural formulas represent constitutional isomers:

$CH_3CHCH_2CH(CH_3)CH_3$ (with CH_3 on C-2 and C-4) and $CH_3CH_2CHCHCH_3$ (with CH_3 on C-3 and C-2)

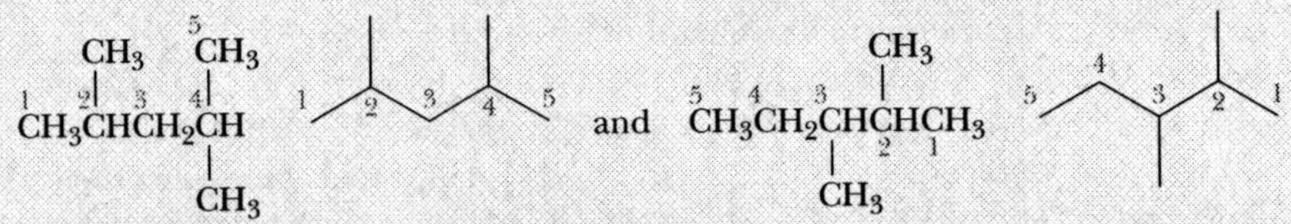

Practice Problem 3.1

Do the structural formulas in each pair represent the same compound or constitutional isomers?

(a) 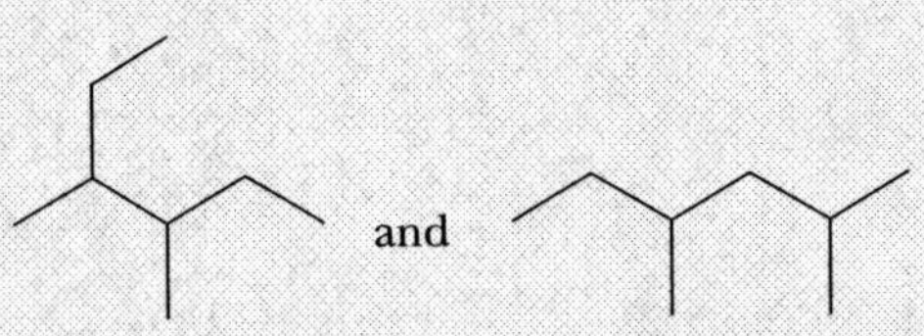 and

(b) 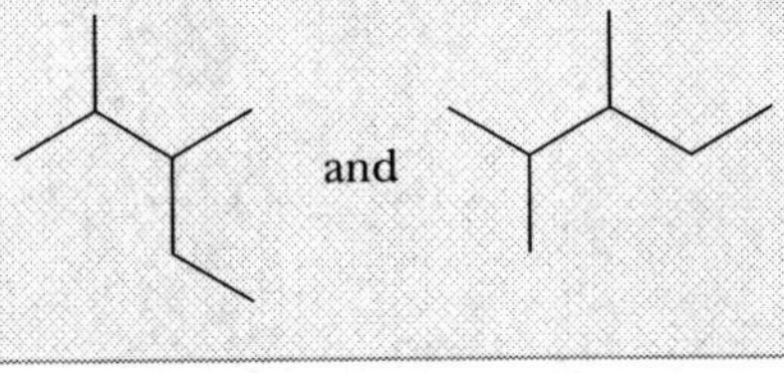 and

EXAMPLE 3.2

Draw structural formulas for the five constitutional isomers with molecular formula C_6H_{14}.

SOLUTION

In solving problems of this type, you should devise a strategy and then follow it. Here is one such strategy: First, draw a line-angle formula for the constitutional isomer with all six carbons in an unbranched chain. Then, draw line-angle formulas for all constitutional isomers with five carbons in a chain and one carbon as a branch on the chain. Finally, draw line-angle formulas for all constitutional isomers with four carbons in a chain and two carbons as branches.

Six carbons in an unbranched chain

Five carbons in a chain; one carbon as a branch

Four carbons in a chain; two carbons as branches

No constitutional isomers with only three carbons in the longest chain are possible for C_6H_{14}.

Practice Problem 3.2

Draw structural formulas for the three constitutional isomers with molecular formula C_5H_{12}.

The ability of carbon atoms to form strong, stable bonds with other carbon atoms results in a staggering number of constitutional isomers. As the following table shows, there are 3 constitutional isomers with molecular formula C_5H_{12}, 75 constitutional isomers for molecular formula $C_{10}H_{22}$, and almost 37 million constitutional isomers with molecular formula $C_{25}H_{52}$:

Carbon Atoms	Constitutional Isomers
1	0
5	3
10	75
15	4,347
25	36,797,588

Thus, for even a small number of carbon and hydrogen atoms, a very large number of constitutional isomers is possible. In fact, the potential for structural and functional group individuality among organic molecules made from just the basic building blocks of carbon, hydrogen, nitrogen, and oxygen is practically limitless.

3.4 NOMENCLATURE OF ALKANES

A. The IUPAC System

Ideally, every organic compound should have a name from which its structural formula can be drawn. For this purpose, chemists have adopted a set of rules established by an organization called the International Union of Pure and Applied Chemistry (IUPAC).

TABLE 3.2 Prefixes Used in the IUPAC System to Show the Presence of 1 to 20 Carbons in an Unbranched Chain

Prefix	Number of Carbon Atoms	Prefix	Number of Carbon Atoms
meth-	1	undec-	11
eth-	2	dodec-	12
prop-	3	tridec-	13
but-	4	tetradec-	14
pent-	5	pentadec-	15
hex-	6	hexadec-	16
hept-	7	heptadec-	17
oct-	8	octadec-	18
non-	9	nonadec-	19
dec-	10	eicos-	20

The IUPAC name of an alkane with an unbranched chain of carbon atoms consists of two parts: (1) a prefix that indicates the number of carbon atoms in the chain and (2) the ending **-ane** to show that the compound is a saturated hydrocarbon. Table 3.2 gives the prefixes used to show the presence of 1 to 20 carbon atoms.

The first four prefixes listed in Table 3.2 were chosen by the IUPAC because they were well established in the language of organic chemistry. In fact, they were well established even before there were hints of the structural theory underlying the discipline. For example, the prefix *but-* appears in the name *butyric acid,* a compound of four carbon atoms formed by the air oxidation of butter fat (Latin: *butyrum,* butter). Prefixes to show five or more carbons are derived from Greek or Latin numbers. (See Table 3.1 for the names, molecular formulas, and condensed structural formulas for the first 20 alkanes with unbranched chains.)

The IUPAC name of an alkane with a branched chain consists of a parent name that indicates the longest chain of carbon atoms in the compound and substituent names that indicate the groups bonded to the parent chain.

substituent

parent chain

$$\overset{1}{C}H_3\overset{2}{C}H_2\overset{3}{C}H_2\overset{4}{C}H(CH_3)\overset{5}{C}H_2\overset{6}{C}H_2\overset{7}{C}H_2\overset{8}{C}H_3$$

4-Methyloctane

Alkyl group A group derived by removing a hydrogen from an alkane; given the symbol R-.

R- A symbol used to represent an alkyl group.

A substituent group derived from an alkane by the removal of a hydrogen atom is called an **alkyl group** and is commonly represented by the symbol **R-**. We name alkyl groups by dropping the *-ane* from the name of the parent alkane and adding the suffix *-yl.* Table 3.3 gives names and structural formulas for eight of the most common alkyl groups. The prefix *sec-* is an abbreviation for secondary, meaning a carbon bonded to two other carbons. The prefix *tert-* is an abbreviation for tertiary, meaning a carbon bonded to three other carbons. Note that when these two prefixes are part of a name, they are always italicized.

The rules of the IUPAC system for naming alkanes are as follows:

1. The name for an alkane with an unbranched chain of carbon atoms consists of a prefix showing the number of carbon atoms in the chain and the ending *-ane.*
2. For branched-chain alkanes, take the longest chain of carbon atoms as the parent chain, and its name becomes the root name.

TABLE 3.3 Names of the Most Common Alkyl Groups

Name	Condensed Structural Formula	Name	Condensed Structural Formula
methyl	$-CH_3$	isobutyl	$-CH_2CHCH_3$ $\vert$ CH_3
ethyl	$-CH_2CH_3$		
propyl	$-CH_2CH_2CH_3$	*sec*-butyl	$-CHCH_2CH_3$ $\vert$ CH_3
isopropyl	$-CHCH_3$ $\vert$ CH_3		
butyl	$-CH_2CH_2CH_2CH_3$	*tert*-butyl	CH_3 $\vert$ $-CCH_3$ $\vert$ CH_3

3. Give each substituent on the parent chain a name and a number. The number shows the carbon atom of the parent chain to which the substituent is bonded. Use a hyphen to connect the number to the name:

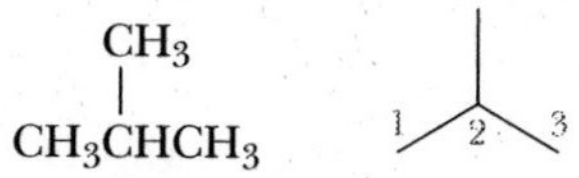

$$\begin{array}{c} CH_3 \\ | \\ CH_3CHCH_3 \end{array}$$

2-Methylpropane

4. If there is one substituent, number the parent chain from the end that gives it the lower number:

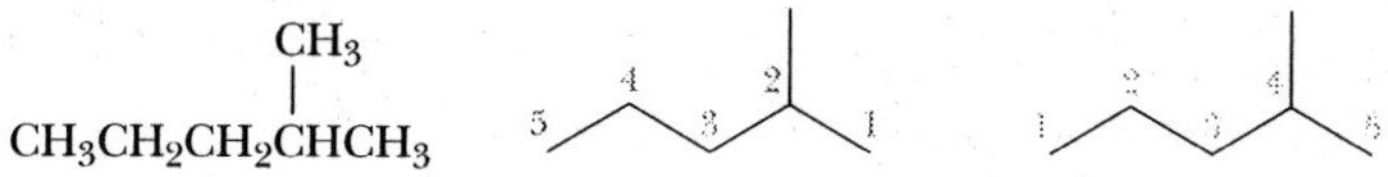

$$\begin{array}{l} \qquad\qquad\quad CH_3 \\ \qquad\qquad\quad | \\ CH_3CH_2CH_2CHCH_3 \end{array}$$

2-Methylpentane (not 4-methylpentane)

5. If there are two or more identical substituents, number the parent chain from the end that gives the lower number to the substituent encountered first. The number of times the substituent occurs is indicated by the prefix *di-*, *tri-*, *tetra-*, *penta-*, *hexa-*, and so on. A comma is used to separate position numbers:

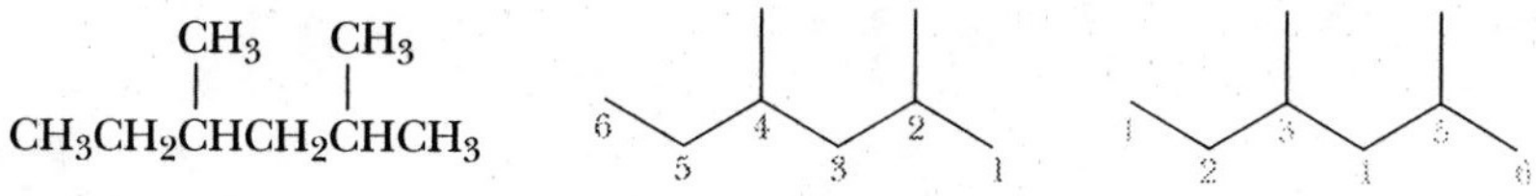

$$\begin{array}{l} \qquad\quad CH_3 \quad CH_3 \\ \qquad\quad | \qquad\quad | \\ CH_3CH_2CHCH_2CHCH_3 \end{array}$$

2,4-Dimethylhexane (not 3,5-dimethylhexane)

6. If there are two or more different substituents, list them in alphabetical order, and number the chain from the end that gives the lower number to the substituent encountered first. If there are different substituents in equivalent positions on opposite ends of the parent chain, the substituent of lower alphabetical order is given the lower number:

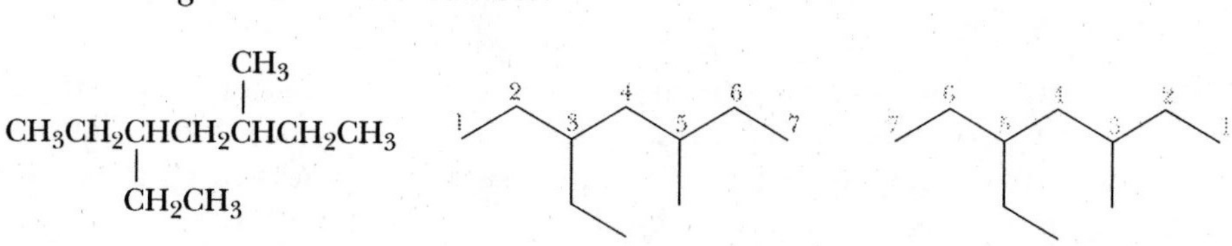

$$\begin{array}{l} \qquad\qquad\qquad\quad CH_3 \\ \qquad\qquad\qquad\quad | \\ CH_3CH_2CHCH_2CHCH_2CH_3 \\ \qquad\qquad | \\ \qquad\quad CH_2CH_3 \end{array}$$

3-Ethyl-5-methylheptane (not 3-methyl-5-ethylheptane)

7. The prefixes di-, tri-, tetra-, and so on are not included in alphabetizing. Neither are the hyphenated prefixes *sec-* and *tert-*. Alphabetize the names of the substituents first, and then insert the prefix. In the following example, the alphabetizing parts are ethyl and methyl, not ethyl and dimethyl:

$CH_3CCH_2CHCH_2CH_3$ (with CH_3 groups on C-2 and CH_2CH_3 on C-4)

4-Ethyl-2,2-dimethylhexane
(not 2,2-dimethyl-4-ethylhexane)

EXAMPLE 3.3

Write IUPAC names for these alkanes:

(a) (b)

SOLUTION

Number the longest chain in each compound from the end of the chain toward the substituent encountered first (rule 4). List the substituents in (b) in alphabetical order (rule 6).

(a) 2-Methylbutane

(b) 4-Isopropyl-2-methylheptane

Practice Problem 3.3

Write IUPAC names for these alkanes:

(a) (b)

B. Common Names

In the older system of common nomenclature, the total number of carbon atoms in an alkane, regardless of their arrangement, determines the name. The first three alkanes are methane, ethane, and propane. All alkanes of formula C_4H_{10} are called butanes, all those of formula C_5H_{12} are called pentanes, and all those of formula C_6H_{14} are called hexanes. For alkanes beyond propane, *iso* indicates that one end of an otherwise unbranched chain terminates in a $(CH_3)_2CH-$group. Following are examples of common names:

$CH_3CH_2CH_2CH_3$	$CH_3CH(CH_3)CH_3$	$CH_3CH_2CH_2CH_2CH_3$	$CH_3CH_2CH(CH_3)CH_3$
Butane	Isobutane	Pentane	Isopentane

This system of common names has no good way of handling other branching patterns, so, for more complex alkanes, it is necessary to use the more flexible IUPAC system of nomenclature.

In this text, we concentrate on IUPAC names. However, we also use common names, especially when the common name is used almost exclusively in the everyday discussions of chemists and biochemists. When both IUPAC and common names are given in the text, we always give the IUPAC name first, followed by the common name in parentheses. In this way, you should have no doubt about which name is which.

C. Classification of Carbon and Hydrogen Atoms

We classify a carbon atom as primary (1°), secondary (2°), tertiary (3°), or quaternary (4°), depending on the number of carbon atoms bonded to it. A carbon bonded to one carbon atom is a primary carbon; a carbon bonded to two carbon atoms is a secondary carbon, and so forth. For example, propane contains two primary carbons and one secondary carbon, 2-methylpropane contains three primary carbons and one tertiary carbon, and 2,2,4-trimethylpentane contains five primary carbons, one secondary carbon, one tertiary carbon, and one quaternary carbon:

two 1° carbons

$CH_3-CH_2-CH_3$

a 2° carbon

Propane

a 3° carbon

$CH_3-CH(CH_3)-CH_3$

2-Methylpropane

a 4° carbon

$CH_3-C(CH_3)_2-CH_2-CH(CH_3)-CH_3$

2,2,4-Trimethylpentane

Similarly, hydrogens are also classified as primary, secondary, or tertiary, depending on the type of carbon to which each is bonded. Those bonded to a primary carbon are classified as primary hydrogens, those on a secondary carbon are secondary hydrogens, and those on a tertiary carbon are tertiary hydrogens.

3.5 CYCLOALKANES

A hydrocarbon that contains carbon atoms joined to form a ring is called a *cyclic hydrocarbon*. When all carbons of the ring are saturated, we call the hydrocarbon a **cycloalkane**. Cycloalkanes of ring sizes ranging from 3 to over 30 abound in nature, and, in principle, there is no limit to ring size. Five-membered (cyclopentane) and six-membered (cyclohexane) rings are especially abundant in nature and have received special attention.

Cycloalkane A saturated hydrocarbon that contains carbon atoms joined to form a ring.

Figure 3.3 shows the structural formulas of cyclobutane, cyclopentane, and cyclohexane. When writing structural formulas for cycloalkanes, chemists rarely show all carbons and hydrogens. Rather, they use line-angle formulas to represent cycloalkane rings. Each ring is represented by a regular polygon having the same number of sides as there are carbon atoms in the ring. For example, chemists represent cyclobutane by a square, cyclopentane by a pentagon, and cyclohexane by a hexagon.

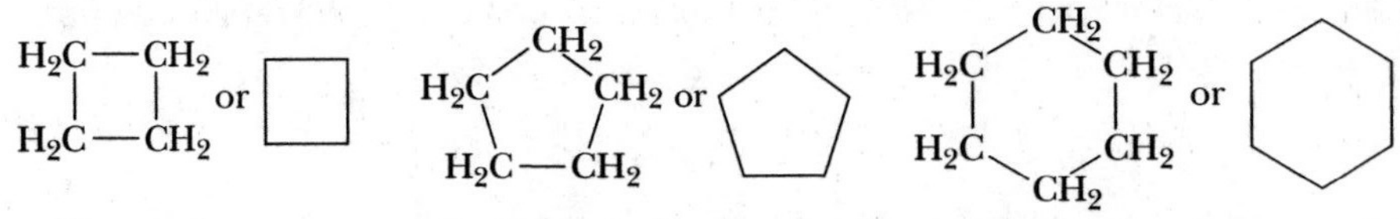

Figure 3.3
Examples of cycloalkanes.

Cyclobutane Cyclopentane Cyclohexane

Cycloalkanes contain two fewer hydrogen atoms than an alkane with the same number of carbon atoms. For instance, compare the molecular formulas of cyclohexane (C_6H_{12}) and hexane (C_6H_{14}). The general formula of a cycloalkane is C_nH_{2n}.

To name a cycloalkane, prefix the name of the corresponding open-chain hydrocarbon with *cyclo-*, and name each substituent on the ring. If there is only one substituent, there is no need to give it a number. If there are two substituents, number the ring by beginning with the substituent of lower alphabetical order. If there are three or more substituents, number the ring so as to give them the lowest set of numbers, and then list the substituents in alphabetical order.

EXAMPLE 3.4

Write the molecular formula and IUPAC name for each cycloalkane:

(a) 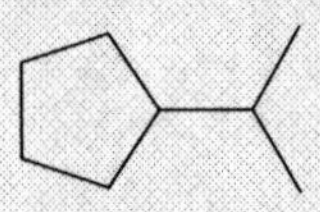 (b)

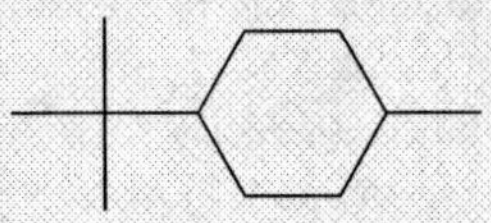

SOLUTION

(a) The molecular formula of this cycloalkane is C_8H_{16}. Because there is only one substituent on the ring, there is no need to number the atoms of the ring. The IUPAC name of this compound is isopropylcyclopentane.

(b) Number the atoms of the cyclohexane ring by beginning with *tert*-butyl, the substituent of lower alphabetical order. The compound's name is 1-*tert*-butyl-4-methylcyclohexane, and its molecular formula is $C_{11}H_{22}$.

Practice Problem 3.4

Write the molecular formula and IUPAC name for each cycloalkane:

(a) 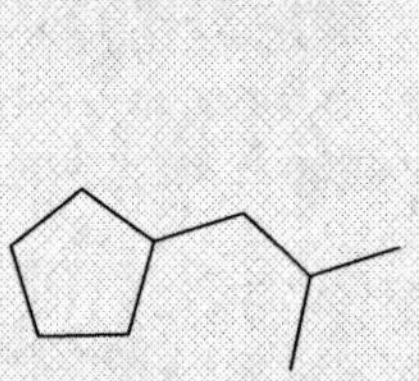 (b) 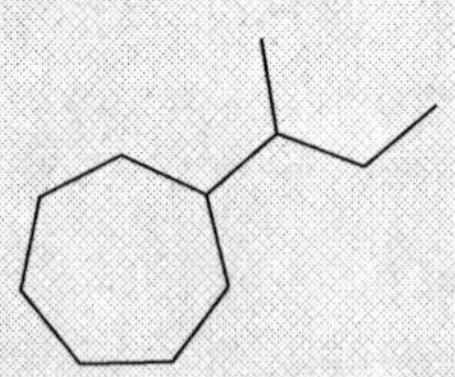 (c)

3.6 THE IUPAC SYSTEM: A GENERAL SYSTEM OF NOMENCLATURE

The naming of alkanes and cycloalkanes in Sections 3.4 and 3.5 illustrates the application of the IUPAC system of nomenclature to two specific classes of organic compounds. Now let us describe the general approach of the IUPAC system. The name we give to any compound with a chain of carbon atoms consists of three parts: a prefix, an infix (a modifying element inserted into a word), and a suffix. Each part provides specific information about the structural formula of the compound.

1. The prefix shows the number of carbon atoms in the parent chain. Prefixes that show the presence of 1 to 20 carbon atoms in a chain were given in Table 3.2.

2. The infix shows the nature of the carbon–carbon bonds in the parent chain;

Infix	Nature of Carbon–Carbon Bonds in the Parent Chain
-an-	all single bonds
-en-	one or more double bonds
-yn-	one or more triple bonds

3. The suffix shows the class of compound to which the substance belongs;

Suffix	Class of Compound
-e	hydrocarbon
-ol	alcohol
-al	aldehyde
-one	ketone
-oic acid	carboxylic acid

EXAMPLE 3.5

Following are IUPAC names and structural formulas for four compounds:

(a) $CH_2{=}CHCH_3$ Propene (b) CH_3CH_2OH Ethanol (c) $CH_3CH_2CH_2CH_2\overset{O}{\overset{\|}{C}}OH$ Pentanoic acid (d) $HC{\equiv}CH$ Ethyne

Divide each name into a prefix, an infix, and a suffix, and specify the information about the structural formula that is contained in each part of the name.

SOLUTION

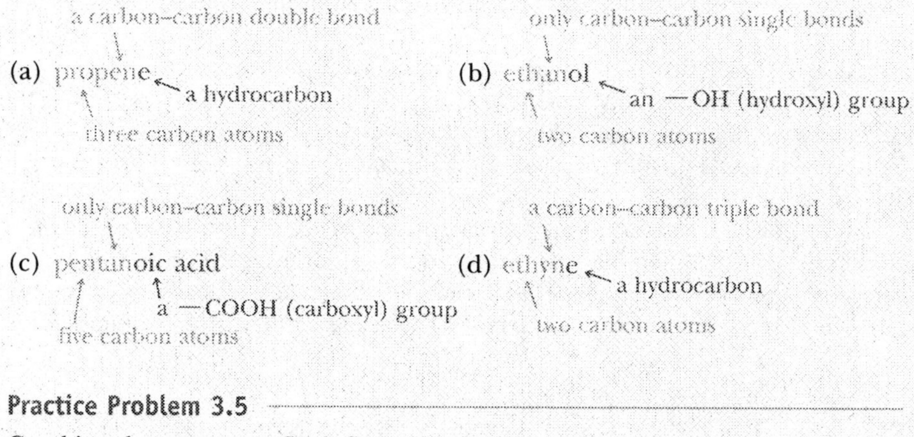

Practice Problem 3.5

Combine the proper prefix, infix, and suffix, and write the IUPAC name for each compound:

(a) $CH_3\overset{O}{\overset{\|}{C}}CH_3$ (b) $CH_3CH_2CH_2CH_2\overset{O}{\overset{\|}{C}}H$ (c) 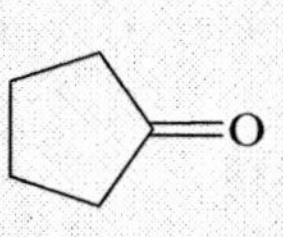 (d)

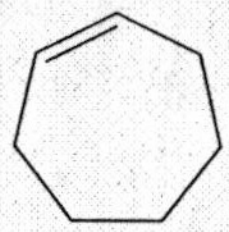

3.7 CONFORMATIONS OF ALKANES AND CYCLOALKANES

Even though structural formulas are useful for showing the order of attachment of atoms, they do not show three-dimensional shapes. As chemists try to understand more and more about the relationships between structure and the chemical and physical properties of molecules, it becomes increasingly important to know more about the three-dimensional shapes of molecules.

In this section, we concentrate on ways to visualize molecules as three-dimensional objects and to visualize not only bond angles within molecules, but also distances between various atoms and groups of atoms not bonded to each other. We also describe strain, which we divide into three types: torsional strain, angle strain, and steric strain. We urge you to build models and to study and manipulate them. Organic molecules are three-dimensional objects, and it is essential that you become comfortable in dealing with them as such.

A. Alkanes

Conformation Any three-dimensional arrangement of atoms in a molecule that results by rotation about a single bond.

Staggered conformation A conformation about a carbon–carbon single bond in which the atoms on one carbon are as far apart as possible from the atoms on the adjacent carbon.

Newman projection A way to view a molecule by looking along a carbon–carbon bond.

Eclipsed conformation A conformation about a carbon–carbon single bond in which the atoms on one carbon are as close as possible to the atoms on the adjacent carbon.

Alkanes of two or more carbons can be twisted into a number of different three-dimensional arrangements of their atoms by rotating about one or more carbon–carbon bonds. Any three-dimensional arrangement of atoms that results from rotation about a single bond is called a **conformation.** Figure 3.4(a) shows a ball-and-stick model of a **staggered conformation** of ethane. In this conformation, the three C—H bonds on one carbon are as far apart as possible from the three C—H bonds on the adjacent carbon. Figure 3.4(b), called a Newman projection, is a shorthand way of representing the staggered conformation of ethane. In a **Newman projection,** we view a molecule along the axis of a C—C bond. The three atoms or groups of atoms nearer your eye appear on lines extending from the center of the circle at angles of 120°. The three atoms or groups of atoms on the carbon farther from your eye appear on lines extending from the circumference of the circle at angles of 120°. Remember that bond angles about each carbon in ethane are approximately 109.5° and not 120°, as this Newman projection might suggest.

Figure 3.5 shows a ball-and-stick model and a Newman projection of an **eclipsed conformation** of ethane. In this conformation, the three C—H bonds on one carbon are as close as possible to the three C—H bonds on the adjacent carbon. In other words, hydrogen atoms on the back carbon are eclipsed by the hydrogen atoms on the front carbon.

For a long time, chemists believed that rotation about the C—C single bond in ethane was completely free. Studies of ethane and other molecules, however, have shown that a potential energy difference exists between its staggered and eclipsed conformations and that rotation is not completely free. In ethane, the potential energy of the eclipsed conformation is a maximum and that of the staggered conformation is a minimum. The difference in potential energy between these two conformations is approximately 3.0 kcal/mol (12.6 kJ/mol), which means that, at room temperature,

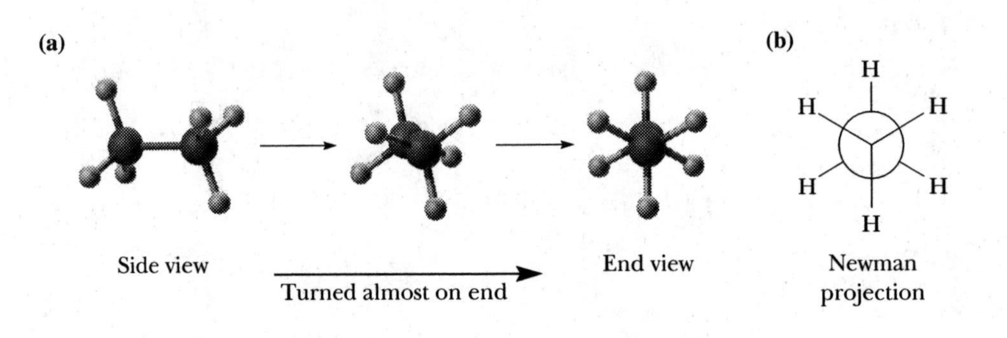

Figure 3.4
A staggered conformation of ethane. (a) Ball-and-stick model and (b) Newman projection.

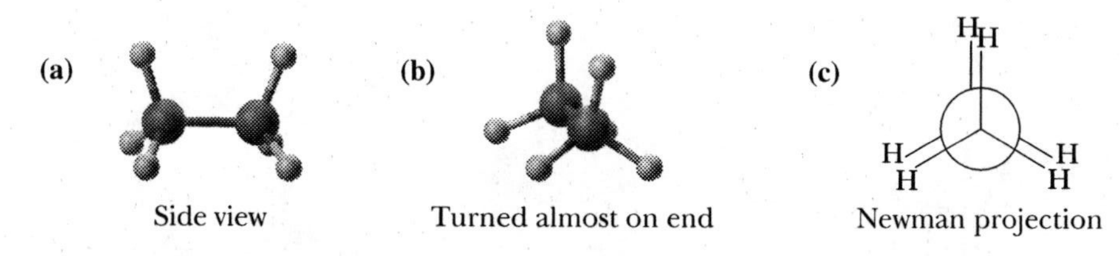

Figure 3.5 An eclipsed conformation of ethane. (a, b) Ball-and-stick models and (c) Newman projection.

the ratio of ethane molecules in a staggered conformation to those in an eclipsed conformation is approximately 100 to 1.

The strain induced in the eclipsed conformation of ethane is an example of torsional strain. **Torsional strain** (also called eclipsed interaction strain) is strain that arises when nonbonded atoms separated by three bonds are forced from a staggered conformation to an eclipsed conformation.

Torsional strain (also called eclipsed interaction strain) Strain that arises when atoms separated by 3 bonds are forced from a staggered conformation to an eclipsed conformation.

Example 3.6

Draw Newman projections for one staggered conformation and one eclipsed conformation of propane.

Solution

Following are Newman projections and ball-and-stick models of these conformations:

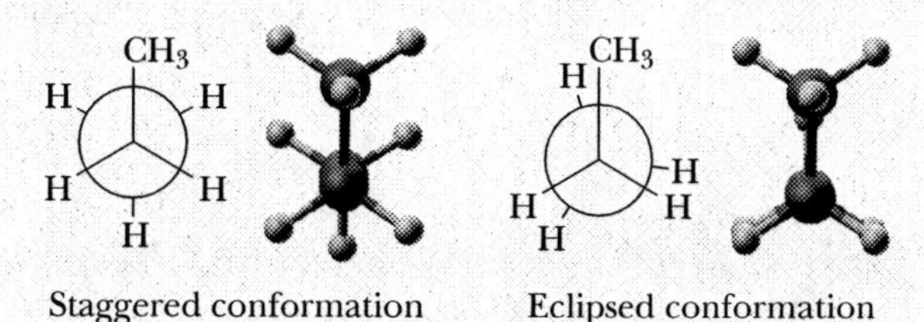

Practice Problem 3.6

Draw Newman projections for two staggered and two eclipsed conformations of 1,2-dichloroethane.

B. Cycloalkanes

We limit our discussion to the conformations of cyclopentanes and cyclohexanes, because these are the most common carbon rings in the molecules of nature.

Cyclopentane

We can draw cyclopentane [Figure 3.6(a)] as a planar conformation with all C—C—C bond angles equal to 108° [Figure 3.6(b)]. This angle differs only slightly from the tetrahedral angle of 109.5°; consequently, there is little angle strain in the planar conformation of cyclopentane. **Angle strain** results when a bond

Angle strain The strain that arises when a bond angle is either compressed or expanded compared with its optimal value.

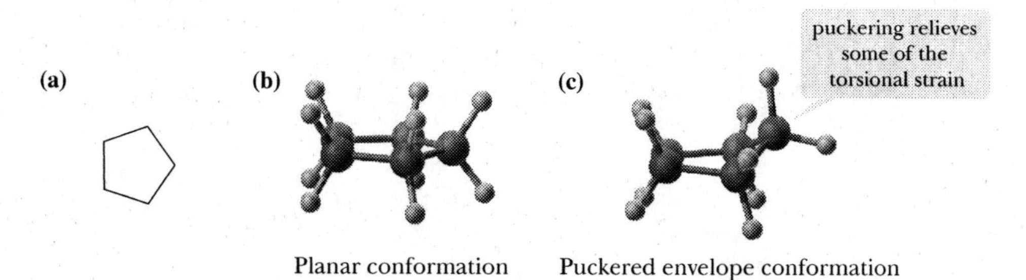

Figure 3.6 Cyclopentane. (a) Structural formula. (b) In the planar conformation, there are 10 pairs of eclipsed C—H interactions. (c) The most stable conformation is a puckered "envelope" conformation.

Figure 3.7
Cyclohexane. The most stable conformation is the puckered "chair" conformation.

Skeletal model

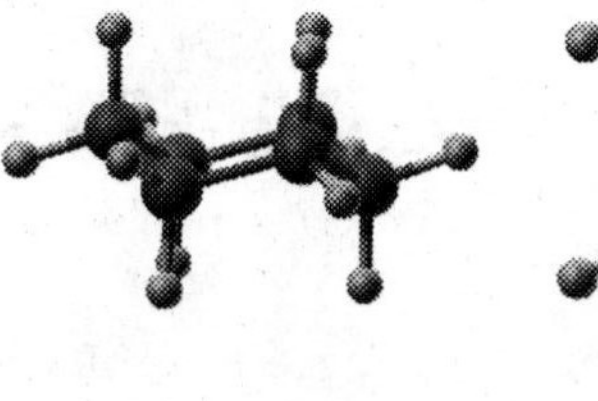

Ball-and-stick model viewed from the side

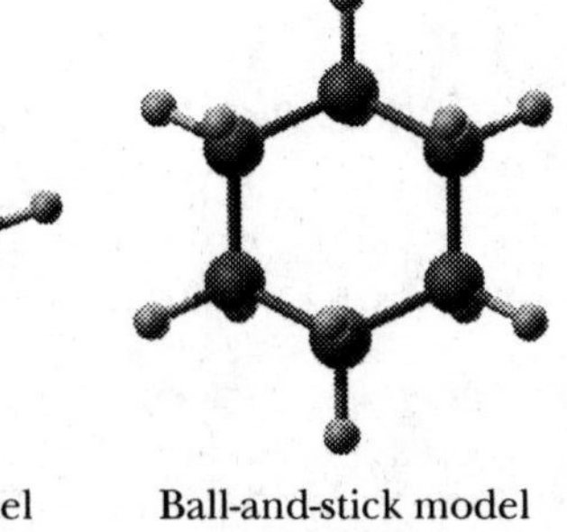

Ball-and-stick model viewed from above

angle in a molecule is either expanded or compressed compared with its optimal values. There are 10 fully eclipsed C—H bonds creating a torsional strain of approximately 10 kcal/mol (42 kJ/mol). To relieve at least a part of this strain, the atoms of the ring twist into the "envelope" conformation [Figure 3.6(c)]. In this conformation, four carbon atoms are in a plane, and the fifth is bent out of the plane, rather like an envelope with its flap bent upward.

In the envelope conformation, the number of eclipsed hydrogen interactions is reduced, thereby decreasing torsional strain. The C—C—C bond angles, however, are also reduced, which increases angle strain. The observed C—C—C bond angles in cyclopentane are 105°, indicating that, in its conformation of lowest energy, cyclopentane is slightly puckered. The strain energy in cyclopentane is approximately 5.6 kcal/mol (23.4 kJ/mol).

Cyclohexane

Cyclohexane adopts a number of puckered conformations, the most stable of which is a **chair conformation**. In this conformation (Figure 3.7), all C—C—C bond angles are 109.5° (minimizing angle strain), and hydrogens on adjacent carbons are staggered with respect to one another (minimizing torsional strain). Thus, there is very little strain in a chair conformation of cyclohexane.

Chair conformation The most stable puckered conformation of a cyclohexane ring; all bond angles are approximately 109.5°, and bonds to all adjacent carbons are staggered.

Equatorial bond A bond on a chair conformation of a cyclohexane ring that extends from the ring roughly perpendicular to the imaginary axis of the ring.

Axial bond A bond on a chair conformation of a cyclohexane ring that extends from the ring parallel to the imaginary axis of the ring.

In a chair conformation, the C—H bonds are arranged in two different orientations. Six C—H bonds are called **equatorial bonds**, and the other six are called **axial bonds**. One way to visualize the difference between these two types of bonds is to imagine an axis through the center of the chair, perpendicular to the floor [Figure 3.8(a)]. Equatorial bonds are approximately perpendicular to our imaginary axis and alternate first slightly up and then slightly down as you move from one carbon of the ring to the next. Axial bonds are parallel to the imaginary axis. Three axial bonds point up; the other three point down. Notice that axial bonds alternate also, first up and then down as you move from one carbon of the ring to the next. Notice further that if the axial

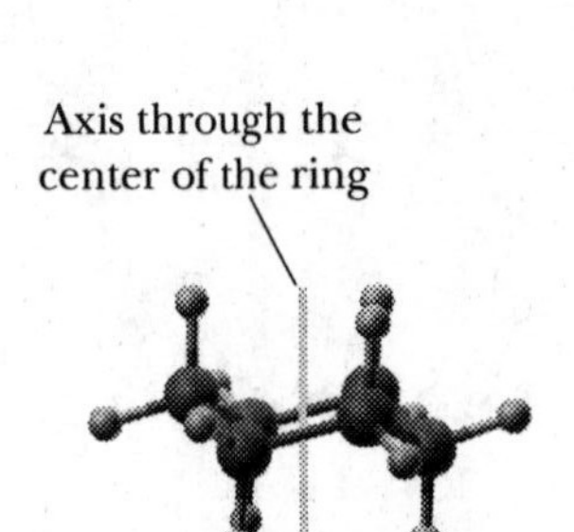

Figure 3.8
Chair conformation of cyclohexane, showing axial and equatorial C—H bonds.

(a) Ball-and-stick model showing all 12 hydrogens

(b) The six equatorial C—H bonds shown in red

(c) The six axial C—H bonds shown in blue

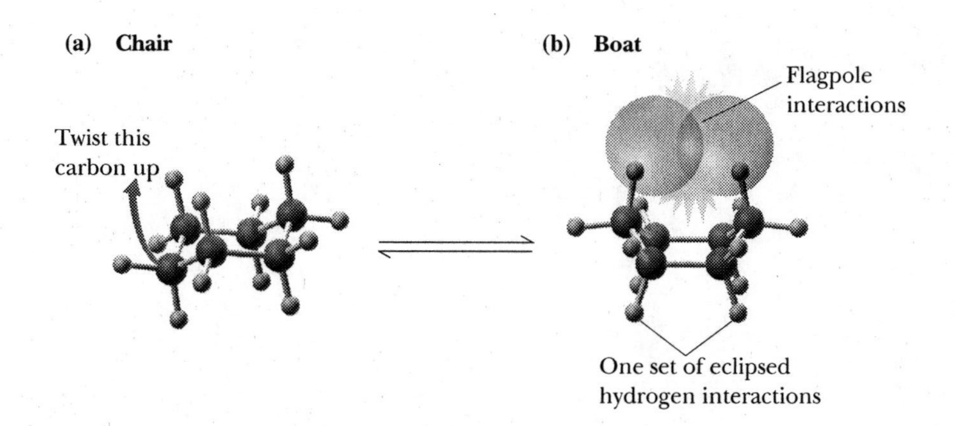

Figure 3.9 Conversion of (a) a chair conformation to (b) a boat conformation. In the boat conformation, there is torsional strain due to the four sets of eclipsed hydrogen interactions and steric strain due to the one set of flagpole interactions. A chair conformation is more stable than a boat conformation.

bond on a carbon points upward, then the equatorial bond on that carbon points slightly downward. Conversely, if the axial bond on a particular carbon points downward, then the equatorial bond on that carbon points slightly upward.

There are many other nonplanar conformations of cyclohexane, one of which is the **boat conformation**. You can visualize the interconversion of a chair conformation to a boat conformation by twisting the ring as illustrated in Figure 3.9. A boat conformation is considerably less stable than a chair conformation. In a boat conformation, torsional strain is created by four sets of eclipsed hydrogen interactions, and steric strain is created by the one set of flagpole interactions. **Steric strain** (also called nonbonded interaction strain) results when nonbonded atoms separated by four or more bonds are forced abnormally close to each other—that is, when they are forced closer than their atomic (contact) radii allow. The difference in potential energy between chair and boat conformations is approximately 6.5 kcal/mol (27 kJ/mol), which means that, at room temperature, approximately 99.99% of all cyclohexane molecules are in the chair conformation.

Boat conformation A puckered conformation of a cyclohexane ring in which carbons 1 and 4 of the ring are bent toward each other.

Steric strain The strain that arises when atoms separated by four or more bonds are forced abnormally close to one another.

For cyclohexane, the two equivalent chair conformations can interconvert by one chair twisting first into a boat and then into the other chair. When one chair is converted to the other, a change occurs in the relative orientations in space of the hydrogen atoms bonded to each carbon: All hydrogen atoms equatorial in one chair become axial in the other, and vice versa (Figure 3.10). The interconversion of one chair conformation of cyclohexane to the other occurs rapidly at room temperature.

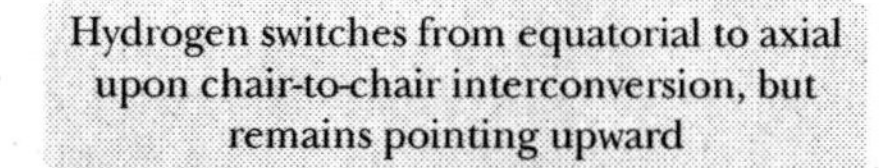

Hydrogen switches from equatorial to axial upon chair-to-chair interconversion, but remains pointing upward

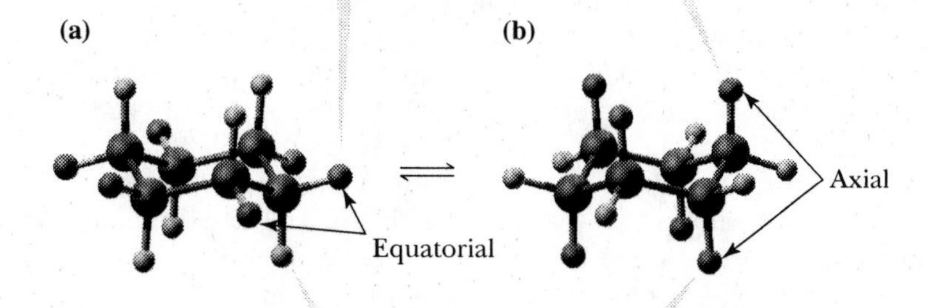

Figure 3.10 Interconversion of chair cyclohexanes. All C—H bonds that are equatorial in one chair are axial in the alternative chair, and vice versa.

EXAMPLE 3.7

Following is a chair conformation of cyclohexane showing a methyl group and one hydrogen:

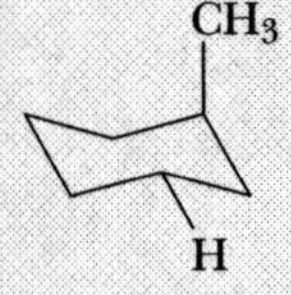

(a) Indicate by a label whether each group is equatorial or axial.
(b) Draw the other chair conformation, and again label each group as equatorial or axial.

SOLUTION

CH_3 (axial) ⇌ CH_3 (equatorial)

H (equatorial) H (axial)

(a) (b)

Practice Problem 3.7

Following is a chair conformation of cyclohexane with carbon atoms numbered 1 through 6:

5 6 1
4 3 2

(a) Draw hydrogen atoms that are above the plane of the ring on carbons 1 and 2 and below the plane of the ring on carbon 4.
(b) Which of these hydrogens are equatorial? Which are axial?
(c) Draw the other chair conformation. Now which hydrogens are equatorial? Which are axial? Which are above the plane of the ring, and which are below it?

If we replace a hydrogen atom of cyclohexane by an alkyl group, the group occupies an equatorial position in one chair and an axial position in the other chair. This means that the two chairs are no longer equivalent and no longer of equal stability.

Diaxial interactions Interactions between groups in parallel axial positions on the same side of a chair conformation of a cyclohexane ring.

A convenient way to describe the relative stabilities of chair conformations with equatorial or axial substituents is in terms of a type of steric strain called **axial–axial (diaxial) interaction**. *Axial–axial interaction* refers to the steric strain existing between an axial substituent and an axial hydrogen (or other group) on the same side of the ring. Consider methylcyclohexane (Figure 3.11). When the $—CH_3$ is equatorial, it is staggered with respect to all other groups on its adjacent carbon atoms. When the $—CH_3$ is axial, it is parallel to the axial C—H bonds on carbons 3 and 5. Thus, for axial methylcyclohexane, there are two unfavorable methyl–hydrogen axial–axial interactions. For methylcyclohexane, the equatorial methyl conformation is favored over the axial methyl conformation by approximately 1.74 kcal/mol (7.28 kJ/mol). At equilibrium at room temperature, approximately 95% of all methylcyclohexane molecules have their methyl group equatorial, and less than 5% have their methyl group axial.

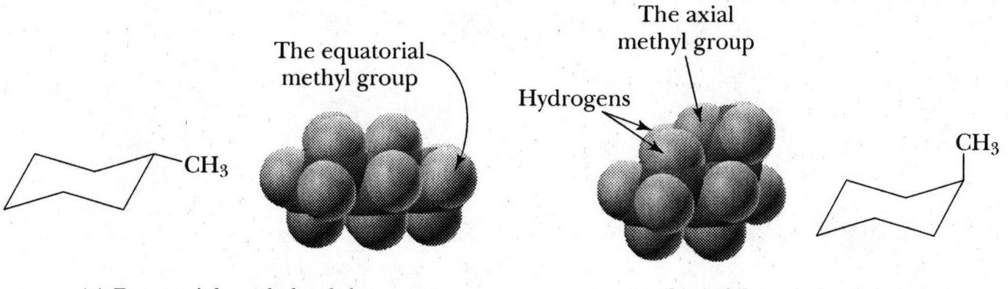

Figure 3.11 Two chair conformations of methylcyclohexane. The two axial–axial interactions (steric strain) make conformation (b) less stable than conformation (a) by approximately 1.74 kcal/mol (7.28 kJ/mol).

As the size of the substituent increases, the conformations with the group equatorial increases. When the group is as large as *tert*-butyl, the equatorial conformation is approximately 4,000 times more abundant at room temperature than the axial conformation, and, in effect, the ring is "locked" into a chair conformation with the *tert*-butyl group equatorial.

EXAMPLE 3.8

Label all axial–axial interactions in the following chair conformation:

SOLUTION

There are four axial–axial interactions: Each axial methyl group has two sets of axial–axial interactions with parallel hydrogen atoms on the same side of the ring. The equatorial methyl group has no axial–axial interactions.

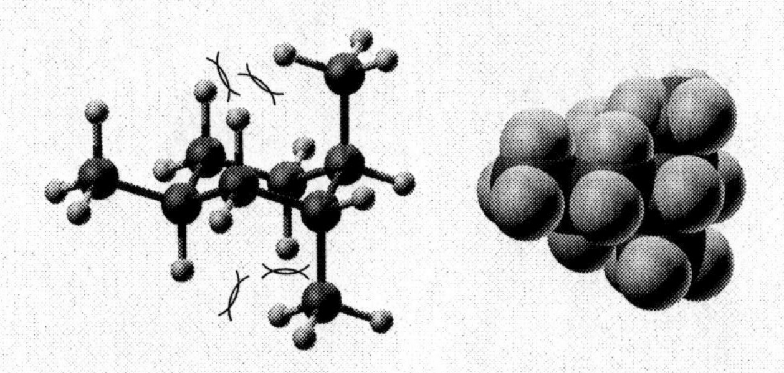

Practice Problem 3.8

The conformational equilibria for methyl-, ethyl-, and isopropylcyclohexane are all about 95% in favor of the equatorial conformation, but the conformational equilibrium for *tert*-butylcyclohexane is almost completely on the equatorial side. Explain why the conformational equilibria for the first three compounds are comparable, but that for *tert*-butylcyclohexane lies considerably farther toward the equatorial conformation.

CHEMICAL CONNECTIONS 3A

The Poisonous Puffer Fish

Nature is by no means limited to carbon in six-membered rings. Tetrodotoxin, one of the most potent toxins known, is composed of a set of interconnected six-membered rings, each in a chair conformation. All but one of these rings have atoms other than carbon in them. Tetrodotoxin is produced in the liver and ovaries of many species of *Tetraodontidae*, especially the puffer fish, so called because it inflates itself to an almost spherical spiny ball when alarmed. The puffer is evidently a species that is highly preoccupied with defense, but the Japanese are not put off. They regard the puffer, called *fugu* in Japanese, as a delicacy. To serve it in a public restaurant, a chef must be registered as sufficiently skilled in removing the toxic organs so as to make the flesh safe to eat.

Symptoms of tetrodotoxin poisoning begin with attacks of severe weakness, progressing to complete paralysis and eventual death. Tetrodotoxin exerts its severe poisonous effect by blocking Na^+ ion channels in excitable membranes. The $=NH_2^+$ end of tetrodotoxin lodges in the mouth of a Na^+ ion channel, thus blocking further transport of Na^+ ions through the channel.

A puffer fish inflated. *(Tim Rock/Animals Animals)*

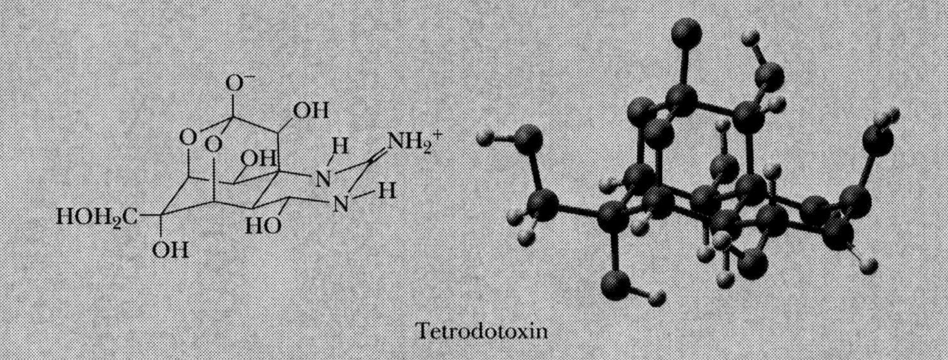

Tetrodotoxin

3.8 CIS–TRANS ISOMERISM IN CYCLOALKANES

Cis–trans isomers Isomers that have the same order of attachment of their atoms, but a different arrangement of their atoms in space, due to the presence of either a ring or a carbon–carbon double bond.

Cis A prefix meaning "on the same side."

Trans A prefix meaning "across from."

Cycloalkanes with substituents on two or more carbons of the ring show a type of isomerism called **cis–trans isomerism**. Cis-trans isomers have (1) the same molecular formula, (2) the same order of attachment of atoms, and (3) an arrangement of atoms that cannot be interchanged by rotation about sigma bonds under ordinary conditions. By way of comparison, the potential energy difference between conformations is so small that they can be interconverted easily at or near room temperature by rotation about single bonds.

We can illustrate cis–trans isomerism in cycloalkanes using 1,2-dimethylcyclopentane as an example. In the following structural formula, the cyclopentane ring is drawn as a planar pentagon viewed edge on (in determining the number of cis–trans isomers in a substituted cycloalkane, it is adequate to draw the cycloalkane ring as a planar polygon):

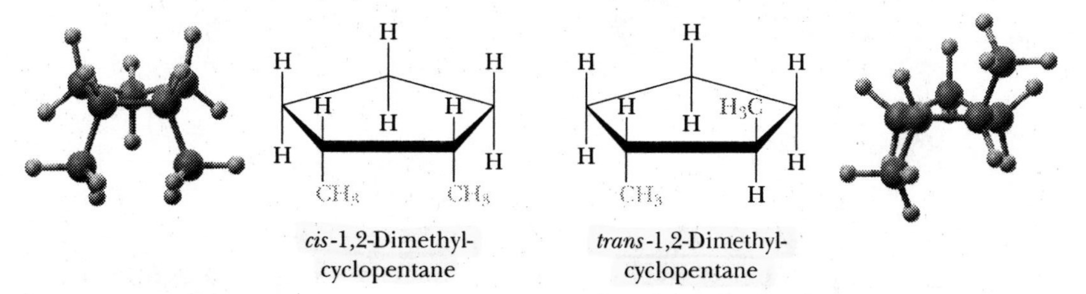

cis-1,2-Dimethyl-cyclopentane *trans*-1,2-Dimethyl-cyclopentane

Carbon–carbon bonds of the ring that project forward are shown as heavy lines. When viewed from this perspective, substituents bonded to the cyclopentane ring project above and below the plane of the ring. In one isomer of 1,2-dimethylcyclopentane, the methyl groups are on the same side of the ring (either both above or both below the plane of the ring); in the other isomer, they are on opposite sides of the ring (one above and one below the plane of the ring).

Alternatively, the cyclopentane ring can be viewed from above, with the ring in the plane of the paper. Substituents on the ring then either project toward you (that is, they project up above the page) and are shown by solid wedges, or they project away from you (they project down below the page) and are shown by broken wedges. In the following structural formulas, only the two methyl groups are shown (hydrogen atoms of the ring are not shown):

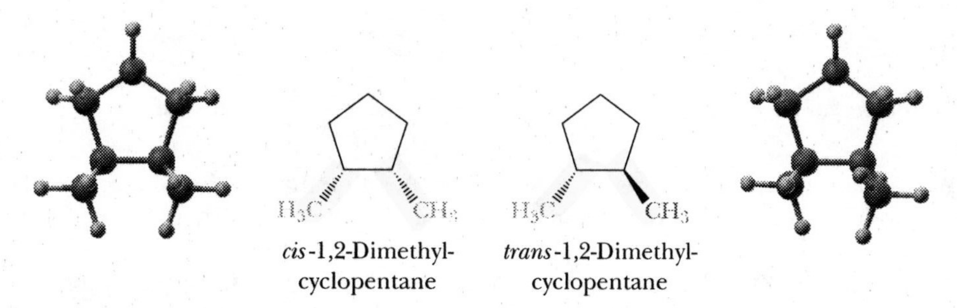

cis-1,2-Dimethyl-cyclopentane *trans*-1,2-Dimethyl-cyclopentane

EXAMPLE 3.9

Which cycloalkanes show cis–trans isomerism? For each that does, draw both isomers.

(a) Methylcyclopentane (b) 1,1-Dimethylcyclobutane
(c) 1,3-Dimethylcyclobutane

SOLUTION

(a) Methylcyclopentane does not show cis–trans isomerism: It has only one substituent on the ring.

(b) 1,1-Dimethylcyclobutane does not show cis–trans isomerism: Only one arrangement is possible for the two methyl groups on the ring, and they must be trans.

(c) 1,3-Dimethylcyclobutane shows cis–trans isomerism. Note that, in these structural formulas, we show only the hydrogens on carbons bearing the methyl groups.

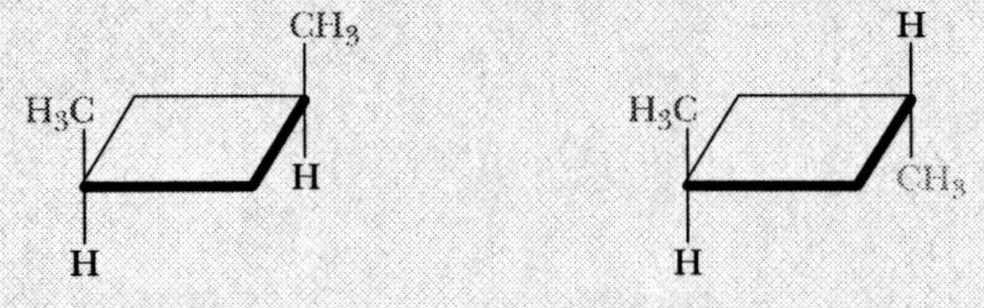

cis-1,3-Dimethylcyclobutane *trans*-1,3-Dimethylcyclobutane

Practice Problem 3.9

Which cycloalkanes show cis–trans isomerism? For each that does, draw both isomers.

(a) 1,3-Dimethylcyclopentane (b) Ethylcyclopentane
(c) 1-Ethyl-2-methylcyclobutane

Two cis–trans isomers exist for 1,4-dimethylcyclohexane. For the purposes of determining the number of cis–trans isomers in substituted cycloalkanes, it is adequate to draw the cycloalkane ring as a planar polygon, as is done in the following disubstituted cyclohexanes:

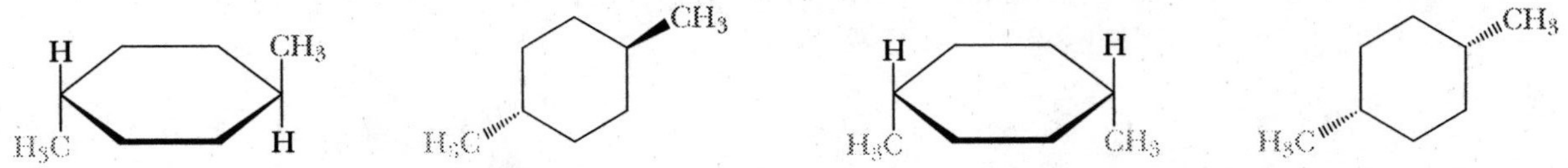

trans-1,4-Dimethylcyclohexane *cis*-1,4-Dimethylcyclohexane

We can also draw the cis and trans isomers of 1,4-dimethylcyclohexane as nonplanar chair conformations. In working with alternative chair conformations, it is helpful to remember that all groups axial on one chair are equatorial in the alternative chair, and vice versa. In one chair conformation of *trans*-1,4-dimethylcyclohexane, the two methyl groups are axial; in the alternative chair conformation, they are equatorial. Of these chair conformations, the one with both methyls equatorial is considerably more stable.

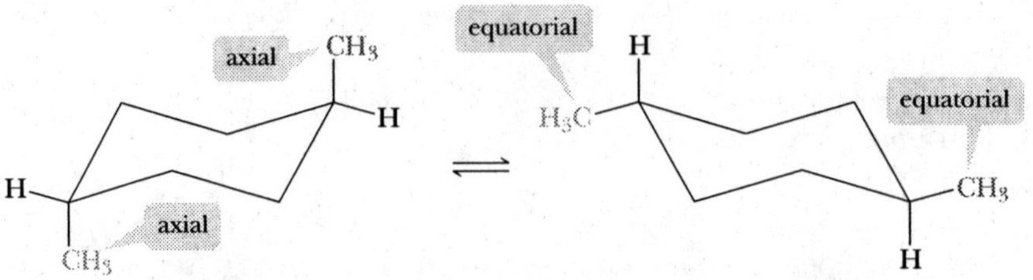

(less stable) (more stable)

trans-1,4-Dimethylcyclohexane

The alternative chair conformations of *cis*-1,4-dimethylcyclohexane are of equal energy. In each chair conformation, one methyl group is equatorial and the other is axial.

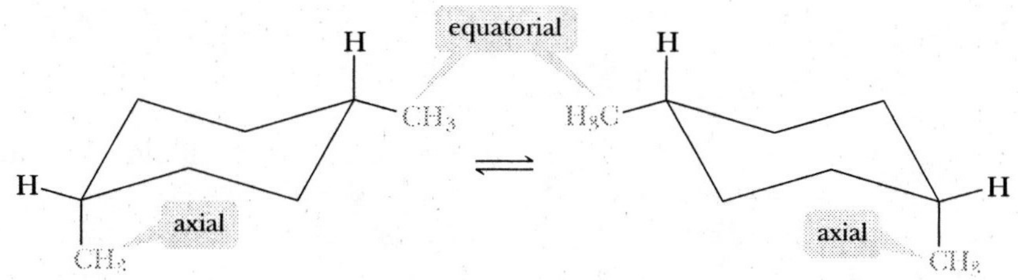

$$cis\text{-1,4-Dimethylcyclohexane} \rightleftharpoons$$

cis-1,4-Dimethylcyclohexane
(these conformations are of equal stability)

EXAMPLE 3.10

Following is a chair conformation of 1,3-dimethylcyclohexane:

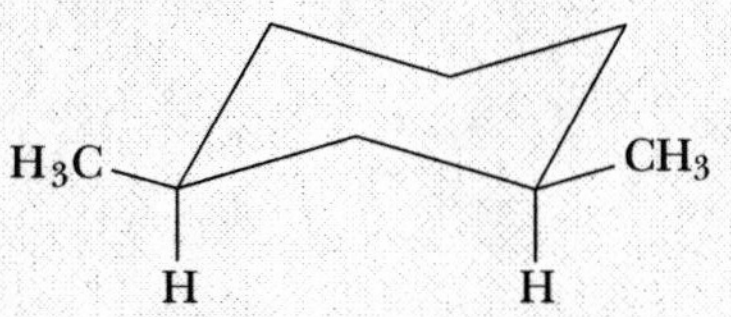

(a) Is this a chair conformation of *cis*-1,3-dimethylcyclohexane or of *trans*-1,3-dimethylcyclohexane?
(b) Draw the alternative chair conformation. Of the two chair conformations, which is the more stable?
(c) Draw a planar hexagon representation of the isomer shown in this example.

SOLUTION

(a) The isomer shown is *cis*-1,3-dimethylcyclohexane; the two methyl groups are on the same side of the ring.

(b) (more stable) $\rightleftharpoons$ (less stable)

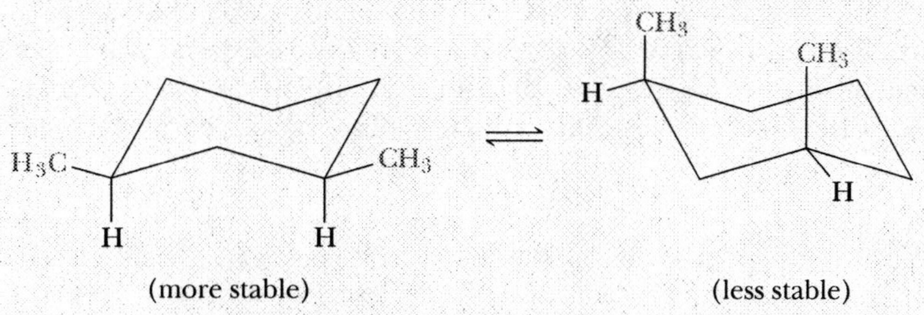

(c) or

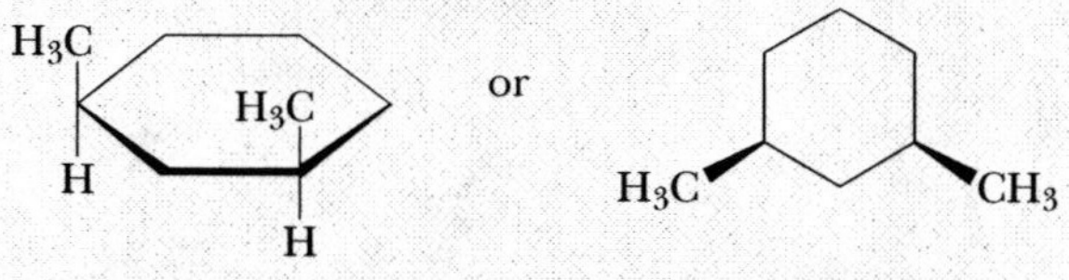

Practice Problem 3.10

Following is a planar hexagon representation of one isomer of 1,2,4-trimethylcyclohexane:

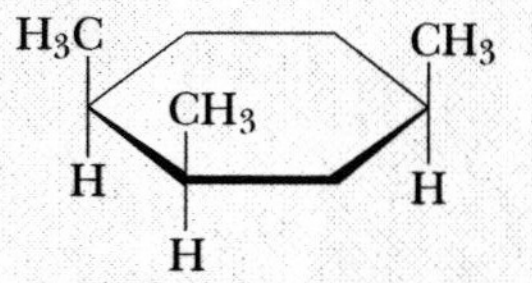

Draw the alternative chair conformations of this compound, and state which is the more stable.

3.9 PHYSICAL PROPERTIES OF ALKANES AND CYCLOALKANES

The most important property of alkanes and cycloalkanes is their almost complete lack of polarity. As we saw in Section 1.3C, the difference in electronegativity between carbon and hydrogen is $2.5 - 2.1 = 0.4$ on the Pauling scale, and given this small difference, we classify a C—H bond as nonpolar covalent. Therefore, alkanes are nonpolar compounds, and there is only weak interaction between their molecules.

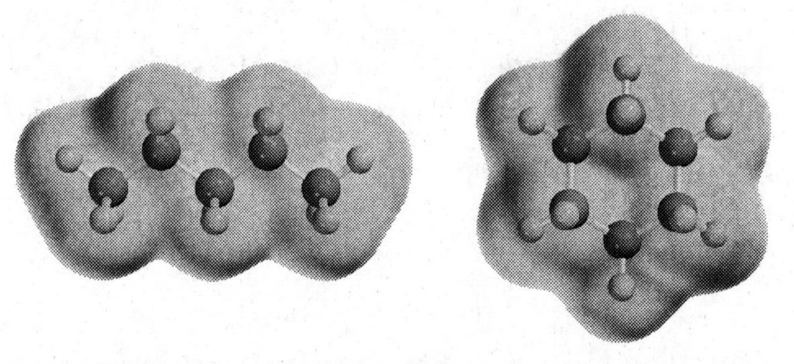

Pentane and cyclohexane. The electron density models show no evidence of any polarity in alkanes and cycloalkanes.

A. Boiling Points

The boiling points of alkanes are lower than those of almost any other type of compound of the same molecular weight. In general, both boiling and melting points of alkanes increase with increasing molecular weight (Table 3.4).

Alkanes containing 1 to 4 carbons are gases at room temperature, and those containing 5 to 17 carbons are colorless liquids. High-molecular-weight alkanes (those with 18 or more carbons) are white, waxy solids. Several plant waxes are high-molecular-weight alkanes. The wax found in apple skins, for example, is an unbranched alkane with molecular formula $C_{27}H_{56}$. Paraffin wax, a mixture of

TABLE 3.4 Physical Properties of Some Unbranched Alkanes

Name	Condensed Structural Formula	Melting Point (°C)	Boiling Point (°C)	*Density of Liquid (g/mL at 0°C)
methane	CH_4	−182	−164	(a gas)
ethane	CH_3CH_3	−183	−88	(a gas)
propane	$CH_3CH_2CH_3$	−190	−42	(a gas)
butane	$CH_3(CH_2)_2CH_3$	−138	0	(a gas)
pentane	$CH_3(CH_2)_3CH_3$	−130	36	0.626
hexane	$CH_3(CH_2)_4CH_3$	−95	69	0.659
heptane	$CH_3(CH_2)_5CH_3$	−90	98	0.684
octane	$CH_3(CH_2)_6CH_3$	−57	126	0.703
nonane	$CH_3(CH_2)_7CH_3$	−51	151	0.718
decane	$CH_3(CH_2)_8CH_3$	−30	174	0.730

*For comparison, the density of H_2O is 1 g/mL at 4°C.

high-molecular-weight alkanes, is used for wax candles, in lubricants, and to seal home canned jams, jellies, and other preserves. Petrolatum, so named because it is derived from petroleum refining, is a liquid mixture of high-molecular-weight alkanes. Sold as mineral oil and Vaseline, petrolatum is used as an ointment base in pharmaceuticals and cosmetics and as a lubricant and rust preventative.

B. Dispersion Forces and Interactions between Alkane Molecules

Methane is a gas at room temperature and atmospheric pressure. It can be converted to a liquid if cooled to −164°C and to a solid if further cooled to −182°C. The fact that methane (or any other compound, for that matter) can exist as a liquid or a solid depends on the existence of forces of attraction between particles of the pure compound. Although the forces of attraction between particles are all electrostatic in nature, they vary widely in their relative strengths. The strongest attractive forces are between ions—for example, between Na^+ and Cl^- in NaCl (188 kcal/mol, 787 kJ/mol). Hydrogen bonding is a weaker attractive force (2–10 kcal/mol, 8–42 kJ/mol). We will have more to say about hydrogen bonding in Chapter 8 when we discuss the physical properties of alcohols—compounds containing polar O—H groups.

Dispersion forces (0.02–2 kcal/mol, 0.08–8 kJ/mol) are the weakest intermolecular attractive forces. It is the existence of dispersion forces that accounts for the fact that low-molecular-weight, nonpolar substances such as methane can be liquefied. When we convert methane from a liquid to a gas at −164°C, for example, the process of separating its molecules requires only enough energy to overcome the very weak dispersion forces.

To visualize the origin of these forces, it is necessary to think in terms of instantaneous distributions of electron density rather than average distributions. Over time, the distribution of electron density in a methane molecule is symmetrical [Figure 3.12(a)], and there is no separation of charge. However, at any instant, there is a nonzero probability that the electron density is polarized (shifted) more toward one part of a methane molecule than toward another. This temporary polarization creates temporary partial positive and partial negative charges, which in turn induce temporary partial positive and negative charges in adjacent methane molecules [Figure 3.12(b)]. **Dispersion forces** are weak electrostatic attractive forces that occur between temporary partial positive and partial negative charges in adjacent atoms or molecules.

Dispersion forces Very weak intermolecular forces of attraction resulting from the interaction of temporary induced dipoles.

Because interactions between alkane molecules consist only of these very weak dispersion forces, the boiling points of alkanes are lower than those of almost any other type of compound of the same molecular weight. As the number of atoms and the molecular weight of an alkane increase, the strength of the dispersion forces among alkane molecules increases, and consequently, boiling points increase.

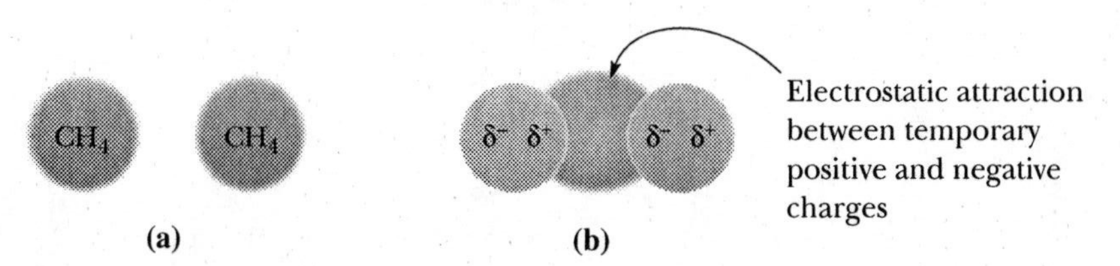

Figure 3.12
Dispersion forces. (a) The average distribution of electron density in a methane molecule is symmetrical, and there is no polarity. (b) Temporary polarization of one molecule induces temporary polarization in an adjacent molecule. Electrostatic attractions between temporary partial positive and partial negative charges are called *dispersion forces.*

C. Melting Point and Density

The melting points of alkanes increase with increasing molecular weight. The increase, however, is not as regular as that observed for boiling points, because the ability of molecules to pack into ordered patterns of solids changes as the molecular size and shape change.

The average density of the alkanes listed in Table 3.4 is about 0.7 g/mL; that of higher-molecular-weight alkanes is about 0.8 g/mL. All liquid and solid alkanes are less dense than water (1.0 g/mL); therefore, they float on water.

D. Constitutional Isomers Have Different Physical Properties

Alkanes that are constitutional isomers are different compounds and have different physical properties. Table 3.5 lists the boiling points, melting points, and densities of the five constitutional isomers with molecular formula C_6H_{14}. The boiling point of each of its branched-chain isomers is lower than that of hexane itself, and the more branching there is, the lower is the boiling point. These differences in boiling points are related to molecular shape in the following way: The only forces of attraction between alkane molecules are dispersion forces. As branching increases, the shape of an alkane molecule becomes more compact, and its surface area decreases. As the surface area decreases, the strength of the dispersion forces decreases, and boiling points also decrease. Thus, for any group of alkane constitutional isomers, it is usually observed that the least-branched isomer has the highest boiling point and the most-branched isomer has the lowest boiling point. The trend in melting points is less obvious, but, as previously mentioned, it correlates with a molecule's ability to pack into ordered patterns of solids.

TABLE 3.5 Physical Properties of the Isomeric Alkanes with Molecular Formula C_6H_{14}

Name	Melting Point (°C)	Boiling Point (°C)	Density (g/mL)
hexane	−95	69	0.659
3-methylpentane	−6	64	0.664
2-methylpentane	−23	62	0.653
2,3-dimethylbutane	−129	58	0.662
2,2-dimethylbutane	−100	50	0.649

more surface area, an increase in dispersion forces, and a higher boiling point

Hexane

smaller surface area, a decrease in dispersion forces, and a lower boiling point

2,2-Dimethylbutane

EXAMPLE 3.11

Arrange the alkanes in each set in order of increasing boiling point:

(a) Butane, decane, and hexane
(b) 2-Methylheptane, octane, and 2,2,4-trimethylpentane

SOLUTION

(a) All of the compounds are unbranched alkanes. As the number of carbon atoms in the chain increases, the dispersion forces among molecules increase, and the boiling points increase. Decane has the highest boiling point, butane the lowest:

Butane (bp −0.5°C) Hexane (bp 69°C) Decane (bp 174°C)

(b) These three alkanes are constitutional isomers with molecular formula C_8H_{18}. Their relative boiling points depend on the degree of branching. 2,2,4-Trimethylpentane, the most highly branched isomer, has the smallest surface area and the lowest boiling point. Octane, the unbranched isomer, has the largest surface area and the highest boiling point.

2,2,4-Trimethylpentane (bp 99°C) 2-Methylheptane (bp 118°C) Octane (bp 125°C)

Practice Problem 3.11

Arrange the alkanes in each set in order of increasing boiling point:

(a) 2-Methylbutane, 2,2-dimethylpropane, and pentane
(b) 3,3-Dimethylheptane, 2,2,4-trimethylhexane, and nonane

3.10 REACTIONS OF ALKANES

The most important chemical property of alkanes and cycloalkanes is their inertness. They are quite unreactive toward most reagents, a behavior consistent with the fact that they are nonpolar compounds containing only strong sigma bonds. Under certain conditions, however, alkanes and cycloalkanes do react, with oxygen, O_2. By far their most important reaction with oxygen is oxidation (combustion) to form carbon dioxide and water. The oxidation of saturated hydrocarbons is the basis for their use as energy sources for heat [natural gas, liquefied petroleum gas (LPG), and fuel oil] and power (gasoline, diesel fuel, and aviation fuel). Following are balanced equations for the complete combustion of methane, the major component of natural gas, and for propane, the major component of LPG:

$$\underset{\text{Methane}}{CH_4} + 2O_2 \longrightarrow CO_2 + 2H_2O \qquad \Delta H° = -212 \text{ kcal/mol } (-886 \text{ kJ/mol})$$

$$\underset{\text{Propane}}{CH_3CH_2CH_3} + 5O_2 \longrightarrow 3CO_2 + 4H_2O \qquad \Delta H° = -530 \text{ kcal/mol } (-2{,}220 \text{ kJ/mol})$$

3.11 SOURCES OF ALKANES

The three major sources of alkanes throughout the world are the fossil fuels: natural gas, petroleum, and coal. Fossil fuels account for approximately 90% of the total energy consumed in the United States. Nuclear electric power and hydroelectric power make up most of the remaining 10%. In addition, fossil fuels provide the bulk of the raw material for the organic chemicals consumed worldwide.

A. Natural Gas

Natural gas consists of approximately 90–95% methane, 5–10% ethane, and a mixture of other relatively low-boiling alkanes—chiefly propane, butane, and 2-methylpropane. The current widespread use of ethylene as the organic chemical industry's most important building block is largely the result of the ease with which ethane can be separated from natural gas and cracked into ethylene. Cracking is a process whereby a saturated hydrocarbon is converted into an unsaturated hydrocarbon plus H_2. Ethane is cracked by heating it in a furnace at 800 to 900°C for a fraction of a second. The production of ethylene in the United States in 1997 was 51.1 billion pounds, making it the number-one organic compound produced by the U.S. chemical industry, on a weight basis. The bulk of the ethylene produced is used to create organic polymers, as described in Chapter 17.

$$\underset{\text{Ethane}}{CH_3CH_3} \xrightarrow[\text{(thermal cracking)}]{800\text{–}900^\circ C} \underset{\text{Ethylene}}{CH_2{=}CH_2} + H_2$$

B. Petroleum

A petroleum refinery.
(K. Straiton/Photo Researchers, Inc.)

Petroleum is a thick, viscous liquid mixture of literally thousands of compounds, most of them hydrocarbons, formed from the decomposition of marine plants and animals. Petroleum and petroleum-derived products fuel automobiles, aircraft, and trains. They provide most of the greases and lubricants required for the machinery of our highly industrialized society. Furthermore, petroleum, along with natural gas, provides close to 90% of the organic raw materials used in the synthesis and manufacture of synthetic fibers, plastics, detergents, drugs, dyes, and a multitude of other products.

It is the task of a petroleum refinery to produce usable products, with a minimum of waste, from the thousands of different hydrocarbons in this liquid mixture. The various physical and chemical processes for this purpose fall into two broad categories: separation processes, which separate the complex mixture into various fractions, and re-forming processes, which alter the molecular structure of the hydrocarbon components themselves.

The fundamental separation process utilized in refining petroleum is fractional distillation (Figure 3.13). Practically all crude oil that enters a refinery goes to distillation units, where it is heated to temperatures as high as 370 to 425°C and separated into fractions. Each fraction contains a mixture of hydrocarbons that boils within a particular range:

1. Gases boiling below 20°C are taken off at the top of the distillation column. This fraction is a mixture of low-molecular-weight hydrocarbons, predominantly propane, butane, and 2-methylpropane, substances that can be liquefied under pressure at room temperature. The liquefied mixture, known as liquefied petroleum gas (LPG), can be stored and shipped in metal tanks and is a convenient source of gaseous fuel for home heating and cooking.

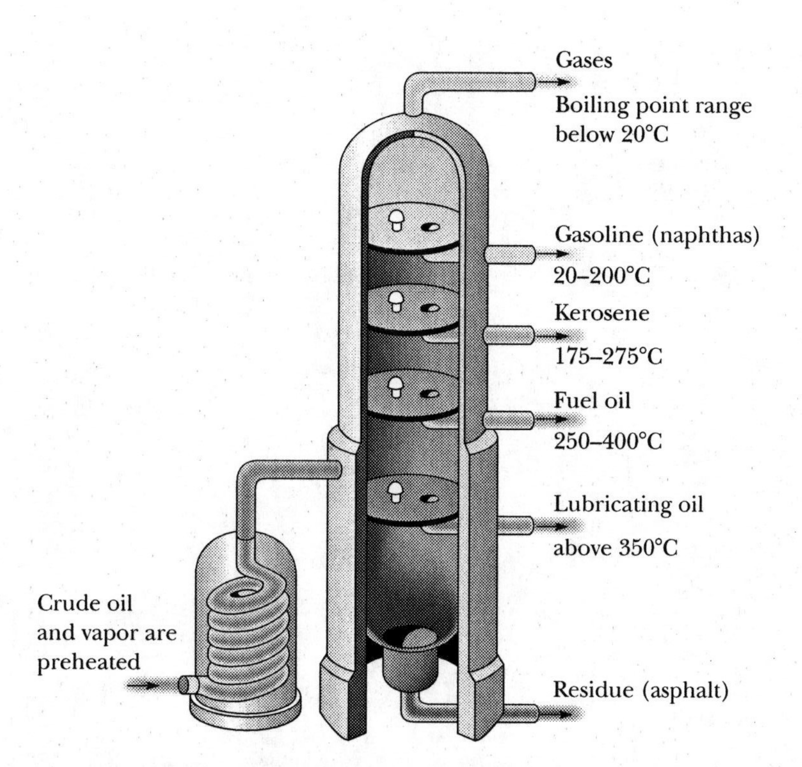

Figure 3.13
Fractional distillation of petroleum. The lighter, more volatile fractions are removed from higher up the column and the heavier, less volatile fractions from lower down.

2. Naphthas, bp 20 to 200°C, are a mixture of C_5 to C_{12} alkanes and cycloalkanes. Naphthas also contain small amounts of benzene, toluene, xylene, and other aromatic hydrocarbons (Chapter 9). The light naphtha fraction, bp 20 to 150°C, is the source of straight-run gasoline and averages approximately 25% of crude petroleum. In a sense, naphthas are the most valuable distillation fractions, because they are useful not only as fuel, but also as sources of raw materials for the organic chemical industry.
3. Kerosene, bp 175 to 275°C, is a mixture of C_9 to C_{15} hydrocarbons.
4. Fuel oil, bp 250 to 400°C, is a mixture of C_{15} to C_{18} hydrocarbons. Diesel fuel is obtained from this fraction.
5. Lubricating oil and heavy fuel oil distill from the column at temperatures above 350°C.
6. Asphalt is the black, tarry residue remaining after the removal of the other volatile fractions.

The two most common re-forming processes are cracking, illustrated by the thermal conversion of ethane to ethylene (Section 3.11A), and catalytic re-forming, illustrated by the conversion of hexane first to cyclohexane and then to benzene:

$$CH_3CH_2CH_2CH_2CH_2CH_3 \xrightarrow[-H_2]{\text{catalyst}} \text{Cyclohexane} \xrightarrow[-3H_2]{\text{catalyst}} \text{Benzene}$$

Hexane Cyclohexane Benzene

C. Coal

To understand how coal can be used as a raw material for the production of organic compounds, it is necessary to discuss synthesis gas. Synthesis gas is a mixture of carbon monoxide and hydrogen in varying proportions, depending on the means

CHEMICAL CONNECTIONS 3B

Octane Rating: What Those Numbers at the Pump Mean

Gasoline is a complex mixture of C_6 to C_{12} hydrocarbons. The quality of gasoline as a fuel for internal combustion engines is expressed in terms of an *octane rating*. Engine knocking occurs when a portion of the air–fuel mixture explodes prematurely (usually as a result of heat developed during compression) and independently of ignition by the spark plug. Two compounds were selected as reference fuels. One of these, 2,2,4-trimethylpentane (isooctane), has very good antiknock properties (the fuel–air mixture burns smoothly in the combustion chamber) and was assigned an octane rating of 100. (The name *isooctane* is a trivial name; its only relation to the name 2,2,4-trimethylpentane is that both compounds show eight carbon atoms.) Heptane, the other reference compound, has poor antiknock properties and was assigned an octane rating of 0.

2,2,4-Trimethylpentane (octane rating 100)

Heptane (octane rating 0)

The octane rating of a particular gasoline is that percentage of isooctane in a mixture of isooctane and heptane that has antiknock properties equivalent to those of the gasoline. For example, the antiknock properties of 2-methylhexane are the same as those of a mixture of 42% isooctane and 58% heptane; therefore, the octane rating of 2-methylhexane is 42. Octane itself has an octane rating of −20, which means that it produces even more engine knocking than heptane. Ethanol, the additive to gasohol, has an octane rating of 105. Benzene and toluene have octane ratings of 106 and 120, respectively.

Typical octane ratings of commonly available gasolines. *(Charles D. Winters)*

by which it is manufactured. Synthesis gas is prepared by passing steam over coal. It is also prepared by the partial oxidation of methane with oxygen.

$$\underset{\text{Coal}}{C} + H_2O \xrightarrow{\text{heat}} CO + H_2$$

$$\underset{\text{Methane}}{CH_4} + \frac{1}{2}O_2 \xrightarrow{\text{catalyst}} CO + 2H_2$$

Two important organic compounds produced today almost exclusively from carbon monoxide and hydrogen are methanol and acetic acid. In the production of methanol, the ratio of carbon monoxide to hydrogen is adjusted to 1:2, and the mixture is passed over a catalyst at elevated temperature and pressure:

$$CO + 2H_2 \xrightarrow{\text{catalyst}} \underset{\text{methanol}}{CH_3OH}$$

The treatment of methanol, in turn, with carbon monoxide over a different catalyst gives acetic acid:

$$\underset{\text{Methanol}}{CH_3OH} + CO \xrightarrow{\text{catalyst}} \underset{\text{Acetic acid}}{CH_3\overset{\overset{\large O}{\|}}{C}OH}$$

Because the processes for making methanol and acetic acid directly from carbon monoxide are commercially proven, it is likely that the decades ahead will see the development of routes to other organic chemicals from coal via methanol.

SUMMARY

A **hydrocarbon** is a compound that contains only carbon and hydrogen. A **saturated hydrocarbon** contains only single bonds. Alkanes have the general formula C_nH_{2n+2}. **Constitutional isomers** (Section 3.3) have the same molecular formula, but a different connectivity (a different order of attachment of their atoms). Alkanes are named according to a set of rules developed by the **International Union of Pure and Applied Chemistry (IUPAC**; Section 3.4A). A carbon atom is classified as **primary (1°), secondary (2°), tertiary (3°),** or **quaternary (4°)**, depending on the number of carbon atoms bonded to it (Section 3.4C). A hydrogen atom is classified as primary (1°), secondary (2°), or tertiary (3°), depending on the type of carbon to which it is bonded.

An alkane that contains carbon atoms bonded to form a ring is called a **cycloalkane** (Section 3.5). To name a cycloalkane, prefix the name of the open-chain hydrocarbon with "*cyclo-*." Five-membered rings (cyclopentanes) and six-membered rings (cyclohexanes) are especially abundant in the biological world.

The IUPAC system is a general system of nomenclature (Section 3.6). The IUPAC name of a compound consists of three parts: (1) a **prefix** that indicates the number of carbon atoms in the parent chain, (2) an **infix** that indicates the nature of the carbon–carbon bonds in the parent chain, and (3) a **suffix** that indicates the class to which the compound belongs. Substituents derived from alkanes by the removal of a hydrogen atom are called **alkyl groups** and are given the symbol **R**. The name of an alkyl group is formed by dropping the suffix *-ane* from the name of the parent alkane and adding *-yl* in its place.

A **conformation** is any three-dimensional arrangement of the atoms of a molecule that results from rotation about a single bond (Section 3.7). One convention for showing conformations is the **Newman projection**. Staggered conformations are lower in energy (more stable) than eclipsed conformations.

Molecular strain (Section 3.7) is of three types:

1. **torsional strain** (also called eclipsed interaction strain) results when nonbonded atoms separated by three bonds are forced from a staggered conformation to an eclipsed conformation,
2. **angle strain** results when a bond angle in a molecule is either expanded or compressed compared with its optimal values, and
3. **steric strain** (also called **nonbonded interaction strain**) results when nonbonded atoms separated by four or more bonds are forced abnormally close to each other—that is, when they are forced closer than their atomic (contact) radii would otherwise allow.

Cyclopentanes, cyclohexanes, and all larger cycloalkanes exist in dynamic equilibrium between a set of puckered conformations. The lowest energy conformation of cyclopentane is an envelope conformation. The lowest energy conformations of cyclohexane are two interconvertible **chair conformations** (Section 3.7B). In a chair conformation, six bonds are axial and six are **equatorial.** Bonds axial in one chair are equatorial in the alternative chair, and vice versa. A **boat conformation** is higher in energy than chair conformations. The more stable conformation of a substituted cyclohexane is the one that minimizes **axial–axial interactions.**

Cis–trans isomers (Section 3.8) have the same molecular formula and the same order of attachment of atoms, but arrangements of atoms in space that cannot be interconverted by rotation about single bonds. **Cis** means that substituents are on the same side of the ring; **trans** means that they are on opposite sides of the ring. Most cycloalkanes with substituents on two or more carbons of the ring show cis–trans isomerism.

Alkanes are nonpolar compounds, and the only forces of attraction between their molecules are **dispersion forces** (Section 3.9), which are weak electrostatic interactions between temporary partial positive and negative charges of atoms or molecules. Low-molecular-weight alkanes, such as methane, ethane, and propane, are gases at room temperature and atmospheric pressure. Higher-molecular-weight alkanes, such as those in gasoline and kerosene, are liquids. Very high-molecular-weight alkanes, such as those in paraffin wax, are solids. Among a set of alkane constitutional isomers, the least branched isomer generally has the highest boiling point; the most branched isomer generally has the lowest boiling point.

Natural gas (Section 3.11A) consists of 90–95% methane, with lesser amounts of ethane and other lower-molecular-weight hydrocarbons. **Petroleum** (Section 3.11B) is a liquid mixture of literally thousands of different hydrocarbons. **Synthesis gas**, a mixture of carbon monoxide and hydrogen, can be derived from natural gas and coal (Section 3.11C).

KEY REACTIONS

1. Oxidation of Alkanes (Section 3.10)

The oxidation of alkanes to carbon dioxide and water is the basis for their use as energy sources of heat and power:

$$CH_3CH_2CH_3 + 5O_2 \longrightarrow 3CO_2 + 4H_2O + \text{energy}$$

PROBLEMS

A problem number set in red indicates an applied "real-world" problem.

Structure of Alkanes

3.12 For each condensed structural formula, write a line-angle formula:

(a) 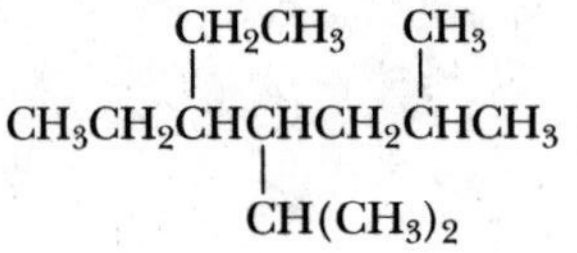

(b) 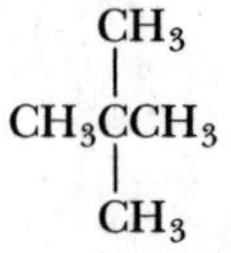

(c) $(CH_3)_2CHCH(CH_3)_2$

(d) 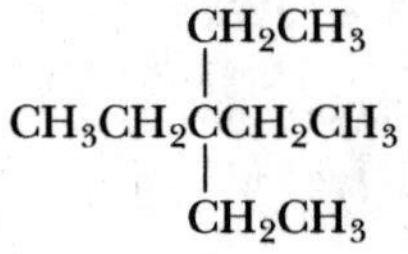

(e) $(CH_3)_3CH$

(f) $CH_3(CH_2)_3CH(CH_3)_2$

3.13 Write a condensed structural formula and the molecular formula of each alkane:

(a) 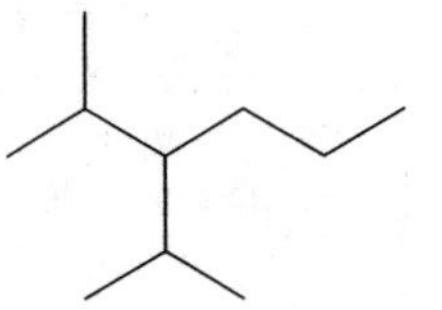

(b) 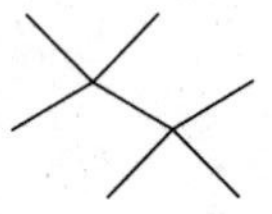

(c) 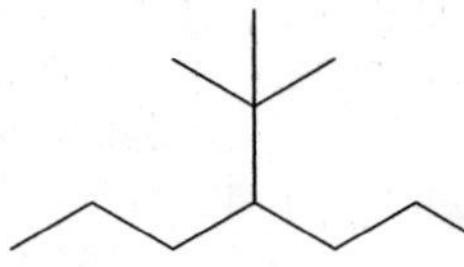

3.14 For each of the following condensed structural formulas, provide an even more abbreviated formula, using parentheses and subscripts:

(a) 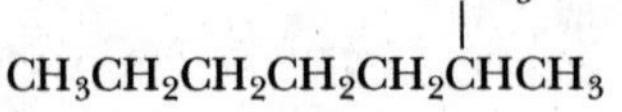

(b) 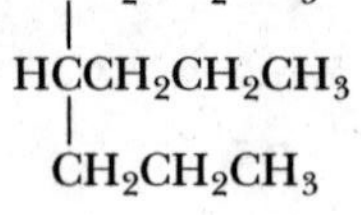

(c) 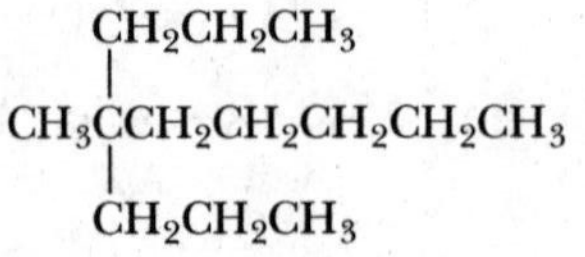

Constitutional Isomerism

3.15 Which statements are true about constitutional isomers?

(a) They have the same molecular formula.
(b) They have the same molecular weight.
(c) They have the same order of attachment of atoms.
(d) They have the same physical properties.

3.16 Each member of the following set of compounds is an alcohol; that is, each contains an —OH (hydroxyl group, Section 1.8A):

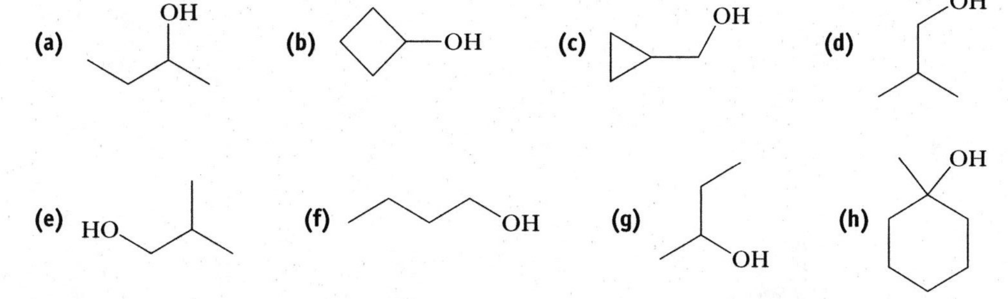

Which structural formulas represent (1) the same compound, (2) different compounds that are constitutional isomers, or (3) different compounds that are not constitutional isomers?

3.17 Each member of the following set of compounds is an amine; that is, each contains a nitrogen bonded to one, two, or three carbon groups (Section 1.8B):

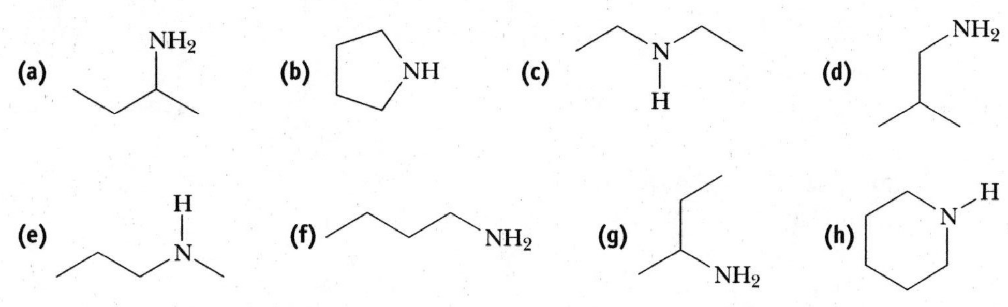

Which structural formulas represent (1) the same compound, (2) different compounds that are constitutional isomers, or (3) different compounds that are not constitutional isomers?

3.18 Each member of the following set of compounds is either an aldehyde or a ketone (Section 1.8C):

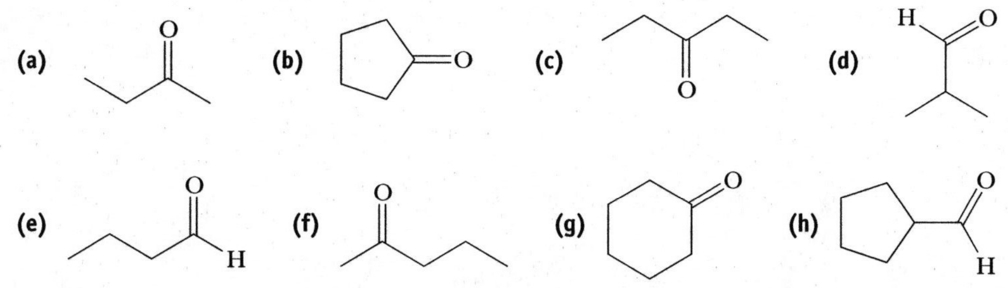

Which structural formulas represent (1) the same compound, (2) different compounds that are constitutional isomers, or (3) different compounds that are not constitutional isomers?

3.19 For each pair of compounds, tell whether the structural formulas shown represent

(1) the same compound,

(2) different compounds that are constitutional isomers, or

(3) different compounds that are not constitutional isomers:

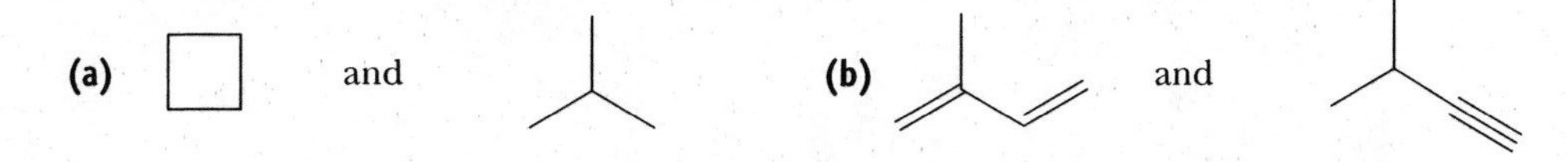

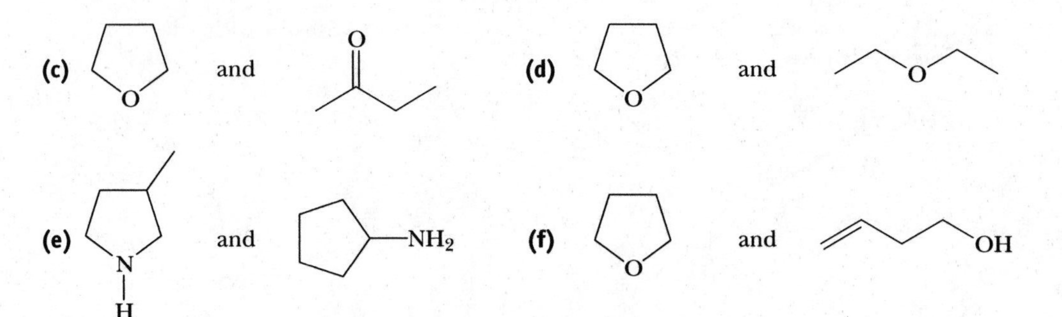

3.20 Name and draw line-angle formulas for the nine constitutional isomers with molecular formula C_7H_{16}.

3.21 Tell whether the compounds in each set are constitutional isomers:

(a) CH_3CH_2OH and CH_3OCH_3
(b) $CH_3\overset{O}{\overset{\|}{C}}CH_3$ and $CH_3CH_2\overset{O}{\overset{\|}{C}}H$
(c) $CH_3\overset{O}{\overset{\|}{C}}OCH_3$ and $CH_3CH_2\overset{O}{\overset{\|}{C}}OH$
(d) $CH_3\overset{OH}{\overset{|}{C}}HCH_2CH_3$ and $CH_3\overset{O}{\overset{\|}{C}}CH_2CH_3$
(e) (cyclopentane) and $CH_3CH_2CH_2CH_2CH_3$
(f) (cyclopentane) and $CH_2{=}CHCH_2CH_2CH_3$

3.22 Draw line-angle formulas for

(a) The four alcohols with molecular formula $C_4H_{10}O$.
(b) The two aldehydes with molecular formula C_4H_8O.
(c) The one ketone with molecular formula C_4H_8O.
(d) The three ketones with molecular formula $C_5H_{10}O$.
(e) The four carboxylic acids with molecular formula $C_5H_{10}O_2$.

Nomenclature of Alkanes and Cycloalkanes

3.23 Write IUPAC names for these alkanes and cycloalkanes:

(a) $CH_3\underset{CH_3}{\underset{|}{C}}HCH_2CH_2CH_3$
(b) $CH_3\underset{CH_3}{\underset{|}{C}}HCH_2CH_2\underset{CH_3}{\underset{|}{C}}HCH_3$
(c) $CH_3(CH_2)_4\underset{CH_2CH_3}{\underset{|}{C}}HCH_2CH_3$
(d) **(e)** **(f)**

3.24 Write line-angle formulas for these alkanes:

(a) 2,2,4-Trimethylhexane
(b) 2,2-Dimethylpropane
(c) 3-Ethyl-2,4,5-trimethyloctane
(d) 5-Butyl-2,2-dimethylnonane
(e) 4-Isopropyloctane
(f) 3,3-Dimethylpentane
(g) *trans*-1,3-Dimethylcyclopentane
(h) *cis*-1,2-Diethylcyclobutane

3.25 Explain why each of the following names is an incorrect IUPAC name and write the correct IUPAC name for the intended compound:

(a) 1,3-Dimethylbutane
(b) 4-Methylpentane
(c) 2,2-Diethylbutane
(d) 2-Ethyl-3-methylpentane
(e) 2-Propylpentane
(f) 2,2-Diethylheptane
(g) 2,2-Dimethylcyclopropane
(h) 1-Ethyl-5-methylcyclohexane

3.26 Draw a structural formula for each compound:

(a) Ethanol
(b) Ethanal
(c) Ethanoic acid
(d) Butanone
(e) Butanal
(f) Butanoic acid

(g) Propanal **(h)** Cyclopropanol **(i)** Cyclopentanol
(j) Cyclopentene **(k)** Cyclopentanone

3.27 Write the IUPAC name for each compound:

(a) 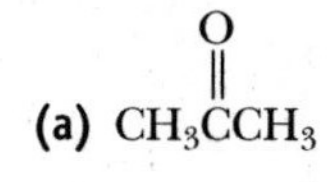

(b) $CH_3(CH_2)_3\overset{O}{\overset{\|}{C}}H$

(c) $CH_3(CH_2)_8\overset{O}{\overset{\|}{C}}OH$

(d) 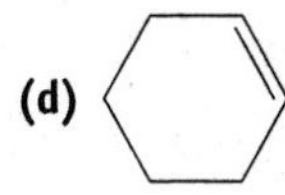

(e) (cyclohexane ring)=O

(f) (cyclobutane ring)—OH

Conformations of Alkanes and Cycloalkanes

3.28 How many different staggered conformations are there for 2-methylpropane? How many different eclipsed conformations are there?

3.29 Looking along the bond between carbons 2 and 3 of butane, there are two different staggered conformations and two different eclipsed conformations. Draw Newman projections of each, and arrange them in order from the most stable conformation to the least stable conformation.

3.30 Explain why each of the following Newman projections might not represent the most stable conformation of that molecule:

(a) 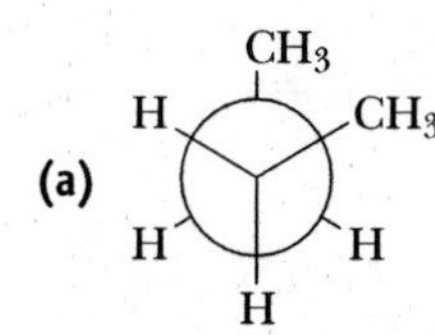

(b) 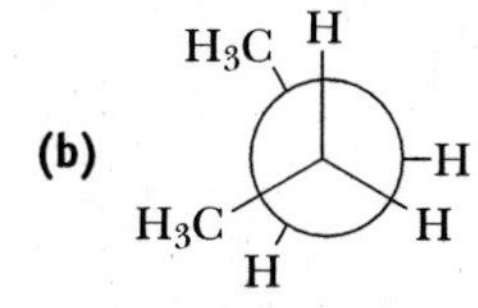

(c) 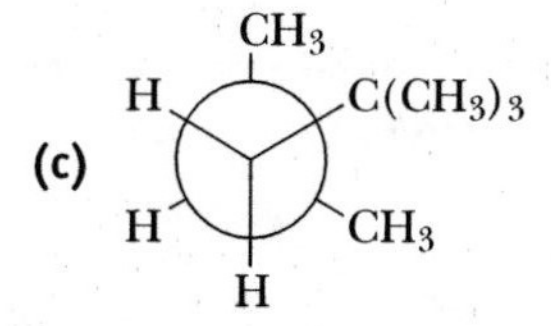

3.31 Explain why the following are not different conformations of 3-hexene:

H H H H

3.32 Which of the following two conformations is the more stable? (*Hint:* Use molecular models to compare structures):

H H CH3 H3C CH3 CH3 — rotate by 180° about the indicated bond → H H CH3 H3C CH3 CH3

(a) (b)

3.33 Determine whether the following pairs of structures in each set represent the same molecule or constitutional isomers, and if they are the same molecule, determine whether they are in the same or different conformations:

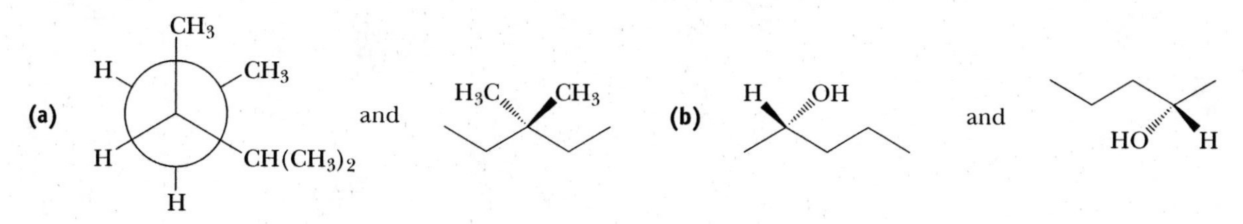

(c) CH_3, H_3C and H_3C, CH_3 **(d)** H_3C, $C(CH_3)_3$, H, H, H, H and

Cis–trans Isomerism in Cycloalkanes

3.34 What structural feature of cycloalkanes makes cis–trans isomerism in them possible?

3.35 Is cis–trans isomerism possible in alkanes?

3.36 Name and draw structural formulas for the cis and trans isomers of 1,2-dimethylcyclopropane.

3.37 Name and draw structural formulas for all cycloalkanes with molecular formula C_5H_{10}. Be certain to include cis–trans isomers, as well as constitutional isomers.

3.38 Using a planar pentagon representation for the cyclopentane ring, draw structural formulas for the cis and trans isomers of

(a) 1,2-Dimethylcyclopentane **(b)** 1,3-Dimethylcyclopentane

3.39 Draw the alternative chair conformations for the cis and trans isomers of 1,2-dimethylcyclohexane, 1,3-dimethylcyclohexane, and 1,4-dimethylcyclohexane.

(a) Indicate by a label whether each methyl group is axial or equatorial.

(b) For which isomer(s) are the alternative chair conformations of equal stability?

(c) For which isomer(s) is one chair conformation more stable than the other?

3.40 Use your answers from Problem 3.39 to complete the following table, showing correlations between cis, trans isomers and axial, equatorial positions for disubstituted derivatives of cyclohexane:

Position of Substitution	cis	trans
1,4-	a,e or e,a	e,e or a,a
1,3-	____ or ____	____ or ____
1,2-	____ or ____	____ or ____

3.41 There are four cis–trans isomers of 2-isopropyl-5-methylcyclohexanol:

OH, 1, 5, 2

2-Isopropyl-5-methylcyclohexanol

(a) Using a planar hexagon representation for the cyclohexane ring, draw structural formulas for these four isomers.

(b) Draw the more stable chair conformation for each of your answers in Part (a).

(c) Of the four cis–trans isomers, which is the most stable? If you answered this part correctly, you picked the isomer found in nature and given the name menthol.

Peppermint plant (*Mentha piperita*), a source of menthol, is a perennial herb with aromatic qualities used in candies, gums, hot and cold beverages, and garnish for punch and fruit. (*John Kaprielian/ Photo Researchers, Inc.*)

3.42 Draw alternative chair conformations for each substituted cyclohexane, and state which chair is the more stable:

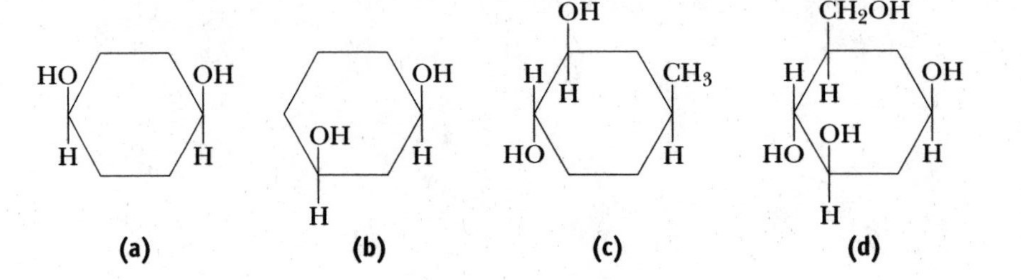

3.43 What kinds of conformations do the six-membered rings exhibit in adamantane?

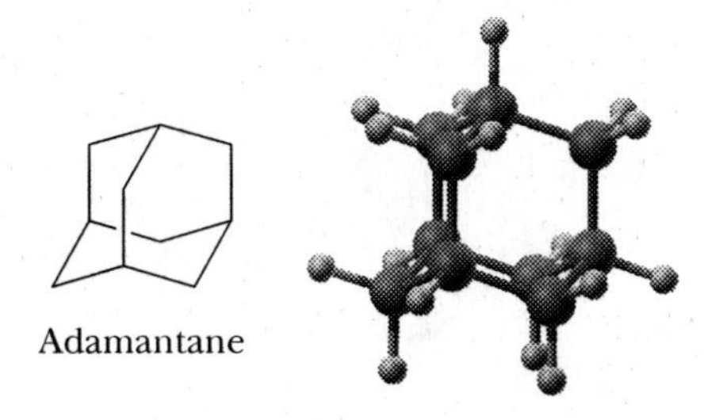
Adamantane

Physical Properties of Alkanes and Cycloalkanes

3.44 In Problem 3.20, you drew structural formulas for all constitutional isomers with molecular formula C_7H_{16}. Predict which isomer has the lowest boiling point and which has the highest.

3.45 What generalizations can you make about the densities of alkanes relative to that of water?

3.46 What unbranched alkane has about the same boiling point as water? (See Table 3.4.) Calculate the molecular weight of this alkane, and compare it with that of water.

3.47 As you can see from Table 3.4, each CH_2 group added to the carbon chain of an alkane increases the boiling point of the alkane. The increase is greater going from CH_4 to C_2H_6 and from C_2H_6 to C_3H_8 than it is from C_8H_{18} to C_9H_{20} or from C_9H_{20} to $C_{10}H_{22}$. What do you think is the reason for this trend?

3.48 Dodecane, $C_{12}H_{26}$, is an unbranched alkane. Predict the following:

(a) Will it dissolve in water?
(b) Will it dissolve in hexane?
(c) Will it burn when ignited?
(d) Is it a liquid, solid, or gas at room temperature and atmospheric pressure?
(e) Is it more or less dense than water?

3.49 As stated in Section 3.9A, the wax found in apple skins is an unbranched alkane with molecular formula $C_{27}H_{56}$. Explain how the presence of this alkane prevents the loss of moisture from within an apple.

Reactions of Alkanes

3.50 Write balanced equations for the combustion of each hydrocarbon. Assume that each is converted completely to carbon dioxide and water.

(a) Hexane **(b)** Cyclohexane **(c)** 2-Methylpentane

3.51 Following are heats of combustion of methane and propane:

Hydrocarbon	Component of	ΔH° [kcal/mol (kJ/mol)]
CH_4	natural gas	−212 (−886)
$CH_3CH_2CH_3$	LPG	−530 (−2220)

On a gram-for-gram basis, which of these hydrocarbons is the better source of heat energy?

3.52 When ethanol is added to gasoline to produce gasohol, the ethanol promotes more complete combustion of the gasoline and is an octane booster (Section 3.11B). Compare the heats of combustion of 2,2,4-trimethylpentane (1304 kcal/mol) and ethanol (327 kcal/mol). Which has the higher heat of combustion in kcal/mol? in kcal/g?

Looking Ahead

3.53 Explain why 1,2-dimethylcyclohexane can exist as cis–trans isomers, while 1,2-dimethylcyclododecane cannot.

3.54 On the left is a representation of the glucose molecule (we discuss the structure and chemistry of glucose in Chapter 18):

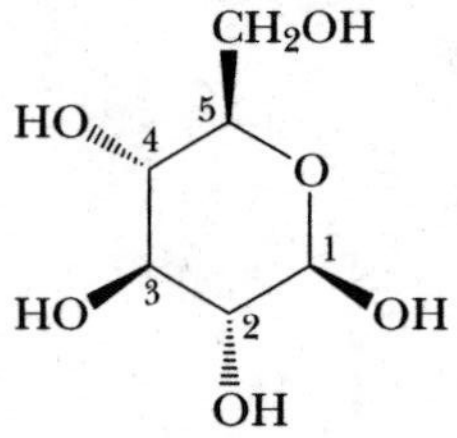

CH_2OH HO 4 5 O 1 HO 3 2 OH OH

Glucose

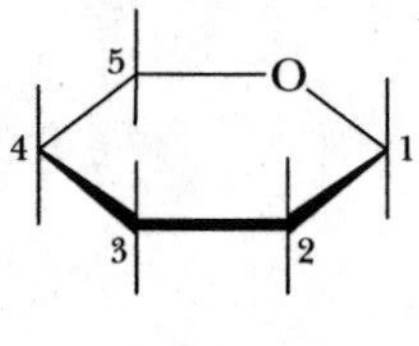

(a)

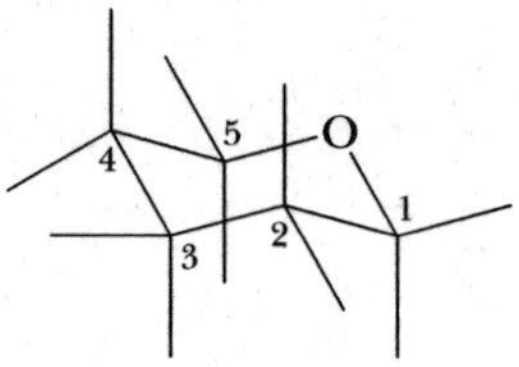

(b)

(a) Convert this representation to a planar hexagon representation.

(b) Convert this representation to a chair conformation. Which substituent groups in the chair conformation are equatorial? Which are axial?

3.55 Following is the structural formula of cholic acid (Section 21.5A), a component of human bile whose function is to aid in the absorption and digestion of dietary fats:

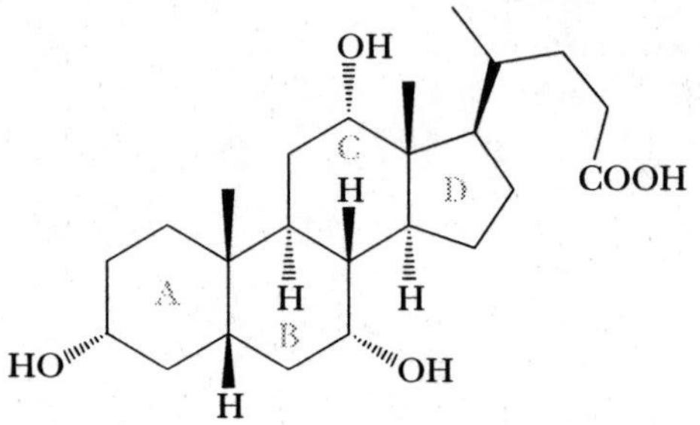

Cholic acid

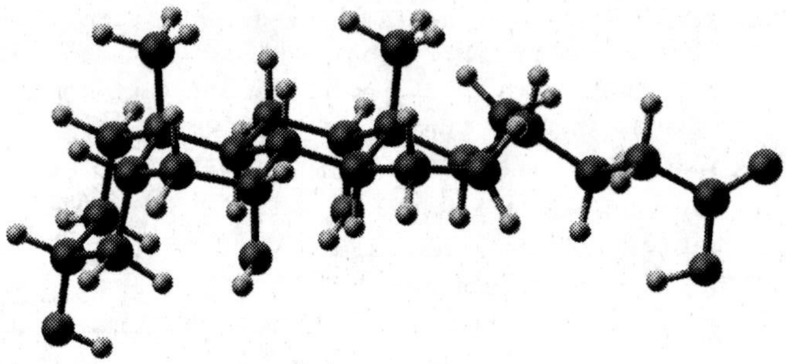

(a) What is the conformation of ring A? of ring B? of ring C? of ring D?

(b) There are hydroxyl groups on rings A, B, and C. Tell whether each is axial or equatorial.

(c) Is the methyl group at the junction of rings A and B axial or equatorial to ring A? Is it axial or equatorial to ring B?

(d) Is the methyl group at the junction of rings C and D axial or equatorial to ring C?

3.56 Following is the structural formula and ball-and-stick model of cholestanol:

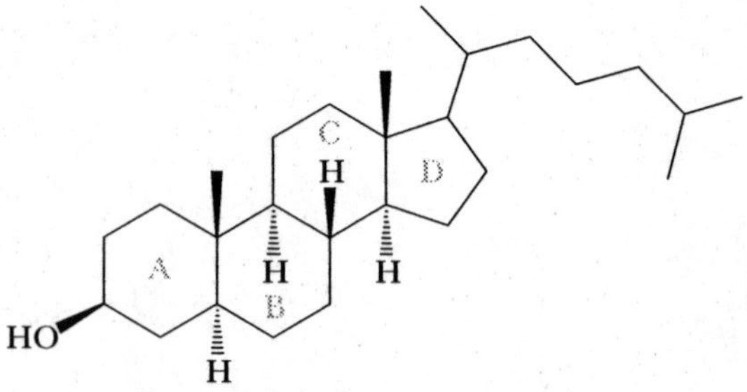

Cholestanol

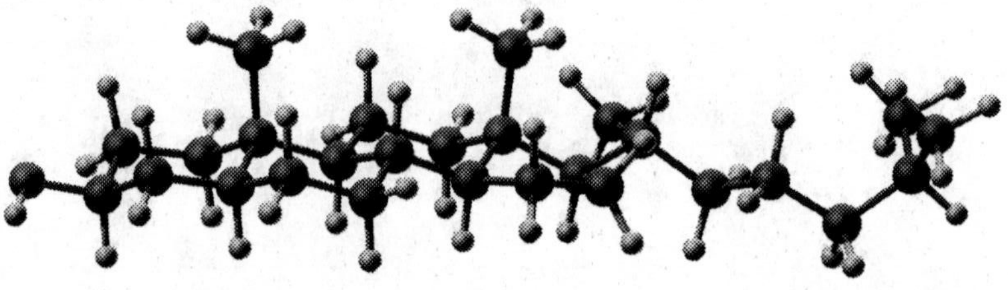

The only difference between this compound and cholesterol (Section 21.5A) is that cholesterol has a carbon–carbon double bond in ring B.

(a) Describe the conformation of rings A, B, C, and D in cholestanol.

(b) Is the hydroxyl group on ring A axial or equatorial?

(c) Consider the methyl group at the junction of rings A and B. Is it axial or equatorial to ring A? Is it axial or equatorial to ring B?

(d) Is the methyl group at the junction of rings C and D axial or equatorial to ring C?

3.57 As we have seen in Section 3.5, the IUPAC system divides the name of a compound into a prefix (showing the number of carbon atoms), an infix (showing the presence of carbon–carbon single, double, or triple bonds), and a suffix (showing the presence of an alcohol, amine, aldehyde, ketone, or carboxylic acid). Assume for the purposes of this problem that, to be alcohol (-ol) or amine (-amine), the hydroxyl or amino group must be bonded to a tetrahedral (sp^3 hybridized) carbon atom.

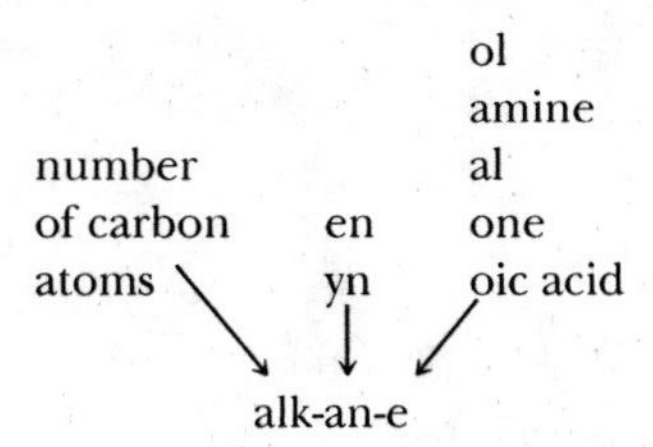

Given this information, write the structural formula of a compound with an unbranched chain of four carbon atoms that is an:

(a) alkane
(b) alkene
(c) alkyne
(d) alkanol
(e) alkenol
(f) alkynol
(g) alkanamine
(h) alkenamine
(i) alkynamine
(j) alkanal
(k) alkenal
(l) alkynal
(m) alkanone
(n) alkenone
(o) alkynone
(p) alkanoic acid
(q) alkenoic acid
(r) alkynoic acid

(Note: There is only one structural formula possible for some parts of this problem. For other parts, two or more structural formulas are possible. Where two are more are possible, we will deal with how the IUPAC system distinguishes among them when we come to the chapters on those particular functional groups.)

4 Alkenes and Alkynes

Carotene and carotene-like molecules are partnered with chlorophyll in nature to assist in the harvest of sunlight. In autumn, green chlorophyll molecules are destroyed and the yellows and reds of carotene and related molecules become visible. The red color of tomatoes comes from lycopene, a molecule closely related to carotene. See Problems 4.33 and 4.34. Inset: A model of β-carotene. *(Charles D. Winters)*

4.1 INTRODUCTION

In this chapter, we begin our study of unsaturated hydrocarbons. A hydrocarbon is unsaturated when it has fewer hydrogens bonded to carbon than an alkane has. There are three classes of unsaturated hydrocarbons: alkenes, alkynes, and arenes. **Alkenes** contain one or more carbon–carbon double bonds, and **alkynes** contain one or more carbon–carbon triple bonds. Ethene (ethylene) is the simplest alkene, and ethyne (acetylene) is the simplest alkyne:

Alkene An unsaturated hydrocarbon that contains a carbon–carbon double bond.

Alkyne An unsaturated hydrocarbon that contains a carbon–carbon triple bond.

$H_2C{=}CH_2$ — Ethene (an alkene)

$H{-}C{\equiv}C{-}H$ — Ethyne (an alkyne)

CHEMICAL CONNECTIONS 4A

Ethylene, a Plant Growth Regulator

As we have noted, ethylene occurs only in trace amounts in nature. Still, scientists have discovered that this small molecule is a natural ripening agent for fruits. Thanks to this knowledge, fruit growers can pick fruit while it is green and less susceptible to bruising. Then, when they are ready to pack the fruit for shipment, the growers can treat it with ethylene gas to induce ripening. Alternatively, the fruit can be treated with ethephon (Ethrel), which slowly releases ethylene and initiates ripening.

Ethephon $Cl-CH_2-CH_2-P(=O)(OH)-OH$

The next time you see ripe bananas in the market, you might wonder when they were picked and whether their ripening was artificially induced.

Arenes are the third class of unsaturated hydrocarbons. The simplest arene is benzene:

Arene A compound containing one or more benzene rings.

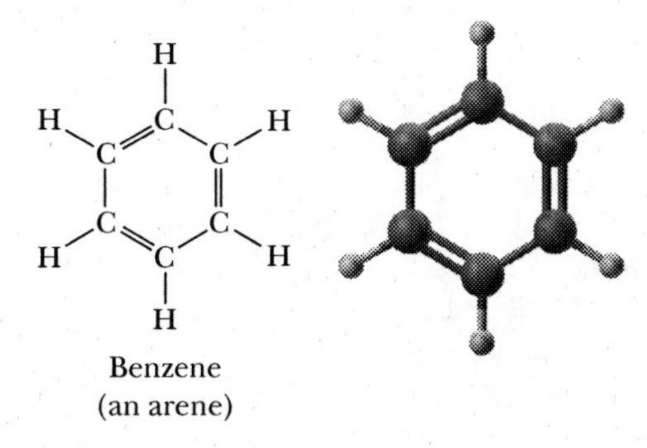

Benzene
(an arene)

The chemistry of benzene and its derivatives is quite different from that of alkenes and alkynes. Even though we do not study the chemistry of arenes until Chapter 9, we will show structural formulas of compounds containing benzene rings in earlier chapters. What you need to remember at this point is that a benzene ring is not chemically reactive under any of the conditions we describe in Chapters 4–8.

Compounds containing carbon–carbon double bonds are especially widespread in nature. Furthermore, several low-molecular-weight alkenes, including ethylene and propene, have enormous commercial importance in our modern, industrialized society. The organic chemical industry produces more pounds of ethylene worldwide than any other chemical. Annual production in the United States alone exceeds 55 billion pounds.

What is unusual about ethylene is that it occurs only in trace amounts in nature. The enormous amounts of it required to meet the needs of the chemical industry are derived the world over by thermal cracking of hydrocarbons. In the United States and other areas of the world with vast reserves of natural gas, the major process for the production of ethylene is thermal cracking of the small quantities of ethane extracted from natural gas (Section 3.11A):

$$\underset{\text{Ethane}}{CH_3CH_3} \xrightarrow[\text{(thermal cracking)}]{800-900^\circ C} \underset{\text{Ethylene}}{CH_2{=}CH_2} + H_2$$

Europe, Japan, and other areas of the world with limited supplies of natural gas depend almost entirely on thermal cracking of petroleum for their ethylene.

The crucial point to recognize is that ethylene and all of the commercial and industrial products made from it are derived from either natural gas or petroleum—both nonrenewable natural resources!

4.2 STRUCTURE

A. Shapes of Alkenes

Using the valence-shell electron-pair repulsion model (Section 1.4), we predict a value of 120° for the bond angles about each carbon in a double bond. The observed H—C—C bond angle in ethylene is 121.7°, a value close to that predicted by this model. In other alkenes, deviations from the predicted angle of 120° may be somewhat larger as a result of strain between groups bonded to one or both carbons of the double bond. The C—C—C bond angle in propene, for example, is 124.7°:

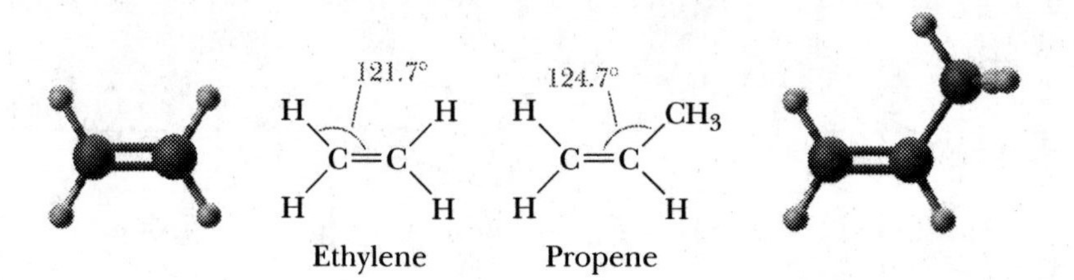

B. Orbital Overlap Model of a Carbon–Carbon Double Bond

In Section 1.7D, we described the formation of a carbon–carbon double bond in terms of the overlap of atomic orbitals. A carbon–carbon double bond consists of one sigma bond and one pi bond. Each carbon of the double bond uses its three sp^2 hybrid orbitals to form sigma bonds to three atoms. The unhybridized $2p$ atomic orbitals, which lie perpendicular to the plane created by the axes of the three sp^2 hybrid orbitals, combine to form the pi bond of the carbon–carbon double bond.

It takes approximately 63 kcal/mol (264 kJ/mol) to break the pi bond in ethylene; that is, to rotate one carbon by 90° with respect to the other so that no overlap occurs between $2p$ orbitals on adjacent carbons (Figure 4.1). This energy is considerably greater than the thermal energy available at room temperature, and, as a consequence, rotation about a carbon–carbon double bond is severely restricted. You might compare rotation about a carbon–carbon double bond, such as the bond in

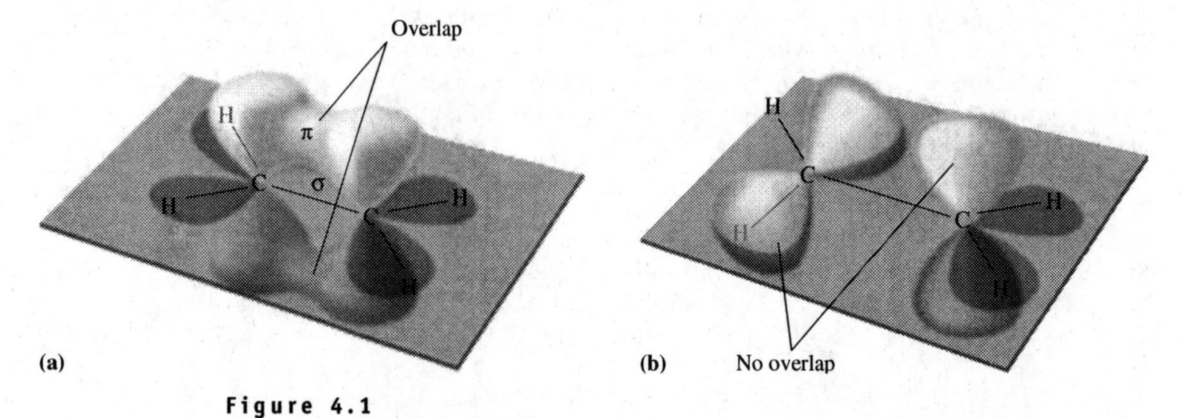

Figure 4.1
Restricted rotation about the carbon–carbon double bond in ethylene. (a) Orbital overlap model showing the pi bond. (b) The pi bond is broken by rotating the plane of one H—C—H group by 90° with respect to the plane of the other H—C—H group.

ethylene, with that about a carbon–carbon single bond, such as the bond in ethane (Section 3.7A), as follows: Whereas rotation about the carbon–carbon single bond in ethane is relatively free (the energy barrier is approximately 3 kcal/mol), rotation about the carbon–carbon double bond in ethylene is restricted (the energy barrier is approximately 63 kcal/mol).

C. Cis–Trans Isomerism in Alkenes

Cis–trans isomerism Isomers that have the same order of attachment of their atoms, but a different arrangement of their atoms in space due to the presence of either a ring (Chapter 3) or a carbon–carbon double bond (Chapter 4).

Because of restricted rotation about a carbon–carbon double bond, an alkene in which each carbon of the double bond has two different groups bonded to it shows cis–trans isomerism. Consider, for example, 2-butene: In *cis*-2-butene, the two methyl groups are on the same side of the double bond; in *trans*-2-butene, the two methyl groups are on opposite sides of the double bond:

cis-2-Butene
mp −139°C, bp 4°C

trans-2-Butene
mp −106°C, bp 1°C

These two compounds cannot be converted into one another at room temperature because of the restricted rotation about the double bond; they are different compounds, with different physical and chemical properties.

Cis alkenes are less stable than their trans isomers because of nonbonded interaction strain between alkyl substituents on the same side of the double bond in the cis isomer, as can be seen in space-filling models of the cis and trans isomers of 2-butene. This is the same type of steric strain that results in the preference for equatorial methylcyclohexane over axial methylcyclohexane (Section 3.7B).

The combustion of acetylene yields energy that produces the very hot temperatures of an oxyacetylene torch. *(Charles D. Winters)*

D. Shapes of Alkynes

The functional group of an alkyne is a **carbon–carbon triple bond**. The simplest alkyne is ethyne, C_2H_2. Ethyne is a linear molecule; all of its bond angles are 180° (Figure 1.12).

According to the valence bond model (Section 1.7E), a triple bond is described in terms of the overlap of *sp* hybrid orbitals of adjacent carbons to form a sigma bond, the overlap of parallel $2p_y$ orbitals to form one pi bond, and the overlap of parallel $2p_z$ orbitals to form the second pi bond. In ethyne, each carbon forms a bond to a hydrogen by the overlap of an *sp* hybrid orbital of carbon with a 1*s* atomic orbital of hydrogen.

4.3 NOMENCLATURE

Alkenes are named using the IUPAC system, but, as we shall see, some are still referred to by their common names.

A. IUPAC Names

We form IUPAC names of alkenes by changing the **-an-** infix of the parent alkane to **-en-** (Section 3.6). Hence, $CH_2{=}CH_2$ is named ethene, and $CH_3CH{=}CH_2$ is named propene. In higher alkenes, where isomers exist that differ in the location of the double bond, we use a numbering system. We number the longest carbon chain that contains the double bond in the direction that gives the carbon atoms of the double bond the lower set of numbers. We then use the number of the first carbon of the double bond to show its location. We name branched or substituted alkenes in a manner similar to the way we name alkanes. We number the carbon atoms, locate the double bond, locate and name substituent groups, and name the main chain.

$\overset{6}{C}H_3\overset{5}{C}H_2\overset{4}{C}H_2\overset{3}{C}H_2\overset{2}{C}H=\overset{1}{C}H_2$ — 1-Hexene

$\overset{6}{C}H_3\overset{5}{C}H_2\overset{4}{C}H(CH_3)\overset{3}{C}H_2\overset{2}{C}H=\overset{1}{C}H_2$ — 4-Methyl-1-hexene

$\overset{5}{C}H_3\overset{4}{C}H_2\overset{3}{C}H(CH_3)\overset{2}{C}(CH_2CH_3)=\overset{1}{C}H_2$ — 2-Ethyl-3-methyl-1-pentene

Note that there is a six-carbon chain in 2-ethyl-3-methyl-1-pentene. However, because the longest chain that contains the double bond has only five carbons, the parent hydrocarbon is pentane, and we name the molecule as a disubstituted 1-pentene.

We form IUPAC names of alkynes by changing the **-an-** infix of the parent alkane to **-yn-** (Section 3.6). Thus, $HC{\equiv}CH$ is named ethyne, and $CH_3C{\equiv}CH$ is named propyne. The IUPAC system retains the name *acetylene*; therefore, there are two acceptable names for $HC{\equiv}CH$: *ethyne* and *acetylene*. Of these two names, *acetylene* is used much more frequently. For larger molecules, we number the longest carbon chain that contains the triple bond from the end that gives the triply bonded carbons the lower set of numbers. We indicate the location of the triple bond by the number of the first carbon of the triple bond.

$\overset{4}{C}H_3\overset{3}{C}H(CH_3)\overset{2}{C}{\equiv}\overset{1}{C}H$ — 3-Methyl-1-butyne

$\overset{1}{C}H_3\overset{2}{C}H_2\overset{3}{C}{\equiv}\overset{4}{C}\overset{5}{C}H_2\overset{6}{C}(CH_3)_2\overset{7}{C}H_3$ — 6,6-Dimethyl-3-heptyne

EXAMPLE 4.1

Write the IUPAC name of each unsaturated hydrocarbon:

(a) $CH_2{=}CH(CH_2)_5CH_3$

(b) $(H_3C)_2C{=}C(CH_3)H$

(c) $CH_3(CH_2)_2C{\equiv}CCH_3$

SOLUTION

(a) 1-Octene (b) 2-Methyl-2-butene (c) 2-Hexyne

Practice Problem 4.1

Write the IUPAC name of each unsaturated hydrocarbon:

(a) (b) (c)

B. Common Names

Despite the precision and universal acceptance of IUPAC nomenclature, some alkenes—particularly those of low molecular weight—are known almost exclusively by their common names, as illustrated by the common names of these alkenes:

	$CH_2{=}CH_2$	$CH_3CH{=}CH_2$	$CH_3C(CH_3){=}CH_2$
IUPAC name:	Ethene	Propene	2-Methylpropene
Common name:	Ethylene	Propylene	Isobutylene

C. Systems for Designating Configuration in Alkenes

The Cis–Trans System

The most common method for specifying the configuration of a disubstituted alkene uses the prefixes *cis* and *trans*. In this system, the orientation of the atoms of the parent chain determines whether the alkene is cis or trans. Following is a structural formula for the cis isomer of 4-methyl-2-pentene:

cis-4-Methyl-2-pentene

In this example, carbon atoms of the main chain (carbons 1 and 4) are on the same side of the double bond; therefore, the configuration of this alkene is cis.

EXAMPLE 4.2

Name each alkene, and, using the cis–trans system, show the configuration about each double bond:

SOLUTION

(a) The chain contains seven carbon atoms and is numbered from the end that gives the lower number to the first carbon of the double bond. The carbon atoms of the parent chain are on opposite sides of the double bond. The compound's name is *trans*-3-heptene.

(b) The longest chain contains seven carbon atoms and is numbered from the right, so that the first carbon of the double bond is carbon 3 of the chain. The carbon atoms of the parent chain are on the same side of the double bond. The compound's name is *cis*-6-methyl-3-heptene.

Practice Problem 4.2

Name each alkene, and, using the cis–trans system, specify its configuration:

E,Z system A system used to specify the configuration of groups about a carbon–carbon double bond.

Z From the German *zusammen*, together; specifies that groups of higher priority on the carbons of a double bond are on the same side.

E From the German *entgegen*, opposite; specifies that groups of higher priority on the carbons of a double bond are on opposite sides.

E,Z : for 3 + 4 substituted alkenes

The E,Z System

The **E,Z system** must be used for tri- and tetrasubstituted alkenes. This system uses a set of rules to assign priorities to the substituents on each carbon of a double bond. If the groups of higher priority are on the same side of the double bond, the configuration of the alkene is **Z** (German: *zusammen*, together). If the groups of higher priority are on opposite sides of the double bond, the configuration is **E** (German: *entgegen*, opposite).

higher, higher / C=C / lower, lower — **Z** (zusammen)

higher, lower / C=C / lower, higher — **E** (entgegen)

The first step in assigning an E or a Z configuration to a double bond is to label the two groups bonded to each carbon in order of priority.

Priority Rules

1. Priority is based on atomic number: The higher the atomic number, the higher is the priority. Following are several substituents arranged in order of increasing priority (the atomic number of the atom determining priority is shown in parentheses):

$$\underset{(1)}{-H},\ \underset{(6)}{-CH_3},\ \underset{(7)}{-NH_2},\ \underset{(8)}{-OH},\ \underset{(16)}{-SH},\ \underset{(17)}{-Cl},\ \underset{(35)}{-Br},\ \underset{(53)}{-I}$$

Increasing priority →

2. If priority cannot be assigned on the basis of the atoms that are bonded directly to the double bond, look at the next set of atoms, and continue until a priority can be assigned. Priority is assigned at the first point of difference. Following is a series of groups, arranged in order of increasing priority (again, numbers in parentheses give the atomic number of the atom on which the assignment of priority is based):

$$-CH_2-\underset{(1)}{H}\quad -CH_2-\underset{(6)}{CH_3}\quad -CH_2-\underset{(7)}{NH_2}\quad -CH_2-\underset{(8)}{OH}\quad -CH_2-\underset{(17)}{Cl}$$

Increasing priority →

3. In order to compare carbons that are not sp^3 hybridized, the carbons must be manipulated in a way that allows us to maximize the number of groups bonded to them. Thus, we treat atoms participating in a double or triple bond as if they are bonded to an equivalent number of similar atoms by single bonds; that is, atoms of a double bond are replicated. Accordingly,

$$-CH{=}CH_2 \xrightarrow{\text{is treated as}} -\overset{C}{\overset{|}{C}H}-\overset{C}{\overset{|}{C}H_2} \quad \text{and} \quad -\overset{O}{\overset{\|}{C}H} \xrightarrow{\text{is treated as}} -\overset{O-C}{\overset{|}{\underset{H}{\underset{|}{C}}}}-O$$

EXAMPLE 4.3

Assign priorities to the groups in each set:

(a) $-\overset{\text{O}}{\overset{\|}{\text{C}}}\text{OH}$ and $-\overset{\text{O}}{\overset{\|}{\text{C}}}\text{H}$ (b) $-CH_2NH_2$ and $-\overset{\text{O}}{\overset{\|}{\text{C}}}\text{OH}$

SOLUTION

(a) The first point of difference is the O of the —OH in the carboxyl group, compared with the —H in the aldehyde group. The carboxyl group is higher in priority:

$$-\overset{\text{O}}{\overset{\|}{\text{C}}}-\text{O}-\text{H} \qquad -\overset{\text{O}}{\overset{\|}{\text{C}}}-\text{H}$$

Carboxyl group (higher priority) Aldehyde group (lower priority)

(b) Oxygen has a higher priority (higher atomic number) than nitrogen. Therefore, the carboxyl group has a higher priority than the primary amino group:

$$-CH_2NH_2 \qquad -\overset{\text{O}}{\overset{\|}{\text{C}}}\text{OH}$$

lower priority higher priority

EXAMPLE 4.4

Name each alkene and specify its configuration by the E,Z system:

(a) $(H)(H_3C)C{=}C(CH_3)(CH(CH_3)_2)$ (b) $(Cl)(H_3C)C{=}C(H)(CH_2CH_3)$

SOLUTION

(a) The group of higher priority on carbon 2 is methyl; that of higher priority on carbon 3 is isopropyl. Because the groups of higher priority are on the same side of the carbon–carbon double bond, the alkene has the Z configuration. Its name is (*Z*)-3,4-dimethyl-2-pentene.

(b) Groups of higher priority on carbons 2 and 3 are —Cl and $-CH_2CH_3$. Because these groups are on opposite sides of the double bond, the configuration of this alkene is E, and its name is (*E*)-2-chloro-2-pentene.

Practice Problem 4.3

Name each alkene and specify its configuration by the E,Z system:

(a) Cl (b) Cl Br (c)

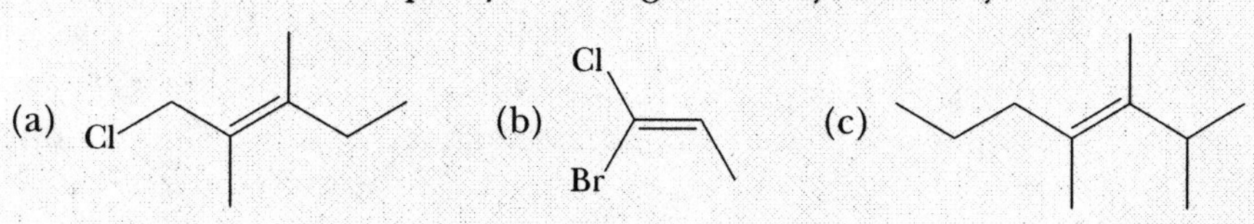

D. Cycloalkenes

In naming **cycloalkenes**, we number the carbon atoms of the ring double bond 1 and 2 in the direction that gives the substituent encountered first the smaller number. We name and locate substituents and list them in alphabetical order, as in the following compounds:

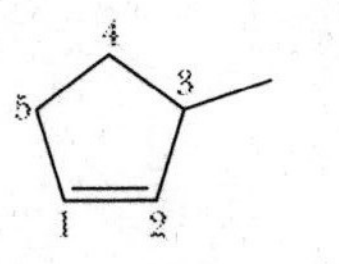

3-Methylcyclopentene (not 5-methylcyclopentene)

4-Ethyl-1-methylcyclohexene (not 5-ethyl-2-methylcyclohexene)

EXAMPLE 4.5

Write the IUPAC name for each cycloalkene:

(a) (b) (c)

SOLUTION

(a) 3,3-Dimethylcyclohexene
(b) 1,2-Dimethylcyclopentene
(c) 4-Isopropyl-1-methylcyclohexene

Practice Problem 4.4

Write the IUPAC name for each cycloalkene:

(a) (b) (c)

E. Cis–Trans Isomerism in Cycloalkenes

Following are structural formulas for four cycloalkenes:

Cyclopentene Cyclohexene Cycloheptene Cyclooctene

In these representations, the configuration about each double bond is cis. Because of angle strain, it is not possible to have a trans configuration in cycloalkenes of seven or fewer carbons. To date, *trans*-cyclooctene is the smallest *trans*-cycloalkene that has been prepared in pure form and is stable at room temperature. Yet, even in this *trans*-

cycloalkene, there is considerable intramolecular strain. *cis*-Cyclooctene is more stable than its trans isomer by 9.1 kcal/mol (38 kJ/mol).

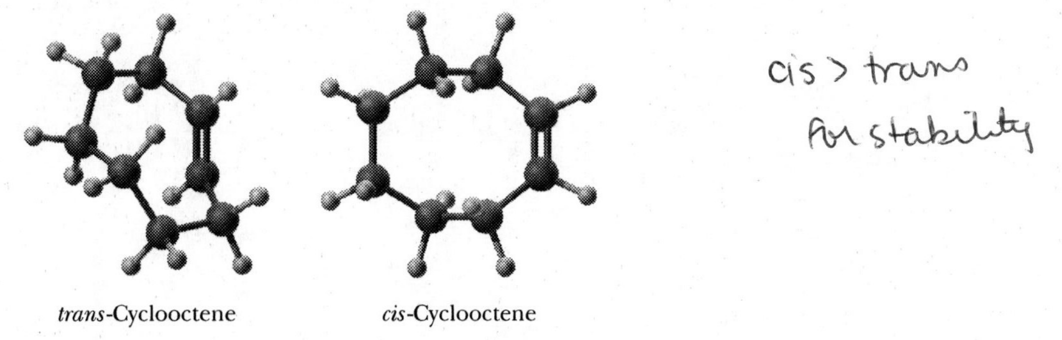

trans-Cyclooctene *cis*-Cyclooctene

F. Dienes, Trienes, and Polyenes

We name alkenes that contain more than one double bond as alkadienes, alkatrienes, and so on. We refer to those that contain several double bonds more generally as polyenes (Greek: *poly*, many). Following are three examples of dienes:

$CH_2{=}CHCH_2CH{=}CH_2$
1,4-Pentadiene

$CH_2{=}C(CH_3)CH{=}CH_2$
2-Methyl-1,3-butadiene
(Isoprene)

1,3-Cyclopentadiene

G. Cis–Trans Isomerism in Dienes, Trienes, and Polyenes

Thus far, we have considered cis–trans isomerism in alkenes containing only one carbon–carbon double bond. For an alkene with one carbon–carbon double bond that can show cis–trans isomerism, two cis–trans isomers are possible. For an alkene with ***n*** carbon–carbon double bonds, each of which can show cis–trans isomerism, $\mathbf{2^n}$ cis–trans isomers are possible.

EXAMPLE 4.6

How many cis–trans isomers are possible for 2,4-heptadiene?

SOLUTION

This molecule has two carbon–carbon double bonds, each of which exhibits cis–trans isomerism. As the following table shows, $2^2 = 4$ cis–trans isomers are possible (to the right of the table are line angle-formulas for two of these isomers):

Double bond	
$C_2{-}C_3$	$C_4{-}C_5$
trans	trans
trans	cis
cis	trans
cis	cis

trans,trans-2,4-Heptadiene *trans,cis*-2,4-Heptadiene

Practice Problem 4.5

Draw structural formulas for the other two cis–trans isomers of 2,4-heptadiene.

EXAMPLE 4.7

Draw all possible cis–trans isomers for the following unsaturated alcohol:

$$\begin{array}{c} \quad CH_3 \qquad\qquad\quad CH_3 \\ \quad | \qquad\qquad\qquad | \\ CH_3C{=}CHCH_2CH_2C{=}CHCH_2OH \end{array}$$

SOLUTION

Cis–trans isomerism is possible only about the double bond between carbons 2 and 3 of the chain. It is not possible for the other double bond, because carbon 7 has two identical groups on it. Thus, $2^1 = 2$ cis–trans isomers are possible. The trans isomer of this alcohol, named geraniol, is a major component of the oils of rose, citronella, and lemongrass.

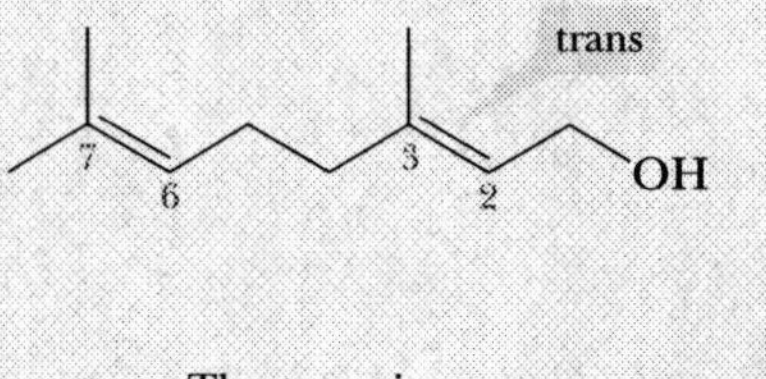

trans 7 6 3 2 OH cis 3 2 OH

The trans isomer 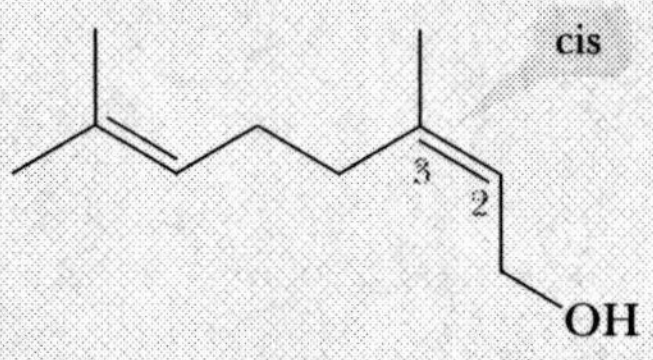 The cis isomer

Practice Problem 4.6

How many cis–trans isomers are possible for the following unsaturated alcohol?

$$\begin{array}{c} \quad CH_3 \qquad\qquad\quad CH_3 \qquad\qquad\quad CH_3 \\ \quad | \qquad\qquad\qquad | \qquad\qquad\qquad | \\ CH_3C{=}CHCH_2CH_2C{=}CHCH_2CH_2C{=}CHCH_2OH \end{array}$$

Vitamin A is an example of a biologically important compound for which a number of cis–trans isomers are possible. There are four carbon–carbon double bonds in the chain of carbon atoms bonded to the substituted cyclohexene ring, and each has the potential for cis–trans isomerism. Thus, $2^4 = 16$ cis–trans isomers are possible for this structural formula. Vitamin A is the all-trans isomer. The enzyme-catalyzed oxidation of vitamin A converts the primary hydroxyl group to an aldehyde group, to give retinal, the biologically active form of the vitamin:

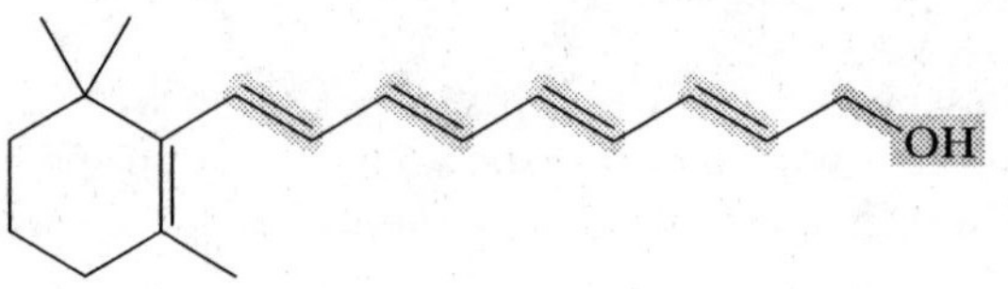

OH

Vitamin A (retinol)

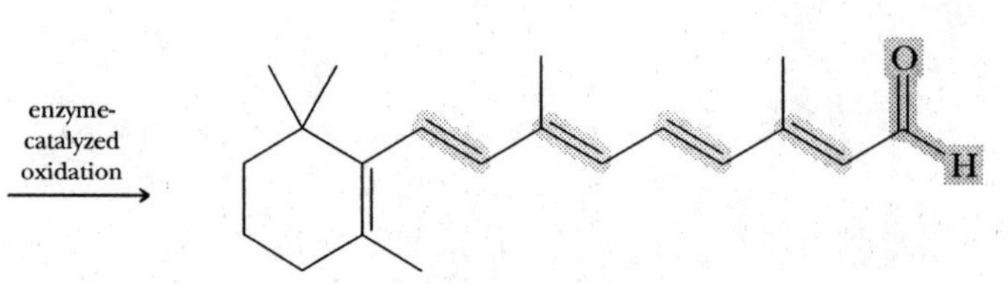

enzyme-catalyzed oxidation → O H

Vitamin A aldehyde (retinal)

CHEMICAL CONNECTIONS 4B

Cis–Trans Isomerism in Vision

The retina—the light-detecting layer in the back of our eyes—contains reddish compounds called *visual pigments*. Their name, *rhodopsin*, is derived from the Greek word meaning "rose colored." Each rhodopsin molecule is a combination of one molecule of a protein called opsin and one molecule of 11-*cis*-retinal, a derivative of vitamin A in which the CH_2OH group of the vitamin is converted to an aldehyde group, —CHO, and the double bond between carbons 11 and 12 of the side chain is in the less stable cis configuration. When rhodopsin absorbs light energy, the less stable 11-cis double bond is converted to the more stable 11-trans double bond. This isomerization changes the shape of the rhodopsin molecule, which in turn causes the neurons of the optic nerve to fire and produce a visual image:

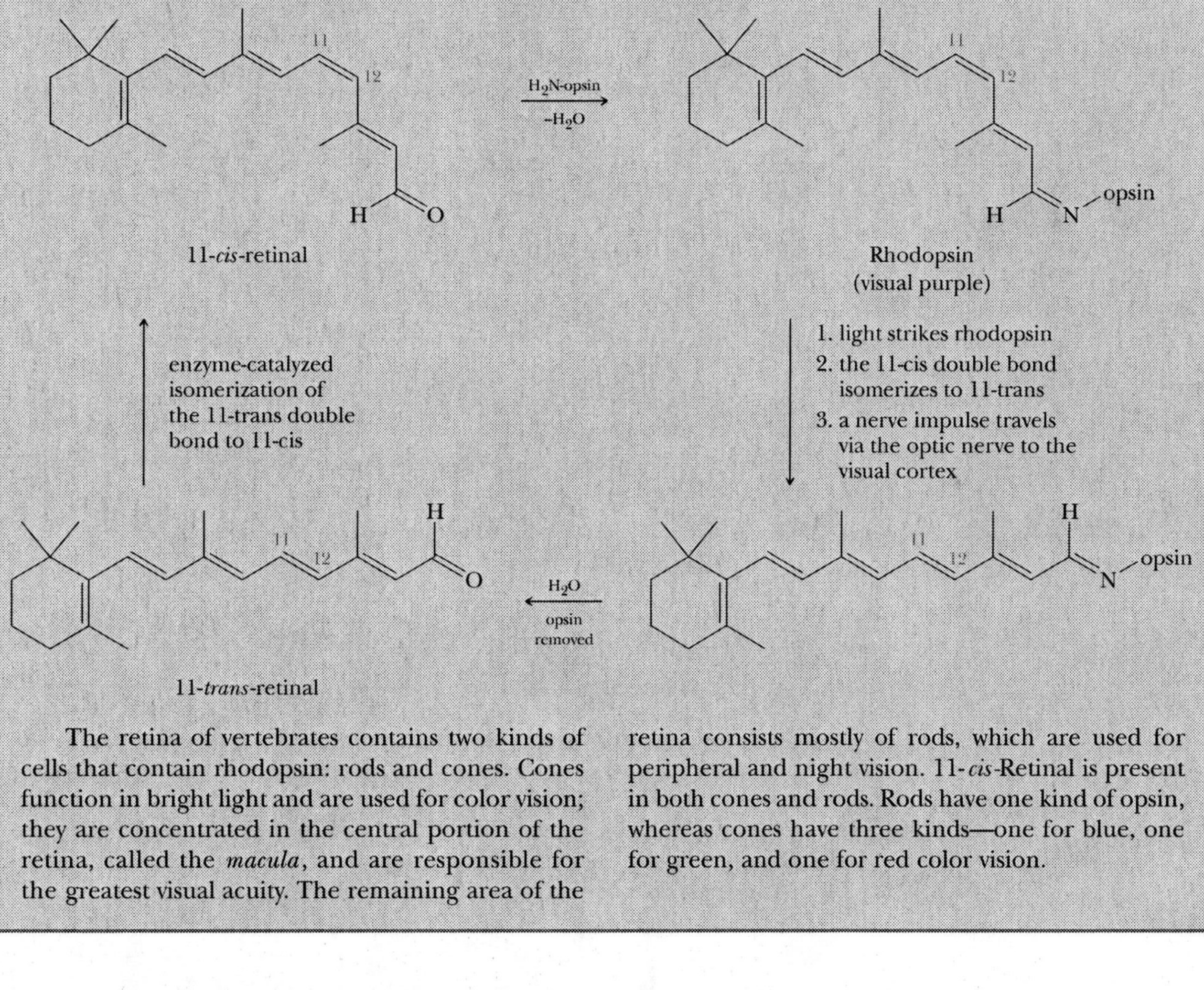

11-*cis*-retinal → (H_2N-opsin, $-H_2O$) → Rhodopsin (visual purple)

Rhodopsin → 1. light strikes rhodopsin 2. the 11-cis double bond isomerizes to 11-trans 3. a nerve impulse travels via the optic nerve to the visual cortex

→ (H_2O, opsin removed) → 11-*trans*-retinal

enzyme-catalyzed isomerization of the 11-trans double bond to 11-cis

The retina of vertebrates contains two kinds of cells that contain rhodopsin: rods and cones. Cones function in bright light and are used for color vision; they are concentrated in the central portion of the retina, called the *macula*, and are responsible for the greatest visual acuity. The remaining area of the retina consists mostly of rods, which are used for peripheral and night vision. 11-*cis*-Retinal is present in both cones and rods. Rods have one kind of opsin, whereas cones have three kinds—one for blue, one for green, and one for red color vision.

4.4 PHYSICAL PROPERTIES

Alkenes and alkynes are nonpolar compounds, and the only attractive forces between their molecules are dispersion forces (Section 3.9B). Therefore, their physical properties are similar to those of alkanes (Section 3.9) with the same carbon skeletons. Alkenes and alkynes that are liquid at room temperature have densities less than 1.0 g/mL. Thus, they are less dense than water. Like alkanes, alkenes and alkynes are nonpolar and are soluble in each other. Because of their contrasting polarity with water, they do not dissolve in water. Instead, they form two layers when mixed with water or another polar organic liquid such as ethanol.

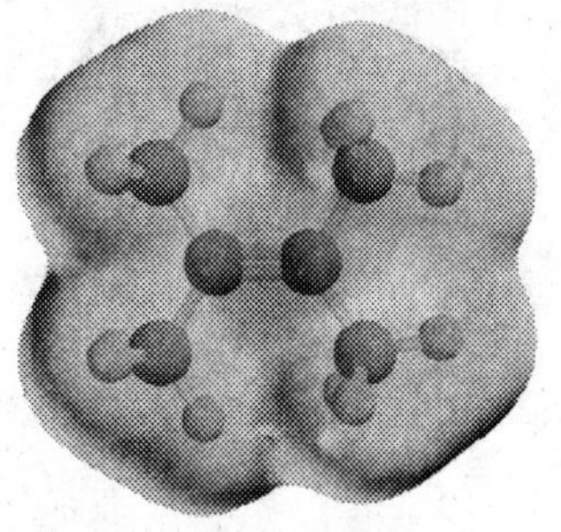
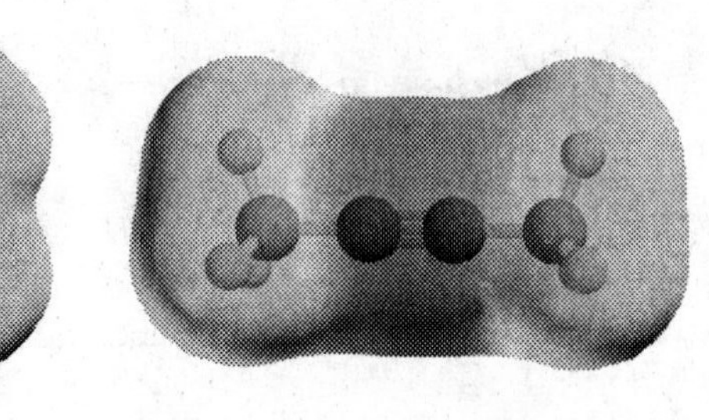

Tetramethylethylene and dimethylacetylene. Both a carbon double bond and a carbon-carbon triple bond are sites of high electron density and, therefore, sites of chemical reactivity.

EXAMPLE 4.8

Describe what will happen when 1-nonene is added to the following compounds:

(a) Water (b) 8-Methyl-1-nonyne

SOLUTION

(a) 1-Nonene is an alkene and, therefore, nonpolar. It will not dissolve in a polar solvent such as water. Water and 1-nonene will form two layers; water, which has the higher density will be the lower layer, and 1-nonene will be the upper layer.

(b) Because alkenes and alkynes are both nonpolar, they will dissolve in one another.

4.5 NATURALLY OCCURRING ALKENES: THE TERPENES

Terpene A compound whose carbon skeleton can be divided into two or more units identical with the carbon skeleton of isoprene.

Among the compounds found in the essential oils of plants is a group of substances called **terpenes,** all of which have in common the property that their carbon skeletons can be divided into two or more carbon units that are identical with the carbon skeleton of isoprene. Carbon 1 of an **isoprene unit** is called the head, and carbon 4 is called the tail. A terpene is a compound in which the tail of one isoprene unit becomes bonded to the head of another isoprene unit.

$$CH_2{=}C(CH_3){-}CH{=}CH_2$$

2-Methyl-1,3-butadiene (Isoprene)

$$\underset{1}{C}-\underset{2}{C}(C)-\underset{3}{C}-\underset{4}{C}$$

head → C1 ... C4 ← tail

Isoprene unit

Terpenes are among the most widely distributed compounds in the biological world, and a study of their structure provides a glimpse of the wondrous diversity that nature can generate from a simple carbon skeleton. Terpenes also illustrate an impor-

CHEMICAL CONNECTIONS 4C

Why Plants Emit Isoprene

Names like Virginia's *Blue Ridge*, Jamaica's *Blue Mountain Peak*, and Australia's *Blue Mountains* remind us of the bluish haze that hangs over wooded hills in the summertime. In the 1950s, it was discovered that this haze is rich in isoprene, which means that isoprene is far more abundant in the atmosphere than anyone thought. The haze is caused by the scattering of light from an aerosol produced by the photooxidation of isoprene and other hydrocarbons. Scientists now estimate that the global emission of isoprene by plants is 3×10^{11} kg/yr (3.3×10^{8} ton/yr), which represents approximately 2% of all carbon fixed by photosynthesis. A recent study of hydrocarbon emissions in the Atlanta area revealed that plants are by far the largest emitters of hydrocarbons, with plant-derived isoprene accounting for almost 60% of the total.

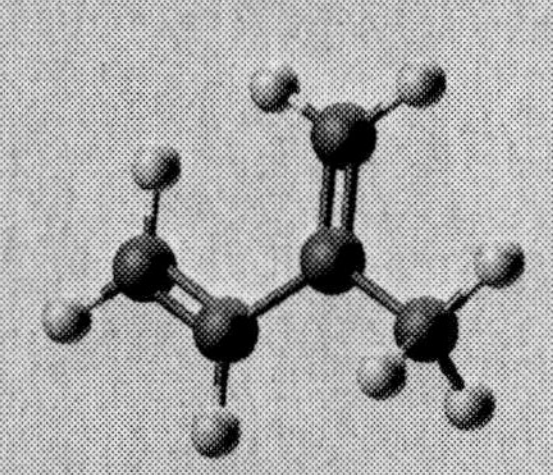

Why do plants emit so much isoprene into the atmosphere rather than use it for the synthesis of terpenes and other natural products? Tom Sharkey, a University of Wisconsin plant physiologist, found that the emission of isoprene is extremely sensitive to temperature. Plants grown at 20°C do not emit isoprene, but they begin to emit it when the temperature of their leaves increases to 30°C. In certain plants, isoprene emission can increase as much as tenfold for a 10°C increase in leaf temperature. Sharkey studied the relationship between temperature-induced leaf damage and isoprene concentration in leaves of the kudzu plant, a nonnative invasive vine. He discovered that leaf damage, as measured by the destruction of chlorophyll, begins to occur at 37.5°C in the absence of isoprene, but not until 45°C in its presence. Sharkey speculates that isoprene dissolves in leaf membranes and in some way increases their tolerance to heat stress. Because isoprene is made rapidly and also lost rapidly, its concentration correlates with temperature throughout the day.

The haze of the Smoky Mountains is caused by light-scattering from the aerosol produced by the photooxidation of isoprene and other hydrocarbons. See the Chemistry in Action box "Why Plants Emit Isoprene". Inset: A model of isoprene *(Digital Vision)*

tant principle of the molecular logic of living systems, namely, that, in building large molecules, small subunits are strung together enzymatically by an iterative process and are then chemically modified by precise enzyme-catalyzed reactions. Chemists use the same principles in the laboratory, but our methods do not have the precision and selectivity of the enzyme-catalyzed reactions of cellular systems.

Probably the terpenes most familiar to you, at least by odor, are components of the so-called essential oils obtained by steam distillation or ether extraction of various parts of plants. Essential oils contain the relatively low-molecular-weight substances that are in large part responsible for characteristic plant fragrances. Many essential oils, particularly those from flowers, are used in perfumes.

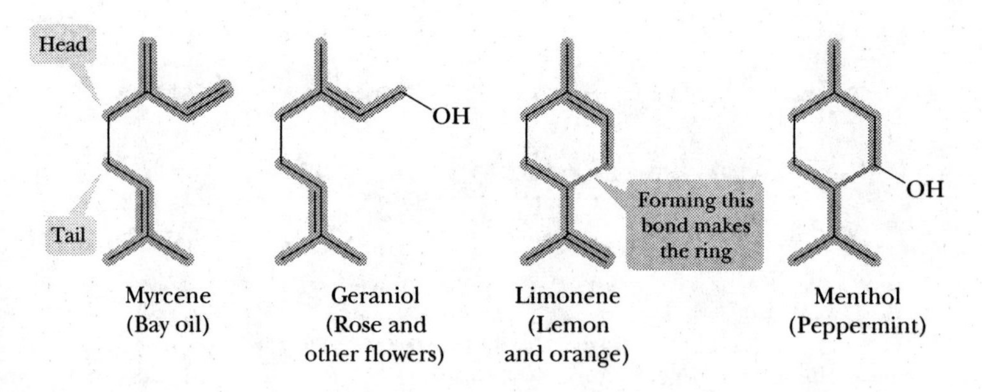

Figure 4.2
Four terpenes, each derived from two isoprene units (highlighted) bonded from the tail of the first unit to the head of the second unit. In limonene and menthol, he formation of an additional carbon–carbon bond creates a six-membered ring.

One example of a terpene obtained from an essential oil is myrcene, $C_{10}H_{16}$, a component of bayberry wax and oils of bay and verbena. Myrcene is a triene with a parent chain of eight carbon atoms and two one-carbon branches (Figure 4.2).

Farnesol, a terpene with molecular formula $C_{15}H_{26}O$, includes three isoprene units:

OH

Farnesol
(Lily-of-the-valley)

Derivatives of both farnesol and geraniol are intermediates in the biosynthesis of cholesterol (Section 21.5B).

Vitamin A (Section 4.3G), a terpene with molecular formula $C_{20}H_{30}O$, consists of four isoprene units linked head-to-tail and cross-linked at one point to form a six-membered ring.

SUMMARY

An **alkene** is an unsaturated hydrocarbon that contains a carbon–carbon double bond. Alkenes have the general formula C_nH_{2n}. An **alkyne** is an unsaturated hydrocarbon that contains a carbon–carbon triple bond. Alkynes have the general formula C_nH_{2n-2}. According to the **orbital overlap model** (Section 4.2B), a carbon–carbon double bond consists of one sigma bond formed by the overlap of sp^2 hybrid orbitals and one pi bond formed by the overlap of parallel $2p$ atomic orbitals. It takes approximately 63 kcal/mol (264 kJ/mol) to break the pi bond in ethylene. A carbon–carbon triple bond consists of one sigma bond formed by the overlap of sp hybrid orbitals and two pi bonds formed by the overlap of pairs of parallel $2p$ orbitals.

The structural feature that makes **cis–trans isomerism** possible in alkenes is restricted rotation about the two carbons of the double bond (Section 4.2C). To date, *trans*-cyclooctene is the smallest trans cycloalkene that has been prepared in pure form and is stable at room temperature.

According to the IUPAC system (Section 4.3A), we show the presence of a **carbon–carbon double bond** by changing the infix of the parent hydrocarbon from **-an-** to **-en-**. We show the presence of a **carbon–carbon triple bond** by changing the infix of the parent alkane from **-an-** to **-yn-**.

The orientation of the carbon atoms of the parent chain about the double bond determines whether an alkene is cis or trans (Section 4.3C). If atoms of the parent are on the same side of the double bond, the configuration of the alkene is cis; if they are on opposite sides, the configuration is trans. Using a set of priority rules, we can also specify the configuration of a carbon–carbon double bond by the **E,Z system** (Section 4.3C). If the two groups of higher priority are on the same side of the double bond, the configuration of the alkene is **Z** (German: *zusammen*, together); if they are on opposite sides, it is **E** (German: *entgegen*, opposite).

To name an alkene containing two or more double bonds, we change the infix to *-adien-*, *-atrien-*, and so on (Section 4.3F). Compounds containing several double bonds are called polyenes.

Alkenes and alkynes are nonpolar compounds, and the only interactions between their molecules are **dispersion forces**. The physical properties of alkenes and alkynes are similar to those of alkanes (Section 4.4).

The characteristic structural feature of a **terpene** (Section 4.5) is a carbon skeleton that can be divided into two or more **isoprene units**. Terpenes illustrate an important principle of the molecular logic of living systems, namely, that, in building large molecules, small subunits are strung together by an iterative process and are then chemically modified by precise enzyme-catalyzed reactions.

PROBLEMS

A problem number set in red indicates an applied real-world" problem.

Structure of Alkenes and Alkynes

4.7 Each carbon atom in ethane and in ethylene is surrounded by eight valence electrons and has four bonds to it. Explain how the VSEPR model (Section 1.4) predicts a bond angle of 109.5° about each carbon in ethane, but an angle of 120° about each carbon in ethylene.

4.8 Use the valence-shell electron-pair repulsion (VSEPR) model to predict all bond angles about each of the following highlighted carbon atoms:

(a) 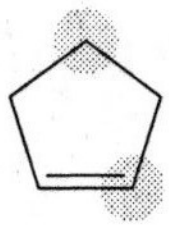 **(b)** 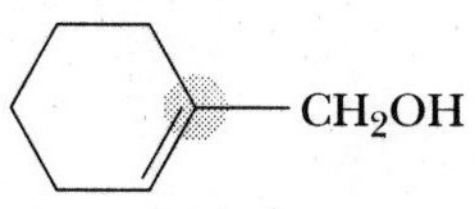 CH_2OH **(c)** $HC{\equiv}C{-}CH{=}CH_2$ **(d)**

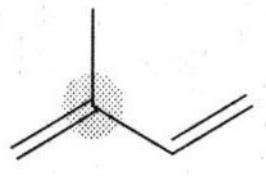

4.9 For each highlighted carbon atom in Problem 4.8, identify which orbitals are used to form each sigma bond and which are used to form each pi bond.

4.10 Predict all bond angles about each highlighted carbon atom:

(a) 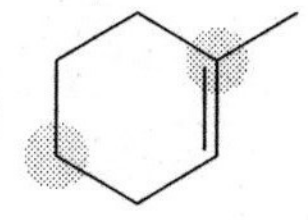 **(b)** 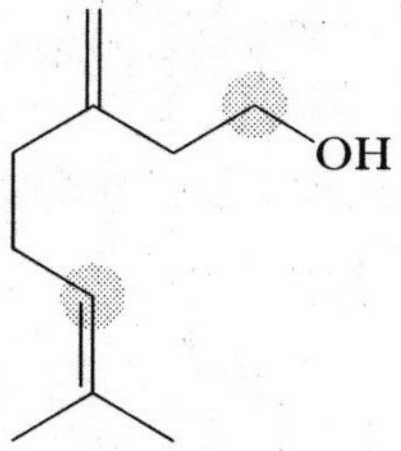 OH **(c)** 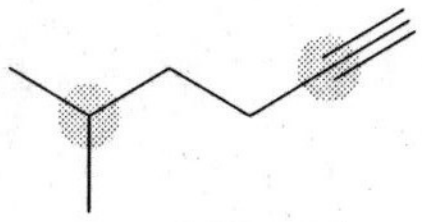 **(d)** 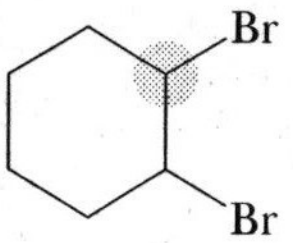 Br Br

4.11 For each highlighted carbon atom in Problem 4.10, identify which orbitals are used to form each sigma bond and which are used to form each pi bond.

4.12 Following is the structure of 1,2-propadiene (allene):

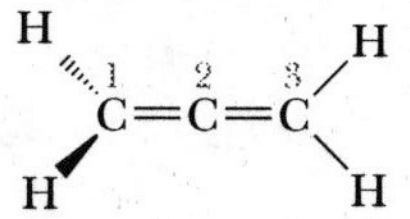

H H 1 2 3 C=C=C H H

1,2-Propadiene (Allene)

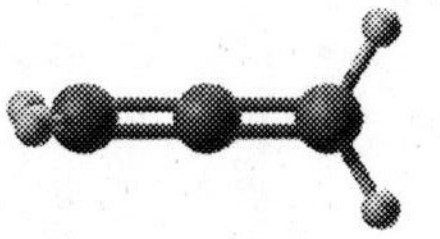

Ball-and-stick model

The plane created by H—C—H of carbon 1 is perpendicular to that created by H—C—H of carbon 3.

(a) State the orbital hybridization of each carbon in allene.

(b) Account for the molecular geometry of allene in terms of the orbital overlap model. Specifically, explain why all four hydrogen atoms are not in the same plane.

Nomenclature of Alkenes and Alkynes

4.13 Draw a structural formula for each compound:

(a) *trans*-2-Methyl-3-hexene **(b)** 2-Methyl-3-hexyne
(c) 2-Methyl-1-butene **(d)** 3-Ethyl-3-methyl-1-pentyne
(e) 2,3-Dimethyl-2-butene **(f)** *cis*-2-Pentene
(g) (*Z*)-1-Chloropropene **(h)** 3-Methylcyclohexene

4.14 Draw a structural formula for each compound:

(a) 1-Isopropyl-4-methylcyclohexene **(b)** (6*E*)-2,6-Dimethyl-2,6-octadiene
(c) *trans*-1,2-Diisopropylcyclopropane **(d)** 2-Methyl-3-hexyne
(e) 2-Chloropropene **(f)** Tetrachloroethylene

4.15 Write the IUPAC name for each compound:

(a) (b) (c) (d)

(e) (f) 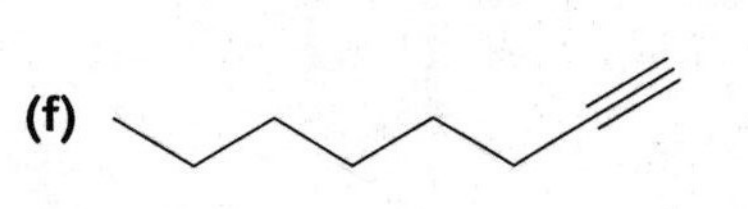 (g) 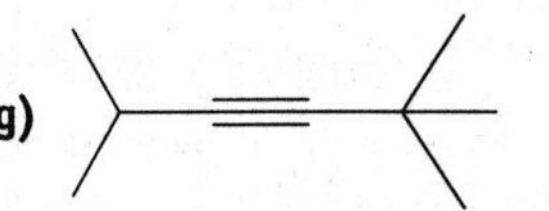 (h)

4.16 Explain why each name is incorrect, and then write a correct name for the intended compound:

(a) 1-Methylpropene **(b)** 3-Pentene
(c) 2-Methylcyclohexene **(d)** 3,3-Dimethylpentene
(e) 4-Hexyne **(f)** 2-Isopropyl-2-butene

4.17 Explain why each name is incorrect, and then write a correct name for the intended compound:

(a) 2-Ethyl-1-propene **(b)** 5-Isopropylcyclohexene
(c) 4-Methyl-4-hexene **(d)** 2-*sec*-Butyl-1-butene
(e) 6,6-Dimethylcyclohexene **(f)** 2-Ethyl-2-hexene

Cis–Trans Isomerism in Alkenes and Cycloalkenes

4.18 Which of these alkenes show cis–trans isomerism? For each that does, draw structural formulas for both isomers.

(a) 1-Hexene **(b)** 2-Hexene
(c) 3-Hexene **(d)** 2-Methyl-2-hexene
(e) 3-Methyl-2-hexene **(f)** 2,3-Dimethyl-2-hexene

4.19 Which of these alkenes show cis–trans isomerism? For each that does, draw structural formulas for both isomers.

(a) 1-Pentene **(b)** 2-Pentene
(c) 3-Ethyl-2-pentene **(d)** 2,3-Dimethyl-2-pentene
(e) 2-Methyl-2-pentene **(f)** 2,4-Dimethyl-2-pentene

4.20 Which alkenes can exist as pairs of cis-trans isomers? For each alkene that does, draw the trans isomer.

(a) $CH_2{=}CHBr$ **(b)** $CH_3CH{=}CHBr$
(c) $(CH_3)_2C{=}CHCH_3$ **(d)** $(CH_3)_2CHCH{=}CHCH_3$

4.21 There are three compounds with molecular formula $C_2H_2Br_2$. Two of these compounds have a dipole greater than zero, and one has no dipole. Draw structural formulas for the three compounds, and explain why two have dipole moments but the third one has none.

4.22 Name and draw structural formulas for all alkenes with molecular formula C_5H_{10}. As you draw these alkenes, remember that cis and trans isomers are different compounds and must be counted separately.

4.23 Name and draw structural formulas for all alkenes with molecular formula C_6H_{12} that have the following carbon skeletons (remember cis and trans isomers):

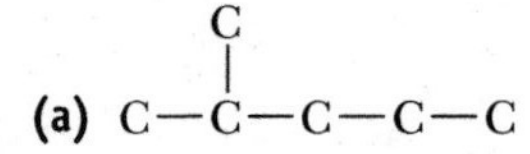

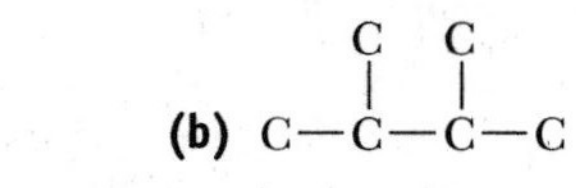

(c) C—C—C—C with two C on the second carbon

4.24 Arrange the groups in each set in order of increasing priority:
(a) $-CH_3$, $-Br$, $-CH_2CH_3$
(b) $-OCH_3$, $-CH(CH_3)_2$, $-CH_2CH_2NH_2$
(c) $-CH_2OH$, $-COOH$, $-OH$
(d) $-CH{=}CH_2$, $-CH{=}O$, $-CH(CH_3)_2$

4.25 Draw the structural formula for at least one bromoalkene with molecular formula C_5H_9Br that **(a)** shows E,Z isomerism and **(b)** does not show E,Z isomerism.

4.26 For each molecule that shows cis–trans isomerism, draw the cis isomer:

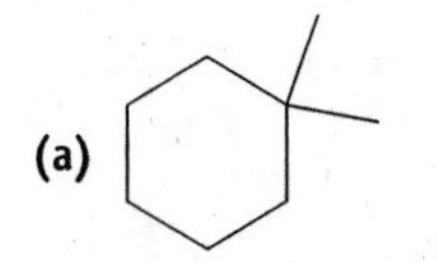

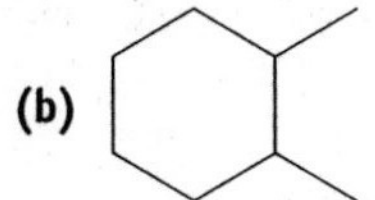

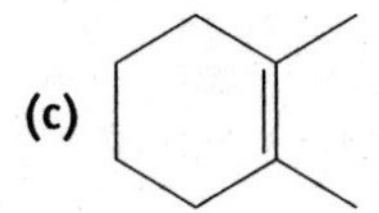

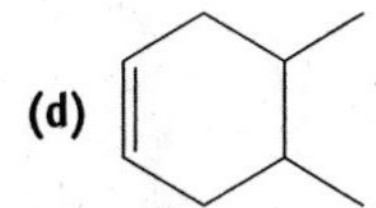

4.27 Explain why each name is incorrect or incomplete, and then write a correct name:
(a) (*Z*)-2-Methyl-1-pentene
(b) (*E*)-3,4-Diethyl-3-hexene
(c) *trans*-2,3-Dimethyl-2-hexene
(d) (1*Z*,3*Z*)-2,3-Dimethyl-1,3-butadiene

4.28 Draw structural formulas for all compounds with molecular formula C_5H_{10} that are
(a) Alkenes that do not show cis–trans isomerism.
(b) Alkenes that do show cis–trans isomerism.
(c) Cycloalkanes that do not show cis–trans isomerism.
(d) Cycloalkanes that do show cis–trans isomerism.

4.29 β-Ocimene, a triene found in the fragrance of cotton blossoms and several essential oils, has the IUPAC name (3*Z*)-3,7-dimethyl-1,3,6-octatriene. Draw a structural formula for β-ocimene.

4.30 Oleic acid and elaidic acid are, respectively, the cis and trans isomers of 9-octadecenoic acid. One of these fatty acids, a colorless liquid that solidifies at 4°C, is a major component of butterfat. The other, a white solid with a melting point of 44–45°C, is a major component of partially hydrogenated vegetable oils. Which of these two fatty acids is the cis isomer and which is the trans isomer?

4.31 Determine whether the structures in each set represent the same molecule, cis–trans isomers, or constitutional isomers. If they are the same molecule, determine whether they are in the same or different conformations.

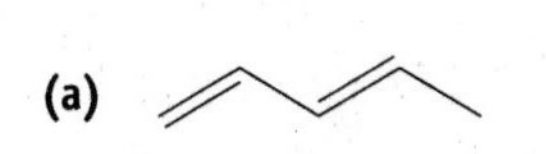
and
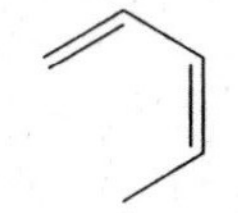

and
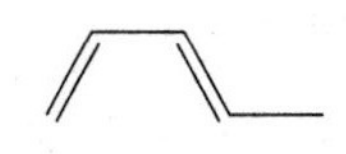

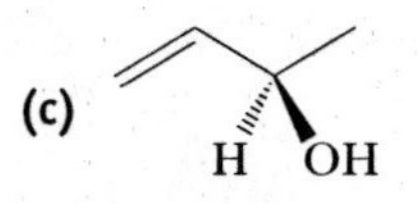

and

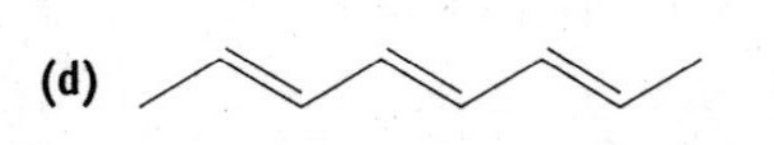

and
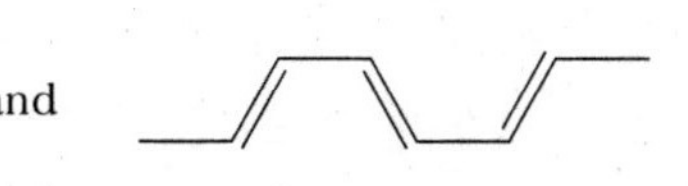

Terpenes

4.32 Show that the structural formula of vitamin A (Section 4.3G) can be divided into four isoprene units joined by head-to-tail linkages and cross-linked at one point to form the six-membered ring.

4.33 Following is the structural formula of lycopene, a deep-red compound that is partially responsible for the red color of ripe fruits, especially tomatoes:

Lycopene

Approximately 20 mg of lycopene can be isolated from 1 kg of fresh, ripe tomatoes.

(a) Show that lycopene is a terpene; that is, show that lycopene's carbon skeleton can be divided into two sets of four isoprene units with the units in each set joined head-to-tail.

(b) How many of the carbon–carbon double bonds in lycopene have the possibility for cis–trans isomerism? Lycopene is the all-trans isomer.

4.34 As you might suspect, β-carotene, a precursor of vitamin A, was first isolated from carrots. Dilute solutions of β-carotene are yellow—hence its use as a food coloring. In plants, it is almost always present in combination with chlorophyll to assist in the harvesting of the energy of sunlight. As tree leaves die in the fall, the green of their chlorophyll molecules is replaced by the yellows and reds of carotene and carotene-related molecules.

β-Carotene

(a) Compare the carbon skeletons of β-carotene and lycopene. What are the similarities? What are the differences?

(b) Show that β-carotene is a terpene.

4.35 α-Santonin, isolated from the flower heads of certain species of artemisia, is an anthelmintic—that is, a drug used to rid the body of worms (helminths). It has been estimated that over one-third of the world's population is infested with these parasites. Farnesol is an alcohol with a florid odor:

Santonin Farnesol

Locate the three isoprene units in santonin, and show how the carbon skeleton of farnesol might be coiled and then cross-linked to give santonin. Two different coiling patterns of the carbon skeleton of farnesol can lead to santonin. Try to find them both.

4.36 Periplanone is a pheromone (a chemical sex attractant) isolated from a species of cockroach. Show that the carbon skeleton of periplanone classifies it as a terpene:

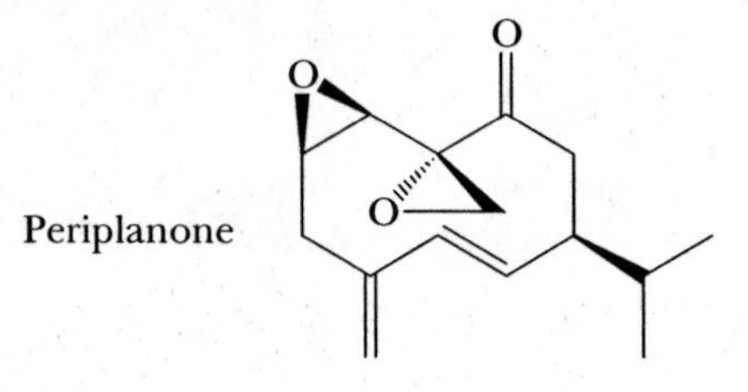

Periplanone

4.37 Gossypol, a compound found in the seeds of cotton plants, has been used as a male contraceptive in overpopulated countries such as China. Show that gossypol is a terpene:

Gossypol

4.38 In many parts of South America, extracts of the leaves and twigs of *Montanoa tomentosa* are used as a contraceptive, to stimulate menstruation, to facilitate labor, and as an abortifacient. The compound responsible for these effects is zoapatanol:

Zoapatanol

(a) Show that the carbon skeleton of zoapatanol can be divided into four isoprene units bonded head-to-tail and then cross-linked in one point along the chain.

(b) Specify the configuration about the carbon–carbon double bond to the seven-membered ring, according to the E,Z system.

(c) How many cis–trans isomers are possible for zoapatanol? Consider the possibilities for cis–trans isomerism in cyclic compounds and about carbon–carbon double bonds.

4.39 Pyrethrin II and pyrethrosin are natural products isolated from plants of the chrysanthemum family:

Chrysanthemum blossoms. *(Scott Camazine/Photo Researchers, Inc.)*

Pyrethrin II   Pyrethrosin

Pyrethrin II is a natural insecticide and is marketed as such.

(a) Label all carbon–carbon double bonds in each about which cis–trans isomerism is possible.

(b) Why are cis–trans isomers possible about the three-membered ring in pyrethrin II, but not about its five-membered ring?

(c) Show that the ring system of pyrethrosin is composed of three isoprene units.

4.40 Cuparene and herbertene are naturally occurring compounds isolated from various species of lichen:

Cuparene Herbertene

Determine whether one or both of these compounds can be classified as terpenes.

Looking Ahead

4.41 Explain why the $=C-C$ single bond in 1,3-butadiene is slightly shorter than the $=C-C$ single bond in 1-butene:

1.47 Å 1.51 Å

1,3-butadiene 1-butene

4.42 What effect might the ring size in the following cycloalkenes have on the reactivity of the $C=C$ double bond in each?

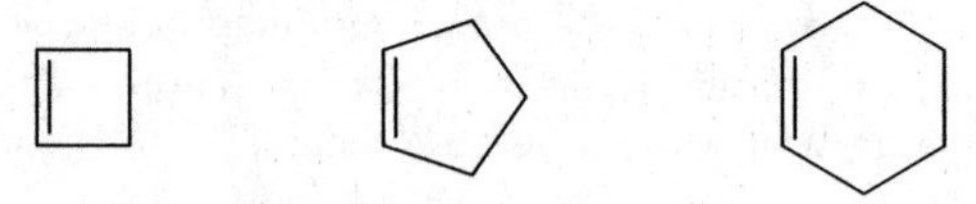

4.43 What effect might each substituent have on the electron density surrounding the alkene $C=C$ bond?

(a) OCH_3 **(b)** CN **(c)** $Si(CH_3)_3$

4.44 In Section 21.1 on the biochemistry of fatty acids, we will study the following three long-chain unsaturated carboxylic acids:

Oleic acid $CH_3(CH_2)_7CH{=}CH(CH_2)_7COOH$

Linoleic acid $CH_3(CH_2)_4CH{=}CHCH_2CH{=}CH(CH_2)_7COOH$

Linolenic acid $CH_3CH_2CH{=}CHCH_2CH{=}CHCH_2CH{=}CH(CH_2)_7COOH$

Each has 18 carbons and is a component of animal fats, vegetable oils, and biological membranes. Because of their presence in animal fats, they are called fatty acids.

(a) How many cis–trans isomers are possible for each fatty acid?

(b) These three fatty acids occur in biological membranes almost exclusively in the cis configuration. Draw line-angle formulas for each fatty acid, showing the cis configuration about each carbon–carbon double bond.

4.45 Assign an E or a Z configuration and a cis or a trans configuration to these carboxylic acids, each of which is an intermediate in the citric acid cycle (Section 22.7; under each is given its common name):

(a) H, COOH / C=C / HOOC, H

Fumaric acid

(b) HOOC, COOH / C=C / H, CH_2COOH

Aconitic acid

5 Reactions of Alkenes

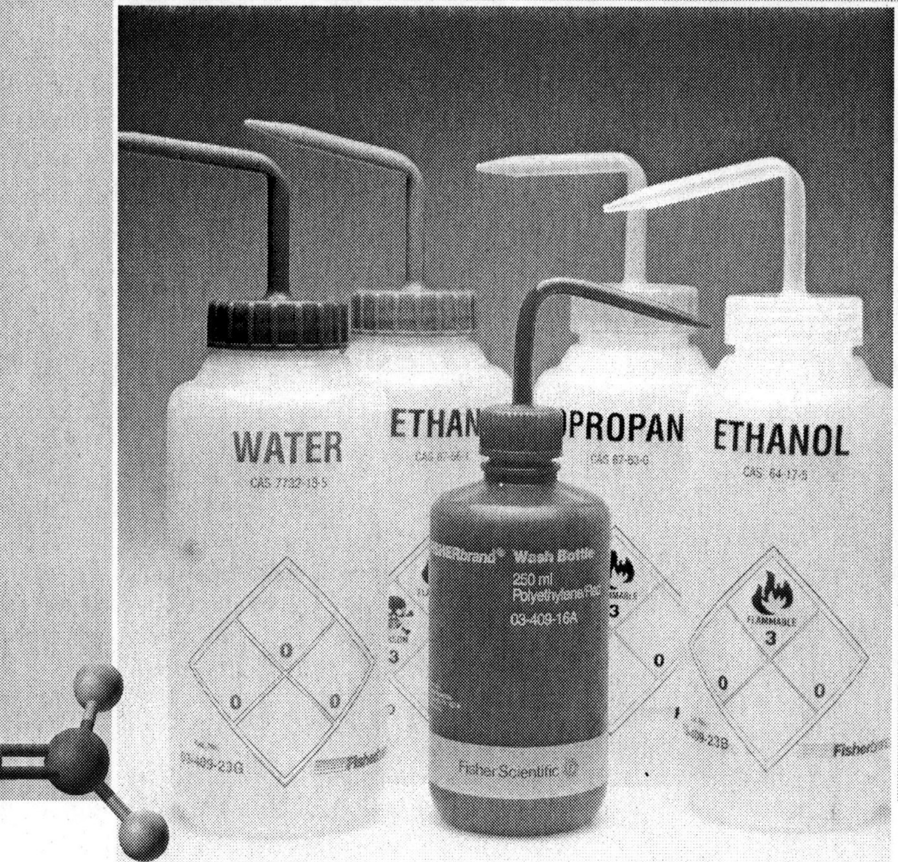

These wash bottles are made of polyethylene. Inset: A model of ethylene. *(Charles D. Winters)*

In this chapter, we begin our systematic study of reaction mechanisms, one of the most important unifying concepts in organic chemistry. We use the reactions of alkenes as the vehicle to introduce this concept.

5.1 INTRODUCTION

The most characteristic reaction of alkenes is **addition to the carbon–carbon double bond** in such a way that the pi bond is broken and, in its place, sigma bonds are formed to two new atoms or groups of atoms. Several examples of reactions at the carbon–carbon double bond are shown in Table 5.1, along with the descriptive name(s) associated with each.

TABLE 5.1 Characteristic Addition Reactions of Alkenes

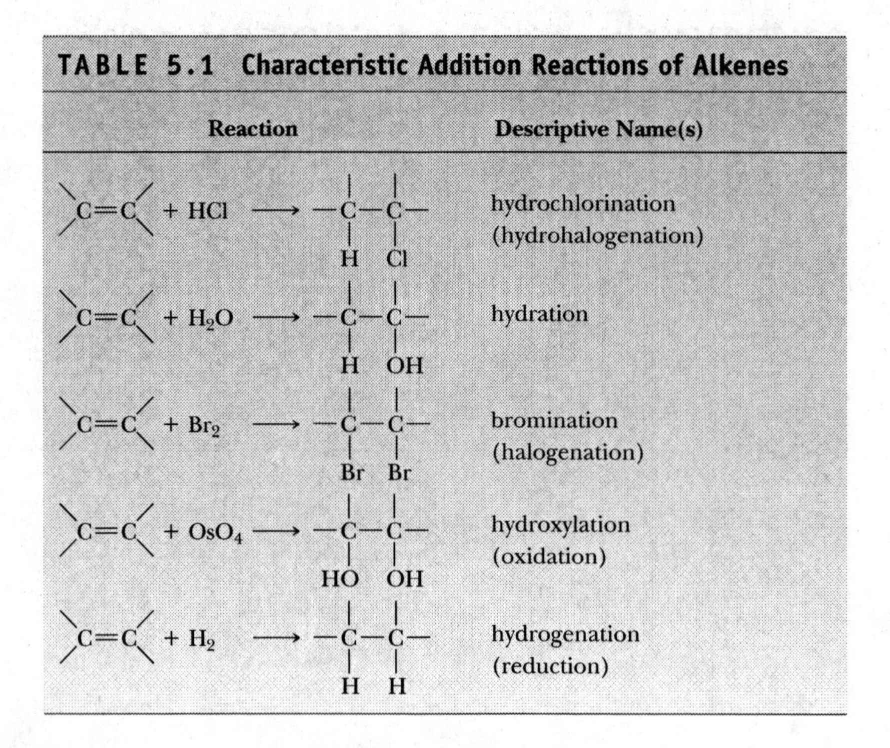

Reaction	Descriptive Name(s)
$>C=C< + HCl \longrightarrow -C(H)-C(Cl)-$	hydrochlorination (hydrohalogenation)
$>C=C< + H_2O \longrightarrow -C(H)-C(OH)-$	hydration
$>C=C< + Br_2 \longrightarrow -C(Br)-C(Br)-$	bromination (halogenation)
$>C=C< + OsO_4 \longrightarrow -C(HO)-C(OH)-$	hydroxylation (oxidation)
$>C=C< + H_2 \longrightarrow -C(H)-C(H)-$	hydrogenation (reduction)

From the perspective of the chemical industry, the single most important reaction of ethylene and other low-molecular-weight alkenes is the production of **chain-growth polymers** (Greek: *poly*, many, and *meros*, part). In the presence of certain catalysts called *initiators*, many alkenes form polymers by the addition of **monomers** (Greek: *mono*, one, and *meros*, part) to a growing polymer chain, as illustrated by the formation of polyethylene from ethylene:

$$n\mathrm{CH_2} = \mathrm{CH_2} \xrightarrow{\text{initiator}} \left(\mathrm{CH_2CH_2}\right)_n$$

In alkene polymers of industrial and commercial importance, n is a large number, typically several thousand. We discuss this alkene reaction in Chapter 17.

5.2 REACTION MECHANISMS

Reaction mechanism A step-by-step description of how a chemical reaction occurs.

A **reaction mechanism** describes in detail how a chemical reaction occurs. It describes which bonds break and which new ones form, as well as the order and relative rates of the various bond-breaking and bond-forming steps. If the reaction takes place in solution, the reaction mechanism describes the role of the solvent; if the reaction involves a catalyst, the reaction mechanism describes the role of the catalyst.

A. Energy Diagrams and Transition States

Energy diagram A graph showing the changes in energy that occur during a chemical reaction; energy is plotted on the y-axis, and the progress of the reaction is plotted on the x-axis.

Reaction coordinate A measure of the progress of a reaction, plotted on the x-axis in an energy diagram.

To understand the relationship between a chemical reaction and energy, think of a chemical bond as a spring. As a spring is stretched from its resting position, its energy increases. As it returns to its resting position, its energy decreases. Similarly, during a chemical reaction, bond breaking corresponds to an increase in energy, and bond forming corresponds to a decrease in energy. We use an **energy diagram** to show the changes in energy that occur in going from reactants to products. Energy is measured along the vertical axis, and the change in position of the atoms during a reaction is measured on the horizontal axis, called the **reaction coordinate**. The reaction coordinate indicates how far the reaction has progressed, from no reaction to a completed reaction.

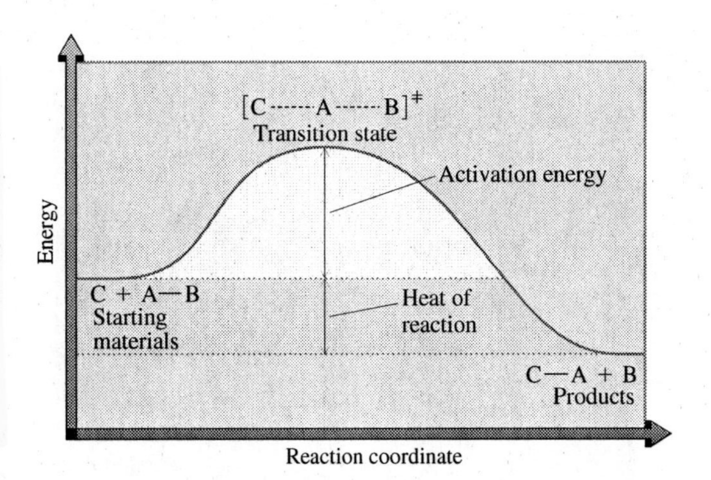

Figure 5.1
An energy diagram for a one-step reaction between C and A—B. The dashed lines indicate that, in the transition state, the new C—A bond is partially formed and the A—B bond is partially broken. The energy of the reactants is higher than that of the products.

Figure 5.1 shows an energy diagram for the reaction of C + A—B to form C—A + B. This reaction occurs in one step, meaning that bond breaking in reactants and bond forming in products occur simultaneously.

The difference in energy between the reactants and products is called the **heat of reaction, ΔH**. If the energy of the products is lower than that of the reactants, heat is released and the reaction is called **exothermic**. If the energy of the products is higher than that of the reactants, heat is absorbed and the reaction is called **endothermic**. The one-step reaction shown in Figure 5.1 is exothermic.

Heat of reaction The difference in energy between reactants and products.

Exothermic reaction A reaction in which the energy of the products is lower than the energy of the reactants; a reaction in which heat is liberated.

Endothermic reaction A reaction in which the energy of the products is higher than the energy of the reactants; a reaction in which heat is absorbed.

Transition state An unstable species of maximum energy formed during the course of a reaction; a maximum on an energy diagram.

A **transition state** is the point on the reaction coordinate at which the energy is at a maximum. At the transition state, sufficient energy has become concentrated in the proper bonds so that bonds in the reactants break. As they break, energy is redistributed and new bonds form, giving products. Once the transition state is reached, the reaction proceeds to give products, with the release of energy.

A transition state has a definite geometry, a definite arrangement of bonding and nonbonding electrons, and a definite distribution of electron density and charge. Because a transition state is at an energy maximum on an energy diagram, we cannot isolate it and we cannot determine its structure experimentally. Its lifetime is on the order of a picosecond (the duration of a single bond vibration). As we will see, however, even though we cannot observe a transition state directly by any experimental means, we can often infer a great deal about its probable structure from other experimental observations.

For the reaction shown in Figure 5.1, we use dashed lines to show the partial bonding in the transition state. As C begins to form a new covalent bond with A (as shown by the dashed line), the covalent bond between A and B begins to break (also shown by a dashed line). Upon completion of the reaction, the A—B bond is fully broken and the C—A bond is fully formed.

The difference in energy between the reactants and the transition state is called the **activation energy**. The activation energy is the minimum energy required for a reaction to occur; it can be considered an energy barrier for the reaction. The activation energy determines the rate of a reaction—that is, how fast the reaction occurs. If the activation energy is large, only a very few molecular collisions occur with sufficient energy to reach the transition state, and the reaction is slow. If the activation energy is small, many collisions generate sufficient energy to reach the transition state, and the reaction is fast.

Activation energy The difference in energy between reactants and the transition state.

In a reaction that occurs in two or more steps, each step has its own transition state and activation energy. Shown in Figure 5.2 is an energy diagram for the conversion of reactants to products in two steps. A **reaction intermediate** corresponds to an energy minimum between two transition states, in this case transition States 1 and 2. Note that because the energies of the reaction intermediates we describe are higher than the energies of either the reactants or the products, these intermediates are highly reactive, and rarely, if ever, can one be isolated.

Reaction intermediate An unstable species that lies in an energy minimum between two transition states.

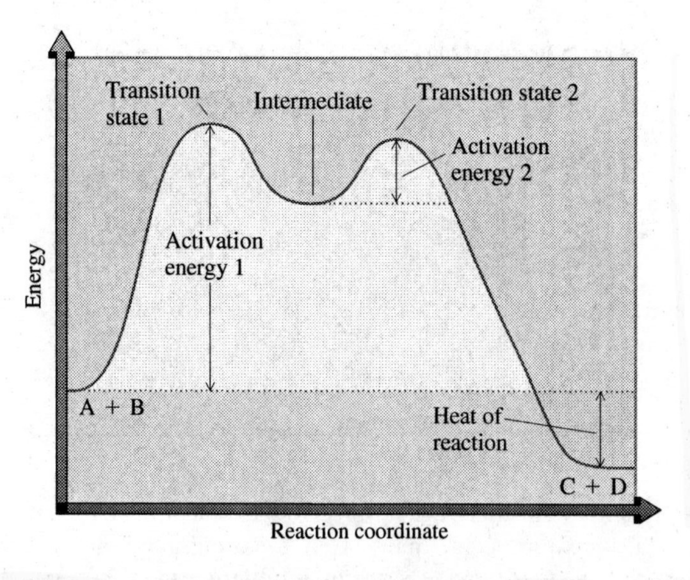

Figure 5.2
Energy diagram for a two-step reaction involving the formation of an intermediate. The energy of the reactants is higher than that of the products, and energy is released in the conversion of A + B to C + D.

Rate-determining step The step in a reaction sequence that crosses the highest energy barrier; the slowest step in a multistep reaction.

The slowest step in a multistep reaction, called the **rate-determining step**, is the step that crosses the highest energy barrier. In the two-step reaction shown in Figure 5.2, Step 1 crosses the higher energy barrier and is, therefore, the rate-determining step.

EXAMPLE 5.1

Draw an energy diagram for a two-step exothermic reaction in which the second step is rate determining.

SOLUTION

A two-step reaction involves the formation of an intermediate. In order for the reaction to be exothermic, the products must be lower in energy than the reactants. In order for the second step to be rate determining, it must cross the higher energy barrier.

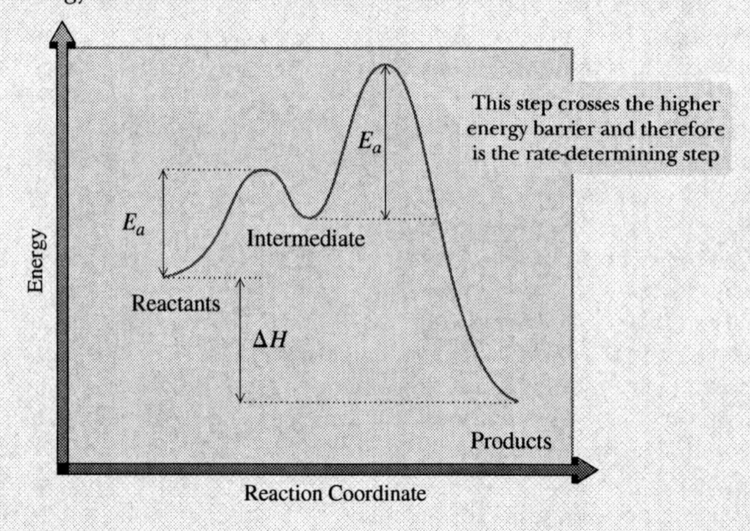

Practice Problem 5.1

In what way would the energy diagram drawn in Example 5.1 change if the reaction were endothermic?

B. Developing a Reaction Mechanism

To develop a reaction mechanism, chemists begin by designing experiments that will reveal details of a particular chemical reaction. Next, through a combination of experience and intuition, they propose several sets of steps or mechanisms, each of which might account for the overall chemical transformation. Finally, they test each proposed mechanism against the experimental observations to exclude those mechanisms which are not consistent with the facts.

A mechanism becomes generally established by excluding reasonable alternatives and by showing that it is consistent with every test that can be devised. This, of course, does not mean that a generally accepted mechanism is a completely accurate description of the chemical events, but only that it is the best chemists have been able to devise. It is important to keep in mind that, as new experimental evidence is obtained, it may be necessary to modify a generally accepted mechanism or possibly even discard it and start all over again.

Before we go on to consider reactions and reaction mechanisms, we might ask why it is worth the trouble to establish them and your time to learn about them. One reason is very practical: Mechanisms provide a theoretical framework within which to organize a great deal of descriptive chemistry. For example, with insight into how reagents add to particular alkenes, it is possible to make generalizations and then predict how the same reagents might add to other alkenes. A second reason lies in the intellectual satisfaction derived from constructing models that accurately reflect the behavior of chemical systems. Finally, to a creative scientist, a mechanism is a tool to be used in the search for new knowledge and new understanding. A mechanism consistent with all that is known about a reaction can be used to make predictions about chemical interactions as yet unexplored, and experiments can be designed to test these predictions. Thus, reaction mechanisms provide a way not only to organize knowledge, but also to extend it.

5.3 ELECTROPHILIC ADDITION REACTIONS

We begin our introduction to the chemistry of alkenes with an examination of three types of addition reactions: the addition of hydrogen halides (HCl, HBr, and HI), water (H_2O), and halogens (Br_2, Cl_2). We first study some of the experimental observations about each addition reaction and then its mechanism. By examining these particular reactions, we develop a general understanding of how alkenes undergo addition reactions.

A. Addition of Hydrogen Halides

The hydrogen halides HCl, HBr, and HI add to alkenes to give haloalkanes (alkyl halides). These additions may be carried out either with the pure reagents or in the presence of a polar solvent such as acetic acid. The addition of HCl to ethylene gives chloroethane (ethyl chloride):

$$\underset{\text{Ethylene}}{CH_2{=}CH_2} + HCl \longrightarrow \underset{\text{Chloroethane}}{\overset{H}{\overset{|}{C}H_2}-\overset{Cl}{\overset{|}{C}H_2}}$$

The addition of HCl to propene gives 2-chloropropane (isopropyl chloride); hydrogen adds to carbon 1 of propene and chlorine adds to carbon 2. If the orientation of addition were reversed, 1-chloropropane (propyl chloride) would be formed.

The observed result is that 2-chloropropane is formed to the virtual exclusion of 1-chloropropane:

$$CH_3CH{=}CH_2 + HCl \longrightarrow CH_3\underset{|}{\overset{Cl}{C}}H{-}\overset{H}{C}H_2 + CH_3\overset{H}{C}H{-}\overset{Cl}{C}H_2$$

Propene — 2-Chloropropane — 1-Chloropropane (not observed)

We say that the addition of HCl to propene is highly regioselective and that 2-chloropropane is the major product of the reaction. A **regioselective reaction** is a reaction in which one direction of bond forming or breaking occurs in preference to all other directions.

Vladimir Markovnikov observed this regioselectivity and made the generalization, known as **Markovnikov's rule**, that, in the addition of HX to an alkene, hydrogen adds to the doubly bonded carbon that has the greater number of hydrogens already bonded to it. Although Markovnikov's rule provides a way to predict the product of many alkene addition reactions, it does not explain why one product predominates over other possible products.

Regioselective reaction A reaction in which one direction of bond forming or bond breaking occurs in preference to all other directions.

Markovnikov's rule In the addition of HX or H_2O to an alkene, hydrogen adds to the carbon of the double bond having the greater number of hydrogens.

'H-rich get richer'

EXAMPLE 5.2

Name and draw a structural formula for the major product of each alkene addition reaction:

(a) $CH_3\overset{CH_3}{\overset{|}{C}}{=}CH_2 + HI \longrightarrow$ (b) 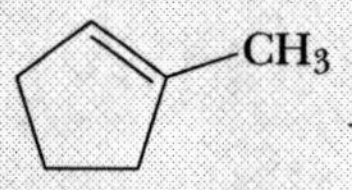 1-methylcyclopentene (CH_3 on ring double-bond carbon) $+ HCl \longrightarrow$

SOLUTION

Using Markovnikov's rule, we predict that 2-iodo-2-methylpropane is the product in (a) and 1-chloro-1-methylcyclopentane is the product in (b):

(a) $CH_3\overset{CH_3}{\overset{|}{C}}CH_3$ with I on the central carbon

2-Iodo-2-methylpropane

(b) 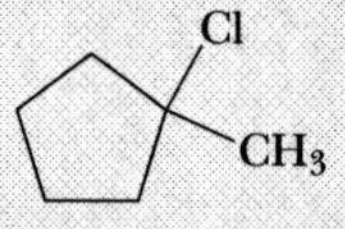 cyclopentane ring bearing Cl and CH_3 on the same carbon

1-Chloro-1-methylcyclopentane

Practice Problem 5.2

Name and draw a structural formula for the major product of each alkene addition reaction:

(a) $CH_3CH{=}CH_2 + HI \longrightarrow$ (b) 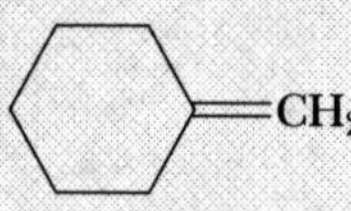 cyclohexane ring $={CH_2} + HI \longrightarrow$

Chemists account for the addition of HX to an alkene by a two-step mechanism, which we illustrate by the reaction of 2-butene with hydrogen chloride to give 2-chlorobutane. Let us first look at this two-step mechanism in general and then go back and study each step in detail.

Mechanism: Electrophilic Addition of HCl to 2-Butene

Step 1: The reaction begins with the transfer of a proton from HCl to 2-butene, as shown by the two curved arrows on the left side of Step 1:

$$CH_3CH{=}CHCH_3 + H{-}\overset{\delta-}{Cl}: \xrightleftharpoons[\text{determining}]{\text{slow, rate}} CH_3\overset{+}{C}H{-}\overset{H}{\overset{|}{C}}HCH_3 + :\ddot{Cl}:^-$$

sec-Butyl cation (a 2° carbocation intermediate)

The first curved arrow shows the breaking of the pi bond of the alkene and its electron pair now forming a new covalent bond with the hydrogen atom of HCl. The second curved arrow shows the breaking of the polar covalent bond in HCl and this electron pair being given entirely to chlorine, forming chloride ion. Step 1 in this mechanism results in the formation of an organic cation and chloride ion.

Step 2: The reaction of the *sec*-butyl cation (a Lewis acid) with chloride ion (a Lewis base) completes the valence shell of carbon and gives 2-chlorobutane:

$$:\ddot{Cl}:^- + CH_3\overset{+}{C}HCH_2CH_3 \xrightarrow{\text{fast}} CH_3\overset{:\ddot{Cl}:}{\overset{|}{C}}HCH_2CH_3$$

Chloride ion (a Lewis base) — *sec*-Butyl cation (a Lewis acid) — 2-Chlorobutane

Now let us go back and look at the individual steps in more detail. There is a great deal of important organic chemistry embedded in these two steps, and it is crucial that you understand it now.

Step 1 results in the formation of an organic cation. One carbon atom in this cation has only six electrons in its valence shell and carries a charge of +1. A species containing a positively charged carbon atom is called a **carbocation** (*carbon* + *cation*). Carbocations are classified as primary (1°), secondary (2°), or tertiary (3°), depending on the number of carbon atoms bonded to the carbon bearing the positive charge. All carbocations are Lewis acids (Section 2.7). They are also electrophiles. The term **electrophile** quite literally means "electron lover."

Carbocation A species containing a carbon atom with only three bonds to it and bearing a positive charge.

Electrophile Any molecule or ion that can accept a pair of electrons to form a new covalent bond; a Lewis acid.

In a carbocation, the carbon bearing the positive charge is bonded to three other atoms, and, as predicted by the valence-shell electron-pair repulsion (VSEPR) model, the three bonds about that carbon are coplanar and form bond angles of approximately 120°. According to the orbital overlap model, the electron-deficient carbon of a carbocation uses its sp^2 hybrid orbitals to form sigma bonds to the three attached groups. The unhybridized $2p$ orbital lies perpendicular to the sigma bond framework and contains no electrons. A Lewis structure and an orbital overlap diagram for the *tert*-butyl cation are shown in Figure 5.3.

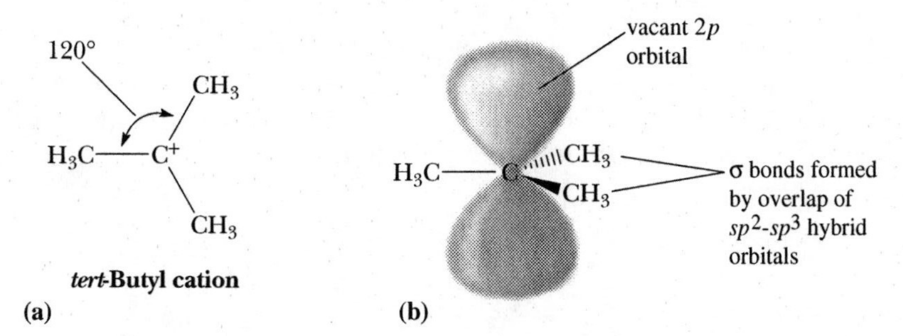

Figure 5.3 The structure of the *tert*-butyl cation. (a) Lewis structure and (b) an orbital picture.

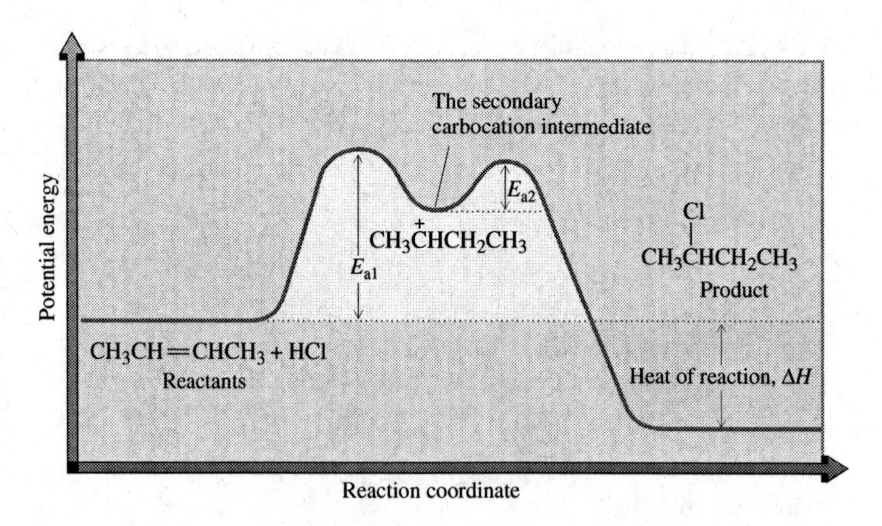

Figure 5.4
Energy diagram for the two-step addition of HCl to 2-butene. The reaction is exothermic.

Figure 5.4 shows an energy diagram for the two-step reaction of 2-butene with HCl. The slower, rate-determining step (the one that crosses the higher energy barrier) is Step 1, which leads to the formation of the 2° carbocation intermediate. This intermediate lies in an energy minimum between the transition states for Steps 1 and 2. As soon as the carbocation intermediate (a Lewis acid) forms, it reacts with chloride ion (a Lewis base) in a Lewis acid–base reaction to give 2-chlorobutane. Note that the energy level for 2-chlorobutane (the product) is lower than the energy level for 2-butene and HCl (the reactants). Thus, in this alkene addition reaction, heat is released; the reaction is, accordingly, exothermic.

Relative Stabilities of Carbocations: Regioselectivity and Markovnikov's Rule

The reaction of HX and an alkene can, at least in principle, give two different carbocation intermediates, depending on which of the doubly bonded carbon atoms forms a bond with H^+, as illustrated by the reaction of HCl with propene:

$$\underset{\text{Propene}}{CH_3CH{=}CH_2} + H{-}\ddot{\underset{..}{Cl}}: \longrightarrow \underset{\substack{\text{Propyl cation} \\ \text{(a 1° carbocation)}}}{CH_3CH_2\overset{+}{C}H_2} \xrightarrow{:\ddot{Cl}:^-} \underset{\substack{\text{1-Chloropropane} \\ \text{(not formed)}}}{CH_3CH_2CH_2\ddot{\underset{..}{Cl}}:}$$

$$\underset{\text{Propene}}{CH_3CH{=}CH_2} + H{-}\ddot{\underset{..}{Cl}}: \longrightarrow \underset{\substack{\text{Isopropyl cation} \\ \text{(a 2° carbocation)}}}{CH_3\overset{+}{C}HCH_3} \xrightarrow{:\ddot{Cl}:^-} \underset{\substack{\text{2-Chloropropane} \\ \text{(product formed)}}}{CH_3\overset{\overset{:\ddot{Cl}:}{|}}{C}HCH_3}$$

The observed product is 2-chloropropane. Because carbocations react very quickly with chloride ions, the absence of 1-chloropropane as a product tells us that the 2° carbocation is formed in preference to the 1° carbocation.

Similarly, in the reaction of HCl with 2-methylpropene, the transfer of a proton to the carbon–carbon double bond might form either the isobutyl cation (a 1° carbocation) or the *tert*-butyl cation (a 3° carbocation):

$$\underset{\text{2-Methylpropene}}{CH_3\overset{\overset{CH_3}{|}}{C}{=}CH_2} + H{-}\ddot{\underset{..}{Cl}}: \longrightarrow \underset{\substack{\text{Isobutyl cation} \\ \text{(a 1° carbocation)}}}{CH_3\overset{\overset{CH_3}{|}}{C}H\overset{+}{C}H_2} \xrightarrow{:\ddot{Cl}:^-} \underset{\substack{\text{1-Chloro-} \\ \text{2-methylpropane} \\ \text{(not formed)}}}{CH_3\overset{\overset{CH_3}{|}}{C}HCH_2\ddot{\underset{..}{Cl}}:}$$

$$CH_3C(CH_3){=}CH_2 + H{-}Cl \longrightarrow CH_3\overset{+}{C}(CH_3)CH_3 \;\; Cl^- \longrightarrow CH_3C(CH_3)(Cl)CH_3$$

2-Methylpropene; *tert*-Butyl cation (a 3° carbocation); 2-Chloro-2-methylpropane (product formed)

In this reaction, the observed product is 2-chloro-2-methylpropane, indicating that the 3° carbocation forms in preference to the 1° carbocation.

From such experiments and a great amount of other experimental evidence, we learn that a 3° carbocation is more stable and requires a lower activation energy for its formation than a 2° carbocation. A 2° carbocation, in turn, is more stable and requires a lower activation energy for its formation than a 1° carbocation. It follows that a more stable carbocation intermediate forms faster than a less stable carbocation intermediate. Following is the order of stability of four types of alkyl carbocations:

$$H{-}C^+H_2 \qquad H_3C{-}C^+H_2 \qquad H_3C{-}C^+H(CH_3) \qquad H_3C{-}C^+(CH_3)_2$$

Methyl cation (methyl) — Ethyl cation (1°) — Isopropyl cation (2°) — *tert*-Butyl cation (3°)

Order of increasing carbocation stability →

Although the concept of the relative stabilities of carbocations had not been developed in Markovnikov's time, their relative stabilities is the underlying basis for his rule; that is, the proton of H—X adds to the less substituted carbon of a double bond because this mode of addition produces the more stable carbocation intermediate.

Now that we know the order of stability of carbocations, how do we account for it? The principles of physics teach us that a system bearing a charge (either positive or negative) is more stable if the charge is delocalized. Using this principle, we can explain the order of stability of carbocations if we assume that alkyl groups bonded to a positively charged carbon release electrons toward the cationic carbon and thereby help delocalize the charge on the cation. The electron-releasing ability of alkyl groups bonded to a cationic carbon is accounted for by the **inductive effect** (Section 2.6C).

The inductive effect operates in the following way: The electron deficiency of the carbon atom bearing a positive charge exerts an electron-withdrawing inductive effect that polarizes electrons from adjacent sigma bonds toward it. Thus, the positive charge of the cation is not localized on the trivalent carbon, but rather is delocalized over nearby atoms. The larger the volume over which the positive charge is delocalized, the greater is the stability of the cation. Thus, as the number of alkyl groups bonded to the cationic carbon increases, the stability of the cation increases as well. Figure 5.5 illustrates the electron-withdrawing inductive effect of the positively charged carbon and the resulting delocalization of charge. According to quantum mechanical calculations, the charge on carbon in the methyl cation is approximately +0.645, and the charge on each of the hydrogen atoms is +0.118. Thus, even in the methyl cation, the positive charge is not localized on carbon. Rather, it is delocalized over the volume of space occupied by the entire ion. The polarization of electron density and the delocalization of charge are even more extensive in the *tert*-butyl cation.

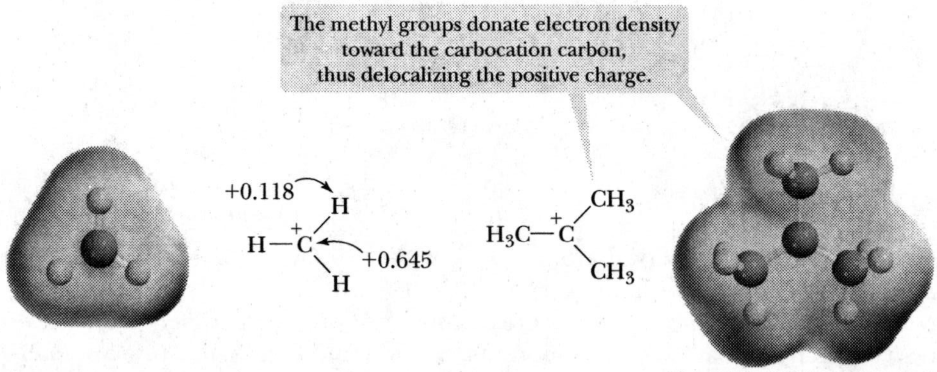

Figure 5.5
Methyl and *tert*-butyl cations. Delocalization of positive charge by the electron-withdrawing inductive effect of the trivalent, positively charged carbon according to molecular orbital calculations.

The methyl groups donate electron density toward the carbocation carbon, thus delocalizing the positive charge.

+0.118 H–C⁺(H)(H) +0.645

$H_3C-\overset{+}{C}(CH_3)_2$

EXAMPLE 5.3

Arrange these carbocations in order of increasing stability:

(a) (b) (c)

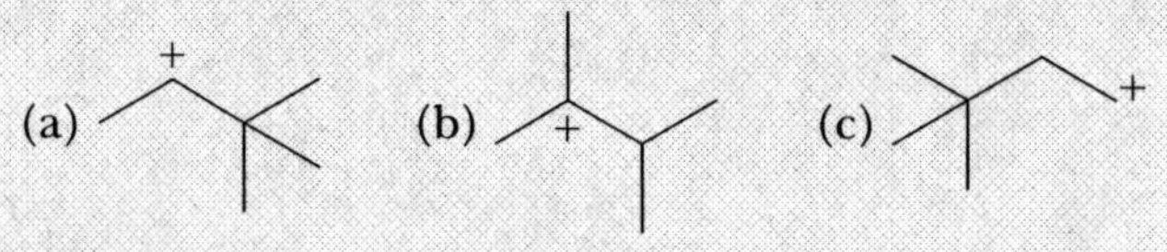

SOLUTION

Carbocation (a) is secondary, (b) is tertiary, and (c) is primary. In order of increasing stability, they are $c < a < b$.

Practice Problem 5.3

Arrange these carbocations in order of increasing stability:

(a) $-CH_3$ (b) $-CH_3$ (c) $-\overset{+}{C}H_2$

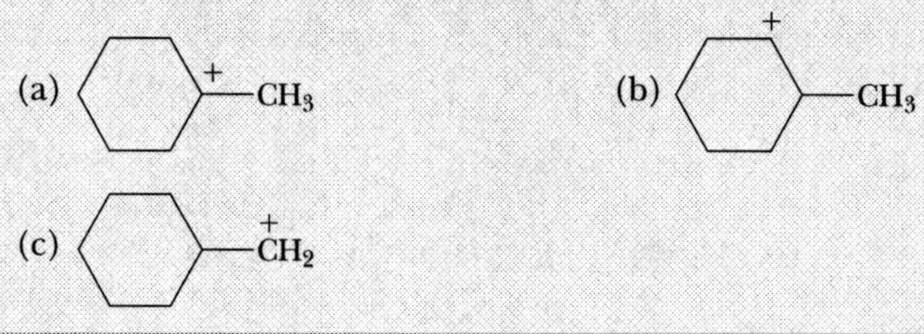

EXAMPLE 5.4

Propose a mechanism for the addition of HI to methylenecyclohexane to give 1-iodo-1-methylcyclohexane:

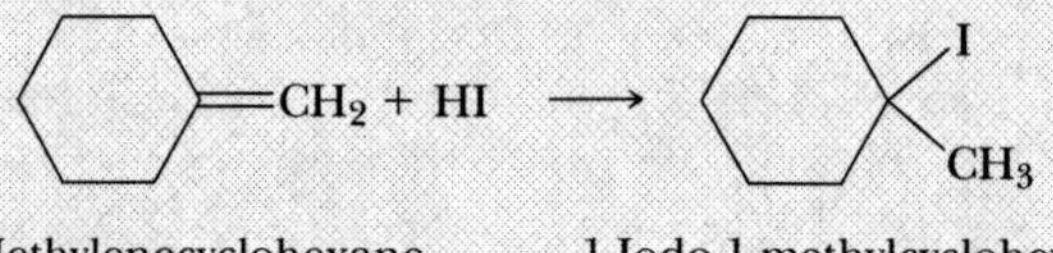

$=CH_2 + HI \longrightarrow$ I, CH_3

Methylenecyclohexane 1-Iodo-1-methylcyclohexane

Which step in your mechanism is rate determining?

SOLUTION

Propose a two-step mechanism similar to that proposed for the addition of HCl to propene.

Step 1: A rate-determining proton transfer from HI to the carbon–carbon double bond gives a 3° carbocation intermediate:

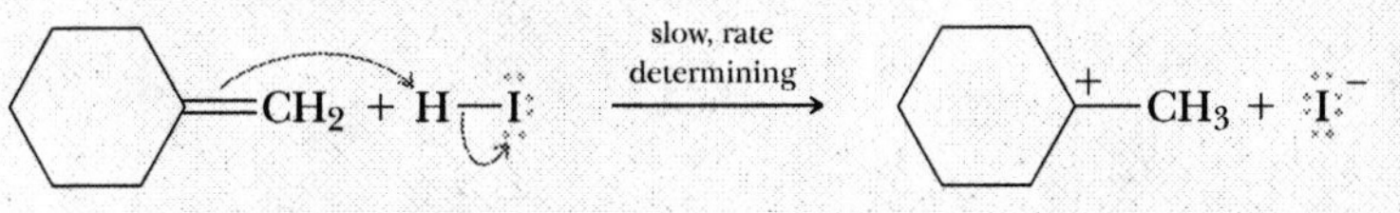

$$\text{Methylenecyclohexane}=CH_2 + H-\ddot{I}: \xrightarrow{\text{slow, rate determining}} \overset{+}{C}-CH_3 + :\ddot{I}:^-$$

Methylenecyclohexane — A 3° carbocation intermediate

Step 2: Reaction of the 3° carbocation intermediate (a Lewis acid) with iodide ion (a Lewis base) completes the valence shell of carbon and gives the product:

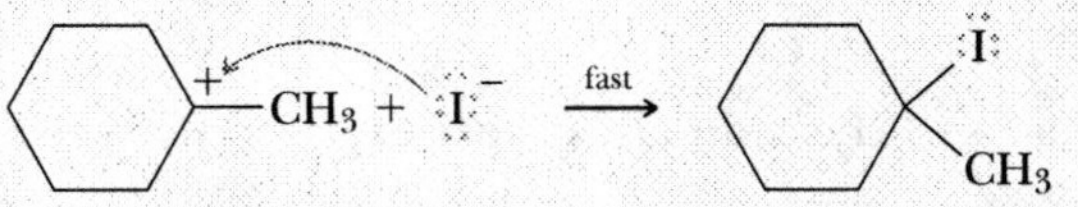

$$\overset{+}{C}-CH_3 + :\ddot{I}:^- \xrightarrow{\text{fast}} \text{1-Iodo-1-methylcyclohexane}$$

1-Iodo-1-methylcyclohexane

Practice Problem 5.4

Propose a mechanism for the addition of HI to 1-methylcyclohexene to give 1-iodo-1-methylcyclohexane. Which step in your mechanism is rate determining?

B. Addition of Water: Acid-Catalyzed Hydration

In the presence of an acid catalyst—most commonly, concentrated sulfuric acid—water adds to the carbon–carbon double bond of an alkene to give an alcohol. The addition of water is called **hydration**. In the case of simple alkenes, H adds to the carbon of the double bond with the greater number of hydrogens and OH adds to the carbon with the lesser number of hydrogens. Thus, H—OH adds to alkenes in accordance with Markovnikov's rule:

Hydration Addition of water.

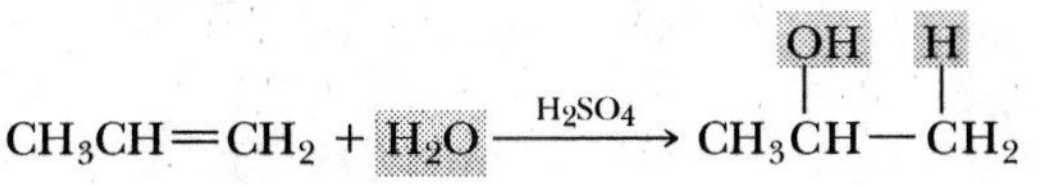

$$CH_3CH{=}CH_2 + H_2O \xrightarrow{H_2SO_4} CH_3\underset{OH}{CH}-\underset{H}{CH_2}$$

Propene — 2-Propanol

$$CH_3\overset{CH_3}{C}{=}CH_2 + H_2O \xrightarrow{H_2SO_4} CH_3\overset{CH_3}{\underset{HO}{C}}-\underset{H}{CH_2}$$

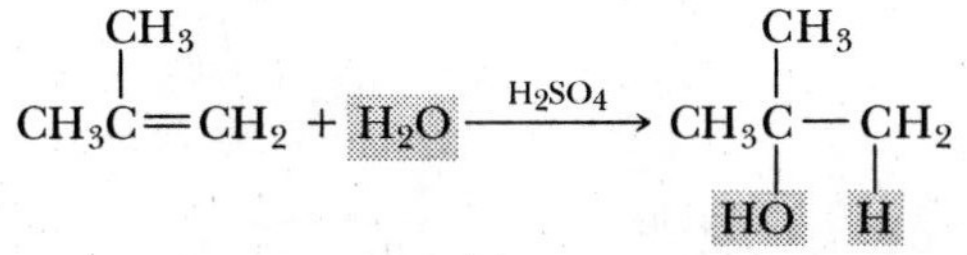

2-Methylpropene — 2-Methyl-2-propanol

EXAMPLE 5.5

Draw a structural formula for the product of the acid-catalyzed hydration of 1-methylcyclohexene.

SOLUTION

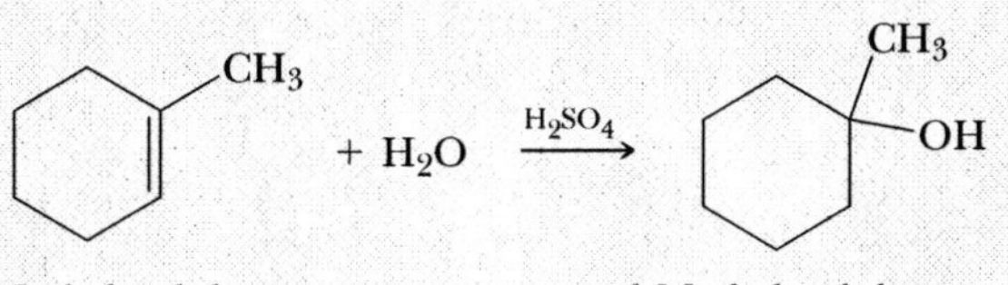

$$\text{1-Methylcyclohexene} + H_2O \xrightarrow{H_2SO_4} \text{1-Methylcyclohexanol}$$

1-Methylcyclohexene — 1-Methylcyclohexanol

Practice Problem 5.5

Draw a structural formula for the product of each alkene hydration reaction:

(a) + H_2O $\xrightarrow{H_2SO_4}$ (b) + H_2O $\xrightarrow{H_2SO_4}$

The mechanism for the acid-catalyzed hydration of alkenes is quite similar to what we have already proposed for the addition of HCl, HBr, and HI to alkenes and is illustrated by the hydration of propene to 2-propanol. This mechanism is consistent with the fact that acid is a catalyst. An H_3O^+ is consumed in Step 1, but another is generated in Step 3.

Mechanism: Acid-Catalyzed Hydration of Propene

Step 1: Proton transfer from the acid catalyst to propene gives a 2° carbocation intermediate (a Lewis acid):

$$CH_3CH{=}CH_2 + H{-}\ddot{O}^+{-}H\ (\text{with } H) \underset{\text{determining}}{\overset{\text{slow, rate}}{\rightleftharpoons}} CH_3\overset{+}{C}HCH_3 + :\ddot{O}{-}H\ (\text{with } H)$$

A 2° carbocation intermediate

Oxonium ion An ion in which oxygen is bonded to three other atoms and bears a positive charge.

Step 2: Reaction of the carbocation intermediate (a Lewis acid) with water (a Lewis base) completes the valence shell of carbon and gives an **oxonium ion**:

$$CH_3\overset{+}{C}HCH_3 + :\ddot{O}{-}H\ (\text{with } H) \overset{\text{fast}}{\rightleftharpoons} CH_3CHCH_3\ (\text{with } {-}\ddot{O}^+H_2)$$

An oxonium ion

Step 3: Proton transfer from the oxonium ion to water gives the alcohol and generates a new molecule of the catalyst:

$$CH_3CHCH_3\ (\text{with } {-}\ddot{O}^+H_2) + H{-}\ddot{O}{-}H \overset{\text{fast}}{\rightleftharpoons} CH_3CHCH_3\ (\text{with } {-}\ddot{O}{-}H) + H{-}\ddot{O}^+{-}H\ (\text{with } H)$$

EXAMPLE 5.6

Propose a mechanism for the acid-catalyzed hydration of methylenecyclohexane to give 1-methylcyclohexanol. Which step in your mechanism is rate determining?

SOLUTION

Propose a three-step mechanism similar to that for the acid-catalyzed hydration of propene. The formation of the 3° carbocation intermediate in Step 1 is rate determining.

Step 1: Proton transfer from the acid catalyst to the alkene gives a 3° carbocation intermediate (a Lewis acid):

$$C_6H_{10}{=}CH_2 + H{-}\ddot{O}^+{-}H\ (\text{with } H) \underset{\text{determining}}{\overset{\text{slow, rate}}{\rightleftharpoons}} C_6H_{10}^+{-}CH_3 + :\ddot{O}{-}H\ (\text{with } H)$$

A 3° carbocation intermediate

Step 2: Reaction of the carbocation intermediate (a Lewis acid) with water (a Lewis base) completes the valence shell of carbon and gives an oxonium ion:

$$\text{(1-methylcyclohexyl cation)}^{+}\!-CH_3 + :\ddot{O}H_2 \overset{\text{fast}}{\rightleftharpoons} \text{(cyclohexane)}(CH_3)\!-\!\overset{+}{O}H_2$$

An oxonium ion

Step 3: Proton transfer from the oxonium ion to water gives the alcohol and generates a new molecule of the catalyst:

$$\text{(cyclohexane)}(CH_3)\!-\!\overset{+}{O}H_2 + :\ddot{O}H_2 \overset{\text{fast}}{\rightleftharpoons} \text{(cyclohexane)}(CH_3)\!-\!\ddot{O}H + H\!-\!\overset{+}{O}H_2$$

Practice Problem 5.6

Propose a mechanism for the acid-catalyzed hydration of 1-methylcyclohexene to give 1-methylcyclohexanol. Which step in your mechanism is rate determining?

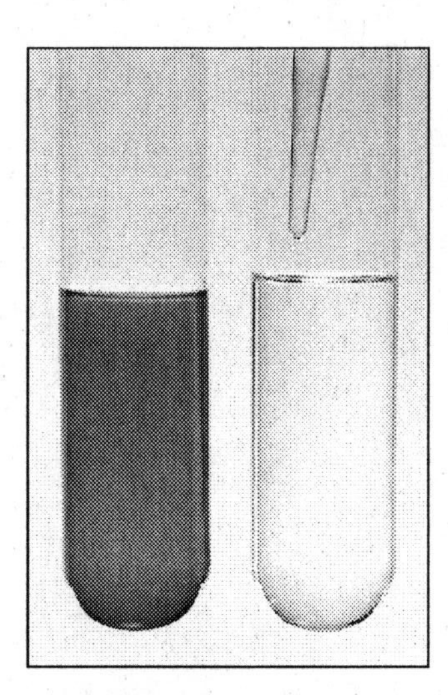

A solution of bromine in dichloromethane is red. Add a few drops of an alkene and the red color disappears. *(Charles D. Winters)*

C. Addition of Bromine and Chlorine

Chlorine (Cl_2) and bromine (Br_2) react with alkenes at room temperature by the addition of halogen atoms to the two carbon atoms of the double bond, forming two new carbon–halogen bonds:

$$\underset{\text{2-Butene}}{CH_3CH{=}CHCH_3} + Br_2 \xrightarrow{CH_2Cl_2} \underset{\text{2,3-Dibromobutane}}{CH_3\overset{Br}{\overset{|}{C}}H-\overset{Br}{\overset{|}{C}}HCH_3}$$

Fluorine, F_2, also adds to alkenes, but because its reactions are very fast and difficult to control, addition of fluorine is not a useful laboratory reaction. Iodine, I_2, also adds, but the reaction is not preparatively useful.

The addition of bromine and chlorine to a cycloalkene gives a trans dihalocycloalkane. For example, the addition of bromine to cyclohexene gives *trans*-1,2-dibromocyclohexane; the cis isomer is not formed. Thus, the addition of a halogen to a cycloalkene is stereoselective. A **stereoselective reaction** is a reaction in which one stereoisomer is formed or destroyed in preference to all others that might be formed or destroyed. Addition of bromine to a cycloalkene is highly stereoselective; the halogen atoms always add trans to each other:

Stereoselective reaction A reaction in which one stereoisomer is formed or destroyed in preference to all others that might be formed or destroyed.

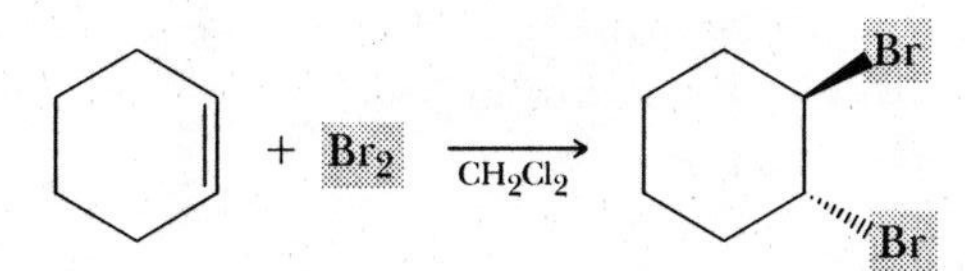

Cyclohexene *trans*-1,2-Dibromocyclohexane

The reaction of bromine with an alkene is a particularly useful qualitative test for the presence of a carbon–carbon double bond. If we dissolve bromine in dichloromethane, the solution turns red. Both alkenes and dibromoalkanes are colorless. If we now mix a few drops of the bromine solution with an alkene, a dibromoalkane is formed, and the solution becomes colorless.

Example 5.7

Complete these reactions, showing the relative orientation of the substituents in the product:

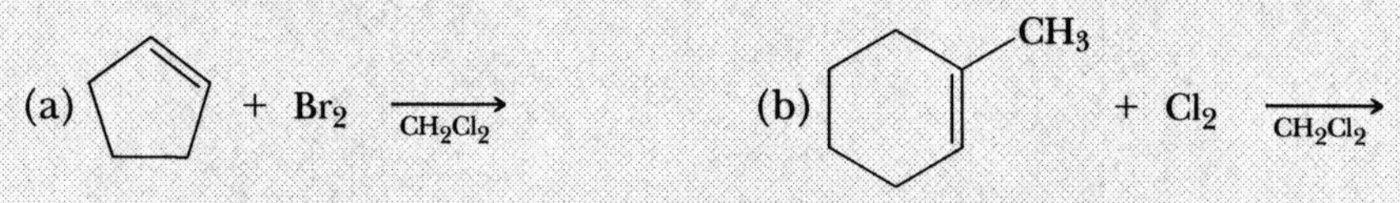

(a) cyclopentene + Br_2 $\xrightarrow{CH_2Cl_2}$ (b) 1-methylcyclohexene + Cl_2 $\xrightarrow{CH_2Cl_2}$

SOLUTION

The halogen atoms are trans to each other in each product:

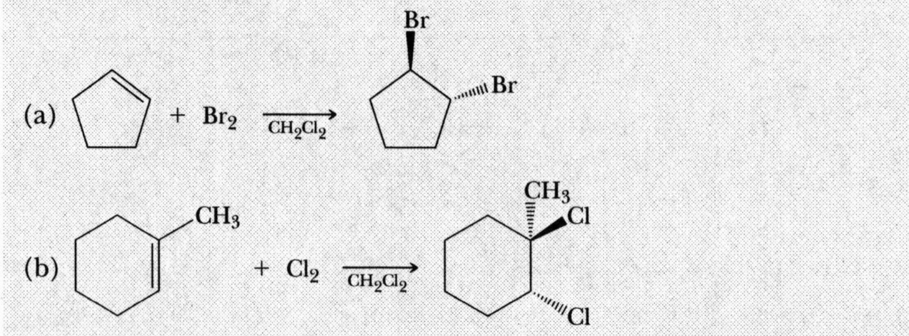

(a) cyclopentene + Br_2 $\xrightarrow{CH_2Cl_2}$ trans-1,2-dibromocyclopentane (Br, Br)

(b) 1-methylcyclohexene + Cl_2 $\xrightarrow{CH_2Cl_2}$ trans-1,2-dichloro-1-methylcyclohexane (CH_3, Cl, Cl)

Practice Problem 5.7

Complete these reactions:

(a) $CH_3C(CH_3)_2CH{=}CH_2 + Br_2 \xrightarrow{CH_2Cl_2}$ (b) methylenecyclohexane ($=CH_2$) + Cl_2 $\xrightarrow{CH_2Cl_2}$

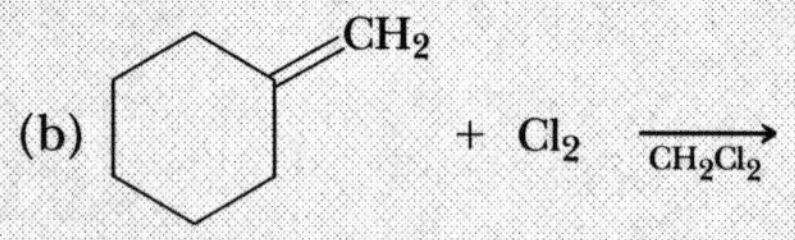

Anti Selectivity and Bridged Halonium Ion Intermediates

We explain the addition of bromine and chlorine to cycloalkenes, as well as their selectivity (they always add trans to each other), by a two-step mechanism that involves a halogen atom bearing a positive charge, called a **halonium ion**. The cyclic structure of which this ion is a part is called a **bridged halonium ion.** The bridged bromonium ion shown in the mechanism may look odd to you, but it is an acceptable Lewis structure. A calculation of formal charge places a positive charge on bromine. Then, in Step 2, a bromide ion reacts with the bridged intermediate from the side opposite that occupied by the bromine atom, giving the dibromoalkane. Thus, bromine atoms add from opposite faces of the carbon–carbon double bond. We say that this addition occurs with **anti selectivity**. Alternatively, we say that the addition of halogens is stereoselective involving anti addition of the halogen atoms.

Halonium ion An ion in which a halogen atom bears a positive charge.

Anti stereoselectivity Addition of atoms or groups of atoms from opposite sides or faces of a carbon–carbon double bond.

Mechanism: Addition of Bromine with Anti Selectivity

Step 1: Reaction of the pi electrons of the carbon–carbon double bond with bromine forms a bridged bromonium ion intermediate in which bromine bears a positive formal charge:

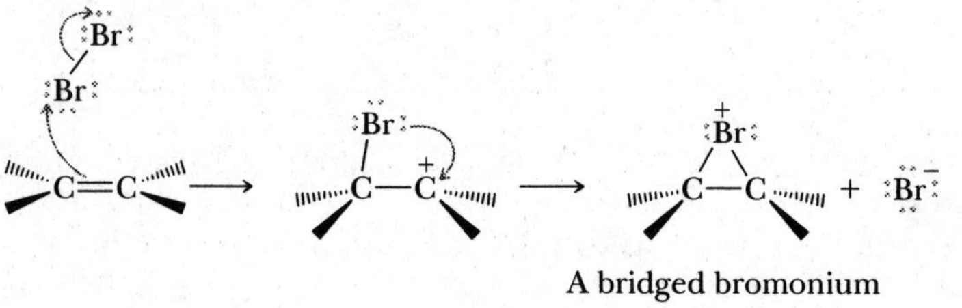

Br–Br + C=C ⟶ Br–C–C^+ ⟶ Br^+ bridged C–C + Br^-

A bridged bromonium ion intermediate

Step 2: A bromide ion (a Lewis base) attacks carbon (a Lewis acid) from the side opposite the bridged bromonium ion, opening the three-membered ring:

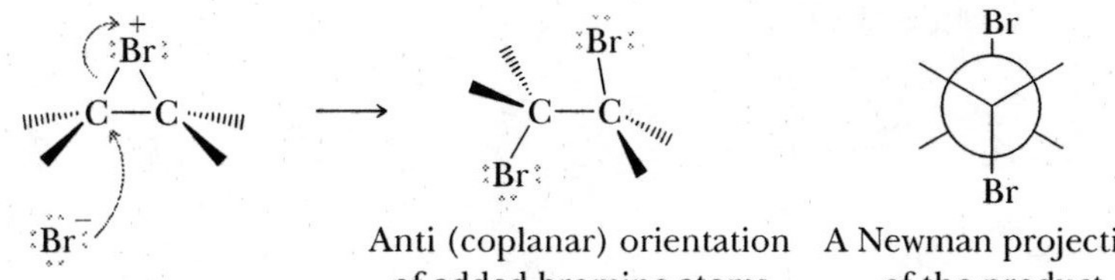

Anti (coplanar) orientation of added bromine atoms

A Newman projection of the product

The addition of chlorine or bromine to cyclohexene and its derivatives gives a trans diaxial product because only axial positions on adjacent atoms of a cyclohexane ring are anti and coplanar. The initial trans diaxial conformation of the product is in equilibrium with the trans diequatorial conformation, and, in simple derivatives of cyclohexane, the latter is more stable and predominates.

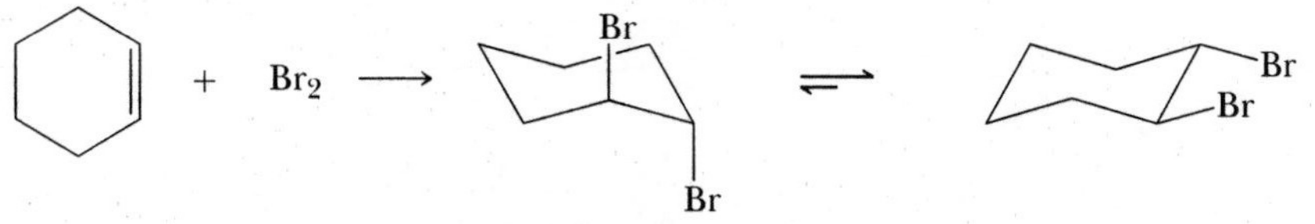

trans Diaxial

trans Diequatorial (more stable)

5.4 OXIDATION OF ALKENES: FORMATION OF GLYCOLS

Recall from your course in general chemistry that oxidation and reduction can be defined in terms of the loss or gain of oxygens or hydrogens by a compound. For organic compounds, we define oxidation and reduction as follows:

oxidation: addition of O to, or removal of H from, a carbon atom

reduction: removal of O from, or addition of H to, a carbon atom

Osmium tetroxide, OsO_4, and certain other transition metal oxides are effective oxidizing agents for the conversion of an alkene to a **glycol**—a compound with two hydroxyl groups on adjacent carbons. The oxidation of an alkene by osmium tetroxide is stereoselective, involving **syn addition** (addition from the same side) of —OH groups to the carbons of the double bond. For example, the oxidation of cyclopentene by OsO_4 gives *cis*-1,2-cyclopentanediol, a cis glycol:

Glycol A compound with two hydroxyl (—OH) groups on adjacent carbons.

Syn addition Addition of atoms or groups of atoms from the same side or face of a carbon–carbon double bond.

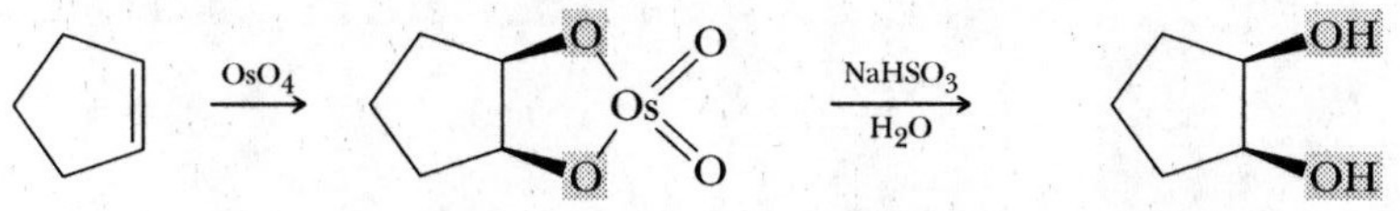

A cyclic osmate

cis-1,2-Cyclopentanediol (a cis glycol)

Note that both cis and trans isomers are possible for this glycol, but only the cis isomer is formed.

The syn selectivity of the osmium tetroxide oxidation of alkenes is accounted for by the formation of a cyclic osmate in which oxygen atoms of OsO_4 form new covalent bonds with each carbon of the double bond in such a way that the five-membered osmium-containing ring fuses cis to the original alkene. Osmates can be isolated and characterized. Usually, however, they are treated directly with a reducing agent, such as $NaHSO_3$, which cleaves osmium–oxygen bonds to give a cis glycol and reduced forms of osmium.

The drawbacks of OsO_4 are that it is both expensive and highly toxic. One strategy to circumvent the high cost of OsO_4 is to use it in catalytic amounts along with stoichiometric amounts of another oxidizing agent, which reoxidizes the reduced forms of osmium and thus recycles the osmium reagent. Oxidizing agents commonly used for this purpose are hydrogen peroxide (H_2O_2) and *tert*-butyl hydroperoxide ($(CH_3)_3COOH$). When this procedure is used, there is no need for a reducing step using $NaHSO_3$.

5.5 REDUCTION OF ALKENES: FORMATION OF ALKANES

A. Catalytic Reduction

Most alkenes react quantitatively with molecular hydrogen, H_2, in the presence of a transition metal catalyst to give alkanes. Commonly used transition metal catalysts include platinum, palladium, ruthenium, and nickel. Yields are usually quantitative or nearly so. Because the conversion of an alkene to an alkane involves reduction by hydrogen in the presence of a catalyst, the process is called **catalytic reduction** or, alternatively, **catalytic hydrogenation**.

$$\text{Cyclohexene} + H_2 \xrightarrow[25°C,\ 3\ atm]{Pd} \text{Cyclohexane}$$

The metal catalyst is used as a finely powdered solid, which may be supported on some inert material such as powdered charcoal or alumina. The reaction is carried out by dissolving the alkene in ethanol or another nonreacting organic solvent, adding the solid catalyst, and exposing the mixture to hydrogen gas at pressures from 1 to 100 atm. Alternatively, the metal may be chelated with certain organic molecules and used in the form of a soluble complex.

Catalytic reduction is stereoselective, the most common pattern being the **syn addition** of hydrogens to the carbon–carbon double bond. The catalytic reduction of 1,2-dimethylcyclohexene, for example, yields *cis*-1,2-dimethylcyclohexane:

$$\text{1,2-Dimethylcyclohexene} + H_2 \xrightarrow{Pt} \textit{cis}\text{-1,2-Dimethylcyclohexane}$$

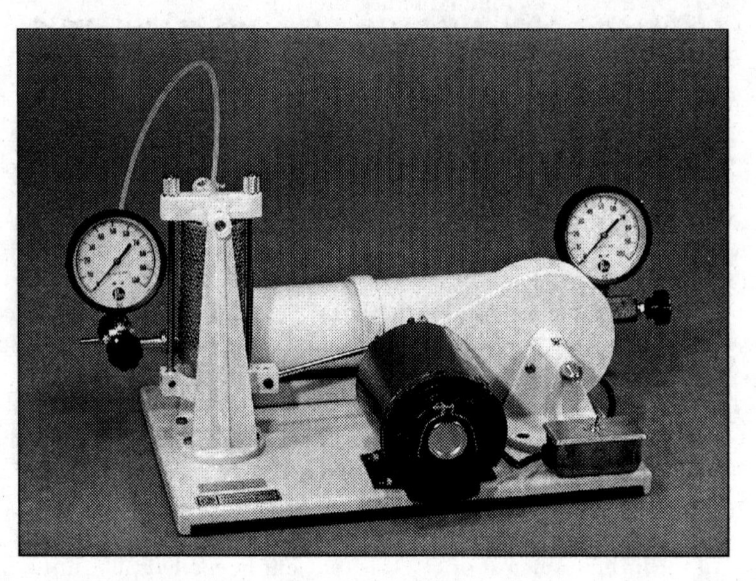

A Paar shaker-type hydrogenation apparatus. *(Paar Instrument Co., Moline, IL.)*

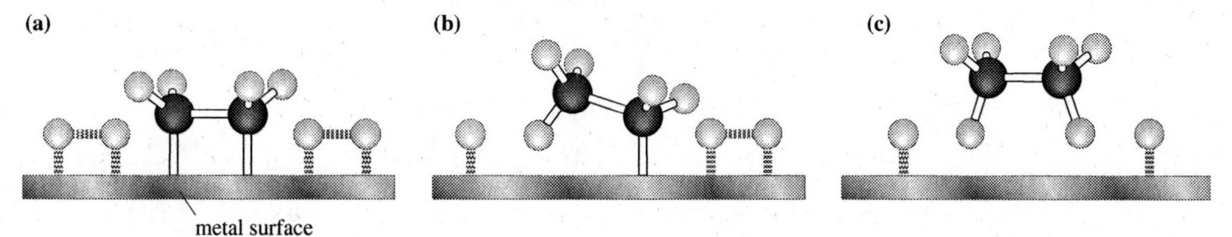

Figure 5.6 Syn addition of hydrogen to an alkene involving a transition metal catalyst. (a) Hydrogen and the alkene are adsorbed on the metal surface, and (b) one hydrogen atom is transferred to the alkene, forming a new C—H bond. The other carbon remains adsorbed on the metal surface. (c) A second C—H bond forms and the alkane is desorbed.

The transition metals used in catalytic reduction are able to adsorb large quantities of hydrogen onto their surfaces, probably by forming metal–hydrogen sigma bonds. Similarly, these transition metals adsorb alkenes on their surfaces, with the formation of carbon–metal bonds [Figure 5.6(a)]. Hydrogen atoms are added to the alkene in two steps.

B. Heats of Hydrogenation and the Relative Stabilities of Alkenes

The **heat of hydrogenation** of an alkene is defined as its heat of reaction, ΔH°, with hydrogen, to form an alkane. Table 5.2 lists the heats of hydrogenation of several alkenes.

Three important points follow from the information given in the table:

TABLE 5.2 Heats of Hydrogenation of Several Alkenes

Name	Structural Formula	ΔH° kJ/mol (kcal/mol)	
Ethylene	$CH_2{=}CH_2$	−137 (−32.8)	
Propene	$CH_3CH{=}CH_2$	−126 (−30.1)	
1-Butene	$CH_3CH_2CH{=}CH_2$	−127 (−30.3)	Ethylene
cis-2-Butene	$\mathrm{(H_3C)(H)C{=}C(H)(CH_3)}$ (both CH_3 on same side)	−120 (−28.6)	
trans-2-Butene	$\mathrm{(H_3C)(H)C{=}C(CH_3)(H)}$ (CH_3 groups on opposite sides)	−115 (−27.6)	*trans*-2-Butene
2-Methyl-2-butene	$\mathrm{(H_3C)_2C{=}C(H)CH_3}$	−113 (−26.9)	
2,3-Dimethyl-2-butene	$\mathrm{(H_3C)_2C{=}C(CH_3)_2}$	−111 (−26.6)	2,3-Dimethyl-2-butene

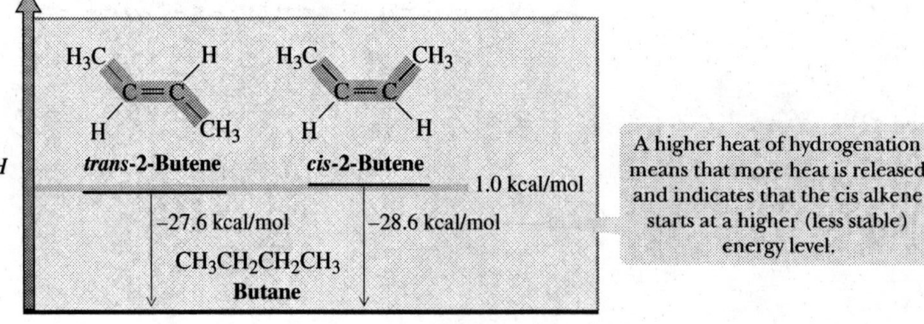

Figure 5.7
Heats of hydrogenation of *cis*-2-butene and *trans*-2-butene. *Trans*-2-Butene is more stable than *cis*-2-butene by 1.0 kcal/mol (4.2 kJ/mol).

1. The reduction of an alkene to an alkane is an exothermic process. This observation is consistent with the fact that, during hydrogenation, there is net conversion of a weaker pi bond to a stronger sigma bond; that is, one sigma bond (H—H) and one pi bond (C=C) are broken, and two new sigma bonds (C—H) are formed.
2. The heats of hydrogenation depend on the degree of substitution of the carbon–carbon double bond: The greater the substitution, the lower is the heat of hydrogenation. Compare, for example, the heats of hydrogenation of ethylene (no substituents), propene (one substituent), 1-butene (one substituent), and the cis and trans isomers of 2-butene (two substituents each).
3. The heat of hydrogenation of a trans alkene is lower than that of the isomeric cis alkene. Compare, for example, the heats of hydrogenation of *cis*-2-butene and *trans*-2-butene. Because the reduction of each alkene gives butane, any difference in their heats of hydrogenation must be due to a difference in relative energy between the two alkenes (Figure 5.7). The alkene with the lower (less negative) value of ΔH° is the more stable alkene.

We explain the greater stability of trans alkenes relative to cis alkenes in terms of nonbonded interaction strain. In *cis*-2-butene, the two $-CH_3$ groups are sufficiently close to each other that there is repulsion between their electron clouds. This repulsion is reflected in the larger heat of hydrogenation (decreased stability) of *cis*-2-butene compared with that of *trans*-2-butene (approximately 1.0 kcal/mol).

SUMMARY

A **reaction mechanism** (Section 5.2) is a description of (1) how and why a chemical reaction occurs, (2) which bonds break and which new ones form, (3) the order and relative rates with which the various bond-breaking and bond-forming steps take place, and (4) the role of the catalyst if the reaction involves one. **Transition state theory** (Section 5.2A) provides a model for understanding the relationships among reaction rates, molecular structure, and energetics. A key postulate of transition state theory is that a **transition state** is formed. The difference in energy between the reactants and the transition state is called the **activation energy**. An **intermediate** is an energy minimum between two transition states. The slowest step in a multistep reaction, called the **rate-determining step**, is the one that crosses the highest energy barrier.

A characteristic reaction of alkenes is **addition**, during which a pi bond is broken and sigma bonds to two new atoms or groups of atoms are formed.

An **electrophile** (Section 5.3A) is any molecule or ion that can accept a pair of electrons to form a new covalent bond. All electrophiles are Lewis acids. The rate-determining step in **electrophilic addition** to an alkene is the reaction of an electrophile with a carbon–carbon double bond to form a **carbocation**—an ion that contains a carbon with only six electrons in its valence shell and has a positive charge. Carbocations are planar, with bond angles of 120° about the positive carbon. The order of stability of carbocations is $3^\circ > 2^\circ > 1^\circ >$ methyl (Section 5.3A).

KEY REACTIONS

1. Addition of H—X (Section 5.3A)

Addition of H—X is regioselective and follows Markovnikov's rule. Reaction occurs in two steps and involves the formation of a carbocation intermediate:

$$\text{(1-methylcyclopentene)}\;CH_3 + HCl \longrightarrow \text{(1-chloro-1-methylcyclopentane)}\;Cl,\ CH_3$$

2. Acid-Catalyzed Hydration (Section 5.3B)

Hydration is regioselective and follows Markovnikov's rule. Reaction occurs in two steps and involves the formation of a carbocation intermediate:

$$CH_3\overset{CH_3}{\overset{|}{C}}{=}CH_2 + H_2O \xrightarrow{H_2SO_4} CH_3\underset{OH}{\underset{|}{\overset{CH_3}{\overset{|}{C}}}}CH_3$$

3. Addition of Bromine and Chlorine (Section 5.3C)

Addition occurs in two steps, and involves anti addition by way of a bridged bromonium or chloronium ion intermediate:

$$\text{(cyclohexene)} + Br_2 \xrightarrow{CH_2Cl_2} \text{(trans-1,2-dibromocyclohexane)}\;Br,\ Br$$

4. Oxidation: Formation of Glycols (Section 5.4B)

Oxidation occurs by the syn addition of —OH groups to the double bond via a cyclic osmate:

$$\text{(cyclopentene)} \xrightarrow[ROOH]{OsO_4} \text{(cis-1,2-cyclopentanediol)}\;OH,\ OH$$

5. Reduction: Formation of Alkanes (Section 5.5)

Catalytic reduction involves predominantly the syn addition of hydrogen:

$$\text{(1,2-dimethylcyclohexene)}\;CH_3,\ CH_3 + H_2 \xrightarrow{\text{transition metal catalyst}} \text{(cis-1,2-dimethylcyclohexane)}\;H,\ CH_3,\ CH_3,\ H$$

PROBLEMS

A problem number set in red indicates an applied "real-world" problem.

Energy Diagrams

5.8 Describe the differences between a transition state and a reaction intermediate.

5.9 Sketch an energy diagram for a one-step reaction that is very slow and only slightly exothermic. How many transition states are present in this reaction? How many intermediates are present?

5.10 Sketch an energy diagram for a two-step reaction that is endothermic in the first step, exothermic in the second step, and exothermic overall. How many transition states are present in this two-step reaction? How many intermediates are present?

5.11 Determine whether each of the following statements is true or false, and provide a rationale for your decision:

(a) A transition state can never be lower in energy than the reactants from which it was formed.

(b) An endothermic reaction cannot have more than one intermediate.

(c) An exothermic reaction cannot have more than one intermediate.

Electrophilic Additions

5.12 From each pair, select the more stable carbocation:

(a) $CH_3CH_2CH_2^+$ or $CH_3\overset{+}{C}HCH_3$ **(b)** $CH_3\underset{+}{C}H(CH_3)CHCH_3$ or $CH_3\underset{+}{C}(CH_3)CH_2CH_3$

5.13 From each pair, select the more stable carbocation:

(a) + —CH_3 or +CH_3 **(b)** + —CH_3 or —CH_2^+

5.14 Draw structural formulas for the isomeric carbocation intermediates formed by the reaction of each alkene with HCl. Label each carbocation as primary, secondary, or tertiary, and state which, if either, of the isomeric carbocations is formed more readily.

(a) **(b)** **(c)** CH_3 **(d)** $=CH_2$

5.15 From each pair of compounds, select the one that reacts more rapidly with HI, draw the structural formula of the major product formed in each case, and explain the basis for your ranking:

(a) and **(b)** and

5.16 Complete these equations by predicting the major product formed in each reaction:

(a) + HCl ⟶ **(b)** + $H_2O \xrightarrow{H_2SO_4}$

(c) + HI ⟶ **(d)** + HCl ⟶

(e) + $H_2O \xrightarrow{H_2SO_4}$ **(f)** + $H_2O \xrightarrow{H_2SO_4}$

5.17 The reaction of 2-methyl-2-pentene with each reagent is regioselective. Draw a structural formula for the product of each reaction, and account for the observed regioselectivity.

(a) HI **(b)** H_2O in the presence of H_2SO_4

5.18 The addition of bromine and chlorine to cycloalkenes is stereoselective. Predict the stereochemistry of the product formed in each reaction:

(a) 1-Methylcyclohexene + Br_2 **(b)** 1,2-Dimethylcyclopentene + Cl_2

5.19 Draw a structural formula for an alkene with the indicated molecular formula that gives the compound shown as the major product. Note that more than one alkene may give the same compound as the major product.

(a) $C_5H_{10} + H_2O \xrightarrow{H_2SO_4}$ OH **(b)** $C_5H_{10} + Br_2 \longrightarrow$ Br Br

(c) $C_7H_{12} + HCl \longrightarrow$ Cl

5.20 Draw the structural formula for an alkene with molecular formula C_5H_{10} that reacts with Br_2 to give each product:

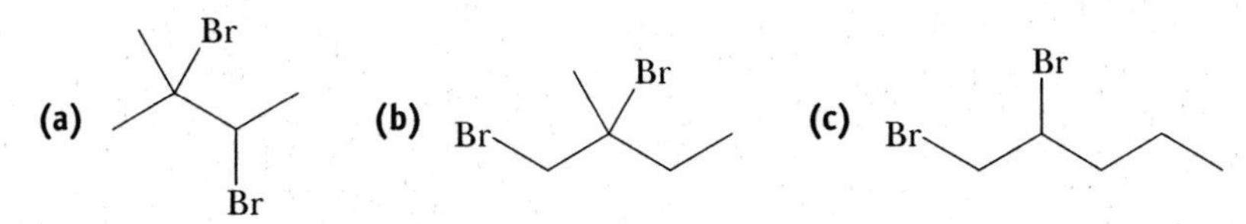

5.21 Draw the structural formula for a cycloalkene with molecular formula C_6H_{10} that reacts with Cl_2 to give each compound:

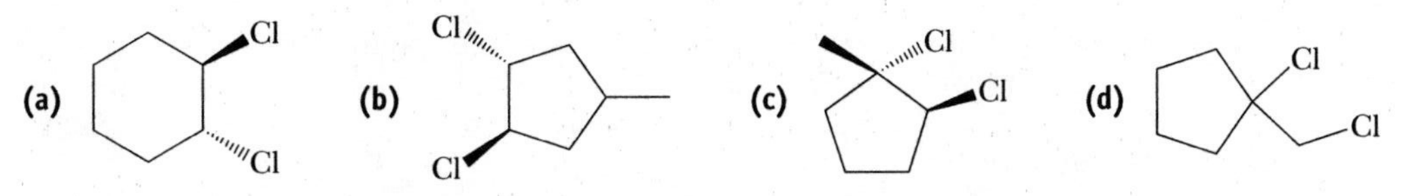

5.22 Draw the structural formula for an alkene with molecular formula C_5H_{10} that reacts with HCl to give the indicated chloroalkane as the major product:

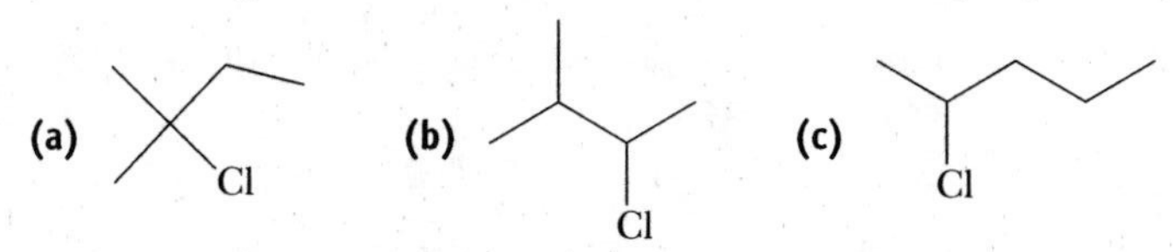

5.23 Draw the structural formula of an alkene that undergoes acid-catalyzed hydration to give the indicated alcohol as the major product. More than one alkene may give each compound as the major product.

(a) 3-Hexanol **(b)** 1-Methylcyclobutanol
(c) 2-Methyl-2-butanol **(d)** 2-Propanol

5.24 Draw the structural formula of an alkene that undergoes acid-catalyzed hydration to give each alcohol as the major product. More than one alkene may give each compound as the major product.

(a) Cyclohexanol **(b)** 1,2-Dimethylcyclopentanol
(c) 1-Methylcyclohexanol **(d)** 1-Isopropyl-4-methylcyclohexanol

5.25 Terpin is prepared commercially by the acid-catalyzed hydration of limonene:

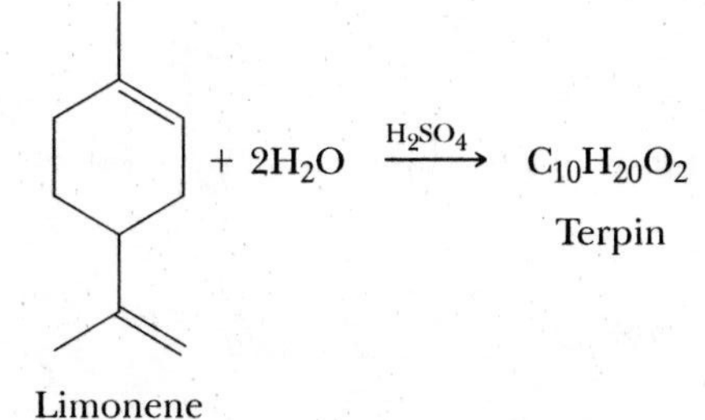

$+ 2H_2O \xrightarrow{H_2SO_4} C_{10}H_{20}O_2$
Terpin

(a) Propose a structural formula for terpin and a mechanism for its formation.
(b) How many cis–trans isomers are possible for the structural formula you propose?
(c) Terpin hydrate, the isomer in terpin in which the one-carbon and three-carbon substituents are cis to each other, is used as an expectorant in cough medicines. Draw the alternative chair conformations for terpin hydrate, and state which of the two is the more stable.

5.26 The treatment of 2-methylpropene with methanol in the presence of a sulfuric acid catalyst gives *tert*-butyl methyl ether:

$$\underset{\text{2-Methylpropene}}{CH_3\overset{\overset{CH_3}{|}}{C}{=}CH_2} + \underset{\text{Methanol}}{CH_3OH} \xrightarrow{H_2SO_4} \underset{\textit{tert}\text{-Butyl methyl ether}}{CH_3\overset{\overset{CH_3}{|}}{\underset{\underset{CH_3}{|}}{C}}{-}OCH_3}$$

Propose a mechanism for the formation of this ether.

5.27 The treatment of 1-methylcyclohexene with methanol in the presence of a sulfuric acid catalyst gives a compound with molecular formula $C_8H_{16}O$:

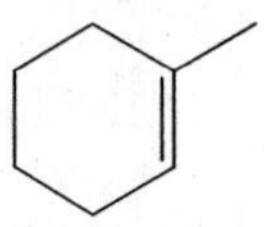

$$\text{1-Methylcyclohexene} + \underset{\text{Methanol}}{CH_3OH} \xrightarrow{H_2SO_4} C_8H_{16}O$$

Propose a structural formula for this compound and a mechanism for its formation.

5.28 *cis*-3-Hexene and *trans*-3-hexene are different compounds and have different physical and chemical properties. Yet, when treated with H_2O/H_2SO_4, each gives the same alcohol. What is the alcohol, and how do you account for the fact that each alkene gives the same one?

Oxidation–Reduction

5.29 Which of these transformations involve oxidation, which involve reduction, and which involve neither oxidation nor reduction?

(a) $CH_3\overset{\overset{OH}{|}}{C}HCH_3 \longrightarrow CH_3\overset{\overset{O}{\|}}{C}CH_3$ **(b)** $CH_3\overset{\overset{OH}{|}}{C}HCH_3 \longrightarrow CH_3CH{=}CH_2$

(c) $CH_3CH{=}CH_2 \longrightarrow CH_3CH_2CH_3$

5.30 Write a balanced equation for the combustion of 2-methylpropene in air to give carbon dioxide and water. The oxidizing agent is O_2, which makes up approximately 20% of air.

5.31 Draw the product formed by treating each alkene with aqueous $OsO_4/ROOH$:
(a) 1-Methylcyclopentene **(b)** 1-Cyclohexylethylene **(c)** *cis*-2-Pentene

5.32 What alkene, when treated with $OsO_4/ROOH$, gives each glycol?

(a) 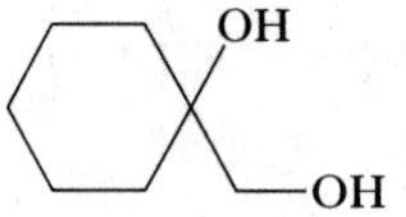 OH, OH **(b)** 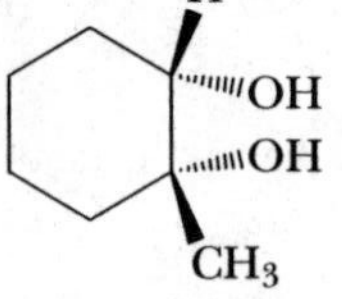 H, OH, OH, CH_3 **(c)** 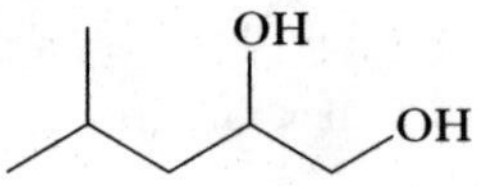 OH, OH

5.33 Draw the product formed by treating each alkene with H_2/Ni:

(a)  **(b)** 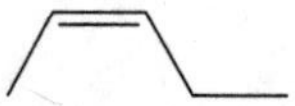 **(c)** 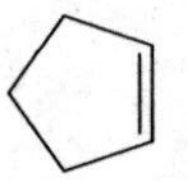 **(d)**

5.34 Hydrocarbon A, C_5H_8, reacts with 2 moles of Br_2 to give 1,2,3,4-tetrabromo-2-methylbutane. What is the structure of hydrocarbon A?

5.35 Two alkenes, A and B, each have the formula C_5H_{10}. Both react with H_2/Pt and with HBr to give identical products. What are the structures of A and B?

Synthesis

5.36 Show how to convert ethylene into these compounds:

(a) Ethane **(b)** Ethanol **(c)** Bromoethane
(d) 1,2-Dibromoethane **(e)** 1,2-Ethanediol **(f)** Chloroethane

5.37 Show how to convert cyclopentene into these compounds:

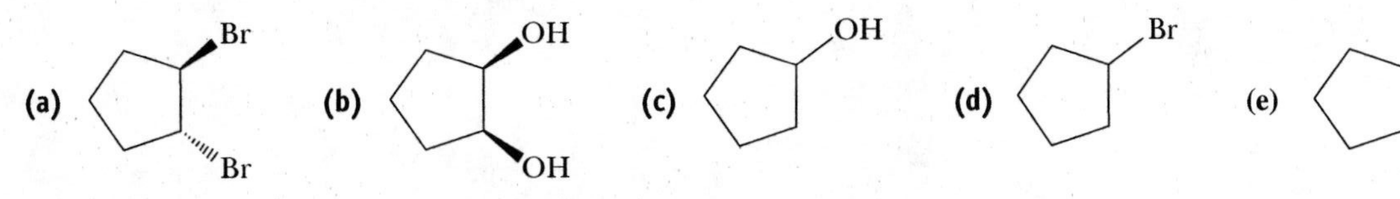

5.38 Show how to convert 1-butene into these compounds:

(a) Butane **(b)** 2-Butanol
(c) 2-Bromobutane **(d)** 1,2-Dibromobutane

5.39 Show how the following compounds can be synthesized in good yields from an alkene:

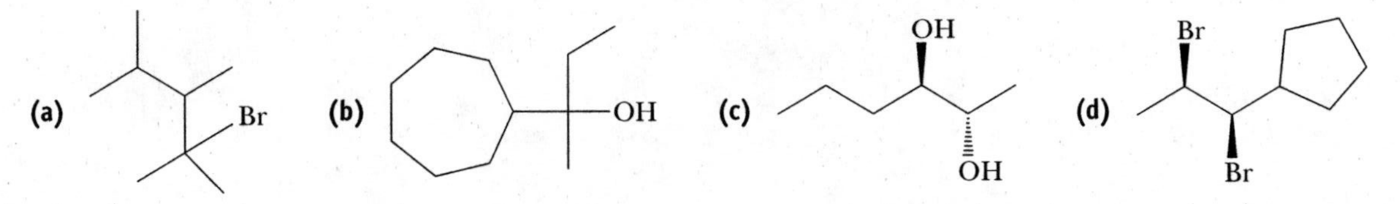

Looking Ahead

5.40 Each of the following 2° carbocations is more stable than the tertiary carbocation shown:

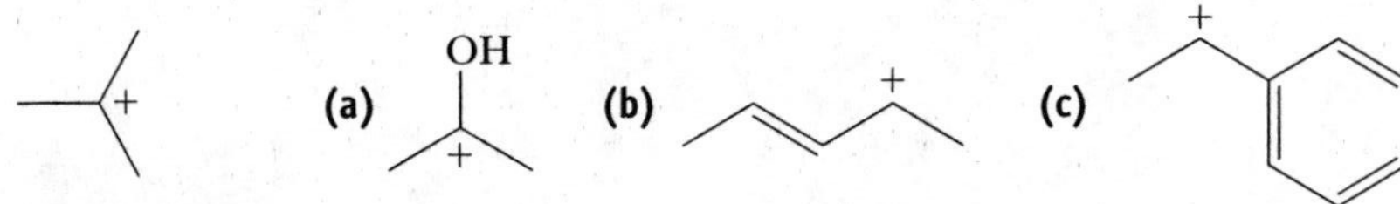

Provide an explanation for each cation's enhanced stability.

5.41 Recall that an alkene possesses a π cloud of electrons above and below the plane of the C=C bond. Any reagent can therefore react with either face of the double bond. Determine whether the reaction of each of the given reagents with the top face of *cis*-2-butene will produce the same product as the reaction of the same reagent with the bottom face. (*Hint:* Build molecular models of the products and compare them.)

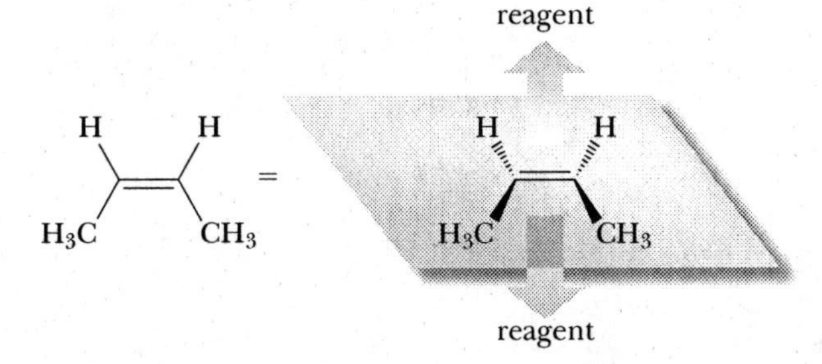

(a) H_2/Pt **(b)** $OsO_4/ROOH$ **(c)** Br_2/CH_2Cl_2

5.42 Each of the following reactions yields two products in differing amounts:

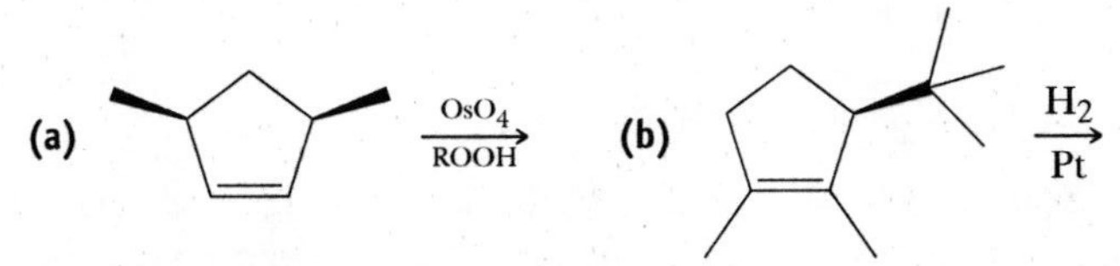

Draw the products of each reaction and determine which product is favored.

7 Haloalkanes

Compact disks are made from poly(vinyl chloride), which is in turn made from 1,2-dichloroethane. Inset: A molecule of vinyl chloride. *(Charles D. Winter)*

7.1 INTRODUCTION

Compounds containing a halogen atom covalently bonded to an sp^3 hybridized carbon atom are named *haloalkanes* or, in the common system of nomenclature, *alkyl halides.* The general symbol for an **alkyl halide** is R—X, where X may be F, Cl, Br, or I:

Alkyl halide A compound containing a halogen atom covalently bonded to an alkyl group; given the symbol RX.

$$R-\ddot{\underset{\cdot\cdot}{X}}:$$

A haloalkane (An alkyl halide)

In this chapter, we study two characteristic reactions of haloalkanes: nucleophilic substitution and β-elimination. By these reactions, haloalkanes can be converted to alcohols, ethers, thiols, amines, and alkenes and are thus versatile molecules. Indeed, haloalkanes are often used as starting materials for the synthesis of many useful compounds encountered in all walks of life, such as medicine, food chemistry, and agriculture (to name a few).

7.2 NOMENCLATURE

A. IUPAC Names

IUPAC names for haloalkanes are derived by naming the parent alkane according to the rules given in Section 3.4A:

- Locate and number the parent chain from the direction that gives the substituent encountered first the lower number.
- Show halogen substituents by the prefixes *fluoro-*, *chloro-*, *bromo-*, and *iodo-*, and list them in alphabetical order along with other substituents.
- Use a number preceding the name of the halogen to locate each halogen on the parent chain.
- In haloalkenes, the location of the double bond determines the numbering of the parent hydrocarbon. In molecules containing functional groups designated by a suffix (for example, *-ol*, *-al*, *-one*, *-oic acid*), the location of the functional group indicated by the suffix determines the numbering:

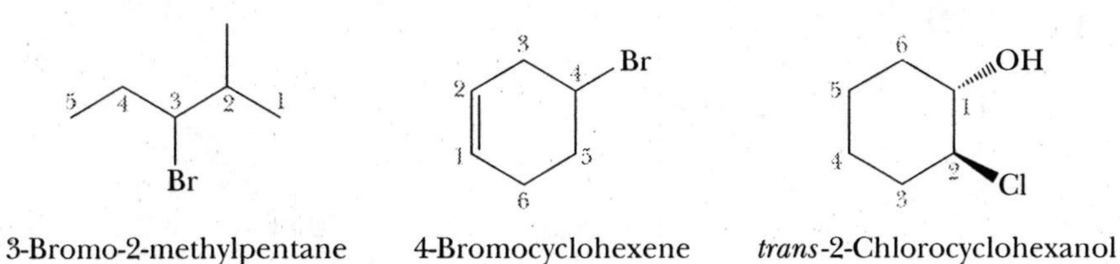

3-Bromo-2-methylpentane 4-Bromocyclohexene *trans*-2-Chlorocyclohexanol

B. Common Names

Common names of haloalkanes consist of the common name of the alkyl group, followed by the name of the halide as a separate word. Hence, the name **alkyl halide** is a common name for this class of compounds. In the following examples, the IUPAC name of the compound is given first, followed by its common name, in parentheses:

$CH_3CH(F)CH_2CH_3$	$CH_2{=}CHCl$
2-Fluorobutane	Chloroethene
(*sec*-Butyl fluoride)	(Vinyl chloride)

Several of the polyhalomethanes are common solvents and are generally referred to by their common, or trivial, names. Dichloromethane (methylene chloride) is the most widely used haloalkane solvent. Compounds of the type CHX_3 are called **haloforms**. The common name for $CHCl_3$, for example, is *chloroform*. The common name for CH_3CCl_3 is *methyl chloroform*. Methyl chloroform and trichloroethylene are solvents for commercial dry cleaning.

CH_2Cl_2	$CHCl_3$	CH_3CCl_3	$CCl_2{=}CHCl$
Dichloromethane	Trichloromethane	1,1,1-Trichloroethane	Trichloroethylene
(Methylene chloride)	(Chloroform)	(Methyl chloroform)	(Trichlor)

EXAMPLE 7.1

Write the IUPAC name for each compound:

(a) Br (b) Br (c) H Br

SOLUTION

(a) 1-Bromo-2-methylpropane. Its common name is isobutyl bromide.
(b) (*E*)-4-Bromo-3-methyl-2-pentene or *trans*-4-bromo-3-methyl-2-pentene.
(c) (*S*)-2-Bromohexane.

Practice Problem 7.1

Write the IUPAC name for each compound:

(a) Cl (b) CH_3 Br

(c) $CH_3CH(Cl)CH_2Cl$ (d) Cl

Of all the haloalkanes, the **chlorofluorocarbons (CFCs)** manufactured under the trade name Freon® are the most widely known. CFCs are nontoxic, nonflammable, odorless, and noncorrosive. Originally, they seemed to be ideal replacements for the hazardous compounds such as ammonia and sulfur dioxide formerly used as heat-transfer agents in refrigeration systems. Among the CFCs most widely used for this purpose were trichlorofluoromethane (CCl_3F, Freon-11) and dichlorodifluoromethane (CCl_2F_2, Freon-12). The CFCs also found wide use as industrial cleaning solvents to prepare surfaces for coatings, to remove cutting oils and waxes from millings, and to remove protective coatings. In addition, they were employed as propellants in aerosol sprays.

7.3 NUCLEOPHILIC ALIPHATIC SUBSTITUTION AND β-ELIMINATION

Nucleophile An atom or a group of atoms that donates a pair of electrons to another atom or group of atoms to form a new covalent bond.

Nucleophilic substitution A reaction in which one nucleophile is substituted for another.

A **nucleophile** (nucleus-loving reagent) is any reagent that donates an unshared pair of electrons to form a new covalent bond. **Nucleophilic substitution** is any reaction in which one nucleophile is substituted for another. In the following general equations, $Nu:^-$ is the nucleophile, X is the leaving group, and substitution takes place on an sp^3 hybridized carbon atom:

$$\underset{\text{Nucleophile}}{Nu:^-} + -\overset{|}{\underset{|}{C}}-X \xrightarrow{\text{nucleophilic substitution}} -\overset{|}{\underset{|}{C}}-Nu + :X^-$$

(leaving group: X)

Halide ions are among the best and most important leaving groups.

CHEMICAL CONNECTIONS

The Environmental Impact of Chlorofluorocarbons

Concern about the environmental impact of CFCs arose in the 1970s when researchers found that more than 4.5×10^5 kg/yr of these compounds were being emitted into the atmosphere. In 1974, Sherwood Rowland and Mario Molina announced their theory, which has since been amply confirmed, that CFCs catalyze the destruction of the stratospheric ozone layer. When released into the air, CFCs escape to the lower atmosphere. Because of their inertness, however, they do not decompose there. Slowly, they find their way to the stratosphere, where they absorb ultraviolet radiation from the sun and then decompose. As they do so, they set up a chemical reaction that leads to the destruction of the stratospheric ozone layer, which shields the earth against short-wavelength ultraviolet radiation from the sun. An increase in short-wavelength ultraviolet radiation reaching the earth is believed to promote the destruction of certain crops and agricultural species and even to increase the incidence of skin cancer in light-skinned individuals.

The concern about CFCs prompted two conventions, one in Vienna in 1985 and one in Montreal in 1987, held by the United Nations Environmental Program. The 1987 meeting produced the Montreal Protocol, which set limits on the production and use of ozone-depleting CFCs and urged the complete phaseout of their production by the year 1996. This phaseout resulted in enormous costs for manufacturers and is not yet complete in developing countries.

Rowland, Molina, and Paul Crutzen (a Dutch chemist at the Max Planck Institute for Chemistry in Germany) were awarded the 1995 Nobel prize for chemistry. As the Royal Swedish Academy of Sciences noted in awarding the prize, "By explaining the chemical mechanisms that affect the thickness of the ozone layer, these three researchers have contributed to our salvation from a global environmental problem that could have catastrophic consequences."

The chemical industry responded to the crisis by developing replacement refrigerants that have a much lower ozone-depleting potential. The most prominent replacements are the hydrofluorocarbons (HFCs) and hydrochlorofluorocarbons (HCFCs), such as the following:

```
    F  F            H  Cl
    |  |            |  |
 F—C—C—H        H—C—C—F
    |  |            |  |
    F  H            H  Cl
  HFC-134a        HCFC-141b
```

These compounds are much more chemically reactive in the atmosphere than the Freons are and are destroyed before they reach the stratosphere. However, they cannot be used in air conditioners in 1994 and earlier model cars.

Because all nucleophiles are also bases, nucleophilic substitution and base-promoted **β-elimination** are competing reactions. The ethoxide ion, for example, is both a nucleophile and a base. With bromocyclohexane, it reacts as a nucleophile (pathway shown in red) to give ethoxycyclohexane (cyclohexyl ethyl ether) and as a base (pathway shown in blue) to give cyclohexene and ethanol:

β-Elimination reaction The removal of atoms or groups of atoms from two adjacent carbon atoms, as for example, the removal of H and X from an alkyl halide or H and OH from an alcohol to form a carbon–carbon double bond.

as a nucleophile, ethoxide ion attacks this carbon

Bromocyclohexane (Br, H) + $CH_3CH_2O^-Na^+$ (a nucleophile and a base)

as a base, ethoxide ion attacks this hydrogen

nucleophilic substitution, ethanol → cyclohexyl–OCH_2CH_3 + Na^+Br^-

β-elimination, ethanol → cyclohexene + CH_3CH_2OH + Na^+Br^-

In this chapter, we study both of these organic reactions. Using them, we can convert haloalkanes to compounds with other functional groups including alcohols, ethers, thiols, sulfides, amines, nitriles, alkenes, and alkynes. Thus, an understanding of nucleophilic substitution and β-elimination opens entirely new areas of organic chemistry.

7.4 NUCLEOPHILIC ALIPHATIC SUBSTITUTION

Nucleophilic substitution is one of the most important reactions of haloalkanes and can lead to a wide variety of new functional groups, several of which are illustrated in Table 7.1. As you study the entries in this table, note the following points:

1. If the nucleophile is negatively charged, as, for example, OH^- and RS^-, then the atom donating the pair of electrons in the substitution reaction becomes neutral in the product.
2. If the nucleophile is uncharged, as, for example, NH_3 and CH_3OH, then the atom donating the pair of electrons in the substitution reaction becomes positively charged in the product. The products then often undergo a second step involving proton transfer to yield a neutral substitution product.

Nu = negative → neutral product
Nu = neutral → positive product.
↳ 2nd step using proton transfer to get neutral product.

TABLE 7.1 Some Nucleophilic Substitution Reactions

Reaction: $Nu^- + CH_3X \longrightarrow CH_3Nu + X^-$

Nucleophile		Product	Class of Compound Formed
HO^-	⟶	CH_3OH	An alcohol
RO^-	⟶	CH_3OR	An ether
HS^-	⟶	CH_3SH	A thiol (a mercaptan)
RS^-	⟶	CH_3SR	A sulfide (a thioether)
I^-	⟶	CH_3I	An alkyl iodide
NH_3	⟶	$CH_3NH_3^+$	An alkylammonium ion
HOH	⟶	$CH_3O^+(H)-H$	An alcohol (after proton transfer)
CH_3OH	⟶	$CH_3O^+(H)-CH_3$	An ether (after proton transfer)

EXAMPLE 7.2

Complete these nucleophilic substitution reactions:

(a) 1-bromobutane (Br) $+ Na^+OH^- \longrightarrow$ (b) 1-chlorobutane (Cl) $+ NH_3 \longrightarrow$

SOLUTION

(a) Hydroxide ion is the nucleophile and bromine is the leaving group:

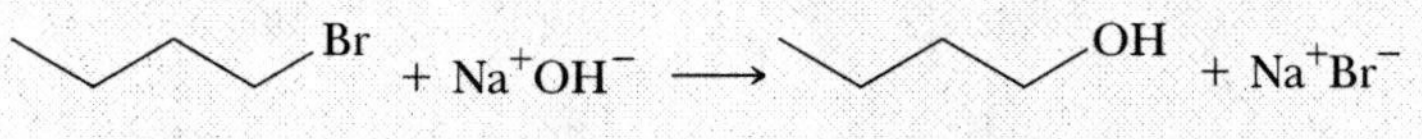

1-Bromobutane + Sodium hydroxide (Na^+OH^-) ⟶ 1-Butanol + Sodium bromide (Na^+Br^-)

(b) Ammonia is the nucleophile and chlorine is the leaving group:

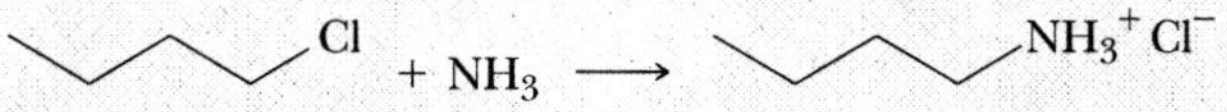

1-Chlorobutane + Ammonia (NH_3) ⟶ Butylammonium chloride ($NH_3^+\ Cl^-$)

Practice Problem 7.2

Complete these nucleophilic substitution reactions:

(a) 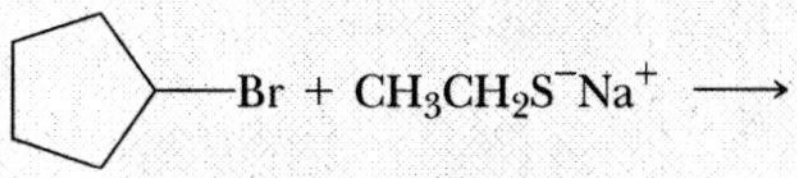 Bromocyclopentane + $CH_3CH_2S^-Na^+$ ⟶

(b) 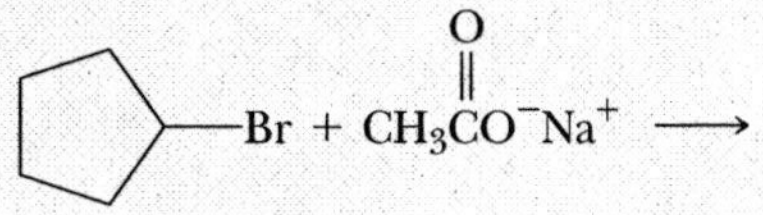 Bromocyclopentane + $CH_3CO^-Na^+$ (with C=O) ⟶

7.5 MECHANISMS OF NUCLEOPHILIC ALIPHATIC SUBSTITUTION

On the basis of a wealth of experimental observations developed over a 70-year period, chemists have proposed two limiting mechanisms for nucleophilic substitutions. A fundamental difference between them is the timing of bond breaking between carbon and the leaving group and of bond forming between carbon and the nucleophile.

A. S_N2 Mechanism

At one extreme, the two processes are *concerted*, meaning that bond breaking and bond forming occur simultaneously. Thus, the departure of the leaving group is assisted by the incoming nucleophile. This mechanism is designated **S_N2**, where *S* stands for *S*ubstitution, *N* for *N*ucleophilic, and 2 for a ***bi*molecular reaction**. This type of substitution reaction is classified as bimolecular because both the haloalkane and the nucleophile are involved in the rate-determining step. That is, both species contribute to the rate law of the reaction:

Bimolecular reaction A reaction in which two species are involved in the reaction leading to the transition state of the rate-determining step.

$$\text{Rate} = k[\text{haloalkane}][\text{nucleophile}]$$

Following is an S_N2 mechanism for the reaction of hydroxide ion and bromomethane to form methanol and bromide ion:

Mechanism: An S_N2 Reaction

The nucleophile attacks the reactive center from the side opposite the leaving group; that is, an S_N2 reaction involves a backside attack by the nucleophile.

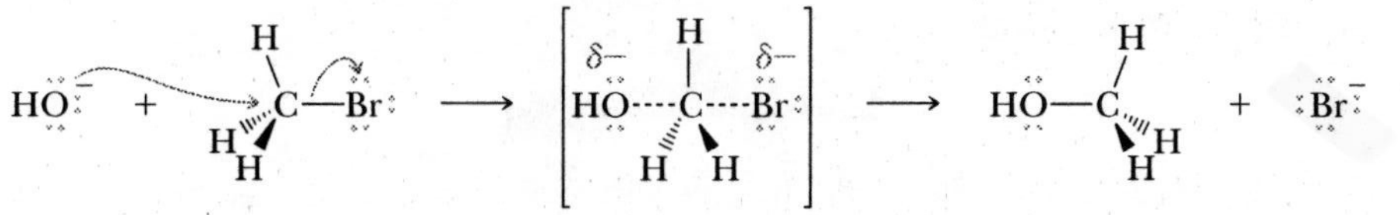

HO^- + CH_3—Br ⟶ [$HO^{\delta-}$····C····$Br^{\delta-}$] ⟶ HO—CH_3 + Br^-

Reactants — Transition state with simultaneous bond breaking and bond forming — Products

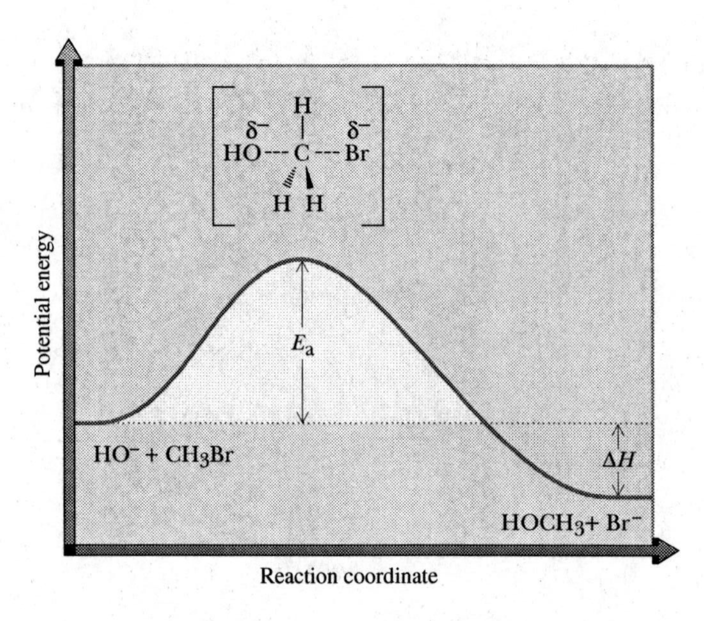

Figure 7.1
An energy diagram for an S_N2 reaction. There is one transition state and no reactive intermediate.

Figure 7.1 shows an energy diagram for an S_N2 reaction. There is a single transition state and no reactive intermediate.

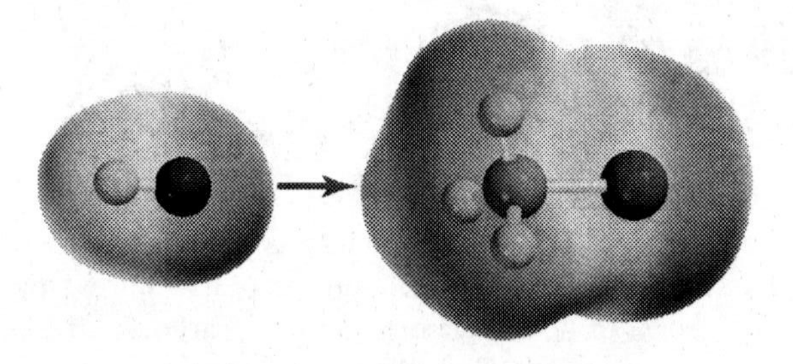

Nucleophilic attack from the side opposite the leaving group.

An S_N2 reaction is driven by the attraction between the negative charge of the nucleophile (in this case the negatively charged oxygen of the hydroxide ion) and the center of positive charge of the electrophile (in this case the partial positive charge on the carbon bearing the chlorine leaving group).

B. S_N1 Mechanism

In the other limiting mechanism, called S_N1, bond breaking between carbon and the leaving group is completed before bond forming with the nucleophile begins. In the designation **S_N1**, *S* stands for *S*ubstitution, *N* stands for *N*ucleophilic, and *1* stands for a ***unimolecular* reaction**. This type of substitution is classified as unimolecular because only the haloalkane is involved in the rate-determining step; that is, only the haloalkane contributes to the rate law governing the rate-determining step:

Unimolecular reaction A reaction in which only one species is involved in the reaction leading to the transition state of the rate-determining step.

$$\text{Rate} = k[\text{haloalkane}]$$

An S_N1 reaction is illustrated by the **solvolysis** reaction of 2-bromo-2-methylpropane (*tert*-butyl bromide) in methanol to form 2-methoxy-2-methylpropane (*tert*-butyl methyl ether).

Solvolysis A nucleophilic substitution reaction in which the solvent is the nucleophile.

*because its uncharged

Mechanism: An S_N1 Reaction

Step 1: The ionization of a C—X bond forms a 3° carbocation intermediate:

$$(CH_3)_3C\text{—}\ddot{Br}: \xrightarrow{\text{slow, rate determining}} (CH_3)_3C^+ + :\ddot{Br}:^-$$

A carbocation intermediate; carbon is trigonal planar

neutral reactant + Nu
↓
positive intermediate
↓
[proton transfer to the solvent]
extra step

Step 2: Reaction of methanol from either face of the planar carbocation intermediate gives an oxonium ion:

$$CH_3\ddot{O}H + (CH_3)_3C^+ + H\ddot{O}CH_3 \xrightarrow{\text{fast}} CH_3O^+(H)\text{—}C(CH_3)_3 + (CH_3)_3C\text{—}O^+(H)CH_3$$

Step 3: Proton transfer from the oxonium ion to methanol (the solvent) completes the reaction and gives *tert*-butyl methyl ether:

$$(CH_3)_3C\text{—}O^+(H)CH_3 + H\ddot{O}CH_3 \overset{\text{fast}}{\rightleftharpoons} (CH_3)_3C\text{—}\ddot{O}\text{—}CH_3 + H\text{—}O^+(H)CH_3$$

Figure 7.2 shows an energy diagram for the S_N1 reaction of 2-bromo-2-methylpropane and methanol. There is one transition state leading to formation of the carbocation intermediate in Step 1 and a second transition state for reaction of the carbocation intermediate with methanol in Step 2 to give the oxonium ion. The reaction leading to formation of the carbocation intermediate crosses the higher energy barrier and is, therefore, the rate-determining step.

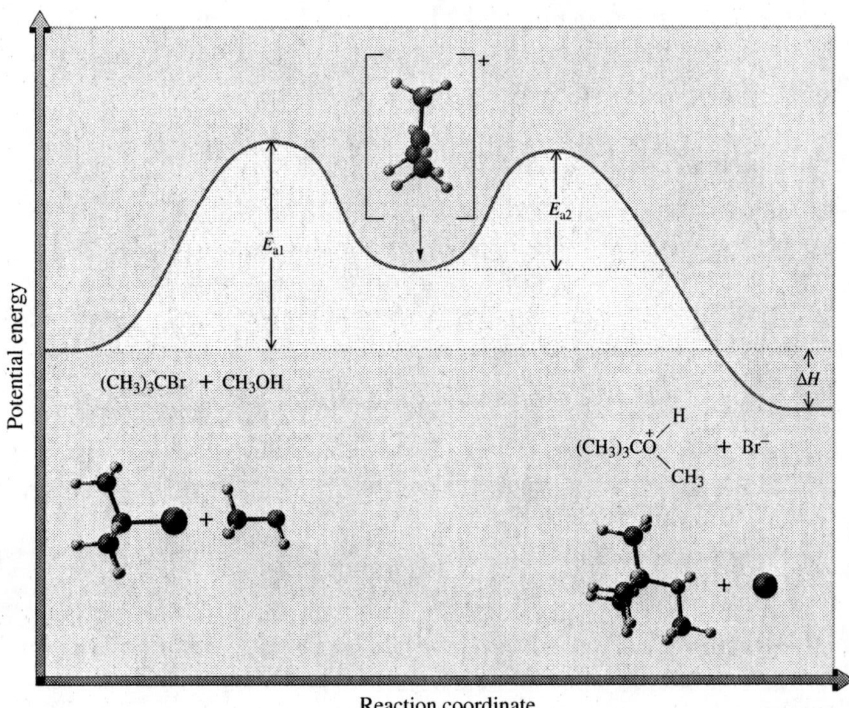

Figure 7.2
An energy diagram for the S_N1 reaction of 2-bromo-2-methylpropane and methanol. There is one transition state leading to formation of the carbocation intermediate in Step 1 and a second transition state for the reaction of the carbocation intermediate with methanol in Step 2. Step 1 crosses the higher energy barrier and is, therefore, rate determining.

If an S_N1 reaction is carried out at a tetrahedral stereocenter, the major product is a racemic mixture. We can illustrate this result with the following example: Upon ionization, the R enantiomer forms an achiral carbocation intermediate. Attack by the nucleophile from the left face of the carbocation intermediate gives the S enantiomer; attack from the right face gives the R enantiomer. Because attack by the nucleophile occurs with equal probability from either face of the planar carbocation intermediate, the R and S enantiomers are formed in equal amounts, and the product is a racemic mixture.

Ionization $-Cl^-$; 1) CH_3OH 2) Proton transfer $(-H^+)$

R enantiomer — Planar carbocation (achiral) — S enantiomer, R enantiomer: A racemic mixture

7.6 EXPERIMENTAL EVIDENCE FOR S_N1 AND S_N2 MECHANISMS

Let us now examine some of the experimental evidence on which these two contrasting mechanisms are based. As we do, we consider the following questions:

1. What effect does the structure of the nucleophile have on the rate of reaction?
2. What effect does the structure of the haloalkane have on the rate of reaction?
3. What effect does the structure of the leaving group have on the rate of reaction?
4. What is the role of the solvent?

A. Structure of the Nucleophile

Nucleophilicity is a kinetic property, which we measure by relative rates of reaction. We can establish the relative nucleophilicities for a series of nucleophiles by measuring the rate at which each displaces a leaving group from a haloalkane—for example, the rate at which each displaces bromide ion from bromoethane in ethanol at 25°C:

$$CH_3CH_2Br + NH_3 \longrightarrow CH_3CH_2NH_3^+ + Br^-$$

From these studies, we can then make correlations between the structure of the nucleophile and its **relative nucleophilicity**. Table 7.2 lists the types of nucleophiles we deal with most commonly in this text.

Relative nucleophilicity The relative rates at which a nucleophile reacts in a reference nucleophilic substitution reaction.

Because the nucleophile participates in the rate-determining step in an S_N2 reaction, the better the nucleophile, the more likely it is that the reaction will occur by that mechanism. The nucleophile does not participate in the rate-determining step for an S_N1 reaction. Thus, an S_N1 reaction can, in principle, occur at approximately the same rate with any of the common nucleophiles, regardless of their relative nucleophilicities.

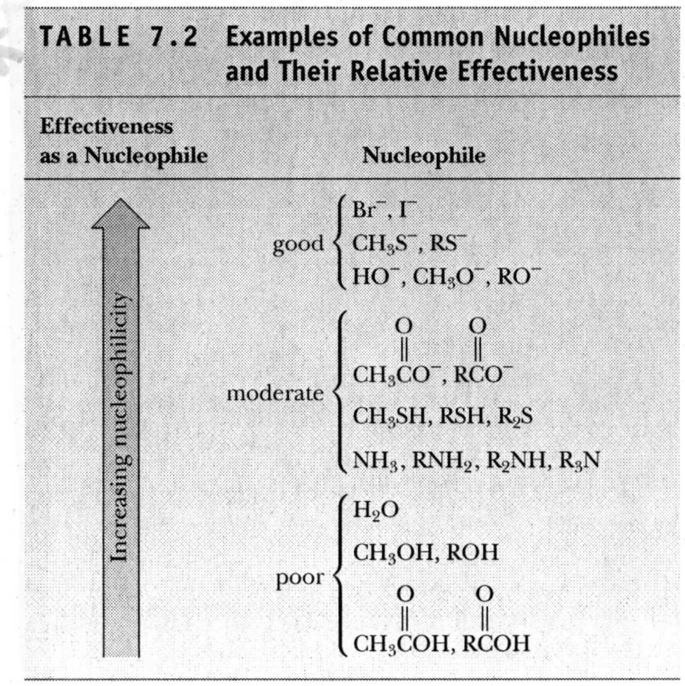

TABLE 7.2 Examples of Common Nucleophiles and Their Relative Effectiveness

Effectiveness as a Nucleophile	Nucleophile
good	Br^-, I^-
	CH_3S^-, RS^-
	HO^-, CH_3O^-, RO^-
moderate	$CH_3C(=O)O^-$, $RC(=O)O^-$
	CH_3SH, RSH, R_2S
	NH_3, RNH_2, R_2NH, R_3N
poor	H_2O
	CH_3OH, ROH
	$CH_3C(=O)OH$, $RC(=O)OH$

Increasing nucleophilicity (↑)

B. Structure of the Haloalkane

S_N1 reactions are governed mainly by **electronic factors**, namely, the relative stabilities of carbocation intermediates. S_N2 reactions, by contrast, are governed mainly by **steric factors**, and their transition states are particularly sensitive to crowding about the site of reaction. The distinction is as follows:

Steric hindrance The ability of groups, because of their size, to hinder access to a reaction site within a molecule.

1. *Relative stabilities of carbocations.* As we learned in Section 5.3A, 3° carbocations are the most stable carbocations, requiring the lowest activation energy for their formation, whereas 1° carbocations are the least stable requiring the highest activation energy for their formation. In fact, 1° carbocations are so unstable that they have never been observed in solution. Therefore, 3° haloalkanes are most likely to react by carbocation formation; 2° haloalkanes are less likely to react in this manner, and methyl and 1° haloalkanes never react in that manner.
2. *Steric hindrance.* To complete a substitution reaction, the nucleophile must approach the substitution center and begin to form a new covalent bond to it. If we compare the ease of approach by the nucleophile to the substitution center of a 1° haloalkane with that of a 3° haloalkane, we see that the approach is considerably easier in the case of the 1° haloalkane. Two hydrogen atoms and one alkyl group screen the backside of the substitution center of a 1° haloalkane. In contrast, three alkyl groups screen the backside of the substitution center of a 3° haloalkane. This center in bromoethane is easily accessible to a nucleophile, while there is extreme crowding around it in 2-bromo-2-methylpropane:

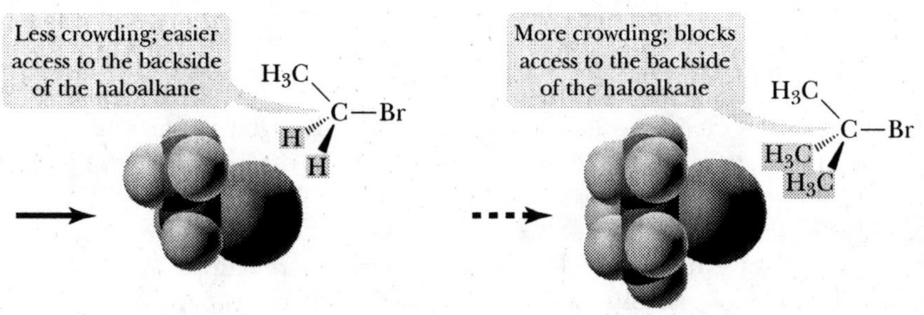

Bromoethane (Ethyl bromide) 2-Bromo-2-methylpropane (*tert*-Butyl bromide)

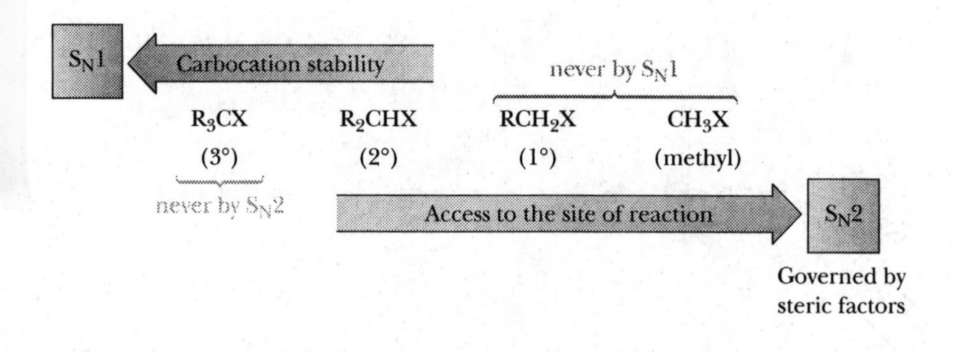

Figure 7.3
Effect of electronic and steric factors in competition between S_N1 and S_N2 reactions of haloalkanes.

Given the competition between electronic and steric factors, we find that 3° haloalkanes react by an S_N1 mechanism because 3° carbocation intermediates are particularly stable and because the backside approach of a nucleophile to the substitution center in a 3° haloalkane is hindered by the three groups surrounding it; 3° haloalkanes never react by an S_N2 mechanism. Halomethanes and 1° haloalkanes have little crowding around the substitution center and react by an S_N2 mechanism; they never react by an S_N1 mechanism, because methyl and primary carbocations are so unstable. Secondary haloalkanes may react by either an S_N1 or an S_N2 mechanism, depending on the nucleophile and solvent. The competition between electronic and steric factors and their effects on relative rates of nucleophilic substitution reactions of haloalkanes are summarized in Figure 7.3.

C. The Leaving Group

In the transition state for nucleophilic substitution on a haloalkane, the leaving group develops a partial negative charge in both S_N1 and S_N2 reactions; therefore, the ability of a group to function as a leaving group is related to how stable it is as an anion. The most stable anions and the best leaving groups are the conjugate bases of strong acids. Thus, we can use the information on the relative strengths of organic and inorganic acids in Table 2.1 to determine which anions are the best leaving groups:

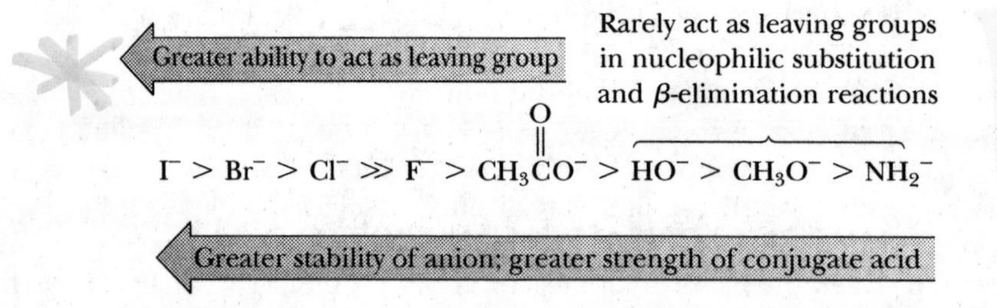

The best leaving groups in this series are the halogens I^-, Br^-, and Cl^-. Hydroxide ion (OH^-), methoxide ion (CH_3O^-), and amide ion (NH_2^-) are such poor leaving groups that they rarely, if ever, are displaced in nucleophilic substitution reactions.

D. The Solvent

Solvents provide the medium in which reactants are dissolved and in which nucleophilic substitution reactions take place. Common solvents for these reactions are divided into two groups: **protic** and **aprotic**.

Protic solvent A hydrogen bond donor solvent, as for example water, ethanol, and acetic acid.

Protic solvents contain —OH groups and are hydrogen-bond donors. Common protic solvents for nucleophilic substitution reactions are water, low-molecular-weight alcohols, and low-molecular-weight carboxylic acids (Table 7.3). Each is able to solvate both the anionic and cationic components of ionic compounds

TABLE 7.3 Common Protic Solvents

Protic Solvent	Structure	Polarity of Solvent	Notes
Water	H_2O	↑ Increasing	These solvents favor S_N1 reactions. The greater the polarity of the solvent, the easier it is to form carbocations in it, because both the carbocation and the negatively charged leaving group can be solvated.
Formic acid	$HCOOH$		
Methanol	CH_3OH		
Ethanol	CH_3CH_2OH		
Acetic acid	CH_3COOH		

by electrostatic interaction between its partially negatively charged oxygen(s) and the cation and between its partially positively charged hydrogen and the anion. These same properties aid in the ionization of C—X bonds to give an X^- anion and a carbocation; thus, protic solvents are good solvents in which to carry out S_N1 reactions.

Aprotic solvents do not contain —OH groups and cannot function as hydrogen-bond donors. Table 7.4 lists the aprotic solvents most commonly used for nucleophilic substitution reactions. Dimethyl sulfoxide and acetone are polar aprotic solvents; dichloromethane and diethyl ether are nonpolar aprotic solvents. The aprotic solvents listed in the table are particularly good ones in which to carry out S_N2 reactions. Polar aprotic solvents are able to solvate only cations; they are not able to solvate anions; therefore, they allow for "naked" and highly reactive anions as nucleophiles.

Aprotic solvent A solvent that cannot serve as a hydrogen bond donor, as for example, acetone, diethyl ether, and dichloromethane.

TABLE 7.4 Common Aprotic Solvents

Aprotic Solvent	Structure	Polarity of Solvent	Notes
Dimethyl sulfoxide (DMSO)	$CH_3\overset{O}{\overset{\|\|}{S}}CH_3$	↑ Increasing	These solvents favor S_N2 reactions. Although solvents at the top of this list are polar, the formation of carbocations in them is far more difficult than in protic solvents, because the anionic leaving group cannot be solvated by these solvents.
Acetone	$CH_3\overset{O}{\overset{\|\|}{C}}CH_3$		
Dichloromethane	CH_2Cl_2		
Diethyl ether	$(CH_3CH_2)_2O$		

Table 7.5 summarizes the factors favoring S_N1 or S_N2 reactions; it also shows the change in configuration when nucleophilic substitution takes place at a stereocenter.

7.7 ANALYSIS OF SEVERAL NUCLEOPHILIC SUBSTITUTION REACTIONS

Predictions about the mechanism for a particular nucleophilic substitution reaction must be based on considerations of the structure of the haloalkane, the nucleophile, the leaving group, and the solvent. Following are analyses of three such reactions:

TABLE 7.5 Summary of S_N1 versus S_N2 Reactions of Haloalkanes

Type of Haloalkane	S_N2	S_N1
Methyl CH_3X	S_N2 is favored.	S_N1 does not occur. The methyl cation is so unstable that it is never observed in solution.
Primary RCH_2X	S_N2 is favored.	S_N1 does not occur. Primary carbocations are so unstable that they are not observed in solution.
Secondary R_2CHX	S_N2 is favored in aprotic solvents with good nucleophiles.	S_N1 is favored in protic solvents with poor nucleophiles.
Tertiary R_3CX	S_N2 does not occur, because of steric hindrance around the substitution center.	S_N1 is favored because of the ease of formation of tertiary carbocations.
Substitution at a stereocenter	**Inversion of configuration.** The nucleophile attacks the stereocenter from the side opposite the leaving group.	**Racemization.** The carbocation intermediate is planar, and an attack by the nucleophile occurs with equal probability from either side.

Nucleophilic Substitution 1

2-chlorobutane (Cl) + CH_3OH ⟶ 2-methoxybutane (OCH_3) + HCl

R enantiomer

Methanol is a polar protic solvent and a good one in which to form carbocations. 2-Chlorobutane ionizes in methanol to form a 2° carbocation intermediate. Methanol is a weak nucleophile. From this analysis, we predict that reaction is by an S_N1 mechanism. The 2° carbocation intermediate (an electrophile) then reacts with methanol (a nucleophile) followed by proton transfer to give the observed product. The product is formed as a 50:50 mixture of R and S configurations; that is, it is formed as a racemic mixture.

Nucleophilic Substitution 2

Br + Na^+I^- $\xrightarrow{\text{DMSO}}$ I + Na^+Br^-

This is a 1° bromoalkane in the presence of iodide ion, a good nucleophile. Because 1° carbocations are so unstable, they never form in solution, and an S_N1 reaction is not possible. Dimethyl sulfoxide (DMSO), a polar aprotic solvent, is a good solvent in which to carry out S_N2 reactions. From this analysis, we predict that reaction is by an S_N2 mechanism.

Nucleophilic Substitution 3

Br + $CH_3S^-Na^+$ $\xrightarrow{\text{acetone}}$ SCH_3 + Na^+Br^-

S enantiomer

Bromine ion is a good leaving group on a 2° carbon. The methylsulfide ion is a good nucleophile. Acetone, a polar aprotic solvent, is a good medium in which to carry out S_N2 reactions, but a poor medium in which to carry out S_N1 reactions. We predict that reaction is by an S_N2 mechanism and that the product formed has the R configuration.

EXAMPLE 7.3

Write the expected product for each nucleophilic substitution reaction, and predict the mechanism by which the product is formed:

(a) 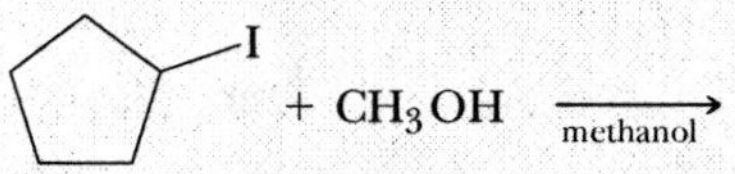 I (on cyclopentane) + CH_3OH $\xrightarrow{\text{methanol}}$

(b) 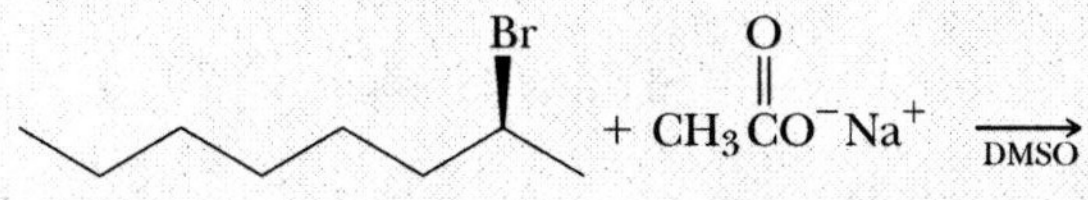 Br + $CH_3\overset{O}{\overset{\|}{C}}O^-Na^+$ $\xrightarrow{\text{DMSO}}$

SOLUTION

(a) Methanol is a poor nucleophile. It is also a polar protic solvent that is able to solvate carbocations. Ionization of the carbon–iodine bond forms a 2° carbocation intermediate. We predict an S_N1 mechanism:

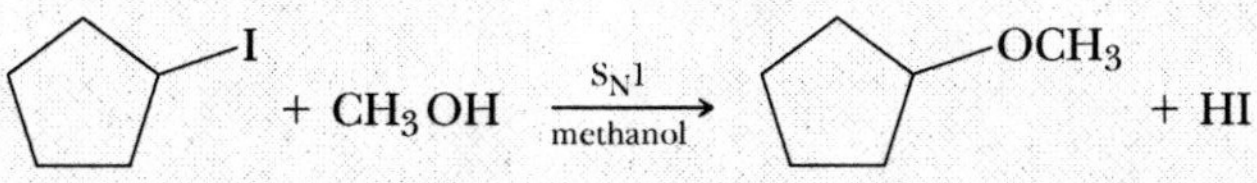

I (on cyclopentane) + CH_3OH $\xrightarrow[\text{methanol}]{S_N1}$ OCH_3 (on cyclopentane) + HI

(b) Bromide is a good leaving group on a 2° carbon. Acetate ion is a moderate nucleophile. DMSO is a particularly good solvent for S_N2 reactions. We predict substitution by an S_N2 mechanism with inversion of configuration at the stereocenter:

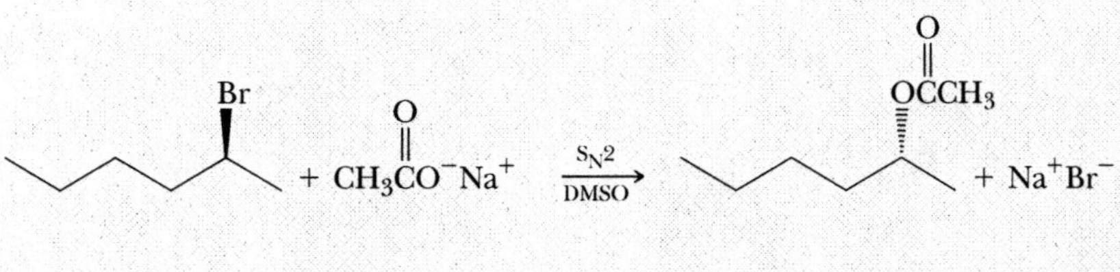

Br + $CH_3\overset{O}{\overset{\|}{C}}O^-Na^+$ $\xrightarrow[\text{DMSO}]{S_N2}$ $\overset{O}{\overset{\|}{O}}CCH_3$ + Na^+Br^-

Practice Problem 7.3

Write the expected product for each nucleophilic substitution reaction, and predict the mechanism by which the product is formed:

(a) 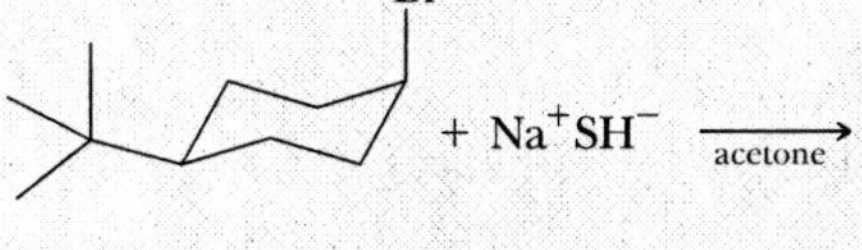 Br + Na^+SH^- $\xrightarrow{\text{acetone}}$

(b) 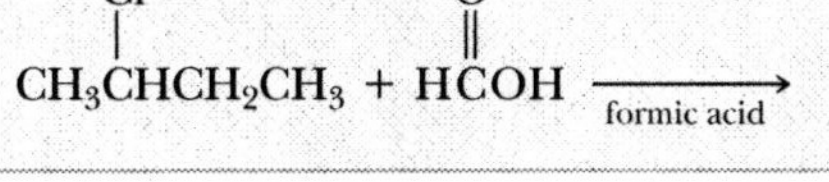 $CH_3\overset{Cl}{\overset{|}{C}}HCH_2CH_3$ + $H\overset{O}{\overset{\|}{C}}OH$ $\xrightarrow{\text{formic acid}}$

7.8 β-ELIMINATION

Dehydrohalogenation Removal of —H and —X from adjacent carbons; a type of β-elimination.

In this section, we study a type of β-elimination called **dehydrohalogenation**. In the presence of a strong base, such as hydroxide ion or ethoxide ion, halogen can be removed from one carbon of a haloalkane and hydrogen from an adjacent carbon to form a carbon–carbon double bond:

$$-\overset{|}{\underset{H}{C}}{}^{\beta}-\overset{|}{\underset{X}{C}}{}^{\alpha}- + CH_3CH_2O^-Na^+ \xrightarrow{CH_3CH_2OH} >C=C< + CH_3CH_2OH + Na^+X^-$$

A haloalkane | Base | An alkene

As the equation shows, we call the carbon bearing the halogen the *α-carbon* and the adjacent carbon the *β-carbon.*

Because most nucleophiles can also act as bases and vice versa, it is important to keep in mind that β-elimination and nucleophilic substitution are competing reactions. In this section, we concentrate on β-elimination. In Section 7.10, we examine the results of competition between the two.

Common strong bases used for β-elimination are OH^-, OR^-, and NH_2^-. Following are three examples of base-promoted β-elimination reactions:

1-Bromooctane (β, α) Br + t-BuO^-K^+ (Potassium *tert*-butoxide) ⟶ 1-Octene + t-BuOH + Na^+Br^-

2-Bromo-2-methylbutane $\xrightarrow[CH_3CH_2OH]{CH_3CH_2O^-Na^+}$ 2-Methyl-2-butene (major product) + 2-Methyl-1-butene

1-Bromo-1-methylcyclopentane $\xrightarrow[CH_3OH]{CH_3O^-Na^+}$ 1-Methylcyclopentene (major product) + Methylenecyclopentane ($=CH_2$)

Zaitsev's rule A rule stating that the major product from a β-elimination reaction is the most stable alkene; that is, the major product is the alkene with the greatest number of substituents on the carbon–carbon double bond.

In the first example, the base is shown as a reactant. In the second and third examples, the base is a reactant, but is shown over the reaction arrow. Also in the second and third examples, there are nonequivalent β-carbons, each bearing a hydrogen; therefore, two alkenes are possible from each β-elimination reaction. In each case, the major product of these and most other β-elimination reactions is the more substituted (and therefore the more stable—see Section 5.5B) alkene. We say that each reaction follows **Zaitsev's rule** or, alternatively, that each undergoes Zaitsev elimination, to honor the chemist who first made this generalization.

EXAMPLE 7.4

Predict the β-elimination product(s) formed when each bromoalkane is treated with sodium ethoxide in ethanol (if two might be formed, predict which is the major product):

(a) 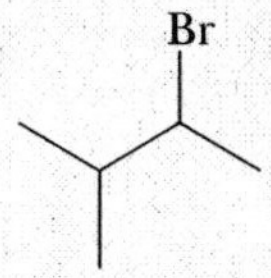 Br (b) 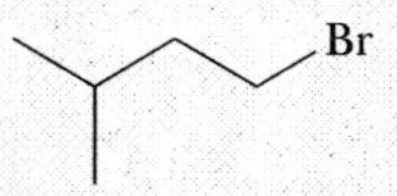 Br

SOLUTION

(a) There are two nonequivalent β-carbons in this bromoalkane, and two alkenes are possible. 2-Methyl-2-butene, the more substituted alkene, is the major product:

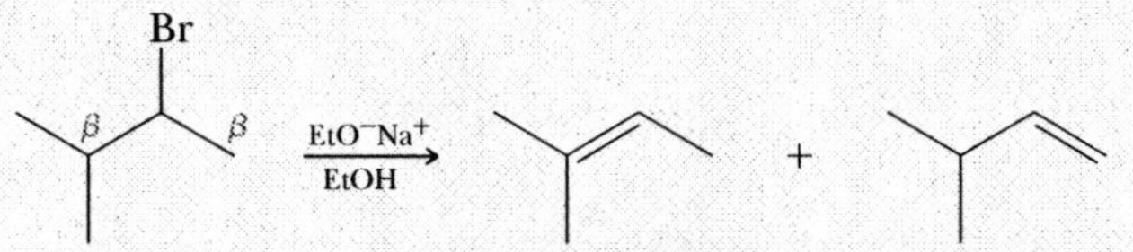

Br β β $\xrightarrow[EtOH]{EtO^-Na^+}$ +

2-Methyl-2-butene (major product) 3-Methyl-1-butene

(b) There is only one β-carbon in this bromoalkane, and only one alkene is possible:

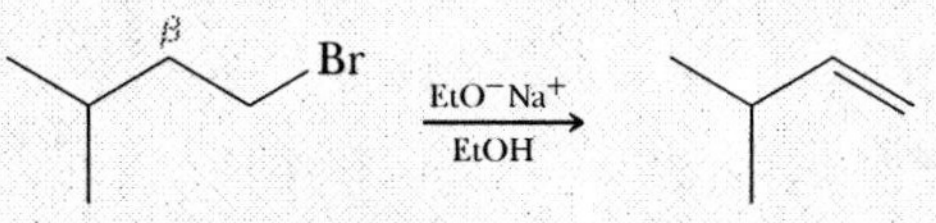

β Br $\xrightarrow[EtOH]{EtO^-Na^+}$

3-Methyl-1-butene

Practice Problem 7.4

Predict the β-elimination products formed when each chloroalkane is treated with sodium ethoxide in ethanol (if two products might be formed, predict which is the major product):

(a) 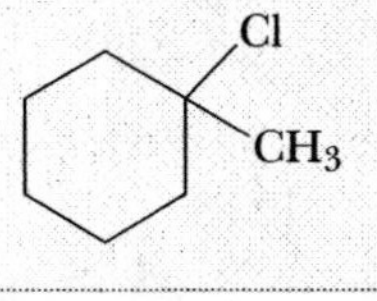 Cl CH_3 (b) 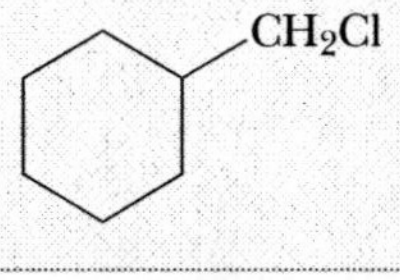 CH_2Cl (c) 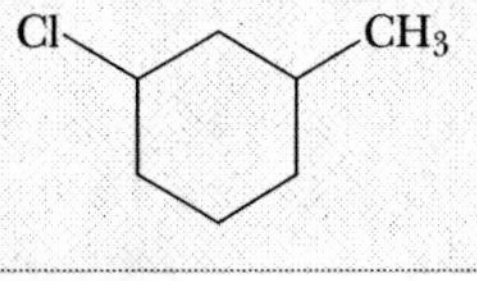 Cl CH_3

7.9 MECHANISMS OF β-ELIMINATION

There are two limiting mechanisms of β-elimination reactions. A fundamental difference between them is the timing of the bond-breaking and bond-forming steps. Recall that we made this same statement about the two limiting mechanisms for nucleophilic substitution reactions in Section 7.5.

A. E1 Mechanism

At one extreme, breaking of the C—X bond is complete before any reaction occurs with base to lose a hydrogen and before the carbon–carbon double bond is formed. This mechanism is designated **E1**, where *E* stands for *e*limination and *1* stands for a *uni*molecular reaction; only *one* species, in this case the haloalkane, is involved in the rate-determining step. The rate law for an E1 reaction has the same form as that for an S_N1 reaction:

$$\text{Rate} = k[\text{haloalkane}]$$

The mechanism for an E1 reaction is illustrated by the reaction of 2-bromo-2-methylpropane to form 2-methylpropene. In this two-step mechanism, the rate-determining step is the ionization of the carbon–halogen bond to form a carbocation intermediate (just as it is in an S_N1 mechanism).

Mechanism: E1 Reaction of 2-Bromo-2-methylpropane

Step 1: Rate-determining ionization of the C—Br bond gives a carbocation intermediate:

$$CH_3-\underset{Br}{\overset{CH_3}{C}}-CH_3 \xrightarrow[\text{determining}]{\text{slow, rate}} CH_3-\underset{+}{\overset{CH_3}{C}}-CH_3 + :\ddot{Br}:^-$$

A carbocation intermediate

Step 2: Proton transfer from the carbocation intermediate to methanol (which in this instance is both the solvent and a reactant) gives the alkene:

$$H_3C-\ddot{O}-H + H-CH_2-\underset{+}{\overset{CH_3}{C}}-CH_3 \xrightarrow{\text{fast}} H_3C-\overset{+}{O}(H)-H + CH_2=\overset{CH_3}{C}-CH_3$$

B. E2 Mechanism

At the other extreme is a concerted process. In an **E2** reaction, *E* stands for *e*limination, and *2* stands for *bi*molecular. Because the base removes a β-hydrogen at the same time the C—X bond is broken to form a halide ion, the rate law for the rate-determining step is dependent on both the haloalkane and the base:

$$\text{Rate} = k[\text{haloalkane}][\text{base}]$$

The stronger the base, the more likely it is that the E2 mechanism will be in operation. We illustrate an E2 mechanism by the reaction of 1-bromopropane with sodium ethoxide.

Mechanism: E2 Reaction of 1-Bromopropane

In this mechanism, proton transfer to the base, formation of the carbon–carbon double bond, and the ejection of bromide ion occur simultaneously; that is, all bond-forming and bond-breaking steps occur at the same time.

$$CH_3CH_2\ddot{O}:^- + H-\overset{CH_3}{CH}-CH_2-\ddot{Br}: \longrightarrow CH_3CH_2\ddot{O}-H + CH_3CH=CH_2 + :\ddot{Br}:^-$$

For both E1 and E2 reactions, the major product is that formed in accordance with Zaitsev's rule as illustrated by this E2 reaction:

$$\text{2-Bromohexane} \xrightarrow[CH_3OH]{CH_3O^-Na^+} \text{2-Hexene (74\%)} + \text{1-Hexene (26\%)}$$

Table 7.6 summarizes these generalizations about β-elimination reactions of haloalkanes.

TABLE 7.6 Summary of E1 versus E2 Reactions of Haloalkanes

Haloalkane	E1	E2
Primary RCH_2X	E1 does not occur. Primary carbocations are so unstable that they are never observed in solution.	E2 is favored.
Secondary R_2CHX	Main reaction with weak bases such as H_2O and ROH.	Main reaction with strong bases such as OH^- and OR^-.
Tertiary R_3CX	Main reaction with weak bases such as H_2O and ROH.	Main reaction with strong bases such as OH^- and OR^-.

EXAMPLE 7.5

Predict whether each β-elimination reaction proceeds predominantly by an E1 or E2 mechanism, and write a structural formula for the major organic product:

(a) $CH_3C(CH_3)(Cl)CH_2CH_3 + NaOH \xrightarrow[H_2O]{80°C}$ (b) $CH_3C(CH_3)(Cl)CH_2CH_3 \xrightarrow{CH_3COOH}$

SOLUTION

(a) A 3° chloroalkane is heated with a strong base. Elimination by an E2 reaction predominates, giving 2-methyl-2-butene as the major product:

$$CH_3C(CH_3)(Cl)CH_2CH_3 + NaOH \xrightarrow[H_2O]{80°C} CH_3C(CH_3){=}CHCH_3 + NaCl + H_2O$$

(b) A 3° chloroalkane dissolved in acetic acid, a solvent that promotes the formation of carbocations, forms a 3° carbocation that then loses a proton to give 2-methyl-2-butene as the major product. The reaction is by an E1 mechanism:

$$CH_3C(CH_3)(Cl)CH_2CH_3 \xrightarrow{CH_3COOH} CH_3C(CH_3){=}CHCH_3 + HCl$$

Practice Problem 7.5

Predict whether each elimination reaction proceeds predominantly by an E1 or E2 mechanism, and write a structural formula for the major organic product:

(a) Br (3-bromopentane, line structure) $+ CH_3O^-Na^+ \xrightarrow{\text{methanol}}$

(b) Cl (cyclohexane ring with Cl and methyl, line structure) $+ CH_3CH_2O^-Na^+ \xrightarrow{\text{ethanol}}$

7.10 SUBSTITUTION VERSUS ELIMINATION

Thus far, we have considered two types of reactions of haloalkanes: nucleophilic substitution and β-elimination. Many of the nucleophiles we have examined—for example, hydroxide ion and alkoxide ions—are also strong bases. Accordingly, nucleophilic substitution and β-elimination often compete with each other, and the ratio of products formed by these reactions depends on the relative rates of the two reactions:

$$\text{H}-\overset{|}{\underset{|}{\text{C}}}-\overset{|}{\underset{|}{\text{C}}}-\text{X} + :\text{Nu}^- \begin{cases} \xrightarrow{\text{nucleophilic substitution}} \text{H}-\overset{|}{\underset{|}{\text{C}}}-\overset{|}{\underset{|}{\text{C}}}-\text{Nu} + :\text{X}^- \\ \xrightarrow{\beta\text{-elimination}} \text{C}=\text{C} + \text{H-Nu} + :\text{X}^- \end{cases}$$

SN1 + E1

Step 1: formation of carbocation

Step 2:
E1 → loss of H = alkene
SN1 → rxn w/solvent = sub. product

SN2 + E2

A. S_N1-versus-E1 Reactions

Reactions of secondary and tertiary haloalkanes in polar protic solvents give mixtures of substitution and elimination products. In both reactions, Step 1 is the formation of a carbocation intermediate. This step is then followed by either (1) the loss of a hydrogen to give an alkene (E1) or (2) reaction with solvent to give a substitution product (S_N1). In polar protic solvents, the products formed depend only on the structure of the particular carbocation. For example, *tert*-butyl chloride and *tert*-butyl iodide in 80% aqueous ethanol both react with solvent, giving the same mixture of substitution and elimination products:

$$(CH_3)_3C-I \xrightarrow{-I^-} (CH_3)_3C^+ \qquad (CH_3)_3C-Cl \xrightarrow{-Cl^-} (CH_3)_3C^+$$

$$(CH_3)_3C^+ \begin{cases} \xrightarrow{\text{E1}} CH_2=C(CH_3)_2 + H^+ \\ \xrightarrow[H_2O]{S_N1} (CH_3)_3C-OH + H^+ \\ \xrightarrow[CH_3CH_2OH]{S_N1} (CH_3)_3C-OCH_2CH_3 + H^+ \end{cases}$$

Because iodide ion is a better leaving group than chloride ion, *tert*-butyl iodide reacts over 100 times faster than *tert*-butyl chloride. Yet the ratio of products is the same.

B. S_N2-versus-E2 Reactions

It is considerably easier to predict the ratio of substitution to elimination products for reactions of haloalkanes with reagents that act as both nucleophiles and bases. The guiding principles are as follows:

1. Branching at the α-carbon or β-carbon(s) increases steric hindrance about the α-carbon and significantly retards S_N2 reactions. By contrast, branching at the α-carbon or β-carbon(s) increases the rate of E2 reactions because of the increased stability of the alkene product.

2. The greater the nucleophilicity of the attacking reagent, the greater is the S_N2-to-E2 ratio. Conversely, the greater the basicity of the attacking reagent, the greater is the E2-to-S_N2 ratio.

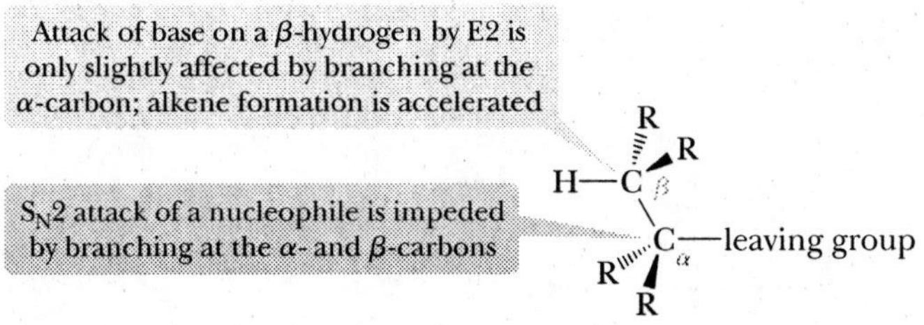

Attack of base on a β-hydrogen by E2 is only slightly affected by branching at the α-carbon; alkene formation is accelerated

S_N2 attack of a nucleophile is impeded by branching at the α- and β-carbons

R R H—C β R C α —leaving group R

Primary halides react with bases/nucleophiles to give predominantly substitution products. With strong bases, such as hydroxide ion and ethoxide ion, a percentage of the product is formed by an E2 reaction, but it is generally small compared with that formed by an S_N2 reaction. With strong, bulky bases, such as *tert*-butoxide ion, the E2 product becomes the major product. Tertiary halides react with all strong bases/good nucleophiles to give only elimination products.

Secondary halides are borderline, and substitution or elimination may be favored, depending on the particular base/nucleophile, solvent, and temperature at which the reaction is carried out. Elimination is favored with strong bases/good nucleophiles—for example, hydroxide ion and ethoxide ion. Substitution is favored with weak bases/poor nucleophiles—for example, acetate ion. Table 7.7 summarizes these generalizations about substitution versus elimination reactions of haloalkanes.

TABLE 7.7 Summary of Substitution versus Elimination Reactions of Haloalkanes

Halide	Reaction	Comments
Methyl CH_3X	S_N2	The only substitution reactions observed.
	~~S_N1~~	S_N1 reactions of methyl halides are never observed. The methyl cation is so unstable that it is never formed in solution.
Primary RCH_2X	S_N2	The main reaction with strong bases such as OH^- and EtO^-. Also, the main reaction with good nucleophiles/weak bases, such as I^- and CH_3COO^-.
	E2	The main reaction with strong, bulky bases, such as potassium *tert*-butoxide.
	~~S_N1/E1~~	Primary cations are never formed in solution; therefore, S_N1 and E1 reactions of primary halides are never observed.
Secondary R_2CHX	S_N2	The main reaction with weak bases/good nucleophiles, such as I^- and CH_3COO^-.
	E2	The main reaction with strong bases/good nucleophiles, such as OH^- and $CH_3CH_2O^-$.
	S_N1/E1	Common in reactions with weak nucleophiles in polar protic solvents, such as water, methanol, and ethanol.
Tertiary R_3CX	~~S_N2~~	S_N2 reactions of tertiary halides are never observed because of the extreme crowding around the 3° carbon.
	E2	Main reaction with strong bases, such as HO^- and RO^-.
	S_N1/E1	Main reactions with poor nucleophiles/weak bases.

EXAMPLE 7.6

Predict whether each reaction proceeds predominantly by substitution (S_N1 or S_N2) or elimination (E1 or E2) or whether the two compete, and write structural formulas for the major organic product(s):

(a) Cl + NaOH $\xrightarrow[H_2O]{80°C}$ (b) Br + $(C_2H_5)_3N$ $\xrightarrow[CH_2Cl_2]{30°C}$

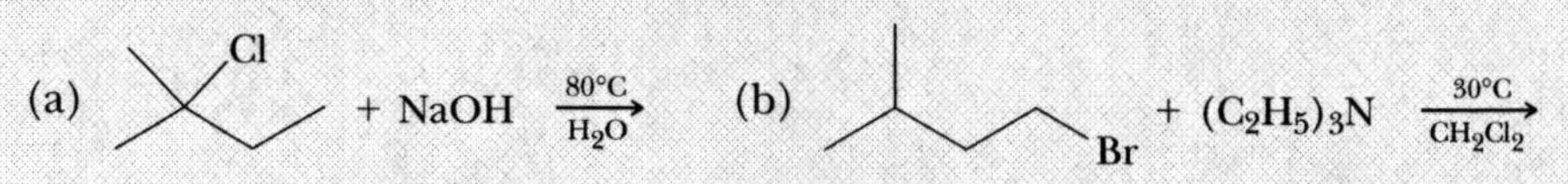

SOLUTION

(a) A 3° halide is heated with a strong base/good nucleophile. Elimination by an E2 reaction predominates to give 2-methyl-2-butene as the major product:

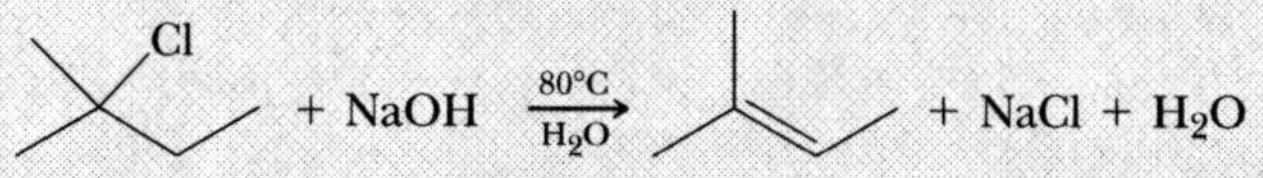

Cl + NaOH $\xrightarrow[H_2O]{80°C}$ + NaCl + H_2O

(b) Reaction of a 1° halide with triethylamine, a moderate nucleophile/weak base, gives substitution by an S_N2 reaction:

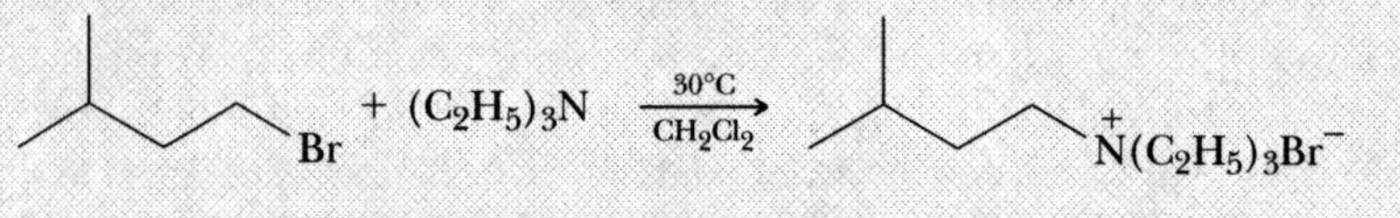

Br + $(C_2H_5)_3N$ $\xrightarrow[CH_2Cl_2]{30°C}$ $\overset{+}{N}(C_2H_5)_3Br^-$

Practice Problem 7.6

Predict whether each reaction proceeds predominantly by substitution (S_N1 or S_N2) or elimination (E1 or E2) or whether the two compete, and write structural formulas for the major organic product(s):

(a) Br + $CH_3O^- Na^+$ $\xrightarrow[methanol]{}$

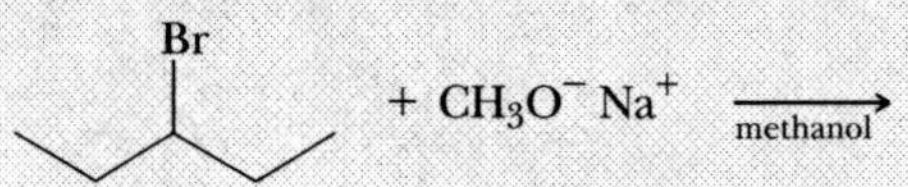

(b) Cl + $Na^+ I^-$ $\xrightarrow[acetone]{}$

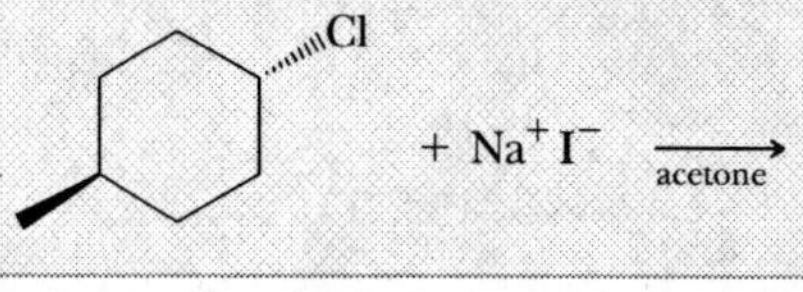

SUMMARY

Haloalkanes contain a halogen covalently bonded to an sp^3-hybridized carbon (Section 7.1). In the IUPAC system, halogen atoms are named as fluoro-, chloro-, bromo-, or iodo- substituents and are listed in alphabetical order with other substituents (Section 7.2A). In the common system, haloalkanes are named **alkyl halides.** Common names are derived by naming the alkyl group, followed by the name of the halide as a separate word (Section 7.2B). Compounds of the type CHX_3 are called **haloforms.**

A **nucleophile** (Section 7.3) is any molecule or ion with an unshared pair of electrons that can be donated to another atom or ion to form a new covalent bond; alternatively, a nucleophile is a Lewis base. An **S_N2 reaction** (Section 7.5A) occurs in one step. The departure of the leaving group is assisted by the incoming nucleophile, and both nucleophile and leaving group are involved in the transition state. S_N2 reactions are stereoselective; reaction at a stereocenter proceeds with inversion of configuration.

An **S_N1 reaction** occurs in two steps (Section 7.5B). Step 1 is a slow, rate-determining ionization of the C–X bond to form a carbocation intermediate, followed in Step 2 by its rapid reaction with a nucleophile to complete the substitution. For S_N1 reactions taking place at a stereocenter, the major reaction occurs with racemization.

The **nucleophilicity** of a reagent is measured by the rate of its reaction in a reference nucleophilic substitution (Section 7.6A). S_N1 reactions are governed by **electronic factors**, namely, the relative stabilities of carbocation intermediates. S_N2 reactions are governed by **steric factors**, namely, the degree of crowding around the site of substitution.

The ability of a group to function as a leaving group is related to its stability as an anion (Section 7.6C). The most stable anions and the best leaving groups are the conjugate bases of strong acids.

Protic solvents contain —OH groups (Section 7.6D). Protic solvents interact strongly with polar molecules and ions and are good solvents in which to form carbocations. Protic solvents favor S_N1 reactions. **Aprotic solvents** do not contain —OH groups. Common aprotic solvents are dimethyl sulfoxide, acetone, diethyl ether, and dichloromethane. Aprotic solvents do not interact as strongly with polar molecules and ions, and carbocations are less likely to form in them. Aprotic solvents favor S_N2 reactions.

Dehydrohalogenation, a type of **β-elimination reaction**, is the removal of H and X from adjacent carbon atoms (Section 7.8). A β-elimination that gives the most highly substituted alkene is called **Zaitsev elimination**. An **E1 reaction** occurs in two steps: breaking the C–X bond to form a carbocation intermediate, followed by the loss of an H^+ to form an alkene. An **E2 reaction** occurs in one step: reaction with base to remove an H^+, formation of the alkene, and departure of the leaving group, all occurring simultaneously.

KEY REACTIONS

1. Nucleophilic Aliphatic Substitution: S_N2 (Section 7.5A)

S_N2 reactions occur in one step, and both the nucleophile and the leaving group are involved in the transition state of the rate-determining step. The nucleophile may be negatively charged or neutral. S_N2 reactions result in an inversion of configuration at the reaction center. They are accelerated in polar aprotic solvents, compared with polar protic solvents. S_N2 reactions are governed by steric factors, namely, the degree of crowding around the site of reaction.

$$I^- + (CH_3CH_2)(H)(H_3C)C{-}Cl \longrightarrow I{-}C(CH_2CH_3)(H)(CH_3) + Cl^-$$

$$(CH_3)_3N + (CH_3CH_2)(H)(H_3C)C{-}Cl \longrightarrow (CH_3)_3\overset{+}{N}{-}C(CH_2CH_3)(H)(CH_3) + Cl^-$$

2. Nucleophilic Aliphatic Substitution: S_N1 (Section 7.5B)

An S_N1 reaction occurs in two steps. Step 1 is a slow, rate-determining ionization of the C–X bond to form a carbocation intermediate, followed in Step 2 by its rapid reaction with a nucleophile to complete the substitution. Reaction at a stereocenter gives a racemic product. S_N1 reactions are governed by electronic factors, namely, the relative stabilities of carbocation intermediates:

$$\text{1-chloro-1-methylcyclohexane (}CH_3\text{, Cl)} + CH_3CH_2OH \longrightarrow \text{1-ethoxy-1-methylcyclohexane (}CH_3\text{, }OCH_2CH_3\text{)} + HCl$$

3. β-Elimination: E1 (Section 7.9A)

E1 reactions involve the elimination of atoms or groups of atoms from adjacent carbons. Reaction occurs in two steps and involves the formation of a carbocation intermediate:

$$\text{(Cl-substituted alkane)} \xrightarrow[CH_3COOH]{E1} \text{(trisubstituted alkene)} + HCl$$

4. β-Elimination: E2 (Section 7.9B)
An E2 reaction occurs in one step: reaction with base to remove a hydrogen, formation of the alkene, and departure of the leaving group, all occurring simultaneously:

Br; $CH_3O^-Na^+$ / CH_3OH → (74%) + (26%)

PROBLEMS

A problem number set in red indicates an applied "real-world" problem.

Nomenclature

7.7 Write the IUPAC name for each compound:

(a) $CH_2{=}CF_2$ **(b)** Br **(c)** Cl

(d) $Cl(CH_2)_6Cl$ **(e)** CF_2Cl_2 **(f)** Br

7.8 Write the IUPAC name for each compound (be certain to include a designation of configuration, where appropriate, in your answer):

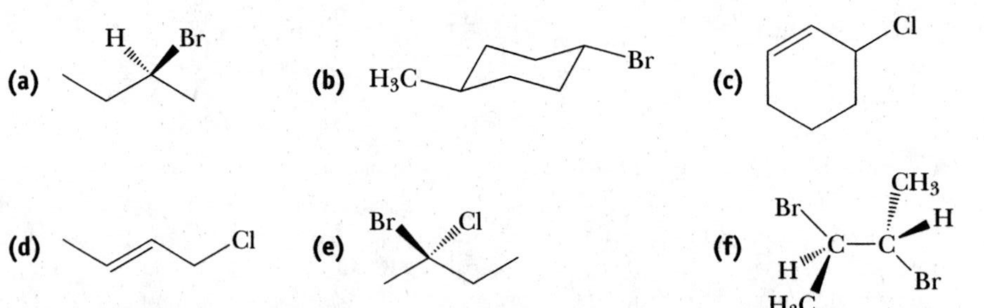

7.9 Draw a structural formula for each compound (given are IUPAC names):
(a) 3-Bromopropene **(b)** (*R*)-2-Chloropentane
(c) meso-3,4-Dibromohexane **(d)** *trans*-1-Bromo-3-isopropylcyclohexane
(e) 1,2-Dichloroethane **(f)** Bromocyclobutane

7.10 Draw a structural formula for each compound (given are common names):
(a) Isopropyl chloride **(b)** *sec*-Butyl bromide **(c)** Allyl iodide
(d) Methylene chloride **(e)** Chloroform **(f)** *tert*-Butyl chloride
(g) Isobutyl chloride

7.11 Which compounds are 2° alkyl halides?
(a) Isobutyl chloride **(b)** 2-Iodooctane **(c)** *trans*-1-Chloro-4-methylcyclohexane

Synthesis of Alkyl Halides

7.12 What alkene or alkenes and reaction conditions give each alkyl halide in good yield? (*Hint:* Review Chapter 5.)

(a) 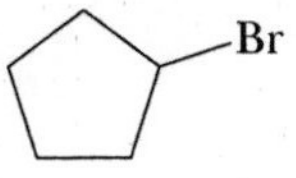 Bromocyclopentane (cyclopentane–Br) **(b)** $CH_3C(CH_3)(Br)CH_2CH_2CH_3$ **(c)** 1-chloro-1-methylcyclohexane (cyclohexane with CH_3 and Cl)

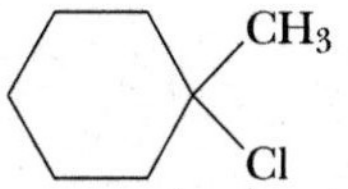

7.13 Show reagents and conditions that bring about these conversions:

(a) 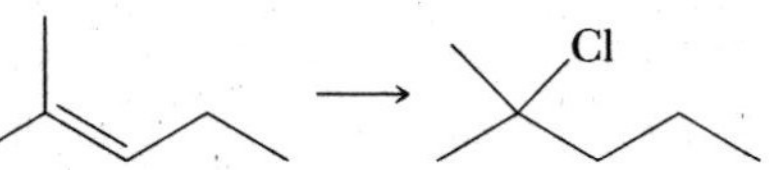 2-methyl-2-pentene ⟶ 2-chloro-2-methylpentane (Cl) **(b)** $CH_3CH_2CH{=}CH_2 \longrightarrow CH_3CH_2CH(I)CH_3$

(c) $CH_3CH{=}CHCH_3 \longrightarrow CH_3CH(Cl)CH_2CH_3$ **(d)** 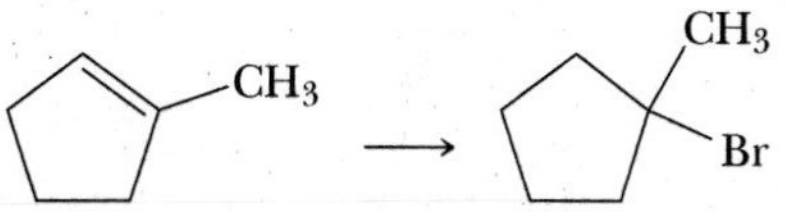 1-methylcyclopentene (CH_3) ⟶ 1-bromo-1-methylcyclopentane (CH_3, Br)

Nucleophilic Aliphatic Substitution

7.14 Write structural formulas for these common organic solvents:
(a) Dichloromethane **(b)** Acetone **(c)** Ethanol
(d) Diethyl ether **(e)** Dimethyl sulfoxide

7.15 Arrange these protic solvents in order of increasing polarity:
(a) H_2O **(b)** CH_3CH_2OH **(c)** CH_3OH

7.16 Arrange these aprotic solvents in order of increasing polarity:
(a) Acetone **(b)** Pentane **(c)** Diethyl ether

7.17 From each pair, select the better nucleophile:
(a) H_2O or OH^- **(b)** CH_3COO^- or OH^- **(c)** CH_3SH or CH_3S^-

7.18 Which statements are true for S_N2 reactions of haloalkanes?
(a) Both the haloalkane and the nucleophile are involved in the transition state.
(b) The reaction proceeds with inversion of configuration at the substitution center.
(c) The reaction proceeds with retention of optical activity.
(d) The order of reactivity is 3° > 2° > 1° > methyl.
(e) The nucleophile must have an unshared pair of electrons and bear a negative charge.
(f) The greater the nucleophilicity of the nucleophile, the greater is the rate of reaction.

7.19 Complete these S_N2 reactions:

(a) $Na^+I^- + CH_3CH_2CH_2Cl \xrightarrow[\text{acetone}]{}$ **(b)** NH_3 + bromocyclohexane (cyclohexane–Br) $\xrightarrow[\text{ethanol}]{}$

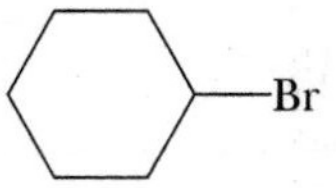

(c) $CH_3CH_2O^-Na^+ + CH_2{=}CHCH_2Cl \xrightarrow[\text{ethanol}]{}$

7.20 Complete these S_N2 reactions:

(a) 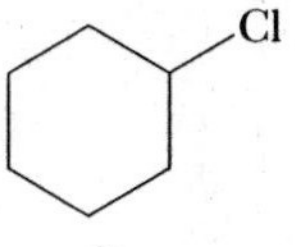 Chlorocyclohexane (cyclohexane–Cl) $+ CH_3C(=O)O^-\,Na^+ \xrightarrow[\text{ethanol}]{}$

(b) $CH_3CH(I)CH_2CH_3 + CH_3CH_2S^-Na^+ \xrightarrow[\text{acetone}]{}$

(c) $CH_3CH(CH_3)CH_2CH_2Br + Na^+I^- \xrightarrow[\text{acetone}]{}$

(d) $(CH_3)_3N + CH_3I \xrightarrow[\text{acetone}]{}$

(e) 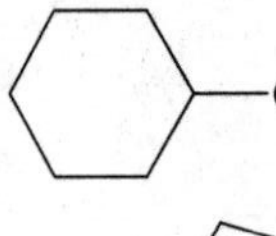 (cyclohexyl)$-CH_2Br + CH_3O^-Na^+ \xrightarrow[\text{methanol}]{}$

(f) H_3C–(cyclohexane)–Cl $+ CH_3S^-Na^+ \xrightarrow[\text{ethanol}]{}$

(g) 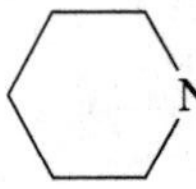 (piperidine)NH $+ CH_3(CH_2)_6CH_2Cl \xrightarrow[\text{ethanol}]{}$

(h) 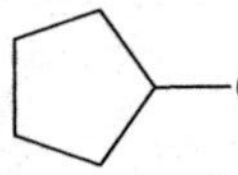 (cyclopentyl)$-CH_2Cl + NH_3 \xrightarrow[\text{ethanol}]{}$

7.21 You were told that each reaction in Problem 7.20 proceeds by an S_N2 mechanism. Suppose you were not told the mechanism. Describe how you could conclude, from the structure of the haloalkane, the nucleophile, and the solvent, that each reaction is in fact an S_N2 reaction.

7.22 In the following reactions, a haloalkane is treated with a compound that has two nucleophilic sites. Select the more nucleophilic site in each part, and show the product of each S_N2 reaction:

(a) $HOCH_2CH_2NH_2 + CH_3I \xrightarrow[\text{ethanol}]{}$

(b) 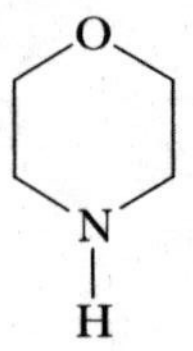 (morpholine, O and N–H) $+ CH_3I \xrightarrow[\text{ethanol}]{}$

(c) $HOCH_2CH_2SH + CH_3I \xrightarrow[\text{ethanol}]{}$

7.23 Which statements are true for S_N1 reactions of haloalkanes?

(a) Both the haloalkane and the nucleophile are involved in the transition state of the rate-determining step.
(b) The reaction at a stereocenter proceeds with retention of configuration.
(c) The reaction at a stereocenter proceeds with loss of optical activity.
(d) The order of reactivity is 3° > 2° > 1° > methyl.
(e) The greater the steric crowding around the reactive center, the lower is the rate of reaction.
(f) The rate of reaction is greater with good nucleophiles compared with poor nucleophiles.

7.24 Draw a structural formula for the product of each S_N1 reaction:

(a) $CH_3CH(Cl)CH_2CH_3 + CH_3CH_2OH \xrightarrow[\text{ethanol}]{}$
S enantiomer

(b) 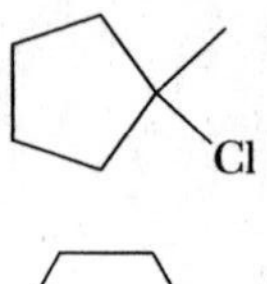 (1-chloro-1-methylcyclopentane) $+ CH_3OH \xrightarrow[\text{methanol}]{}$

(c) $CH_3C(CH_3)_2Cl + CH_3COOH \xrightarrow[\text{acetic acid}]{}$

(d) 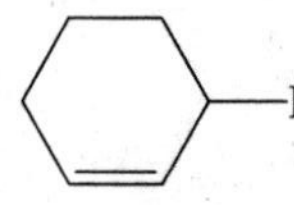 (cyclohex-2-enyl)$-Br + CH_3OH \xrightarrow[\text{methanol}]{}$

7.25 You were told that each substitution reaction in Problem 7.24 proceeds by an S_N1 mechanism. Suppose that you were not told the mechanism. Describe how you could conclude, from the structure of the haloalkane, the nucleophile, and the solvent, that each reaction is in fact an S_N1 reaction.

7.26 Select the member of each pair that undergoes nucleophilic substitution in aqueous ethanol more rapidly:

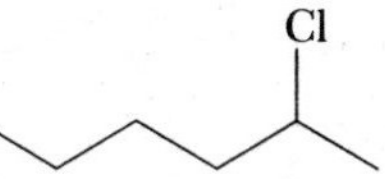

(a)  Cl or Cl **(b)** 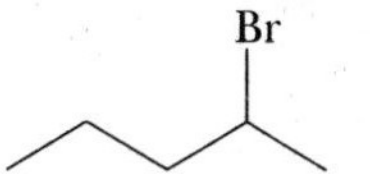 Br or Br

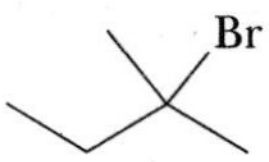

(c) 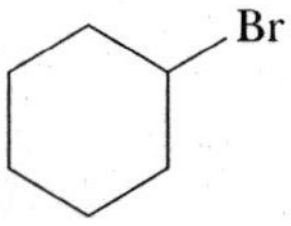 Br or CH_3 Br

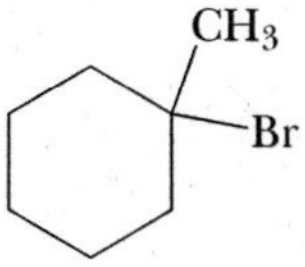

7.27 Propose a mechanism for the formation of the products (but not their relative percentages) in this reaction:

$$(CH_3)_3CCl \xrightarrow[25°C]{20\%H_2O,\ 80\%CH_3CH_2OH} \underbrace{(CH_3)_3COCH_2CH_3 + (CH_3)_3COH}_{85\%} + \underset{15\%}{CH_3C(CH_3){=}CH_2} + HCl$$

7.28 The rate of reaction in Problem 7.27 increases by 140 times when carried out in 80% water to 20% ethanol, compared with 40% water to 60% ethanol. Account for this difference.

7.29 Select the member of each pair that shows the greater rate of S_N2 reaction with KI in acetone:

(a)  Cl or Cl **(b)**  Cl or Br

(c) 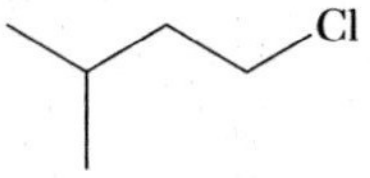 Cl or Cl **(d)** 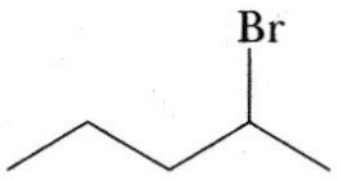 Br or Br

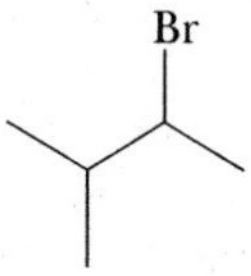

7.30 What hybridization best describes the reacting carbon in the S_N2 transition state?

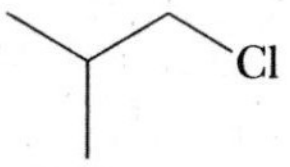

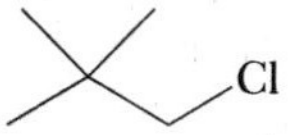

7.31 Haloalkenes such as vinyl bromide, $CH_2{=}CHBr$, undergo neither S_N1 nor S_N2 reactions. What factors account for this lack of reactivity?

7.32 Show how you might synthesize the following compounds from a haloalkane and a nucleophile:

(a) 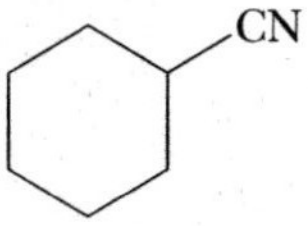 CN **(b)** 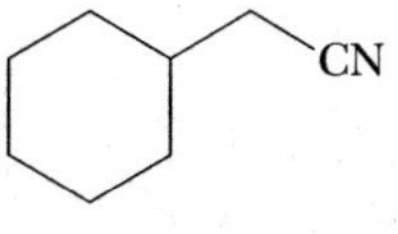 CN **(c)** 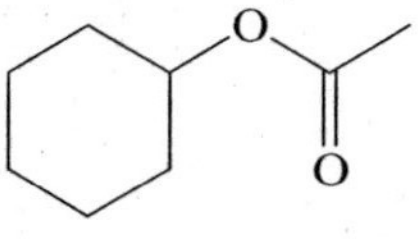 O O **(d)** 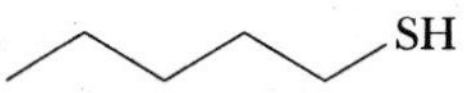 SH

(e)  OCH_3 **(f)** 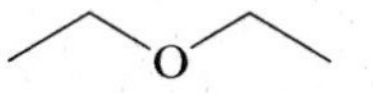 O **(g)** 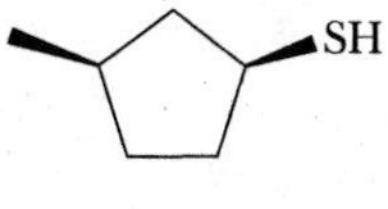 SH

7.33 Show how you might synthesize each compound from a haloalkane and a nucleophile:

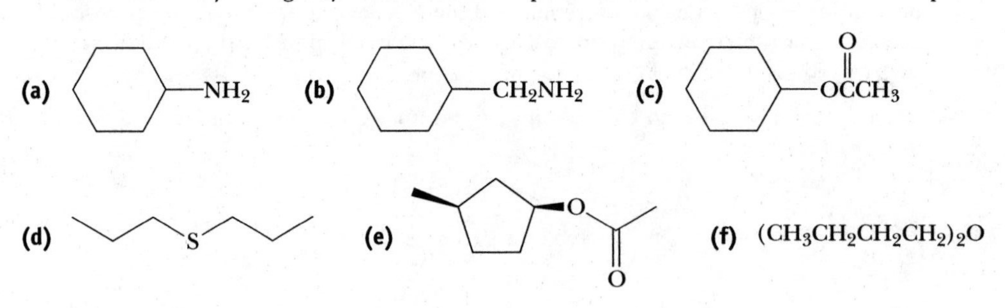

β-Eliminations

7.34 Draw structural formulas for the alkene(s) formed by treating each of the following haloalkanes with sodium ethoxide in ethanol. Assume that elimination is by an E2 mechanism. Where two alkenes are possible, use Zaitsev's rule to predict which alkene is the major product:

(a) Br **(b)** Cl **(c)** Cl **(d)** Br

7.35 Which of the following haloalkanes undergo dehydrohalogenation to give alkenes that do not show cis–trans isomerism?

(a) 2-Chloropentane **(b)** 2-Chlorobutane
(c) Chlorocyclohexane **(d)** Isobutyl chloride

7.36 How many isomers, including cis–trans isomers, are possible for the major product of dehydrohalogenation of each of the following haloalkanes?

(a) 3-Chloro-3-methylhexane **(b)** 3-Bromohexane

7.37 What haloalkane might you use as a starting material to produce each of the following alkenes in high yield and uncontaminated by isomeric alkenes?

(a) $=CH_2$ **(b)** $CH_3\underset{|}{C}HCH_2CH{=}CH_2$ (with CH_3 on the second carbon)

7.38 For each of the following alkenes, draw structural formulas of all chloroalkanes that undergo dehydrohalogenation when treated with KOH to give that alkene as the major product (for some parts, only one chloroalkane gives the desired alkene as the major product; for other parts, two chloroalkanes may work):

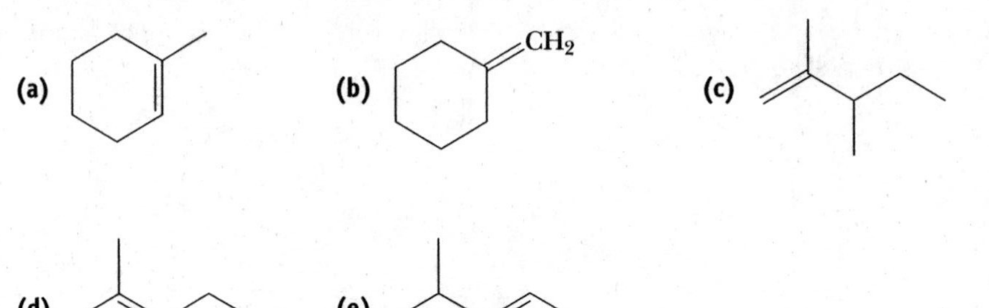

7.39 When *cis*-4-chlorocyclohexanol is treated with sodium hydroxide in ethanol, it gives only the substitution product *trans*-1,4-cyclohexanediol (1). Under the same experimental conditions, *trans*-4-chlorocyclohexanol gives 3-cyclohexenol (2) and the bicyclic ether (3):

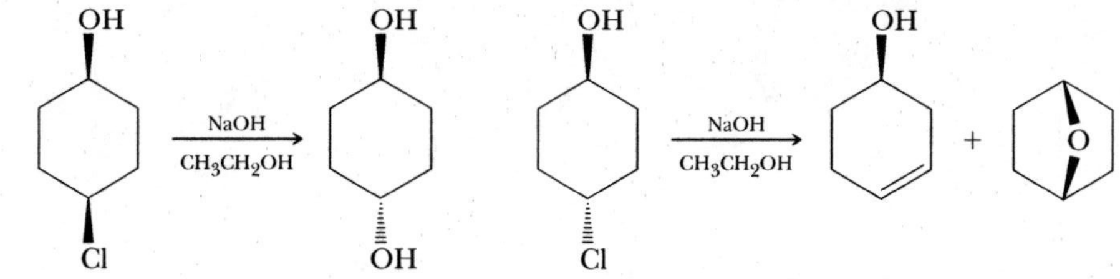

OH, Cl; NaOH / CH_3CH_2OH → OH, OH; OH, Cl; NaOH / CH_3CH_2OH → OH + O

cis-4-Chloro-cyclohexanol (1) *trans*-4-Chloro-cyclohexanol (2) (3)

(a) Propose a mechanism for the formation of product (1), and account for its configuration.

(b) Propose a mechanism for the formation of product (2).

(c) Account for the fact that the bicyclic ether (3) is formed from the trans isomer, but not from the cis isomer.

Synthesis

7.40 Show how to convert the given starting material into the desired product (note that some syntheses require only one step, whereas others require two or more steps):

(a) 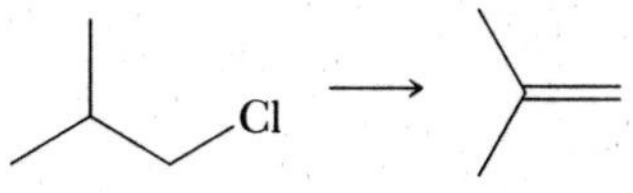 Cl →

(b) 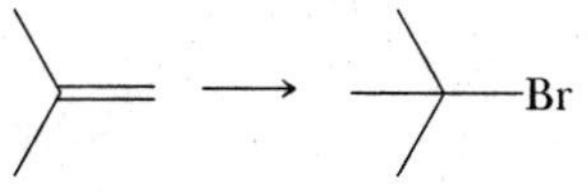 → Br

(c) 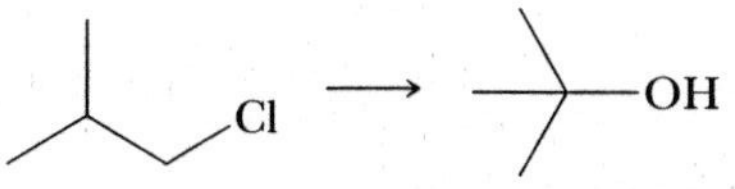 Cl → OH

(d) 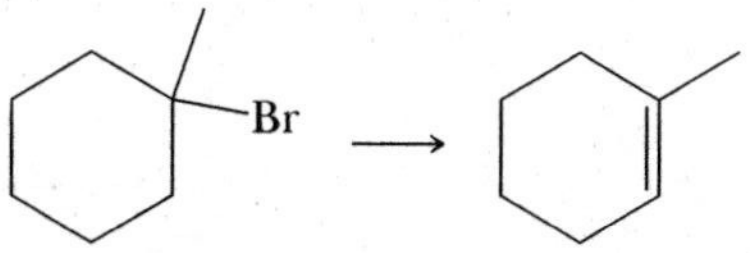 Br →

(e) 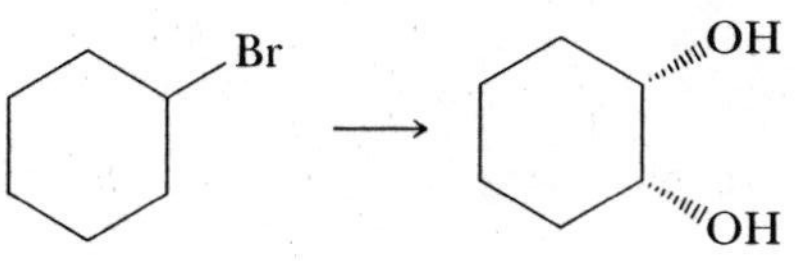 Br → OH, OH

(f) 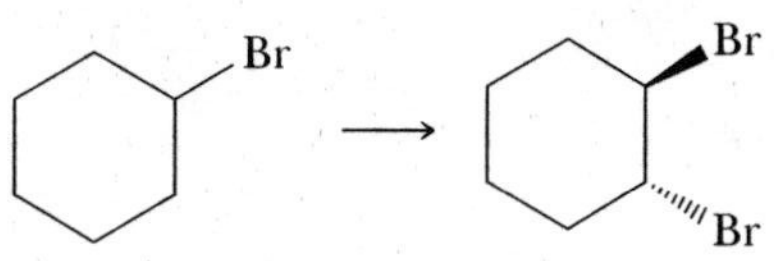 Br → Br, Br

Looking Ahead

7.41 The Williamson ether synthesis involves treating a haloalkane with a metal alkoxide. Following are two reactions intended to give benzyl *tert*-butyl ether. One reaction gives the ether in good yield, the other does not. Which reaction gives the ether? What is the product of the other reaction, and how do you account for its formation?

(a) 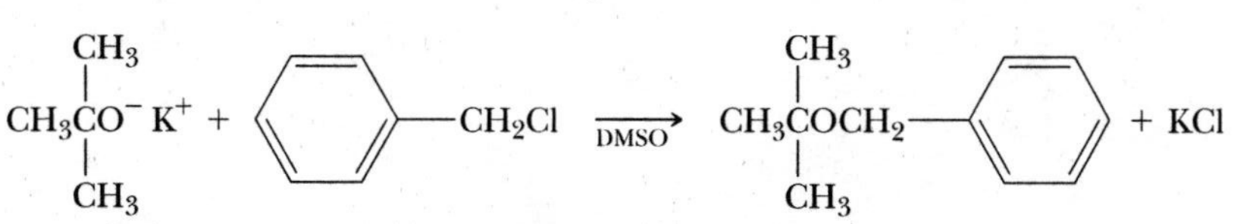 $CH_3C(CH_3)_2O^- K^+ + C_6H_5CH_2Cl \xrightarrow{DMSO} CH_3C(CH_3)_2OCH_2C_6H_5 + KCl$

(b) 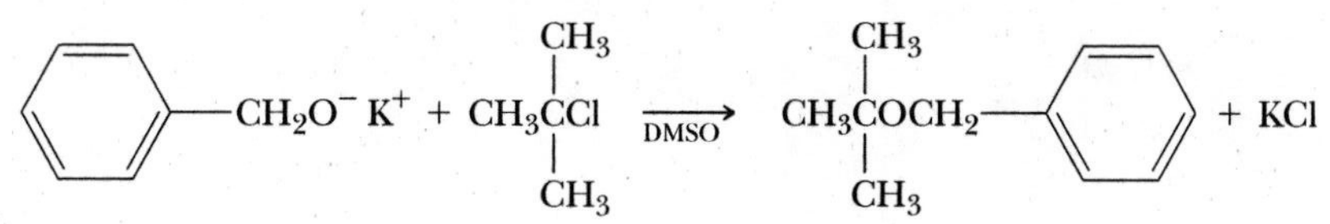 $C_6H_5CH_2O^- K^+ + CH_3C(CH_3)_2Cl \xrightarrow{DMSO} CH_3C(CH_3)_2OCH_2C_6H_5 + KCl$

7.42 The following ethers can, in principle, be synthesized by two different combinations of haloalkane or halocycloalkane and metal alkoxide. Show one combination that forms ether bond (1) and another that forms ether bond (2). Which combination gives the higher yield of ether?

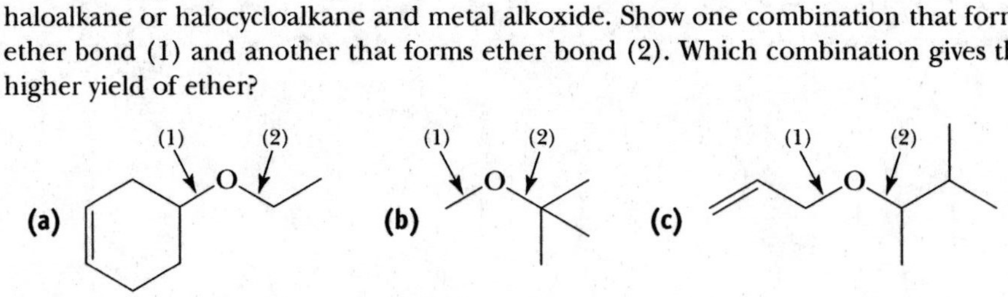

(1) (2) (1) (2) (1) (2)

(a) **(b)** **(c)**

7.43 Propose a mechanism for this reaction:

$$Cl-CH_2-CH_2-OH \xrightarrow{Na_2CO_3,\ H_2O} H_2C\overset{O}{-}CH_2$$

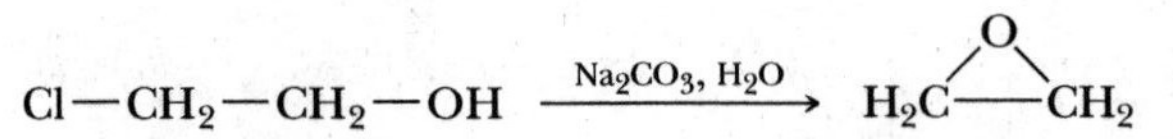

2-Chloroethanol Ethylene oxide

7.44 An OH group is a poor leaving group, and yet substitution occurs readily in the following reaction. Propose a mechanism for this reaction that shows how OH overcomes its limitation of being a poor leaving group.

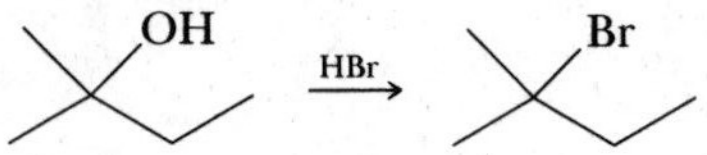

OH $\xrightarrow{HBr}$ Br

7.45 Explain why (*S*)-2-bromobutane becomes optically inactive when treated with sodium bromide in DMSO:

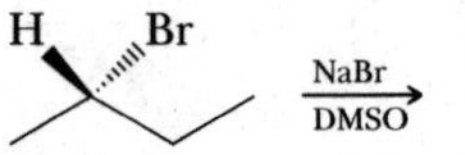

H Br $\xrightarrow[DMSO]{NaBr}$ optically inactive

optically active

7.46 Explain why phenoxide is a much poorer nucleophile than cyclohexoxide:

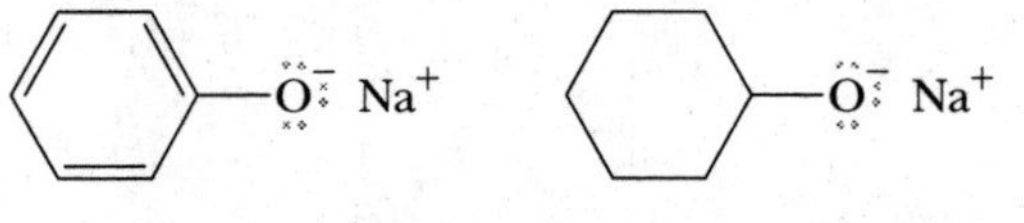

O^- Na^+ O^- Na^+

Sodium phenoxide Sodium cyclohexoxide

7.47 In ethers, each side of the oxygen is essentially an OR group and is thus a poor leaving group. Epoxides are three-membered ring ethers. Explain why an epoxide reacts readily with a nucleophile despite being an ether.

$$R-O-R \ + \ :Nu^- \longrightarrow \text{no reaction}$$

An ether

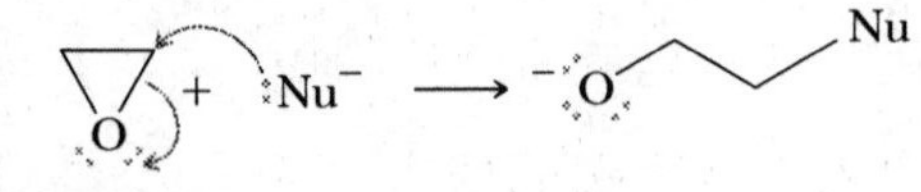

$+ \ :Nu^- \longrightarrow {}^-O$ Nu

An epoxide

8 Alcohols, Ethers, and Thiols

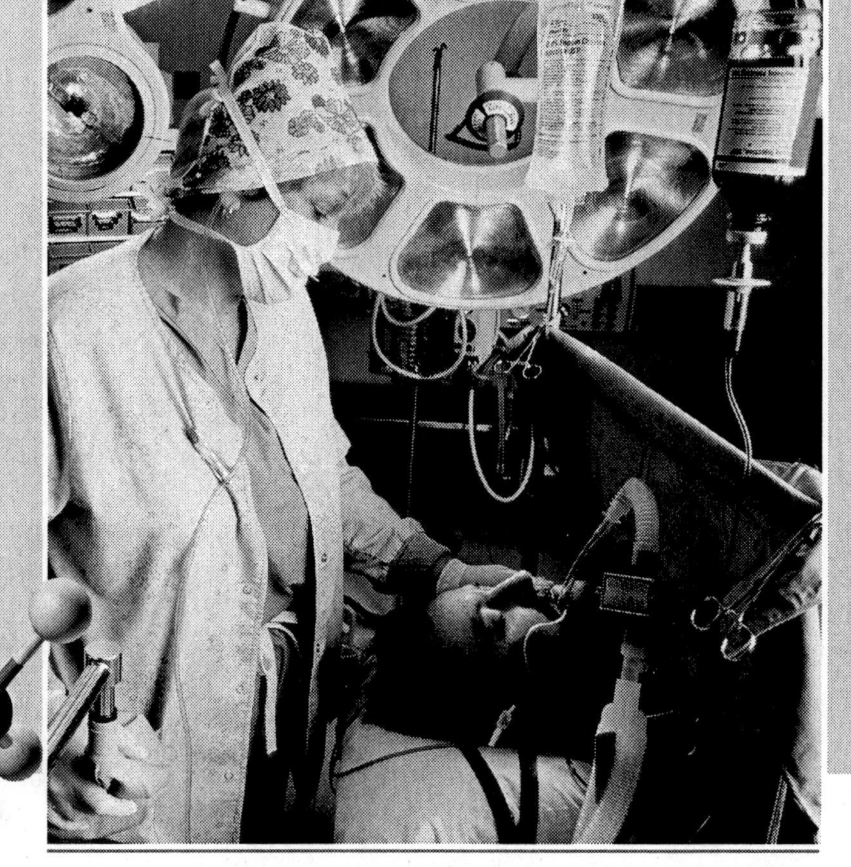

The discovery that inhaling ethers could make a patient insensitive to pain revolutionized the practice of medicine. Inset: A model of isoflurane, $CF_3CHClOCHF_2$, a halogenated ether widely used as an inhalation anesthetic in both human and veterinary medicine. *(Allan Levenson/Stone/Getty Images)*

8.1 INTRODUCTION

In this chapter, we study the physical and chemical properties of alcohols and ethers, two classes of oxygen-containing compounds. We also study thiols, a class of sulfur-containing compounds. Thiols are like alcohols in structure, except that they contain an —SH group rather than an —OH group.

CH_3CH_2OH	$CH_3CH_2OCH_2CH_3$	CH_3CH_2SH
Ethanol (an alcohol)	Diethyl ether (an ether)	Ethanethiol (a thiol)

These three compounds are certainly familiar to you. Ethanol is the fuel additive in gasohol, the alcohol in alcoholic beverages, and an important industrial and laboratory solvent. Diethyl ether was the first inhalation anesthetic used in general surgery. It is also an important industrial and laboratory solvent. Ethanethiol, like other low-molecular-weight thiols, has a stench. Smells such as those from skunks, rotten eggs, and sewage are caused by thiols.

Alcohols are particularly important in both laboratory and biochemical transformations of organic compounds. They can be converted into other types of compounds, such as alkenes, haloalkanes, aldehydes, ketones, carboxylic acids, and esters. Not only can alcohols be converted to these compounds, but they also can be prepared from them. Thus, alcohols play a central role in the interconversion of organic functional groups.

8.2 ALCOHOLS

A. Structure

Alcohol A compound containing an —OH (hydroxyl) group bonded to an sp^3 hybridized carbon.

The functional group of an **alcohol** is an **—OH (hydroxyl) group** bonded to an sp^3 hybridized carbon atom (Section 1.8A). The oxygen atom of an alcohol is also sp^3 hybridized. Two sp^3 hybrid orbitals of oxygen form sigma bonds to atoms of carbon and hydrogen. The other two sp^3 hybrid orbitals of oxygen each contain an unshared pair of electrons. Figure 8.1 shows a Lewis structure and ball-and-stick model of methanol, CH_3OH, the simplest alcohol.

(a)

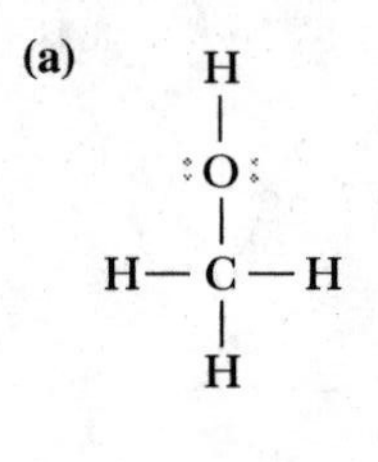

(b)

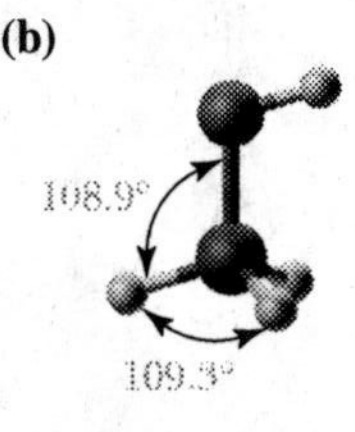

Figure 8.1
Methanol, CH_3OH. (a) Lewis structure and (b) ball-and-stick model. The measured H—C—O bond angle in methanol is 108.6°, very close to the tetrahedral angle of 109.5°.

B. Nomenclature

We derive the IUPAC names for alcohols in the same manner as those for alkanes, with the exception that the ending of the parent alkane is changed from *-e* to *-ol*. The ending *-ol* tells us that the compound is an alcohol.

1. Select, as the parent alkane, the longest chain of carbon atoms that contains the —OH, and number that chain from the end closer to the —OH group. In numbering the parent chain, the location of the —OH group takes precedence over alkyl groups and halogens.
2. Change the suffix of the parent alkane from *-e* to *-ol* (Section 3.6), and use a number to show the location of the —OH group. For cyclic alcohols, numbering begins at the carbon bearing the —OH group.
3. Name and number substituents and list them in alphabetical order.

To derive common names for alcohols, we name the alkyl group bonded to —OH and then add the word *alcohol*. Following are the IUPAC names and, in parentheses, the common names of eight low-molecular-weight alcohols:

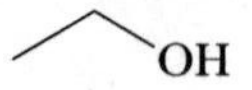

Ethanol (Ethyl alcohol)

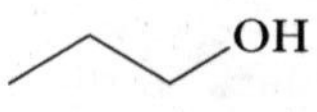

1-Propanol (Propyl alcohol)

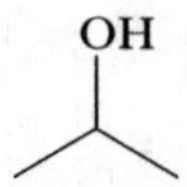

2-Propanol (Isopropyl alcohol)

1-Butanol (Butyl alcohol)

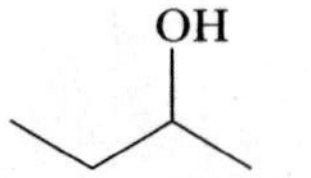

2-Butanol (*sec*-Butyl alcohol)

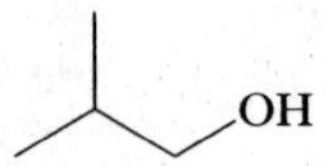

2-Methyl-1-propanol (Isobutyl alcohol)

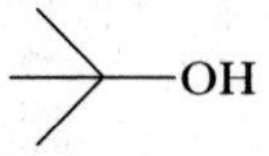

2-Methyl-2-propanol (*tert*-Butyl alcohol)

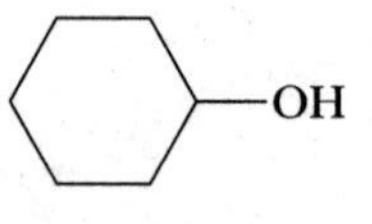

Cyclohexanol (Cyclohexyl alcohol)

Example 8.1

Write the IUPAC name for each alcohol:

(a) $CH_3(CH_2)_6CH_2OH$ (b) 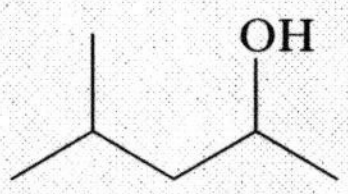 (c)

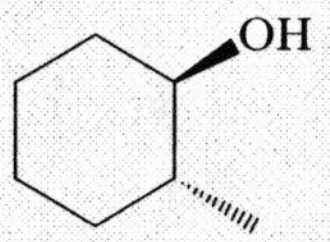

SOLUTION

(a) 1-Octanol (b) 4-Methyl-2-pentanol
(c) *trans*-2-Methylcyclohexanol or (1*R*,2*R*)-2-Methylcyclohexanol

Practice Problem 8.1

Write the IUPAC name for each alcohol:

(a) 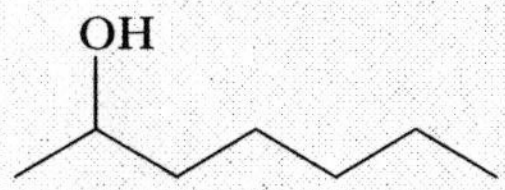 (b) 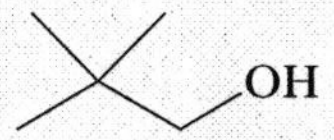 (c)

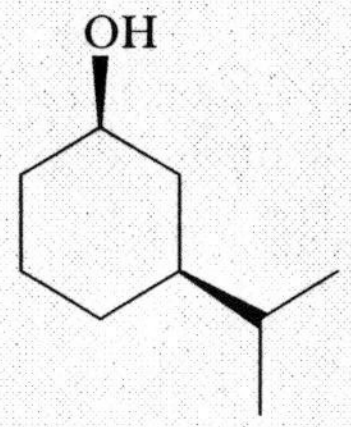

We classify alcohols as **primary (1°)**, **secondary (2°)**, or **tertiary (3°)**, depending on whether the —OH group is on a primary, secondary, or tertiary carbon (Section 1.8A).

Example 8.2

Classify each alcohol as primary, secondary, or tertiary:

(a) 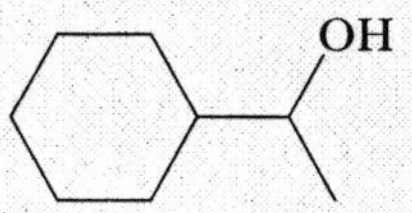 (b) $CH_3\overset{CH_3}{\underset{CH_3}{C}}OH$ (c)

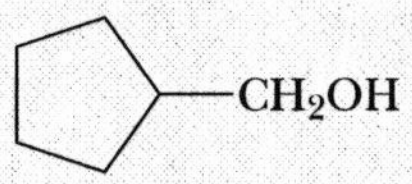

SOLUTION

(a) Secondary (2°) (b) Tertiary (3°) (c) Primary (1°)

Practice Problem 8.2

Classify each alcohol as primary, secondary, or tertiary:

(a) 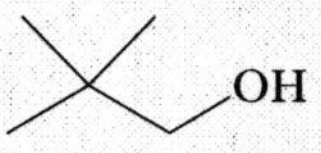 (b)

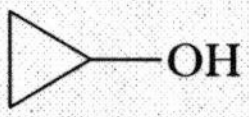

(c) $CH_2{=}CHCH_2OH$ (d)

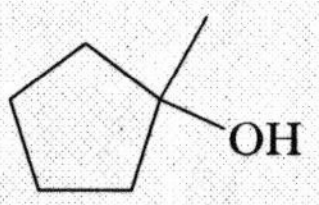

In the IUPAC system, a compound containing two hydroxyl groups is named as a **diol**, one containing three hydroxyl groups is named as a **triol**, and so on. In IUPAC names for diols, triols, and so on, the final *-e* of the parent alkane name is retained, as for example, in 1,2-ethanediol.

Glycol A compound with two hydroxyl (—OH) groups on adjacent carbons.

As with many organic compounds, common names for certain diols and triols have persisted. Compounds containing two hydroxyl groups on adjacent carbons are often referred to as **glycols** (Section 5.4). Ethylene glycol and propylene glycol are synthesized from ethylene and propylene, respectively—hence their common names:

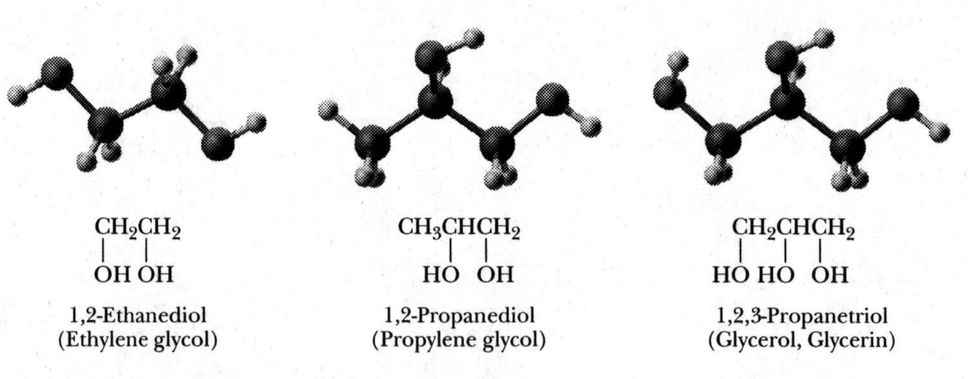

CH_2CH_2 (OH OH)	CH_3CHCH_2 (HO OH)	CH_2CHCH_2 (HO HO OH)
1,2-Ethanediol (Ethylene glycol)	1,2-Propanediol (Propylene glycol)	1,2,3-Propanetriol (Glycerol, Glycerin)

Ethylene glycol is a polar molecule and dissolves readily in water, a polar solvent. *(Charles D. Winter)*

We often refer to compounds containing —OH and C=C groups as unsaturated alcohols. To name an unsaturated alcohol,

1. Number the parent alkane so as to give the —OH group the lowest possible number.
2. Show the double bond by changing the infix of the parent alkane from *-an-* to *-en-* (Section 3.6), and show the alcohol by changing the suffix of the parent alkane from *-e* to *-ol*.
3. Use numbers to show the location of both the carbon–carbon double bond and the hydroxyl group.

Example 8.3

Write the IUPAC name for each unsaturated alcohol:

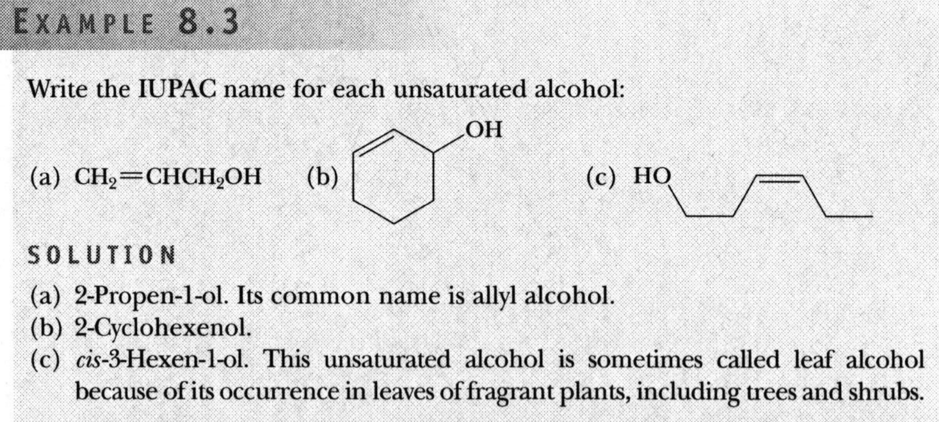

(a) $CH_2{=}CHCH_2OH$ (b) (c)

SOLUTION

(a) 2-Propen-1-ol. Its common name is allyl alcohol.
(b) 2-Cyclohexenol.
(c) *cis*-3-Hexen-1-ol. This unsaturated alcohol is sometimes called leaf alcohol because of its occurrence in leaves of fragrant plants, including trees and shrubs.

Practice Problem 8.3

Write the IUPAC name for each unsaturated alcohol:

(a) (b)

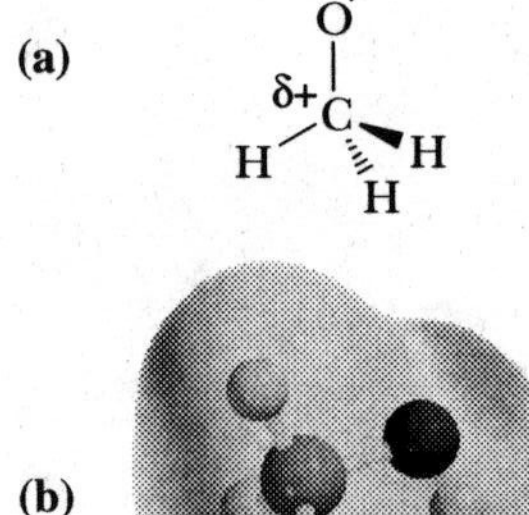

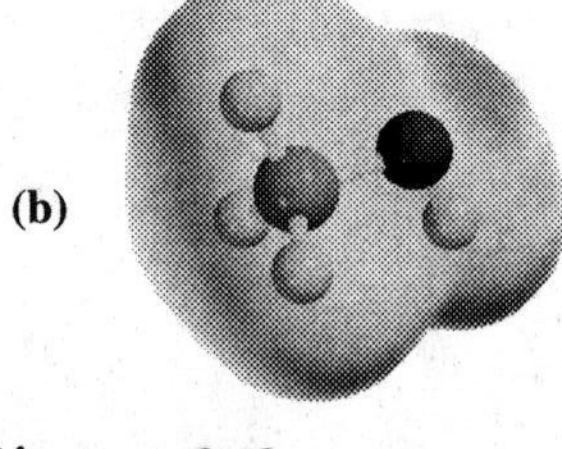

Figure 8.2
Polarity of the C—O—H bond in methanol.
(a) There are partial positive charges on carbon and hydrogen and a partial negative charge on oxygen.
(b) An electron density map showing the partial negative charge (in red) around oxygen and a partial positive charge (in blue) around hydrogen of the —OH group.

C. Physical Properties

The most important physical property of alcohols is the polarity of their —OH groups. Because of the large difference in electronegativity (Table 1.5) between oxygen and carbon $(3.5 - 2.5 = 1.0)$ and between oxygen and hydrogen $(3.5 - 2.1 = 1.4)$, both the C—O and O—H bonds of an alcohol are polar covalent, and alcohols are polar molecules, as illustrated in Figure 8.2 for methanol.

Chemical Connections 8A

Nitroglycerin: An Explosive and a Drug

In 1847, Ascanio Sobrero (1812–1888) discovered that 1,2,3-propanetriol, more commonly named glycerin, reacts with nitric acid in the presence of sulfuric acid to give a pale yellow, oily liquid called nitroglycerin:

$$\begin{array}{l} CH_2-OH \\ | \\ CH-OH \\ | \\ CH_2-OH \end{array} + 3\,HNO_3 \xrightarrow{H_2SO_4} \begin{array}{l} CH_2-ONO_2 \\ | \\ CH-ONO_2 \\ | \\ CH_2-ONO_2 \end{array} + 3\,H_2O$$

1,2,3-Propanetriol (Glycerol, Glycerin) — 1,2,3-Propanetriol trinitrate (Nitroglycerin)

Sobrero also discovered the explosive properties of the compound: When he heated a small quantity of it, it exploded! Soon, nitroglycerin became widely used for blasting in the construction of canals, tunnels, roads, and mines and, of course, for warfare.

One problem with the use of nitroglycerin was soon recognized: It was difficult to handle safely, and accidental explosions occurred frequently. The Swedish chemist Alfred Nobel (1833–1896) solved the problem: He discovered that a claylike substance called diatomaceous earth absorbs nitroglycerin so that it will not explode without a fuse. He gave the name *dynamite* to this mixture of nitroglycerine, diatomaceous earth, and sodium carbonate.

The fortune of Alfred Nobel, 1833–1896, built on the manufacture of dynamite, now funds the Nobel Prizes. *(Bettmann/Corbis)*

Surprising as it may seem, nitroglycerin is used in medicine to treat angina pectoris, the symptoms of which are sharp chest pains caused by a reduced flow of blood in the coronary artery. Nitroglycerin, which is available in liquid (diluted with alcohol to render it nonexplosive), tablet, or paste form, relaxes the smooth muscles of blood vessels, causing dilation of the coronary artery. This dilation, in turn, allows more blood to reach the heart.

When Nobel became ill with heart disease, his physicians advised him to take nitroglycerin to relieve his chest pains. He refused, saying he could not understand how the explosive could relieve chest pains. It took science more than 100 years to find the answer. We now know that it is nitric oxide, NO, derived from the nitro groups of nitroglycerin, that relieves the pain.

Table 8.1 lists the boiling points and solubilities in water for five groups of alcohols and alkanes of similar molecular weight. Notice that, of the compounds compared in each group, the alcohol has the higher boiling point and is the more soluble in water.

Alcohols have higher boiling points than alkanes of similar molecular weight, because alcohols are polar molecules and can associate in the liquid state by a type of intermolecular attraction called **hydrogen bonding** (Figure 8.3). The strength of hydrogen bonding between alcohol molecules is approximately 2 to 5 kcal/mol (8.4 to 21 kJ/mol). For comparison, the strength of the O—H covalent bond in an alcohol molecule is approximately 110 kcal/mol (460 kJ/mol). As we see by comparing these numbers, an O----H hydrogen bond is considerably weaker than an O—H covalent bond. Nonetheless, it is sufficient to have a dramatic effect on the physical properties of alcohols.

Hydrogen bonding The attractive force between a partial positive charge on hydrogen and partial negative charge on a nearby oxygen, nitrogen, or fluorine atom.

Because of hydrogen bonding between alcohol molecules in the liquid state, extra energy is required to separate each hydrogen-bonded alcohol molecule from its neighbors—hence the relatively high boiling points of alcohols compared with those of alkanes. The presence of additional hydroxyl groups in a molecule further increases the extent of hydrogen bonding, as can be seen by comparing the boiling points of 1-pentanol (138°C) and 1,4-butanediol (230°C), both of which have approximately the same molecular weight.

TABLE 8.1 Boiling Points and Solubilities in Water of Five Groups of Alcohols and Alkanes of Similar Molecular Weight

Structural Formula	Name	Molecular Weight	Boiling Point (°C)	Solubility in Water
CH_3OH	methanol	32	65	infinite
CH_3CH_3	ethane	30	−89	insoluble
CH_3CH_2OH	ethanol	46	78	infinite
$CH_3CH_2CH_3$	propane	44	−42	insoluble
$CH_3CH_2CH_2OH$	1-propanol	60	97	infinite
$CH_3CH_2CH_2CH_3$	butane	58	0	insoluble
$CH_3CH_2CH_2CH_2OH$	1-butanol	74	117	8 g/100 g
$CH_3CH_2CH_2CH_2CH_3$	pentane	72	36	insoluble
$CH_3CH_2CH_2CH_2CH_2OH$	1-pentanol	88	138	2.3 g/100 g
$HOCH_2CH_2CH_2CH_2OH$	1,4-butanediol	90	230	infinite
$CH_3CH_2CH_2CH_2CH_2CH_3$	hexane	86	69	insoluble

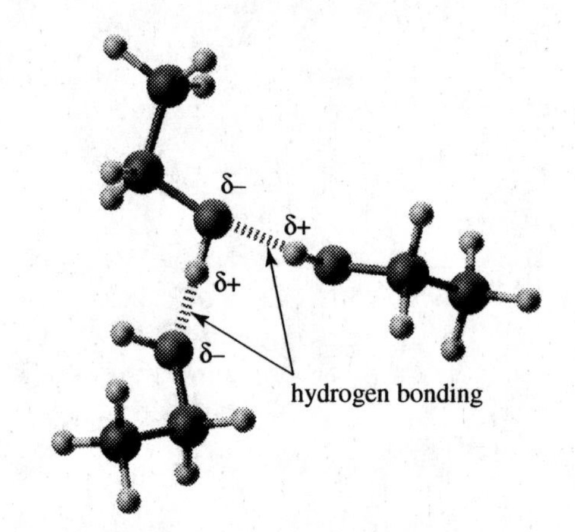

Figure 8.3
The association of ethanol molecules in the liquid state. Each O—H can participate in up to three hydrogen bonds (one through hydrogen and two through oxygen). Only two of these three possible hydrogen bonds per molecule are shown in the figure.

Because of increased dispersion forces (Section 3.9B) between larger molecules, boiling points of all types of compounds, including alcohols, increase with increasing molecular weight. (Compare, for example, the boiling points of ethanol, 1-propanol, 1-butanol, and 1-pentanol.)

Alcohols are much more soluble in water than are alkanes, alkenes, and alkynes of comparable molecular weight. Their increased solubility is due to hydrogen bonding between alcohol molecules and water. Methanol, ethanol, and 1-propanol are soluble in water in all proportions. As molecular weight increases, the physical properties of alcohols become more like those of hydrocarbons of comparable molecular weight. Alcohols of higher molecular weight are much less soluble in water because of the increase in size of the hydrocarbon portion of their molecules.

8.3 REACTIONS OF ALCOHOLS

In this section, we study the acidity and basicity of alcohols, their dehydration to alkenes, their conversion to haloalkanes, and their oxidation to aldehydes, ketones, or carboxylic acids.

A. Acidity of Alcohols

Alcohols have about the same pK_a values as water (15.7), which means that aqueous solutions of alcohols have about the same pH as that of pure water. The pK_a of methanol, for example, is 15.5:

$$CH_3\ddot{O}{-}H + :\ddot{O}{-}H \rightleftharpoons CH_3\ddot{O}:^- + H{-}\overset{+}{\ddot{O}}{-}H \qquad K_a = \frac{[CH_3O^-][H_3O^+]}{[CH_3OH]} = 3.2 \times 10^{-16}$$

(with H below each of the two water-derived O atoms)

$$pK_a = 15.5$$

Table 8.2 gives the acid ionization constants for several low-molecular-weight alcohols. Methanol and ethanol are about as acidic as water. Higher-molecular-weight, water-soluble alcohols are slightly weaker acids than water. Even though alcohols have some slight acidity, they are not strong enough acids to react with weak bases such as sodium bicarbonate or sodium carbonate. (At this point, it would be worthwhile to review Section 2.5 and the discussion of the position of equilibrium in acid–base reactions.) Note that, although acetic acid is a "weak acid" compared with acids such as HCl, it is still 10^{10} times stronger as an acid than alcohols are.

TABLE 8.2 pK_a Values for Selected Alcohols in Dilute Aqueous Solution*

Compound	Structural Formula	pK_a	
hydrogen chloride	HCl	−7	Stronger acid
acetic acid	CH_3COOH	4.8	↑
methanol	CH_3OH	15.5	
water	H_2O	15.7	
ethanol	CH_3CH_2OH	15.9	
2-propanol	$(CH_3)_2CHOH$	17	
2-methyl-2-propanol	$(CH_3)_3COH$	18	Weaker acid

* Also given for comparison are pK_a values for water, acetic acid, and hydrogen chloride.

B. Basicity of Alcohols

In the presence of strong acids, the oxygen atom of an alcohol is a weak base and reacts with an acid by proton transfer to form an oxonium ion:

$$\underset{\text{Ethanol}}{CH_3CH_2{-}\ddot{O}{-}H} + \underset{\substack{\text{Hydronium ion}\\(pK_a - 1.7)}}{H{-}\overset{+}{\ddot{O}}{-}H} \xrightleftharpoons{H_2SO_4} \underset{\substack{\text{Ethyloxonium ion}\\(pK_a - 2.4)}}{CH_3CH_2{-}\overset{+}{\ddot{O}}{-}H} + :\ddot{O}{-}H$$

(with H below each of the hydronium O, ethyloxonium O, and water O atoms)

Thus, alcohols can function as both weak acids and weak bases.

Methanol reacts with sodium metal with the evolution of hydrogen gas. *(Charles D. Winters)*

C. Reaction with Active Metals

Like water, alcohols react with Li, Na, K, Mg, and other active metals to liberate hydrogen and to form metal alkoxides. In the following oxidation–reduction reaction, Na is oxidized to Na^+ and H^+ is reduced to H_2:

$$2\,CH_3OH + 2\,Na \longrightarrow \underset{\text{Sodium methoxide}}{2\,CH_3O^-Na^+} + H_2$$

To name a metal alkoxide, name the cation first, followed by the name of the anion. The name of an alkoxide ion is derived from a prefix showing the number of carbon atoms and their arrangement (*meth-*, *eth-*, *isoprop-*, tert-*but-*, and so on) followed by the suffix *-oxide*.

Alkoxide ions are somewhat stronger bases than is the hydroxide ion. In addition to sodium methoxide, the following metal salts of alcohols are commonly used in organic reactions requiring a strong base in a nonaqueous solvent; sodium ethoxide in ethanol and potassium *tert*-butoxide in 2-methyl-2-propanol (*tert*-butyl alcohol):

$$CH_3CH_2O^-Na^+ \qquad CH_3\overset{CH_3}{\underset{CH_3}{C}}O^-K^+$$

Sodium ethoxide Potassium *tert*-butoxide

As we saw in Chapter 7, alkoxide ions can also be used as nucleophiles in substitution reactions.

EXAMPLE 8.4

Write a balanced equation for the reaction of cyclohexanol with sodium metal.

SOLUTION

$$2\,C_6H_{11}\text{—}OH + 2\,Na \longrightarrow 2\,C_6H_{11}\text{—}O^-Na^+ + H_2$$

Cyclohexanol Sodium cyclohexoxide

Practice Problem 8.4

Predict the position of equilibrium for the following acid–base reaction. (*Hint:* Review Section 2.5.)

$$CH_3CH_2O^-Na^+ + CH_3\overset{O}{\overset{\|}{C}}OH \rightleftharpoons CH_3CH_2OH + CH_3\overset{O}{\overset{\|}{C}}O^-Na^+$$

D. Conversion to Haloalkanes

The conversion of an alcohol to an alkyl halide involves substituting halogen for —OH at a saturated carbon. The most common reagents for this conversion are the halogen acids and $SOCl_2$.

Reaction with HCl, HBr, and HI

Water-soluble tertiary alcohols react very rapidly with HCl, HBr, and HI. Mixing a tertiary alcohol with concentrated hydrochloric acid for a few minutes at room temperature converts the alcohol to a water-insoluble chloroalkane that separates from the aqueous layer.

$$CH_3\overset{CH_3}{\underset{CH_3}{C}}OH + HCl \xrightarrow{25°C} CH_3\overset{CH_3}{\underset{CH_3}{C}}Cl + H_2O$$

2-Methyl-2-propanol 2-Chloro-2-methylpropane

Low-molecular-weight, water-soluble primary and secondary alcohols do not react under these conditions.

Water-insoluble tertiary alcohols are converted to tertiary halides by bubbling gaseous HX through a solution of the alcohol dissolved in diethyl ether or tetrahydrofuran (THF):

$$\text{1-Methylcyclohexanol} + HCl \xrightarrow[\text{ether}]{0°C} \text{1-Chloro-1-methylcyclohexane} + H_2O$$

1-Methyl-cyclohexanol 1-Chloro-1-methyl cyclohexane

Water-insoluble primary and secondary alcohols react only slowly under these conditions.

Primary and secondary alcohols are converted to bromoalkanes and iodoalkanes by treatment with concentrated hydrobromic and hydroiodic acids. For example, heating 1-butanol with concentrated HBr gives 1-bromobutane:

$$\text{1-Butanol} + HBr \longrightarrow \text{1-Bromobutane (Butyl bromide)} + H_2O$$

On the basis of observations of the relative ease of reaction of alcohols with HX (3° > 2° > 1°), it has been proposed that the conversion of tertiary and secondary alcohols to haloalkanes by concentrated HX occurs by an S_N1 mechanism and involves the formation of a carbocation intermediate.

Mechanism: Reaction of a Tertiary Alcohol with HCl: An S_N1 Reaction

Step 1: Rapid and reversible proton transfer from the acid to the OH group gives an oxonium ion. The result of this proton transfer is to convert the leaving group from OH^-, a poor leaving group, to H_2O, a better leaving group:

$$(CH_3)_3C{-}\ddot{O}{-}H + H{-}\overset{+}{O}H_2 \underset{\text{reversible}}{\overset{\text{rapid and}}{\rightleftharpoons}} (CH_3)_3C{-}\overset{+}{O}H_2 + :\ddot{O}H_2$$

2-Methyl-2-propanol (*tert*-Butyl alcohol) An oxonium ion

Step 2: Loss of water from the oxonium ion gives a 3° carbocation intermediate:

$$(CH_3)_3C{-}\overset{+}{O}H_2 \xrightarrow[S_N1]{\text{slow, rate determining}} (CH_3)_3C^+ + :\ddot{O}H_2$$

An oxonium ion A 3° carbocation intermediate

Step 3: Reaction of the 3° carbocation intermediate (an electrophile) with chloride ion (a nucleophile) gives the product:

$$(CH_3)_3C^+ + :\ddot{Cl}:^- \xrightarrow{\text{fast}} (CH_3)_3C{-}\ddot{Cl}:$$

2-Chloro-2-methylpropane (*tert*-Butyl chloride)

Primary alcohols react with HX by an S_N2 mechanism. In the rate-determining step, the halide ion displaces H_2O from the carbon bearing the oxonium ion. The displacement of H_2O and the formation of the C—X bond are simultaneous.

Mechanism: Reaction of a Primary Alcohol with HBr: An S_N2 Reaction

Step 1: Rapid and reversible proton transfer to the OH group which converts the leaving group from OH^-, a poor leaving group, to H_2O, a better leaving group:

$$CH_3CH_2CH_2CH_2-\ddot{O}H + H-\overset{+}{O}(H)-H \underset{\text{reversible}}{\overset{\text{rapid and}}{\rightleftharpoons}} CH_3CH_2CH_2CH_2-\overset{+}{O}H_2 + :\ddot{O}(H)-H$$

An oxonium ion

Step 2: The nucleophilic displacement of H_2O by Br^- gives the bromoalkane:

$$:\ddot{Br}:^- + CH_3CH_2CH_2CH_2-\overset{+}{O}H_2 \xrightarrow[S_N2]{\text{slow, rate determining}} CH_3CH_2CH_2CH_2-\ddot{Br}: + :\ddot{O}H_2$$

Why do tertiary alcohols react with HX by formation of carbocation intermediates, whereas primary alcohols react by direct displacement of —OH (more accurately, by displacement of $-OH_2^+$)? The answer is a combination of the same two factors involved in nucleophilic substitution reactions of haloalkanes (Section 7.6B):

1. *Electronic factors* Tertiary carbocations are the most stable (require the lowest activation energy for their formation), whereas primary carbocations are the least stable (require the highest activation energy for their formation). Therefore, tertiary alcohols are most likely to react by carbocation formation; secondary alcohols are intermediate, and primary alcohols rarely, if ever, react by carbocation formation.
2. *Steric factors* To form a new carbon–halogen bond, halide ion must approach the substitution center and begin to form a new covalent bond to it. If we compare the ease of approach to the substitution center of a primary oxonium ion with that of a tertiary oxonium ion, we see that approach is considerably easier in the case of a primary oxonium ion. Two hydrogen atoms and one alkyl group screen the back side of the substitution center of a primary oxonium ion, whereas three alkyl groups screen the back side of the substitution center of a tertiary oxonium ion.

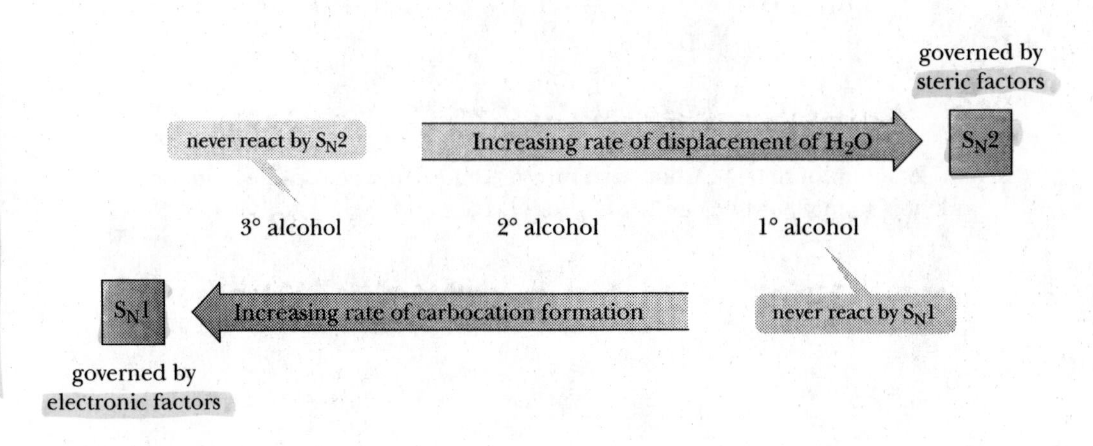

Reaction with Thionyl Chloride

The most widely used reagent for the conversion of primary and secondary alcohols to alkyl chlorides is thionyl chloride, $SOCl_2$. The by-products of this nucleophilic substitution reaction are HCl and SO_2, both given off as gases. Often, an organic base such as pyridine (Section 10.2) is added to react with and neutralize the HCl by-product:

$$\text{1-Heptanol (}\sim\!\!\sim\!\!\sim\text{OH)} + \underset{\text{Thionyl chloride}}{SOCl_2} \xrightarrow{\text{pyridine}} \text{1-Chloroheptane (}\sim\!\!\sim\!\!\sim\text{Cl)} + SO_2 + HCl$$

1-Heptanol | Thionyl chloride | 1-Chloroheptane

E. Acid-Catalyzed Dehydration to Alkenes

Dehydration Elimination of a molecule of water from a compound.

An alcohol can be converted to an alkene by **dehydration**—that is, by the elimination of a molecule of water from adjacent carbon atoms. In the laboratory, the dehydration of an alcohol is most often brought about by heating it with either 85% phosphoric acid or concentrated sulfuric acid. Primary alcohols are the most difficult to dehydrate and generally require heating in concentrated sulfuric acid at temperatures as high as 180°C. Secondary alcohols undergo acid-catalyzed dehydration at somewhat lower temperatures. The acid-catalyzed dehydration of tertiary alcohols often requires temperatures only slightly above room temperature:

$$CH_3CH_2OH \xrightarrow[180^\circ C]{H_2SO_4} CH_2{=}CH_2 + H_2O$$

$$\text{Cyclohexanol} \xrightarrow[140^\circ C]{H_2SO_4} \text{Cyclohexene} + H_2O$$

Cyclohexanol | Cyclohexene

$$CH_3\overset{CH_3}{\underset{CH_3}{C}}OH \xrightarrow[50^\circ C]{H_2SO_4} CH_3\overset{CH_3}{C}{=}CH_2 + H_2O$$

2-Methyl-2-propanol (*tert*-Butyl alcohol) | 2-Methylpropene (Isobutylene)

Thus, the ease of acid-catalyzed dehydration of alcohols occurs in this order:

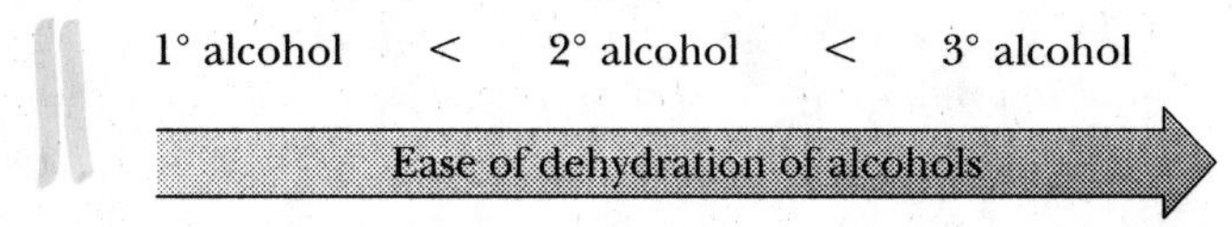

When isomeric alkenes are obtained in the acid-catalyzed dehydration of an alcohol, the more stable alkene (the one with the greater number of substituents on the double bond; see Section 5.5B) generally predominates; that is, the acid-catalyzed dehydration of alcohols follows Zaitsev's rule (Section 7.8):

$$\underset{\text{2-Butanol}}{CH_3CH_2\overset{OH}{C}HCH_3} \xrightarrow[\text{heat}]{85\%\ H_3PO_4} \underset{\substack{\text{2-Butene}\\(80\%)}}{CH_3CH{=}CHCH_3} + \underset{\substack{\text{1-Butene}\\(20\%)}}{CH_3CH_2CH{=}CH_2}$$

EXAMPLE 8.5

For each of the following alcohols, draw structural formulas for the alkenes that form upon acid-catalyzed dehydration of that alcohol, and predict which alkene is the major product:

(a) 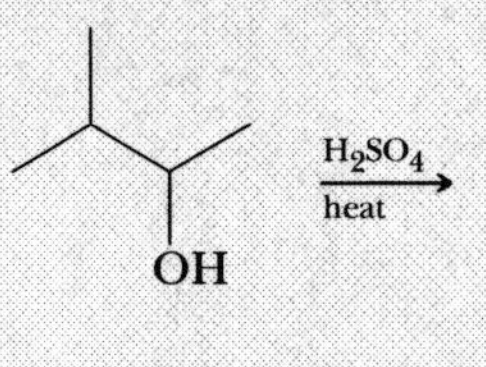 OH $\xrightarrow[\text{heat}]{H_2SO_4}$ (b) 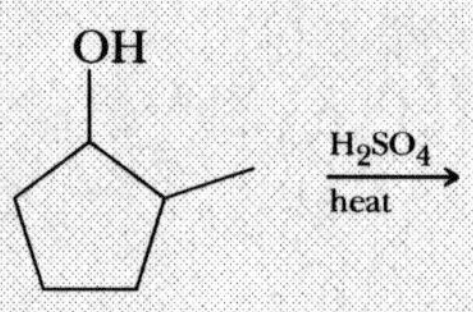 OH $\xrightarrow[\text{heat}]{H_2SO_4}$

SOLUTION

(a) The elimination of H_2O from carbons 2 and 3 gives 2-methyl-2-butene; the elimination of H_2O from carbons 1 and 2 gives 3-methyl-1-butene. 2-Methyl-2-butene, with three alkyl groups (three methyl groups) on the double bond, is the major product. 3-Methyl-1-butene, with only one alkyl group (an isopropyl group) on the double bond, is the minor product:

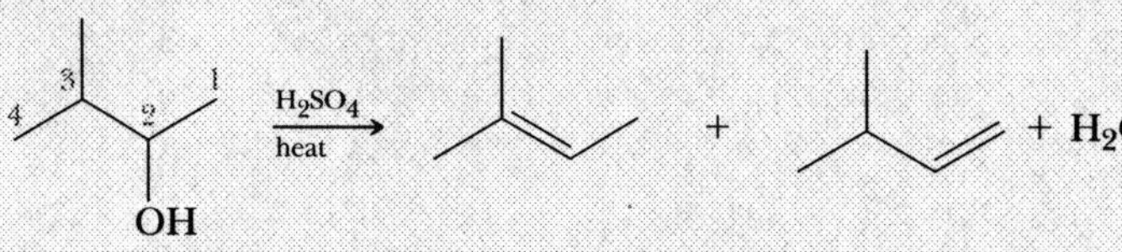

4 3 2 1 OH $\xrightarrow[\text{heat}]{H_2SO_4}$ + + H_2O

3-Methyl-2-butanol 2-Methyl-2-butene (major product) 3-Methyl-1-butene

(b) The major product, 1-methylcyclopentene, has three alkyl substituents on the double bond. The minor product, 3-methylcyclopentene, only two alkyl substituents on the double bond:

OH $\xrightarrow[\text{heat}]{H_2SO_4}$ +

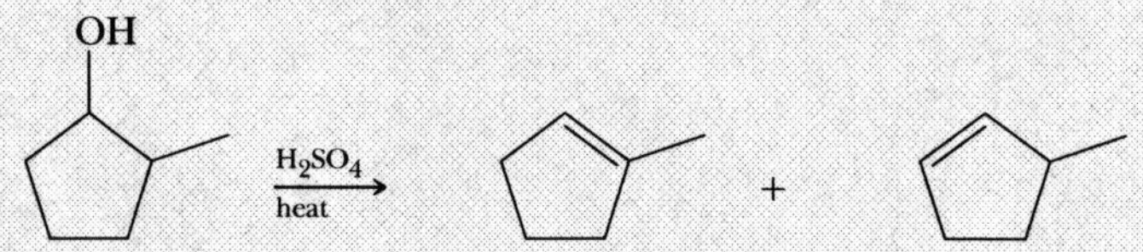

2-Methylcyclopentanol 1-Methylcyclopentene (major product) 3-Methylcyclopentene

Practice Problem 8.5

For each of the following alcohols, draw structural formulas for the alkenes that form upon acid-catalyzed dehydration of that alcohol, and predict which alkene is the major product from each alcohol:

(a) 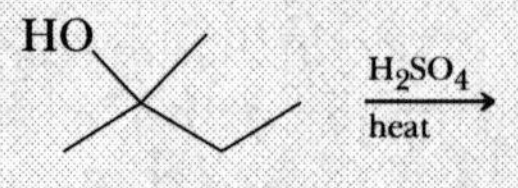 HO $\xrightarrow[\text{heat}]{H_2SO_4}$ (b) 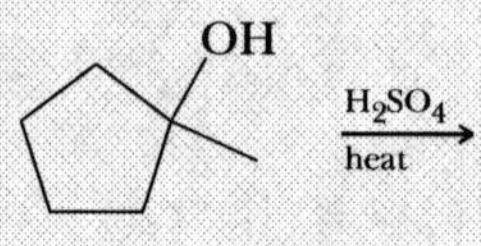 OH $\xrightarrow[\text{heat}]{H_2SO_4}$

On the basis of the relative ease of dehydration of alcohols (3° > 2° > 1°), chemists propose a three-step mechanism for the acid-catalyzed dehydration of secondary and tertiary alcohols. This mechanism involves the formation of a carbocation intermediate in the rate-determining step and therefore is an E1 mechanism.

Mechanism: Acid-Catalyzed Dehydration of 2-Butanol: An E1 Mechanism

Step 1: Proton transfer from H_3O^+ to the OH group of the alcohol gives an oxonium ion. A result of this step is to convert OH^-, a poor leaving group, into H_2O, a better leaving group:

$$CH_3CH(OH)CH_2CH_3 + H-\overset{+}{O}(H)-H \xrightleftharpoons[\text{reversible}]{\text{rapid and}} CH_3CH(\overset{+}{O}H_2)CH_2CH_3 + :\ddot{O}(H)-H$$

An oxonium ion

Step 2: Breaking of the C—O bond gives a 2° carbocation intermediate and H_2O:

H₂O is a good leaving group

$$CH_3CH(\overset{+}{O}H_2)CH_2CH_3 \xrightleftharpoons[\text{determining}]{\text{slow, rate}} CH_3\overset{+}{C}HCH_2CH_3 + H_2\ddot{O}:$$

A 2° carbocation intermediate

Step 3: Proton transfer from the carbon adjacent to the positively charged carbon to H_2O gives the alkene and regenerates the catalyst. The sigma electrons of a C—H bond become the pi electrons of the carbon–carbon double bond:

$$CH_3-\overset{+}{C}H-CH(H)-CH_3 + :\ddot{O}(H)-H \xrightarrow{\text{rapid}} CH_3-CH=CH-CH_3 + H-\overset{+}{O}(H)-H$$

Because the rate-determining step in the acid-catalyzed dehydration of secondary and tertiary alcohols is the formation of a carbocation intermediate, the relative ease of dehydration of these alcohols parallels the ease of formation of carbocations.

Primary alcohols react by the following two-step mechanism, in which Step 2 is the rate-determining step:

Mechanism: Acid-Catalyzed Dehydration of a Primary Alcohol: An E2 Mechanism

Step 1: Proton transfer from H_3O^+ to the OH group of the alcohol gives an oxonium ion:

$$CH_3CH_2-\ddot{O}-H + H-\overset{+}{O}(H)-H \xrightleftharpoons[\text{reversible}]{\text{rapid and}} CH_3CH_2-\overset{+}{O}H_2 + :\ddot{O}(H)-H$$

Step 2: Simultaneous proton transfer to solvent and loss of H_2O gives the alkene:

$$H-\ddot{O}(H): + H-C(H)(H)-CH_2-\overset{+}{O}H_2 \xrightarrow[\text{E2}]{\text{slow, rate determining}} H-\overset{+}{O}(H)-H + H_2C=CH_2 + :\ddot{O}(H)-H$$

In Section 5.3B, we discussed the acid-catalyzed hydration of alkenes to give alcohols. In the current section, we discussed the acid-catalyzed dehydration of alcohols to give alkenes. In fact, hydration–dehydration reactions are reversible.

Alkene hydration and alcohol dehydration are competing reactions, and the following equilibrium exists:

$$\gt C{=}C\lt + H_2O \underset{}{\overset{\text{acid catalyst}}{\rightleftharpoons}} -\underset{H}{\overset{|}{C}}-\underset{OH}{\overset{|}{C}}-$$

An alkene — An alcohol

How, then, do we control which product will predominate? Recall that LeChâtelier's principle states that a system in equilibrium will respond to a stress in the equilibrium by counteracting that stress. This response allows us to control these two reactions to give the desired product. Large amounts of water (achieved with the use of dilute aqueous acid) favor alcohol formation, whereas a scarcity of water (achieved with the use of concentrated acid) or experimental conditions by which water is removed (for example, heating the reaction mixture above 100°C) favor alkene formation. Thus, depending on the experimental conditions, it is possible to use the hydration–dehydration equilibrium to prepare either alcohols or alkenes, each in high yields.

F. Oxidation of Primary and Secondary Alcohols

The oxidation of a primary alcohol gives an aldehyde or a carboxylic acid, depending on the experimental conditions. Secondary alcohols are oxidized to ketones. Tertiary alcohols are not oxidized. Following is a series of transformations in which a primary alcohol is oxidized first to an aldehyde and then to a carboxylic acid. The fact that each transformation involves oxidation is indicated by the symbol O in brackets over the reaction arrow:

$$CH_3-\underset{H}{\overset{OH}{C}}-H \xrightarrow{[O]} CH_3-\overset{O}{\overset{\|}{C}}-H \xrightarrow{[O]} CH_3-\overset{O}{\overset{\|}{C}}-OH$$

A primary alcohol — An aldehyde — A carboxylic acid

The reagent most commonly used in the laboratory for the oxidation of a primary alcohol to a carboxylic acid and a secondary alcohol to a ketone is chromic acid, H_2CrO_4. Chromic acid is prepared by dissolving either chromium(VI) oxide or potassium dichromate in aqueous sulfuric acid:

$$CrO_3 + H_2O \xrightarrow{H_2SO_4} H_2CrO_4$$

Chromium(VI) oxide — Chromic acid

$$K_2Cr_2O_7 \xrightarrow{H_2SO_4} H_2Cr_2O_7 \xrightarrow{H_2O} 2\ H_2CrO_4$$

Potassisum dichromate — Chromic acid

The oxidation of 1-octanol by chromic acid in aqueous sulfuric acid gives octanoic acid in high yield. These experimental conditions are more than sufficient to oxidize the intermediate aldehyde to a carboxylic acid:

$$CH_3(CH_2)_6CH_2OH \xrightarrow[H_2SO_4,\ H_2O]{CrO_3} \left[CH_3(CH_2)_6\overset{O}{\overset{\|}{C}}H \right] \longrightarrow CH_3(CH_2)_6\overset{O}{\overset{\|}{C}}OH$$

1-Octanol — Octanal (not isolated) — Octanoic acid

The form of Cr(VI) commonly used for the oxidation of a primary alcohol to an aldehyde is prepared by dissolving CrO_3 in aqueous HCl and adding pyridine to precipitate **pyridinium chlorochromate (PCC)** as a solid. PCC oxidations are carried out in aprotic solvents, most commonly dichloromethane, CH_2Cl_2:

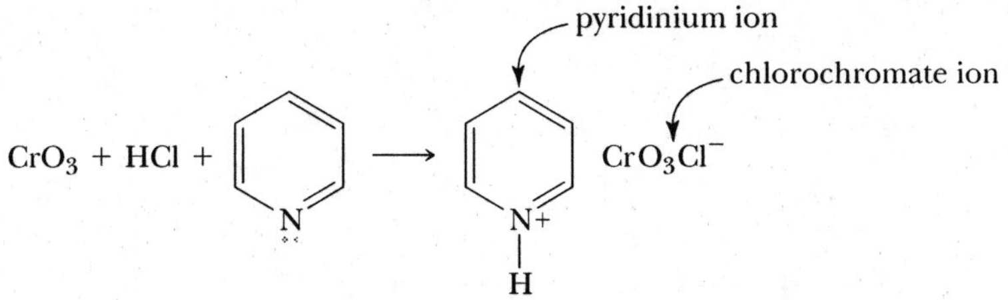

pyridinium ion

chlorochromate ion

CrO_3 + HCl + [Pyridine] ⟶ [Pyridinium] CrO_3Cl^-

Pyridine

Pyridinium chlorochromate (PCC)

This reagent is not only selective for the oxidation of primary alcohols to aldehydes, but also has little effect on carbon–carbon double bonds or other easily oxidized functional groups. In the following example, geraniol is oxidized to geranial without affecting either carbon–carbon double bond:

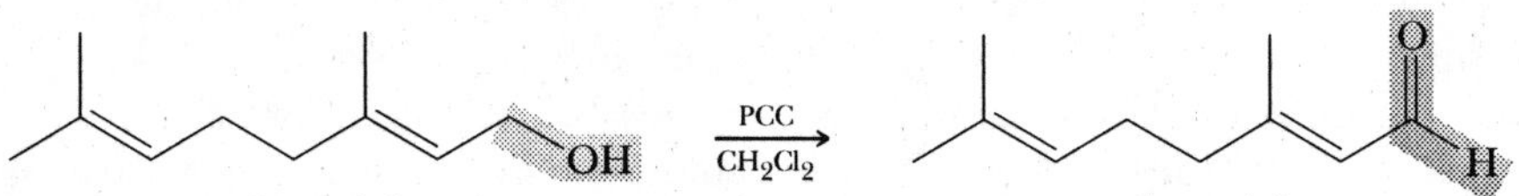

Geraniol —PCC, CH_2Cl_2⟶ Geranial

Secondary alcohols are oxidized to ketones by both chromic acid and PCC:

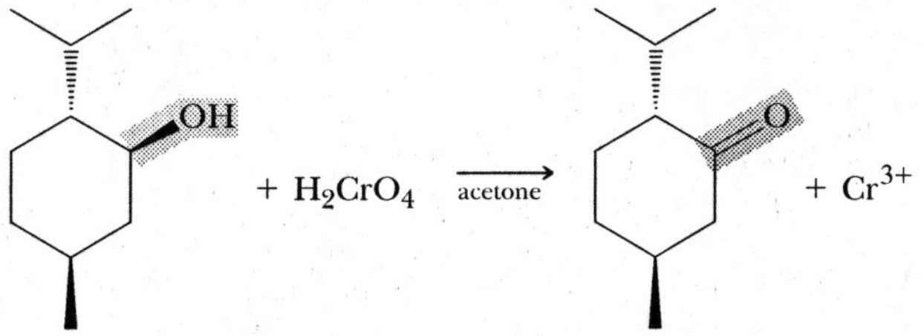

2-Isopropyl-5-methylcyclohexanol (Menthol) + H_2CrO_4 —acetone⟶ 2-Isopropyl-5-methylcyclohexanone (Menthone) + Cr^{3+}

Tertiary alcohols are resistant to oxidation, because the carbon bearing the —OH is bonded to three carbon atoms and therefore cannot form a carbon–oxygen double bond:

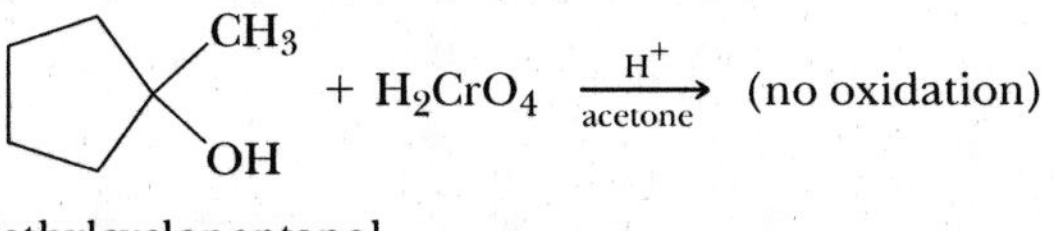

1-Methylcyclopentanol + H_2CrO_4 —H^+, acetone⟶ (no oxidation)

Note that the essential feature of the oxidation of an alcohol is the presence of at least one hydrogen on the carbon bearing the OH group. Tertiary alcohols lack such a hydrogen; therefore, they are not oxidized.

EXAMPLE 8.6

Draw the product of the treatment of each of the following alcohols with PCC:

(a) 1-Hexanol (b) 2-Hexanol (c) Cyclohexanol

This painting by Robert Hinckley shows the first use of diethyl ether as an anesthetic in 1846. Dr. Robert John Collins was removing a tumor from the patient's neck, and the dentist W. T. G. Morton - who discovered its anesthetic properties - administered the ether. *(Boston Medical Library in the Francis A. Courtney Library of Medicine)*

SOLUTION

1-Hexanol, a primary alcohol, is oxidized to hexanal. 2-Hexanol, a secondary alcohol, is oxidized to 2-hexanone. Cyclohexanol, a secondary alcohol, is oxidized to cyclohexanone.

(a) Hexanal (b) 2-Hexanone (c) Cyclohexanone

Practice Problem 8.6

Draw the product of the treatment of each alcohol in Example 8.6 with chromic acid.

8.4 ETHERS

A. Structure

Ether A compound containing an oxygen atom bonded to two carbon atoms.

The functional group of an **ether** is an atom of oxygen bonded to two carbon atoms. Figure 8.4 shows a Lewis structure and ball-and-stick model of dimethyl ether, CH_3OCH_3, the simplest ether. In dimethyl ether, two sp^3 hybrid orbitals of oxygen form sigma bonds to carbon atoms. The other two sp^3 hybrid orbitals of oxygen each contain an unshared pair of electrons. The C—O—C bond angle in dimethyl ether is 110.3°, close to the predicted tetrahedral angle of 109.5°.

In ethyl vinyl ether, the ether oxygen is bonded to one sp^3 hybridized carbon and one sp^2 hybridized carbon:

$CH_3CH_2—O—CH=CH_2$

Ethyl vinyl ether

B. Nomenclature

Alkoxy group An —OR group, where R is an alkyl group.

In the IUPAC system, ethers are named by selecting the longest carbon chain as the parent alkane and naming the —OR group bonded to it as an **alkoxy** (*alk*yl + *oxy*gen) group. Common names are derived by listing the alkyl groups bonded to oxygen in alphabetical order and adding the word *ether*.

$CH_3CH_2OCH_2CH_3$	$CH_3OC(CH_3)_2CH_3$	
Ethoxyethane (Diethyl ether)	2-Methoxy-2-methylpropane (methyl *tert*-butyl ether, MTBE)	*trans*-2-Ethoxycyclohexanol

Chemists almost invariably use common names for low-molecular-weight ethers. For example, although ethoxyethane is the IUPAC name for $CH_3CH_2OCH_2CH_3$, it is rarely called that, but rather is called diethyl ether, ethyl ether, or, even more commonly, simply ether. The abbreviation for *tert*-butyl methyl ether, used at one time as an octane-improving additive to gasolines, is *MTBE*, after the common name of methyl *tert*-butyl ether.

(a) H—C(H)(H)—Ö—C(H)(H)—H

(b)

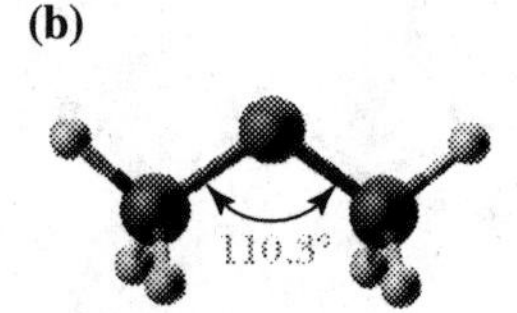

Figure 8.4 Dimethyl ether, CH_3OCH_3. (a) Lewis structure and (b) ball-and-stick model.

CHEMICAL CONNECTIONS 8B

Blood Alcohol Screening

Potassium dichromate oxidation of ethanol to acetic acid is the basis for the original breath alcohol screening test used by law enforcement agencies to determine a person's blood alcohol content. The test is based on the difference in color between the dichromate ion (reddish orange) in the reagent and the chromium(III) ion (green) in the product. Thus, color change can be used as a measure of the quantity of ethanol present in a breath sample:

$$\underset{\text{Ethanol}}{CH_3CH_2OH} + \underset{\substack{\text{Dichromate ion}\\\text{(reddish orange)}}}{Cr_2O_7^{2-}} \xrightarrow[H_2O]{H_2SO_4} \underset{\text{Acetic acid}}{CH_3\overset{O}{\overset{\|}{C}}OH} + \underset{\substack{\text{Chromium(III)}\\\text{ion (green)}}}{Cr^{3+}}$$

In its simplest form, a breath alcohol screening test consists of a sealed glass tube containing a potassium dichromate–sulfuric acid reagent impregnated on silica gel. To administer the test, the ends of the tube are broken off, a mouthpiece is fitted to one end, and the other end is inserted into the neck of a plastic bag. The person being tested then blows into the mouthpiece until the plastic bag is inflated.

As breath containing ethanol vapor passes through the tube, reddish orange dichromate ion is reduced to green chromium(III) ion. The concentration of ethanol in the breath is then estimated by measuring how far the green color extends along the length of the tube. When it extends beyond the halfway point, the person is judged as having a sufficiently high blood alcohol content to warrant further, more precise testing.

The Breathalyzer, a more precise testing device, operates on the same principle as the simplified screening test. In a Breathalyzer test, a measured volume of breath is bubbled through a solution of potassium dichromate in aqueous sulfuric acid, and the color change is measured spectrophotometrically.

Both tests measure alcohol in the breath. The legal definition of being under the influence of alcohol is based on *blood* alcohol content, not breath alcohol content. The chemical correlation between these two measurements is that air deep within the lungs is in equilibrium with blood passing through the pulmonary arteries, and an equilibrium is established between blood alcohol and breath alcohol. It has been determined by tests in persons drinking alcohol that 2100 mL of breath contains the same amount of ethanol as 1.00 mL of blood.

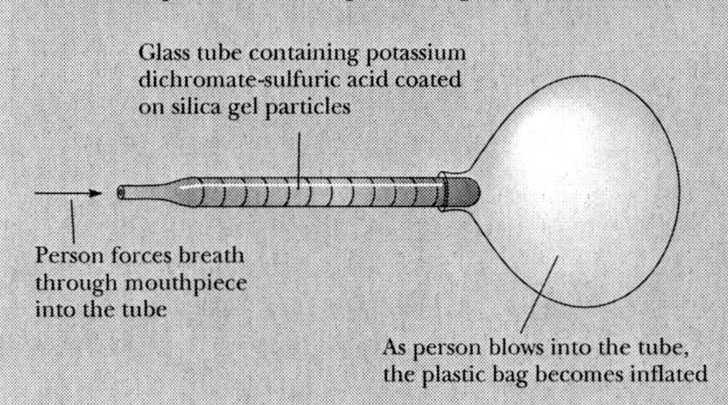
Glass tube containing potassium dichromate-sulfuric acid coated on silica gel particles
Person forces breath through mouthpiece into the tube
As person blows into the tube, the plastic bag becomes inflated

FINAL CALL
Alcohol Breath Tester
Don't Drive Drunk!

A device for testing the breath for the presence of ethanol. When ethanol is oxidized by potassium dichromate, the reddish-orange color of dichromate ion turns to green as it is reduced to chromium(III) ion. *(Charles D. Winters)*

Cyclic ethers are heterocyclic compounds in which the ether oxygen is one of the atoms in a ring. These ethers are generally known by their common names:

Cyclic ether An ether in which the oxygen is one of the atoms of a ring.

Ethylene oxide

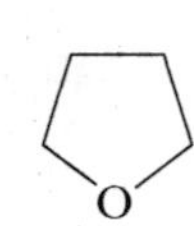
Tetrahydrofuran (THF)

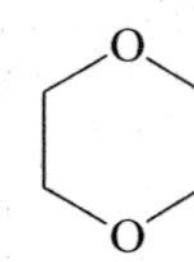
1,4-Dioxane

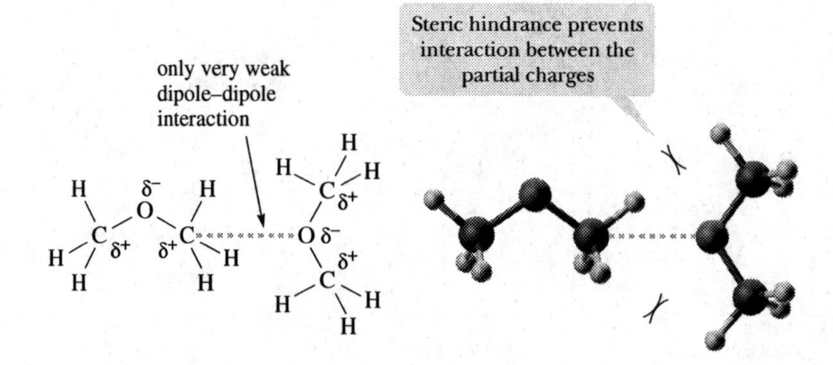

Figure 8.5
Ethers are polar molecules, but because of steric hindrance, only weak attractive interactions exist between their molecules in the pure liquid.

C. Physical Properties

Ethers are polar compounds in which oxygen bears a partial negative charge and each carbon bonded to it bears a partial positive charge (Figure 8.5). Because of steric hindrance, however, only weak forces of attraction exist between ether molecules in the pure liquid. Consequently, boiling points of ethers are much lower than those of alcohols of comparable molecular weight (Table 8.3). Boiling points of ethers are close to those of hydrocarbons of comparable molecular weight (compare Tables 3.4 and 8.3).

TABLE 8.3 Boiling Points and Solubilities in Water of Some Alcohols and Ethers of Comparable Molecular Weight

Structural Formula	Name	Molecular Weight	Boiling Point (°C)	Solubility in Water
CH_3CH_2OH	ethanol	46	78	infinite
CH_3OCH_3	dimethyl ether	46	−24	7.8 g/100 g
$CH_3CH_2CH_2CH_2OH$	1-butanol	74	117	7.4 g/100 g
$CH_3CH_2OCH_2CH_3$	diethyl ether	74	35	8 g/100 g
$CH_3CH_2CH_2CH_2CH_2OH$	1-pentanol	88	138	2.3 g/100 g
$HOCH_2CH_2CH_2CH_2OH$	1,4-butanediol	90	230	infinite
$CH_3CH_2CH_2CH_2OCH_3$	butyl methyl ether	88	71	slight
$CH_3OCH_2CH_2OCH_3$	ethylene glycol dimethyl ether	90	84	infinite

Because the oxygen atom of an ether carries a partial negative charge, ethers form hydrogen bonds with water (Figure 8.6) and are more soluble in water than are hydrocarbons of comparable molecular weight and shape (compare data in Tables 3.4 and 8.3).

The effect of hydrogen bonding is illustrated dramatically by comparing the boiling points of ethanol (78°C) and its constitutional isomer dimethyl ether (−24°C). The difference in boiling points between these two compounds is due to the polar O—H group in the alcohol, which is capable of forming intermolecular hydrogen bonds. This hydrogen bonding increases the attractive force between molecules of ethanol; thus, ethanol has a higher boiling point than dimethyl ether:

CH_3CH_2OH	CH_3OCH_3
Ethanol	Dimethyl ether
bp 78°C	bp −24°C

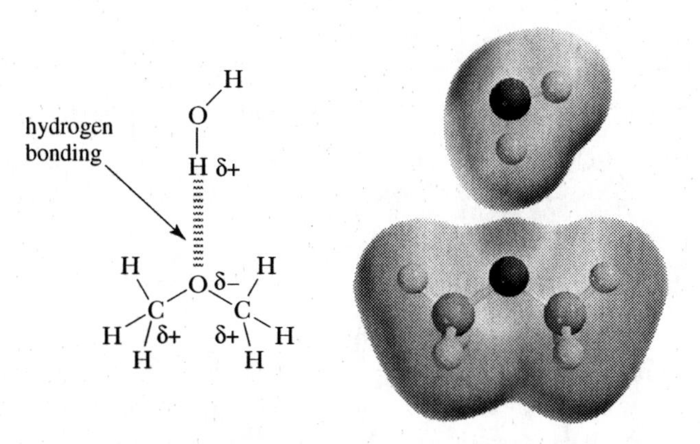

Dimethyl ether in water. The partially negative oxygen of the ether is the hydrogen bond acceptor, and a partially positive hydrogen of a water molecule is the hydrogen bond donor.

Figure 8.6
Ethers are hydrogen-bond acceptors only. They are not hydrogen-bond donors.

EXAMPLE 8.7

Write the IUPAC and common names for each ether:

(a) $CH_3C(CH_3)_2OCH_2CH_3$ (structure drawn with two CH_3 groups on the central C)

(b) cyclohexyl—O—cyclohexyl (two cyclohexane rings joined through O)

SOLUTION

(a) 2-Ethoxy-2-methylpropane. Its common name is *tert*-butyl ethyl ether.
(b) Cyclohexoxycyclohexane. Its common name is dicyclohexyl ether.

Practice Problem 8.7

Write the IUPAC and common names for each ether:

(a) $CH_3CH(CH_3)CH_2OCH_2CH_3$ (structure drawn with CH_3 on the second C)

(b) cyclopentyl—OCH_3

EXAMPLE 8.8

Arrange these compounds in order of increasing solubility in water:

$CH_3OCH_2CH_2OCH_3$	$CH_3CH_2OCH_2CH_3$	$CH_3CH_2CH_2CH_2CH_2CH_3$
Ethylene glycol dimethyl ether	Diethyl ether	Hexane

SOLUTION

Water is a polar solvent. Hexane, a nonpolar hydrocarbon, has the lowest solubility in water. Both diethyl ether and ethylene glycol dimethyl ether are polar compounds, due to the presence of their polar C—O—C groups, and each interacts with water as a hydrogen-bond acceptor. Because ethylene glycol dimethyl ether has more sites within its molecules for hydrogen bonding, it is more soluble in water than diethyl ether:

$CH_3CH_2CH_2CH_2CH_2CH_3$	$CH_3CH_2OCH_2CH_3$	$CH_3OCH_2CH_2OCH_3$
Insoluble	8 g/100 g water	Soluble in all proportions

Practice Problem 8.8

Arrange these compounds in order of increasing boiling point:

$CH_3OCH_2CH_2OCH_3$ $\quad$ $HOCH_2CH_2OH$ $\quad$ $CH_3OCH_2CH_2OH$

D. Reactions of Ethers

Ethers, R—O—R, resemble hydrocarbons in their resistance to chemical reaction. They do not react with oxidizing agents, such as potassium dichromate or potassium permanganate. They are not affected by most acids or bases at moderate temperatures. Because of their good solvent properties and general inertness to chemical reaction, ethers are excellent solvents in which to carry out many organic reactions.

8.5 EPOXIDES

A. Structure and Nomenclature

Epoxide A cyclic ether in which oxygen is one atom of a three-membered ring.

An **epoxide** is a cyclic ether in which oxygen is one atom of a three-membered ring:

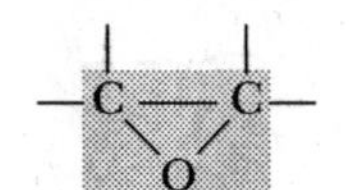

Functional group of an epoxide

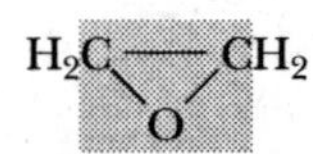

Ethylene oxide

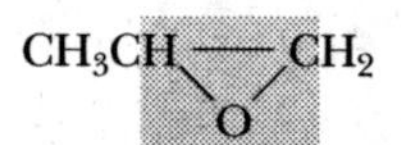

Propylene oxide

Although epoxides are technically classed as ethers, we discuss them separately because of their exceptional chemical reactivity compared with other ethers.

Common names for epoxides are derived by giving the common name of the alkene from which the epoxide might have been derived, followed by the word *oxide*; an example is ethylene oxide.

B. Synthesis from Alkenes

Ethylene oxide, one of the few epoxides manufactured on an industrial scale, is prepared by passing a mixture of ethylene and air (or oxygen) over a silver catalyst:

$$CH_2{=}CH_2 + O_2 \xrightarrow[\text{heat}]{\text{Ag}} H_2C\text{—}CH_2 \text{ (bridged by O)}$$

Ethylene $\qquad$ Ethylene oxide

In the United States, the annual production of ethylene oxide by this method is approximately 10^9 kg.

The most common laboratory method for the synthesis of epoxides from alkenes is oxidation with a peroxycarboxylic acid (a peracid), RCO_3H. One peracid used for this purpose is peroxyacetic acid:

$$CH_3\overset{\overset{\displaystyle O}{\|}}{C}OOH$$

Peroxyacetic acid
(Peracetic acid)

Following is a balanced equation for the epoxidation of cyclohexene by a peroxycarboxylic acid. In the process, the peroxycarboxylic acid is reduced to a carboxylic acid:

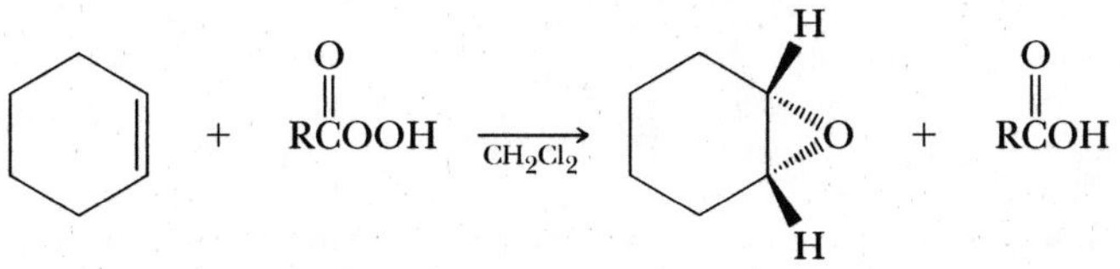

Cyclohexene + A peroxycarboxylic acid (RCOOH) $\xrightarrow{CH_2Cl_2}$ 1,2-Epoxycyclohexane (Cyclohexene oxide) + A carboxylic acid (RCOH)

The epoxidation of an alkene is stereoselective. The epoxidation of *cis*-2-butene, for example, yields only *cis*-2-butene oxide:

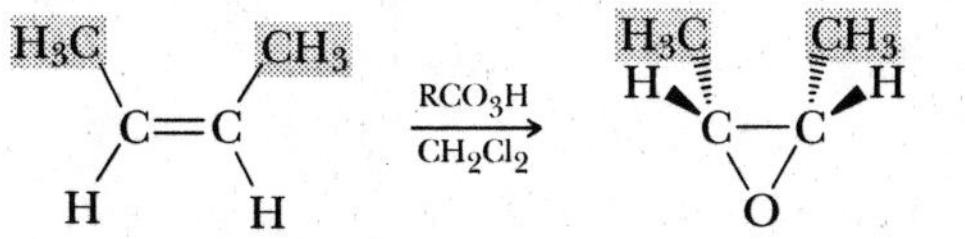

cis-2-Butene $\xrightarrow[CH_2Cl_2]{RCO_3H}$ *cis*-2-Butene oxide

EXAMPLE 8.9

Draw a structural formula of the epoxide formed by treating *trans*-2-butene with a peroxycarboxylic acid.

SOLUTION

The oxygen of the epoxide ring is added by forming both carbon–oxygen bonds from the same side of the carbon–carbon double bond:

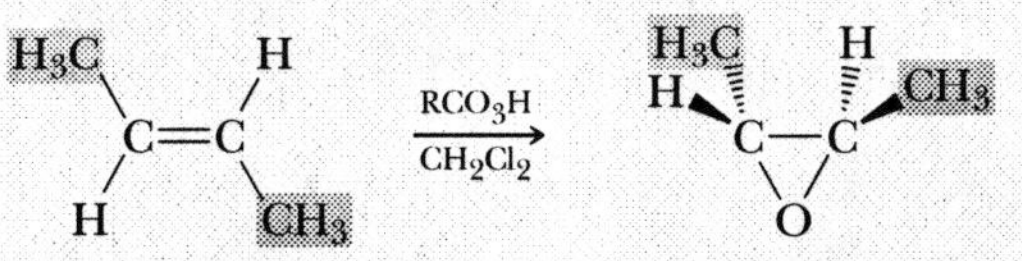

trans-2-Butene $\xrightarrow[CH_2Cl_2]{RCO_3H}$ *trans*-2-Butene oxide

Practice Problem 8.9

Draw the structural formula of the epoxide formed by treating 1,2-dimethylcyclopentene with a peroxycarboxylic acid.

C. Ring-Opening Reactions

Ethers are not normally susceptible to reaction with aqueous acid (Section 8.4D). Epoxides, however, are especially reactive because of the angle strain in the three-membered ring. The normal bond angle about an sp^3 hybridized carbon or oxygen atom is 109.5°. Because of the strain associated with the compression of bond angles in the three-membered epoxide ring from the normal 109.5° to 60°, epoxides undergo ring-opening reactions with a variety of reagents.

In the presence of an acid catalyst—most commonly, perchloric acid—epoxides are hydrolyzed to glycols. As an example, the acid-catalyzed hydrolysis of ethylene oxide gives 1,2-ethanediol:

$$CH_2\text{—}CH_2 (\text{—O—}) + H_2O \xrightarrow{H^+} HOCH_2CH_2OH$$

Ethylene oxide — 1,2-Ethanediol (Ethylene glycol)

Annual production of ethylene glycol in the United States is approximately 10^{10} kg. Two of its largest uses are in automotive antifreeze and as one of the two starting materials for the production of polyethylene terephthalate (PET), which is fabricated into such consumer products as Dacron® polyester, Mylar®, and packaging films (Section 17.5B).

The acid-catalyzed ring opening of epoxides shows a stereoselectivity typical of S_N2 reactions: The nucleophile attacks anti to the leaving hydroxyl group, and the —OH groups in the glycol thus formed are anti. As a result, the hydrolysis of an epoxycycloalkane yields a *trans*-1,2-cycloalkanediol:

1,2-Epoxycyclopentane (Cyclopentene oxide) + H_2O $\xrightarrow{H^+}$ *trans*-1,2-Cyclopentanediol

At this point, let us compare the stereochemistry of the glycol formed by the acid-catalyzed hydrolysis of an epoxide with that formed by the OsO_4 oxidation of an alkene (Section 5.4). Each reaction sequence is stereoselective, but gives a different stereoisomer. The acid-catalyzed hydrolysis of cyclopentene oxide gives *trans*-1,2-cyclopentanediol, whereas the osmium tetroxide oxidation of cyclopentene gives *cis*-1,2-cyclopentanediol. Thus, a cycloalkene can be converted to either a cis glycol or a trans glycol by the proper choice of reagents.

Cyclopentene $\xrightarrow{RCO_3H}$ cyclopentene oxide $\xrightarrow[H_2O]{H^+}$ *trans*-1,2-Cyclopentanediol

Cyclopentene $\xrightarrow{OsO_4,\ t\text{-BuOOH}}$ *cis*-1,2-Cyclopentanediol

EXAMPLE 8.10

Draw the structural formula of the product formed by treating cyclohexene oxide with aqueous acid. Be certain to show the stereochemistry of the product.

SOLUTION

The acid-catalyzed hydrolysis of the three-membered epoxide ring gives a trans glycol:

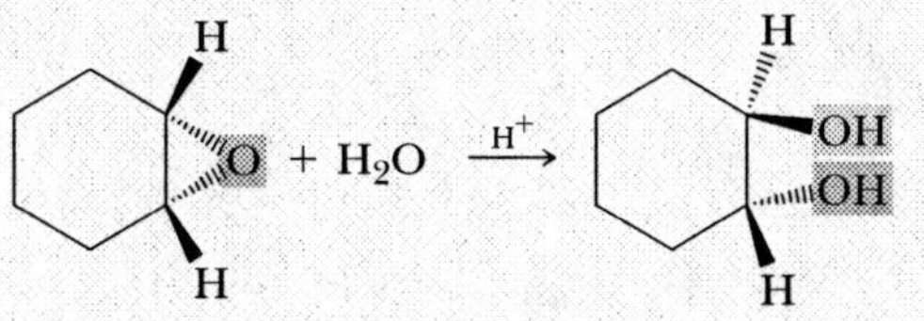

$$\text{cyclohexene oxide} + H_2O \xrightarrow{H^+} \textit{trans}\text{-1,2-cyclohexanediol}$$

trans-1,2-cyclohexanediol

Practice Problem 8.10

Show how to convert cyclohexene to *cis*-1,2-cyclohexanediol.

Just as ethers are not normally susceptible to reaction with electrophiles, neither are they normally susceptible to reaction with nucleophiles. Because of the strain associated with the three-membered ring, however, epoxides undergo ring-opening reactions with good nucleophiles such as ammonia and amines (Chapter 10), alkoxide ions, and thiols and their anions (Section 8.7). Good nucleophiles attack the ring by an S_N2 mechanism and show a stereoselectivity for attack of the nucleophile at the less hindered carbon of the three-membered ring. An illustration is the reaction of cyclohexene oxide with ammonia to give *trans*-2-aminocyclohexanol:

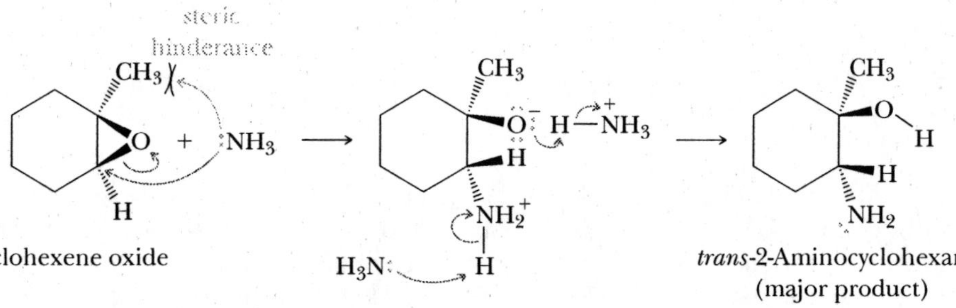

steric hinderance

CH_3 + $:NH_3$ ⟶ $H-NH_3^+$ ⟶ NH_2

Cyclohexene oxide

$H_3N:$

trans-2-Aminocyclohexanol (major product)

The value of epoxides lies in the number of nucleophiles that bring about ring opening and the combinations of functional groups that can be prepared from them. The following chart summarizes the three most important of these nucleophilic ring-opening reactions (the characteristic structural feature of each ring-opening product is shown in color):

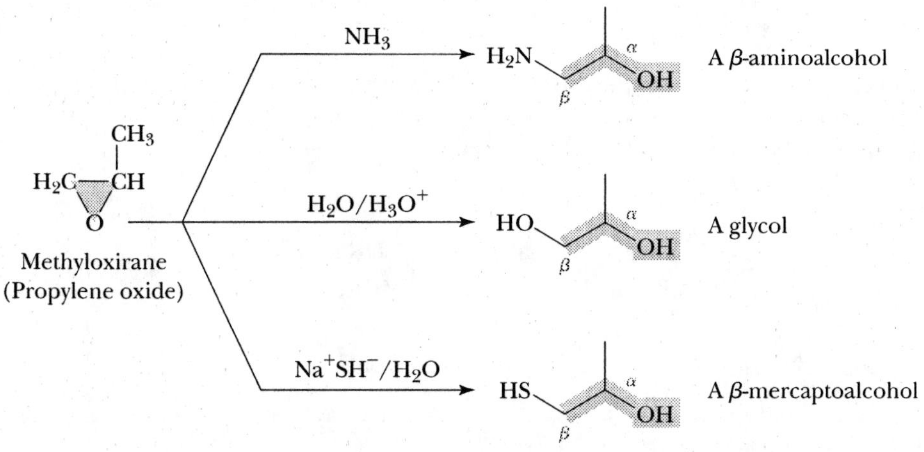

Methyloxirane (Propylene oxide)

NH_3 ⟶ H_2N–CH$_2$–CH(CH$_3$)–OH — A β-aminoalcohol

H_2O/H_3O^+ ⟶ HO–CH$_2$–CH(CH$_3$)–OH — A glycol

Na^+SH^-/H_2O ⟶ HS–CH$_2$–CH(CH$_3$)–OH — A β-mercaptoalcohol

CHEMICAL CONNECTIONS 8C

Ethylene Oxide: A Chemical Sterilant

Because ethylene oxide is such a highly strained molecule, it reacts with the types of nucleophilic groups present in biological materials. At sufficiently high concentrations, ethylene oxide reacts with enough molecules in cells to cause the death of microorganisms. This toxic property is the basis for using ethylene oxide as a chemical sterilant. In hospitals, surgical instruments and other items that cannot be made disposable are now sterilized by exposure to ethylene oxide.

Ethylene oxide and substituted ethylene oxides are valuable building blocks for the synthesis of larger organic molecules. Following are structural formulas for two common drugs, each synthesized in part from ethylene oxide:

Procaine (Novocaine)

Diphenhydramine (Benadryl)

Novocaine was the first injectable local anesthetic. Benadryl was the first synthetic antihistamine. The portion of the carbon skeleton of each that is derived from the reaction of ethylene oxide with a nitrogen–nucleophile is shown in color.

In later chapters, after we have developed the chemistry of more functional groups, we will show how to synthesize Novocaine and Benadryl from readily available starting materials. For the moment, however, it is sufficient to recognize that the unit —O—C—C—Nu can be derived by nucleophilic opening of ethylene oxide or a substituted ethylene oxide.

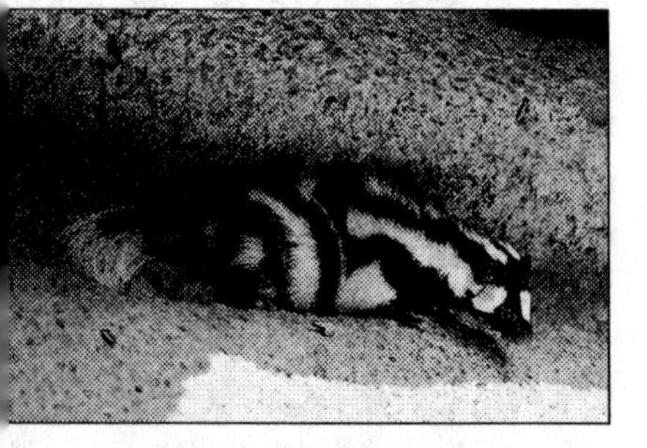

The scent of skunks is a mixture of two thiols, 3-methyl-1-butanethiol and 2-butene-1-thiol. *(Stephen J. Krasemann/Photo Researchers, Inc.)*

8.6 THIOLS

The most outstanding property of low-molecular-weight thiols is their stench. They are responsible for unpleasant odors such as those from skunks, rotten eggs, and sewage. The scent of skunks is due primarily to two thiols:

$CH_3CH{=}CHCH_2SH$ — 2-Butene-1-thiol

$CH_3CH(CH_3)CH_2CH_2SH$ — 3-Methyl-1-butanethiol

A. Structure

Thiol A compound containing an —SH (sulfhydryl) group.

The functional group of a **thiol** is an —SH (sulfhydryl) group. Figure 8.7 shows a Lewis structure and a ball-and-stick model of methanethiol, CH_3SH, the simplest thiol.

$$\begin{array}{c} H \\ | \\ H-C-\ddot{\underset{..}{S}}-H \\ | \\ H \end{array}$$

(a) (b)

Figure 8.7
Methanethiol, CH_3SH.
(a) Lewis structure and
(b) ball-and-stick model.
The C—S—H bond angle is 100.9°, somewhat smaller than the tetrahedral angle of 109.5°.

Methanethiol. The electronegativities of carbon and sulfur are virtually identical (2.5 each), while sulfur is slightly more electronegative than hydrogen (2.5 versus 2.1). The electron density model shows some slight partial positive charge on hydrogen of the S—H group and some slight partial negative charge on sulfur.

B. Nomenclature

The sulfur analog of an alcohol is called a thiol (thi- from the Greek: *theion*, sulfur) or, in the older literature, a **mercaptan**, which literally means "mercury capturing." Thiols react with Hg^{2+} in aqueous solution to give sulfide salts as insoluble precipitates. Thiophenol, C_6H_5SH, for example, gives $(C_6H_5S)_2Hg$.

Mercaptan A common name for any molecule containing an —SH group.

In the IUPAC system, thiols are named by selecting as the parent alkane the longest chain of carbon atoms that contains the —SH group. To show that the compound is a thiol, we add *-thiol* to the name of the parent alkane and number the parent chain in the direction that gives the —SH group the lower number.

Common names for simple thiols are derived by naming the alkyl group bonded to —SH and adding the word *mercaptan*. In compounds containing other functional groups, the presence of an —SH group is indicated by the prefix **mercapto-**. According to the IUPAC system, —OH takes precedence over —SH in both numbering and naming:

CH_3CH_2SH	$CH_3CH(CH_3)CH_2SH$	$HSCH_2CH_2OH$
Ethanethiol (Ethyl mercaptan)	2-Methyl-1-propanethiol (Isobutyl mercaptan)	2-Mercaptoethanol

Sulfur analogs of ethers are named by using the word *sulfide* to show the presence of the —S— group. Following are common names of two sulfides:

CH_3SCH_3	$CH_3CH_2SCH(CH_3)CH_3$
Dimethyl sulfide	Ethyl isopropyl sulfide

Mushrooms, onions, garlic, and coffee all contain sulfur compounds. One of these present in coffee is

[2-furanthiol structure: furan ring (O) with SH substituent]

(Charles D. Winters)

EXAMPLE 8.11

Write the IUPAC name for each thiol:

(a) [line-angle structure: five-carbon chain ending in SH] (b) [line-angle structure: four-carbon chain with SH on C-2]

SOLUTION

(a) The parent alkane is pentane. We show the presence of the —SH group by adding *thiol* to the name of the parent alkane. The IUPAC name of this thiol is 1-pentanethiol. Its common name is pentyl mercaptan.
(b) The parent alkane is butane. The IUPAC name of this thiol is 2-butanethiol. Its common name is *sec*-butyl mercaptan.

Practice Problem 8.11

Write the IUPAC name for each thiol:

(a) SH

(b) SH

C. Physical Properties

Because of the small difference in electronegativity between sulfur and hydrogen ($2.5 - 2.1 = 0.4$), we classify the S—H bond as nonpolar covalent. Because of this lack of polarity, thiols show little association by hydrogen bonding. Consequently, they have lower boiling points and are less soluble in water and other polar solvents than are alcohols of similar molecular weight. Table 8.4 gives the boiling points of three low-molecular-weight thiols. For comparison, the table also gives the boiling points of alcohols with the same number of carbon atoms.

TABLE 8.4 Boiling Points of Three Thiols and Three Alcohols with the Same Number of Carbon Atoms

Thiol	Boiling Point (°C)	Alcohol	Boiling Point (°C)
methanethiol	6	methanol	65
ethanethiol	35	ethanol	78
1-butanethiol	98	1-butanol	117

Earlier, we illustrated the importance of hydrogen bonding in alcohols by comparing the boiling points of ethanol (78°C) and its constitutional isomer dimethyl ether (24°C). By comparison, the boiling point of ethanethiol is 35°C, and that of its constitutional isomer dimethyl sulfide is 37°C:

CH_3CH_2SH	CH_3SCH_3
Ethanethiol	Dimethyl sulfide
bp 35°C	bp 37°C

The fact that the boiling points of these constitutional isomers are almost identical indicates that little or no association by hydrogen bonding occurs between thiol molecules.

8.7 REACTIONS OF THIOLS

In this section, we discuss the acidity of thiols and their reaction with strong bases, such as sodium hydroxide, and with molecular oxygen.

A. Acidity

Hydrogen sulfide is a stronger acid than water:

$$H_2O + H_2O \rightleftharpoons HO^- + H_3O^+ \qquad pK_a = 15.7$$

$$H_2S + H_2O \rightleftharpoons HS^- + H_3O^+ \qquad pK_a = 7.0$$

Similarly, thiols are stronger acids than alcohols. Compare, for example, the pK_a's of ethanol and ethanethiol in dilute aqueous solution:

$$CH_3CH_2OH + H_2O \rightleftharpoons CH_3CH_2O^- + H_3O^+ \quad pK_a = 15.9$$

$$CH_3CH_2SH + H_2O \rightleftharpoons CH_3CH_2S^- + H_3O^+ \quad pK_a = 8.5$$

Thiols are sufficiently strong acids that, when dissolved in aqueous sodium hydroxide, they are converted completely to alkylsulfide salts:

$$CH_3CH_2SH + Na^+OH^- \longrightarrow CH_3CH_2S^-Na^+ + H_2O$$

pK_a 8.5			pK_a15.7
Stronger acid	Stronger base	Weaker base	Weaker acid

To name salts of thiols, give the name of the cation first, followed by the name of the alkyl group to which the suffix *-sulfide* is added. For example, the sodium salt derived from ethanethiol is named sodium ethylsulfide.

B. Oxidation to Disulfides

Many of the chemical properties of thiols stem from the fact that the sulfur atom of a thiol is oxidized easily to several higher oxidation states. The most common reaction of thiols in biological systems is their oxidation to disulfides, the functional group of which is a **disulfide** (—S—S—) bond. Thiols are readily oxidized to disulfides by molecular oxygen. In fact, they are so susceptible to oxidation that they must be protected from contact with air during storage. Disulfides, in turn, are easily reduced to thiols by several reagents. This easy interconversion between thiols and disulfides is very important in protein chemistry, as we will see in Chapter 20:

$$2\ HOCH_2CH_2SH \underset{\text{reduction}}{\overset{\text{oxidation}}{\rightleftharpoons}} HOCH_2CH_2S{-}SCH_2CH_2OH$$

A thiol — A disulfide

We derive common names of simple disulfides by listing the names of the groups bonded to sulfur and adding the word *disulfide*, as, for example, $CH_3S{-}SCH_3$, which is named dimethyldisulfide.

SUMMARY

The functional group of an **alcohol** (Section 8.2A) is an **—OH (hydroxyl)** group bonded to an sp^3 hybridized carbon. Alcohols are classified as **1°, 2°,** or **3°** (Section 8.2A), depending on whether the —OH group is bonded to a primary, secondary, or tertiary carbon. IUPAC names of alcohols (Section 8.2B) are derived by changing the suffix of the parent alkane from *-e* to *-ol*. The chain is numbered to give the carbon bearing —OH the lower number. Common names for alcohols are derived by naming the alkyl group bonded to —OH and adding the word *alcohol*.

Alcohols are polar compounds (Section 8.2C) with oxygen bearing a partial negative charge and both the carbon and hydrogen bonded to it bearing partial positive charges. Because of intermolecular association by **hydrogen bonding**, the boiling points of alcohols are higher than those of hydrocarbons of comparable molecular weight. Because of increased dispersion forces, the boiling points of alcohols increase with increasing molecular weight. Alcohols interact with water by hydrogen bonding and therefore are more soluble in water than are hydrocarbons of comparable molecular weight.

The functional group of an **ether** is an atom of oxygen bonded to two carbon atoms (Section 8.4A). In the IUPAC name of an ether (Section 8.4B), the parent alkane is named, and then the —OR group is named as an alkoxy substituent. Common names are derived by naming the two groups bonded to oxygen, followed by the word *ether*. Ethers are weakly polar compounds (Section 8.4C). Their boiling points are close to those of hydrocarbons of comparable molecular weight. Because ethers are hydrogen-bond acceptors, they are more soluble in water than are hydrocarbons of comparable molecular weight. An epoxide is a cyclic ether in which oxygen is one of the atoms of the three-membered ring (Section 8.5A).

A **thiol** (Section 8.6A) is the sulfur analog of an alcohol; it contains an —**SH (sulfhydryl)** group in place of an —OH group. Thiols are named in the same manner as alcohols, but the suffix *-e* is retained, and **-thiol** is added (Section 8.6B). Common names for thiols are derived by naming the alkyl group bonded to —SH and adding the word *mercaptan*. In compounds containing functional groups of higher precedence, the presence of —SH is indicated by the prefix **mercapto-**. For **thioethers**, name the two groups bonded to sulfur, followed by the word *sulfide*. The S—H bond is nonpolar, and the physical properties of thiols are more like those of hydrocarbons of comparable molecular weight (Section 8.6C).

KEY REACTIONS

1. Acidity of alcohols (Section 8.3A)

In dilute aqueous solution, methanol and ethanol are comparable in acidity to water. Secondary and tertiary alcohols are weaker acids than water.

$$CH_3OH + H_2O \rightleftharpoons CH_3O^- + H_3O^+ \qquad pK_a = 15.5$$

2. Reaction of alcohols with Active Metals (Section 8.3C)

Alcohols react with Li, Na, K, and other active metals to form metal alkoxides, which are somewhat stronger bases than NaOH and KOH:

$$2\,CH_3CH_2OH + 2\,Na \longrightarrow 2\,CH_3CH_2O^-Na^+ + H_2$$

3. Reaction of alcohols with HCl, HBr, and HI (Section 8.3D)

Primary alcohols react with HBr and HI by an S_N2 mechanism:

$$CH_3CH_2CH_2CH_2OH + HBr \longrightarrow CH_3CH_2CH_2CH_2Br + H_2O$$

Tertiary alcohols react with HCl, HBr, and HI by an S_N1 mechanism, with the formation of a carbocation intermediate:

$$CH_3\overset{\displaystyle CH_3}{\underset{\displaystyle CH_3}{C}}OH + HCl \xrightarrow{25°C} CH_3\overset{\displaystyle CH_3}{\underset{\displaystyle CH_3}{C}}Cl + H_2O$$

Secondary alcohols may react with HCl, HBr, and HI by an S_N2 or an S_N1 mechanism, depending on the alcohol and experimental conditions.

4. Reaction of alcohols with $SOCl_2$ (Section 8.3D)

This is often the method of choice for converting an alcohol to an alkyl chloride:

$$CH_3(CH_2)_5OH + SOCl_2 \longrightarrow CH_3(CH_2)_5Cl + SO_2 + HCl$$

5. Acid-catalyzed dehydration of alcohols (Section 8.3E)

When isomeric alkenes are possible, the major product is generally the more substituted alkene (Zaitsev's rule):

$$CH_3CH_2\overset{\displaystyle OH}{C}HCH_3 \xrightarrow[\text{heat}]{H_3PO_4} \underset{\text{Major product}}{CH_3CH{=}CHCH_3} + CH_3CH_2CH{=}CH_2 + H_2O$$

6. Oxidation of a Primary Alcohol to an Aldehyde (Section 8.3F)

This oxidation is most conveniently carried out by using pyridinium chlorochromate (PCC):

$$\text{C}_5\text{H}_9\text{—}CH_2OH \xrightarrow[CH_2Cl_2]{PCC} \text{C}_5\text{H}_9\text{—}\overset{\displaystyle O}{\overset{\|}{C}}H$$

7. Oxidation of a Primary Alcohol to a Carboxylic Acid (Section 8.3F)
A primary alcohol is oxidized to a carboxylic acid by chromic acid:

$$CH_3(CH_2)_4CH_2OH + H_2CrO_4 \xrightarrow[\text{acetone}]{H_2O} CH_3(CH_2)_4\overset{\overset{O}{\|}}{C}OH + Cr^{3+}$$

8. Oxidation of a Secondary Alcohol to a Ketone (Section 8.3F)
A secondary alcohol is oxidized to a ketone by chromic acid and by PCC:

$$CH_3(CH_2)_4\overset{\overset{OH}{|}}{C}HCH_3 + H_2CrO_4 \longrightarrow CH_3(CH_2)_4\overset{\overset{O}{\|}}{C}CH_3 + Cr^{3+}$$

9. Oxidation of an alkene to an epoxide (Section 8.5B)
The most common method for the synthesis of an epoxide from an alkene is oxidation with a peroxycarboxylic acid, such as peroxyacetic acid:

$$\text{cyclohexene} + R\overset{\overset{O}{\|}}{C}OOH \longrightarrow \text{cyclohexene oxide (H, H)} + R\overset{\overset{O}{\|}}{C}OH$$

10. Acid-catalyzed hydrolysis of epoxides (Section 8.5C)
Acid-catalyzed hydrolysis of an epoxide derived from a cycloalkene gives a trans glycol (hydrolysis of cycloalkene oxide is stereoselective, giving the trans glycol):

$$\text{cyclopentene oxide (H, H)} + H_2O \xrightarrow{H^+} \text{trans-1,2-cyclopentanediol (OH, OH)}$$

11. Nucleophilic ring opening of epoxides (Section 8.5C)
Good nucleophiles, such as ammonia and amines, open the highly strained epoxide ring by an S_N2 mechanism and show a stereoselectivity for attack of the nucleophile at the less hindered carbon of the three-membered ring:

$$\text{O} + NH_3 \longrightarrow \text{OH}, NH_2$$

Cyclohexene oxide — *trans*-2-Aminocyclohexanol

12. Acidity of thiols (Section 8.7A)
Thiols are weak acids, pK_a 8–9, but are considerably stronger acids than alcohols, pK_a 16–18.

$$CH_3CH_2SH + H_2O \rightleftharpoons CH_3CH_2S^- + H_3O^+ \qquad pK_a = 8.5$$

13. Oxidation to Disulfides (Section 8.7B)
Oxidation of a thiol by O_2 gives a disulfide:

$$2\,RSH + \tfrac{1}{2}O_2 \longrightarrow RS\text{-}SR + H_2O$$

PROBLEMS

A problem number set in red indicates an applied "real-world" problem.

Structure and Nomenclature

8.12 Which of the following compounds are secondary alcohols?

(a) 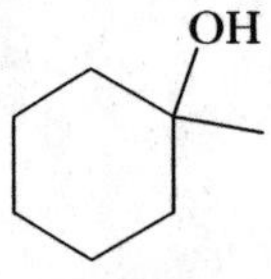 OH

(b) $(CH_3)_3COH$

(c) 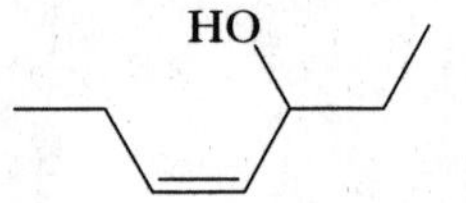 HO

(d) 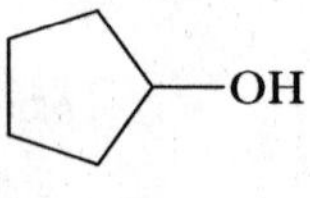 OH

8.13 Name these compounds:

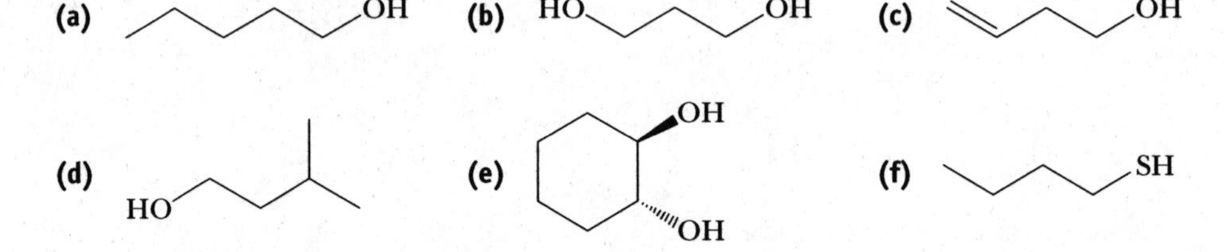

8.14 Draw a structural formula for each alcohol:

(a) Isopropyl alcohol
(b) Propylene glycol
(c) (*R*)-5-Methyl-2-hexanol
(d) 2-Methyl-2-propyl-1,3-propanediol
(e) 2,2-Dimethyl-1-propanol
(f) 2-Mercaptoethanol
(g) 1,4-Butanediol
(h) (*Z*)-5-Methyl-2-hexen-1-ol
(i) *cis*-3-Pentene-1-ol
(j) *trans*-1,4-Cyclohexanediol

8.15 Write names for these ethers:

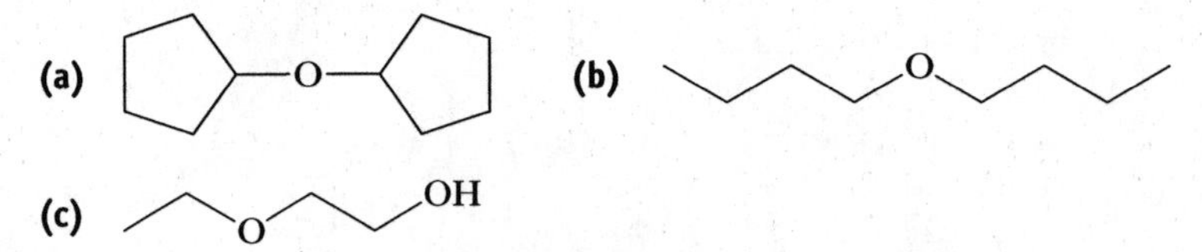

8.16 Name and draw structural formulas for the eight isomeric alcohols with molecular formula $C_5H_{12}O$. Which are chiral?

Physical Properties

8.17 Arrange these compounds in order of increasing boiling point (values in °C are −42, 78, 117, and 198):

(a) $CH_3CH_2CH_2CH_2OH$
(b) CH_3CH_2OH
(c) $HOCH_2CH_2OH$
(d) $CH_3CH_2CH_3$

8.18 Arrange these compounds in order of increasing boiling point (values in °C are −42, −24, 78, and 118):

(a) CH_3CH_2OH
(b) CH_3OCH_3
(c) $CH_3CH_2CH_3$
(d) CH_3COOH

8.19 Propanoic acid and methyl acetate are constitutional isomers, and both are liquids at room temperature:

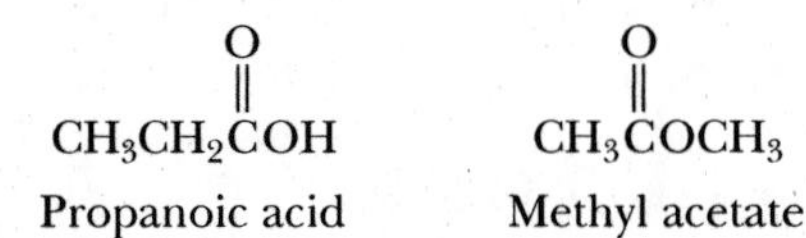

$CH_3CH_2\overset{O}{\overset{\|}{C}}OH$ — Propanoic acid

$CH_3\overset{O}{\overset{\|}{C}}OCH_3$ — Methyl acetate

One of these compounds has a boiling point of 141°C; the other has a boiling point of 57°C. Which compound has which boiling point?

8.20 Draw all possible staggered conformations of ethylene glycol ($HOCH_2CH_2OH$). Can you explain why the gauche conformation is more stable than the anti conformation by approximately 1 kcal/mol?

8.21 Following are structural formulas for 1-butanol and 1-butanethiol:

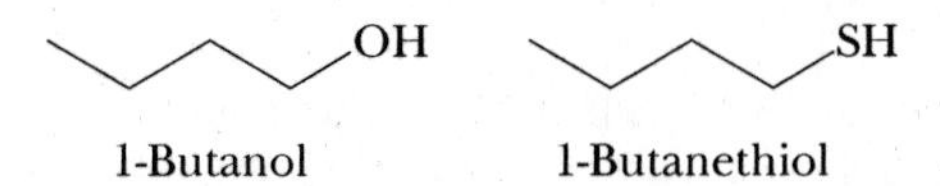

1-Butanol 1-Butanethiol

One of these compounds has a boiling point of 98.5°C; the other has a boiling point of 117°C. Which compound has which boiling point?

8.22 From each pair of compounds, select the one that is more soluble in water:

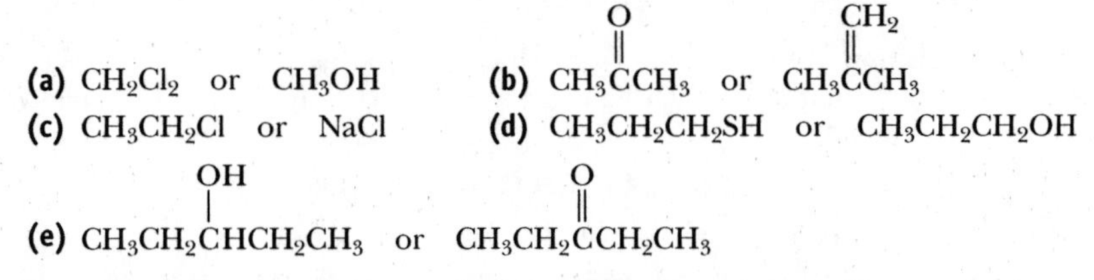

(a) CH_2Cl_2 or CH_3OH **(b)** $CH_3\overset{O}{\overset{\|}{C}}CH_3$ or $CH_3\overset{CH_2}{\overset{\|}{C}}CH_3$

(c) CH_3CH_2Cl or NaCl **(d)** $CH_3CH_2CH_2SH$ or $CH_3CH_2CH_2OH$

(e) $CH_3CH_2\overset{OH}{\overset{|}{C}}HCH_2CH_3$ or $CH_3CH_2\overset{O}{\overset{\|}{C}}CH_2CH_3$

8.23 Arrange the compounds in each set in order of decreasing solubility in water:

(a) Ethanol; butane; diethyl ether **(b)** 1-Hexanol; 1,2-hexanediol; hexane

8.24 Each of the following compounds is a common organic solvent:

(a) CH_2Cl_2 or CH_3CH_2OH

(b) $CH_3CH_2OCH_2CH_3$ or CH_3CH_2OH

(c) $CH_3\overset{O}{\overset{\|}{C}}CH_3$ or $CH_3CH_2OCH_2CH_3$

(d) $CH_3CH_2OCH_2CH_3$ or $CH_3(CH_2)_3CH_3$

From each pair of compounds, select the solvent with the greater solubility in water.

Synthesis of Alcohols

8.25 Give the structural formula of an alkene or alkenes from which each alcohol or glycol can be prepared:

(a) 2-Butanol **(b)** 1-Methylcyclohexanol

(c) 3-Hexanol **(d)** 2-Methyl-2-pentanol

(e) Cyclopentanol **(f)** 1,2-Propanediol

8.26 The addition of bromine to cyclopentene and the acid-catalyzed hydrolysis of cyclopentene oxide are both stereoselective; each gives a trans product. Compare the mechanisms of these two reactions, and show how each mechanism accounts for the formation of the trans product.

Acidity of Alcohols and Thiols

8.27 From each pair, select the stronger acid, and, for each stronger acid, write a structural formula for its conjugate base:

(a) H_2O or H_2CO_3 **(b)** CH_3OH or CH_3COOH
(c) CH_3COOH or CH_3CH_2SH

8.28 Arrange these compounds in order of increasing acidity (from weakest to strongest):

$$CH_3CH_2CH_2OH \qquad CH_3CH_2\overset{\overset{O}{\|}}{C}OH \qquad CH_3CH_2CH_2SH$$

8.29 From each pair, select the stronger base, and, for each stronger base, write the structural formula of its conjugate acid:

(a) OH^- or CH_3O^- **(b)** $CH_3CH_2S^-$ or $CH_3CH_2O^-$
(c) $CH_3CH_2O^-$ or NH_2^-

8.30 Label the stronger acid, stronger base, weaker acid, and weaker base in each of the following equilibria, and then predict the position of each equilibrium (for pK_a values, see Table 2.1):

(a) $CH_3CH_2O^- + HCl \rightleftharpoons CH_3CH_2OH + Cl^-$

(b) $CH_3\overset{\overset{O}{\|}}{C}OH + CH_3CH_2O^- \rightleftharpoons CH_3\overset{\overset{O}{\|}}{C}O^- + CH_3CH_2OH$

8.31 Predict the position of equilibrium for each acid–base reaction; that is, does each lie considerably to the left, does each lie considerably to the right, or are the concentrations evenly balanced?

(a) $CH_3CH_2OH + Na^+OH^- \rightleftharpoons CH_3CH_2O^-Na^+ + H_2O$

(b) $CH_3CH_2SH + Na^+OH^- \rightleftharpoons CH_3CH_2S^-Na^+ + H_2O$

(c) $CH_3CH_2OH + CH_3CH_2S^-Na^+ \rightleftharpoons CH_3CH_2O^-Na^+ + CH_3CH_2SH$

(d) $CH_3CH_2S^-Na^+ + CH_3\overset{\overset{O}{\|}}{C}OH \rightleftharpoons CH_3CH_2SH + CH_3\overset{\overset{O}{\|}}{C}O^-Na^+$

Reactions of Alcohols

8.32 Show how to distinguish between cyclohexanol and cyclohexene by a simple chemical test. (*Hint:* Treat each with Br_2 in CCl_4 and watch what happens.)

8.33 Write equations for the reaction of 1-butanol, a primary alcohol, with these reagents:

(a) Na metal **(b)** HBr, heat
(c) $K_2Cr_2O_7$, H_2SO_4, heat **(d)** $SOCl_2$
(e) Pyridinium chlorochromate (PCC)

8.34 Write equations for the reaction of 2-butanol, a secondary alcohol, with these reagents:

(a) Na metal **(b)** H_2SO_4, heat
(c) HBr, heat **(d)** $K_2Cr_2O_7$, H_2SO_2, heat
(e) $SOCl_2$ **(f)** Pyridinium chlorochromate (PCC)

8.35 When (*R*)-2-butanol is left standing in aqueous acid, it slowly loses its optical activity. When the organic material is recovered from the aqueous solution, only 2-butanol is found. Account for the observed loss of optical activity.

8.36 What is the most likely mechanism of the following reaction?

OH + HCl ⟶ Cl + H_2O

Draw a structural formula for the intermediate(s) formed during the reaction.

8.37 Complete the equations for these reactions:

(a) OH $+ H_2CrO_4 \longrightarrow$ **(b)** OH $+ SOCl_2 \longrightarrow$

(c) OH $+$ HCl $\longrightarrow$ **(d)** HO OH $+$ HBr (excess) $\longrightarrow$

(e) OH $+ H_2CrO_4 \longrightarrow$ **(f)** $+ OsO_4, H_2O_2 \longrightarrow$

8.38 In the commercial synthesis of methyl *tert*-butyl ether (MTBE), once used as an anti-knock, octane-improving gasoline additive, 2-methylpropene and methanol are passed over an acid catalyst to give the ether:

$$CH_3\overset{\displaystyle CH_3}{\overset{|}{C}}{=}CH_2 + CH_3OH \xrightarrow[\text{catalyst}]{\text{acid}} CH_3\underset{\underset{\displaystyle CH_3}{|}}{\overset{\overset{\displaystyle CH_3}{|}}{C}}OCH_3$$

2-Methylpropene (Isobutylene) Methanol 2-Methoxy-2-methyl-propane (Methyl *tert*-butyl ether, MTBE)

Propose a mechanism for this reaction.

8.39 Cyclic bromoalcohols, upon treatment with base, can sometimes undergo intramolecular S_N2 reactions to form bicyclic ethers. Determine whether each of the following compounds is capable of forming a bicyclic ether, and draw the product for those which can:

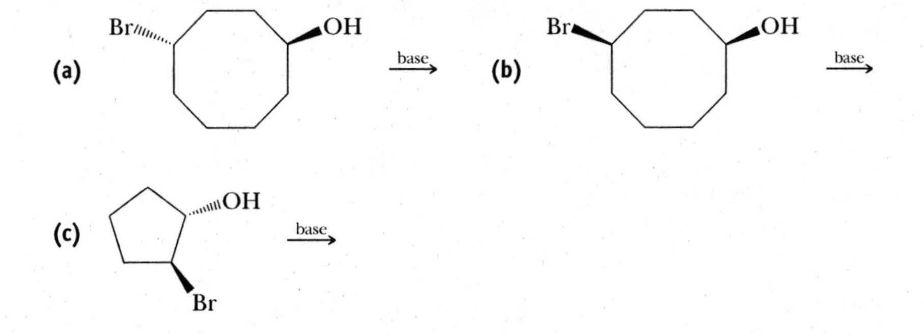

Syntheses

8.40 Show how to convert

(a) 1-Propanol to 2-propanol in two steps.
(b) Cyclohexene to cyclohexanone in two steps.
(c) Cyclohexanol to *cis*-1,2-cyclohexanediol in two steps.
(d) Propene to propanone (acetone) in two steps.

8.41 Show how to convert cyclohexanol to these compounds:

(a) Cyclohexene **(b)** Cyclohexane **(c)** Cyclohexanone

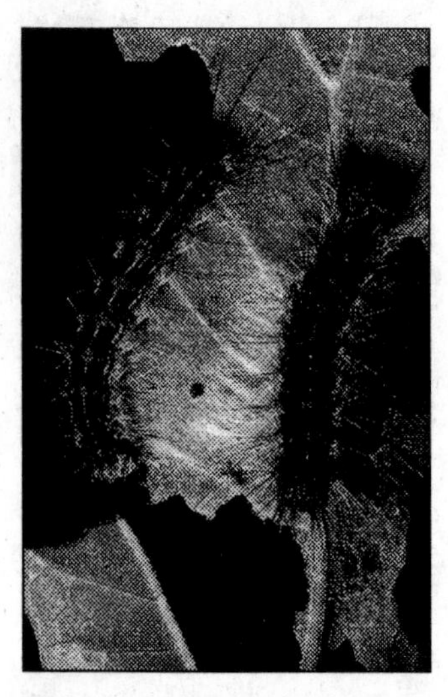

Gypsy moth caterpillars. *(William D. Griffin/Animals, Animals)*

8.42 Show reagents and experimental conditions that can be used to synthesize these compounds from 1-propanol (any derivative of 1-propanol prepared in an earlier part of this problem may be used for a later synthesis):

(a) Propanal **(b)** Propanoic acid
(c) Propene **(d)** 2-Propanol
(e) 2-Bromopropane **(f)** 1-Chloropropane
(g) Propanone **(h)** 1,2-Propanediol

8.43 Show how to prepare each compound from 2-methyl-1-propanol (isobutyl alcohol):

(a) $CH_3C(CH_3){=}CH_2$ **(b)** $CH_3C(CH_3)(OH)CH_3$ **(c)** $CH_3C(CH_3)(OH){-}CH_2OH$ **(d)** $CH_3CH(CH_3)COOH$

For any preparation involving more than one step, show each intermediate compound formed.

8.44 Show how to prepare each compound from 2-methylcyclohexanol:

(a) **(b)** OH **(c)** O

(d) O **(e)** OH OH **(f)** OH OH

For any preparation involving more than one step, show each intermediate compound formed.

8.45 Show how to convert the alcohol on the left to compounds (a), (b), and (c).

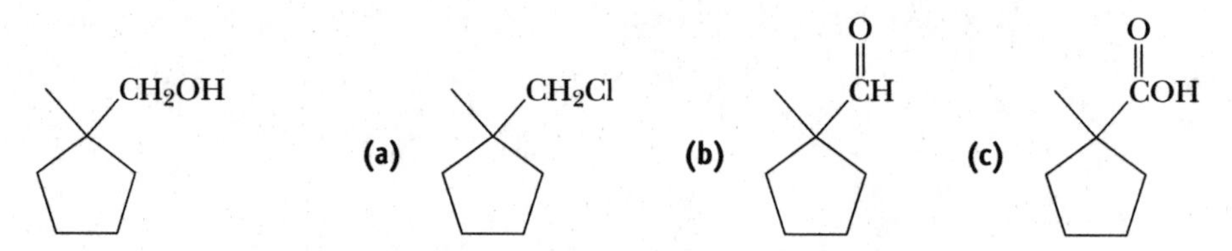

8.46 Disparlure, a sex attractant of the gypsy moth (*Porthetria dispar*), has been synthesized in the laboratory from the following (*Z*)-alkene:

H H → H O H

(*Z*)-2-Methyl-7-octadecene Disparlure

(a) How might the (*Z*)-alkene be converted to disparlure?

(b) How many stereoisomers are possible for disparlure? How many are formed in the sequence you chose?

8.47 The chemical name for bombykol, the sex pheromone secreted by the female silkworm moth to attract male silkworm moths, is *trans*-10-*cis*-12-hexadecadien-1-ol. (The compound has one hydroxyl group and two carbon–carbon double bonds in a 16-carbon chain.)

(a) Draw a structural formula for bombykol, showing the correct configuration about each carbon–carbon double bond.

(b) How many cis–trans isomers are possible for the structural formula you drew in part (a)? All possible cis–trans isomers have been synthesized in the laboratory, but only the one named bombykol is produced by the female silkworm moth, and only it attracts male silkworm moths.

Looking Ahead

8.48 Compounds that contain an N—H group associate by hydrogen bonding.

(a) Do you expect this association to be stronger or weaker than that between compounds containing an O—H group?

(b) Based on your answer to part (a), which would you predict to have the higher boiling point, 1-butanol or 1-butanamine?

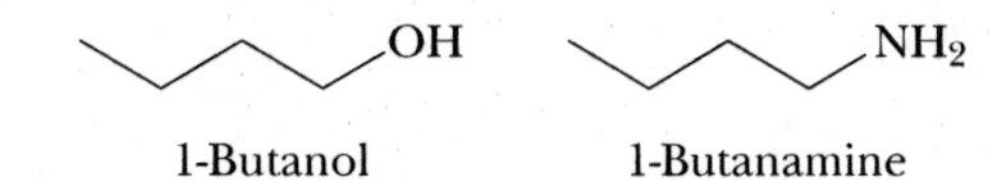

1-Butanol 1-Butanamine

8.49 Write balanced equations for the reactions of phenol and cyclohexanol, with NaOH:

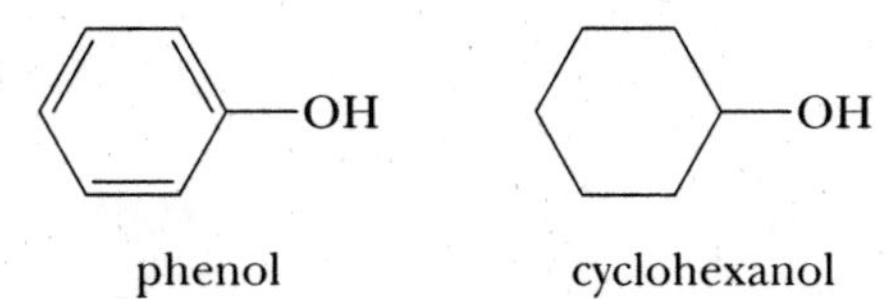

phenol cyclohexanol

(a) Which compound is more acidic? (See Table 2.2.)

(b) Which conjugate base is more nucleophilic?

8.50 Draw a resonance structure for each of the following compounds in which the heteroatom (O or S) is positively charged:

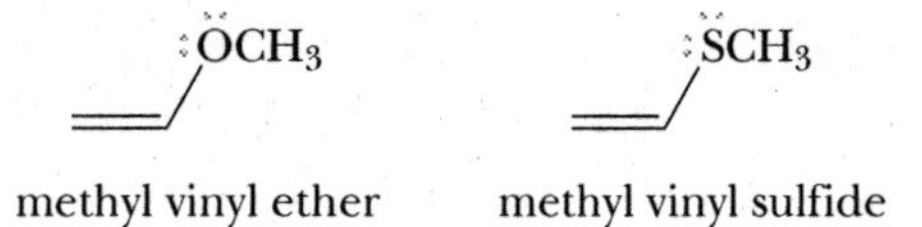

methyl vinyl ether methyl vinyl sulfide

(a) Compared with ethylene, how does each resonance structure influence the reactivity of the alkene towards an electrophile?

(b) Peracids are known to be electrophilic reagents. Based on the resonance picture and your knowledge of periodic properties of the elements, would an epoxide be more likely to form with methyl vinyl ether or methyl vinyl sulfide?

(c) Would your answer to part (b) above be the same or different if only inductive effects were taken into consideration?

8.51 Rank the members in each set of reagents from most to least nucleophilic:

(a)

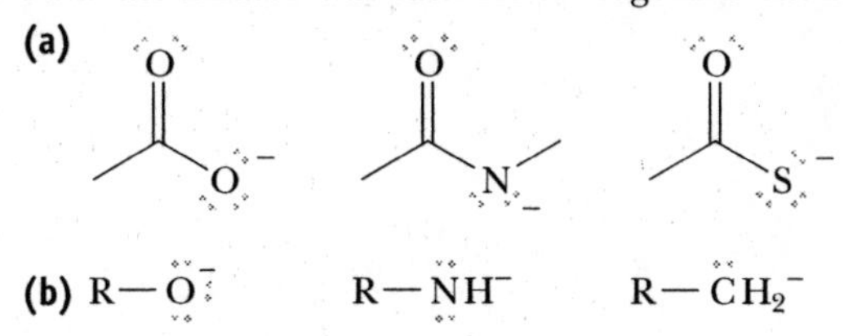

(b) $R—\ddot{O}^{-}$ $R—\ddot{N}H^{-}$ $R—\ddot{C}H_2^{-}$

8.52 Which of the following compounds is more basic?

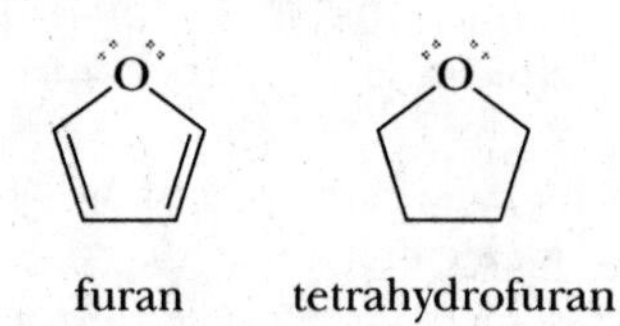

8.53 In Chapter 15 we will see that the reactivity of the following carbonyl compounds is directly proportional to the stability of the leaving group. Rank the order of reactivity of these carbonyl compounds from most reactive to least reactive based on the stability of the leaving group.

$RC(=O)OCH_3$ A $RC(=O)NH_2$ B $RC(=O)Cl$ C

9 Benzene and Its Derivatives

Peppers of the capsicum family. See Chemical Connections "Capsaicin - For Those Who Like it Hot". Inset: A model of capsaicin. *(Douglas Brown)*

9.1 INTRODUCTION

Benzene, a colorless liquid, was first isolated by Michael Faraday in 1825 from the oily residue that collected in the illuminating gas lines of London. Benzene's molecular formula, C_6H_6, suggests a high degree of unsaturation. For comparison, an alkane with six carbons has a molecular formula of C_6H_{14}, and a cycloalkane with six carbons has a molecular formula of C_6H_{12}. Considering benzene's high degree of unsaturation, it might be expected to show many of the reactions characteristic of alkenes. Yet, benzene is remarkably *un*reactive! It does not undergo the addition, oxidation, and reduction reactions characteristic of alkenes. For example, benzene does not react with bromine, hydrogen chloride, or other reagents that usually add to carbon–carbon double bonds. Nor is benzene oxidized by chromic acid or osmium tetroxide under conditions that readily oxidize alkenes. When benzene reacts, it does so by substitution in which a hydrogen atom is replaced by another atom or a group of atoms.

The term "aromatic" was originally used to classify benzene and its derivatives because many of them have distinctive odors. It became clear, however, that a sounder classification for these compounds would be one based on structure and chemical reactivity, not aroma. As it is now used, the term **aromatic** refers instead to the fact that benzene and its derivatives are highly unsaturated compounds which are unexpectedly stable toward reagents that react with alkenes.

Aromatic compound A term used to classify benzene and its derivatives.

Arene An aromatic hydrocarbon.

We use the term **arene** to describe aromatic hydrocarbons, by analogy with alkane and alkene. Benzene is the parent arene. Just as we call a group derived by the removal of an H from an alkane an alkyl group and give it the symbol R—, we call a group derived by the removal of an H from an arene an **aryl group** and give it the symbol **Ar**—.

Aryl group A group derived from an aromatic compound (an arene) by the removal of an H; given the symbol Ar—.

Ar— The symbol used for an aryl group, by analogy with R— for an alkyl group.

9.2 THE STRUCTURE OF BENZENE

Let us imagine ourselves in the mid-19th century and examine the evidence on which chemists attempted to build a model for the structure of benzene. First, because the molecular formula of benzene is C_6H_6, it seemed clear that the molecule must be highly unsaturated. Yet benzene does not show the chemical properties of alkenes, the only unsaturated hydrocarbons known at that time. Benzene does undergo chemical reactions, but its characteristic reaction is substitution rather than addition. When benzene is treated with bromine in the presence of ferric chloride as a catalyst, for example, only one compound with molecular formula C_6H_5Br forms:

$$\underset{\text{Benzene}}{C_6H_6} + Br_2 \xrightarrow{FeCl_3} \underset{\text{Bromobenzene}}{C_6H_5Br} + HBr$$

Chemists concluded, therefore, that all six carbons and all six hydrogens of benzene must be equivalent. When bromobenzene is treated with bromine in the presence of ferric chloride, three isomeric dibromobenzenes are formed:

$$\underset{\text{Bromobenzene}}{C_6H_5Br} + Br_2 \xrightarrow{FeCl_3} \underset{\text{Dibromobenzene}}{C_6H_4Br_2} + HBr$$

(formed as a mixture of three constitutional isomers)

For chemists in the mid-19th century, the problem was to incorporate these observations, along with the accepted tetravalence of carbon, into a structural formula for benzene. Before we examine their proposals, we should note that the problem of the structure of benzene and other aromatic hydrocarbons has occupied the efforts of chemists for over a century. It was not until 1930s that chemists developed a general understanding of the unique chemical properties of benzene and its derivatives.

A. Kekulé's Model of Benzene

The first structure for benzene, proposed by August Kekulé in 1872, consisted of a six-membered ring with alternating single and double bonds and with one hydrogen bonded to each carbon. Kekulé further proposed that the ring contains three double bonds which shift back and forth so rapidly that the two forms cannot be separated. Each structure has become known as a **Kekulé structure**.

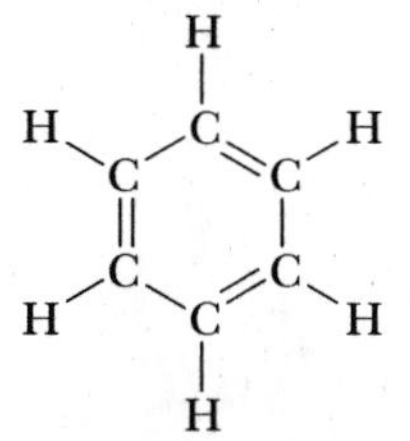

A Kekulé structure, showing all atoms

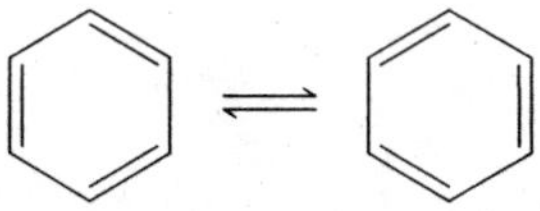

Kekulé structures as line-angle formulas

Because all of the carbons and hydrogens of Kekulé's structure are equivalent, substituting bromine for any one of the hydrogens gives the same compound. Thus, Kekulé's proposed structure was consistent with the fact that treating benzene with bromine in the presence of ferric chloride gives only one compound with molecular formula C_6H_5Br.

His proposal also accounted for the fact that the bromination of bromobenzene gives three (and only three) isomeric dibromobenzenes:

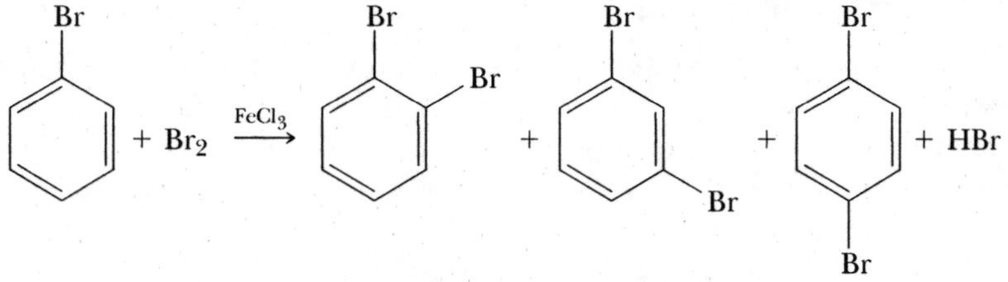

The three isomeric dibromobenzenes

Although Kekulé's proposal was consistent with many experimental observations, it was contested for years. The major objection was that it did not account for the unusual chemical behavior of benzene. If benzene contains three double bonds, why, his critics asked, doesn't it show the reactions typical of alkenes? Why doesn't it add three moles of bromine to form 1,2,3,4,5,6-hexabromocyclohexane? Why, instead, does benzene react by substitution rather than addition?

B. The Orbital Overlap Model of Benzene

The concepts of the **hybridization of atomic orbitals** and the **theory of resonance**, developed by Linus Pauling in the 1930s, provided the first adequate description of the structure of benzene. The carbon skeleton of benzene forms a regular hexagon with C—C—C and H—C—C bond angles of 120°. For this type of bonding, carbon uses sp^2 hybrid orbitals (Section 1.7E). Each carbon forms sigma bonds to two adjacent carbons by the overlap of sp^2–sp^2 hybrid orbitals and one sigma bond to hydrogen by the overlap of sp^2–$1s$ orbitals. As determined experimentally, all carbon–carbon bonds in benzene are the same length, 1.39 Å, a value almost midway between the length of a single bond between sp^3 hybridized carbons (1.54 Å) and that of a double bond between sp^2 hybridized carbons (1.33 Å):

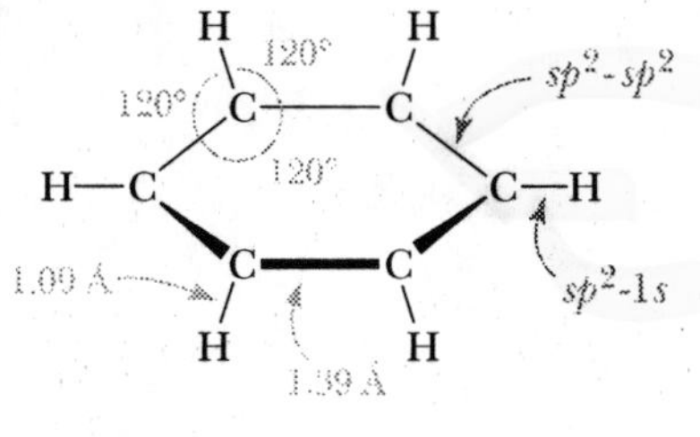

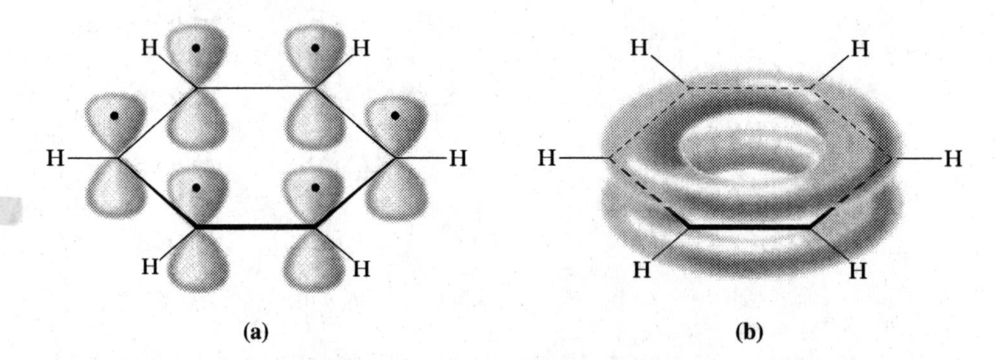

Figure 9.1
Orbital overlap model of bonding in benzene. (a) The carbon, hydrogen framework. The six $2p$ orbitals, each with one electron, are shown uncombined. (b) The overlap of parallel $2p$ orbitals forms a continuous pi cloud, shown by one torus above the plane of the ring and a second below the plane of the ring.

Each carbon also has a single unhybridized $2p$ orbital that contains one electron. These six $2p$ orbitals lie perpendicular to the plane of the ring and overlap to form a continuous pi cloud encompassing all six carbons. The electron density of the pi system of a benzene ring lies in one torus (a doughnut-shaped region) above the plane of the ring and a second torus below the plane (Figure 9.1).

C. The Resonance Model of Benzene

One of the postulates of resonance theory is that, if we can represent a molecule or ion by two or more contributing structures, then that molecule cannot be adequately represented by any single contributing structure. We represent benzene as a hybrid of two equivalent contributing structures, often referred to as *Kekulé structures*:

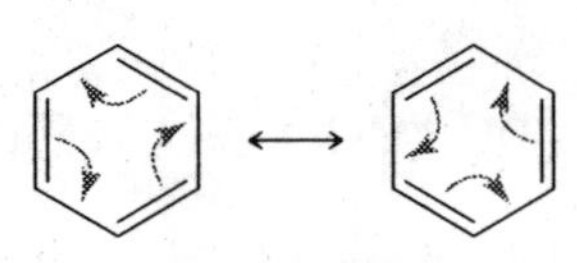

Benzene as a hybrid of two equivalent contributing structures

Each Kekulé structure makes an equal contribution to the hybrid; thus, the C—C bonds are neither single nor double bonds, but something intermediate. We recognize that neither of these contributing structures exists (they are merely alternative ways to pair $2p$ orbitals, with no reason to prefer one over the other) and that the actual structure is a superposition of both. Nevertheless, chemists continue to use a single contributing structure to represent this molecule because it is as close as we can come to an accurate structure within the limitations of classical valence bond structures and the tetravalence of carbon.

D. The Resonance Energy of Benzene

Resonance energy The difference in energy between a resonance hybrid and the most stable of its hypothetical contributing structures.

Resonance energy is the difference in energy between a resonance hybrid and its most stable hypothetical contributing structure. One way to estimate the resonance energy of benzene is to compare the heats of hydrogenation of cyclohexene and benzene. In the presence of a transition metal catalyst, hydrogen readily reduces cyclohexene to cyclohexane (Section 5.5):

$$\text{cyclohexene} + H_2 \xrightarrow[\text{1–2 atm}]{\text{Ni}} \text{cyclohexane} \qquad \Delta H^0 = -28.6 \text{ kcal/mol} \; (-120 \text{ kJ/mol})$$

By contrast, benzene is reduced only very slowly to cyclohexane under these conditions. It is reduced more rapidly when heated and under a pressure of several hundred atmospheres of hydrogen:

$$\text{benzene} + 3\,H_2 \xrightarrow[200\text{–}300\ \text{atm}]{\text{Ni}} \text{cyclohexane} \qquad \Delta H^0 = -49.8\ \text{kcal/mol}\ (-208\ \text{kJ/mol})$$

The catalytic reduction of an alkene is an exothermic reaction (Section 5.5B). The heat of hydrogenation per double bond varies somewhat with the degree of substitution of the double bond; for cyclohexene, $\Delta H^0 = -28.6$ kcal/mol (−120 kJ/mol). If we consider benzene to be 1,3,5-cyclohexatriene, a hypothetical compound with alternating single and double bonds, we might expect its heat of hydrogenation to be $3 \times -28.6 = -85.8$ kcal/mol (−359 kJ/mol). Instead, the heat of hydrogenation of benzene is only −49.8 kcal/mol (−208 kJ/mol). The difference of 36.0 kcal/mol (151 kJ/mol) between the expected value and the experimentally observed value is the **resonance energy of benzene**. Figure 9.2 shows these experimental results in the form of a graph.

For comparison, the strength of a carbon–carbon single bond is approximately 80–100 kcal/mol (333–418 kJ/mol), and that of hydrogen bonding in water and low-molecular-weight alcohols is approximately 2–5 kcal/mol (8.4–21 kJ/mol). Thus, although the resonance energy of benzene is less than the strength of a carbon–carbon single bond, it is considerably greater than the strength of hydrogen bonding in water and alcohols. In Section 8.2C, we saw that hydrogen bonding has a dramatic effect on the physical properties of alcohols compared with those of alkanes. In this chapter, we see that the resonance energy of benzene and other aromatic hydrocarbons has a dramatic effect on their chemical reactivity.

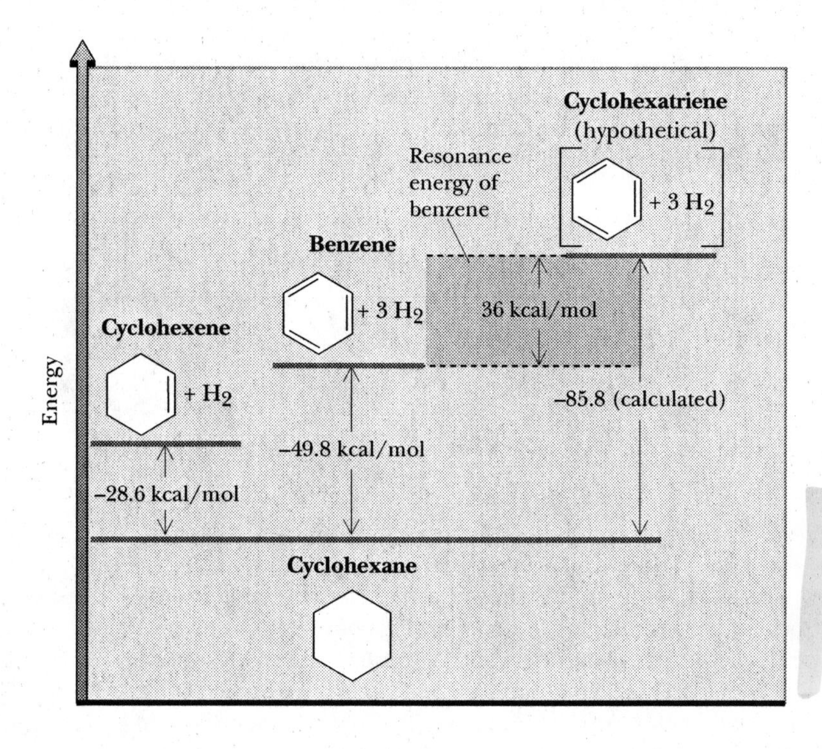

Figure 9.2
The resonance energy of benzene, as determined by a comparison of the heats of hydrogenation of cyclohexene, benzene, and the hypothetical 1,3,5-cyclohexatriene.

Following are resonance energies for benzene and several other aromatic hydrocarbons:

Resonance energy [kcal/mol (kJ/mol)]	Benzene	Naphthalene	Anthracene	Phenanthrene
	36 (151)	61 (255)	83 (347)	91 (381)

9.3 THE CONCEPT OF AROMATICITY

Many other types of molecules besides benzene and its derivatives show aromatic character; that is, they contain high degrees of unsaturation, yet fail to undergo characteristic alkene addition and oxidation–reduction reactions. What chemists had long sought to understand were the principles underlying aromatic character. The German chemical physicist Eric Hückel solved this problem in the 1930s.

Hückel's criteria are summarized as follows. To be aromatic, a ring must

1. Have one $2p$ orbital on each of its atoms.
2. Be planar or nearly planar, so that there is continuous overlap or nearly continuous overlap of all $2p$ orbitals of the ring.
3. Have 2, 6, 10, 14, 18, and so forth pi electrons in the cyclic arrangement of $2p$ orbitals.

Benzene meets these criteria. It is cyclic, planar, has one $2p$ orbital on each carbon atom of the ring, and has 6 pi electrons (an aromatic sextet) in the cyclic arrangement of its $2p$ orbitals.

Heterocyclic compound An organic compound that contains one or more atoms other than carbon in its ring.

Let us apply these criteria to several **heterocyclic compounds**, all of which are aromatic. Pyridine and pyrimidine are heterocyclic analogs of benzene. In pyridine, one CH group of benzene is replaced by a nitrogen atom, and in pyrimidine, two CH groups are replaced by nitrogen atoms:

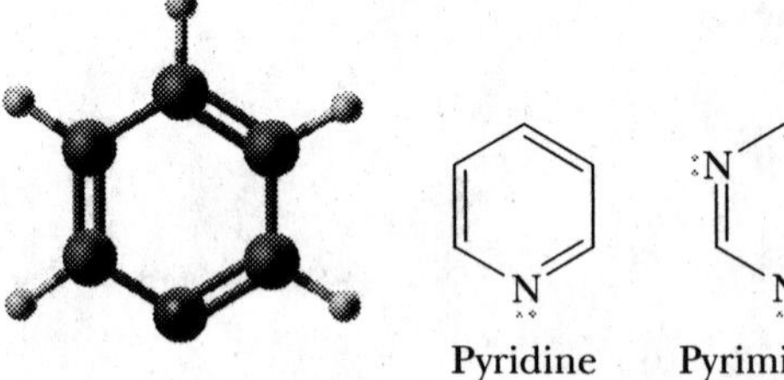

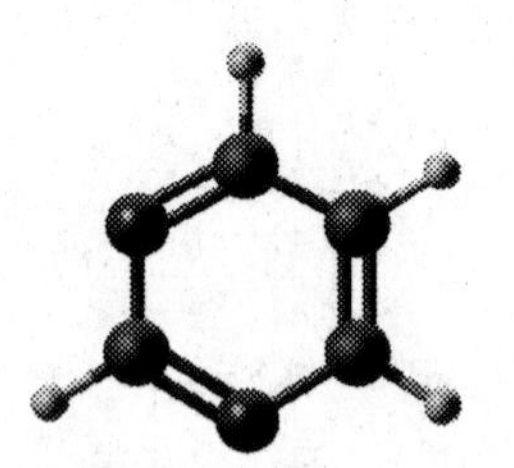

Pyridine Pyrimidine

Each molecule meets the Hückel criteria for aromaticity: Each is cyclic and planar, has one $2p$ orbital on each atom of the ring, and has six electrons in the pi system. In pyridine, nitrogen is sp^2 hybridized, and its unshared pair of electrons occupies an sp^2 orbital perpendicular to the $2p$ orbitals of the pi system and thus is not a part of the pi system. In pyrimidine, neither unshared pair of electrons of nitrogen is part of the pi system. The resonance energy of pyridine is 32 kcal/mol (134 kJ/mol), slightly less than that of benzene. The resonance energy of pyrimidine is 26 kcal/mol (109 kJ/mol).

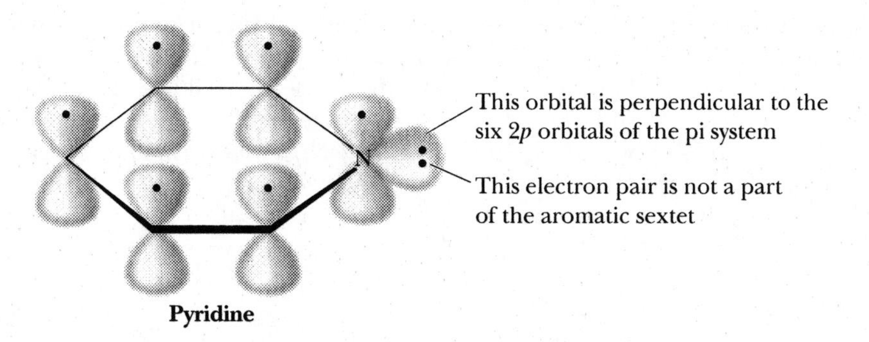
This orbital is perpendicular to the six 2p orbitals of the pi system
N
This electron pair is not a part of the aromatic sextet
Pyridine

The five-membered-ring compounds furan, pyrrole, and imidazole are also aromatic:

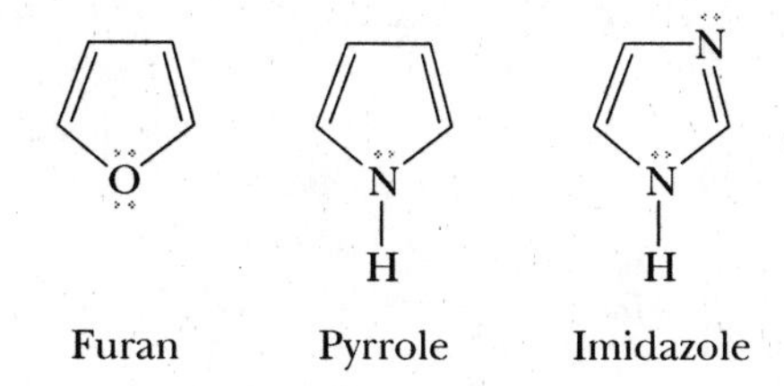
O
N
H
N
N
H

Furan Pyrrole Imidazole

In these planar compounds, each heteroatom is sp^2 hybridized, and its unhybridized $2p$ orbital is part of a continuous cycle of five $2p$ orbitals. In furan, one unshared pair of electrons of the heteroatom lies in the unhybridized $2p$ orbital and is a part of the pi system (Figure 9.3). The other unshared pair of electrons lies in an sp^2 hybrid orbital, perpendicular to the $2p$ orbitals, and is not a part of the pi system. In pyrrole, the unshared pair of electrons on nitrogen is part of the aromatic sextet. In imidazole, the unshared pair of electrons on one nitrogen is part of the aromatic sextet; the unshared pair on the other nitrogen is not.

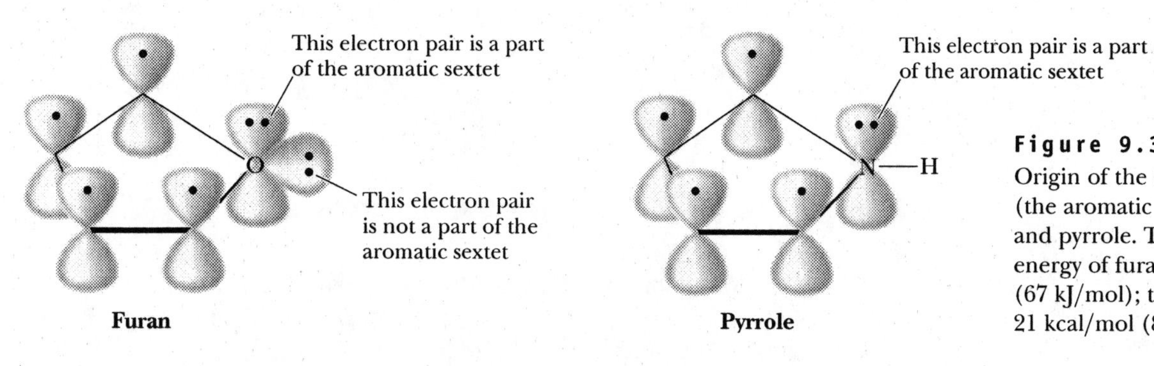
This electron pair is a part of the aromatic sextet
O
This electron pair is not a part of the aromatic sextet
Furan
This electron pair is a part of the aromatic sextet
N—H
Pyrrole

Figure 9.3
Origin of the 6 pi electrons (the aromatic sextet) in furan and pyrrole. The resonance energy of furan is 16 kcal/mol (67 kJ/mol); that of pyrrole is 21 kcal/mol (88 kJ/mol).

Nature abounds with compounds having a heterocyclic ring fused to one or more other rings. Two such compounds especially important in the biological world are indole and purine:

Indole Serotonin (a neurotransmitter) Purine Adenine

Indole contains a pyrrole ring fused with a benzene ring. Compounds derived from indole include the amino acid L-tryptophan (Section 19.2A) and the neurotransmitter serotonin. Purine contains a six-membered pyrimidine ring fused with a five-membered imidazole ring. Adenine is one of the building blocks of deoxyribonucleic acids (DNA) and ribonucleic acids (RNA), as described in Chapter 20. It is also a component of the biological oxidizing agent nicotinamide adenine dinucleotide, abbreviated NAD^+ (Section 22.2B).

9.4 NOMENCLATURE

A. Monosubstituted Benzenes

Monosubstituted alkylbenzenes are named as derivatives of benzene; an example is ethylbenzene. The IUPAC system retains certain common names for several of the simpler monosubstituted alkylbenzenes. Examples are **toluene** (rather than methylbenzene) and **styrene** (rather than phenylethylene):

CH_2CH_3 CH_3 $CH{=}CH_2$

Benzene Ethylbenzene Toluene Styrene

The common names **phenol, aniline, benzaldehyde, benzoic acid,** and **anisole** are also retained by the IUPAC system:

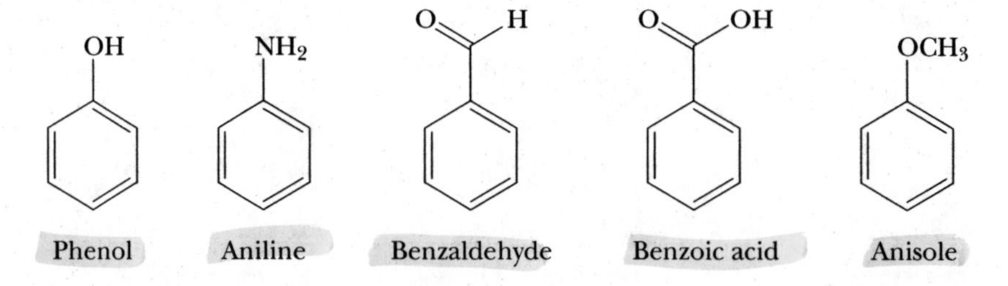

Phenyl group C_6H_5—, the aryl group derived by removing a hydrogen from benzene.

Benzyl group $C_6H_5CH_2$—, the alkyl group derived by removing a hydrogen from the methyl group of toluene.

As noted in the introduction to Chapter 5, the substituent group derived by the loss of an H from benzene is a **phenyl** group (Ph); that derived by the loss of an H from the methyl group of toluene is a **benzyl group** (Bn):

—H —CH_3 —CH_2—

Benzene Phenyl group (Ph) Toluene Benzyl group (Bn)

In molecules containing other functional groups, phenyl groups and benzyl groups are often named as substituents:

CH_3, H, C=C, Ph, CH_3 — (*Z*)-2-Phenyl-2-butene

$PhCH_2CH_2OH$ — 2-Phenylethanol

$PhCH_2Cl$ — Benzyl chloride

B. Disubstituted Benzenes

When two substituents occur on a benzene ring, three constitutional isomers are possible. We locate substituents either by numbering the atoms of the ring or by using the locators **ortho**, **meta**, and **para**. The numbers 1,2- are equivalent to *ortho* (Greek: straight); 1,3- to *meta* (Greek: after); and 1,4- to *para* (Greek: beyond).

Ortho (*o*) Refers to groups occupying positions 1 and 2 on a benzene ring.

Meta (*m*) Refers to groups occupying positions 1 and 3 on a benzene ring.

Para (*p*) Refers to groups occupying positions 1 and 4 on a benzene ring.

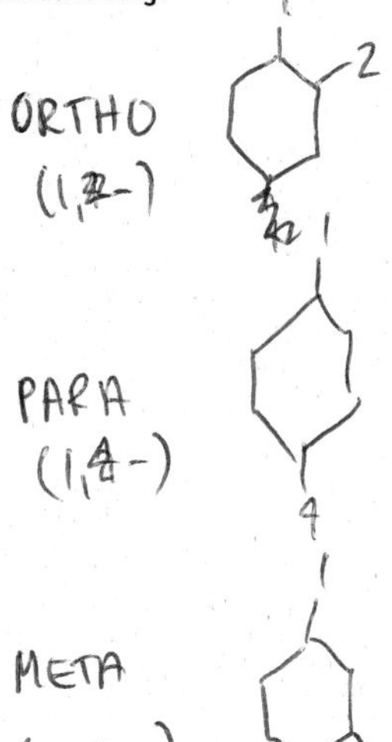

When one of the two substituents on the ring imparts a special name to the compound, as, for example, toluene, phenol, and aniline, then we name the compound as a derivative of that parent molecule. In this case, the special substituent occupies ring position number 1. The IUPAC system retains the common name **xylene** for the three isomeric dimethylbenzenes. When neither group imparts a special name, we locate the two substituents and list them in alphabetical order before the ending *-benzene*. The carbon of the benzene ring with the substituent of lower alphabetical ranking is numbered C-1.

4-Bromotoluene (*p*-Bromotoluene)

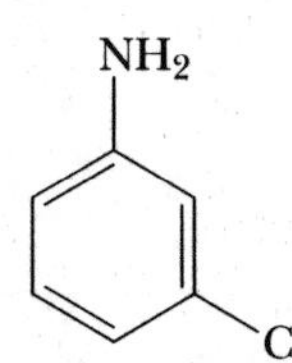

3-Chloroaniline (*m*-Chloroaniline)

1,3-Dimethylbenzene (*m*-Xylene)

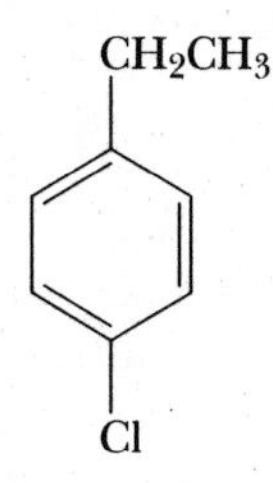

1-Chloro-4-ethylbenzene (*p*-Chloroethylbenzene)

C. Polysubstituted Benzenes

When three or more substituents are present on a ring, we specify their locations by numbers. If one of the substituents imparts a special name, then the molecule is named as a derivative of that parent molecule. If none of the substituents imparts a special name, we number them to give the smallest set of numbers and list them in alphabetical order before the ending *-benzene*. In the following examples, the first compound is a derivative of toluene, and the second is a derivative of phenol. Because there is no special name for the third compound, we list its three substituents in alphabetical order, followed by the word *benzene*:

4-Chloro-2-nitrotoluene

2,4,6-Tribromophenol

2-Bromo-1-ethyl-4-nitrobenzene

3+ substituted ≠ omp

EXAMPLE 9.1

Write names for these compounds:

(a) (b) (c) (d)

SOLUTION

(a) 3-Iodotoluene or *m*-iodotoluene (b) 3,5-Dibromobenzoic acid
(c) 1-Chloro-2,4-dinitrobenzene (d) 3-Phenylpropene

Practice Problem 9.1

Write names for these compounds:

(a) (b) (c)

Polynuclear aromatic hydrocarbon A hydrocarbon containing two or more fused aromatic rings.

Polynuclear aromatic hydrocarbons (PAHs) contain two or more aromatic rings, each pair of which shares two ring carbon atoms. Naphthalene, anthracene, and phenanthrene, the most common PAHs, and substances derived from them are found in coal tar and high-boiling petroleum residues. At one time, naphthalene was used as a moth repellent and insecticide in preserving woolens and furs, but its use has decreased due to the introduction of chlorinated hydrocarbons such as *p*-dichlorobenzene. Also found in coal tar are lesser amounts of benzo[a]pyrene. This compound is found as well in the exhausts of gasoline-powered internal combustion engines (for example, automobile engines) and in cigarette smoke. Benzo[a]pyrene is a very potent carcinogen and mutagen.

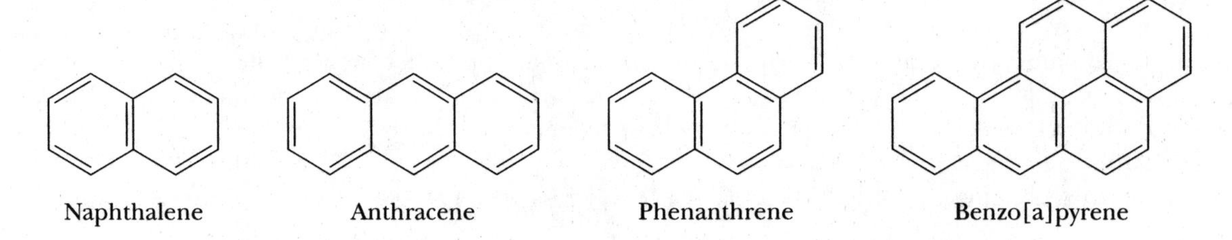

9.5 REACTIONS OF BENZENE: OXIDATION AT A BENZYLIC POSITION

As we have mentioned, benzene's aromaticity causes it to resist many of the reactions that alkenes typically undergo. However, chemists have been able to react benzene in other ways. This is fortunate, because benzene rings are abundant in many of the compounds that society depends upon, including various medications, plastics, and preservatives for food. We begin our discussion of benzene reactions with processes that take place not on the ring itself, but at the carbon immediately attached to the benzene ring. This is known as the **benzylic position.**

Benzylic carbon An sp^3 hybridized carbon bonded to a benzene ring.

Benzene is unaffected by strong oxidizing agents, such as H_2CrO_4 and $KMnO_4$. When we treat toluene with these oxidizing agents under vigorous conditions, the side-chain methyl group is oxidized to a carboxyl group to give benzoic acid:

$$\text{Toluene } (C_6H_5CH_3) + H_2CrO_4 \longrightarrow \text{Benzoic acid } (C_6H_5COOH) + Cr^{3+}$$

CHEMICAL CONNECTIONS 9A

Carcinogenic Polynuclear Aromatics and Smoking

A **carcinogen** is a compound that causes cancer. The first carcinogens to be identified were a group of polynuclear aromatic hydrocarbons, all of which have at least four aromatic rings. Among them is benzo[a]pyrene, one of the most carcinogenic of the aromatic hydrocarbons. It forms whenever there is incomplete combustion of organic compounds. Benzo[a]pyrene is found, for example, in cigarette smoke, automobile exhaust, and charcoal-broiled meats.

Benzo[a]pyrene causes cancer in the following way: Once it is absorbed or ingested, the body attempts to convert it into a more soluble compound that can be excreted easily. To this end, a series of enzyme-catalyzed reactions transforms benzo[a]pyrene into a **diol epoxide**, a compound that can bind to DNA by reacting with one of its amino groups, thereby altering the structure of DNA and producing a cancer-causing mutation:

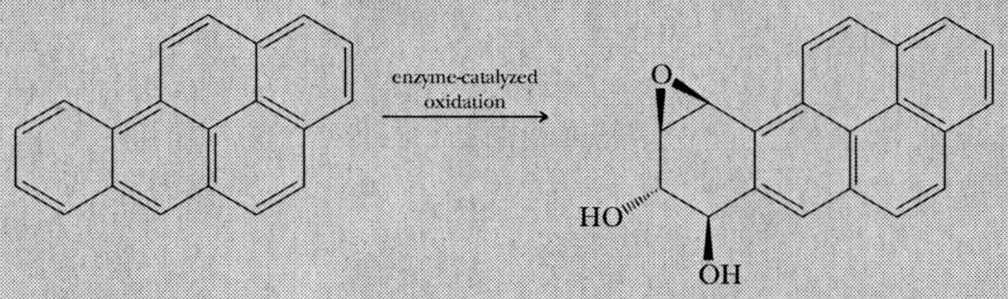

enzyme-catalyzed oxidation

O HO OH

Benzo[a]pyrene A diol epoxide

The fact that the side-chain methyl group is oxidized, but the aromatic ring is unchanged, illustrates the remarkable chemical stability of the aromatic ring. Halogen and nitro substituents on an aromatic ring are unaffected by these oxidations. For example, chromic acid oxidizes 2-chloro-4-nitrotoluene to 2-chloro-4-nitrobenzoic acid. Notice that in this oxidation, the nitro and chloro groups remain unaffected:

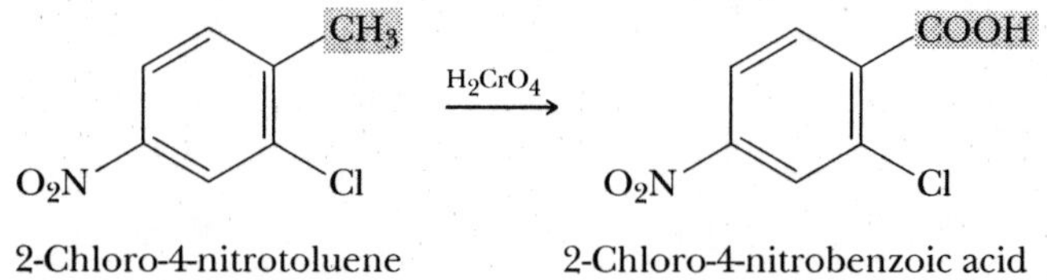

CH_3 $\xrightarrow{H_2CrO_4}$ COOH

O_2N Cl O_2N Cl

2-Chloro-4-nitrotoluene 2-Chloro-4-nitrobenzoic acid

Ethylbenzene and isopropylbenzene are also oxidized to benzoic acid under these conditions. The side chain of *tert*-butylbenzene, which has no benzylic hydrogen, is not affected by these oxidizing conditions.

From these observations, we conclude that, if a benzylic hydrogen exists, then the benzylic carbon (Section 9.4A) is oxidized to a carboxyl group and all other carbons of the side chain are removed. If no benzylic hydrogen exists, as in the case of *tert*-butylbenzene, then the side chain is not oxidized.

If more than one alkyl side chain exists, each is oxidized to —COOH. Oxidation of *m*-xylene gives 1,3-benzenedicarboxylic acid, more commonly named isophthalic acid:

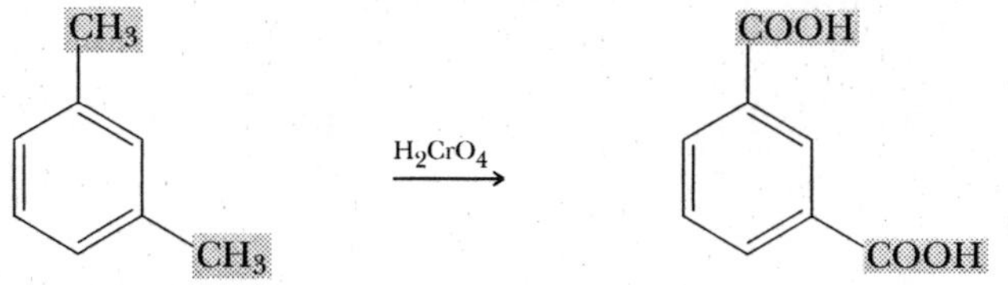

CH_3 COOH

$\xrightarrow{H_2CrO_4}$

CH_3 COOH

m-Xylene 1,3-Benzenedicarboxylic acid (Isophthalic acid)

EXAMPLE 9.2

Draw a structural formula for the product of vigorous oxidation of 1,4-dimethylbenzene (*p*-xylene) by H_2CrO_4.

SOLUTION

Chromic acid oxidizes both alkyl groups to —COOH groups, and the product is terephthalic acid, one of two monomers required for the synthesis of Dacron® polyester and Mylar® (Section 17.5B):

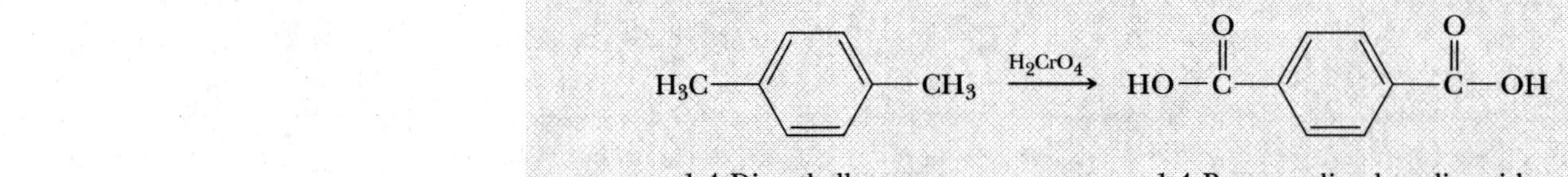

$$H_3C-C_6H_4-CH_3 \xrightarrow{H_2CrO_4} HO-\overset{O}{\overset{\|}{C}}-C_6H_4-\overset{O}{\overset{\|}{C}}-OH$$

1,4-Dimethylbenzene (*p*-Xylene) — 1,4-Benzenedicarboxylic acid (Terephthalic acid)

Practice Problem 9.2

Predict the products resulting from vigorous oxidation of each compound by H_2CrO_4:

(a) 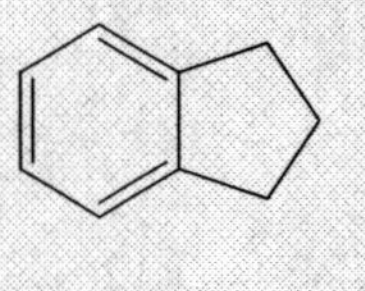 (b) 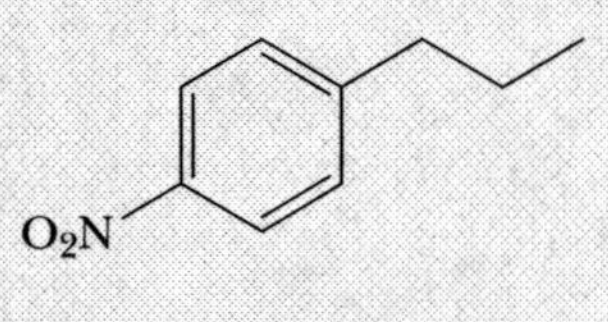 O_2N

9.6 REACTIONS OF BENZENE: ELECTROPHILIC AROMATIC SUBSTITUTION

By far the most characteristic reaction of aromatic compounds is substitution at a ring carbon. Some groups that can be introduced directly onto the ring are the halogens, the nitro ($-NO_2$) group, the sulfonic acid ($-SO_3H$) group, alkyl ($-R$) groups, and acyl ($RCO-$) groups.

Halogenation:

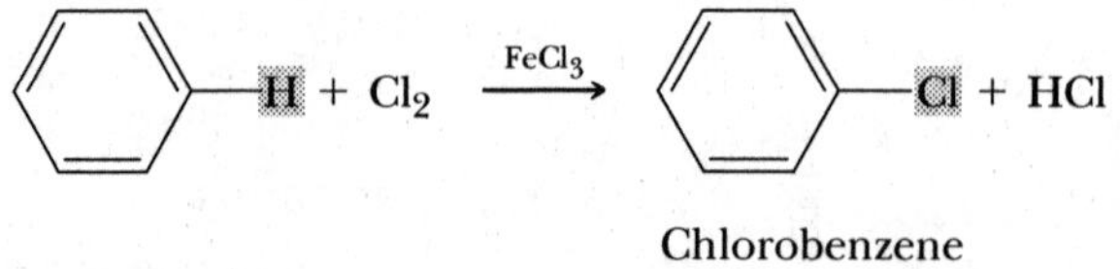

$$C_6H_5-H + Cl_2 \xrightarrow{FeCl_3} C_6H_5-Cl + HCl$$

Chlorobenzene

Nitration:

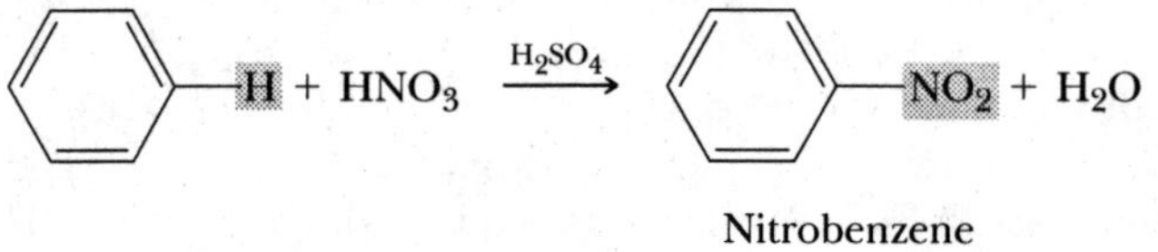

$$C_6H_5-H + HNO_3 \xrightarrow{H_2SO_4} C_6H_5-NO_2 + H_2O$$

Nitrobenzene

Sulfonation:

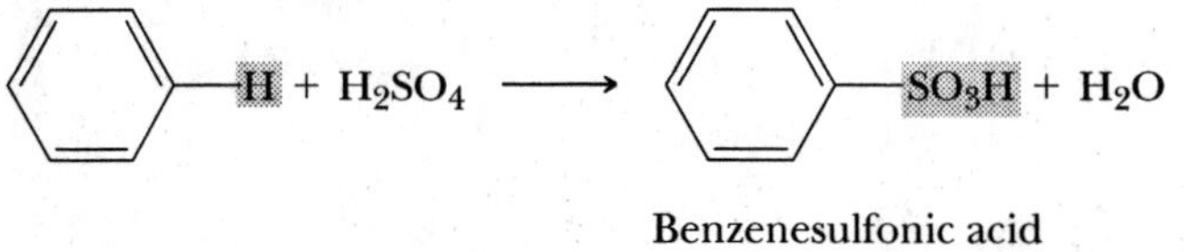

$$C_6H_5-H + H_2SO_4 \longrightarrow C_6H_5-SO_3H + H_2O$$

Benzenesulfonic acid

Alkylation:

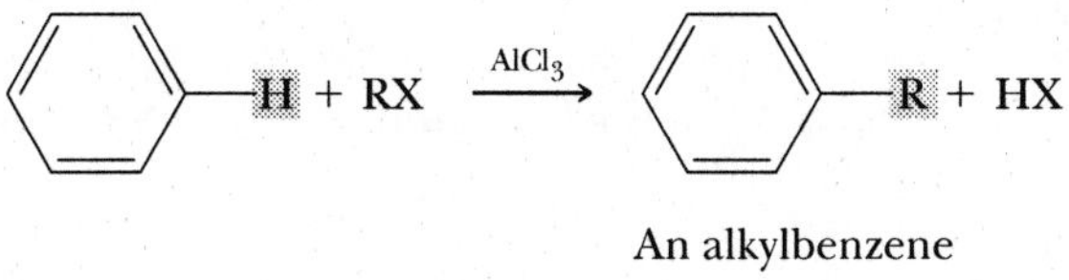

$$C_6H_5\text{—H} + RX \xrightarrow{AlCl_3} C_6H_5\text{—R} + HX$$

An alkylbenzene

Acylation:

$$C_6H_5\text{—H} + R\text{—}\overset{O}{\overset{\|}{C}}\text{—}X \xrightarrow{AlCl_3} C_6H_5\text{—}\overset{O}{\overset{\|}{C}}R + HX$$

An acyl halide — An acylbenzene

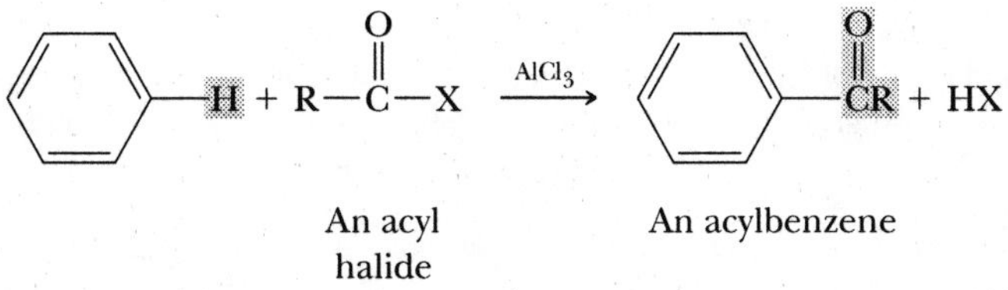

9.7 MECHANISM OF ELECTROPHILIC AROMATIC SUBSTITUTION

In this section, we study several types of **electrophilic aromatic substitution** reactions—that is, reactions in which a hydrogen of an aromatic ring is replaced by an electrophile, E^+. The mechanisms of these reactions are actually very similar. In fact, they can be broken down into three common steps:

Electrophilic aromatic substitution A reaction in which an electrophile, E^+, substitutes for a hydrogen on an aromatic ring.

Step 1: Generation of the electrophile:

$$\text{Reagent(s)} \longrightarrow E^+$$

Step 2: Attack of the electrophile on the aromatic ring to give a resonance-stabilized cation intermediate:

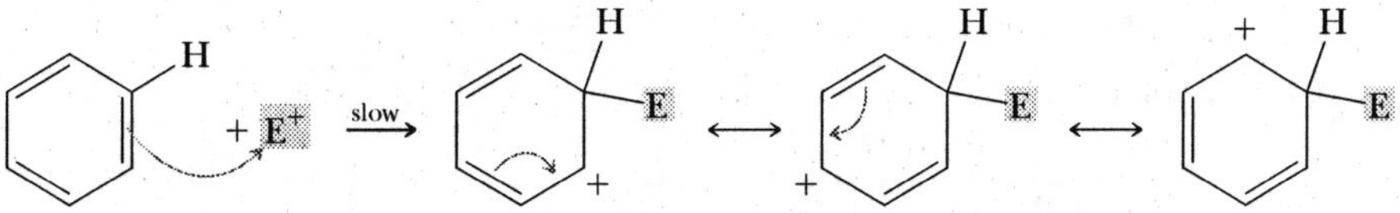

$$C_6H_5\text{—H} + E^+ \xrightarrow{\text{slow}} [\text{cation}] \longleftrightarrow [\text{cation}] \longleftrightarrow [\text{cation}]$$

Resonance-stabilized cation intermediate

Step 3: Proton transfer to a base to regenerate the aromatic ring:

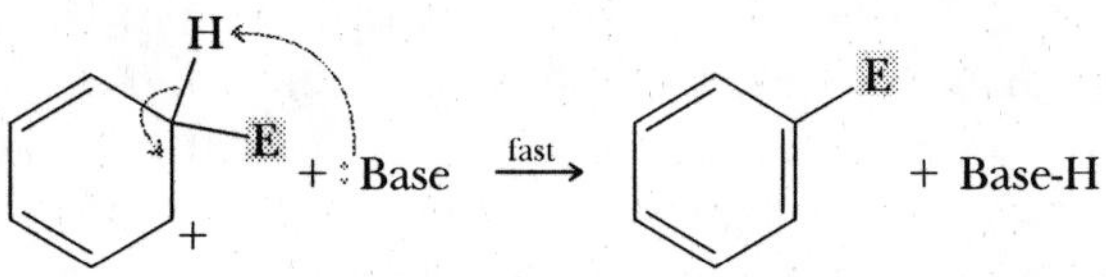

$$[\text{cation}] + :\text{Base} \xrightarrow{\text{fast}} C_6H_5\text{—E} + \text{Base-H}$$

The reactions we are about to study differ only in the way the electrophile is generated and in the base that removes the proton to re-form the aromatic ring. You should keep this principle in mind as we explore the details of each reaction.

A. Chlorination and Bromination

Chlorine alone does not react with benzene, in contrast to its instantaneous addition to cyclohexene (Section 5.3C). However, in the presence of a Lewis acid catalyst, such as ferric chloride or aluminum chloride, chlorine reacts to give chlorobenzene and HCl. Chemists account for this type of electrophilic aromatic substitution by the following three-step mechanism:

Mechanism: Electrophilic Aromatic Substitution—Chlorination

Step 1: *Formation of the Electrophile:* Reaction between chlorine (a Lewis base) and $FeCl_3$ (a Lewis acid) gives an ion pair containing a chloronium ion (an electrophile):

Chlorine (a Lewis base) + Ferric chloride (a Lewis acid) ⇌ A molecular complex with a positive charge on chlorine and a negative charge on iron ⇌ An ion pair containing a chloronium ion

Step 2: *Attack of the Electrophile on the Ring:* Reaction of the Cl_2–$FeCl_3$ ion pair with the pi electron cloud of the aromatic ring forms a resonance-stabilized cation intermediate, represented here as a hybrid of three contributing structures:

slow, rate determining

Resonance-stabilized cation intermediate

The positive charge on the resonance-stabilized intermediate is distributed approximately equally on the carbon atoms 2, 4, and 6 of the ring relative to the point of substitution.

Step 3: *Proton Transfer:* Proton transfer from the cation intermediate to $FeCl_4^-$ forms HCl, regenerates the Lewis acid catalyst, and gives chlorobenzene:

Cation intermediate —fast→ Chlorobenzene + H—Cl + $FeCl_3$

Treatment of benzene with bromine in the presence of ferric chloride or aluminum chloride gives bromobenzene and HBr. The mechanism for this reaction is the same as that for chlorination of benzene.

The major difference between the addition of halogen to an alkene and substitution by halogen on an aromatic ring is the fate of the cation intermediate formed in the first step of each reaction. Recall from Section 5.3C that the addition of chlorine to an alkene is a two-step process, the first and slower step of which is the formation of a bridged chloronium ion intermediate. This intermediate then reacts with chloride ion to complete the addition. With aromatic compounds, the cation intermediate loses H^+ to regenerate the aromatic ring and regain its large resonance stabilization. There is no such resonance stabilization to be regained in the case of an alkene.

B. Nitration and Sulfonation

The sequence of steps for the nitration and sulfonation of benzene is similar to that for chlorination and bromination. For nitration, the electrophile is the **nitronium ion**, NO_2^+, generated by the reaction of nitric acid with sulfuric acid. In the following equations nitric acid is written $HONO_2$ to show more clearly the origin of the nitronium ion.

Mechanism: Formation of the Nitronium Ion

Step 1: Proton transfer from sulfuric acid to the OH group of nitric acid gives the conjugate acid of nitric acid:

$$H-\ddot{O}-NO_2 + H-\ddot{O}-SO_3H \rightleftharpoons H-\overset{H}{\overset{|}{\overset{+}{O}}}-NO_2 + HSO_4^-$$

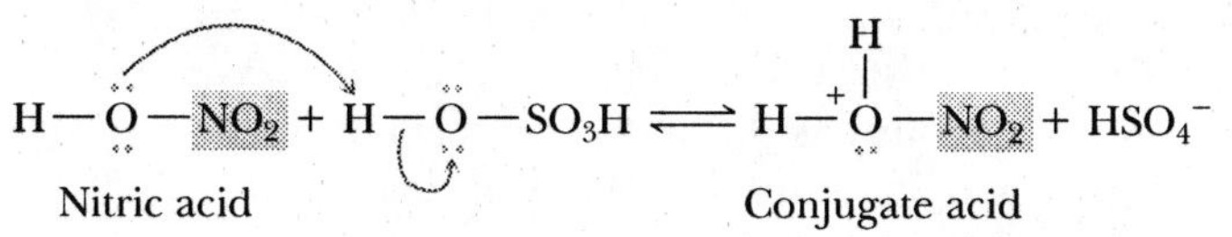

Nitric acid — Conjugate acid of nitric acid

Step 2: Loss of water from this conjugate acid gives the nitronium ion, NO_2^+:

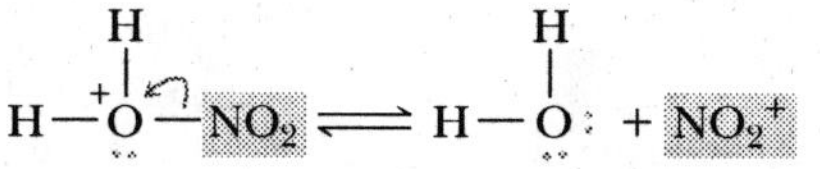

$$H-\overset{H}{\overset{|}{\overset{+}{O}}}-NO_2 \rightleftharpoons H-\overset{H}{\overset{|}{\ddot{O}}}: + NO_2^+$$

The nitronium ion

The sulfonation of benzene is carried out using hot, concentrated sulfuric acid. The electrophile under these conditions is either SO_3 or HSO_3^+, depending on the experimental conditions. The HSO_3^+ electrophile is formed from sulfuric acid in the following way:

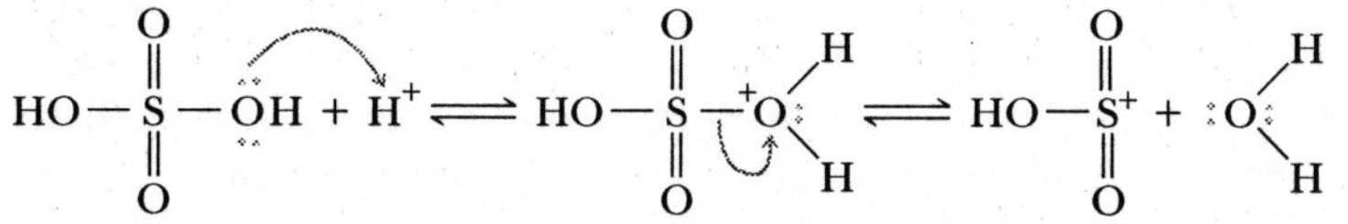

$$HO-\overset{O}{\overset{\|}{\underset{\underset{O}{\|}}{S}}}-\ddot{O}H + H^+ \rightleftharpoons HO-\overset{O}{\overset{\|}{\underset{\underset{O}{\|}}{S}}}-\overset{+}{O}H_2 \rightleftharpoons HO-\overset{O}{\overset{\|}{\underset{\underset{O}{\|}}{S^+}}} + :\ddot{O}H_2$$

Sulfuric acid — The electrophile

EXAMPLE 9.3

Write a stepwise mechanism for the nitration of benzene.

SOLUTION

Step 1: Reaction of the nitronium ion (an electrophile) with the benzene ring (a nucleophile) gives a resonance-stabilized cation intermediate.

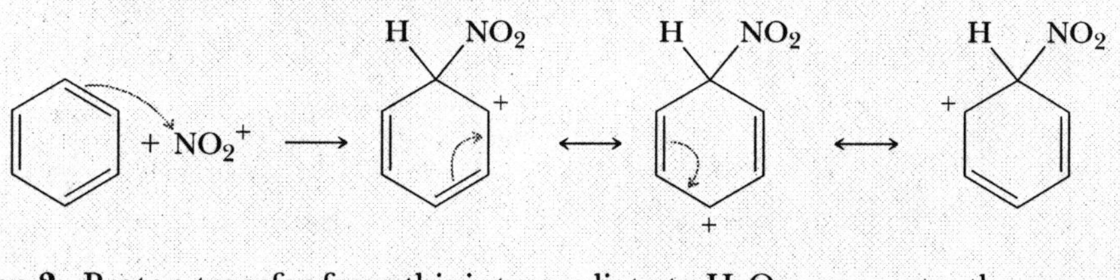

$$C_6H_6 + NO_2^+ \longrightarrow [C_6H_6NO_2]^+ \longleftrightarrow [C_6H_6NO_2]^+ \longleftrightarrow [C_6H_6NO_2]^+$$

Step 2: Proton transfer from this intermediate to H_2O regenerates the aromatic ring and gives nitrobenzene:

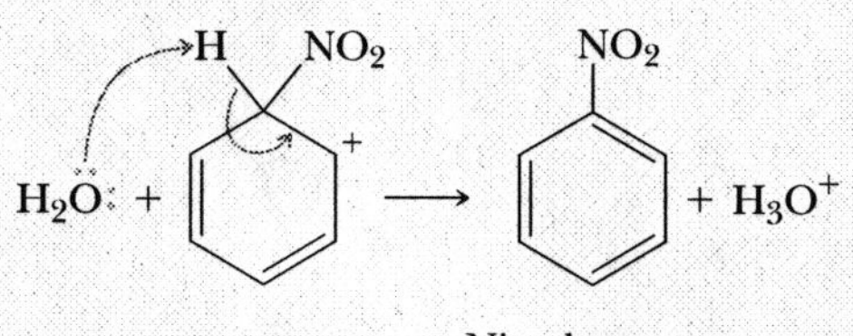

$$H_2\ddot{O}: + [C_6H_6NO_2]^+ \longrightarrow C_6H_5NO_2 + H_3O^+$$

Nitrobenzene

Practice Problem 9.3

Write a stepwise mechanism for the sulfonation of benzene. Use HSO_3^+ as the electrophile.

C. Friedel–Crafts Alkylation

Alkylation of aromatic hydrocarbons was discovered in 1877 by the French chemist Charles Friedel and a visiting American chemist, James Crafts. They discovered that mixing benzene, a haloalkane, and $AlCl_3$ results in the formation of an alkylbenzene and HX. **Friedel–Crafts alkylation** forms a new carbon–carbon bond between benzene and an alkyl group, as illustrated by reaction of benzene with 2-chloropropane in the presence of aluminum chloride:

Benzene + 2-Chloropropane (Isopropyl chloride) $\xrightarrow{AlCl_3}$ Isopropylbenzene (Cumene) + HCl

Friedel–Crafts alkylation is among the most important methods for forming new carbon–carbon bonds to aromatic rings.

Mechanism: Friedel–Crafts Alkylation

Step 1: Reaction of a haloalkane (a Lewis base) with aluminum chloride (a Lewis acid) gives a molecular complex in which aluminum has a negative formal charge and the halogen of the haloalkane has a positive formal charge. Redistribution of electrons in this complex then gives an alkyl carbocation as part of an ion pair:

$$R{-}Cl + AlCl_3 \rightleftharpoons R{-}\overset{+}{Cl}{-}\overset{-}{Al}Cl_3 \rightleftharpoons R^+ \; Cl{-}\overset{-}{Al}Cl_3$$

the electrophile

A molecular complex with a positive charge on chlorine and a negative charge on aluminum

An ion pair containing a carbocation

Step 2: Reaction of the alkyl carbocation with the pi electrons of the aromatic ring gives a resonance-stabilized cation intermediate:

Benzene + R^+ ⟶ (cation intermediate, H and R on same carbon) ⟷ ⟷

The positive charge is delocalized onto three atoms of the ring

Step 3: Proton transfer regenerates the aromatic character of the ring and the Lewis acid catalyst:

$$\text{(cation intermediate)} + Cl{-}\overset{-}{Al}Cl_3 \longrightarrow C_6H_5{-}R + AlCl_3 + H{-}Cl$$

There are two major limitations on Friedel–Crafts alkylations. The first is that it is practical only with stable carbocations, such as 3° and 2° carbocations. The reasons for this limitation are beyond the scope of this text.

The second limitation on Friedel–Crafts alkylation is that it fails altogether on benzene rings bearing one or more strongly electron-withdrawing groups. The following table shows some of these groups:

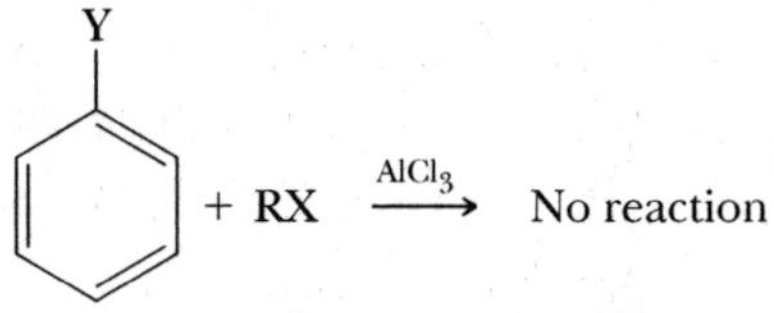

$$C_6H_5Y + RX \xrightarrow{AlCl_3} \text{No reaction}$$

When Y Equals Any of These Groups, the Benzene Ring Does Not Undergo Friedel–Crafts Alkylation

$-CHO$	$-COR$	$-COOH$	$-COOR$	$-CONH_2$
$-SO_3H$	$-C{\equiv}N$	$-NO_2$	$-NR_3^+$	
$-CF_3$	$-CCl_3$			

A common characteristic of the groups listed in the preceding table is that each has either a full or partial positive charge on the atom bonded to the benzene ring. For carbonyl-containing compounds, this partial positive charge arises because of the difference in electronegativity between the carbonyl oxygen and carbon. For $-CF_3$ and $-CCl_3$ groups, the partial positive charge on carbon arises because of the difference in electronegativity between carbon and the halogens bonded to it. In both the nitro group and the trialkylamonium group, there is a positive charge on nitrogen:

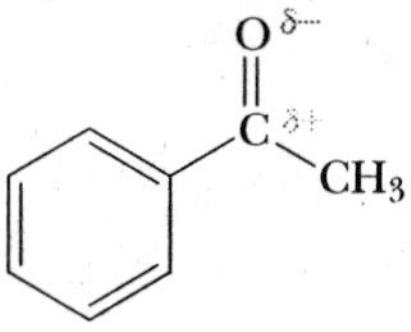

The carbonyl group of a ketone 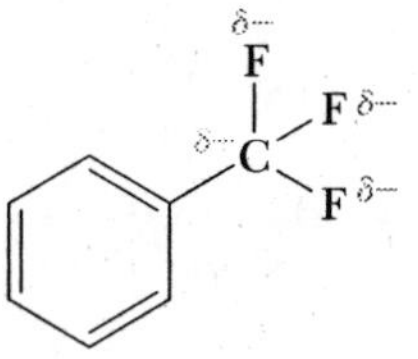 | A trifluoromethyl group 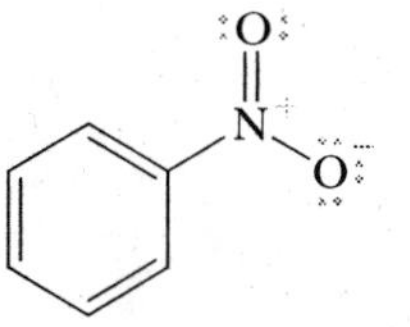 | A nitro group 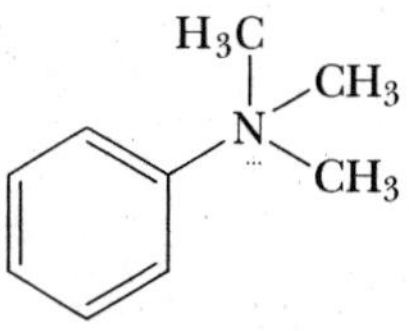 | A trimethylammonium group

D. Friedel–Crafts Acylation

Friedel and Crafts also discovered that treating an aromatic hydrocarbon with an acyl halide (Section 15.2A) in the presence of aluminum chloride gives a ketone. An **acyl halide** is a derivative of a carboxylic acid in which the $-OH$ of the carboxyl group is replaced by a halogen, most commonly chlorine. Acyl halides are also referred to as acid halides. An $RCO-$ group is known as an acyl group; hence, the reaction of an acyl halide with an aromatic hydrocarbon is known as **Friedel–Crafts acylation**, as illustrated by the reaction of benzene and acetyl chloride in the presence of aluminum chloride to give acetophenone:

Acyl halide A derivative of a carboxylic acid in which the $-OH$ of the carboxyl group is replaced by a halogen—most commonly, chlorine.

$$C_6H_6 + CH_3CCl(=O) \xrightarrow{AlCl_3} C_6H_5CCH_3(=O) + HCl$$

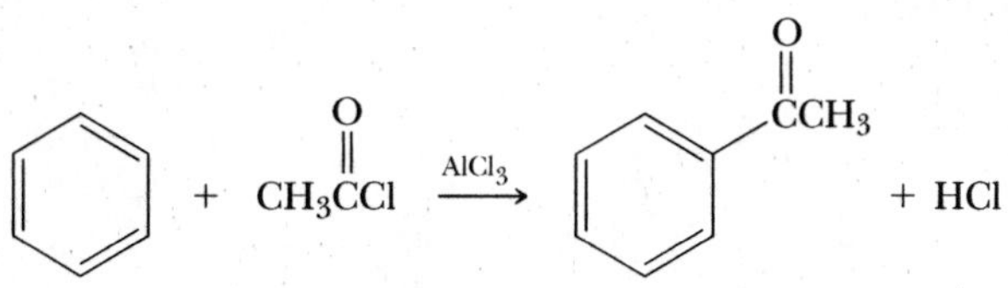

Benzene | Acetyl chloride (an acyl halide) | Acetophenone (a ketone)

In Friedel–Crafts acylations, the electrophile is an acylium ion, generated in the following way:

Mechanism: Friedel–Crafts Acylation—Generation of an Acylium Ion

Reaction between the halogen atom of the acyl chloride (a Lewis base) and aluminum chloride (a Lewis acid) gives a molecular complex. The redistribution of valence electrons in turn gives an ion pair containing an acylium ion:

$$R{-}\overset{O}{\overset{\|}{C}}{-}\ddot{\underset{\cdot\cdot}{Cl}}{:} + Al(Cl)_3 \rightleftharpoons R{-}\overset{O}{\overset{\|}{C}}{-}\overset{+}{Cl}{-}\overset{-}{Al}Cl_3 \rightleftharpoons R{-}\overset{O}{\overset{\|}{C^+}}\ \ {:}\ddot{Cl}{-}\overset{-}{Al}Cl_3$$

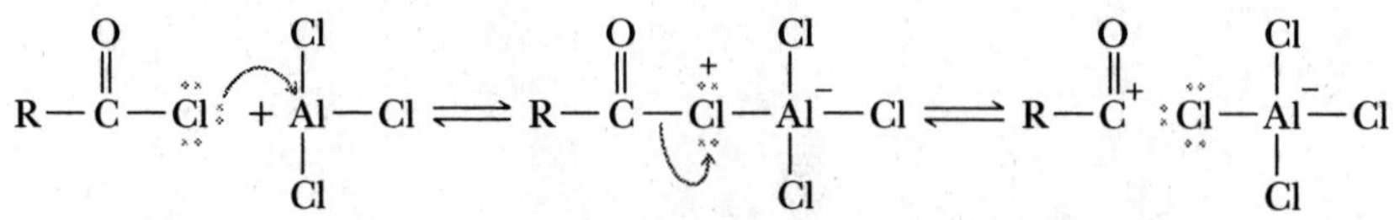

An acyl chloride (a Lewis base) — Aluminum chloride (a Lewis acid) — A molecular complex with a positive charge on chlorine and a negative charge on aluminum — An ion pair containing an acylium ion

EXAMPLE 9.4

Write a structural formula for the product formed by Friedel–Crafts alkylation or acylation of benzene with

(a) $C_6H_5CH_2Cl$
Benzyl chloride

(b) $C_6H_5\overset{O}{\overset{\|}{C}}Cl$
Benzoyl chloride

SOLUTION

(a) Treatment of benzyl chloride with aluminum chloride gives the resonance-stabilized benzyl cation. Reaction of this cation with benzene, followed by loss of H^+, gives diphenylmethane:

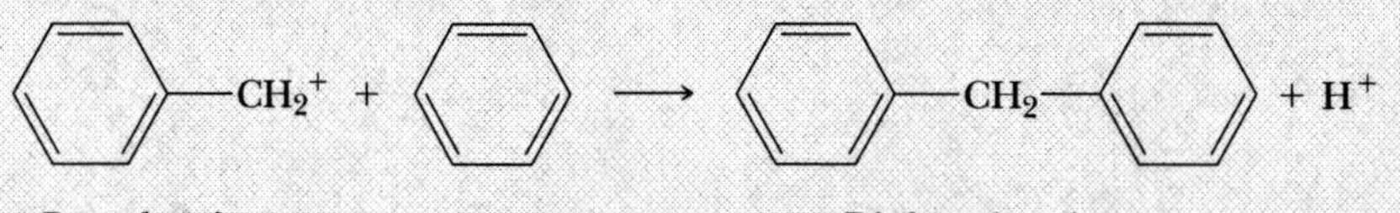

$$C_6H_5{-}CH_2^+ + C_6H_6 \longrightarrow C_6H_5{-}CH_2{-}C_6H_5 + H^+$$

Benzyl cation — Diphenylmethane

(b) Treatment of benzoyl chloride with aluminum chloride gives an acyl cation. Reaction of this cation with benzene, followed by loss of H^+, gives benzophenone:

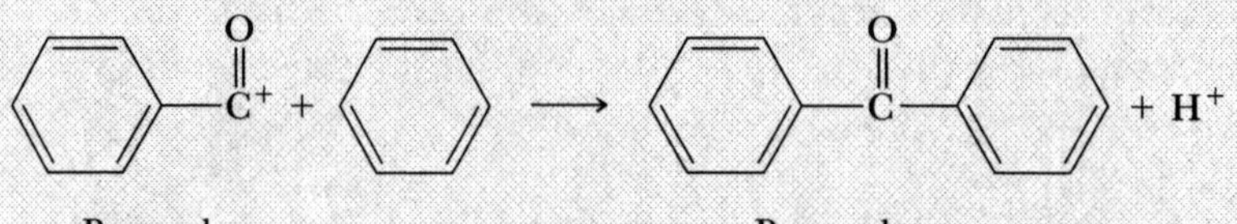

$$C_6H_5{-}\overset{O}{\overset{\|}{C^+}} + C_6H_6 \longrightarrow C_6H_5{-}\overset{O}{\overset{\|}{C}}{-}C_6H_5 + H^+$$

Benzoyl cation — Benzophenone

Practice Problem 9.4

Write a structural formula for the product formed from Friedel–Crafts alkylation or acylation of benzene with

(a) 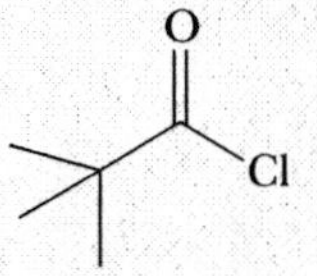 (b) 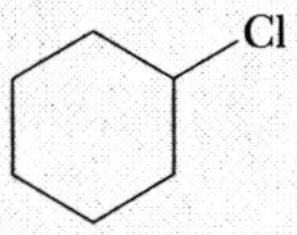 (c)

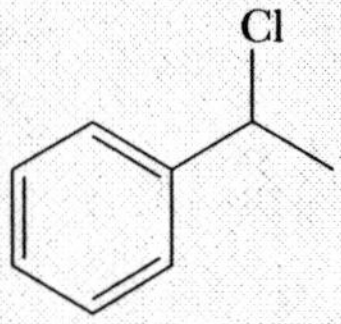

E. Other Electrophilic Aromatic Alkylations

Once it was discovered that Friedel–Crafts alkylations and acylations involve cationic intermediates, chemists realized that other combinations of reagents and catalysts could give the same products. We study two of these reactions in this section: the generation of carbocations from alkenes and from alcohols.

As we saw in Section 5.3A, treatment of an alkene with a strong acid, most commonly H_2SO_4 or H_3PO_4, generates a carbocation. Isopropylbenzene is synthesized industrially by reacting benzene with propene in the presence of an acid catalyst:

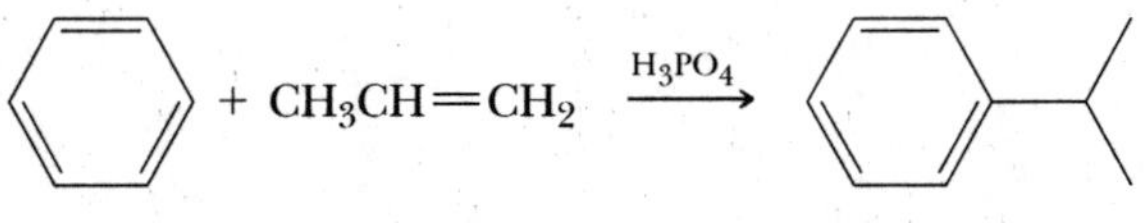

$$\text{Benzene} + CH_3CH{=}CH_2 \xrightarrow{H_3PO_4} \text{Isopropylbenzene}$$

Benzene Propene Isopropylbenzene (Cumene)

Carbocations are also generated by treating an alcohol with H_2SO_4 or H_3PO_4 (Section 8.3E):

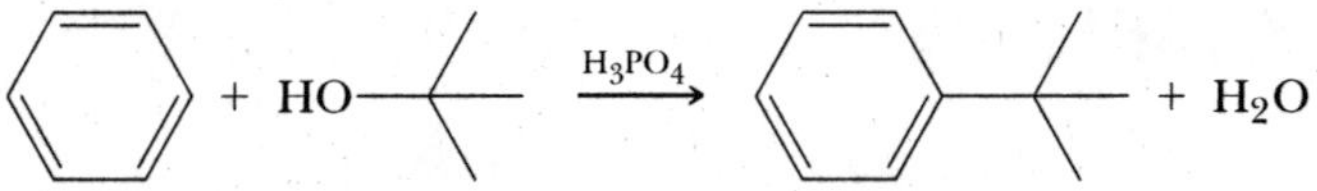

$$\text{Benzene} + HO{-}C(CH_3)_3 \xrightarrow{H_3PO_4} C_6H_5{-}C(CH_3)_3 + H_2O$$

Benzene 2-Methyl-2-phenylpropane (*tert*-Butylbenzene)

EXAMPLE 9.5

Write a mechanism for the formation of isopropylbenzene from benzene and propene in the presence of phosphoric acid.

SOLUTION

Step 1: Proton transfer from phosphoric acid to propene gives the isopropyl cation:

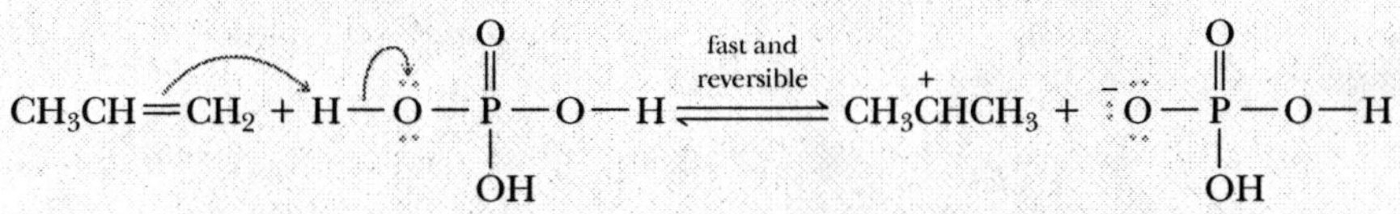

$$CH_3CH{=}CH_2 + H{-}O{-}P(=O)(OH){-}O{-}H \underset{\text{reversible}}{\overset{\text{fast and}}{\rightleftharpoons}} CH_3\overset{+}{C}HCH_3 + {}^{-}O{-}P(=O)(OH){-}O{-}H$$

Step 2: Reaction of the isopropyl cation with benzene gives a resonance-stabilized carbocation intermediate:

$$C_6H_6 + {}^{+}CH(CH_3)_2 \xrightleftharpoons[\text{limiting}]{\text{slow, rate}} [C_6H_6\text{–}H(CH(CH_3)_2)]^{+}$$

Step 3: Proton transfer from this intermediate to dihydrogen phosphate ion gives isopropylbenzene:

$$[C_6H_5(H)CH(CH_3)_2]^{+} + {}^{-}O\text{–}P(=O)(OH)\text{–}O\text{–}H \xrightarrow{\text{fast}} C_6H_5\text{–}CH(CH_3)_2 + H\text{–}O\text{–}P(=O)(OH)\text{–}O\text{–}H$$

Isopropylbenzene

Practice Problem 9.5

Write a mechanism for the formation of *tert*-butylbenzene from benzene and *tert*-butyl alcohol in the presence of phosphoric acid.

9.8 DISUBSTITUTION AND POLYSUBSTITUTION

A. Effects of a Substituent Group on Further Substitution

In the electrophilic aromatic substitution of a monosubstituted benzene, three isomeric products are possible: The new group may be oriented ortho, meta, or para to the existing group. On the basis of a wealth of experimental observations, chemists have made the following generalizations about the manner in which an existing substituent influences further electrophilic aromatic substitution:

1. *Substituents affect the orientation of new groups.* Certain substituents direct a second substituent preferentially to the ortho and para positions; other substituents direct it preferentially to a meta position. In other words, we can classify substituents on a benzene ring as **ortho–para directing** or **meta directing**.
2. *Substituents affect the rate of further substitution.* Certain substituents cause the rate of a second substitution to be greater than that of benzene itself, whereas other substituents cause the rate of a second substitution to be lower than that of benzene. In other words, we can classify groups on a benzene ring as **activating** or **deactivating** toward further substitution.

Ortho–para director Any substituent on a benzene ring that directs electrophilic aromatic substitution preferentially to ortho and para positions.

Meta director Any substituent on a benzene ring that directs electrophilic aromatic substitution preferentially to a meta position.

Activating group Any substituent on a benzene ring that causes the rate of electrophilic aromatic substitution to be greater than that for benzene.

Deactivating group Any substituent on a benzene ring that causes the rate of electrophilic aromatic substitution to be lower than that for benzene.

To see the operation of these directing and activating–deactivating effects, compare, for example, the products and rates of bromination of anisole and nitrobenzene. Bromination of anisole proceeds at a rate considerably greater than that of bromination of benzene (the methoxy group is activating), and the product is a mixture of *o*-bromoanisole and *p*-bromoanisole (the methoxy group is ortho–para directing):

$$\text{Anisole} + Br_2 \xrightarrow{CH_3COOH} o\text{-Bromoanisole } (4\%) + p\text{-Bromoanisole } (96\%) + HBr$$

We see quite another situation in the nitration of nitrobenzene, which proceeds much more slowly than the nitration of benzene itself. (A nitro group is strongly deactivating.) Also, the product consists of approximately 93% of the meta isomer and less than 7% of the ortho and para isomers combined (the nitro group is meta directing):

$$\text{Nitrobenzene} + HNO_3 \xrightarrow[100^\circ C]{H_2SO_4} m\text{-Dinitrobenzene } (93\%) + o\text{-Dinitrobenzene} + p\text{-Dinitrobenzene} + H_2O$$

(o-Dinitrobenzene and p-Dinitrobenzene: Less than 7% combined)

Table 9.1 lists the directing and activating–deactivating effects for the major functional groups with which we are concerned in this text.

TABLE 9.1 Effects of Substituents on Further Electrophilic Aromatic Substitution

Ortho–Para Directing	strongly activating	$-NH_2$	$-NHR$	$-NR_2$	$-OH$	$-OR$	
Ortho–Para Directing	moderately activating	$-NHC(=O)R$	$-NHC(=O)Ar$	$-OC(=O)R$	$-OC(=O)Ar$		
Ortho–Para Directing	weakly activating	$-R$	$-C_6H_5$ (phenyl)				
Ortho–Para Directing	weakly deactivating	$-F$	$-Cl$	$-Br$	$-I$		
Meta Directing	moderately deactivating	$-C(=O)H$	$-C(=O)R$	$-C(=O)OH$	$-C(=O)OR$	$-C(=O)NH_2$	$-S(=O)_2OH$
Meta Directing	strongly deactivating	$-NO_2$	$-NH_3^+$	$-CF_3$	$-CCl_3$		

Relative importance in directing further substitution (increases upward)

If we compare these ortho–para and meta directors for structural similarities and differences, we can make the following generalizations:

1. Alkyl groups, phenyl groups, and substituents in which the atom bonded to the ring has an unshared pair of electrons are ortho–para directing. All other substituents are meta directing.
2. Except for the halogens, all ortho–para directing groups are activating toward further substitution. The halogens are weakly deactivating.
3. All meta directing groups carry either a partial or full positive charge on the atom bonded to the ring.

We can illustrate the usefulness of these generalizations by considering the synthesis of two different disubstituted derivatives of benzene. Suppose we wish to prepare *m*-bromonitrobenzene from benzene. This conversion can be carried out in two steps: nitration and bromination. If the steps are carried out in just that order, the major product is indeed *m*-bromonitrobenzene. The nitro group is a meta director and directs bromination to a meta position:

NO_2 NO_2

$\xrightarrow[H_2SO_4]{HNO_3}$ $\xrightarrow[FeCl_3]{Br_2}$

Br

Nitrobenzene *m*-Bromonitrobenzene

If, however, we reverse the order of the steps and first form bromobenzene, we now have an ortho–para directing group on the ring. Nitration of bromobenzene then takes place preferentially at the ortho and para positions, with the para product predominating:

Br Br Br

NO_2

$\xrightarrow[FeCl_3]{Br_2}$ $\xrightarrow[H_2SO_4]{HNO_3}$ +

NO_2

Bromobenzene *o*-Bromonitrobenzene *p*-Bromonitrobenzene

As another example of the importance of order in electrophilic aromatic substitutions, consider the conversion of toluene to nitrobenzoic acid. The nitro group can be introduced with a nitrating mixture of nitric and sulfuric acids. The carboxyl group can be produced by oxidation of the methyl group (Section 9.5).

CH_3 COOH

$\xrightarrow[H_2SO_4]{HNO_3}$ $\xrightarrow[H_2SO_4]{K_2Cr_2O_7}$

CH_3

NO_2 NO_2

4-Nitrotoluene 4-Nitrobenzoic acid

COOH COOH

Toluene

$\xrightarrow[H_2SO_4]{K_2Cr_2O_7}$ $\xrightarrow[H_2SO_4]{HNO_3}$

NO_2

Benzoic acid 3-Nitrobenzoic acid

Nitration of toluene yields a product with the two substituents para to each other, whereas nitration of benzoic acid yields a product with the substituents meta to each other. Again, we see that the order in which the reactions are performed is critical.

Note that, in this last example, we show nitration of toluene producing only the para isomer. In practice because methyl is an ortho–para directing group, both ortho and para isomers are formed. In problems in which we ask you to prepare one or the other of these isomers, we assume that both form and that there are physical methods by which you can separate them and obtain the desired isomer.

EXAMPLE 9.6

Complete the following electrophilic aromatic substitution reactions. Where you predict meta substitution, show only the meta product. Where you predict ortho–para substitution, show both products:

(a) OCH_3-benzene + 2-chloropropane $\xrightarrow{AlCl_3}$ (b) SO_3H-benzene + HNO_3 $\xrightarrow[H_2SO_4]{}$

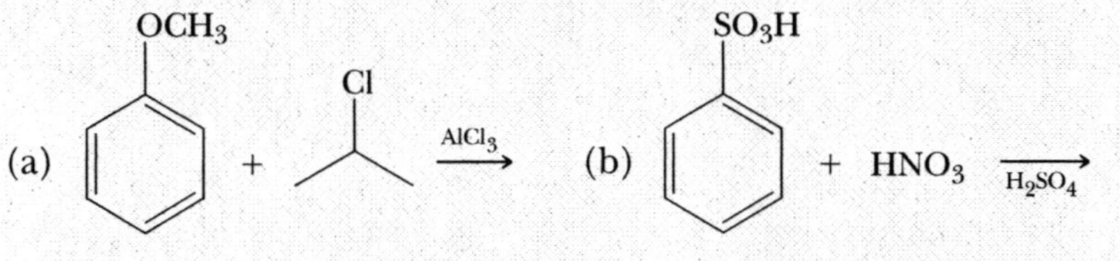

SOLUTION

The methoxyl group in (a) is ortho–para directing and strongly activating. The sulfonic acid group in (b) is meta directing and moderately deactivating:

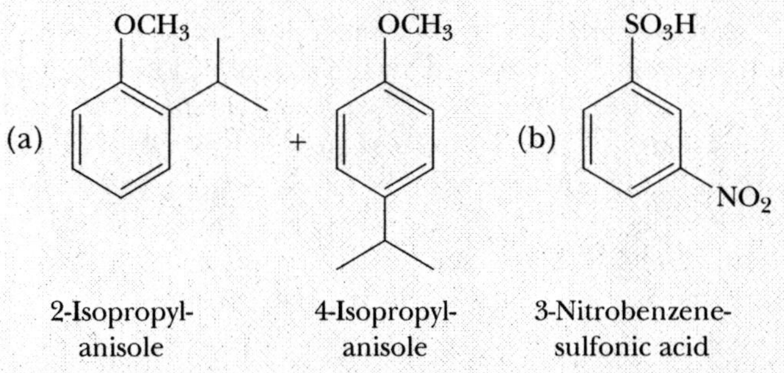

(a) 2-Isopropylanisole + 4-Isopropylanisole (b) 3-Nitrobenzenesulfonic acid

Practice Problem 9.6

Complete the following electrophilic aromatic substitution reactions. Where you predict meta substitution, show only the meta product. Where you predict ortho–para substitution, show both products:

(a) methyl benzoate + HNO_3 $\xrightarrow{H_2SO_4}$

(b) phenyl acetate + HNO_3 $\xrightarrow{H_2SO_4}$

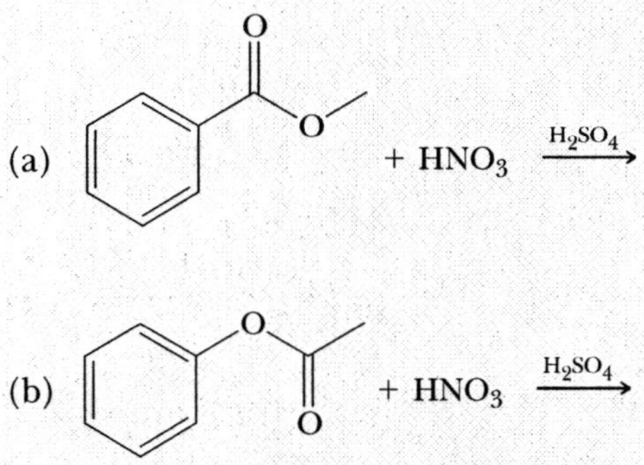

B. Theory of Directing Effects

As we have just seen, a group on an aromatic ring exerts a major effect on the patterns of further substitution. We can make these three generalizations:

1. If there is a lone pair of electrons on the atom bonded to the ring, the group is an ortho–para director.
2. If there is a full or partial positive charge on the atom bonded to the ring, the group is a meta director.
3. Alkyl groups are ortho–para directors.

We account for these patterns by means of the general mechanism for electrophilic aromatic substitution first presented in Section 9.6. Let us extend that mechanism to consider how a group already present on the ring might affect the relative stabilities of cation intermediates formed during a second substitution reaction.

We begin with the fact that the rate of electrophilic aromatic substitution is determined by the slowest step in the mechanism, which, in almost every reaction of an electrophile with the aromatic ring, is attack of the electrophile on the ring to give a resonance-stabilized cation intermediate. Thus, we must determine which of the alternative carbocation intermediates (that for ortho–para substitution or that for meta substitution) is the more stable. That is, which of the alternative cationic intermediates has the lower activation energy for its formation.

Nitration of Anisole

The rate-determining step in nitration is reaction of the nitronium ion with the aromatic ring to produce a resonance-stabilized cation intermediate. Figure 9.4 shows the cation intermediate formed by reaction meta to the methoxy group. The figure also shows the cationic intermediate formed by reaction para to the methoxy group. The intermediate formed by reaction at a meta position is a hybrid of three major contributing structures: (a), (b), and (c). These three are the only important contributing structures we can draw for reaction at a meta position.

The cationic intermediate formed by reaction at the para position is a hybrid of four major contributing structures: (d), (e), (f), and (g). What is important about structure (f) is that all atoms in it have complete octets, which means that this structure contributes more to the hybrid than structures (d), (e), or (g). Because the cation formed by reaction at an ortho or para position on anisole has a greater resonance stabilization and, hence, a lower activation energy for its formation, nitration of anisole occurs preferentially in the ortho and para positions.

Figure 9.4
Nitration of anisole. Reaction of the electrophile meta and para to a methoxy group. Regeneration of the aromatic ring is shown from the rightmost contributing structure in each case.

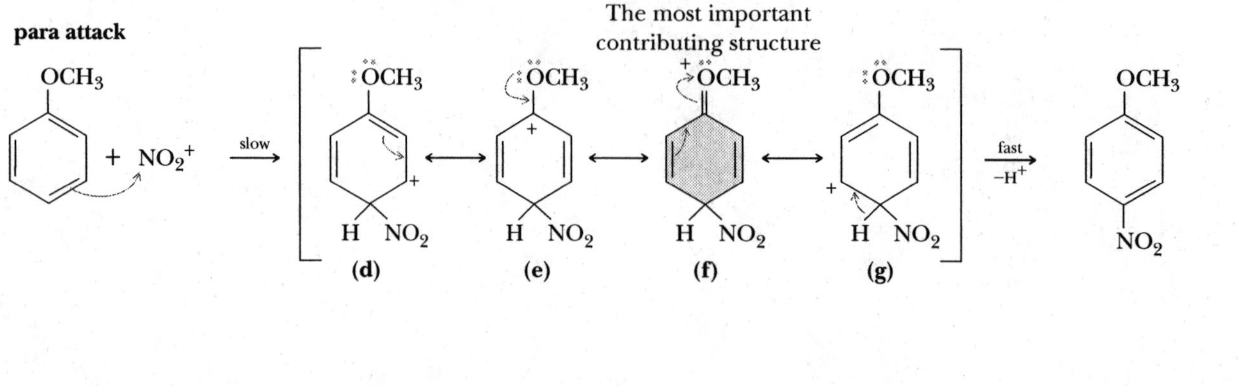

meta attack

OCH_3 + NO_2^+ —slow→ [(a) ⟷ (b) ⟷ (c)] —fast, $-H^+$→ (product with OCH_3 and NO_2 meta)

para attack

The most important contributing structure

OCH_3 + NO_2^+ —slow→ [(d) ⟷ (e) ⟷ (f) ⟷ (g)] —fast, $-H^+$→ (product with OCH_3 and NO_2 para)

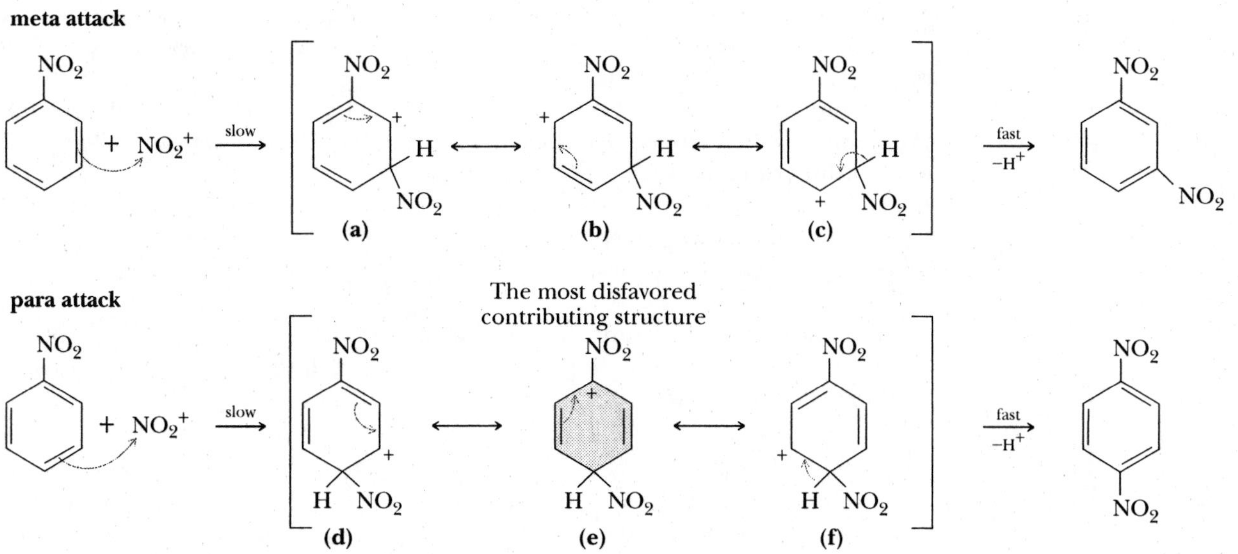

meta attack

NO_2-benzene + NO_2^+ —slow→ [(a) ⟷ (b) ⟷ (c)] —fast, $-H^+$→ m-dinitrobenzene

para attack

The most disfavored contributing structure

NO_2-benzene + NO_2^+ —slow→ [(d) ⟷ (e) ⟷ (f)] —fast, $-H^+$→ p-dinitrobenzene

Figure 9.5
Nitration of nitrobenzene. Reaction of the electrophile meta and para to a nitro group. Regeneration of the aromatic ring is shown from the rightmost contributing structure in each case.

Nitration of Nitrobenzene

Figure 9.5 shows the resonance-stabilized cation intermediates formed by reaction of the nitronium ion meta to the nitro group and also para to it.

Each cation in the figure is a hybrid of three contributing structures; no additional ones can be drawn. Now we must compare the relative resonance stabilizations of each hybrid. If we draw a Lewis structure for the nitro group showing the positive formal charge on nitrogen, we see that contributing structure (e) places positive charges on adjacent atoms:

positive charges on adjacent atoms destabilizes the intermediate

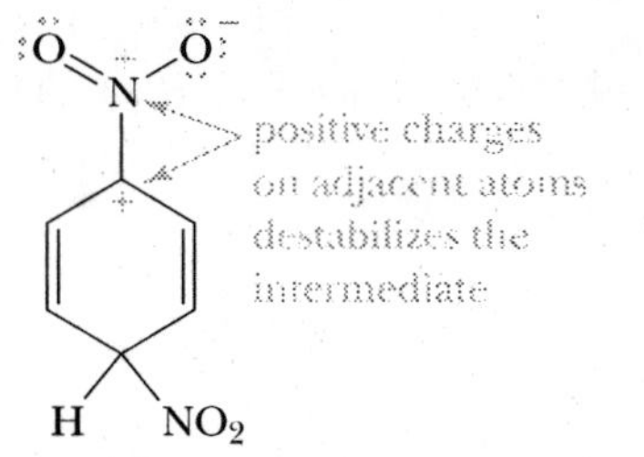

(e)

Because of the electrostatic repulsion thus generated, structure (e) makes only a negligible contribution to the hybrid. None of the contributing structures for reaction at a meta position places positive charges on adjacent atoms. As a consequence, resonance stabilization of the cation formed by reaction at a meta position is greater than that for the cation formed by reaction at a para (or ortho) position. Stated alternatively, the activation energy for reaction at a meta position is less than that for reaction at a para position.

A comparison of the entries in Table 9.1 shows that almost all ortho–para directing groups have an unshared pair of electrons on the atom bonded to the aromatic ring. Thus, the directing effect of most of these groups is due primarily to the ability of the atom bonded to the ring to delocalize further the positive charge on the cation intermediate.

The fact that alkyl groups are also ortho–para directing indicates that they, too, help to stabilize the cation intermediate. In Section 5.3A, we saw that alkyl groups stabilize carbocation intermediates and that the order of stability of carbocations is

3° > 2° > 1° > methyl. Just as alkyl groups stabilize the cation intermediates formed in reactions of alkenes, they also stabilize the carbocation intermediates formed in electrophilic aromatic substitutions.

To summarize, any substituent on an aromatic ring that further stabilizes the cation intermediate directs ortho–para, and any group that destabilizes the cation intermediate directs meta.

EXAMPLE 9.7

Draw contributing structures formed during the para nitration of chlorobenzene, and show how chlorine participates in directing the incoming nitronium ion to ortho–para positions.

SOLUTION

Contributing structures (a), (b), and (d) place the positive charge on atoms of the ring, while contributing structure (c) places it on chlorine and thus creates additional resonance stabilization for the cation intermediate:

(a) (b) (c) (d)

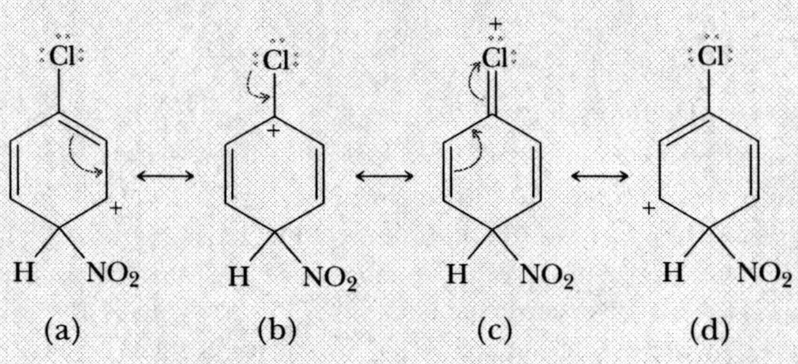

Practice Problem 9.7

Because the electronegativity of oxygen is greater than that of carbon, the carbon of a carbonyl group bears a partial positive charge, and its oxygen bears a partial negative charge. Using this information, show that a carbonyl group is meta directing:

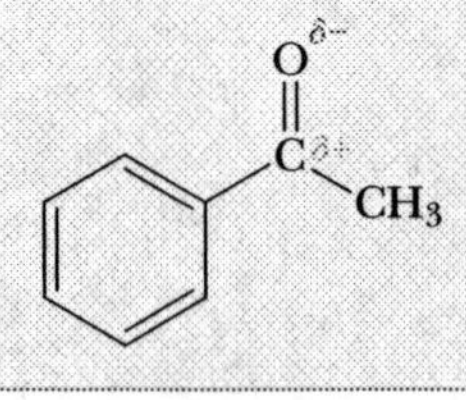

C. Theory of Activating–Deactivating Effects

We account for the activating–deactivating effects of substituent groups by a combination of resonance and inductive effects:

1. Any resonance effect, such as that of $-NH_2$, $-OH$, and $-OR$, which delocalizes the positive charge of the cation intermediate lowers the activation energy for its formation and is activating toward further electrophilic aromatic substitution. That is, these groups increase the rate of electrophilic aromatic substitution, compared with the rate at which benzene itself reacts.

2. Any resonance or inductive effect, such as that of $—NO_2$, $—C{=}O$, $—SO_3H$, $—NR_3^+$, $—CCl_3$, and $—CF_3$, which decreases electron density on the ring deactivates the ring to further substitution. That is, these groups decrease the rate of further electrophilic aromatic substitution, compared with the rate at which benzene itself reacts.
3. Any inductive effect (such as that of $—CH_3$ or another alkyl group) which releases electron density toward the ring activates the ring toward further substitution.

In the case of the halogens, the resonance and inductive effects operate in opposite directions. As Table 9.1 shows, the halogens are ortho–para directing, but, unlike other ortho–para directors listed in the table, the halogens are weakly deactivating. These observations can be accounted for in the following way.

1. *The inductive effect of halogens.* The halogens are more electronegative than carbon and have an electron-withdrawing inductive effect. Aryl halides, therefore, react more slowly in electrophilic aromatic substitution than benzene does.
2. *The resonance effect of halogens.* A halogen ortho or para to the site of electrophilic attack stabilizes the cation intermediate by delocalization of the positive charge:

Cl + E⁺ ⟶ [Cl (+) H E ⟷ +Cl= H E]

EXAMPLE 9.8

Predict the product of each electrophilic aromatic substitution:

(a) OH, NO₂ + Br_2 ⟶ (b) COOH, CH₃ + HNO_3 $\xrightarrow{H_2SO_4}$

SOLUTION

The key to predicting the orientation of further substitution on a disubstituted arene is that ortho–para directing groups activate the ring toward further substitution, whereas meta directing groups deactivate it. This means that, when there is competition between ortho–para directing and meta directing groups, the ortho–para group wins.

(a) The ortho–para directing and activating $—OH$ group determines the position of bromination. Bromination between the $—OH$ and $—NO_2$ groups is only a minor product, because of steric hindrance to attack of bromine at this position:

OH, NO₂ + Br_2 ⟶ OH, NO₂, Br + Br, OH, NO₂ + HBr

(b) The ortho–para directing and activating methyl group determines the position of nitration:

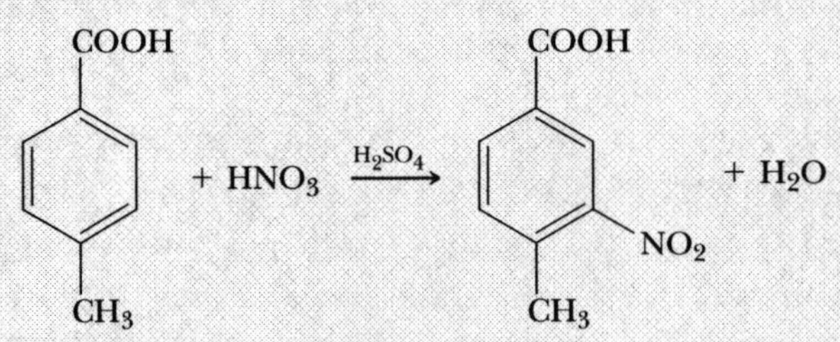

COOH, CH_3 (benzene ring) + HNO_3 $\xrightarrow{H_2SO_4}$ COOH, NO_2, CH_3 (benzene ring) + H_2O

Practice Problem 9.8

Predict the product of treating each compound with HNO_3/H_2SO_4:

(a) CH_3, Cl (benzene ring)

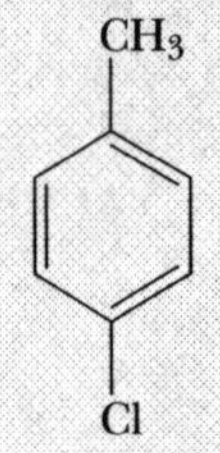

(b) COOH, NO_2 (benzene ring)

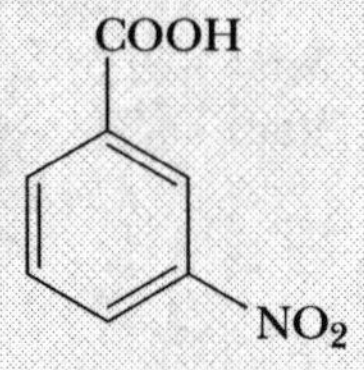

9.9 PHENOLS

A. Structure and Nomenclature

Phenol A compound that contains an —OH bonded to a benzene ring.

The functional group of a **phenol** is a hydroxyl group bonded to a benzene ring. We name substituted phenols either as derivatives of phenol or by common names:

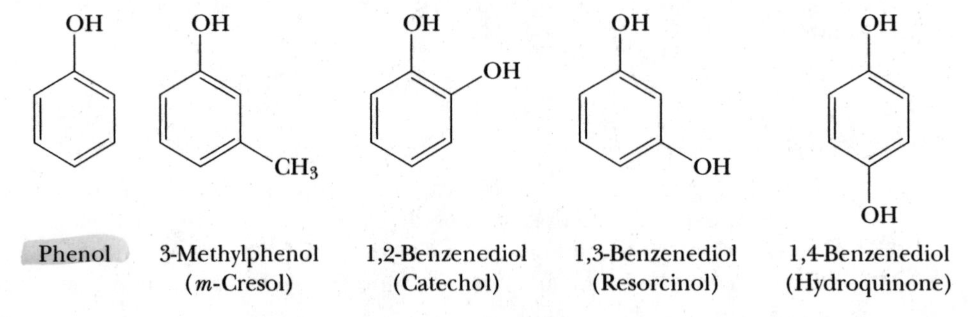

Phenol | 3-Methylphenol (*m*-Cresol) | 1,2-Benzenediol (Catechol) | 1,3-Benzenediol (Resorcinol) | 1,4-Benzenediol (Hydroquinone)

Phenols are widely distributed in nature. Phenol itself and the isomeric cresols (*o*-, *m*-, and *p*-cresol) are found in coal tar. Thymol and vanillin are important constituents of thyme and vanilla beans, respectively:

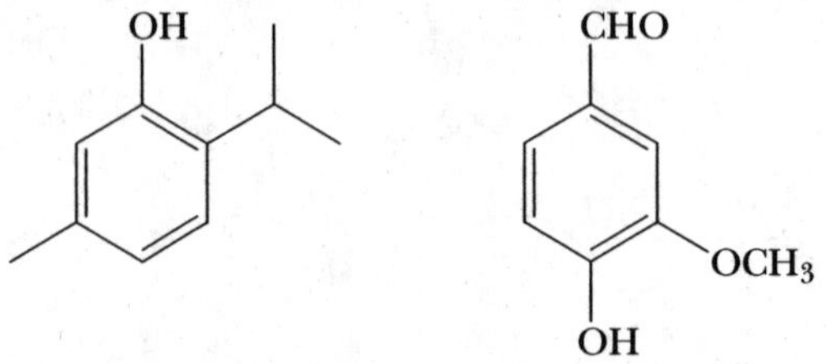

OH (thymol structure) — CHO, OCH_3, OH (vanillin structure)

2-Isopropyl-5-methylphenol (Thymol) | 4-Hydroxy-3-methoxybenzaldehyde (Vanillin)

Thymol is a constituent of garden thyme, Thymus vulgaris. *(Wally Eberhart/ Visuals Unlimited)*

CHEMICAL CONNECTIONS 9B

Capsaicin, for Those Who Like It Hot

Capsaicin, the pungent principle from the fruit of various peppers (*Capsicum* and Solanaceae), was isolated in 1876, and its structure was determined in 1919:

Capsaicin
(from various types of peppers)

The inflammatory properties of capsaicin are well known; the human tongue can detect as little as one drop of it in 5 L of water. Many of us are familiar with the burning sensation in the mouth and sudden tearing in the eyes caused by a good dose of hot chili peppers. Capsaicin-containing extracts from these flaming foods are also used in sprays to ward off dogs or other animals that might nip at your heels while you are running or cycling.

Ironically, capsaicin is able to cause pain and relieve it as well. Currently, two capsaicin-containing creams, Mioton and Zostrix®, are prescribed to treat the burning pain associated with postherpetic neuralgia, a complication of shingles. They are also prescribed for diabetics, to relieve persistent foot and leg pain.

The mechanism by which capsaicin relieves pain is not fully understood. It has been suggested that, after it is applied, the nerve endings in the area responsible for the transmission of pain remain temporarily numb. Capsaicin remains bound to specific receptor sites on these pain-transmitting neurons, blocking them from further action. Eventually, capsaicin is removed from the receptor sites, but in the meantime, its presence provides needed relief from pain.

Phenol, or carbolic acid, as it was once called, is a low-melting solid that is only slightly soluble in water. In sufficiently high concentrations, it is corrosive to all kinds of cells. In dilute solutions, phenol has some antiseptic properties and was introduced into the practice of surgery by Joseph Lister, who demonstrated his technique of aseptic surgery in the surgical theater of the University of Glasgow School of Medicine in 1865. Nowadays, phenol has been replaced by antiseptics that are both more powerful and have fewer undesirable side effects. Among these is hexylresorcinol, which is widely used in nonprescription preparations as a mild antiseptic and disinfectant.

Poison Ivy. *(Charles D. Winters)*

Hexylresorcinol   Eugenol   Urushiol

Eugenol, which can be isolated from the flower buds (cloves) of *Eugenia aromatica*, is used as a dental antiseptic and analgesic. Urushiol is the main component in the irritating oil of poison ivy.

B. Acidity of Phenols

Phenols and alcohols both contain an —OH group. We group phenols as a separate class of compounds, however, because their chemical properties are quite different from those of alcohols. One of the most important of these differences is that phenols are significantly more acidic than are alcohols. Indeed, the acid ionization constant for phenol is 10^6 times larger than that of ethanol!

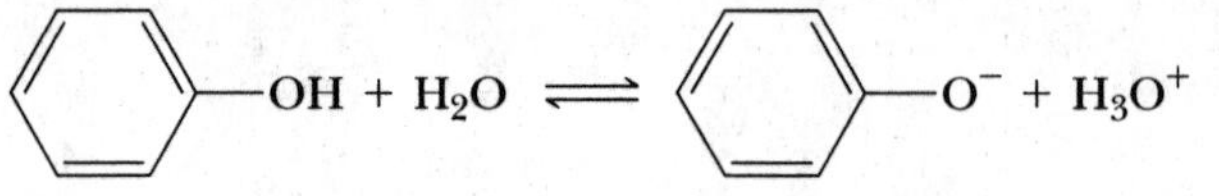

$$C_6H_5\text{—}OH + H_2O \rightleftharpoons C_6H_5\text{—}O^- + H_3O^+ \quad K_a = 1.1 \times 10^{-10} \quad pK_a = 9.95$$

Phenol — Phenoxide ion

$$CH_3CH_2OH + H_2O \rightleftharpoons CH_3CH_2O^- + H_3O^+ \quad K_a = 1.3 \times 10^{-16} \quad pK_a = 15.9$$

Ethanol — Ethoxide ion

Another way to compare the relative acid strengths of ethanol and phenol is to look at the hydrogen ion concentration and pH of a 0.1-M aqueous solution of each (Table 9.2). For comparison, the hydrogen ion concentration and pH of 0.1 M HCl are also included.

TABLE 9.2 Relative Acidities of 0.1-M Solutions of Ethanol, Phenol, and HCl

Acid Ionization Equation	$[H^+]$	pH
$CH_3CH_2OH + H_2O \rightleftharpoons CH_3CH_2O^- + H_3O^+$	1×10^{-7}	7.0
$C_6H_5OH + H_2O \rightleftharpoons C_6H_5O^- + H_3O^+$	3.3×10^{-6}	5.4
$HCl + H_2O \rightleftharpoons Cl^- + H_3O^+$	0.1	1.0

In aqueous solution, alcohols are neutral substances, and the hydrogen ion concentration of 0.1 M ethanol is the same as that of pure water. A 0.1-M solution of phenol is slightly acidic and has a pH of 5.4. By contrast, 0.1 M HCl, a strong acid (completely ionized in aqueous solution), has a pH of 1.0.

The greater acidity of phenols compared with alcohols results from the greater stability of the phenoxide ion compared with an alkoxide ion. The negative charge on the phenoxide ion is delocalized by resonance. The two contributing structures on the left for the phenoxide ion place the negative charge on oxygen, while the three on the right place the negative charge on the ortho and para positions of the ring. Thus, in the resonance hybrid, the negative charge of the phenoxide ion is delocalized over four atoms, which stabilizes the phenoxide ion realtive to an alkoxide ion, for which no delocalization is possible:

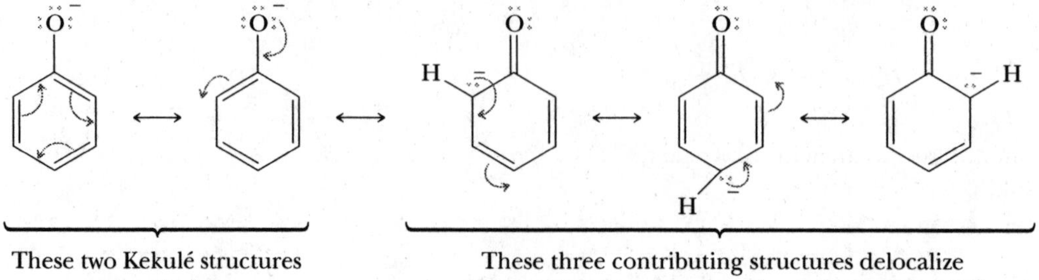

These two Kekulé structures are equivalent

These three contributing structures delocalize the negative charge onto carbon atoms of the ring

Note that, although the resonance model gives us a way of understanding why phenol is a stronger acid than ethanol, it does not provide us with any quantitative means of predicting just how much stronger an acid it might be. To find out how much stronger one acid is than another, we must determine their pK_a values experimentally and compare them.

Ring substituents, particularly halogen and nitro groups, have marked effects on the acidities of phenols through a combination of inductive and resonance effects. Because the halogens are more electronegative than carbon, they withdraw electron density from the aromatic ring, weaken the O—H bond, and stabilize the phenoxide ion. Nitro groups also withdraw electron density from the ring, weaken the O—H bond, stabilize the phenoxide ion, and thus increase acidity.

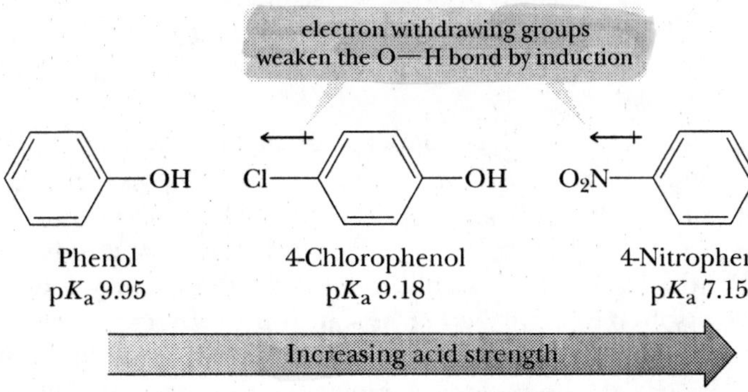

electron withdrawing groups weaken the O—H bond by induction

Phenol	4-Chlorophenol	4-Nitrophenol
pK_a 9.95	pK_a 9.18	pK_a 7.15

Increasing acid strength →

EXAMPLE 9.9

Arrange these compounds in order of increasing acidity: 2,4-dinitrophenol, phenol, and benzyl alcohol.

SOLUTION

Benzyl alcohol, a primary alcohol, has a pK_a of approximately 16–18 (Section 8.3A). The pK_a of phenol is 9.95. Nitro groups are electron withdrawing and increase the acidity of the phenolic —OH group. In order of increasing acidity, these compounds are:

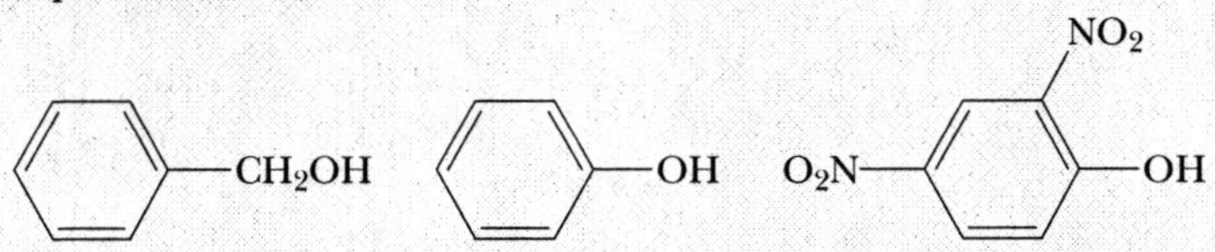

Benzyl alcohol	Phenol	2,4-Dinitrophenol
pK_a 16–18	pK_a 9.95	pK_a 3.96

Practice Problem 9.9

Arrange these compounds in order of increasing acidity: 2,4-dichlorophenol, phenol, cyclohexanol.

C. Acid–Base Reactions of Phenols

Phenols are weak acids and react with strong bases, such as NaOH, to form water-soluble salts:

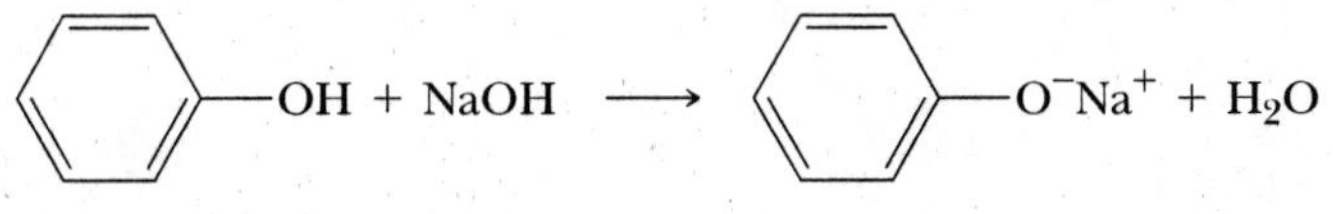

$$C_6H_5OH + NaOH \longrightarrow C_6H_5O^-Na^+ + H_2O$$

Phenol	Sodium hydroxide	Sodium phenoxide	Water
pK_a 9.95			pK_a 15.7
(stronger acid)	(stronger base)	(weaker base)	(weaker acid)

Most phenols do not react with weaker bases, such as sodium bicarbonate; and do not dissolve in aqueous sodium bicarbonate. Carbonic acid is a stronger acid than most phenols and, consequently, the equilibrium for their reaction with bicarbonate ion lies far to the left (see Section 2.5):

$$C_6H_5\text{-}OH + NaHCO_3 \rightleftharpoons C_6H_5\text{-}O^-Na^+ + H_2CO_3$$

Phenol	Sodium	Sodium	Carbonic acid
pK_a 9.95	bicarbonate	phenoxide	pK_a 6.36
(weaker acid)	(weaker base)	(stronger base)	(stronger acid)

The fact that phenols are weakly acidic, whereas alcohols are neutral, provides a convenient way to separate phenols from water-insoluble alcohols. Suppose that we want to separate 4-methylphenol from cyclohexanol. Each is only slightly soluble in water; therefore, they cannot be separated on the basis of their water solubility. They can be separated, however, on the basis of their difference in acidity. First, the mixture of the two is dissolved in diethyl ether or some other water-immiscible solvent. Next, the ether solution is placed in a separatory funnel and shaken with dilute aqueous NaOH. Under these conditions, 4-methylphenol reacts with NaOH to give sodium 4-methylphenoxide, a water-soluble salt. The upper layer in the separatory funnel is now diethyl ether (density 0.74 g/cm^3), containing only dissolved cyclohexanol. The lower aqueous layer contains dissolved sodium 4-methylphenoxide. The layers are separated, and distillation of the ether (bp 35°C) leaves pure cyclohexanol (bp 161°C). Acidification of the aqueous phase with 0.1 M HCl or another strong acid converts sodium 4-methylphenoxide to 4-methylphenol, which is insoluble in water and can be extracted with ether and recovered in pure form. The following flowchart summarizes these experimental steps:

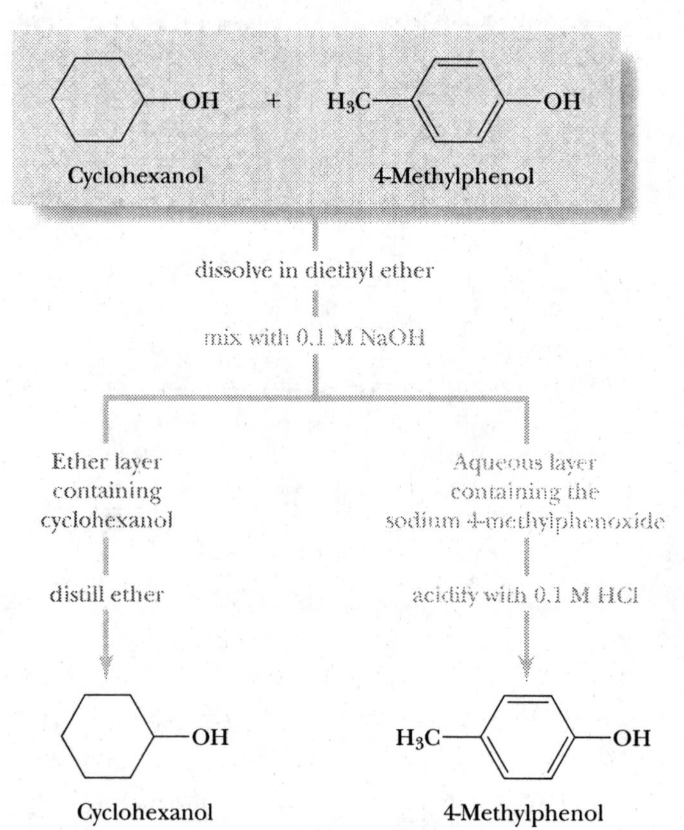

D. Phenols as Antioxidants

An important reaction in living systems, foods, and other materials that contain carbon–carbon double bonds is **autoxidation**—that is, oxidation requiring oxygen and no other reactant. If you open a bottle of cooking oil that has stood for a long time, you will notice a hiss of air entering the bottle. This sound occurs because the consumption of oxygen by autoxidation of the oil creates a negative pressure inside the bottle.

Cooking oils contain esters of polyunsaturated fatty acids. You need not worry now about what esters are; we will discuss them in Chapter 15. The important point here is that all vegetable oils contain fatty acids with long hydrocarbon chains, many of which have one or more carbon–carbon double bonds. (See Problem 4.44 for the structures of three of these fatty acids.) Autoxidation takes place at a carbon adjacent to a double bond—that is, at an **allylic carbon**.

Autoxidation is a radical chain process that converts an R—H group into an R—O—O—H group, called a *hydroperoxide*. The process begins by removing a hydrogen atom, together with one of its electrons (H·), from an allylic carbon. The carbon losing the H· now has only seven electrons in its valence shell, one of which is unpaired. An atom or molecule with an unpaired electron is called a **radical**.

Step 1: *Chain Initiation—Formation of a Radical from a Nonradical Compound* The removal of a hydrogen atom (H·) gives an allylic radical:

$$-CH_2CH{=}CH{-}\underset{}{\overset{H}{\overset{|}{C}}}H- \xrightarrow[\text{or heat}]{\text{light}} -CH_2CH{=}CH{-}\dot{C}H-$$

Section of a fatty acid hydrocarbon chain — An allylic radical

Step 2a: *Chain Propagation—Reaction of a Radical to Form a New Radical* The allylic radical reacts with oxygen, itself a diradical, to form a hydroperoxy radical. The new covalent bond of the hydroperoxy radical forms by the combination of one electron from the allylic radical and one electron from the oxygen diradical:

$$-CH_2CH{=}CH{-}\dot{C}H- + \cdot O{-}O\cdot \longrightarrow -CH_2CH{=}CH{-}\overset{O-O\cdot}{\overset{|}{C}}H-$$

Oxygen is a diradical — A hydroperoxy radical

Step 2b: *Chain Propagation—Reaction of a Radical to Form a New Radical* The hydroperoxy radical removes an allylic hydrogen atom (H·) from a new fatty acid hydrocarbon chain to complete the formation of a hydroperoxide and, at the same time, produce a new allylic radical:

$$-CH_2CH{=}CH{-}\overset{O-O\cdot}{\overset{|}{C}}H- + -CH_2CH{=}CH{-}\overset{H}{\overset{|}{C}}H- \longrightarrow$$

Section of a new fatty acid hydrocarbon chain

$$-CH_2CH{=}CH{-}\overset{O-O-H}{\overset{|}{C}}H{-}CH_2- + -CH_2CH{=}CH{-}\dot{C}H-$$

A hydroperoxide — A new allylic radical

The most important point about the pair of chain propagation steps is that they form a continuous cycle of reactions. The new radical formed in Step 2b next reacts

Butylated hydroxytoluene (BHT) is often used as an antioxidant in baked goods to "retard spoilage." *(Charles D. Winters)*

with another molecule of O_2 in Step 2a to give a new hydroperoxy radical, which then reacts with a new hydrocarbon chain to repeat Step 2b, and so forth. This cycle of propagation steps repeats over and over in a chain reaction. Thus, once a radical is generated in Step 1, the cycle of propagation steps may repeat many thousands of times, generating thousands and thousands of hydroperoxide molecules. The number of times the cycle of chain propagation steps repeats is called the **chain length.**

Hydroperoxides themselves are unstable and, under biological conditions, degrade to short-chain aldehydes and carboxylic acids with unpleasant "rancid" smells. These odors may be familiar to you if you have ever smelled old cooking oil or aged foods that contain polyunsaturated fats or oils. A similar formation of hydroperoxides in the low-density lipoproteins deposited on the walls of arteries leads to cardiovascular disease in humans. In addition, many effects of aging are thought to be the result of the formation and subsequent degradation of hydroperoxides.

Fortunately, nature has developed a series of defenses, including the phenol vitamin E, ascorbic acid (vitamin C), and glutathione, against the formation of destructive, hydroperoxides. The compounds that defend against hydroperoxides are "natures scavengers." Vitamin E, for example, inserts itself into either Step 2a or 2b, donates an H· from its phenolic —OH group to the allylic radical, and converts the radical to its original hydrocarbon chain. Because the vitamin E radical is stable, it breaks the cycle of chain propagation steps, thereby preventing the further formation of destructive hydroperoxides. While some hydroperoxides may form, their numbers are very small and they are easily decomposed to harmless materials by one of several enzyme-catalyzed reactions.

Unfortunately, vitamin E is removed in the processing of many foods and food products. To make up for this loss, phenols such as BHT and BHA are added to foods to "retard [their] spoilage" (as they say on the packages) by autoxidation:

Vitamin E

*B*utylated *h*ydroxy-*t*oluene (BHT)

*B*utylated *h*ydroxy-*a*nisole (BHA)

Similar compounds are added to other materials, such as plastics and rubber, to protect them against autoxidation.

SUMMARY

Benzene and its alkyl derivatives are classified as **aromatic hydrocarbons**, or **arenes**. The concepts of **hybridization of atomic orbitals** and the **theory of resonance** (Section 9.2C), developed in the 1930s, provided the first adequate description of the structure of benzene. The **resonance energy** of benzene is approximately 36 kcal/mol (151 kJ/mol) (Section 9.2D).

According to the Hückel criteria for aromaticity, a five- or six-membered ring is aromatic if it (1) has one *p* orbital on each atom of the ring, (2) is planar, so that overlap of all *p* orbitals of the ring is continuous or nearly so, and (3) has 2, 6, 10, 14 and so forth pi electrons in the overlapping system of *p* orbitals (Section 9.3). A **heterocyclic aromatic compound** contains one or more atoms other than carbon in an aromatic ring.

Aromatic compounds are named by the IUPAC system (Section 9.4). The common names toluene, xylene, styrene, phenol, aniline, benzaldehyde, and benzoic acid are retained. The C_6H_5— group is named **phenyl**, and the $C_6H_5CH_2$— group is named **benzyl**. To locate two substituents on a benzene ring, either number the atoms of the ring or use the locators **ortho (o)**, **meta (m)**, and **para (p)**.

Polynuclear aromatic hydrocarbons (Section 9.4C) contain two or more fused benzene rings. Particularly abundant are naphthalene, anthracene, phenanthrene, and their derivatives.

A characteristic reaction of aromatic compounds is **electrophilic aromatic substitution** (Section 9.6). Substituents on an aromatic ring influence both the site and rate of further substitution (Section 9.8). Substituent groups that direct an incoming group preferentially to the ortho and para positions are called **ortho–para directors**. Those which direct an incoming group preferentially to the meta positions are called **meta directors**. **Activating groups** cause the rate of further substitution to be faster than that for benzene; **deactivating groups** cause it to be slower than that for benzene.

A mechanistic rationale for directing effects is based on the degree of resonance stabilization of the possible cation intermediates formed upon reaction of the aromatic ring and the electrophile (Section 9.8B). Groups that stabilize the cation intermediate are ortho–para directors; groups that destabilize it are deactivators and meta directors.

The functional group of a **phenol** is an —OH group bonded to a benzene ring (Section 9.9A). Phenol and its derivatives are weak acids, with pK_a approximately 10.0, but are considerably stronger acids than alcohols, with pK_a 16–18.

KEY REACTIONS

1. Oxidation at a Benzylic Position (Section 9.5)

A benzylic carbon bonded to at least one hydrogen is oxidized to a carboxyl group:

$$H_3C-C_6H_4-CH(CH_3)_2 \xrightarrow[H_2SO_4]{K_2Cr_2O_7} HOOC-C_6H_4-COOH$$

2. Chlorination and Bromination (Section 9.7A)

The electrophile is a halonium ion, Cl^+ or Br^+, formed by treating Cl_2 or Br_2 with $AlCl_3$ or $FeCl_3$:

$$C_6H_6 + Cl_2 \xrightarrow{AlCl_3} C_6H_5-Cl + HCl$$

3. Nitration (Section 9.7B)

The electrophile is the nitronium ion, NO_2^+, formed by treating nitric acid with sulfuric acid:

$$C_6H_5Br + HNO_3 \xrightarrow{H_2SO_4} o\text{-}BrC_6H_4NO_2 + p\text{-}BrC_6H_4NO_2 + H_2O$$

4. Sulfonation (Section 9.7B)

The electrophile is HSO_3^+:

$$C_6H_6 + H_2SO_4 \longrightarrow C_6H_5-SO_3H + H_2O$$

5. Friedel–Crafts Alkylation (Section 9.7C)

The electrophile is an alkyl carbocation formed by treating an alkyl halide with a Lewis acid:

$$C_6H_6 + (CH_3)_2CHCl \xrightarrow{AlCl_3} C_6H_5-CH(CH_3)_2 + HCl$$

6. Friedel–Crafts Acylation (Section 9.7D)

The electrophile is an acyl cation formed by treating an acyl halide with a Lewis acid:

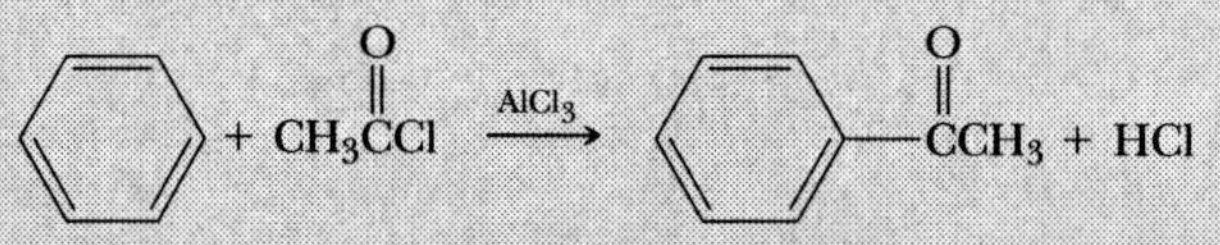

$$C_6H_6 + CH_3CCl(=O) \xrightarrow{AlCl_3} C_6H_5CCH_3(=O) + HCl$$

7. Alkylation Using an Alkene (Section 9.7E)

The electrophile is a carbocation formed by treating an alkene with H_2SO_4 or H_3PO_4:

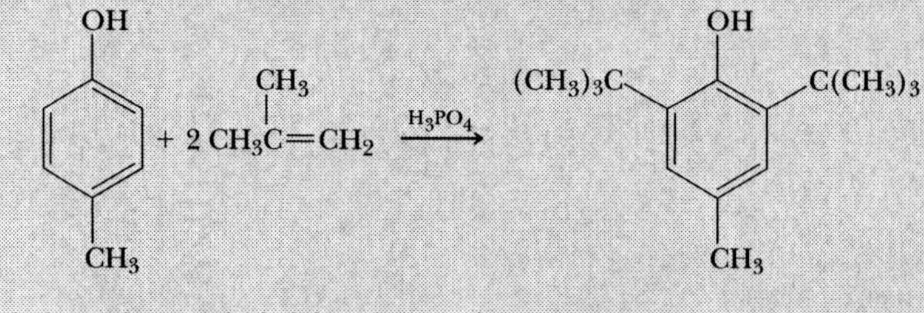

$$\text{4-}CH_3C_6H_4OH + 2\ CH_3C(CH_3){=}CH_2 \xrightarrow{H_3PO_4} \text{2,6-}((CH_3)_3C)_2\text{-4-}CH_3C_6H_2OH$$

8. Alkylation Using an Alcohol (Section 9.7E)

The electrophile is a carbocation formed by treating an alcohol with H_2SO_4 or H_3PO_4:

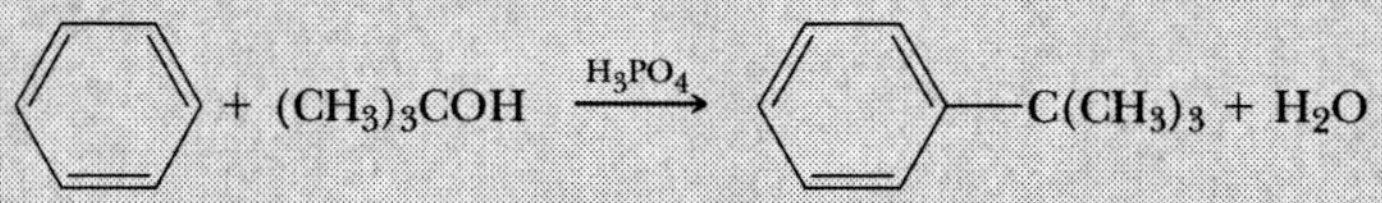

$$C_6H_6 + (CH_3)_3COH \xrightarrow{H_3PO_4} C_6H_5C(CH_3)_3 + H_2O$$

9. Acidity of Phenols (Section 9.9B)

Phenols are weak acids:

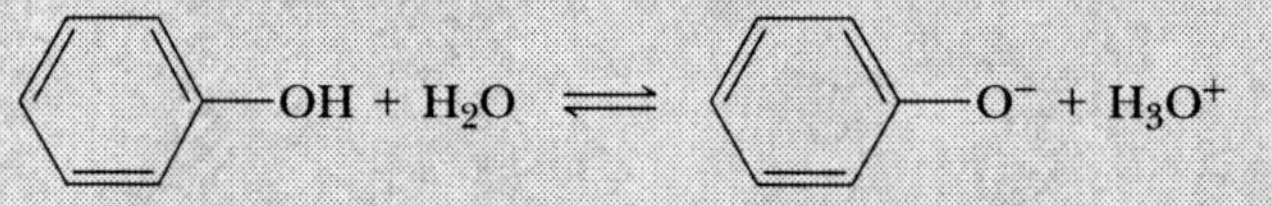

$$C_6H_5OH + H_2O \rightleftharpoons C_6H_5O^- + H_3O^+ \qquad K_a = 1.1 \times 10^{-10}, \quad pK_a = 9.95$$

Phenol — Phenoxide ion

Substitution by electron-withdrawing groups, such as the halogens and the nitro group, increases the acidity of phenols.

10. Reaction of Phenols with Strong Bases (Section 9.9C)

Water-insoluble phenols react quantitatively with strong bases to form water-soluble salts:

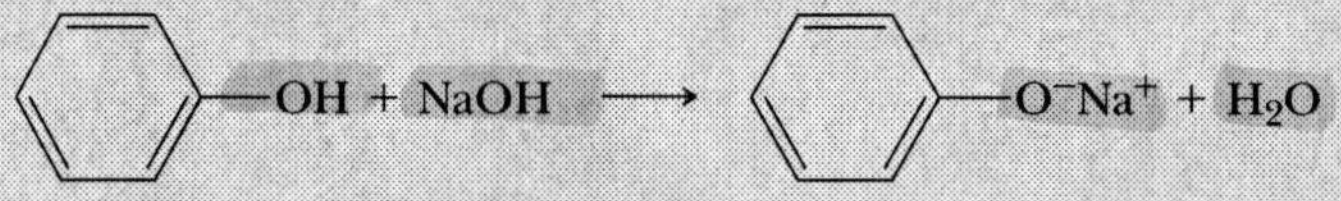

$$C_6H_5OH + NaOH \longrightarrow C_6H_5O^-Na^+ + H_2O$$

Phenol	Sodium hydroxide	Sodium phenoxide	Water
pK_a 9.95			pK_a 15.7
(stronger acid)	(stronger base)	(weaker base)	(weaker acid)

PROBLEMS

A problem number set in red indicates an applied "real-world" problem.

Aromaticity

9.10 Which of the following compounds are aromatic?

(a)

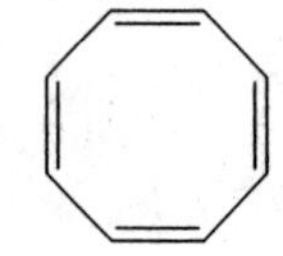

(b)

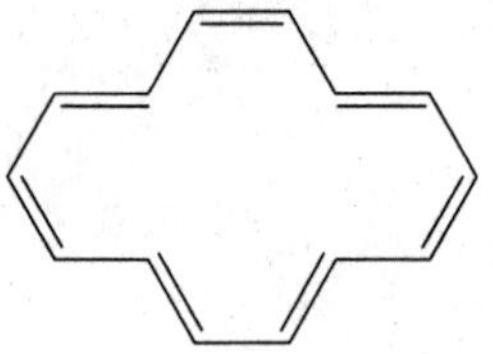

(c) 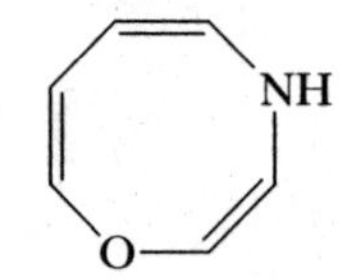 NH O

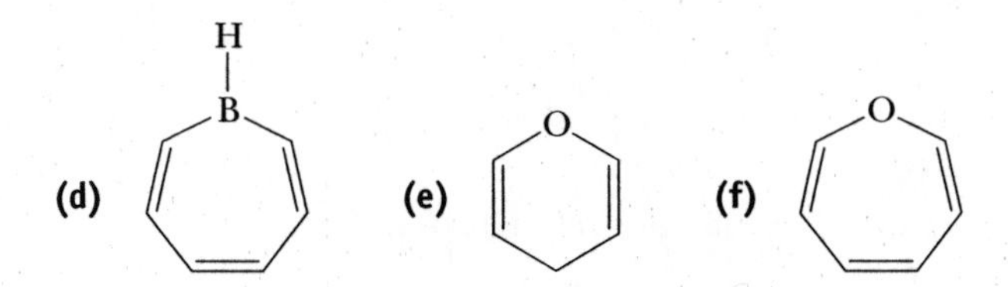

9.11 Explain why cyclopentadiene (pK_a 16) is many orders of magnitude more acidic than cyclopentane (pK_a > 50). (*Hint:* Draw the structural formula for the anion formed by removing one of the protons on the $—CH_2—$ group, and then apply the Hückel criteria for aromaticity.)

Cyclopentadiene Cyclopentane

Nomenclature and Structural Formulas

9.12 Name these compounds:

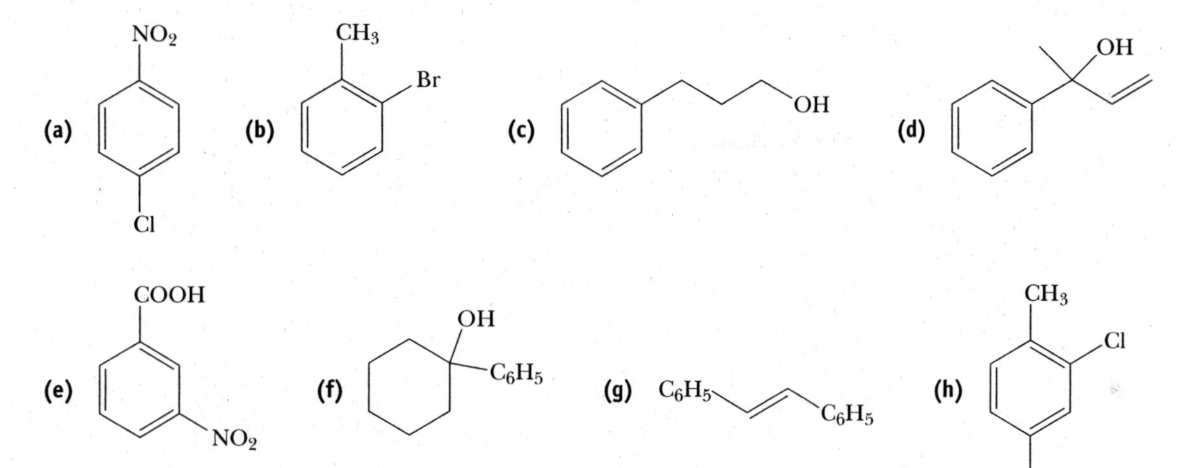

9.13 Draw structural formulas for these compounds:

(a) 1-Bromo-2-chloro-4-ethylbenzene
(b) 4-Iodo-1,2-dimethylbenzene
(c) 2,4,6-Trinitrotoluene
(d) 4-Phenyl-2-pentanol
(e) *p*-Cresol
(f) 2,4-Dichlorophenol
(g) 1-Phenylcyclopropanol
(h) Styrene (phenylethylene)
(i) *m*-Bromophenol
(j) 2,4-Dibromoaniline
(k) Isobutylbenzene
(l) *m*-Xylene

9.14 Show that pyridine can be represented as a hybrid of two equivalent contributing structures.

9.15 Show that naphthalene can be represented as a hybrid of three contributing structures. Show also, by the use of curved arrows, how one contributing structure is converted to the next.

9.16 Draw four contributing structures for anthracene.

Electrophilic Aromatic Substitution: Monosubstitution

9.17 Draw a structural formula for the compound formed by treating benzene with each of the following combinations of reagents:

(a) $CH_3CH_2Cl/AlCl_3$
(b) $CH_2{=}CH_2/H_2SO_4$
(c) CH_3CH_2OH/H_2SO_4

9.18 Show three different combinations of reagents you might use to convert benzene to isopropylbenzene.

9.19 How many monochlorination products are possible when naphthalene is treated with $Cl_2/AlCl_3$?

9.20 Write a stepwise mechanism for the following reaction, using curved arrows to show the flow of electrons in each step:

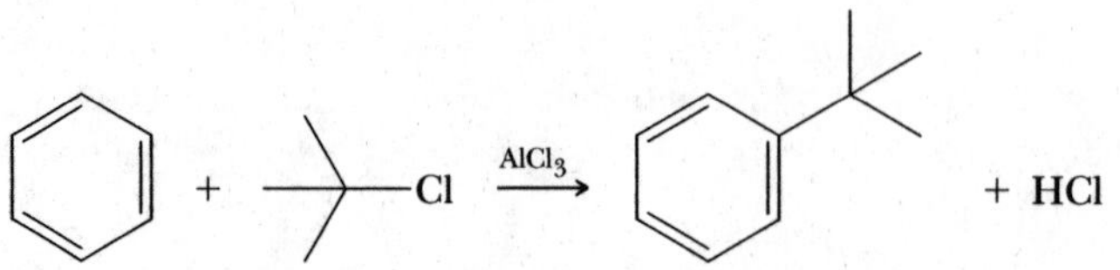

$$C_6H_6 + (CH_3)_3C{-}Cl \xrightarrow{AlCl_3} C_6H_5C(CH_3)_3 + HCl$$

9.21 Write a stepwise mechanism for the preparation of diphenylmethane by treating benzene with dichloromethane in the presence of an aluminum chloride catalyst.

Electrophilic Aromatic Substitution: Disubstitution

9.22 When treated with $Cl_2/AlCl_3$, 1,2-dimethylbenzene (*o*-xylene) gives a mixture of two products. Draw structural formulas for these products.

9.23 How many monosubstitution products are possible when 1,4-dimethylbenzene (*p*-xylene) is treated with $Cl_2/AlCl_3$? When *m*-xylene is treated with $Cl_2/AlCl_3$?

9.24 Draw the structural formula for the major product formed upon treating each compound with $Cl_2/AlCl_3$:

(a) Toluene **(b)** Nitrobenzene **(c)** Chlorobenzene

(d) *tert*-Butylbenzene **(e)** 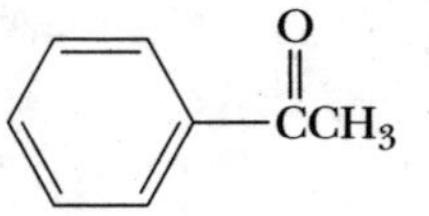 $C_6H_5{-}\overset{O}{\overset{\|}{C}}CH_3$ **(f)** 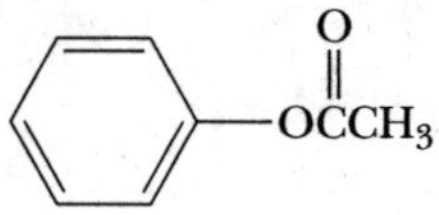 $C_6H_5{-}O\overset{O}{\overset{\|}{C}}CH_3$

(g) 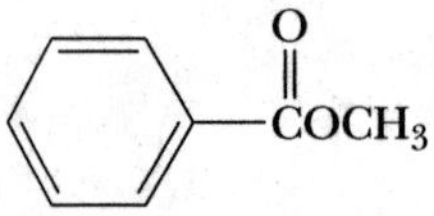 $C_6H_5{-}\overset{O}{\overset{\|}{C}}OCH_3$

9.25 Which compound, chlorobenzene or toluene, undergoes electrophilic aromatic substitution more rapidly when treated with $Cl_2/AlCl_3$? Explain and draw structural formulas for the major product(s) from each reaction.

9.26 Arrange the compounds in each set in order of decreasing reactivity (fastest to slowest) toward electrophilic aromatic substitution:

(a) 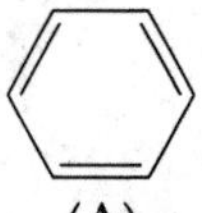 (A) C_6H_6 (B) $C_6H_5{-}O\overset{O}{\overset{\|}{C}}CH_3$ (C) $C_6H_5{-}\overset{O}{\overset{\|}{C}}OCH_3$

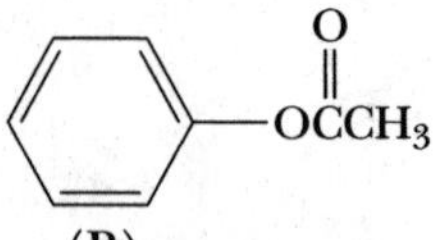

(b) (A) $C_6H_5{-}NO_2$ (B) $C_6H_5{-}COOH$ (C) C_6H_6

(c) (A) $C_6H_5{-}NH_2$ (B) $C_6H_5{-}NH\overset{O}{\overset{\|}{C}}CH_3$ (C) $C_6H_5{-}\overset{O}{\overset{\|}{C}}NHCH_3$

(d) 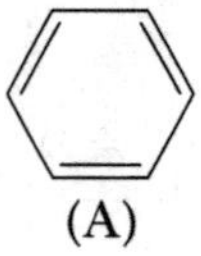 (A) C_6H_6 (B) $C_6H_5{-}CH_3$ (C) $C_6H_5{-}OCH_3$

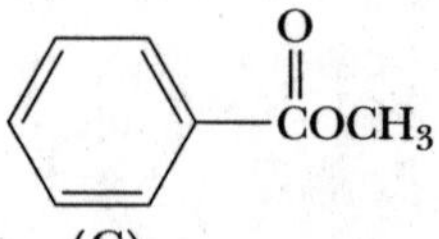

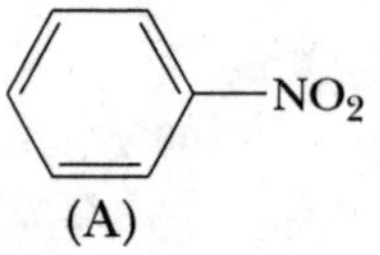

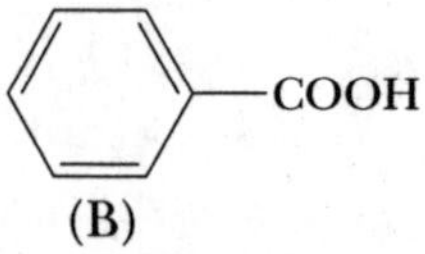

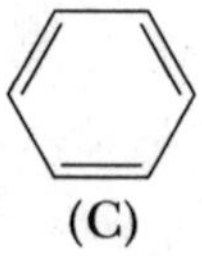

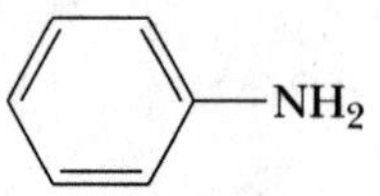

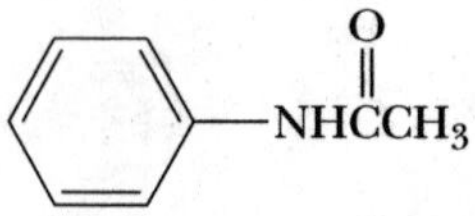

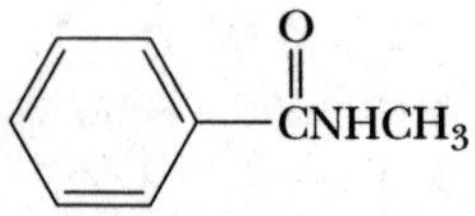

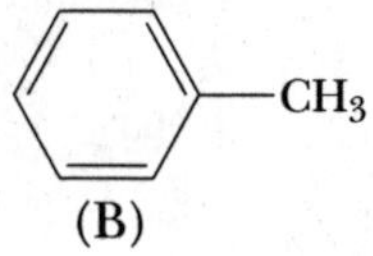

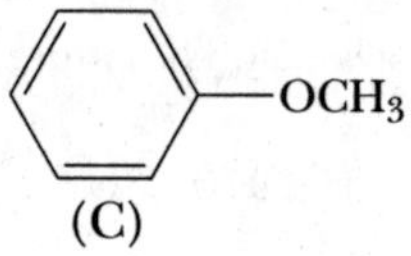

9.27 Account for the observation that the trifluoromethyl group is meta directing, as shown in the following example:

$$C_6H_5CF_3 + HNO_3 \xrightarrow{H_2SO_4} m\text{-}O_2NC_6H_4CF_3 + H_2O$$

9.28 Show how to convert toluene to these carboxylic acids:

(a) 4-Chlorobenzoic acid **(b)** 3-Chlorobenzoic acid

9.29 Show reagents and conditions that can be used to bring about these conversions:

(a) $C_6H_5CH_3 \longrightarrow p\text{-}CH_3C_6H_4C(=O)CH_2CH_3$

(b) $C_6H_5OH \longrightarrow C_6H_5OCH_2CH_3$

(c) $C_6H_5OCH_3 \longrightarrow p\text{-}CH_3OC_6H_4CCH_3$ (C=O)

(d) $C_6H_5NHCCH_3$ (C=O) $\longrightarrow p\text{-}HO_3SC_6H_4NHCCH_3$ (C=O)

9.30 Propose a synthesis of triphenylmethane from benzene as the only source of aromatic rings. Use any other necessary reagents.

9.31 Reaction of phenol with acetone in the presence of an acid catalyst gives bisphenol A, a compound used in the production of polycarbonate and epoxy resins (Sections 17.5C and 17.5E):

$$2\ C_6H_5OH + CH_3\overset{O}{\overset{\|}{C}}CH_3 \xrightarrow{H_3PO_4} HO{-}C_6H_4{-}C(CH_3)_2{-}C_6H_4{-}OH + H_2O$$

Acetone Bisphenol A

Propose a mechanism for the formation of bisphenol A. (*Hint:* The first step is a proton transfer from phosphoric acid to the oxygen of the carbonyl group of acetone.)

9.32 2,6-Di-*tert*-butyl-4-methylphenol, more commonly known as butylated hydroxytoluene, or BHT, is used as an antioxidant in foods to "retard spoilage." BHT is synthesized industrially from 4-methylphenol (*p*-cresol) by reaction with 2-methylpropene in the presence of phosphoric acid:

$$\text{4-Methylphenol} + 2\ \text{2-Methylpropene} \xrightarrow{H_3PO_4} \text{BHT}$$

4-Methylphenol 2-Methylpropene 2,6-Di-*tert*-butyl-4-methylphenol (Butylated hydroxytoluene, BHT)

Propose a mechanism for this reaction.

9.33 The first herbicide widely used for controlling weeds was 2,4-dichlorophenoxyacetic acid (2,4-D). Show how this compound might be synthesized from 2,4-dichlorophenol and chloroacetic acid, $ClCH_2COOH$:

2,4-Dichlorophenol → 2,4-Dichlorophenoxyacetic acid (2,4-D)

Acidity of Phenols

9.34 Use the resonance theory to account for the fact that phenol (pK_a 9.95) is a stronger acid than cyclohexanol (pK_a 18).

9.35 Arrange the compounds in each set in order of increasing acidity (from least acidic to most acidic):

(a) C_6H_5–OH cyclohexyl–OH CH_3COOH

(b) C_6H_5–OH $NaHCO_3$ H_2O

(c) O_2N–C_6H_4–OH C_6H_5–OH C_6H_5–CH_2OH

9.36 From each pair, select the stronger base:

(a) C_6H_5–O^- or OH^- **(b)** C_6H_5–O^- or cyclohexyl–O^-

(c) C_6H_5–O^- or HCO_3^- **(d)** C_6H_5–O^- or CH_3COO^-

9.37 Account for the fact that water-insoluble carboxylic acids (pK_a 4–5) dissolve in 10% sodium bicarbonate with the evolution of a gas, but water-insoluble phenols (pK_a 9.5–10.5) do not show this chemical behavior.

9.38 Describe a procedure for separating a mixture of 1-hexanol and 2-methylphenol (*o*-cresol) and recovering each in pure form. Each is insoluble in water, but soluble in diethyl ether.

Syntheses

9.39 Using styrene, $C_6H_5CH{=}CH_2$, as the only aromatic starting material, show how to synthesize these compounds. In addition to styrene, use any other necessary organic or inorganic chemicals. Any compound synthesized in one part of this problem may be used to make any other compound in the problem:

(a) C_6H_5–COOH **(b)** C_6H_5–CHBrCH$_3$ **(c)** C_6H_5–CH(OH)CH$_3$

(d) C_6H_5–$\overset{O}{\overset{\|}{C}}CH_3$ **(e)** C_6H_5–CH_2CH_3 **(f)** C_6H_5–$\overset{OH}{\overset{|}{C}}HCH_2OH$

9.40 Show how to synthesize these compounds, starting with benzene, toluene, or phenol as the only sources of aromatic rings. Assume that, in all syntheses, you can separate mixtures of ortho–para products to give the desired isomer in pure form:

(a) *m*-Bromonitrobenzene
(b) 1-Bromo-4-nitrobenzene
(c) 2,4,6-Trinitrotoluene (TNT)
(d) *m*-Bromobenzoic acid
(e) *p*-Bromobenzoic acid
(f) *p*-Dichlorobenzene
(g) *m*-Nitrobenzenesulfonic acid
(h) 1-Chloro-3-nitrobenzene

9.41 Show how to synthesize these aromatic ketones, starting with benzene or toluene as the only sources of aromatic rings. Assume that, in all syntheses, mixtures of ortho–para products can be separated to give the desired isomer in pure form:

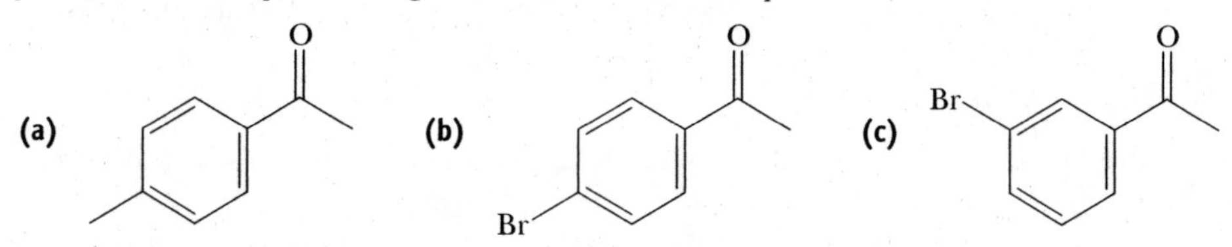

9.42 The following ketone, isolated from the roots of several members of the iris family, has an odor like that of violets and is used as a fragrance in perfumes. Describe the synthesis of this ketone from benzene.

4-Isopropylacetophenone

9.43 The bombardier beetle generates *p*-quinone, an irritating chemical, by the enzyme-catalyzed oxidation of hydroquinone, using hydrogen peroxide as the oxidizing agent. Heat generated in this oxidation produces superheated steam, which is ejected, along with *p*-quinone, with explosive force.

Hydroquinone + H_2O_2 $\xrightarrow{\text{enzyme catalyst}}$ *p*-Quinone + H_2O + heat

(a) Balance the equation.
(b) Show that this reaction of hydroquinone is an oxidation.

9.44 Following is a structural formula for musk ambrette, a synthetic musk used in perfumes to enhance and retain fragrance:

m-Cresol $\xrightarrow{?}$ Musk ambrette

Propose a synthesis for musk ambrette from *m*-cresol.

9.45 1-(3-Chlorophenyl)propanone is a building block in the synthesis of bupropion, the hydrochloride salt of which is the antidepressant Wellbutrin. During clinical trials, researchers discovered that smokers reported a lessening in their craving for tobacco after one to two weeks on the drug. Further clinical trials confirmed this finding, and the drug is also marketed under the trade name Zyban® as an aid in smoking cessation. Propose a synthesis for this building block from benzene. (We will see in Section 13.9 how to complete the synthesis of bupropion.)

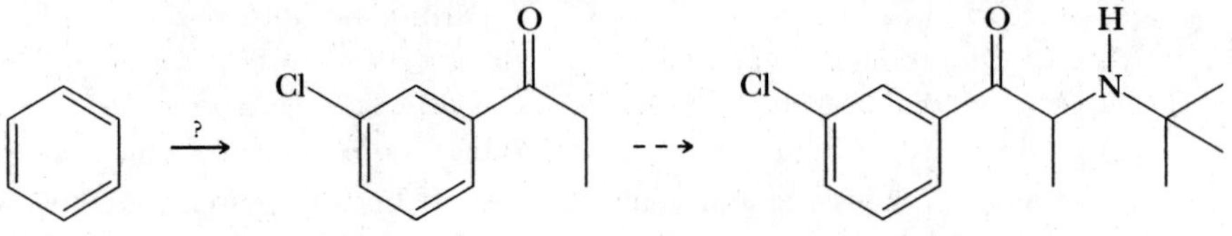

Benzene $\xrightarrow{?}$ 1-(3-Chlorophenyl)-1-propanone - - → Bupropion (Wellbutrin, Zyban)

Looking Ahead

9.46 Which of the following compounds can be made directly by using an electrophilic aromatic substitution reaction?

(a) 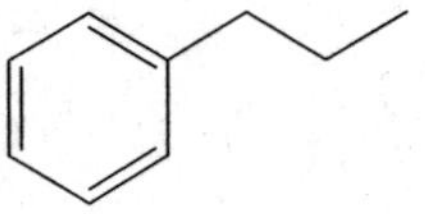 (b) 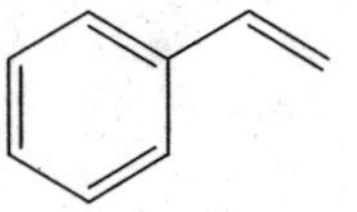 (c) 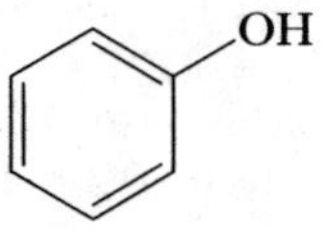 OH (d) 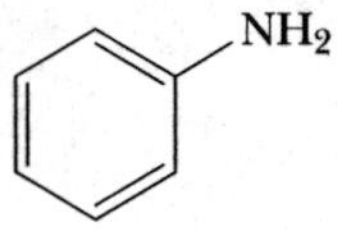 NH_2

9.47 Which compound is a better nucleophile?

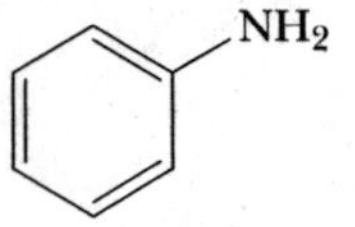

NH_2 or NH_2

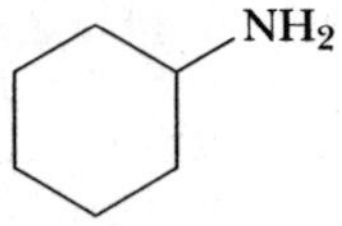

Aniline Cyclohexanamine

9.48 Suggest a reason that the following arenes do not undergo electrophilic aromatic substitution when $AlCl_3$ is used in the reaction:

(a) 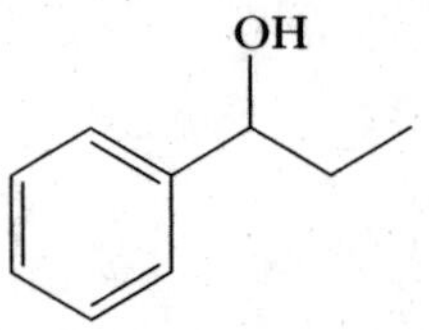 OH (b) 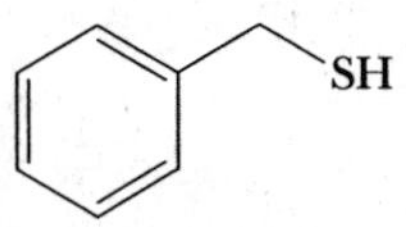 SH (c) 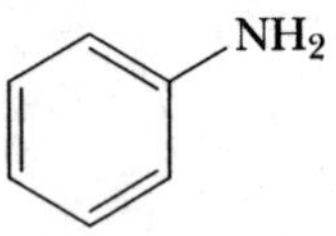 NH_2

9.49 Predict the product of the following acid–base reaction:

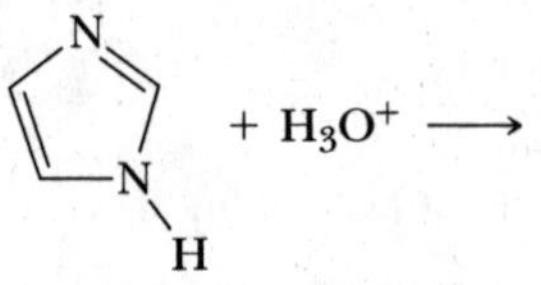

N, N–H + H_3O^+ ⟶

9.50 Which haloalkane reacts faster in an S_N1 reaction?

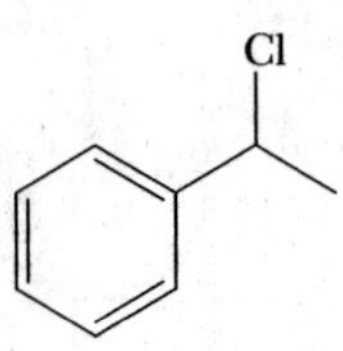

Cl or Cl

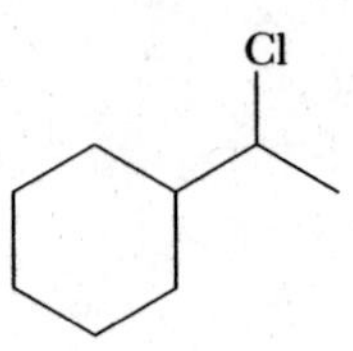

10 Amines

This inhaler delivers puffs of albuterol (Proventil), a potent synthetic bronchodilator whose structure is patterned after that of epinephrine (adrenaline). See Problem 10.13. Inset: A model of morphine. *(Mark Clarke/Photo Researchers, Inc.)*

10.1 INTRODUCTION

Carbon, hydrogen, and oxygen are the three most common elements in organic compounds. Because of the wide distribution of amines in the biological world, nitrogen is the fourth most common component of organic compounds. The most important chemical property of amines is their basicity and their nucleophilicity.

10.2 STRUCTURE AND CLASSIFICATION

Amines are derivatives of ammonia in which one or more hydrogens are replaced by alkyl or aryl groups. Amines are classified as primary (1°), secondary (2°), or

CHEMICAL CONNECTIONS 10A

Morphine as a Clue in the Design and Discovery of Drugs

The analgesic, soporific, and euphoriant properties of the dried juice obtained from unripe seed pods of the opium poppy *Papaver somniferum* have been known for centuries. By the beginning of the 19th century, the active principal, morphine, had been isolated and its structure determined:

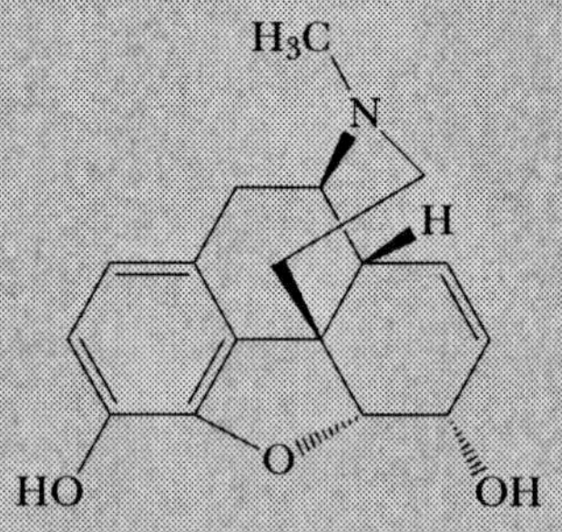

Morphine

Also occurring in the opium poppy is codeine, a monomethyl ether of morphine. Heroin is synthesized by treating morphine with two moles of acetic anhydride:

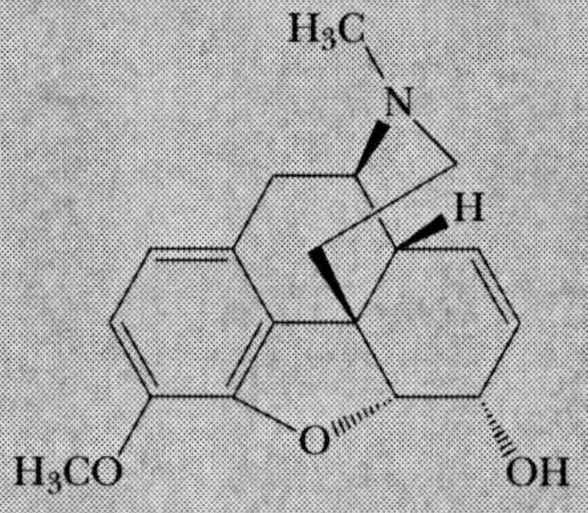

Codeine

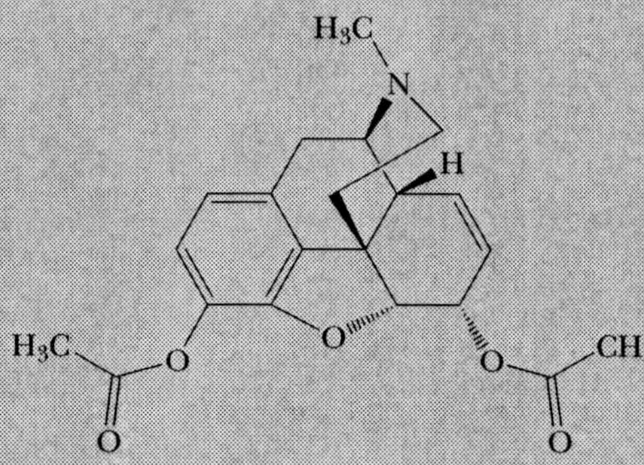

Heroin

Even though morphine is one of modern medicine's most effective painkillers, it has two serious side effects: It is addictive, and it depresses the respiratory control center of the central nervous system. Large doses of morphine (or heroin) can lead to death by respiratory failure. For these reasons, chemists have sought to produce painkillers related in structure to morphine, but without these serious side disadvantages. One strategy in this ongoing research has been to synthesize compounds related in structure to morphine, in the hope that they would be equally effective analgesics, but with diminished side effects. Following are structural formulas for two such compounds that have proven to be clinically useful:

tertiary (3°), depending on the number of hydrogen atoms of ammonia that are replaced by alkyl or aryl groups (Section 1.8B):

$:NH_3$	$CH_3-\ddot{N}H_2$	$CH_3-\ddot{N}H-CH_3$	$CH_3-\ddot{N}(CH_3)-CH_3$
Ammonia	Methylamine (a 1° amine)	Dimethylamine (a 2° amine)	Trimethylamine (a 3° amine)

Aliphatic amine An amine in which nitrogen is bonded only to alkyl groups.

Amines are further divided into aliphatic amines and aromatic amines. In an **aliphatic amine**, all the carbons bonded directly to nitrogen are derived from alkyl

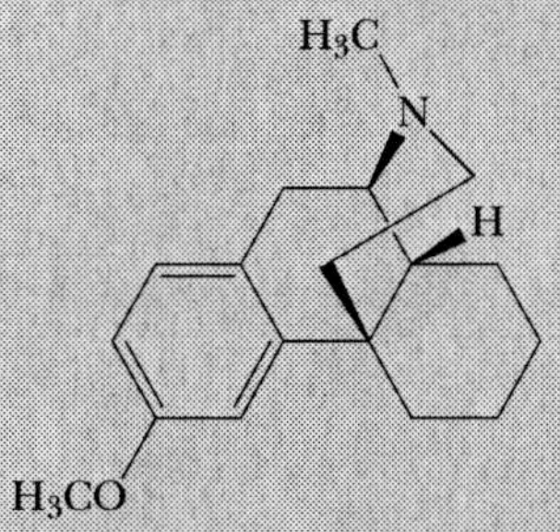

(−)-enantiomer = Levomethorphan
(+)-enantiomer = Dextromethorphan

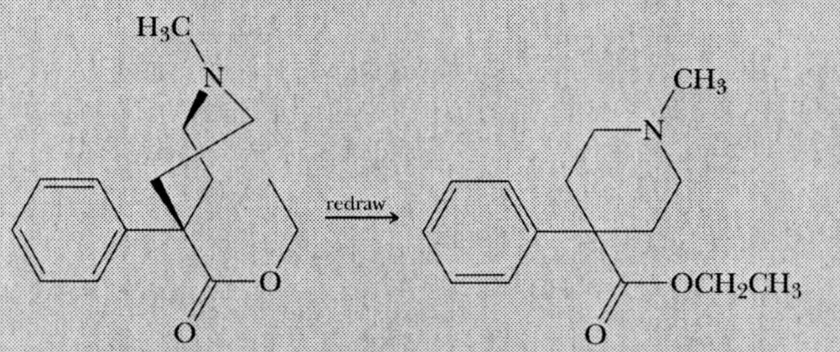

Meperidine
(Demerol)

Levomethorphan is a potent analgesic. Interestingly, its dextrorotatory enantiomer, dextromethorphan, has no analgesic activity. It does, however, show approximately the same cough-suppressing activity as morphine and is used extensively in cough remedies.

It has been discovered that there can be even further simplification in the structure of morphine-like analgesics. One such simplification is represented by meperidine, the hydrochloride salt of which is the widely used analgesic Demerol®.

It was hoped that meperidine and related synthetic drugs would be free of many of the morphine-like undesirable side effects. It is now clear, however, that they are not. Meperidine, for example, is definitely addictive. In spite of much determined research, there are as yet no agents as effective as morphine for the relief of severe pain that are absolutely free of the risk of addiction.

How and in what regions of the brain does morphine act? In 1979, scientists discovered that there are specific receptor sites for morphine and other opiates and that these sites are clustered in the brain's limbic system, the area involved in emotion and the perception of pain. Scientists then asked, Why does the human brain have receptor sites specific for morphine? Could it be that the brain produces its own opiates? In 1974, scientists discovered that opiate-like compounds are indeed present in the brain; in 1975, they isolated a brain opiate that was named *enkephalin*, meaning "in the brain." Scientists have yet to understand the role of these natural brain opiates. Perhaps when we do understand their biochemistry, we will discover clues that will lead to the design and synthesis of more potent, but less addictive, analgesics.

groups; in an **aromatic amine**, one or more of the groups bonded directly to nitrogen are aryl groups:

Aromatic amine An amine in which nitrogen is bonded to one or more aryl groups.

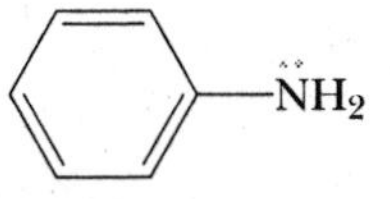

Aniline
(a 1° aromatic amine)

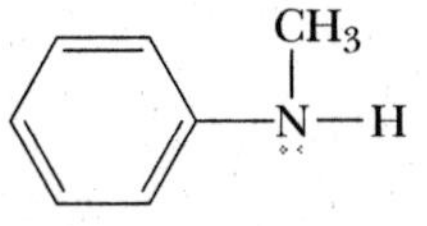

N-Methylaniline
(a 2° aromatic amine)

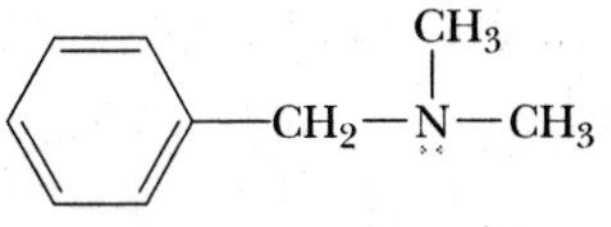

Benzyldimethylamine
(a 3° aliphatic amine)

An amine in which the nitrogen atom is part of a ring is classified as a **heterocyclic amine**. When the nitrogen is part of an aromatic ring (Section 9.3), the

Heterocyclic amine An amine in which nitrogen is one of the atoms of a ring.

Heterocyclic aromatic amine An amine in which nitrogen is one of the atoms of an aromatic ring.

amine is classified as a **heterocyclic aromatic amine**. Following are structural formulas for two heterocyclic aliphatic amines and two heterocyclic aromatic amines:

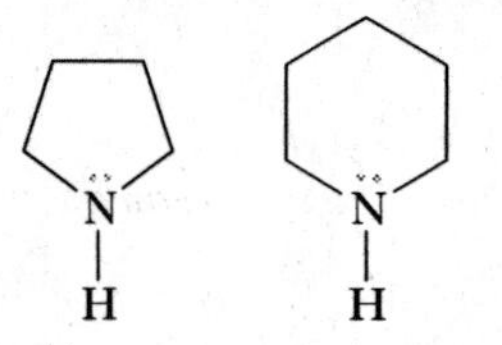

Pyrrolidine Piperidine
(heterocyclic aliphatic amines)

Pyrrole Pyridine
(heterocyclic aromatic amines)

EXAMPLE 10.1

Alkaloids are basic nitrogen-containing compounds of plant origin, many of which have physiological activity when administered to humans. The ingestion of coniine, present in water hemlock, can cause weakness, labored respiration, paralysis, and, eventually, death. Coniine was the toxic substance in "poison hemlock" that caused the death of Socrates. In small doses, nicotine is an addictive stimulant. In larger doses, it causes depression, nausea, and vomiting. In still larger doses, it is a deadly poison. Solutions of nicotine in water are used as insecticides. Cocaine is a central nervous system stimulant obtained from the leaves of the coca plant. Classify each amino group in these alkaloids according to type (that is, primary, secondary, tertiary, heterocyclic, aliphatic, or aromatic):

(a) (*S*)-Coniine

(b) (*S*)-Nicotine

(c) Cocaine

SOLUTION

(a) A secondary heterocyclic aliphatic amine.
(b) One tertiary heterocyclic aliphatic amine and one heterocyclic aromatic amine.
(c) A tertiary heterocyclic aliphatic amine.

Practice Problem 10.1

Identify all carbon stereocenters in coniine, nicotine, and cocaine.

10.3 NOMENCLATURE

A. Systematic Names

Systematic names for aliphatic amines are derived just as they are for alcohols. The suffix *-e* of the parent alkane is dropped and is replaced by *-amine*; that is, they are named alkanamines:

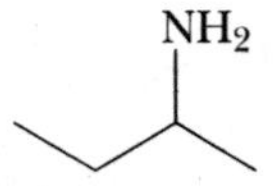

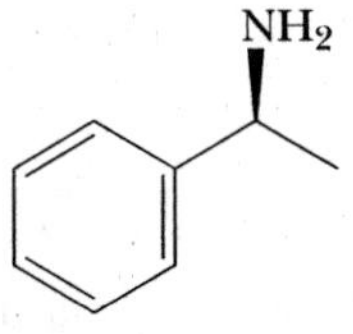

NH_2 NH_2 $H_2N(CH_2)_6NH_2$

2-Butanamine (*S*)-1-Phenylethanamine 1,6-Hexanediamine

EXAMPLE 10.2

Write the IUPAC name for each amine:

(a) 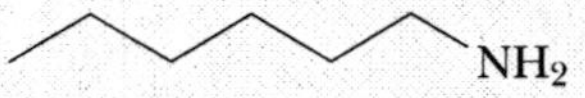 NH_2 (b) 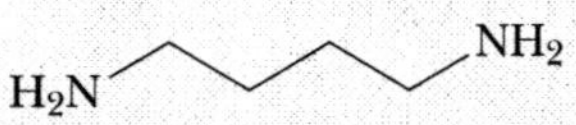 H_2N NH_2

(c) 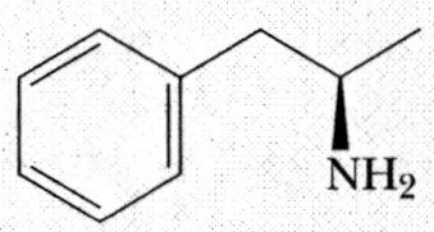 NH_2

SOLUTION

(a) 1-Hexanamine
(b) 1,4-Butanediamine
(c) The systematic name of this compound is (*S*)-1-phenyl-2-propanamine. Its common name is amphetamine. The dextrorotatory isomer of amphetamine (shown here) is a central nervous system stimulant and is manufactured and sold under several trade names. The salt with sulfuric acid is marketed as Dexedrine® sulfate.

Practice Problem 10.2

Write a structural formula for each amine:

(a) 2-Methyl-1-propanamine (b) Cyclohexanamine (c) (*R*)-2-Butanamine

IUPAC nomenclature retains the common name **aniline** for $C_6H_5NH_2$, the simplest aromatic amine. Its simple derivatives are named with the prefixes *o*-, *m*-, and *p*-, or numbers to locate substituents. Several derivatives of aniline have common names that are still widely used. Among these are **toluidine**, for a methyl-substituted aniline, and **anisidine**, for a methoxy-substituted aniline:

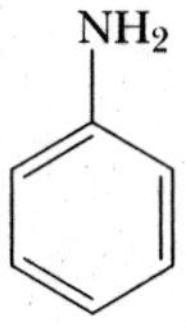

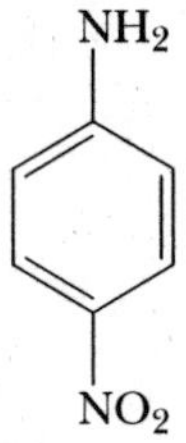

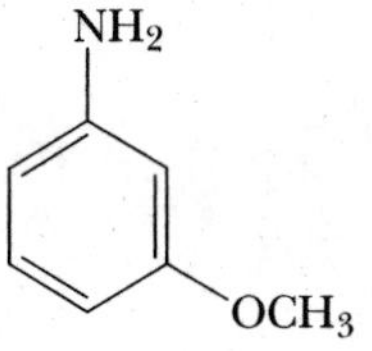

NH_2 NH_2 NH_2 NH_2

OCH_3

NO_2 CH_3

Aniline 4-Nitroaniline (*p*-Nitroaniline) 4-Methylaniline (*p*-Toluidine) 3-Methoxyaniline (*m*-Anisidine)

Secondary and tertiary amines are commonly named as *N*-substituted primary amines. For unsymmetrical amines, the largest group is taken as the parent amine; then the smaller group or groups bonded to nitrogen are named, and their location is indicated by the prefix *N* (indicating that they are attached to nitrogen):

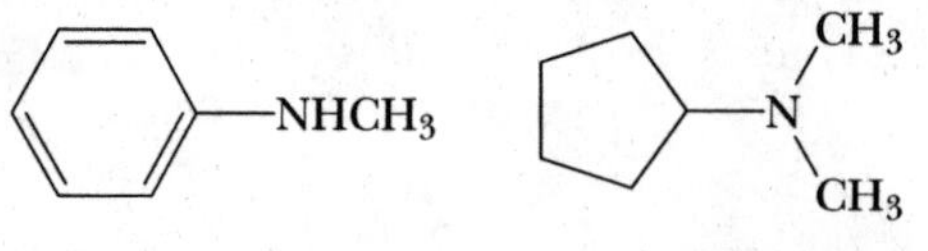

$-NHCH_3$ *N*-Methylaniline

$N(CH_3)_2$ *N,N*-Dimethylcyclopentanamine

Following are names and structural formulas for four heterocyclic aromatic amines, the common names of which have been retained by the IUPAC:

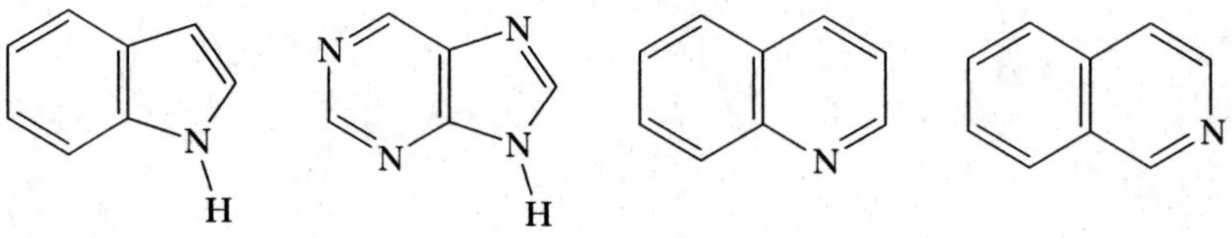

Indole Purine Quinoline Isoquinoline

Among the various functional groups discussed in this text, the $-NH_2$ group has one of the lowest priorities. The following compounds each contain a functional group of higher precedence than the amino group, and, accordingly, the amino group is indicated by the prefix *amino-*:

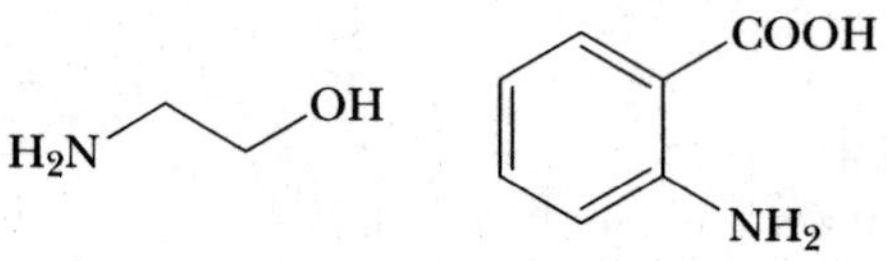

H_2N—OH
2-Aminoethanol
(Ethanolamine)

COOH, NH_2
2-Aminobenzoic acid
(Anthranilic acid)

B. Common Names

Common names for most aliphatic amines are derived by listing the alkyl groups bonded to nitrogen in alphabetical order in one word ending in the suffix *-amine*; that is, they are named as **alkylamines**:

CH_3NH_2 Methylamine

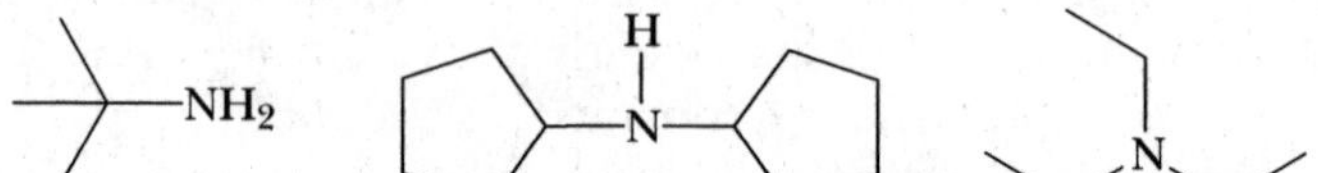

$-NH_2$ *tert*-Butylamine

Dicyclopentylamine

Triethylamine

EXAMPLE 10.3

Write a structural formula for each amine:

(a) Isopropylamine (b) Cyclohexylmethylamine (c) Benzylamine

SOLUTION

(a) $(CH_3)_2CHNH_2$ (b) [cyclohexane ring]—$NHCH_3$ (c) [benzene ring]—CH_2NH_2

Practice Problem 10.3

Write a structural formula for each amine:

(a) Isobutylamine (b) Triphenylamine (c) Diisopropylamine

When four atoms or groups of atoms are bonded to a nitrogen atom, we name the compound as a salt of the corresponding amine. We replace the ending *-amine* (or aniline, pyridine, or the like) by *-ammonium* (or *anilinium, pyridinium*, or the like) and add the name of the anion (chloride, acetate, and so on). Compounds containing such ions have properties characteristic of salts. Following are three examples (cetylpyridinium chloride is used as a topical antiseptic and disinfectant):

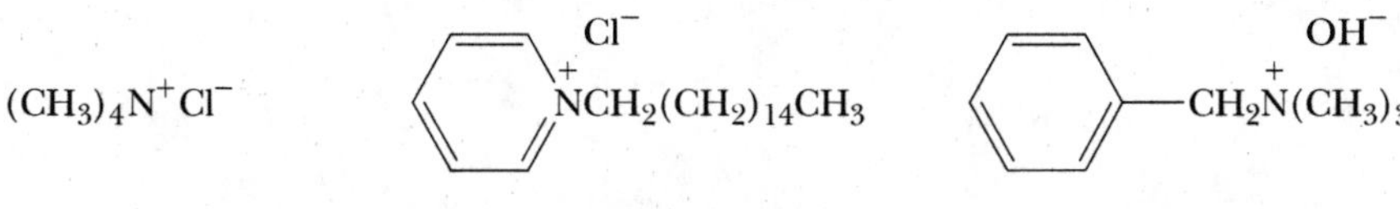
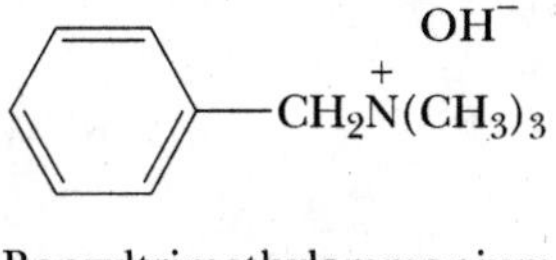

Tetramethylammonium chloride | Hexadecylpyridinium chloride (Cetylpyridinium chloride) | Benzyltrimethylammonium hydroxide

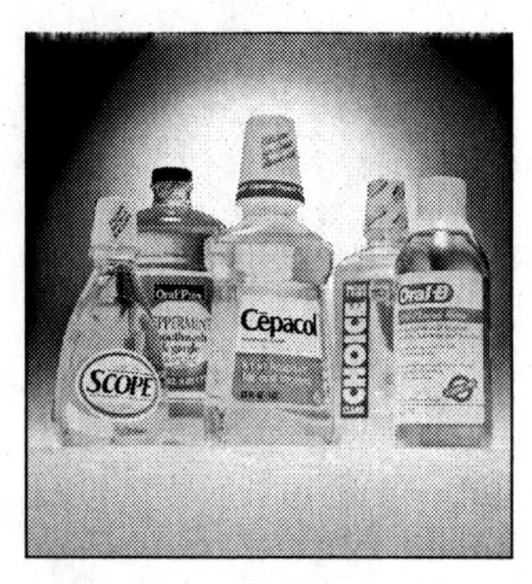

Several over-the-counter mouthwashes contain *N*-alkylatedpyridinium chlorides as an antibacterical agent. *(Charles D. Winters)*

10.4 PHYSICAL PROPERTIES

Amines are polar compounds, and both primary and secondary amines form intermolecular hydrogen bonds (Figure 10.1).

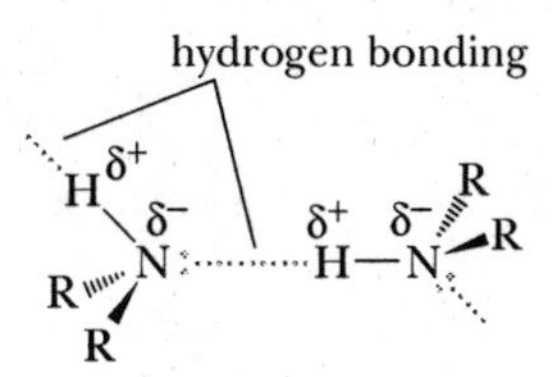

Figure 10.1
Intermolecular association of 1° and 2° amines by hydrogen bonding. Nitrogen is approximately tetrahedral in shape, with the axis of the hydrogen bond along the fourth position of the tetrahedron.

An N—H----N hydrogen bond is weaker than an O—H----O hydrogen bond, because the difference in electronegativity between nitrogen and hydrogen ($3.0 - 2.1 = 0.9$) is less than that between oxygen and hydrogen ($3.5 - 2.1 = 1.4$). We can illustrate the effect of intermolecular hydrogen bonding by comparing the boiling points of methylamine and methanol:

	CH_3NH_2	CH_3OH
molecular weight (g/mol)	31.1	32.0
boiling point (°C)	−6.3	65.0

Both compounds have polar molecules and interact in the pure liquid by hydrogen bonding. Methanol has the higher boiling point because hydrogen bonding between its molecules is stronger than that between molecules of methylamine.

CHEMICAL CONNECTIONS 10B

The Poison Dart Frogs of South America: Lethal Amines

The Noanamá and Embrá peoples of the jungles of western Colombia have used poison blow darts for centuries, perhaps millennia. The poisons are obtained from the skin secretions of several highly colored frogs of the genus *Phyllobates* (*neará* and *kokoi* in the language of the native peoples). A single frog contains enough poison for up to 20 darts. For the most poisonous species (*Phyllobates terribilis*), just rubbing a dart over the frog's back suffices to charge the dart with poison.

Scientists at the National Institutes of Health became interested in studying these poisons when it was discovered that they act on cellular ion channels, which would make them useful tools in basic research on mechanisms of ion transport. A field station was established in western Colombia to collect the relatively common poison dart frogs. From 5,000 frogs, 11 mg of batrachotoxin and batrachotoxinin A was isolated. These names are derived from *batrachos*, the Greek word for frog.

Poison dart frog, *Phyllobates terribilis*. (*Juan M. Renjifo/ Animals, Animals*)

Batrachotoxin and batrachotoxinin A are among the most lethal poisons ever discovered:

Batrachotoxin

Batrachotoxinin A

It is estimated that as little as 200 μg of batrachotoxin is sufficient to induce irreversible cardiac arrest in a human being. It has been determined that they act by causing voltage-gated Na^+ channels in nerve and muscle cells to be blocked in the open position, which leads to a huge influx of Na^+ ions into the affected cell.

The batrachotoxin story illustrates several common themes in the discovery of new drugs. First, information about the kinds of biologically active compounds and their sources are often obtained from the native peoples of a region. Second, tropical rain forests are a rich source of structurally complex, biologically active substances. Third, an entire ecosystem, not only the plants, is a potential source of fascinating organic molecules.

All classes of amines form hydrogen bonds with water and are more soluble in water than are hydrocarbons of comparable molecular weight. Most low-molecular-weight amines are completely soluble in water (Table 10.1). Higher-molecular-weight amines are only moderately soluble or insoluble.

TABLE 10.1 Physical Properties of Selected Amines

Name	Structural Formula	Melting Point (°C)	Boiling Point (°C)	Solubility in Water
Ammonia	NH_3	−78	−33	very soluble
Primary Amines				
methylamine	CH_3NH_2	−95	−6	very soluble
ethylamine	$CH_3CH_2NH_2$	−81	17	very soluble
propylamine	$CH_3CH_2CH_2NH_2$	−83	48	very soluble
butylamine	$CH_3(CH_2)_3NH_2$	−49	78	very soluble
benzylamine	$C_6H_5CH_2NH_2$	10	185	very soluble
cyclohexylamine	$C_6H_{11}NH_2$	−17	135	slightly soluble
Secondary Amines				
dimethylamine	$(CH_3)_2NH$	−93	7	very soluble
diethylamine	$(CH_3CH_2)_2NH$	−48	56	very soluble
Tertiary Amines				
trimethylamine	$(CH_3)_3N$	−117	3	very soluble
triethylamine	$(CH_3CH_2)_3N$	−114	89	slightly soluble
Aromatic Amines				
aniline	$C_6H_5NH_2$	−6	184	slightly soluble
Heterocyclic Aromatic Amines				
pyridine	C_5H_5N	−42	116	very soluble

10.5 BASICITY OF AMINES

Like ammonia, all amines are weak bases, and aqueous solutions of amines are basic. The following acid–base reaction between an amine and water is written using curved arrows to emphasize that, in this proton-transfer reaction, the unshared pair of electrons on nitrogen forms a new covalent bond with hydrogen and displaces hydroxide ion:

$$CH_3-\underset{H}{\overset{H}{|}}\!\!\!N\!: + H-\ddot{O}-H \rightleftharpoons CH_3-\underset{H}{\overset{H}{|}}\!\!\!N^+-H \quad {}^-\!:\ddot{O}-H$$

Methylamine — Methylammonium hydroxide

The equilibrium constant for the reaction of an amine with water, K_{eq}, has the following form, illustrated for the reaction of methylamine with water to give methylammonium hydroxide:

$$K_{eq} = \frac{[CH_3NH_3^+][OH^-]}{[CH_3NH_2][H_2O]}$$

Because the concentration of water in dilute solutions of methylamine in water is essentially a constant ($[H_2O] = 55.5$ mol/L), it is combined with K_{eq} in a new constant called a *base ionization constant*, K_b. The value of K_b for methylamine is 4.37×10^{-4} ($pK_b = 3.36$):

$$K_b = K_{eq}[H_2O] = \frac{[CH_3NH_3^+][OH^-]}{[CH_3NH_2]} = 4.37 \times 10^{-4}$$

It is also common to discuss the basicity of amines by referring to the acid ionization constant of the corresponding conjugate acid, as illustrated for the ionization of the methylammonium ion:

$$CH_3NH_3^+ + H_2O \rightleftharpoons CH_3NH_2 + H_3O^+$$

$$K_a = \frac{[CH_3NH_2][H_3O^+]}{[CH_3NH_3^+]} = 2.29 \times 10^{-11} \qquad pK_a = 10.64$$

Values of pK_a and pK_b for any acid–conjugate base pair are related by the equation

$$pK_a + pK_b = 14.00$$

Values of pK_a and pK_b for selected amines are given in Table 10.2.

TABLE 10.2 Base Strengths (pK_b) of Selected Amines and Acid Strengths (pK_a) of Their Conjugate Acids*

Amine	Structure	pK_b	pK_a
Ammonia	NH_3	4.74	9.26
Primary Amines			
methylamine	CH_3NH_2	3.36	10.64
ethylamine	$CH_3CH_2NH_2$	3.19	10.81
cyclohexylamine	$C_6H_{11}NH_2$	3.34	10.66
Secondary Amines			
dimethylamine	$(CH_3)_2NH$	3.27	10.73
diethylamine	$(CH_3CH_2)_2NH$	3.02	10.98
Tertiary Amines			
trimethylamine	$(CH_3)_3N$	4.19	9.81
triethylamine	$(CH_3CH_2)_3N$	3.25	10.75
Aromatic Amines			
aniline	C_6H_5–NH_2	9.37	4.63
4-methylaniline	H_3C–C_6H_4–NH_2	8.92	5.08
4-chloroaniline	Cl–C_6H_4–NH_2	9.85	4.15
4-nitroaniline	O_2N–C_6H_4–NH_2	13.0	1.0
Heterocyclic Aromatic Amines			
pyridine	(pyridine ring, N)	8.75	5.25
imidazole	(imidazole ring, N, N–H)	7.05	6.95

* For each amine, $pK_a + pK_b = 14.00$.

EXAMPLE 10.4

Predict the position of equilibrium for this acid–base reaction:

$$CH_3NH_2 + CH_3COOH \rightleftharpoons CH_3NH_3^+ + CH_3COO^-$$

SOLUTION

Use the approach we developed in Section 2.5 to predict the position of equilibrium in acid–base reactions. Equilibrium favors reaction of the stronger acid and stronger base to form the weaker acid and the weaker base. Thus, in this reaction, equilibrium favors the formation of methylammonium ion and acetate ion:

$$CH_3NH_2 + CH_3COOH \rightleftharpoons CH_3NH_3^+ + CH_3COO^-$$

CH_3NH_2	CH_3COOH	$CH_3NH_3^+$	CH_3COO^-
	$pK_a = 4.76$	$pK_a = 10.64$	
Stronger base	Stronger acid	Weaker acid	Weaker base

Practice Problem 10.4

Predict the position of equilibrium for this acid–base reaction:

$$CH_3NH_3^+ + H_2O \rightleftharpoons CH_3NH_2 + H_3O^+$$

Given information such as that in Table 10.2, we can make the following generalizations about the acid–base properties of the various classes of amines:

1. All aliphatic amines have about the same base strength, pK_b 3.0–4.0, and are slightly stronger bases than ammonia.
2. Aromatic amines and heterocyclic aromatic amines are considerably weaker bases than are aliphatic amines. Compare, for example, values of pK_b for aniline and cyclohexylamine:

aromatic < aliphatic amines amines base strength

$$C_6H_{11}\text{-}NH_2 + H_2O \rightleftharpoons C_6H_{11}\text{-}NH_3^+OH^- \qquad pK_b = 3.34 \quad K_b = 4.5 \times 10^{-4}$$

Cyclohexylamine — Cyclohexylammonium hydroxide

$$C_6H_5\text{-}NH_2 + H_2O \rightleftharpoons C_6H_5\text{-}NH_3^+OH^- \qquad pK_b = 9.37 \quad K_b = 4.3 \times 10^{-10}$$

Aniline — Anilinium hydroxide

The base ionization constant for aniline is smaller (the larger the value of pK_b, the weaker is the base) than that for cyclohexylamine by a factor of 10^6.

Aromatic amines are weaker bases than are aliphatic amines because of the resonance interaction of the unshared pair on nitrogen with the pi system of the aromatic ring. Because no such resonance interaction is possible for an alkylamine, the electron pair on its nitrogen is more available for reaction with an acid:

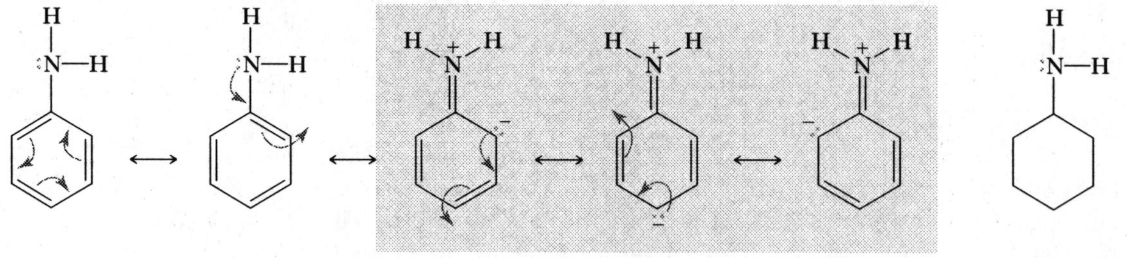

Two Kekulé structures

Interaction of the electron pair on nitrogen with the pi system of the aromatic ring

No resonance is possible with alkylamines

3. Electron-withdrawing groups such as halogen, nitro, and carbonyl decrease the basicity of substituted aromatic amines by decreasing the availability of the electron pair on nitrogen:

e⁻ withdrawing groups ↓ base strength of aromatic amines

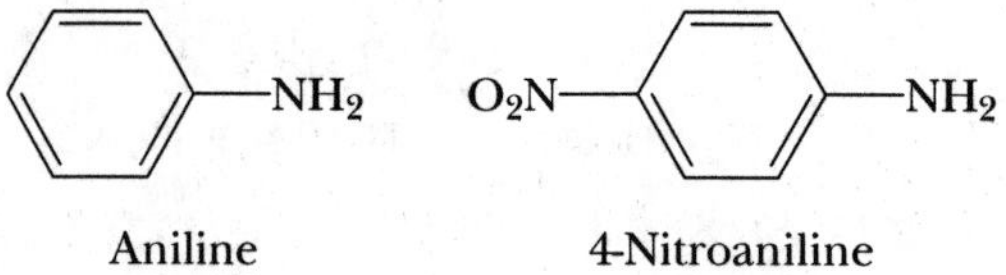

Aniline
pK_b 9.37

4-Nitroaniline
pK_b 13.0

Recall from Section 9.9B that these same substituents increase the acidity of phenols.

EXAMPLE 10.5

Select the stronger base in each pair of amines:

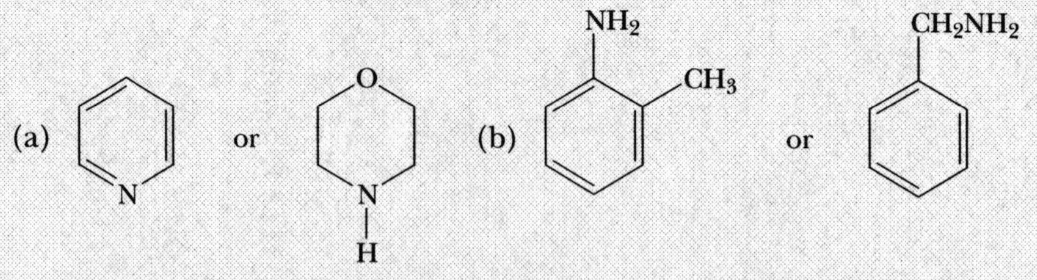

(a) (A) or (B) (b) (C) or (D)

SOLUTION

(a) Morpholine (B) is the stronger base (pK_b 5.79). It has a basicity comparable to that of secondary aliphatic amines. Pyridine (A), a heterocyclic aromatic amine (pK_b 8.75), is considerably less basic than aliphatic amines.
(b) Benzylamine (D), a primary aliphatic amine, is the stronger base (pK_b 3–4). *o*-Toluidine (C), an aromatic amine, is the weaker base (pK_b 9–10).

Practice Problem 10.5

Select the stronger acid from each pair of ions:

(a) $O_2N-C_6H_4-NH_3^+$ (A) or $H_3C-C_6H_4-NH_3^+$ (B)

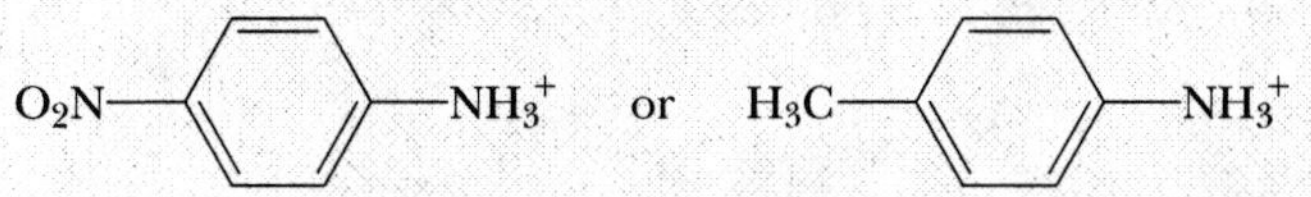

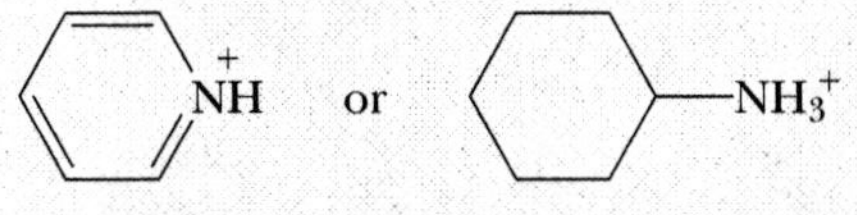

(b) Pyridinium ion, $C_5H_5NH^+$ (C) or cyclohexyl$-NH_3^+$ (D)

Guanidine, with pK_b 0.4, is the strongest base among neutral compounds:

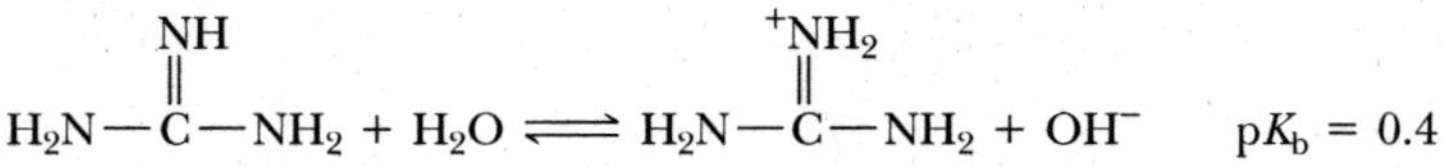

$$H_2N-C(=NH)-NH_2 + H_2O \rightleftharpoons H_2N-C(=\overset{+}{N}H_2)-NH_2 + OH^- \qquad pK_b = 0.4$$

Guanidine Guanidinium ion

The remarkable basicity of guanidine is attributed to the fact that the positive charge on the guanidinium ion is delocalized equally over the three nitrogen atoms, as shown by these three equivalent contributing structures:

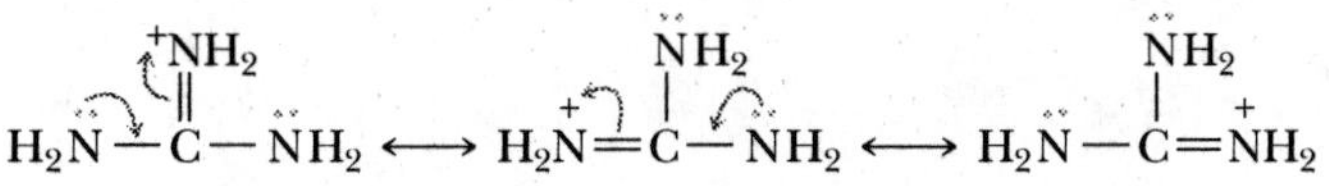

$$H_2\ddot{N}-C(=\overset{+}{N}H_2)-\ddot{N}H_2 \longleftrightarrow H_2\overset{+}{N}=C(-\ddot{N}H_2)-\ddot{N}H_2 \longleftrightarrow H_2\ddot{N}-C(-\ddot{N}H_2)=\overset{+}{N}H_2$$

Three equivalent contributing structures

Hence, the guanidinium ion is a highly stable cation. The presence of a guanidine group on the side chain of the amino acid arginine accounts for the basicity of its side chain (Section 19.2A).

10.6 REACTION WITH ACIDS

Amines, whether soluble or insoluble in water, react quantitatively with strong acids to form water-soluble salts, as illustrated by the reaction of (*R*)-norepinephrine (noradrenaline) with aqueous HCl to form a hydrochloride salt:

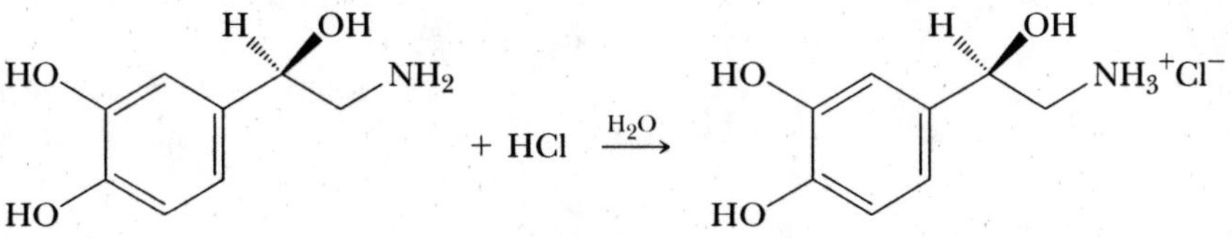

$$\text{(R)-Norepinephrine} + HCl \xrightarrow{H_2O} \text{(R)-Norepinephrine hydrochloride}$$

(*R*)-Norepinephrine (only slightly soluble in water)

(*R*)-Norepinephrine hydrochloride (a water-soluble salt)

Norepinephrine, secreted by the medulla of the adrenal gland, is a neurotransmitter. It has been suggested that it is a neurotransmitter in those areas of the brain which mediate emotional behavior.

EXAMPLE 10.6

Complete each acid–base reaction, and name the salt formed:

(a) $(CH_3CH_2)_2NH + HCl \longrightarrow$ (b) pyridine $+ CH_3COOH \longrightarrow$

SOLUTION

(a) $(CH_3CH_2)_2NH_2^+Cl^-$

Diethylammonium chloride

(b)

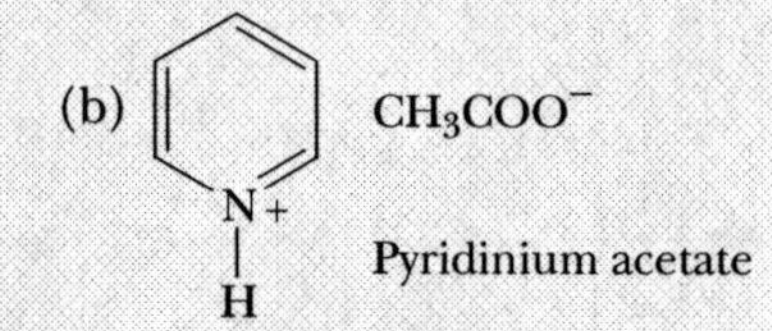

Practice Problem 10.6

Complete each acid–base reaction and name the salt formed:

(a) $(CH_3CH_2)_3N + HCl \longrightarrow$ (b) piperidine (cyclic NH) $+ CH_3COOH \longrightarrow$

The basicity of amines and the solubility of amine salts in water can be used to separate amines from water-insoluble, nonbasic compounds. Shown in Figure 10.2 is a flowchart for the separation of aniline from anisole. Note that aniline is recovered from its salt by treatment with NaOH.

EXAMPLE 10.7

Following are two structural formulas for alanine (2-aminopropanoic acid), one of the building blocks of proteins (Chapter 19):

$$\underset{\text{(A)}}{CH_3\underset{\displaystyle NH_2}{\underset{|}{C}}H\overset{\displaystyle O}{\overset{\|}{C}}OH} \quad \text{or} \quad \underset{\text{(B)}}{CH_3\underset{\displaystyle NH_3^+}{\underset{|}{C}}H\overset{\displaystyle O}{\overset{\|}{C}}O^-}$$

Is alanine better represented by structural formula (A) or structural formula (B)?

SOLUTION

Structural formula (A) contains both an amino group (a base) and a carboxyl group (an acid). Proton transfer from the stronger acid (—COOH) to the stronger base (—NH_2) gives an internal salt; therefore, (B) is the better representation for alanine. Within the field of amino acid chemistry, the internal salt represented by (B) is called a **zwitterion** (Chapter 19).

Practice Problem 10.7

As shown in Example 10.7, alanine is better represented as an internal salt. Suppose that the internal salt is dissolved in water.

(a) In what way would you expect the structure of alanine in aqueous solution to change if concentrated HCl were added to adjust the pH of the solution to 2.0?
(b) In what way would you expect the structure of alanine in aqueous solution to change if concentrated NaOH were added to bring the pH of the solution to 12.0?

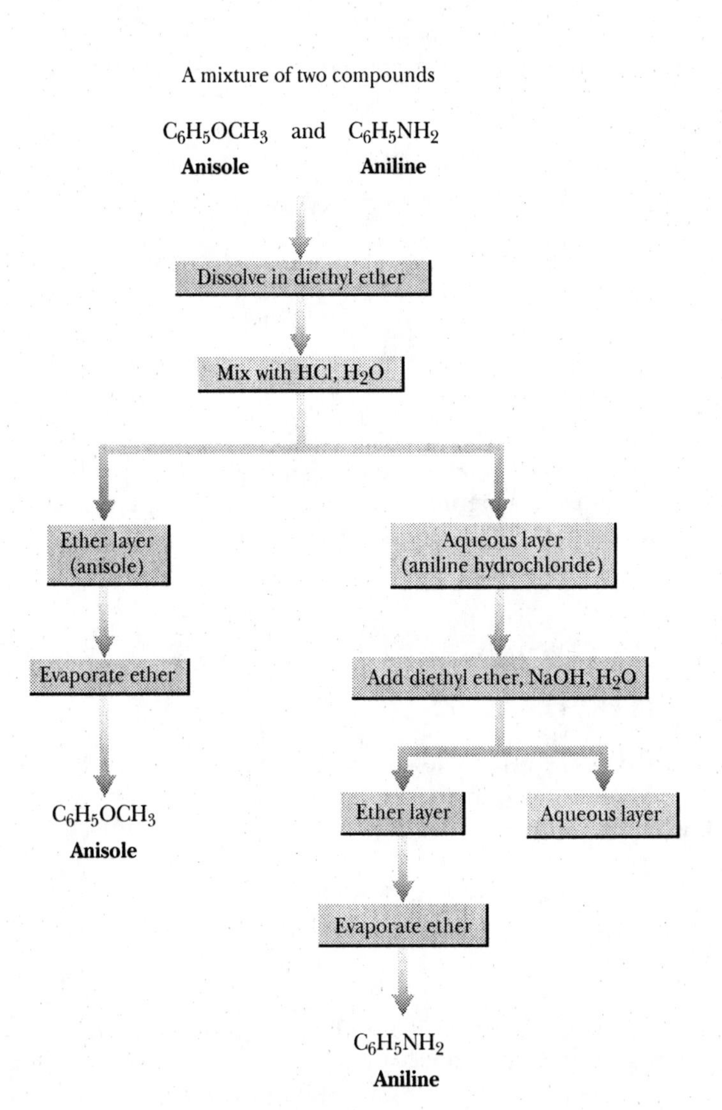

Figure 10.2
Separation and purification of an amine and a neutral compound.

10.7 SYNTHESIS OF ARYLAMINES: REDUCTION OF THE —NO_2 GROUP

As we have already seen (Section 9.7B), the nitration of an aromatic ring introduces a NO_2 group. A particular value of nitration is the fact that the resulting nitro group can be reduced to a primary amino group, —NH_2, by hydrogenation in the presence of a transition metal catalyst such as nickel, palladium, or platinum:

$$\text{3-Nitrobenzoic acid } (m\text{-}O_2NC_6H_4COOH) + 3H_2 \xrightarrow[\text{(3 atm)}]{\text{Ni}} \text{3-Aminobenzoic acid } (m\text{-}H_2NC_6H_4COOH) + 2H_2O$$

This method has the potential disadvantage that other susceptible groups, such as a carbon–carbon double bond, and the carbonyl group of an aldehyde or ketone, may also be reduced. Note that neither the —COOH nor the aromatic ring is reduced under these conditions.

Alternatively, a nitro group can be reduced to a primary amino group by a metal in acid:

$$\text{2,4-Dinitrotoluene} \xrightarrow[C_2H_5OH,\ H_2O]{Fe,\ HCl} CH_3C_6H_3(NH_3^+Cl^-)_2 \xrightarrow{NaOH,\ H_2O} \text{2,4-Diaminotoluene}$$

2,4-Dinitrotoluene   2,4-Diaminotoluene

The most commonly used metal-reducing agents are iron, zinc, and tin in dilute HCl. When reduced by this method, the amine is obtained as a salt, which is then treated with a strong base to liberate the free amine.

10.8 REACTION OF PRIMARY AROMATIC AMINES WITH NITROUS ACID

Nitrous acid, HNO_2, is an unstable compound that is prepared by adding sulfuric or hydrochloric acid to an aqueous solution of sodium nitrite, $NaNO_2$. Nitrous acid is a weak acid and ionizes according to the following equation:

$$HNO_2 + H_2O \rightleftharpoons H_3O^+ + NO_2^- \qquad K_a = 4.26 \times 10^{-4}$$

Nitrous acid $pK_a = 3.37$

Nitrous acid reacts with amines in different ways, depending on whether the amine is primary, secondary, or tertiary and whether it is aliphatic or aromatic. We concentrate on the reaction of nitrous acid with primary aromatic amines, because this reaction is useful in organic synthesis.

Treatment of a primary aromatic amine—for example, aniline—with nitrous acid gives a diazonium salt:

$$C_6H_5{-}NH_2 + NaNO_2 + HCl \xrightarrow[0°C]{H_2O} C_6H_5{-}\overset{+}{N}{\equiv}N\ Cl^- + H_2O$$

Aniline (a 1° aromatic amine)   Sodiun nitrite   Benzenediazonium chloride

We can also write the equation for this reaction in the following more abbreviated form:

$$C_6H_5{-}NH_2 \xrightarrow[0°C]{NaNO_2,\ HCl} C_6H_5{-}N_2^+Cl^-$$

Benzenediazonium chloride

When we warm an aqueous solution of an arenediazonium salt, the $—N_2^+$ group is replaced by an —OH group. This reaction is one of the few methods we have for the synthesis of phenols. It enables us to convert an aromatic amine to a phenol by first forming the arenediazonium salt and then heating the solution. In this manner, we can convert 2-bromo-4-methylaniline to 2-bromo-4-methylphenol:

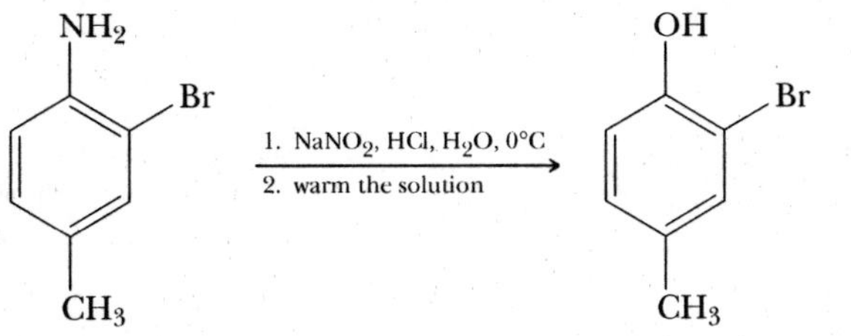

NH_2 Br — 1. $NaNO_2$, HCl, H_2O, 0°C; 2. warm the solution → OH Br
CH_3 CH_3

2-Bromo-4-methylaniline 2-Bromo-4-methylphenol

EXAMPLE 10.8

Show the reagents that will bring about each step in this conversion of toluene to 4-hydroxybenzoic acid:

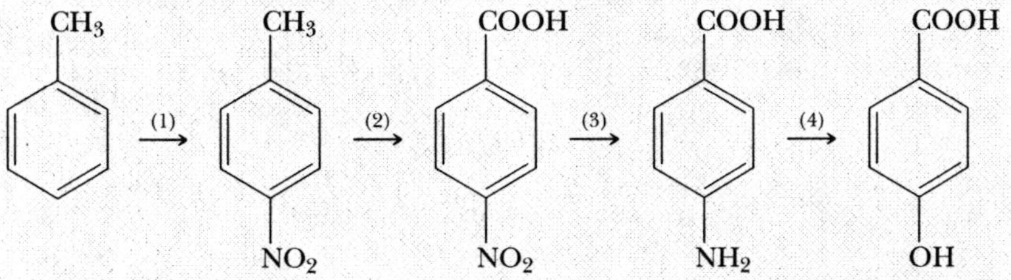

CH_3 (1)→ CH_3 / NO_2 (2)→ COOH / NO_2 (3)→ COOH / NH_2 (4)→ COOH / OH

Toluene 4-Hydroxybenzoic acid

SOLUTION

Step 1: Nitration of toluene, using nitric acid/sulfuric acid (Section 9.7B), followed by separation of the ortho and para isomers.

Step 2: Oxidation of the benzylic carbon, using chromic acid (Section 9.5).

Step 3: Reduction of the nitro group, either using H_2 in the presence of a transition metal catalyst or using Fe, Sn, or Zn in the presence of aqueous HCl (Section 10.8).

Step 4: Treatment of the aromatic amine with $NaNO_2$/HCl to form the diazonium ion salt and then warming the solution.

Practice Problem 10.8

Show how you can use the same set of steps in Example 10.8, but in a different order, to convert toluene to 3-hydroxybenzoic acid.

Treatment of arenediazonium salt with hypophosphorous acid, H_3PO_2, reduces the diazonium group and replaces it with —H, as illustrated by the conversion of aniline to 1,3,5-trichlorobenzene. Recall that the $—NH_2$ group is a powerful activating and ortho-para directing group (Section 9.8A). Treatment of aniline with chlorine requires no catalyst and gives 2,4,6-trichloroaniline (to complete the conversion, we treat the trichloroaniline with nitrous acid followed by hypophosphorous acid):

$$\text{Aniline} \xrightarrow{Cl_2} \text{2,4,6-trichloroaniline} \xrightarrow[0°C]{NaNO_2, HCl} \text{2,4,6-trichlorobenzenediazonium } (N_2^+Cl^-) \xrightarrow{H_3PO_2} \text{1,3,5-Trichloro-aniline}$$

Aniline

1,3,5-Trichloro-aniline

SUMMARY

Amines are classified as **primary**, **secondary**, or **tertiary**, depending on the number of hydrogen atoms of ammonia replaced by alkyl or aryl groups (Section 10.2). In an **aliphatic amine**, all carbon atoms bonded to nitrogen are derived from alkyl groups. In an **aromatic amine**, one or more of the groups bonded to nitrogen are aryl groups. A **heterocyclic amine** is an amine in which the nitrogen atom is part of a ring. A **heterocyclic aromatic amine** is an amine in which the nitrogen atom is part of an aromatic ring.

In systematic nomenclature, aliphatic amines are named **alkanamines** (Section 10.3A). In the common system of nomenclature (Section 10.3B), aliphatic amines are named **alkylamines**; the alkyl groups are listed in alphabetical order in one word ending in the suffix *-amine*. An ion containing nitrogen bonded to four alkyl or aryl groups is named as a **quaternary ammonium ion**.

Amines are polar compounds, and primary and secondary amines associate by intermolecular hydrogen bonding (Section 10.4). Because an N—H----N hydrogen bond is weaker than an O—H----O hydrogen bond, amines have lower boiling points than alcohols of comparable molecular weight and structure. All classes of amines form hydrogen bonds with water and are more soluble in water than are hydrocarbons of comparable molecular weight.

Amines are weak bases, and aqueous solutions of amines are basic (Section 10.5). The base ionization constant for an amine in water is given the symbol K_b. It is also common to discuss the acid–base properties of amines by reference to the acid ionization constant, K_a, for the conjugate acid of the amine. Acid and base ionization constants for an amine in water are related by the equation $pK_a + pK_b = 14.0$.

KEY REACTIONS

1. Basicity of Aliphatic Amines (Section 10.5)

Most aliphatic amines have comparable basicities (pK_b 3.0–4.0) and are slightly stronger bases than ammonia:

$$CH_3NH_2 + H_2O \rightleftharpoons CH_3NH_3^+ + OH^- \qquad pK_b = 3.36$$

2. Basicity of Aromatic Amines (Section 10.5)

Aromatic amines (pK_b 9.0–10.0) are considerably weaker bases than are aliphatic amines. Resonance stabilization from interaction of the unshared electron pair on nitrogen with the pi system of the aromatic ring decreases the availability of that electron pair for reaction with an acid. Substitution on the ring by electron-withdrawing groups decreases the basicity of the —NH_2 group:

$$C_6H_5\text{—}NH_2 + H_2O \rightleftharpoons C_6H_5\text{—}NH_3^+ + OH^- \qquad pK_b = 9.37$$

3. Reaction of Amines with Strong Acids (Section 10.6)

All amines react quantitatively with strong acids to form water-soluble salts:

$$C_6H_5N(CH_3)_2 + HCl \longrightarrow C_6H_5\overset{+}{N}H(CH_3)_2\,Cl^-$$

Insoluble in water — A water-soluble salt

4. Reduction of an Aromatic NO_2 group (Section 10.7)

An NO_2 group on an aromatic ring can be reduced to an amino group by catalytic hydrogenation or by treatment with a metal and hydrochloric acid, followed by a strong base to liberate the free amine:

$$C_6H_5NO_2 + 3H_2 \xrightarrow[(3\ atm)]{Ni} C_6H_5NH_2$$

$$C_6H_4(NO_2)_2 \xrightarrow[C_2H_5OH, H_2O]{Fe,\ HCl} C_6H_4(NH_3^+\,Cl^-)_2 \xrightarrow{NaOH,\ H_2O} C_6H_4(NH_2)_2$$

5. Conversion of a Primary Aromatic Amine to a Phenol (Section 10.8)

Treatment of a primary aromatic amine with nitrous acid gives an arenediazonium salt. Heating the aqueous solution of this salt brings about the evolution of N_2 and forms a phenol:

$$CH_3C_6H_4NH_2 \xrightarrow[0°C]{NaNO_2,\ HCl} CH_3C_6H_4N_2^+\,Cl^- \xrightarrow{heat} CH_3C_6H_4OH$$

6. Reduction of an arenediazonium Salt (Section 10.8)

Treatment of an arenediazonium salt with hypophosphorous acid, H_3PO_2, results in replacement of the N_2^+ group by H:

$$Cl_2C_6H_3NH_2 \xrightarrow[0°C]{NaNO_2,\ HCl} Cl_2C_6H_3N_2^+\,Cl^- \xrightarrow{H_3PO_2} Cl_2C_6H_4$$

PROBLEMS

A problem number set in red indicates an applied "real-world" problem.

Structure and Nomenclature

10.9 Draw a structural formula for each amine:

(a) (*R*)-2-Butanamine
(b) 1-Octanamine
(c) 2,2-Dimethyl-1-propanamine
(d) 1,5-Pentanediamine
(e) 2-Bromoaniline
(f) Tributylamine
(g) *N,N*-Dimethylaniline
(h) Benzylamine
(i) *tert*-Butylamine
(j) *N*-Ethylcyclohexanamine
(k) Diphenylamine
(l) Isobutylamine

10.10 Draw a structural formula for each amine:

(a) 4-Aminobutanoic acid
(b) 2-Aminoethanol (ethanolamine)
(c) 2-Aminobenzoic acid
(d) (*S*)-2-Aminopropanoic acid (alanine)
(e) 4-Aminobutanal
(f) 4-Amino-2-butanone

10.11 Draw examples of 1°, 2°, and 3° amines that contain at least four sp^3 hybridized carbon atoms. Using the same criterion, provide examples of 1°, 2°, and 3° alcohols. How does the classification system differ between the two functional groups?

10.12 Classify each amino group as primary, secondary, or tertiary and as aliphatic or aromatic:

(a) Benzocaine
(a topical anesthetic)

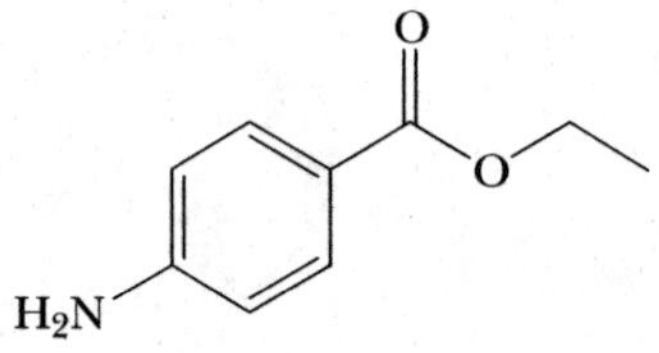

(b) Chloroquine
(a drug for the treatment of malaria)

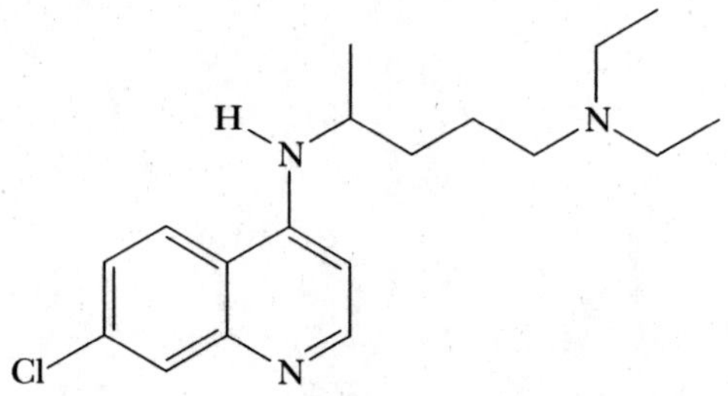

10.13 Epinephrine is a hormone secreted by the adrenal medulla. Among epinephrine's actions, it is a bronchodilator. Albuterol, sold under several trade names, including Proventil® and Salbumol®, is one of the most effective and widely prescribed anti-asthma drugs. The R enantiomer of albuterol is 68 times more effective in the treatment of asthma than the S enantiomer.

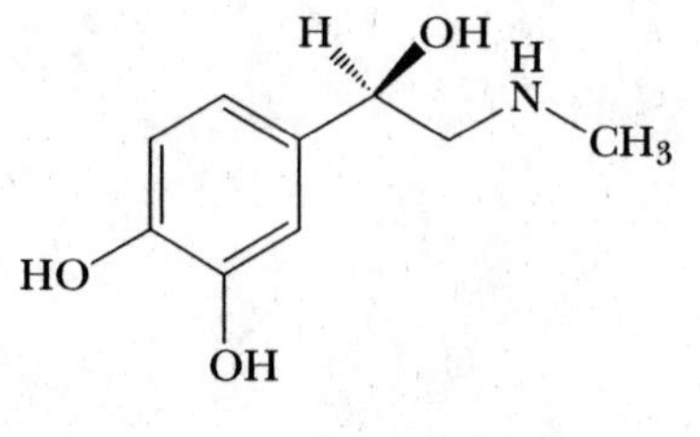

(*R*)-Epinephrine
(Adrenaline)

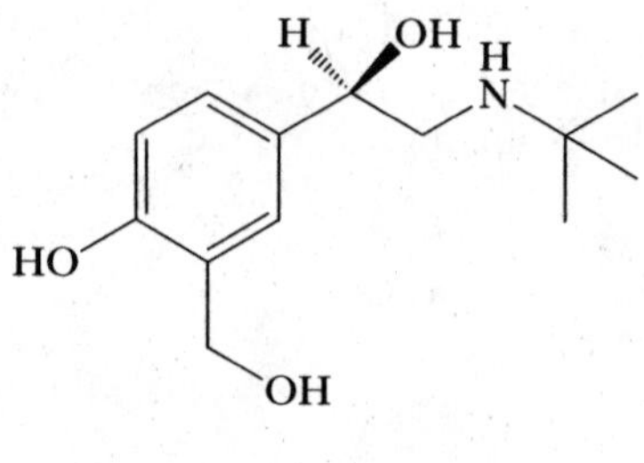

(*R*)-Albuterol

(a) Classify each amino group as primary, secondary, or tertiary.
(b) List the similarities and differences between the structural formulas of these compounds.

10.14 There are eight constitutional isomers with molecular formula $C_4H_{11}N$. Name and draw structural formulas for each. Classify each amine as primary, secondary, or tertiary.

10.15 Draw a structural formula for each compound with the given molecular formula:

(a) A 2° arylamine, C_7H_9N **(b)** A 3° arylamine, $C_8H_{11}N$
(c) A 1° aliphatic amine, C_7H_9N **(d)** A chiral 1° amine, $C_4H_{11}N$
(e) A 3° heterocyclic amine, $C_5H_{11}N$ **(f)** A trisubstituted 1° arylamine, $C_9H_{13}N$
(g) A chiral quaternary ammonium salt, $C_9H_{22}NCl$

Physical Properties

10.16 Propylamine, ethylmethylamine, and trimethylamine are constitutional isomers with molecular formula C_3H_9N:

$CH_3CH_2CH_2NH_2$	$CH_3CH_2NHCH_3$	$(CH_3)_3N$
bp 48°C	bp 37°C	bp 3°C
Propylamine	Ethylmethylamine	Trimethylamine

Account for the fact that trimethylamine has the lowest boiling point of the three, and propylamine has the highest.

10.17 Account for the fact that 1-butanamine has a lower boiling point than 1-butanol:

NH$_2$ OH
bp 78°C bp 117°C
1-Butanamine 1-Butanol

10.18 Account for the fact that putrescine, a foul-smelling compound produced by rotting flesh, ceases to smell upon treatment with two equivalents of HCl:

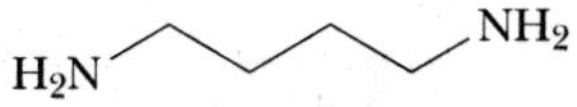

H$_2$N NH$_2$
1,4-Butanediamine
(Putrescine)

Basicity of Amines

10.19 Account for the fact that amines are more basic than alcohols.

10.20 From each pair of compounds, select the stronger base:

(a) N H or N **(b)** N(CH$_3$)$_2$ or N(CH$_3$)$_2$

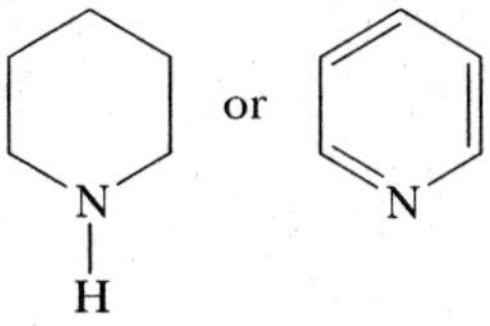

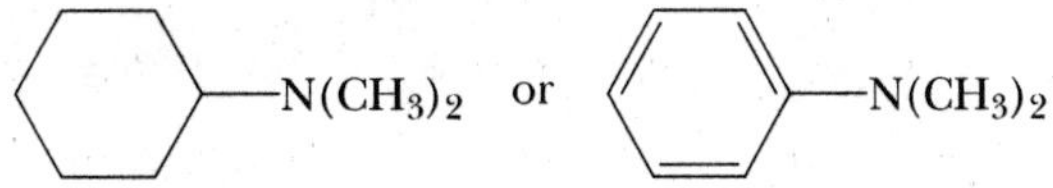

(c) NH$_2$ CH$_3$ or CH$_2$NH$_2$ **(d)** NH$_2$ O$_2$N or NH$_2$ H$_3$C

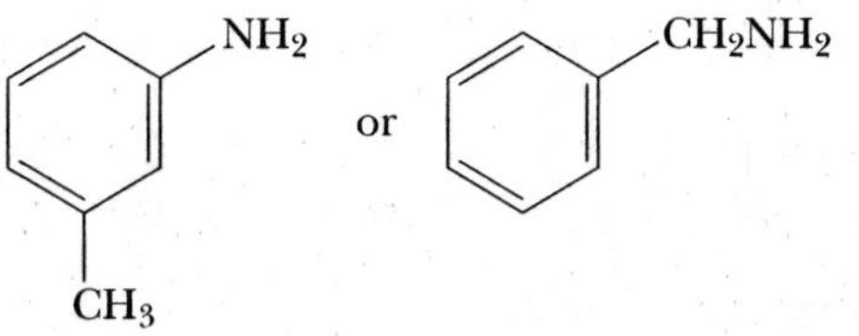

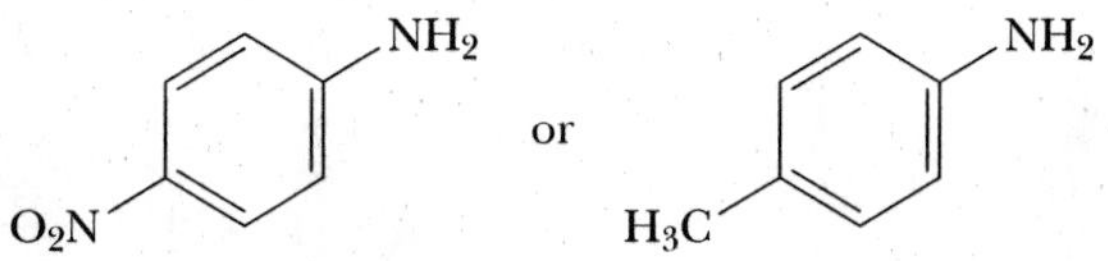

10.21 Account for the fact that substitution of a nitro group makes an aromatic amine a weaker base, but makes a phenol a stronger acid. For example, 4-nitroaniline is a weaker base than aniline, but 4-nitrophenol is a stronger acid than phenol.

10.22 Select the stronger base in this pair of compounds:

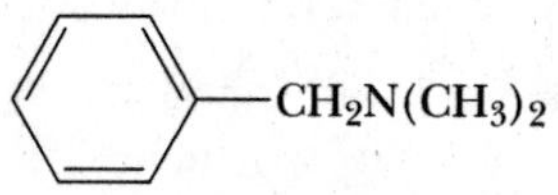

$C_6H_5-CH_2N(CH_3)_2$ or $C_6H_5-CH_2\overset{+}{N}(CH_3)_3\ OH^-$

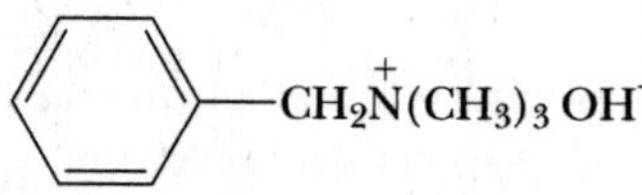

10.23 Complete the following acid–base reactions and predict the position of equilibrium for each. Justify your prediction by citing values of pK_a for the stronger and weaker acid in each equilibrium. For values of acid ionization constants, consult Table 2.2 (pK_a's of some inorganic and organic acids), Table 8.2 (pK_a's of alcohols), Section 9.9B (acidity of phenols), and Table 10.2 (base strengths of amines). Where no ionization constants are given, make the best estimate from aforementioned tables and section.

(a) 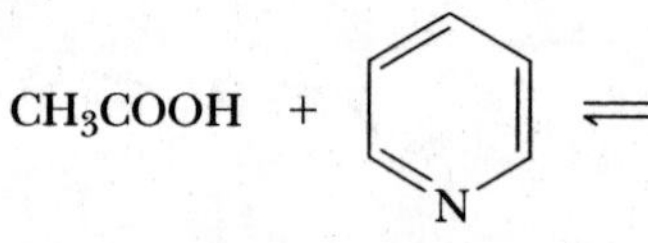 CH_3COOH + Pyridine $\rightleftharpoons$

Acetic acid   Pyridine

(b) 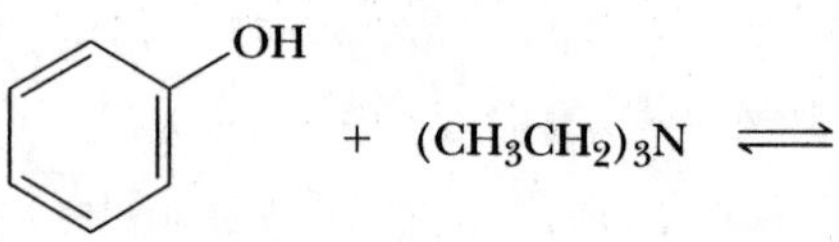 Phenol (OH) + $(CH_3CH_2)_3N$ $\rightleftharpoons$

Phenol   Triethylamine

(c) $PhCH_2CH(CH_3)NH_2 + CH_3CH(OH)COOH \rightleftharpoons$

1-Phenyl-2-propanamine (Amphetamine)   2-Hydroxypropanoic acid (Lactic acid)

(d) $PhCH_2CH(CH_3)NHCH_3 + CH_3COOH \rightleftharpoons$

Methamphetamine   Acetic acid

10.24 The pK_a of the morpholinium ion is 8.33:

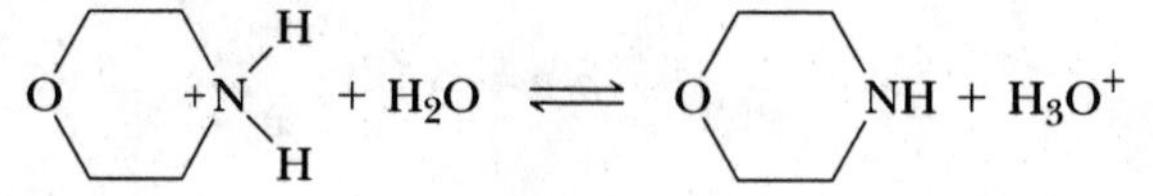

$$\text{Morpholinium ion } (O(CH_2CH_2)_2\overset{+}{N}H_2) + H_2O \rightleftharpoons \text{Morpholine } (O(CH_2CH_2)_2NH) + H_3O^+ \qquad pK_a = 8.33$$

Morpholinium ion   Morpholine

(a) Calculate the ratio of morpholine to morpholinium ion in aqueous solution at pH 7.0.

(b) At what pH are the concentrations of morpholine and morpholinium ion equal?

10.25 The pK_b of amphetamine (Example 10.2) is approximately 3.2. Calculate the ratio of amphetamine to its conjugate acid at pH 7.4, the pH of blood plasma.

10.26 Calculate the ratio of amphetamine to its conjugate acid at pH 1.0, such as might be present in stomach acid.

10.27 Following is a structural formula of pyridoxamine, one form of vitamin B_6:

CH_2NH_2 HO CH_2OH H_3C N

Pyridoxamine
(Vitamin B_6)

(a) Which nitrogen atom of pyridoxamine is the stronger base?

(b) Draw the structural formula of the hydrochloride salt formed when pyridoxamine is treated with one mole of HCl.

10.28 Epibatidine, a colorless oil isolated from the skin of the Ecuadorian poison frog *Epipedobates tricolor*, has several times the analgesic potency of morphine. It is the first chlorine-containing, nonopioid (nonmorphine-like in structure) analgesic ever isolated from a natural source:

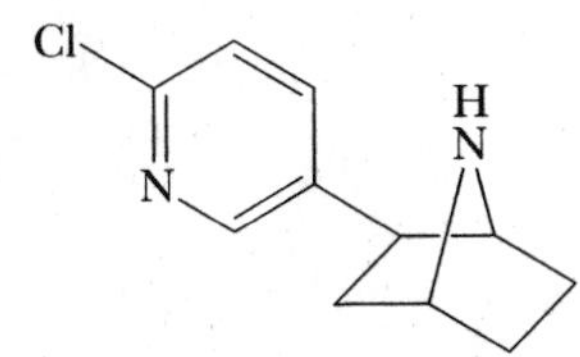

Epibatidine

Poison arrow frog. *(Stephen J. Krasemann/ Photo Researchers, Inc.)*

(a) Which of the two nitrogen atoms of epibatidine is the more basic?

(b) Mark all stereocenters in this molecule.

10.29 Procaine was one of the first local anesthetics for infiltration and regional anesthesia:

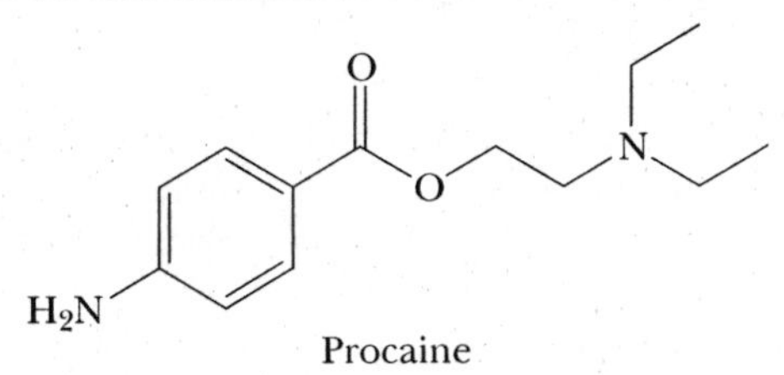

Procaine

The hydrochloride salt of procaine is marketed as Novocaine®.

(a) Which nitrogen atom of procaine is the stronger base?

(b) Draw the formula of the salt formed by treating procaine with one mole of HCl.

(c) Is procaine chiral? Would a solution of Novocaine in water be optically active or optically inactive?

10.30 Treatment of trimethylamine with 2-chloroethyl acetate gives the neurotransmitter acetylcholine as its chloride salt:

$$(CH_3)_3N + CH_3\overset{\overset{\displaystyle O}{\|}}{C}OCH_2CH_2Cl \longrightarrow C_7H_{16}ClNO_2$$

Acetylcholine chloride

Propose a structural formula for this quaternary ammonium salt and a mechanism for its formation.

10.31 Aniline is prepared by the catalytic reduction of nitrobenzene:

$$C_6H_5{-}NO_2 \xrightarrow[Ni]{H_2} C_6H_5{-}NH_2$$

Devise a chemical procedure based on the basicity of aniline to separate it from any unreacted nitrobenzene.

10.32 Suppose that you have a mixture of the following three compounds:

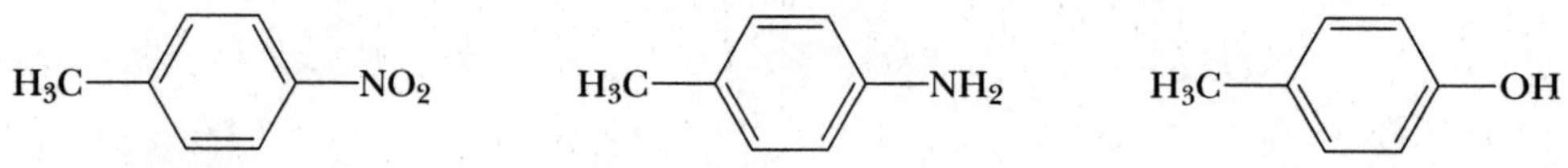

$H_3C-C_6H_4-NO_2$ — 4-Nitrotoluene (*p*-Nitrotoluene)

$H_3C-C_6H_4-NH_2$ — 4-Methylaniline (*p*-Toluidine)

$H_3C-C_6H_4-OH$ — 4-Methylphenol (*p*-Cresol)

Devise a chemical procedure based on their relative acidity or basicity to separate and isolate each in pure form.

10.33 Following is a structural formula for metformin, the hydrochloride salt of which is marketed as the antidiabetic Glucophage®:

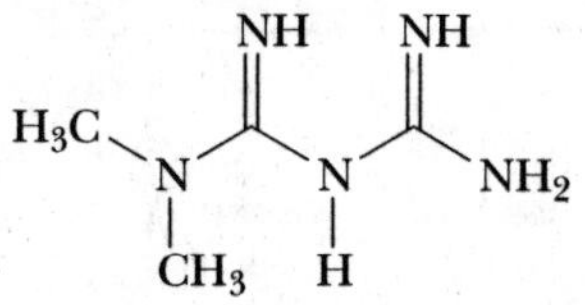

$(CH_3)_2N-C(=NH)-NH-C(=NH)-NH_2$

Metformin

Metformin was introduced into clinical medicine in the United States in 1995 for the treatment of type 2 diabetes. More than 25 million prescriptions for this drug were written in 2000, making it the most commonly prescribed brand-name diabetes medication in the nation.

(a) Draw the structural formula for Glucophage®.

(b) Would you predict Glucophage® to be soluble or insoluble in water? Soluble or insoluble in blood plasma? Would you predict it to be soluble or insoluble in diethyl ether? In dichloromethane? Explain your reasoning.

Synthesis

10.34 4-Aminophenol is a building block in the synthesis of the analgesic acetaminophen. Show how this building block can be synthesized in two steps from phenol (in Chapter 15, we will see how to complete the synthesis of acetaminophen):

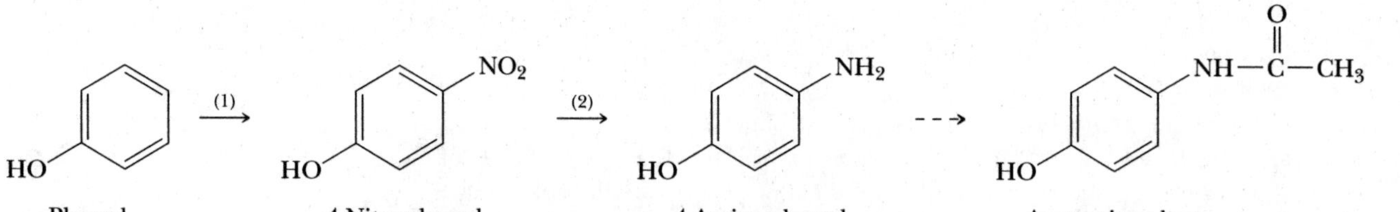

Phenol $\xrightarrow{(1)}$ 4-Nitrophenol $\xrightarrow{(2)}$ 4-Aminophenol $\dashrightarrow$ Acetaminophen ($HO-C_6H_4-NH-C(=O)-CH_3$)

10.35 4-Aminobenzoic acid is a building block in the synthesis of the topical anesthetic benzocaine. Show how this building block can be synthesized in three steps from toluene (in Chapter 15, we will see how to complete the synthesis of benzocaine):

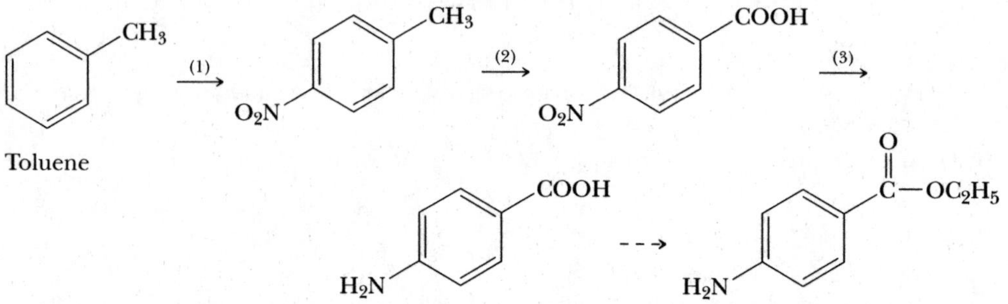

Toluene $\xrightarrow{(1)}$ $O_2N-C_6H_4-CH_3$ $\xrightarrow{(2)}$ $O_2N-C_6H_4-COOH$ $\xrightarrow{(3)}$ 4-Aminobenzoic acid ($H_2N-C_6H_4-COOH$) $\dashrightarrow$ Ethyl 4-aminobenzoate (Benzocaine) ($H_2N-C_6H_4-C(=O)-OC_2H_5$)

10.36 The compound 4-aminosalicylic acid is one of the building blocks needed for the synthesis of propoxycaine, one of the family of "caine" anesthetics. Some other members of this family of local anesthetics are procaine (Novocaine®), lidocaine (Xylocaine®), and mepivicaine (Carbocaine®). 4-Aminosalicylic acid is synthesized from salicylic acid in five steps (in Chapter 15, we will see how to complete the synthesis of propoxycaine):

OH COOH (1)→ O_2N OH COOH (2)→ H_2N OH COOH (3)→

Salicylic acid

O_2N H_2N OH COOH (4)→ O_2N OH COOH

(5)→ H_2N OH COOH - - → H_2N O O N O

4-Aminosalicylic acid Propoxycaine

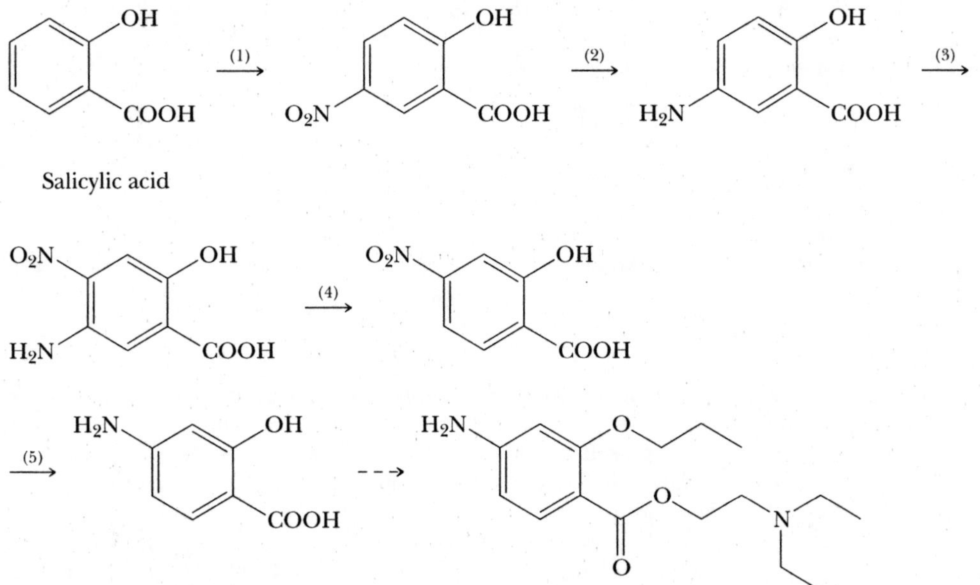

Show reagents that will bring about the synthesis of 4-aminosalicylic acid.

10.37 A second building block for the synthesis of propoxycaine is 2-diethylaminoethanol:

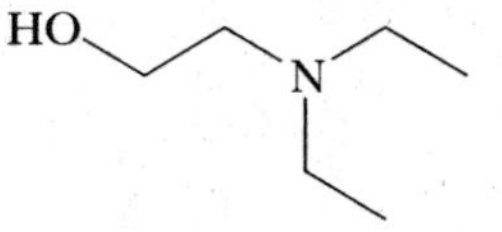

HO N

2-Diethylaminoethanol

Show how this compound can be prepared from ethylene oxide and diethylamine.

10.38 Following is a two-step synthesis of the antihypertensive drug propranolol, a so-called beta-blocker with vasodilating action:

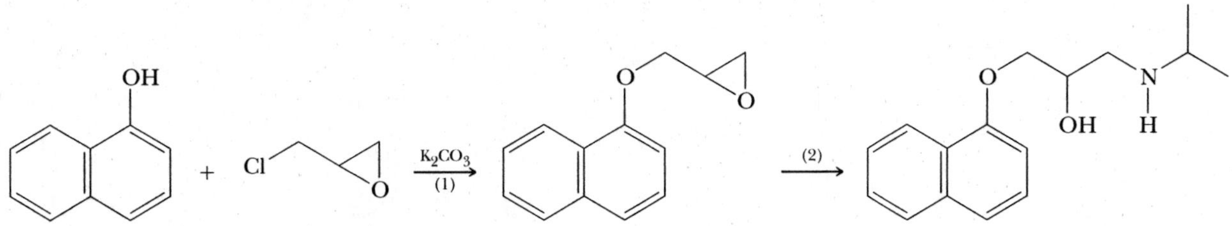

OH + Cl O $\xrightarrow[(1)]{K_2CO_3}$ O O (2)→ O OH N H

1-Naphthol Epichlorohydrin Propranolol (Cardinol)

Propranolol and other beta blockers have received enormous clinical attention because of their effectiveness in treating hypertension (high blood pressure), migraine headaches, glaucoma, ischemic heart disease, and certain cardiac arrhythmias. The hydrochloride salt of propranolol has been marketed under at least 30 brand names, one of which is Cardinol®. (Note the "card-" part of the name, after *cardiac*.)

(a) What is the function of potassium carbonate, K_2CO_3, in Step 1? Propose a mechanism for the formation of the new oxygen–carbon bond in this step.

(b) Name the amine used to bring about Step 2, and propose a mechanism for this step.

(c) Is propranolol chiral? If so, how many stereoisomers are possible for it?

10.39 The compound 4-ethoxyaniline, a building block of the over-the-counter analgesic phenacetin, is synthesized in three steps from phenol:

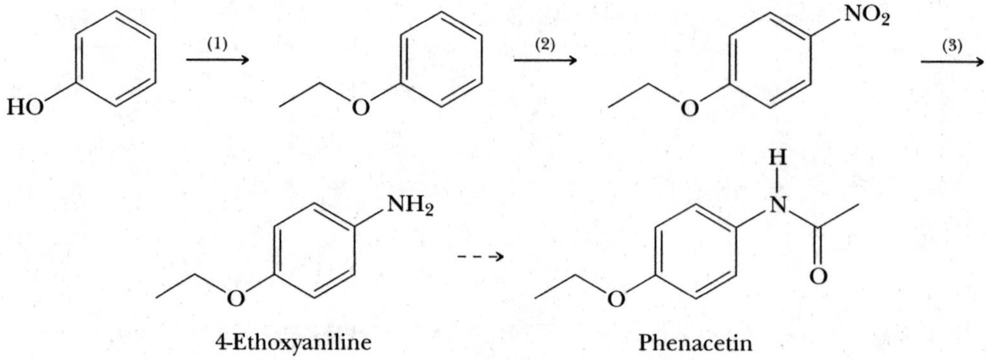

4-Ethoxyaniline

Phenacetin

Show reagents for each step of the synthesis of 4-ethoxyaniline. (In Chapter 15, we will see how to complete this synthesis.)

10.40 Radiopaque imaging agents are substances administered either orally or intravenously that absorb X rays more strongly than body material does. One of the best known of these agents is barium sulfate, the key ingredient in the "barium cocktail" used for imaging of the gastrointestinal tract. Among other X-ray imaging agents are the so-called triiodoaromatics. You can get some idea of the kinds of imaging for which they are used from the following selection of trade names: Angiografin®, Gastrografin, Cardiografin, Cholografin, Renografin, and Urografin®. The most common of the triiodiaromatics are derivatives of these three triiodobenzenecarboxylic acids:

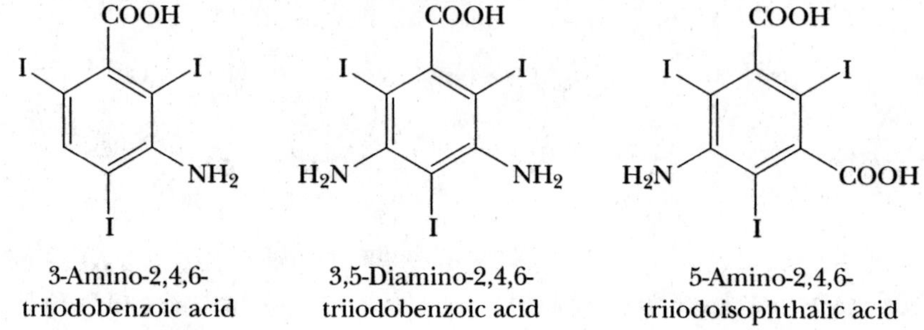

3-Amino-2,4,6-triiodobenzoic acid

3,5-Diamino-2,4,6-triiodobenzoic acid

5-Amino-2,4,6-triiodoisophthalic acid

3-Amino-2,4,6-triiodobenzoic acid is synthesized from benzoic acid in three steps:

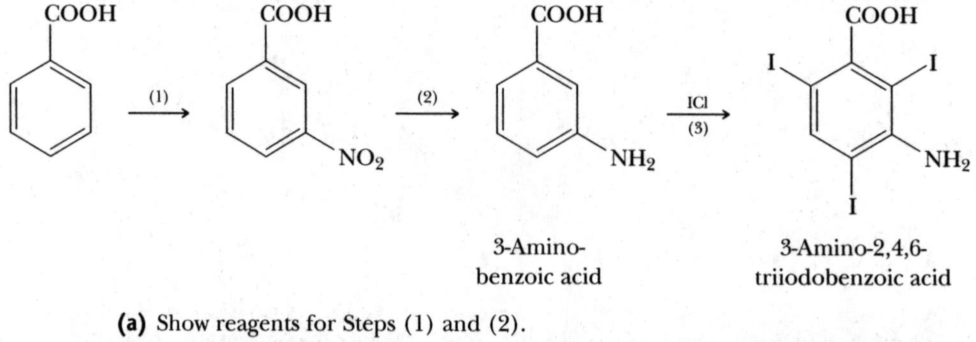

3-Amino-benzoic acid

3-Amino-2,4,6-triiodobenzoic acid

(a) Show reagents for Steps (1) and (2).

(b) Iodine monochloride, ICl, a black crystalline solid with a melting point of 27.2°C and a boiling point of 97°C, is prepared by mixing equimolar amounts of I_2 and Cl_2. Propose a mechanism for the iodination of 3-aminobenzoic acid by this reagent.

(c) Show how to prepare 3,5-diamino-2,4,6-triiodobenzoic acid from benzoic acid.

(d) Show how to prepare 5-amino-2,4,6-triiodoisophthalic acid from isophthalic acid (1,3-benzenedicarboxylic acid).

10.41 The intravenous anesthetic propofol is synthesized in four steps from phenol:

Phenol → (1) → (2) → (3) → (4) → Propofol

Show reagents to bring about each step.

Looking Ahead

10.42 State the hybridization of the nitrogen atom in each of the following compounds:

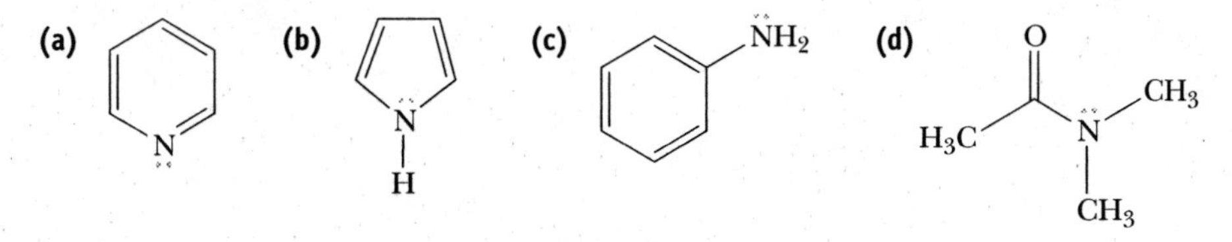

10.43 Amines can act as nucleophiles. For each of the following molecules, circle the most likely atom that would be attacked by the nitrogen of an amine:

(a) (b) O, OCH_3 (c) Cl, Br

10.44 Draw a Lewis structure for a molecule with formula C_3H_7N that does not contain a ring or an alkene (a carbon–carbon double bond).

10.45 Rank the following leaving groups in order from best to worst:

R—Cl R—O—C(=O)—R R—OCH_3 R—$N(CH_3)_2$

13 Aldehydes and Ketones

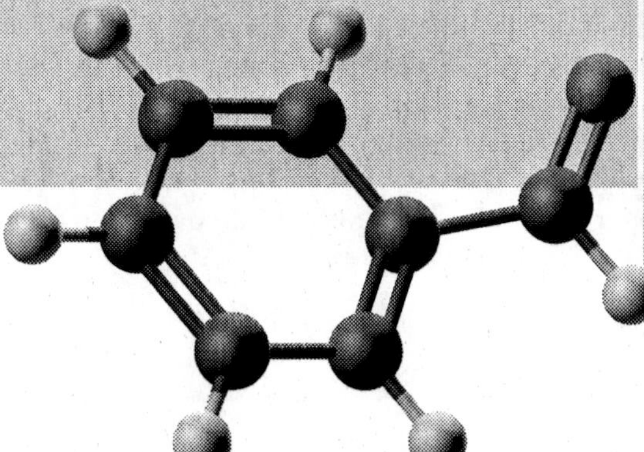

Benzaldehyde is found in the kernels of bitter almonds. Cinnamaldehyde is found in Ceylonese and Chinese cinnamon oils. Inset: A model of benzaldehyde. *(Charles D. Winters)*

13.1 INTRODUCTION

In this and several of the following chapters, we study the physical and chemical properties of compounds containing the carbonyl group, C═O. Because this group is the functional group of aldehydes, ketones, and carboxylic acids and their derivatives, it is one of the most important functional groups in organic chemistry. The chemical properties of the carbonyl group are straightforward, and an understanding of its characteristic reaction themes leads very quickly to an understanding of a wide variety of organic reactions.

13.2 STRUCTURE AND BONDING

Aldehyde A compound containing a carbonyl group bonded to hydrogen (a CHO group).

Ketone A compound containing a carbonyl group bonded to two carbons.

The functional group of an **aldehyde** is a carbonyl group bonded to a hydrogen atom (Section 1.8C). In methanal, the simplest aldehyde, the carbonyl group is bonded to two hydrogen atoms. In other aldehydes, it is bonded to one hydrogen atom and one carbon atom. The functional group of a **ketone** is a carbonyl group bonded to two carbon atoms. Following are Lewis structures for the aldehydes methanal and ethanal and a Lewis structure for propanone, the simplest ketone (the common names of each are in parentheses underneath):

$$\overset{O}{\overset{\|}{HCH}} \qquad \overset{O}{\overset{\|}{CH_3CH}} \qquad \overset{O}{\overset{\|}{CH_3CCH_3}}$$

Methanal (Formaldehyde) — Ethanal (Acetaldehyde) — Propanone (Acetone)

Dihydroxyacetone is the active ingredient in several artificial tanning preparations.
(Andy Washnik)

A carbon–oxygen double bond consists of one sigma bond formed by the overlap of sp^2 hybrid orbitals of carbon and oxygen and one pi bond formed by the overlap of parallel $2p$ orbitals. The two nonbonding pairs of electrons on oxygen lie in the two remaining sp^2 hybrid orbitals (Figure 1.20).

13.3 NOMENCLATURE

A. IUPAC Nomenclature

The IUPAC system of nomenclature for aldehydes and ketones follows the familiar pattern of selecting the longest chain of carbon atoms that contains the functional group as the parent alkane. We show the aldehyde group by changing the suffix *-e* of the parent alkane to *-al*, as in methanal (Section 3.6). Because the carbonyl group of an aldehyde can appear only at the end of a parent chain and numbering must start with that group as carbon-1, its position is unambiguous; there is no need to use a number to locate it.

For **unsaturated aldehydes**, the presence of a carbon–carbon double bond is indicated by the infix *-en-*. As with other molecules with both an infix and a suffix, the location of the suffix determines the numbering pattern.

3-Methylbutanal

2-Propenal (Acrolein)

(2*E*)-3,7-Dimethyl-2,6-octadienal (Geranial)

For cyclic molecules in which —CHO is bonded directly to the ring, we name the molecule by adding the suffix *-carbaldehyde* to the name of the ring. We number the atom of the ring bearing the aldehyde group as number 1:

Cyclopentane-carbaldehyde

trans-4-Hydroxycyclo-hexanecarbaldehyde

Among the aldehydes for which the IUPAC system retains common names are benzaldehyde and cinnamaldehyde:

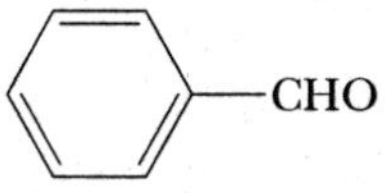

-CHO

Benzaldehyde

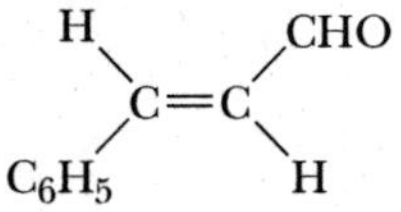

H CHO
C=C
C_6H_5 H

trans-3-Phenyl-2-propenal
(Cinnamaldehyde)

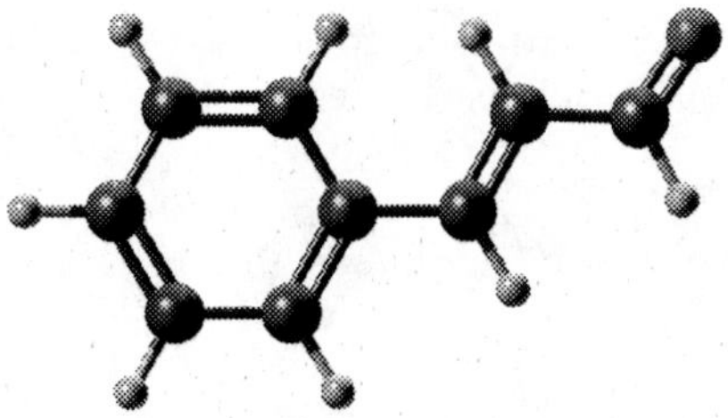

Note here the alternative ways of writing the phenyl group. In benzaldehyde, it is written as a line-angle formula and is abbreviated C_6H_5— in cinnamaldehyde. Two other aldehydes whose common names are retained in the IUPAC system are formaldehyde and acetaldehyde.

In the IUPAC system, we name ketones by selecting the longest chain that contains the carbonyl group and making that chain the parent alkane. We indicate the presence of the ketone by changing the suffix from *-e* to *-one* (Section 3.6). We number the parent chain from the direction that gives the carbonyl carbon the smaller number. The IUPAC system retains the common names acetophenone and benzophenone:

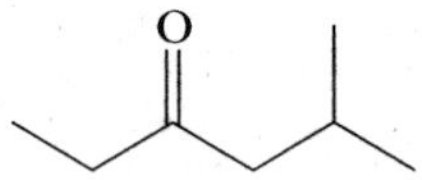

5-Methyl-3-hexanone

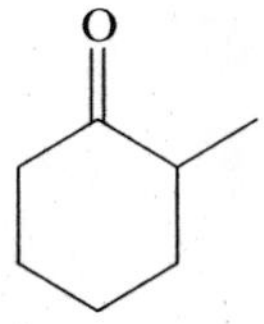

2-Methyl-
cyclohexanone

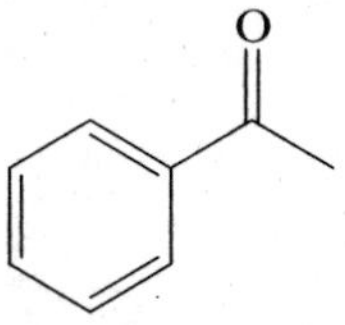

Acetophenone

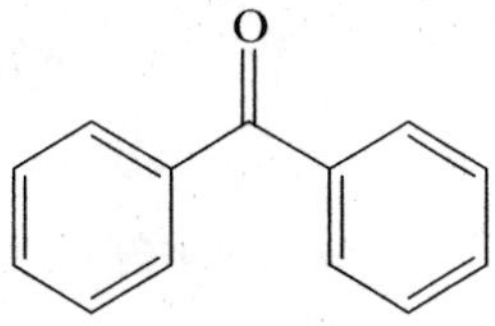

Benzophenone

EXAMPLE 13.1

Write the IUPAC name for each compound:

(a) 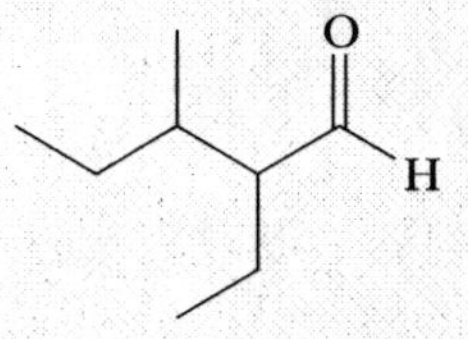 (b) 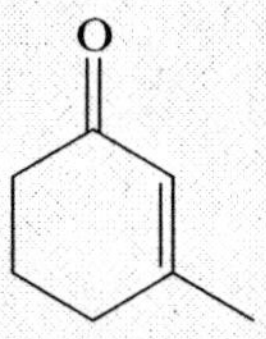 (c)

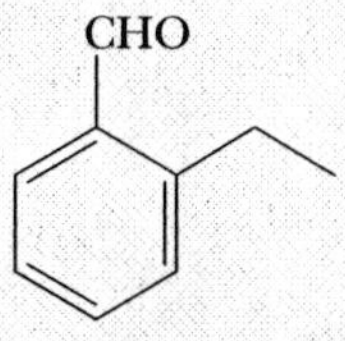

SOLUTION

(a) The longest chain has six carbons, but the longest chain that contains the carbonyl group has five carbons. The IUPAC name of this compound is 2-ethyl-3-methylpentanal.

(b) Number the six-membered ring beginning with the carbonyl carbon. The IUPAC name of this compound is 3-methyl-2-cyclohexenone.

(c) This molecule is derived from benzaldehyde. Its IUPAC name is 2-ethylbenzaldehyde.

Practice Problem 13.1

Write the IUPAC name for each compound, and specify the configuration of (c):

(a) 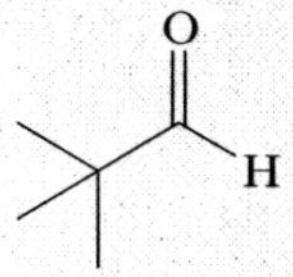 (b) 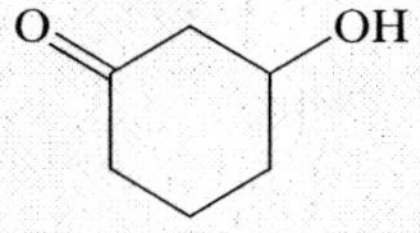 (c)

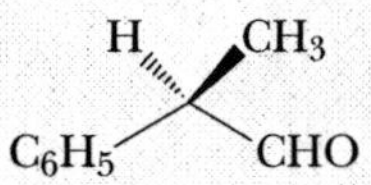

H CH_3
C_6H_5 CHO

EXAMPLE 13.2

Write structural formulas for all ketones with molecular formula $C_6H_{12}O$, and give each its IUPAC name. Which of these ketones are chiral?

SOLUTION

Following are line-angle formulas and IUPAC names for the six ketones with the given molecular formula:

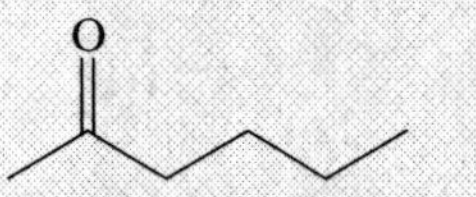

2-Hexanone 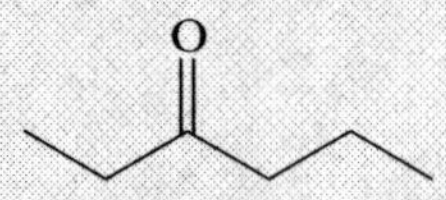 3-Hexanone 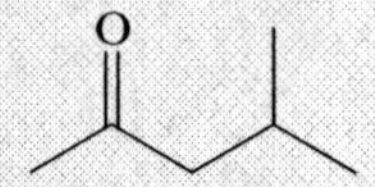 4-Methyl-2-pentanone

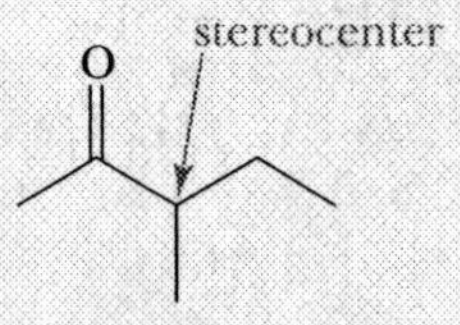

stereocenter

3-Methyl-2-pentanone 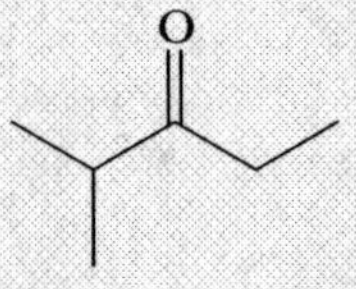 2-Methyl-3-pentanone 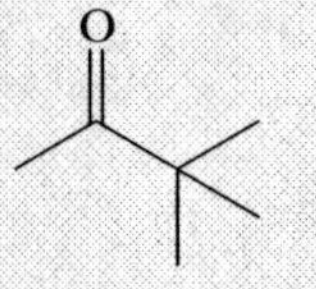 3,3-Dimethyl-2-butanone

Only 3-methyl-2-pentanone has a stereocenter and is chiral.

Practice Problem 13.2

Write structural formulas for all aldehydes with molecular formula $C_6H_{12}O$, and give each its IUPAC name. Which of these aldehydes are chiral?

Order of precedence of functional groups A system for ranking functional groups in order of priority for the purposes of IUPAC nomenclature.

B. IUPAC Names for More Complex Aldehydes and Ketones

In naming compounds that contain more than one functional group, the IUPAC has established an **order of precedence of functional groups**. Table 13.1 gives the order of precedence for the functional groups we have studied so far.

TABLE 13.1 Increasing Order of Precedence of Six Functional Groups

Functional Group	Suffix	Prefix	Example of When the Functional Group Has Lower Priority
Carboxyl group	-oic acid	—	
Aldehyde group	-al	oxo-	3-Oxopropanoic acid 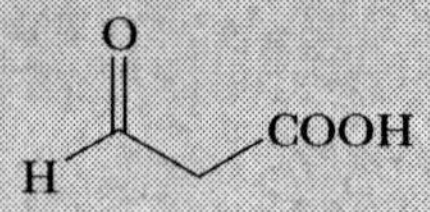
Ketone group	-one	oxo-	3-Oxobutanoic acid 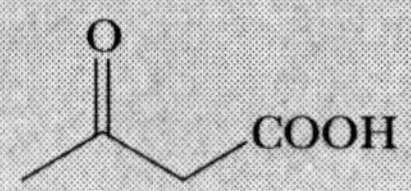
Alcohol group	-ol	hydroxy-	4-Hydroxybutanoic acid 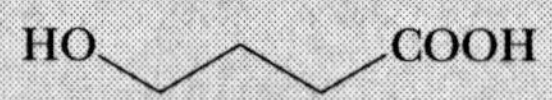
Amino group	-amine	amino-	3-Aminobutanoic acid 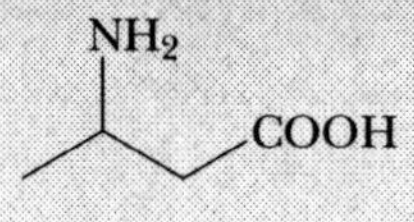
Sulfhydryl	-thiol	mercapto-	2-Mercaptoethanol 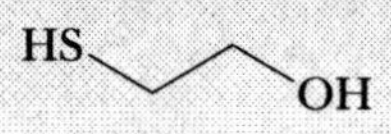

EXAMPLE 13.3

Write the IUPAC name for each compound:

(a) 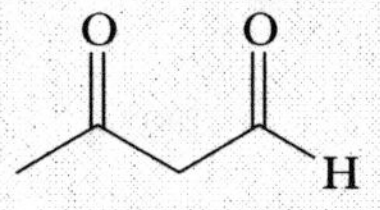 [structure] (b) 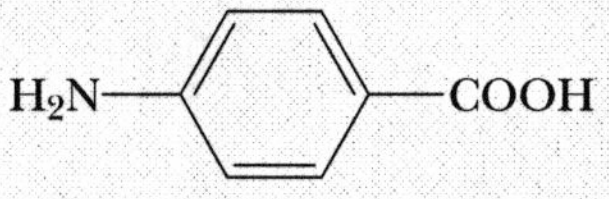 H_2N—[benzene ring]—COOH

(c) 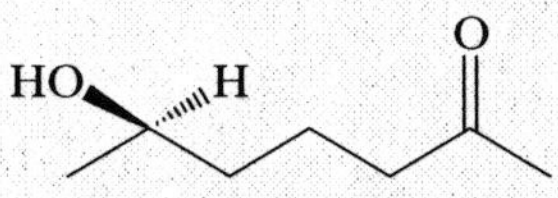 [structure: HO, H, O]

SOLUTION

(a) An aldehyde has higher precedence than a ketone, so we indicate the presence of the carbonyl group of the ketone by the prefix *oxo-*. The IUPAC name of this compound is 3-oxobutanal.

(b) The carboxyl group has higher precedence, so we indicate the presence of the amino group by the prefix *amino-*. The IUPAC name is 4-aminobenzoic acid. Alternatively, the compound may be named *p*-aminobenzoic acid, abbreviated PABA. PABA, a growth factor of microorganisms, is required for the synthesis of folic acid.

(c) The C=O group has higher precedence than the —OH group, so we indicate the —OH group by the prefix *hydroxy-*. The IUPAC name of this compound is (*R*)-6-hydroxy-2-heptanone.

Practice Problem 13.3

Write IUPAC names for these compounds, each of which is important in intermediary metabolism:

(a) $CH_3CH(OH)COOH$ Lactic acid (b) $CH_3C(=O)COOH$ Pyruvic acid (c) 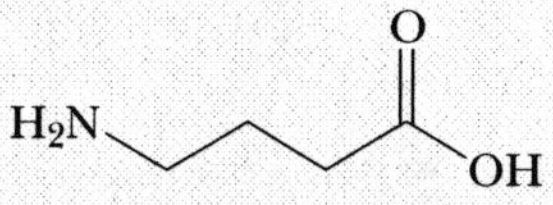 H_2N [structure] OH γ-Aminobutyric acid

The name shown is the one by which the compound is more commonly known in the biological sciences.

C. Common Names

The common name for an aldehyde is derived from the common name of the corresponding carboxylic acid by dropping the word *acid* and changing the suffix *-ic* or *-oic* to *-aldehyde*. Because we have not yet studied common names for carboxylic acids, we are not in a position to discuss common names for aldehydes. We can, however, illustrate how they are derived by reference to two common names of carboxylic acids with which you are familiar. The name formaldehyde is derived from formic acid, the name acetaldehyde from acetic acid:

$$\underset{\text{Formaldehyde}}{HC(=O)H} \qquad \underset{\text{Formic acid}}{HC(=O)OH} \qquad \underset{\text{Acetaldehyde}}{CH_3C(=O)H} \qquad \underset{\text{Acetic acid}}{CH_3C(=O)OH}$$

Common names for ketones are derived by naming each alkyl or aryl group bonded to the carbonyl group as a separate word, followed by the word *ketone*.

Groups are generally listed in order of increasing atomic weight. (Methyl ethyl ketone, abbreviated MEK, is a common solvent for varnishes and lacquers):

Methyl ethyl ketone (MEK) — Diethyl ketone — Dicyclohexyl ketone

13.4 PHYSICAL PROPERTIES

Oxygen is more electronegative than carbon (3.5 compared with 2.5; Table 1.5); therefore, a carbon–oxygen double bond is polar, with oxygen bearing a partial negative charge and carbon bearing a partial positive charge:

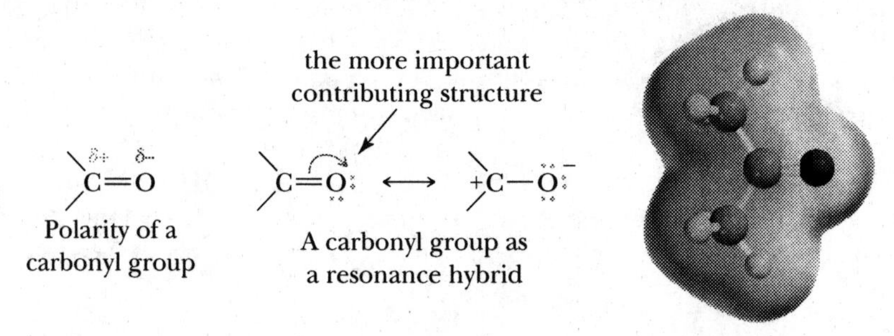

The electron density model shows that the partial positive charge on an acetone molecule is distributed both on the carbonyl carbon and on the two attached methyl groups as well.

In addition, the resonance structure on the right emphasizes that, in reactions of a carbonyl group, carbon acts as an electrophile and a Lewis acid. The carbonyl oxygen, by contrast, acts as a nucleophile and a Lewis base.

Because of the polarity of the carbonyl group, aldehydes and ketones are polar compounds and interact in the liquid state by dipole–dipole interactions. As a result, aldehydes and ketones have higher boiling points than those of nonpolar compounds of comparable molecular weight. Table 13.2 lists the boiling points of six compounds of comparable molecular weight.

Boiling Points
ether < alkyl < aldehyde < ketone < alcohol < carboxylic acid.

TABLE 13.2 Boiling Points of Six Compounds of Comparable Molecular Weight

Name	Structural Formula	Molecular Weight	Boiling Point (°C)
Diethyl ether	$CH_3CH_2OCH_2CH_3$	74	34
Pentane	$CH_3CH_2CH_2CH_2CH_3$	72	36
Butanal	$CH_3CH_2CH_2CHO$	72	76
2-Butanone	$CH_3CH_2COCH_3$	72	80
1-Butanol	$CH_3CH_2CH_2CH_2OH$	74	117
Propanoic acid	CH_3CH_2COOH	72	141

Pentane and diethyl ether have the lowest boiling points of these six compounds. Both butanal and 2-butanone are polar compounds, and because of the intermolecular attraction between carbonyl groups, their boiling points are higher than those of pentane and diethyl ether. Alcohols (Section 8.2C) and carboxylic acids (Section 14.4) are polar compounds, and their molecules associate by hydrogen bonding; their boiling points are higher than those of butanal and 2-butanone, compounds whose molecules cannot associate in that manner.

Because the carbonyl groups of aldehydes and ketones interact with water molecules by hydrogen bonding, low-molecular-weight aldehydes and ketones are more soluble in water than are nonpolar compounds of comparable molecular weight. Table 13.3 lists the boiling points and solubilities in water of several low-molecular-weight aldehydes and ketones.

TABLE 13.3 Physical Properties of Selected Aldehydes and Ketones

IUPAC Name	Common Name	Structural Formula	Boiling Point (°C)	Solubility (g/100 g water)
Methanal	Formaldehyde	HCHO	−21	infinite
Ethanal	Acetaldehyde	CH_3CHO	20	infinite
Propanal	Propionaldehyde	CH_3CH_2CHO	49	16
Butanal	Butyraldehyde	$CH_3CH_2CH_2CHO$	76	7
Hexanal	Caproaldehyde	$CH_3(CH_2)_4CHO$	129	slight
Propanone	Acetone	CH_3COCH_3	56	infinite
2-Butanone	Methyl ethyl ketone	$CH_3COCH_2CH_3$	80	26
3-Pentanone	Diethyl ketone	$CH_3CH_2COCH_2CH_3$	101	5

13.5 REACTIONS

The most common reaction theme of the carbonyl group is the addition of a nucleophile to form a **tetrahedral carbonyl addition intermediate**. In the following general reaction, the nucleophilic reagent is written as Nu:⁻ to emphasize the presence of its unshared pair of electrons:

$$\text{Nu:}^- + R_2C{=}\ddot{O}: \longrightarrow \text{Nu}{-}CR_2{-}\ddot{O}:^-$$

Tetrahedral carbonyl addition intermediate

13.6 ADDITION OF GRIGNARD REAGENTS

From the perspective of the organic chemist, the addition of carbon nucleophiles is the most important type of nucleophilic addition to a carbonyl group, because these reactions form new carbon–carbon bonds. In this section, we describe the preparation and reactions of Grignard reagents and their reaction with aldehydes and ketones.

A. Formation and Structure of Organomagnesium Compounds

Alkyl, aryl, and vinylic halides react with Group I, Group II, and certain other metals to form **organometallic compounds**. Within the range of organometallic compounds, organomagnesium compounds are among the most readily available,

Organometallic compound A compound containing a carbon–metal bond.

Grignard reagent An organomagnesium compound of the type RMgX or ArMgX.

easily prepared, and easily handled. They are commonly named **Grignard reagents,** after Victor Grignard, who was awarded a 1912 Nobel prize in chemistry for their discovery and their application to organic synthesis.

Grignard reagents are typically prepared by the slow addition of a halide to a stirred suspension of magnesium metal in an ether solvent, most commonly diethyl ether or tetrahydrofuran (THF). Organoiodides and bromides generally react rapidly under these conditions, whereas chlorides react more slowly. Butylmagnesium bromide, for example, is prepared by adding 1-bromobutane to an ether suspension of magnesium metal. Aryl Grignards, such as phenylmagnesium bromide, are prepared in a similar manner:

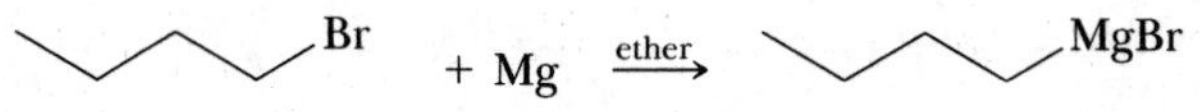

Br + Mg $\xrightarrow{\text{ether}}$ MgBr

1-Bromobutane — Butylmagnesium bromide

—Br + Mg $\xrightarrow{\text{ether}}$ —MgBr

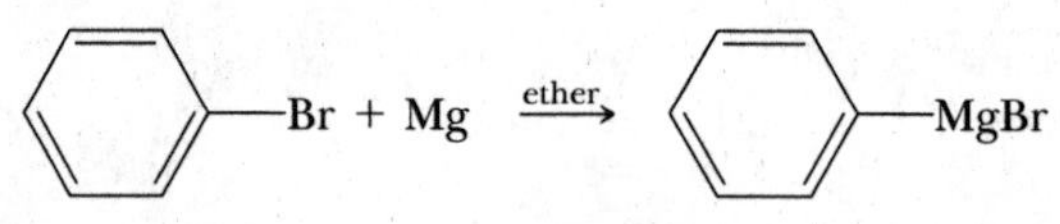

Bromobenzene — Phenylmagnesium bromide

Given that the difference in electronegativity between carbon and magnesium is 1.3 units (2.5 − 1.2), the carbon–magnesium bond is best described as polar covalent, with carbon bearing a partial negative charge and magnesium bearing a partial positive charge. In the structural formula on the right, the carbon–magnesium bond is shown as ionic to emphasize its nucleophilic character. Note that although we can write a Grignard reagent as a **carbanion**, a more accurate representation shows it as a polar covalent compound:

Carbanion An anion in which carbon has an unshared pair of electrons and bears a negative charge.

$$CH_3(CH_2)_2CH_2\overset{\delta-}{}-\overset{\delta+}{MgBr} \qquad CH_3(CH_2)_2CH_2{:}^- \ \ {}^{+}MgBr$$

carbon is a nucleophile

The feature that makes Grignard reagents so valuable in organic synthesis is that the carbon bearing the halogen is now transformed into a nucleophile.

B. Reaction with Protic Acids

Grignard reagents are very strong bases and react readily with a wide variety of acids (proton donors) to form alkanes. Ethylmagnesium bromide, for example, reacts instantly with water to give ethane and magnesium salts. This reaction is an example of a stronger acid and a stronger base reacting to give a weaker acid and a weaker base (Section 2.5):

$$\overset{\delta-}{CH_3CH_2}-\overset{\delta+}{MgBr} + H-OH \longrightarrow CH_3CH_2-H + Mg^{2+} + OH^- + Br^-$$

Stronger base	pK_a 15.7 Stronger acid	pK_a 51 Weaker acid	Weaker base

Any compound containing an O—H, N—H, or S—H bond will react with a Grignard reagent by proton transfer. Following are examples of compounds containing those functional groups:

HOH	ROH	ArOH	RCOOH	RNH_2	RSH
Water	Alcohols	Phenols	Carboxylic acids	Amines	Thiols

Because Grignard reagents react so rapidly with these proton acids, Grignard reagents cannot be made from any halogen-containing compounds that also contain them.

EXAMPLE 13.4

Write an equation for the acid–base reaction between ethylmagnesium iodide and an alcohol. Use curved arrows to show the flow of electrons in this reaction. In addition, show that the reaction is an example of a stronger acid and stronger base reacting to form a weaker acid and weaker base.

SOLUTION

The alcohol is the stronger acid and ethyl carbanion is the stronger base:

$$CH_3CH_2{-}MgI + H{-}\ddot{O}R \longrightarrow CH_3CH_2{-}H + R\ddot{O}{:}^{-}MgI^{+}$$

Ethylmagnesium iodide	An alcohol pK_a 16–18	Ethane pK_a 51	A magnesium alkoxide
(stronger base)	(stronger acid)	(weaker acid)	(weaker base)

Practice Problem 13.4

Explain how these Grignard reagents react with molecules of their own kind to "self-destruct":

(a) HO—C_6H_4—MgBr (b) $HOC(=O)CH_2CH_2CH_2CH_2MgBr$

C. Addition of Grignard Reagents to Aldehydes and Ketones

The special value of Grignard reagents is that they provide excellent ways to form new carbon–carbon bonds. In their reactions, Grignard reagents behave as carbanions. A carbanion is a good nucleophile and adds to the carbonyl group of an aldehyde or a ketone to form a tetrahedral carbonyl addition compound. The driving force for these reactions is the attraction of the partial negative charge on the carbon of the organometallic compound to the partial positive charge of the carbonyl carbon. In the examples that follow, the magnesium–oxygen bond is written $-O^{-}[MgBr]^{+}$ to emphasize its ionic character. The alkoxide ions formed in Grignard reactions are strong bases (Section 8.3C) and form alcohols when treated with an aqueous acid such as HCl or aqueous NH_4Cl during workup.

Addition to Formaldehyde Gives a 1° Alcohol

Treatment of a Grignard reagent with formaldehyde, followed by hydrolysis in aqueous acid, gives a primary alcohol:

$$CH_3CH_2{-}MgBr + H{-}\overset{\overset{\ddot{O}}{\|}}{C}{-}H \xrightarrow{\text{ether}} CH_3CH_2{-}\overset{\overset{:\ddot{O}:^{-}[MgBr]^{+}}{|}}{C}H_2 \xrightarrow[H_2O]{HCl} CH_3CH_2{-}\overset{\overset{:\ddot{O}H}{|}}{C}H_2 + Mg^{2+}$$

Formaldehyde — A magnesium alkoxide — 1-Propanol (a 1° alcohol)

Addition to an Aldehyde (Except Formaldehyde) Gives a 2° Alcohol

Treatment of a Grignard reagent with any aldehyde other than formaldehyde, followed by hydrolysis in aqueous acid, gives a secondary alcohol:

$$\text{C}_6\text{H}_{11}\text{—MgBr} + \text{CH}_3\text{—}\overset{\overset{\Large O}{\|}}{\text{C}}\text{—H} \xrightarrow{\text{ether}} \text{C}_6\text{H}_{11}\text{—}\overset{\overset{O^-\,[MgBr]^+}{|}}{\text{C}}\text{HCH}_3 \xrightarrow[\text{H}_2\text{O}]{\text{HCl}} \text{C}_6\text{H}_{11}\text{—}\overset{\overset{OH}{|}}{\text{C}}\text{HCH}_3 + \text{Mg}^{2+}$$

Acetaldehyde — A magnesium alkoxide — 1-Cyclohexylethanol (a 2° alcohol)

Addition to a Ketone Gives a 3° Alcohol

Treatment of a Grignard reagent with a ketone, followed by hydrolysis in aqueous acid, gives a tertiary alcohol:

$$\text{C}_6\text{H}_5\text{—MgBr} + \text{CH}_3\text{—}\overset{\overset{\Large O}{\|}}{\text{C}}\text{—CH}_3 \xrightarrow{\text{ether}} \text{C}_6\text{H}_5\text{—}\overset{\overset{O^-\,[MgBr]^+}{|}}{\underset{\underset{CH_3}{|}}{\text{C}}}\text{CH}_3 \xrightarrow[\text{H}_2\text{O}]{\text{HCl}} \text{C}_6\text{H}_5\text{—}\overset{\overset{OH}{|}}{\underset{\underset{CH_3}{|}}{\text{C}}}\text{CH}_3 + \text{Mg}^{2+}$$

Acetone — A magnesium alkoxide — 2-Phenyl-2-propanol (a 3° alcohol)

EXAMPLE 13.5

2-Phenyl-2-butanol can be synthesized by three different combinations of a Grignard reagent and a ketone. Show each combination.

SOLUTION

Curved arrows in each solution show the formation of the new carbon–carbon bond and the alkoxide ion, and labels on the final product show which set of reagents forms each bond:

(a) $\text{C}_6\text{H}_5\text{—MgBr} + \overset{\overset{O}{\|}}{\text{C}}(\text{CH}_3)\text{CH}_2\text{CH}_3$

(b) $\text{C}_6\text{H}_5\text{—}\overset{\overset{O}{\|}}{\text{C}}\text{—CH}_3 + \text{BrMg—CH}_2\text{CH}_3$

(c) $\text{C}_6\text{H}_5\text{—}\overset{\overset{O}{\|}}{\text{C}}\text{CH}_2\text{CH}_3 + \text{BrMg—CH}_3$

$$\longrightarrow \text{C}_6\text{H}_5\text{—}\overset{\overset{O^-\,[MgBr]^+}{|}}{\underset{\underset{CH_3}{|}}{\text{C}}}\text{—CH}_2\text{CH}_3 \xrightarrow[\text{H}_2\text{O}]{\text{HCl}} \text{C}_6\text{H}_5\text{—}\overset{\overset{OH}{|}}{\underset{\underset{CH_3}{|}}{\text{C}}}\text{—CH}_2\text{CH}_3$$

(a) (b) (c)

Practice Problem 13.5

Show how these three compounds can be synthesized from the same Grignard reagent:

(a) 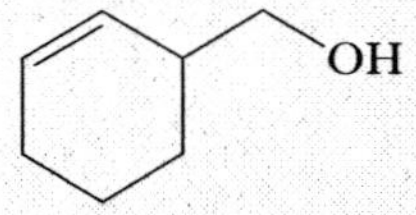 (b) 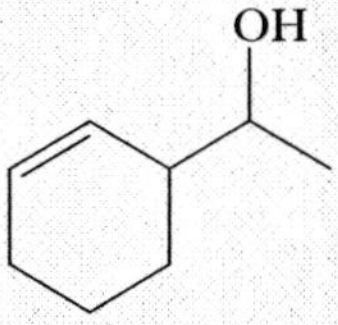 (c)

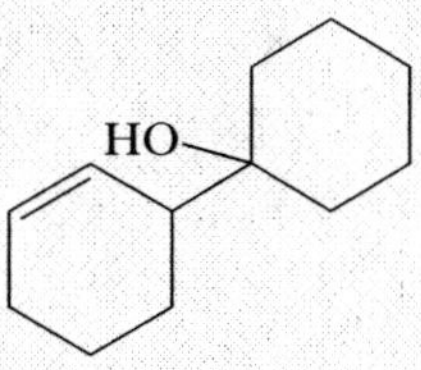

13.7 ADDITION OF ALCOHOLS

A. Formation of Acetals

The addition of a molecule of alcohol to the carbonyl group of an aldehyde or a ketone forms a **hemiacetal** (a half-acetal). This reaction is catalyzed by both acid and base: Oxygen adds to the carbonyl carbon and hydrogen adds to the carbonyl oxygen:

Hemiacetal A molecule containing an —OH and an —OR or —OAr group bonded to the same carbon.

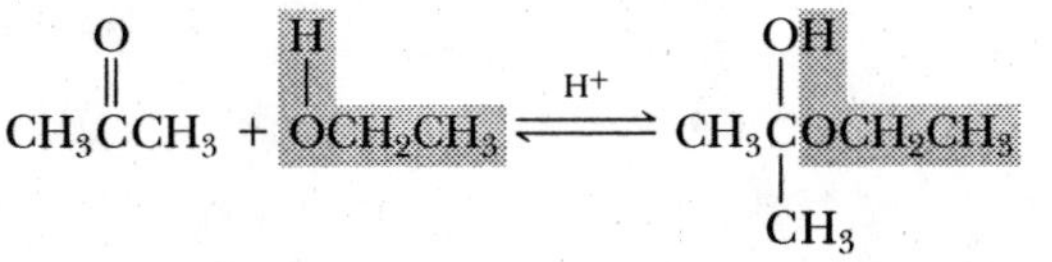

$$CH_3\overset{O}{\overset{\|}{C}}CH_3 + \overset{H}{\overset{|}{O}}CH_2CH_3 \overset{H^+}{\rightleftharpoons} CH_3\underset{CH_3}{\underset{|}{\overset{OH}{\overset{|}{C}}}}OCH_2CH_3$$

A hemiacetal

The functional group of a hemiacetal is a carbon bonded to an —OH group and an —OR or —OAr group:

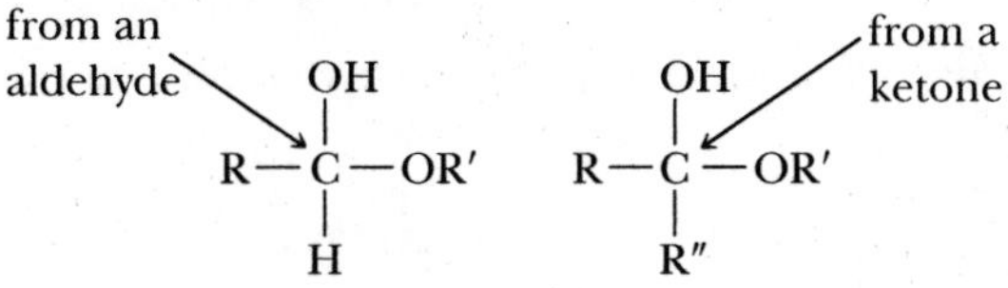

from an aldehyde $\quad R-\underset{H}{\underset{|}{\overset{OH}{\overset{|}{C}}}}-OR' \quad R-\underset{R''}{\underset{|}{\overset{OH}{\overset{|}{C}}}}-OR'$ from a ketone

Hemiacetals

Hemiacetals are generally unstable and are only minor components of an equilibrium mixture, except in one very important type of molecule. When a hydroxyl group is part of the same molecule that contains the carbonyl group, and a five- or six-membered ring can form, the compound exists almost entirely in a cyclic hemiacetal form:

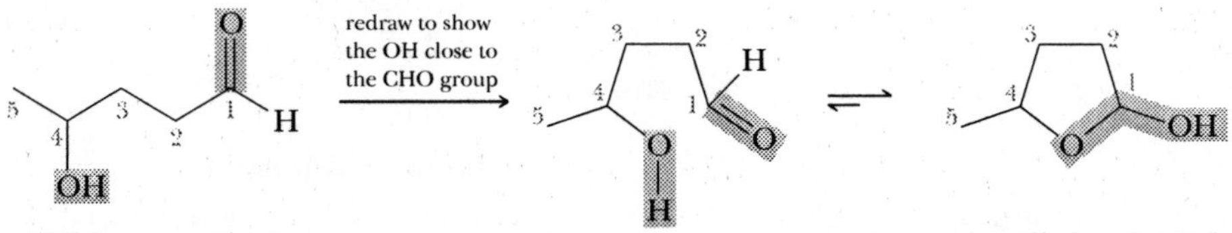

redraw to show the OH close to the CHO group

4-Hydroxypentanal

A cyclic hemiacetal (major form present at equilibrium)

We shall have much more to say about cyclic hemiacetals when we consider the chemistry of carbohydrates in Chapter 18.

Acetal A molecule containing two —OR or —OAr groups bonded to the same carbon.

Hemiacetals can react further with alcohols to form **acetals** plus a molecule of water. This reaction is acid catalyzed:

$$CH_3\overset{\overset{OH}{|}}{\underset{\underset{CH_3}{|}}{C}}OCH_2CH_3 + CH_3CH_2OH \overset{H^+}{\rightleftharpoons} CH_3\overset{\overset{OCH_2CH_3}{|}}{\underset{\underset{CH_3}{|}}{C}}OCH_2CH_3 + H_2O$$

A hemiacetal — A diethyl acetal

The functional group of an acetal is a carbon bonded to two —OR or —OAr groups:

$$R-\overset{\overset{OR'}{|}}{\underset{\underset{H}{|}}{C}}-OR' \qquad R-\overset{\overset{OR'}{|}}{\underset{\underset{R''}{|}}{C}}-OR'$$

from an aldehyde — from a ketone

Acetals

The mechanism for the acid-catalyzed conversion of a hemiacetal to an acetal can be divided into four steps. Note that acid H—A is a true catalyst in this reaction; it is used in Step 1, but a replacement H—A is generated in Step 4.

Mechanism: Acid-Catalyzed Formation of an Acetal

Step 1: Proton transfer from the acid, H—A, to the hemiacetal OH group gives an oxonium ion:

$$R-\overset{\overset{HO:}{|}}{\underset{\underset{H}{|}}{C}}-OCH_3 + H-A \rightleftharpoons R-\overset{\overset{H_2O^+}{|}}{\underset{\underset{H}{|}}{C}}-OCH_3 + A:^-$$

An oxonium ion

Step 2: Loss of water from the oxonium ion gives a resonance-stabilized cation:

$$R-\overset{\overset{H_2O^+}{|}}{\underset{\underset{H}{|}}{C}}-OCH_3 \rightleftharpoons R-\underset{\underset{H}{|}}{C}=\overset{+}{O}CH_3 \longleftrightarrow R-\underset{\underset{H}{|}}{\overset{+}{C}}-OCH_3 + H_2O:$$

A resonance-stabilized cation

Step 3: Reaction of the resonance-stabilized cation (an electrophile) with methanol (a nucleophile) gives the conjugate acid of the acetal:

$$CH_3-\overset{\overset{H}{|}}{O}: + R-\underset{\underset{H}{|}}{C}=\overset{+}{O}CH_3 \rightleftharpoons R-\overset{\overset{H\overset{+}{O}CH_3}{|}}{\underset{\underset{H}{|}}{C}}-OCH_3$$

A protonated acetal

Step 4: Proton transfer from the protonated acetal to A^- gives the acetal and generates a new molecule of H—A, the acid catalyst:

$$A:^- + R-\overset{\overset{H\overset{+}{O}CH_3}{|}}{\underset{\underset{H}{|}}{C}}-OCH_3 \rightleftharpoons HA + R-\overset{\overset{:OCH_3}{|}}{\underset{\underset{H}{|}}{C}}-OCH_3$$

A protonated acetal — An acetal

Formation of acetals is often carried out using the alcohol as a solvent and dissolving either dry HCl (hydrogen chloride) or arenesulfonic acid (Section 9.7B) in the alcohol. Because the alcohol is both a reactant and the solvent, it is present in large molar excess, which drives the reaction to the right and favors acetal formation. Alternatively, the reaction may be driven to the right by the removal of water as it is formed:

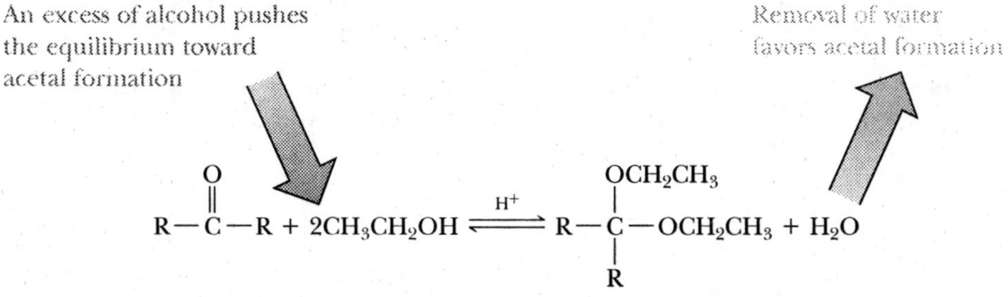

An excess of alcohol pushes the equilibrium toward acetal formation

Removal of water favors acetal formation

$$R-\overset{O}{\overset{\|}{C}}-R + 2CH_3CH_2OH \underset{}{\overset{H^+}{\rightleftharpoons}} R-\underset{R}{\overset{OCH_2CH_3}{C}}-OCH_2CH_3 + H_2O$$

A diethyl acetal

EXAMPLE 13.6

Show the reaction of the carbonyl group of each ketone with one molecule of alcohol to form a hemiacetal and then with a second molecule of alcohol to form an acetal (note that, in part (b), ethylene glycol is a diol, and one molecule of it provides both —OH groups):

(a) 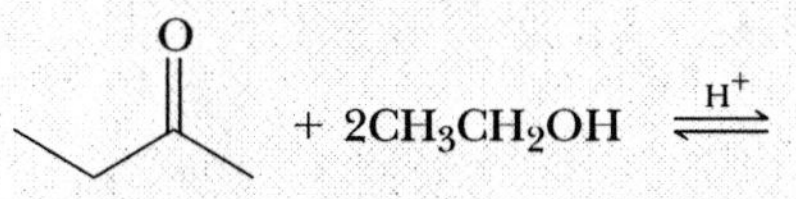 [2-butanone] + $2CH_3CH_2OH \overset{H^+}{\rightleftharpoons}$

(b) 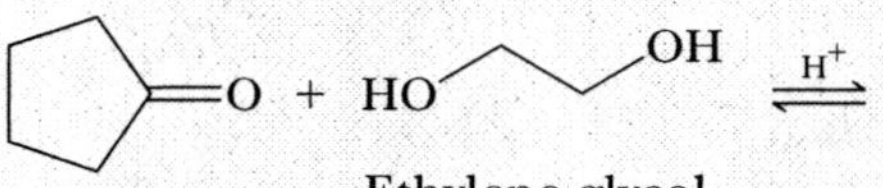 [cyclopentanone]=O + HO~~OH $\overset{H^+}{\rightleftharpoons}$

Ethylene glycol

SOLUTION

Here are structural formulas of the hemiacetal and then the acetal:

(a) 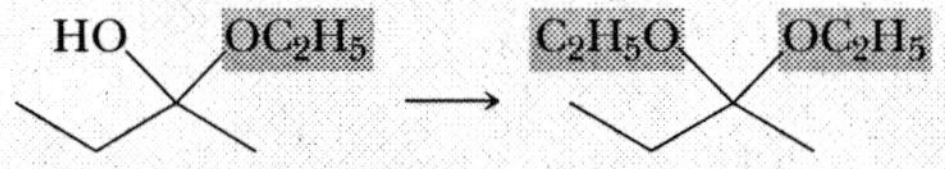 HO, OC_2H_5 → C_2H_5O, OC_2H_5

(b) 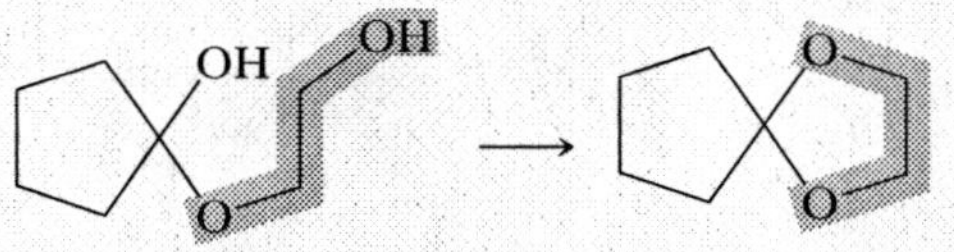 OH, OH, O → O, O

Practice Problem 13.6

The hydrolysis of an acetal forms an aldehyde or a ketone and two molecules of alcohol. Following are structural formulas for three acetals:

(a) 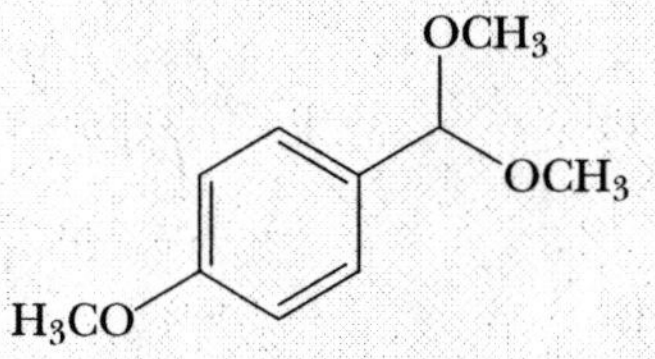 OCH_3, OCH_3, H_3CO (b) 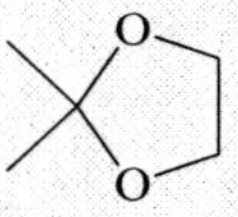 O, O (c) 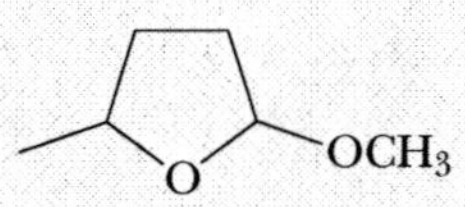 O, OCH_3

Draw the structural formulas for the products of the hydrolysis of each in aqueous acid.

Like ethers, acetals are unreactive to bases, to reducing agents such as H_2/M, to Grignard reagents, and to oxidizing agents (except, of course, those which involve aqueous acid). Because of their lack of reactivity toward these reagents, acetals are often used to protect the carbonyl groups of aldehydes and ketones while reactions are carried out on functional groups in other parts of the molecule.

B. Acetals as Carbonyl-Protecting Groups

The use of acetals as carbonyl-protecting groups is illustrated by the synthesis of 5-hydroxy-5-phenylpentanal from benzaldehyde and 4-bromobutanal:

Benzaldehyde + 4-Bromobutanal $\xrightarrow{??}$ 5-Hydroxy-5-phenylpentanal

One obvious way to form a new carbon–carbon bond between these two molecules is to treat benzaldehyde with the Grignard reagent formed from 4-bromobutanal. This Grignard reagent, however, would react immediately with the carbonyl group of another molecule of 4-bromobutanal, causing it to self-destruct during preparation (Section 13.6B). A way to avoid this problem is to protect the carbonyl group of 4-bromobutanal by converting it to an acetal. Cyclic acetals are often used because they are particularly easy to prepare.

4-Bromobutanal + Ethylene glycol $\underset{}{\overset{H^+}{\rightleftharpoons}}$ A cyclic acetal + H_2O

Treatment of the protected bromoaldehyde with magnesium in diethyl ether, followed by the addition of benzaldehyde, gives a magnesium alkoxide:

A cyclic acetal + Mg $\xrightarrow{\text{ether}}$ A Grignard reagent

Benzaldehyde + BrMg–(acetal) $\longrightarrow$ A magnesium alkoxide

Treatment of the magnesium alkoxide with aqueous acid accomplishes two things. First, protonation of the alkoxide anion gives the desired hydroxyl group, and then, hydrolysis of the cyclic acetal regenerates the aldehyde group:

O^-MgBr^+ (acetal) $\xrightarrow{HCl,\ H_2O}$ 5-hydroxy-5-phenylpentanal + HO–CH$_2$CH$_2$–OH

13.8 ADDITION OF AMMONIA AND AMINES

A. Formation of Imines

Ammonia, primary aliphatic amines (RNH_2), and primary aromatic amines ($ArNH_2$) react with the carbonyl group of aldehydes and ketones in the presence of an acid catalyst to give a product that contains a carbon–nitrogen double bond. A molecule containing a carbon–nitrogen double bond is called an **imine** or, alternatively, a **Schiff base**:

Imine A compound containing a carbon–nitrogen double bond; also called a Schiff base.

Schiff base An alternative name for an imine.

$$CH_3\overset{O}{\overset{\|}{C}}H + H_2N\text{—}C_6H_5 \underset{}{\overset{H^+}{\rightleftharpoons}} CH_3CH{=}N\text{—}C_6H_5 + H_2O$$

Ethanal Aniline An imine (A Schiff base)

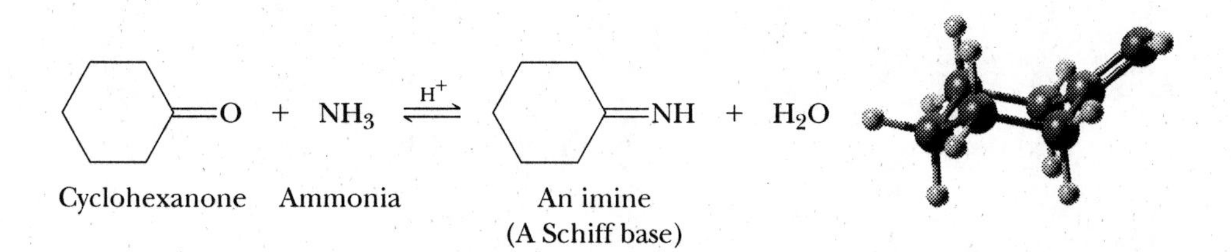

Mechanism: Formation of an Imine from an Aldehyde or a Ketone

Step 1: Addition of the nitrogen atom of ammonia or a primary amine, both good nucleophiles, to the carbonyl carbon, followed by a proton transfer, gives a tetrahedral carbonyl addition intermediate:

$$>C{=}O + H_2\ddot{N}\text{—}R \rightleftharpoons \text{—}\overset{:\ddot{O}:^-}{\underset{|}{\overset{|}{C}}}\text{—}\overset{H}{\underset{H}{N^+}}\text{—}R \rightleftharpoons \text{—}\overset{OH}{\underset{|}{\overset{|}{C}}}\text{—}\underset{H}{\ddot{N}}\text{—}R$$

A tetrahedral carbonyl addition intermediate

Step 2: Protonation of the OH group, followed by loss of water and proton transfer to solvent gives the imine. Notice that the loss of water and the proton transfer have the characteristics of an E2 reaction. Three things happen simultaneously in this dehydration: a base (in this case a water molecule) removes a proton from N, the carbon–nitrogen double bond forms, and the leaving group (in this case, a water molecule) departs:

$$H_3O^+ + \text{—}\overset{OH}{\underset{|}{\overset{|}{C}}}\text{—}\underset{H}{\ddot{N}}\text{—}R \rightleftharpoons \text{—}\overset{H_2O^+}{\underset{|}{\overset{|}{C}}}\text{—}\underset{H\cdots :OH_2}{\ddot{N}}\text{—}R \rightleftharpoons >C{=}\ddot{N}\text{—}R + H_2\ddot{O}: + H\text{—}\overset{H}{\underset{}{\overset{|}{O^+}}}\text{—}H$$

An imine

(The flow of electrons here is similar to that in an E2 reaction.)

To give but one example of the importance of imines in biological systems, the active form of vitamin A aldehyde (retinal) is bound to the protein opsin in the human retina in the form of an imine called *rhodopsin* or *visual purple*. The amino acid lysine (Table 18.1) provides the primary amino group for this reaction:

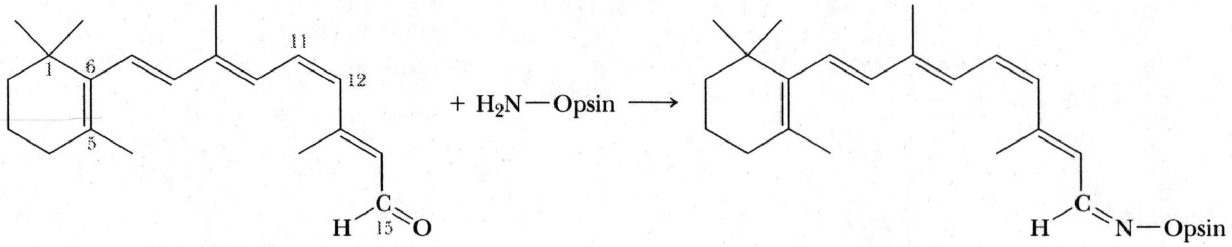

$$\text{11-}cis\text{-Retinal} + H_2N\text{—Opsin} \longrightarrow \text{Rhodopsin (Visual purple)}$$

11-*cis*-Retinal

Rhodopsin
(Visual purple)

EXAMPLE 13.7

Write a structural formula for the imine formed in each reaction:

(a) 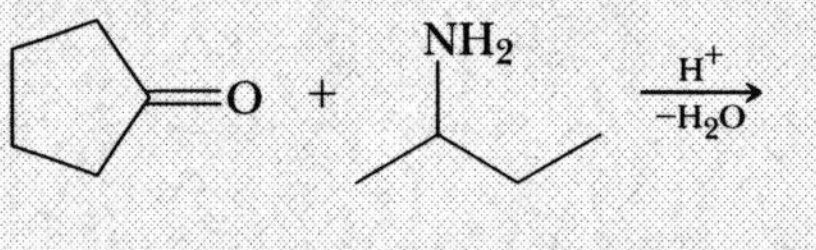 Cyclopentanone + 2-butanamine $\xrightarrow[-H_2O]{H^+}$

(b) 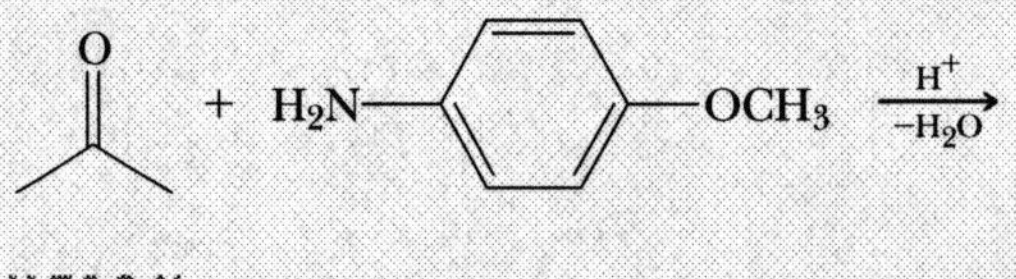 Acetone + H_2N–C_6H_4–OCH_3 $\xrightarrow[-H_2O]{H^+}$

SOLUTION

Here is a structural formula for each imine:

(a) 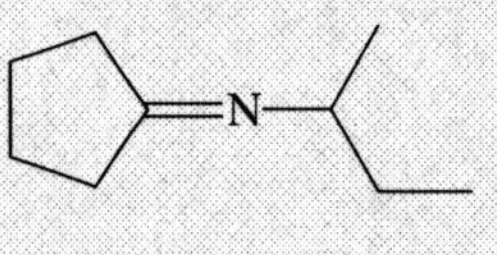 Cyclopentylidene=N–CH(CH_3)CH_2CH_3

(b) 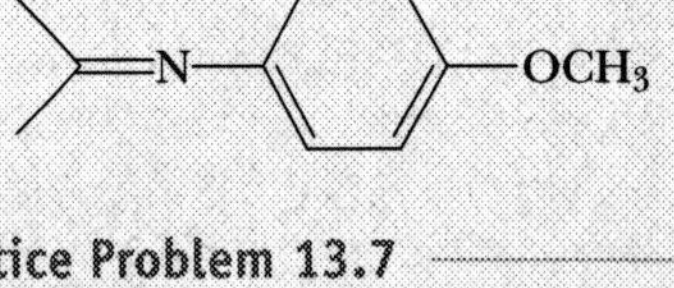 $(CH_3)_2C$=N–C_6H_4–OCH_3

Practice Problem 13.7

Acid-catalyzed hydrolysis of an imine gives an amine and an aldehyde or a ketone. When one equivalent of acid is used, the amine is converted to its ammonium salt. For each of the following imines, write a structural formula for the products of hydrolysis, using one equivalent of HCl:

(a) 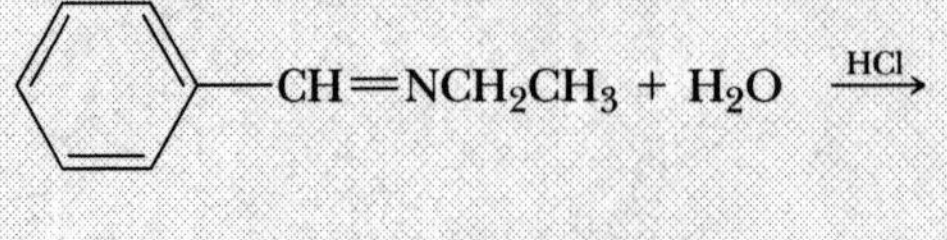 C_6H_5–$CH{=}NCH_2CH_3 + H_2O \xrightarrow{HCl}$

(b) 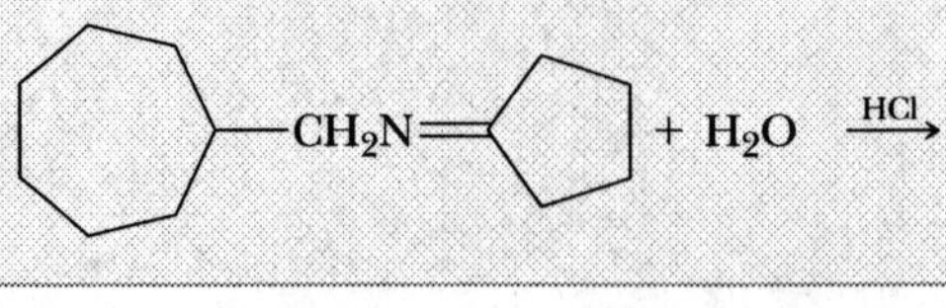 Cycloheptyl–$CH_2N{=}$cyclopentylidene $+ H_2O \xrightarrow{HCl}$

B. Reductive Amination of Aldehydes and Ketones

One of the chief values of imines is that the carbon–nitrogen double bond can be reduced to a carbon–nitrogen single bond by hydrogen in the presence of a nickel or other transition metal catalyst. By this two-step reaction, called **reductive amination**, a primary amine is converted to a secondary amine by way of an imine, as illustrated by the conversion of cyclohexylamine to dicyclohexylamine:

Reductive amination The formation of an imine from an aldehyde or a ketone, followed by the reduction of the imine to an amine.

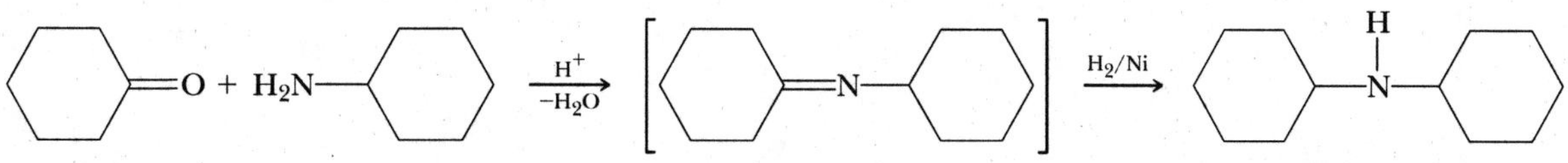

=O + H_2N— $\xrightarrow[-H_2O]{H^+}$ [=N—] $\xrightarrow{H_2/Ni}$ H —N—

Cyclohexanone Cyclohexyl-amine (a 1° amine) (An imine) Dicyclohexylamine (a 2° amine)

Conversion of an aldehyde or a ketone to an amine is generally carried out in one laboratory operation by mixing together the carbonyl-containing compound, the amine or ammonia, hydrogen, and the transition metal catalyst. The imine intermediate is not isolated.

EXAMPLE 13.8

Show how to synthesize each amine by a reductive amination:

(a) 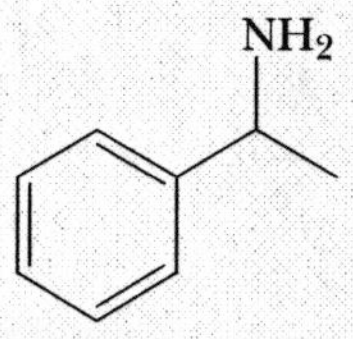 NH_2 (b) 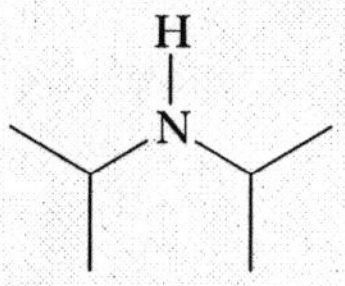 H N

SOLUTION

Treat the appropriate compound, in each case a ketone, with ammonia or an amine in the presence of H_2/Ni:

(a) 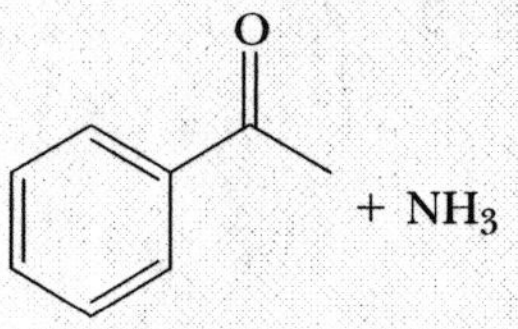 O + NH_3 (b) 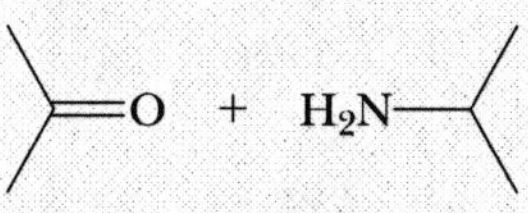 =O + H_2N—

Practice Problem 13.8

Show how to prepare each amine by the reductive amination of an appropriate aldehyde or ketone:

(a) 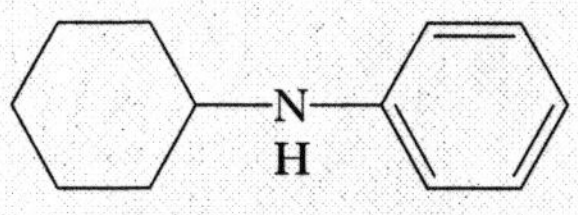 N H (b) 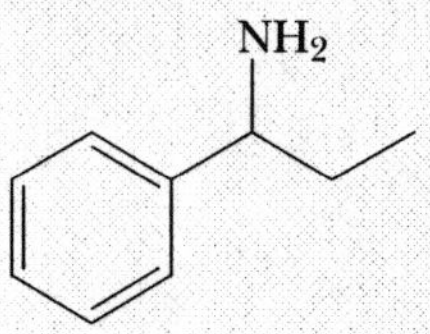 NH_2

13.9 KETO-ENOL TAUTOMERISM

A. Keto and Enol Forms

α-Carbon A carbon atom adjacent to a carbonyl group.

α-Hydrogen A hydrogen on an α-carbon.

A carbon atom adjacent to a carbonyl group is called an **α-carbon,** and any hydrogen atoms bonded to it are called **α-hydrogens:**

α-hydrogens

$$CH_3-\overset{\overset{\displaystyle O}{\|}}{C}-CH_2-CH_3$$

α-carbons

Enol A molecule containing an —OH group bonded to a carbon of a carbon–carbon double bond.

An aldehyde or ketone that has at least one α-hydrogen is in equilibrium with a constitutional isomer called an **enol**. The name *enol* is derived from the IUPAC designation of it as both an alkene (*-en-*) and an alcohol (*-ol*):

$$\underset{\text{(keto form)}}{\underset{\text{Acetone}}{CH_3-\overset{\overset{\displaystyle O}{\|}}{C}-CH_3}} \rightleftharpoons \underset{\text{(enol form)}}{\underset{\text{Acetone}}{CH_3-\overset{\overset{\displaystyle OH}{|}}{C}=CH_2}}$$

Tautomers Constitutional isomers that differ in the location of hydrogen and a double bond relative to O, N, or S.

Keto and enol forms are examples of **tautomers**—constitutional isomers in equilibrium with each other and that differ in the location of a hydrogen atom and a double bond relative to a heteroatom, most commonly O, S, or N. This type of isomerism is called **tautomerism.**

For most simple aldehydes and ketones, the position of the equilibrium in keto–enol tautomerism lies far on the side of the keto form (Table 13.4), because a carbon–oxygen double bond is stronger than a carbon–carbon double bond.

The equilibration of keto and enol forms is catalyzed by acid, as shown in the following two-step mechanism (note that a molecule of H—A is consumed in Step 1, but another is generated in Step 2):

TABLE 13.4 The Position of Keto-Enol Equilibrium for Four Aldehydes and Ketones*

Keto form		Enol form	% Enol at Equilibrium
$CH_3\overset{\overset{O}{\|}}{C}H$	⇌	$CH_2=\overset{\overset{OH}{|}}{C}H$	6×10^{-5}
$CH_3\overset{\overset{O}{\|}}{C}CH_3$	⇌	$CH_3\overset{\overset{OH}{|}}{C}=CH_2$	6×10^{-7}
cyclopentanone	⇌	cyclopent-1-en-1-ol	1×10^{-6}
cyclohexanone	⇌	cyclohex-1-en-1-ol	4×10^{-5}

*Data from J. March, *Advanced Organic Chemistry,* 4th ed. (New York, Wiley Interscience, 1992) p. 70.

Mechanism: Acid-Catalyzed Equilibration of Keto and Enol Tautomers

Step 1: Proton transfer from the acid catalyst, H—A, to the carbonyl oxygen forms the conjugate acid of the aldehyde or ketone:

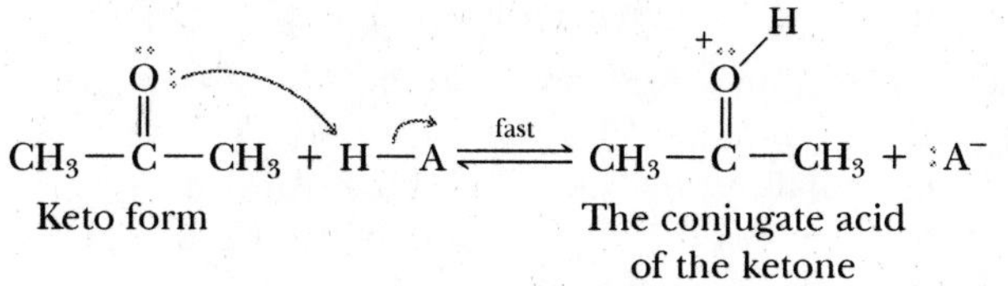

$$CH_3-\overset{\overset{\displaystyle O}{\|}}{C}-CH_3 + H-A \overset{\text{fast}}{\rightleftharpoons} CH_3-\overset{\overset{\displaystyle \overset{+}{O}-H}{\|}}{C}-CH_3 + :A^-$$

Keto form — The conjugate acid of the ketone

Step 2: Proton transfer from the α-carbon to the base, A^-, gives the enol and generates a new molecule of the acid catalyst, H—A:

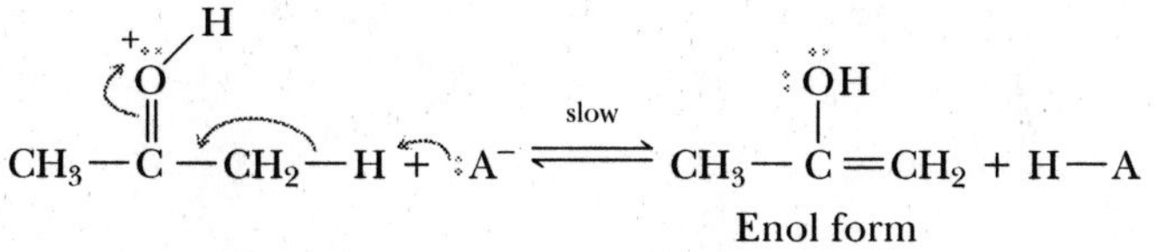

$$CH_3-\overset{\overset{\displaystyle \overset{+}{O}-H}{\|}}{C}-CH_2-H + :A^- \overset{\text{slow}}{\rightleftharpoons} CH_3-\overset{\overset{\displaystyle :\ddot{O}H}{|}}{C}=CH_2 + H-A$$

Enol form

EXAMPLE 13.9

Write two enol forms for each compound, and state which enol of each predominates at equilibrium:

(a) 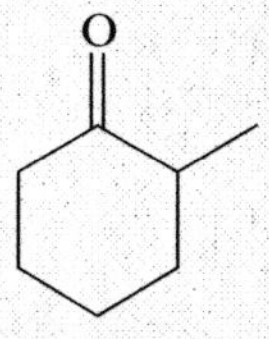 (b)

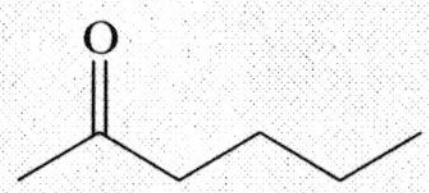

SOLUTION

In each case, the major enol form has the more substituted (the more stable) carbon–carbon double bond:

(a) 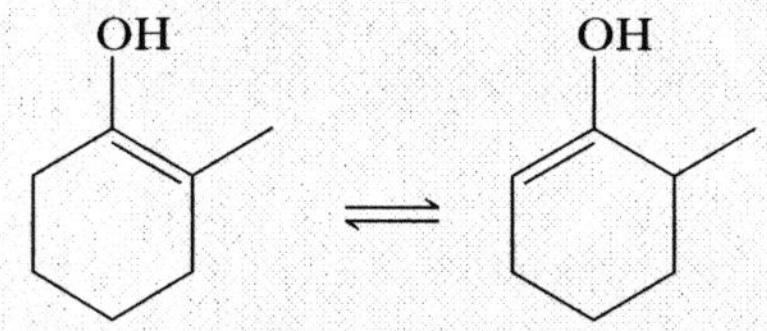 OH OH ⇌

Major enol

(b) 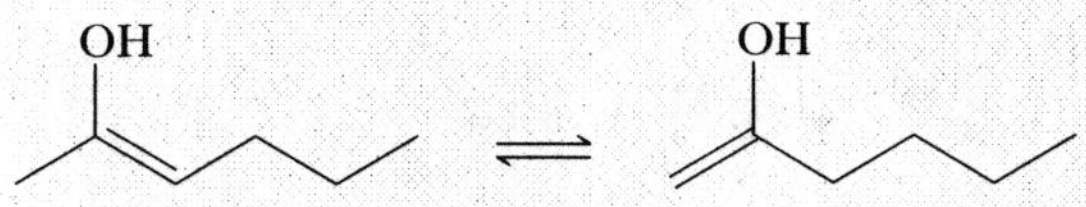 OH OH ⇌

Major enol

Practice Problem 13.9

Draw the structural formula for the keto form of each enol:

(a) 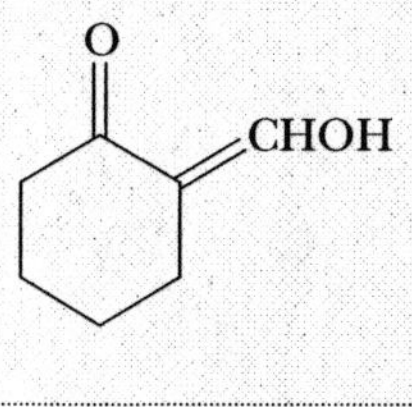 O CHOH (b) 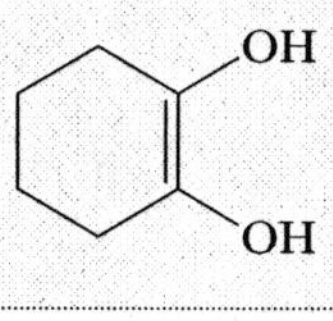 OH OH (c) 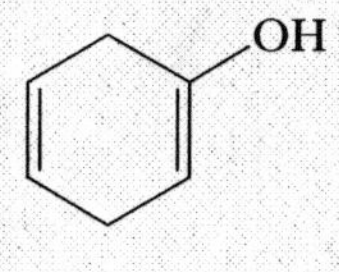 OH

Racemization The conversion of a pure enantiomer into a racemic mixture.

B. Racemization at an α-Carbon

When enantiomerically pure (either *R* or *S*) 3-phenyl-2-butanone is dissolved in ethanol, no change occurs in the optical activity of the solution over time. If, however, a trace of acid (for example, HCl) is added, the optical activity of the solution begins to decrease and gradually drops to zero. When 3-phenyl-2-butanone is isolated from this solution, it is found to be a racemic mixture (Section 6.9C). This observation can be explained by the acid-catalyzed formation of an achiral enol intermediate. Tautomerism of the achiral enol to the chiral keto form generates the *R* and *S* enantiomers with equal probability:

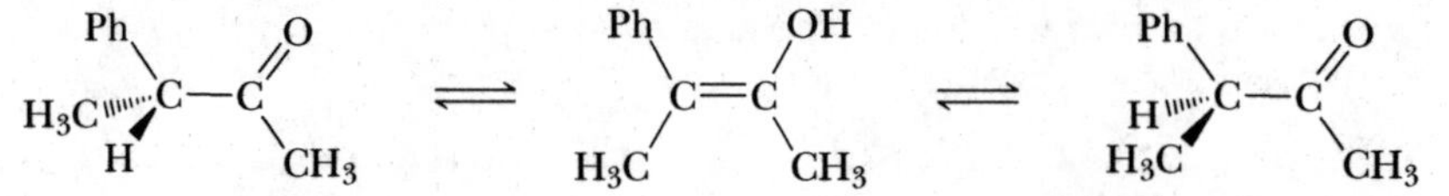

(*R*)-3-Phenyl-2-butanone ⇌ An achiral enol ⇌ (*S*)-3-Phenyl-2-butanone

Racemization by this mechanism occurs only at α-carbon stereocenters with at least one α-hydrogen.

C. α-Halogenation

Aldehydes and ketones with at least one α-hydrogen react with bromine and chlorine at the α-carbon to give an α-haloaldehyde or α-haloketone. Acetophenone, for example, reacts with bromine in acetic acid to give an α-bromoketone:

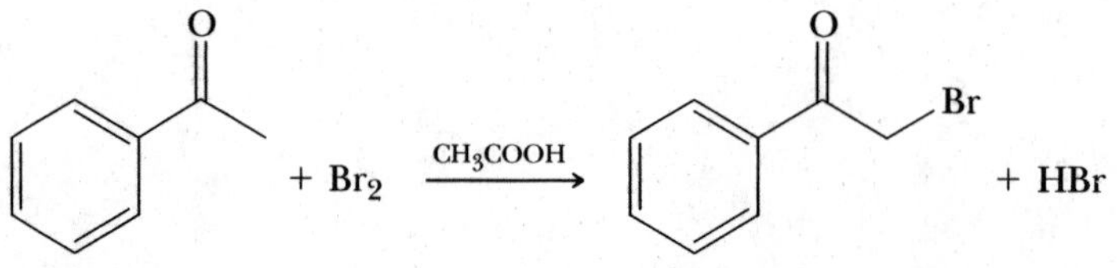

Acetophenone + Br_2 $\xrightarrow{CH_3COOH}$ α-Bromoacetophenone + HBr

α-Halogenation is catalyzed by both acid and base. For acid-catalyzed halogenation, the HBr or HCl generated by the reaction catalyzes further reaction.

Mechanism: Acid-Catalyzed α-Halogenation of a Ketone

Step 1: Acid-catalyzed keto–enol tautomerism gives the enol:

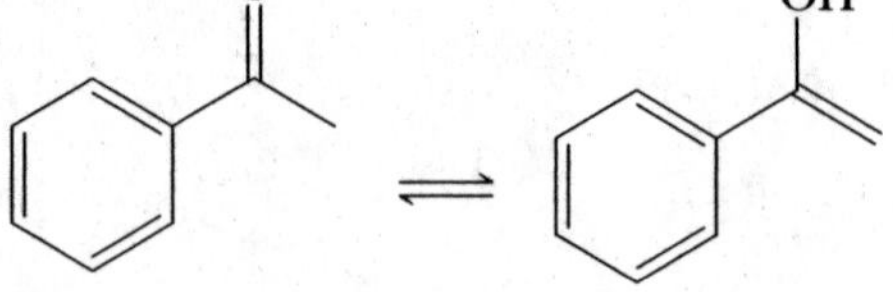

Keto form ⇌ Enol form

Step 2: Nucleophilic attack of the enol on the halogen molecule gives the α-haloketone:

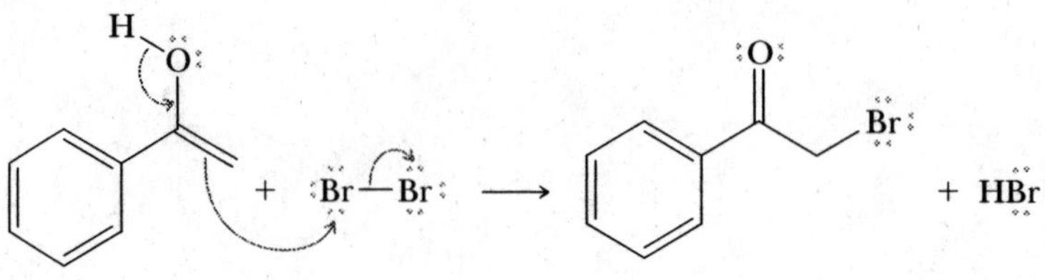

+ Br—Br → + HBr

The value of α-halogenation is that it converts an α-carbon into a center that now has a good leaving group bonded to it and that is therefore susceptible to attack by a variety of good nucleophiles. In the following illustration, diethylamine (a nucleophile) reacts with the α-bromoketone to give an α-diethylaminoketone:

$$\text{PhCOCH}_2\text{Br} + \text{H—N(CH}_2\text{CH}_3)_2 \longrightarrow \text{PhCOCH}_2\text{N(CH}_2\text{CH}_3)_2 + \text{HBr}$$

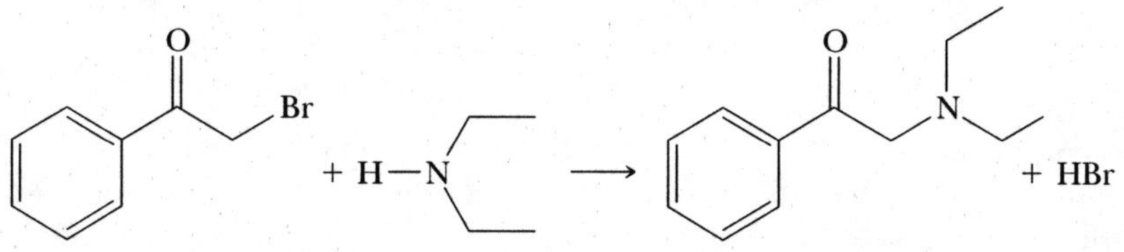

An α-bromoketone — An α-diethylaminoketone

In practice, this type of nucleophilic substitution is generally carried out in the presence of a weak base such as potassium carbonate to neutralize the HX as it is formed.

13.10 OXIDATION

A. Oxidation of Aldehydes to Carboxylic Acids

Aldehydes are oxidized to carboxylic acids by a variety of common oxidizing agents, including chromic acid and molecular oxygen. In fact, aldehydes are one of the most easily oxidized of all functional groups. Oxidation by chromic acid is illustrated by the conversion of hexanal to hexanoic acid:

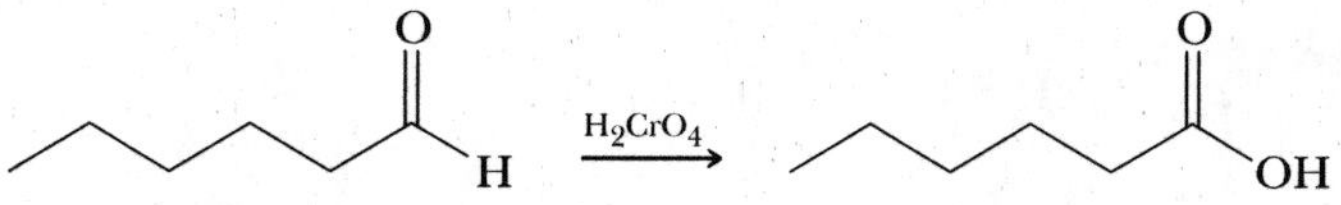

$$\text{CH}_3(\text{CH}_2)_4\text{CHO} \xrightarrow{H_2CrO_4} \text{CH}_3(\text{CH}_2)_4\text{COOH}$$

Hexanal — Hexanoic acid

Aldehydes are also oxidized to carboxylic acids by silver ion. One laboratory procedure is to shake a solution of the aldehyde dissolved in aqueous ethanol or tetrahydrofuran (THF) with a slurry of Ag_2O:

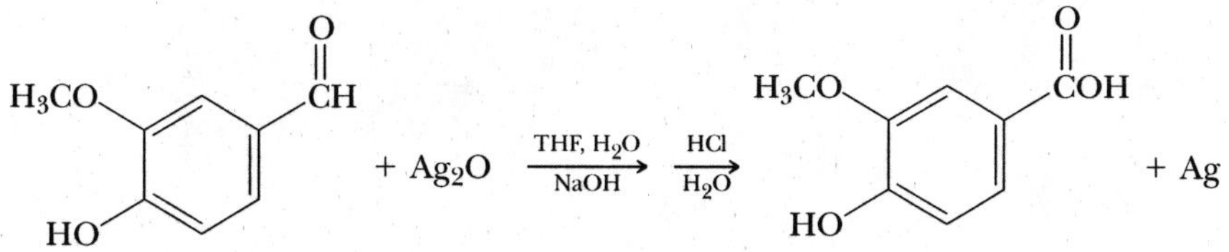

$$\text{Vanillin} + Ag_2O \xrightarrow[NaOH]{THF,\ H_2O} \xrightarrow[H_2O]{HCl} \text{Vanillic acid} + Ag$$

(H$_3$CO, HO-substituted benzaldehyde (CH=O) → H$_3$CO, HO-substituted benzoic acid (COOH))

Vanillin (from vanilla) — Vanillic acid

Tollens' reagent, another form of silver ion, is prepared by dissolving $AgNO_3$ in water, adding sodium hydroxide to precipitate silver ion as Ag_2O, and then adding aqueous ammonia to redissolve silver ion as the silver–ammonia complex ion:

$$Ag^+NO_3^- + 2NH_3 \overset{NH_3,\ H_2O}{\rightleftharpoons} Ag(NH_3)_2{}^+NO_3^-$$

When Tollens' reagent is added to an aldehyde, the aldehyde is oxidized to a carboxylic anion, and Ag^+ is reduced to metallic silver. If this reaction is carried

A silver mirror has been deposited in the inside of this flask by the reaction of an aldehyde with Tollens' reagent. *(Charles D. Winters)*

out properly, silver precipitates as a smooth, mirrorlike deposit—hence the name **silver-mirror test**:

$$\mathrm{R\overset{O}{\overset{\|}{C}}H} + 2\mathrm{Ag(NH_3)_2}^{+} \xrightarrow{\mathrm{NH_3,\ H_2O}} \mathrm{R\overset{O}{\overset{\|}{C}}O^-} + 2\mathrm{Ag} + 4\mathrm{NH_3}$$

Precipitates as silver mirror

Nowadays, Ag^+ is rarely used for the oxidation of aldehydes, because of the cost of silver and because other, more convenient methods exist for this oxidation. The reaction, however, is still used for silvering mirrors. In the process, formaldehyde or glucose is used as the aldehyde to reduce Ag^+.

Aldehydes are also oxidized to carboxylic acids by molecular oxygen and by hydrogen peroxide.

$$2\,\mathrm{C_6H_5\overset{O}{\overset{\|}{C}}H} + \mathrm{O_2} \longrightarrow 2\,\mathrm{C_6H_5\overset{O}{\overset{\|}{C}}OH}$$

Benzaldehyde → Benzoic acid

Molecular oxygen is the least expensive and most readily available of all oxidizing agents, and, on an industrial scale, air oxidation of organic molecules, including aldehydes, is common. Air oxidation of aldehydes can also be a problem: Aldehydes that are liquid at room temperature are so sensitive to oxidation by molecular oxygen that they must be protected from contact with air during storage. Often, this is done by sealing the aldehyde in a container under an atmosphere of nitrogen.

EXAMPLE 13.10

Draw a structural formula for the product formed by treating each compound with Tollens' reagent, followed by acidification with aqueous HCl:

(a) Pentanal (b) Cyclopentanecarbaldehyde

SOLUTION

The aldehyde group in each compound is oxidized to a carboxyl group:

(a) $\mathrm{CH_3CH_2CH_2CH_2\overset{O}{\overset{\|}{C}}OH}$ — Pentanoic acid

(b) $\mathrm{C_5H_9{-}COOH}$ — Cyclopentanecarboxylic acid

Practice Problem 13.10

Complete these oxidations:

(a) 3-Oxobutanal + O_2 $\longrightarrow$

(b) 3-Phenylpropanal + Tollens' reagent $\longrightarrow$

B. Oxidation of Ketones to Carboxylic Acids

Ketones are much more resistant to oxidation than are aldehydes. For example, ketones are not normally oxidized by chromic acid or potassium permanganate. In fact, these reagents are used routinely to oxidize secondary alcohols to ketones in good yield (Section 8.3F).

CHEMICAL CONNECTIONS

A Green Synthesis of Adipic Acid

The current industrial production of adipic acid relies on the oxidation of a mixture of cyclohexanol and cyclohexanone by nitric acid:

$$4\ \text{Cyclohexanol} + 6HNO_3 \longrightarrow 4\ \text{Hexanedioic acid (Adipic acid)} + 3N_2O\ (\text{Nitrous oxide}) + 3H_2O$$

A by-product of this oxidation is nitrous oxide, a gas considered to play a role in global warming and the depletion of the ozone layer in the atmosphere, as well as contributing to acid rain and acid smog. Given the fact that worldwide production of adipic acid is approximately 2.2 billion metric tons per year, the production of nitrous oxide is enormous. In spite of technological advances that allow for the recovery and recycling of nitrous oxide, it is estimated that approximately 400,000 metric tons escapes recovery and is released into the atmosphere each year.

Recently, Ryoji Noyori and coworkers at Nagoya University in Japan developed a "green" route to adipic acid, one that involves the oxidation of cyclohexene by 30% hydrogen peroxide catalyzed by sodium tungstate, Na_2WO_4:

$$\text{Cyclohexene} + 4H_2O_2 \xrightarrow[{[CH_3(C_8H_{17})_3N]HSO_4}]{Na_2WO_4} \text{Hexanedioic acid (Adipic acid)} + 4H_2O$$

In this process, cyclohexene is mixed with aqueous 30% hydrogen peroxide, and sodium tungstate and methyltrioctylammonium hydrogen sulfate are added to the resulting two-phase system. (Cyclohexene is insoluble in water.) Under these conditions, cyclohexene is oxidized to adipic acid in approximately 90% yield.

While this route to adipic acid is environmentally friendly, it is not yet competitive with the nitric acid oxidation route because of the high cost of 30% hydrogen peroxide. What will make it competitive is either a considerable reduction in the cost of hydrogen peroxide or the institution of more stringent limitations on the emission of nitrous oxide into the atmosphere (or a combination of these).

Ketones undergo oxidative cleavage, via their enol form, by potassium dichromate and potassium permanganate at higher temperatures and by higher concentrations of nitric acid, HNO_3. The carbon–carbon double bond of the enol is cleaved to form two carboxyl or ketone groups, depending on the substitution pattern of the original ketone. An important industrial application of this reaction is the oxidation of cyclohexanone to hexanedioic acid (adipic acid), one of the two monomers required for the synthesis of the polymer nylon 66 (Section 17.5A):

$$\text{Cyclohexanone (keto form)} \rightleftharpoons \text{Cyclohexanone (enol form)} \xrightarrow{HNO_3} \text{Hexanedioic acid (Adipic acid)}$$

13.11 REDUCTION

Aldehydes are reduced to primary alcohols and ketones to secondary alcohols:

$$\underset{\text{An aldehyde}}{\overset{\overset{\large O}{\|}}{RCH}} \xrightarrow{\text{reduction}} \underset{\text{A primary alcohol}}{RCH_2OH} \qquad \underset{\text{A ketone}}{\overset{\overset{\large O}{\|}}{RCR'}} \xrightarrow{\text{reduction}} \underset{\text{A secondary alcohol}}{\overset{\overset{OH}{|}}{RCHR'}}$$

A. Catalytic Reduction

The carbonyl group of an aldehyde or a ketone is reduced to a hydroxyl group by hydrogen in the presence of a transition metal catalyst, most commonly finely divided palladium, platinum, nickel, or rhodium. Reductions are generally carried out at temperatures from 25 to 100°C and at pressures of hydrogen from 1 to 5 atm. Under such conditions, cyclohexanone is reduced to cyclohexanol:

$$\text{Cyclohexanone} + H_2 \xrightarrow[25°C,\ 2\ atm]{Pt} \text{Cyclohexanol}$$

Cyclohexanone Cyclohexanol

The catalytic reduction of aldehydes and ketones is simple to carry out, yields are generally very high, and isolation of the final product is very easy. A disadvantage is that some other functional groups (for example, carbon–carbon double bonds) are also reduced under these conditions.

$$\text{trans-2-Butenal} \xrightarrow[Ni]{2H_2} \text{1-Butanol}$$

trans-2-Butenal (Crotonaldehyde) 1-Butanol

B. Metal Hydride Reductions

Hydride ion A hydrogen atom with two electrons in its valence shell; H:⁻.

By far the most common laboratory reagents used to reduce the carbonyl group of an aldehyde or a ketone to a hydroxyl group are sodium borohydride and lithium aluminum hydride. Each of these compounds behaves as a source of **hydride ion**, a very strong nucleophile. The structural formulas drawn here for these reducing agents show formal negative charges on boron and aluminum:

$$\underset{\text{Sodium borohydride}}{Na^+ H{-}\overset{\overset{H}{|}}{\underset{\underset{H}{|}}{B^-}}{-}H} \qquad \underset{\text{Lithium aluminum hydride}}{Li^+ H{-}\overset{\overset{H}{|}}{\underset{\underset{H}{|}}{Al^-}}{-}H} \qquad \underset{\text{Hydride ion}}{H{:}^-}$$

In fact, hydrogen is more electronegative than either boron or aluminum (H = 2.1, Al = 1.5, and B = 2.0), and the formal negative charge in the two reagents resides more on hydrogen than on the metal.

Lithium aluminum hydride is a very powerful reducing agent; it rapidly reduces not only the carbonyl groups of aldehydes and ketones, but also those of carboxylic acids (Section 14.6) and their functional derivatives (Section 15.9). Sodium borohydride is a much more selective reagent, reducing only aldehydes and ketones rapidly.

Reductions using sodium borohydride are most commonly carried out in aqueous methanol, in pure methanol, or in ethanol. The initial product of reduction is a tetraalkyl borate, which is converted to an alcohol and sodium borate salts upon treatment with water. One mole of sodium borohydride reduces 4 moles of aldehyde or ketone:

$$4\mathrm{R\overset{O}{\overset{\|}{C}}H} + \mathrm{NaBH_4} \xrightarrow{\mathrm{CH_3OH}} \underset{\text{A tetraalkyl borate}}{(\mathrm{RCH_2O})_4\mathrm{B^-Na^+}} \xrightarrow{\mathrm{H_2O}} 4\mathrm{RCH_2OH} + \text{borate salts}$$

The key step in the metal hydride reduction of an aldehyde or a ketone is the transfer of a hydride ion from the reducing agent to the carbonyl carbon to form a tetrahedral carbonyl addition compound. In the reduction of an aldehyde or a ketone to an alcohol, only the hydrogen atom attached to carbon comes from the hydride-reducing agent; the hydrogen atom bonded to oxygen comes from the water added to hydrolyze the metal alkoxide salt.

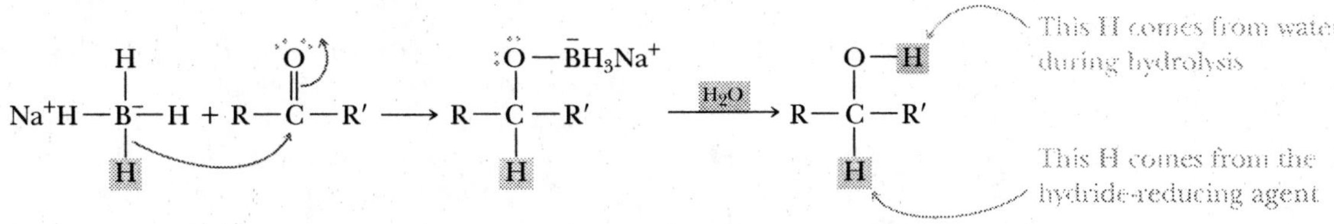

$$\mathrm{Na^+H{-}\overset{H}{\underset{H}{B^-}}{-}H} + \mathrm{R{-}\overset{O}{\overset{\|}{C}}{-}R'} \longrightarrow \mathrm{R{-}\overset{:\ddot{O}-\bar{B}H_3Na^+}{\underset{H}{C}}{-}R'} \xrightarrow{\mathrm{H_2O}} \mathrm{R{-}\overset{O-H}{\underset{H}{C}}{-}R'}$$

This H comes from water during hydrolysis

This H comes from the hydride-reducing agent

The next two equations illustrate the selective reduction of a carbonyl group in the presence of a carbon–carbon double bond and, alternatively, the selective reduction of a carbon–carbon double bond in the presence of a carbonyl group.

Selective reduction of a carbonyl group:

$$\mathrm{RCH{=}CH\overset{O}{\overset{\|}{C}}R'} \xrightarrow[2.\ \mathrm{H_2O}]{1.\ \mathrm{NaBH_4}} \mathrm{RCH{=}CH\overset{OH}{\overset{|}{C}}HR'}$$

Selective reduction of a carbon–carbon double bond:

$$\mathrm{RCH{=}CH\overset{O}{\overset{\|}{C}}R'} + \mathrm{H_2} \xrightarrow{\mathrm{Rh}} \mathrm{RCH_2CH_2\overset{O}{\overset{\|}{C}}R'}$$

Example 13.11

Complete these reductions:

(a) 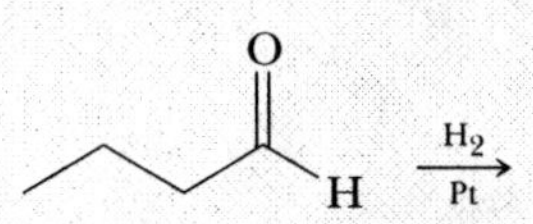 butanal (CH₃CH₂CH₂CHO) $\xrightarrow[\mathrm{Pt}]{\mathrm{H_2}}$ (b) 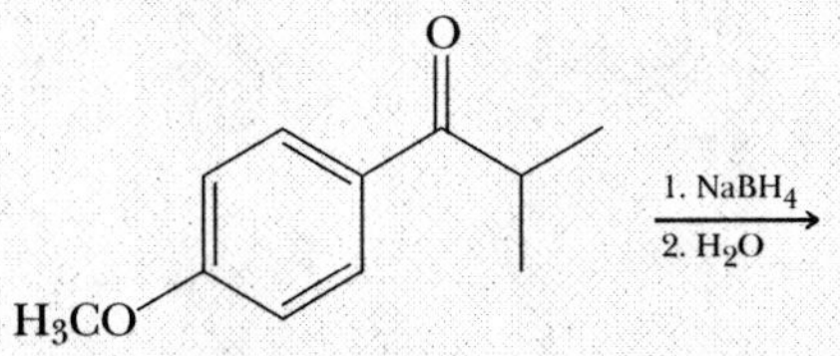 1-(4-methoxyphenyl)-2-methylpropan-1-one (H_3CO–C₆H₄–CO–CH(CH₃)₂) $\xrightarrow[2.\ \mathrm{H_2O}]{1.\ \mathrm{NaBH_4}}$

Solution

The carbonyl group of the aldehyde in (a) is reduced to a primary alcohol, and that of the ketone in (b) is reduced to a secondary alcohol:

(a)  CH₃CH₂CH₂CH₂OH (b) 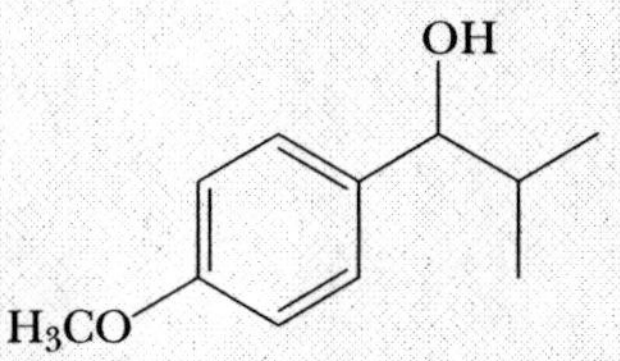 H_3CO–C₆H₄–CH(OH)–CH(CH₃)₂

Practice Problem 13.11

What aldehyde or ketone gives each alcohol upon reduction by $NaBH_4$?

(a) 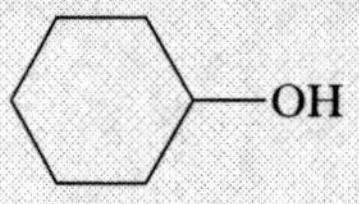 cyclohexane–OH (b) 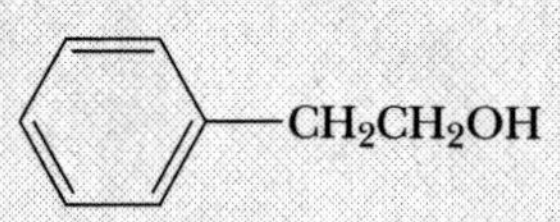 benzene–CH_2CH_2OH

(c) 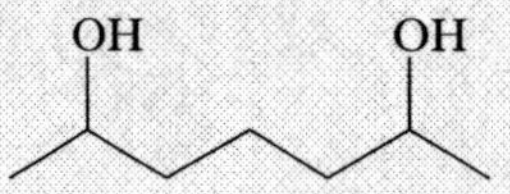 (OH, OH skeletal diol structure)

SUMMARY

An **aldehyde** (Section 13.2) contains a carbonyl group bonded to a hydrogen atom and a carbon atom. A **ketone** contains a carbonyl group bonded to two carbons. An aldehyde is named by changing *-e* of the parent alkane to *-al* (Section 13.3). A CHO group bonded to a ring is indicated by the suffix *-carbaldehyde*. A ketone is named by changing *-e* of the parent alkane to *-one* and using a number to locate the carbonyl group. In naming compounds that contain more than one functional group, the IUPAC system has established an **order of precedence of functional groups** (Section 13.3B). If the carbonyl group of an aldehyde or a ketone is lower in precedence than other functional groups in the molecule, it is indicated by the infix *-oxo-*.

Aldehydes and ketones are polar compounds (Section 13.4) and interact in the pure state by dipole–dipole interactions; they have higher boiling points and are more soluble in water than are nonpolar compounds of comparable molecular weight.

The carbon–metal bond in **Grignard reagents** (Section 13.5) has a high degree of partial ionic character. Grignard reagents behave as carbanions and are both strong bases and good nucleophiles.

A carbon atom adjacent to a carbonyl group is called an **α-carbon** (Section 13.9A), and a hydrogen attached to it is called an **α-hydrogen.**

KEY REACTIONS

1. Reaction with Grignard Reagents (Section 13.6C)

Treatment of formaldehyde with a Grignard reagent, followed by hydrolysis in aqueous acid, gives a primary alcohol. Similar treatment of any other aldehyde gives a secondary alcohol:

$$CH_3\overset{O}{\overset{\|}{C}}H \xrightarrow[2.\ HCl,\ H_2O]{1.\ C_6H_5MgBr} C_6H_5\overset{OH}{\overset{|}{C}}HCH_3$$

Treatment of a ketone with a Grignard reagent gives a tertiary alcohol:

$$CH_3\overset{O}{\overset{\|}{C}}CH_3 \xrightarrow[2.\ HCl,\ H_2O]{1.\ C_6H_5MgBr} C_6H_5\overset{OH}{\overset{|}{C}}(CH_3)_2$$

2. Addition of Alcohols to Form Hemiacetals (Section 13.7)

Hemiacetals are only minor components of an equilibrium mixture of aldehyde or ketone and alcohol, except where the —OH and C=O groups are parts of the same molecule and a five- or six-membered ring can form:

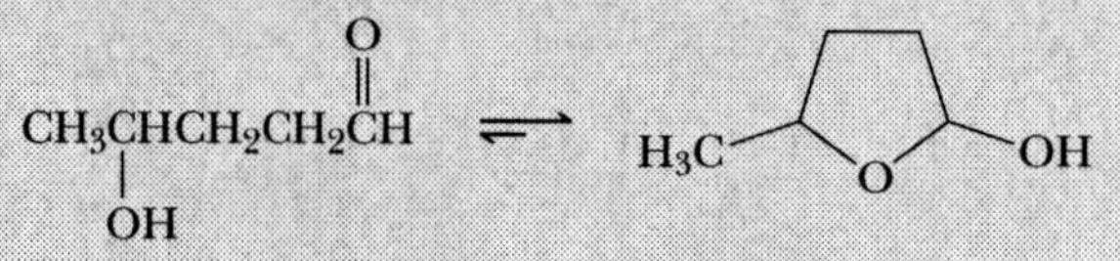

$$CH_3\underset{OH}{\underset{|}{C}}HCH_2CH_2\overset{O}{\overset{\|}{C}}H \rightleftharpoons H_3C\text{–(cyclic, ring O)–}OH$$

4-Hydroxypentanal A cyclic hemiacetal

3. Addition of Alcohols to Form Acetals (Section 13.7)

The formation of acetals is catalyzed by acid:

$$\text{cyclopentanone} + HOCH_2CH_2OH \overset{H^+}{\rightleftharpoons} \text{(cyclic acetal: O–}CH_2\text{–}CH_2\text{–O)} + H_2O$$

4. Addition of Ammonia and Amines (Section 13.8)

The addition of ammonia or a primary amine to the carbonyl group of an aldehyde or a ketone forms a tetrahedral carbonyl addition intermediate. Loss of water from this intermediate gives an imine (a Schiff base):

$$\text{cyclopentanone} + H_2NCH_3 \overset{H^+}{\rightleftharpoons} \text{cyclopentylidene}{=}NCH_3 + H_2O$$

5. Reductive Amination to Amines (Section 13.8B)

The carbon–nitrogen double bond of an imine can be reduced by hydrogen in the presence of a transition metal catalyst to a carbon–nitrogen single bond:

$$\text{cyclohexanone} + H_2N\text{–cyclohexyl} \xrightarrow[-H_2O]{} [\text{cyclohexylidene}{=}N\text{–cyclohexyl}] \xrightarrow{H_2/Ni} \text{cyclohexyl–N(H)–cyclohexyl}$$

6. Keto–Enol Tautomerism (Section 13.9A)

The keto form generally predominates at equilibrium:

$$\underset{\substack{\text{Keto form} \\ \text{(Approx 99.9\%)}}}{CH_3\overset{O}{\overset{\|}{C}}CH_3} \rightleftharpoons \underset{\text{Enol form}}{CH_3\overset{OH}{\overset{|}{C}}{=}CH_2}$$

7. Oxidation of an Aldehyde to a Carboxylic Acid (Section 13.10)

The aldehyde group is among the most easily oxidized functional groups. Oxidizing agents include H_2CrO_4, Tollens' reagent, and O_2:

$$\text{2-hydroxybenzaldehyde (}\overset{O}{\overset{\|}{C}}H\text{, OH)} + Ag_2O \xrightarrow[\text{2. } H_2O,\ HCl]{\text{1. THF, } H_2O,\ NaOH} \text{2-hydroxybenzoic acid (}\overset{O}{\overset{\|}{C}}OH\text{, OH)} + Ag$$

8. Catalytic Reduction (Section 13.11A)

Catalytic reduction of the carbonyl group of an aldehyde or a ketone to a hydroxyl group is simple to carry out and yields of alcohols are high:

$$\text{cyclohexanone} + H_2 \xrightarrow[25°C,\ 2\ atm]{Pt} \text{cyclohexanol (–OH)}$$

9. Metal Hydride Reduction (Section 13.11B)

Both $LiAlH_4$ and $NaBH_4$ reduce the carbonyl group of an aldehyde or a ketone to an hydroxyl group. They are selective in that neither reduces isolated carbon–carbon double bonds:

$$\text{cyclohex-3-enone (=O)} \xrightarrow[\text{2. } H_2O]{\text{1. } NaBH_4} \text{cyclohex-3-enol (–OH)}$$

PROBLEMS

A problem number set in red indicates an applied "real-world" problem.

Preparation of Aldehydes and Ketones (See Chapters 8 and 9)

13.12 Complete these reactions:

(a) 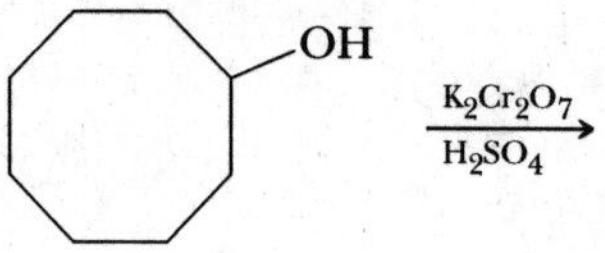 cyclooctanol $\xrightarrow[H_2SO_4]{K_2Cr_2O_7}$ (b) 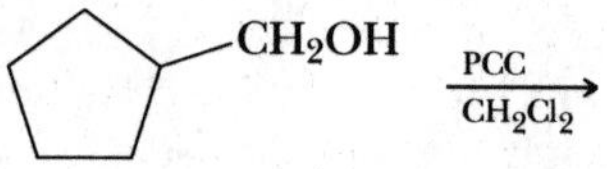 cyclopentyl–CH_2OH $\xrightarrow[CH_2Cl_2]{PCC}$

(c) 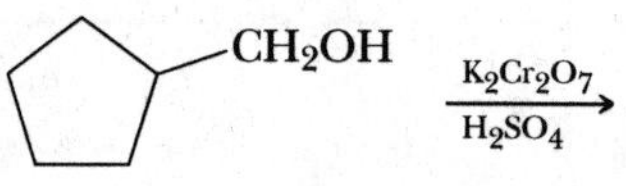 cyclopentyl–CH_2OH $\xrightarrow[H_2SO_4]{K_2Cr_2O_7}$ (d) 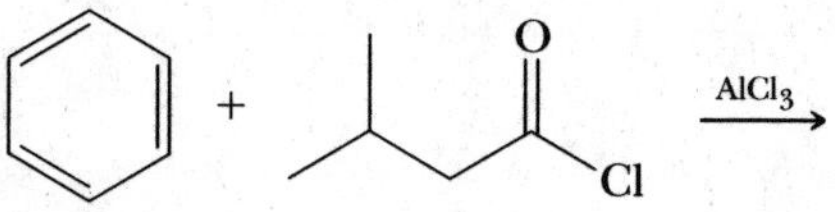 benzene + 3-methylbutanoyl chloride (Cl, O) $\xrightarrow{AlCl_3}$

13.13 Show how you would bring about these conversions:

(a) 1-Pentanol to pentanal
(b) 1-Pentanol to pentanoic acid
(c) 2-Pentanol to 2-pentanone
(d) 1-Pentene to 2-pentanone
(e) Benzene to acetophenone
(f) Styrene to acetophenone
(g) Cyclohexanol to cyclohexanone
(h) Cyclohexene to cyclohexanone

Structure and Nomenclature

13.14 Draw a structural formula for the one ketone with molecular formula C_4H_8O and for the two aldehydes with molecular formula C_4H_8O.

13.15 Draw structural formulas for the four aldehydes with molecular formula $C_5H_{10}O$. Which of these aldehydes are chiral?

13.16 Name these compounds:

(a) 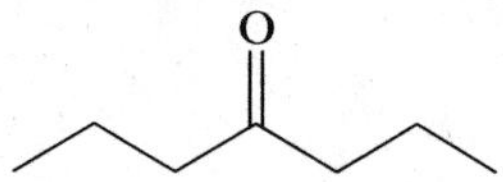 O (b) 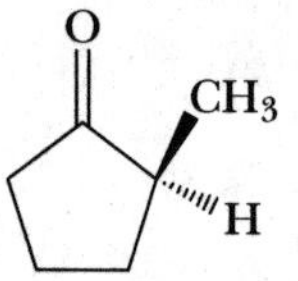 O, CH_3, H (c) 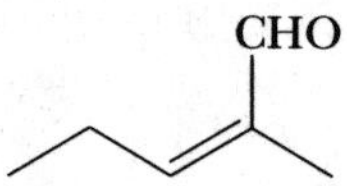 CHO (d) 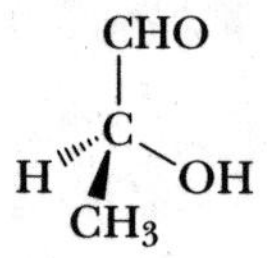 CHO, H, C, OH, CH_3

(e) 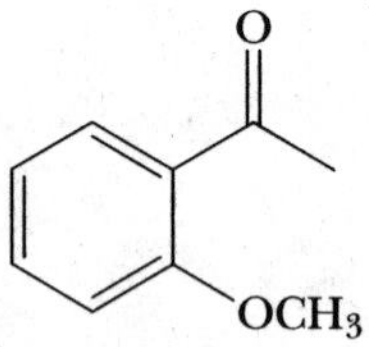 O, OCH_3 (f) 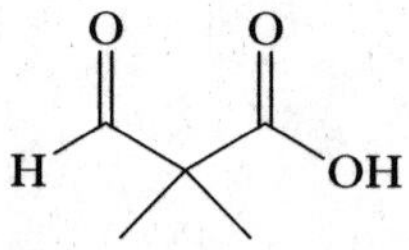 O, O, H, OH (g) 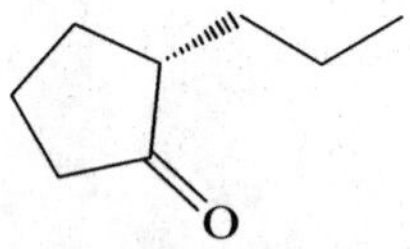 O

13.17 Draw structural formulas for these compounds:

(a) 1-Chloro-2-propanone
(b) 3-Hydroxybutanal
(c) 4-Hydroxy-4-methyl-2-pentanone
(d) 3-Methyl-3-phenylbutanal
(e) (*S*)-3-bromocyclohexanone
(f) 3-Methyl-3-buten-2-one
(g) 5-Oxohexanal
(h) 2,2-Dimethylcyclohexanecarbaldehyde
(i) 3-Oxobutanoic acid

Addition of Carbon Nucleophiles

13.18 Write an equation for the acid–base reaction between phenylmagnesium iodide and a carboxylic acid. Use curved arrows to show the flow of electrons in this reaction. In addition, show that the reaction is an example of a stronger acid and stronger base reacting to form a weaker acid and weaker base.

13.19 Diethyl ether is prepared on an industrial scale by the acid-catalyzed dehydration of ethanol:

$$2CH_3CH_2OH \xrightarrow[180°C]{H_2SO_4} CH_3CH_2OCH_2CH_3 + H_2O$$

Explain why diethyl ether used in the preparation of Grignard reagents must be carefully purified to remove all traces of ethanol and water.

13.20 Draw structural formulas for the product formed by treating each compound with propylmagnesium bromide, followed by hydrolysis in aqueous acid:

(a) CH_2O **(b)** 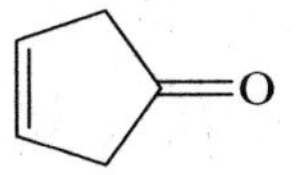 =O **(c)** 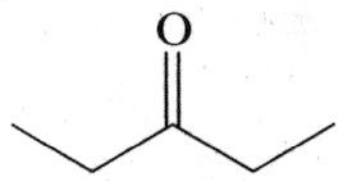 O **(d)** 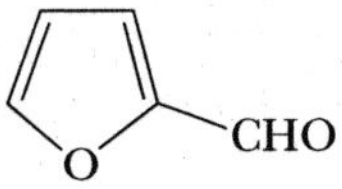 O CHO **(e)** 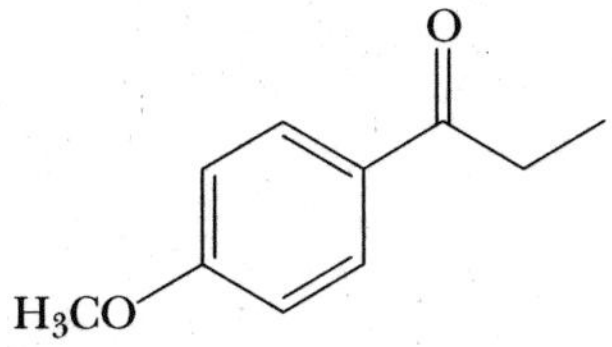 O H_3CO

13.21 Suggest a synthesis for each alcohol, starting from an aldehyde or a ketone and an appropriate Grignard reagent (the number of combinations of Grignard reagent and aldehyde or ketone that might be used is shown in parentheses below each target molecule):

(a) 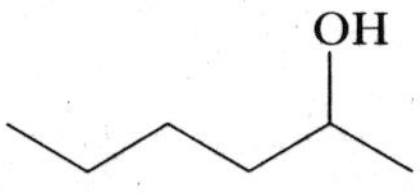 OH (Two combinations) **(b)** 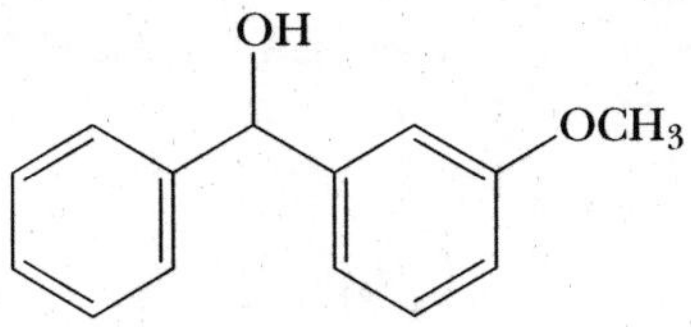 OH OCH_3 (Two combinations) **(c)** 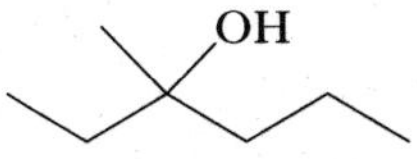 OH (Three combinations)

Addition of Oxygen Nucleophiles

13.22 5-Hydroxyhexanal forms a six-membered cyclic hemiacetal that predominates at equilibrium in aqueous solution:

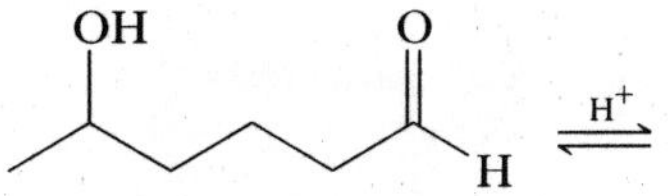

OH O H $\overset{H^+}{\rightleftharpoons}$ a cyclic hemiacetal

5-Hydroxyhexanal

(a) Draw a structural formula for this cyclic hemiacetal.
(b) How many stereoisomers are possible for 5-hydroxyhexanal?
(c) How many stereoisomers are possible for the cyclic hemiacetal?
(d) Draw alternative chair conformations for each stereoisomer.
(e) For each stereoisomer, which alternative chair conformation is the more stable?

13.23 Draw structural formulas for the hemiacetal and then the acetal formed from each pair of reactants in the presence of an acid catalyst:

(a) 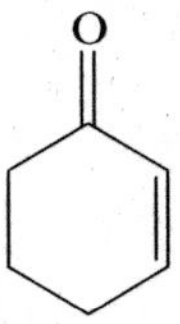 O + CH_3CH_2OH **(b)** 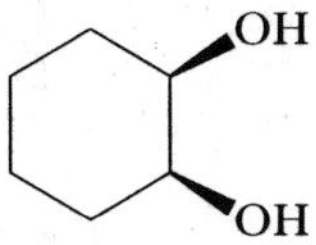 OH OH + $CH_3\overset{O}{\overset{\|}{C}}CH_3$ **(c)** CHO + CH_3OH

13.24 Draw structural formulas for the products of hydrolysis of each acetal in aqueous acid:

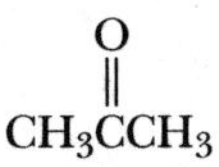

(a) 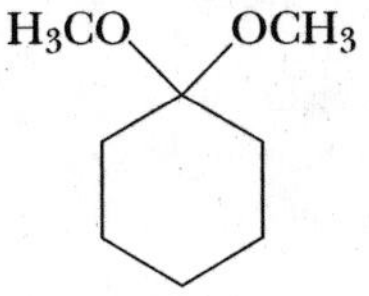 H_3CO OCH_3 **(b)** 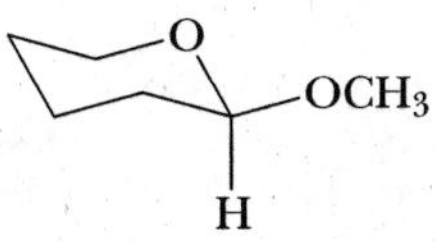 O OCH_3 H **(c)** 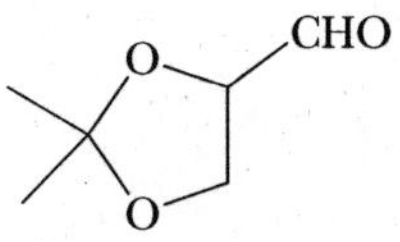 O CHO O

13.25 The following compound is a component of the fragrance of jasmine: From what carbonyl-containing compound and alcohol is the compound derived?

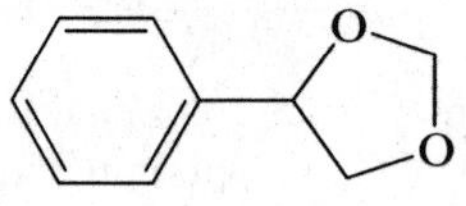

13.26 Propose a mechanism for the formation of the cyclic acetal by treating acetone with ethylene glycol in the presence of an acid catalyst. Make sure that your mechanism is consistent with the fact that the oxygen atom of the water molecule is derived from the carbonyl oxygen of acetone.

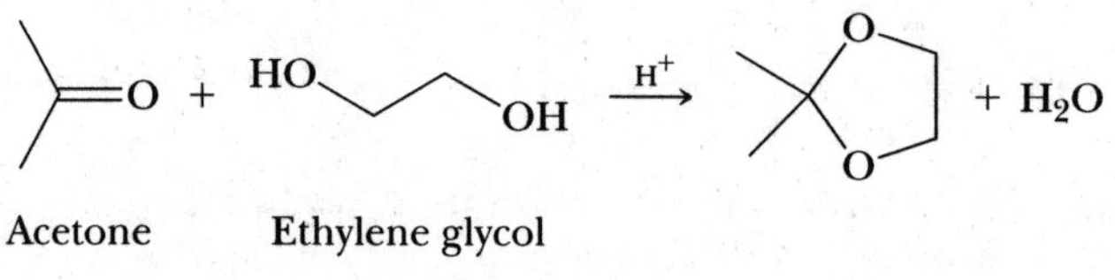

Acetone + Ethylene glycol $\xrightarrow{H^+}$ + H_2O

13.27 Propose a mechanism for the formation of a cyclic acetal from 4-hydroxypentanal and one equivalent of methanol: If the carbonyl oxygen of 4-hydroxypentanal is enriched with oxygen-18, does your mechanism predict that the oxygen label appears in the cyclic acetal or in the water? Explain.

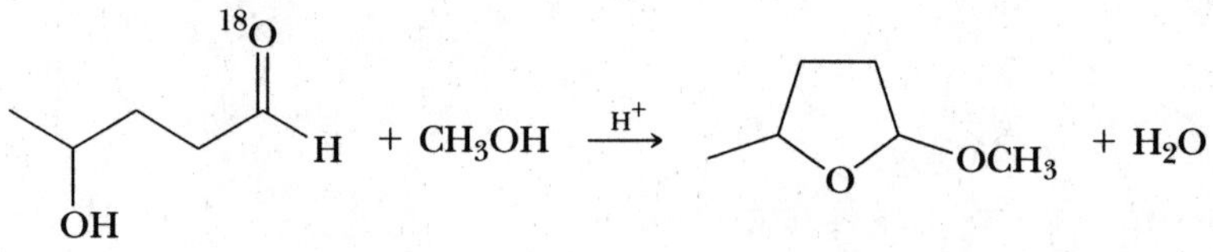

+ CH_3OH $\xrightarrow{H^+}$ OCH_3 + H_2O

Addition of Nitrogen Nucleophiles

13.28 Show how this secondary amine can be prepared by two successive reductive aminations:

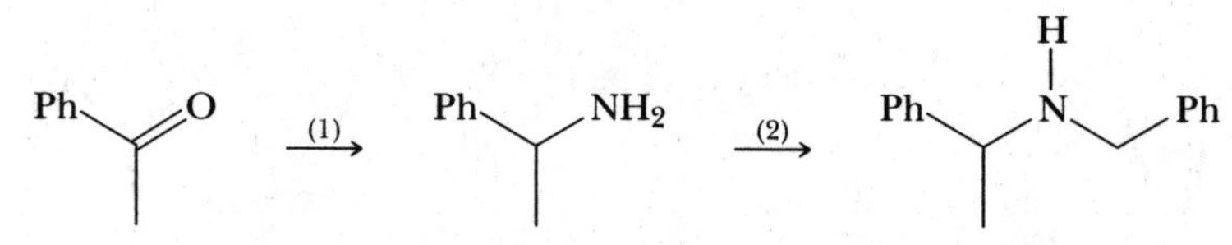

Ph–C(=O)–CH₃ $\xrightarrow{(1)}$ Ph–CH(CH₃)–NH_2 $\xrightarrow{(2)}$ Ph–CH(CH₃)–NH–CH₂–Ph

13.29 Show how to convert cyclohexanone to each of the following amines:

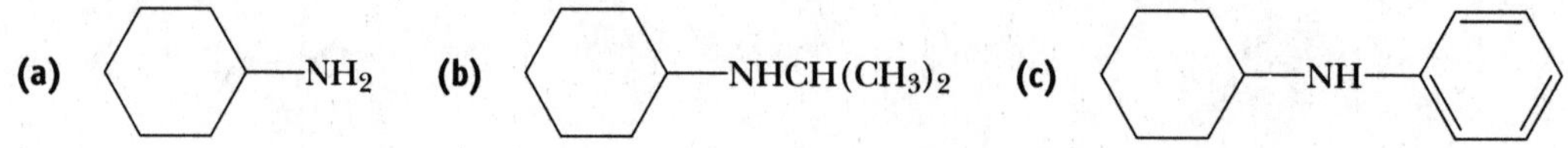

(a) cyclohexyl–NH_2 (b) cyclohexyl–$NHCH(CH_3)_2$ (c) cyclohexyl–NH–phenyl

13.30 Following are structural formulas for amphetamine and methamphetamine:

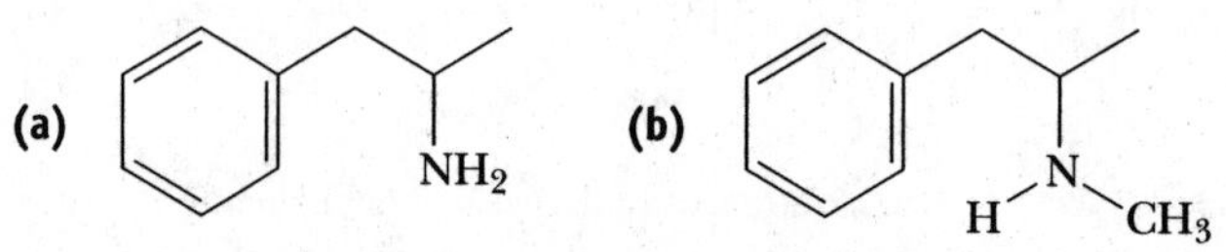

(a) Amphetamine (NH_2) (b) Methamphetamine (H–N–CH_3)

The major central nervous system effects of amphetamine and amphetaminelike drugs are locomotor stimulation, euphoria and excitement, stereotyped behavior, and anorexia. Show how each drug can be synthesized by the reductive amination of an appropriate aldehyde or ketone.

13.31 Rimantadine is effective in preventing infections caused by the influenza A virus and in treating established illness. The drug is thought to exert its antiviral effect by blocking a late stage in the assembly of the virus. Following is the final step in the synthesis of rimantadine:

O ⟶ NH_2

Rimantadine
(an antiviral agent)

(a) Describe experimental conditions to bring about this conversion.

(b) Is rimantadine chiral?

13.32 Methenamine, a product of the reaction of formaldehyde and ammonia, is a *prodrug*—a compound that is inactive by itself, but is converted to an active drug in the body by a biochemical transformation. The strategy behind the use of methenamine as a prodrug is that nearly all bacteria are sensitive to formaldehyde at concentrations of 20 mg/mL or higher. Formaldehyde cannot be used directly in medicine, however, because an effective concentration in plasma cannot be achieved with safe doses. Methenamine is stable at pH 7.4 (the pH of blood plasma), but undergoes acid-catalyzed hydrolysis to formaldehyde and ammonium ion under the acidic conditions of the kidneys and the urinary tract:

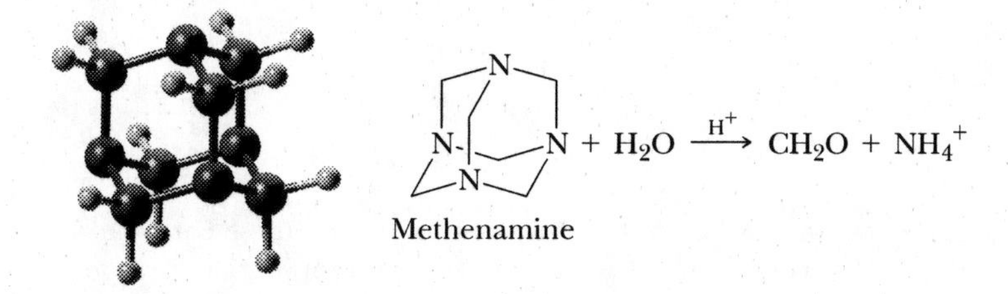

Thus, methenamine can be used as a site-specific drug to treat urinary infections.

(a) Balance the equation for the hydrolysis of methenamine to formaldehyde and ammonium ion.

(b) Does the pH of an aqueous solution of methenamine increase, remain the same, or decrease as a result of the hydrolysis of the compound? Explain.

(c) Explain the meaning of the following statement: The functional group in methenamine is the nitrogen analog of an acetal.

(d) Account for the observation that methenamine is stable in blood plasma, but undergoes hydrolysis in the urinary tract.

Keto–Enol Tautomerism

13.33 The following molecule belongs to a class of compounds called enediols: Each carbon of the double bond carries an —OH group:

HC—OH
‖
α-hydroxyaldehyde ⇌ C—OH ⇌ α-hydroxyketone
|
CH_3

An enediol

Draw structural formulas for the α-hydroxyketone and the α-hydroxyaldehyde with which this enediol is in equilibrium.

13.34 In dilute aqueous acid, (R)-glyceraldehyde is converted into an equilibrium mixture of (R,S)-glyceraldehyde and dihydroxyacetone:

$$\underset{(R)\text{-Glyceraldehyde}}{\begin{array}{c}CHO\\|\\CHOH\\|\\CH_2OH\end{array}} \xrightleftharpoons{H_2O,\ HCl} \underset{(R,S)\text{-Glyceraldehyde}}{\begin{array}{c}CHO\\|\\CHOH\\|\\CH_2OH\end{array}} + \underset{\text{Dihydroxyacetone}}{\begin{array}{c}CH_2OH\\|\\C{=}O\\|\\CH_2OH\end{array}}$$

Propose a mechanism for this isomerization.

Oxidation/Reduction of Aldehydes and Ketones

13.35 Draw a structural formula for the product formed by treating butanal with each of the following sets of reagents:

(a) $LiAlH_4$ followed by H_2O
(b) $NaBH_4$ in CH_3OH/H_2O
(c) H_2/Pt
(d) $Ag(NH_3)_2^+$ in NH_3/H_2O and then HCl/H_2O
(e) H_2CrO_4
(f) $C_6H_5NH_2$ in the presence of H_2/Ni

13.36 Draw a structural formula for the product of the reaction of p-bromoacetophenone with each set of reagents in Problem 13.35.

Synthesis

13.37 Show the reagents and conditions that will bring about the conversion of cyclohexanol to cyclohexanecarbaldehyde:

Cyclohexanol (OH) $\xrightarrow{(1)}$ chlorocyclohexane (Cl) $\xrightarrow{(2)}$ cyclohexylmagnesium chloride (MgCl) $\xrightarrow{(3)}$ cyclohexylmethanol (OH) $\xrightarrow{(4)}$ cyclohexanecarbaldehyde (CHO)

13.38 Starting with cyclohexanone, show how to prepare these compounds (in addition to the given starting material, use any other organic or inorganic reagents, as necessary):

(a) Cyclohexanol
(b) Cyclohexene
(c) *cis*-1,2-Cyclohexanediol
(d) 1-Methylcyclohexanol
(e) 1-Methylcyclohexene
(f) 1-Phenylcyclohexanol
(g) 1-Phenylcyclohexene
(h) Cyclohexene oxide
(i) *trans*-1,2-Cyclohexanediol

13.39 Show how to bring about these conversions (in addition to the given starting material, use any other organic or inorganic reagents, as necessary):

(a) $C_6H_5\overset{O}{\overset{\|}{C}}CH_2CH_3 \longrightarrow C_6H_5\overset{OH}{\overset{|}{C}}HCH_2CH_3 \longrightarrow C_6H_5CH{=}CHCH_3$

(b) Cyclopentanone (=O) $\longrightarrow$ cyclopentanol (OH) $\longrightarrow$ chlorocyclopentane (Cl) $\longrightarrow$ cyclopentylmethanol (CH_2OH)

(c) Cyclopentanone (=O) $\longrightarrow$ 1-cyclopentylcyclopentanol (OH)

(d) Cyclopentanone (=O) $\longrightarrow$ 1-(cyclopent-1-enyl)pyrrolidine (N)

13.40 Many tumors of the breast are estrogen dependent. Drugs that interfere with estrogen binding have antitumor activity and may even help prevent the occurrence of tumors. A widely used antiestrogen drug is tamoxifen:

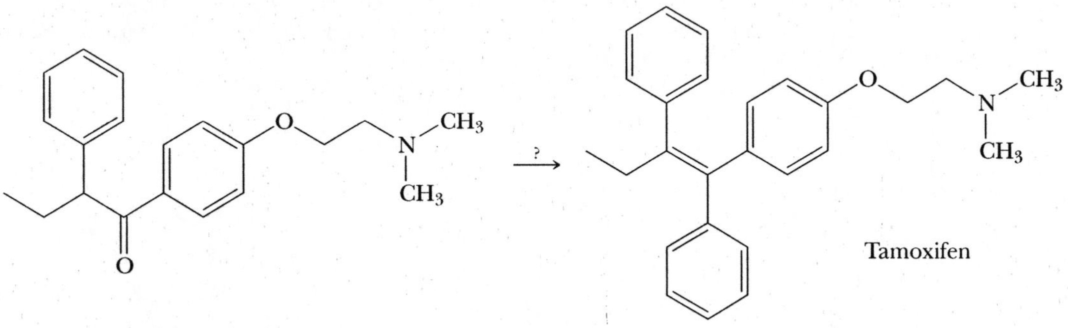

O O N CH_3 CH_3 $\xrightarrow{?}$ O N CH_3 CH_3

Tamoxifen

(a) How many stereoisomers are possible for tamoxifen?
(b) Specify the configuration of the stereoisomer shown here.
(c) Show how tamoxifen can be synthesized from the given ketone using a Grignard reaction, followed by dehydration.

13.41 Following is a possible synthesis of the antidepressant bupropion (Wellbutrin®):

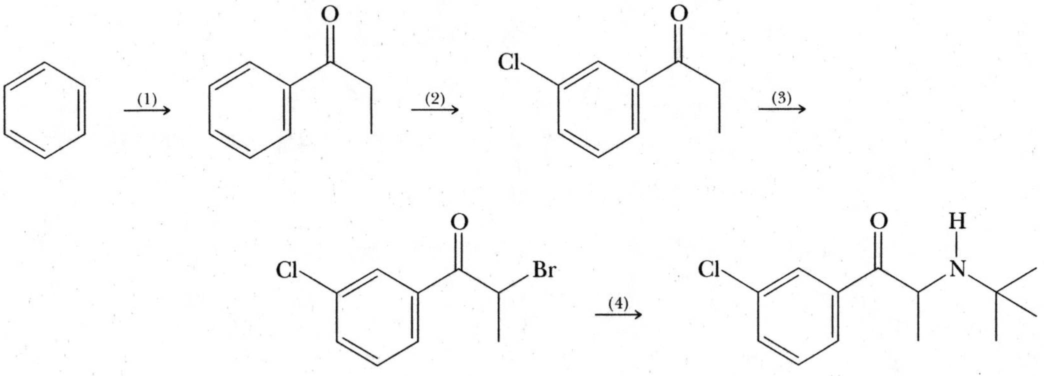

$\xrightarrow{(1)}$ O $\xrightarrow{(2)}$ Cl O $\xrightarrow{(3)}$

Cl O Br $\xrightarrow{(4)}$ Cl O H N

Bupropion
(Wellbutrin®)

Show the reagents that will bring about each step in this synthesis.

13.42 The synthesis of chlorpromazine in the 1950s and the discovery soon thereafter of the drug's antipsychotic activity opened the modern era of biochemical investigations into the pharmacology of the central nervous system. One of the compounds prepared in the search for more effective antipsychotics was amitriptyline.

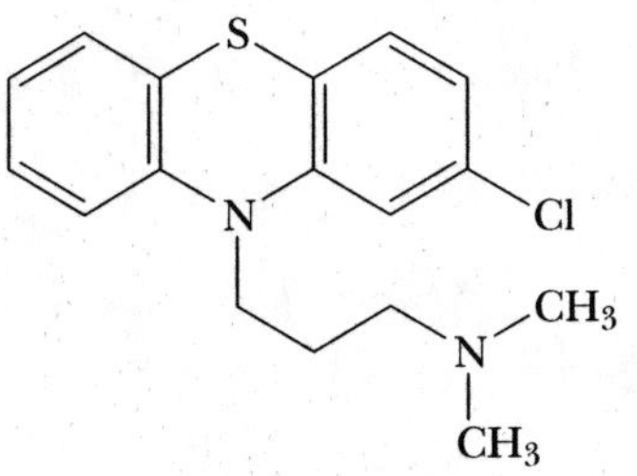

S N Cl N CH_3 CH_3

Chlorpromazine

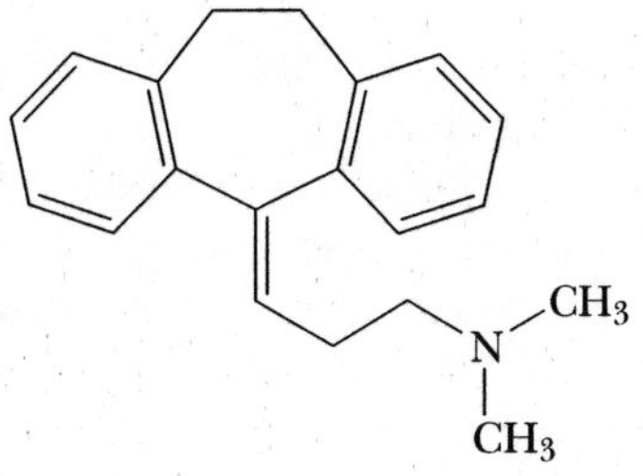

N CH_3 CH_3

Amitriptyline

Surprisingly, amitriptyline shows antidepressant activity rather than antipsychotic activity. It is now known that amitriptyline inhibits the reuptake of norepinephrine and serotonin from the synaptic cleft. Because the reuptake of these neurotransmitters is inhibited, their effects are potentiated. That is, the two neurotransmitters remain available to interact with serotonin and norepinephrine receptor sites longer and continue to cause excitation of serotonin and norepinephrine-mediated neural pathways. The following is a synthesis for amitriptyline:

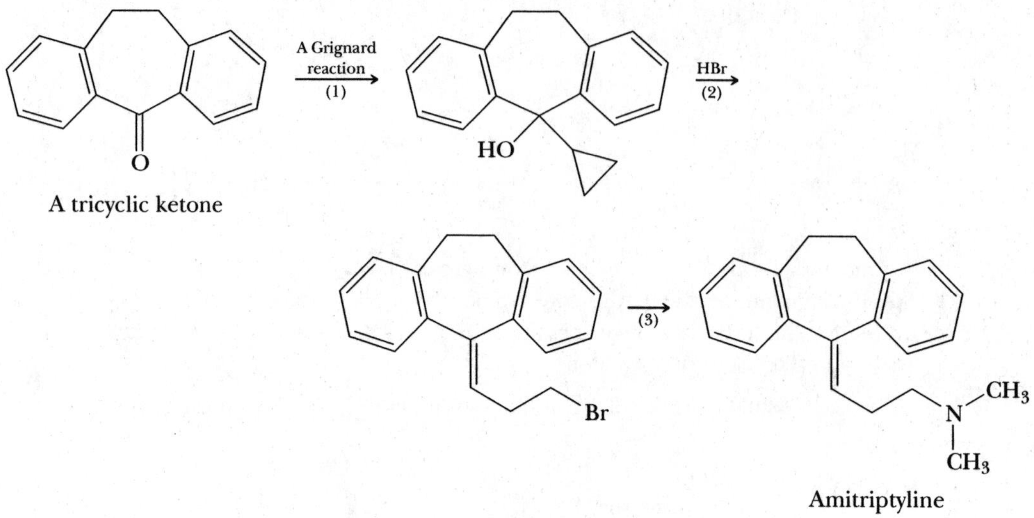

A Grignard reaction (1)

HBr (2)

HO

O

A tricyclic ketone

(3)

Br

CH_3 N CH_3

Amitriptyline

(a) Propose a reagent for Step 1.

(b) Propose a mechanism for Step 2. (*Note*: It is not acceptable to propose a primary carbocation as an intermediate.)

(c) Propose a reagent for Step (3).

13.43 Following is a synthesis for diphenhydramine:

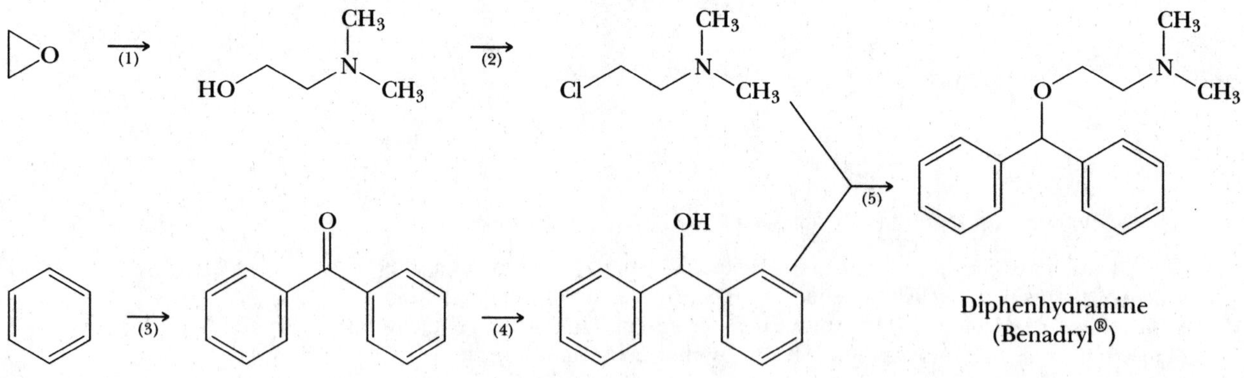

O (1) HO N CH_3 CH_3 (2) Cl N CH_3 CH_3

(3) O (4) OH (5) O N CH_3 CH_3

Diphenhydramine (Benadryl®)

The hydrochloride salt of this compound, best known by its trade name, Benadryl®, is an antihistamine.

(a) Propose reagents for Steps 1 and 2.

(b) Propose reagents for Steps 3 and 4.

(c) Show that Step 5 is an example of nucleophilic aliphatic substitution. What type of mechanism—S_N1 or S_N2—is more likely for this reaction? Explain.

13.44 Following is a synthesis for the antidepressant venlafaxine:

Venlafaxine

(a) Propose a reagent for Step 1, and name the type of reaction that takes place.
(b) Propose reagents for Steps 2 and 3.
(c) Propose reagents for Steps 4 and 5.
(d) Propose a reagent for Step 6, and name the type of reaction that takes place.

Spectroscopy

13.45 Compound A, $C_5H_{10}O$, is used as a flavoring agent for many foods that possess a chocolate or peach flavor. Its common name is isovaleraldehyde and it gives ^{13}C-NMR peaks at δ 202.7, 52.7, 23.6, and 22.6. Provide a structural formula for isovaleraldehyde and give its IUPAC name.

13.46 Following are ^{1}H-NMR and IR spectra of compound B, $C_6H_{12}O_2$:

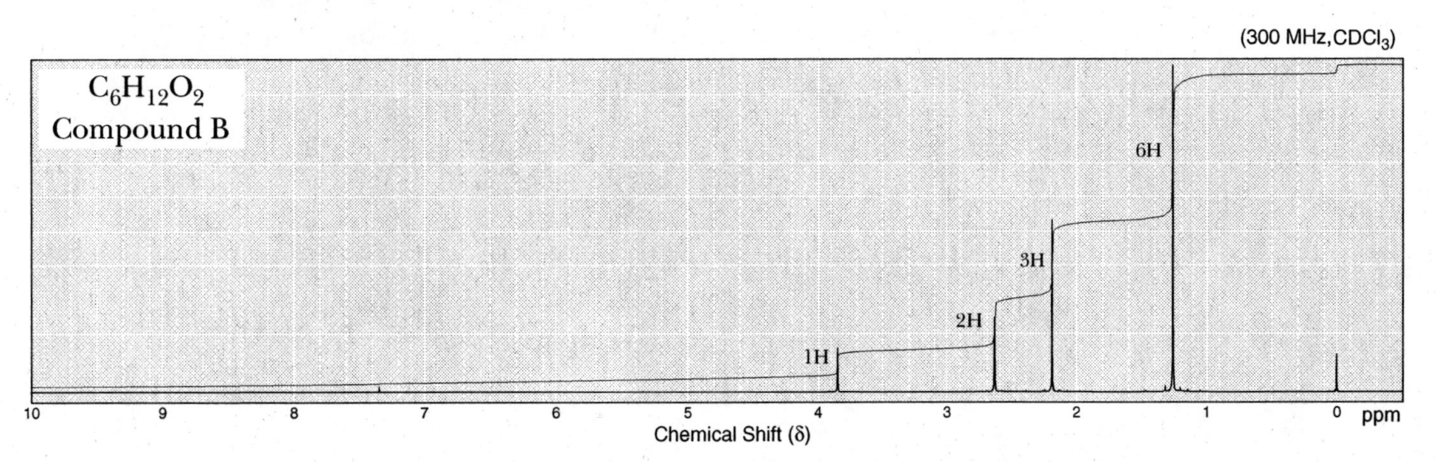

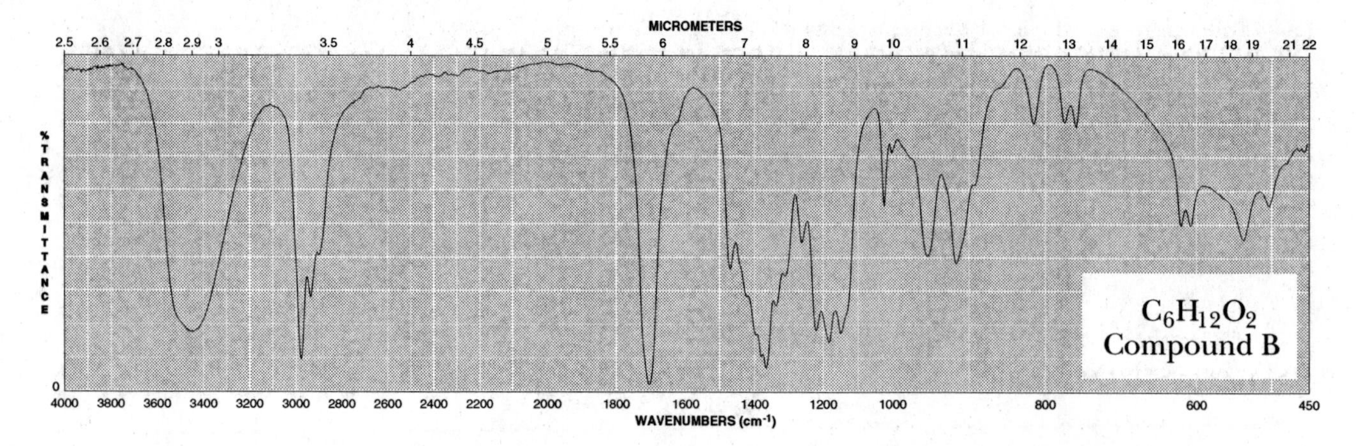

Propose a structural formula for compound B.

13.47 Compound C, $C_9H_{18}O$, is used in the automotive industry to retard the flow of solvent and thus improve the application of paints and coatings. It yields ^{13}C-NMR peaks at δ 210.5, 52.4, 24.5, and 22.6. Provide a structure and an IUPAC name for C.

Looking Ahead

13.48 Reaction of a Grignard reagent with carbon dioxide, followed by treatment with aqueous HCl, gives a carboxylic acid. Propose a structural formula for the bracketed intermediate formed by the reaction of phenylmagnesium bromide with CO_2, and propose a mechanism for the formation of this intermediate:

MgBr + CO_2 ⟶ [intermediate (not isolated)] $\xrightarrow{HCl,\ H_2O}$ C(=O)OH

13.49 Rank the following carbonyls in order of increasing reactivity to nucleophilic attack, and explain your reasoning.

O   O   O
   O   N

13.50 Provide the enol form of this ketone and predict the direction of equilibrium:

O
keto-enol tautomerism ⇌ an enol

13.51 Draw the cyclic hemiacetal formed by reaction of the highlighted —OH group with the aldehyde group:

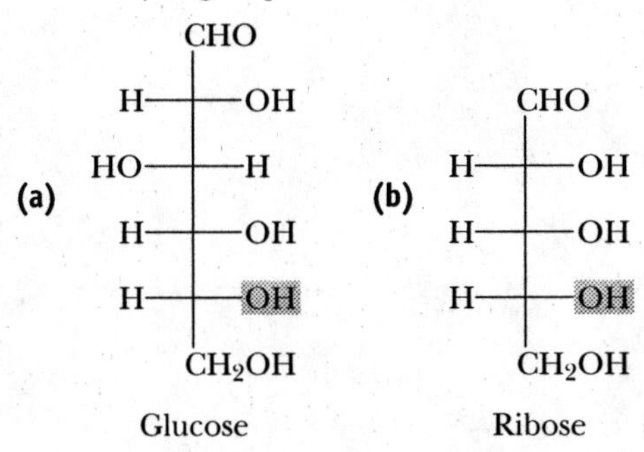

13.52 Propose a mechanism for the acid-catalyzed reaction of the following hemiacetal, with an amine acting as a nucleophile:

$$\xrightarrow[H_2N\text{—}CH_2CH_3]{H_3O^+} \quad NHCH_2CH_3 \quad + H_2O$$

14 Carboxylic Acids

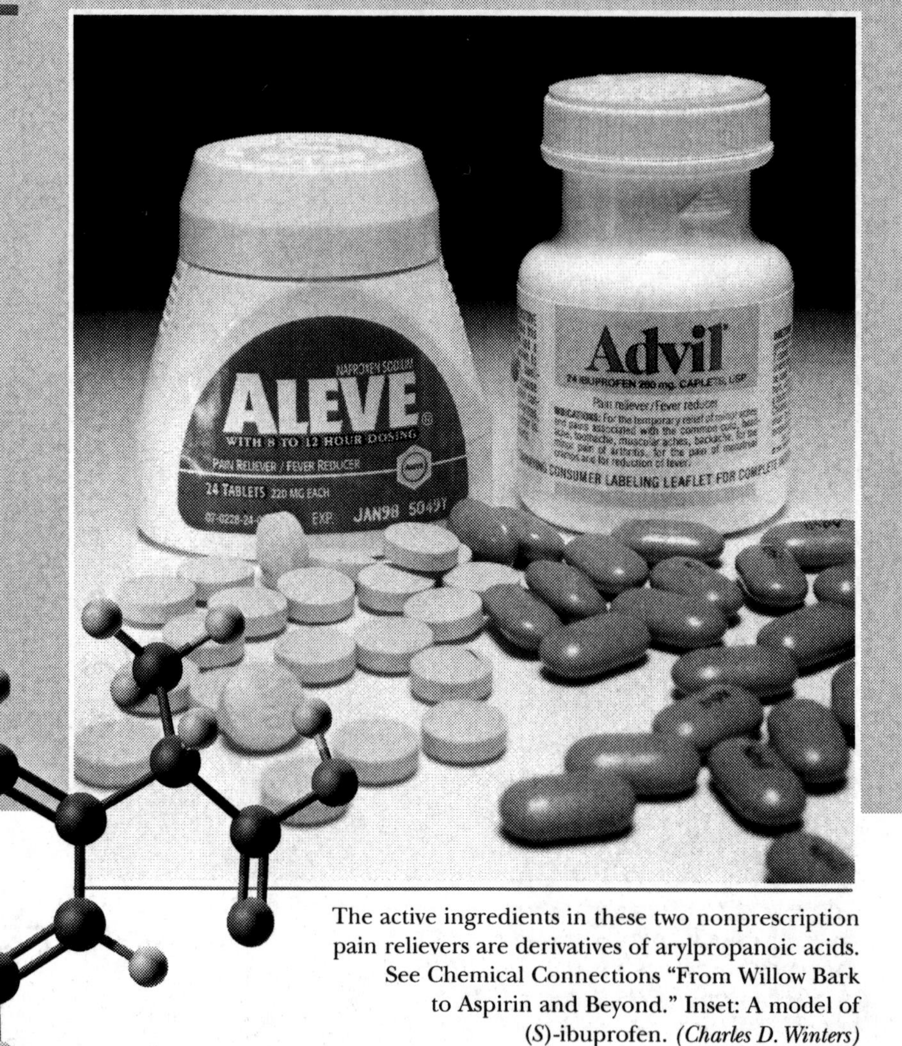

The active ingredients in these two nonprescription pain relievers are derivatives of arylpropanoic acids. See Chemical Connections "From Willow Bark to Aspirin and Beyond." Inset: A model of (*S*)-ibuprofen. *(Charles D. Winters)*

14.1 INTRODUCTION

The most important chemical property of carboxylic acids, another class of organic compounds containing the carbonyl group, is their acidity. Furthermore, carboxylic acids form numerous important derivatives, including esters, amides, anhydrides, and acid halides. In this chapter, we study carboxylic acids themselves; in Chapters 15 and 16, we study their derivatives.

14.2 STRUCTURE

Carboxyl group A —COOH group.

The functional group of a carboxylic acid is a **carboxyl group**, so named because it is made up of a **carb**onyl group and a hydr**oxyl** group (Section 1.8D). Following is a Lewis structure of the carboxyl group, as well as two alternative representations of it:

—C(=Ö:)—Ö—H   —COOH $—CO_2H$

The general formula of an aliphatic carboxylic acid is RCOOH; that of an aromatic carboxylic acid is ArCOOH.

14.3 NOMENCLATURE

A. IUPAC System

We derive the IUPAC name of a carboxylic acid from that of the longest carbon chain which contains the carboxyl group by dropping the final *-e* from the name of the parent alkane and adding the suffix *-oic*, followed by the word *acid* (Section 3.6). We number the chain beginning with the carbon of the carboxyl group. Because the carboxyl carbon is understood to be carbon 1, there is no need to give it a number. If the carboxylic acid contains a carbon–carbon double bond, we change the infix from *-an-* to *-en-* to indicate the presence of the double bond, and we show the location of the double bond by a number. In the following examples, the common name of each acid is given in parentheses:

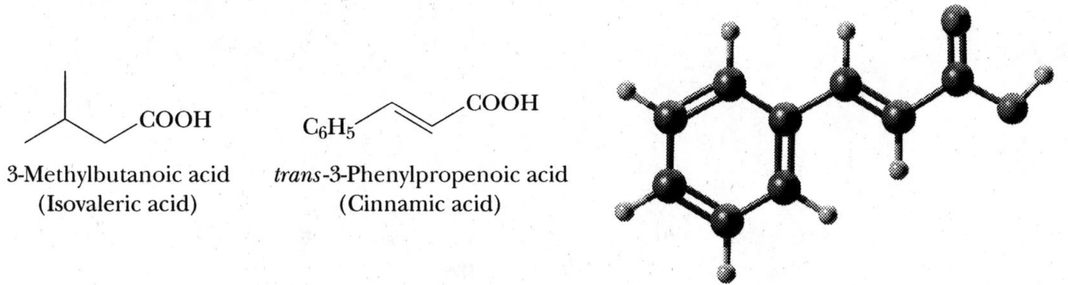

3-Methylbutanoic acid (Isovaleric acid)   *trans*-3-Phenylpropenoic acid (Cinnamic acid)

In the IUPAC system, a carboxyl group takes precedence over most other functional groups (Table 13.1), including hydroxyl and amino groups, as well as the carbonyl groups of aldehydes and ketones. As illustrated in the following examples, an —OH group of an alcohol is indicated by the prefix *hydroxy-*, an $—NH_2$ group of an amine by *amino-*, and an =O group of an aldehyde or ketone by *oxo-*:

OH ... COOH H_2N ... COOH   O ... COOH

5-Hydroxyhexanoic acid   4-Aminobutanoic acid   5-Oxohexanoic acid

Dicarboxylic acids are named by adding the suffix *-dioic*, followed by the word *acid*, to the name of the carbon chain that contains both carboxyl groups. Because the two carboxyl groups can be only at the ends of the parent chain, there is no need to number them. Following are IUPAC names and common names for several important aliphatic dicarboxylic acids:

Ethanedioic acid
(Oxalic acid)

Propanedioic acid
(Malonic acid)

Butanedioic acid
(Succinic acid)

Pentanedioic acid
(Glutaric acid)

Hexanedioic acid
(Adipic acid)

Leaves of the rhubarb plant contain the poison oxalic acid as its potassium and sodium salts. *(Hans Reinhard/OKAPIA/ Photo Researchers, Inc.)*

The name *oxalic acid* is derived from one of its sources in the biological world, namely, plants of the genus *Oxalis*, one of which is rhubarb. Oxalic acid also occurs in human and animal urine, and calcium oxalate (the calcium salt of oxalic acid) is a major component of kidney stones. Adipic acid is one of the two monomers required for the synthesis of the polymer nylon 66. The U.S. chemical industry produces approximately 1.8 billion pounds of adipic acid annually, solely for the synthesis of nylon 66 (Section 17.5A).

A carboxylic acid containing a carboxyl group bonded to a cycloalkane ring is named by giving the name of the ring and adding the suffix *-carboxylic acid*. The atoms of the ring are numbered beginning with the carbon bearing the —COOH group:

2-Cyclohexenecarboxylic acid

trans-1,3-Cyclopentanedicarboxylic acid

The simplest aromatic carboxylic acid is benzoic acid. Derivatives are named by using numbers and prefixes to show the presence and location of substituents relative to the carboxyl group. Certain aromatic carboxylic acids have common names by which they are more usually known. For example, 2-hydroxybenzoic acid is more often called salicylic acid, a name derived from the fact that this aromatic carboxylic acid was first obtained from the bark of the willow, a tree of the genus *Salix*. Aromatic dicarboxylic acids are named by adding the words *dicarboxylic acid* to *benzene*. Examples are 1,2-benzenedicarboxylic acid and 1,4-benzenedicarboxylic acid. Each is more usually known by its common name: phthalic acid and terephthalic acid, respectively. Terephthalic acid is one of the two organic components required for the synthesis of the textile fiber known as Dacron® polyester (Section 17.5B).

Benzoic acid

2-Hydroxybenzoic acid
(Salicylic acid)

1,2-Benzenedicarboxylic acid
(Phthalic acid)

1,4-Benzenedicarboxylic acid
(Terephthalic acid)

B. Common Names

Aliphatic carboxylic acids, many of which were known long before the development of structural theory and IUPAC nomenclature, are named according to their source or for some characteristic property. Table 14.1 lists several of the unbranched aliphatic carboxylic acids found in the biological world, along with the common name of each. Those with 16, 18, and 20 carbon atoms are particularly abundant in fats and oils (Section 21.2) and the phospholipid components of biological membranes (Section 21.4).

TABLE 14.1 Several Aliphatic Carboxylic Acids and Their Common Names

Structure	IUPAC Name	Common Name	Derivation
HCOOH	methanoic acid	formic acid	Latin: *formica*, ant
CH_3COOH	ethanoic acid	acetic acid	Latin: *acetum*, vinegar
CH_3CH_2COOH	propanoic acid	propionic acid	Greek: *propion*, first fat
$CH_3(CH_2)_2COOH$	butanoic acid	butyric acid	Latin: *butyrum*, butter
$CH_3(CH_2)_3COOH$	pentanoic acid	valeric acid	Latin: *valere*, to be strong
$CH_3(CH_2)_4COOH$	hexanoic acid	caproic acid	Latin: *caper*, goat
$CH_3(CH_2)_6COOH$	octanoic acid	caprylic acid	Latin: *caper*, goat
$CH_3(CH_2)_8COOH$	decanoic acid	capric acid	Latin: *caper*, goat
$CH_3(CH_2)_{10}COOH$	dodecanoic acid	lauric acid	Latin: *laurus*, laurel
$CH_3(CH_2)_{12}COOH$	tetradecanoic acid	myristic acid	Greek: *myristikos*, fragrant
$CH_3(CH_2)_{14}COOH$	hexadecanoic acid	palmitic acid	Latin: *palma*, palm tree
$CH_3(CH_2)_{16}COOH$	octadecanoic acid	stearic acid	Greek: *stear*, solid fat
$CH_3(CH_2)_{18}COOH$	eicosanoic acid	arachidic acid	Greek: *arachis*, peanut

Formic acid was first obtained in 1670 from the destructive distillation of ants, whose genus is *Formica*. It is one of the components of the venom of stinging ants. *(Ted Nelson/Dembinsky Photo Associates)*

When common names are used, the Greek letters α, β, γ, δ, and so forth are often added as a prefix to locate substituents. The α-position in a carboxylic acid is the position next to the carboxyl group; an α-substituent in a common name is equivalent to a 2-substituent in an IUPAC name. *GABA*, short for *g*amma-*a*mino*b*utyric *a*cid, is an inhibitory neurotransmitter in the central nervous system of humans:

O, OH, δ 5, γ 4, β 3, α 2, 1

H_2N γ, O, OH

4-Aminobutanoic acid
(γ-Aminobutyric acid, GABA)

In common nomenclature, the prefix *keto-* indicates the presence of a ketone carbonyl in a substituted carboxylic acid (as illustrated by the common name β-ketobutyric acid), and the substituent $CH_3CO—$ is named an **aceto group**:

Aceto group A $CH_3CO—$ group.

O O OH

3-Oxobutanoic acid
(β-Ketobutyric acid;
Acetoacetic acid)

$CH_3\overset{O}{\overset{\|}{C}}—$

Acetyl group
(Aceto group)

An alternative common name for 3-oxobutanoic acid is acetoacetic acid. In deriving this common name, this ketoacid is regarded as a substituted acetic acid, and the CH_3CO— substituent is named an aceto group.

EXAMPLE 14.1

Write the IUPAC name for each carboxylic acid:

(a) 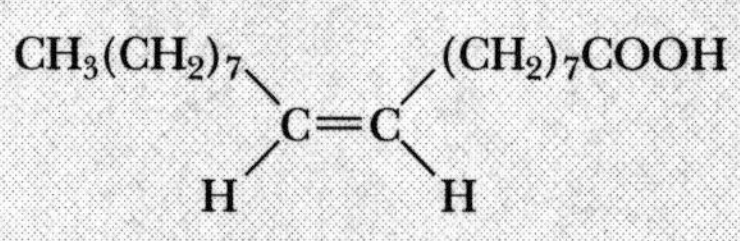 $CH_3(CH_2)_7$(H)C=C(H)$(CH_2)_7COOH$

(b) 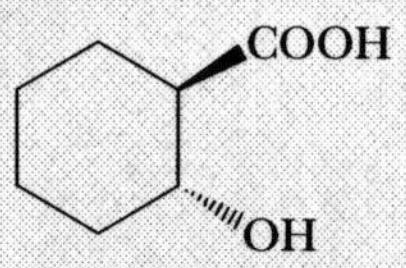 cyclohexane ring with COOH (wedge) and OH (hash)

(c) C with OH, H, COOH, CH_3

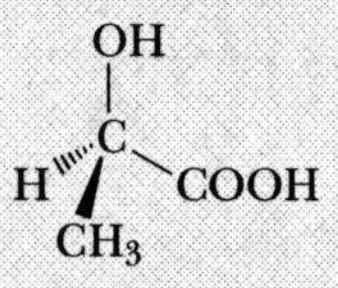

(d) $ClCH_2COOH$

SOLUTION

(a) *cis*-9-Octadecenoic acid (oleic acid)
(b) *trans*-2-Hydroxycyclohexanecarboxylic acid
(c) (*R*)-2-Hydroxypropanoic acid [(*R*)-lactic acid]
(d) Chloroethanoic acid (chloroacetic acid)

Practice Problem 14.1

Each of the following compounds has a well-recognized common name. A derivative of glyceric acid is an intermediate in glycolysis (Section 22.4). Maleic acid is an intermediate in the tricarboxylic acid (TCA) cycle. Mevalonic acid is an intermediate in the biosynthesis of steroids (Section 21.5B).

(a) 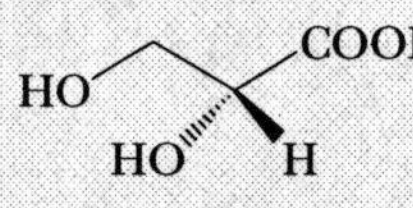 HO, COOH, HO, H — Glyceric acid

(b) 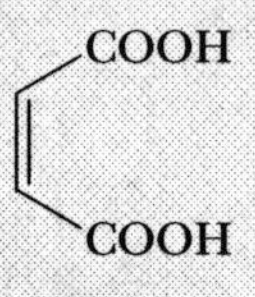 COOH, COOH — Maleic acid

(c) 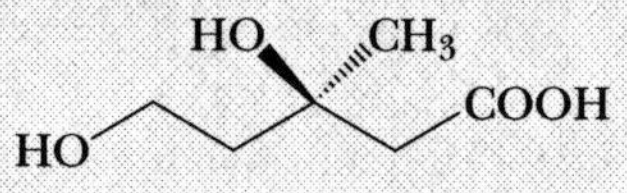 HO, CH_3, COOH, HO — Mevalonic acid

Write the IUPAC name for each compound. Be certain to show the configuration of each.

14.4 PHYSICAL PROPERTIES

In the liquid and solid states, carboxylic acids are associated by intermolecular hydrogen bonding into dimers, as shown for acetic acid:

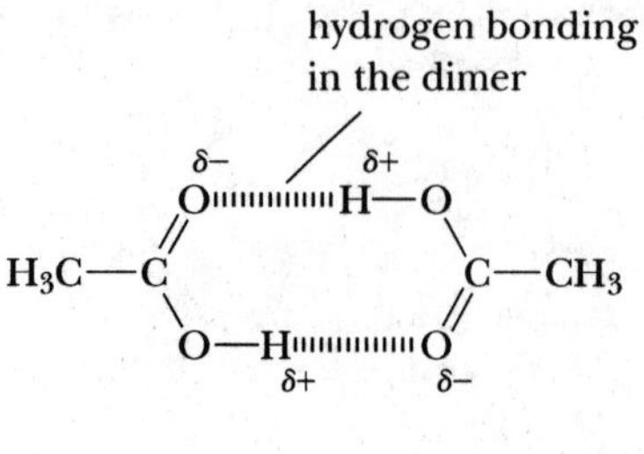

hydrogen bonding in the dimer

δ− δ+
O⋯⋯H—O
H_3C—C C—CH_3
O—H⋯⋯O
δ+ δ−

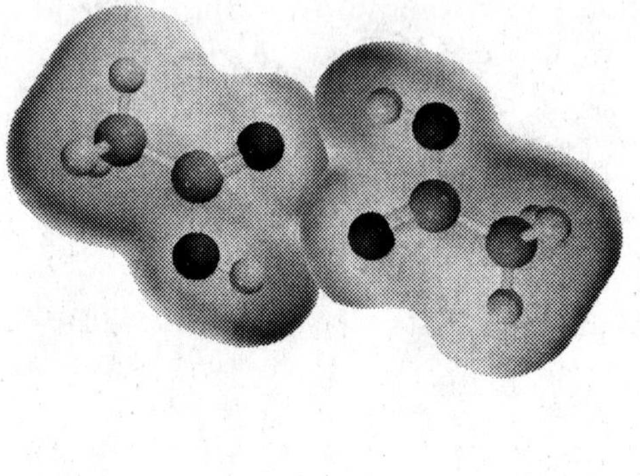

Carboxylic acids have significantly higher boiling points than other types of organic compounds of comparable molecular weight, such as alcohols, aldehydes, and ketones. For example, butanoic acid (Table 14.2) has a higher boiling point than either 1-pentanol or pentanal. The higher boiling points of carboxylic acids result from their polarity and from the fact that they form very strong intermolecular hydrogen bonds.

TABLE 14.2 Boiling Points and Solubilities in Water of Selected Carboxylic Acids, Alcohols, and Aldehydes of Comparable Molecular Weight

Structure	Name	Molecular Weight	Boiling Point (°C)	Solubility (g/100 mL H_2O)
CH_3COOH	acetic acid	60.5	118	infinite
$CH_3CH_2CH_2OH$	1-propanol	60.1	97	infinite
CH_3CH_2CHO	propanal	58.1	48	16
$CH_3(CH_2)_2COOH$	butanoic acid	88.1	163	infinite
$CH_3(CH_2)_3CH_2OH$	1-pentanol	88.1	137	2.3
$CH_3(CH_2)_3CHO$	pentanal	86.1	103	slight
$CH_3(CH_2)_4COOH$	hexanoic acid	116.2	205	1.0
$CH_3(CH_2)_5CH_2OH$	1-heptanol	116.2	176	0.2
$CH_3(CH_2)_5CHO$	heptanal	114.1	153	0.1

Carboxylic acids also interact with water molecules by hydrogen bonding through both their carbonyl and hydroxyl groups. Because of these hydrogen-bonding interactions, carboxylic acids are more soluble in water than are alcohols, ethers, aldehydes, and ketones of comparable molecular weight. The solubility of a carboxylic acid in water decreases as its molecular weight increases. We account for this trend in the following way: A carboxylic acid consists of two regions of different polarity—a polar hydrophilic carboxyl group and, except for formic acid, a nonpolar hydrophobic hydrocarbon chain. The **hydrophilic** carboxyl group increases water solubility; the **hydrophobic** hydrocarbon chain decreases water solubility.

Hydrophilic From the Greek, meaning "water loving."

Hydrophobic From the Greek, meaning "water hating."

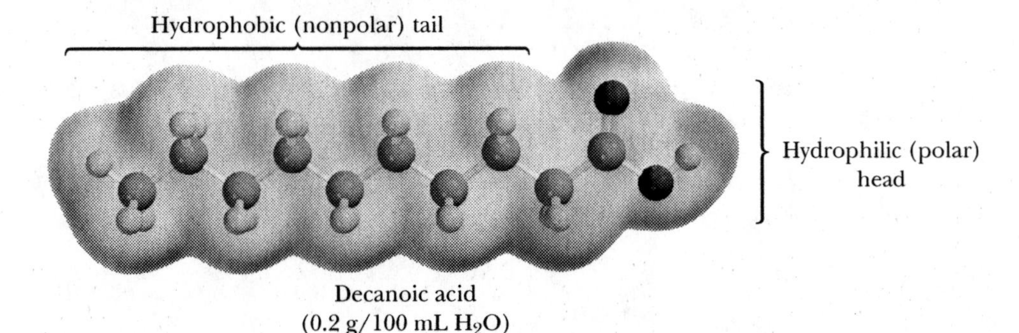

Decanoic acid
(0.2 g/100 mL H_2O)

The first four aliphatic carboxylic acids (formic, acetic, propanoic, and butanoic acids) are infinitely soluble in water because the hydrophilic character of the carboxyl group more than counterbalances the hydrophobic character of the hydrocarbon chain. As the size of the hydrocarbon chain increases relative to the size of the carboxyl group, water solubility decreases. The solubility of hexanoic acid in water is 1.0 g/100 g water; that of decanoic acid is only 0.2 g/100 g water.

One other physical property of carboxylic acids must be mentioned: The liquid carboxylic acids, from propanoic acid to decanoic acid, have extremely foul odors, about as bad as those of thiols, although different. Butanoic acid is found in stale perspiration and is a major component of "locker room odor." Pentanoic acid smells even worse, and goats, which secrete C_6, C_8, and C_{10} acids, are not famous for their pleasant odors.

14.5 ACIDITY

A. Acid Ionization Constants

Carboxylic acids are weak acids. Values of K_a for most unsubstituted aliphatic and aromatic carboxylic acids fall within the range from 10^{-4} to 10^{-5}. The value of K_a for acetic acid, for example, is 1.74×10^{-5}, and the pK_a of acetic acid is 4.76:

$$CH_3COOH + H_2O \rightleftharpoons CH_3COO^- + H_3O^+$$

$$K_a = \frac{[CH_3COO^-][H_3O^+]}{[CH_3COOH]} = 1.74 \times 10^{-5}$$

$$pK_a = 4.76$$

As we discussed in Section 2.6B, carboxylic acids are stronger acids (pK_a 4–5) than alcohols (pK_a 16–18) because resonance stabilizes the carboxylate anion by delocalizing its negative charge. No comparable resonance stabilization exists in alkoxide ions.

Substitution at the α-carbon of an atom or a group of atoms of higher electronegativity than carbon increase the acidity of carboxylic acids, often by several orders of magnitude (Section 2.6C). Compare, for example, the acidities of acetic acid (pK_a 4.76) and chloroacetic acid (pK_a 2.86). A single chlorine substituent on the α-carbon increases acid strength by nearly 100! Both dichloroacetic acid and trichloroacetic acid are stronger acids than phosphoric acid (pK_a 2.1):

Formula:	CH_3COOH	$ClCH_2COOH$	$Cl_2CHCOOH$	Cl_3CCOOH
Name:	Acetic acid	Chloroacetic acid	Dichloroacetic acid	Trichloroacetic acid
pK_a:	4.76	2.86	1.48	0.70

Increasing acid strength →

The acid-strengthening effect of halogen substitution falls off rather rapidly with increasing distance from the carboxyl group. Although the acid ionization constant for 2-chlorobutanoic acid (pK_a 2.83) is 100 times that for butanoic acid, the acid ionization constant for 4-chlorobutanoic acid (pK_a 4.52) is only about twice that for butanoic acid:

2-Chlorobutanoic acid	3-Chlorobutanoic acid	4-Chlorobutanoic acid	Butanoic acid
(pK_a 2.83)	(pK_a 3.98)	(pK_a 4.52)	(pK_a 4.82)

Decreasing acid strength →

EXAMPLE 14.2

Which acid in each set is the stronger?

(a) 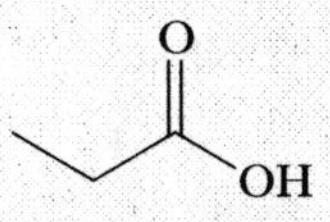 Propanoic acid or 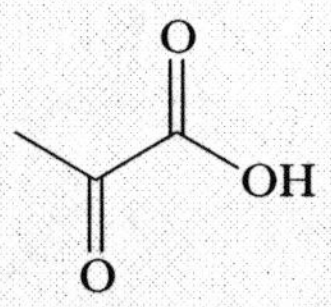 2-Hydroxypropanoic acid (Lactic acid)

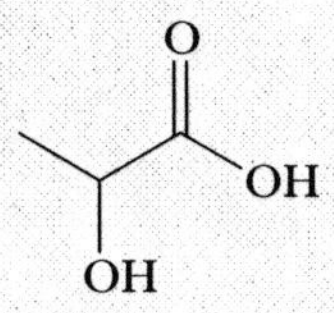

(b) 2-Hydroxypropanoic acid (Lactic acid) or 2-Oxopropanoic acid (Pyruvic acid)

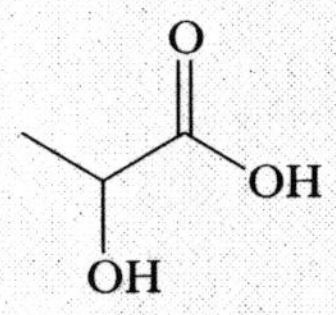

SOLUTION

(a) 2-Hydroxypropanoic acid (pK_a 3.08) is a stronger acid than propanoic acid (pK_a 4.87), because of the electron-withdrawing inductive effect of the hydroxyl oxygen.

(b) 2-Oxopropanoic acid (pK_a 2.06) is a stronger acid than 2-hydroxypropanoic acid (pK_a 3.08), because of the greater electron-withdrawing inductive effect of the carbonyl oxygen compared with that of the hydroxyl oxygen.

Practice Problem 14.2

Match each compound with its appropriate pK_a value:

$CH_3C(CH_3)_2COOH$ — 2,2-Dimethylpropanoic acid

CF_3COOH — Trifluoroacetic acid

$CH_3CH(OH)COOH$ — 2-Hydroxypropanoic acid (Lactic acid)

pK_a values = 5.03, 3.08, and 0.22.

B. Reaction with Bases

All carboxylic acids, whether soluble or insoluble in water, react with NaOH, KOH, and other strong bases to form water-soluble salts:

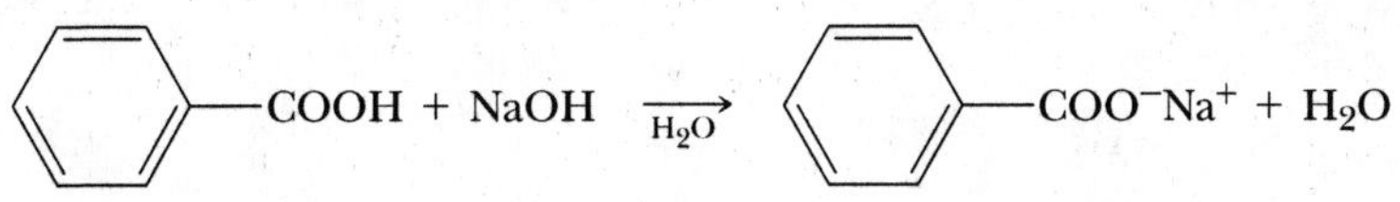

$$C_6H_5COOH + NaOH \xrightarrow{H_2O} C_6H_5COO^-Na^+ + H_2O$$

Benzoic acid (slightly soluble in water) → Sodium benzoate (60 g/100 mL water)

Sodium benzoate, a fungal growth inhibitor, is often added to baked goods "to retard spoilage." Calcium propanoate is used for the same purpose.

Carboxylic acids also form water-soluble salts with ammonia and amines:

$$C_6H_5\text{—}COOH + NH_3 \xrightarrow{H_2O} C_6H_5\text{—}COO^-NH_4^+$$

Benzoic acid (slightly soluble in water) — Ammonium benzoate (20 g/100 mL water)

As described in Section 2.5, carboxylic acids react with sodium bicarbonate and sodium carbonate to form water-soluble sodium salts and carbonic acid (a relatively weak acid). Carbonic acid, in turn, decomposes to give water and carbon dioxide, which evolves as a gas:

$$CH_3COOH + Na^+HCO_3^- \xrightarrow{H_2O} CH_3COO^-Na^+ + H_2CO_3$$

$$H_2CO_3 \longrightarrow CO_2 + H_2O$$

$$CH_3COOH + Na^+HCO_3^- \longrightarrow CH_3COO^-Na^+ + CO_2 + H_2O$$

Salts of carboxylic acids are named in the same manner as are salts of inorganic acids: Name the cation first and then the anion. Derive the name of the anion from the name of the carboxylic acid by dropping the suffix *-ic acid* and adding the suffix *-ate*. For example, the name of $CH_3CH_2COO^-Na^+$ is sodium propanoate, and that of $CH_3(CH_2)_{14}COO^-Na^+$ is sodium hexadecanoate (sodium palmitate).

EXAMPLE 14.3

Complete each acid–base reaction and name the salt formed:

(a) $CH_3CH_2CH_2COOH + NaOH \longrightarrow$

(b) $CH_3CH(OH)COOH + NaHCO_3 \longrightarrow$

SOLUTION

Each carboxylic acid is converted to its sodium salt. In (b), carbonic acid forms and decomposes to carbon dioxide and water:

(a) $CH_3CH_2CH_2COOH + NaOH \longrightarrow CH_3CH_2CH_2COO^-Na^+ + H_2O$

Butanoic acid — Sodium butanoate

(b) $CH_3CH(OH)COOH + NaHCO_3 \longrightarrow CH_3CH(OH)COO^-Na^+ + H_2O + CO_2$

2-Hydroxypropanoic acid (Lactic acid) — Sodium 2-hydroxypropanoate (Sodium lactate)

Practice Problem 14.3

Write an equation for the reaction of each acid in Example 14.3 with ammonia, and name the salt formed.

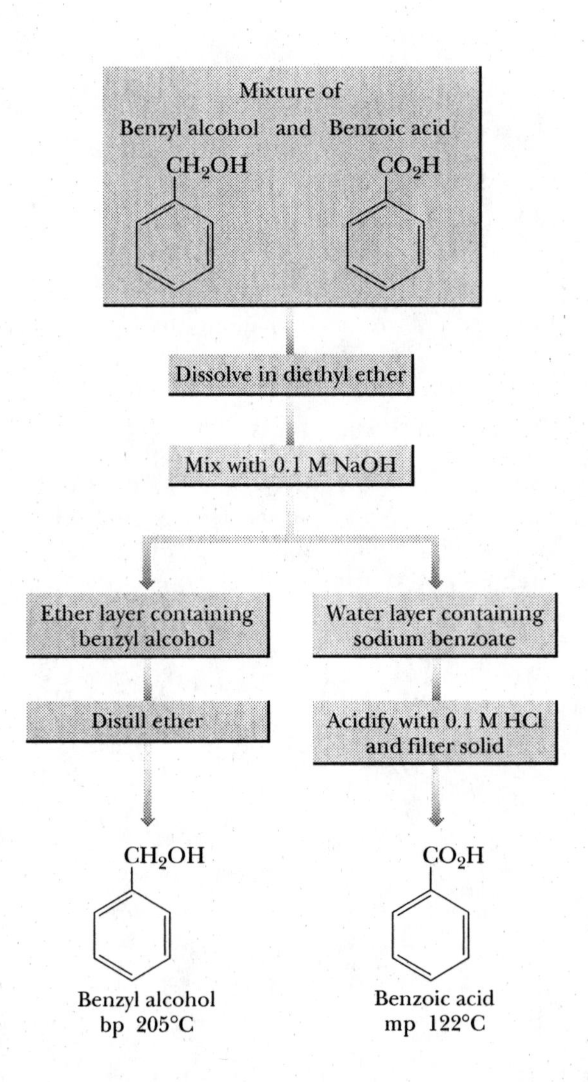

Figure 14.1
Flowchart for separation of benzoic acid from benzyl alcohol.

A consequence of the water solubility of carboxylic acid salts is that we can convert water-insoluble carboxylic acids to water-soluble alkali metal or ammonium salts and then extract them into aqueous solution. In turn, we can transform the salt into the free carboxylic acid by adding HCl, H_2SO_4, or some other strong acid. These reactions allow us to separate water-insoluble carboxylic acids from water-insoluble neutral compounds.

Figure 14.1 shows a flowchart for the separation of benzoic acid, a water-insoluble carboxylic acid, from benzyl alcohol, a water-insoluble nonacidic compound. First, we dissolve the mixture of benzoic acid and benzyl alcohol in diethyl ether. Next, we shake the ether solution with aqueous NaOH to convert benzoic acid to its water-soluble sodium salt. Then we separate the ether from the aqueous phase. Distillation of the ether solution yields first diethyl ether (bp 35°C) and then benzyl alcohol (bp 205°C). When we acidify the aqueous solution with HCl, benzoic acid precipitates as a water-insoluble solid (mp 122°C) and is recovered by filtration.

14.6 REDUCTION

The carboxyl group is one of the organic functional groups that is most resistant to reduction. It is not affected by catalytic reduction under conditions that easily reduce aldehydes and ketones to alcohols and that reduce alkenes to alkanes.

CHEMICAL CONNECTIONS 14A

From Willow Bark to Aspirin and Beyond

The first drug developed for widespread use was aspirin, today's most common pain reliever. Americans alone consume approximately 80 billion tablets of aspirin a year! The story of the development of this modern pain reliever goes back more than 2,000 years: In 400 B.C.E., the Greek physician Hippocrates recommended chewing bark of the willow tree to alleviate the pain of childbirth and to treat eye infections.

The active component of willow bark was found to be salicin, a compound composed of salicyl alcohol joined to a unit of β-D-glucose (Section 18.3). Hydrolysis of salicin in aqueous acid gives salicyl alcohol, which can then be oxidized to salicylic acid, an even more effective reliever of pain, fever, and inflammation than salicin and one without its extremely bitter taste:

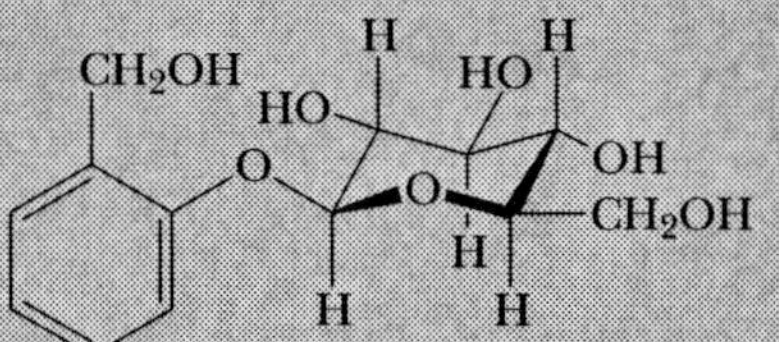

Salicin

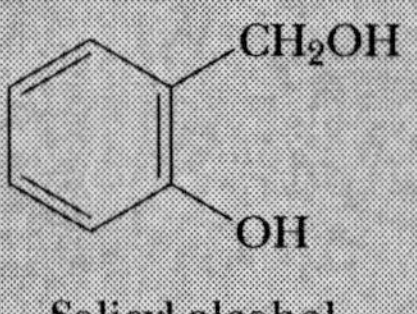

Salicyl alcohol

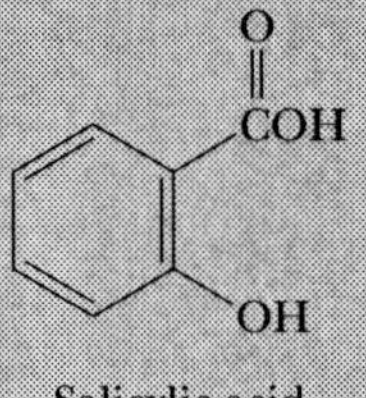

Salicylic acid

Unfortunately, patients quickly recognized salicylic acid's major side effect: It causes severe irritation of the mucous membrane lining the stomach. In the search for less irritating, but still effective, derivatives of salicylic acid, chemists at the Bayer division of I. G. Farben in Germany prepared acetylsalicylic acid in 1883 and gave it the name *aspirin*, a word derived from the German *spirsäure* (salicylic acid), with the initial *a* for the acetyl group:

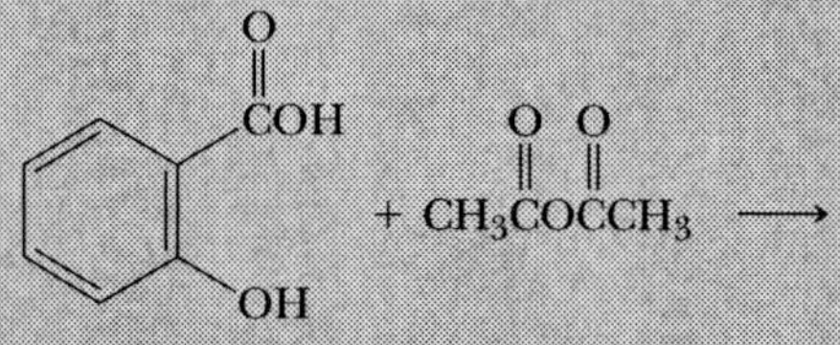

Salicylic acid + Acetic anhydride 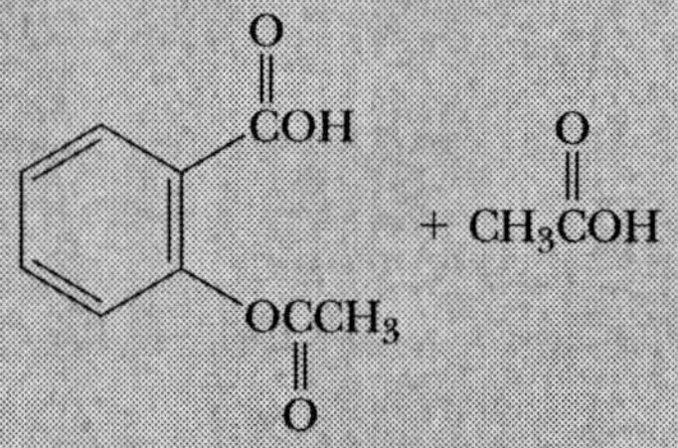 ⟶ Acetyl salicylate (Aspirin) + CH_3COOH

The most common reagent for the reduction of a carboxylic acid to a primary alcohol is the very powerful reducing agent lithium aluminum hydride (Section 13.11B).

A. Reduction of a Carboxyl Group

Lithium aluminum hydride, $LiAlH_4$, reduces a carboxyl group to a primary alcohol in excellent yield. Reduction is most commonly carried out in diethyl ether or tetrahydrofuran (THF). The initial product is an aluminum alkoxide, which is then treated with water to give the primary alcohol and lithium and aluminum hydroxides:

Aspirin proved to be less irritating to the stomach than salicylic acid and also more effective in relieving the pain and inflammation of rheumatoid arthritis. Bayer began large-scale production of aspirin in 1899.

In the 1960s, in a search for even more effective and less irritating analgesics and anti-inflammatory drugs, the Boots Pure Drug Company in England studied compounds related in structure to salicylic acid. They discovered an even more potent compound, which they named ibuprofen, and soon thereafter, Syntex Corporation in the United States developed naproxen and Rhone–Poulenc in France developed ketoprofen:

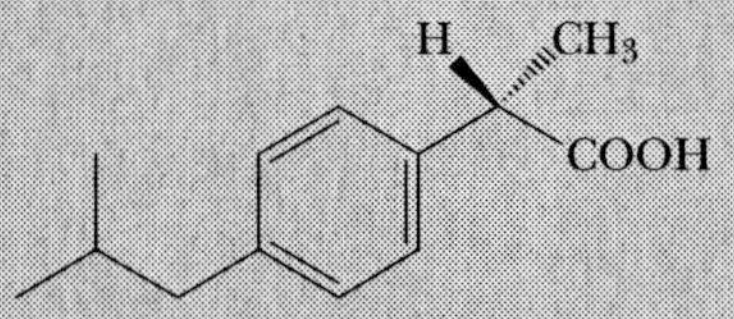

H CH_3 COOH

(*S*)-Ibuprofen

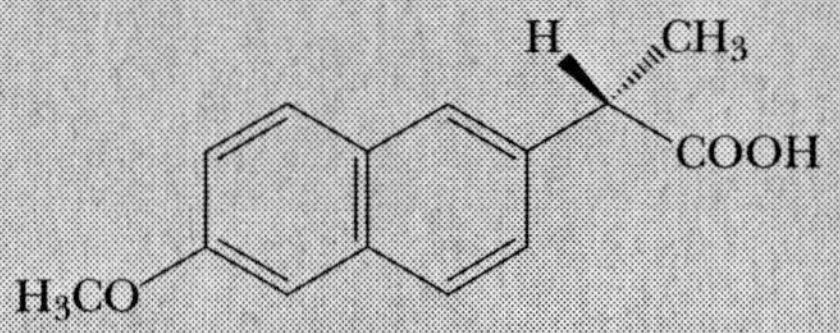

H CH_3 COOH H_3CO

(*S*)-Naproxen

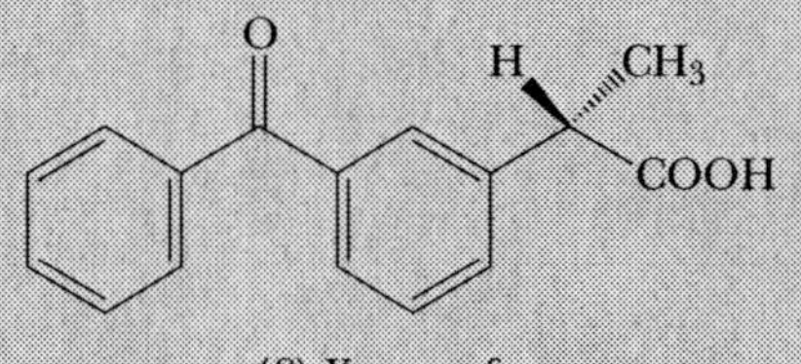

O H CH_3 COOH

(*S*)-Ketoprofen

Notice that each compound has one stereocenter and can exist as a pair of enantiomers. For each drug, the physiologically active form is the *S* enantiomer. Even though the *R* enantiomer of ibuprofen has none of the analgesic or anti-inflammatory activity, it is converted in the body to the active *S* enantiomer.

In the 1960s, scientists discovered that aspirin acts by inhibiting cyclooxygenase (COX), a key enzyme in the conversion of arachidonic acid to prostaglandins (Section 21.6). With this discovery, it became clear why only one enantiomer of ibuprofen, naproxen, and ketoprofen is active: Only the *S* enantiomer of each has the correct handedness to bind to COX and inhibit its activity.

The discovery that these drugs owe their effectiveness to the inhibition of COX opened an entirely new avenue for drug research. If we know more about the structure and function of this key enzyme, might it be possible to design and discover even more effective nonsteroidal anti-inflammatory drugs for the treatment of rheumatoid arthritis and other inflammatory diseases?

And so continues the story that began with the discovery of the beneficial effects of chewing willow bark.

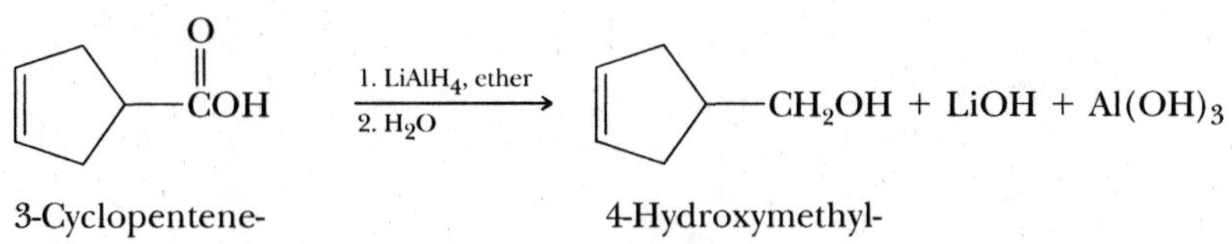

$$\text{3-Cyclopentenecarboxylic acid (COOH)} \xrightarrow[\text{2. } H_2O]{\text{1. } LiAlH_4\text{, ether}} \text{4-Hydroxymethylcyclopentene } (CH_2OH) + LiOH + Al(OH)_3$$

3-Cyclopentene-carboxylic acid

4-Hydroxymethyl-cyclopentene

These hydroxides are insoluble in diethyl ether or THF and are removed by filtration. Evaporation of the solvent yields the primary alcohol.

Alkenes are generally not affected by metal hydride-reducing reagents. These reagents function as hydride ion donors; that is, they function as nucleophiles, and alkenes are not normally attacked by nucleophiles.

B. Selective Reduction of Other Functional Groups

Catalytic hydrogenation does not reduce carboxyl groups, but does reduce alkenes to alkanes. Therefore, we can use H_2/M to reduce this functional group selectively in the presence of a carboxyl group:

$$\text{5-Hexenoic acid} + H_2 \xrightarrow[25°C,\ 2\ atm]{Pt} \text{Hexanoic acid}$$

5-Hexenoic acid — Hexanoic acid

We saw in Section 13.11B that aldehydes and ketones are reduced to alcohols by both $LiAlH_4$ and $NaBH_4$. Only $LiAlH_4$, however, reduces carboxyl groups. Thus, it is possible to reduce an aldehyde or a ketone carbonyl group selectively in the presence of a carboxyl group by using the less reactive $NaBH_4$ as the reducing agent:

$$C_6H_5C(=O)CH_2CH_2CH_2COOH \xrightarrow[2.\ H_2O]{1.\ NaBH_4} C_6H_5CH(OH)CH_2CH_2CH_2COOH$$

5-Oxo-5-phenylpentanoic acid — 5-Hydroxy-5-phenylpentanoic acid

Fischer esterification The process of forming an ester by refluxing a carboxylic acid and an alcohol in the presence of an acid catalyst, commonly sulfuric acid.

These products all contain ethyl acetate as a solvent. *(Charles D. Winter)*

14.7 FISCHER ESTERIFICATION

Treatment of a carboxylic acid with an alcohol in the presence of an acid catalyst—most commonly, concentrated sulfuric acid—gives an ester. This method of forming an ester is given the special name **Fischer esterification** after the German chemist Emil Fischer (1852–1919). As an example of Fischer esterification, treating acetic acid with ethanol in the presence of concentrated sulfuric acid gives ethyl acetate and water:

$$CH_3\overset{O}{\overset{\|}{C}}OH + CH_3CH_2OH \overset{H_2SO_4}{\rightleftharpoons} CH_3\overset{O}{\overset{\|}{C}}OCH_2CH_3 + H_2O$$

Ethanoic acid (Acetic acid)	Ethanol (Ethyl alcohol)	Ethyl ethanoate (Ethyl acetate)

We study the structure, nomenclature, and reactions of esters in detail in Chapter 15. In this chapter, we discuss only their preparation from carboxylic acids.

Acid-catalyzed esterification is reversible, and generally, at equilibrium, the quantities of remaining carboxylic acid and alcohol are appreciable. By controlling the experimental conditions, however, we can use Fischer esterification to prepare esters in high yields. If the alcohol is inexpensive compared with the carboxylic acid, we can use a large excess of the alcohol to drive the equilibrium to the right and achieve a high conversion of carboxylic acid to its ester.

EXAMPLE 14.4

Complete these Fischer esterification reactions:

(a) 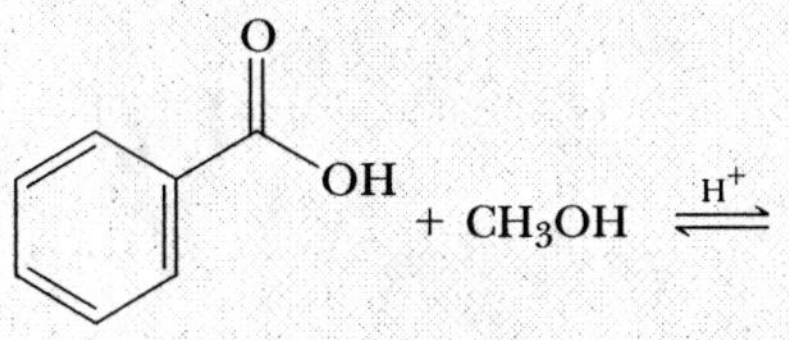 [benzoic acid] + CH_3OH $\xrightleftharpoons{H^+}$

(b) 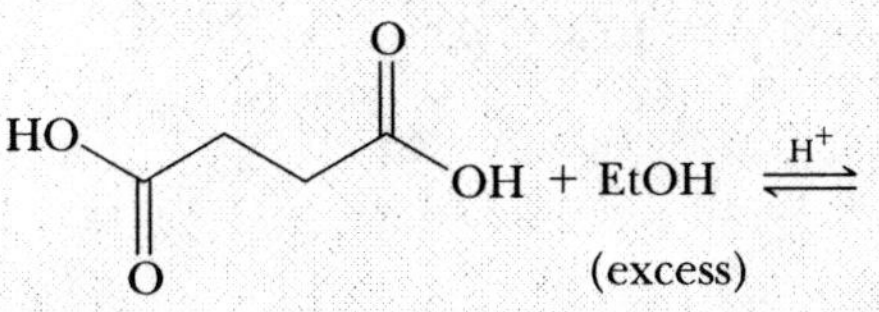 HO(O)C–CH₂CH₂–C(O)OH + EtOH $\xrightleftharpoons{H^+}$
(excess)

SOLUTION

Here is a structural formula for the ester produced in each reaction:

(a) 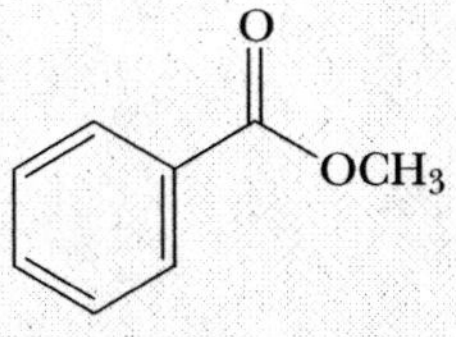 [benzene ring]–C(O)OCH_3

Methyl benzoate

(b) 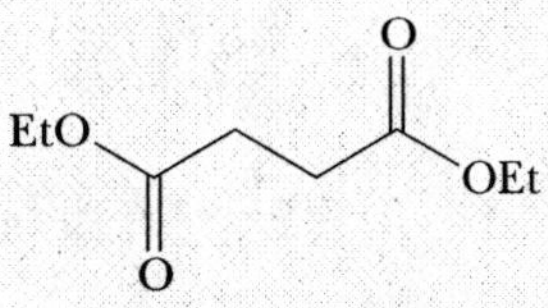 EtO(O)C–CH₂CH₂–C(O)OEt

Diethyl butanedioate
(Diethyl succinate)

Practice Problem 14.4

Complete these Fischer esterification reactions:

(a) 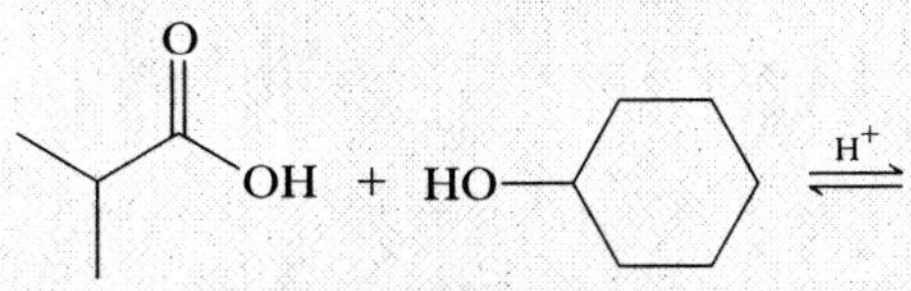 (CH₃)₂CH–C(O)OH + HO–[cyclohexane] $\xrightleftharpoons{H^+}$

(b) 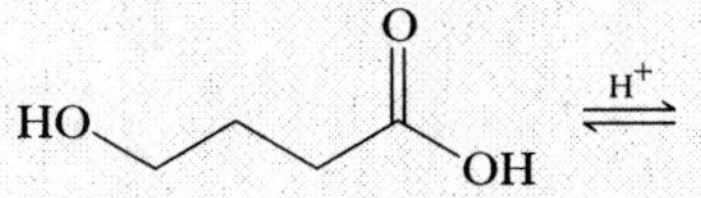 HO–CH₂CH₂CH₂–C(O)OH $\xrightleftharpoons{H^+}$ (a cyclic ester)

Following is a mechanism for Fischer esterification and we urge you to study it carefully. It is important that you understand this mechanism thoroughly, because it is a model for many of the reactions of the functional derivatives of carboxylic acids presented in Chapter 15. Note that, although we show the acid catalyst as H_2SO_4 when we write Fisher esterification reactions, the actual proton-transfer acid that initiates the reaction is the oxonium formed by the transfer of a proton

CHEMICAL CONNECTIONS 14B

Esters as Flavoring Agents

Flavoring agents are the largest class of food additives. At present, over a thousand synthetic and natural flavors are available. The majority of these are concentrates or extracts from the material whose flavor is desired and are often complex mixtures of from tens to hundreds of compounds. A number of ester flavoring agents are synthesized industrially. Many have flavors very close to the target flavor, and adding only one or a few of them is sufficient to make ice cream, soft drinks, or candy taste natural. (Isopentane is the common name for 2-methylbutane.) The table shows the structures of a few of the esters used as flavoring agents:

Structure	Name	Flavor
$HCOOCH_2CH_3$	Ethyl formate	Rum
$CH_3COOCH_2CH_2CH(CH_3)_2$	Isopentyl acetate	Banana
$CH_3COO(CH_2)_7CH_3$	Octyl acetate	Orange
$CH_3CH_2CH_2COOCH_3$	Methyl butanoate	Apple
$CH_3CH_2CH_2COOCH_2CH_3$	Ethyl butanoate	Pineapple
$2\text{-}H_2NC_6H_4COOCH_3$ (NH$_2$)	Methyl 2-aminobenzoate (Methyl anthranilate)	Grape

from H_2SO_4 (the stronger acid) to the alcohol (the stronger base) used in the esterification reaction:

$$CH_3-\ddot{O}-H + H-\ddot{O}-\overset{O}{\underset{O}{\overset{\|}{\underset{\|}{S}}}}-O-H \rightleftharpoons CH_3-\underset{H}{\underset{|}{\overset{+}{\ddot{O}}}}-H + {}^{-}:\ddot{O}-\overset{O}{\underset{O}{\overset{\|}{\underset{\|}{S}}}}-O-H$$

Mechanism: Fischer Esterification

1 Proton transfer from the acid catalyst to the carbonyl oxygen increases the electrophilicity of the carbonyl carbon . . .

2 which is then attacked by the nucleophilic oxygen atom of the alcohol . . .

3 to form an oxonium ion.

4 Proton transfer from the oxonium ion to a second molecule of alcohol . . .

5 gives a tetrahedral carbonyl addition intermediate (TCAI).

6 Proton transfer to one of the —OH groups of the TCAI . . .

7 gives a new oxonium ion.

8 Loss of water from this oxonium ion . . .

9 gives the ester and water, and regenerates the acid catalyst.

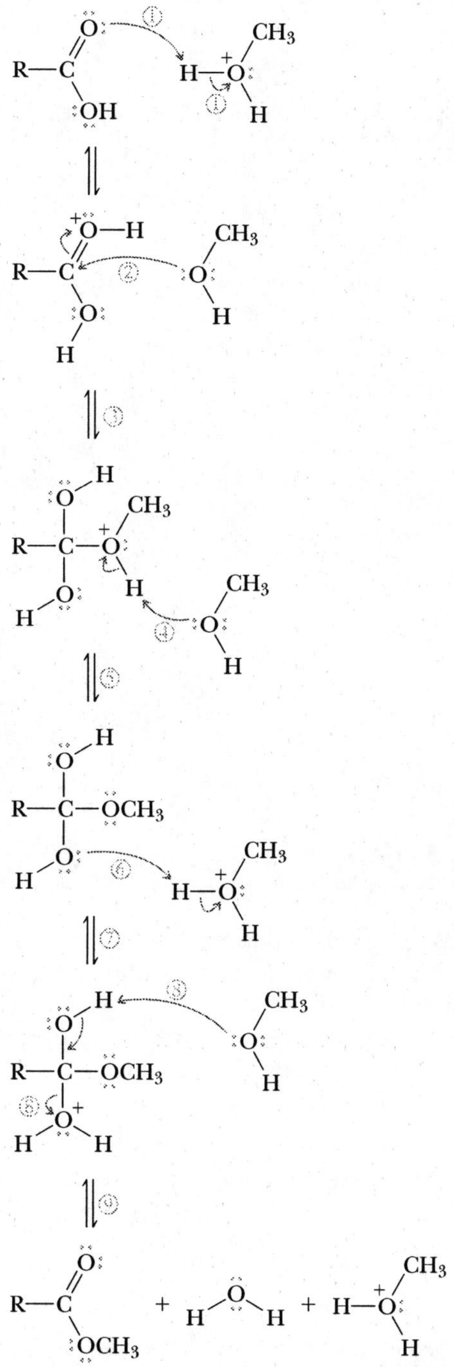

14.8 CONVERSION TO ACID HALIDES

The functional group of an acid halide is a carbonyl group bonded to a halogen atom. Among the acid halides, acid chlorides are the most frequently used in the laboratory and in industrial organic chemistry:

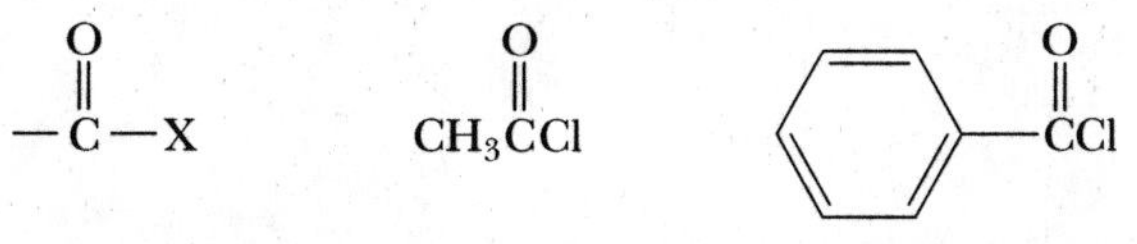

Functional group of an acid halide   Acetyl chloride   Benzoyl chloride

CHEMICAL CONNECTIONS 14C

The Pyrethrins: Natural Insecticides of Plant Origin

Pyrethrum is a natural insecticide obtained from the powdered flower heads of several species of *Chrysanthemum*, particularly *C. cinerariaefolium*. The active substances in pyrethrum—principally, pyrethrins I and II—are contact poisons for insects and cold-blooded vertebrates. Because their concentrations in the pyrethrum powder used in chrysanthemum-based insecticides are nontoxic to plants and higher animals, pyrethrum powder is used in household and livestock sprays, as well as in dusts for edible plants. Natural pyrethrins are esters of chrysanthemic acid.

While pyrethrum powders are effective insecticides, the active substances in them are destroyed rapidly in the environment. In an effort to develop synthetic compounds as effective as these natural insecticides, but with greater biostability, chemists have prepared a series of esters related in structure to chrysanthemic acid. Permethrin is one of the most commonly used synthetic pyrethrinlike compounds in household and agricultural products.

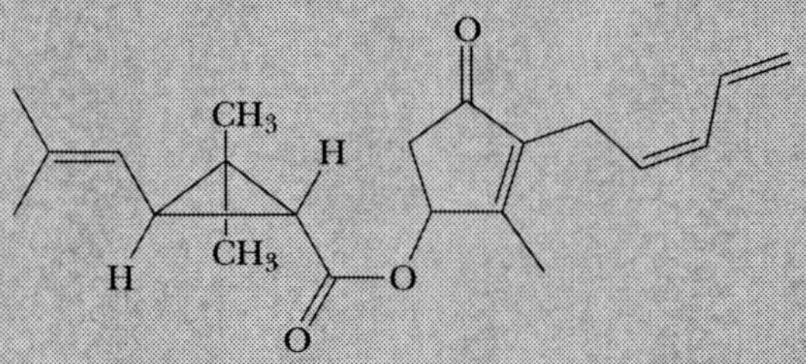

Pyrethrin I

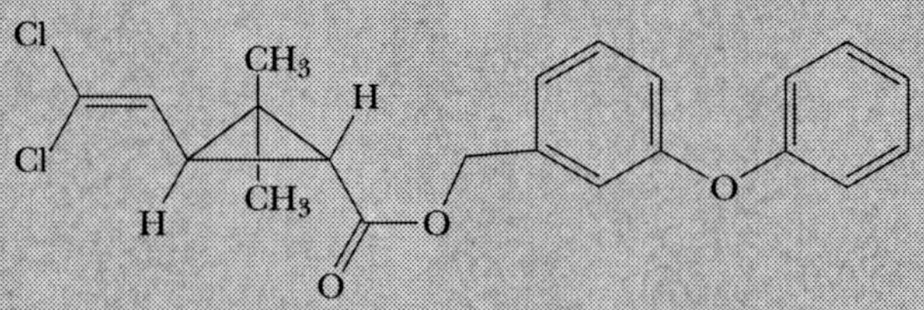

Permethrin

We study the nomenclature, structure, and characteristic reactions of acid halides in Chapter 15. In this chapter, our concern is only with their synthesis from carboxylic acids.

The most common way to prepare an acid chloride is to treat a carboxylic acid with thionyl chloride, the same reagent that converts an alcohol to a chloroalkane (Section 8.3D):

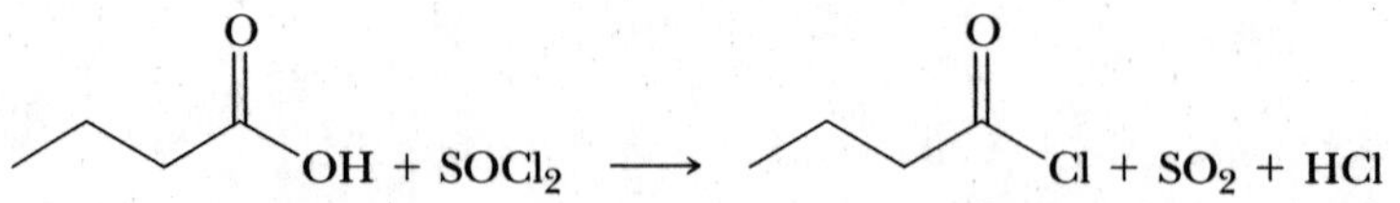

$$\text{Butanoic acid} + SOCl_2 \text{ (Thionyl chloride)} \longrightarrow \text{Butanoyl chloride} + SO_2 + HCl$$

EXAMPLE 14.5

Complete each equation:

(a) 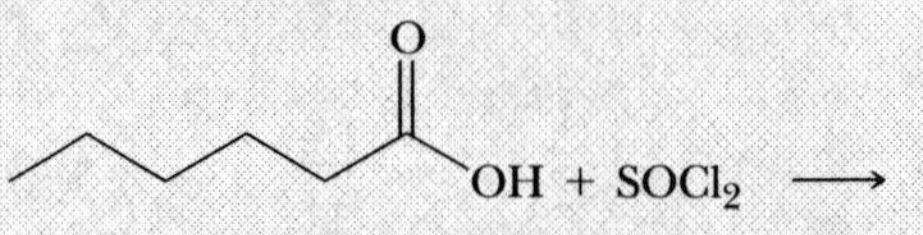 $CH_3(CH_2)_4COOH + SOCl_2 \longrightarrow$

(b) 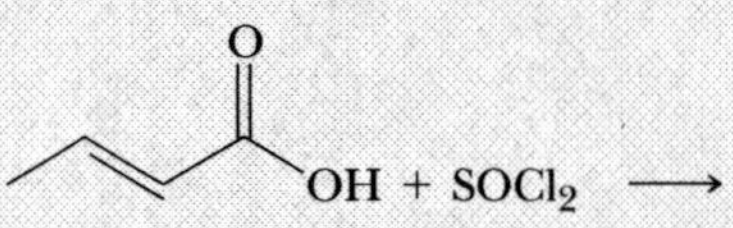 $CH_3CH{=}CHCOOH + SOCl_2 \longrightarrow$

SOLUTION

Following are the products for each reaction:

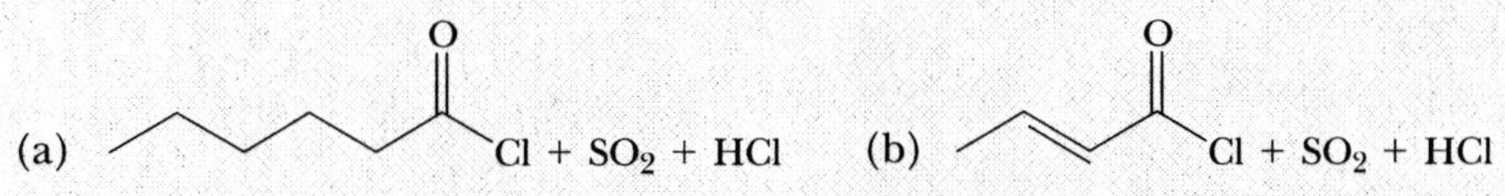

(a) hexanoyl chloride: $CH_3(CH_2)_4COCl + SO_2 + HCl$ (b) but-2-enoyl chloride: $CH_3CH{=}CHCOCl + SO_2 + HCl$

Practice Problem 14.5

Complete each equation:

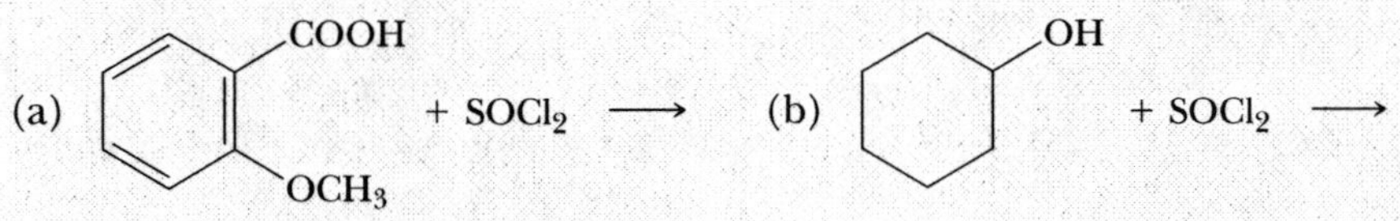

(a) 2-methoxybenzoic acid (COOH, OCH_3) $+ SOCl_2 \longrightarrow$ (b) cyclohexanol (OH) $+ SOCl_2 \longrightarrow$

14.9 DECARBOXYLATION

A. β-Ketoacids

Decarboxylation is the loss of CO_2 from a carboxyl group. Almost any carboxylic acid, heated to a very high temperature, undergoes decarboxylation:

Decarboxylation Loss of CO_2 from a carboxyl group.

$$RCOOH \xrightarrow{\text{decarboxylation}} RH + CO_2$$

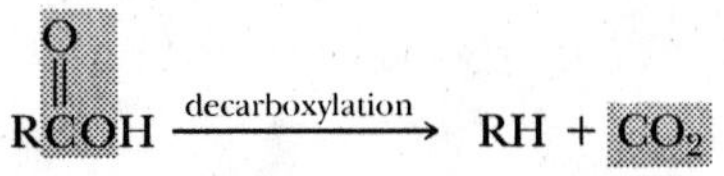

Most carboxylic acids, however, are quite resistant to moderate heat and melt or even boil without decarboxylation. Exceptions are carboxylic acids that have a carbonyl group β to the carboxyl group. This type of carboxylic acid undergoes decarboxylation quite readily on mild heating. For example, when 3-oxobutanoic acid (acetoacetic acid) is heated moderately, it undergoes decarboxylation to give acetone and carbon dioxide:

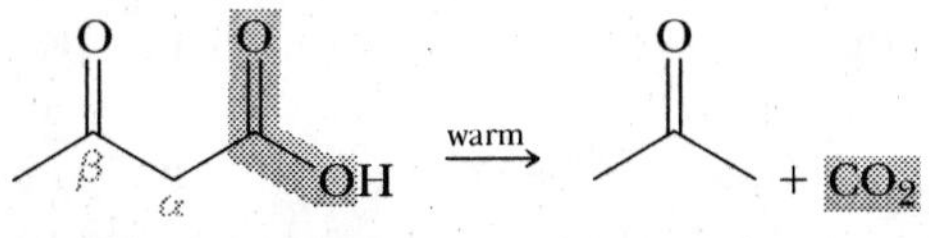

$$CH_3COCH_2COOH \xrightarrow{\text{warm}} CH_3COCH_3 + CO_2$$

3-Oxobutanoic acid (Acetoacetic acid) → Acetone

Decarboxylation on moderate heating is a unique property of 3-oxocarboxylic acids (β-ketoacids) and is not observed with other classes of ketoacids.

Mechanism: Decarboxylation of a β-Ketocarboxylic Acid

Step 1: Redistribution of six electrons in a cyclic six-membered transition state gives carbon dioxide and an enol:

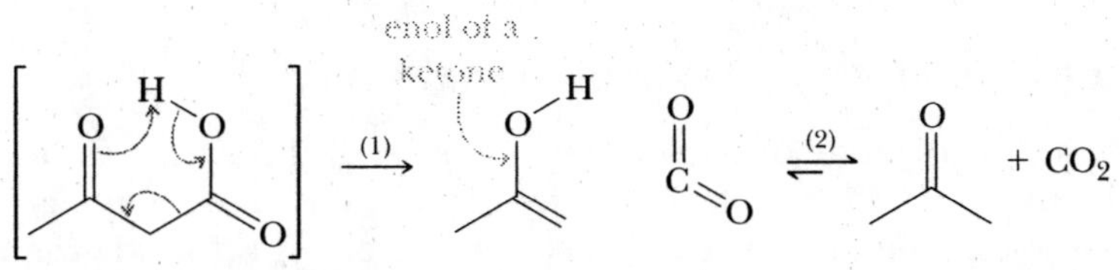

(A cyclic six-membered transition state) $\xrightarrow{(1)}$ enol of a ketone + $O{=}C{=}O$ $\overset{(2)}{\rightleftharpoons}$ acetone $+ CO_2$

CHEMICAL CONNECTIONS 14D

Ketone Bodies and Diabetes

3-Oxobutanoic acid (acetoacetic acid) and its reduction product, 3-hydroxybutanoic acid, are synthesized in the liver from acetyl-CoA, a product of the metabolism of fatty acids (Section 22.6C) and certain amino acids:

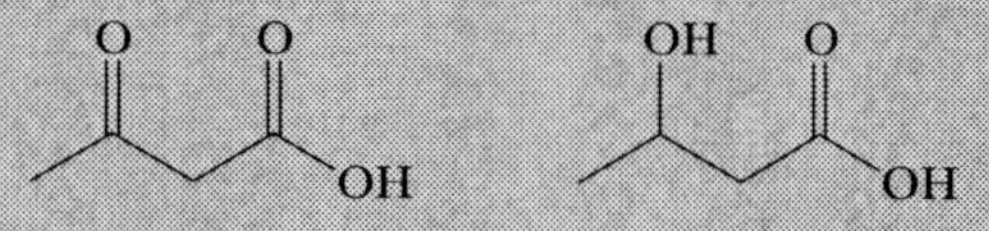

3-Oxobutanoic acid (Acetoacetic acid) 3-Hydroxybutanoic acid (β-Hydroxybutyric acid)

3-Hydroxybutanoic acid and 3-oxobutanoic acid are known collectively as ketone bodies.

The concentration of ketone bodies in the blood of healthy, well-fed humans is approximately 0.01 mM/L. However, in persons suffering from starvation or diabetes mellitus, the concentration of ketone bodies may increase to as much as 500 times normal. Under these conditions, the concentration of acetoacetic acid increases to the point where it undergoes spontaneous decarboxylation to form acetone and carbon dioxide. Acetone is not metabolized by humans and is excreted through the kidneys and the lungs. The odor of acetone is responsible for the characteristic "sweet smell" on the breath of severely diabetic patients.

Step 2: Keto–enol tautomerism (Section 13.9A) of the enol gives the more stable keto form of the product:

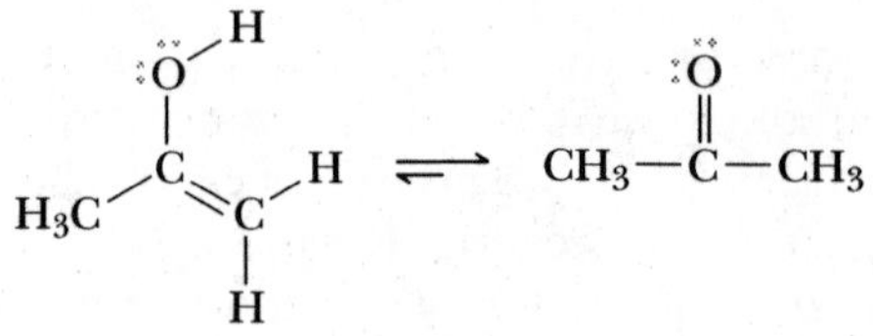

$$H_3C{-}C(OH){=}CH_2 \rightleftharpoons CH_3{-}\overset{O}{\overset{\|}{C}}{-}CH_3$$

An important example of decarboxylation of a β-ketoacid in the biological world occurs during the oxidation of foodstuffs in the tricarboxylic acid (TCA) cycle. Oxalosuccinic acid, one of the intermediates in this cycle, undergoes spontaneous decarboxylation to produce α-ketoglutaric acid. Only one of the three carboxyl groups of oxalosuccinic acid has a carbonyl group in the position β to it, and it is this carboxyl group that is lost as CO_2:

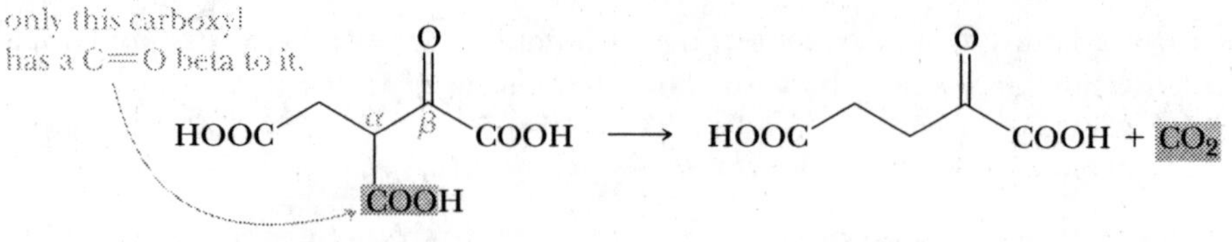

only this carboxyl has a C=O beta to it.

HOOC–CH₂–CH(COOH)–CO–COOH ⟶ HOOC–CH₂–CH₂–CO–COOH + CO_2

Oxalosuccinic acid α-Ketoglutaric acid

B. Malonic Acid and Substituted Malonic Acids

The presence of a ketone or an aldehyde carbonyl group on the carbon β to the carboxyl group is sufficient to facilitate decarboxylation. In the more general reaction, decarboxylation is facilitated by the presence of any carbonyl group on the β carbon, including that of a carboxyl group or ester. Malonic acid and substituted malonic acids, for example, undergo decarboxylation on heating, as illustrated by the decarboxylation of malonic acid when it is heated slightly above its melting point of 135–137°C:

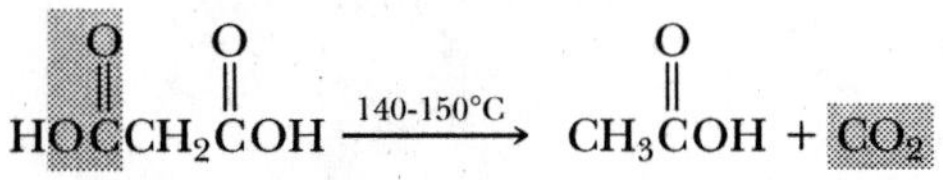

$$HOCCH_2COH \xrightarrow{140\text{-}150°C} CH_3COH + CO_2$$

Propanedioic acid
(Malonic acid)

The mechanism for decarboxylation of malonic acids is similar to what we have just studied for the decarboxylation of β-ketoacids. The formation of a cyclic, six-membered transition state involving a redistribution of three electron pairs gives the enol form of a carboxylic acid, which, in turn, isomerizes to the carboxylic acid.

Mechanism: Decarboxylation of a β-Dicarboxylic Acid

Step 1: Rearrangement of six electrons in a cyclic six-membered transition state gives carbon dioxide and the enol form of a carboxyl group.

Step 2: Keto–enol tautomerism (Section 13.9A) of the enol gives the more stable keto form of the carboxyl group.

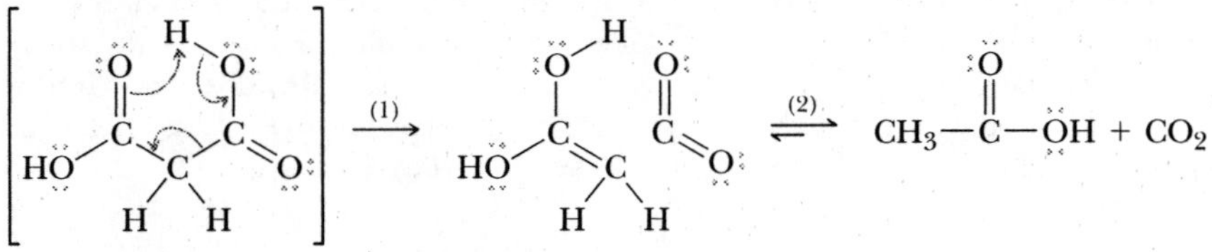

(1) → (2) ⇌ $CH_3—C—OH + CO_2$

A cyclic six-membered transition state

Enol of a carboxyl group

EXAMPLE 14.6

Each of these carboxylic acids undergoes thermal decarboxylation:

(a) 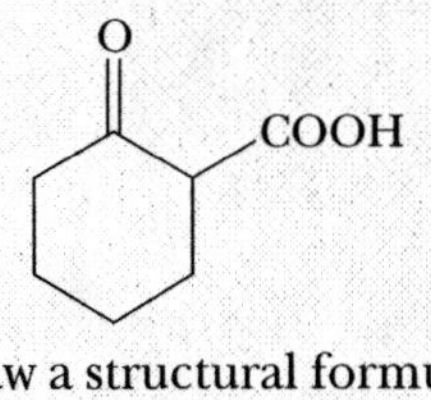 COOH

(b) 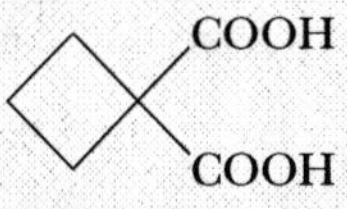 COOH COOH

Draw a structural formula for the enol intermediate and final product formed in each reaction.

SOLUTION

(a) 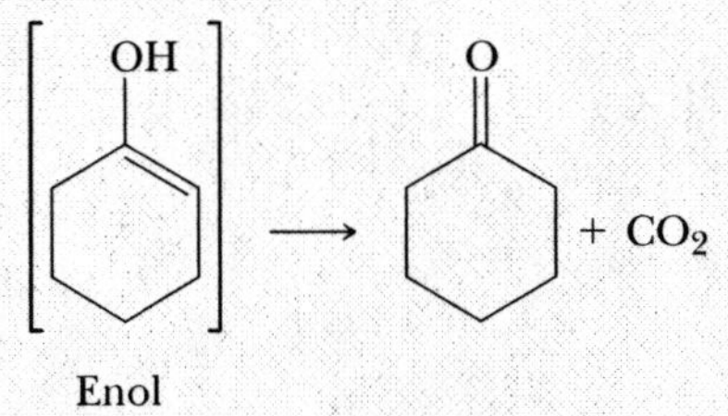 OH → O + CO_2

Enol intermediate

(b) 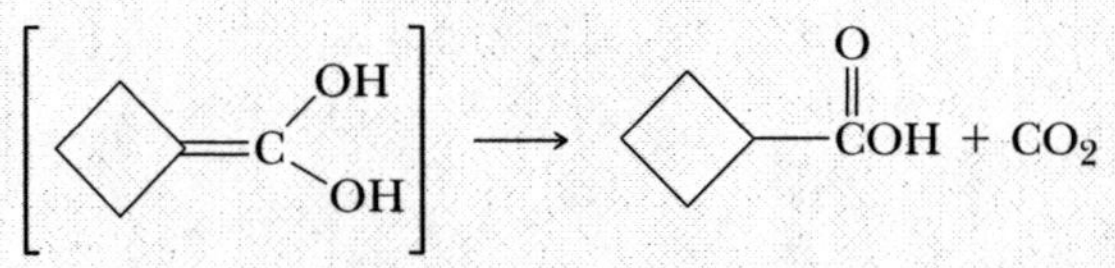 OH C OH → O COH + CO_2

Enol intermediate

Practice Problem 14.6

Draw the structural formula for the indicated β-ketoacid:

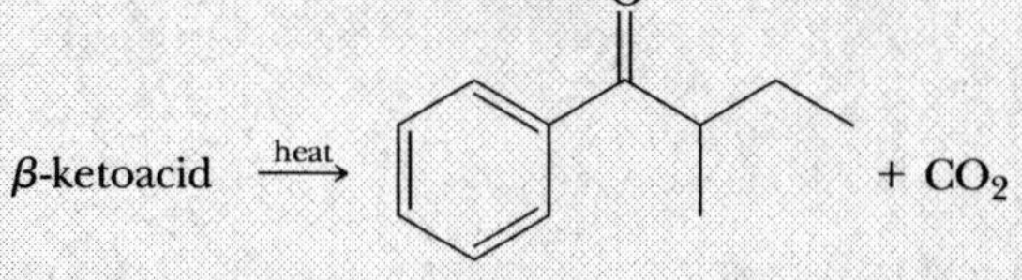

β-ketoacid $\xrightarrow{\text{heat}}$ [1-phenyl-2-methyl-1-butanone structure] + CO_2

SUMMARY

The functional group of a **carboxylic acid** (Section 14.2) is the **carboxyl group**, —**COOH**. IUPAC names of carboxylic acids (Section 14.3) are derived from the parent alkane by dropping the suffix *-e* and adding *-oic acid*. Dicarboxylic acids are named as *-dioic acids*.

Carboxylic acids are polar compounds (Section 14.4) that associate by hydrogen bonding into dimers in the liquid and solid states. Carboxylic acids have higher boiling points and are more soluble in water than alcohols, aldehydes, ketones, and ethers of comparable molecular weight. A carboxylic acid consists of two regions of different polarity; a polar, **hydrophilic** carboxyl group, which increases solubility in water, and a nonpolar, **hydrophobic** hydrocarbon chain, which decreases solubility in water. The first four aliphatic carboxylic acids are infinitely soluble in water, because the hydrophilic carboxyl group more than counterbalances the hydrophobic hydrocarbon chain. As the size of the carbon chain increases, however, the hydrophobic group becomes dominant, and solubility in water decreases.

Values of **pK_a** for aliphatic carboxylic acids are in the range from 4.0 to 5.0 (Section 14.5A). Electron-withdrawing substituents near the carboxyl group increase acidity in both aliphatic and aromatic carboxylic acids.

KEY REACTIONS

1. Acidity of Carboxylic Acids (Section 14.5A)

Values of pK_a for most unsubstituted aliphatic and aromatic carboxylic acids are within the range from 4 to 5:

$$CH_3\overset{O}{\overset{\|}{C}}OH + H_2O \rightleftharpoons CH_3\overset{O}{\overset{\|}{C}}O^- + H_3O^+ \quad pK_a = 4.76$$

Substitution by electron-withdrawing groups decreases pK_a (increases acidity).

2. Reaction of Carboxylic Acids with Bases (Section 14.5B)

Carboxylic acids form water-soluble salts with alkali metal hydroxides, carbonates, and bicarbonates, as well as with ammonia and amines:

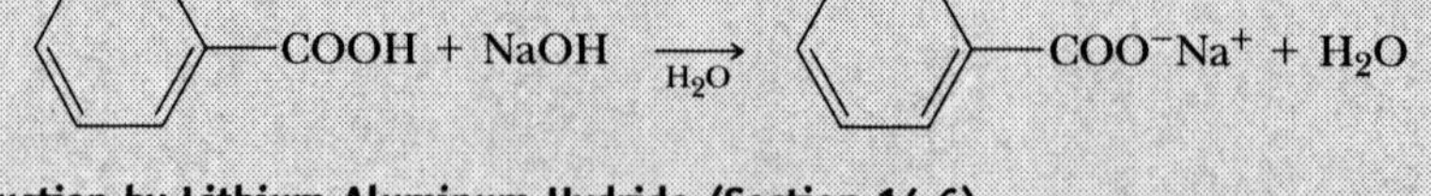

$$C_6H_5\text{—}COOH + NaOH \xrightarrow[H_2O]{} C_6H_5\text{—}COO^-Na^+ + H_2O$$

3. Reduction by Lithium Aluminum Hydride (Section 14.6)

Lithium aluminum hydride reduces a carboxyl group to a primary alcohol:

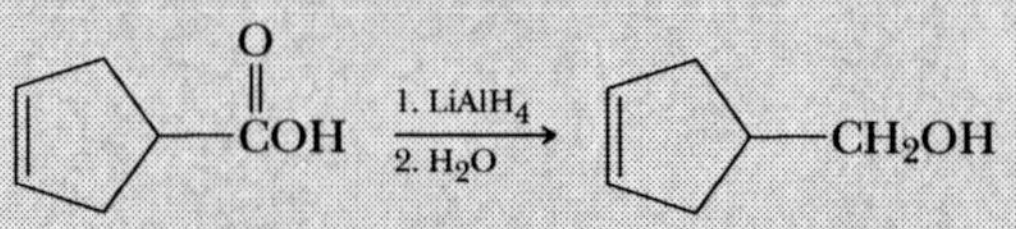

$$\text{(cyclopentenyl)}\text{—}\overset{O}{\overset{\|}{C}}OH \xrightarrow[2.\ H_2O]{1.\ LiAlH_4} \text{(cyclopentenyl)}\text{—}CH_2OH$$

4. Fischer Esterification (Section 14.7)

Fischer esterification is reversible:

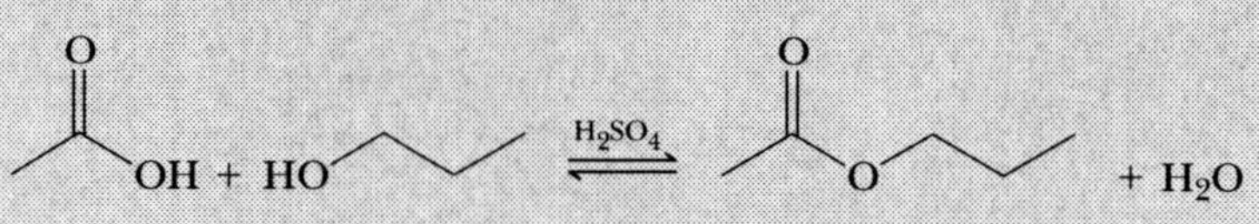

$$CH_3\overset{O}{\overset{\|}{C}}OH + HOCH_2CH_2CH_3 \underset{}{\overset{H_2SO_4}{\rightleftharpoons}} CH_3\overset{O}{\overset{\|}{C}}OCH_2CH_2CH_3 + H_2O$$

One way to force the equilibrium to the right is to use an excess of the alcohol.

5. Conversion to Acid Halides (Section 14.8)

Acid chlorides, the most common and widely used of the acid halides, are prepared by treating carboxylic acids with thionyl chloride:

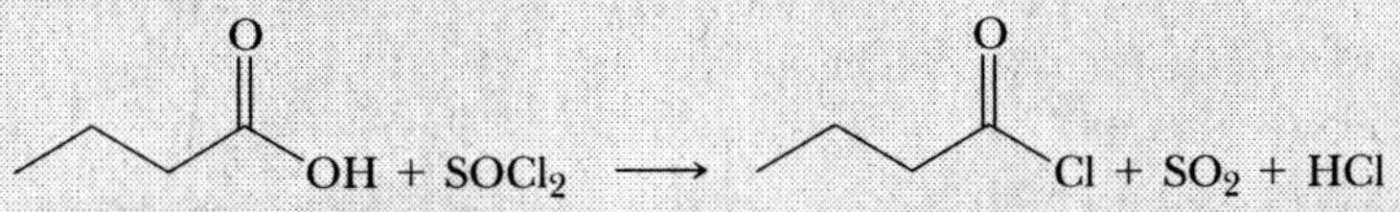

$$CH_3CH_2CH_2COOH + SOCl_2 \longrightarrow CH_3CH_2CH_2COCl + SO_2 + HCl$$

6. Decarboxylation of β-Ketoacids (Section 14.9A)

The mechanism of decarboxylation involves the redistribution of bonding electrons in a cyclic, six-membered transition state:

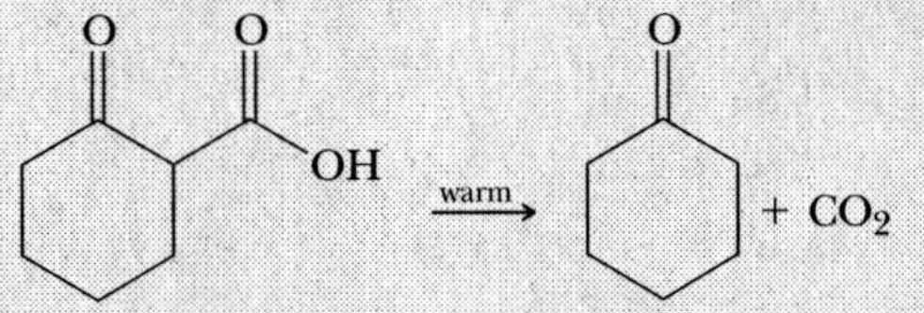

$$\text{2-oxocyclohexanecarboxylic acid} \xrightarrow{\text{warm}} \text{cyclohexanone} + CO_2$$

7. Decarboxylation of β-Dicarboxylic Acids (Section 14.9B)

The mechanism of decarboxylation of a β-dicarboxylic acid is similar to that of decarboxylation of a β-ketoacid:

$$HO\overset{O}{\overset{\|}{C}}CH_2\overset{O}{\overset{\|}{C}}OH \xrightarrow{\text{heat}} CH_3\overset{O}{\overset{\|}{C}}OH + CO_2$$

PROBLEMS

A problem number set in red indicates an applied "real-world" problem.

Structure and Nomenclature

14.7 Name and draw structural formulas for the four carboxylic acids with molecular formula $C_5H_{10}O_2$. Which of these carboxylic acids is chiral?

14.8 Write the IUPAC name for each compound:

(a) 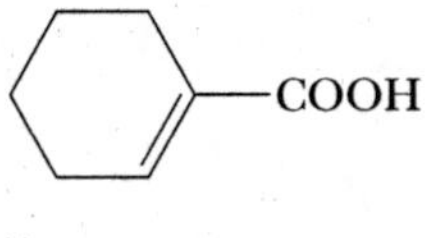 cyclohexene ring with —COOH

(b) 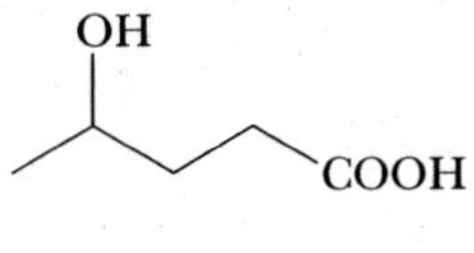 OH-substituted chain with COOH

(c) 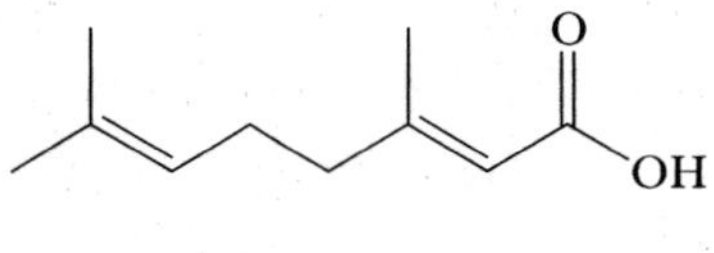 diene chain ending in C(=O)OH

(d) 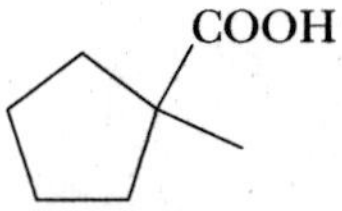 cyclopentane ring with COOH and methyl on the same carbon

(e)  chain ending in $COO^-NH_4^+$

(f) 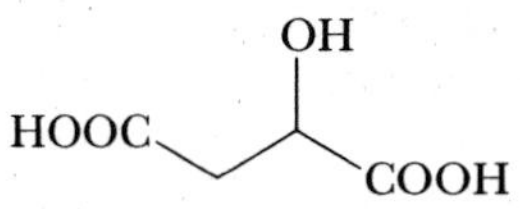 HOOC chain with OH and COOH

14.9 Draw a structural formula for each carboxylic acid:

(a) 4-Nitrophenylacetic acid
(b) 4-Aminopentanoic acid
(c) 3-Chloro-4-phenylbutanoic acid
(d) *cis*-3-Hexenedioic acid
(e) 2,3-Dihydroxypropanoic acid
(f) 3-Oxohexanoic acid
(g) 2-Oxocyclohexanecarboxylic acid
(h) 2,2-Dimethylpropanoic acid

14.10 Megatomoic acid, the sex attractant of the female black carpet beetle, has the structure

$$CH_3(CH_2)_7CH{=}CHCH{=}CHCH_2COOH$$

Megatomoic acid

(a) What is the IUPAC name of megatomoic acid?
(b) State the number of stereoisomers possible for this compound.

14.11 The IUPAC name of ibuprofen is 2-(4-isobutylphenyl)propanoic acid. Draw a structural formula of ibuprofen.

14.12 Draw structural formulas for these salts:

(a) Sodium benzoate
(b) Lithium acetate
(c) Ammonium acetate
(d) Disodium adipate
(e) Sodium salicylate
(f) Calcium butanoate

14.13 The monopotassium salt of oxalic acid is present in certain leafy vegetables, including rhubarb. Both oxalic acid and its salts are poisonous in high concentrations. Draw a structural formula of monopotassium oxalate.

14.14 Potassium sorbate is added as a preservative to certain foods to prevent bacteria and molds from causing spoilage and to extend the foods' shelf life. The IUPAC name of potassium sorbate is potassium (2*E*,4*E*)-2,4-hexadienoate. Draw a structural formula of potassium sorbate.

14.15 Zinc 10-undecenoate, the zinc salt of 10-undecenoic acid, is used to treat certain fungal infections, particularly *tinea pedis* (athlete's foot). Draw a structural formula of this zinc salt.

Physical Properties

14.16 Arrange the compounds in each set in order of increasing boiling point:

(a) $CH_3(CH_2)_5COOH$ $CH_3(CH_2)_6CHO$ $CH_3(CH_2)_6CH_2OH$

(b) CH_3CH_2COOH $CH_3CH_2CH_2CH_2OH$ $CH_3CH_2OCH_2CH_3$

Preparation of Carboxylic Acids

14.17 Draw a structural formula for the product formed by treating each compound with warm chromic acid, H_2CrO_4:

(a) $CH_3(CH_2)_4CH_2OH$

(b)

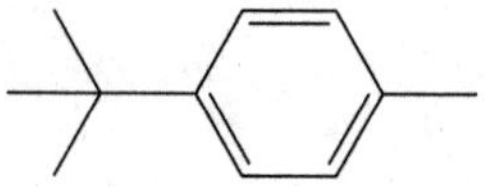

(c) 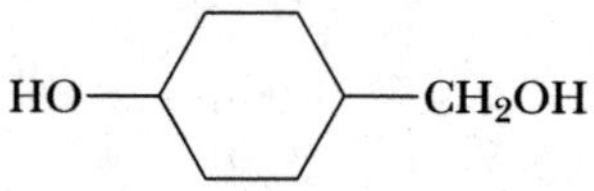 HO— —CH_2OH

14.18 Draw a structural formula for a compound with the given molecular formula that, on oxidation by chromic acid, gives the carboxylic acid or dicarboxylic acid shown:

(a) $C_6H_{14}O \xrightarrow{\text{oxidation}}$ COOH

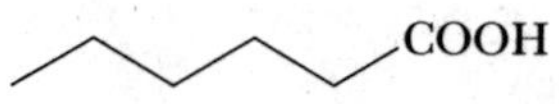

(b) $C_6H_{12}O \xrightarrow{\text{oxidation}}$ COOH

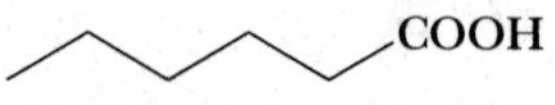

(c) $C_6H_{14}O_2 \xrightarrow{\text{oxidation}}$ HOOC COOH

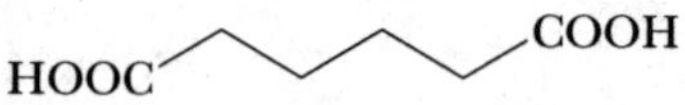

Acidity of Carboxylic Acids

14.19 Which is the stronger acid in each pair?

(a) Phenol (pK_a 9.95) or benzoic acid (pK_a 4.17)

(b) Lactic acid (K_a 8.4×10^{-4}) or ascorbic acid (K_a 7.9×10^{-5})

14.20 Arrange these compounds in order of increasing acidity: benzoic acid, benzyl alcohol, and phenol.

14.21 Assign the acid in each set its appropriate pK_a:

(a) 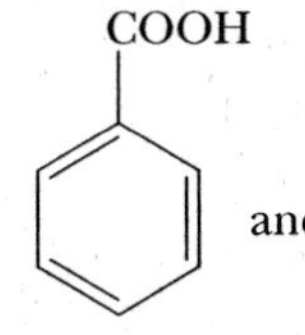 Benzoic acid (C_6H_5COOH) and 4-nitrobenzoic acid ($4\text{-}O_2NC_6H_4COOH$) (pK_a 4.19 and 3.14)

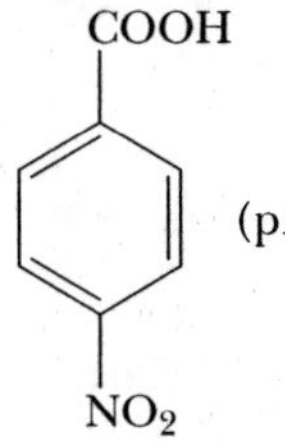

(b) 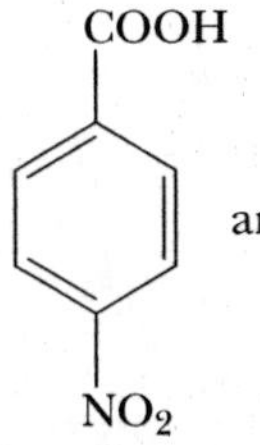 4-nitrobenzoic acid ($4\text{-}O_2NC_6H_4COOH$) and 4-aminobenzoic acid ($4\text{-}H_2NC_6H_4COOH$) (pK_a 4.92 and 3.14)

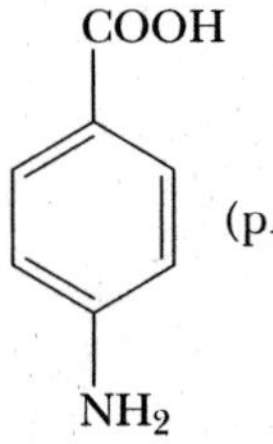

(c) $CH_3\overset{O}{\overset{\|}{C}}CH_2COOH$ and $CH_3\overset{O}{\overset{\|}{C}}COOH$ (pK_a 3.58 and 2.49)

(d) $CH_3\overset{OH}{\overset{|}{C}}HCOOH$ and CH_3CH_2COOH (pK_a 4.78 and 3.08)

14.22 Complete these acid–base reactions:

(a) 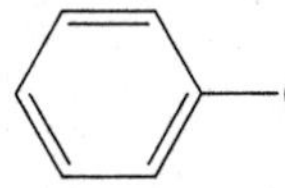 $C_6H_5CH_2COOH + NaOH \longrightarrow$

(b) $CH_3CH{=}CHCH_2COOH + NaHCO_3 \longrightarrow$

(c) 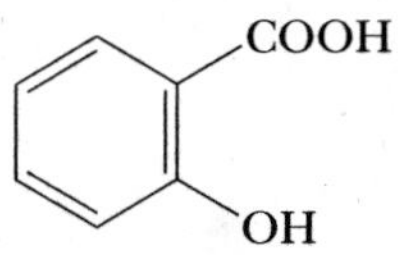 2-hydroxybenzoic acid ($2\text{-}HOC_6H_4COOH$) $+ NaHCO_3 \longrightarrow$

(d) $CH_3\overset{OH}{\overset{|}{C}}HCOOH + H_2NCH_2CH_2OH \longrightarrow$

(e) $CH_3CH{=}CHCH_2COO^-Na^+ + HCl \longrightarrow$

14.23 The normal pH range for blood plasma is 7.35–7.45. Under these conditions, would you expect the carboxyl group of lactic acid (pK_a 4.07) to exist primarily as a carboxyl group or as a carboxylate anion? Explain.

14.24 The pK_a of ascorbic acid (Section 18.7) is 4.76. Would you expect ascorbic acid dissolved in blood plasma (pH 7.35–7.45) to exist primarily as ascorbic acid or as ascorbate anion? Explain.

14.25 Excess ascorbic acid is (pK_a 4.76) excreted in the urine, the pH of which is normally in the range from 4.8 to 8.4. What form of ascorbic acid, ascorbic acid itself or ascorbate anion, would you expect to be present in urine with pH 8.4?

14.26 The pH of human gastric juice is normally in the range from 1.0 to 3.0. What form of lactic acid (pK_a 4.07), lactic acid itself or its anion, would you expect to be present in the stomach?

14.27 Following are two structural formulas for the amino acid alanine (Section 19.2):

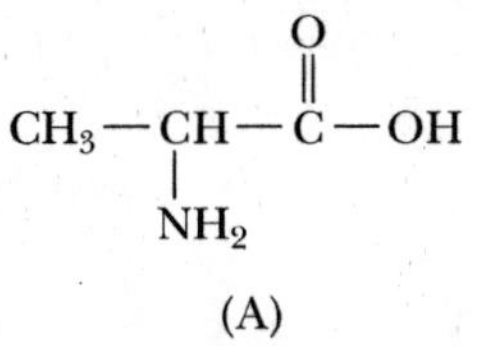

$$\underset{\text{(A)}}{CH_3{-}\underset{NH_2}{\underset{|}{C}H}{-}\overset{O}{\overset{\|}{C}}{-}OH} \qquad \underset{\text{(B)}}{CH_3{-}\underset{NH_3^+}{\underset{|}{C}H}{-}\overset{O}{\overset{\|}{C}}{-}O^-}$$

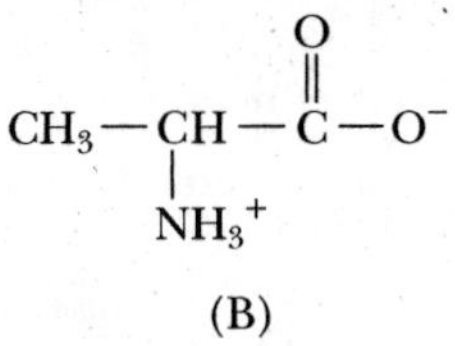

Is alanine better represented by structural formula A or B? Explain.

14.28 In Chapter 19, we discuss a class of compounds called amino acids, so named because they contain both an amino group and a carboxyl group. Following is a structural formula for the amino acid alanine in the form of an internal salt:

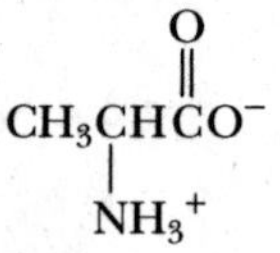

$$CH_3\underset{\displaystyle NH_3^+}{\underset{|}{C}}H\overset{\displaystyle O}{\overset{\|}{C}}O^- \quad \text{Alanine}$$

What would you expect to be the major form of alanine present in aqueous solution at (a) pH 2.0, (b) pH 5–6, and (c) pH 11.0? Explain.

Reactions of Carboxylic Acids

14.29 Give the expected organic products formed when phenylacetic acid, $PhCH_2COOH$, is treated with each of the following reagents:

(a) $SOCl_2$
(b) $NaHCO_3$, H_2O
(c) $NaOH$, H_2O
(d) NH_3, H_2O
(e) $LiAlH_4$, followed by H_2O
(f) $NaBH_4$, followed by H_2O
(g) $CH_3OH + H_2SO_4$ (catalyst)
(h) H_2/Ni at 25°C and 3 atm pressure

14.30 Show how to convert *trans*-3-phenyl-2-propenoic acid (cinnamic acid) to these compounds:

(a) 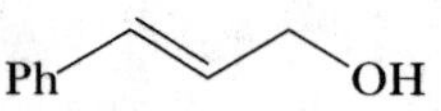 Ph⁄⁀⁀OH (alkene) **(b)**  Ph⁄⁀COOH **(c)**  Ph⁄⁀⁀OH

14.31 Show how to convert 3-oxobutanoic acid (acetoacetic acid) to these compounds:

(a) $CH_3\overset{OH}{\overset{|}{C}}HCH_2COOH$ **(b)** $CH_3\overset{OH}{\overset{|}{C}}HCH_2CH_2OH$ **(c)** $CH_3CH{=}CHCOOH$

14.32 Complete these examples of Fischer esterification (assume an excess of the alcohol):

(a) 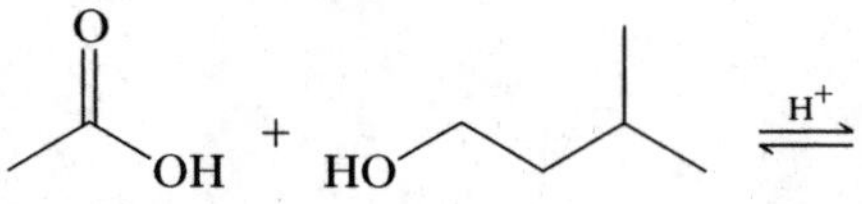 CH_3COOH + HO⁄⁀⁀ (3-methylbutanol) $\overset{H^+}{\rightleftharpoons}$

(b) 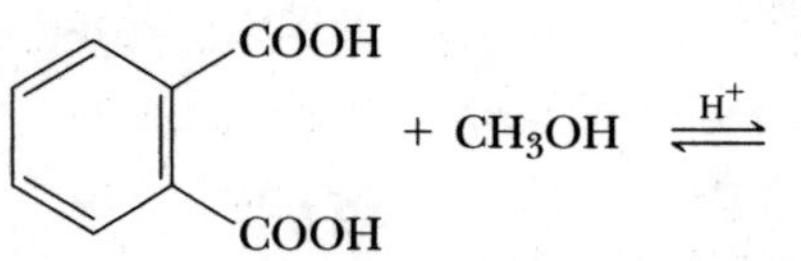 (benzene-1,2-dicarboxylic acid: COOH, COOH) $+ CH_3OH \overset{H^+}{\rightleftharpoons}$

(c) 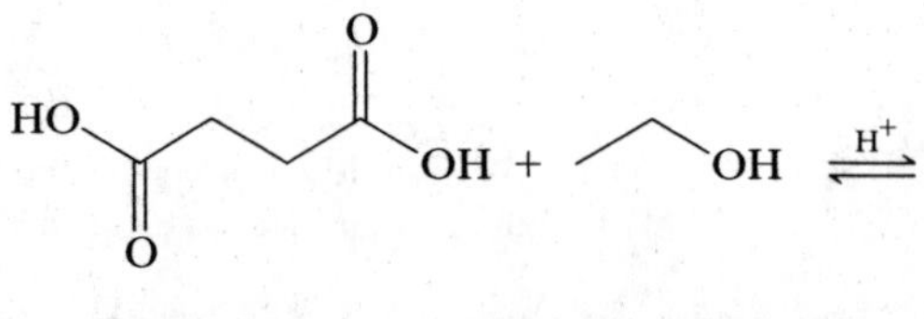 HO(C=O)CH₂CH₂(C=O)OH + ⁄⁀OH $\overset{H^+}{\rightleftharpoons}$

14.33 Formic acid is one of the components responsible for the sting of biting ants and is injected under the skin by bees and wasps. A way to relieve the pain is to rub the area of the sting with a paste of baking soda ($NaHCO_3$) and water, which neutralizes the acid. Write an equation for this reaction.

14.34 Methyl 2-hydroxybenzoate (methyl salicylate) has the odor of oil of wintergreen. This ester is prepared by the Fischer esterification of 2-hydroxybenzoic acid (salicylic acid) with methanol. Draw a structural formula of methyl 2-hydroxybenzoate.

14.35 Benzocaine, a topical anesthetic, is prepared by treating 4-aminobenzoic acid with ethanol in the presence of an acid catalyst, followed by neutralization. Draw a structural formula of benzocaine.

14.36 Examine the structural formulas of pyrethrin and permethrin. (See Chemical Connections 14C.)

(a) Locate the ester groups in each compound.
(b) Is pyrethrin chiral? How many stereoisomers are possible for it?
(c) Is permethrin chiral? How many stereoisomers are possible for it?

14.37 A commercial Clothing & Gear Insect Repellant gives the following information about permethrin, its active ingredient:

Cis/trans ratio: Minimum 35% (+/−) cis and maximum 65% (+/−) trans

(a) To what does the cis/trans ratio refer?

(b) To what does the designation "(+/−)" refer?

14.38 From what carboxylic acid and alcohol is each of the following esters derived?

(a) 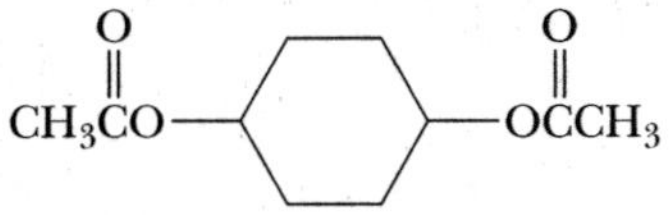 $CH_3C(=O)O$–(cyclohexane-1,4-diyl)–$OC(=O)CH_3$ **(b)** $CH_3OC(=O)CH_2CH_2C(=O)OCH_3$

(c) 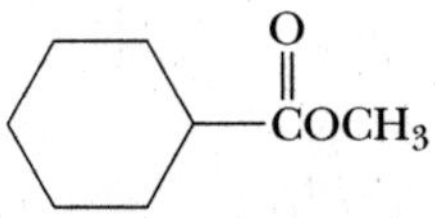 (cyclohexyl)–$C(=O)OCH_3$ **(d)** $CH_3CH_2CH{=}CHC(=O)OCH(CH_3)_2$

14.39 When treated with an acid catalyst, 4-hydroxybutanoic acid forms a cyclic ester (a lactone). Draw the structural formula of this lactone.

14.40 Draw a structural formula for the product formed on thermal decarboxylation of each of the following compounds:

(a) $C_6H_5C(=O)CH_2COOH$ **(b)** $C_6H_5CH_2CH(COOH)COOH$ **(c)** 1-acetylcyclopentane-1-carboxylic acid ($C(=O)CH_3$ and $COOH$ on the same ring carbon)

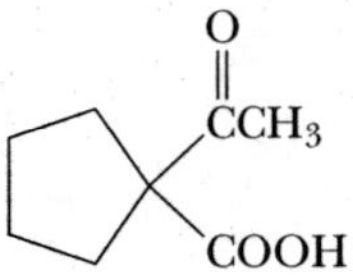

Synthesis

14.41 Methyl 2-aminobenzoate, a flavoring agent with the taste of grapes (see "Chemical Connections 14B"), can be prepared from toluene by the following series of steps:

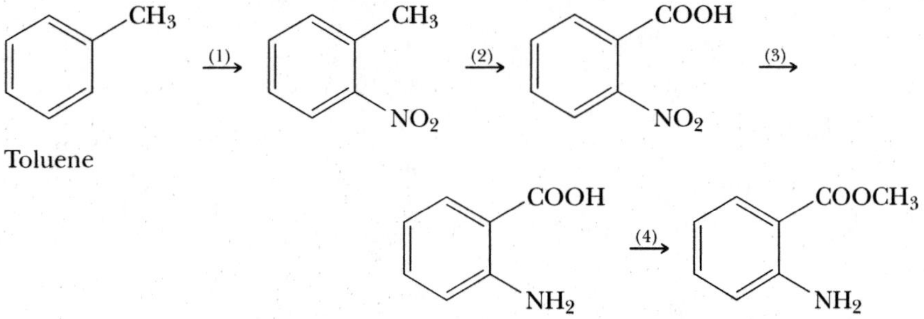

Toluene $\xrightarrow{(1)}$ 2-nitrotoluene (CH_3, NO_2) $\xrightarrow{(2)}$ 2-nitrobenzoic acid ($COOH$, NO_2) $\xrightarrow{(3)}$ 2-aminobenzoic acid ($COOH$, NH_2) $\xrightarrow{(4)}$ ($COOCH_3$, NH_2) Methyl 2-aminobenzoate

Show how you might bring about each step in this synthesis.

14.42 Methylparaben and propylparaben are used as preservatives in foods, beverages, and cosmetics:

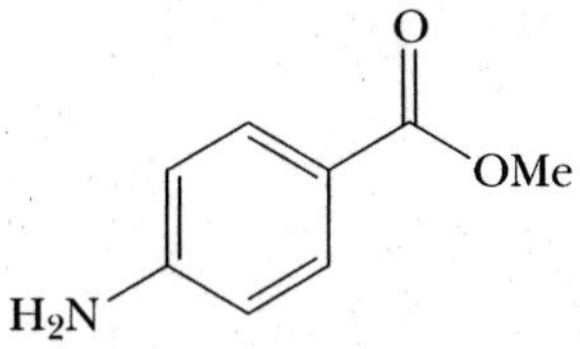

H_2N–C_6H_4–$C(=O)OMe$ — Methyl 4-aminobenzoate (Methylparaben)

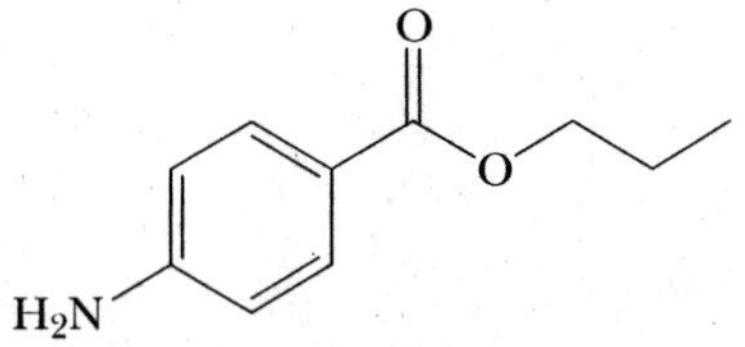

H_2N–C_6H_4–$C(=O)OCH_2CH_2CH_3$ — Propyl 4-aminobenzoate (Propylparaben)

Show how the synthetic scheme in Problem 14.41 can be modified to give each of these compounds.

14.43 Procaine (its hydrochloride is marketed as Novocaine®) was one of the first local anesthetics developed for infiltration and regional anesthesia. It is synthesized by the following Fischer esterification:

H_2N–C₆H₄–COOH + HO–CH₂CH₂–N(CH₂CH₃)₂ —(Fischer esterification)→ Procaine

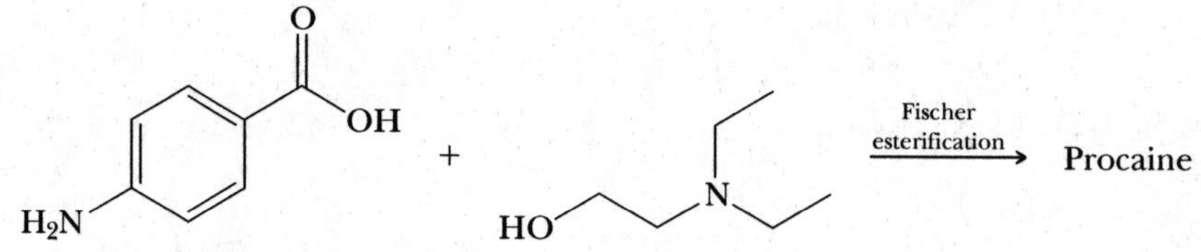

p-Aminobenzoic acid   2-Diethylaminoethanol

Draw a structural formula for procaine.

14.44 Meclizine is an antiemetic: It helps prevent, or at least lessen, the vomiting associated with motion sickness, including seasickness. Among the names of the over-the-counter preparations of meclizine are Bonine®, Sea-Legs, Antivert®, and Navicalm®. Meclizine can be synthesized by the following series of steps:

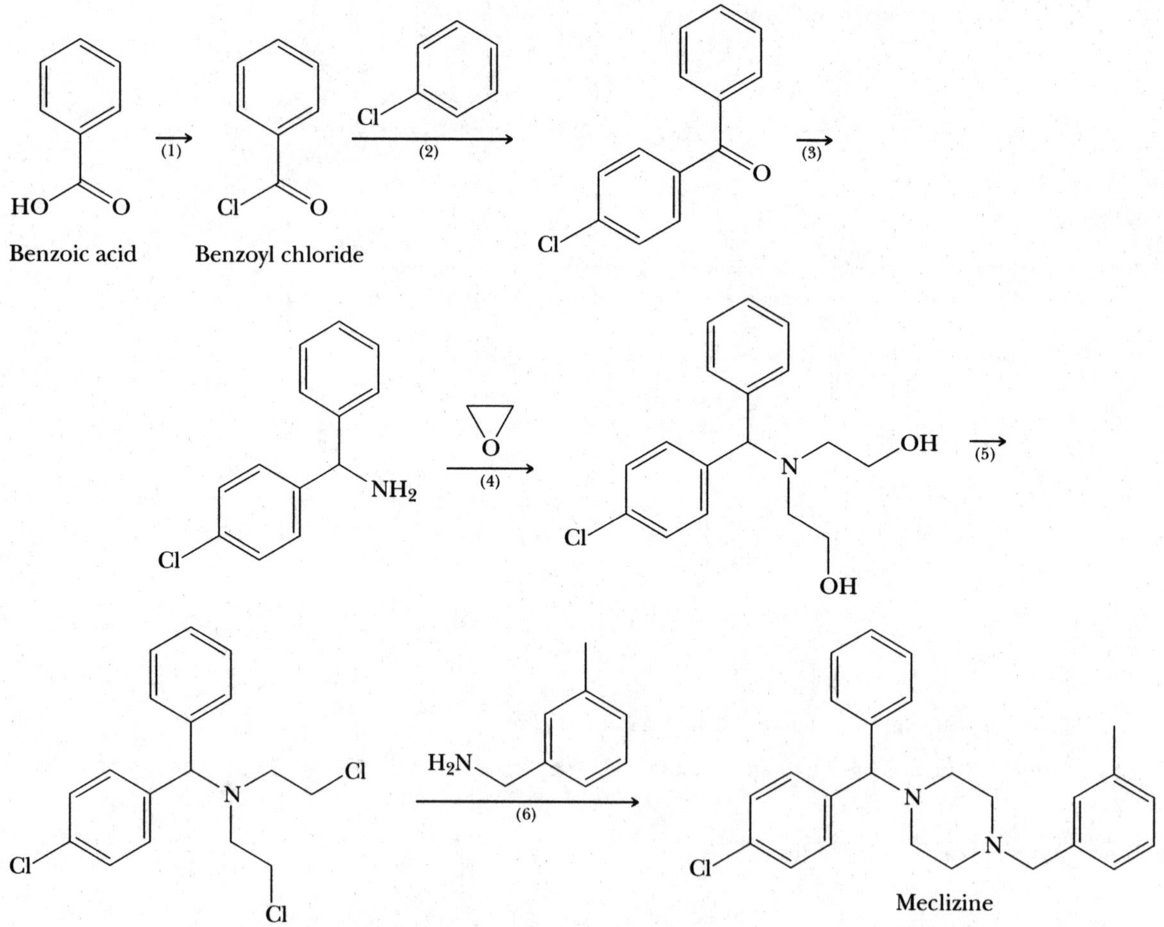

Benzoic acid —(1)→ Benzoyl chloride —(2), chlorobenzene→ 4-chlorobenzophenone —(3)→

(4-chlorophenyl)(phenyl)methylamine (NH_2) —(4), ethylene oxide→ N,N-bis(2-hydroxyethyl) amine (OH, OH) —(5)→

N,N-bis(2-chloroethyl) amine (Cl, Cl) —(6), H_2N–CH₂–(3-methylphenyl)→ Meclizine

(a) Propose a reagent for Step 1.
(b) The catalyst for Step 2 is $AlCl_3$. Name the type of reaction that occurs in Step 2.
(c) Propose reagents for Step 3.
(d) Propose a mechanism for Step 4, and show that it is an example of nucleophilic aliphatic substitution.
(e) Propose a reagent for Step 5.
(f) Show that Step 6 is also an example of nucleophilic aliphatic substitution.

14.45 Chemists have developed several syntheses for the antiasthmatic drug albuterol (Proventil). One of these syntheses starts with salicylic acid, the same acid that is the starting material for the synthesis of aspirin:

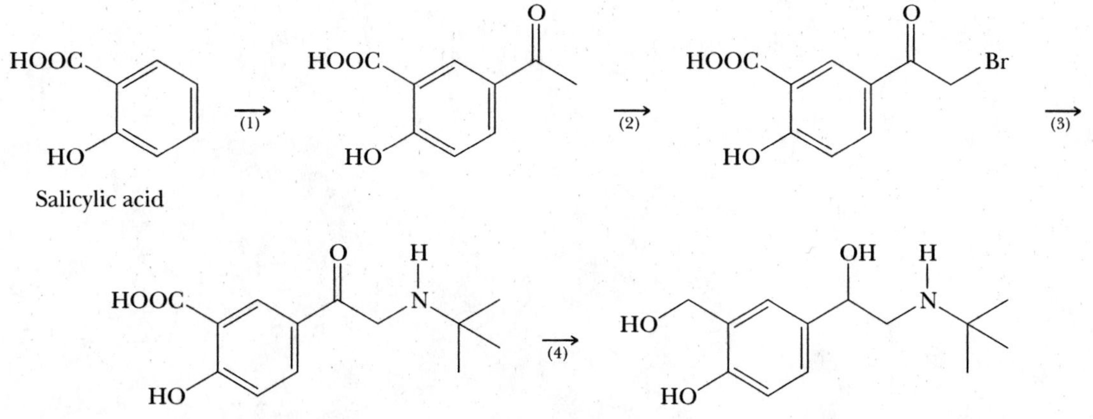

Salicylic acid $\xrightarrow{(1)}$ $\xrightarrow{(2)}$ $\xrightarrow{(3)}$ $\xrightarrow{(4)}$ Albuterol

(a) Propose a reagent and a catalyst for Step 1. What name is given to this type of reaction?

(b) Propose a reagent for Step 2.

(c) Name the amine used to bring about Step 3.

(d) Step 4 is a reduction of two functional groups. Name the functional groups reduced and tell what reagent will accomplish the reduction.

Looking Ahead

14.46 Explain why α-amino acids, the building blocks of proteins (Chapter 20), are nearly a thousand times more acidic than aliphatic carboxylic acids:

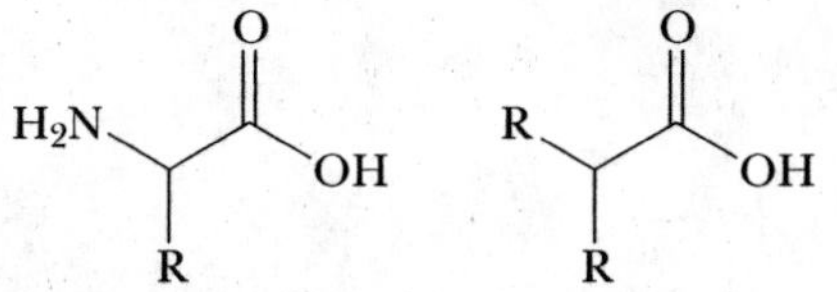

An α-amino acid $pK_a \approx 2$

An aliphatic acid $pK_a \approx 5$

14.47 Which is more difficult to reduce with $LiAlH_4$, a carboxylic acid or a carboxylate ion?

14.48 Show how an ester can react with H^+/H_2O to give a carboxylic acid and an alcohol (*Hint:* This is the reverse of Fischer esterification):

$$\xrightleftharpoons[H_2O]{H^+} + HOR'$$

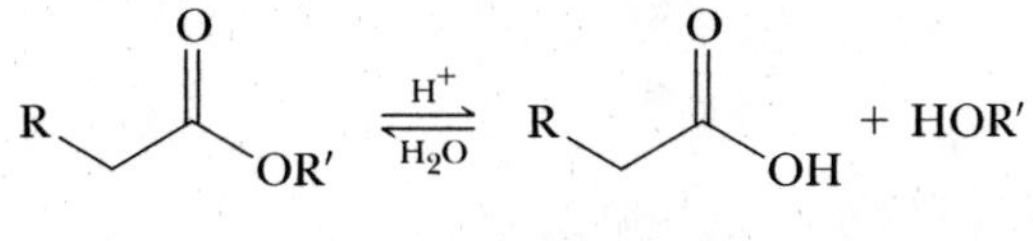

14.49 In Chapter 13, we saw how Grignard reagents readily attack the carbonyl carbon of ketones and aldehydes. Should the same process occur with Grignards and carboxylic acids? With esters?

14.50 In Section 14.7, it was suggested that the mechanism for the Fischer esterification of carboxylic acids would be a model for many of the reactions of the functional derivatives of carboxylic acids. One such reaction, the reaction of an acid halide with water, is the following:

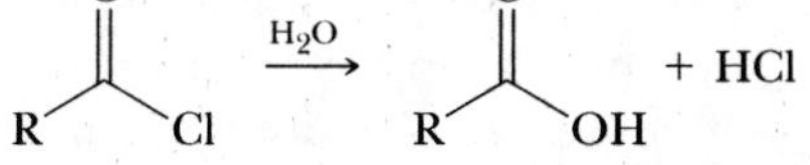

$$\xrightarrow{H_2O} + HCl$$

Suggest a mechanism for this reaction.

17 Organic Polymer Chemistry

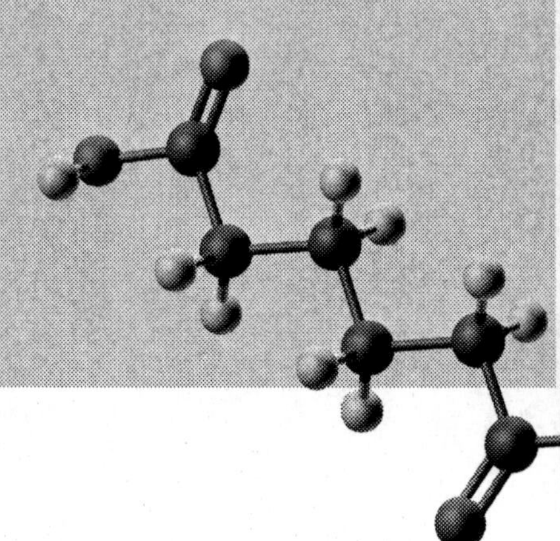

Sea of umbrellas on a rainy day in Shanghai, China. Inset: A model of adipic acid, one of the two monomers from which nylon 66 in made. *(Gavin Hellier/Stone/Getty Images)*

17.1 INTRODUCTION

The technological advancement of any society is inextricably tied to the materials available to it. Indeed, historians have used the emergence of new materials as a way of establishing a time line to mark the development of human civilization. As part of the search to discover new materials, scientists have made increasing use of organic chemistry for the preparation of synthetic materials known as polymers. The versatility afforded by these polymers allows for the creation and fabrication of materials with ranges of properties unattainable using such materials as wood, metals, and ceramics. Deceptively simple changes in the chemical structure of a given polymer, for example, can change its mechanical properties from those of a sandwich bag to those of a bulletproof vest. Furthermore, structural changes can introduce properties never before imagined in organic polymers. For instance, using well-defined organic reactions, chemists can turn one type of polymer into an insulator (e.g., the rubber sheath that surrounds electrical cords). Treated differently, the same type of polymer can be made into an electrical conductor with a conductivity nearly equal to that of metallic copper!

The years since the 1930s have seen extensive research and development in organic polymer chemistry, and an almost explosive growth in plastics, coatings, and rubber technology has created a worldwide multibillion-dollar industry. A few basic characteristics account for this phenomenal growth. First, the raw materials for synthetic polymers are derived mainly from petroleum. With the development of petroleum-refining processes, raw materials for the synthesis of polymers became generally cheap and plentiful. Second, within broad limits, scientists have learned how to tailor polymers to the requirements of the end use. Third, many consumer products can be fabricated more cheaply from synthetic polymers than from such competing materials as wood, ceramics, and metals. For example, polymer technology created the water-based (latex) paints that have revolutionized the coatings industry, and plastic films and foams have done the same for the packaging industry. The list could go on and on as we think of the manufactured items that are everywhere around us in our daily lives.

17.2 THE ARCHITECTURE OF POLYMERS

Polymers (Greek: *poly* + *meros*, many parts) are long-chain molecules synthesized by linking **monomers** (Greek: *mono* + *meros*, single part) through chemical reactions. The molecular weights of polymers are generally high compared with those of common organic compounds and typically range from 10,000 g/mol to more than 1,000,000 g/mol. The architectures of these macromolecules can also be quite diverse: There are polymer architectures with linear and branched chains, as well as those with comb, ladder, and star structures (Figure 17.1). Additional structural variations can be achieved by introducing covalent cross-links between individual polymer chains.

Polymer From the Greek *poly*, many and *meros*, parts; any long-chain molecule synthesized by linking together many single parts called monomers.

Monomer From the Greek *mono*, single and *meros*, part; the simplest nonredundant unit from which a polymer is synthesized.

Linear and branched polymers are often soluble in solvents such as chloroform, benzene, toluene, dimethyl sulfoxide (DMSO), and tetrahydrofuran (THF). In addition, many linear and branched polymers can be melted to form highly viscous liquids. In polymer chemistry, the term **plastic** refers to any polymer that can be molded when hot and that retains its shape when cooled. **Thermoplastics** are polymers which, when melted, become sufficiently fluid that they can be molded into shapes that are retained when they are cooled. **Thermosetting plastics**, or thermosets, can be molded when they are first prepared, but once cooled, they harden irreversibly and cannot be remelted. Because of their very different physical characteristics, thermoplastics and thermosets must be processed differently and are used in very different applications.

Plastic A polymer that can be molded when hot and retains its shape when cooled.

Thermoplastic A polymer that can be melted and molded into a shape that is retained when it is cooled.

Thermosetting plastic A polymer that can be molded when it is first prepared, but, once cooled, hardens irreversibly and cannot be remelted.

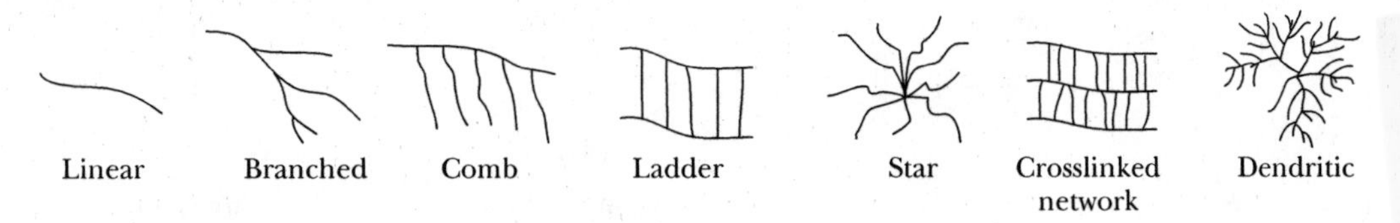

Figure 17.1
Various polymer architectures.

The single most important property of polymers at the molecular level is the size and shape of their chains. A good example of the importance of size is a comparison of paraffin wax, a natural polymer, and polyethylene, a synthetic polymer. These two distinct materials have identical repeat units, namely, $-CH_2-$, but differ greatly in the size of their chains. Paraffin wax has between 25 and 50 carbon atoms per chain, whereas polyethylene has between 1,000 and 3,000 carbons per chain. Paraffin wax, such as that in birthday candles, is soft and brittle, but polyethylene, from which plastic beverage bottles are fabricated, is strong, flexible, and tough. These vastly different properties arise directly from the difference in size and molecular architecture of the individual polymer chains.

17.3 POLYMER NOTATION AND NOMENCLATURE

Average degree of polymerization, *n* A subscript placed outside the parentheses of the simplest nonredundant unit of a polymer to indicate that the unit repeats *n* times in the polymer.

We typically show the structure of a polymer by placing parentheses around the **repeating unit**, which is the smallest molecular fragment that contains all the nonrepeating structural features of the chain. A subscript n placed outside the parentheses indicates that the unit repeats n times. Thus, we can reproduce the structure of an entire polymer chain by repeating the enclosed structure in both directions. An example is polypropylene, which is derived from the polymerization of propylene:

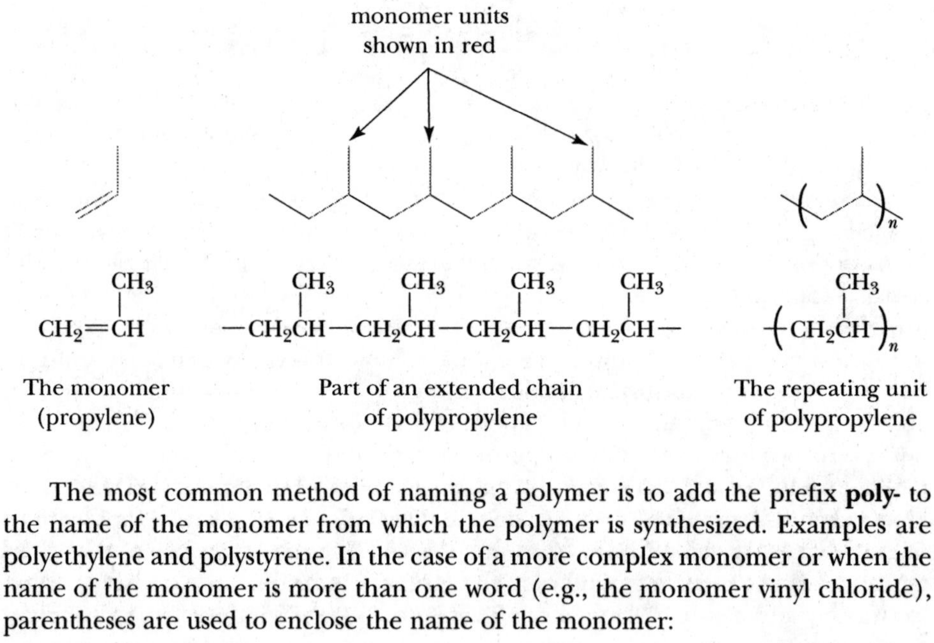

monomer units shown in red

$CH_2{=}CH(CH_3)$

The monomer (propylene)

$-CH_2CH(CH_3)-CH_2CH(CH_3)-CH_2CH(CH_3)-CH_2CH(CH_3)-$

Part of an extended chain of polypropylene

$\left(CH_2CH(CH_3)\right)_n$

The repeating unit of polypropylene

The most common method of naming a polymer is to add the prefix **poly-** to the name of the monomer from which the polymer is synthesized. Examples are polyethylene and polystyrene. In the case of a more complex monomer or when the name of the monomer is more than one word (e.g., the monomer vinyl chloride), parentheses are used to enclose the name of the monomer:

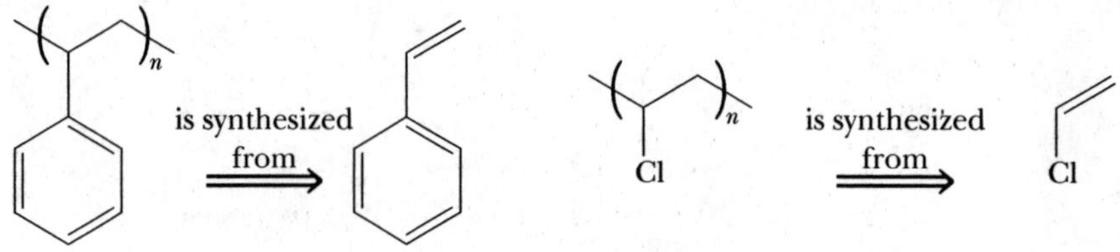

Polystyrene is synthesized from ⟹ Styrene

Poly(vinyl chloride) (PVC) is synthesized from ⟹ Vinyl chloride

EXAMPLE 17.1

Given the following structure, determine the polymer's repeating unit; redraw the structure, using the simplified parenthetical notation; and name the polymer:

F F F F F F F F F F F F

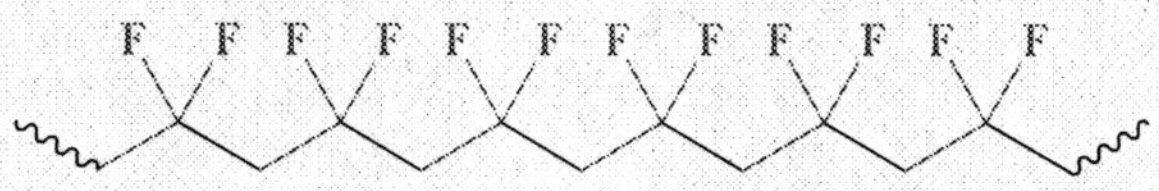

(repeating unit in red)

SOLUTION

The repeating unit is $—CH_2CF_2—$ and the polymer is written $\leftarrow\!(CH_2CF_2)\!\rightarrow_n$. The repeat unit is derived from 1,1-difluoroethylene and the polymer is named poly(1,1-difluoroethylene). This polymer is used in microphone diaphragms.

Practice Problem 17.1

Given the following structure, determine the polymer's repeat unit; redraw the structure, using the simplified parenthetical notation; and name the polymer:

Cl Cl Cl Cl Cl

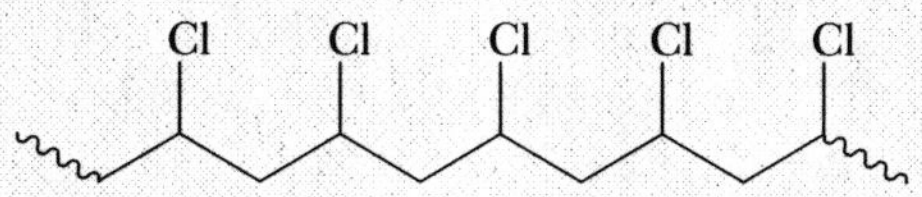

17.4 POLYMER MORPHOLOGY: CRYSTALLINE VERSUS AMORPHOUS MATERIALS

Polymers, like small organic molecules, tend to crystallize upon precipitation or as they are cooled from a melt. Acting to inhibit this tendency are their very large molecules, which tend to inhibit diffusion, and their sometimes complicated or irregular structures, which prevent efficient packing of the chains. The result is that polymers in the solid state tend to be composed of both ordered **crystalline domains** (crystallites) and disordered **amorphous domains**. The relative amounts of crystalline and amorphous domains differ from polymer to polymer and frequently depend upon the manner in which the material is processed.

Crystalline domains Ordered crystalline regions in the solid state of a polymer; also called crystallites.

Amorphous domains Disordered, noncrystalline regions in the solid state of a polymer.

Melt transition temperature, T_m The temperature at which crystalline regions of a polymer melt.

We often find high degrees of crystallinity in polymers with regular, compact structures and strong intermolecular forces, such as hydrogen bonding. The temperature at which crystallites melt corresponds to the **melt transition temperature** (**T_m**) of the polymer. As the degree of crystallinity of a polymer increases, its T_m increases, and the polymer becomes more opaque because of the scattering of light by its crystalline domains. With an increase in crystallinity comes a corresponding increase in strength and stiffness. For example, poly(6-aminohexanoic acid), known more commonly as nylon 6, has a $T_m = 223°C$. At and well above room temperature, this polymer is a hard, durable material that does not undergo any appreciable change in properties, even on a very hot summer afternoon. Its uses range from textile fibers to the heels of shoes.

Amorphous domains have little or no long-range order. Highly amorphous polymers are sometimes referred to as **glassy** polymers. Because they lack crystalline domains that scatter light, amorphous polymers are transparent. In addition, they are typically weak polymers, in terms of both their high flexibility and their low

Glass transition temperature, T_g The temperature at which a polymer undergoes the transition from a hard glass to a rubbery state.

mechanical strength. On being heated, amorphous polymers are transformed from a hard, glassy state to a soft, flexible, rubbery state. The temperature at which this transition occurs is called the **glass transition temperature (T_g)**. Amorphous polystyrene, for example, has a $T_g = 100°C$. At room temperature, it is a rigid solid used for drinking cups, foamed packaging materials, disposable medical wares, tape reels, and so forth. If it is placed in boiling water, it becomes soft and rubbery.

This relationship between a polymer's mechanical properties and degree of crystallinity can be illustrated by poly(ethylene terephthalate) (PET):

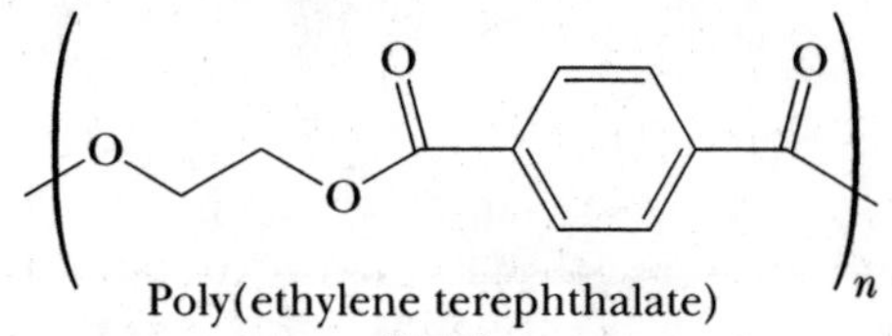

Poly(ethylene terephthalate)
(PET)

PET can be made with a percentage of crystalline domains ranging from 0% to about 55%. Completely amorphous PET is formed by cooling the melt quickly. By prolonging the cooling time, more molecular diffusion occurs and crystallites form as the chains become more ordered. The differences in mechanical properties between these forms of PET are great. PET with a low degree of crystallinity is used for plastic beverage bottles, whereas fibers drawn from highly crystalline PET are used for textile fibers and tire cords.

Elastomer A material that, when stretched or otherwise distorted, returns to its original shape when the distorting force is released.

Rubber materials must have low T_g values in order to behave as **elastomers (elastic polymers)**. If the temperature drops below its T_g value, then the material is converted to a rigid glassy solid and all elastomeric properties are lost. A poor understanding of this behavior of elastomers contributed to the *Challenger* spacecraft disaster in 1985. The elastomeric O-rings used to seal the solid booster rockets had a T_g value around 0°C. When the temperature dropped to an unanticipated low on the morning of the launch of the craft, the O-ring seals dropped below their T_g value and obediently changed from elastomers to rigid glasses, losing any sealing capabilities. The rest is tragic history. The physicist Richard Feynman sorted this out publicly in a famous televised hearing in which he put a *Challenger*-type O-ring in ice water and showed that its elasticity was lost!

17.5 STEP-GROWTH POLYMERS

Step-growth polymerization A polymerization in which chain growth occurs in a stepwise manner between difunctional monomers, for example, between adipic acid and hexamethylenediamine to form nylon 66. Also referred to as Condensation Polymerization.

Polymerizations in which chain growth occurs in a stepwise manner are called **step-growth**, or **condensation, polymerizations**. Step-growth polymers are formed by reaction between difunctional molecules, with each new bond created in a separate step. During polymerization, monomers react to form dimers, dimers react with monomers to form trimers, dimers react with dimers to form tetramers, and so on.

There are two common types of step-growth processes: (1) reaction between A—M—A and B—M—B type monomers to give $\left(A—M—A—B—M—B\right)_n$ polymers and (2) the self-condensation of A—M—B monomers to give $\left(A—M—B\right)_n$ polymers. In this notation, "M" indicates the monomer and "A" and "B" the reactive functional groups on the monomer. In each type of step-growth polymerization, an A functional group reacts exclusively with a B functional group, and a B functional group reacts exclusively with an A functional group. New covalent bonds in step-growth polymerizations are generally formed by polar reactions between A and B functional groups—for example, nucleophilic acyl substitution. In this section, we discuss five types of step-growth polymers: polyamides, polyesters, polycarbonates, polyurethanes, and epoxy resins.

A. Polyamides

In the early 1930s, chemists at E. I. DuPont de Nemours & Company began fundamental research into the reactions between dicarboxylic acids and diamines to form **polyamides**. In 1934, they synthesized the first purely synthetic fiber, nylon-66, so named because it is synthesized from two different monomers, each containing six carbon atoms.

Polyamide A polymer in which each monomer unit is joined to the next by an amide bond, as for example nylon 66.

In the synthesis of nylon-66, hexanedioic acid and 1,6-hexanediamine are dissolved in aqueous ethanol, in which they react to form a one-to-one salt called nylon salt. This salt is then heated in an autoclave to 250°C and an internal pressure of 15 atm. Under these conditions, $-COO^-$ groups from the diacid and $-NH_3^+$ groups from diamine react by the loss of H_2O to form a polyamide. Nylon 66 formed under these conditions melts at 250 to 260°C and has a molecular weight ranging from 10,000 to 20,000 g/mol:

Hexanedioic acid (Adipic acid) + 1,6-Hexanediamine (Hexamethylenediamine)

↓

Nylon salt

heat & pressure ↓ $-H_2O$

Nylon 66

In the first stage of fiber production, crude nylon 66 is melted, spun into fibers, and cooled. Next, the melt-spun fibers are **cold drawn** (drawn at room temperature) to about four times their original length to increase their degree of crystallinity. As the fibers are drawn, individual polymer molecules become oriented in the direction of the fiber axis, and hydrogen bonds form between carbonyl oxygens of one chain and amide hydrogens of another chain (Figure 17.2). The effects of the orientation of polyamide molecules on the physical properties of the fiber are dramatic: Both tensile strength and stiffness are increased markedly. Cold drawing is an important step in the production of most synthetic fibers.

Figure 17.2
The structure of cold-drawn nylon 66. Hydrogen bonds between adjacent polymer chains provide additional tensile strength and stiffness to the fibers.

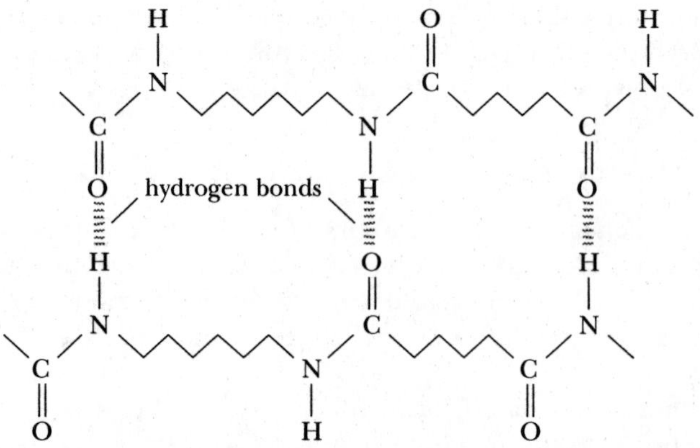

hydrogen bonds

The current raw-material base for the production of adipic acid is benzene, which is derived almost entirely from catalytic cracking and re-forming of petroleum (Section 3.11B). Catalytic reduction of benzene to cyclohexane (Section 9.2D), followed by catalyzed air oxidation, gives a mixture of cyclohexanol and cyclohexanone. Oxidation of this mixture by nitric acid gives adipic acid:

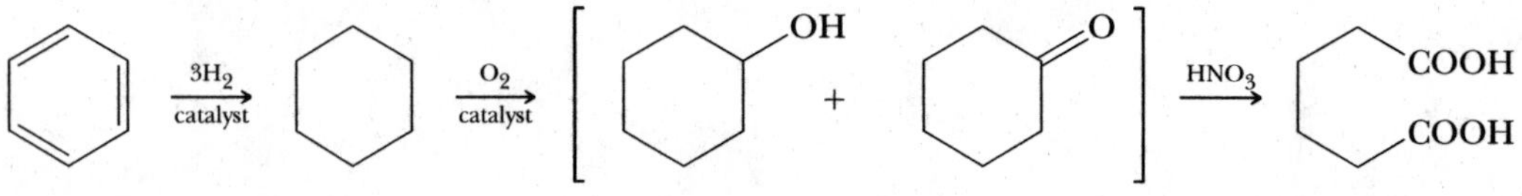

Benzene $\xrightarrow[\text{catalyst}]{3H_2}$ Cyclohexane $\xrightarrow[\text{catalyst}]{O_2}$ [Cyclohexanol + Cyclohexanone] $\xrightarrow{HNO_3}$ Hexanedioic acid (Adipic acid)

Adipic acid, in turn, is a starting material for the synthesis of hexamethylenediamine. Treating adipic acid with ammonia yields an ammonium salt that, when heated, gives adipamide. The catalytic reduction of adipamide then gives hexamethylenediamine:

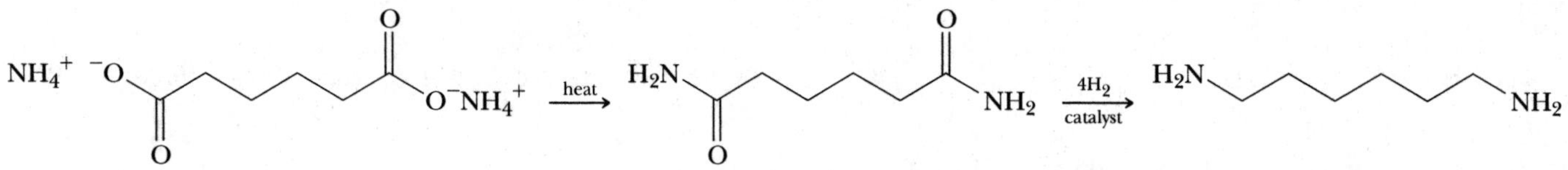

Ammonium hexanedioate (Ammonium adipate) $\xrightarrow{\text{heat}}$ Hexanediamide (Adipamide) $\xrightarrow[\text{catalyst}]{4H_2}$ 1,6-Hexanediamine (Hexamethylenediamine)

Note that carbon sources for the production of nylon 66 are derived entirely from petroleum, which, unfortunately, is not a renewable resource.

The nylons are a family of polymers, the members of which have subtly different properties that suit them to one use or another. The two most widely used members of the family are nylon 66 and nylon 6. Nylon 6 is so named because it is synthesized from caprolactam, a six-carbon monomer. In this synthesis, caprolactam is partially hydrolyzed to 6-aminohexanoic acid and then heated to 250°C to bring about polymerization:

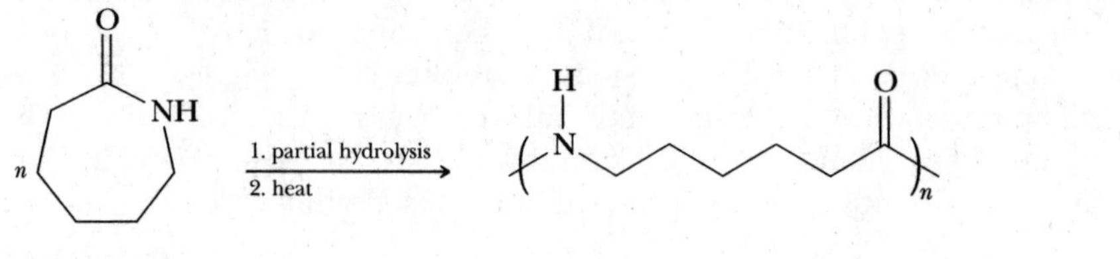

n Caprolactam $\xrightarrow[\text{2. heat}]{\text{1. partial hydrolysis}}$ Nylon 6

Nylon 6 is fabricated into fibers, bristles, rope, high-impact moldings, and tire cords.

Based on extensive research into the relationships between molecular structure and bulk physical properties, scientists at DuPont reasoned that a polyamide containing aromatic rings would be stiffer and stronger than either nylon 66 or nylon 6. In early 1960, DuPont introduced Kevlar, a polyaromatic amide **(aramid)** fiber synthesized from terephthalic acid and *p*-phenylenediamine:

Aramid A poly*aromatic amide*; a polymer in which the monomer units are an aromatic diamine and an aromatic dicarboxylic acid.

$$n\text{HOC(=O)}-C_6H_4-\text{C(=O)OH} + n\text{H}_2\text{N}-C_6H_4-\text{NH}_2 \longrightarrow \left(\text{C(=O)}-C_6H_4-\text{C(=O)NH}-C_6H_4-\text{NH}\right)_n + 2n\text{H}_2\text{O}$$

1,4-Benzenedicarboxylic acid (Terephthalic acid) 1,4-Benzenediamine (*p*-Phenylenediamine) Kevlar

One of the remarkable features of Kevlar is its light weight compared with that of other materials of similar strength. For example, a 7.6-cm (3-in.) cable woven of Kevlar has a strength equal to that of a similarly woven 7.6-cm (3-in.) steel cable. However, whereas the steel cable weighs about 30 kg/m (20 lb/ft), the Kevlar cable weighs only 6 kg/m (4 lb/ft). Kevlar now finds use in such articles as anchor cables for offshore drilling rigs and reinforcement fibers for automobile tires. Kevlar is also woven into a fabric that is so tough that it can be used for bulletproof vests, jackets, and raincoats.

Bulletproof vests have a thick layer of Kevlar. *(Charles D. Winters)*

B. Polyesters

The first **polyester**, developed in the 1940s, involved the polymerization of benzene 1,4-dicarboxylic acid (terephthalic acid) with 1,2-ethanediol (ethylene glycol) to give poly(ethylene terephthalate), abbreviated PET. Virtually all PET is now made from the dimethyl ester of terephthalic acid by the following transesterification reaction (Section 15.5C):

Polyester A polymer in which each monomer unit is joined to the next by an ester bond as, for example, poly(ethylene terephthalate).

$$CH_3O\text{C(=O)}-C_6H_4-\text{C(=O)OCH}_3 + \text{HOCH}_2\text{CH}_2\text{OH} \xrightarrow[-CH_3OH]{\text{heat}} \left(\text{C(=O)}-C_6H_4-\text{C(=O)OCH}_2\text{CH}_2\text{O}\right)_n$$

(remove CH_3OH from OCH_3 and HO)

Dimethyl terephthalate 1,2-Ethanediol (Ethylene glycol) Poly(ethylene terephthalate) (Dacron®, Mylar®)

The crude polyester can be melted, extruded, and then cold drawn to form the textile fiber Dacron® polyester, the outstanding features of which are its stiffness (about four times that of nylon 66), very high strength, and remarkable resistance to creasing and wrinkling. Because the early Dacron® polyester fibers were harsh to the touch, due to their stiffness, they were usually blended with cotton or wool to make acceptable textile fibers. Newly developed fabrication techniques now produce less harsh Dacron® polyester textile fibers. PET is also fabricated into Mylar® films and recyclable plastic beverage containers.

Ethylene glycol for the synthesis of PET is obtained by the air oxidation of ethylene to ethylene oxide (Section 8.5B), followed by hydrolysis to the glycol (Section 8.5C). Ethylene is, in turn, derived entirely from cracking either petroleum

Because Mylar film has very tiny pores, it is used for balloons that can be inflated with helium; the helium atoms diffuse only slowly through the pores of the film. *(Charles D. Winters)*

or ethane derived from natural gas (Section 3.11). Terephthalic acid is obtained by the oxidation of *p*-xylene (Section 9.5), an aromatic hydrocarbon obtained along with benzene and toluene from the catalytic cracking and re-forming of naphtha and other petroleum fractions (Section 3.11B):

$$\underset{\text{Ethylene}}{CH_2{=}CH_2} \xrightarrow[\text{catalyst}]{O_2} \underset{\text{Ethylene oxide}}{H_2C\overset{O}{—}CH_2} \xrightarrow{H^+,\ H_2O} \underset{\substack{\text{1,2-Ethanediol}\\ \text{(Ethylene glycol)}}}{HOCH_2CH_2OH}$$

$$\underset{p\text{-Xylene}}{H_3C\text{—}C_6H_4\text{—}CH_3} \xrightarrow[\text{catalyst}]{O_2} \underset{\text{Terephthalic acid}}{HO\overset{O}{\overset{\|}{C}}\text{—}C_6H_4\text{—}\overset{O}{\overset{\|}{C}}OH}$$

C. Polycarbonates

Polycarbonate A polyester in which the carboxyl groups are derived from carbonic acid.

Polycarbonates, the most familiar of which is Lexan®, are a class of commercially important engineering polyesters. Lexan® forms by the reaction between the disodium salt of bisphenol A (Problem 9.31) and phosgene:

remove Na^+Cl^-

$$\underset{\text{Disodium salt of bisphenol A}}{{}^+Na^-O\text{—}C_6H_4\text{—}C(CH_3)_2\text{—}C_6H_4\text{—}O^-Na^+} + \underset{\text{Phosgene}}{Cl\overset{O}{\overset{\|}{C}}Cl} \longrightarrow \underset{\substack{\text{Lexan®}\\ \text{(a polycarbonate)}}}{\left(C_6H_4\text{—}C(CH_3)_2\text{—}C_6H_4\text{—}O\overset{O}{\overset{\|}{C}}O\right)_n} + NaCl$$

Note that phosgene is the diacid chloride (Section 15.2A) of carbonic acid; hydrolysis of phosgene gives H_2CO_3 and 2HCl.

Lexan® is a tough, transparent polymer with high impact and tensile strengths that retains its properties over a wide temperature range. It is used in sporting equipment (for helmets and face masks), to make light, impact-resistant housings for household appliances, and in the manufacture of safety glass and unbreakable windows.

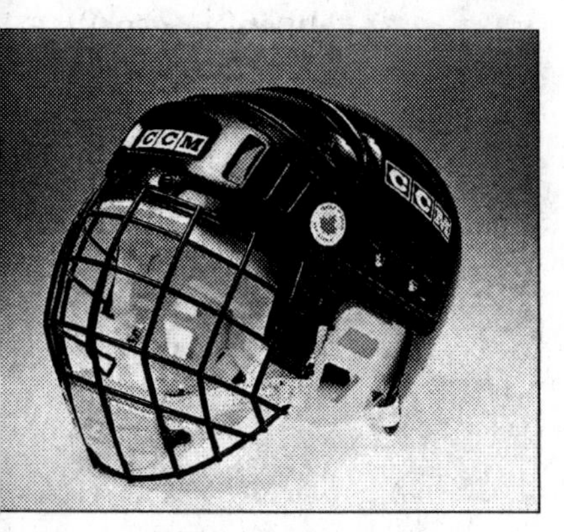

A polycarbonate hockey mask. *(Charles D. Winters)*

D. Polyurethanes

A urethane, or carbamate, is an ester of carbamic acid, H_2NCOOH. Carbamates are most commonly prepared by treating an isocyanate with an alcohol. In this reaction, the H and OR′ of the alcohol add to the C=N bond in a reaction comparable to the addition of an alcohol to a C=O bond:

$$\underset{\text{An isocyanate}}{RN{=}C{=}O} + R'OH \longrightarrow \underset{\text{A carbamate}}{RNH\overset{O}{\overset{\|}{C}}OR'}$$

Polyurethanes consist of flexible polyester or polyether units (blocks) alternating with rigid urethane units (blocks) derived from a diisocyanate, commonly a mixture of 2,4- and 2,6-toluene diisocyanate:

Polyurethane A polymer containing the —NHCOO— group as a repeating unit.

2,6-Toluene diisocyanate + nHO-polymer-OH ⟶ A polyurethane

Low-molecular-weight polyester or polyether with OH groups at each end of the chain

The more flexible blocks are derived from low-molecular-weight (MW 1,000 to 4,000) polyesters or polyethers with —OH groups at each end of their chains. Polyurethane fibers are fairly soft and elastic and have found use as spandex and Lycra®, the "stretch" fabrics used in bathing suits, leotards, and undergarments.

Polyurethane foams for upholstery and insulating materials are made by adding small amounts of water during polymerization. Water reacts with isocyanate groups to form a carbamic acid that undergoes spontaneous decarboxylation to produce gaseous carbon dioxide, which then acts as the foaming agent:

$$RN{=}C{=}O + H_2O \longrightarrow \left[RNH{-}\overset{O}{\overset{\|}{C}}{-}OH \right] \longrightarrow RNH_2 + CO_2$$

An isocyanate — A carbamic acid (unstable)

Epoxy resin A material prepared by a polymerization in which one monomer contains at least two epoxy groups.

E. Epoxy Resins

Epoxy resins are materials prepared by a polymerization in which one monomer contains at least two epoxy groups. Within this range, a large number of polymeric materials are possible, and epoxy resins are produced in forms ranging from low-viscosity liquids to high-melting solids. The most widely used epoxide monomer is the diepoxide prepared by treating 1 mole of bisphenol A (Problem 9.31) with 2 moles of epichlorohydrin:

Epichlorohydrin + Disodium salt of Bisphenol A + Epichlorohydrin ⟶ A diepoxide + 2NaCl

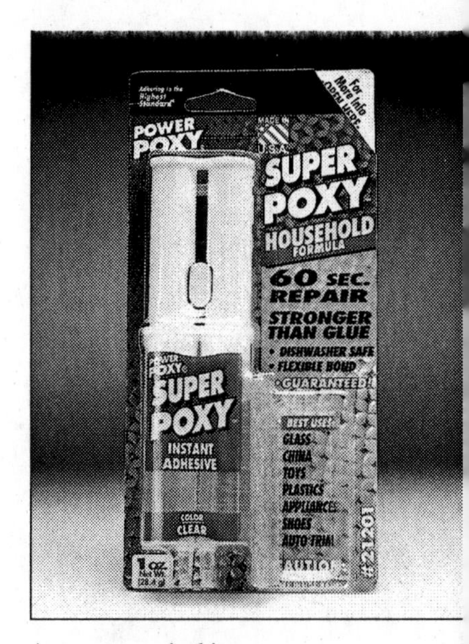

An epoxy resin kit. *(Charles D. Winters)*

To prepare the following epoxy resin, the diepoxide monomer is treated with 1,2-ethanediamine (ethylene diamine):

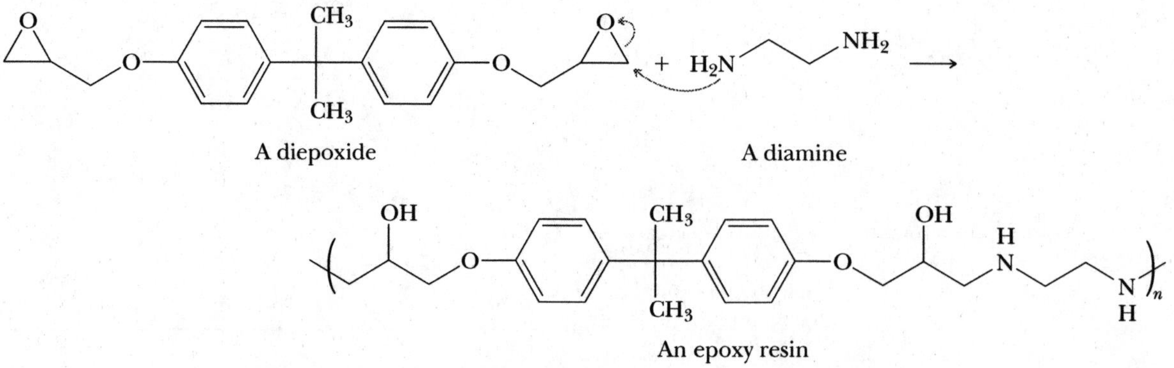

A diepoxide

A diamine

An epoxy resin

Ethylene diamine is usually called the catalyst in the two-component formulations that you buy in hardware or craft stores; it is also the component with the acrid smell. The preceding reaction corresponds to nucleophilic opening of the highly strained three-membered epoxide ring (Section 8.5C).

Epoxy resins are widely used as adhesives and insulating surface coatings. They have good electrical insulating properties, which lead to their use in encapsulating electrical components ranging from integrated circuit boards to switch coils and insulators for power transmission systems. Epoxy resins are also used as composites with other materials, such as glass fiber, paper, metal foils, and other synthetic fibers, to create structural components for jet aircraft, rocket motor casings, and so on.

EXAMPLE 17.2

By what type of mechanism does the reaction between the disodium salt of bisphenol A and epichlorohydrin take place?

SOLUTION

The mechanism is an S_N2 mechanism. The phenoxide ion of bisphenol A is a good nucleophile, and chlorine on the primary carbon of epichlorohydrin is the leaving group.

Practice Problem 17.2

Write the repeating unit of the epoxy resin formed from the following reaction:

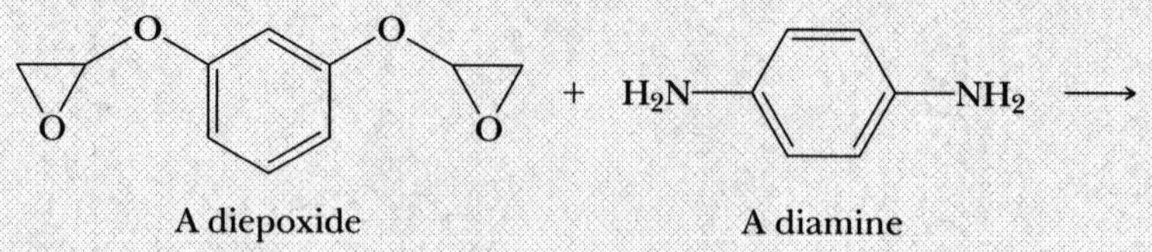

A diepoxide

A diamine

17.6 CHAIN-GROWTH POLYMERS

Chain-growth polymerization A polymerization that involves sequential addition reactions, either to unsaturated monomers or to monomers possessing other reactive functional groups.

From the perspective of the chemical industry, the single most important reaction of alkenes is **chain-growth polymerization**, a type of polymerization in which monomer units are joined together without the loss of atoms. An example is the formation of polyethylene from ethylene:

$$nCH_2{=}CH_2 \xrightarrow{\text{catalyst}} \text{-}(CH_2CH_2)\text{-}_n$$

Ethylene   Polyethylene

CHEMICAL CONNECTIONS 17A

Stitches That Dissolve

As the technological capabilities of medicine have grown, the demand for synthetic materials that can be used inside the body has increased as well. Polymers have many of the characteristics of an ideal biomaterial: They are lightweight and strong, are inert or biodegradable (depending on their chemical structure), and have physical properties (softness, rigidity, elasticity) that are easily tailored to match those of natural tissues. Carbon–carbon backbone polymers are resistant to degradation and are used widely in permanent organ and tissue replacements.

Even though most medical uses of polymeric materials require biostability, applications have been developed that use the biodegradable nature of some macromolecules. An example is the use of glycolic acid/lactic acid copolymers as absorbable sutures:

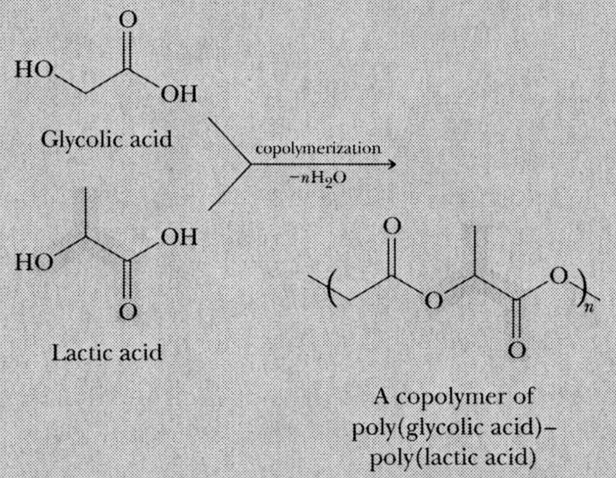

Glycolic acid + Lactic acid $\xrightarrow[-nH_2O]{\text{copolymerization}}$ A copolymer of poly(glycolic acid)–poly(lactic acid)

Traditional suture materials such as catgut must be removed by a health-care specialist after they have served their purpose. Stitches of these hydroxyester polymers, however, are hydrolyzed slowly over a period of approximately two weeks, and by the time the torn tissues have fully healed, the stitches are fully degraded and the sutures need not be removed. Glycolic and lactic acids formed during hydrolysis of the stitches are metabolized and excreted by existing biochemical pathways.

The mechanisms of chain-growth polymerization differ greatly from the mechanism of step-growth polymerizations. In the latter, all monomers plus the polymer end groups possess equally reactive functional groups, allowing for all possible combinations of reactions to occur, including monomer with monomer, dimer with dimer, monomer with tetramer, and so forth. In contrast, chain-growth polymerizations involve end groups possessing reactive intermediates that react only with a monomer. The reactive intermediates used in chain-growth polymerizations include radicals, carbanions, carbocations, and organometallic complexes.

The number of monomers that undergo chain-growth polymerization is large and includes such compounds as alkenes, alkynes, allenes, isocyanates, and cyclic compounds such as lactones, lactams, ethers, and epoxides. We concentrate on the chain-growth polymerizations of ethylene and substituted ethylenes and show how these compounds can be polymerized by radical and organometallic-mediated mechanisms.

Table 17.1 lists several important polymers derived from ethylene and substituted ethylenes, along with their common names and most important uses.

A. Radical Chain-Growth Polymerization

The first commercial polymerizations of ethylene were initiated by radicals formed by thermal decomposition of organic peroxides, such as benzoyl peroxide. A **radical** is any molecule that contains one or more unpaired electrons. Radicals can be

Radical Any molecule that contains one or more unpaired electrons.

CHEMICAL CONNECTIONS 17B

Paper or Plastic?

Any audiophile will tell you that the quality of any sound system is highly dependent upon its speakers. Speakers create sound by moving a diaphragm in and out to displace air. Most diaphragms are in the shape of a cone, traditionally made of paper. Paper cones are inexpensive, lightweight, rigid, and nonresonant. One disadvantage is their susceptibility to damage by water and humidity. Over time and with exposure, paper cones become weakened, losing their fidelity of sound. Many of the speakers that are available today are made of polypropylene, which is also inexpensive, lightweight, rigid, and nonresonant. Furthermore, not only are polypropylene cones immune to water and humidity, but also, their performance is less influenced by heat or cold. Moreover, their added strength makes them less prone to splitting than paper. They last longer and can be displaced more frequently and for longer distances, creating deeper bass notes and higher high notes.

(Photo Courtesy of Crutchfield.com)

TABLE 17.1 Polymers Derived from Ethylene and Substituted Ethylenes

Monomer Formula	Common Name	Polymer Name(s) and Common Uses
$CH_2{=}CH_2$	ethylene	polyethylene, Polythene; break-resistant containers and packaging materials
$CH_2{=}CHCH_3$	propylene	polypropylene, Herculon; textile and carpet fibers
$CH_2{=}CHCl$	vinyl chloride	poly(vinyl chloride), PVC; construction tubing
$CH_2{=}CCl_2$	1,1-dichloroethylene	poly(1,1-dichloroethylene); Saran Wrap® is a copolymer with vinyl chloride
$CH_2{=}CHCN$	acrylonitrile	polyacrylonitrile, Orlon®; acrylics and acrylates
$CF_2{=}CF_2$	tetrafluoroethylene	polytetrafluoroethylene, PTFE; Teflon®, nonstick coatings
$CH_2{=}CHC_6H_5$	styrene	polystyrene, Styrofoam™; insulating materials
$CH_2{=}CHCOOCH_2CH_3$	ethyl acrylate	poly(ethyl acrylate); latex paints
$CH_2{=}C(CH_3)COOCH_3$	methyl methacrylate	poly(methyl methacrylate), Lucite®, Plexiglas®; glass substitutes

formed by the cleavage of a bond in such a way that each atom or fragment participating in the bond retains one electron. In the following equation, **fishhook arrows** are used to show the change in position of single electrons:

Fishhook arrow A single-barbed, curved arrow used to show the change in position of a single electron.

Benzoyl peroxide $\xrightarrow{\text{heat}}$ Benzoyloxy radicals

Radical polymerization of ethylene and substituted ethylenes involves three steps: (1) chain initiation, (2) chain propagation, and (3) chain termination. We show these steps here and then discuss each separately in turn.

Mechanism: Radical Polymerization of Ethylene

Step 1: Chain initiation—formation of radicals from nonradical compounds:

$$\text{In—In} \xrightarrow[\text{or light}]{\text{heat}} 2\text{In}\cdot$$

In this equation, In-In represents an initiator which, when heated or irradiated with radiation of a suitable wavelength, cleaves to give two radicals (In·).

Chain initiation In radical polymerization, the formation of radicals from molecules containing only paired electrons.

Step 2: Chain propagation—reaction of a radical and a molecule to form a new radical:

Chain propagation In radical polymerization, a reaction of a radical and a molecule to give a new radical.

Step 3: Chain termination—destruction of radicals:

radical coupling

Chain termination In radical polymerization, a reaction in which two radicals combine to form a covalent bond.

The characteristic feature of a chain-initiation step is the formation of radicals from a molecule with only paired electrons. In the case of peroxide-initiated polymerizations of alkenes, chain initiation is by (1) heat cleavage of the O—O bond of a peroxide to give two alkoxy radicals and (2) reaction of an alkoxy radical with a molecule of alkene to give an alkyl radical. In the general mechanism shown, the initiating catalyst is given the symbol In-In and its radical is given the symbol In·.

The structure and geometry of carbon radicals are similar to those of alkyl carbocations. They are planar or nearly so, with bond angles of approximately 120° about the carbon with the unpaired electron. The relative stabilities of alkyl radicals are similar to those of alkyl carbocations:

methyl < 1° < 2° < 3°

Increasing stability of alkyl radicals

The characteristic feature of a chain-propagation step is the reaction of a radical and a molecule to give a new radical. Propagation steps repeat over and over (propagate), with the radical formed in one step reacting with a monomer to produce a new radical, and so on. The number of times a cycle of chain-propagation steps repeats is called the **chain length** and is given the symbol ***n***. In the polymerization of ethylene, chain-lengthening reactions occur at a very high rate, often as fast as thousands of additions per second, depending on the experimental conditions.

Radical polymerizations of substituted ethylenes almost always give the more stable (more substituted) radical. Because additions are biased in this fashion, the polymerizations of substituted ethylene monomers tend to yield polymers with monomer units joined by the head (carbon 1) of one unit to the tail (carbon 2) of the next unit:

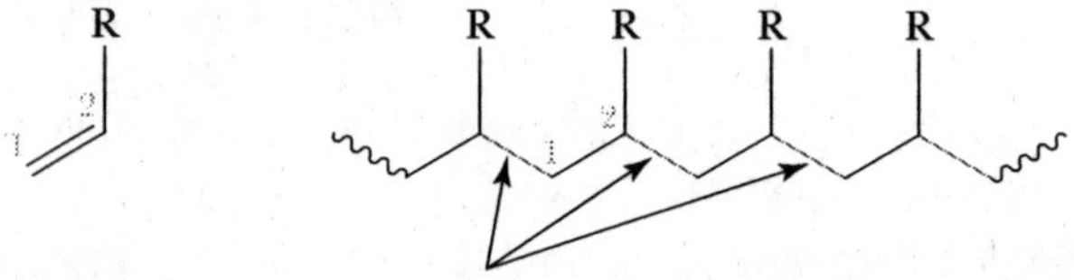

Substituted ethylene monomer   head-to-tail linkages

In principle, chain-propagation steps can continue until all starting materials are consumed. In practice, they continue only until two radicals react with each other to terminate the process. The characteristic feature of a chain-termination step is the destruction of radicals. In the mechanism shown for radical polymerization of the substituted ethylene, chain termination occurs by the coupling of two radicals to form a new carbon–carbon single bond.

The first commercial process for ethylene polymerization used peroxide catalysts at temperatures of 500°C and pressures of 1,000 atm and produced a soft, tough polymer known as low-density polyethylene (LDPE) with a density of between 0.91 and 0.94 g/cm^3 and a melt transition temperature (T_m) of about 115°C. Because LDPE's melting point is only slightly above 100°C, it cannot be used for products that will be exposed to boiling water. At the molecular level, chains of LDPE are highly branched.

The branching on chains of low-density polyethylene results from a "back-biting" reaction in which the radical end group abstracts a hydrogen from the fourth carbon back (the fifth carbon in the chain). Abstraction of this hydrogen is particularly facile because the transition state associated with the process can adopt a conformation like that of a chair cyclohexane. In addition, the less stable 1° radical is converted to a more stable 2° radical. This side reaction is called a **chain-transfer reaction**, because the activity of the end group is "transferred" from one chain to another. Continued polymerization of monomer from this new radical center leads to a branch four carbons long:

Chain-transfer reaction In radical polymerization, the transfer of reactivity of an end group from one chain to another during a polymerization.

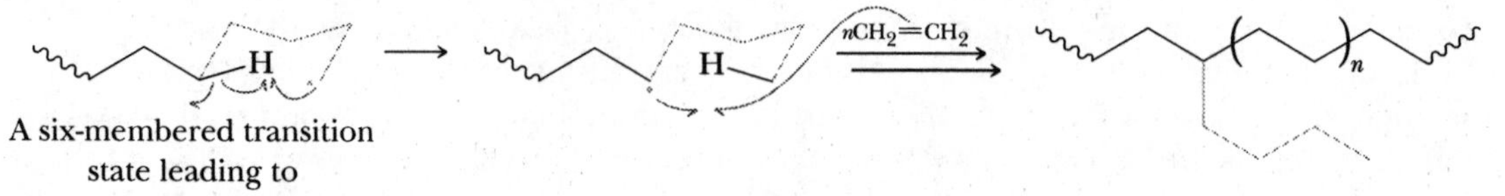

$nCH_2{=}CH_2$

A six-membered transition state leading to 1,5-hydrogen abstraction

Approximately 65% of all LDPE is used for the manufacture of films by a blow-molding technique illustrated in Figure 17.3. LDPE film is inexpensive, which makes it ideal for packaging such consumer items as baked goods, vegetables and other produce and for trash bags.

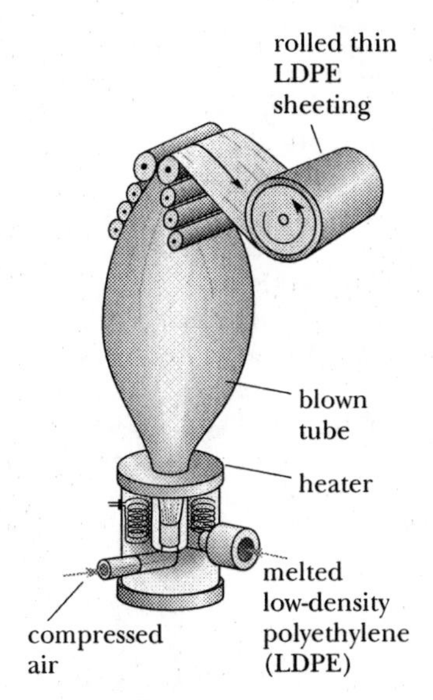

Figure 17.3
Fabrication of an LDPE film. A tube of melted LDPE along with a jet of compressed air is forced through an opening and blown into a giant, thin-walled bubble. The film is then cooled and taken up onto a roller. This double-walled film can be slit down the side to give LDPE film, or it can be sealed at points along its length to make LDPE bags.

B. Ziegler–Natta Chain-Growth Polymerization

In the 1950s, Karl Ziegler of Germany and Giulio Natta of Italy developed an alternative method for the polymerization of alkenes, work for which they shared the Nobel prize in chemistry in 1963. The early Ziegler–Natta catalysts were highly active, heterogeneous materials composed of an $MgCl_2$ support, a Group 4B transition metal halide such as $TiCl_4$, and an alkylaluminum compound—for example, diethylaluminum chloride, $Al(CH_2CH_3)_2Cl$. These catalysts bring about the polymerization of ethylene and propylene at 1–4 atm and at temperatures as low as 60°C.

The catalyst in a Ziegler–Natta polymerization is an alkyltitanium compound formed by reaction between $Al(CH_2CH_3)_2Cl$ and the titanium halide on the surface of a $MgCl_2/TiCl_4$ particle. Once formed, this alkyltitanium species repeatedly inserts ethylene units into the titanium–carbon bond to yield polyethylene.

Mechanism: Ziegler–Natta Catalysis of Ethylene Polymerization

Step 1: Formation of a titanium–ethyl bond:

$$\text{Ti—Cl} + Al(CH_2CH_3)_2Cl \longrightarrow \text{Ti—}CH_2CH_3 + Al(CH_2CH_3)Cl_2$$

Step 2: Insertion of ethylene into the titanium–carbon bond:

$$\text{Ti—}CH_2CH_3 + CH_2{=}CH_2 \longrightarrow \text{Ti—}CH_2CH_2CH_2CH_3$$

Over 60 billion pounds of polyethylene are produced worldwide every year with Ziegler–Natta catalysts. Polyethylene from Ziegler–Natta systems, termed **high-density polyethylene (HDPE)**, has a higher density (0.96 g/cm^3) and melt transition temperature (133°C) than low-density polyethylene, is 3 to 10 times stronger, and is opaque rather than transparent. The added strength and opacity are due to a much lower degree of chain branching and a resulting higher degree of crystallinity of HDPE compared with LDPE. Approximately 45% of all HDPE used in the United States is blow molded (Figure 17.4).

Polyethylene films are produced by extruding the molten plastic through a ring-like gap and inflating the film into a balloon. *(Brownie Harris/Corbis)*

Even greater improvements in properties of HDPE can be realized through special processing techniques. In the melt state, HDPE chains have random coiled conformations similar to those of cooked spaghetti. Engineers have developed extrusion techniques that force the individual polymer chains of HDPE to uncoil into linear conformations. These linear chains then align with one another to form highly crystalline materials. HDPE processed in this fashion is stiffer than steel and has approximately four times its tensile strength! Because the density of polyethylene (≈1.0 g/cm^3) is considerably less than that of steel (8.0 g/cm^3), these comparisons of strength and stiffness are even more favorable if they are made on a weight basis.

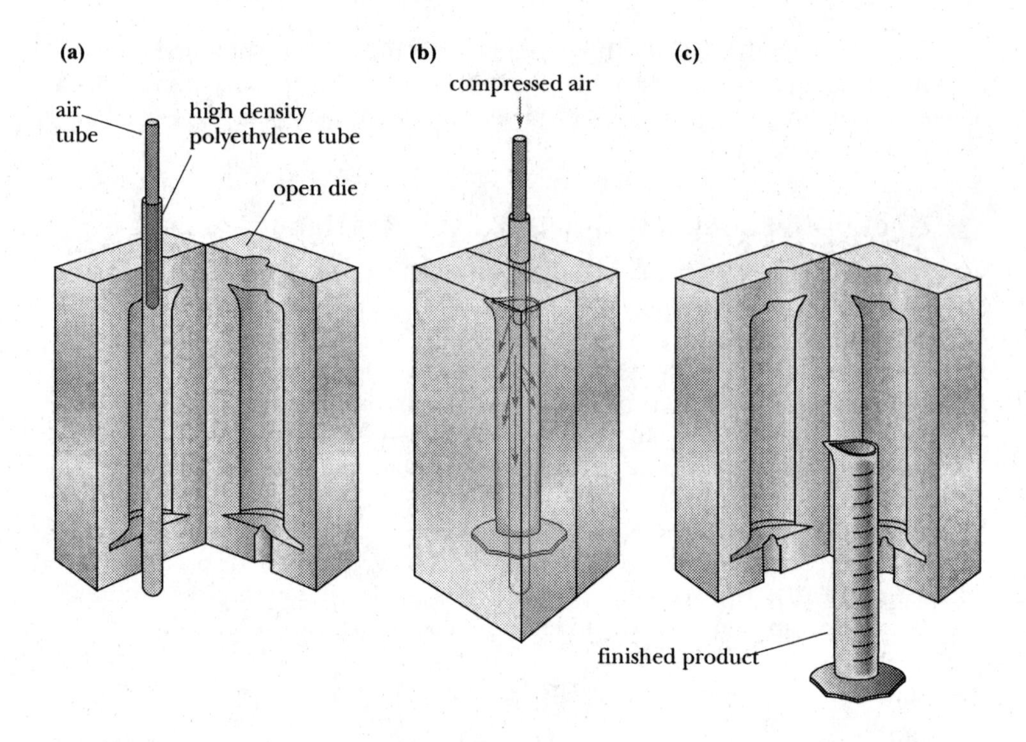

Figure 17.4
Blow molding of an HDPE container. (a) A short length of HDPE tubing is placed in an open die, and the die is closed, sealing the bottom of the tube. (b) Compressed air is forced into the hot polyethylene–die assembly, and the tubing is literally blown up to take the shape of the mold. (c) After cooling, the die is opened, and there is the container!

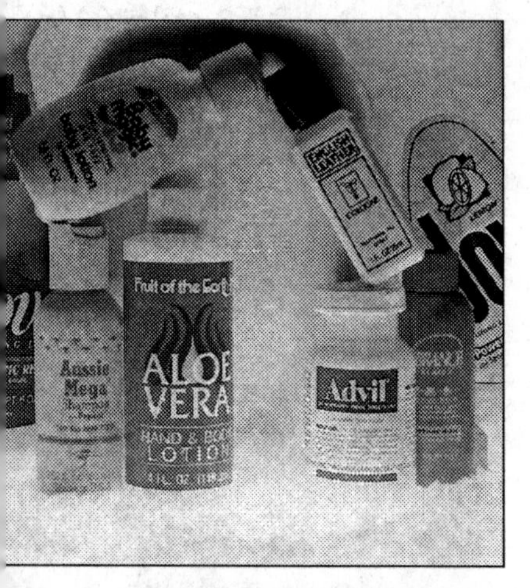

Some common products packaged in high-density-polyethylene containers. *(Charles D. Winters)*

17.7 RECYCLING PLASTICS

Polymers in the form of plastics are materials upon which our society is incredibly dependent. Durable and lightweight, plastics are probably the most versatile synthetic materials in existence; in fact, their current production in the United States exceeds that of steel. Plastics have come under criticism, however, for their role in the current trash crisis. They make up 21% of the volume and 8% of the weight of solid waste, most of which is derived from disposable packaging and wrapping. Of the 1.5×10^8 kg of thermoplastic materials produced in the United States per year, less than 2% is recycled.

If the durability and chemical inertness of most plastics make them ideally suited for reuse, why aren't more plastics being recycled? The answer to this question has more to do with economics and consumer habits than with technological obstacles. Because curbside pickup and centralized drop-off stations for recyclables are just now becoming common, the amount of used material available for reprocessing has traditionally been small. This limitation, combined with the need for an additional sorting and separation step, rendered the use of recycled plastics in manufacturing expensive compared with virgin materials. The increase in environmental awareness over the last decade, however, has resulted in a greater demand for recycled products. As manufacturers adapt to satisfy this new market, the recycling of plastics will eventually catch up with that of other materials, such as glass and aluminum.

Six types of plastics are commonly used for packaging applications. In 1988, manufacturers adopted recycling code numbers developed by the Society of the Plastics Industry (Table 17.2). Because the plastics recycling industry still is not fully developed, only PET and HDPE are currently being recycled in large quantities. LDPE, which accounts for about 40% of plastic trash, has been slow in finding acceptance with recyclers. Facilities for the reprocessing of poly(vinyl chloride) (PVC), polypropylene (PP), and polystyrene (PS) exist, but are rare.

The process for the recycling of most plastics is simple, with separation of the desired plastics from other contaminants the most labor-intensive step. For example, PET soft-drink bottles usually have a paper label and adhesive that must be removed

TABLE 17.2 Recycling Codes for Plastics

Recycling Code	Polymer	Common Uses	Uses of Recycled Polymer
1 PET	poly(ethylene terephthalate)	soft-drink bottles, household chemical bottles, films, textile fibers	soft-drink bottles, household chemical bottles, films, textile fibers
2 HDPE	high-density polyethylene	milk and water jugs, grocery bags, bottles	bottles, molded containers
3 V	poly(vinyl chloride), PVC	shampoo bottles, pipes, shower curtains, vinyl siding, wire insulation, floor tiles, credit cards	plastic floor mats
4 LDPE	low-density polyethylene	shrink wrap, trash and grocery bags, sandwich bags, squeeze bottles	trash bags and grocery bags
5 PP	polypropylene	plastic lids, clothing fibers, bottle caps, toys, diaper linings	mixed-plastic components
6 PS	polystyrene	Styrofoam™ cups, egg cartons, disposable utensils, packaging materials, appliances	molded items such as cafeteria trays, rulers, Frisbees™, trash cans, videocasettes
7	all other plastics and mixed plastics	various	plastic lumber, playground equipment, road reflectors

before the PET can be reused. The recycling process begins with hand or machine sorting, after which the bottles are shredded into small chips. An air cyclone then removes paper and other lightweight materials. Any remaining labels and adhesives are eliminated with a detergent wash, and the PET chips are then dried. PET produced by this method is 99.9% free of contaminants and sells for about half the price of the virgin material. Unfortunately, plastics with similar densities cannot be separated with this technology, nor can plastics composed of several polymers be broken down into pure components. However, recycled mixed plastics can be molded into plastic lumber that is strong, durable, and resistant to graffiti.

An alternative to the foregoing process, which uses only physical methods of purification, is chemical recycling. Eastman Kodak salvages large amounts of its PET film scrap by a transesterification reaction. The scrap is treated with methanol in the presence of an acid catalyst to give ethylene glycol and dimethyl terephthalate, monomers that are purified by distillation or recrystallization and used as feedstocks for the production of more PET film:

$$\text{Poly(ethylene terephthalate) (PET)} \xrightarrow[H^+]{CH_3OH} \underset{\text{Ethylene glycol}}{HO\!-\!CH_2CH_2\!-\!OH} + \underset{\text{Dimethyl terephthalate}}{H_3CO\!-\!C(=O)\!-\!C_6H_4\!-\!C(=O)\!-\!OCH_3}$$

SUMMARY

Polymerization is the process of joining together many small **monomers** into large, high-molecular-weight **polymers** (Section 17.1). The properties of polymeric materials depend on the structure of the repeat unit, as well as on the chain architecture and morphology of the material (Section 17.4).

Step-growth polymerizations involve the stepwise reaction of difunctional monomers (Section 17.5). Important commercial polymers synthesized through step-growth processes include polyamides, polyesters, polycarbonates, polyurethanes, and epoxy resins.

Chain-growth polymerization proceeds by the sequential addition of monomer units to an active chain end group (Section 17.6). **Radical chain-growth polymerization** (Section 17.6A) consists of three stages: chain initiation, chain propagation, and chain termination. In **chain initiation**, radicals are formed from nonradical molecules. In **chain propagation**, a radical and a monomer react to give a new radical. The **chain length** is the number of times a cycle of chain-propagation steps repeats. In **chain termination**, radicals are destroyed. Alkyl radicals are planar or almost so, with bond angles of 120° about the carbon with the unpaired electron. **Ziegler–Natta chain-growth polymerizations** involve the formation of an alkyl-transition metal compound and then the repeated insertion of alkene monomers into the transition metal-to-carbon bond to yield a saturated polymer chain (Section 17.6B).

KEY REACTIONS

1. Step-growth polymerization of a dicarboxylic acid and a diamine gives a polyamide (Section 17.5A)

In this equation, M and M′ indicate the remainder of each monomer unit:

$$HO\overset{O}{\overset{\|}{C}}-M-\overset{O}{\overset{\|}{C}}OH + H_2N-M'-NH_2 \xrightarrow{\text{heat}} \left(\overset{O}{\overset{\|}{C}}-M-\overset{O}{\overset{\|}{C}}-\underset{H}{N}-M'-\underset{H}{N} \right)_n + 2nH_2O$$

2. Step-growth polymerization of a dicarboxylic acid and a diol gives a polyester (Section 17.4B)

$$HO\overset{O}{\overset{\|}{C}}-M-\overset{O}{\overset{\|}{C}}OH + HO-M'-OH \xrightarrow[\text{catalyst}]{\text{acid}} \left(\overset{O}{\overset{\|}{C}}-M-\overset{O}{\overset{\|}{C}}-O-M'-O \right)_n + 2nH_2O$$

3. Step-growth polymerization of a phosgene and a diol gives a polycarbonate (Section 17.5C)

$$Cl-\overset{O}{\overset{\|}{C}}-Cl + HO-M-OH \longrightarrow \left(O-\overset{O}{\overset{\|}{C}}-O-M \right)_n + 2nHCl$$

4. Step-growth polymerization of a diisocyanate and a diol gives a polyurethane (Section 17.5D)

$$O=C=N-M-N=C=O + HO-M'-OH \longrightarrow \left(\overset{O}{\overset{\|}{C}}-\underset{H}{N}-M-\underset{H}{N}-\overset{O}{\overset{\|}{C}}-O-M'-O \right)_n$$

5. Step-growth polymerization of a diepoxide and a diamine gives an epoxy resin (Section 17.5E)

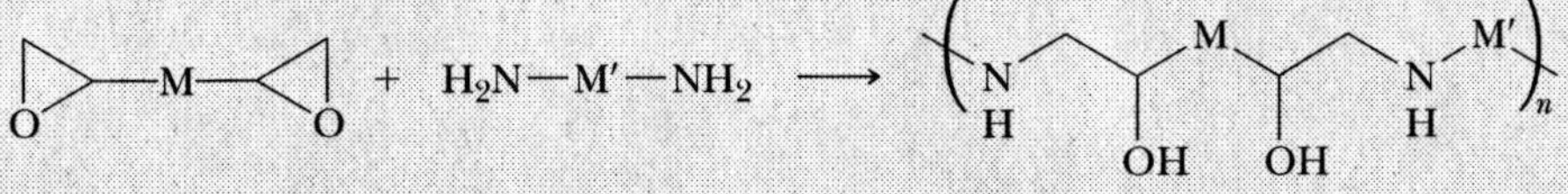

$$\text{diepoxide (O-ring–M–O-ring)} + H_2N{-}M'{-}NH_2 \longrightarrow \left(N(H){-}CH_2{-}CH(OH){-}M{-}CH(OH){-}CH_2{-}N(H){-}M' \right)_n$$

6. Radical chain-growth polymerization of ethylene and substituted ethylenes (Section 17.6A)

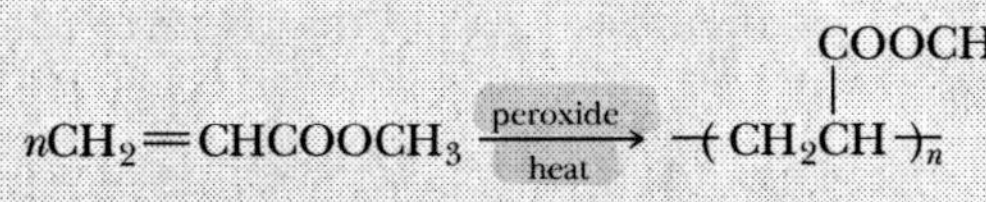

$$nCH_2{=}CHCOOCH_3 \xrightarrow[\text{heat}]{\text{peroxide}} \left(CH_2CH(COOCH_3) \right)_n$$

7. Ziegler–Natta chain-growth polymerization of ethylene and substituted ethylenes (Section 17.6B)

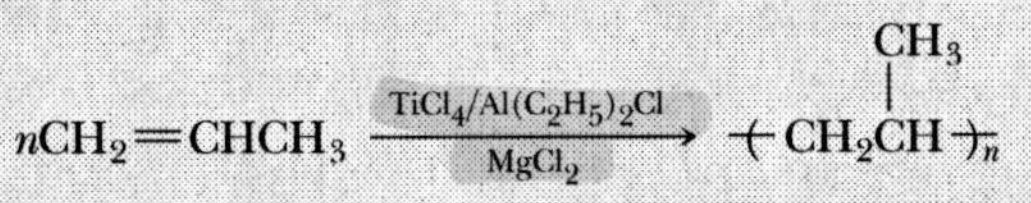

$$nCH_2{=}CHCH_3 \xrightarrow[\text{MgCl}_2]{\text{TiCl}_4/\text{Al}(C_2H_5)_2\text{Cl}} \left(CH_2CH(CH_3) \right)_n$$

PROBLEMS

A problem number set in red indicates an applied "real-world" problem.

Step-Growth Polymers

17.3 Identify the monomers required for the synthesis of each step-growth polymer:

(a) $\left(C(=O){-}C_6H_4{-}C(=O){-}OCH_2{-}C_6H_{10}{-}CH_2O \right)_n$

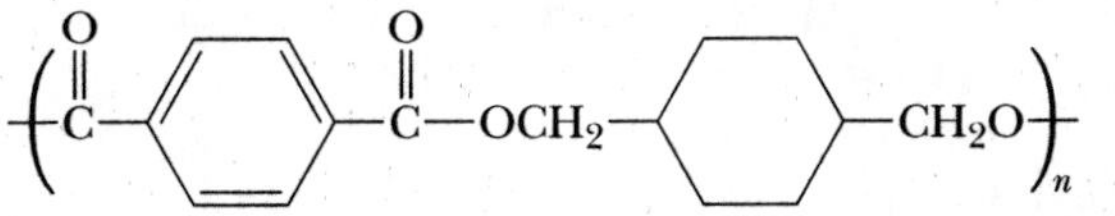

Kodel™
(a polyester)

(b) $\left(C(=O)(CH_2)_6C(=O){-}NH{-}C_6H_{10}{-}CH_2{-}C_6H_{10}{-}NH \right)_n$

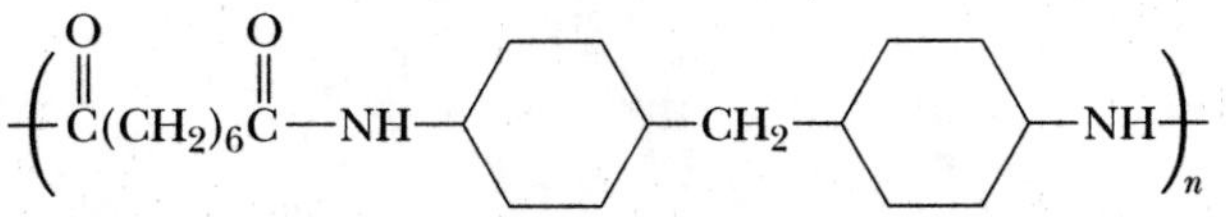

Quiana™
(a polyamide)

(c) $\left(O{-}(CH_2)_6{-}O{-}C(=O){-}C_6H_4{-}C(=O) \right)_n$

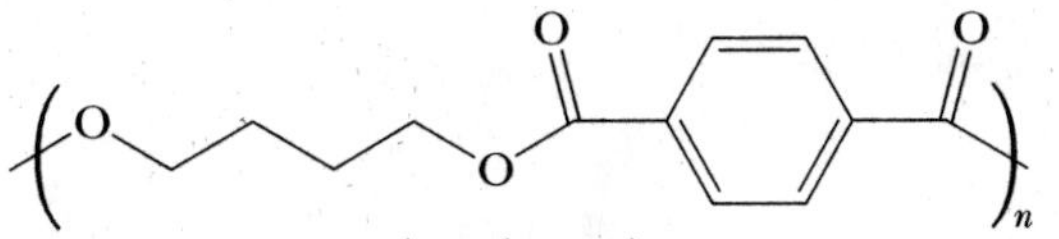

(a polyester)

(d) $\left(C(=O){-}(CH_2)_8{-}C(=O){-}NH{-}(CH_2)_6{-}NH \right)_n$

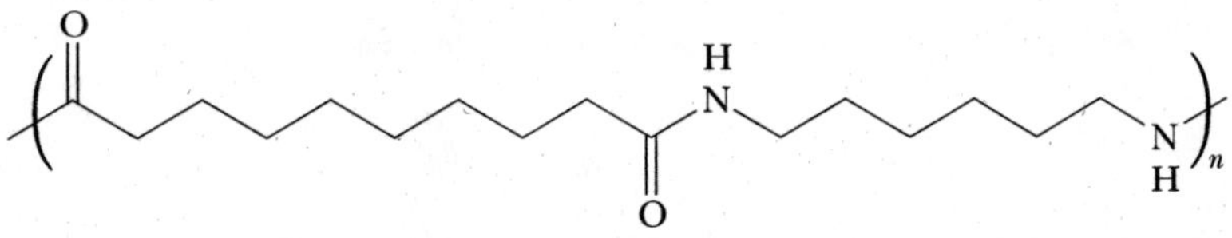

Nylon 6,10
(a polyamide)

17.4 Poly(ethylene terephthalate) (PET) can be prepared by the following reaction:

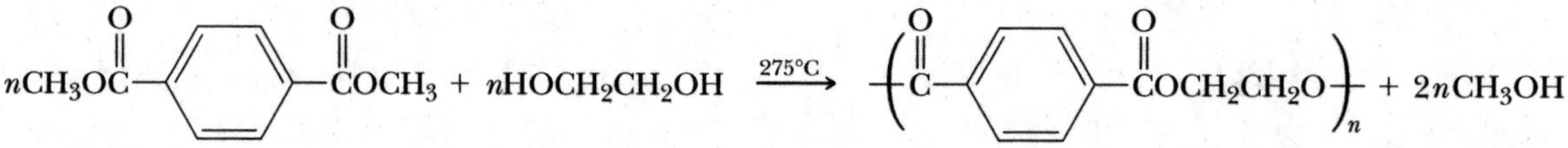

$$n\text{CH}_3\text{O}\overset{\text{O}}{\overset{\|}{\text{C}}}-\text{C}_6\text{H}_4-\overset{\text{O}}{\overset{\|}{\text{C}}}\text{OCH}_3 + n\text{HOCH}_2\text{CH}_2\text{OH} \xrightarrow{275°\text{C}} \left(\overset{\text{O}}{\overset{\|}{\text{C}}}-\text{C}_6\text{H}_4-\overset{\text{O}}{\overset{\|}{\text{C}}}\text{OCH}_2\text{CH}_2\text{O}\right)_n + 2n\text{CH}_3\text{OH}$$

Dimethyl terephthalate | Ethylene glycol | Poly(ethylene terephthalate) | Methanol

Propose a mechanism for the step-growth reaction in this polymerization.

17.5 Currently, about 30% of PET soft-drink bottles are being recycled. In one recycling process, scrap PET is heated with methanol in the presence of an acid catalyst. The methanol reacts with the polymer, liberating ethylene glycol and dimethyl terephthalate. These monomers are then used as feedstock for the production of new PET products. Write an equation for the reaction of PET with methanol to give ethylene glycol and dimethyl terephthalate.

17.6 Nomex® is an aromatic polyamide (aramid) prepared from the polymerization of 1,3-benzenediamine and the acid chloride of 1,3-benzenedicarboxylic acid:

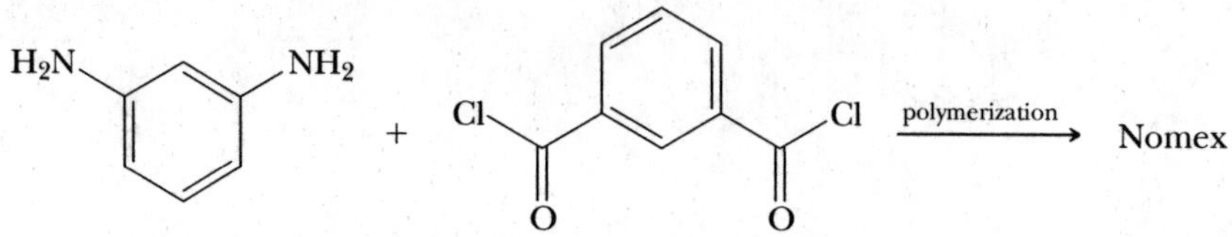

$$\text{H}_2\text{N}-\text{C}_6\text{H}_4-\text{NH}_2 + \text{Cl}\overset{\text{O}}{\overset{\|}{\text{C}}}-\text{C}_6\text{H}_4-\overset{\text{O}}{\overset{\|}{\text{C}}}\text{Cl} \xrightarrow{\text{polymerization}} \text{Nomex}$$

1,3-Benzenediamine | 1,3-Benzenedicarbonyl chloride

The physical properties of the polymer make it suitable for high-strength, high-temperature applications such as parachute cords and jet aircraft tires. Draw a structural formula for the repeating unit of Nomex.

17.7 Nylon 6,10 [Problem 17.3(d)] can be prepared by reacting a diamine and a diacid chloride. Draw the structural formula of each reactant.

Chain-Growth Polymerization

17.8 Following is the structural formula of a section of polypropylene derived from three units of propylene monomer:

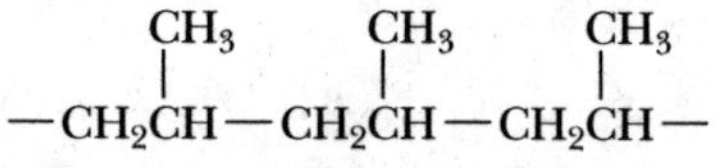

$$-\text{CH}_2\underset{}{\overset{\text{CH}_3}{\overset{|}{\text{C}}}}\text{H}-\text{CH}_2\overset{\text{CH}_3}{\overset{|}{\text{C}}}\text{H}-\text{CH}_2\overset{\text{CH}_3}{\overset{|}{\text{C}}}\text{H}-$$

Polypropylene

Draw a structural formula for a comparable section of

(a) Poly(vinyl chloride)
(b) Polytetrafluoroethylene (PTFE)
(c) Poly(methyl methacrylate)

17.9 Following are structural formulas for sections of two polymers:

(a) 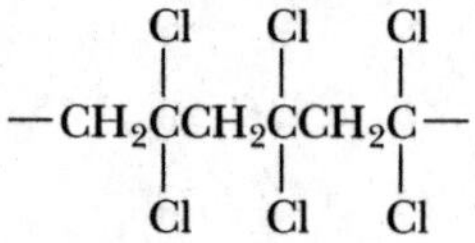 $-\text{CH}_2\underset{\text{Cl}}{\underset{|}{\overset{\text{Cl}}{\overset{|}{\text{C}}}}}\text{CH}_2\underset{\text{Cl}}{\underset{|}{\overset{\text{Cl}}{\overset{|}{\text{C}}}}}\text{CH}_2\underset{\text{Cl}}{\underset{|}{\overset{\text{Cl}}{\overset{|}{\text{C}}}}}-$

(b) 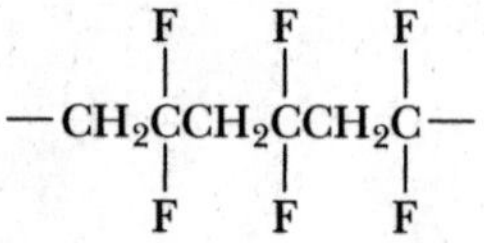 $-\text{CH}_2\underset{\text{F}}{\underset{|}{\overset{\text{F}}{\overset{|}{\text{C}}}}}\text{CH}_2\underset{\text{F}}{\underset{|}{\overset{\text{F}}{\overset{|}{\text{C}}}}}\text{CH}_2\underset{\text{F}}{\underset{|}{\overset{\text{F}}{\overset{|}{\text{C}}}}}-$

From what alkene monomer is each polymer derived?

17.10 Draw the structure of the alkene monomer used to make each chain-growth polymer:

(a) 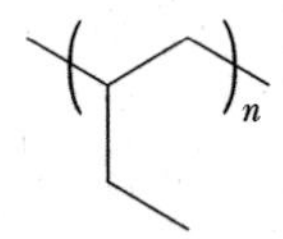 (b) 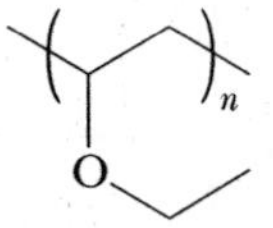 (c) 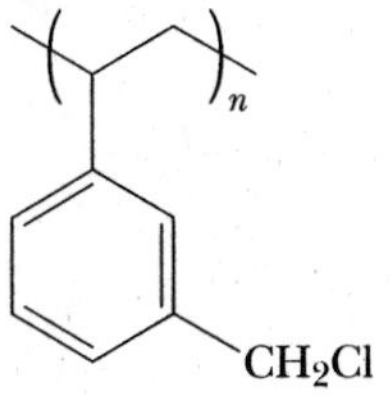 (d)

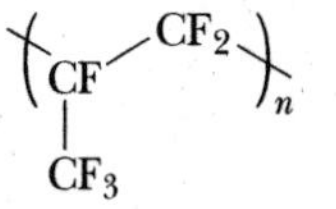

CH_2Cl

CF_2 CF CF_3 n

17.11 LDPE has a higher degree of chain branching than HDPE. Explain the relationship between chain branching and density.

17.12 Compare the densities of LDPE and HDPE with the densities of the liquid alkanes listed in Table 3.4. How might you account for the differences between them?

17.13 The polymerization of vinyl acetate gives poly(vinyl acetate). Hydrolysis of this polymer in aqueous sodium hydroxide gives poly(vinyl alcohol). Draw the repeat units of both poly(vinyl acetate) and poly(vinyl alcohol):

$$\text{Vinyl acetate} \quad CH_3-\overset{\overset{\displaystyle O}{\|}}{C}-O-CH=CH_2$$

17.14 As seen in the previous problem, poly(vinyl alcohol) is made by the polymerization of vinyl acetate, followed by hydrolysis in aqueous sodium hydroxide. Why is poly(vinyl alcohol) not made instead by the polymerization of vinyl alcohol, $CH_2=CHOH$?

17.15 As you know, the shape of a polymer chain affects its properties. Consider the following three polymers:

A Cl Cl Cl Cl Cl Cl Cl

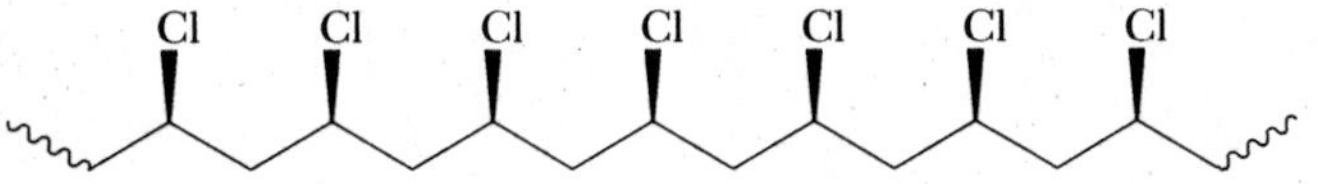

B Cl Cl Cl Cl Cl Cl Cl

C Cl Cl Cl Cl Cl Cl Cl

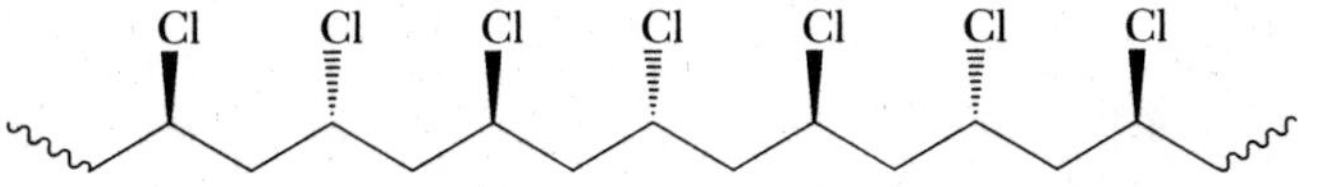

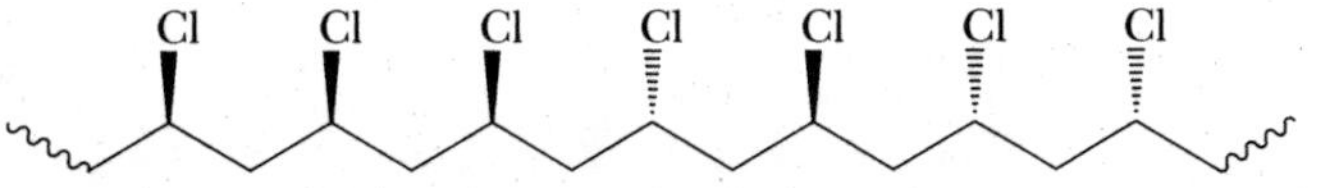

Which do you expect to be the most rigid? Which do you expect to be the most transparent? (Assume the same molecular weights.)

Looking Ahead

17.16 Cellulose, the principle component of cotton, is a polymer of D-glucose in which the monomer unit repeats at the indicated atoms:

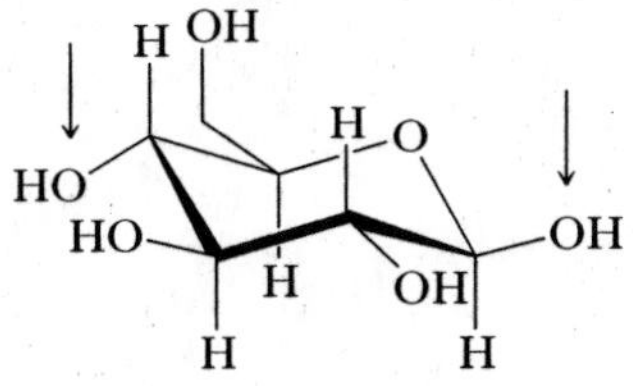

H OH H O HO HO OH H OH H H

D-Glucose

Draw a three-unit section of cellulose.

17.17 Is a repeating unit a requirement for a compound to be called a polymer?

17.18 Proteins are polymers of naturally occurring monomers called amino acids:

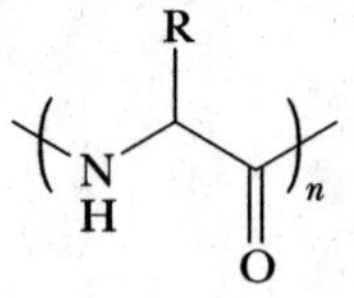

$$\left[\mathrm{NH{-}CH(R){-}C({=}O)} \right]_n$$

a protein

Amino acids differ in the types of R groups available in nature. Explain how the following properties of a protein might be affected upon changing the R groups from $-CH_2CH(CH_3)_2$ to $-CH_2OH$:

(a) solubility in water

(b) T_m

(c) crystallinity

(d) elasticity

INTRODUCTION TO PHYSICAL POLYMER SCIENCE

FOURTH EDITION

L.H. Sperling

Lehigh University
Bethlehem, Pennsylvania

Published by John Wiley & Sons, Inc.,. Hoboken, New Jersey
Published simultaneously in Canada

For general information on our other products and services or for technical support, please contact our Customer Care Department within the United States at (800) 762-2974, outside the United States at (317) 572-3993 or fax (317) 572-4002.

Wiley also publishes its books in a variety of electronic formats. Some content that appears in print may not be available in electronic formats. For more information about Wiley products, visit our web site at www.wiley.com.

Library of Congress Cataloging-in-Publication Data:
Sperling, L. H. (Leslie Howard), 1932–
Introduction to physical polymer science / L.H. Sperling.—4th ed.
p. cm.
Includes index.
ISBN-13 978-0-471-70606-9 (cloth)
ISBN-10 0-471-70606-X (cloth)
1. Polymers. 2. Polymerization. I. Title.
QD381.S635 2006
668.9—dc22

2005021351

Printed in the United States of America
10 9 8 7 6 5 4 3 2 1

1

INTRODUCTION TO POLYMER SCIENCE

Polymer science was born in the great industrial laboratories of the world of the need to make and understand new kinds of plastics, rubber, adhesives, fibers, and coatings. Only much later did polymer science come to academic life. Perhaps because of its origins, polymer science tends to be more interdisciplinary than most sciences, combining chemistry, chemical engineering, materials, and other fields as well.

Chemically, polymers are long-chain molecules of very high molecular weight, often measured in the hundreds of thousands. For this reason, the term "macromolecules" is frequently used when referring to polymeric materials. The trade literature sometimes refers to polymers as resins, an old term that goes back before the chemical structure of the long chains was understood.

The first polymers used were natural products, especially cotton, starch, proteins, and wool. Beginning early in the twentieth century, synthetic polymers were made. The first polymers of importance, Bakelite and nylon, showed the tremendous possibilities of the new materials. However, the scientists of that day realized that they did not understand many of the relationships between the chemical structures and the physical properties that resulted. The research that ensued forms the basis for physical polymer science.

This book develops the subject of physical polymer science, describing the interrelationships among polymer structure, morphology, and physical and mechanical behavior. Key aspects include molecular weight and molecular weight distribution, and the organization of the atoms down the polymer chain. Many polymers crystallize, and the size, shape, and organization of the

Introduction to Physical Polymer Science, by L.H. Sperling
ISBN 0-471-70606-X

crystallites depend on how the polymer was crystallized. Such effects as annealing are very important, as they have a profound influence on the final state of molecular organization.

Other polymers are amorphous, often because their chains are too irregular to permit regular packing. The onset of chain molecular motion heralds the glass transition and softening of the polymer from the glassy (plastic) state to the rubbery state. Mechanical behavior includes such basic aspects as modulus, stress relaxation, and elongation to break. Each of these is relatable to the polymer's basic molecular structure and history.

This chapter provides the student with a brief introduction to the broader field of polymer science. Although physical polymer science does not include polymer synthesis, some knowledge of how polymers are made is helpful in understanding configurational aspects, such as tacticity, which are concerned with how the atoms are organized along the chain. Similarly polymer molecular weights and distributions are controlled by the synthetic detail. This chapter starts at the beginning of polymer science, and it assumes no prior knowledge of the field.

1.1 FROM LITTLE MOLECULES TO BIG MOLECULES

The behavior of polymers represents a continuation of the behavior of smaller molecules at the limit of very high molecular weight. As a simple example, consider the normal alkane hydrocarbon series

```
    H           H   H           H   H   H
    |           |   |           |   |   |
H — C — H   H — C — C — H   H — C — C — C — H      (1.1)
    |           |   |           |   |   |
    H           H   H           H   H   H
 Methane        Ethane            Propane
```

These compounds have the general structure

$$H{+}CH_2{+}_nH \tag{1.2}$$

where the number of —CH_2— groups, n, is allowed to increase up to several thousand. The progression of their state and properties is shown in Table 1.1.

At room temperature, the first four members of the series are gases. n-Pentane boils at 36.1°C and is a low-viscosity liquid. As the molecular weight of the series increases, the viscosity of the members increases. Although commercial gasolines contain many branched-chain materials and aromatics as well as straight-chain alkanes, the viscosity of gasoline is markedly lower than that of kerosene, motor oil, and grease because of its lower average chain length.

These latter materials are usually mixtures of several molecular species, although they are easily separable and identifiable. This point is important

Table 1.1 Properties of the alkane/polyethylene series

Number of Carbons in Chain	State and Properties of Material	Applications
1–4	Simple gas	Bottled gas for cooking
5–11	Simple liquid	Gasoline
9–16	Medium-viscosity liquid	Kerosene
16–25	High-viscosity liquid	Oil and grease
25–50	Crystalline solid	Paraffin wax candles
50–1000	Semicrystalline solid	Milk carton adhesives and coatings
1000–5000	Tough plastic solid	Polyethylene bottles and containers
$3–6 \times 10^5$	Fibers	Surgical gloves, bullet-proof vests

because most polymers are also "mixtures"; that is, they have a molecular weight distribution. In high polymers, however, it becomes difficult to separate each of the molecular species, and people talk about molecular weight averages.

Compositions of normal alkanes averaging more than about 20 to 25 carbon atoms are crystalline at room temperature. These are simple solids known as wax. It must be emphasized that at up to 50 carbon atoms the material is far from being polymeric in the ordinary sense of the term.

The polymeric alkanes with no side groups that contain 1000 to 3000 carbon atoms are known as polyethylenes. Polyethylene has the chemical structure

$$\left(-CH_2-CH_2-\right)_n \tag{1.3}$$

which originates from the structure of the monomer ethylene, $CH_2{=}CH_2$. The quantity n is the number of mers—or monomeric units in the chain. In some places the structure is written

$$\left(-CH_2-\right)_{n'} \tag{1.4}$$

or polymethylene. (Then $n' = 2n$.) The relationship of the latter structure to the alkane series is clearer. While true alkanes have CH_3— as end groups, most polyethylenes have initiator residues.

Even at a chain length of thousands of carbons, the melting point of polyethylene is still slightly molecular-weight-dependent, but most linear polyethylenes have melting or fusion temperatures, T_f, near 140°C. The approach to the theoretical asymptote of about 145°C at infinite molecular weight (1) is illustrated schematically in Figure 1.1.

The greatest differences between polyethylene and wax lie in their mechanical behavior, however. While wax is a brittle solid, polyethylene is a tough plastic. Comparing resistance to break of a child's birthday candle with a wash bottle tip, both of about the same diameter, shows that the wash bottle tip can be repeatedly bent whereas the candle breaks on the first deformation.

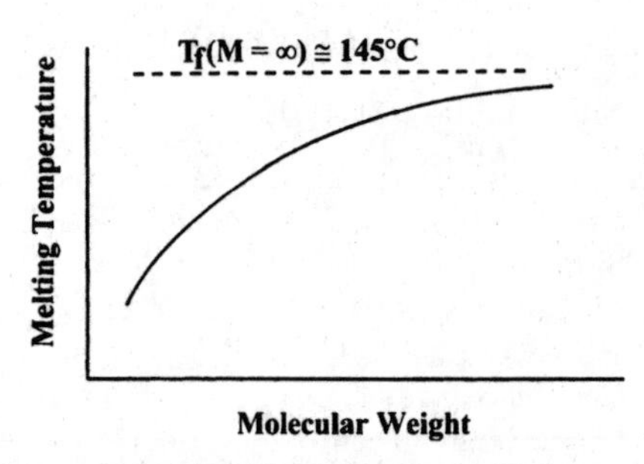

Figure 1.1 The molecular weight-melting temperature relationship for the alkane series. An asymptotic value of about 145°C is reached for very high molecular weight linear polyethylenes.

Polyethylene is a tough plastic solid because its chains are long enough to connect individual stems together within a lamellar crystallite by chain folding (see Figure 1.2). The chains also wander between lamellae, connecting several of them together. These effects add strong covalent bond connections both within the lamellae and between them. On the other hand, only weak van der Waals forces hold the chains together in wax.

In addition a certain portion of polyethylene is amorphous. The chains in this portion are rubbery, imparting flexibility to the entire material. Wax is 100% crystalline, by difference.

The long chain length allows for entanglement (see Figure 1.3). The entanglements help hold the whole material together under stress. In the melt state, chain entanglements cause the viscosity to be raised very significantly also.

The long chains shown in Figure 1.3 also illustrate the coiling of polymer chains in the amorphous state. One of the most powerful theories in polymer science (2) states that the conformations of amorphous chains in space are random coils; that is, the directions of the chain portions are statistically determined.

1.2 MOLECULAR WEIGHT AND MOLECULAR WEIGHT DISTRIBUTIONS

While the exact molecular weight required for a substance to be called a polymer is a subject of continued debate, often polymer scientists put the number at about 25,000 g/mol. This is the minimum molecular weight required for good physical and mechanical properties for many important polymers. This molecular weight is also near the onset of entanglement.

1.2.1 Effect on Tensile Strength

The tensile strength of any material is defined as the stress at break during elongation, where stress has the units of Pa, dyn/cm^2, or lb/in^2; see Chapter 11.

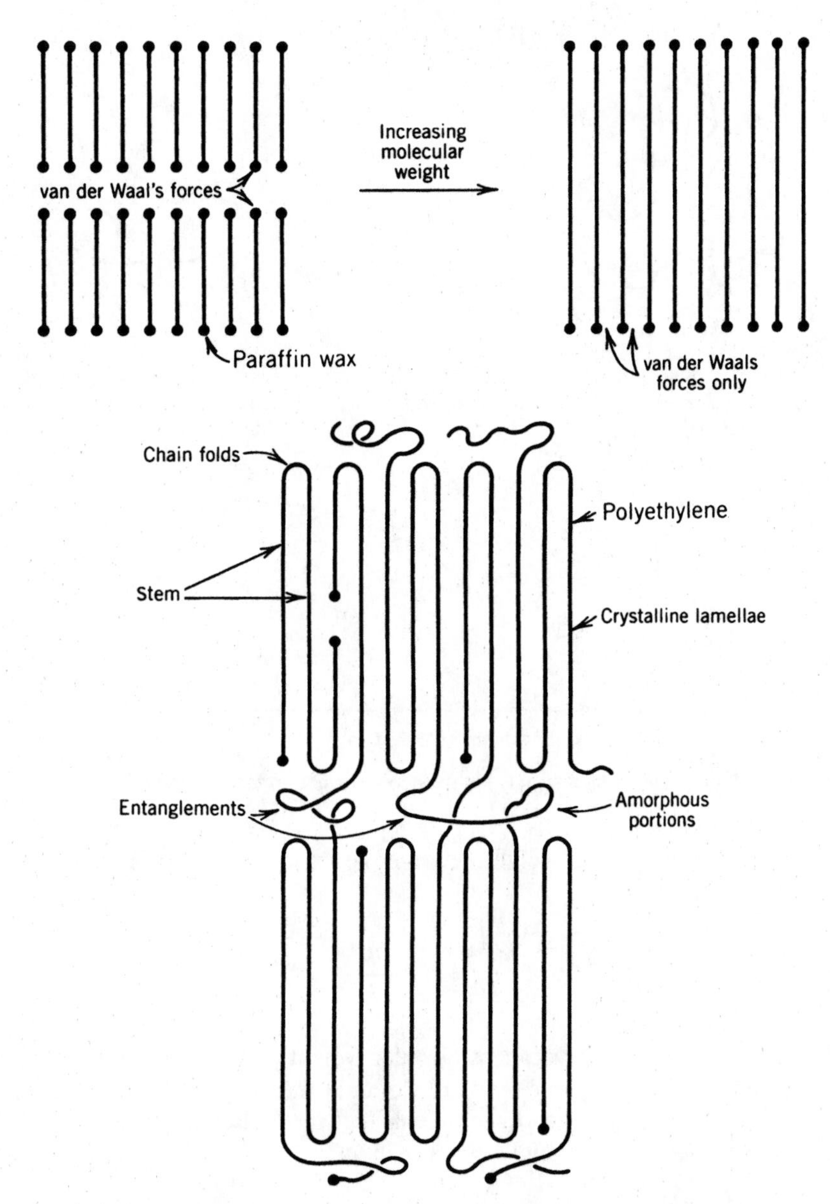

Figure 1.2 Comparison of wax and polyethylene structure and morphology.

The effect of molecular weight on the tensile strength of polymers is illustrated in Figure 1.4. At very low molecular weights the tensile stress to break, σ_b, is near zero. As the molecular weight increases, the tensile strength increases rapidly, and then gradually levels off. Since a major point of weakness at the molecular level involves the chain ends, which do not transmit the covalent bond strength, it is predicted that the tensile strength reaches an asymptotic

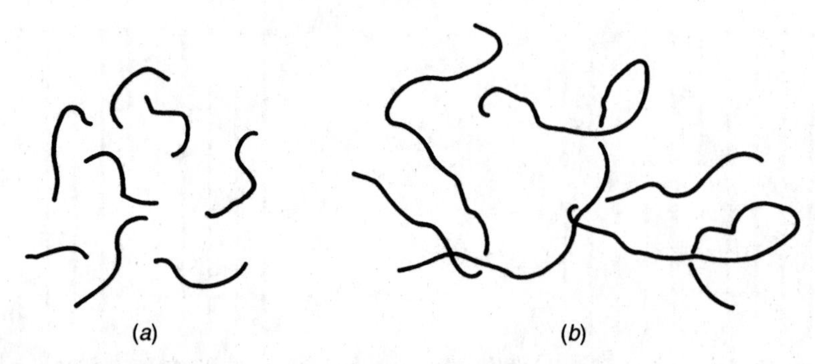

Figure 1.3 Entanglement of polymer chains. (*a*) Low molecular weight, no entanglement. (*b*) High molecular weight, chains are entangled. The transition between the two is often at about 600 backbone chain atoms.

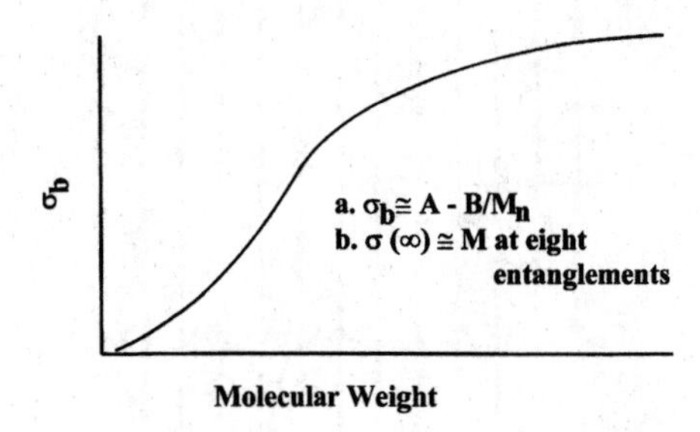

Figure 1.4 Effect of polymer molecular weight on tensile strength.

value at infinite molecular weight. A large part of the curve in Figure 1.4 can be expressed (3,4)

$$\sigma_b = A - \frac{B}{M_n} \tag{1.5}$$

where M_n is the number-average molecular weight (see below) and A and B are constants. Newer theories by Wool (3) and others suggest that more than 90% of tensile strength and other mechanical properties are attained when the chain reaches eight entanglements in length.

1.2.2 Molecular Weight Averages

The same polymer from different sources may have different molecular weights. Thus polyethylene from source A may have a molecular weight of 150,000 g/mol, whereas polyethylene from source B may have a molecular weight of 400,000 g/mol (see Figure 1.5). To compound the difficulty, all common synthetic polymers and most natural polymers (except proteins) have a distribution in molecular weights. That is, some molecules in a given sample

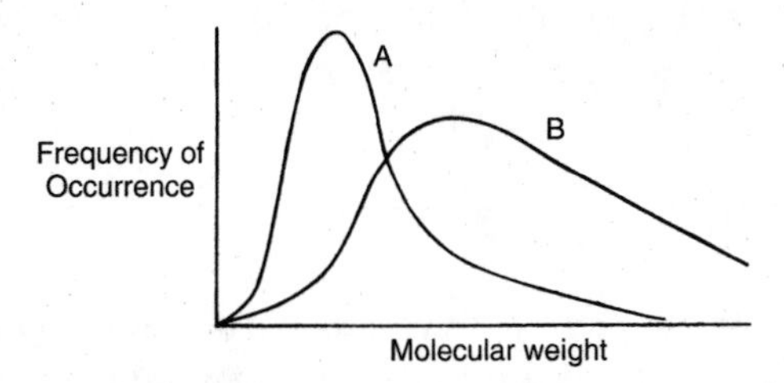

Figure 1.5 Molecular weight distributions of the same polymer from two different sources, A and B.

of polyethylene are larger than others. The differences result directly from the kinetics of polymerization.

However, these facts led to much confusion for chemists early in the twentieth century. At that time chemists were able to understand and characterize small molecules. Compounds such as hexane all have six carbon atoms. If polyethylene with 2430 carbon atoms were declared to be "polyethylene," how could that component having 5280 carbon atoms also be polyethylene? How could two sources of the material having different average molecular weights both be polyethylene, noting A and B in Figure 1.5?

The answer to these questions lies in defining average molecular weights and molecular weight distributions (5,6). The two most important molecular weight averages are the number-average molecular weight, M_n,

$$M_n = \frac{\sum_i N_i M_i}{\sum_i N_i} \tag{1.6}$$

where N_i is the number of molecules of molecular weight M_i, and the weight-average molecular weight, M_w,

$$M_w = \frac{\sum_i N_i M_i^2}{\sum_i N_i M_i} \tag{1.7}$$

For single-peaked distributions, M_n is usually near the peak. The weight-average molecular weight is always larger. For simple distributions, M_w may be 1.5 to 2.0 times M_n. The ratio M_w/M_n, sometimes called the polydispersity index, provides a simple definition of the molecular weight distribution. Thus all compositions of $\text{+}CH_2\text{—}CH_2\text{+}_n$ are called polyethylene, the molecular weights being specified for each specimen.

For many polymers a narrower molecular distribution yields better properties. The low end of the distribution may act as a plasticizer, softening the material. Certainly it does not contribute as much to the tensile strength. The high-molecular-weight tail increases processing difficulties, because of its enormous contribution to the melt viscosity. For these reasons, great emphasis is placed on characterizing polymer molecular weights.

1.3 MAJOR POLYMER TRANSITIONS

Polymer crystallinity and melting were discussed previously. Crystallization is an example of a first-order transition, in this case liquid to solid. Most small molecules crystallize, an example being water to ice. Thus this transition is very familiar.

A less classical transition is the glass–rubber transition in polymers. At the glass transition temperature, T_g, the amorphous portions of a polymer soften. The most familiar example is ordinary window glass, which softens and flows at elevated temperatures. Yet glass is not crystalline, but rather it is an amorphous solid. It should be pointed out that many polymers are totally amorphous. Carried out under ideal conditions, the glass transition is a type of second-order transition.

The basis for the glass transition is the onset of coordinated molecular motion is the polymer chain. At low temperatures, only vibrational motions are possible, and the polymer is hard and glassy (Figure 1.6, region 1) (7). In the glass transition region, region 2, the polymer softens, the modulus drops three orders of magnitude, and the material becomes rubbery. Regions 3, 4, and 5 are called the rubbery plateau, the rubbery flow, and the viscous flow regions, respectively. Examples of each region are shown in Table 1.2.

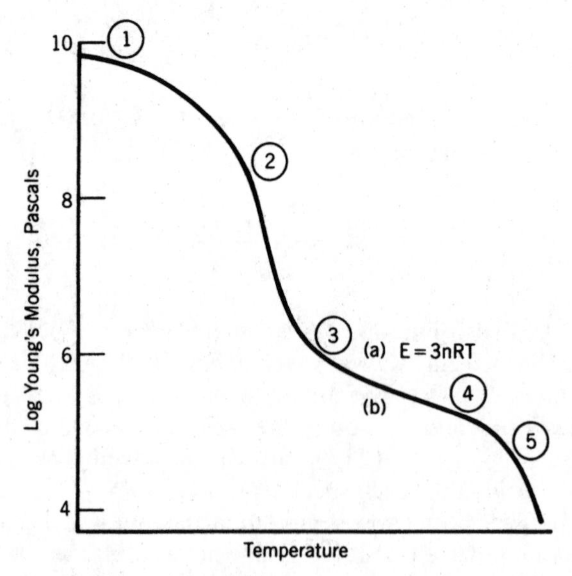

Figure 1.6 Idealized modulus–temperature behavior of an amorphous polymer. Young's modulus, stress/strain, is a measure of stiffness.

Table 1.2 Typical polymer viscoelastic behavior at room temperature (7a)

Region	Polymer	Application
Glassy	Poly(methyl methacrylate)	Plastic
Glass transition	Poly(vinyl acetate)	Latex paint
Rubbery plateau	*Cross*-poly(butadiene–*stat*–styrene)	Rubber bands
Rubbery flow	Chicle[a]	Chewing gum
Viscous flow	Poly(dimethylsiloxane)	Lubricant

[a] From the latex of *Achras sapota*, a mixture of *cis*- and *trans*-polyisoprene plus polysaccharides.

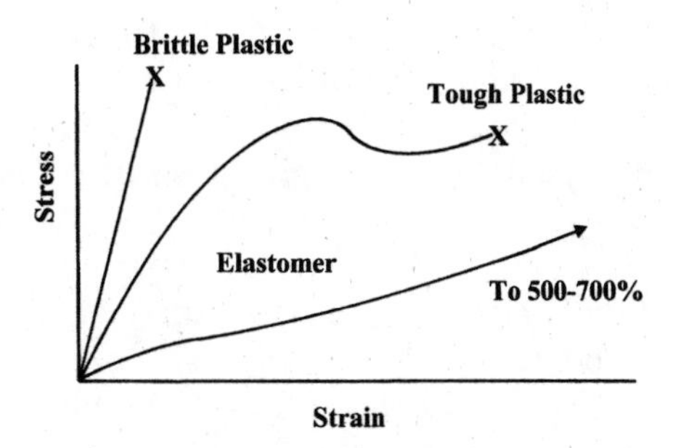

Figure 1.7 Stress–strain behavior of various polymers. While the initial slope yields the modulus, the area under the curve provides the energy to fracture.

Depending on the region of viscoelastic behavior, the mechanical properties of polymers differ greatly. Model stress–strain behavior is illustrated in Figure 1.7 for regions 1, 2, and 3. Glassy polymers are stiff and often brittle, breaking after only a few percent extension. Polymers in the glass transition region are more extensible, sometimes exhibiting a yield point (the hump in the tough plastic stress–strain curve). If the polymer is above its brittle–ductile transition, Section 11.2.3, rubber-toughened, Chapter 13, or semicrystalline with its amorphous portions above T_g, tough plastic behavior will also be observed. Polymers in the rubbery plateau region are highly elastic, often stretching to 500% or more. Regions 1, 2, and 3 will be discussed further in Chapters 8 and 9. Regions 4 and 5 flow to increasing extents under stress; see Chapter 10.

Cross-linked amorphous polymers above their glass transition temperature behave rubbery. Examples are rubber bands and automotive tire rubber. In general, Young's modulus of elastomers in the rubbery-plateau region is higher than the corresponding linear polymers, and is governed by the relation $E = 3nRT$, in Figure 1.6 (line not shown); the linear polymer behavior is illustrated by the line (b). Here, n represents the number of chain segments bound at both ends in a network, per unit volume. The quantities R and T are the gas constant and the absolute temperature, respectively.

Polymers may also be partly crystalline. The remaining portion of the polymer, the amorphous material, may be above or below its glass transition

Table 1.3 Examples of polymers at room temperature by transition behavior

	Crystalline	Amorphous
Above T_g	Polyethylene	Natural rubber
Below T_g	Cellulose	Poly(methyl methacrylate)

temperature, creating four subclasses of materials. Table 1.3 gives a common example of each. While polyethylene and natural rubber need no further introduction, common names for processed cellulose are rayon and cellophane. Cotton is nearly pure cellulose, and wood pulp for paper is 80 to 90% cellulose. A well-known trade name for poly(methyl methacrylate) is Plexiglas®. The modulus–temperature behavior of polymers in either the rubbery-plateau region or in the semicrystalline region are illustrated further in Figure 8.2, Chapter 8.

Actually there are two regions of modulus for semicrystalline polymers. If the amorphous portion is above T_g, then the modulus is generally between rubbery and glassy. If the amorphous portion is glassy, then the polymer will be actually be a bit stiffer than expected for a 100% glassy polymer.

1.4 POLYMER SYNTHESIS AND STRUCTURE

1.4.1 Chain Polymerization

Polymers may be synthesized by two major kinetic schemes, chain and stepwise polymerization. The most important of the chain polymerization methods is called free radical polymerization.

1.4.1.1 Free Radical Polymerization The synthesis of poly(ethyl acrylate) will be used as an example of free radical polymerization. Benzoyl peroxide is a common initiator. Free radical polymerization has three major kinetic steps—initiation, propagation, and termination.

1.4.1.2 Initiation On heating, benzoyl peroxide decomposes to give two free radicals:

$$\mathrm{C_6H_5{-}\overset{\overset{\displaystyle O}{\|}}{C}{-}O{:}O{-}\overset{\overset{\displaystyle O}{\|}}{C}{-}C_6H_5 \xrightarrow{\Delta} 2\,C_6H_5{-}\overset{\overset{\displaystyle O}{\|}}{C}{-}O\cdot} \tag{1.8}$$

Benzoyl peroxide — Free radical, R·

In this reaction the electrons in the oxygen–oxygen bond are unpaired and become the active site. With R representing a generalized organic chemical group, the free radical can be written R·. (It should be pointed out that hydrogen peroxide undergoes the same reaction on a wound, giving a burning sensation as the free radicals "kill the germs.")

The initiation step usually includes the addition of the first monomer molecule:

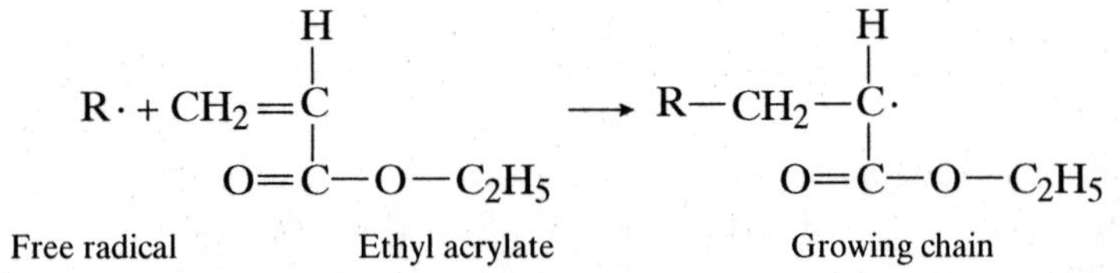

$$\underset{\text{Free radical}}{R\cdot} + \underset{\text{Ethyl acrylate}}{CH_2{=}\overset{\overset{H}{|}}{\underset{\underset{O=C-O-C_2H_5}{|}}{C}}} \longrightarrow \underset{\text{Growing chain}}{R-CH_2-\overset{\overset{H}{|}}{\underset{\underset{O=C-O-C_2H_5}{|}}{C}}\cdot} \tag{1.9}$$

In this reaction the free radical attacks the monomer and adds to it. The double bond is broken open, and the free radical reappears at the far end.

1.4.1.3 Propagation After initiation reactions (1.8) and (1.9), many monomer molecules are added rapidly, perhaps in a fraction of a second:

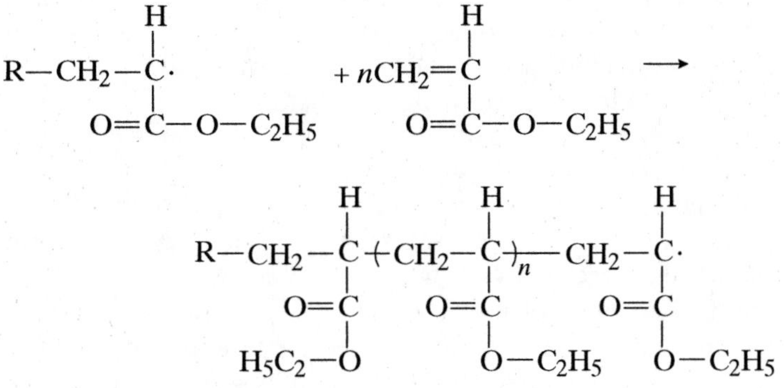

$$R-CH_2-\overset{\overset{H}{|}}{\underset{\underset{O=C-O-C_2H_5}{|}}{C}}\cdot + nCH_2{=}\overset{\overset{H}{|}}{\underset{\underset{O=C-O-C_2H_5}{|}}{C}} \longrightarrow R-CH_2-\overset{\overset{H}{|}}{\underset{\underset{\underset{H_5C_2-O}{|}}{O=C}}{|}}{C}}\!\left(CH_2-\overset{\overset{H}{|}}{\underset{\underset{\underset{O-C_2H_5}{|}}{O=C}}{|}}{C}}\right)_n\!-CH_2-\overset{\overset{H}{|}}{\underset{\underset{\underset{O-C_2H_5}{|}}{O=C}}{|}}{C}}\cdot \tag{1.10}$$

On the addition of each monomer, the free radical moves to the end of the chain.

1.4.1.4 Termination In the termination reaction, two free radicals react with each other. Termination is either by combination,

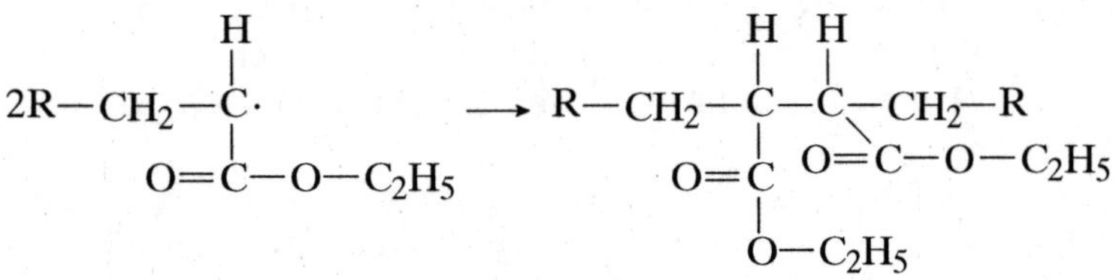

$$2R-CH_2-\overset{\overset{H}{|}}{\underset{\underset{O=C-O-C_2H_5}{|}}{C}}\cdot \longrightarrow R-CH_2-\overset{\overset{H}{|}}{\underset{\underset{\underset{O-C_2H_5}{|}}{O=C}}{|}}{C}}-\overset{\overset{H}{|}}{\underset{O=C-O-C_2H_5}{C}}-CH_2-R \tag{1.11}$$

where R now represents a long-chain portion, or by disproportionation, where a hydrogen is transferred from one chain to the other. This latter result produces in two final chains. While the normal mode of addition is a head-to-tail reaction (1.10), this termination step is normally head-to-head.

As a homopolymer, poly(ethyl acrylate) is widely used as an elastomer or adhesive, being a polymer with a low T_g, –22°C. As a copolymer with other acrylics it is used as a latex paint.

1.4.1.5 Structure and Nomenclature The principal method of polymerizing monomers by the chain kinetic scheme involves the opening of double bonds to form a linear molecule. In a reacting mixture, monomer, fully reacted polymer, and only a small amount of rapidly reacting species are present. Once the polymer terminates, it is "dead" and cannot react further by the synthesis scheme outlined previously.

Polymers are named by rules laid out by the IUPAC Nomenclature Committee (8,9). For many simple polymers the source-based name utilizes the monomer name prefixed by "poly." If the monomer name has two or more words, parentheses are placed around the monomer name. Thus, in the above, the monomer ethyl acrylate is polymerized to make poly(ethyl acrylate). Source-based and IUPAC names are compared in Appendix 1.1.

Table 1.4 provides a selected list of common chain polymer structures and names along with comments as to how the polymers are used. The "vinyl" monomers are characterized by the general structure $CH_2{=}CHR$, where R represents any side group. One of the best-known vinyl polymers is poly(vinyl chloride), where R is —Cl.

Polyethylene and polypropylene are the major members of the class of polymers known as *polyolefins*; see Section 14.1. The term *olefin* derives from the double-bond characteristic of the alkene series.

A slight dichotomy exists in the writing of vinyl polymer structures. From a correct nomenclature point of view, the pendant moiety appears on the left-hand carbon. Thus poly(vinyl chloride) should be written $\leftarrow\!CHCl{-}CH_2\!\rightarrow_n$. However, from a synthesis point of view, the structure is written $\leftarrow\!CH_2{-}CHCl\!\rightarrow_n$, because the free radical is borne on the pendant moiety carbon. Thus both forms appear in the literature.

The diene monomer has the general structure $CH_2{=}CR{-}CH{=}CH_2$, where on polymerization one of the double bonds forms the chain bonds, and the other goes to the central position. The vinylidenes have two groups on one carbon. Table 1.4 also lists some common copolymers, which are formed by reacting two or more monomers together. In general, the polymer structure most closely resembling the monomer structure will be presented herein.

Today, recycling of plastics has become paramount in preserving the environment. On the bottom of plastic bottles and other plastic items is an identification number and letters; see Table 1.5. This information serves to help in separation of the plastics prior to recycling. Observation of the properties of the plastic such as modulus, together with the identification, will help

Table 1.4 Selected chain polymer structures and nomenclature

Structure	Name	Where Used
$-\!\!\left[CH_2-CH(R)\right]_n-$	"Vinyl" class	
R = —H	Polyethylene	Plastic
R = $-CH_3$	Polypropylene	Rope
R 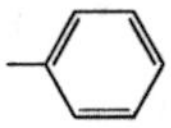 = $-C_6H_5$ (phenyl)	Polystyrene	Drinking cups
R = —Cl	Poly(vinyl chloride)	"Vinyl," water pipes
R = $-O-C(=O)-CH_3$	Poly(vinyl acetate)	Latex paints
R = —OH	Poly(vinyl alcohol)	Fiber
$-\!\!\left[CH_2-C(X)(C(=O)-O-R)\right]_n-$	X = —H, acrylics X = $-CH_3$, methacrylics	
X = —H, R = $-C_2H_5$	Poly(ethyl acrylate)	Latex paints
X = $-CH_3$, R = $-CH_3$	Poly(methyl methacrylate)	Plexiglas®
X = $-CH_3$, R = $-C_2H_5$	Poly(ethyl methacrylate)	Adhesives
$-\!\!\left[CH_2-C(H)(C\equiv N)\right]_n-$	Polyacrylonitrile[a]	Orlon®
$-\!\!\left[CH_2-C(R)=CH-CH_2\right]_n-$ [a']	"Diene" class	
R = —H	Polybutadiene	Tires
R = $-CH_3$	Polyisoprene	Natural rubber
R = —Cl	Polychloroprene	Neoprene
$-\!\!\left[CX_2-CR_2\right]_n-$	Vinylidenes	
X = —H, R = —F	Poly(vinylidene fluoride)	Plastic
X = —H, R = —F	Polytetrafluoroethylene	Teflon®
X = —H, R = $-CH_3$	Polyisobutene[b]	Elastomer
	Common Copolymers	
EPDM	Ethylene–propylene–diene–monomer	Elastomer
SBR	Styrene–butadiene–rubber Poly(styrene–*stat*–butadiene)[c]	Tire rubber
NBR	Acrylonitrile–butadiene–rubber Poly(acrylonitrile–*stat*–butadiene)	Elastomer
ABS	Acrylonitrile–butadiene–styrene[d]	Plastic

[a] Polyacrylonitrile is technically a number of the acrylic class because it forms acrylic acid on hydrolysis.

[a'] IUPAC recommends $-\!\!\left[C(R)=CH-CH_2-CH_2\right]_n-$

[b] Also called polyisobutylene. The 2% copolymer with isoprene, after vulcanization, is called butyl rubber.

[c] The term–*stat*–means statistical copolymer, as explained in Chapter 2.

[d] ABS is actually a blend or graft of two random copolymers, poly(acrylonitrile–*stat*–butadiene) and poly(acrylonitrile–*stat*–styrene).

Table 1.5 The plastics identification code

Code	Letter I.D.	Polymer Name
1	PETE	Poly(ethylene terephthalate)
2	HDPE	High-density polyethylene
3	V	Poly(vinyl chloride)
4	LDPE	Low-density polyethylene
5	PP	Polypropylene
6	PS	Polystyrene
7	Other	Different polymers

Source: From the *Plastic Container Code System*, The Plastic Bottle Information Bureau, Washington, DC.

the student understand the kinds and properties of the plastics in common service.

1.4.2 Step Polymerization

1.4.2.1 A Polyester Condensation Reaction The second important kinetic scheme is step polymerization. As an example of a step polymerization, the synthesis of a polyester is given.

The general reaction to form esters starts with an acid and an alcohol:

$$\underset{\text{Ethyl alcohol}}{CH_3-CH_2OH} + \underset{\text{Acetic acid}}{CH_3-\overset{\overset{\displaystyle O}{\|}}{C}-OH} \rightarrow \underset{\text{Ethyl acetate}}{CH_3-CH_2-O-\overset{\overset{\displaystyle O}{\|}}{C}-CH_3} + \underset{\text{Water}}{H_2O} \tag{1.12}$$

where the ester group is $-O-\overset{\overset{\displaystyle O}{\|}}{C}-$, and water is eliminated.

The chemicals above cannot form a polyester because they have only one functional group each. When the two reactants each have bifunctionality, a linear polymer is formed:

$$n\mathrm{HO{-}CH_2{-}CH_2{-}OH} + n\mathrm{HO{-}\overset{O}{\overset{\|}{C}}{-}C_6H_4{-}\overset{O}{\overset{\|}{C}}{-}OH} \rightarrow$$

Ethylene glycol Terephthalic acid

$$\mathrm{H}\left[\mathrm{O{-}CH_2{-}CH_2{-}O{-}\overset{O}{\overset{\|}{C}}{-}C_6H_4{-}\overset{O}{\overset{\|}{C}}}\right]_n\mathrm{OH} + (2n-1)\mathrm{H_2O} \quad (1.13)$$

Poly(ethylene terephthalate)

In the stepwise reaction scheme, monomers, dimers, trimers, and so on, may all react together. All that is required is that the appropriate functional groups meet in space. Thus the molecular weight slowly climbs as the small molecule water is eliminated. Industrially, $\mathrm{{-}\overset{O}{\overset{\|}{C}}{-}OH}$ is replaced by $\mathrm{{-}\overset{O}{\overset{\|}{C}}{-}O{-}CH_3}$. Then, the reaction is an ester interchange, releasing methanol.

Poly(ethylene terephthalate) is widely known as the fiber Dacron®. It is highly crystalline, with a melting temperature of about +265°C.

Another well-known series of polymers made by step polymerization reactions is the polyamides, known widely as the nylons. In fact there are two series of nylons. In the first series, the monomer has an amine at one end of the molecule and a carboxyl at the other. For example,

$$n\mathrm{H_2N{-}CH_2{-}CH_2{-}CH_2{-}\overset{O}{\overset{\|}{C}}{-}OH} \rightarrow \mathrm{H}\left(\mathrm{\overset{H}{\overset{|}{N}}{-}CH_2{-}CH_2{-}CH_2{-}\overset{O}{\overset{\|}{C}}}\right)_n\mathrm{OH} + (n-1)\mathrm{H_2O} \quad (1.14)$$

which is known as nylon 4. The number 4 indicates the number of carbon atoms in the mer.

In the second series, a dicarboxylic acid is reacted with a diamine:

$$n\mathrm{H_2N(CH_2)_4NH_2} + n\mathrm{H{-}O{-}\overset{O}{\overset{\|}{C}}(CH_2)_6\overset{O}{\overset{\|}{C}}{-}OH} \longrightarrow \mathrm{H}\left[\mathrm{\overset{H}{\overset{|}{N}}{-}(CH_2)_4{-}\overset{H}{\overset{|}{N}}{-}\overset{O}{\overset{\|}{C}}{-}(CH_2)_6\overset{O}{\overset{\|}{C}}}\right]_n\mathrm{OH} + (2n-1)\mathrm{H_2O} \quad (1.15)$$

which is named nylon 48. Note that the amine carbon number is written first, and the acid carbon number second. For reaction purposes, acyl chlorides are frequently substituted for the carboxyl groups. An excellent demonstration experiment is described by Morgan and Kwolek (10), called the nylon rope trick.

1.4.2.2 Stepwise Nomenclature and Structures Table 1.6 names some of the more important stepwise polymers. The polyesters have already been mentioned. The nylons are known technically as polyamides. There are two important subseries of nylons, where amine and the carboxylic acid are on different monomer molecules (thus requiring both monomers to make the polymer) or one each on the ends of the same monomer molecule. These are numbered by the number of carbons present in the monomer species. It must be mentioned that the proteins are also polyamides.

Other classes of polymers mentioned in Table 1.6 include the polyurethanes, widely used as elastomers; the silicones, also elastomeric; and the cellulosics, used in fibers and plastics. Cellulose is a natural product.

Another class of polymers are the polyethers, prepared by ring-opening reactions. The most important member of this series is poly(ethylene oxide),

$$\text{+}CH_2\text{—}CH_2\text{—}O\text{+}_n$$

Because of the oxygen atom, poly(ethylene oxide) is water soluble.

To summarize the material in Table 1.6, the major stepwise polymer classes contain the following identifying groups:

Polyesters	$\text{—}\overset{\overset{\displaystyle O}{\|}}{C}\text{—}O\text{—}$
Polyamides	$\text{—}\overset{H}{N}\text{—}\overset{\overset{\displaystyle O}{\|}}{C}\text{—}$
Polyurethanes	$\text{—}\overset{H}{N}\text{—}\overset{\overset{\displaystyle O}{\|}}{C}\text{—}O\text{—}$
Silicones	$\text{—}\underset{\underset{\displaystyle CH_3}{\|}}{\overset{\overset{\displaystyle CH_3}{\|}}{Si}}\text{—}O\text{—}$
Epoxy resins	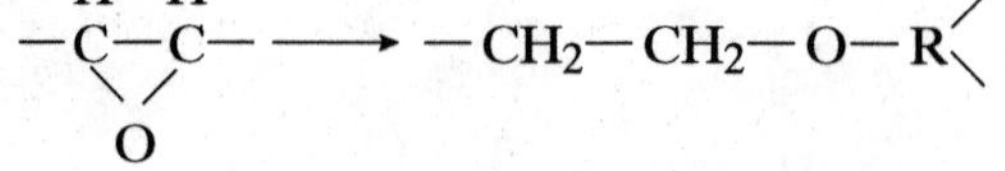 $\text{—}\overset{H}{C}\text{—}\overset{H}{C}\text{—}$ (bridged by O) $\longrightarrow \text{—}CH_2\text{—}CH_2\text{—}O\text{—}R\langle$
Polyethers	$\text{—}O\text{—}$

1.4.2.3 Natural Product Polymers Living organisms make many polymers, nature's best. Most such natural polymers strongly resemble step-polymerized materials. However, living organisms make their polymers enzymatically, the structure ultimately being controlled by DNA, itself a polymer.

Table 1.6 Selected stepwise structures and nomenclature

Structure[a]	Name	Where Known
$\left(O-CH_2-CH_2-O-\overset{O}{\overset{\|}{C}}-C_6H_4-\overset{O}{\overset{\|}{C}}\right)_n$	Poly(ethylene terephthalate)	Dacron®
$\left[\overset{H}{N}\left(CH_2\right)_6\overset{H}{N}-\overset{O}{\overset{\|}{C}}\left(CH_2\right)_8\overset{O}{\overset{\|}{C}}\right]_n$	Poly(hexamethylene sebacamide)	Polyamide 610[b]
$\left[\overset{H}{N}-\overset{O}{\overset{\|}{C}}\left(CH_2\right)_5\right]_n$	Polycaprolactam	Polyamide 6
$\left(O-CH_2\right)_n$	Polyoxymethylene	Polyacetal
$\left[O\left(CH_2\right)_4\right]_n$	Polytetrahydrofuran	Polyether
$\left[\left(O\left(CH_2\right)_4\right)_m\underset{H}{N}-\overset{O}{\overset{\|}{C}}\right]_n$	Polyurethane[c]	Spandex Lycra®
$\left(O-\underset{CH_3}{\overset{CH_3}{Si}}\right)_n$	Poly(dimethyl siloxane)	Silicone rubber
$\left(O-C_6H_4-\overset{CH_3}{\underset{\|}{C}}-C_6H_4-O-\overset{O}{\overset{\|}{C}}\right)_n$	Polycarbonate	Lexan®
(glucose ring repeat unit: OH, CH_2, H, O, OH, H, H, OH; O links)$_n$	Cellulose	Cotton
$H_2\overset{O}{C-CH}-R-\overset{O}{CH-CH_2}\longrightarrow$ $-R'-O\left(CH_2-\overset{OH}{CH}-R-CH_2-CH_2-O\right)_n R''-$	Epoxy resins	Epon®

[a] Some people see the mer structure in the third row more clearly with

$$\left(\overset{H}{N}\left(CH_2\right)_5\overset{O}{\overset{\|}{C}}\right)_n$$

Some other step polymerization mers can also be drawn in two or more different ways. The student should learn to recognize the structures in different ways.

[b] The "6" refers to the number of carbons in the diamine portion, and the "10" to the number of carbons in the diacid. An old name is nylon 610.

[c] The urethane group usually links polyether or polyester low molecular weight polymers together.

The cross-linking of rubber with sulfur is called vulcanization. Cross-linking bonds the chains together to form a network. The resulting product is called a thermoset, because it does not flow on heating.

Plasticizers are small molecules added to soften a polymer by lowering its glass transition temperature or reducing its crystallinity or melting temperature. The most widely plasticized polymer is poly(vinyl chloride). The distinctive odor of new "vinyl" shower curtains is caused by the plasticizer, for example.

Fillers may be of two types, reinforcing and nonreinforcing. Common reinforcing fillers are the silicas and carbon blacks. The latter are most widely used in automotive tires to improve wear characteristics such as abrasion resistance. Nonreinforcing fillers, such as calcium carbonate, may provide color or opacity or may merely lower the price of the final product.

1.6 THE MACROMOLECULAR HYPOTHESIS

In the nineteenth century, the structure of polymers was almost entirely unknown. The Germans called it *Schmierenchemie*, meaning grease chemistry (11), but a better translation might be "the gunk at the bottom of the flask," that portion of an organic reaction that did not result in characterizable products. In the nineteenth century and early twentieth century the field of polymers and the field of colloids were considered integral parts of the same field. Wolfgang Ostwald declared in 1917 (12):

> All those sticky, mucilaginous, resinous, tarry masses which refuse to crystallize, and which are the abomination of the normal organic chemist; those substances which he carefully sets toward the back of his cupboard . . . , just these are the substances which are the delight of the colloid chemist.

Indeed, those old organic colloids (now polymers) and inorganic colloids such as soap micelles and silver or sulfur sols have much in common (11):

1. Both types of particles are relatively small, 10^{-6} to 10^{-4} mm, and visible via ultramicroscopy† as dancing light flashes, that is, Brownian motion.
2. The elemental composition does not change with the size of the particle.

Thus, soap micelles (true aggregates) and polymer chains (which repeat the same structure but are covalently bonded) appeared the same in those days. Partial valences (see Section 6.12) seemed to explain the bonding in both types.

† Ultramicroscopy is an old method used to study very small particles dispersed in a fluid for examination, and below normal resolution. Although invisible in ordinary light, colloidal particles become visible when intensely side-illuminated against a dark background.

In 1920 Herman Staudinger (13,14) enunciated the *Macromolecular Hypothesis*. It states that certain kinds of these colloids actually consist of very long-chained molecules. These came to be called polymers because many (but not all) were composed of the same repeating unit, or mer. In 1953 Staudinger won the Nobel prize in chemistry for his discoveries in the chemistry of macromolecular substances (15). The Macromolecular Hypothesis is the origin of modern polymer science, leading to our current understanding of how and why such materials as plastics and rubber have the properties they do.

1.7 HISTORICAL DEVELOPMENT OF INDUSTRIAL POLYMERS

Like most other technological developments, polymers were first used on an empirical basis, with only a very incomplete understanding of the relationships between structure and properties. The first polymers used were natural products that date back to antiquity, including wood, leather, cotton, various grasses for fibers, papermaking, and construction, wool, and protein animal products boiled down to make glues and related material.

Then came several semisynthetic polymers, which were natural polymers modified in some way. One of the first to attain commercial importance was cellulose nitrate plasticized with camphor, popular around 1885 for stiff collars and cuffs as celluloid, later most notably used in Thomas Edison's motion picture film (11). Cellulose nitrates were also sold as lacquers, used to coat wooden staircases, and so on. The problem was the terrible fire hazard existing with the nitrates, which were later replaced by the acetates.

Other early polymer materials included Chardonnet's artificial silk, made by regenerating and spinning cellulose nitrate solutions, eventually leading to the viscose process for making rayon (see Section 6.10) still in use today.

The first truly synthetic polymer was a densely cross-linked material based on the reaction of phenol and formaldehyde; see Section 14.2. The product, called Bakelite, was manufactured from 1910 onward for applications ranging from electrical appliances to phonograph records (16,17). Another early material was the General Electric Company's Glyptal, based on the condensation reaction of glycerol and phthalic anhydride (18), which followed shortly after Bakelite. However, very little was known about the actual chemical structure of these polymers until after Staudinger enunciated the Macromolecular Hypothesis in 1920.

All of these materials were made on a more or less empirical basis; trial and error have been the basis for very many advances in history, including polymers. However, in the late 1920s and 1930s, a DuPont chemist by the name of Wallace Carothers succeeded in establishing the reality of the Macromolecular Hypothesis by bringing the organic-structural approach back to the study of polymers, resulting in the discovery of nylon and neoprene. Actually the first polymers that Carothers discovered were polyesters (19). He reasoned that if the Macromolecular Hypothesis was correct, then if one mixed a molecule with dihydroxide end groups with a another molecule with diacid end

Table 1.7 Some natural product polymers

Name	Source	Application
Cellulose	Wood, cotton	Paper, clothing, rayon, cellophane
Starch	Potatoes, corn	Food, thickener
Wool	Sheep	Clothing
Silk	Silkworm	Clothing
Natural rubber	Rubber tree	Tires
Pitch	Oil deposits	Coating, roads

Some of the more important commercial natural polymers are shown in Table 1.7. People sometimes refer to these polymers as natural products or renewable resources.

Wool and silk are both proteins. All proteins are actually copolymers of polyamide-2 (or nylon-2, old terminology). As made by plants and animals, however, the copolymers are highly ordered, and they have monodisperse molecular weights, meaning that all the chains have the same molecular weights.

Cellulose and starch are both polysaccharides, being composed of chains of glucose-based rings but bonded differently. Their structures are discussed further in Appendix 2.1.

Natural rubber, the hydrocarbon polyisoprene, more closely resembles chain polymerized materials. In fact synthetic polyisoprene can be made either by free radical polymerization or anionic polymerization. The natural and synthetic products compete commercially with each other.

Pitch, a decomposition product, usually contains a variety of aliphatic and aromatic hydrocarbons, some of very high molecular weight.

1.5 CROSS-LINKING, PLASTICIZERS, AND FILLERS

The above provides a brief introduction to simple homopolymers, as made pure. Only a few of these are finally sold as "pure" polymers, such as polystyrene drinking cups and polyethylene films. Much more often, polymers are sold with various additives. That the student may better recognize the polymers, the most important additives are briefly discussed.

On heating, linear polymers flow and are termed thermoplastics. To prevent flow, polymers are sometimes cross-linked (•):

(1.16)

Table 1.8 Commercialization dates of selected synthetic polymers (20)

Year	Polymer	Producer
1909	Poly(phenol–*co*–formaldehyde)	General Bakelite Corporation
1927	Poly(vinyl chloride)	B.F. Goodrich
1929	Poly(styrene–*stat*–butadiene)	I.G. Farben
1930	Polystyrene	I.G. Farben/Dow
1936	Poly(methyl methacrylate)	Rohm and Haas
1936	Nylon 66 (Polyamide 66)	DuPont
1936	Neoprene (chloroprene)	DuPont
1939	Polyethylene	ICI
1943	Poly(dimethylsiloxane)	Dow Corning
1954	Poly(ethylene terephthalate)	ICI
1960	Poly(*p*-phenylene terephthalamide)[a]	DuPont
1982	Polyetherimide	GEC

[a] Kevlar; see Chapter 7.

groups and allowed them to react, a long, linear chain should result if the stoichiometry was one-to-one.

The problem with the aliphatic polyesters made at that time was their low melting point, making them unsuitable for clothing fibers because of hot water washes and ironing. When the ester groups were replaced with the higher melting amide groups, the nylon series was born. In the same time frame, Carothers discovered neoprene, which was a chain-polymerized product of an isoprene-like monomer with a chlorine replacing the methyl group.

Bakelite was a thermoset; that is, it did not flow after the synthesis was complete (20). The first synthetic thermoplastics, materials that could flow on heating, were poly(vinyl chloride), poly(styrene–*stat*–butadiene), polystyrene, and polyamide 66; see Table 1.8 (20). Other breakthrough polymers have included the very high modulus aromatic polyamides, known as Kevlar™ (see Section 7.4), and a host of high temperature polymers.

Further items on the history of polymer science can be found in Appendix 5.1, and Sections 6.1.1 and 6.1.2.

1.8 MOLECULAR ENGINEERING

The discussion above shows that polymer science is an admixture of pure and applied science. The structure, molecular weight, and shape of the polymer molecule are all closely tied to the physical and mechanical properties of the final material.

This book emphasizes physical polymer science, the science of the interrelationships between polymer structure and properties. Although much of the material (except the polymer syntheses) is developed in greater detail in the remaining chapters, the intent of this chapter is to provide an overview of the subject and a simple recognition of polymers as encountered in everyday life. In addition to the books in the General Reading section, a listing of handbooks, encyclopedias, and websites is given at the end of this chapter.

REFERENCES

1. L. Mandelkern and G. M. Stack, *Macromolecules*, **17**, 87 (1984).
2. P. J. Flory, *Principles of Polymer Chemistry*, Cornell University, Ithaca, NY, 1953.
3. R. P. Wool, *Polymer Interfaces: Structure and Stength*, Hanser, Munich, 1995.
4. L. E. Nielsen and R. F. Landel, *Mechanical Properties of Polymers*, Reinhold, New York, 1994.
5. H. Pasch and B. Trathnigg, *HPLC of Polymers*, Springer, Berlin, 1997.
6. T. C. Ward, *J. Chem. Ed.*, **58**, 867 (1981).
7. L. H. Sperling et al., *J. Chem. Ed.*, **62**, 780, 1030 (1985).

7a. M. S. Alger, *Polymer Science Dictionary*, Elsevier, New York, 1989.

8. A. D. Jenkins, in *Chemical Nomenclature*, K. J. Thurlow, ed., Kluwer Academic Publishers, Dordrecht, 1998.
9. (a) E. S. Wilks, *Polym. Prepr.*, **40**(2), 6 (1999); (b) N. A. Platé and I. M. Papisov, *Pure Appl. Chem.*, **61**, 243 (1989).
10. P. W. Morgan and S. L. Kwolek, *J. Chem. Ed.*, **36**, 182, 530 (1959).
11. Y. Furukawa, *Inventing Polymer Science*, University of Pennsylvania Press, Philadelphia, 1998.
12. W. Ostwald, *An Introduction to Theoretical and Applied Colloid Chemistry: The World of Neglected Dimensions*, Dresden and Leipzig, Verlag von Theodor Steinkopff, 1917.
13. H. Staudinger, *Ber.*, **53**, 1073 (1920).
14. H. Staudinger, *Die Hochmolecular Organischen Verbindung*, Springer, Berlin, 1932; reprinted 1960.
15. E. Farber, *Nobel Prize Winners in Chemistry, 1901–1961*, rev. ed., Abelard-Schuman, London, 1963.
16. H. Morawitz, *Polymers: The Origins and Growth of a Science*, Wiley-Interscience, New York, 1985.
17. L. H. Sperling, *Polymer News*, **132**, 332 (1987).
18. R. H. Kienle and C. S. Ferguson, *Ind. Eng. Chem.*, **21**, 349 (1929).
19. D. A. Hounshell and J. K. Smith, *Science and Corporate Strategy: DuPont R&D, 1902–1980*, Cambridge University Press, Cambridge, 1988.
20. L. A. Utracki, *Polymer Alloys and Blends*, Hanser, New York, 1990.

GENERAL READING

H. R. Allcock, F. W. Lampe, and J. E. Mark, *Contemporary Polymer Chemistry*, 3rd ed., Pearson Prentice-Hall, Upper Saddle River, NJ, 2003.

P. Bahadur and N. V. Sastry, *Principles of Polymer Science*, CRC Press, Boca Raton, FL, 2002.

D. I. Bower, *An Introduction to Polymer Physics*, Cambridge University Press, Cambridge, U.K., 2002.

I. M. Campbell, *Introduction to Synthetic Polymers*, Oxford University Press, Oxford, England, 2000.

C. E. Carraher Jr., *Giant Molecules: Essential Materials for Everyday Living and Problem Solving*, 2nd ed., Wiley-Interscience, Hoboken, NJ, 2003.

C. E. Carraher Jr., *Seymour/Carraher's Polymer Chemistry: An Introduction*, 6th ed., Dekker, New York, 2004.

M. Doi, *Introduction to Polymer Physics*, Oxford Science, Clarendon Press, Wiley, New York, 1996.

R. O. Ebewele, *Polymer Science and Technology,* CRC Press, Boca Raton, FL, 2000.

U. Eisele, *Introduction to Polymer Physics*, Springer, Berlin, 1990.

H. G. Elias, *An Introduction to Polymer Science*, VCH, Weinheim, 1997.

J. R. Fried, *Polymer Science and Technology*, 2nd ed., Prentice-Hall, Upper Saddle River, NJ, 2003.

U. W. Gedde, *Polymer Physics*, Chapman and Hall, London, 1995.

A. Yu. Grosberg and A. R. Khokhlov, *Giant Molecules*, Academic Press, San Diego, 1997.

A. Kumar and R. K. Gupta, *Fundamentals of Polymers*, McGraw-Hill, New York, 1998.

J. E. Mark, H. R. Allcock, and R. West, *Inorganic Polymers*, Prentice-Hall, Englewood Cliffs, NJ, 1992.

J. E. Mark, A. Eisenberg, W. W. Graessley, L. Mandelkern, E. T. Samulski, J. L. Koenig, and G. D. Wignall, *Physical Properties of Polymers*, 2nd ed., American Chemical Society, Washington, DC, 1993.

N. G. McCrum, C. P. Buckley, and C. B. Bucknall, *Principles of Polymer Engineering*, 2nd ed., Oxford Science, Oxford, England, 1997.

P. Munk and T. M. Aminabhavi, *Introduction to Macromolecular Science,* 2nd ed., Wiley-Interscience, Hoboken, NJ, 2002.

P. C. Painter and M. M. Coleman, *Fundamentals of Polymer Science: An Introductory Text*, 2nd ed., Technomic, Lancaster, 1997.

J. Perez, *Physics and Mechanics of Amorphous Polymers*, Balkema, Rotterdam, 1998.

A. Ram, *Fundamentals of Polymer Engineering*, Plenum Press, New York, 1997.

A. Ravve, *Principles of Polymer Chemistry*, 2nd ed., Kluwer, Norwell, MA, 2000.

F. Rodriguez, C. Cohen, C. K. Ober, and L. Archer, *Principles of Polymer Systems*, 5th ed., Taylor and Francis, Washington, DC, 2003.

M. Rubinstein, *Polymer Physics,* Oxford University Press, Oxford, 2003.

A. Rudin, *The Elements of Polymer Science and Engineering*, 2nd ed., Academic Press, San Diego, 1999.

M. P. Stevens, *Polymer Chemistry: An Introduction*, 3rd ed., Oxford University Press, New York, 1999.

G. R. Strobl, *The Physics of Polymers*, 2nd ed., Springer, Berlin, 1997.

A. B. Strong, *Plastics Materials and Processing*, 2nd ed., Prentice Hall, Upper Saddle River, NJ, 2000.

HANDBOOKS, ENCYCLOPEDIAS, AND DICTIONARIES

M. Alger, *Polymer Science Dictionary*, 2nd ed., Chapman and Hall, London, 1997.

G. Allen, ed., *Comprehensive Polymer Science*, Pergamon, Oxford, 1989.

Compendium of Macromolecular Nomenclature, IUPAC, CRC Press, Boca Raton, FL, 1991.

ASM, *Engineered Materials Handbook, Volume 2: Engineering Plastics*, ASM International, Metals Park, OH, 1988.

D. Bashford, ed., *Thermoplastics: Directory and Databook*, Chapman and Hall, London, 1997.

J. Brandrup, E. H. Immergut, and E. A. Grulke, eds., *Polymer Handbook*, 4th ed., Wiley-Interscience, New York, 1999.

S. H. Goodman, *Handbook of Thermoset Plastics*, 2nd ed., Noyes Publishers, Westwood, NJ, 1999.

C. A. Harper, ed., *Handbook of Plastics, Elastomers, and Composites*, McGraw-Hill, New York, 2002.

W. A. Kaplan, ed., *Modern Plastics World Encyclopedia*, McGraw-Hill, New York, 2004 (published annually).

H. G. Karian, ed., *Handbook of Polypropylene and Polypropylene Composites*, 2nd ed., Marcel Dekker, New York, 2003.

J. I. Kroschwitz ed., *Encyclopedia of Polymer Science and Engineering*, 3rd ed., Wiley, Hoboken, NJ, 2004.

J. E. Mark, ed., *Polymer Data Handbook*, Oxford University Press, New York, 1999.

J. E. Mark, ed., *Physical Properties of Polymers Handbook*, Springer, New York, 1996.

H. S. Nalwa, *Encyclopedia of Nanoscience and Nanotechnology*, 10 Vol., American Scientific Publications, Stevenson Ranch, CA, 2004.

O. Olabisi, ed., *Handbook of Thermoplastics*, Marcel Dekker, New York, 1997.

D. V. Rosato, *Rosato's Plastics Encyclopedia and Dictionary*, Hanser Publishers, Munich, 1993.

J. C. Salamone, ed., *Polymer Materials Encyclopedia*, CRC Press, Boca Raton, FL, 1996.

D. W. Van Krevelen, *Properties of Polymers*, 3rd ed., Elsevier, Amsterdam, 1997.

C. Vasile, ed., *Handbook of Polyolefins*, 2nd ed., Marcel Dekker, New York, 2000.

T. Whelen, *Polymer Technology Dictionary*, Chapman and Hall, London, 1992.

E. S. Wilks, ed., *Industrial Polymers Handbook, Vol. 1–4*, Wiley-VCH, Weinheim, 2001.

G. Wypych, *Handbook of Fillers*, 2nd ed., William Anderson, Norwich, NY, 1999.

WEB SITES

Case-Western Reserve University, Department of Macromolecular Chemistry: *http://abalone.cwru.edu/tutorial/enhanced/main.htm*

Chemical Abstracts: *http://www.cas.org/EO/polymers.pdf*

Conducts classroom teachers polymer workshops: *http://www.polymerambassadors.org*

Educational materials about polymers: *http://matse1.mse.uiuc.edu/~tw/polymers/polymers.html*

History of polymers, activities, and tutorials: *http://www.chemheritage.org/EducationalServies/faces/poly/home.htm*

Online courses in polymer science and engineering: *http://agpa.uakron.edu*

Pennsylvania College of Technology, Pennsylvania State University, and University of Massachusetts at Lowell: *http://www.pct.edu/prep/*

Polymer education at the K-12 level: *http://www.uwsp.edu//chemistry/ipec.htm*

Recycling of plastics: *http://www.plasticbag.com/environmental/pop.html*

Teacher's workshops in materials and polymers: *http://matse1.mse.uiuc.edu/~tw*

Teaching of plastics and science: *http://www.teachingplastics.org*

The American Chemical Society Polymer Education Committee site: *http://www.polyed.org*

The National Plastics Center & Museum main page; museum, polymer education, PlastiVan: *http://www.plasticsmuseum.org*

The Society of Plastics Engineers main page; training and education, scholarships: *http://www.4spe.org*

The Society of the Plastics Industry main page; information about plastics, environmental issues: *http://www.plasticsindustry.org/outreach/environment/index.htm*

University of Southern Mississippi, Dept. of Polymer Science, *The Macrogalleria: http://www.psrc.usm.edu/macrog/index.html*

World Wide Web sites for polymer activities and information: *http://www.polymerambassadors.org/WWWsites2.htm*

STUDY PROBLEMS

1. Polymers are obviously different from small molecules. How does polyethylene differ from oil, grease, and wax, all of these materials being essentially $-CH_2-$?

2. Write chemical structures for polyethylene, polyproplyene, poly(vinyl chloride), polystyrene, and polyamide 66.

3. Name the following polymers:

$$\text{(a)}\quad -\!\!\left[CH_2-\underset{\substack{|\\ O=C-O-CH_3}}{\overset{\substack{H\\ |}}{C}}\right]\!\!-_n$$

$$\text{(b)}\quad -\!\!\left[CH_2-\underset{\substack{|\\ O=C-O-C_2H_5}}{\overset{\substack{CH_3\\ |}}{C}}\right]\!\!-_n$$

$$\text{(c)}\quad -\!\!\left[CH_2-\underset{\substack{|\\ O-\underset{\substack{\|\\ O}}{C}-CH_3}}{\overset{\substack{H\\ |}}{C}}\right]\!\!-_n$$

$$\text{(d)}\quad -\!\!\left[CH_2-CF_2\right]\!\!-_n$$

4. What molecular characteristics are required for good mechanical properties? Distinguish between amorphous and crystalline polymers.

5. Show the synthesis of polyamide 610 from the monomers.

6. Name some commercial polymer materials by chemical name that are (a) amorphous, cross-linked, and above T_g; (b) crystalline at ambient temperatures.

7. Take any 10 books off a shelf and note the last page number. What are the number-average and weight-average number of pages of these books? Why is the weight-average number of pages greater than the number-average? What is the polydispersity index? Can it ever be unity?

8. Draw a log modulus–temperature plot for an amorphous polymer. What are the five regions of viscoelasticity, and where do they fit? To which regions do the following belong at room temperature: chewing gum, rubber bands, Plexiglas®?

9. Define the terms: Young's modulus, tensile strength, chain entanglements, and glass–rubber transition.

10. A cube 1 cm on a side is made up of one giant polyethylene molecule, having a density of 1.0 g/cm^3. (a) What is the molecular weight of this molecule? (b) Assuming an all trans conformation, what is the contour length of the chain (length of the chain stretched out)? Hint: The mer length is 0.254 nm.

APPENDIX 1.1 NAMES FOR POLYMERS

The IUPAC Macromolecular Nomenclature Commission has developed a systematic nomenclature for polymers (A1, A2). The Commission recognized, however, that a number of common polymers have semisystematic or trivial names that are well established by usage. For the reader's convenience, the recommended trivial name (or the source-based name) of the polymer is given under the polymer structure, and then the structure-based name is given. For example, the trivial name, polystyrene, is a source-based name, literally "the polymer made from styrene." The structure-based name, poly(1-phenylethylene), is useful both in addressing people who may not be familiar with the structure of polystyrene and in cases where the polymer is not well known. This book uses a source-based nomenclature, unless otherwise specified. The following structures are IUPAC recommended.

$-\!\!(CH_2CH_2)\!\!-_n$

polyethylene
poly(methylene)

$-\!\!(CH=CHCH_2CH_2)\!\!-_n$

polybutadiene[a]
poly(1-butenylene)

$-\!\!(\underset{\displaystyle CH_3}{\underset{|}{C}}HCH_2)\!\!-_n$

polypropylene
poly(1-methylethylene)

$-\!\!(\underset{\displaystyle CH_3}{\underset{|}{C}}=CHCH_2CH_2)\!\!-_n$

polyisoprene[b]
poly(1-methyl-1-butenylene)

$-\!\!(CH_2-\overset{\displaystyle CH_3}{\overset{|}{\underset{\displaystyle CH_3}{\underset{|}{C}}}})\!\!-_n$

polyisobutylene
poly(1,1-dimethylethylene)

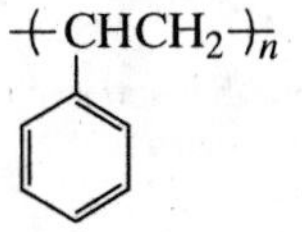

$-\!\!(\underset{\displaystyle C_6H_5}{\underset{|}{C}}HCH_2)\!\!-_n$

polystyrene
poly(1-phenylethylene)

$-\!\!(\underset{\displaystyle CN}{\underset{|}{C}}HCH_2)\!\!-_n$

polyacrylonitrile
poly(1-cyanoethylene)

$-\!\!(\underset{\displaystyle OH}{\underset{|}{C}}HCH_2)\!\!-_n$

poly(vinyl alcohol)
poly(1-hydroxyethylene)

$-\!\!(\underset{\displaystyle OOCCH_3}{\underset{|}{C}}HCH_2)\!\!-_n$

poly(vinyl acetate)
poly(1-acetoxyethylene)

$-\!\!(\underset{\displaystyle Cl}{\underset{|}{C}}HCH_2)\!\!-_n$

poly(vinyl chloride)
poly(1-chloroethylene)

$-\!\!(\overset{\displaystyle F}{\overset{|}{\underset{\displaystyle F}{\underset{|}{C}}}}CH_2)\!\!-_n$

poly(vinylidene Fluoride)
poly(1,1-difluoroethylene)

---

[a]Polybutadiene is usually written $-\!\!(CH_2CH=CHCH_2)\!\!-_n$, that is, with the double bond in the center. The structure-based name is given.

[b]Polyisoprene is usually written $-\!\!(CH_2\overset{\displaystyle CH_3}{\overset{|}{C}}=CHCH_2)\!\!-_n$.

$\text{-(}CF_2CF_2\text{)-}_n$ poly(tetrafluoroethylene) poly(difluoromethylene)	-(1,3-dioxane ring$\text{-}CH_2\text{)-}_n$ with C_3H_7 at C-2 poly(vinyl butyral) poly[(2-propyl-1,3-dioxane-4,6-diyl)methylene]
$\text{-(}CH(COOCH_3)CH_2\text{)-}_n$ poly(methyl acrylate) poly[1-(methoxycarbonyl)ethylene]	$\text{-(}C(CH_3)(COOCH_3)\text{-}CH_2\text{)-}_n$ poly(methyl methacrylate) poly[1-(methoxycarbonyl)-1-methylethylene]
$\text{-(}OCH_2\text{)-}_n$ polyformaldehyde poly(oxymethylene)	$\text{-(}O\text{-}C_6H_4\text{)-}_n$ poly(phenylene oxide) poly(oxy-1,4-phenylene)
$\text{-(}NH(CH_2)_6NHCO(CH_2)_4CO\text{)-}_n$ polyamide 66[a] poly(hexamethylene adipamide) poly(iminohexamethyleneiminoadipoyl)	$\text{-(}OCH_2CH_2\text{)-}_n$ poly(ethylene oxide) poly(oxyethylene)
$\text{-(}OCH_2CH_2OOC\text{-}C_6H_4\text{-}CO\text{)-}_n$ poly(ethylene terephthalate) poly(oxyethyleneoxyterephthaloyl)	$\text{-(}NHCO(CH_2)_5\text{)-}_n$ polyamide 6[b] poly(ε-caprolactam) poly[imino(1-oxohexamethylene)]

[a] Common name. Other ways this is named include nylon 6,6, 66-nylon, 6,6-nylon, and nylon 66.
[b] Common name.

REFERENCE

A1. E. S. Wilks, *Polym. Prepr.*, **40(2)**, 6 (1999).

A2. N. A. Platé and I. M. Papisov, *Pure Appl. Chem.*, **61**, 243 (1989).

5

THE AMORPHOUS STATE

The bulk state, sometimes called the condensed or solid state, includes both amorphous and crystalline polymers. As opposed to polymer solutions, generally there is no solvent present. This state comprises polymers as ordinarily observed, such as plastics, elastomers, fibers, adhesives, and coatings.

While amorphous polymers do not contain any crystalline regions, "crystalline" polymers generally are only semicrystalline, containing appreciable amounts of amorphous material. When a crystalline polymer is melted, the melt is amorphous. In treating the kinetics and thermodynamics of crystallization, the transformation from the amorphous state to the crystalline state and back again is constantly being considered. The subjects of amorphous and crystalline polymers are treated in the next two chapters. This will be followed by a discussion of liquid crystalline polymers, Chapter 7. Although polymers in the bulk state may contain plasticizers, fillers, and other components, this chapter emphasizes the polymer molecular organization itself.

A few definitions are in order. Depending on temperature and structure, amorphous polymers exhibit widely different physical and mechanical behavior patterns. At low temperatures, amorphous polymers are glassy, hard, and brittle. As the temperature is raised, they go through the glass–rubber transition. The glass transition temperature (T_g) is defined as the temperature at which the polymer softens because of the onset of long-range coordinated molecular motion. This is the subject of Chapter 8.

Above T_g, cross-linked amorphous polymers exhibit rubber elasticity. An example is styrene–butadiene rubber (SBR), widely used in materials ranging

Introduction to Physical Polymer Science, by L.H. Sperling
ISBN 0-471-70606-X

from rubber bands to automotive tires. Rubber elasticity is treated in Chapter 9. Linear amorphous polymers flow above T_g.

Polymers that cannot crystallize usually have some irregularity in their structure. Examples include the atactic vinyl polymers and statistical copolymers.

5.1 THE AMORPHOUS POLYMER STATE

5.1.1 Solids and Liquids

An amorphous polymer does not exhibit a crystalline X-ray diffraction pattern, and it does not have a first-order melting transition. If the structure of crystalline polymers is taken to be regular or ordered, then by difference, the structure of amorphous polymers contains greater or lesser amounts of disorder.

The older literature often referred to the amorphous state as a liquid state. Water is a noncrystalline (amorphous) condensed substance and is surely a liquid. However, polymers such as polystyrene or poly(methyl methacrylate) at room temperature are glassy, taking months or years for significant creep or flow. By contrast, skyscrapers are also undergoing creep (or flow), becoming measurably shorter as the years pass, as the steel girders creep (or flow). Today, amorphous polymers in the glassy state are better called amorphous solids.

Above the glass transition temperature, if the polymer is amorphous and linear, it will flow, albeit the viscosity may be very high. Such materials are liquids in the modern sense of the term. It should be noted that the glass transition itself is named after the softening of ordinary glass, an amorphous inorganic polymer. If the polymer is crystalline, the melting temperature is always above the glass transition temperature.

5.1.2 Possible Residual Order in Amorphous Polymers?

As a point of focus, the evidence for and against partial order in amorphous polymers is presented. On the simplest level, the structure of bulk amorphous polymers has been likened to a pot of spaghetti, where the spaghetti strands weave randomly in and out among each other. The model would be better if the strands of spaghetti were much longer, because by ratio of length to diameter, spaghetti more resembles wax chain lengths than it does high polymers.

The spaghetti model provides an entry into the question of residual order in amorphous polymers. An examination of relative positions of adjacent strands shows that they have short regions where they appear to lie more or less parallel. One group of experiments finds that oligomeric polymers also exhibit similar parallel regions (1,2). Accordingly, the chains appear to lie parallel for short runs because of space-filling requirements, permitting a higher density (3). This point, the subject of much debate, is discussed further later.

Questions of interest to amorphous state studies include the design of critical experiments concerning the shape of the polymer chain, the estimation of type and extent of order or disorder, and the development of models suitable for physical and mechanical applications. It must be emphasized that our knowledge of the amorphous state remains very incomplete, and that this and other areas of polymer science are the subjects of intensive research at this time. Pechhold and Grossmann (4) capture the spirit of the times exactly:

> Our current knowledge about the level of order in amorphous polymers should stimulate further development of competing molecular models, by making their suppositions more precise in order to provide a bridge between their microscopic structure description and the understanding of macroscopic properties, thereby predicting effects which might be proved experimentally.

The subject of structure in amorphous polymers has been entensively reviewed (5–11) and has been the subject of two published symposia (12,13).

5.2 EXPERIMENTAL EVIDENCE REGARDING AMORPHOUS POLYMERS

The experimental methods used to characterize amorphous polymers may be divided into those that measure relatively short-range interactions (nonrandom versus random chain positions) (14), below about 20 Å, and those that measure longer-range interactions. In the following paragraphs the role of these several techniques will be explored. The information obtainable from these methods is summarized in Table 5.1.

5.2.1 Short-Range Interactions in Amorphous Polymers

Methods that measure short-range interactions can be divided into two groups: those that measure the orientation or correlation of the mers along the *axial* direction of a chain, and those that measure the order between chains, in the *radial* direction. Figure 5.1 illustrates the two types of measurements.

There are several measures of the axial direction in the literature. Two of the more frequently used are the *Kuhn segment length* (see Section 5.3.1.2) and the *persistence length* (15). The persistence length, $p = C_\infty/6$, is a measure of chain dimensions particularly consided to be the length that a segment of chain "remembers" the direction the first mer was pointed. C_∞ is defined in Section 5.3.1.1. For example, the value of the persistence length for polyethylene is 5.75 Å, comprising only a few mers.

One of the most powerful experimental methods of determining short-range order in polymers utilizes birefringence (6). Birefringence measures orientation in the axial direction. The birefringence of a sample is defined by

Table 5.1 Selected studies of the amorphous state

Method	Information Obtainable	Principal Findings	Reference
	A. Short-Range Interactions		
Stress–optical coefficient	Orientation of segments in isolated chain	Orientation limited to 5–10 Å	(a)
Depolarized light-scattering	Segmental orientation correlation	2–3 —CH_2— units along chain correlated	(b)
Magnetic birefringence	Segmental orientation correlation	Orientation correlations very small	(b)
Raman scattering	Trans and gauche populations	Little or no modification in chain conformation initiated by intermolecular forces	(b)
NMR relaxation	Relaxation times	Small fluctuating bundles in the melt	(c)
Small-angle X-ray scattering, SAXS	Density variations	Amorphous polymers highly homogeneous; thermal fluctuations predominate	(d)
Birefringence	$n_1 - n_2$	Orientation	
	B. Long-Range Interactions		
Small-angle neutron scattering	Conformation of single chains	Radius of gyration the same in melt as in θ-solvents	(e, f)
Electron microscopy	Surface inhomogeneities	Nodular structures of 50–200 Å in diameter	(g, h)
Electron diffraction and wide-angle X-ray diffraction	Amorphous halos	Bundles of radial dimension = 25 Å and axial dimension = 50 Å, but order may extend to only one or two adjacent chains	(i, j)
	C. General		
Enthalpy relaxation	Deviations from equilibrium state	Changes not related to formation of structure	(k)
Density	Packing of chains	Density in the amorphous state is about 0.9 times the density in the crystalline state	(l, m)

References: (a) R. S. Stein and S. D. Hong, *J. Macromol. Sci. Phys.*, **B12** (11), 125 (1976). (b) E. W. Fischer, G. R. Strobl, M. Dettenmaier, M. Stamm, and N. Steidle, *Faraday Discuss. Chem. Soc.*, **68**, 26 (1979). (c) W. L. F. Golz and H. G. Zachmann, *Makromol. Chem.*, **176**, 2721 (1975). (d) D. R. Uhlmann, *Faraday Discuss. Chem. Soc.*, **68**, 87 (1979). (e) H. Benoit, *J. Macromol. Sci. Phys.*, **B12** (1), 27 (1976). (f) G. D. Wignall, D. G. H. Ballard, and J. Schelten, *J. Macromol. Sci. Phys.*, **B12** (1), 75 (1976). (g) G. S. Y. Yeh, *Crit. Rev. Macromol. Sci.*, **1**, 173 (1972). (h) R. Lam and P. H. Gell, *J. Macromol. Sci. Phys.*, **B20** (1), 37 (1981). (i) Yu. K. Ovchinnikov, G. S. Markova, and V. A. Kargin, *Vysokomol. Soedin.* **AII** (2), 329 (1969). (j) R. Lovell, G. R. Mitchell, and A. H. Windle, *Faraday Discuss. Chem. Soc.*, **68**, 46 (1979). (k) S. E. B. Petrie, *J. Macromol. Sci. Phys.*, **B12** (2), 225 (1976). (l) R. E. Robertson, *J. Phys. Chem.*, **69**, 1575 (1965). (m) R. F. Boyer, *J. Macromol. Sci. Phys.*, **B12**, 253 (1976).

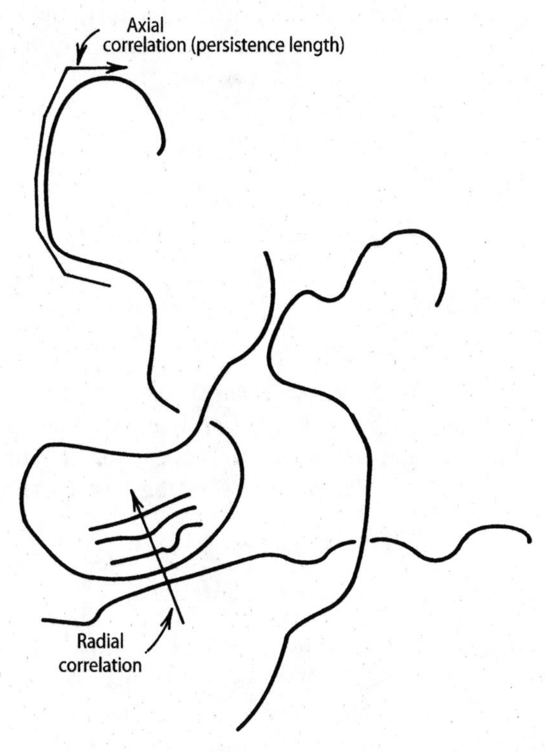

Figure 5.1 Schematic diagram illustrating the axial and radial correlation directions.

$$\Delta n = n_1 - n_2 \tag{5.1}$$

where n_1 and n_2 are the refractive indexes for light polarized in two directions 90° apart. If a polymer sample is stretched, n_1 and n_2 are taken as the refractive indexes for light polarized parallel and perpendicular to the stretching direction.

The anisotropy of refractive index of the stretched polymer can be demonstrated by placing a thin film between crossed polaroids. The field of view is dark before stretching, but vivid colors develop as orientation is imposed. For stretching at 45° to the polarization directions, the fraction of light transmitted is given by (6)

$$\mathbf{T} = \sin^2\left(\frac{\pi' d\, \Delta n}{\lambda_0}\right) \tag{5.2}$$

where d represents the thickness, λ_0 represents the wavelength of light in vacuum, and $\pi' = 3.14$.

By measuring the transmitted light quantitatively, the birefringence is obtained. The birefringence is related to the orientation of molecular units such as mers, crystals, or even chemical bonds by

$$\Delta n = \frac{2}{9}\pi \frac{(\bar{n}^2 + 2)^2}{\bar{n}} \sum_i (b_1 - b_2)_i f_i \tag{5.3}$$

where f_i is an orientation function of such units given by

$$f_i = \frac{3\cos^2\theta_i - 1}{2} \tag{5.4}$$

where θ_i is the angle that the symmetry axis of the unit makes with respect to the stretching direction, $\bar{n}$ is the average refractive index, and b_1 and b_2 are the polarizabilities along and perpendicular to the axes of such units.

Equation (5.4) contains two important solutions for fibers and films:

$$\theta = 0°, \text{ perfect orientation}$$

$$\theta = 54°, \text{ zero orientation}$$

Many commercial fibers such as nylon or rayon will have θ equal to about 5°.

The stress–optical coefficient (SOC) is a measure of the change in birefringence on stretching a sample under a stress σ (16)

$$\text{SOC} = \frac{\Delta n}{\sigma} \tag{5.5}$$

If the polymer is assumed to obey rubbery elasticity relations (see Chapter 9), then

$$\text{SOC} = \frac{\Delta n}{\sigma} = \frac{2\pi_1}{45kT} \frac{(\bar{n}^2 + 2)^2}{\bar{n}} (b_1 - b_2) \tag{5.6}$$

where $\pi_1 = 3.14$, and $\bar{n}$ represents the average refractive index. The change in birefringence that occurs when an amorphous polymer is deformed yields important information concerning the state of order in the amorphous solid. It should be emphasized that the theory expressed in equation (5.6) involves the orientation of segments within a single isolated chain. From an experimental point of view, it has been found that the strain–optical coefficient (STOC) is independent of the extension (17) but that the SOC is not.

The anisotropy of a segment is given by $b_1 - b_2$. Experiments carried out by Stein and Hong (16) on this quantity as a function of swelling and extension show no appreciable changes, leading to the conclusion that the order within a chain (axial correlation) does not change beyond a range of 5 to

10 Å, comparable with the range of ordering found for low-molecular-weight liquids.

Depolarized light-scattering (DPS) is a related technique whereby the intensity of scattered light is measured when the sample is irradiated by visible light. During this experiment, the sample is held between crossed Nicols. Studies on DPS on *n*-alkane liquids (13) reveal that there is a critical chain length of 8 to 9 carbons, below which there is no order in the melt. For longer chains, only 2 to 3 —CH_2— units in one chain are correlated with regard to their orientation, indicating an extremely weak orientational correlation.

Other electromagnetic radiation interactions with polymers useful for the study of short-range interactions in polymers include:

1. *Rayleigh scattering*: elastically scattered light, usually measured as a function of scattering angle.
2. *Brillouin scattering*: in essence a Doppler effect, which yields small frequency shifts.
3. *Raman scattering*: an inelastic process with a shift in wavelength due to chemical absorption or emission.

Results of measurements utilizing SOC, DPS, and other short-range experimental methods such as magnetic birefringence, Raman scattering, Brillouin scattering, NMR relaxation, and small-angle X-ray scattering are summarized in Table 5.1. The basic conclusion is that intramolecular orientation is little affected by the presence of other chains in the bulk amorphous state. The extent of order indicated by these techniques is limited to at most a few tens of angstroms, approximately that which was found in ordinary low-molecular-weight liquids.

5.2.2 Long-Range Interactions in Amorphous Polymers

5.2.2.1 Small-Angle Neutron Scattering The long-range interactions are more interesting from a polymer conformation and structure point of view. The most powerful of the methods now available is small-angle neutron scattering (SANS). For these experiments, the de Broglie wave nature of neutrons is utilized. Applied to polymers, SANS techniques can be used to determine the actual chain radius of gyration in the bulk state (18–24).

The basic theory of SANS follows the development of light-scattering (see Section 3.6.1). For small-angle neutron scattering, the weight-average molecular weight, M_w, and the z-average radius of gyration, R_g, may be determined (18–19):

$$\frac{Hc}{R(\theta)-R(\text{solvent})}=\frac{1}{M_w P(\theta)}+2A_2c \qquad (5.7)$$

Table 5.2 Evolution of SANS instrumentation

Method	Location	Comments
Long flight path	(a) ILL, Grenoble (b) Oak Ridge (c) Jülich	Inverse square distance law means long experimental times
Long wavelength	NIST	Neutrons cooled via liquid He or H_2
Time-of-flight[a] (TOF)	Los Alamos Nat. Lab	"White" neutrons, liquid H_2, pulsed source

[a] Pulsed neutrons are separated via TOF according to wavelength. TOF is to long flight path as FT-IR is to IR.

where $R(\theta)$ is the scattering intensity known as the "Rayleigh ratio,"

$$R(\theta) = \frac{I_\theta \omega^2}{I_0 V_s} \tag{5.8}$$

where ω represents the sample-detector distance, V_s is the scattering volume, and I_θ/I_0 is the ratio of scattered radiation intensity to the initial intensity (20–22). The quantity $P(\theta)$ is the scattering form factor, identical to the form factor used in light-scattering formulations [see equation (3.47)]. The formulation for $P(\theta)$, originally derived by Peter Debye (25), forms one of the mainspring relationships between physical measurements in both the dilute solution and solid states and in the interpretation of the data. For very small particles or molecules, $P(\theta)$ equals unity. In both equation (5.7) (explicit) and equation (5.8) (implicit), the scattering intensity of the solvent or background must be subtracted.

In SANS experiments of the type of interest here, a deuterated polymer is dissolved in an ordinary hydrogen-bearing polymer of the same type (or vice versa). The calculations are simplified if the two polymers have the same molecular weight. The background to be subtracted originates from the scattering of the protonated species, and the coherent scattering of interest originates from the dissolved deuterated species. The quantity H in equation (5.7) was already defined in equation (3.56) for neutron scattering. SANS instrumentation has evolved through several generations, as delineated in Table 5.2.

As currently used (26–27), the coherent intensity in a SANS experiment is described by the cross section, $d\Sigma/d\Omega$, which is the probability that a neutron will be scattered in a solid angle, Ω, per unit volume of the sample. This cross section, which is normally used to express the neutron scattering power of a sample, is identical with the quantity R defined in equation (5.8).

Then it is convenient to express equation (5.7) as

$$\frac{C_N}{d\Sigma/d\Omega} = \frac{1}{M_w P(\theta)} \tag{5.9}$$

Table 5.3 Scattering lengths of elements (20,21)

Element	Coherent Scattering Length[a] $b \times 10^{12}$ cm
Carbon, ^{12}C	0.665
Oxygen, ^{16}O	0.580
Hydrogen, ^{1}H	−0.374
Deuterium, ^{2}H	0.667
Fluorine, ^{19}F	0.560
Sulfur, ^{32}S	0.280

[a] Here $a = \Sigma_i b_i$.

where C_N, the analogue of H, may be expressed

$$C_N = \frac{(a_{\mathrm{H}} - a_{\mathrm{D}})^2 N_a \rho (1-n)n}{M_P^2} \tag{5.10}$$

The quantities a_{H} and a_{D} are the scattering length of a normal protonated and deuterated (labeled) structural unit (mer), and n is the mole fraction of labeled chains. Thus C_N contains the concentration term as well as the "optical" constants. The quantities a_{H} and a_{D} are calculated by adding up the scattering lengths of each atom in the mer (see Table 5.3). In the case of high dilution, the quantity $(1 - n)n$ reduces to the concentration c, as in equation (5.7).

After rearranging, equation (5.7) becomes

$$\left[\frac{d\Sigma}{d\Omega}\right]^{-1} = \frac{1}{C_N M_w}\left(1 + \frac{K^2 R_g^2}{3} + \cdots\right) \tag{5.11}$$

where K is the wave vector. Thus the mean square radius of gyration, R_g^2, and the polymer molecular weight, M_w, may be obtained from the ratio of the slope to the intercept and the intercept, respectively, of a plot of $[d\Sigma/d\Omega]^{-1}$ versus K^2. If $A_2 = 0$, this result is satisfactory. For finite A_2 values, a second extrapolation to zero concentration is required (see below).

A problem in neutron scattering and light-scattering alike stems from the fact that R_g is a z-average quantity, whereas the molecular weight is a weight-average quantity. The preferred solution has been to work with nearly monodisperse polymer samples, such as prepared by anionic polymerization. If the molecular weight distribution is known, an approximate correction can be made.

Typical data for polyprotostyrene dissolved in polydeuterostyrene are shown in Figure 5.2 (28). Use is made of the Zimm plot, which allows simultaneous plotting of both concentration and angular functions for a more compact representation of the data (29).

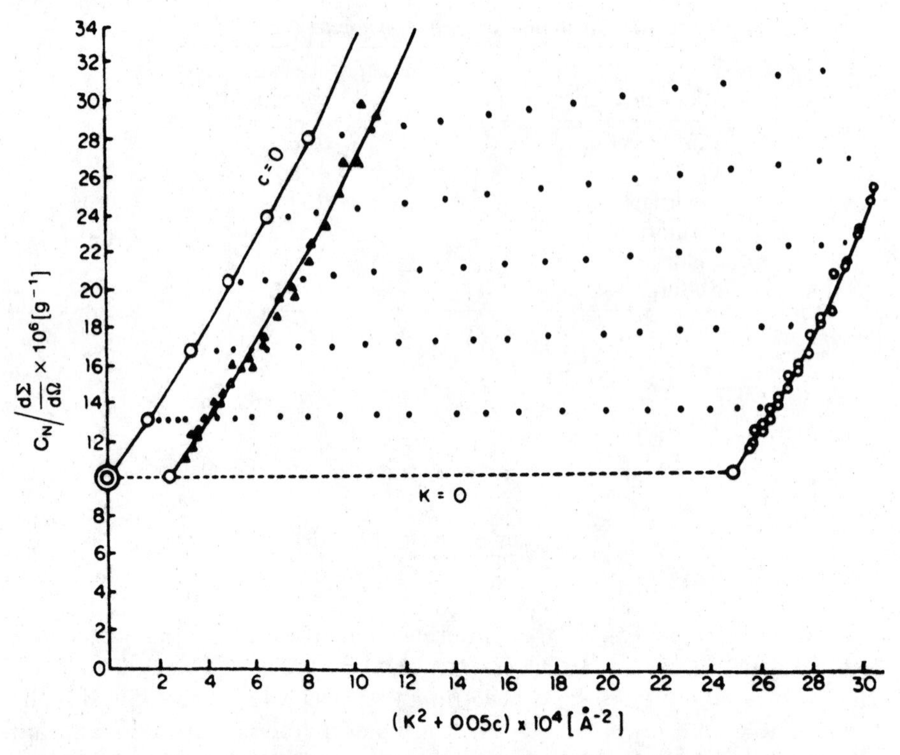

Figure 5.2 Small-angle neutron scattering of polyprotostyrene dissolved in polydeuterostyrene. A Zimm plot with extrapolations to both zero angle and zero concentration. Note that the second virial coefficient is zero, because polystyrene is essentially dissolved in polystyrene (28).

From data such as presented in Figure 5.2, both R_g and M_w may be calculated. The results are tabulated in Table 5.4. Also shown in Table 5.4 are corresponding data obtained by light-scattering in Flory θ-solvents, where the conformation of the chain is unperturbed because the free energies of solvent–polymer and polymer–polymer interactions are all the same (see Section 3.3).

Values of $(R_g^2/M)^{1/2}$ are shown in Table 5.4 because this quantity is independent of the molecular weight when the chain is unperturbed (30), being a constant characteristic of each polymer. An examination of Table 5.4 reveals that the values in θ-solvents and in the bulk state are identical within experimental error. This important finding confirms earlier theories (30) that these two quantities ought to be equal, since under these conditions the polymer chain theoretically is unable to distinguish between a solvent molecule and a polymer segment with which it may be in contact.† Since it was believed that

†This finding was predicted in 1953 by P. J. Flory, 20 years before it was confirmed experimentally.

Table 5.4 Molecular dimensions in bulk polymer samples (20)

		$(R_g^2/M_w)^{1/2}\ \frac{\text{Å}\cdot\text{mol}^{1/2}}{\text{g}^{1/2}}$			
Polymer	State of Bulk	SANS Bulk	Light-Scattering θ-Solvent	SAXS	Reference
Polystyrene	Glass	0.275	0.275	0.27 (i)	(a)
Polystyrene	Glass	0.28	0.275	—	(b)
Polyethylene	Melt	0.46	0.45	—	(c)
Polyethylene	Melt	0.45	0.45	—	(d)
Poly(methyl methacrylate)	Glass	0.31	0.30	—	(e)
Poly(ethylene oxide)	Melt	0.343	—	—	(f)
Poly(vinyl chloride)	Glass	0.30	0.37	—	(g)
Polycarbonate	Glass	0.457	—	—	(h)

References: (a) J. P. Cotton, D. Decker, H. Benoit, B. Farnoux, J. Higgins, G. Jannink, R. Ober, C. Picot, and J. desCloizeaux, *Macromolecules*, **7**, 863 (1974). (b) G. D. Wignall, D. G. Ballard, and J. Schelten, *Eur. Polym. J.*, **10**, 861 (1974). (c) J. Schelten, D. G. H. Ballard, G. Wignall, G. Longman, and W. Schmatz, *Polymer*, **17**, 751 (1976). (d) G. Lieser, E. W. Fischer, and K. Ibel, *J. Polym. Sci. Polym. Lett. Ed.*, **13**, 39 (1975). (e) R. G. Kirste, W. A. Kruse, and K. Ibel, *Polymer*, **16**, 120 (1975). (f) G. Allen, *Proc. R. Soc. Lond., Ser. A*, **351**, 381 (1976). (g) P. Herchenroeder and M. Dettenmaier, Unpublished manuscript (1977). (h) D. G. H. Ballard, A. N. Burgess, P. Cheshire, E. W. Janke, A. Nevin, and J. Schelten, *Polymer*, **22**, 1353 (1981). (i) H. Hayashi, F. Hamada, and A. Nakajima, *Macromolecules*, **9**, 543 (1976). (i) G. J. Fleer, M. A. C. Stuart, J. M. H. M. Scheutjens, T. Cosgrove, and B. Vincent, *Polymers at Interfaces*, Chapman and Hall, London, 1993.

polymer chains in dilute solution were random coils, this finding provided powerful evidence that random coils also existed in the bulk amorphous state.

The reader should note the similarities between $R_g = KM^{1/2}$ and the Brownian motion relationship, $X = k't^{1/2}$, where X is the average distance traversed. For random coils, the end-to-end distance $r = 6^{1/2}R_g$.

5.2.2.2 Electron and X-Ray Diffraction

Under various conditions, crystalline substances diffract X-rays and electrons to give spots or rings. According to Bragg's law,† these can be interpreted as interplanar spacings. Amorphous materials, including ordinary liquids, also diffract X-rays and electrons, but the diffraction is much more diffuse, sometimes called halos. For low-molecular-weight liquids, the diffuse halos have long been interpreted to mean that the nearest-neighbor spacings are slightly irregular and that after two or three molecular spacings all sense of order is lost. The situation is complicated in the case of polymers because of the presence of long chains. Questions to be resolved center about whether or not chains lie parallel for some distance, and if so, to what extent (31–34).

†See Section 6.2.2. Bragg's law: $n\lambda = 2d \sin\theta$, where $n = 1$ here, d is the distance between chains, and θ is the angle of diffraction.

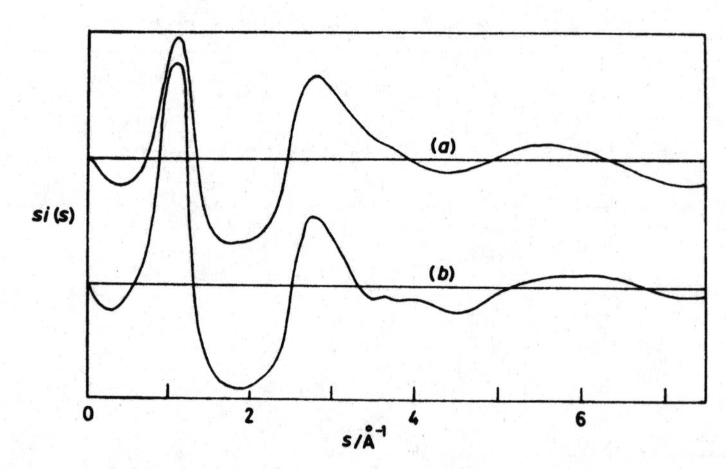

Figure 5.3 WAXS data on polytetrafluorethylene: (*a*) experimental data and (*b*) theory. Model is based on a disordered helix arranged with fivefold packing in a 24-Å diameter cylinder.

X-ray diffraction studies are frequently called wide-angle X-ray scattering, or WAXS. Typical data are illustrated in Figure 5.3 for polytetrafluorethylene (33). The first scattering maximum indicates the chain spacing distance. Maxima at larger values of *s* indicate other, shorter spacings. The (33) reduced intensity data are plotted as a function of angle,† $s = 4\pi_1(\sin\theta)/\lambda$, which is sometimes called inverse space because the dimensions are $Å^{-1}$. The diffracted intensity is plotted in the *y*-axis multiplied by the quantity *s* to permit the features to be more evenly weighted. Lovell et al. (33) fitted the experimental data with various theoretical models, also illustrated in Figure 5.3.

In analyzing WAXS data, the two different molecular directions must be borne in mind: (a) conformational orientation in the axial direction, which is a measure of how ordered or straight a given chain might be, and (b) organization in the radial direction, which is a direct measure of intermolecular order. WAXS measures both parameters. Lovell et al. (33) concluded from their study that the axial direction of molten polyethylene could be described by a chain with three rotational states, 0° and ±120°, with an average *trans* sequence length of three to four backbone bonds. The best radial packing model consisted of flexible chains arranged in a random manner (see Figure 5.3*b*).

Polytetrafluoroethylene, on the other hand, was found to have more or less straight chains in the axial direction for distances of at least 24 Å (30). Many other studies have also shown that this polymer has extraordinarily stiff chains, because of its extensive substitution. In the radial direction, a model of parallel straight-chain segments was strongly supported by the WAXS data,

†Variously, *K* is used by SANS experimenters for the angular function. The quantity *s* is called the scattering vector.

Table 5.5 Interchain spacing in selected amorphuos polymers

Polymer	Spacing, Å	Reference
Polyethylene	5.5	(a)
Silicone rubber	9.0	(a)
Polystyrene	10.0	(b)
Polycarbonate	4.8	(c)

References: (a) Y. K. Ovchinnikov, G. S. Markova, and V. A. Kargin, *Vysokomol. Soyed.*, **A11**, 329 (1969). (b) A. Bjornhaug, O. Ellefsen, and B. A. Tonnesen, *J. Polym. Sci.*, **12**, 621 (1954). (c) A. Siegmann and P. H. Geil, *J. Macromol. Sci.* (*Phys.*), **4** (2), 239 (1970).

although the exact nature of the packing and the extent of chain disorder are still the subject of current research. Poly(methyl methacrylate) and polystyrene were found to have a level of order intermediate between polyethylene and polytetrafluoroethylene.

The first interchain spacing of typical amorphous polymers is shown in Table 5.5. The greater interchain spacing of polystyrene and silicone rubbers is in part caused by bulky side groups compared with polyethylene.

The interpretation of diffraction data on amorphous polymers is currently a subject of debate. Ovchinnikov et al. (31,34) interpreted their electron diffraction data to show considerable order in the bulk amorphous state, even for polyethylene. Miller and co-workers (35,36) found that spacings increase with the size of the side groups, supporting the idea of local order in amorphous polymers. Fischer et al. (32), on the other hand, found that little or no order fits their data best. Schubach et al. (37) take an intermediate position, finding that they were able to characterize first- and second-neighbor spacings for polystyrene and polycarbonate, but no further.

5.2.2.3 General Properties Two of the most important general properties of the amorphous polymers are the density and the excess free energy due to nonattainment of equilibrium. The latter shows mostly smooth changes on relaxation and annealing (38) and is not suggestive of any particular order. Changes in enthalpy on relaxation and annealing are touched on in Chapter 8.

However, many polymer scientists have been highly concerned with the density of polymers (3,32). For many common polymers the density of the amorphous phase is approximately 0.85 to 0.95 that of the crystalline phase (3,10). Returning to the spaghetti model, some scientists think that the polymer chains have to be organized more or less parallel over short distances, or the experimental densities cannot be attained. Others (32) have pointed out that different statistical methods of calculation lead, in fact, to satisfactory agreement between the experimental densities and a more random arrangement of the chains.

Using computer simulation of polymer molecular packing, Weber and Helfand (39) studied the relative alignment of polyethylene chains expected from certain models. They calculated the angle between pairs of chords of chains (from the center of one bond to the center of the next), which showed a small but clear tendency toward alignment between closely situated molecular segments, and registered well-developed first and second density peaks. However, no long-range order was observed.

Table 5.1 summarized the several experiments designed to obtain information about the organization of polymer chains in the bulk amorphous state, both for short- and long-range order and for the general properties. Table 5.6 outlines some of the major order–disorder arguments. Some of these are discussed below. The next section is concerned with the development of molecular models that best fit the data and understanding obtained to date.

Table 5.6 Major order–disorder arguments in amorphous polymers

Order	Disorder
Conceptual difficulties in dense packing without order (a, b)	Rubber elasticity of polymer networks (e, f)
Appearance of nodules (b, c)	Absence of anomalous thermodynamic dilution effects (g)
Amorphous halos intensifying on equatorial plane during extension (d)	Radii of gyration the same in bulk as in θ-solvents (h)
Nonzero Mooney–Rivlin C_2 constants (a)	Fit of $P(\theta)$ for random coil model to scattering data (i)
Electron diffraction (l) lateral order to 15–20 Å	Rayleigh–Brillouin scattering, X-ray diffraction (j, k), stress–optical coefficient, etc. studies showing only modest (if any) short-range order (j, k, m, n)

References: (a) R. F. Boyer, *J. Macromol. Sci. Phys.*, **B12** (2), 253 (1976). (b) G. S. Y. Yeh, *Crit. Rev. Macromol. Sci.*, **1**, 173 (1972). (c) P. H. Geil, *Faraday Discuss. Chem. Soc.*, **68**, 141 (1979); but see S. W. Lee, H. Miyaji, and P. H. Geil, *J. Macromol. Sci., Phys.*, **B22** (3), 489 (1983); and D. R. Uhlmann, *Faraday Discuss. Chem. Soc.*, **68**, 87 (1979). (d) S. Krimm and A. V. Tobolsky, *Text. Res. J.*, **21**, 805 (1951). (e) P. J. Flory, *J. Macromol. Sci. Phys.*, **B12** (1), 1 (1976). (f) P. J. Flory, *Faraday Discuss. Chem. Soc.*, **68**, 14 (1979). (g) P. J. Flory, *Principles of Polymer Chemistry*, Cornell University Press, Ithaca, NY, 1953. (h) J. S. Higgins and R. S. Stein, *J. Appl. Crystallog.* **11**, 346 (1978). (i) H. Hayashi, F. Hamada, and A. Nakajima, *Macromolecules*, **9**, 543 (1976). (j) E. W. Fischer, J. H. Wendorff, M. Dettenmaier, G. Leiser, and I. Voigt-Martin, *J. Macromol. Sci. Phys.*, **B12** (1), 41 (1976). (k) D. R. Uhlmann, *Faraday Discuss. Chem. Soc.*, **68**, 87 (1979). (l) Yu. K. Ovchinnikov, G. S. Markova, and V. A. Kargin, *Vysokomol. Soyed*, **A11** (2), 329 (1969); *Polym. Sci. USSR*, **11**, 369 (1969). (m) R. S. Stein and S. O. Hong, *J. Macromol. Sci. Phys.*, **B12** (1), 125 (1976). (n) R. E. Robertson, *J. Phys. Chem.*, **69**, 1575 (1965).

5.3 CONFORMATION OF THE POLYMER CHAIN

One of the great classic problems in polymer science has been the determination of the conformation of the polymer chain in space. The data in Table 5.4 show that the radius of gyration divided by the square root of the molecular weight is a constant for any given polymer in the Flory θ-state, or in the bulk state. However, the detailed arrangement in space must be determined by other experiments and, in particular, by modeling. The resulting models are important in deriving equations for viscosity, diffusion, rubbery elasticity, and mechanical behavior.

5.3.1 Models and Ideas

5.3.1.1 The Freely Jointed Chain The simplest mathematical model of a polymer chain in space is the freely jointed chain. It has n links, each of length l, joined in a linear sequence with no restrictions on the angles between successive bonds (see Figure 5.4). By analogy and Brownian motion statistics, the root-mean-square end-to-end distance is given by (38–42)

$$\left(\overline{r_f^2}\right)^{1/2} = ln^{1/2} \tag{5.12}$$

where the subscript f indicates free rotation.

A more general equation yielding the average end-to-end distance of a random coil, r_0, is given by

$$r_0^2 = l^2 n \frac{(1-\cos\theta)(1+\cos\phi)}{(1+\cos\theta)(1-\cos\phi)} \tag{5.13}$$

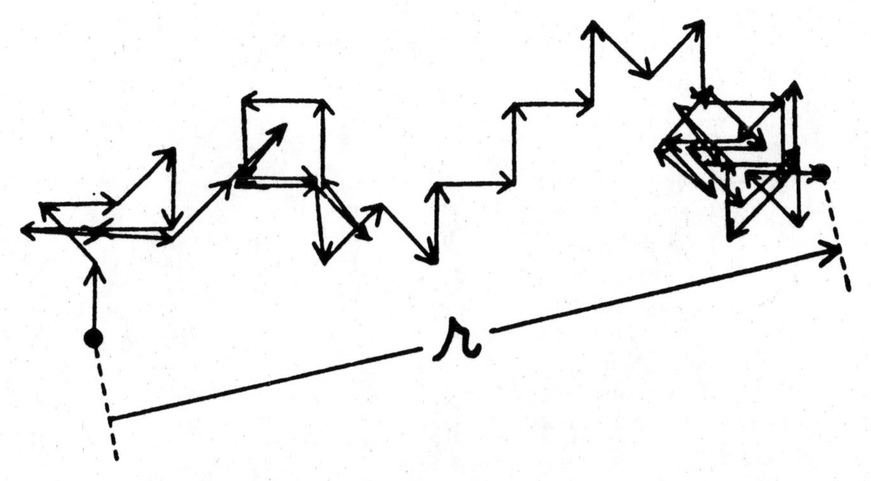

Figure 5.4 A vectorial representation of a freely jointed chain in two dimensions. A random walk of 50 steps (42).

where θ is the bond angle between atoms, and ϕ is the conformation angle. This latter is as the angle of rotation for each bond, defined as the angle each bond makes from the plane delineated by the previous two bonds.

However, equation (5.13) still underestimates the end-to-end distance of the polymer chain, omitting such factors as excluded volume. This last arises from the fact that a chain cannot cross itself in space.

The placement of regular bond angles (109° between carbon atoms) expands the chain by a factor of $[(1-\cos\theta)/(1+\cos\theta)]^{1/2} = \sqrt{2}$, and the three major positions of successive placement obtained by rotation about the previous band (two gauche and one trans positions) results in the chain being extended still further. Other short-range interactions include steric hindrances. Long-range interactions include excluded volume, which eliminates conformations in which two widely separated segments would occupy the same space. The total expansion is represented by a constant, C_∞, after squaring both sides of equation (5.12):

$$r^2 = nl^2 C_\infty \tag{5.14}$$

The characteristic ratio $C_\infty = r^2/l^2 n$ varies from about 5 to about 10, depending on the foliage present on the individual chains (see Table 5.7). The values of $l^2 n$ can be calculated by a direct consideration of the bond angles and energies of the various states and a consideration of longer-range interactions between portions of the chain (40).

5.3.1.2 Kuhn Segments There are several approaches for dividing the polymer chain into specified lengths for conceptual or analytic purposes. For example, Section 4.2 introduced the blob, useful for semidilute solution

Table 5.7 Typical values of the characteristic ratio C_∞

Polymer	C_∞
Polyethylene	6.8
Polystyrene	9.85
it-Polyproplyene	5.5
Poly(ethylene oxide)	4.1
Polyamide 66	5.9
Polybutadiene, 98% *cis*	4.75
cis-polyisoprene	4.75
trans-polyisoprene	7.4
Polycarbonate	2.4
Poly(methyl methacrylate)	8.2
Poly(vinyl acetate)	9.4

Source: J. Brandrup and E. H. Immergut, eds., *Polymer Handbook*, 2nd ed., Wiley-Interscience, New York, 1975, and R. Wool, *Polymer Interfaces*, Hanser, Munich, 1995.

calculations. In the bulk state, the *Kuhn segment* serves a similar purpose. The Kuhn segment length, b, depends on the chain's end-to-end distance under Flory θ-conditions, or its equivalent in the unoriented, amorphous bulk state, r_θ,

$$b = \frac{r_\theta^2}{L} \tag{5.15}$$

where L represents the chain contour length, nl (43). The Kuhn segment length is the basic scale for specifying the size of a chain segment, providing a quantitative basis for evaluating the axial correlation length of Section 5.2.1. For flexible polymers, the Kuhn segment size varies between six and 12 mers (44), having a value of eight mers for polystyrene, and six for poly(methyl methacrylate). The Kuhn segment also expresses the idea of how far one must travel along a chain until all memory of the starting direction is lost, similar to the axial correlation distance, Figure 5.1.

The quantity b is also related to the persistence length, p (45), see Section 5.2.1:

$$2p = b \tag{5.15a}$$

5.3.2 The Random Coil

The term "random coil" is often used to describe the unperturbed shape of the polymer chains in both dilute solutions and in the bulk amorphous state. In dilute solutions the random coil dimensions are present under Flory θ-solvent conditions, where the polymer–solvent interactions and the excluded volume terms just cancel each other. In the bulk amorphous state the mers are surrounded entirely by identical mers, and the sum of all the interactions is zero. Considering mer–mer contacts, the interaction between two distant mers on the same chain is the same as the interaction between two mers on different chains. The same is true for longer chain segments.

In the limit of high molecular weight, the end-to-end distance of the random coil divided by the square root of 6 yields the radius of gyration (Section 3.6.2). Since the n links are proportional to the molecular weight, these relations lead directly to the result that $R_g/M^{1/2}$ is constant.

Of course, there is a distribution in end-to-end distances for random coils, even of the same molecular weight. The distribution of end-to-end distances can be treated by Gaussian distribution functions (see Chapter 9). The most important result is that, for relaxed random coils, there is a well-defined maximum in the frequency of the end-to-end distances, this distance is designated as r_0.

Appendix 5.1 describes the historical development of the random coil.

5.3.3 Models of Polymer Chains in the Bulk Amorphous State

Ever since Hermann Staudinger developed the macromolecular hypothesis in the 1920s (41), polymer scientists have wondered about the spatial arrangement of polymer chains, both in dilute solution and in the bulk. The earliest models included both rods and bedspring-like coils. X-ray and mechanical studies led to the development of the random coil model. In this model the polymer chains are permitted to wander about in a space-filling way as long as they do not pass through themselves or another chain (excluded-volume theory).

The development of the random coil model by Mark, and the many further developments by Flory (5–8,42), led to a description of the conformation of chains in the bulk amorphous state. Neutron-scattering studies found the conformation in the bulk to be close to that found in the θ-solvents, strengthening the random coil model. On the other hand, some workers suggested that the chains have various degrees of either local or long-range order (46–49).

Some of the better-developed models are described in Table 5.8. They range from the random coil model of Mark and Flory (37) to the highly organized meander model of Pechhold et al. (49). Several of the models have taken an intermediate position of suggesting some type of tighter than random coiling, or various extents of chain folding in the amorphous state (47–49). A collage of the most different models is illustrated in Figure 5.5.

The most important reasons why some polymer scientists are suggesting nonrandom chain conformations in the bulk state include the high amorphous/crystalline density ratio, and electron and X-ray diffraction studies, which suggest lateral order (see Table 5.6). Experiments that most favor the random coil model include small-angle neutron scattering and a host of short-range interaction experiments that suggest little or no order at the local level. Both the random coil proponents and the order-favoring proponents claim points in the area of rubber elasticity, which is examined further in Chapter 9.

The SANS experiments bear further development. As shown above, the radius of gyration (R_g) of the chains is the same in the bulk amorphous state as it is in θ-solvents. However, virtually the same values of R_g are also obtained in rapidly crystallized polymers (50–54), where significant order is known to exist. This finding at first appeared to support the possibility of short-range order of the type suggested by the appearance of X-ray halos. Two points need to be mentioned. (a) A more sensitive indication of random chains is the Debye scattering form factor for random coils [see equations (3.47) and (5.11)]. Plots of $P(\theta)$ versus $\sin^2(\theta/2)$ follow the experimental data over surprisingly long ranges of θ, including regions where the Guinier approximation, implicit in equation (5.11), no longer holds (55,56). (b) The cases where R_g is the same in the melt as in the crystallized polymer appear to be in crystallization regime III, where chain folding is significantly reduced. (See Section 6.6.2.5.)

Table 5.8 Major models of the amorphous polymer state

Principals	Description of Model	Reference
H. Mark and P. J. Flory	Random coil model; chains mutually penetrable and of the same dimension as in θ-solvents	(a, b)
B. Vollmert	Individual cell structure model, close-packed structure of individual chains	(c, d)
P. H. Lindenmeyer	Highly coiled or irregularly folded conformational model, limited chain interpenetration	(e)
T. G. F. Schoon	Pearl necklace model of spherical structural units	(f)
V. A. Kargin	Bundle model, aggregates of molecules exist in parallel alignment	(g)
W. Pechhold	Meander model, with defective bundle structure, with meander-like folds	(h, i)
G. S. Y. Yeh	Folded-chain fringed-micellar grain model. Contains two elements: grain (ordered) domain of quasi-parallel chains, and intergrain region of randomly packed chains	(j)
V. P. Privalko and Y. S. Lipatov	Conformation having folded structures with R_g equaling the unperturbed dimension	(k)
R. Hosemann	Paracrystalline model with disorder within the lamellae (see Figure 6.38)	(l, m, n)
S. A. Arzhakov	Folded fibril model, with folded chains perpendicular to fibrillar axis	(o)

References: (a) P. J. Flory, *Principles of Polymer Chemistry*, Cornell University Press, Ithaca, NY, 1953. (b) P. J. Flory, *Faraday Discuss. Chem. Soc.*, **68**, 14 (1979). (c) B. Vollmert, *Polymer Chemistry*, Springer-Verlag, Berlin, 1973, p. 552. (d) B. Vollmert and H. Stuty, in *Colloidal and Morphological Behavior of Block and Graft Copolymers*, G. E. Molau, ed., Plenum, New York, 1970. (e) P. H. Lindenmeyer, *J. Macromol. Sci. Phys.*, **8**, 361 (1973). (f) T. G. F. Schoon and G. Rieber, *Angew. Makromol. Chem.*, **15**, 263 (1971). (g) Y. K. Ovchinnikov, G. S. Markova, and V. A. Kargin, *Polym. Sci. USSR* (Eng. Transl.), **11**, 369 (1969); V. A. Kargin, A. I. Kitajgorodskij, and G. L. Slonimskii, *Kolloid-Zh.*, **19**, 131 (1957). (h) W. Pechhold, M. E. T. Hauber, and E. Liska, *Kolloid Z. Z. Polym.*, **251**, 818 (1973). (i) W. R. Pechhold and H. P. Grossmann, *Faraday Discuss. Chem. Soc.*, **68**, 58 (1979). (j) G. S. Y. Yeh, *J. Macromol. Sci. Phys.*, **6**, 451 (1972). (k) V. P. Privalko and Yu. S. Lipatov, *Makromol. Chem.*, **175**, 641 (1974). (l) R. Hosemann, *J. Polym. Sci.*, **C20**, 1 (1967). (m) R. Hosemann, *Colloid Polym. Sci.*, **260**, 864 (1982). (n) R. Hosemann, *CRC Crit. Rev. Macromol. Sci.*, **1**, 351 (1972). (o) S. A. Arzhakov, N. F. Bakeyev, and V. A. Kabanov, *Vysokomol. Soyed.*, **A15** (5), 1154 (1973).

A major advantage of the random coil model, interestingly, is its simplicity. By not assuming any particular order, the random coil has become amenable to extensive mathematical development. Thus, detailed theories have been developed including rubber elasticity (Chapter 9) and viscosity behavior (Section 3.8), which predict polymer behavior quite well. By difference, little or no analytical development of the other models has taken place, so few properties can be quantitatively predicted. Until such developments have taken place, their absence alone is a strong driving force for the use of the random coil model.

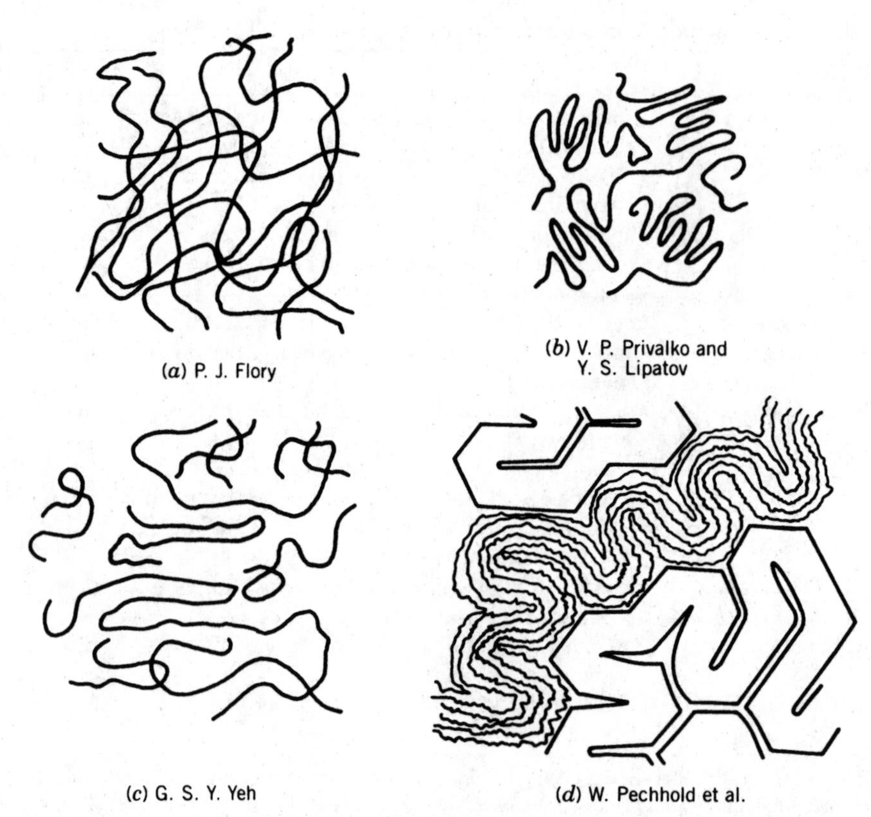

Figure 5.5 Models of the amorphous state in pictorial form. (*a*) Flory's random coil model; the (*b*) Privalko and Lipatov randomly folded chain conformations; (*c*) Yeh's folded-chain fringed-micellar model; and (*d*) Pechhold's meander model. Models increase in degree of order from (*a*) to (*d*). ***References***: (*a*) P. J. Flory, ***Principles of Polymer Chemistry***, Cornell University Press, Ithaca, NY, 1953. (*b*) V. P. Privalko and Y. S. Lipatov, *Makromol. Chem.*, **175**, 641 (1972). (*c*) G. S. Y. Yeh, *J. Makoromol. Sci. Phys.*, **6**, 451 (1972). (*d*) W. Pechhold, M. E. T. Hauber, and E. Liska, *Kolloid Z. Z. Polym.*, **251**, 818 (1973). W. Pechhold, IUPAC Preprints, 789 (1971).

Some of the models may not be quite as far apart as first imagined, however. Privalko and Lipatov have pointed out some of the possible relationships between the random, Gaussian coil, and their own folded-chain model (49) (see Figure 5.5*b*). As a result of thermal motion, they suggest that both the size and location of regions of short-range order in amorphous polymers depend on the time of observation, assuming that the polymers are above T_g and in rapid motion. The instantaneous conformation of the polymer corresponds to a loosely folded chain. However, when the time of observation is long relative to the time required for molecular motion (see Section 5.4), the various chain conformations will be averaged out in time, yielding radii of gyration more like the unperturbed random coil. For polymers in the glassy state, a similar argument holds, because the very many different chains and their respective conformations replace the argument of a single chain varying its conformation with time.

Clearly, the issue of the conformation of polymer chains in the bulk amorphous state is not yet settled; indeed it remains an area of current research. The vast bulk of research to date strongly suggests that the random coil must be at least close to the truth for many polymers of interest. Points such as the extent of local order await further developments. Thus, this book will expound the Mark–Flory theory of the random coil, except where specifically mentioned to the contrary.

5.4 MACROMOLECULAR DYNAMICS

Since the basic notions of chain motion in the bulk state are required to understand much of physical polymer science, a brief introduction is given here. Applications include chain crystallization (to be considered beginning in Chapter 6), the onset of motions in the glass transition region (Chapter 8), and the extension and relaxation of elastomers (Chapters 9 and 10).

Small molecules move primarily by translation. A simple case is of a gas molecule moving in space, following a straight line until hitting another molecule or a wall. In the liquid state, small molecules also move primarily by translation, although the path length is usually only of the order of molecular dimensions.

Polymer motion can take two forms: (a) the chain can change its overall conformation, as in relaxation after strain, or (b) it can move relative to its neighbors. Both motions can be considered in terms of self-diffusion. All such diffusion is a subcase of Brownian motion, being induced by random thermal processes. For center-of-mass diffusion, the center-of-mass distance diffused depends on the square root of time. For high enough temperatures, an Arrhenius temperature dependence is found.

Polymer chains find it almost impossible to move "sideways" by simple translation, for such motion is exceedingly slow for long, entangled chains. This is because the surrounding chains that block sideways diffusion are also long and entangled, and sideways diffusion can only occur by many cooperative motions. Thus polymer chain diffusion demands separate theoretical treatment.

5.4.1 The Rouse–Bueche Theory

The first molecular theories concerned with polymer chain motion were developed by Rouse (57) and Bueche (58), and modified by Peticolas (59). This theory begins with the notion that a polymer chain may be considered as a succession of equal submolecules, each long enough to obey the Gaussian distribution function; that is, they are random coils in their own right. These submolecules are replaced by a series of beads of mass M connected by springs

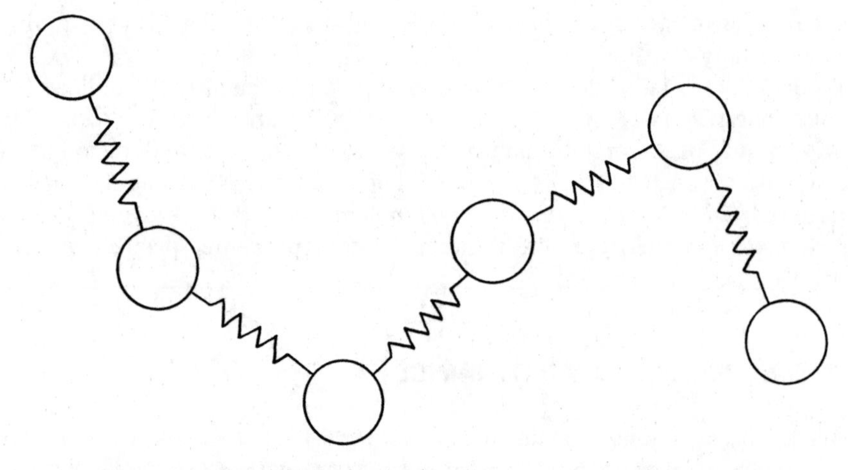

Figure 5.6 Rouse–Bueche bead and spring model of a polymer chain.

with the proper Hooke's force constant; see Figure 5.6. The beads also act as universal joints. This model resembles a one-dimensional crystal.

In the development of rubber elasticity theory (Section 9.7.1), it will be shown that the restoring force, f, on a chain or chain portion large enough to be Gaussian, is given by

$$f = \frac{3kT\Delta x}{\overline{r^2}} \tag{5.16}$$

where Δx is the displacement and r is the end-to-end distance of the chain or Gaussian segment. Thus the springs (but not the whole mass) have an equivalent modulus.

If the beads in Figure 5.6 are numbered 1, 2, 3 . . . , z, so that there are z springs and $z + 1$ beads, the restoring force on the ith bead may be written

$$f_i = \frac{-3kT}{r^2}(-x_{i-1} + 2x_i - x_{i+1}), \qquad 1 \le i \le z-1 \tag{5.17}$$

where f_i represents the force on the ith bead in the x direction, x_i represents the amount by which the bead i has been displaced from its equilibrium position, and r is the end-to-end distance of the segment.

The segments move through a viscous medium (other polymer chains and segments) in which they are immersed. This viscous medium exerts a drag force on the system, damping out the motions. It is assumed that the force is proportional to the velocity of the beads, which is equivalent to assuming that the bead behaves exactly as if it were a macroscopic bead in a continuous viscous medium. The viscous force on the ith bead is given by

$$f_i = \rho\left(\frac{dx_i}{dt}\right) \tag{5.18}$$

where ρ is the friction factor.

Zimm (60) advanced the theory by introducing the concepts of Brownian motion and hydrodynamic shielding into the system. One advantage is that the friction factor is replaced by the macroscopic viscosity of the medium. This leads to a matrix algebra solution with relaxation times of

$$\tau_{p,i} = \frac{6\eta_0 M_i^2}{\pi^2 cRTM_w p^2} \tag{5.19}$$

where η_0 is the bulk-melt viscosity, p is a running index, and c is the polymer concentration.

The Rouse–Bueche theory is useful especially below 1% concentration. However, only poor agreement is obtained on studies of the bulk melt. The theory describes the relaxation of deformed polymer chains, leading to advances in creep and stress relaxation. While it does not speak about the center-of-mass diffusional motions of the polymer chains, the theory is important because it serves as a precursor to the de Gennes reptation theory, described next.

5.4.2 Reptation and Chain Motion

5.4.2.1 The de Gennes Reptation Theory While the Rouse–Bueche theory was highly successful in establishing the idea that chain motion was responsible for creep, relaxation, and viscosity, quantitative agreement with experiment was generally unsatisfactory. More recently, de Gennes (61) introduced his theory of reptation of polymer chains. His model consisted of a single polymeric chain, P, trapped inside a three-dimensional network, G, such as a polymeric gel. The gel itself may be reduced to a set of fixed obstacles—$O_1, O_2, \ldots, O_n \ldots$ His model is illustrated in Figure 5.7 (61). The chain P is not allowed to cross any of the obstacles; however, it may move in a snakelike fashion among them (62).

The snakelike motion is called reptation. The chain is assumed to have certain "defects," each with stored length, b (see Figure 5.8) (61). These defects migrate along the chain in a type of defect current. When the defects move, the chain progresses, as shown in Figure 5.9 (61). The tubes are made up of the surrounding chains. The velocity of the nth mer is related to the defect current J_n by

$$\frac{d\vec{r}_n}{dt} = bJ_n \tag{5.20}$$

where $\vec{r}_n$ represents the position vector of the nth mer.

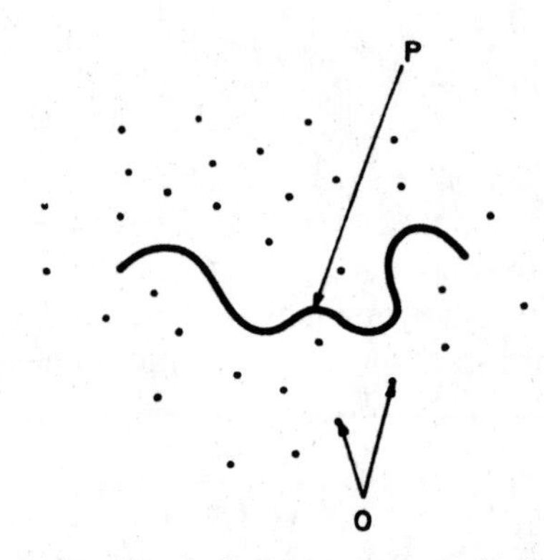

Figure 5.7 A model for reptation. The chain *P* moves among the fixed obstacles, *O*, but cannot cross any of them (61).

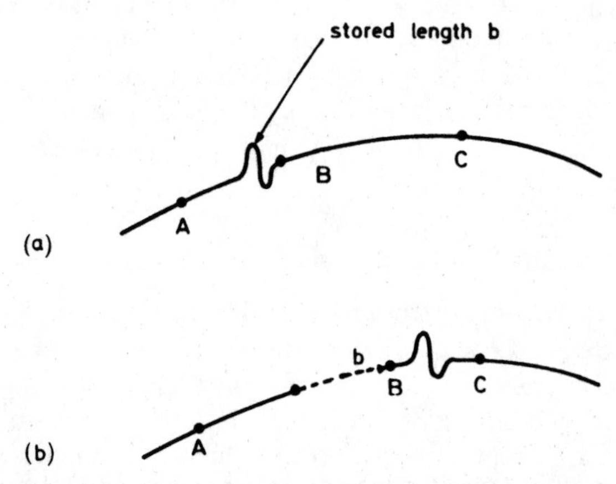

Figure 5.8 Reptation as a motion of defects. (*a*) The stored length *b* moves from A toward C along the chain. (*b*) When the defect crosses mer B, it is displaced by an amount *b* (61).

The reptation motion yields forward motion when a defect leaves the chain at the extremity. The end of the chain may assume various new orientations. In zoological terms, the head of the snake must decide which direction it will go through the bushes. De Gennes assumes that this choice is at random.

Using scaling concepts, see Appendix 4.1, de Gennes (61) found that the self-diffusion coefficient, D, of a chain in the gel depends on the molecular weight M as

$$D \propto M^{-2} \tag{5.21}$$

Numerical values of the diffusion coefficient in bulk systems range from 10^{-12} to 10^{-6} cm^2/s. In data reviewed by Tirrell (63), polyethylene of $1 \times$

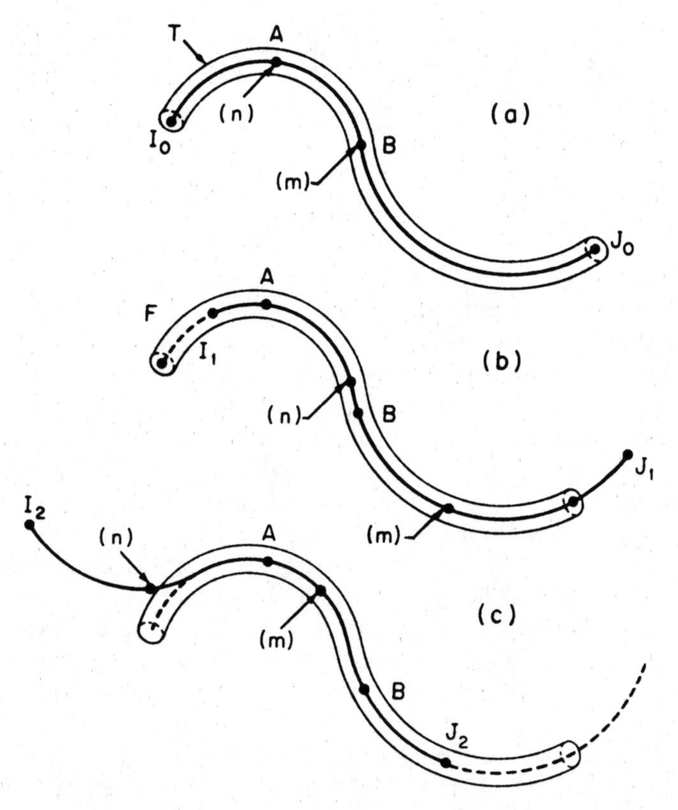

Figure 5.9 The chain is considered within a tube. (*a*) Initial position. (*b*) The chain has moved to the right by reptation. (*c*) The chain has moved to the left, the extremity choosing another path, I_2J_2. A certain fraction of the chain, I_1J_2, remains trapped within the tube at this stage (61). The tubes are made up of the surrounding chains.

10^4 g/mol at 176°C has a value of D near 1×10^{-8} cm^2/s. Polystyrene of 1×10^5 g/mol has a diffusion coefficient of about 1×10^{-12} cm^2/s at 175°C. The inverse second-power molecular weight relationship holds. The temperature dependence can be determined either through activation energies ($E_a = 90$ kJ/mol for polystyrene and 23 kJ/mol for polyethylene) or through the WLF equation (Chapter 8). Calculations using the diffusion coefficient are illustrated in Appendix 5.2.

The reptation time, T_r, depends on the molecular weight as

$$T_r \propto M^3 \tag{5.22}$$

Continuing these theoretical developments, Doi and Edwards (64) developed the relationship of the dynamics of reptating chains to mechanical properties. In brief, expressions for the rubbery plateau shear modulus, G_N^0, steady-state viscosity, η_0, and the steady-state recoverable compliance, J_e^0, were found to be related to the molecular weight as follows:

$$G_N^0 \propto M^0 \tag{5.23}$$

$$\eta_0 \propto M^3 \tag{5.24}$$

$$J_e^0 \propto M^0 \tag{5.25}$$

An important reason why the modulus and the compliance are independent of the molecular weight (above about 8 M_c) is that the number of entanglements (contacts between the reptating chain and the gel) are large for each chain and occur at roughly constant intervals. Experimentally, the viscosity is found to depend on the molecular weight to the 3.4 power (Chapter 10), higher than predicted. The de Gennes theory of reptation as a mechanism for diffusion has also seen applications in the dissolution of polymers, termination by combination in free radical polymerizations, and polymer–polymer welding (63). Graessley (65) reviewed the theory of reptation recently.

Dr. Pierre de Gennes was awarded the 1991 Nobel Prize in Physics for his work in polymers and liquid crystals, as detailed in part above. Several sections following also discuss his contributions. The field of polymer science and engineering has, in fact, several Nobel Prize winners, as delineated in Appendix 5.3.

5.4.2.2 Fickian and Non-Fickian Diffusion

The three-dimensional self-diffusion coefficient, D, of a polymer chain in a melt is given by

$$X = (6Dt)^{1/2} \tag{5.26}$$

where X is the center-of-mass distance traversed in three dimensions, and t represents the time (66). For one-dimensional diffusion in a particular direction, the six is replaced by a two. This is a simple case of Fickian diffusion, noting the time to the one-half dependence. The reptation model of de Gennes supports the $t^{1/2}$ dependence, also supplying the molecular weight dependence and leading to other important analytical features, as discussed above. For one-dimensional diffusion, say in the vertical direction away from an interface, the six in equation (5.26) is replaced by a two.

According to the scaling laws for interdiffusion at a polymer–polymer interface, see Section 11.5.3, the initial diffusion rate as the chain leaves the tube goes as $t^{1/4}$, representing a case of non-Fickian diffusion. An important example involves the interdiffusion of the chains in latex particles to form a film; see Section 5.4.4.

5.4.3 Nonlinear Chains

The discussion above models the motion of *linear* chains in a tube. Physical entanglements define a tube of some 50 Å diameter. This permits the easy

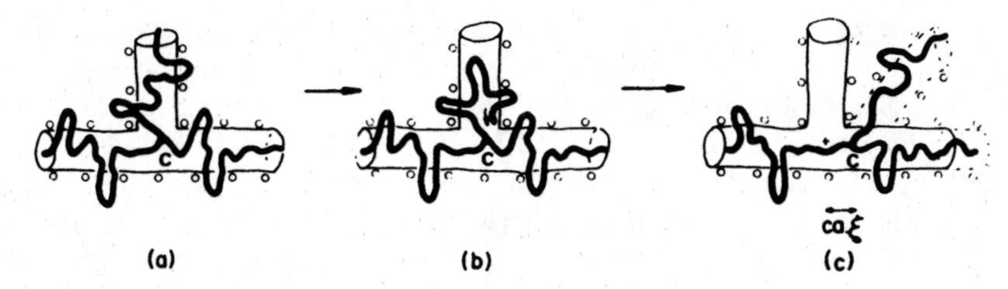

Figure 5.10 The basic diffusion steps for a branched polymer. Note motion of mer C, which requires a fully retracted branch before it can take a step into a new topological environment (68).

passage of defects but effectively prevents sideways motion of the chain. According to the reptation theory, the chains wiggle onward randomly from their ends. How do branched, star, and cyclic polymers diffuse?

Two possibilities exist for translational motion in branched polymers. First, one end may move forward, pulling the other end and the branch into the same tube. This process is strongly resisted by the chains as it requires a considerable decrease in entropy to cause a substantial portion of a branch to lie parallel to the main chain in an adjacent tube (67).

Instead, it is energetically cheaper for an entangled branched-chain polymer to renew its conformation by retracting a branch so that it retraces its path along the confining tube to the position of the center mer. Then it may extend outward again, adopting a new conformation at random; see Figure 5.10 (68). The basic requirement is that the branch not loop around another chain in the process, or it must drag it along also.

De Gennes (65) calculated the probability P_1 of an arm of n-mers folding back on itself as

$$P_1 = \exp\left(\frac{-\alpha n}{n_c}\right) \tag{5.27}$$

where n_c is the critical number of mers between physical entanglements and α is a constant.

The result is that diffusion in branched-chain polymers is much slower than in linear chains. For rings, diffusion is even more sluggish, because the ring is forced to collapse into a quasilinear conformation in order to have center-of-mass motion. Since many commercial polymers are branched or star-shaped, the self-diffusion of the polymer is correspondingly decreased, and the melt viscosity increased.

5.4.4 Experimental Methods of Determining Diffusion Coefficients

Two general methods exist for determining the translational diffusion coefficient, D, in polymer melts: (a) by measuring the broadening of concentration gradients as a function of time (in such cases, two portions of polymer, which differ in some identifiable mode, are placed in juxtaposition and the two polymer portions are allowed to interdiffuse) and (b) by measuring the translation of molecules directly using local probes such as NMR.

Diffusion broadening has been the more widely reported. Small-angle neutron scattering (70), forward recoil spectrometry (71), radioactive labeling (72), infrared absorption methods (73), secondary mass spectroscopy (SIMS) (74) and dynamic mechanial spectroscopy (75) have been employed, among others. A generalized scheme of sample preparation and analysis is shown in Figure 5.11 (73). Of course, pure labeled polymers may be used as well as the blend. The objective is to measure some characteristic change at the interface.

Today, the self-diffusion coefficients, D, of many polymers are known (76), having been measured by a number of investigators using a variety of techniques. Of course, the diffusion coefficient depends on the inverse square of the molecular weight. Above the glass transition temperature, the temperature dependence can be estimated either through activation energies and the Arrhenius equation (Section 8.5) or through the WLF equation (see Section 8.6). Below the glass transition temperature, Fickian diffusion of polymer chains is substantially absent and vibrational modes of motion dominate.

On normalizing the data to 150,000 g/mol and 135°C, the average values of the diffusion coefficients for a few common polymers are

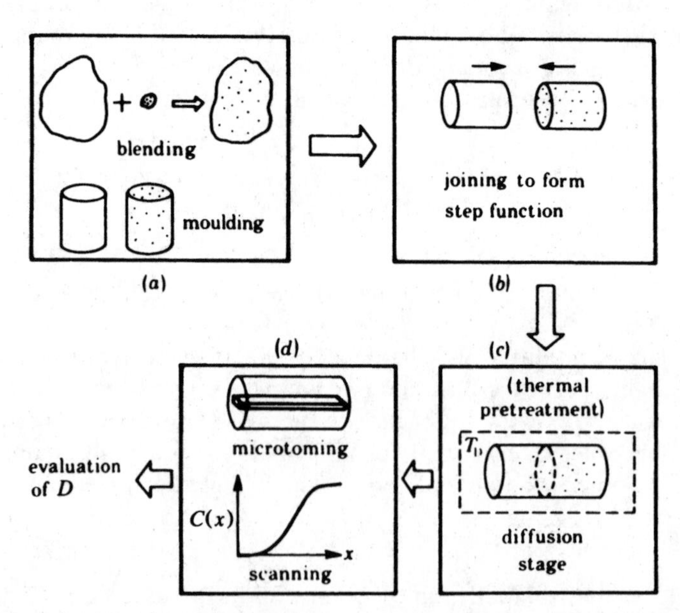

Figure 5.11 Schematic flowchart showing the stages involved in an interdiffusion experiment (68). The slices are scanned in an IR microdensitometer to obtain the broadened concentration profile, from which D is evaluated.

Polystyrene	1.2×10^{-15} cm^2/s
Poly(methyl methacrylate)	6.9×10^{-17} cm^2/s
Poly(n-butyl methacrylate)	8.0×10^{-11} cm^2/s

The much larger value of D for poly(n-butyl methacrylate) is due to its much lower glass transition. For polystyrenes and poly(methyl methacrylate)s as used in many applications, the molecules would diffuse at the rate of a few Ångstroms a minute at 135°C. This is convenient for scientific research. However, commercial molding usually takes place at significantly higher temperatures, as delineated in Appendix 5.2.

Bartels et al. (77) reacted H_2 or D_2 with monodisperse polybutadiene to saturate the double bond, making the equivalent of polyethylene. Films with up to 25 alternating layers of HPB and DPB for the diffusion studies were prepared with layer thicknesses ranging from 3 to 15 μm, chosen to keep homogenization times in the order of a few hours. SANS measurements were made with the incident beam perpendicular to the film surface. The resulting diffusion coefficient on the linear polymer is compared to values obtained on three-armed stars of the same molecular weight in Table 5.9. The stars diffuse almost three orders of magnitude slower than linear polymer, illustrating the detriment that long side chains present to reptation.

Mixtures of deuterated and protonated polystyrene latexes have also been studied by SANS (78,79). The main advantage is the relatively large surface area presented for interdiffusion, since the latex particles are relatively small. Figure 5.12 (79) illustrates the time dependence of the interdiffusion process. The data follow the non-Fickian $t^{1/4}$ relationship up to a diffusion distance of about $0.4R_g$. For diffusion from one side only, Wool (80) derived the break point as $0.8R_g$. However, equal diffusion from both sides, the present case, yields half that value as intuitively expected, noting that the diffusion distance is measured from the original interface plane rather than the true diffusion distance.

Table 5.9 Diffusion coefficients of hydrogenated polybutadienes†

Method	Shape	T, °C	M_w, g/mol	D, cm^2/s	Reference
SANS	Linear	125	7.3×10^4	4.8×10^{-11}	(a)
Forward recoil spectrometry	3-arm	125	7.5×10^4	2.4×10^{-14}	(b)
SANS	3-arm	165	7.5×10^4	1.4×10^{-13}	(c)

References: (a) C. R. Bartels, B. Crist, and W. W. Graessley, *Macromolecules*, **17**, 2702 (1984). (b) B. Crist, P. F. Green, R. A. L. Jones, and E. J. Kramer, *Macromolecules*, **22**, 2857 (1989). (c) C. R. Bartels, B. Crist Jr., L. J. Fetters, and W. W. Graessley, *Macromolecules*, **19**, 785 (1986).

† Hydrogenated polybutadiene makes polyethylene. In this case, a narrow polydispersity polymer not available otherwise.

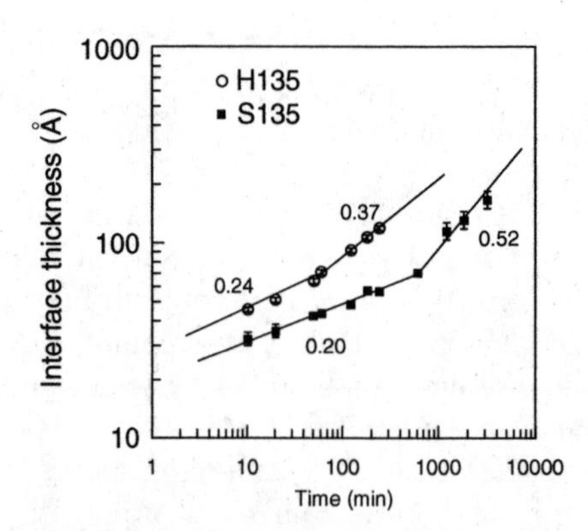

Figure 5.12 Scaling law time dependence for two polystyrene latexes. The H135 material was composed of latexes of 325,000 g/mol polymer having hydrogen end groups. The S135 material had —SO_3 groups at the end of each molecule, with a molecular weight of 325,000 g/mol also. Particle sizes by transmission electron microscopy were in the range of 100–120 nm in all cases. Measurements were by SANS on samples annealed for interdiffusion at 135°C (79).

Beyond a diffusion distance of $0.4R_g$, the normal Fickian diffusion coefficient of $t^{1/2}$ kicks in, note the break point in Figure 5.12. Of course, at this point, considering diffusion from both sides, a total interdiffusion of $0.8R_g$ has taken place. This means that the film is substantially completely interdiffused, and its tensile strength will be approaching its full value, see Section 11.5.4. Note that while the slopes in Figure 5.12 theoretically should have been exactly 0.25 and 0.50 before and after the break point, respectively, some experimental error was observed due to the paucity of experimental points.

Of course, there are other methods for measuring interdiffusion in latex based films. More recently Winnik et al. (81–84) employed a direct nonradiative energy transfer, DET, fluorescence technique to measure diffusion. In this method, latexes are prepared in two different batches. In one batch, the chains contain a "donor" group, while in the other, an "acceptor" group is attached. When the two groups are close to one another, the excitation energy of the donor molecules may be transferred by the resonance dipole–dipole interaction mechanism known as DET, if the emission spectrum of the "donor" overlaps the absorption spectrum of the "acceptor." The ratio of emission intensities to the acceptor intensities changes with interdiffusion, permitting the calculation of self-diffusion coefficients, and hence the interdiffusion depth. For poly(methyl methacrylate)-based latexes, self-diffusion coefficients of approximately 10^{-15} cm^2/s were calculated, similar to the results obtained by Yoo et al. (79). The field has been reviewed by Morawetz (85), who pointed out that changes in the emission spectrum in nonradiative energy transfer between fluorescent labels has been used to characterize polymer miscibility, interpenetration of chain molecules in solution, micelle formation in graft copolymers, and other important effects.

5.5 CONCLUDING REMARKS

The amorphous state is defined as a condensed, noncrystalline state of matter. Many polymers are amorphous under ordinary use conditions, including polystyrene, poly(methyl methacrylate), and poly(vinyl acetate). Crystalline polymers such as polyethylene, polypropylene, and nylon become amorphous above their melting temperatures.

In the amorphous state the position of one chain segment relative to its neighbors is relatively disordered. In the relaxed condition, the polymer chains making up the amorphous state form random coils. The chains are highly entangled with one another, with physical cross-links appearing at about every 600 backbone atoms.

While the amorphous polymer state is "liquid-like" in the classical sense, if the polymer is glassy, a better term would be "amorphous solid," since measurable flow takes years or centuries. Ordinary glass, an inorganic polymer, is such a glassy polymer. The chains are rigidly interlocked in glassy polymers, motion being restricted to vibrational modes.

Above the glass transition, the polymer may flow if it is not in network form. On a submicroscopic scale, the chains interdiffuse with one another with a reptating motion. Common values of the diffusion coefficient vary from 10^{-10} to 10^{-17} cm^2/s, depending inversely on the square of the molecular weight.

REFERENCES

1. E. W. Fischer, G. R. Strobl, M. Dettenmaier, M. Stamm, and N. Steidle, *Faraday Discuss. Chem. Soc.*, **68**, 26 (1979).
2. F. J. Balta-Calleja, K. D. Berling, H. Cackovic, R. Hosemann, and J. Loboda-Cackovic, *J. Macromol. Sci. Phys.*, **B12**, 383 (1976).
3. R. E. Robertson, *J. Phys. Chem.*, **69**, 1575 (1965).
4. W. R. Pechhold and H. P. Grossmann, *Faraday Discuss. Chem. Soc.*, **68**, 58 (1979).
5. P. J. Flory, *J. Macromol. Sci. Phys.*, **B12** (1), 1 (1976).
6. R. S. Stein, *J. Chem. Ed.*, **50**, 748 (1973).
7. P. J. Flory, *Faraday Discuss. Chem. Soc.*, **68**, 15 (1979).
8. P. J. Flory, *Pure Appl. Chem. Macromol. Chem.*, **8**, 1 (1972); reprinted in *Rubber Chem. Tech.*, **48**, 513 (1975).
9. G. S. Y. Yeh, *Crit. Rev. Macromol. Sci.*, **1**, 173 (1972).
10. R. F. Boyer, *J. Macromol. Sci. Phys.*, **B12**, 253 (1976).
11. V. P. Privalko, Yu. S. Lipatov, and A. P. Lobodina, *J. Macromol. Sci. Phys.*, **B11** (4), 441 (1975).
12. Symposium on "Physical Structure of the Amorphous State," *J. Macromol. Sci. Phys.*, **B12** (1976).
13. "Organization of Macromolecules in the Condensed Phase," *Faraday Discuss. Chem. Soc.*, **68** (1979).

14. E. W. Fischer, G. R. Strobl, M. Dettenmaier, M. Stamm, and N. Steidle, *Faraday Discuss. Chem. Soc.*, **68**, 26 (1979).
15. B. Erman, P. J. Flory and J. P. Hummel, *Macromolecules*, **13**, 484 (1980).
16. R. S. Stein and S. D. Hong, *J. Macromol. Sci. Phys.*, **B12** (1), 125 (1976).
17. G. M. Estes, R. W. Seymour, D. S. Huh, and S. L. Cooper, *Polym. Eng. Sci.*, **9**, 383 (1969).
18. R. G. Kirste, W. A. Kruse, and K. Ibel, *Polymer*, **16**, 120 (1975).
19. D. G. H. Ballard, A. N. Burgess, P. Cheshire, E. W. Janke, A. Nevin, and J. Schelten, *Polymer*, **22**, 1353 (1981).
20. J. S. Higgins and R. S. Stein, *J. Appl. Cryst.*, **11**, 346 (1978).
21. A. Maconnachie and R. W. Richards, *Polymer*, **19**, 739 (1978).
22. L. H. Sperling, *Polym. Eng. Sci.*, **24**, 1 (1984).
23. H. Benoit, *J. Macromol. Sci. Phys.*, **B12** (1), 27 (1976).
24. G. D. Wignall, D. G. H. Ballard, and J. Schelten, *J. Macromol. Sci. Phys.*, **B12** (1), 75 (1976).
25. P. Debye, *J. Phys. Coll. Chem.*, **51**, 18 (1947). See also B. H. Zimm, R. S. Stein, and P. Doty, *Polym. Bull.*, **1**, 90 (1945).
26. *National Center for Small-Angle Neutron Scattering Research User's Guide*, Oak Ridge National Laboratory, 1980. Solid State Divisions, Oak Ridge National Laboratory, Oak Ridge, TN 37830.
27. W. C. Koehler, R. W. Hendricks, H. R. Child, S. P. King, J. S. Lin, and G. D. Wignall, in *Proceedings of NATO Advanced Study Institute on Scattering Techniques Applied to Supramolecular and Nonequilibrium Systems*, Vol. 73, S. H. Chen, B. Chu, and R. Nossal, eds., Plenum, New York, 1981.
28. G. D. Wignall, D. G. H. Ballard, and J. Schelten, *Eur. Polym. J.*, **10**, 861 (1974); reprinted in *J. Macromol. Sci. Phys.*, **B12** (1), 75 (1976).
29. B. H. Zimm, *J. Chem. Phys.* **16**, 1098 (1948).
30. P. J. Flory, *Principles of Polymer Chemistry*, Cornell University Press, Ithaca, NY, 1953.
31. K. C. Honnell, J. D. McCoy, J. G. Curro, K. C. Schweizer, A. H. Narten, and A. Habenshuss, *J. Chem. Phys.*, **94**, 4659 (1991).
32. E. W. Fischer, J. H. Wendorff, M. Dettenmaier, G. Lieser, and I. Voigt-Martin, *J. Macromol. Sci. Phys.*, **B12** (1), 41 (1976).
33. R. Lovell, G. R. Mitchell, and A. H. Windle, *Faraday Discuss. Chem. Soc.*, **68**, 46 (1979).
34. Yu. K. Ovchinnikov, Ye. M. Antipov, and G. S. Markova, *Polymer Sci. USSR*, **17**, 2081 (1975).
35. R. L. Miller, R. F. Boyer, and J. Heijboer, *J. Polym. Sci. Polym. Phys. Ed.*, **22**, 2021 (1984).
36. R. L. Miller and R. F. Boyer, *J. Polym. Sci. Polym. Phys. Ed.*, **22**, 2043 (1984).
37. H. R. Schubach, E. Nagy, and B. Heise, *Coll. Polym. Sci.* (*Koll. Z.z. Polym*), **259** (8), 789 (1981).
38. S. E. B. Petrie, *J. Macromol. Sci. Phys.*, **B12**, 225 (1976).
39. T. A. Weber and E. Helfand, *J. Chem. Phys.*, **71**, 4760 (1979).

40. P. J. Flory, *Statistical Mechanics of Chain Molecules*, Interscience, New York, 1969.
41. H. Staudinger, *Die Hochmolekularen Organischen Verbindung*, Springer, Berlin, 1932.
42. P. J. Flory, *Principles of Polymer Chemistry*, Cornell University Press, Ithaca, NY, 1953.
43. H. Fujita, *Polymer Solutions*, Elsevier, Amsterdam, 1990.
44. V. A. Bershtein, V. M. Egorov, L. M. Egorova, and V. A. Ryzhov, *Thermochim. Acta*, **238**, 41 (1994).
45. A. V. Zubkov, in *Polyamic Acids and Polyimides*, M. I. Bessonov and V. A. Zubkov, eds., CRC Press, Boca Raton, FL, 1993.
46. W. Pechhold, M. E. T. Hauber, and E. Liska, *Kolloid Z.z. Polym.*, **251**, 818 (1973).
47. B. Vollmert, *Polymer Chemistry*, Springer, Berlin, 1973, p. 552.
48. P. H. Lindenmeyer, *J. Macromol. Sci. Phys.*, **8**, 361 (1973).
49. V. P. Privalko and Yu. S. Lipatov, *Makromol. Chem.*, **175**, 641 (1974).
50. J. Schelten, D. G. H. Ballard, G. Wignall, G. Longman, and W. Schmatz, *Polymer*, **17**, 751 (1976).
51. J. Schelten, G. D. Wignall, D. G. H. Ballard, and G. W. Longman, *Polymer*, **18**, 1111 (1977).
52. D. G. H. Ballard, P. Cheshire, G. W. Longman, and J. Schelten, *Polymer*, **19**, 379 (1978).
53. E. W. Fischer, M. Stamm, M. Dettenmaier, and P. Herschenraeder, *Polym. Prepr. Am. Chem. Soc. Div. Polym. Chem.*, **20** (1), 219 (1979).
54. J. M. Guenet, *Polymer*, **22**, 313 (1981).
55. F. S. Bates, C. V. Berney, R. E. Cohen, and G. D. Wignall, *Polymer*, **24**, 519 (1983).
56. A. M. Fernandez, J. M. Widmaier, G. D. Wignall, and L. H. Sperling, *Polymer*, **25**, 1718 (1984).
57. P. E. Rouse, *J. Chem. Phys.*, **21**, 1272 (1953).
58. F. Bueche, *J. Chem. Phys.*, **22**, 1570 (1954).
59. W. L. Peticolas, *Rubber Chem. Tech.*, **36**, 1422 (1963).
60. B. H. Zimm, *J. Chem. Phys.*, **24**, 269 (1956).
61. P. G. de Gennes, *J. Chem. Phys.*, **55**, 572 (1971).
62. P. G. de Gennes, *Phys. Today*, **36** (6), 33 (1983).
63. M. Tirrell, *Rubber Chem. Tech.*, **57**, 523 (1984).
64. M. Doi and S. F. Edwards, *J. Chem. Soc. Faraday Trans.* 2, **74**, 1789, 1802, 1818 (1978); **75**, 38 (1979).
65. W. W. Graessley, *Adv. Polym. Sci.*, **47**, 67 (1982).
66. K. Binder and H. Sillescu, *Encyclopedia of Polymer Science and Engineering, Supplementary Volume*, J. I. Kroschwitz, ed., Wiley, New York, 1989.
67. J. Klein, in *Encyclopedia of Polymer Science and Engineering*, 2nd ed., Vol. 9, J. I. Kroschwitz, ed., Wiley, New York, 1987.
68. J. Klein, *Macromolecules*, **19**, 105 (1986).
69. P. G. de Gennes, *J. Phys.* (*Les Ulis, Fr.*), **36**, 1199 (1975).

70. C. R. Bartels, B. Crist, Jr., L. J. Fetters, and W. W. Graessley, *Macromolecules*, **19**, 785 (1986).
71. H. Yokoyama, E. J. Kramer, D. A. Hajduk, and F. S. Bates, *Macromolecules*, **32**, 3353 (1999).
72. F. Bueche, W. M. Cashin, and P. Debye, *J. Chem. Phys.*, **20**, 1956 (1952).
73. J. Klein and B. J. Briscoe, *Proc. R. Soc. Lond.* A, **365**, 53 (1979).
74. S. J. Whitlow and R. P. Wool, *Macromolecules*, **24**, 5926 (1991).
75. H. Qiu and M. Bousmina, *J. Rheology*, **43**, 551 (1999).
76. L. H. Sperling, A. Klein, M. Sambasivam, and K. D. Kim, *Polym. Adv. Technol.*, **5**, 453 (1994).
77. C. R. Bartels, B. Crist, and W. W. Graessley, *Macromolecules*, **20**, 2702 (1984).
78. J. H. Jou and J. E. Anderson, *Macromolecules*, **20**, 1544 (1987).
79. S. D. Kim, A. Klein, and L. H. Sperling, *Macromolecules*, **33**, 8334 (2000).
80. R. P. Wool, *Polymer Interfaces*, Hanser Publisher, New York, 1995.
81. O. Pekan, M. A. Winnik, and M. D. Croucher, *Macromolecules*, **23**, 2673 (1990).
82. C. L. Zhao, W. C. Wang, Z. Hruska, and M. A. Winnik, *Macromolecules*, **23**, 4082 (1990).
83. Y. C. Wang and M. A. Winnik, *Macromolecules*, **23**, 4731 (1990).
84. Y. Wang and M. A. Winnik, *J. Phys. Chem.*, **97**, 2507 (1993).
85. H. Morawetz, *Science*, **240**, 172 (1988).

GENERAL READING

D. Campbell, R. A. Pethrick, and J. R. White, *Polymer Characterization: Physical Techniques*, 2nd ed., Stanley Thornes, Cheltenham, England, 2000.

M. Doi and S. F. Edwards, *The Theory of Polymer Dynamics*, Oxford University Press, New York, 1986.

P. G. de Gennes, *Scaling Concepts in Polymer Physics*, Cornell University Press, Ithaca, NY, 1979.

J. S. Higgins and H. C. Benoit, *Polymers and Neutron Scattering*, Oxford University Press, Oxford, 1994.

S. E. Keinath, R. E. Miller, and J. K. Reike, eds., *Order in the Amorphous State of Polymers*, Plenum, New York, 1987.

R.-J. Roe, *Methods of X-Ray and Neutron Scattering in Polymer Science*, Oxford University Press, New York, 2000.

R.-J. Roe, *Methods of X-Ray and Neutron Scattering in Polymer Science*, Oxford University Press, Oxford, 2000.

J. Scheirs and D. Priddy, eds., *Modern Styrenic Polymers*, John Wiley & Sons, Chichester, England, 2003.

STUDY PROBLEMS

1. Why is the radius of gyration of a polymer in the bulk state essentially the same as measured in a θ-solvent but not the same as in other solvents?

2. Estimate the radius of gyration and end-to-end distance of a polystyrene sample having $M_w = 1 \times 10^5$ g/mol, in the bulk state.

3. In an actual kitchen experiment, one quart of cooked spaghetti was measured out level with cold water so that the spaghetti strands just break the water surface. Nine ounces of water were drained.
   (a) Assuming the spaghetti strands were polymer chains, what is the ratio of the specific volumes of the perfectly packed state to the actual disordered state? Assume a hexagonal close pack array. [*Hint:* Allow for water between perfectly aligned spaghetti strands.]
   (b) Calculate the average angle between strands, $\theta/2$, given by the ratio of the specific volumes (specific volumes is the reciprocal of the density),†

$$\frac{v_c}{v_a} = \left(\frac{3}{2}\right)^3 \left\{ \left[\frac{1-\cos^3(\theta/2)}{\sin^3(\theta/2)} + 1 \right]^2 \left(1 - \cos^3\left(\frac{\theta}{2}\right)\right) \right\}^{-1}$$

   (c) Interpret $\theta/2$ in terms of intermolecular orientation and the randomness of the "amorphous state."

4. Compare the Rouse–Bueche theory with the de Gennes theory. How do they model molecular motion?

5. What is the Kuhn segment length of 1×10^5 g/mol polystyrene? What does the result suggest about the chain conformation?

6. With the advent of small-angle neutron scattering, molecular dimensions can now be determined in the bulk state. A polymer scientist determined the following data on a new deuterated polymer dissolved in a sample of (protonated) polymer:

$\left[\frac{d\Sigma}{d\Omega}\right]^{-1}$ (cm)	0.50	0.72	1.20
$K^2 \times 10^4 (\text{Å}^{-2})$	1.00	3.70	10.1

The constant C_N for this system was determined to be 10.0×10^{-5} mol/g·cm. What is the weight-average molecular weight and the z-average radius of gyration of the deuterated polymer? What third quantity is implicit in this experiment, and what is its probable numerical value?

7. What is the activation energy for the three-armed star's diffusion coefficient in Table 5.9, assuming an Arrhenius relationship? How do you interpret this result?

8. Calculate the first interchain radial spacing for polytetrafluoroethylene from the data given in Figure 5.3*a*. How are these data best interpreted? [*Hint:* Use Bragg's law.]

† R. E. Robertson, *J. Phys. Chem.*, **69**, 1575 (1965).

9. Based on the data shown in Figure 5.12, what values of x do you find for xR_g at the break point? How does that compare to theory?

APPENDIX 5.1 HISTORY OF THE RANDOM COIL MODEL FOR POLYMER CHAINS†

Introduction

Advances in science and engineering have never been completely uniform nor followed an orderly pattern. The truth is that science advances by fits and jerks, with ideas propounded by individuals who see the world in a different light. Frequently they face adversity when putting their ideas forward.

Before polymer science came to be, people had the concept of colloids. There were both inorganic and organic colloids, but they shared certain facts. They both were large compared to ordinary molecules, and both were of irregular sizes and shapes. While this concept "explained" certain simple experimental results, it left much to be desired in the way of understanding the properties of rubber and plastics, which were then considered to be colloids.

In 1920 Herman Staudinger formulated the macromolecular hypothesis: there was a special class of organic colloids of high viscosity that were composed of long chains (A1,A2). This revolutionary idea was argued throughout important areas of chemistry (A3). One of the most important experiments was provided by Herman Mark, who showed that crystalline polymers that had cells of ordinary sizes had only a few mers in each cell, but that the mers were connected to those in the next cell. Eventually the idea of long-chained molecules formed one of the most important cornerstones in the development of modern polymer science.

Early Ideas of Polymer Chain Shape

If one accepts the idea of long-chain macromolecules, the next obvious question relates to their conformation or shape in space. This was especially important since it was early thought that the physical and mechanical properties of the material were determined by the spatial arrangement of the long chains. Staudinger himself thought that most amorphous high polymers such as polystyrene were rod-shaped, and when in solution, the rods lay parallel to each other (A2).

Rubbery materials were different, however. According to early scientists (A4), elastomers were coils or spirals resembling bedsprings (see Figure A5.1.1). Staudinger himself described the idea as follows (A2):

† L. H. Sperling, in *Pioneers in Polymer Science*, R. B. Seymour, ed., Kluwer Academic Publishers, Dordrecht, Germany, 1989.

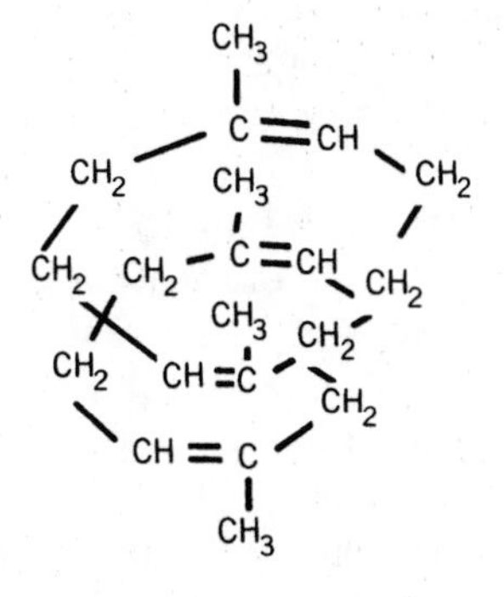

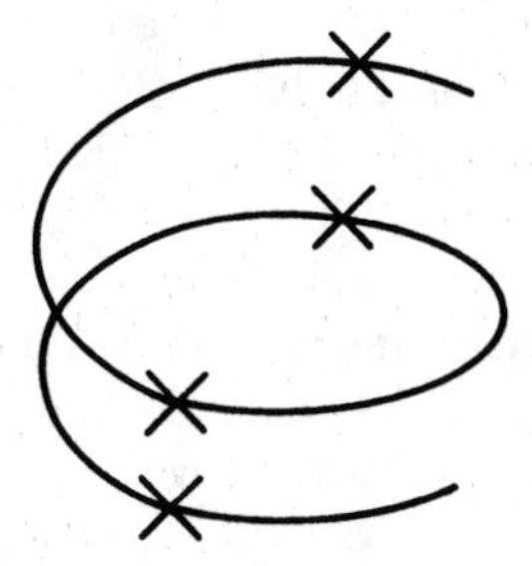

Figure A5.1.1 The spiral structure of natural rubber proposed to explain long-range elasticity. X indicates double bond locations.

> In order to clarify the elasticity of rubber, several investigators have stated that long molecules form spirals, and to be sure the spiral form of the molecules is promoted through the double bonds. By this arrangement, the secondary valences of the double bonds can be satisfied. The elasticity of rubber depends upon the extensibility of such spirals.

The Random Coil

According to H. Mark (A5), the story of the development of the random coil began with the X-ray work of Katz on natural rubber in 1925 (A6–A9). Katz studied the X-ray patterns of rubber both in the relaxed state and the extended or stretched state. In the stretched state, Katz found a characteristic fiber diagram, with many strong and clear diffraction spots, indicating a crystalline material. This contrasted with the diffuse halo found in the relaxed state, indicating that the chains were amorphous under that condition. The fiber periodicity of the elementary cell was found to be about 9 Å, which could only accommodate a few isoprene units. Since the question of how a long chain could fit into a small elementary cell is fundamental to the macromolecular hypothesis, Hauser and Mark repeated the "Katz effect" experiment and, on the basis of improved diagrams and X-ray techniques, established the exact size of the elementary cell (A10). Of course, the answer to the question of the cell size is that the cell actually accommodates the mer, or repeat unit, rather than the whole chain.

The "Katz effect" was particularly important because it established the first relationship between mechanical deformation and concomitant molecular events in polymers (A5). This led Mark and Valko (A11) to carry out stress–strain studies over a wide temperature range together with X-ray studies in order to analyze the phenomenon of rubber reinforcement. This paper contains the first clear statement that the contraction of rubber is not caused by an increase in energy but by the decrease in entropy on elongation.

This finding can be explained by assuming that the rubber chains are in the form of flexible coils (see Figure A5.1.2) (A12). These flexible coils have a high conformational entropy, but they lose their conformational entropy on being straightened out. The fully extended chain, which is rod-shaped, can have only one conformation, and its entropy is zero. This concept was extended to all elastic polymers by Meyer et al. (A13) in 1932. These ideas are developed quantitatively in Chapter 9. Although thermal motion and free chain rotation are required for rubber elasticity, the idea of the random coil was later adopted for glassy polymers such as polystyrene as well.

The main quantitative developments began in 1934 with the work of Guth and Mark (A14) and Kuhn (A15). Guth and Mark chose to study the entropic origin of the rubber elastic forces, whereas Kuhn was more interested in explaining the high viscosity of polymeric solutions. Using the concept of free rotation of the carbon–carbon bond, Guth and Mark developed the idea of the "random walk" or "random flight" of the polymer chain. This led to the familiar Gaussian statistics of today, and eventually to the famous relationship between the end-to-end distance of the chain and the square root of the molecular weight. Three stages in the development of the random coil model have been described by H. Mark (see Table A5.1.1) (A16). Like many great ideas, it apparently occurred to several people nearly simultaneously. It is also clear from Table A5.1.1 that Mark played a central role in the development of the random coil.

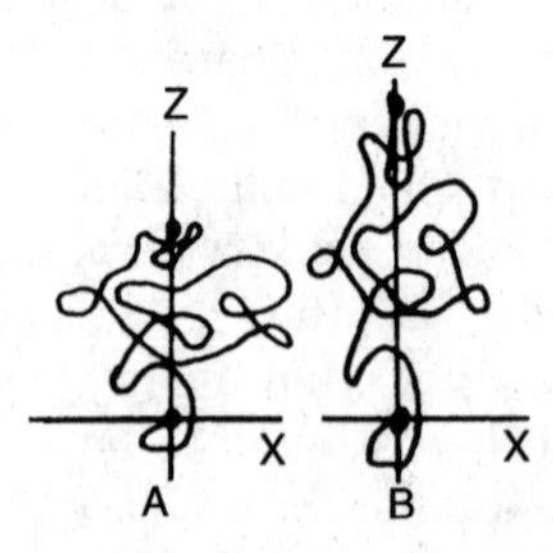

Figure A5.1.2 Early drawing of the random coil. (*A*) Relaxed state. (*B*) Effect of deformation in the *Z* direction. After W. Kuhn (A12).

Table A5.1.1 Early references to randomly coiled flexible macromolecules (A16)

Phase 1	Hypothetic and accidental remarks:
	E. Wachlisch, *Z. S. Biol.*, **85**, 206 (1927)
	H. Mark and E. Valko, *Kautschuk*, **6**, 210 (1930)
	W. Haller, *Koll. Z. S.*, **56**, 257 (1931)
Phase 2	Elaborate *qualitative* interpretation:
	K. H. Meyer, G. von Susich, and E. Valko, *Koll. Z. S.*, **59**, 208 (1932)
Phase 3	*Quantitative* mathematical treatment:
	E. Guth and H. Mark, *Monatschefte Chemie*, **65**, 93 (1934)
	H. Mark, IX Congress for Chemistry, Madrid 1934, Vol. **4**, p. 197 (1934)
	W. Kuhn, *Koll. Z. S.*, **68**, 2 (1934)
	H. Mark, *Der Feste Koerper*, 1937, p. 65

Source: H. Mark, Private communication, May 2, 1983.

The random coil model has remained essentially the same until today (A17,A18), although many mathematical treatments have refined its exact definition. Its main values are twofold: by all experiments, it appears to be the best model for amorphous polymers, and it is the only model that has been extensively treated mathematically. It is interesting to note that by its very randomness, the random coil model is easier to understand quantitatively and analytically than models introducing modest amounts of order (A19).

The random coil model has been supported by many experiments over the years. The most important of these has been light-scattering from dilute solutions and, more recently, small-angle neutron scattering from the bulk state (see earlier sections of Chapter 5). Both of these experiments support the famous relationships between the square root of the molecular weight and the end-to-end distance. The random coil model has been used to explain not only rubber elasticity and dilute solution viscosities but a host of other physical and mechanical phenomena, such as melt rheology, diffusion, and the equilibrium swelling of cross-linked polymers. Some important reviews include the works of Flory (A17), Treloar (A18), Staverman (A21), and Guth and Mark (A22).

REFERENCES

A1. H. Staudinger, *Ber. Dtsch. Chem. Ges.*, **53**, 1074 (1920).

A2. H. Staudinger, *Die Hochmolekularen Organischen Verbingdung*, Springer, Berlin, 1932, at pp. 116–123. Reprinted, 1960.

A3. G. A. Stahl, ed., "*Polymer Science: Overview A Tribute to Herman F. Mark*," ACS Symposium No. 175, American Chemical Society, Washington, DC, 1981.

A4. F. Kirchhof, *Kautschuk*, **6**, 31 (1930).

A5. H. Mark, unpublished, 1982.

A6. J. R. Katz, *Naturwissenschaften*, **13**, 410 (1925).

A7. J. R. Katz, *Chem. Ztg.*, **19**, 353 (1925).
A8. J. R. Katz, *Kolloid Z.*, **36**, 300 (1925).
A9. J. R. Katz, *Kolloid Z.*, **37**, 19 (1925).
A10. E. A. Hauser and H. Mark, *Koll. Chem. Beih.*, **22**, 63; **23**, 64 (1929).
A11. H. Mark and E. Valko, *Kautschuk*, **6**, 210 (1930).
A12. W. Kuhn, *Angew. Chem.*, **49**, 858 (1936).
A13. K. H. Meyer, G. V. Susich, and E. Valko, *Koll. Z.*, **41**, 208 (1932).
A14. E. Guth and H. Mark, *Monatsh. Chem.*, **65**, 93 (1934).
A15. W. Kuhn, *Kolloid. Z.*, **68**, 2 (1934).
A16. H. Mark, private communication, May 2, 1983.
A17. P. J. Flory, *Principles of Polymer Chemistry*, Cornell University Press, Ithaca, NY, 1953.
A18. L. R. G. Treloar, *The Physics of Rubber Elasticity*, 3rd ed., Clarendon Press, Oxford, 1975.
A19. R. F. Boyer, *J. Macromol. Sci. Phys.*, **B12**, 253 (1976).
A20. P. J. Flory, *Statistical Mechanics of Chain Molecules*, Wiley, New York, 1969.
A21. A. J. Staverman, *J. Polym. Sci.*, Symposium No. 51, 45 (1975).
A22. E. Guth and H. F. Mark, *J. Polym. Sci. B Polym. Phys.*, **29**, 627 (1991).

APPENDIX 5.2 CALCULATIONS USING THE DIFFUSION COEFFICIENT

The diffusion of polystyrene of various molecular weights in high-molecular-weight polystyrene ($M_w = 2 \times 10^7$ g/mol) was measured by Mills et al. (B1) using forward recoil spectrometry. At 170°C, the diffusion coefficient was found to depend on the weight-average molecular weight as

$$D = 8 \times 10^{-3} M_w^{-2} \text{ cm}^2/\text{s} \qquad \text{(A5.2.1)}$$

If the concentration gradient of the diffusing species is given in concentration per unit distance, the unit

$$\frac{\text{mol/cm}^3}{\text{cm}} \times \frac{\text{cm}^2}{\text{s}}$$

yields the flux in mol/(cm$^2 \cdot$s)—that is, the number of moles of polystyrene crossing a square centimeter of area per second.

Consider a polymer having a weight-average molecular weight of 1×10^6 g/mol. With a density of 1.05 g/cm^3, a bulk concentration of about 1×10^{-6} mol/cm^3 is obtained. Consider a diffusion over a 100-Å distance, or 1×10^{-6} cm from the bulk concentration to a zero concentration. In 1 sec, $8 \times 10^{-15} \cong 1 \times 10^{-14}$ mol will diffuse through a 1-cm^2 area. This is a significant part of the polymer that is lying on the surface of the hypothetical 1 cm^3 under consideration.

Bulk polymeric materials being pressed together under molding conditions require diffusion of the order of 100 Å to produce a significant number of entanglements, thereby fusing the interfacial boundary. In the commercial molding of polystyrene at 170°C, times of the order of 2 minutes might be employed. The largest portion of this time is actually required for heat transfer to be complete and for a uniform temperature to be achieved. Since the weight-average molecular weight of many polystyrenes is about 1×10^5 g/mol, the original boundary will be obliterated in a few seconds under these conditions. Thus the values obtained via polymer physics research confirm the values used in practice.

REFERENCE

B1. P. J. Mills, P. F. Green, C. J. Palmstrom, J. W. Mayer, and E. J. Kramer, *Appl. Phys. Lett.*, **45** (9), 957 (1984).

APPENDIX 5.3 NOBEL PRIZE WINNERS IN POLYMER SCIENCE AND ENGINEERING

From the time of Hermann Staudinger's enunciation of the macromolecular hypothesis in 1920, polymer science and engineering has had many fundamental advances leading to the understanding of plastics, rubber, adhesives, coatings, and fibers of today. The discoveries that these Nobel Prize winners made are summarized in Table C5.3.1 (C1,C2). These people have revolutionized life in the modern world.

Table C5.3.1 Nobel Prize winners for advances in polymer science and engineering

Scientist	Year	Field	Research and Discovery
Hermann Staudinger	1953	Chemistry	Macromolecular Hypothesis
Karl Ziegler and Giulio Natta	1963	Chemistry	Ziegler–Natta catalysts and resulting stereospecific polymers like isotactic polypropylene
Paul J. Flory	1974	Chemistry	Random coil and organization of polymer chains
Pierre G. de Gennes	1991	Physics	Reptation in polymers and polymer structures at interfaces
A. J. Heeger, A. G. MacDiarmid and H. Shirakawa	2000	Chemistry	Discovery and development of conductive polymers

REFERENCES

C1. P. Canning, ed., *Who's Who in Science and Engineering*, 4th ed., Marquis Who's Who, New Providence, NJ, 1998–9.

C2. F. N. Magill, ed., *The Nobel Prize Winners in Chemistry*, Salem Press, Pasadena, CA, 1990.

6

THE CRYSTALLINE STATE

6.1 GENERAL CONSIDERATIONS

In the previous chapter the structure of amorphous polymers was examined. In this chapter the study of crystalline polymers is undertaken. The crystalline state is defined as one that diffracts X-rays and exhibits the *first-order* transition known as melting.

A first-order transition normally has a discontinuity in the volume–temperature dependence, as well as a heat of transition, ΔH_f, also called the enthalpy of fusion or melting. The most important *second-order* transition is the glass transition, Chapter 8, in which the volume–temperature dependence undergoes a change in slope, and only the derivative of the expansion coefficient, dV/dT, undergoes a discontinuity. There is no heat of transition at T_g, but rather a change in the heat capacity, ΔC_p.

Polymers crystallized in the bulk, however, are never totally crystalline, a consequence of their long-chain nature and subsequent entanglements. The melting (fusion) temperature of the polymer, T_f, is always higher than the glass transition temperature, T_g. Thus the polymer may be either hard and rigid or flexible. An example of the latter is ordinary polyethylene, which has a T_g of about –80°C and a melting temperature of about +139°C. At room temperature it forms a leathery product as a result.

The development of crystallinity in polymers depends on the regularity of structure in the polymer (see Chapter 2). Thus isotactic and syndiotactic polymers usually crystallize, whereas atactic polymers, with a few exceptions

Introduction to Physical Polymer Science, by L.H. Sperling
ISBN 0-471-70606-X

Table 6.1 Properties of selected crystalline polymers (1)

Polymer	T_f, °C	$\Delta H_f, \frac{kJ}{mol}$
Polyethylene	139	7.87[a]
Poly(ethylene oxide)	66	8.29
it-Polystyrene	240	8.37
Poly(vinyl chloride)	212	3.28
Poly(ethylene terephthalate)[b]	265	24.1
Poly(hexamethylene adipamide)[c]	265	46.5
Cellulose tributyrate	207	12.6
cis-Polyisoprene[d]	28	4.40
Polytetrafluoroethylene[e]	330	5.74
it-Polypropylene	171	8.79
Poly(oxymethylene)[f]	182	10.6

[a] Per —CH_2—CH_2—. Note that values for —CH_2— alone are sometimes reported.
[b] Dacron.
[c] Nylon 66 or Polyamide 66.
[d] Natural rubber.
[e] Teflon.
[f] Delrin.

(where the side groups are small or highly polar), do not. Regular structures also appear in the polyamides (nylons), polyesters, and so on, and these polymers make excellent fibers.

Nonregularity of structure first decreases the melting temperature and finally prevents crystallinity. Mers of incorrect tacticity (see Chapter 2) tend to destroy crystallinity, as does copolymerization. Thus statistical copolymers are generally amorphous. Blends of isotactic and atactic polymers show reduced crystallinity, with only the isotactic portion crystallizing. Under some circumstances block copolymers containing a crystallizable block will crystallize; again, only the crystallizable block crystallizes.

Factors that control the melting temperature include polarity and hydrogen bonding as well as packing capability. Table 6.1 lists some important crystalline polymers and their melting temperatures (1).

6.1.1 Historical Aspects

Historically the study of crystallinity of polymers was important in the proof of the Macromolecular Hypothesis, developed originally by Staudinger. In the early 1900s when X-ray studies were first applied to crystal structures, scientists found that the cell size of crystalline polymers was of normal size (about several Ångstoms on a side). This was long before they developed an understanding of the chain nature of polymers required to completely characterize the cell contents. In the case of ordinary sized organic molecules, each cell was found to contain only a few molecules. If polymers were composed of long chains, they asked Staudinger, how could they fit into the small unit cells? The density of such a material would have to be 50 times lead! The answer, developed by Mark and co-workers (2–7), and others, was that the unit cell contains only a few mers that are repeated in adjacent unit cells. The molecule continues in the adjacent axial directions. Mark showed that the bond distances both within a cell and between cells were consistent with covalent bond distances (1.54 Å for carbon–carbon bonds) and inconsistent with the formation of discrete small molecules. For these and many other advances (see Section 3.8), and a life-long leadership in polymers (he died at age 96 in 1992), Herman Mark was called the *Father of Polymer Science.*

One of the first structures to be determined was the natural polysaccharide cellulose. In this case the repeat unit is cellobiose, composed of two glucoside rings. In the 1980s, ^{13}C NMR experiments established that native cellulose is actually a composite of a triclinic parallel-packed unit cell called cellulose I_α, and a monoclinic parallel-packed unit cell called cellulose I_β. Experimentally, the structures are only difficultly distinguishable *via* X-ray analysis (7a, 7b). Figure 6.1 (3) illustrates the general form of the cellulose unit cell.

When vinyl, acrylic, and polyolefin polymers were first synthesized, the only microstructure then known was atactic. Most of these materials were amorphous. Scientific advancement waited until Ziegler's work on novel catalysts (8), together with Natta's work on X-ray characterization of the stereospecific polymers subsequently synthesized (9), when isotactic and syndiotactic crystalline polymers became known. For this great pioneering work, Ziegler and Natta were jointly awarded the 1963 Nobel Prize in Chemistry (see Appendix 5.3). The general class of these catalysts are known today as Ziegler–Natta catalysts.

Of course, crystalline polymers constitute many of the plastics and fibers of commerce. Polyethylene is used in films to cover dry-cleaned clothes, and as water and solvent containers (e.g., wash bottles). Polypropylene makes a highly extensible rope, finding particularly important applications in the marine industry. Polyamides (nylons) and polyesters are used as both plastics and fibers. Their use in clothing is world famous. Cellulose, mentioned above, is used in clothing in both its native state (cotton) and its regenerated state (rayon). The film is called cellophane.

6.1.2 Melting Phenomena

The melting of polymers may be observed by any of several experiments. For linear or branched polymers, the sample becomes liquid and flows. However, there are several possible complications to this experiment, which may make interpretation difficult. First of all, simple liquid behavior may not be immediately apparent because of the polymer's high viscosity. If the polymer is cross-linked, it may not flow at all. It must also be noted that amorphous polymers soften at their glass transition temperature, T_g, which is emphatically not

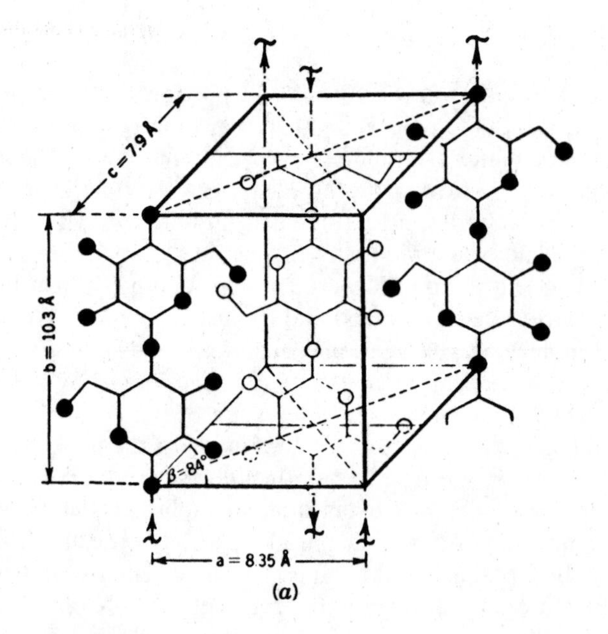

(*a*)

Parameter	Cellulose I[a]	Cellulose II[a]	Cellulose III[b]	Cellulose IV[a] Cellulose *x*
a-axis (Å)	8.35	8.02	7.74	8.12
b-axis (Å)	10.30	10.30	10.30	10.30
c-axis Å)	7.90	9.03	9.96	7.99
β (degrees)	83.3	62.8	58	90
Density (g/cm^3)	1.625	1.62	1.61	1.61

[a] Ø. Ellefsen, J. G. Ønnes, and N. Norman, *Acta Chem. Scand.*, **13**, 853 (1959).
[b] C. Legrand, *J. Polym. Sci.*, **7**, 333 (1951).

(*b*)

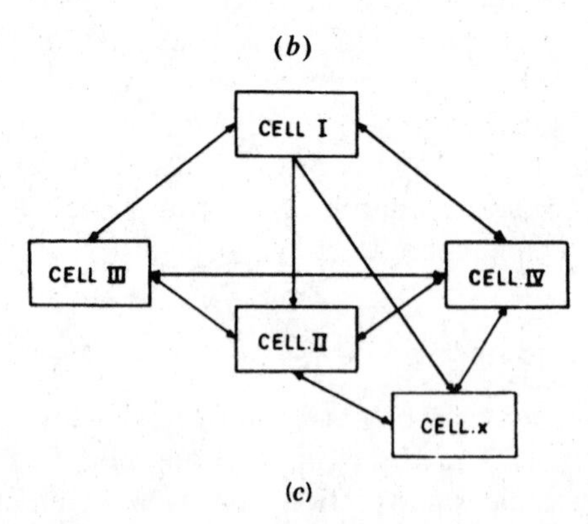

(*c*)

Figure 6.1 The crystalline structure of cellulose. (*a*) The unit cell of native cellulose, or cellulose I, as determined by X-ray analysis (2,10,11). Cellobiose units are not shown on all diagonals for clarity. The volume of the cell is given by $V = abc \sin\beta$. (*b*) Unit cell dimensions of the four forms of cellulose. (*c*) Known pathways to change the crystalline structure of cellulose. The lesser known form of cellulose *x* is also included (10).

a melting temperature but may resemble one, especially to the novice (see Chapter 8). If the sample does not contain colorants, it is usually hazy in the crystalline state because of the difference in refractive index between the amorphous and crystalline portions. On melting, the sample becomes clear, or more transparent.

The disappearance of crystallinity may also be observed in a microscope, for example, between crossed Nicols. The sharp X-ray pattern characteristic of crystalline materials gives way to amorphous halos at the melting temperature, providing one of the best experiments.

Another important way of observing the melting point is to observe the changes in specific volume with temperature. Since melting constitutes a first-order phase change, a discontinuity in the volume is expected. Ideally, the melting temperature should give a discontinuity in the volume, with a concomitant sharp melting point. In fact, because of the very small size of the crystallites in bulk crystallized polymers (or alternatively, their imperfections), most polymers melt over a range of several degrees (see Figure 6.2) (12). The melting temperature is usually taken as the temperature at which the last trace of crystallinity disappears. This is the temperature at which the largest and/or

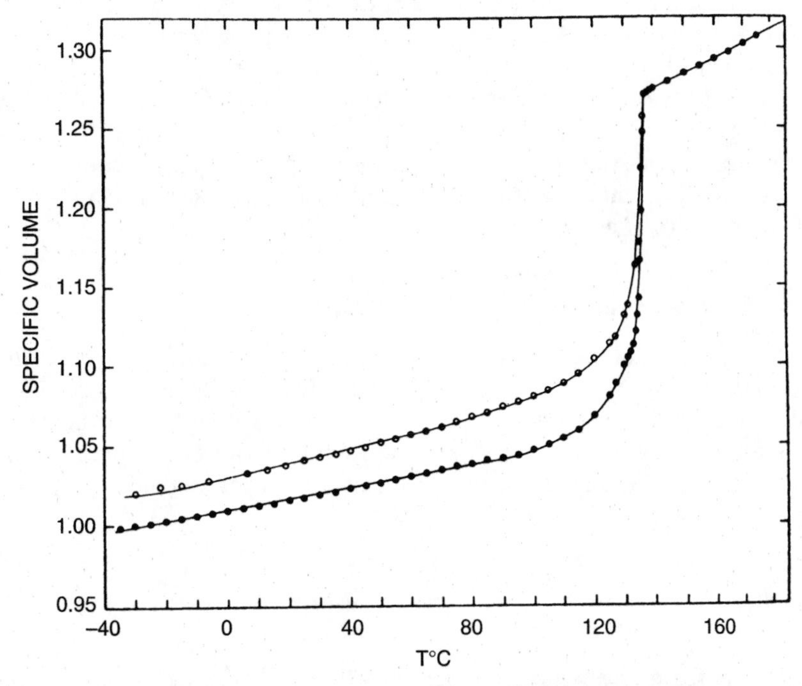

Figure 6.2 Specific volume–temperature relations for linear polyethylene. Open circles, specimen cooled relatively rapidly from the melt to room temperature before fusion experiments; solid circles, specimen crystallized at 130°C for 40 days, then cooled to room temperature prior to fusion (12).

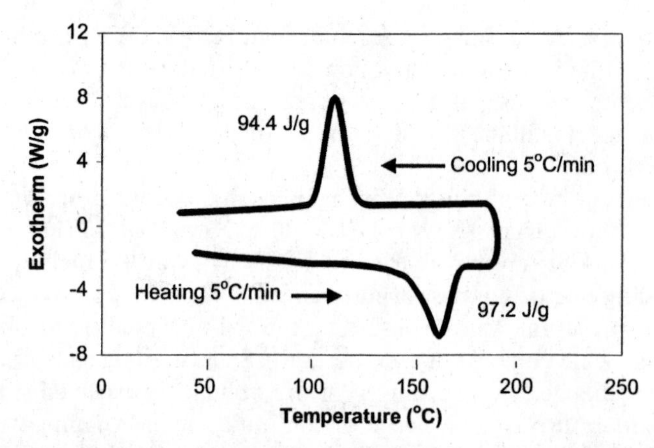

Figure 6.3 Differential scanning calorimetry of a commercial isotactic polypropylene sample, generously provided by Dr. S. J. Han of the Exxon Research and Engineering Company. Note the supercooling effect on crystallization, but the equal and opposite heats of melting and crystallization (13). Experiment by S. D. Kim.

most perfect crystals are melting. The volumetric coefficients of expansion in Figure 6.2 can be calculated from $\alpha = (1/V)\ (dV/dT)_p$.

Alternatively, the melting temperature can be determined thermally. Today, the differential scanning calorimeter (DSC) is popular, since it gives the heat of fusion as well as the melting temperature. Such an experiment is illustrated in Figure 6.3 for *it*-polypropylene (13). The heat of fusion, ΔH_f, is given by the area under the peak.

Further general studies of polymer fusion are presented in Sections 6.8 and 6.9, after the introduction of crystallographic concepts and the kinetics and thermodynamics of crystallization.

6.1.3 Example Calculation of Percent Crystallinity

Exactly how crystalline is the *it*-polypropylene in Figure 6.3? The heat of fusion, ΔH_f, of the whole sample, amorphous plus crystalline parts, is 97.2 J/g, determined by the area under the melting curve. Table 6.1 gives the heat of fusion for *it*-polypropylene as 8.79 kJ/mol. This latter value is for the crystalline component only. Noting that *it*-polypropylene has a mer molecular weight of 42 g/mol,

$$\frac{8790\ \text{J/mol}}{42\ \text{g/mol}} = 209\ \text{J/g}$$

Then the percent crystallinity is given by

$$\frac{97.2\ \text{J/g}}{209\ \text{J/g}} \times 100 = 46\%\ \text{crystallinity}$$

Many semicrystalline polymers are between 40% and 75% crystalline. (See Section 6.5.4 for further information.)

6.2 METHODS OF DETERMINING CRYSTAL STRUCTURE

6.2.1 A Review of Crystal Structure

Before beginning the study of the structure of crystalline polymers, the subject of crystallography and molecular order in crystalline substances is reviewed. Long before X-ray analysis was available, scientists had already deduced a great deal about the atomic order within crystals.

The science of geometric crystallography was concerned with the outward spatial arrangement of crystal planes and the geometric shape of crystals. Workers of that day arrived at three fundamental laws: (a) the law of constancy of interfacial angles, (b) the law of rationality of indexes, and (c) the law of symmetry (14).

Briefly, the law of constancy of interfacial angles states that for a given substance, corresponding faces or planes that form the external surface of a crystal always intersect at a definite angle. This angle remains constant, independent of the sizes of the individual faces.

The law of rationality of indexes states that for any crystal a set of three coordinate axes can be chosen such that all the faces of the crystal will either intercept these axes at definite distances from the origin or be parallel to some of the axes. In 1784 Hauy showed that it was possible to choose among the three coordinate axes unit distances (a, b, c) of not necessarily the same length. Furthermore, Hauy showed that it was possible to choose three coefficients for these three axes—m, n, and p—that are either integral whole numbers, infinity, or fractions of whole numbers such that the ratio of the three intercepts of any plane in the crystal is given by $(ma\text{: } nb\text{: } pc)$. The numbers m, n, and p are known as the Weiss indexes of the plane in question. The Weiss indexes have been replaced by the Miller indexes, which are obtained by taking the reciprocals of the Weiss coefficients and multiplying through by the smallest number that will express the reciprocals as integers. For example, if a plane in the Weiss notation is given by $a{:}\infty b{:}\frac{1}{4}c$, the Miller indexes become $a{:}0b{:}4c$, thus more simply written (104), which is the modern way of expressing the indexes in the Miller system of crystal face notation.

The third law of crystallography states that all crystals of the same compound possess the same elements of symmetry. There are three types of symmetry: a plane of symmetry, a line of symmetry, and a center of symmetry (14). A plane of symmetry passes through the center of the crystal and divides it into two equal portions, each of which is the mirror image of the other. If it is possible to draw an imaginary line through the center of the crystal and then revolve the crystal about this line in such a way as to cause the crystal to appear unchanged two, three, four, or six times in 360° of revolution, then the crystal is said to possess a line of symmetry. Similarly, a crystal possesses a center of symmetry if every face has an identical atom at an equal distance on the opposite side of this center. On the basis of the total number of plane, line, and center symmetries, it is possible to classify the crystal types into six crystal systems, which may in turn be grouped into 32 classes and finally into 230 crystal forms.

The scientists of the pre–X-ray period postulated that any macroscopic crystal was built up by repetition of a fundamental structural unit composed of atoms, molecules, or ions, called the unit crystal lattice or space group. This unit crystal lattice has the same geometric shape as the macroscopic crystal. This line of reasoning led to the 14 basic arrangements of atoms in space, called space lattices. Among these are the familiar simple cubic, hexagonal, and triclinic lattices.

There are four basic methods in wide use for the study of polymer crystallinity: X-ray diffraction, electron diffraction, infrared absorption, and Raman spectra. The first two methods constitute the fundamental basis for crystal cell size and form, and the latter two methods provide a wealth of supporting data such as bond distances and intermolecular attractive forces. These several methods are now briefly described.

6.2.2 X-Ray Methods

In 1895 X-rays were discovered by Roentgen. The new X-rays were first applied to crystalline substances in 1912 and 1913, following the suggestion by Von Laue that crystalline substances ought to act as a three-dimensional diffraction grating for X-rays.

By considering crystals as reflection gratings for X-rays, Bragg (15) derived his now famous equation for the distance d between successive identical planes of atoms in the crystal:

$$d = \frac{n\lambda}{2\sin\theta} \tag{6.1}$$

where λ is the X-ray wavelength, θ is the angle between the X-ray beam and these atomic planes, and n represents the order of diffraction, a whole number. It turns out that both the X-ray wavelength and the distance between crystal planes, d, are of the order of 1 Å. Such an analysis from a single crystal produces a series of spots.

However, not every crystalline substance can be obtained in the form of macroscopic crystals. This led to the Debye–Scherrer (16) method of analysis for powdered crystalline solids or polycrystalline specimens. The crystals are oriented at random so the spots become cones of diffracted beams that can be recorded either as circles on a flat photographic plate or as arcs on a strip of film encircling the specimen (see Figure 6.4) (17). The latter method permits the study of back reflections as well as forward reflections.

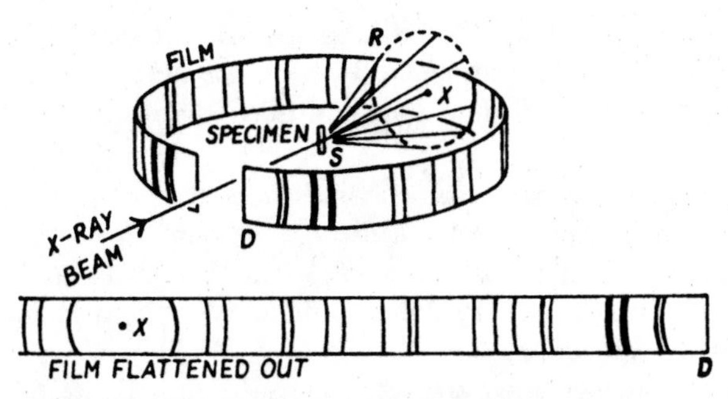

Figure 6.4 The Debye–Scherrer method for taking powder photographs. The angle RSX is 2θ, where θ is the angle of incidence on a set of crystal planes (17).

Basically the intensity of the diffraction spot or line depends on the scattering power of the individual atoms, which in turn depends on the number of electrons in the atom. Other quantities of importance include the arrangement of the atoms with regard to the crystal planes, the angle of reflection, the number of crystallographically equivalent sets of planes contributing, and the amplitude of the thermal vibrations of the atoms. Both the intensities of the spots or arcs and their positions are required to calculate the crystal lattice, plus lots of imagination and hard work. The subject of X-ray analysis of crystalline materials has been widely reviewed (14,17).

6.2.3 Electron Diffraction of Single Crystals

Electron microscopy provides a wealth of information about the very small, including a view of the actual crystal cell size and shape. In another mode of use, the electrons can be made to diffract, using their wavelike properties. In this regard they are made to behave like the neutron scattering considered earlier.

In the case of X-ray studies, the polymer samples are usually uniaxially oriented and yield fiber diagrams that correspond to single-crystal rotation photographs. Electron diffraction studies utilize single crystals.

Since the polymer chains in single crystals are most often oriented perpendicular to their large flat surface, diffraction patterns perpendicular to the 001 plane are common. Tilting of the sample yields diffraction from other planes. The interpretation of the spots obtained utilizes Bragg's law in a manner identical to that of X-rays.

6.2.4 Infrared Absorption

Tadokoro (18) summarized some of the specialized information that infrared absorption spectra yield about crystallinity:

1. Infrared spectra of semicrystalline polymers include "crystallization-sensitive bands." The intensities of these bands vary with the degree of crystallinity and have been used as a measure of the crystallinity.

2. By measuring the polarized infrared spectra of oriented semicrystalline polymers, information about both the molecular and crystal structure can be obtained. Both uniaxially and biaxially oriented samples can be studied.
3. The regular arrangement of polymer molecules in a crystalline region can be treated theoretically, utilizing the symmetry properties of the chain or crystal. With the advent of modern computers, the normal modes of vibrations of crystalline polymers can be calculated and compared with experiment.
4. Deuteration of specific groups yields information about the extent of the contribution of a given group to specific spectral bands. This aids in the assignment of the bands as well as the identification of bands owing to the crystalline and amorphous regions.

6.2.5 Raman Spectra

Although Raman spectra have been known since 1928, studies on high polymers and other materials became popular only after the development of efficient laser sources. According to Tadokoro (18), some of the advantages of Raman spectra are the following:

1. Since the selection rules for Raman and infrared spectra are different, Raman spectra yield information complementary to the infrared spectra. For example, the S—S linkages in vulcanized rubber and the C═C bonds yield strong Raman spectra but are very weak or unobservable in infrared spectra.
2. Since the Raman spectrum is a scattering phenomenon, whereas the infrared methods depend on transmission, small bulk, powdered, or turbid samples can be employed.
3. On analysis, the Raman spectra provide information equivalent to very low-frequency measurements, even lower than $10\,cm^{-1}$. Such low-frequency studies provide information on lattice vibrations.
4. Polarization measurements can be made on oriented samples.

Of course, much of the above is widely practiced by spectroscopists on small molecules as well as big ones. Again, it must be emphasized that polymer chains are ordinary molecules that have been grown long in one direction.

6.3 THE UNIT CELL OF CRYSTALLINE POLYMERS

When polymers are crystallized in the bulk state, the individual crystallites are microscopic or even submicroscopic in size. They are an integral part of the solids and cannot be isolated. Hence studies on crystalline polymers in the bulk were limited to powder diagrams of the Debye-Scherrer type, or fiber diagrams of oriented materials.

It was only in 1957 that Keller (19) and others discovered a method of preparing single crystals from very dilute solutions by slow precipitation. These too were microscopic in size (see Section 6.4). However, X-ray studies could now be carried out on single crystals, with concomitant increases in detail obtainable.

Of course a major difference between polymers and low-molecular-weight compounds relates to the very existence of the macromolecule's long chains. These long chains traverse many unit cells. Their initial entangled nature impedes their motion, however, and leaves regions that are amorphous. Even the crystalline portions may be less than perfectly ordered.

This section describes the structure of the unit cell in polymers, principally as determined by X-ray analysis. The following sections describe the structure and morphology of single crystals, bulk crystallized crystallites, and spherulites and develops the kinetics and thermodynamics of crystallization.

6.3.1 Polyethylene

One of the most important polymers to be studied is polyethylene. It is the simplest of the polyolefins, those polymers consisting only of carbon and hydrogen, and polymerized through a double bond. Because of its simple structure, it has served as a model polymer in many laboratories. Also polyethylene's great commercial importance as a crystalline plastic has made the results immediately usable. It has been investigated both in the bulk and in the single-crystal state.

The unit cell structure of polyethylene was first investigated by Bunn (20). A number of experiments were reviewed by Natta and Corradini (21). The unit cell is orthorhombic, with cell dimensions of $a = 7.40$, $b = 4.93$, and $c = 2.534$ Å. The unit cell contains two mers (see Figure 6.5) (22). Not unexpectedly, the unit cell dimensions are substantially the same as those found for the normal paraffins of molecular weights in the range 300 to 600 g/mol. The chains are in the extended zigzag form; that is, the carbon–carbon bonds are *trans* rather than *gauche*. The zigzag form may also be viewed as a twofold screw axis.

The single-crystal electron diffraction pattern shown in Figure 6.5 was obtained by viewing the crystal along the *c*-axis. Also shown is the single-crystal structure of polyethylene, which is typically diamond-shaped (see below). The unit cell is viewed from the *c*-axis direction, perpendicular to the diamonds.

6.3.2 Other Polyolefin Polymers

Because of the need for regularity along the chain, only those vinyl polymers that are either isotactic or syndiotactic will crystallize. Thus isotactic

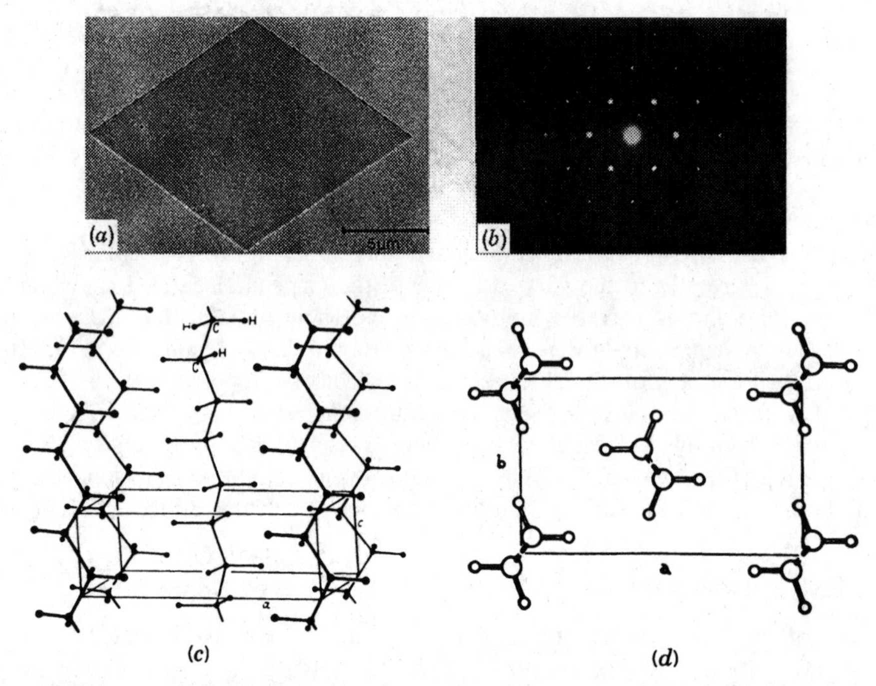

Figure 6.5 A study of polyethylene single-crystal structure. (*a*) A single crystal of polyethylene, precipitated from xylene, as seen by electron microscopy. (*b*) Electron diffraction of the same crystal, with identical orientation. (*c*) Perspective view of the unit cell of polyethylene, after Bunn. (*d*) View along chain axis. This latter corresponds to the crystal and diffraction orientation in (*a*) and (*b*) (22). Courtesy of A. Keller and Sally Argon.

polypropylene crystallizes well and is a good fiber former, whereas atactic polypropylene is essentially amorphous.

The idea of a screw axis along the individual extended chains needs developing. Such chains may be viewed as having an n/p-fold helix, where n is the number of mer units and p is the number of pitches within the identity period. Of course, n/p will be a rational number. Some of the possible types of helices for isotactic polymers are illustrated in Figure 6.6 (23). Group I of Figure 6.6 has a helix that makes one complete turn for every three mer units, so $n = 3$ and $p = 1$. Group II shows seven mer units in two turns, so $n = 7$ and $p = 2$. Group III shows four mer units per turn.

Of course, both left- and right-handed helices are possible. The isotactic hydrocarbon polymers in question occur as enantiomorphic pairs that face each other, a closer packing being realized through the operation of a glide plane with translation parallel to the fiber axis. The enantiomorphic crystal structure of polybutene-1 is illustrated in Figure 6.7 (21). Note that better packing is achieved through the chains' having the opposite sense of helical twist.

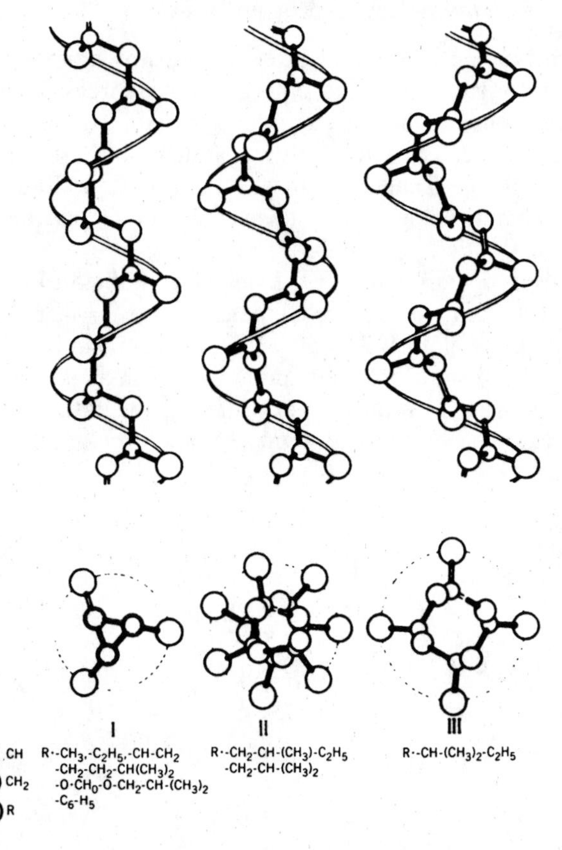

Figure 6.6 Possible types of helices for isotactic chains, with various lateral group (23).

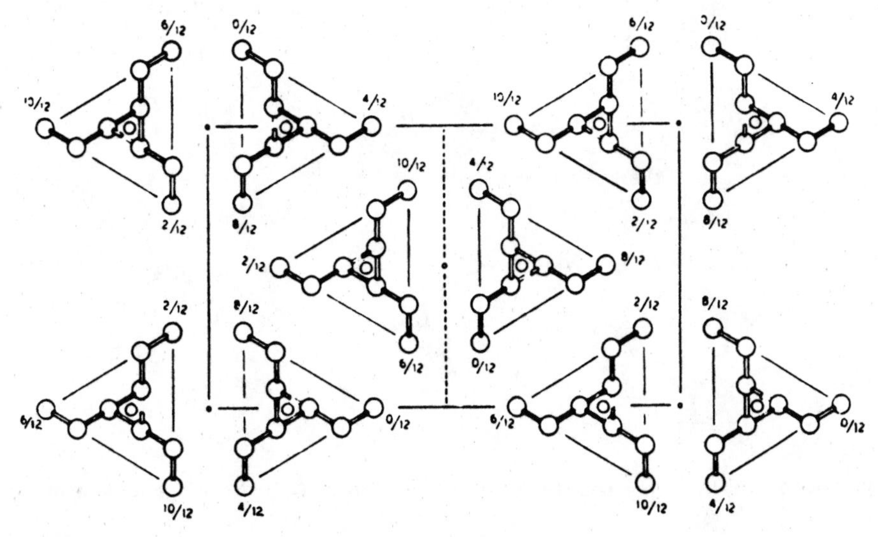

Figure 6.7 Enantiomorphous mode of packing polybutene-1 chains in a crystal (21).

6.3.3 Polar Polymers and Hydrogen Bonding

The hydrocarbon polymers illustrated above are nonpolar, being bonded together only by van der Waals–type attractive forces. When the polymers possess polar groups or hydrogen bonding capability, the most energetically favored crystal structures will tend to capitalize on these features. Figure 6.8 (24) illustrates the molecular organization within crystallites of a polyamide, known as polyamide 66 or nylon 66. The chains in the crystallites are found to occur as fully extended, planar zigzag structures.

X-ray analysis reveals that poly(ethylene terephthalate) (Dacron®) belongs to the triclinic system (25). The cell dimensions are $a = 4.56$, $b = 5.94$, $c = 10.75$ Å, with the angles being $\alpha = 98.5°$, $\beta = 118°$, $\gamma = 112°$. Both the polyamides and the aromatic polyesters are high melting polymers because of hydrogen bonding in the former case and chain stiffness in the latter case (see Table 6.1). As is well known, both of these polymers make excellent fibers and plastics.

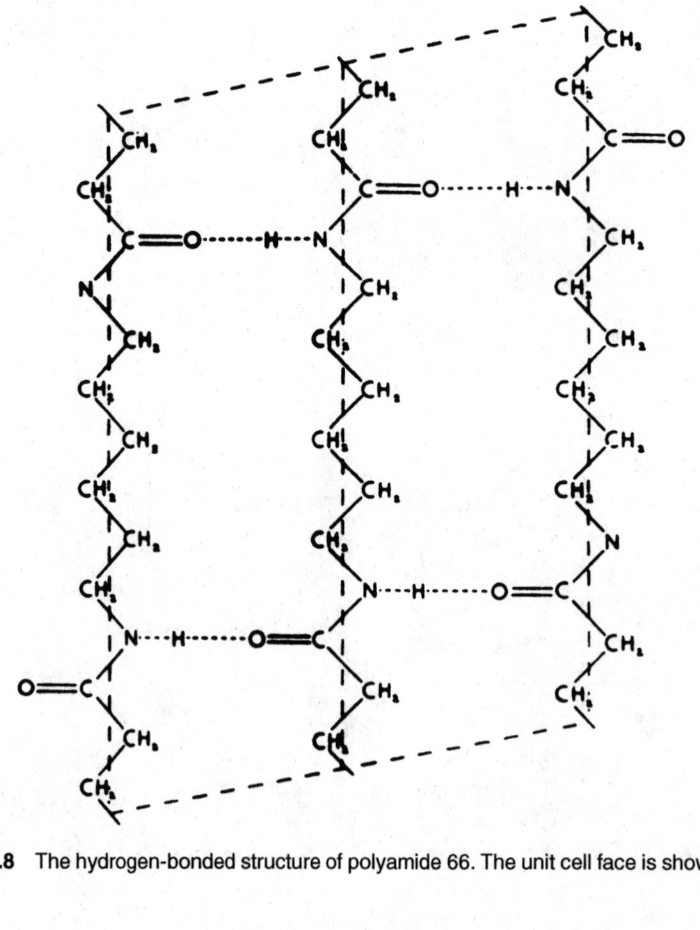

Figure 6.8 The hydrogen-bonded structure of polyamide 66. The unit cell face is shown dotted (24).

The polyethers are a less polar group of polymers. Poly(ethylene oxide) will be taken as an example (26). Four (7/2) helical molecules pass through a unit cell with parameters $a = 8.05$, $b = 13.04$, $c = 19.48$ Å, and $\beta = 125.4°$ with the space group $P2_1/a - C_{2h}$. [Space groups are discussed by Tadokoro (27,28). These symbols represent the particular one of the 230 possible space groups to which poly(ethylene oxide) belongs; see Section 6.2.]

Table 6.2 (27) summarizes the crystallographic data of some important polymers. Several polymers have more than one crystallographic form. Polytetrafluoroethylene (Teflon®), for example, undergoes a first-order crystal–crystal transition at 19°C.

6.3.4 Polymorphic Forms of Cellulose

A few words have already been said about the crystalline structure of cellulose (see Figure 6.1). The monoclinic unit cell structure illustrated for cellulose I was postulated many years ago by Meyer et al. (3–6) and has been confirmed many times. The *b* axis is the fiber direction, and the cell belongs to the space group P2, containing four glucose residues.

The triclinic structure is also known. Note that the structure shown in Figure 6.1*a* illustrates the chains running in alternating directions, up and down, now not thought true for most celluloses. The assumption of all one way or of alternating opposite directions is very important in the development of crystalline models (see below).

As shown further in Figure 6.1*b*, *c*, four different polymorphic forms of crystalline cellulose exist. Cellulose I is native cellulose, the kind found in wood and cotton. Cellulose II is made either by soaking cellulose I in strong alkali solutions (e.g., making mercerized cotton) or by dissolving it in the viscose process, which makes the labile but soluble cellulose xanthate. The regenerated cellulose II products are known as rayon for the fiber form and cellophane for the film form. Cellulose III can be made by treating cellulose with ethylamine. Cellulose IV may be obtained by treatment with glycerol or alkali at high temperatures (29,30).

Going from the polymorphic forms of cellulose back to cellulose I is difficult but can be accomplished by partial hydrolysis. The subject of the polymorphic forms of cellulose has been reviewed (10,11).

6.3.5 Principles of Crystal Structure Determination

Through the use of X-ray analysis, electron diffraction, and the supporting experiments of infrared absorption and Raman spectroscopy, much information has been collected on crystalline polymers. Today, the data are analyzed through the use of computer techniques. Somewhere along the line, however, the investigator is required to use intuition to propose models of crystal structure. The proposed models are then compared to experiment. The models are gradually refined to produce the structures given above. It must be

Table 6.2 Selected crystallographic data (27)

Polymer	Crystal System, Space Group, Lattice Constants, and Number of Chains per Unit Cell[a]	Molecular Conformation	Crystal Density (g/cm³)
Polyethylene $-(CH_2-CH_2)_n-$	Stable form, orthorhombic, $Pnam\text{-}D_{2h}^{16}$, $a = 7.417$ Å, $b = 4.945$ Å, $c = 2.547$ Å, $N = 2$	Planar zigzag (2/1)	1.00
	Metastable form, monoclinic, $C2/m - C_{2h}^{3}$, $a = 8.09$ Å, b (f.a.) = 2.53 Å, $c = 4.79$ Å, $\beta = 107.9°$, $N = 2$	Planar zigzag (2/1)	0.998
	High-pressure form, orthohexagonal (assumed), $a = 8.42$ Å, $b = 4.56$ Å, c (f.a.) has not been determined		
Polytetrafluoroethylene $-(CF_2-CF_2)_n-$	Below 19°C, pseudohexagonal (triclinic), $a' = b' = 5.59$ Å, $c = 16.88$ Å, $\gamma' = 119.3°$, $N = 1$	Helix (13/6) 163.5°	2.35
	Above 19°C, trigonal, $a = 5.66$ Å, $c = 19.50$ Å, $N = 1$	Helix (15/7) 165.8°	2.30
	High-pressure form I[12], orthorhombic, $Pnam\text{-}D_{2h}^{16}$, $a = 8.73$ Å, $b = 5.69$ Å, $c = 2.62$ Å, $N = 2$	Planar zigzag (2/1)	2.55
	High-pressure form II,[193] monoclinic, $B2/m\text{-}C_{2h}{}^{3}$, $a = 9.50$ Å, $b = 5.05$ Å, $c = 2.62$ Å, $\gamma = 105.5°$, $N = 2$	Planar zigzag (2/1)	2.74
it-Polypropylene $-(CH(CH_3)-CH_2)_n-$	α-Form, monoclinic, $C2/c\text{-}C_{2h}^{6}$ of $Cc\text{-}C_{s}^{4}$, $a = 6.65$ Å, $b = 20.96$ Å, $c = 6.50$ Å, $\beta = 99°20'$, $N = 4$	Helix (3/1) $(TG)_3$	0.936
	β-Form, hexagonal, $a = 19.08$ Å, $c = 6.49$Å, $N = 9$	Helix (3/1) $(TG)_3$	0.922
	γ-Form, trigonal, $P3_121\text{-}D_3^4$ or $P3_221\text{-}D_3^6$, $a = 6.38$ Å, $c = 6.33$ Å, $N = 1$	Helix (3/1) $(TG)_3$	0.939
it-Polystyrene $-(CH(C_6H_5)-CH_2)_n-$	Trigonal, $R3c\text{-}C_{3v}^{6}$ or $R\bar{3}c\text{-}D_{3d}^{6}$, $a = 21.90$ Å, $c = 6.65$ Å, $N = 6$	Helix (3/1) $(TG)_3$	1.13

cis-1,4-Polyisoprene $-\!\!\!+CH_2-C(CH_3)=CH-CH_2+\!\!\!-_n$	Monoclinic, $P2_1/a\text{-}C_{2h}^5$, $a = 12.46$ Å, $b = 8.89$ Å, $c = 8.10$ Å, $\beta = 92°$, $N = 4$	*cis*-$ST\bar{S}$-cis-$\bar{S}TS$, (2/0)	1.02
Poly(vinyl chloride) $-\!\!\!+CHCl-CH_2+\!\!\!-_n$	Orthorhombic, $Pcam\text{-}D_{2h}^{11}$, $a = 10.6$ Å, $b = 5.4$ Å, $c = 5.1$ Å, $N = 2$	Planar zigzag	1.42
Polytetrahydrofuran $-\!\!\!+(CH_2)_4-O+\!\!\!-_n$	Monoclinic, $C2/c\text{-}C_{2h}^6$, $a = 5.59$ Å, $b = 8.90$ Å, $c = 12.07$ Å, $\beta = 134.2°$, $N = 2$	Planar zigzag (2/1)	1.11
Polyamide 6 $-\!\!\!+(CH_2)_5-CONH+\!\!\!-_n$	α-Form, monoclinic, $P2_1\text{-}C_2^2$, $a = 9.56$ Å, b (f.a.) = 17.2 Å, $c = 8.01$ Å, $\beta = 67.5°$, $N = 4$	Planar zigzag (2/1)	1.23
	γ-Form, monoclinic, $P2_1/a\text{-}C_{2h}^5$, $a = 9.33$ Å, b (f.a.) = 16.88 Å, $c = 4.78$ Å, $\beta = 121°$, $N = 2$	Helix (2/1) $(T_4ST\bar{S})^2$	1.17
Polyamide 66 $-\!\!\!+NH-(CH_2)_6NHCO-(CH_2)_4-CO+\!\!\!-_n$	α-Form, triclinic, $P\bar{1}\text{-}C_i^1$, $a = 4.9$ Å, $b = 5.4$ Å, $c = 17.2$ Å, $a = 48.5°$, $\beta = 77°$, $\gamma = 63.5°$, $N = 1$	Planar zigzag (1/0)	1.24
	β-Form, triclinic, $P\bar{1}\text{-}C_i^1$, $a = 4.9$ Å, $b = 8.0$ Å, $c = 17.2$ Å, $\alpha = 90°$, $\beta = 77°$, $\gamma = 67°$, $N = 2$	Planar zigzag (1/0)	1.248
Poly(ethylene oxide) $-\!\!\!+CH_2-CH_2-O+\!\!\!-_n$	Form I, monoclinic, $P2_1/a\text{-}C_{2h}^5$, $a = 8.05$ Å, $b = 13.04$ Å, $c = 19.48$ Å, $\beta = 125.4°$, $N = 4$	Helix (7/2)	1.228
	Form II, triclinic, $P\bar{1}\text{-}C_i^1$, $a = 4.17$ Å, $b = 4.44$ Å, $c = 7.12$ Å, $\alpha = 62.8°$, $\beta = 93.2°$, $\gamma = 111.4°$, $N = 1$	Planar zigzag (2/1)	1.197
	β-Form, triclinic, $P\bar{1}\text{-}C_i^1$, $a = 4.9$ Å, $b = 8.0$ Å, $c = 22.4$ Å, $\alpha = 90°$, $\beta = 77°$, $\gamma = 67°$, $N = 2$	Planar zigzag (1/0)	1.196
Poly(ethylene terephthalate) $-\!\!\!+O-(CH_2)_2-O-CO-C_6H_4-CO+\!\!\!-_n$	Triclinic, $P\bar{1}\text{-}C_i^1$, $a = 4.56$ Å, $b = 5.94$ Å, $c = 10.75$ Å, $\alpha = 98.5°$, $\beta = 112°$, $N = 1$	Nearly planar	1.455

emphasized, that the experiments do not yield the crystal structure; only researchers' imagination and hard work yield that.

However, it is possible to simplify the task. Natta and Corradini (23) postulated three principles for the determination of crystal structures, which introduce considerable order into the procedure. These are:

1. *The Equivalence Postulate.* It is possible to assume that all mer units in a crystal occupy geometrically equivalent positions with respect to the chain axis.
2. *The Minimum Energy Postulate.* The conformation of the chain in a crystal may be assumed to approach the conformation of minimum potential energy for an isolated chain oriented along an axis.
3. *The Packing Postulate.* As many elements of symmetry of isolated chain as possible are maintained in the lattice, so equivalent atoms of different mer units along an axis tend to assume equivalent positions with respect to atoms of neighboring chains.

The equivalence postulate is seen in the structures given in Figure 6.6. Here, the chain mers repeat their structure in the next unit cell.

Energy calculations made for both single molecules and their unit cells serve three purposes: (a) they clarify the factors governing the crystal and molecular structure already tentatively arrived at experimentally, (b) they suggest the most stable molecular conformation and its crystal packing starting from the individual mer chemical structure, and (c) they provide a collection of reliable potential functions and parameters for both intra- and intermolecular interactions based on well-defined crystal structures (27). An example of intermolecular interactions is hydrogen bonding in the polyamide structures described in Figure 6.8.

The packing postulate is seen at work in Figure 6.7, where enantiomorphic structures pack closer together in space than if the chains had the same sense of helical twist.

Last, one should not neglect the very simple but all important density. The crystalline cell is usually about 10% more dense than the bulk amorphous polymer. Significant deviations from this density must mean an incorrect model.

Because of the importance of polymer crystallinity generally, and the unit cell in particular, the subject has been reviewed many times (17,21,27,31–41).

6.4 STRUCTURE OF CRYSTALLINE POLYMERS

6.4.1 The Fringed Micelle Model

Very early studies on bulk materials showed that some polymers were partly crystalline. X-ray line broadening indicated that the crystals were either very

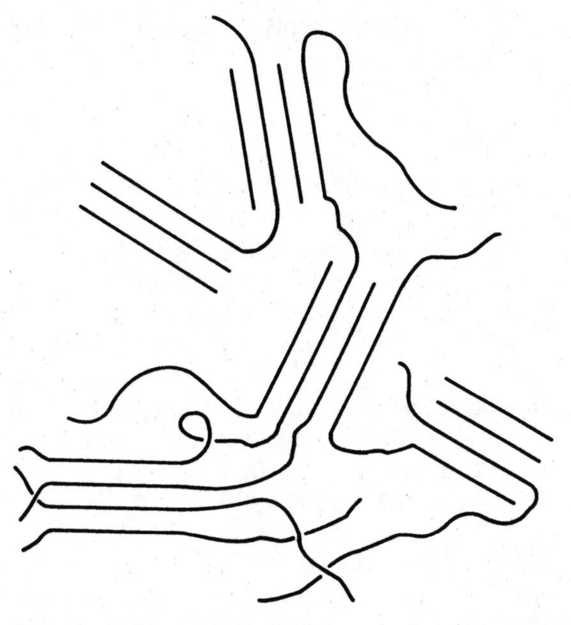

Figure 6.9 The fringed micelle model. Each chain meanders from crystallite to crystallite, binding the whole mass together.

imperfect or very small (42). Assuming the latter, in 1928 Hengstenberg and Mark (43) estimated that the crystallites of ramie, a form of native cellulose, were about 55 Å wide and over 600 Å long by this method. It had already been established that the polymer chain passed through many unit cells. Because of the known high molecular weight, the polymer chain was calculated to be even longer than the crystallites. Hence it was reasoned that they passed in and out of many crystallites (32,44,45). Their findings led to the fringed micelle model.

According to the fringed micelle model, the crystallites are about 100 Å long (Figure 6.9). The disordered regions separating the crystallites are amorphous. The chains wander from the amorphous region through a crystallite, and back out into the amorphous region. The chains are long enough to pass through several crystallites, binding them together.

The fringe micelle model was used with great success to explain a wide range of behavior in semicrystalline plastics, and also in fibers. The amorphous regions, if glassy, yielded a stiff plastic. However, if they were above T_g, then they were rubbery and were held together by the hard crystallites. This model explains the leathery behavior of ordinary polyethylene plastics, for example. The greater tensile strength of polyethylene over that of low-molecular-weight hydrocarbon waxes was attributed to amorphous chains wandering from crystallite to crystallite, holding them together by primary bonds. The flexible nature of fibers was explained similarly; however, the chains were oriented along the fiber axis (see Section 6.3). The exact stiffness of the plastic or fiber was related to the degree of crystallinity, or fraction of the polymer that was crystallized.

6.4.2 Polymer Single Crystals

Ideas about polymer crystallinity underwent an abrupt change in 1957 when Keller (19) succeeded in preparing single crystals of polyethylene. These were made by precipitation from extremely dilute solutions of hot xylene. These crystals tended to be diamond-shaped and of the order of 100 to 200 Å thick (see Figure 6.5) (21). Amazingly electron diffraction analysis showed that the polymer chain axes in the crystal body were essentially perpendicular to the large, flat faces of the crystal. Since the chains were known to have contour lengths of about 2000 Å and the thickness of the single crystals was in the vicinity of 110 to 140 Å, Keller concluded that the polymer molecules in the crystals had to be folded upon themselves. These observations were immediately confirmed by Fischer (47) and Till (48).†

6.4.2.1 The Folded-Chain Model This led to the folded-chain model, illustrated in Figure 6.10 (41). Ideally the molecules fold back and forth with hairpin turns. While adjacent reentry has been generally confirmed by small-angle neutron scattering and infrared studies for single crystals, the present understanding of bulk crystallized polymers indicates a much more complex situation (see below).

Figure 6.10 uses polyethylene as the model material. The orthorhombic cell structure and the *a*- and *b*-axes are illustrated. The *c*-axis runs parallel to the chains. The dimension ℓ is the thickness of the crystal. The predominant fold plane in polyethylene solution-grown crystals is along the (110) plane. Chain folding is also supported by NMR studies (see Section 6.7) (49–51).

For many polymers, the single crystals are not simple flat structures. The crystals often occur in the form of hollow pyramids, which collapse on drying. If the polymer solution is slightly more concentrated, or if the crystallization rate is increased, the polymers will crystallize in the form of various twins, spirals, and dendritic structures, which are multilayered (see Figure 6.11) (34). These latter form a preliminary basis for understanding polymer crystallization from bulk systems.

Simple homopolymers are not the only polymeric materials capable of forming single crystals. Block copolymers of poly(ethylene oxide) crystallize in the presence of considerable weight fractions of amorphous polystyrene (see Figure 6.12) (52). In this case square-shaped crystals with some spirals are

† The early literature reveals signficant premonitions of this discovery. K. H. Storks, *J. Am. Chem. Soc.*, **60**, 1753 (1938) suggested that the macromolecules in crystalline gutta percha are folded back and forth upon themselves in such a way that adjacent sections remain parallel. R. Jaccodine, *Nature (London)*, **176**, 305 (1955) showed the spiral growth of polythylene single crystals by a dislocation mechanism.

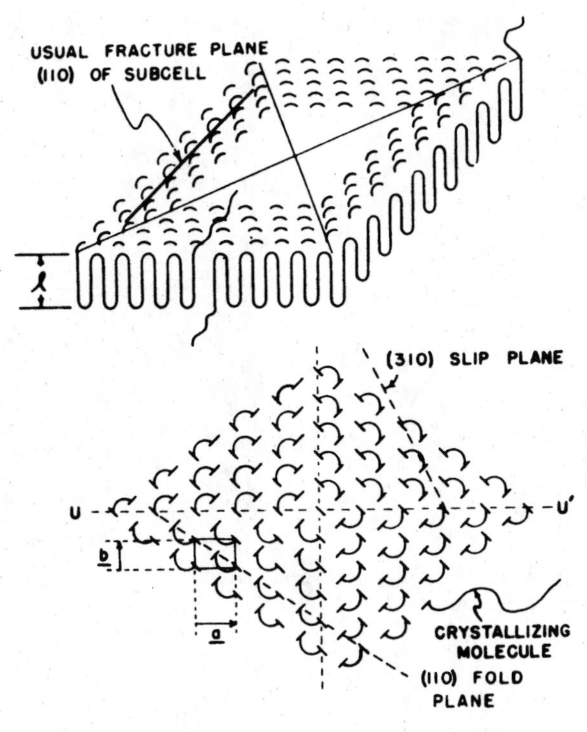

Figure 6.10 Schematic view of a polyethylene single crystal exhibiting adjacent reentry. The orthorhombic subcell with dimensions *a* and *b*, typical of many *n*-paraffins, is illustrated below (41).

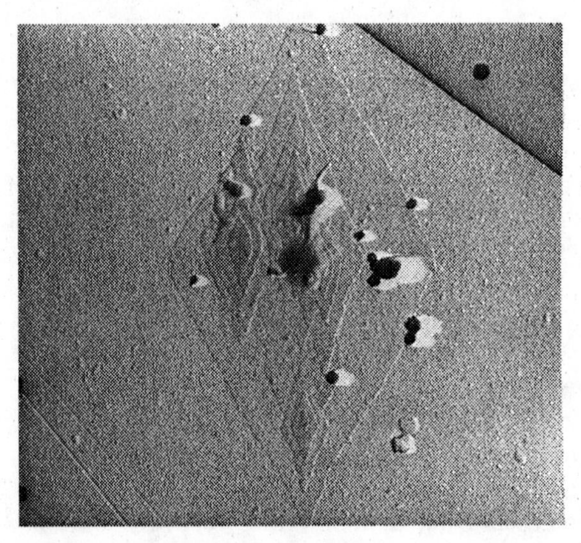

Figure 6.11 Single crystal of polyamide 6 precipitated from a glycerol solution. The lamellae are about 60 Å thick. Black marks indicate 1 μm (34).

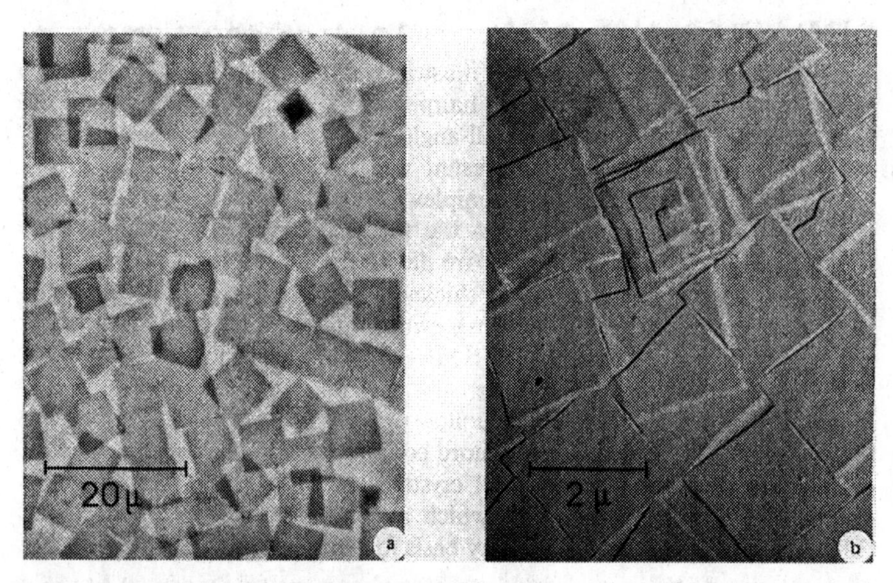

Figure 6.12 Single crystals of poly(ethylene oxide)–*block*–polystyrene diblock copolymers. (*a*) Optical micrograph. (*b*) Electron micrograph. M_n (PS) = 7.3 × 10^3 g/mol; M_n (PEO) = 10.9 × 10^3; weight fraction polystyrene is 0.34 (52).

seen. The crystals reject the amorphous portion (polystyrene), which appears on the surfaces of the crystals.

Amorphous material also appears on the surfaces of homopolymer single crystals. As will be developed below, causes of this amorphous material range from chain-end cilia to irregular folding.

6.4.2.2 The Switchboard Model In the switchboard model the chains do not have a reentry into the lamellae by regular folding; they rather reenter more or less randomly (53). The model more or less resembles and old-time telephone switchboard. Of course, both the perfectly folded chain and switchboard models represent limiting cases. Real systems may combine elements of both. For bulk systems, this aspect is discussed in Section 6.7.

6.5 CRYSTALLIZATION FROM THE MELT

6.5.1 Spherulitic Morphology

In the previous sections it was observed that when polymers are crystallized from dilute solutions, they form lamellar-shaped single crystals. These crystals exhibit a folded-chain habit and are of the order of 100 to 200 Å thick. From somewhat more concentrated solutions, various multilayered dendritic structures are observed.

Figure 6.13 Spherulites of low-density polyethylene, observed through crossed polarizers. Note characteristic Maltese cross pattern (34).

When polymers crystallize from the melt, they usually *supercool* to greater or lesser extents; see Figure 6.3. Thus, the crystallization temperature may be 10 to 20°C lower than the melting temperature. Supercooling arises from the extra free energy required to align chain segments, common in the crystallization of many complex organic compounds as well as polymers.

When polymer samples are crystallized from the melt, the most obvious of the observed structures are the spherulites (33). As the name implies, spherulites are sphere-shaped crystalline structures that form in the bulk (see Figure 6.13) (34). One of the more important problems to be addressed concerns the form of the lamellae within the spherulite.

Spherulites are remarkably easy to grow and observe in the laboratory (36). Simple cooling of a thin section between crossed polarizers is sufficient, although controlled experiments are obviously more demanding. It is observed that each spherulite exhibits an extinction cross, sometimes called a Maltese cross. This extinction is centered at the origin of the spherulite, and the arms of the cross are oriented parallel to the vibration directions of the microscope polarizer and analyzer.

Usually the spherulites are really spherical in shape only during the initial stages of crystallization. During the latter stages of crystallization, the

Figure 6.14 Surface replica of polyoxymethylene fractured at liquid nitrogen temperatures. Lamellae at lower left are oriented at an angle to the fracture surface. Lamellae elsewhere are nearly parallel to the fracture surface, being stacked up like cards or dishes in the bulk state. These structures closely resemble stacks of single crystals, and they have led to ideas about chain folding in bulk materials (34).

spherulites impinge on their neighbors. When the spherulites are nucleated simultaneously, the boundaries between them are straight. However, when the spherulites have been nucleated at different times, so that they are different in size when impinging on one another, their boundaries form hyperbolas. Finally, the spherulites form structures that pervade the entire mass of the material. The kinetics of spherulite crystallization are considered in Section 6.7.

Electron microscopy examination of the spherulitic structure shows that the spherulites are composed of individual lamellar crystalline plates (see Figure 6.14) (34). The lamellar structures sometimes resemble staircases, being composed of nearly parallel (but slightly diverging) lamellae of equal thickness. Amorphous material usually exists between the staircase lamellae.

X-ray microdiffraction (37) and electron diffraction (38) examination of the spherulites indicates that the *c*-axis of the crystals is normal to the radial (growth) direction of the spherulites. Thus the *c*-axis is perpendicular to the lamellae flat surfaces, showing the resemblance to single-crystal structures.

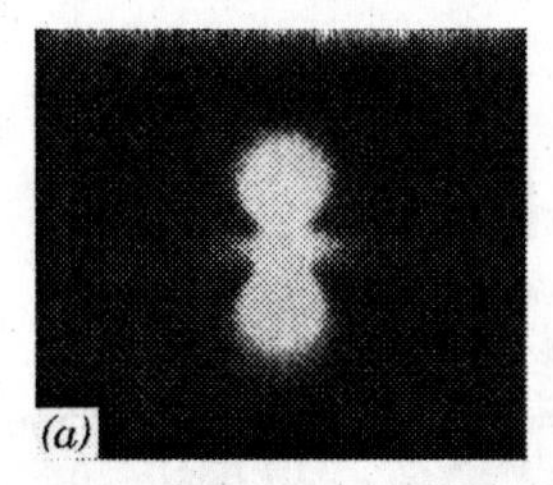

Figure 6.15 Different types of light-scattering patterns are obtained from spherulitic polyethylene using (*a*) V_v and (*b*) H_v polarization (54). Note the twofold symmetry of the V_v pattern, and the fourfold symmetry of the H_v pattern. This provided direct experimental evidence that spherulites were anisotropic.

For some polymers, such as polyethylene (37), it was shown that their lamellae have a screwlike twist along their unit cell *b* axis, on the spherulite radius. The distance corresponding to one-half of the pitch of the lamellar screw is just in accordance with the extinction ring interval visible on some photographs.

The growth and structure of spherulites may also be studied by small-angle light scattering (39,40). The sample is placed between polarizers, a monochromatic or laser light beam is passed through, and the resultant scattered beam is photographed. Two types of scattering patterns are obtained, depending on polarization conditions (54). When the polarization of the incident beam and that of the analyzer are both vertical, it is called a V_v type of pattern. When the incident radiation is vertical in polarization but the analyzer is horizontal (polarizers crossed), an H_v pattern is obtained.

The two types of scattering patterns are illustrated in Figure 6.15 (54). These patterns arise from the spherulitic structure of the polymer, which is optically anisotropic, with the radial and tangential refractive indexes being different.

The scattering pattern can be used to calculate the size of the spherulites (40) (see Figure 6.16). The maximum that occurs in the radial direction, *U*, is related to *R*, the radius of the spherulite by

$$U_{\max} = \left(\frac{4\pi R}{\lambda}\right)\sin\left(\frac{\theta_{\max}}{2}\right) = 4.1 \tag{6.2}$$

where $\theta_{\max}$ is the angle at which the intensity maximum occurs and λ is the wavelength. As the spherulites get larger, the maximum in intensity occurs at smaller angles.

Conversely, Stein (54) points out that in very rapidly crystallized polymers, spherulites are often not observed. The smaller amount of scattering observed results entirely from local structure. These structures are highly disordered.

Mandelkern recently drew a morphological map for polyethylene (55). He showed that the supermolecular structures become less ordered as the molecular weight is increased or the temperature of crystallization is decreased.

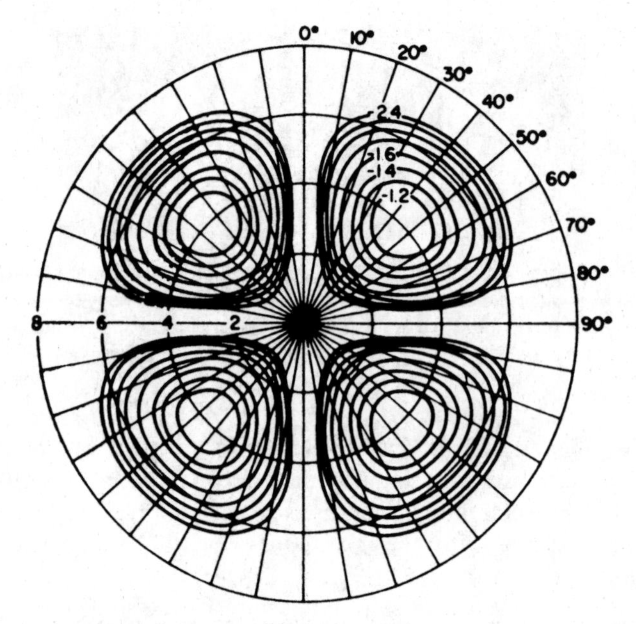

Figure 6.16 The calculated H_v light-scattering pattern for an idealized spherulite (40). Compare with the actual result, Figure 6.15*b*.

A model of the spherulite structure is illustrated in Figure 6.17 (41). The chain direction in the bulk crystallized lamellae is perpendicular to the broad plane of the structure, just like the dilute solution crystallized material.

The spherulite lamellae also contain low-angle branch points, where new lamellar structures are initiated. The new lamellae tend to keep the spacing between the crystallites constant.

While the lamellar structures in the spherulites are the analogue of the single crystals, the folding of the chains is much more irregular, as will be developed further in Section 6.6.2.3. In between the lamellar structures lies amorphous material. This portion is rich in components such as atactic polymers, low-molecular-weight material, or impurities of various kinds.

The individual lamellae in the spherulites are bonded together by tie molecules, which lie partly in one crystallite and partly in another. Sometimes these tie molecules are actually in the form of what are called intercrystalline links (56–59), which are long, threadlike crystalline structures with the *c* axis along their long dimension. These intercrystalline links are thought to be important in the development of the great toughness characteristic of semicrystalline polymers. They serve to tie the entire structure together by crystalline regions and/or primary chain bonds.

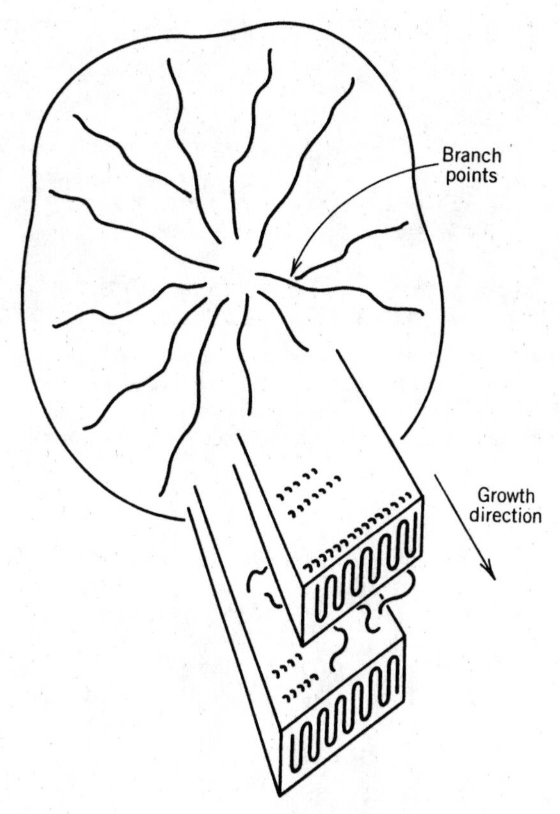

Figure 6.17 Model of spherulitic structure. Note the growth directions and lamellar branch points that fill the space uniformly with crystalline material. After J. D. Hoffman et al. (41).

6.5.2 Mechanism of Spherulite Formation

On cooling from the melt, the first structure that forms is the single crystal. These rapidly degenerate into sheaflike structures during the early stages of the growth of polymer spherulites (see Figure 6.18) (33). These sheaflike structures have been variously called axialites (60) or hedrites (61). These transitional, multilayered structures represent an intermediate stage in the formation of spherulites (62). It is evident from Figure 6.18 that as growth proceeds, the lamellae develop on either side of a central reference plane. The lamellae fan out progressively and grow away from the plane as the structure begins to mature.

The sheaflike structures illustrated in Figure 6.18 are modeled in Figure 6.19 (33). As in Figure 6.18, both edge-on and flat-on views are illustrated. Figure 6.18*a* is modeled by Figure 6.19, row *a*, column III, and Figure 6.18*b* is modeled by row *b*, column III. Gradually the lamellae in the hedrites diverge or fan outward in a splaying motion. Repeated splaying, perhaps aided by lamellae

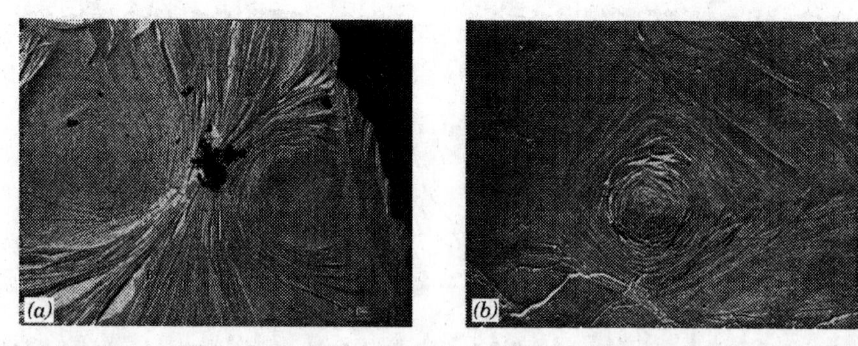

Figure 6.18 Electron micrographs of replicas of hedrites formed in the same melt-crystallized thin film of poly(4-methylpentene-1). (*a*) An edge-on view of a hedrite. Note the distinctly lamellar character and the "sheaflike" arrangement of the lamellae. (*b*) A flat-on view of a hedrite. Note the degenerate overall square outline of the object, whose lamellar texture is evident (33).

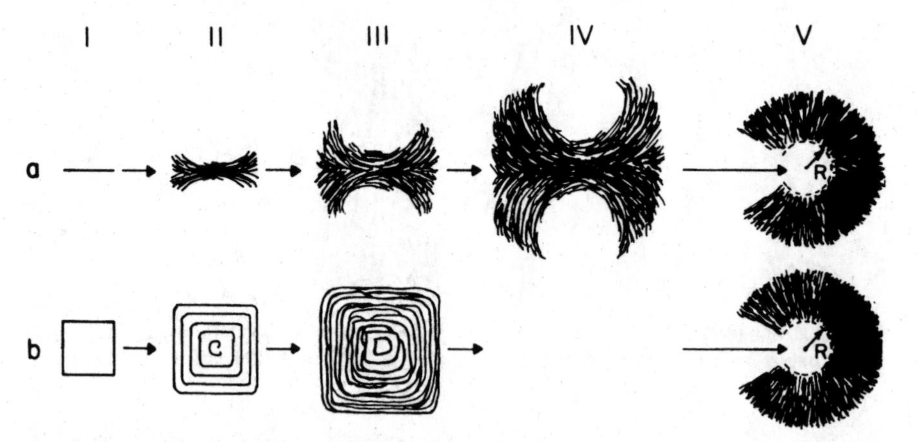

Figure 6.19 Schematic development of a spherulite from a chain-folded precursor crystal. Rows (*a*) and (*b*) represent, respectively, edge-on and flat-on views of the evolution of the spherulite (33).

that are intrinsically curved, eventually leads to the spherical shape characteristic of the spherulite.

6.5.3 Spherulites in Polymer Blends and Block Copolymers

There are two cases to be considered. Either the two polymers composing the blend may be miscible and form one phase in the melt, or they are immiscible and form two phases. Martuscelli (63) pointed out that if the glass transition of the miscible noncrystallizing component is lower than that of the crystallizing component (i.e., its melt viscosity will be lower, other things being equal), then the spherulites will actually grow faster, although the system is diluted. Usually, crystallizable polymers containing low-molecular-weight fractions (which are not incorporated in the spherulite) crystallize faster.

Martuscelli also pointed out that the inverse was also true, especially if the noncrystallizing polymer was glassy at the temperature of crystallization.

The crystallization behavior is quite different if the two polymers are immiscible in the melt. Figure 6.20 (64) shows droplets of polyisobutylene dispersed in isotactic polypropylene. On spherulite formation, the droplets, which are noncrystallizing, become ordered within the growing arms of the crystallizing component.

Block copolymers also form spherulites (65–67). The morphology develops on a finer scale, however, because the domains are constrained to be of the order of the size of the individual blocks. In the case of a triblock copolymer, for example, the chain may be modeled as wandering from one lamella to another through an amorphous phase consisting of the center block (see Figure 6.21) (67).

Figure 6.22 (66) illustrates the spherulite morphology for poly(ethylene oxide)–*block*–polystyrene. Two points should be made. First, the glass

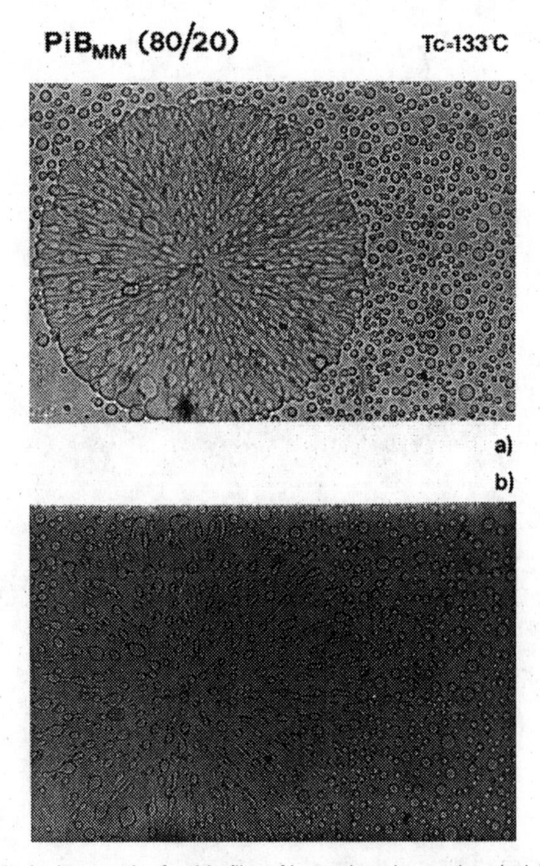

Figure 6.20 Optical micrograph of a thin film of isotactic polypropylene/polyisobutylene blend. (*a*) *it*-PP spherulite (T_c = 133°C), surrounded by melt blend at the early stage of crystallization. (*b*) The same region of film after melting of spherulite. Note the multiphase morphology common to many blends (63).

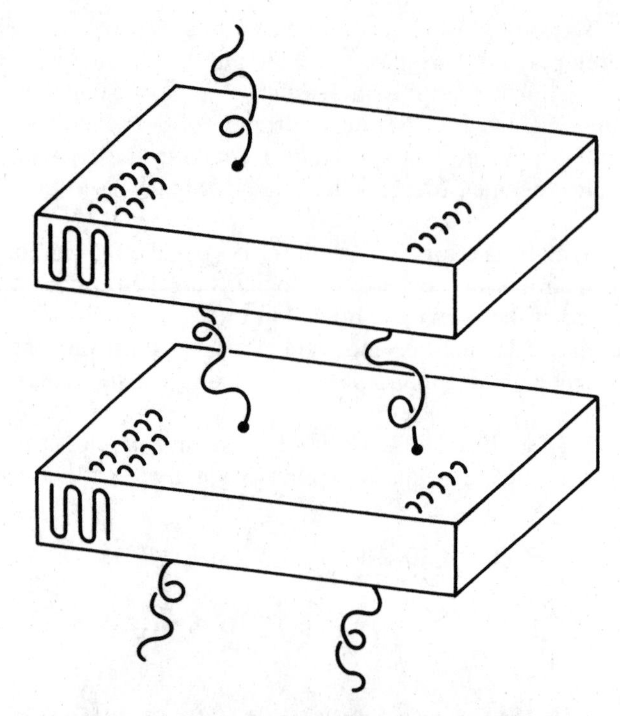

Figure 6.21 Model of crystallizable triblock copolymer thermoplastic elastomer. The center block, amorphous, is rubbery, whereas the end blocks are crystalline (64).

Figure 6.22 An optical micrograph of a poly(ethylene oxide)–*block*–polystyrene copolymer containing 19.6% polystyrene (66). Crystals cast from chloroform and observed through crossed polarizers. White markers are 250 μm apart.

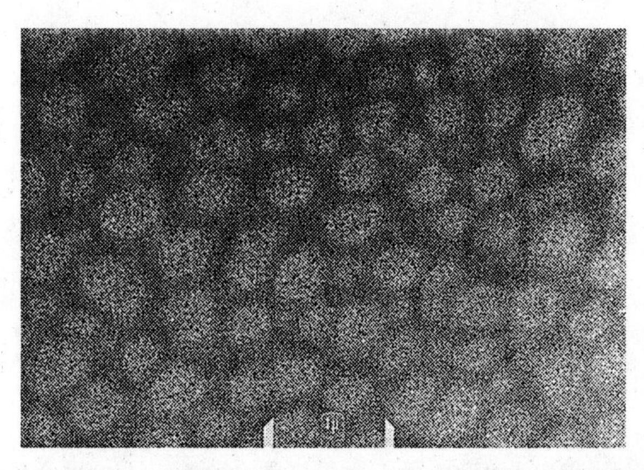

Figure 6.23 A transmission electron micrograph of poly(ethylene oxide)–*block*–polystyrene, containing 70% polystyrene. The polystyrene phase is stained with OsO_4 (66).

transition temperature of the polystyrene component is higher than the temperature of crystallization (65); this particular sample was made by casting from chloroform. Second, the amount of polystyrene is small, only 19.6%. When the polystyrene component is increased, it disturbs the ordering process (see Figure 6.23) (66). This figure shows spheres rich in poly(ethylene oxide) lamellae but containing some polystyrene segments (dark spots) embedded in the poly(ethylene oxide) spherulites, as well as forming the more continuous phase. The size of the fine structure is of the order of a few hundred angstroms, because the two blocks must remain attached, even though they are in different phases.

In the case where the polymer forms a triblock copolymer of the structure A—B—A, or a multiblock copolymer $\text{+}A\text{—}B\text{+}_m$, where A is crystalline at use temperature and B is rubbery (above T_g), then a thermoplastic elastomer is formed (see Figure 6.21). The material exhibits some degree of rubber elasticity at use temperature, the crystallites serving as cross-links. Above the melting temperature of the crystalline phase, the material is capable of flowing: that is, it is thermoplastic. It must be pointed out that a very important kind of thermoplastic elastomer is when the A polymer is glassy rather than crystalline (see Section 9.16).

Two important kinds of $\text{+}A\text{—}B\text{+}_n$ block copolymers, where A is amorphous and above T_g, and B is crystalline are the segmented polyurethanes and poly(ester–ether) materials. Both are fiber formers; see Chapter 13. These fibers tend to be soft and elastic.

6.5.4 Percent Crystallinity in Polymers

As suggested above, most crystallizing polymers are semicrystalline; that is, a certain fraction of the material is amorphous, while the remainder is crys-

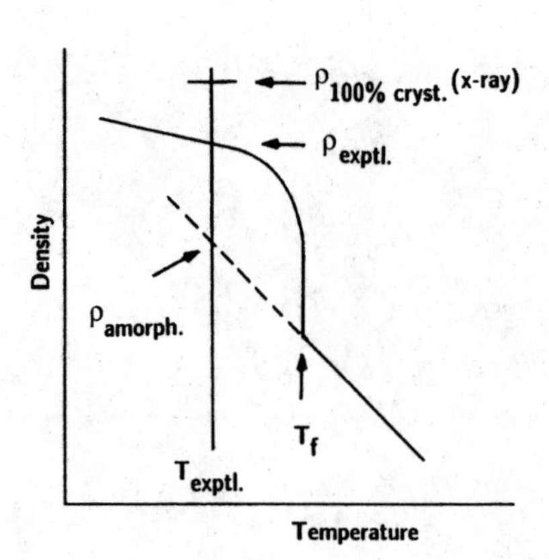

Figure 6.24 The experimental determination of the extent of polymer crystallinity using the density method.

talline. The reason why polymers fail to attain 100% crystallinity is kinetic, resulting from the inability of the polymer chains to completely disentangle and line up properly in a finite period of cooling or annealing.

There are several methods for determining the percent crystallinity in such polymers. The first involves the determination of the heat of fusion of the whole sample by calorimetric methods such as DSC; see Figure 6.3. The heat of fusion per mole of crystalline material can be estimated independently by melting point depression experiments; see Section 6.8.

A second method involves the determination of the density of the crystalline portion via X-ray analysis of the crystal structure, and determining the theoretical density of a 100% crystalline material. The density of the amorphous material can be determined from an extrapolation of the density from the melt to the temperature of interest; see Figure 6.24. Then the percent crystallinity is given by

$$\% \text{ Crystallinity} = \left[\frac{\rho_{\text{exptl}} - \rho_{\text{amorph}}}{\rho_{100\% \text{ cryst}} - \rho_{\text{amorph}}} \right] \times 100 \tag{6.3}$$

where ρ_{exptl} represents the experimental density, and ρ_{amorph} and $\rho_{100\%\text{cryst}}$ are the densities of the amorphous and crystalline portions, respectively.

A third method stems from the fact that the intensity of X-ray diffraction depends on the number of electrons involved and is thus proportional to the density. Besides Bragg diffraction lines for the crystalline portion, there is an amorphous halo caused by the amorphous portion of the polymer. This last occurs at a slightly smaller angle than the corresponding crystalline peak, because the atomic spacings are larger. The amorphous halo is broader than the corresponding crystalline peak, because of the molecular disorder.

This third method, sometimes called wide-angle X-ray scattering (WAXS), can be quantified by the crystallinity index (68), CI,

$$\mathrm{CI} = \frac{A_c}{A_a + A_c} \tag{6.4}$$

where A_c and A_a represent the area under the Bragg diffraction line (or equivalent crystalline Debye-Scherrer diffraction line; see Figure 6.4) and corresponding amorphous halo, respectively.

Naturally these methods will not yield the same answer for a given sample, but surprisingly good agreement is obtained. For many semicrystalline polymers the crystallinity is in the range of 40% to 75%. Polymers such as polytetrafluoroethylene achieve 90% crystallinity, while poly(vinyl chloride) is often down around 15% crystallinity. The latter polymer is largely atactic, but short syndiotactic segments contribute greatly to its crystallinity. Of course, annealing usually increases crystallinity, as does orienting the polymer in fiber or film formation.

6.6 KINETICS OF CRYSTALLIZATION

During crystallization from the bulk, polymers form lamellae, which in turn are organized into spherulites or their predecessor structures, hedrites. This section is concerned with the rates of crystallization under various conditions of temperature, molecular weight, structure, and so on, and the theories that provide not only an insight into the molecular mechanisms but considerable predictive power.

6.6.1 Experimental Observations of Crystallization Kinetics

It has already been pointed out that the volume changes on melting, usually increasing (see Figure 6.2). This phenomenon may be used to study the kinetics of crystallization. Figure 6.25 (69) illustrates the isothermal crystallization of poly(ethylene oxide) as determined dilatometrically. From Table 6.1 the melting temperature of poly(ethylene oxide) is 66°C, where the rate of crystallization is zero. The rate of crystallization increases as the temperature is decreased. This follows from the fact that the driving force increases as the sample is supercooled.

Crystallization rates may also be observed microscopically, by measuring the growth of the spherulites as a function of time. This may be done by optical microscopy, as has been done by Keith and Padden (70,71), or by transmission electron microscopy of thin sections (72). The isothermal radial growth of the spherulites is usually observed to be linear (see Figure 6.26) (71). This implies that the concentration of impurity at the growing tips of the lamellae remains constant through the growth process. The more impurity, the slower is the growth rate (Table 6.3) (71). However, linearity of growth rate is maintained. In such a steady state, the radial diffusion of rejected impurities is outstripped by the more rapidly growing lamellae so that impurities diffuse aside and are

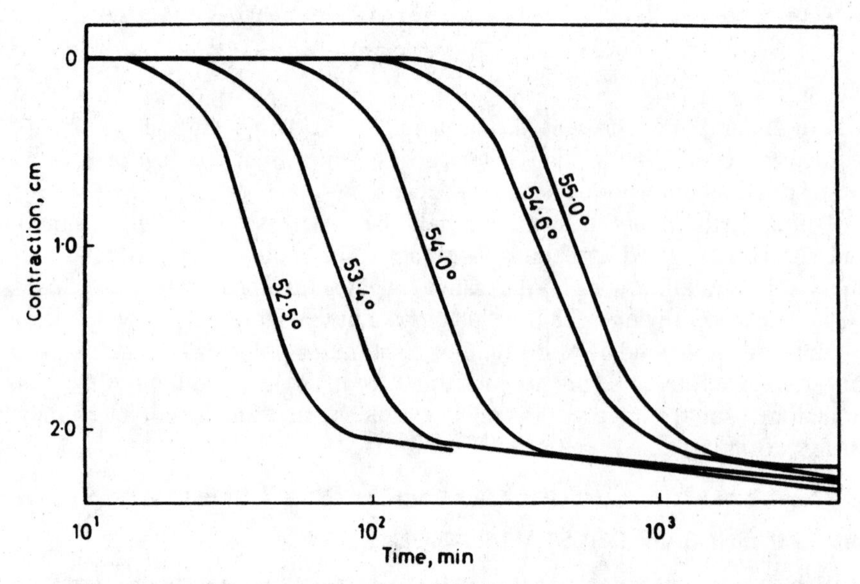

Figure 6.25 Dilatometric crystallization isotherms for poly(ethylene oxide), $M = 20{,}000$ g/mol. The Avrami exponent n falls 4.0 to 2.0 as crystallization proceeds (69).

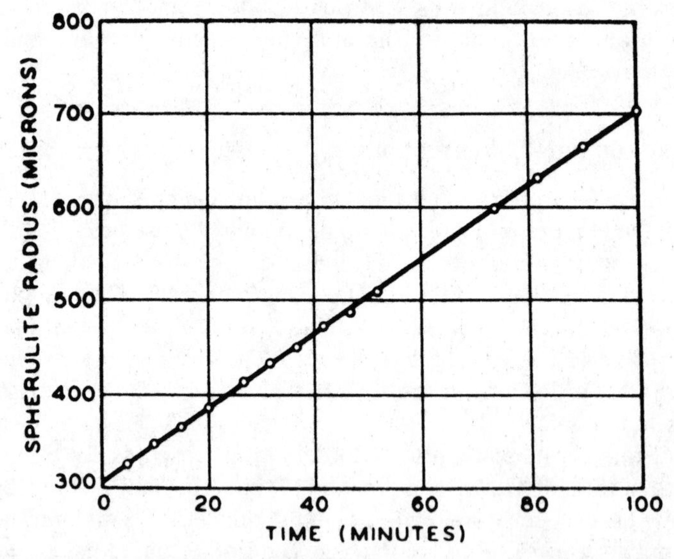

Figure 6.26 Spherulite radius as a function of time, grown isothermally (at 125°C) in a blend of 20% isotactic and 80% atactic ($M = 2600$) polypropylene. Note the linear behavior (71).

Table 6.3 Blends of unextracted Isotactic polypropylene with atactic polypropylene (71)

Temperature (°C)	Radial Growth Rates (μm/min) for Various Compositions: 100% Isotactic	90% Isotactic	80% Isotactic	60% Isotactic	40% Isotactic
120	29.4	29.4	26.4	22.8	21.2
125	13.0	12.0	11.0	8.90	8.57
131	3.88	3.60	3.03	2.37	2.40
135	1.63	1.57	1.35	1.18	1.12
Melting point (°C)	171	169	167	165	162

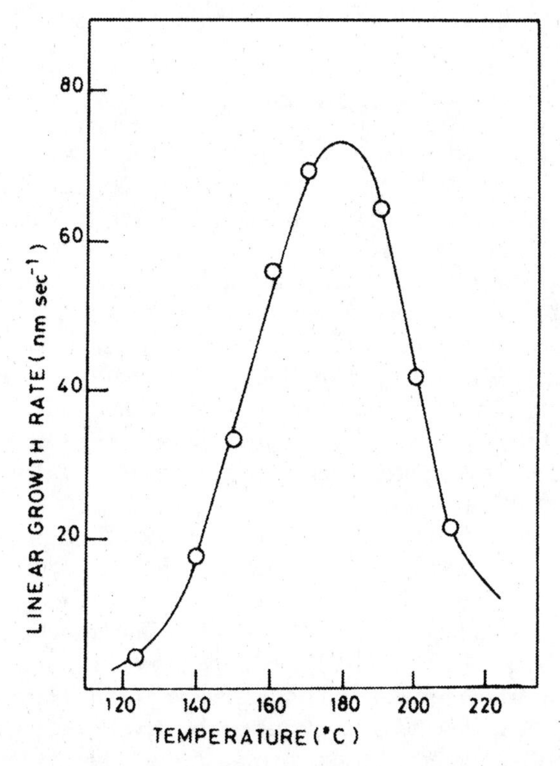

Figure 6.27 Plot of linear growth rate versus crystallization temperature for poly(ethylene terephthalate) (72). T_f = 265°C, and T_g = 67°C, at which points the rates of crystallization are theoretically zero.

trapped in interlamellar channels. In the case illustrated in Figure 6.26, the main "impurity" is low-molecular-weight atactic polypropylene.

When the radial growth rate is plotted as a function of crystallization temperature, a maximum is observed (see Figure 6.27) (72). As mentioned earlier, the increase in rate of crystallization as the temperature is lowered is controlled by the increase in the driving force. As the temperature is lowered still further, molecular motion becomes sluggish as the glass transition is approached, and the crystallization rate decreases again. Below T_g, molecular motion is so sluggish that the rate of crystallization effectively becomes zero. This is another example of supercooling, as illustrated earlier in Figure 6.3.

There is an interesting rule-of-thumb for determining a good temperature to crystallize a polymer, if the melting temperature is known. It is called the *eight-ninths temperature-of-fusion* rule, where T_f is in absolute temperature. At $(8/9)T_f$ the polymer is supposed to crystallize readily. Starting with $T_f = 265°C$, this yields 205°C for the poly(ethylene terephthalate) shown in Figure 6.27. While this result is somewhat higher than the maximum rate of crystallization temperature, it remains a handy rule if no other data are available.

6.6.2 Theories of Crystallization Kinetics

Section 6.4.2 described Keller's early preparations of single crystals from dilute solutions. Since the crystals were only about 100 Å thick and the chains were oriented perpendicular to the flat faces, Keller postulated that the chains had to be folded back and forth.

Similar structures, called lamellae, exist in the bulk state. While their folding is now thought to be much less regular, their proposed molecular organization remains similar. In the bulk state, however, these crystals are organized into the larger structures known as spherulites. Section 6.6.1 showed that the rate of radial growth of the spherulites was linear in time and that the rate of growth goes through a maximum as the temperature of crystallization is lowered. These several experimental findings form the basis for three theories of polymer crystallization kinetics.

The first of these theories is based on the work of Avrami (73,74), which adapts formulations intended for metallurgy to the needs of polymer science. The second theory was developed by Keith and Padden (70,71), providing a qualitative understanding of the rates of spherulitic growth. More recently Hoffman and co-workers (41,77–80) developed the kinetic nucleation theory of chain folding, which provides an understanding of how lamellar structures form from the melt. This theory continues to be developed even as this material is being written. Together, these theories provide insight into the kinetics not only of crystallization but also of the several molecular mechanisms taking part.

6.6.2.1 The Avrami Equation The original derivations by Avrami (73–75) have been simplified by Evans (81) and put into polymer context by Meares (82) and Hay (83). In the following, it is helpful to imagine raindrops falling in a puddle. These drops produce expanding circles of waves that intersect and cover the whole surface. The drops may fall sporadically or all at once. In either case they must strike the puddle surface at random points. The expanding circles of waves, of course, are the growth fronts of the spherulites, and the points of impact are the crystallite nuclei.

The probability p_x that a point P is crossed by x fronts of growing spherulites is given by an equation originally derived by Poisson (84):

$$p_x = \frac{e^{-E}E^x}{x!} \tag{6.5}$$

where E represents the average number of fronts of all such points in the system. The probability that P will not have been crossed by any of the fronts, and is still amorphous, is given by

$$p_0 = e^{-E} \tag{6.6}$$

since E^0 and $0!$ are both unity. Of course, p_0 is equal to $1 - X_t$, where X_t is the volume fraction of crystalline material, known widely as the degree of crystallinity. Equation (6.6) may be written

$$1 - X_t = e^{-E} \tag{6.7}$$

which for low degrees of crystallinity yield the useful approximation

$$X_t \cong E \tag{6.8}$$

For the bulk crystallization of polymers, X_t (in the exponent) may be considered related to the volume of crystallization material, V_t:

$$1 - X_t = e^{-V_t} \tag{6.9}$$

The problem now resides on the evaluation of V_t. There are two cases to be considered: (a) the nuclei are predetermined—that is, they all develop at once on cooling the polymer to the temperature of crystallization—and (b) there is sporadic nucleation of the spheres.

For case (a), L spherical nuclei, randomly placed, are considered to be growing at a constant rate, g. The volume increase in crystallinity in the time period t to $t + dt$ is

$$dV_t = 4\pi r^2 L\, dr \tag{6.10}$$

where r represents the radius of the spheres at time t; that is,

$$r = gt \tag{6.11}$$

and

$$V_t = \int_0^1 4\pi g^2 t^2 Lg \, dt \tag{6.12}$$

Upon integration

$$V_t = \tfrac{4}{3}\pi g^3 L t^3 \tag{6.13}$$

For sporadic nucleation the argument above is followed, but the number of spherical nuclei is allowed to increase linearly with time at a rate l. Then spheres nucleated at time t_i will produce a volume increase of

$$dV_t = 4\pi g^2 (t - t_i)^2 \, ltg \, dt \tag{6.14}$$

Upon integration

$$V_t = \tfrac{3}{2}\pi g^3 l t^4 \tag{6.15}$$

The quantities on the right of equations (6.12) and (6.14) can be substituted into equation (6.8) to produce the familiar form of the Avrami equation:

$$1 - X_t = e^{-Zt^n} \tag{6.16}$$

which is often written in the logarithmic form:

$$\ln(1 - X_t) = -Zt^n \tag{6.17}$$

The quantity Z is replaced by K in some books (82).

The above derivation suggests that the quantity n in equations (6.16) or (6.17) should be either 3 or 4. [If rates of crystallization are diffusion controlled, which occurs in the presence of high concentrations of noncrystallizable impurities (70,71) $r = gt^{1/2}$, leading to half-order values of n.]

Both Z and n are diagnostic of the crystallization mechanism. The equation has been derived for spheres, discs, and rods, representing three-, two-, and one-dimensional forms of growth. The constants are summarized in Table 6.4 (83). It must be emphasized that the approximation given in equation (6.8) limits the equations to low degrees of crystallinity. In practice, the quantity n frequently decreases as the crystallization proceeds. Values for typical polymers are summarized in Table 6.5 (83).

The Avrami equation represents only the initial portions of polymer crystallization correctly. The spherulites grow outward with a constant radial growth rate until impingement takes place when they stop growth at the intersection, as illustrated in Figures 6.13 and 6.22. Then a secondary crystallization process is often observed after the initial spherulite growth in the amorphous interstices (85).

Table 6.4 The Avrami parameters for crystallization of polymers (83)

	Crystallization Mechanism	Avrami Constants		Restrictions
		Z	n	
Spheres	Sporadic	$2/3\pi g^3 l$	4.0	3 dimensions
	Predetermined	$4/3\pi g^3 L$	3.0	3 dimensions
Discs[a]	Sporadic	$\pi/3g^2 ld$	3.0	2 dimensions
	Predetermined	$\pi g^3 Ld$	2.0	2 dimensions
Rods[b]	Sporadic	$\pi/4gld^2$	2.0	1 dimension
	Predetermined	$\frac{1}{2}\pi gLd^2$	1.0	1 dimension

[a] Constant thickness *d*.
[b] Constant radius *d*.

Table 6.5 Range of the Avrami constant for typical polymers (83)

Polymer	Range of n	Reference
Polyethylene	2.6–4.0	(a)
Poly(ethylene oxide)	2.0–4.0	(b, c)
Polypropylene	2.8–4.1	(d)
Poly(decamethylene terephthalate)	2.7–4.0	(e)
it-Polystyrene	2.0–4.0	(f, g)

References: (a) W. Banks, M. Gordon, and A. Sharples, *Polymer*, **4**, 61, 289 (1963). (b) J. N. Hay, M. Sabin, and R. L. T. Stevens, *Polymer*, **10**, 187 (1969). (c) W. Banks and A. Sharples, *Makromol. Chem.*, **59**, 283 (1963). (d) P. Parrini and G. Corrieri, *Makromol. Chem.*, **62**, 83 (1963). (e) A. Sharples and F. L. Swinton, *Ploymer*, **4**, 119 (1963). (f) I. H. Hillier, *J. Polym. Sci.*, **A-2** (4), 1 (1966). (g) J. N. Hay, *J. Polym. Sci.*, **A-3**, 433 (1965).

As an example, miscible blends of high-density (an higher melting temperature) polyethylene (HDPE) and low-density polyethylene (LDPE), which contains both short and long branches, and a lower melting temperature (see Chapter 14), are often utilized commercially. On cooling from the melt, the HDPE portion crystallizes first, forming the spherulites, while the LDPE tends to be preferentially located in the amorphous interlamellar regions (86,87), and partly crystallize later in the remaining space.

If the system is considered as two-phased, then the volume of the amorphous phase is V_a and the volume of the crystalline phase is V_c. The total volume, V, is given by

$$V = X_t V_c + (1 - X_t) V_a \tag{6.18}$$

Then

$$1 - X_t = \frac{V - V_c}{V_a - V_c} \tag{6.19}$$

or for dilatometric experiments,

$$1 - X_t = \frac{h_0 - h_t}{h_0 - h_\infty} \tag{6.20}$$

where h_0, h_t, and h_∞ represent capillary dilatometric heights at time zero, time t, and the final dilatometric reading. Substitution of equation (6.20) into (6.17) yields a method of determining the constants Z and n experimentally (e.g., see Figure 6.25).

6.6.2.2 Keith–Padden Kinetics of Spherulitic Crystallization Although the Avrami equation provides useful data on the overall kinetics of crystallization, it provides little insight as to the molecular organization of the crystalline regions, structure of the spherulites, and so on.

Section 6.5 described how the spherulites are composed of lamellar structures that grow outward radially. The individual chains are folded back and forth tangentially to the growing spherical surface of the spherulite (see Figure 6.28) (71). Normally, the rate of growth in the radial direction is constant until the spherulites meet (see Figure 6.26). As the spherulites grow, the individual lamellae branch. Impurities, atactic components, and so on, become trapped in the interlamellar regions.

The first theory to address the kinetics of spherulitic growth in crystallizing polymers directly was developed by Keith and Padden (70,71,88). According to Keith and Padden (70), a parameter of major significance is the quantity

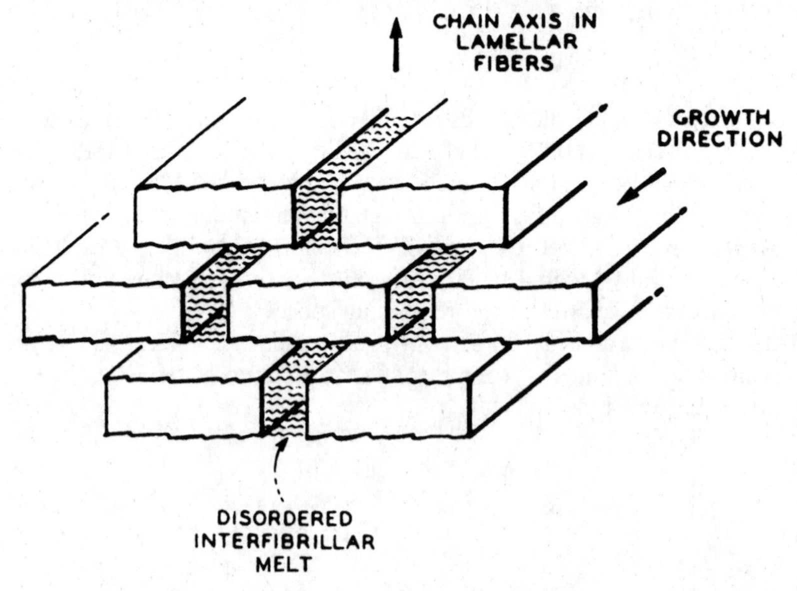

Figure 6.28 Schematic representation of the distribution of residual melt and disordered material among lamellae in a spherulite (71).

$$\delta = \frac{D}{G} \tag{6.21}$$

where D is the diffusion coefficient for impurity in the melt and G represents the radial growth rate of a spherulite. The quantity δ, whose dimension is that of length, determines the lateral dimensions of the lamellae, and that non-crystallographic branching should be observed when δ becomes small enough to be commensurate with the dimensions of the disordered regions on their surfaces. Thus δ is a measure of the internal structure of the spherulite, or its coarseness.

By logarithmic differentiation of equation (6.21),

$$\frac{1}{\delta}\left(\frac{d\delta}{dT}\right) = \frac{1}{D}\left(\frac{dD}{dT}\right) - \frac{1}{G}\frac{dG}{dT} \tag{6.22}$$

The derivative dD/dT always has a positive value. However, dG/dT may be positive or negative (see Figure 6.27). The coarseness of the spherulites depends on which of the two terms on the right of equation (6.22) is the larger. If the quantity on the right-hand side of the equation is positive, an increase in coarseness is expected as the temperature is increased.

The radial growth rate, G, may be described by the equation

$$G = G_0 e^{\Delta E/RT} e^{-\Delta F^*/RT} \tag{6.23}$$

where ΔF^* is the free energy of formation of a surface nucleus of critical size, and ΔE is the free energy of activation for a chain crossing the barrier to the crystal. Equation (6.23) allows the temperature dependence of spherulite growth rates to be understood in terms of two competing processes. Opposing one another are the rate of molecular transport in the melt, which increases with increasing temperature, and the rate of nucleation, which decreases with increasing temperature (see Figure 6.27). According to Keith and Padden, diffusion is the controlling factor at low temperatures, whereas at higher temperatures the rate of nucleation dominates. Between these two extremes the growth rate passes through a maximum where the two factors are approximately equal in magnitude.

6.6.2.3 Hoffman's Nucleation Theory The major shortcoming of the Keith–Padden theory resides in its qualitative nature. Although great insight into the morphology of spherulites was attained, little detail was given concerning growth mechanisms, particularly the thermodynamics and kinetics of the phenomenon.

More recently Hoffman and co-workers attacked the kinetics of polymer crystallization anew (41,76–80). Hoffman began with the assumption that chain folding and lamellar formation are kinetically controlled, the resulting crystals being metastable. The thermodynamically stable form is the extended chain crystal, obtainable by crystallizing under pressure (89).

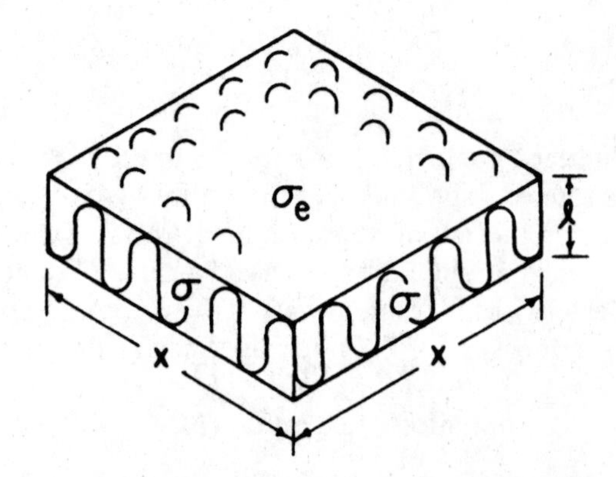

Figure 6.29 Thin chain-folded crystal showing σ and σ_e (schematic) (41).

The basic model is illustrated in Figure 6.29 (41), where ℓ is the thin dimension of the crystal, x the large dimension, σ_e the fold surface interfacial free energy, and σ the lateral surface interfacial free energy, a single chain making up the entire crystal. The free energy of formation of a single chain-folded crystal may be set down in the manner of Gibbs as

$$\Delta G_{\text{crystal}} = 4x\ell\sigma + 2x^2\sigma_e - x^2\ell(\Delta f) \tag{6.24}$$

where the quantity Δf represents the bulk free energy of fusion, which can be approximated from the entropy of fusion, ΔS_f, by assuming that the heat of fusion, Δh_f, is independent of the temperature:

$$\Delta f = \Delta h_f - T\Delta S_f = \Delta h_f - \frac{T\Delta h_f}{T_f^0} = \frac{\Delta h_f(\Delta T)}{T_f^0} \tag{6.25}$$

At the melting temperature of the crystal, the free energy of formation is zero. For $x \gg \ell$,

$$T_f = T_f^0\left[1 - \frac{2\sigma_e}{\Delta h_f \ell}\right] \tag{6.26}$$

Equation (6.26) yields the melting point depression in terms of fundamental parameters. The quantity σ_e may be interpreted in terms of the fold structure. For the actual crystallization, chains are added from the melt or solution to the surface of the crystal defined by the area $x\ell$, and T_f^0 represents the large crystal limit.

6.6.2.4 Example Calculation of the Fold Surface Free Energy Find the fold surface interfacial free energy of a new polymer with the following characteristics: the actual melting temperature is 350°C, while after extended annealing, it melts at 360°C. X-ray analysis shows its lamellae are 150 Å thick. The heat of fusion is 12.0 kJ/mol. (see Table 6.1.)

Equation (6.26) can be rewritten

$$\sigma_e = \frac{\Delta h_f \ell}{2}\left(1 - \frac{T_f}{T_f^0}\right)$$

Substituting the given values obtains

$$\sigma_e = \frac{12.0\,\text{kJ/mol} \times 1.50 \times 10^{-6}\,\text{cm}}{2}\left(1 - \frac{623\,\text{K}}{633\,\text{K}}\right)$$

$$= 1.42 \times 10^{-7}\,\text{kJ}\cdot\text{cm/mol}$$

This polymer has a mer molecular weight of 100 g/mol and a density of 1.0 g/cm^3. Then

$$100\,\text{g/mol} \times 1\,\text{cm}^3/g = 100\,\text{cm}^3/\text{mol}$$

and

$$\sigma_e = \frac{1.42 \times 10^{-7}\,\text{kJ}\cdot\text{cm/mol}}{100\,\text{cm}^3/\text{mol}} = 1.42 \times 10^{-9}\,\text{kJ/cm}^2$$

6.6.2.5 Three Regimes of Crystallization Kinetics Hoffman defined three regimes of crystallization kinetics from the melt, which differ according to the rate that the chains are deposited on the crystal surface.

Regime I. One surface nucleus causes the completion of the entire substrate of length L (see Figure 6.30) (77); that is, one chain is crystallizing at a time. Many molecules may be required to complete L. The term "surface nucleus" refers to a segment of a chain sitting down on a preexisting crystalline lamellar structure, as opposed to the nucleus, which initiates the lamellae from the melt in the first place.

These nuclei are deposited sporadically in time on the substrate at a rate i per unit length in a manner that is highly dependent on the temperature. Substrate completion at a rate g begins at the energetically favorable niche that occurs on either side of the surface nucleus, chain folding assumed.

As illustrated in Figure 6.30, the overall growth rate is given by G, and g is the substrate completion rate. The quantities a_0 and b_0 refer to the molecular width and layer thickness, respectively. The quantity l_g^* refers to the initial fold thickness of the lamellae. The portion of chain occupying the length l_g^* is called a stem.

Again noting Figure 6.30, the "reeling in" or reptation rate r (see Section 5.4) is given by

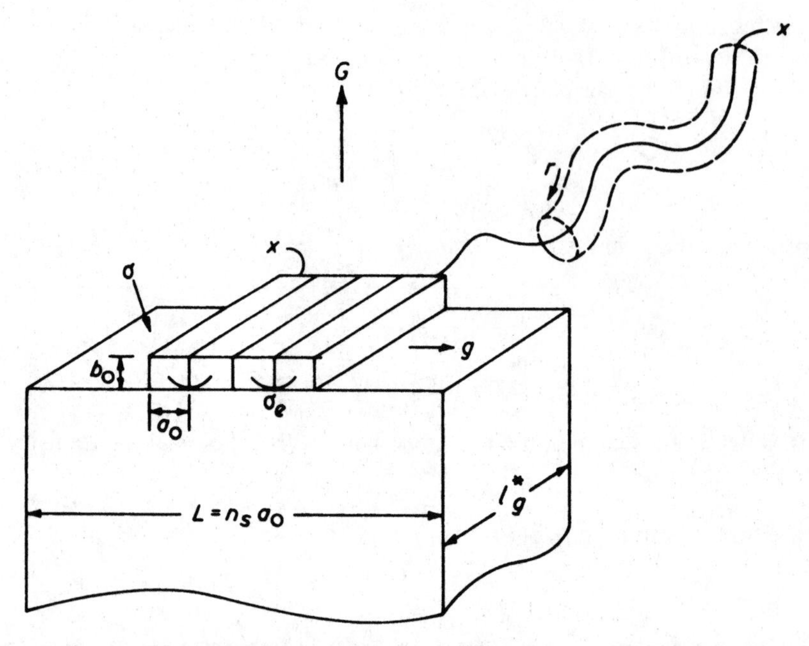

Figure 6.30 Surface nucleation and substrate completion with reptation in regime I, where one surface nucleus deposited at rate I causes completion of substrate of length L, giving overall growth rate $G_I = b_0 iL$. Multiple surface nuclei occur in regime II (not shown) and lead to $G_{II} = b_0(2ig)^{1/2}$, where g is the substarte completion rate. The substrate completion rate, g, is associated with a "reeling in" rate $r = (l_g^*/a_0)g$ for the case of adjacent reentry (77).

$$r = \left(\frac{\ell_g^*}{a_0}\right)g \tag{6.27}$$

A significant question, still being debated in the literature, is whether or not reptation type diffusion is sufficiently rapid to supply chain portions as required (90). An alternate theory, proposed by Yoon and Flory (90) suggests that disengagement of the macromolecule from its entanglements with other chains in the melt is necessary for regular folding. In this theory, the 100 to 200 skeletal bonds corresponding to one traversal of the crystal lamella readily undergo the conformational rearrangements required for their deposition in the growth layer. On the other hand, Hoffman (77) concludes that the reptation rate characteristic of the melt is fast enough to allow a significant degree of adjacent reentry "regular" folding during crystallization.

According to Hoffman, the overall growth rate is given by

$$G_I = b_0 iL = b_0 a_0 n_s i \tag{6.28}$$

where G_I is the growth rate in regime I, n_s represents the number of stems of width a_0 that make up this length, and i is the rate of deposition of surface nuclei.

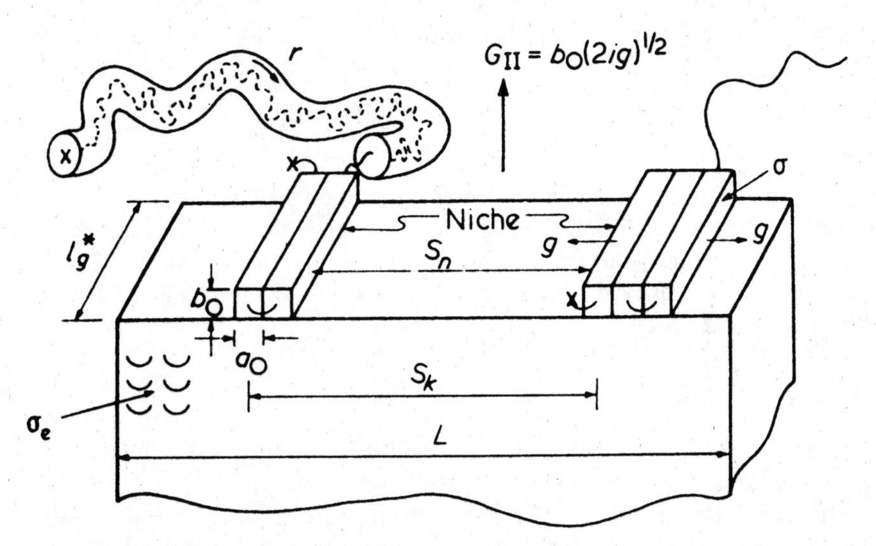

Figure 6.31 Model for regime II growth showing multiple nucleation. The quantity S_k represents the mean separation between the primary nuclei, and S_n denotes the mean distance between the associated niches. The primary nucleation rate is i, and the substrate completion rate is g. The overall observable growth rate is G_{II}. Reptation tube contains molecule being reeled at rate r onto substrate (75).

The free energy of crystallizing v stems and $v_f = v - 1$ folds can be generalized from equation (6.24), using Figure 6.30:

$$\Delta G_v = 2b_o \ell\sigma + 2v_f a_0 b_0 \sigma_e - v a_0 b_0 \ell \, \Delta f \tag{6.29}$$

which for v large becomes

$$\Delta G_v = 2b\ell_g^* \sigma + v a_0 b_0 (2\sigma_e - \ell_g^* \Delta f) \tag{6.30}$$

Regime II. In this regime multiple surface nuclei occur on the same crystallizing surface, because the rate of nucleation is larger than the crystallization rate of each molecule. This, in turn, results from the larger undercooling necessary to reach regime II. As in regime I each molecule is assumed to fold back and forth to give adjacent reentry (see Figure 6.31) (76). An important parameter in this regime is the niche separation distance, S_n.

The rate of growth of the lamellae is

$$G_{\mathrm{II}} = b_0 (2ig)^{1/2} \tag{6.31}$$

The number of nucleation sites per centimeter in regime II is given by

$$N_k = \left(\frac{i}{2g}\right)^{1/2} \tag{6.32}$$

Then the mean separation distance between the sites is

$$S_k = \frac{1}{N_k} = \left(\frac{2g}{i}\right)^{1/2} \tag{6.33}$$

Regime III. This regime becomes important when the niche separation characteristic of the substrate in regime II approaches the width of a stem, producing a snow storm effect. Regime III is very important industially, where rapid cooling is employed. In this regime, the crystallization rate is very rapid. The growth rate is given by

$$G_{\text{III}} = b_0 iL = b_0 i' n'_s a_0 \tag{6.34}$$

where n'_s is the mean number of stems laid down in the niche adjacent to the newly nucleated stem, and i' is the nucleation rate, nuclei/sec-cm.

In regime III the chains do not undergo repeated adjacent reentry into the lamellae but have only a few folds before re-entering the amorphous phase. Then they are free to reenter the same lamella via a type of switchboard model, or go on to the next lamella.

As the temperature is lowered through regimes I, II, and III, substrate completion rates per chain decrease. However, more chains are crystallizing simultaneously. At temperatures approaching T_g, de Gennes reptation is severely limited, as illustrated in Figure 6.27.

6.6.3 Analysis of the Three Crystallization Regimes

Recently, there have been significant advances in both theory and experiment toward understanding both the advantages and limitations of using the three crystallization regimes (91,92).

6.6.3.1 Kinetics of Crystallization The free energy barrier to a stem laying down properly on a growing polymer lamella is given by ΔF. The quantity ΔF is proportional to the length of the stem, l. The growth rate of the crystal, G, depends exponentially on $\Delta F/kT$. Then,

$$G = G_0 \exp(-K_1 l / T_c) \tag{6.34a}$$

where K_1 is a constant and T_c represents the temperature during crystallization. Since the quantity l varies inversely with the relative undercooling, ΔT, from the equilibrium melting temperature, T_f,

$$G \approx G_0 \exp(-K_g / T_c\, \Delta T) \tag{6.34b}$$

The spherulite growth rate in regime I, previously described in equation (6.18), can be expressed as a function of temperature with the relationship

$$G_{\mathrm{I}} = G_{0(\mathrm{I})} \exp(-Q_D^*/kT)\exp(-K_{g(\mathrm{I})}/T\,\Delta T) \quad (6.34c)$$

where Q_D^* is defined as the activation energy of reptational diffusion. Its value is 5736 cal/mol of $-CH_2-$ groups for polyethylene. The quantity K_g represents a nucleation constant, and $G_{0(\mathrm{I})}$ is a preexponential factor, all for regime I as written.

Similarly, for region II equation (6.31) can be expressed as a function of temperature,

$$G_{\mathrm{II}} = G_{0(\mathrm{II})} \exp(-Q_D^*/kT)\exp(-K_{g(\mathrm{II})}/T\,\Delta T) \quad (6.34d)$$

and for region III, equation (6.34) can be written,

$$G_{\mathrm{III}} = G_{0(\mathrm{III})} \exp(-Q_D^*/kT)\exp(-K_{g(\mathrm{III})}/T\,\Delta T) \quad (6.34e)$$

Comparison of the nucleation exponents in equations (6.34c), (6.34d), and (6.34e) show that the relationship among them becomes

$$K_{g(\mathrm{III})} \cong K_{g(\mathrm{I})} = 2K_{g(\mathrm{II})} \quad (6.34f)$$

6.6.3.2 Experimental Data on Regimes I, II, and III Experimental data were obtained by Armistead and Hoffman (92) on fractionated polyethylene with a molecular weight, $M_{nw} = (M_n x M_w)^{1/2}$ of 7.03×10^4 g/mol to examine the above relationships. (Note that many commercial polyethylene materials are made from molecular weights in this range, see Table 1.1.) In polyethylene, the dominant growth front is on the {110} plane, more or less perpendicular to the spherulite radius. Each stem added has a similar orientation with respect to the radius. Growth in the radial direction is caused by successive addition of new layers of thickness b_0 generated by nucleation acts on this {110} substrate.

A polarizing optical microscope equipped with a video camera and a time-lapse VCR was employed, all computer controlled. Data was taken as a function of time for each temperature of undercooling. Regime II and III crystallizations took from under a minute to several hours. Regime I runs took from hours to days. At all temperatures the growth of the spherulites was linear in time up to the point of impingement.

The results are shown in Figure 6.32A. It is evident that all three crystallization regimes are present. Note that the slopes of the data yield values of K_g's fitting the relationship given in equation (6.34f).

An important new concept is that of *perturbed* diffusion (92). Low molecular weight polyethylenes from $M = 15{,}000$ to 39,000 g/mol exhibit near-ideal reptation. Higher molecular weights, from roughly 50,000 to 91,000 g/mol exhibit perturbed diffusion. The perturbation effect is attributed to an increase in the effective mer friction coefficient. The underlying cause of the perturbation results from the increasingly long dangling chains being drawn on to the

A

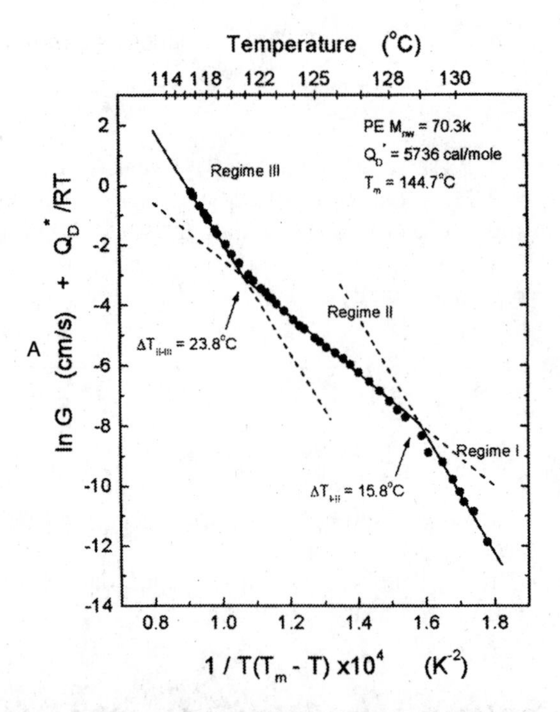

B

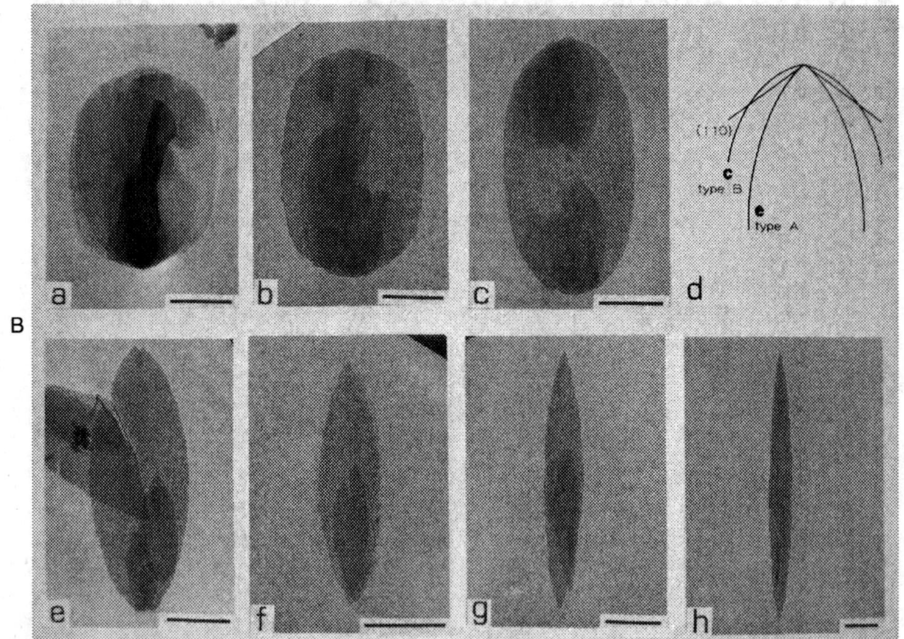

Figure 6.32 Behavior of crystallizing polyethylene. (A) The kinetics of crystallization of $M_{nw} = 7.03 \times 10^4$ g/mol, illustrating the appearance of regimes I, II, and III with decreasing temperature. (B) Transmission electron micrographs of an $M = 3.2 \times 10^4$ g/mol sample, illustrating the change in crystal habit in bulk crystallization from a lenticular shape in regime I to a truncated lozenge with curved edges in regime II: (a) at 125°C, (b) 126°C, (c) 127°C, (e) 128°C, (f) 129°C, (g) 130°C, and (h) 131°C. The bar lines represent 2 μm. (Note that the data are shown in increasing temperature, (a)–(h), with the transition from regime I to regime II taking place between (c) and (e) at 127.4°C.)

substrate by the force of crystallization. These dangling chains form some transient attachments elsewhere in the same (or another) lamella, such as entanglements. This impedes the steady-state *reeling in* reptation process. Some of these attachments become permanent, forming tie chains and loops, and thus lowering the final degree of crystallinity.

The final degree of crystallinity obtained was also molecular weight dependent. For the molecular weight range of 15,000 to 39,000 g/mol, the quasi-equilibrium degree of crystallinity of unoriented polyethylenes is commonly near 0.8. At the molecular weight of 7.03×10^4 g/mol, the degree of crystallinity was 0.65. For very high molecular weights, 6×10^5 to 7×10^5 g/mol, the degree of crystallinity is commonly 0.2 to 0.3. This also has its basis in the larger number of attachments to different lamellae, entanglements, etc.

It should be noted that these quasi-equilibrium degrees of crystallization are for unoriented samples undergoing crystallization. Ultradrawn very high molecular weight polyethylene fibers, see Section 11.2, exhibit much higher degrees of crystallinity. Separate studies suggest that broader molecular weight distributions crystallize in substantially the same way.

At very high molecular weights, above 6.4×10^5 g/mol, distinct spherulitic objects no longer appear. At the highest molecular weights, $(5 - 8) \times 10^6$ g/mol, the low degree of crystallinity bespeaks the presence of a massive fraction of amorphous material between the imperfect microlamellae, some of it consisting of tie chains between different lamellae with much of the remainder being associated with long loops between mostly nonadjacent stems in the same lamella. Such structures in the ultrahigh molecular weight region rule out the *near ideal* or *weakly perturbed* form of steady-state forced reptation as the transport process active for the lower molecular weight materials. Armistead and Hoffman (92) refer to such crystallization as taking place in regime III-A.

6.6.3.3 Changes in Crystal Growth Habit It has now been discovered that polyethylene crystal growth habit in the bulk state differs in regime I and regime II. This was achieved *via* permanganic etching of the polymer, coupled with differential dissolution of the material (93). The key step in this permanganic etching was the use of potassium permanganate in a 2:1 mixture of concentrated sulfuric:orthophosphoric acids. This rather drastic etching process preferentially dissolves the amorphous material, leaving the crystalline portion more or less intact. The etching process was followed by shadowing the samples for electron microscopy with Pt/Pd.

A change in crystal habit for polyethylene was observed at the regime I–regime II transition temperature, 127.4°C, see Figure 6.32B (94). For the 32,000 g/mol sample examined, a lenticular shape elongated structure was observed for regime I, while a truncated lozenge with curved edges in that of regime II. This suggests that there may be a shift in crystal growth direction from the {110} face to the curved {200} face in regime II.

Above, a theory known today as the *surface nucleation theory* was described in Sections 6.6.2.3 to 6.6.3. The surface nucleation theory assumes an ensemble of crystals, each of which grows with constant thickness. This thickness is close to the thickness for which the crystals have the maximum growth rate. While the theory correctly describes a wide range of crystallization kinetics, it has a major shortcoming in being based primarily on enthalpic concepts.

6.6.4 The Entropic Barrier Theory

The *entropic barrier theory* was developed by Sadler and Gilmer (95,96). This latter theory is based upon the interpretation of kinetic Monte Carlo simulations and concomitant rate-theory calculations. The phrase *Monte Carlo* suggests chance events, or in this case, random motion. While individual motions of the molecules are governed by chance, they move according to rules laid out on the computer such as excluded volume considerations and secondary bonding energies and/or repulsive forces.

This theory displays a low-entropy barrier to crystallization that is a consequence of two factors: (a) The tendency of the growth faces of the lamellae to assume a rounded shape, in vertical cross section, with relatively short stems at the outermost positions, and (b) the pinning of a stem at a fixed length because other segments of the stem molecule have been incorporated into the crystal elsewhere, or in another lamella.

Point (a) above needs some explanation. As initiated during crystallization, the leading edge of the lamella is apparently thinner than the equilibrium value. This, at least, is the predicted phenomenon by this and several other computer simulations. Not covered by these theories is long-time annealing of the lamellae, whereby further molecular rearrangements lead to a thickening of the crystal to a more or less uniform thickness.

Figure 6.33 (96) illustrates the geometry of the lamella edge according to the entropic barrier theory. For simplicity, the stems are shown as being vertical and parallel, and most fold segments connecting various stems are omitted. These crystals have a finite degree of surface roughness, not permitted in the surface nucleation theory. An important point of the theory, illustrated in Figure 6.33*b*, is both the addition and removal of the mers are allowed according to the settings on the computer.

Consider an average polymer crystallization taking place (97). Assuming a few Ångstroms for the typical dimension of a crystalline cell (the size of a mer), one obtains a time scale of 10^{-4} s to add one elementary cell to the growth front. The typical time scale of 10^{-10} s characterizes individual motions of the mers in the liquid state on the size scale of the crystal cell. Hence, some 10^{6} motions are necessary just for one elementary step in the growth process! This points out that the rearrangements of the polymer conformations at the growth front are a rather slow process, involving many trial events together with interactions among the chains at the growth front.

An important point in all of the theories developed is that the equilibrium crystal thickness, obtainable by crystallizing the polymer at its equilibrium melting temperature, yields fully extended crystals, as described in Section

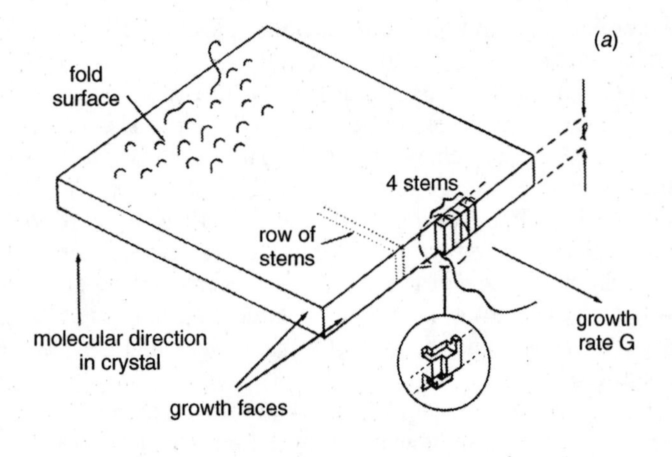

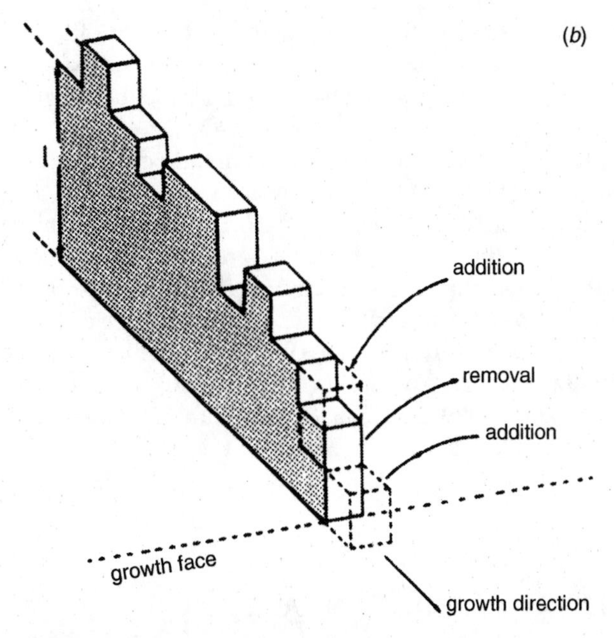

Figure 6.33 Crystallization according to the entropic barrier theory. (*a*) Representation of a lamellar crystal, showing stems (chain direction vertical) and a step in the growth face. The inset provides a description of the step in terms of units that are shorter than the length of the surface nucleation theory (one molecule making up a whole stem). The dotted lines indicate where the row of stems in (*b*) is imagined to occur. (*b*) The basic row of stems model, showing mers along the chains as cubes, chain direction vertical, as in (*a*).

6.7.4. As mentioned above, the equilibrium thickness for crystallizing the polymer at some lower temperature is directly proportional to ΔT.

However, the proportionality is not strictly linear, allowing for significant curvature as the crystallization temperature is lowered. This allows some correspondence to the regime changes in the surface nucleation theory.

The entropic barrier theory has spawned a plethora of computer simulation studies (97–102). For example, simulations by Doye and Frenkel (100) led to the finding that it is unfavorable for a stem to be shorter than l_{min}, the minimum thickness for which the crystal is thermodynamically more stable than the melt. They find instead that the lamellar thickness converges to a value just larger than l_{min} as the crystal grows. This value is at the maximum rate of crystal growth.

It is also unfavorable for a stem to overhang the edge of the previous layer (100). Whenever a stem is significantly longer than l_{min}, the growth of the stem can be terminated by the successful initiation of a new stem.

Chen and Higgs (98) found that although most chain folds are aligned perpendicular to the growth direction (as dictated by the surface nucleation theory), a significant number of chains folding parallel to the growth direction were observed.

Sommer (97) examined the crystallization of bimodal mixtures of oligomers differing in size by a factor of two. Conditions were such that the short chains crystallized only in the fully stretched conformation, while the longer chains folded once in a hairpin conformation. Mixtures of the two chains crystallized slower than the pure components, reaching a minimum near 50:50.

For example, consider the consequences if the longer chains are added in small quantities to the shorter chains. The longer chains act as defects, either in the folded state, where they are easily removed, or in the stretched state, where they are almost always isolated, thus forming cilia which also have a higher rate of removal. As a consequence, the growth rate of the lamella decreases.

6.7 THE REENTRY PROBLEM IN LAMELLAE

In the surface nucleation theory the lamellae were assumed to be formed through regular adjacent reentry, although it was recognized that this was an oversimplification. The concept of the switchboard and folded-chain models were briefly developed in Section 6.4. The question of the molecular organization within polymer single crystals as well as the bulk state has dogged polymer science since the discovery of lamellar-shaped single crystals in 1957 (19). Again, X-ray and other studies show that the chains are perpendicular to the lamellar surface (see Section 6.4.2).

Since the chain length far exceeds the thickness of the crystal, the chains must either reenter the crystal or go elsewhere. However, the relative merits of the switchboard versus folded-chain models remained substantially unresolved for several years for lack of appropriate instrumentation.

6.7.1 Infrared Spectroscopy

Beginning in 1968, Tasumi and Krimm (103) undertook a series of experiments using a mixed crystal infrared spectroscopy technique. Mixed single crystals of protonated and deuterated polymer were made by precipitation from dilute solution. The characteristic crystal field splitting in the infrared spectrum was measured and analyzed to determine the relative locations of the chain stems of one molecule, usually the deuterated portion, in the crystal lattice. The main experiments involved blending protonated and deuterated polyethylenes (104–106).

The main findings were that folding takes place with adjacent reentry along (121) planes for dilute solution-grown crystals. In addition, it was also concluded that there is a high probability for a molecule to fold back along itself on the next adjacent (121) plane.

Melt-crystallized polyethylene was shown to be organized differently, with a much lower (if any) extent of adjacent reentry (106). However, significant undercooling was required to prevent segregation of the deuterated species from the ordinary, hydrogen-bearing species.[†] Since the experimental rate of cooling was estimated to be 1°C per minute down to room temperature, it may be that some crystallization occurred in all three regimes (see Section 6.6.2.3).

6.7.2 Carbon-13 NMR

Additional evidence for chain folding in solution-grown crystals comes from carbon-13 NMR studies of partially epoxidized 1,4-trans-polybutadiene crystals (49,50). This polymer was crystallized from dilute heptane solution and oxidized with *m*-chloroperbenzoic acid. This reaction is thought to epoxidize the amorphous portions present in the folds, while leaving the crystalline stem portions intact.

The result was a type of block copolymer with alternating epoxy and double-bonded segments. NMR analyses showed that for the two samples studied the chain-folded portion was about 2.4 and 3 mers thick, whereas the stems were 15.2 and 40.8 mers thick, respectively. Since the number of mer units to complete the tightest fold in this polymer has been calculated to be about three (51), the NMR study strongly favors a tight adjacent reentry fold model for single crystals.

[†] In fact, one early problem that continues to plague many studies is that deuterated polyethylene tends to phase-separate from ordinary, protonated polymer, even though they are chemically identical. The cause has been related to slightly different crystallization rates owing to polydeuteroethylene melting 6°C lower than ordinary polyethylene.

Broad line proton NMR (107) and Raman analyses (108) of polyethylenes indicated three major regions for crystalline polymers: the crystalline region, the interfacial or interzonal region, and the amorphous or liquidlike region. For molecular weight of 250,000 g/mol, the three regions were 75, 10, and 15% of the total, respectively. The presence of the interfacial regions reduces the requirements for chain folding (109).

6.7.3 Small-Angle Neutron Scattering

6.7.3.1 Single-Crystal Studies With the advent of small-angle neutron scattering (SANS), the several possible modes of chain reentry could be put to the test anew (79, 109–121). Sadler and Keller (119–121) prepared blends of deuterated and normal (protonated) polyethylene and crystallized them from dilute solution. The radius of gyration, R_g, was determined as a function of molecular weight.

The several models possible are illustrated in Figure 6.34 (122). For adjacent reentry (Figure 6.34*c*), R_g should vary as $M^{1.0}$ for high enough molecular weights, since a type of rod would be generated. For the switchboard model (Figure 6.34*d*), R_g is expected to vary close to $M^{0.5}$, since the chains would be expected to be nearly Gaussian in conformation. The several possible relationships between R_g and M are set out in Table 6.6 (123).

For solution-grown crystals, Sadler and Keller (120) found that R_g depended on M only to the 0.1 power. Such a situation could arise only if the stems folded up on themselves beyond a certain number of entries (see Figure 6.33), called superfolding. However, the 0.1 power dependence appears to hold only for intermediate molecular weight ranges. For low enough M, there should not be superfolding. For high enough molecular weight, a square plate with a 0.5 power dependence would be generated.

Recent quantitative calculations of the absolute scattering intensities expected from various crystallite models for single crystals by Keller (121) and Yoon and Flory (125–127) (Figure 6.34) on polyethylene suggested that the model for adjacent reentry does not correlate with experiment. Rather, Yoon and Flory put forward a model requiring a stem dilution by a factor of 2–3. The calculated scattering functions are shown in Figure 6.34; this leads to the

Table 6.6 Relationships among geometric shape, R_g, and M (123)

Geometric Shape	R_g Equals	Molecular Weight Dependence
Sphere	$D/\sqrt{20/3}$	$M^{1/3}$
Rod	$L/\sqrt{12}$	M^{1}
Random coil	$r/\sqrt{6}$	$M^{0.5}$
Rectangular plate	$(b^2+l^2)^{1/2}/\sqrt{12}$	M^{variable}
Square plate	$A^{1/2}/\sqrt{6}$	$M^{0.5}$

Symbols: D, diameter; L, length of rod; r, end-to-end distance; A, area; b, width of plate; l, length of plate.

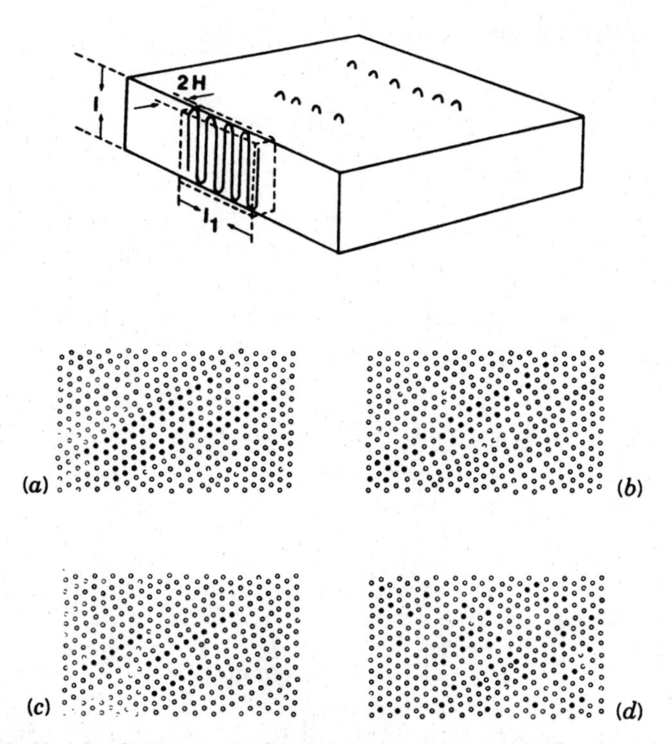

Figure 6.34 Models of stem reentry for chain sequences in a lamellar-shaped crystal. (*a*) Regular reentry with superfolding; (*b*) partial nonadjacency (stem dilution) as required by closer matching of the experimental data in accord with Yoon and Flory; (*c*) adjacent stem positions without superfolding; (*d*) the switchboard model. All reentry is along the (121) plane; superfolding is along adjacent (121) planes. View is from the (001) plane, indicated by dots (122).

model in Figure 6.34*b*. This last suggests a type of skip mechanism, with two or three chains participating.

6.7.3.2 Melt-Crystallized Polymers Upon crystallization from the melt, an entirely different result emerges. Experiments by Sadler and Keller (120–122) showed that nearly random stem reentry was most likely; that is, some type of switchboard model was correct. Quantitative calculations by Yoon and Flory (125–127) and by Dettenmaier et al. (128,129) on melt-crystallized polyethylene (130) and isotactic polypropylene (131) also showed that adjacent reentry should occur only infrequently on cooling from the melt.

Three regions of space were defined by Yoon (126): a crystalline lamellar region about 100 Å thick, an interfacial region about 5 to 15 Å thick, and an amorphous region about 50 Å thick. A nearby reentry model constrains the chain within the interfacial layer during the irregular folding process. The calculated reentry dimensions for solution and melt-crystallized polyethylene are summarized in Table 6.7 (126). A major problem, of course, was that the samples had to be severely undercooled to prevent segregation of the two

Table 6.7 Polyethylene lamellar reentry dimensions (125)

	Reentry Statistics	
Case	Probability of Reentering Same Crystal	Average Displacement on Reentry
Solution-crystallized	1.0	10–15 Å
Melt-crystallized	0.7	25–30 Å

Table 6.8 Comparison of molecular dimensions in molten and crystallized polymers (113)

		$R_g/M_w^{1/2}$ Å/(g/mol)$^{1/2}$	
Polymer	Method of Crystallization	Melt	Crystallized
Polyethylene	Rapidly quenched from melt	0.46	0.46
it-Polypropylene	Rapidly quenched	0.35	0.34
	Isothermally crystallized at 139°C	0.35	0.38
	Rapidly quenched from melt and subsequently annealed at 137°C	0.35	0.36
Poly(ethylene oxide)	Slowly cooled	0.42	0.52
it-Polystyrene	Crystallized at 140°C (5 h)	0.26–0.28[a]	0.24–0.27
	Crystallized at 140°C (5 h) then at 180°C (50 min)		0.26
	Crystallized at 200° (1 h)	0.22[a]	0.24–0.29

[a] Dimensions in the melt were not available. The values quoted are for atactic polystyrene annealed in the same way as the crystalline material.

species during crystallization. Undoubtedly, large portions of the crystallization took place in regime III.

Crist et al. (132) deuterated or protonated a slightly branched polybutadiene to produce a type of polyethylene. Blends of these two materials were used in SANS studies. The scattering curves indicated identical dimensions (R_g values) for both the melt and melt-crystallized materials; these were the same as expected for Flory θ-solvent values for polyethylene. Using wide-angle neutron scattering, Wignall et al. (113) concluded that the number of stems that could be regularly folded had an upper limit of about four.

Of course, other crystallizable polymers have been studied (114–119). These include polypropylene (114–117), poly(ethylene oxide) (118), and isotactic polystyrene (119). Wignall et al. (113) have summarized the values of $R_g/M_w^{1/2}$ in both the melt and crystalline states (see Table 6.8). Most interestingly, the dimensions in the crystalline state and in the melt state are virtually identical for all these polymers. None of these data show a decrease in R_g on crystallization. That the R_g values in the melt and in the crystallized material are the same all but rules out regular folding. In fact, the data for poly(ethylene oxide) (118) shows a slight increase, if anything. By way of summarizing the above

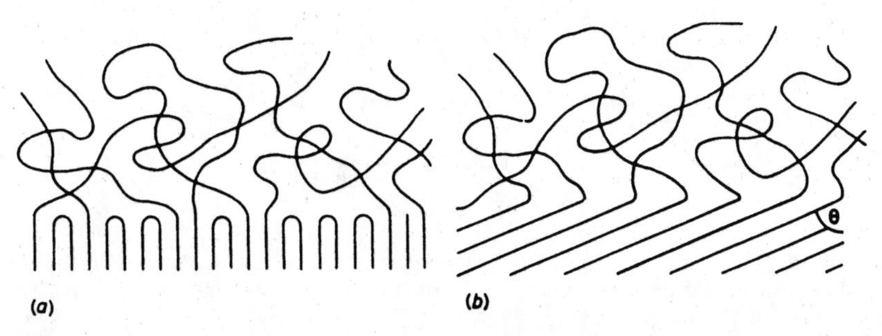

Figure 6.35 Alternative resolutions of the density paradox $(1 - p - 2\ell/L) \cos \theta < 3/10$ (116): (*a*) increased chain folding beyond critical value and (*b*) oblique angle crystalline stems to reduce amorphous chain density at interface.

studies on melt-crystallized polymers, it seems that adjacent reentry occurs much less than in solution-crystallized polymers. Some experiments suggest very little adjacent reentry.

Hoffman (76) and Frank (133) point out, however, that some folding is required. Alternatively, the crystals must have the chains at an oblique angle to the crystal surface. If neither condition is met, a serious density anomaly at the crystalline-amorphous interface is predicted: the density is too high. These workers point out that the difficulty can be mitigated by interspersing some tight folds between (or among) longer loops in the amorphous phase (see Figure 6.35) (133). Frank (133) derived a general equation that combines the probability of back folding, p, and the obliquity angle, θ, with the crystalline stem length, l, and the contour length of the chain, L, to yield the minimum conditions to prevent an anomalous density in the amorphous region:

$$\left(1 - p - \frac{2l}{L}\right)\cos\theta \leqq \frac{3}{10} \tag{6.35}$$

The findings above led to two different models. In 1980 Dettenmaier et al. (128) proposed their solidification model, whereby it was assumed that crystallization occurred by a straightening out of short coil sequences without a long-range diffusion process. Thus these sequences of chains crystallized where they stood, following a modified type of switchboard model (53). This was the first model to illustrate how R_g values could remain virtually unchanged during crystallization.

On the other hand, Hoffman (76) showed that the density of the amorphous phase is better accounted for by having at least about 2/3 adjacent reentries, which he calls the variable cluster model. An illustration of how a chain can crystallize with a few folds in one lamella, then move on through an amorphous region to another lamella, where it folds a few more times and so on, is illustrated in Figure 6.36 (124). Thus a regime III crystallization according to the variable cluster model will substantially retain its melt value of R_g.

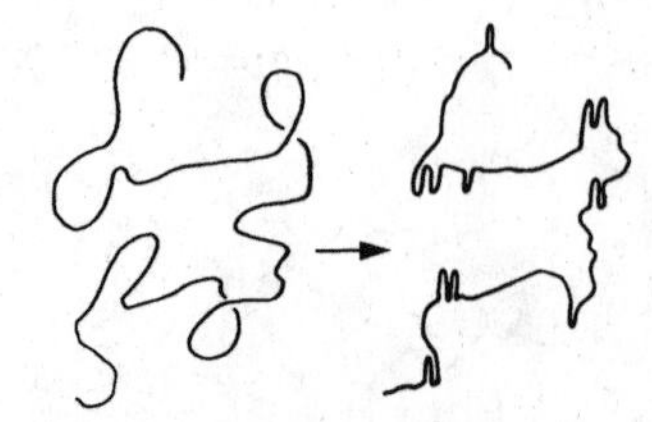

Figure 6.36 The variable cluster model, showing how a chain can crystallize from the melt with some folding and some amorphous portions and retain, substantially, its original dimensions and its melt radius of gyration (124).

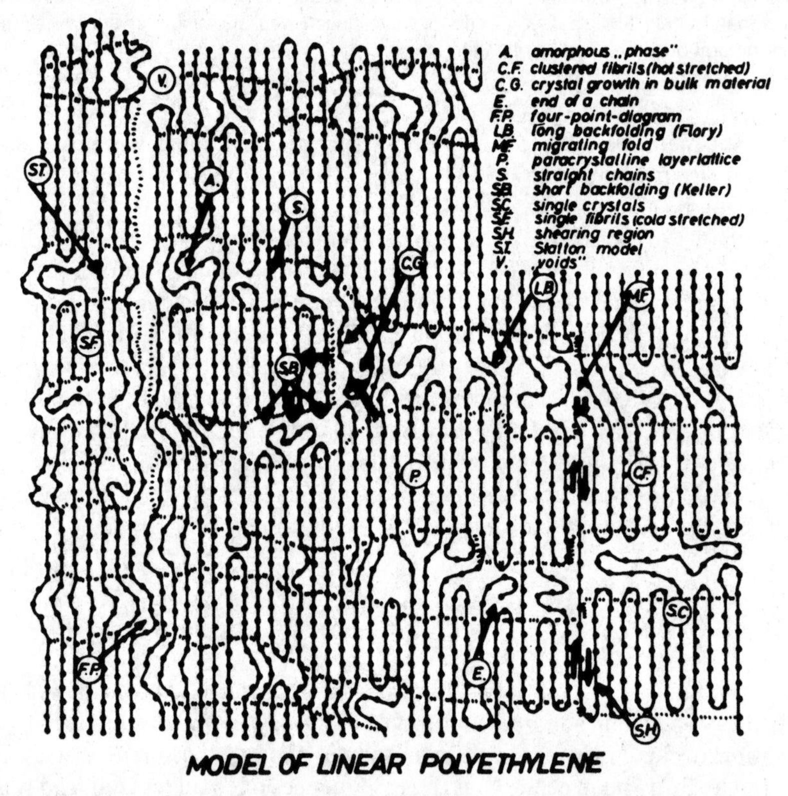

Figure 6.37 The paracrystalline model of Hosemann (134). Amorphous structures are illustrated in terms of defects. A radius of gyration approaching amorphous materials might be expected.

The general conclusion from these studies is that the fringed micelle model fails, because it predicts that the density of the amorphous polymer at both ends of the crystal will be higher than that of the crystal itself. While the chains could be laid over at a sharp angle (Figure 6.35*b*), the most viable alternative is to introduce a significant amount of chain folding.

It is of interest to compare the results of this modern research with Hosemann's paracrystalline model, first published in 1962. As illustrated in Figure 6.37 (134), this model emphasizes lattice imperfections and disorder, as might be expected from regime III crystallization. This model also serves as a bridge between the concepts of crystalline and amorphous polymers (see Figure 5.3). More recent research by Hosemann has continued to examine the partially ordered state (135,136).

By way of summary, for dilute solution-grown crystals a modified regular reentry model fits best, with the same molecule forming a new stem either after immediate reentry or after skipping over one or two nearest-neighbor sites. For melt-formed crystals the concept of folded chains is considerably modified. Since active research in this area is now in progress, perhaps a more definitive set of conclusions will be forthcoming.

It must be remembered that the formation of lamellae, whether with adjacently folded chains or with a switchboardlike structure, is kinetically controlled by the degree of undercooling and finite rates of molecular motion. The thermodynamically most stable crystal form is thought to have extended chains.

6.7.4 Extended Chain Crystals

Wunderlich (137,138) pointed out that the thermodynamic equilibrium crystalline state has an extended chain macroconformation when the crystallization is carried out under great hydrostatic pressure. Thus polyethylene forms extended chain single crystals at pressures approaching 5 kbar (139,140). These crystals form long needlelike structures, which may be several μm in length. In the discussion above it was pointed out that polymer chains fold during crystallization at atmospheric pressure, in significant measure because of kinetic circumstances. It is now thought that the appearance of folded chain instead of extended chain alternative phase variants depends on a complex interaction between thermodynamics and kinetics.

The development of a pressure–temperature phase diagram (141) for polyethylene showed that *orthorhombic* (folded chain), o, and *hexagonal* (extended chain), h, crystal domains were placed in such a way that on cooling from the melt above about 4 kbar, first the hexagonal crystal structure was encountered and then the orthorhombic. The surprising conclusion was that at room temperature and one atmosphere, the hexagonal structure was metastable.

The thermodynamic stability of a crystal depends on its dimensions, and in particular, on its thickness (Section 6.6.2.3). Extended crystals can be considered the limit of thick crystals, with molecules adding end-on-end as well as side-to-side. From a study of a phase stability diagram (Figure 6.38), it was shown that the relative stability of the hexagonal and orthorhombic phases can invert with size; that is, above a certain thickness the folded chain is more stable (142).

Figure 6.38 shows the relation between reciprocal lamellar crystal thickness, $1/l$, and temperature under isobaric conditions (142). There are several regions of isothermal growth possible, indicated by horizontal arrows pointing toward $1/l = 0$ (i.e., infinite thickness) chosen to lie in the two principal temperature

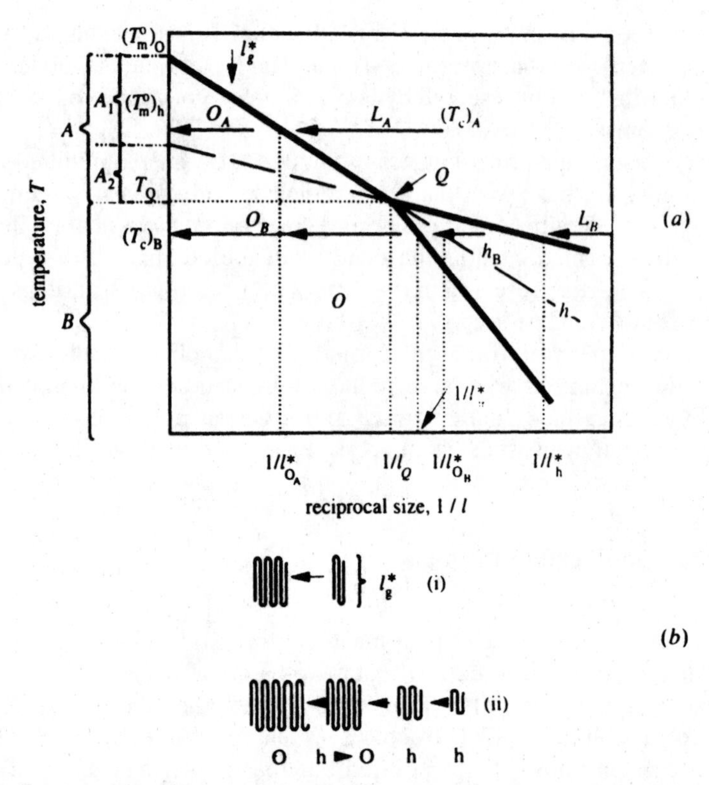

Figure 6.38 Variation of lamellar crystal thickness *vs.* temperature. (*a*) Phase growth in terms of a *phase stability* diagram. Notation *h* and *o* refer to crystal forms in polyethylene, and L stands for the liquid melt. The two sets of horizontal arrows pointing toward 1/l = 0 denote isothermal growth pathways at the two selected crystallization temperatures, T_c, chosen to be in the two temperature regimes *A* and *B*. (*b*) Schematic representation of chain folded polymer crystal grown: (*i*) Region *A* leading to lamellae of a specific thickness l_g, in which they continue to grow laterally through direct growth in phase *o*, (*ii*) Region *B* where crystals arise in the *h* phase and develop by simultaneous lateral and thickening growth with the latter stopping (or slowing down) on the *h* to *o* transformation and/or impingement. T_m: melting temperature.

regions, above and below T_Q, denoted by A and B, respectively. While in the liquid, L, stability region, any crystallization is transient until the size corresponding to a phase line at a specified temperature is reached. At this point the new crystal phase will become stable and capable of continued growth. In region A the first crystal to appear is thought to be in the o phase, but growth will proceed only up to a limited value of l, l_g^*, after which there is an $h \to o$ transformation, and further growth will be in the lateral direction with constant l. Thus it substantially passes straight into the region of ultimate stability, the orthorhombic phase structure, o.

At a lower temperature, region B, first the hexagonal crystal structure forms, h. The lamellar thickness increases in the course of continuing crystal growth. This hexagonal form remains only stable within a limited size range at modest pressures, being metastable for larger dimensions, with $1/l_{tr}^*$ representing the boundary between the two phase regimes in $T - 1/l$ space.

For very high pressures, $P > P_Q$ (143), the h phase remains stable even for infinite l, However, from a kinetic point of view, lateral growth in folded-chain crystals is thought to be faster than thickening growth, favoring the appearance of folded-chain morphologies under most circumstances. The o phase is much less mobile than the h phase, slowing down growth of the h phase in the presence of the o phase.

There is a continuing connection with kinetics of crystallization in this argument. When region A is small, the polymer must be close to $(T_m^o)_o$ throughout the region, hence at small supercoolings. Consequently crystallization in this region will be slow, and may be unrealizable. In this case, on cooling, crystallization will take place in region B, resulting in hexagonal crystallization. If region A is wide, however, the crystallization may start at the ultimately more stable orthorhombic state before it reaches T_Q.

6.8 THERMODYNAMICS OF FUSION

In the previous sections it was shown that the formation of lamellae with folded chains was essentially a kinetically controlled phenomenon. This section treats the free energy of polymer crystallization and melting point depression.

Melting is a first-order transition, ordinarily accompanied by discontinuities of such functions as the volume and the enthalpy. Ideal and real melting in polymers is illustrated in Figure 6.39. Ideally polymers should exhibit the behavior shown in Figure 6.39*a*, where the volume increases a finite amount exactly at the melting (fusion) temperature, T_f. (The subscript M, for melting, is also in wide use. In this text, M represents mixing.) Note that the coefficient of thermal expansion also increases above T_f. Owing to the range of crystallite sizes and degrees of perfection in the real case, a range of melting temperatures is usually encountered, as shown experimentally in Figure 6.2. The

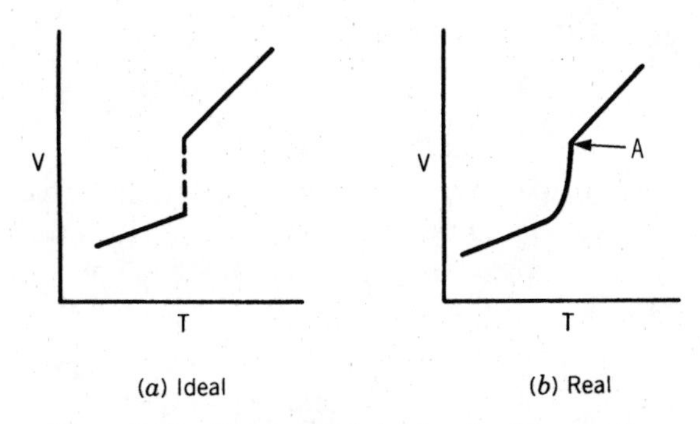

Figure 6.39 Dilatometric behavior of polymer melting.

classic melting temperature is usually taken where the last trace of crystallinity disappears, point A in Figure 6.39b.

The free energy of fusion, ΔG_f, is given by the usual equation,

$$\Delta G_f = \Delta H_f - T\,\Delta S_f \tag{6.36}$$

where ΔH_f and ΔS_f represent the molar enthalpy and entropy of fusion. At the melting temperature, ΔG_f equals zero, and

$$T_f = \frac{\Delta H_f}{\Delta S_f} \tag{6.37}$$

Thus a smaller entropy or a larger enthalpy term raises T_f. Thus, the relative changes in ΔH_f and ΔS_f in going from the amorphous state to the crystalline state determine the melting temperature of the polymer.

6.8.1 Theory of Melting Point Depression

The melting point depression in crystalline substances from the pure state T_f^0 is given by the general equation (143)

$$\frac{1}{T_f} - \frac{1}{T_f^0} = -\frac{R}{\Delta H_f}\ln a \tag{6.38}$$

where a represents the activity of the crystal in the presence of the impurity.

The thermodynamics of melting in polymers was developed by Flory and his co-workers (145–147). To a first approximation, the melting point depression depends on the mole fraction of impurity, X_B, the mole fraction of crystallizable polymer being X_A. Substituting X_A for a in equation (6.38),

$$\frac{1}{T_f} - \frac{1}{T_f^0} = -\frac{R}{\Delta H_f}\ln X_A \tag{6.39}$$

For small values of X_B,

$$-\ln X_A = -\ln(1 - X_B) \cong X_B \tag{6.40}$$

In the following discussion, ΔH_f is the heat of fusion per mole of crystalline mers. There are three important cases in which the melting temperature may be depressed. If X_B represents the mole fraction of noncrystallizable comonomer incorporated in the chain,

$$\frac{1}{T_f} - \frac{1}{T_f^0} = \frac{R}{\Delta H_f}X_B \tag{6.41}$$

The mer unit at the end of the chain must always have a different chemical structure from those of the mers along the chain. Thus end mers constitute a special type of impurity, and the melting point depends on the molecular weight. If M_0 is the molecular weight of the end mer (and assuming that both ends are identical), the mole fraction of the chain ends is given approximately by $2M_0/M_n$. Thus

$$\frac{1}{T_f} - \frac{1}{T_f^0} = \frac{R}{\Delta H_f} \frac{2M_0}{M_n} \tag{6.42}$$

which predicts that the highest possible melting temperature will occur at infinite molecular weight.

If a solvent or plasticizer is added, the case is slightly more complicated. Here, the molar volume of the solvent, V_1, and the molar volume of the polymer repeat unit, V_u, cannot be assumed to be equal. Also, the interaction between the polymer and the solvent needs to be taken into account. The result may be written (147)

$$\frac{1}{T_f} - \frac{1}{T_f^0} = \frac{R}{\Delta H_f} \frac{V_u}{V_1} (v_1) \tag{6.43}$$

where v_1 represents the volume fraction of diluent.

The quantity χ_1 has been interpreted in several ways (145,146). Principally it is a function of the energy of mixing per unit volume. For calculations involving plasticizers, the form using the solubility parameters δ_1 and δ_2 is particularly easy to use (148) (see Section 3.3.2):

$$\chi_1 = \frac{(\delta_1 - \delta_2)^2 V_1}{RT} \tag{6.44}$$

Corresponding relations for the depression of the glass transition temperature, T_g, by plasticizer are given in Section 8.8.1.

Corresponding equations for the dependence of the melting point on pressure were derived by Karasz and Jones (149). For a pressure P_f,

$$P_f - P_f^0 = \frac{RT_f}{\Delta V_f} \frac{V_u}{V_1} (v_1) \tag{6.45}$$

where ΔV_f is the volume change on fusion.

6.8.2 Example Calculation of Melting Point Depression

Suppose that we swell poly(ethylene oxide) with 10% of benzene. What will be the new melting temperature, if it melted at 66°C dry?

Equation (6.43) can be solved, with the aid of Tables 3.1, 3.2, 3.4, and 6.1. First, the molar volumes of the polymer and the solvent must be computed: for benzene, with six carbons and six hydrogens, M = 78 g/mol. Its density is 0.878 g/cm^3. Division yields 88 cm^3/mol. For poly(ethylene oxide), the mer molecular weight is 44 g/mol, and its density is 1.20 g/cm^3, yielding a molar volume of 36.6 cm^3/mol. Equation (6.43), with appropriate numerical values, is

$$\frac{1}{T_f} = -\frac{R}{\Delta H_f}\left(\frac{V_u}{V_1}\right)(v_1) + \frac{1}{T_f^0}$$

$$\frac{1}{T_f} = +\frac{0.0083\,\text{kJ/mol}\cdot\text{K}}{8.29\,\text{kJ/mol}}\left(\frac{36.6\,\text{cm}^3/\text{mol}}{88\,\text{cm}^3/\text{mol}}\right)(0.1) + \frac{1}{339\,\text{K}}$$

$$T_f = 334.3\,\text{K or } 61.4^\circ\text{C}$$

If room temperature is about 25°C, then the crystallinity will not be destroyed by the benzene plasticizer.

6.8.3 Experimental Thermodynamic Parameters

The quantity T_f^0 may be determined either directly on the pure polymer or by a plot of $1/T_f$ versus v_1. The latter is very useful in the case where the polymer decomposes below its melting temperature.

Once T_f^0 is determined, equations (6.43) and (6.44) permit the calculation of both the Flory interaction parameter and the heat of fusion of the polymer from the slope and intercept of a plot of $(1/T_f) - (1/T_f^0)$ versus $(1/T_f)$ (146). The heat of fusion determined in this way measures only the crystalline portion. If heat of fusion data are compared with corresponding data obtained by DSC (see Figure 6.3), which measures the heat of fusion for the whole polymer, the percent crystallinity may be obtained.

6.8.4 Entropy of Melting

In classical thermodynamics, the change in Gibbs' free energy is zero at the melting point,

$$\Delta G_f = \Delta H_f - T\,\Delta S_f = 0 \tag{6.46}$$

where $T = T_f$. For polyethylene, per —CH_2— group, with ΔH_f = 3.94 kJ/mol; see Table 6.1. Then

$$\Delta S_f = \Delta H_f / T_f = 9.61\,\text{J/mol}\cdot\text{K} \tag{6.47}$$

at a T_f of 410 K.

Statistical thermodynamics asks how many conformational changes are involved in the melting process. If the polyethylene is crystallized in the all *trans* conformation, and two *gauche* plus the *trans* are possible in the melt (see Section 2.1.2), then the polymer goes from one possible conformation to three on melting (150,151). From equation (3.13),

$$\Delta S = R \ln \Omega = R \ln 3 = 9.13 \text{ J/mol}\cdot\text{K} \tag{6.48}$$

The agreement between the classical and statistical results, equations (6.47) and (6.48), is seen to be excellent, noting the approximations involved.

A more general statistical thermodynamic theory can be obtained with the quasi-lattice models; see Figure 3.3. If a coordination number z of the lattice is assumed, then there are $z - 1$ choices of where to put the next bond in the chain. (This is a little smaller with excluded volume considered.) Then the entropy varies with dimensionality of the quasi-lattice according to

$$\text{3-D:} \quad \Delta S_f = R \ln 5 = 13.4 \text{ J/mol}\cdot\text{K} \tag{6.49}$$

$$\text{2-D:} \quad \Delta S_f = R \ln 3 = 9.13 \text{ J/mol}\cdot\text{K} \tag{6.50}$$

$$\text{1-D:} \quad \Delta S_f = R \ln 1 = 0 \tag{6.51}$$

If more than one group needs to be considered, the values are multiplicative, rather than additive. Taking two —CH_2— groups, for example, yields nine conformations, rather than six. The general equation can be written

$$\Delta S = R \ln(A^a B^b C^c \cdots) \tag{6.52}$$

where A, B, C, . . . are the number of ways the various moieties can be arranged in space, and a, b, c, . . . are the number of appearances of the moiety in each mer.

Values of entropy of fusion are shown in Table 6.9 (146). While the entropies per mer varied widely, as might be expected from the enormous differences in the sizes and structures of the units, values divided by the number of chain bonds about which free rotation is permitted gave more nearly uniform values. According to Flory (152), the configurational entropy of fusion per segment should be $R \ln(Z' - 1)$, where Z' is the coordination number of the lattice. Values of E.u./No. bonds permitting rotation in Table 6.9 are in rough agreement with Flory's calculation. Based on equation (6.37), it is easy to understand why large heats of fusion produce high melting polymers (see Table 6.1). The quantity E.u. represents entropy units per mer.

In experiments such as the above, heating and cooling are usually done very slowly. Therefore regime I structures may predominate. The melting temperature is higher under these conditions than when cooling or heating is rapid, in which case regime II and III kinetics apply.

Table 6.9 Entropies of fusion for various polymers (146)

Polymer	Repeating Unit	Entropy of fusion, J/mol·K: E.u./mol of Repeating Unit	E.u./No. Bonds Permitting Rotation
Polyethylene	$—CH_2—$	8.37	8.37
Cellulose tributyrate	$—C_{18}H_{28}O_{18}—$	25.9	13.0
Poly(decamethylene sebacate) $—O—(CH_2)_{10}—O—CO—(CH_2)_8CO—$		145	6.3
Poly(*N,N'*-sebacoylpiperazine)	$—N(CH_2CH_2)_2N—CO(CH_2)_8CO—$	57.3	5.0

6.8.5 The Hoffman–Weeks Equilibrium Melting Temperature

There are several definitions of the equilibrium melting temperature currently in use. According to equation (6.42) the highest melting temperature (and presumably the equilibrium melting temperature) is reached at infinite molecular weight. Another definition assumes infinitely thick crystalline lamellae (153); see Section 6.7.4.

According to Hoffman and Weeks (154) the equilibrium melting temperature of a polymer, T_f^*, is defined as the melting point of an assembly of crystals, each of which is so large that surface effects are negligible and that each such large crystal is in equilibrium with the normal polymer liquid. Furthermore the crystals at the melting temperature must have the equilibrium degree of perfection consistent with the minimum free energy at T_f^*.

While this definition holds for most pure compounds, polymers as ordinarily crystallized tend to melt below T_f^* because the crystals are small and all too imperfect. Thus the temperature of crystallization, usually still lower because of supercooling (see Figure 6.3), has an important influence on the experimentally observed melting point. Hoffman and Weeks (154) found the following relation to hold:

$$T_f^* - T_f = \phi'(T_f^* - T_c) \tag{6.53}$$

where ϕ' represents a stability parameter that depends on crystal size and perfection. The quantity ϕ' may assume all values between 0 and 1, where $\phi' = 0$ implies that $T_f = T_f^*$, whereas $\phi' = 1$ implies that $T_f = T_c$. Therefore crystals are most stable at $\phi' = 0$ and inherently unstable at $\phi' = 1$. Values of ϕ' near $\frac{1}{2}$ are common.

The experiment generally involves rapidly cooling the polymer from the melt to some lower temperature, T_c, where it is then crystallized isothermally.

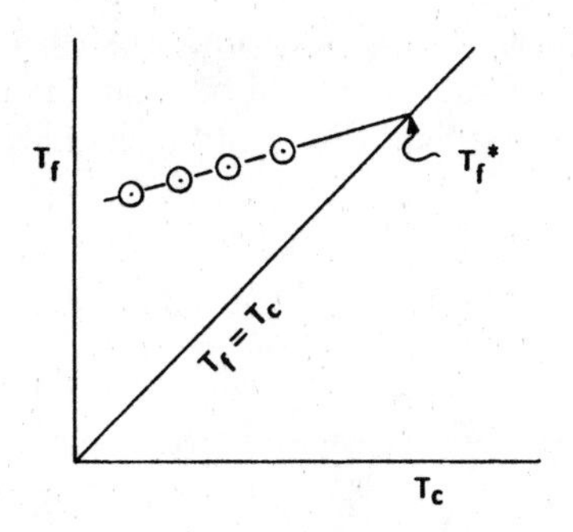

Figure 6.40 Idealized Hoffman-Weeks plot, showing the extrapolation to T_f^*.

For higher crystallization temperatures the polymer forms more perfect crystals; that is, it is better annealed. Hence its melting temperature increases.

To determine T_f^*, a plot of T_c versus T_f is prepared (154) (Figure 6.40). A line is drawn where $T_c = T_f$. The experimental data are extrapolated to the intersection with the line. The temperature of intersection is T_f^*.

6.9 EFFECT OF CHEMICAL STRUCTURE ON THE MELTING TEMPERATURE

The actual values of the enthalpy and entropy of fusion are, of course, controlled by the chemical structure of the polymer. The most important inter- and intramolecular structural characteristics include structural regularity, bond flexibility, close packing ability, and interchain attraction (155,156). In general, high melting points are associated with highly regular structures, rigid molecules, close packing capability, strong interchain attraction, or several of these factors combined.

The effect of structural irregularities can be illustrated by a study of polyesters (156) having the general structure

$$\left[\overset{\overset{\displaystyle O}{\|}}{C} - \langle\bigcirc\rangle - \overset{\overset{\displaystyle O}{\|}}{C} - O - R - O \right]_n \qquad (6.54)$$

when R is —CH_2—CH_2—, the structure is Dacron®. The melting temperature depends on the regularity of the group R. For aliphatic groups, the size and regularity of R are both important:

R	T_f, °C
$-CH_2-CH_2-$	265
$-(CH_2)_3-$	220
$-\underset{\displaystyle CH_3}{\underset{\|}{CH}}-CH_2-$	Noncrystalline

The irregularity of the atactic propylene unit destroys crystallinity entirely.

The effect of bond flexibility may also be examined utilizing the polyester structure (156). In this case substitutions in the rigid aromatic group are made:

$$\left(\!-O-\overset{\overset{\large O}{\|}}{C}-R'-\overset{\overset{\large O}{\|}}{C}-O-CH_2-CH_2-\!\right)_n \tag{6.55}$$

R′	T_f, °C
 $-C_6H_4-$ (p-phenylene)	265
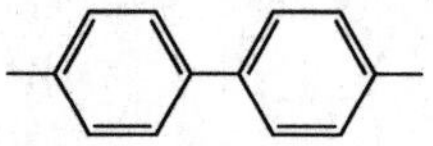 $-C_6H_4-C_6H_4-$ (biphenylene)	355
$\left(\!-CH_2-\!\right)_4$	50

The flexible aliphatic group, although having dimensions similar to those of the phenyl group, has a much lower melting temperature. It should be pointed out that the aliphatic polyesters, first synthesized by Carothers (157), failed as clothing fibers because they melted during washing or ironing. Aromatic polyesters (158) as well as aliphatic nylons achieved the necessary high melting temperatures; see Chapter 7.

Interchain forces can be illustrated by the following substitutions (156) of increasingly polar groups:

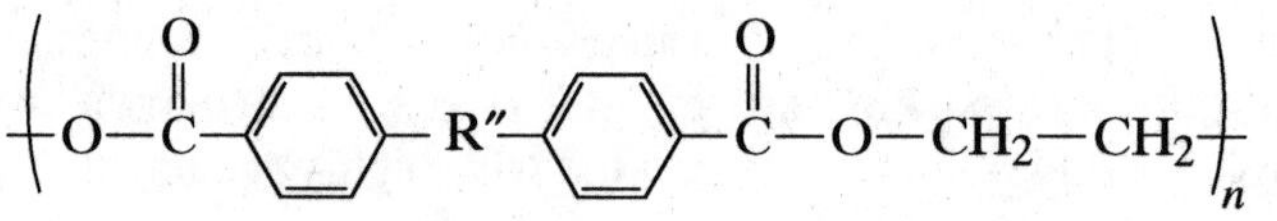

$$\left(\!-O-\overset{\overset{\large O}{\|}}{C}-C_6H_4-R''-C_6H_4-\overset{\overset{\large O}{\|}}{C}-O-CH_2-CH_2-\!\right)_n \tag{6.56}$$

R″	T_f, °C
$\left(\!-CH_2-\!\right)_4$	170
$-O-CH_2-CH_2-O-$	240
$-NH-CH_2-CH_2-NH-$	273

Similar effects are caused by bulky substituents and by odd or even numbers of carbon atoms in hydrocarbon segments of the chain. Generally, bulky groups lower T_f, because they separate the chains. The odd or even number of carbon atoms affects the regularity of packing. Of course, the frequency of occurrence of polar groups is very important. As the length of the aliphatic group R in equation (6.54) is increased, the melting point gradually approaches that of polyethylene, or about 139°C.

In each of the cases above, of course, the crystal structure is governed by the principles laid down in Section 6.3.5 and elsewhere. Generally, those structures that are most tightly bonded, fit the most closely together, and are held in place the most rigidly will have the highest melting temperatures.

6.10 FIBER FORMATION AND STRUCTURE

The synthetic fibers of today, the polyamides, polyesters, rayons, and so on, are manufactured by a process called spinning. Spinning involves extrusion through fine holes known as spinnerets. Immediately after the spinning process, the polymer is oriented by stretching or drawing. This both increases polymer chain orientation and degree of crystallinity. As a result the modulus and tensile strength of the fiber are increased.

Fiber manufacture is subdivided into three basic methods, melt spinning, dry spinning, and wet spinning; see Table 6.10 (159). Melt spinning is the simplest but requires that the polymer be stable above its melting temperature. Polyamide 66 is a typical example. Basically, the polymer is melted and forced through spinnerets, which may have from 50 to 500 holes. The fiber diameter immediately after the hole and before attenuation begins is larger than the hole diameter. This is called die swell, which is due to a relaxation of the viscoelastic stress-induced orientation in the hole; see Section 5.4 and Figure 10.20.

During the cooling process the fiber is subjected to a draw-down force, which introduces the orientation. Additional orientation may be introduced later by stretching the fiber to a higher draw ratio.

In dry spinning, the polymer is dissolved in a solvent. A typical example is polyacrylonitrile dissolved in dimethylformamide to 30% concentration. The polymer solution is extruded through the spinnerets, after which the solvent is rapidly evaporated (Figure 6.41) (159). After the solvent is evaporated, the fiber is drawn as before.

In wet spinning, the polymer solution is spun into a coagulant bath. An example is a 7% aqueous solution of sodium cellulose xanthate (viscose), which is spun into a dilute sulfuric acid bath, also containing sodium sulfate and zinc sulfate (160). The zinc ions form temporary ionic cross-links between the xanthate groups, holding the chains together while the sulfuric acid, in turn,

Table 6.10 Spinning processes (159)

	Solution Spinning		
		Wet Spinning	
Melt Spinning	Dry Spinning	Coagulation	Regeneration
Polyamide	Cellulose acetate		Viscose rayon
Polyester	Cellulose triacetate		Cupro
Polyethylene	Acrylic	Acrylic	
	Modacrylic	Modacrylic	
Polypropylene	Aramid	Aramid	
PVDC	Elastane	Elastane	
	PVC	PVC	
	Vinylal		

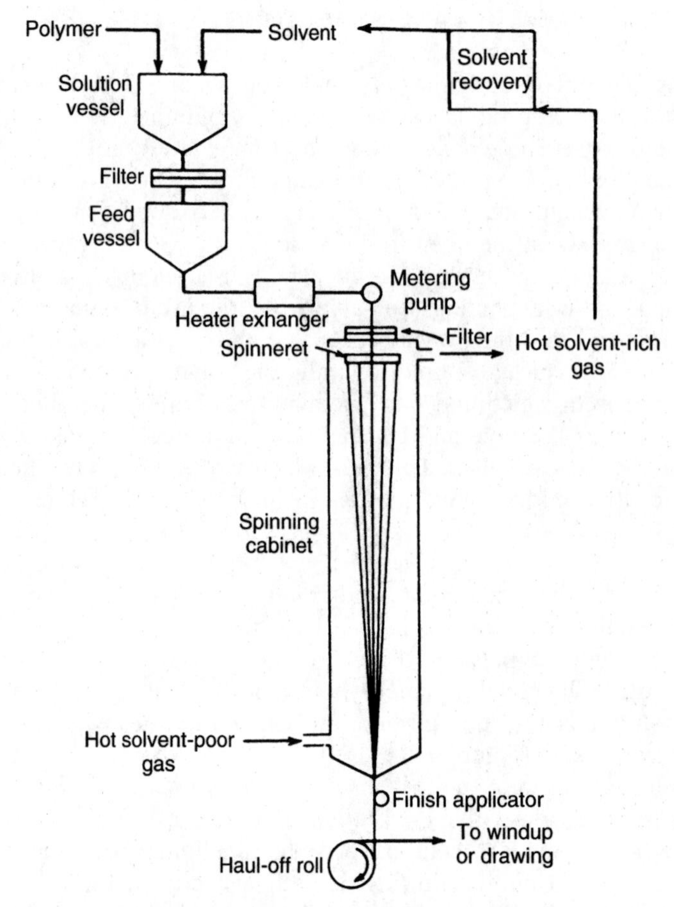

Figure 6.41 In typical dry-spinning operations, hot gas is used to evaporate the solvent in the spinning cabinet. The fibers are simultaneously oriented (159).

removes the xanthate groups, thus precipitating the polymer. After orientation, and so on, the final product is known as rayon.

6.10.1 X-Ray Fiber Diagrams

For the purpose of X-ray analyses, the samples should be as highly oriented and crystalline as possible. Since these are also the conditions required for strong, high-modulus fibers, basic characterization and engineering requirements are almost identical.

Figure 6.42 (27,161,162) illustrates a typical X-ray fiber diagram, for polyallene, $\mathrm{+CH_2{-}C({=}CH_2)+_n}$. The actual fiber orientation is vertical. The most intense diffractions are on the equitorial plane; note the 110 and 200 reflections. Note the rather intense amorphous halo, appearing inside the 011 reflection.

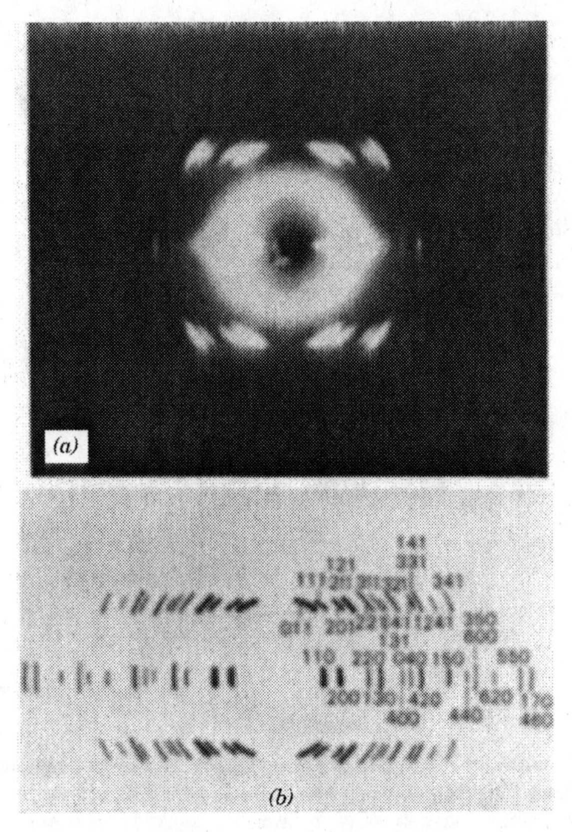

Figure 6.42 X-ray fiber diagram of polyallene (*a*), and its indexing (*b*) (27).

Because of the imperfect orientation of the polymer in the fiber, arcs are seen, rather than spots. The variation in the intensity over the arcs can be used, however, to calculate the average orientation.

A further complication in the interpretation of the fiber diagram arises because it actually is a "full rotation photograph." In these ways fiber diagrams differ from those of single crystals. Vibrational analyses via infrared and Raman spectroscopy studies play an important role in the selection of the molecular model, as described above.

6.10.2 Natural Fibers

Natural fibers were used long before the discovery of the synthetics in the twentieth century. Natural fibers are usually composed of either cellulose or protein, as shown in Table 6.11. Animal hair fibers belong to a class of proteins known as keratin, which serve as the protective outer layer of the higher vertebrates. The silks are partly crystalline protein fibers. The crystalline por-

Table 6.11 Chemical nature of natural fibers

Cellulose	Protein
Cotton	Wool
Tracheid (wood)	Hair
Flax	Silk
Hemp	Spider webs
Coir	
Ramie	
Jute	

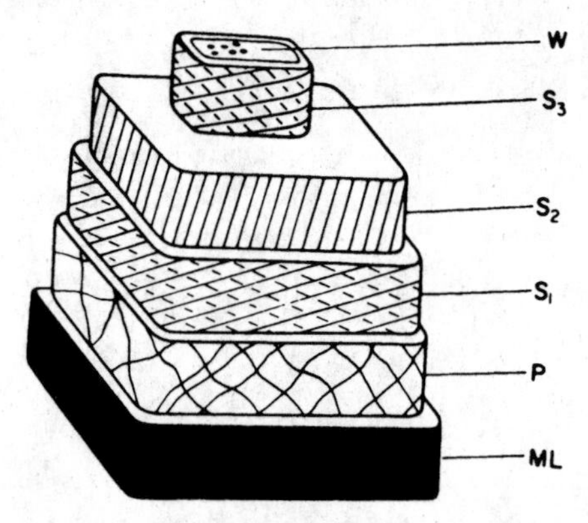

Figure 6.43 The cell walls of a tracheid or wood fiber have several layers, each with a different orientation of microfibrils (164). *ML*, middle lamella, composed of lignin; *P*, primary wall; S_1, S_2, S_3, layers of the secondary wall; *W*, warty layer. The lumen in the interior of the warty layer is used to transport water.

tions of these macromolecules are arranged in antiparallel pleated sheets, a form of the folded-chain lamellae (163).

The morphology of the natural fibers is often quite complex; see Figure 6.43 (164). The cellulose making up these trachieds is a polysaccharide; see Table 1.4. The crystalline portion of the cellulose making up the trachieds is highly oriented, following the various patterns indicated in Figure 6.43. The winding angles of the cellulose form the basis for a natural composite of great strength and resilience. A similar morphology exists in cotton cellulose.

The fibrous proteins (keratin) are likewise highly organized; see Figure 6.44 (165). Proteins are actually polyamide derivatives, a copolymeric form of polyamide 2, where the mers are amino acids. For example, the structure of the amino acid phenylalanine in a protein may be written

$$-\underset{\displaystyle |}{\overset{\mathrm{H}}{\overset{|}{\mathrm{N}}}}-\overset{\mathrm{H}}{\overset{|}{\underset{\underset{\displaystyle \mathrm{CH_2}-\mathrm{C_6H_5}}{|}}{\mathrm{C}}}}-\overset{\mathrm{O}}{\overset{\|}{\mathrm{C}}}- \qquad (6.57)$$

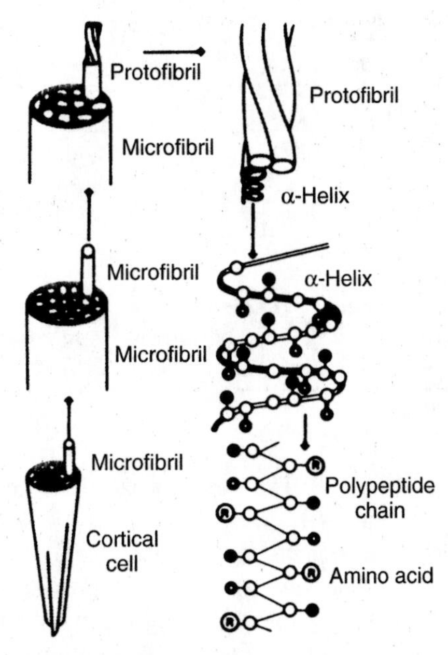

Figure 6.44 A wool-fiber cortical cell is complex structure, being composed ultimately of proteins (165)

There are some 20 amino acids in nature, see Figure 14.29. These are organized into an α-helix in the fibrous proteins, which in turn are combined to form protofibrils as shown. In addition to being crystalline, the fibrous proteins are cross-linked though disulfide bonds contained in the cystine amino acid mer, which is especially high in keratin. Animal tendons, composed of collagen, another fibrous protein, have also been shown to have a surprisingly complex hierarchical structure (166).

6.11 THE HIERARCHICAL STRUCTURE OF POLYMERIC MATERIALS

While all polymers are composed of long-chain molecules, it is the organization of such materials at higher and higher levels that progressively determines their properties and ultimately determines their applications. Table 6.12 summarizes the hierarchical structure of the several classes of polymers according to size range. The largest synthetic crystalline polymer structures are the spherulites, while the largest structures of the natural polymers are the walls of the whole (living) cell.

Of course, there are many kinds of natural polymers. Starch and bread are discussed in Section 14.3, and silk fibers in Section 14.4; both are semicrystalline materials. The hierarchical structure of polymers has been reviewed by

Table 6.12 Hierarchical structure of polymers

	Synthetic		Natural	
Size Range	Amorphous	Crystalline	Cellulosic	Protein
Ångstroms	Mer	Mer	Glucose	Amino acid
Nanometers	Chain coil Physical entanglement Cross-link	Stem	Stem	α-Helix
Microns	Multicomponent morphologies	Lamellae spherulite	Lamellae microfibril	Protofibril microfibril
100 Microns	—	—	Fiber (cell)	Microfibril cell

Baer and co-workers (167,168). They emphasize that the occurrence of crystallinity in synthetic polymers requires a sufficiently stereoregular chemical structure, so that the chain molecules can pack closely in parallel orientation. Structure at the nanometer level is generally determined by the pronounced tendency of these chains to crystallize by folding back and forth within crystals of thin lamellar habit. Chain folding itself is a kinetic phenomenon. Given sufficient annealing time, the lamellar thickness increases until the chains are all straight at thermodynamic equilibrium.

When crystallization occurs in flowing solutions or melts, fine fibrous crystals may be produced that consist of highly extended chains oriented axially. This will be recognized as the basis for the supermolecular structure in fibers. Even after prolonged crystallization from the melt, there always remains an appreciable fraction of a disordered phase known as amorphous polymer. If the polymer can crystallize, the appearance of amorphous material is also a kinetic phenomenon, which at equilibrium should disappear. However, the physical entanglements and already existing crystalline structure may slow further crystallization down to substantially zero.

While amorphous materials are significantly less organized than their crystalline counterparts, entanglements, branches, and cross-links significantly control their properties. When two polymers are blended, they generally phase-separate to create a super-molecular morphology, which may be of the order of tens of micrometers, while graft and block copolymers exhibit morphologies of the order of hundreds of angstroms.

6.12 HOW DO YOU KNOW IT'S A POLYMER?

Suppose that a polymer scientist watches a demonstration. The contents of two bottles are mixed, and a white precipitate appears. "Here, I have created a new polymer for you," the demonstrator announces. The polymer scientist examines the two bottles, which are unlabeled. The demonstrator mysteriously leaves the building. What experiments should be performed to determine if the white precipitate is polymeric? (A separate, but equally important, question is: What is its structure?)

In 1920 such questions were far from trivial; note Appendix 5.1. Even today, questions of this nature need to be addressed. For example, the structure of ordinary (window) glass is considered polymeric by some, and not by others. It surely exhibits a glass transition (see Chapter 8) and has a high viscosity, but also apparently has a time variable structure.

There are several types of experiments that should be performed in order to ascertain that the material is (or is not) polymeric. A key experiment is the determination of the molecular weight. If the molecular weight is above about 25,000 g/mol, most scientists will consider that evidence in favor of a polymeric structure. However, it may be colloidal, as a sulfur or silver sol, and not polymeric. If the material is semicrystalline, the unit cell should be established. Hence it will become known if a chain running from cell to cell is likely to occur.

A next series of experiments involves ordinary chemistry, as outlined in Section 2.2. Elemental analysis, characterization of degradation products, end-group analysis, and determination of the probable mer structure are all important in solving the puzzle.

Perhaps the material can be reacted or degraded to form a soluble compound, which can then be characterized. An example is cellulose, a natural polymer that was part of the earliest discussions. Insoluble itself, it can be acetylated to form soluble cellulose acetate for molecular weight determination, by osmometry at the time. Then it was subjected to degradation to glucose and cellobiose, which were determined to be the monomer and dimer, respectively.

The history of polymer science (169,170), takes the reader back to before the macromolecular hypothesis. Natural rubber, *cis*-polyisoprene, was investigated around the turn of the twentieth century. Morawetz (169) writes of the destructive distillation of rubber to isoprene, and of the ozonolysis of rubber to levulinic aldehyde. These studies led to the dimer ring structure of dimethylcyclooctadiene for *cis*-polyisoprene:

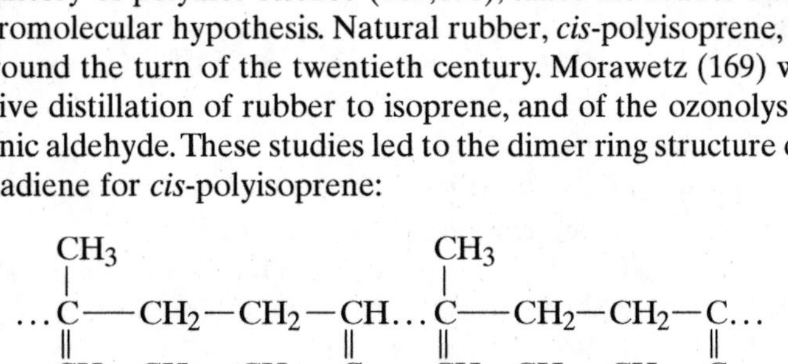

$$\begin{array}{lllll}
\mathrm{CH_3} & & \mathrm{CH_3} & & \\
\quad| & & \quad| & & \\
\ldots\mathrm{C}-\mathrm{CH_2}-\mathrm{CH_2}-\mathrm{CH}\ldots & & \mathrm{C}-\mathrm{CH_2}-\mathrm{CH_2}-\mathrm{C}\ldots & & \\
\quad\| \qquad\qquad\qquad\quad \| & & \| \qquad\qquad\qquad\quad \| & & \\
\ldots\mathrm{CH}-\mathrm{CH_2}-\mathrm{CH_2}-\mathrm{C} & & \ldots\mathrm{CH}-\mathrm{CH_2}-\mathrm{CH_2}-\mathrm{C}\ldots & & \\
\qquad\qquad\qquad\qquad\quad | & & \qquad\qquad\qquad\quad | & & \\
\qquad\qquad\qquad\quad \mathrm{CH_3} & & \qquad\qquad\qquad \mathrm{CH_3} & &
\end{array} \tag{6.58}$$

The strength of the affinity of the carbon–carbon double bonds was supposed to hold the material together with partial valences. Only much later was it determined that the structure was actually a long chain of isoprene mers, connected in the 1,4-positions, with all the double bonds in the *cis*-configuration.

Such structures imply the confusion then reigning between colloids and polymers. Both were large structures. What was the difference between a sulfur colloidal dispersion and a rubber solution? The answer was found after many years of research.

Another class of experiments that is important includes FT-IR, Raman spectroscopy, and NMR, which provide physical evidence as to where each atom is located and what moieties are present; see Section 2.2. In combination with a known high molecular weight and chemical analyses, large parts of the puzzle can be put together.

If the material is crystalline, X-ray and electron diffraction experiments can help decide the relative location of large sections of the chain. Are the presumed "mers" lined up in such a fashion that they can be reasonably bonded together? What is the relation between one cell and its neighbors?

Returning to our mysterious demonstration, the material could have been polyamide 610, synthesized via interfacial polymerization. Then elemental analysis would show the percent nitrogen, carbon, oxygen, and hydrogen, and each of the methods described above would be brought to play their roles. If, however, it was concentrated hydrochloric acid and concentrated sodium hydroxide that were mixed, the result would eventually be shown to be sodium chloride, ordinary salt! Philosophically, of course, conclusions are the product of reasoning power. It must be remembered that experiments provide evidence, and nothing else.

REFERENCES

1. J. Brandrup and E. H. Immergut, eds., *Polymer Handbook*, 2nd ed., Wiley-Interscience, New York, 1975.
2. K. H. Meyer and H. Mark, *Ber. Deutsch. Chem. Ges.*, **61**, 593 (1928).
3. H. Mark and K. H. Meyer, *Z. Phys. Chem.*, **B2**, 115 (1929).
4. K. H. Meyer and H. Mark, *Z. Phys. Chem.*, **B2**, 115 (1929).
5. K. H. Meyer and L. Misch, *Ber. Deutsch. Chem. Ges., B.*, **70B**, 266 (1937).
6. K. H. Meyer and L. Misch, *Helv. Chim. Acta*, **20**, 232 (1937).
7. H. Mark, presented at the American Chemical Society meeting, Seattle, Washington, March 1983.

7a. O. L. van der Hart and R. H. Atalla, *Macromolecules*, **17**, 1465 (1984).

7b. S. Neyertz, A. Pizzi, A. Merlin, B. Maigret, D. Brown, and X. Deglise, *J. Appl. Polym. Sci.*, **78**, 1939 (2000).

8. K. Ziegler, E. Holtzkamp, H. Briel, and H. Martin, *Angew. Chem.*, **67**, 426, 541 (1955).
9. G. Natta, P. Pino and G. Mazzanti, *Gazz. Chim. Ital.*, **87**, 528 (1957).
10. O. Ellefsen and B. A. Tonnesen, in *Cellulose and Cellulose Derivatives*, Part IV, N. M. Bikales and L. Segal, eds., Wiley-Interscience, New York, 1971, p. 151.

11. J. A. Hawsman and W. A. Sisson, in *Cellulose and Cellulose Derivatives*, Part I, E. Ott, H. M. Spurlin, and M. W. Grafflin, eds., Wiley-Interscience, New York, 1954, p. 231.
12. L. Mandelkern, *Rubber Chem. Tech.*, **32**, 1392 (1959).
13. S. D. Kim and L. H. Sperling, Unpublished.
14. P. W. Atkins, *Physical Chemistry*, 6th ed., Oxford University Press, Oxford, 1998.
15. W. L. Bragg, *Proc. Camb. Philos. Soc.*, **17**, 43 (1913).
16. P. Debye and P. Scherrer, *Phys. Z.*, **17**, 277 (1916).
17. C. W. Bunn, *Chemical Crystallography*, Oxford University Press, London, 1945, p. 109.
18. H. Tadokoro, *Structure of Crystalline Polymers*, Wiley-Interscience, New York, 1979, Chap. 5.
19. A. Keller, *Philos. Mag.*, **2**, 1171 (1957).
20. C. W. Bunn, *Trans. Faraday Soc.*, **35**, 482 (1939).
21. G. Natta and P. Corradini, *Rubber Chem. Tech.*, **33**, 703 (1960).
22. Courtesy of Dr. A. Keller and Sally Argon.
23. G. Natta and P. Corradini, *J. Polym. Sci.*, **39**, 29 (1959).
24. D. R. Holmes, C. W. Bunn, and D. J. Smith, *J. Polym. Sci.*, **17**, 159 (1955).
25. R. de P. Daubeny, C. W. Bunn, and C. J. Brown, *Proc. R. Soc. (Lord.)*, **A226**, 531 (1954).
26. Y. Takahashi and H. Tadokoro, *Macromolecules*, **6**, 672 (1973).
27. H. Tadokoro, *Structure of Crystalline Polymers*, Wiley-Interscience, New York, 1979.
28. H. Tadokoro, Y. Chatani, T. Yoshihara, S. Tahara, and S. Murahashi, *Makromol. Chem.*, **73**, 109 (1964).
29. B. C. Rånby, *Acta Chem. Scand.*, **6**, 101, 116 (1952).
30. L. Loeb and L. Segal, *J. Polym. Sci.*, **14**, 121 (1954).
31. J. W. S. Hearle, *Polymers and Their Properties*, Vol. I, *Fundamentals of Structure and Mechanics*, Ellis Horwood Ltd., Chichester, England, 1982.
32. J. W. S. Hearle and R. H. Peters, *Fiber Structure*, Textile Institute, Manchester, England, 1963.
33. F. Khoury and E. Passaglia, in *Treatise on Solid State Chemistry*, Vol. 3, *Crystalline and Noncrystalline Solids*, N. B. Hannay, ed., Plenum, New York, 1976, Ch. 6.
34. P. H. Geil, *Polymer Single Crystals*, Interscience, New York, 1963.
35. D. W. Van Krevelen, *Properties of Polymers*, 3rd ed., Elsevier, Amsterdam, 1990, Chap. 19.
36. J. E. Mark, A. Eisenberg, W. W. Graessley, L. Mandelkern, E. T. Samulski, J. L. Koenig, and G. D. Wignall, *Physical Properties of Polymers*, 2nd ed., American Chemical Society, Washington, DC, 1993, Chap. 4.
37. A. Keller, M. Warner, A. H. Windle, eds., *Self-order and Form in Polymers*, Chapman and Hall, London, 1995.
38. E. J. Roche, R. S. Stein, and E. L. Thomas, *J. Polym. Sci. Polym. Phys. Ed.*, **18**, 1145 (1980).

39. R. S. Stein and A. Misra, *J. Polym Sci. Polym. Phys. Ed.*, **18**, 327 (1980).
40. R. S. Stein, in *Rheology, Theory and Applications*, Vol. 5, F. R. Eirich, ed., Academic, New York 1969, Chap. 6.
41. J. D. Hoffman, G. T. Davis, and J. I. Lauritzen Jr., in *Treatise on Solid State Chemistry*, Vol. 3, *Crystalline and Noncrystalline Solids*, N. B. Hannay, ed., Plenum, New York, 1976, Chap. 7.
42. W. A. Sisson, in *Cellulose and Cellulose Derivatives*, Interscience, New York, 1943, pp. 203–285.
43. J. Hengstenberg and J. Mark, *Z. Kristallogr.*, **69**, 271 (1928).
44. W. O. Statton, *J. Polym. Sci.*, **20C**, 117 (1967).
45. K. Herrmann and O. Gerngross, *Kautschuk*, **8**, 181 (1932).
46. K. Herrmann, O. Gerngross, and W. Abitz, *Z. Phys. Chem.*, **10**, 371 (1930).
47. E. W. Fischer, *Z. Naturforsch.*, **12a**, 753 (1957).
48. P. H. Till, Jr., *J. Polym. Sci.*, **24**, 301 (1957).
49. F. A. Bovey, *Org. Coat. Appl. Polym. Sci Prepr.*, **48** (1), 76 (1983).
50. F. C. Schilling, F. A. Bovey, S. Tseng, and A. E. Woodward, *Macromolecules*, **16**, 808 (1983).
51. T. Oyama, K. Shiokawa, and Y. Murata, *Polym. J.*, **6**, 549 (1974).
52. A. J. Kovacs, J. A. Manson, and D. Levy, *Kolloid Z.*, **214**, 1 (1966).
53. P. J. Flory, *J. Am. Chem. Soc.*, **84**, 2857 (1962).
54. R. S. Stein, *J. Chem. Ed.*, **50**, 748 (1973).
55. L. Mandelkern, in *Physical Properties of Polymers*, J. E. Mark, A. Eisenberg, W. W. Graessley, L. Mandelkern, and J. L. Koenig, eds., American Chemical Society, Washington, DC, 1984.
56. H. D. Keith, F. J. Padden, Jr., and R. G. Vadimsky, *J. Polym. Sci.*, *A-2*, **4**, 267 (1966).
57. H. D. Keith, F. J. Padden, and R. G. Vadimsky, *J. Appl. Phys.*, **42**, 4585 (1971).
58. Y. Hase and P. H. Geil, *Polym. J.* (*Jpn.*), **2**, 560, 581 (1971).
59. F. Rybnikar and P. H. Geil, *J. Macromol. Sci. Phys.*, **B7**, 1 (1973).
60. D. C. Bassett, A. Keller, and S. Mitsuhashi, *J. Polym. Sci.*, **A1**, 763 (1963).
61. P. H. Geil, in *Growth and Perfection of Crystals*, R. H. Doremus, B. W. Roberts, and D. Turnbull, eds., Wiley, New York, 1958, pp. 579–585.
62. H. D. Keith, *J. Polym. Sci.*, **A2**, 4339 (1964).
63. E. Martuscelli, Multicomponent Polymer Blends Symposium, Capri, Italy, May 1983.
64. E. Martuscelli, *Polym. Eng. Sci.*, **24**, 563 (1984).
65. A. J. Kovacs, *Chim. Ind. Genie Chim.*, **97**, 315 (1967).
66. R. G. Crystal, P. F. Erhardt, and J. J. O'Malley, in *Block Copolymers*, S. L. Aggarwal, ed., Plenum, New York, 1970.
67. K. E. Hardenstine, C. J. Murphy, R. B. Jones, L. H. Sperling, and G. E. Manser, *J. Appl. Polym. Sci.*, **30**, 2051 (1985).
68. H. Cornélis, R. G. Kander, and J. P. Martin, *Polymer*, **37**, 4573 (1996).
69. J. N. Hay and M. Sabin, *Polymer*, **10** (3), 203 (1969).
70. H. D. Keith and F. J. Padden Jr., *J. Appl. Phys.*, **35**, 1270 (1964).

71. H. D. Keith and F. J. Padden Jr., *J. Appl. Phys.*, **35**, 1286 (1964).
72. L. H. Palys and P. J. Philips., *J. Polym. Sci. Polym. Phys. Ed.*, **18**, 829 (1980).
73. M. Avrami, *J. Chem. Phys.*, **7**, 1103 (1939).
74. M. Avrami, *J. Chem. Phys.*, **8**, 212 (1940).
75. M. Avrami, *J. Chem. Phys.*, **9**, 177 (1941).
76. J. D. Hoffman, *Polymer*, **24**, 3 (1983).
77. J. D. Hoffman, *Polymer*, **23**, 656 (1982).
78. E. A. DiMarzio, C. M. Guttman, and J. D. Hoffman, *Faraday Discuss. Chem. Soc.*, **68**, 210 (1979).
79. C. M. Guttman, J. D. Hoffman, and E. A. DiMarzio, *Faraday Discuss. Chem. Soc.*, **68**, 297 (1979).
80. J. D. Hoffman, C. M. Guttman, and E. A. DiMarzio, *Faraday Discuss. Chem. Soc.*, **68**, 177 (1979).
81. U. R. Evans, *Trans. Faraday Soc.*, **41**, 365 (1945).
82. P. Meares, *Polymers: Structure and Bulk Properties*, Van Nostrand, New York, 1965, Chap. 5.
83. J. N. Hay, *Br. Polym. J.*, **3**, 74 (1971).
84. S. D. Poisson, *Recherches sur la Probabilite des Judgements en Matiere Criminelle et en Matiere Civile*, Bachelier, Paris, 1837, p. 206.
85. W. Liu, B. S. Hsiao, and R. S. Stein, *Polym. Mater. Sci. Eng. (Prepr.)*, **81**, 363 (1999).
86. R. G. Alamo, J. D. Londono, L. Mandelkern, F. C. Stehling, and G. D. Wignall, *Macromolecules*, **27**, 411 (1994).
87. G. D. Wignall, J. D. Londono, J. S. Lin, R. G. Alamo, M. J. Galante, and L. Mandelkern, *Macromolecules*, **28**, 3156 (1995).
88. H. D. Keith and F. J. Padden Jr., *J. Appl. Phys.*, **34**, 2409 (1963).
89. B. Wunderlich and L. Melillo, *Makromol. Chem.*, **118**, 250 (1968).
90. D. O. Yoon and P. J. Flory, *Faraday Discuss. Chem. Soc.*, **68**, 288 (1979).
91. J. D. Hoffman and R. L. Miller, *Polymer*, **38**, 3151 (1997).
92. J. P. Armistead and J. D. Hoffman, *Macromolecules*, **35**, 3895 (2002).
93. D. C. Bassett, R. H. Olley, and I. A. M. Al Raheil, *Polymer*, **29**, 1539 (1988).
94. A. Toda, *Colloid Polym. Sci.*, **270**, 667 (1992).
95. D. M. Sadler, *Nature*, **326**, 174 (1987).
96. D. M. Sadler and G. H. Gilmer, *Phys. Rev. B*, **38**, 5684 (1988).
97. J.-W. Sommer, *Polymer*, **43**, 929 (2002).
98. C.-M. Chen and P. G. Higgs, *J. Chem. Phys.*, **108**, 4305 (1998).
99. J. P. K. Doye and D. Frenkel, *Polymer*, **41**, 1519 (2000).
100. J. P. K. Doye and D. Frenkel, *J. Chem. Phys.*, **110**, 7073 (1999).
101. S. Z. D. Cheng and B. Lotz, *Phil. Trans. R. Soc. Lond. A*, **361**, 517 (2003).
102. G. Goldbeck-Wood, *J. Polym. Sci.: B: Polym. Phys.*, **31**, 61 (1993).
103. M. Tasumi and S. Krimm, *J. Polym. Sci.*, *A-2*, **6**, 995 (1968).
104. M. I. Bank and S. Krimm, *J. Polym. Sci.*, *A-2*, **7**, 1785 (1969).
105. S. Krimm and T. C. Cheam, *Faraday Discuss. Chem. Soc.*, **68**, 244 (1979).

106. X. Jing and S. Krimm, *Polym. Lett.*, **21**, 123 (1983).
107. R. Kitamarn, F. Horii, and S. H. Hyon, *J. Polym. Sci. Polym. Phys. Ed.*, **15**, 821 (1977).
108. M. Glotin and L. Mandelkern, *Colloid Polym. Sci.*, **260**, 182 (1982).
109. L. Mandelkern, in *Physical Properties of Polymers*, J. E. Mark, A. Eisenberg, W. W. Graessley, L. Mandelkern, and J. L. Koenig, eds. American Chemical Society, Washington, DC, 1984.
110. M. Stamm, E. W. Fischer, M. Dettenmaier, and P. Convert, *Faraday Discuss. Chem. Soc.*, **68**, 263 (1979).
111. D. G. H. Ballard, A. N. Burgess, T. L. Crawley, G. W. Longman, and J. Schelten, *Faraday Discuss. Chem. Soc.*, **68**, 279 (1979).
112. M. Stamm, *J. Polym. Sci. Polym. Phys. Ed.*, **20**, 235 (1982).
113. G. D. Wignall, L. Mandelkern, C. Edwards, and M. Glotin, *J. Polym. Sci. Polym. Phys. Ed.*, **20**, 245 (1982).
114. D. M. Sadler and R. Harris, *J. Polym. Sci. Polym. Phys. Ed.*, **20**, 561 (1982).
115. J. Schelten, G. D. Wignall, D. G. H. Ballard, and G. W. Longman, *Polymer*, **18**, 1111 (1977).
116. J. Schelten, A. Zinken, and D. G. H. Ballard, *Colloid Polym. Sci.*, **259**, 260 (1981).
117. D. G. H. Ballard, P. Cheshire, G. W. Longman, and J. Schelten, *Polymer*, **19**, 379 (1978).
118. E. W. Fischer, M. Stamm, M. Dettenmaier, and P. Herschenraeder, *Polym. Prepr. Am. Chem. Soc. Div. Polym. Chem.*, **20** (1), 219 (1979).
119. J. M. Guenet, *Polymer*, **22**, 313 (1981).
120. D. M. Sadler and A. Keller, *Macromolecules*, **10**, 1128 (1977).
121. D. M. Sadler and A. Keller, *Science*, **203**, 263 (1979).
122. A. Keller, *Faraday Discuss. Chem. Soc.*, **68**, 145 (1979).
123. D. M. Sadler and A. Keller, *Polymer*, **17**, 37 (1976).
124. L. H. Sperling, *Polym. Eng. Sci.*, **24**, 1 (1984).
125. D. Y. Yoon and P. J. Flory, *Polymer*, **18**, 509 (1977).
126. D. Y. Yoon, *J. Appl. Cryst.*, **11**, 531 (1978).
127. D. Y. Yoon and P. J. Flory, *Faraday Discuss. Chem. Soc.*, **68**, 289 (1979).
128. M. Dettenmaier, E. W. Fischer, and M. Stamm, *Colloid Polym. Sci.*, **258**, 343 (1979).
129. M. Stamm, E. W. Fischer, and M. Dettenmaier, *Faraday Discuss. Chem. Soc.*, **68**, 263 (1979).
130. J. Schelten, D. G. H. Ballard, G. D. Wignall, G. W. Longman, and W. Schmatz, *Polymer*, **17**, 751 (1976).
131. D. G. H. Ballard, P. Cheshire, G. W. Longman, and J. Schelten, *Polymer*, **19**, 379 (1978).
132. B. Crist, W. W. Graessley, and G. D. Wignall, *Polymer*, **23**, 1561 (1982).
133. F. C. Frank, *Faraday Discuss. Chem. Soc.*, **68**, 7 (1979).
134. R. Hosemann, *Polymer*, **3**, 349 (1962).
135. F. J. Balta-Calleja and R. Hosemann, *J. Appl. Crystallogr.*, **13**, 521 (1980).
136. R. Hosemann, *Colloid Polym. Sci.*, **260**, 864 (1982).

137. B. Wunderlich, in *Macromolecular Physics*, Vol. I, Academic Press, Orlando, FL, 1973.
138. B. Wunderlich and L. Melillo, *Makromol. Chem.*, **118**, 250 (1968).
139. E. Hellmuth 2nd, and B. Wunderlich, *J. Appl. Phys.*, **36**, 3039 (1965).
140. B. Wunderlich, *J. Polym. Sci. Symp.*, **43**, 29 (1973).
141. D. C. Bassett and B. Turner, *Phil. Mag.*, **29**, 925 (1974).
142. A. Keller, M. Hikosaka, S. Rastogi, A. Toda, P. J. Barham, and G. Goldbeck-Wood, in *Self-order and Form in Polymeric Materials*, A. Keller, M. Warner, and A. H. Windle, eds., Chapman and Hall, London, 1995.
143. A. Keller, M. Hikosaka, S. Rastogi, A. Toda, P. J. Barham, and G. Goldbeck-Wood, *J. Mater. Sci.*, **29**, 2579 (1994).
144. W. J. Moore, *Physical Chemistry*, 4th ed., Prentice-Hall, Englewood Cliffs, NJ, 1972, p. 134.
145. P. J. Flory, *J. Chem. Phys.*, **17**, 223 (1949).
146. L. Mandelkern and P. J. Flory, *J. Am. Chem. Soc.*, **73**, 3206 (1951).
147. L. Mandelkern, R. R. Garrett, and P. J. Flory, *J. Am. Chem. Soc.*, **74**, 3949 (1952).
148. G. M. Bristow and W. F. Watson, *Trans. Faraday Soc.*, **54**, 1731 (1958).
149. F. E. Karasz and L. D. Jones, *J. Phys. Chem.*, **71**, 2234 (1967).
150. A. V. Tobolsky, *Properties and Structure of Polymers*, Wiley, New York, 1960.
151. M. Warner, in *Side-Chain Liquid Crystals*, C. B. McArdle, ed., Blackie, Glasgow, 1989.
152. P. J. Flory, *J. Chem. Phys.*, **10**, 51 (1942).
153. B. Wunderlich, *Macromolecular Physics*, Vol. 1, Academic Press, Orlando, FL, 1973.
154. J. D. Hoffman and J. J. Weeks, *J. Res. Natl. Bur. Stand.*, **66A**, 13 (1962).
155. R. E. Wilfong, *J. Polym. Sci.*, **54**, 385 (1961).
156. R. W. Lenz, *Organic Chemistry of Synthetic High Polymers*, Interscience, New York, 1967, pp. 91–95.
157. W. H. Carothers, *J. Am. Chem. Soc.*, **51**, 2548, 2560 (1929).
158. J. R. Whinfield, *Nature*, **158**, 930 (1946).
159. J. E. McIntire and M. J. Denton, in *Encyclopedia of Polymer Science and Engineering*, Vol. 6, J. I. Kroschwitz, ed., Wiley, New York, 1986.
160. J. W. Schappel and G. C. Bockno, in *Cellulose and Cellulose Derivatives*, N. M. Bikales and L. Segal, eds., Vol. 5, Part 5, High Polymer Series, Wiley-Interscience, New York, 1971.
161. H. Tadokoro, Y. Takahasi, S. Otsuka, K. Mori, and F. Imaizumi, *J. Polym. Sci.*, *3B*, **3B**, 697 (1965).
162. H. Tadokoro, M. Kobayaski, K. Mori, Y. Takahashi, and S. Taniyama, *J. Polym. Sci. C*, **22**, 1031 (1969).
163. L. H. Sperling and C. E. Carraher, in *Encyclopedia of Polymer Science and Engineering*, 2nd ed., Vol. 12, J. I. Kroschwitz, ed., Wiley, New York, 1988.
164. G. Tsoumis, *Wood as a Raw Material*, Pergamon, New York, 1968.

165. W. S. Boston, in *Encylopedia of Textiles, Fibers, and Nonwoven Fabrics*, M. Grayson, ed., Wiley-Interscience, New York, 1984.

166. K. Kastelic, A. Galeski, and E. Baer, *Conn. Tiss. Res.*, **6**, 11 (1978).

167. E. Baer, A. Hiltner, and H. D. Keith, *Science*, **235**, 1015 (1987).

168. E. Baer, *Sci. Am.* **254** (10), 179 (1986).

169. H. Morawetz, *Polymers: The Origins and Growth of a Science*, Wiley, New York, 1985.

170. Y. Furukawa, *Inventing Polymer Science*, University of Pennsylvania Press, Philadelphia, 1998.

GENERAL READING

M. Dosiere, ed., *Crystallization of Polymers*, Kluwer, Dordrecht, 1993.

J. D. Hoffman and R. L. Miller, *Polymer*, **38**, 3151 (1997). (General review of polymer crystallization kinetics.)

A. Keller, M. Warner, and A. H. Windle, eds., *Self-order and Form in Polymeric Materials*, Chapman and Hall, London, 1995.

R. S. Porter and L. H. Wang, *Rev. Macromol. Chem. Phys.*, **C35** (1), 63 (1995).

J. M. Schultz, *Polymer Crystallization: The Development of Crystalline Order in Thermoplastic Polymers*, Oxford U. Press, Oxford, England, 2001.

H. Tadokoro, *Structure of Crystalline Polymers*, Wiley-Interscience, New York, 1979.

STUDY PROBLEMS

1. Based on the unit cell structure for cellulose I, calculate its theoretical crystal density. (See Figure 6.1.)

2. A firm uses crosslinked polyethylene to manufacture outdoor electrical cable insulation. They feel their products might be improved if they understood the kinetics of crystallization and the morphology of the final products better. Typical cooling rates approach 100°C/min.

How will you approach the problem? Since a good researcher usually has one or more possible experimental outcomes in mind to test against, what do you think your data will show?

3. A difficultly crystallizable high-molecular-weight polymer was finally crystallized in regime I. Compare and contrast the properties of the crystallized polymer and the amorphous polymer at the same temperature and pressure. Specifically, how do the densities, radii of gyration, and morphology via optical microscopy differ?

4. Wood and wool are based on renewable-resources, called "green" materials today. What is happening in either one of these fields today, bearing on polymer science? Look up one or more papers or patents from 2004–

present. Please provide the full reference. Write a brief summary of your findings.

5. Polymers are supposed to consist of long chains, yet the unit cell, by X-ray studies, is about the same size as those of ordinary molecules, containing only relatively few atoms. How can this be?

6. Compare and contrast the Avrami, Keith–Padden, and Hoffman theories of crystallization.

7. Note the volume–temperature data for polyethylene in Figure 6.2. What are the volume coefficients of expansion of the melt and two crystalline samples? Why are they different?

8. Given the unit cell structure of polyethylene (Figure 6.5), compute the theoretical density of the 100% crystalline product. [*Hint:* see Table 6.2.]

9. Equation (6.43) shows corrections to the melting point depressing due to mismatch of molar volumes of solvent and mer. Should the corresponding equations for co-polymers and finite molecular weight be corrected similarly? See equations (6.41) and (6.42). If so, derive suitable relations.

10. Poly(decamethylene adipate), $\left[\!-CO-(CH_2)_4-CO-O-(CH_2)_{10}-O-\!\right]_n$, density = 0.99 g/cm^3, was mixed with various quantities of dimethylformamide, $(CH_3)_2NCHO$, d = 0.9445 g/cm^3 and the melting temperatures observed:[†]

v_1	T_f, °C
0.078	72.5
0.202	66.5
0.422	61.5
0.603	57.5

(a) What is the melting temperature of the pure polymer? (b) What is the heat of fusion of poly(decamethylene adipate)?

11. What spherulite radius can be calculated from Figure 6.16?

12. Devise an NMR experiment to study chain folding in (a) cellulose triacetate, (b) isotactic polystyrene, and (c) transpolyisoprene (Gutta percha). [*Hint:* What chemical modifications, if any, are required?]

13. Compare infrared, NMR, and SANS results on chain folding in single crystals. Can you devise a new experiment to investigate the problem?

14. Read an original paper published in the last 12 months on crystalline polymer behavior or theory, and write a brief report on it in your own words. Cite the authors and exact reference. Does it support the present text? Add new ideas or data? Contradict present theories or ideas?

[†] L. Mandelkern, R. R. Garrett, and P. J. Flory, *J. Am. Chem. Soc.*, **74**, 3949 (1952).

15. Single crystals of polyethylene are grown from different molecular weight materials from $M = 2000$ to 5×10^7 g/mol. The crystals are all 150 Å thick, with adjacent reentry and superfolding after each 20 stems. How does R_g depend on M in this region? Plot the results. What dependence of R_g on M is predicted as M goes to infinity?

16. The lattice constants of orthorhombic polyethylene have been determined as a function of temperature:†

	Lattice Constants, Å		
T, K	a	b	c
4	7.121	4.851	2.548
77	7.155	4.899	2.5473
293	7.399	4.946	2.543
303	7.414	4.942	2.5473

(a) What is the theoretical volume coefficient of expansion of 100% crystalline polyethylene? (The volume coefficient of expansion is given by

$$\alpha = \frac{1}{V}\left(\frac{\partial V}{\partial T}\right)_P$$

where the change in volume V is measured as a function of temperature T at constant pressure P.) (b) Why are the c-axis lattice constants substantially independent of the temperature?

17. You were handed bottles of dimethylcyclooctadiene and *cis*-polyisoprene, but they became mixed up. What experiments would you perform to identify them?

18. Toothbrushes are available with soft, medium, and hard bristles, generally made of polyamides. (a) Propose at least two distinctly different ways by which the bristles could be controlled to provide soft, medium, or hard performance. (b) If you had bristles from different toothbrushes, what tests or experiments would you perform to determine the ways that were actually used to control the hardness? (c) Why polyamide?

19. You are handed a glass fiber reinforced sample. "Do not damage this very valuable material," your boss asks, "but we need to know what is in it!" What nondestructive experiments would you perform in situ to determine the chemical structure, crystallinity (if any), orientation, and so on, in the material?

20. Both experiment and computer simulation analyses show that polymer chains like to crystallize in lamella. However, the directions of the chain

† H. Tadokoro, *Structure of Crystalline Polymers*, Wiley-Interscience, New York, 1979, p. 375.

folds appear to be irregular, as is the crystal's top and bottom surfaces. Using your imagination, (a) invent at least one different molecular model for lamella formation or growth, and (b) suggest an experiment to examine your novel idea.

8

GLASS–RUBBER TRANSITION BEHAVIOR

The state of a polymer depends on the temperature and on the time allotted to the experiment. While this is equally true for semicrystalline and amorphous polymers, although in different ways, the discussion in this chapter centers on amorphous materials.

At low enough temperatures, all amorphous polymers are stiff and glassy. This is the glassy state, sometimes called the vitreous state, especially for inorganic materials. On warming, the polymers soften in a characteristic temperature range known as the glass–rubber transition region. Here, the polymers behave in a leathery manner. The importance of the glass transition in polymer science was stated by Eisenberg:[†] "The glass transition is perhaps the most important single parameter that determines the application of many non-crystalline polymers now available."

The glass transition is named after the softening of ordinary glass. On a molecular basis, the glass transition involves the onset of long-range coordinated molecular motion, the beginning of reptation. The glass transition is a second-order transition. Rather than dicontinuities in enthalpy and volume, their temperature derivatives, heat capacity, and coefficients of expansion shift. By difference, melting and boiling are first-order transitions, exhibiting discontinuities in enthalpy and volume, with heats of transition.

[†] In J. E. Mark, A. Eisenberg, W. W. Graessley, L. Mandelkern, E. T. Samulski, J. L. Koenig, and G. D. Wignall, *Physical Properties of Polymers*, 2nd ed., American Chemical Society, Washington, DC, 1993.

Introduction to Physical Polymer Science, by L.H. Sperling
ISBN 0-471-70606-X

For amorphous polymers, the glass transition temperature, T_g, constitutes their most important mechanical property. In fact, upon synthesis of a new polymer, the glass transition temperature is among the first properties measured. This chapter describes the behavior of amorphous polymers in the glass transition range, emphasizing the onset of molecular motions associated with the transition. Before beginning the main topic, two introductory sections are presented. The first defines a number of mechanical terms that will be needed, and the second describes the mechanical spectrum encountered as a polymer's temperature is raised.

8.1 SIMPLE MECHANICAL RELATIONSHIPS

Terms such as "glassy," "rubbery," and "viscous" imply a knowledge of simple material mechanical relationships. Although such information is usually obtained by the student in elementary courses in physics or mechanics, the basic relationships are reviewed here because they are used throughout the text. More detailed treatments are available (1–4).

8.1.1 Modulus

8.1.1.1 Young's Modulus Hook's law assumes perfect elasticity in a material body. Young's modulus, E, may be written

$$E = \sigma / \varepsilon \tag{8.1}$$

where σ and ε represent the tensile (normal) stress and strain, respectively. Young's modulus is a fundamental measure of the stiffness of the material. The higher its value, the more resistant the material is to being stretched.

The tensile stress is defined in terms of force per unit area. If the sample's initial length is L_0 and its final length is L, then the strain is $\varepsilon = (L - L_0)/L_0$[†] (see Figure 8.1).

The forces and subsequent work terms have some simple examples. Consider a postage stamp, which weighs about 1 g and is about 1 cm × 1 cm in size. It requires about 1 dyne of force to lift it from the horizontal position to the vertical position, as in turning it over. One dyne of force through 1 cm gives 1 erg, the amount of work done. Modulus is usually reported in dynes/cm^2, in terms of force per unit area. Frequently the Pascal unit of modulus is used, 10 dynes/cm^2 = 1 Pascal.

8.1.1.2 Shear Modulus Instead of elongating (or compressing!) a sample, it may be subjected to various shearing or twisting motions (see Figure 8.1).

[†] In more graphic language, the amount of stress applied is measured by the amount of grunting the investigator does, and the strain is measured by the sample's groaning.

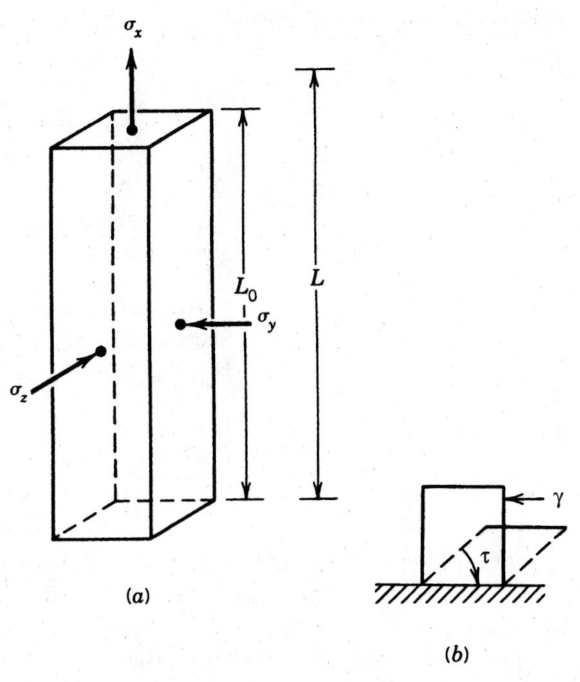

Figure 8.1 Mechanical deformation of solid bodies. (*a*) Triaxial stresses on a material body undergoing elongation. (*b*) Simple shear deformation.

The ratio of the shear stress, τ, to the shear strain, γ, defines the shear modulus, G:

$$G = \frac{\tau}{\gamma} \tag{8.2}$$

These and other mechanical terms are summarized in Table 8.1. (Note that different authors often use different symbols.)

8.1.2 Newton's Law

The equation for a perfect liquid exhibiting a shear viscosity, η, may be written

$$\eta = \tau/(d\gamma/dt) \tag{8.3}$$

where τ and γ represent the shear stress and strain, respectively, and t is the time. For simple liquids such as water or toluene, equation (8.3) reasonably describes their viscosity, especially at low shear rates. For larger values of η, flow is slower at constant shear stress. While neither equation (8.1) nor equation (8.3) accurately describes polymer behavior, they represent two important limiting cases.

Table 8.1 Some mechanical terms

Term	Definition
σ	Normal stress
ε	Normal strain
γ	Shear strain
τ	Shear stress
E	Young's modulus
G	Shear modulus
B	Bulk modulus
D	Tensile compliance
J	Shear compliance
v	Poisson's ratio
η	Shear viscosity
λ	Elongational viscosity
β	Compressibility

The basic definition of viscosity should be considered in terms of equation (8.3). Consider two 1-cm^2 planes 1 cm apart imbedded in a liquid. If it takes 1 dyne of force to move one of the planes 1 cm/s relative to the other in a shearing motion, the liquid has a viscosity of 1 poise. Viscosity is also expressed in pascal-seconds, with $1\,\text{Pa}\cdot\text{s} = 0.1$ poise.

8.1.3 Poisson's Ratio

When a material body is elongated (or undergoes other modes of deformation), in general, the volume changes, usually increasing as elongational (normal) strains are applied. Poisson's ratio, v, is defined as

$$-v\varepsilon_x = \varepsilon_y = \varepsilon_z \tag{8.4}$$

for very small strains, where the strain ε_x is applied in the x direction and the strains ε_y and ε_z are responses in the y and z directions, respectively (see Figure 8.1). Table 8.2 summarizes the behavior of v under several circumstances.

For analytical purposes, Poisson's ratio is defined on the differential scale. If V represents the volume,

$$V = xyz \tag{8.5}$$

Then

$$\frac{d\ln V}{d\ln x} = \frac{d\ln x}{d\ln x} + \frac{d\ln y}{d\ln x} + \frac{d\ln z}{d\ln x} \tag{8.6}$$

and

Table 8.2 Values of Poisson's ratio

Value	Interpretation
0.5	No volume change during stretch
0.0	No lateral contraction
0.49–0.499	Typical values for elastomers
0.20–0.40	Typical values for plastics

$$-\frac{d\ln y}{d\ln x} = -\frac{d\ln z}{d\ln x} = v = -\frac{dy}{y_0}\bigg/\frac{dx}{x_0} \tag{8.7}$$

Since $d\ln x/d\ln x = 1$, for no volume change $v = 0.5$ (see Table 8.2).

On extension, plastics exhibit considerable volume increases, as illustrated by the values of v in Table 8.2. The physical separation of atoms provides a major mechanism for energy storage and short-range elasticity.

Poisson's ratio is only useful for very small strains. Poisson's ratio was originally developed for calculations involving metals, concrete, and other materials with limited extensibility. The approximations built into the theory make rubber elasticity results unrealistic. Glazebrook (5) presents a more general treatment of Poisson's ratio.

8.1.4 The Bulk Modulus and Compressibility

The bulk modulus, B, is defined as

$$B = -V\left(\frac{\partial P}{\partial V}\right)_T \tag{8.8}$$

where P is the hydrostatic pressure. Normally a body shrinks in volume on being exposed to increasing external pressures, so the term $(\partial P/\partial V)_T$ is negative.

The inverse of the bulk modulus is the compressibility, β,

$$\beta = \frac{1}{B} \tag{8.9}$$

which is strictly true only for a solid or liquid in which there is no time-dependent response. Bulk compression usually does not involve long-range conformational changes but rather a forcing together of the chain atoms. Of course, materials ordinarily exist under a hydrostatic pressure of 1 atm (1 bar) at sea level.

8.1.5 Relationships among *E*, *G*, *B*, and *v*

A three-way equation may be written relating the four basic mechanical properties:

$$E = 3B(1-2v) = 2(1+v)G \tag{8.10}$$

Any two of these properties may be varied independently, and conversely, knowledge of any two defines the other two. As an especially important relationship, when $v \cong 0.5$,

$$E \cong 3G \tag{8.11}$$

which defines the relationship between E and G to a good approximation for elastomers.

Equation (8.10) can also be used to evaluate Poisson's ratio for elastomers. Rearranging the two left-hand portions, we have

$$1-2v = \frac{E}{3B} = \frac{\beta E}{3} \tag{8.12}$$

Because the quantity $1 - 2v$ is close to zero for elastomers (but cannot be exactly so), exact evaluation of v depends on the evaluation of the right-hand side of equation (8.12). Values in the literature for elastomers vary from 0.49 to 0.49996 (2).

Thus, in contrast to plastics, separation of the atoms in elastomers plays only a small role in the internal storage of energy. Instead, conformational changes in the chains come to the fore, the main subject of Chapter 9.

8.1.6 Compliance versus Modulus

If the modulus is a measure of the stiffness or hardness of an object, its compliance is a measure of softness. In regions far from transitions, the elongational compliance, D, is defined as

$$D \simeq \frac{1}{E} \tag{8.13}$$

For regions in or near transitions, the relationship is more complex. Likewise, the shear compliance, J, is defined as $1/G$. Ferry has reviewed this topic (2).

8.1.7 Numerical Values for *E*

Before proceeding with the description of the temperature behavior of polymers, it is of interest to establish some numerical values for Young's modulus (see Table 8.3). Polystyrene represents a typical glassy polymer at room temperature. It is about 40 times as soft as elemental copper, however. Soft rubber, exemplified by such materials as rubber bands, is nearly 1000 times softer still. Perhaps the most important observation from Table 8.3 is that the modulus

Table 8.3 Numerical values of Young's modulus

Material	E (dyne/cm^2)	E (Pa)
Copper	1.2×10^{12}	1.2×10^{11}
Polystyrene	3×10^{10}	3×10^{9}
Soft rubber	2×10^{7}	2×10^{6}

varies over wide ranges, leading to the wide use of logarithmic plots to describe the variation of modulus with temperature or time.

8.1.8 Storage and Loss Moduli

The quantities E and G refer to quasistatic measurements. When cyclical or repetitive motions of stress and strain are involved, it is more convenient to talk about dynamic mechanical moduli. The complex Young's modulus has the formal definition

$$E^* = E' + iE'' \tag{8.14}$$

where E' is the storage modulus and E'' is the loss modulus. Note that $E = |E^*|$. The quantity i represents the square root of minus one. The storage modulus is a measure of the energy stored elastically during deformation, and the loss modulus is a measure of the energy converted to heat. Similar definitions hold for G^*, D^*, J^*, and other mechanical quantities.

8.1.9 Elongational Viscosity

The elongational viscosity, λ, sometimes called the extension viscosity, refers to the thinning down of a column of liquid as it is being stretched. A simple example is the narrowing of a stream of water falling under gravity from a faucet. For a Newtonian fluid, $\lambda = 3\eta$ (6). For entangled polymer melts, λ may be many times η. An important example in polymer science and engineering is in the spinning of fibers; see Section 6.10.

8.2 FIVE REGIONS OF VISCOELASTIC BEHAVIOR

Viscoelastic materials simultaneously exhibit a combination of elastic and viscous behavior. While all substances are viscoelastic to some degree, this behavior is especially prominent in polymers. Generally, viscoelasticity refers to both the time and temperature dependence of mechanical behavior.

The states of matter of low-molecular-weight compounds are well known: crystalline, liquid, and gaseous. The first-order transitions that separate these states are equally well known: melting and boiling. Another well-known first-order transition is the crystalline–crystalline transition, in which a compound changes from one crystalline form to another.

By contrast, no high-molecular-weight polymer vaporizes to a gaseous state; all decompose before the boiling point. In addition no high-molecular-weight polymer attains a totally crystalline structure, except in the single-crystal state (see Section 6.4.2).

In fact many important polymers do not crystallize at all but form glasses at low temperatures. At higher temperatures they form viscous liquids. The transition that separates the glassy state from the viscous state is known as the glass–rubber transition. According to theories to be developed later, this transition attains the properties of a second-order transition at very slow rates of heating or cooling.

Before entering into a detailed discussion of the glass transition, the five regions of viscoelastic behavior are briefly discussed to provide a broader picture of the temperature dependence of polymer properties. In the following, quasi-static measurements of the modulus at constant time, perhaps 10 or 100 s, and the temperature being raised 1°C/min will be assumed.

8.2.1 The Glassy Region

The five regions of viscoelastic behavior for linear amorphous polymers (3,7–9) are shown in Figure 8.2. In region 1 the polymer is glassy and frequently brittle. Typical examples at room temperature include polystyrene (plastic) drinking cups and poly(methyl methacrylate) (Plexiglas® sheets).

Young's modulus for glassy polymers just below the glass transition temperature is surprisingly constant over a wide range of polymers, having the

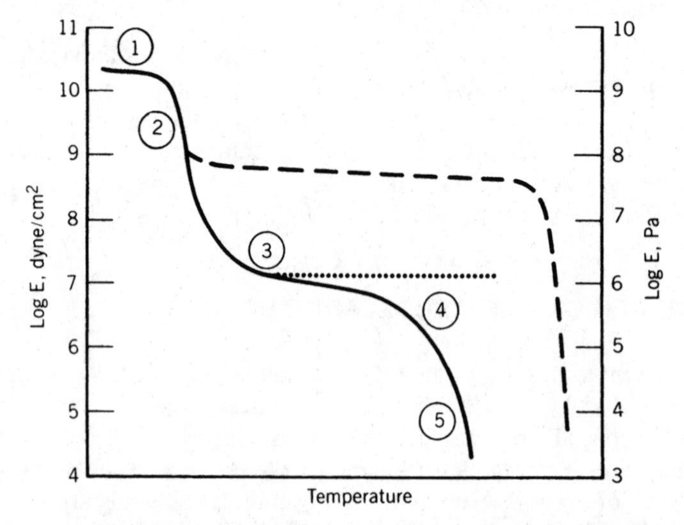

Figure 8.2 Five regions of viscoelastic behavior for a linear, amorphous polymer. Also illustrated are effects of crystallinity (dashed line) and cross-linking (dotted line).

value of approximately 3×10^{10} dynes/cm^2 (3×10^9 Pa). In the glassy state, molecular motions are largely restricted to vibrations and short-range rotational motions.

A classical explanation for the constant modulus values in the glassy state starts with the Lennard-Jones potential, describing the energy of interaction between a pair of isolated molecules. From this, the molar lattice energy between polymer segments was calculated. The bulk modulus, B, was calculated in terms of the cohesive energy density, CED, which represents the energy theoretically required to move a detached segment into the vapor phase. This, in turn, is related to the square of the solubility parameter:

$$B = 8.04(\text{CED}) = 8.04\delta^2 \tag{8.15}$$

The factor 8.04 arises from Lennard-Jones considerations (2,3).

Using polystyrene as an example, its CED is 83 cal/cm^3 (3.47×10^9 ergs/cm^3). Its Poisson ratio is approximately 0.30. From equation (8.10), $E \simeq 1.2B$, and hence

$$E \simeq 9.6(\text{CED}) \tag{8.16}$$

For polystyrene, a value of $E = 3.3 \times 10^{10}$ dynes/cm^2 (3.3×10^9 Pa) is calculated, which is surprisingly close to the value obtained via modulus measurements, 3×10^9 Pa. It should be noted that many hydrocarbon and not-too-polar polymers have CED values within a factor of 2 of the value for polystyrene.

A more modern explanation of the glassy modulus of polymers starts with a consideration of the carbon–carbon bonding force fields adopted for the explanation of vibrational frequencies (10). For perfectly oriented polyethylene, a value of $E = 3.4 \times 10^{11}$ Pa was calculated, with a slightly lower value, 1.6×10^{11} Pa for polytetrafluoroethylene.

Using scanning force microscopy, Du et al. (4) determined Young's modulus of polyethylene single crystals in the chain direction to be 1.7×10^{11} Pa. Ultradrawn polyethylene fibers also have an experimental Young's modulus of 1.7×10^{11} Pa; see Table 11.2. These slightly lower values may be attributed to less than perfect orientation of the chains.

For unoriented, amorphous glassy polymers, the theoretical values should be divided by three. Further consideration is required for the somewhat greater mer cross-sectional area of polymers such as polystyrene or poly(methyl methacrylate), since the side groups should not contribute to the modulus according to this theory. Thus values of approximately 3×10^9 Pa may again be estimated.

Interestingly, both the CED and the vibrational force field theories of glassy polymer modulus arrive at very similar values, although their approaches are entirely different. Further reductions in the experimental modulus may be imagined as caused by free volume, end groups, and so on. It may be that the two theoretical values should better be considered as *additive*, since intermolecular attractive forces and carbon–carbon covalent bonding forces both play roles. This is yet to be determined experimentally.

8.2.2 The Glass Transition Region

Region 2 in Figure 8.2 is the glass transition region. Typically the modulus drops a factor of about 1000 in a 20 to 30°C range. The behavior of polymers in this region is best described as leathery, although a few degrees of temperature change will obviously affect the stiffness of the leather.

For quasi-static measurements such as illustrated in Figure 8.2, the glass transition temperature, T_g, is often taken at the maximum rate of turndown of the modulus at the elbow, where $E \cong 10^9$ Pa. Often the glass transition temperature is defined as the temperature where the thermal expansion coefficient (Section 8.3) undergoes a discontinuity. (Enthalpic and dynamic definitions are given in Section 8.2.9. Other, more precise definitions are given in Section 8.5.)

Qualitatively, the glass transition region can be interpreted as the onset of long-range, coordinated molecular motion. While only 1 to 4 chain atoms are involved in motions below the glass transition temperature, some 10 to 50 chain atoms attain sufficient thermal energy to move in a coordinated manner in the glass transition region (9,11–14) (see Table 8.4) (9,15). The number of chain atoms, 10–50, involved in the coordinated motions was deduced by observing the dependence of T_g on the molecular weight between cross-links, M_c. When T_g became relatively independent of M_c in a plot to T_g versus M_c, the number of chain atoms was counted. It should be emphasized that these results are tenuous at best.

The glass transition temperature itself varies widely with structure and other parameters, as will be discussed later. A few glass transition temperatures are shown in Table 8.4. Interestingly, the idealized map of polymer behavior shown in Figure 8.2 can be made to fit any of these polymers merely by moving the curve to the right or left, so that the glass transition temperature appears in the right place.

8.2.3 The Rubbery Plateau Region

Region 3 in Figure 8.2 is the rubbery plateau region. After the sharp drop that the modulus takes in the glass transition region, it becomes almost constant

Table 8.4 Glass transition parameters (9,15)

Polymer	T_g, °C	Number of Chain Atoms Involved
Poly(dimethyl siloxane)	−127	40
Poly(ethylene glycol)	−41	30
Polystyrene	+100	40–100
Polyisoprene	−73	30–40

again in the rubbery plateau region, with typical values of 2×10^7 dynes/cm^2 (2×10^6 Pa). In the rubbery plateau region, polymers exhibit long-range rubber elasticity, which means that the elastomer can be stretched, perhaps several hundred percent, and snap back to substantially its original length on being released.

Two cases in region 3 need to be distinguished:

1. The polymer is linear. In this case the modulus will drop off slowly, as indicated in Figure 8.2. The width of the plateau is governed primarily by the molecular weight of the polymer; the higher the molecular weight, the longer is the plateau (see Figure 8.3) (16).

   An interesting example of such a material is unvulcanized natural rubber. When Columbus came to America (17), he found the American Indians playing ball with natural rubber. This product, a linear polymer of very high molecular weight, retains its shape for short durations of time. However, on standing overnight, it creeps, first forming a flat spot on the bottom, and eventually flattening out like a pancake. (See Section 9.2 and Chapter 10.)
2. The polymer is cross-linked. In this case the dotted line in Figure 8.2 is followed, and improved rubber elasticity is observed, with the creep portion suppressed. The dotted line follows the equation $E = 3nRT$, where n is the number of active chain segments in the network and RT represents the gas constant times the temperature; see equation (9.36). An example of a cross-linked polymer above its glass transition temperature obeying this relationship is the ordinary rubber band. Cross-linked elastomers and rubber elasticity relationships are the primary subjects of Chapter 9.

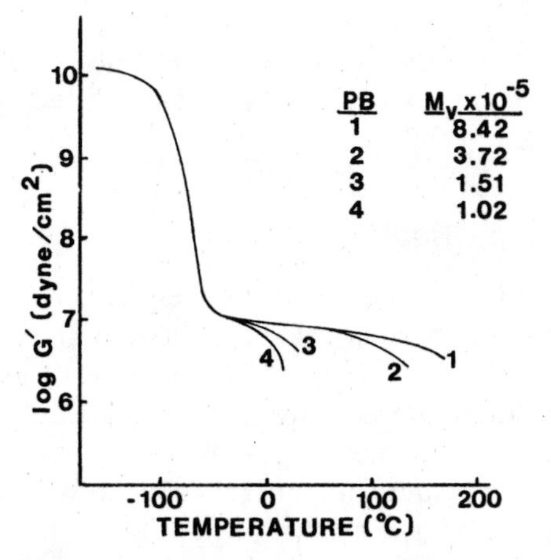

Figure 8.3 Effect of molecular weight on length of plateau (16). PB = polybutadiene.

The rapid, coordinated molecular motion in this region is governed by the principles of reptation and diffusion laid down in Section 5.4. Thus, when the elastomer is stretched, the chains deform with a series of rapid motions of the de Gennes type. The model must be altered slightly for cross-linked systems, for then the chain ends are bound at the cross-links. The motion is thought to become a more complex affair involving the several chain segments that are bound together.

So far the discussion has been limited to amorphous polymers. If a polymer is semicrystalline, the dashed line in Figure 8.2 is followed. The height of the plateau is governed by the degree of crystallinity. This is so because of two reasons: first, the crystalline regions tend to behave as a filler phase, and second, because the crystalline regions also behave as a type of physical cross-link, tying the chains together.

The crystalline plateau extends until the melting point of the polymer. The melting temperature, T_f, is always higher than T_g, T_g being from one-half to two-thirds of T_f on the absolute temperature scale (see Section 8.9.3 for further details).

8.2.4 The Rubbery Flow Region

As the temperature is raised past the rubbery plateau region for linear amorphous polymers, the rubbery flow region is reached—region 4. In this region the polymer is marked by both rubber elasticity and flow properties, depending on the time scale of the experiment. For short time scale experiments, the physical entanglements are not able to relax, and the material still behaves rubbery. For longer times, the increased molecular motion imparted by the increased temperature permits assemblies of chains to move in a coordinated manner (depending on the molecular weight), and hence to flow (see Figure 8.3) (16). An example of a material in the rubbery flow region is Silly Putty®, which can be bounced like a ball (short-time experiment) or pulled out like taffy (a much slower experiment).

It must be emphasized that region 4 does not occur for cross-linked polymers. In that case, region 3 remains in effect up to the decomposition temperature of the polymer (Figure 8.2).

8.2.5 The Liquid Flow Region

At still higher temperatures, the liquid flow region is reached—region 5. The polymer flows readily, often behaving like molasses. In this region, as an idealized limit, equation (8.3) is obeyed. The increased energy allotted to the chains permits them to reptate out through entanglements rapidly and flow as individual molecules.

For semicrystalline polymers, the modulus depends on the degree of crystallinity. The amorphous portions go through the glass transition, but the crystalline portion remains hard. Thus a composite modulus is found. The melting temperature is always above the glass transition temperature (see below). At the melting temperature the modulus drops rapidly to that of the corresponding amorphous material, now in the liquid flow region. It must be mentioned that modulus and viscosity are related through the molecular relaxation time, also discussed below.

8.2.6 Effect of Plasticizers

Polymers are frequently plasticized to "soften" them. These plasticizers are usually small, relatively nonvolatile molecules that dissolve in the polymer, separating the chains from each other and hence making reptation easier. In the context of Figure 8.2, the glass transition temperature is lowered, and the rubbery plateau modulus is lowered. If the polymer is semicrystalline, the plasticizer reduces the melting temperature and/or reduces the extent of crystallinity.

An example is poly(vinyl chloride), which has a T_g of +80°C. Properly plasticized, it has a T_g of about +20°C or lower, forming the familiar "vinyl." A typical plasticizer is dioctyl phthalate, with a solubility parameter of 8.7 $(\text{cal/cm}^3)^{1/2}$, fairly close to that of poly(vinyl chloride), 9.6 $(\text{cal/cm}^3)^{1/2}$; see Table 3.2. A significant parameter in this case is thought to be hydrogen bonding between the hydrogen on the same carbon as the chlorine and the ester group on the dioctyl phthalate. Other factors influencing the glass transition temperature are considered in Sections 8.6.3.2, 8.7, and 8.8.

8.2.7 Definitions of the Terms "Transition," "Relaxation," and "Dispersion"

The term "transition" refers to a change of state induced by changing the temperature or pressure.

The term "relaxation" refers to the time required to respond to a change in temperature or pressure. It also implies some measure of the molecular motion, especially near a transition condition. Frequently an external stress is present, permitting the relaxation to be measured. For example, one could state that $1/e$ (0.367) of the polymer chains respond to an applied stress in 10 s at the glass transition temperature, providing a simple molecular definition.

The term "dispersion" refers to the emission or absorption of energy—that is, a loss peak—at a transition. In practice, the literature sometimes uses these terms somewhat interchangeably.

8.2.8 Melt Viscosity Relationships near T_g

The discussion above emphasizes changes in the modulus with temperature. Equally large changes also take place in the viscosity of the polymer. In fact the term "glass transition" refers to the temperature in which ordinary glass softens and flows (18). (Glass is an inorganic polymer, held together with both

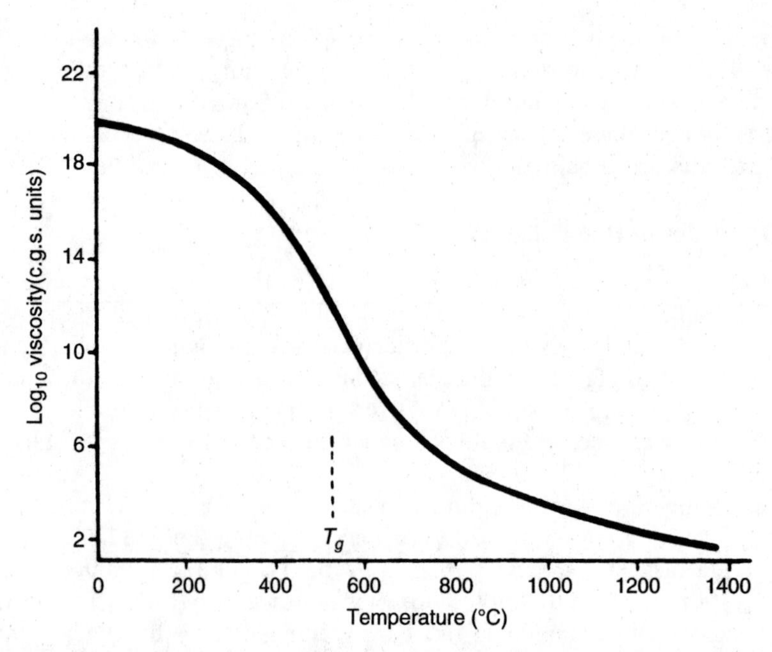

Figure 8.4 Viscosity–temperature relation of a soda–lime–silica glass (19). Soda–lime–silica glass is one of the commonly used glasses for windows and other items.

covalent —Si—O—Si— bonds and ionic bonds.) In this case viscoelastic region 3 is virtually absent.

The viscosity–temperature relationship of glass is shown in Figure 8.4 (19). A criterion sometimes used for T_g for both inorganic and organic polymers is the temperature at which the melt viscosity reaches a value of 1×10^{13} poises (1×10^{12} Pa · s) (20) on cooling.

8.2.9 Dynamic Mechanical Behavior through the Five Regions

The change in the modulus with temperature has already been introduced (see Section 8.2.2). More detail about the transitions is available through dynamic mechanical measurements, sometimes called dynamic mechanical spectroscopy (DMS) (see Section 8.1.8).

The shear storage modulus, G', and the shear loss modulus, G'', are the shear counter parts of E' and E''. While the temperature dependence of G' is similar to that of G, the quantity G'' behaves quite differently (see Figure 8.5) (9). The loss quantities behave somewhat like the absorption spectra in infrared spectroscopy, where the energy of the electromagnetic radiation is just sufficient to cause a portion of a molecule to go to a higher energy state. (Infrared spectrometry is usually carried out by varying the frequency of the radiation at constant temperature.) This exact analogue is frequently carried out for poly-

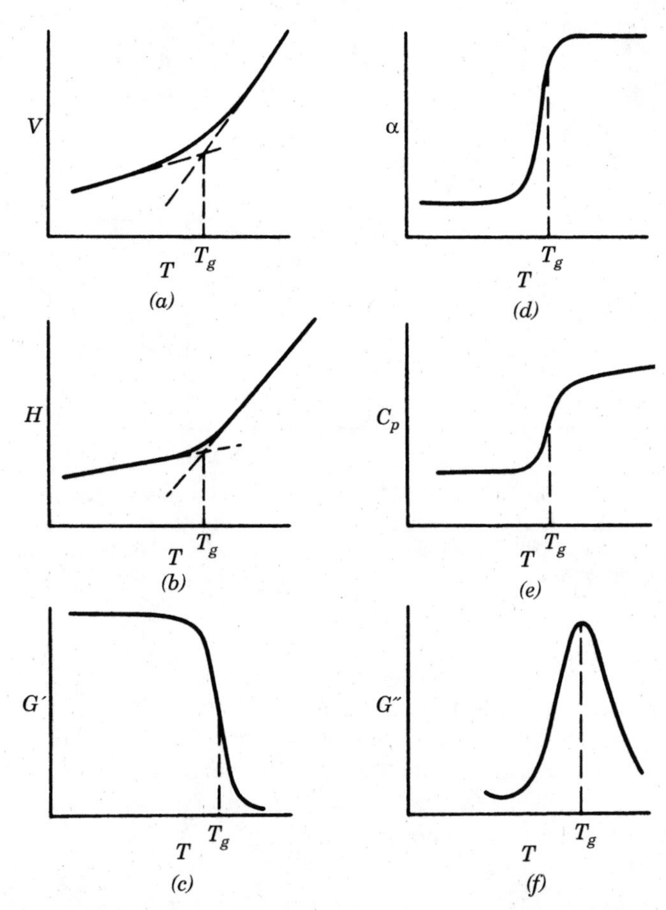

Figure 8.5 Idealized variations in volume, V, enthalpy, H, and storage shear modulus, G' as a function of temperature. Also shown are α, the volume coefficient of expansion, and C_p, the heat capacity, which are, respectively, the first derivatives of V and H with respect to temperature, and the loss shear modulus, G'' (9).

mers also (see below). Of course, in DMS, the energy is imparted by mechanical waves. The subject has been reviewed (18).

The reader will note that in Figure 8.5 there is no discontinuity in the V–T or H–T plots, only a change in slope. This is characteristic of a *second-order transition*. In metallurgy, a second-order transition of note is the Curie temperature of iron at 1043 K, where it looses its ferromagnetic capability (21). *First-order transitions* include melting, boiling, and changes in crystalline structure with temperature. Ordinary melting and boiling of water are common examples. These are characterized by discontinuities in V–T and H–T plots. (Note that Figure 6.2 illustrates such melting.)

Measurements by DMS refer to any one of several methods where the sample undergoes repeated small-amplitude strains in a cyclic manner.

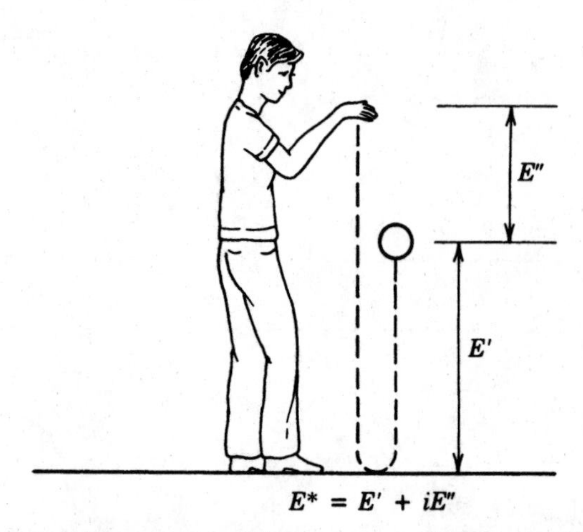

Figure 8.6 Simplified definition of E' and E'' (22). When a viscoelastic ball is dropped onto a perfectly elastic floor, it bounces back to a height E', a measure of the energy stored elastically during the collision between the ball and the floor. The quantity E'' represents the energy lost as heat during the collision.

Molecules perturbed in this way store a portion of the imparted energy elastically and dissipate a portion in the form of heat (1–4). The quantity E', Young's storage modulus, is a measure of the energy stored elastically, whereas E'', Young's loss modulus, is a measure of the energy lost as heat (see Figure 8.6) (22).

Another equation in wide use is

$$\frac{E''}{E'} = \tan\delta \tag{8.17}$$

where $\tan\delta$ is called the loss tangent, δ being the angle between the in-phase and out-of-phase components in the cyclic motion. Tan δ also goes through a series of maxima. The maxima in E'' and $\tan\delta$ are sometimes used as the definition of T_g. For the glass transition, the portion of the molecule excited may be from 10 to 50 atoms or more (see Table 8.4).

The dynamic mechanical behavior of an ideal polymer is illustrated in Figure 8.7. The storage modulus generally follows the behavior of Young's modulus as shown in Figure 8.2. In detail, it is subject to equation (8.14), so the storage modulus is slightly smaller, depending on the value of E''.

The quantities E'' and $\tan\delta$ display decided maxima at T_g, the $\tan\delta$ maximum appearing several degrees centigrade higher than the E'' peak. Also shown in Figure 8.7 is the β peak, generally involving a smaller number of atoms. The area under the peaks, especially when plotted with a linear y axis, is related to the chemical structure of the polymer (23). The width of the transition and shifts in the peak temperatures of E'' or $\tan\delta$ are sensitive guides

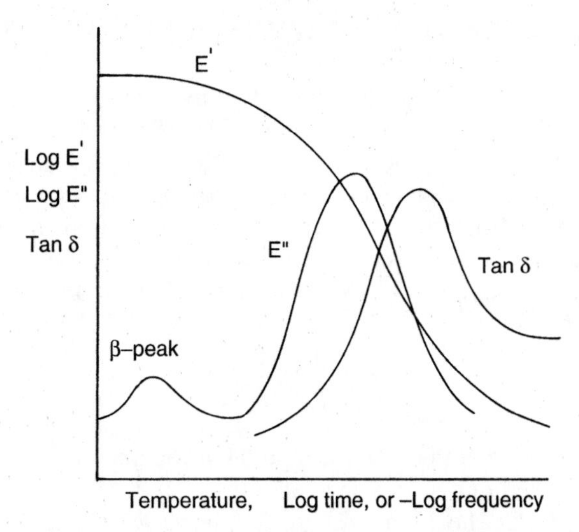

Figure 8.7 Schematic of dynamic mechanical behavior.

to the exact state of the material, molecular mixing in blends, and so on. The smaller transitions, such as the β peak, usually appear at other temperatures, such as the Schatzki crankshaft transition (see Section 8.4.1), or at higher temperatures (see Section 8.4.2).

Also included in the x axis of Figure 8.7 are the log time and the minus log frequency axes. As discussed in Sections 8.5.3 and 8.6.1.2, the minus log frequency dependence of the mechanical behavior takes the same form as temperature. As the imposed frequency on the sample is raised, it goes through the glass transition in much the same way as when the temperature is lowered. The theory of the glass transition will be discussed in Section 8.6.

For small stresses and strains, if both the elastic deformation at equilibrium and the rate of viscous flow are simple functions of the stress, the polymer is said to exhibit linear viscoelasticity. While most of the material in this chapter refers to linear viscoelasticity, much of the material in Chapters 9 and 10 concerns nonlinear behavior.

A principal value of the loss quantities stems from their sensitivity to many other types of transitions besides the glass transition temperature. The maxima in E'', G'', tan δ, and so on, provide a convenient and reproducible measure of each transition's behavior from dynamic experiments. From an engineering point of view, the intensity of the loss quantities can be utilized in mechanical damping problems such as vibration control.

In the following section several types of instrumentation used to measure the glass transition and other transitions are discussed.

8.3 METHODS OF MEASURING TRANSITIONS IN POLYMERS

The glass transition and other transitions in polymers can be observed experimentally by measuring any one of several basic thermodynamic, physical, mechanical, or electrical properties as a function of temperature. Recall that in first-order transitions such as melting and boiling, there is a discontinuity in the volume–temperature plot. For second-order transitions such as the glass transition, a change in slope occurs, as illustrated in Figure 8.5 (9).

The volumetric coefficient of expansion, α, is defined as

$$\alpha = \frac{1}{V}\left(\frac{\partial V}{\partial T}\right)_p \tag{8.18}$$

where V is the volume of the material, and α has the units K^{-1}. While this quantity increases as the temperature increases through T_g, it usually changes over a range of 10–30°C. The elbow shown in Figure 8.5*a* is not sharp. Similar changes occur in the enthalpy, H, and the heat capacity at constant pressure, C_p (Figure 8.5*b* and *e*, respectively).

8.3.1 Dilatometry Studies

There are two ways of characterizing polymers via dilatometry. The most obvious is volume–temperature measurements (24), where the polymer is confined by a liquid and the change in volume is recorded as the temperature is raised (see Figure 8.8). The usual confining liquid is mercury, since it does not swell organic polymers and has no transition of its own through most of the temperature range of interest. In an apparatus such as is shown in Figure 8.8, outgassing is required.

The results may be plotted as specific volume versus temperature (see Figure 8.9) (24). Since the elbow in volume–temperature studies is not sharp (all measurements of T_g show a dispersion of some 20–30°C), the two straight lines below and above the transition are extrapolated until they meet; that point is usually taken as T_g. Dilatometric and other methods of measuring T_g are summarized in Table 8.5.

The straight lines above and below T_g in Figure 8.9, of course, yield the volumetric coefficient of expansion, α. The linear coefficient of expansion, β, can also be employed (25). Note that $\alpha = 3\beta$.

Dilatometric data agree well with modulus–temperature studies, especially if the heating rates and/or length of times between measurements are controlled. (Raising the temperature 1°C/min roughly corresponds to a 10 s mechanical measurement.) Besides being a direct measure of T_g, dilatometric studies provide free volume information, of use in theoretical studies of the glass transition phenomenon (see Section 8.6.1).

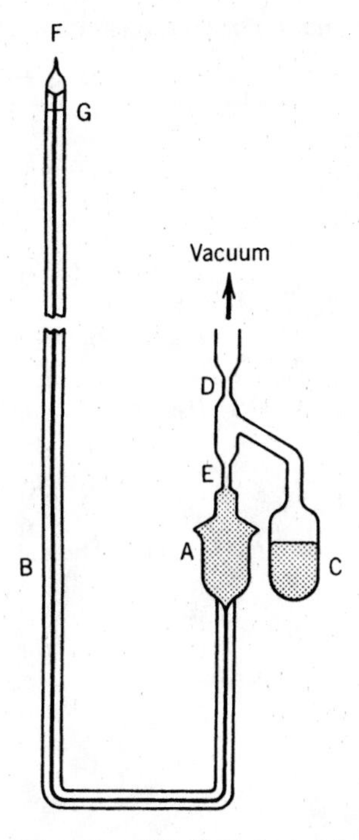

Figure 8.8 A mercury-based dilatometer (24). Bulb *A* contains the polymer (about 1 g), capillary *B* is for recording volume changes (Hg + polymer), *G* is a capillary for calibration, sealed at point *F*. After packing bulb *A*, the inlet is constricted at *E*, *C* contains weighed mercury to fill all dead space, and *D* is a second constriction.

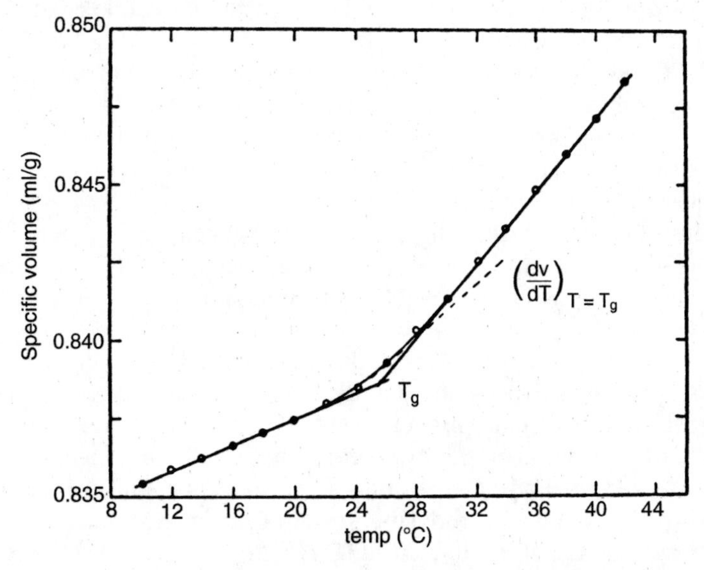

Figure 8.9 Dilatometric studies on branched poly(vinyl acetate) (24).

Table 8.5 Methods of measuring the glass transition and thermal properties

Method	Representative Instrumentation
Dilatometry	
Volume-temperature	Polymer confined by mercury (home made)
Linear expansivity	TMS + computer
Thermal	
DSC	Modulated DSC 2920 (TA Instruments)
DTA	DuPont 900
Calorimetric, C_p	Perkin-Elmer modulated DSC, Pyris 1
Mechanical	
Static	Gehman, Clash-Berg
Dynamic	Rheometrics RDA2 (Rheometrics Scientific)
Torsional braid analysis	TBA (Plastics Analysis Instruments)
Dielectric and magnetic	
Dielectric loss	DuPont Dielectric Analyzer, DEA 2920
Broad-line NMR	Joel JNH 3H60 Spectrometer
Melt viscosity	Weissenberg Rheogoniometer
TGA	HiRes TGA 2950 (TA Instruments)

8.3.2 Thermal Methods

Two closely related methods dominate the field—the older method, differential thermal analysis (DTA), and the newer method, differential scanning calorimetry (DSC). Both methods yield peaks relating to endothermic and exothermic transitions and show changes in heat capacity. The DSC method also yields quantitative information relating to the enthalpic changes in the polymer (26–28) (see Table 8.5).

The DSC method uses a servo system to supply energy at a varying rate to the sample and the reference, so that the temperatures of the two stay equal. The DSC output plots energy supplied against average temperature. By this improved method, the areas under the peaks can be directly related to the enthalpic changes quantitatively.

As illustrated by Figure 8.10 (29), T_g can be taken as the temperature at which one-half of the increase in the heat capacity, ΔC_p, has occurred. The increase in ΔC_p is associated with the increased molecular motion of the polymer.

Figure 8.10 shows a hysteresis peak associated with the glass transition. Such hysteresis peaks appear frequently, but not all the time. They are most often associated with some physical relaxation, such as residual orientation. On repeated measurement, the hysteresis peak usually disappears.

An improved method of separating a transient phenomenon such as the hysteresis peak from the reproducible result of the change in heat capacity is obtained *via* the use of modulated DSC (25,26); see Table 8.5. Here, a sine wave is imposed on the temperature ramp. A real-time computer analysis

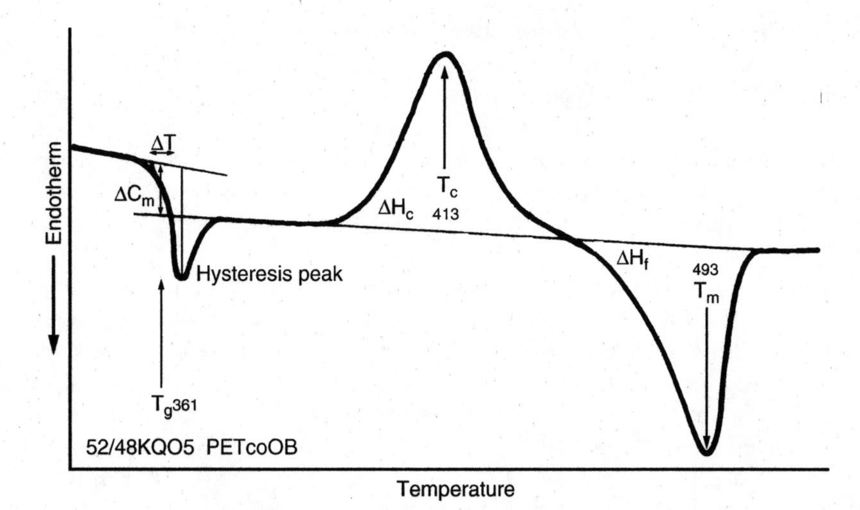

Figure 8.10 Example of differential scanning calorimetry trace of poly(ethylene terephthalate-*stat-p*-oxbenzoate), quenched, reheated, and cooled at 0.5 K/min through the glass transition, and reheated for measurement at 10 K/min (29). T_g is taken at the temperature at which half the increase in heat capacity has occurred. The width of the glass transition is indicated by ΔT. Note that ΔH_c and ΔH_f are equal in magnitude but opposite in sign.

allows a plot of not only the whole data but also its transient and reproducible components.

Figure 8.10 also illustrates the crystalline and melting behavior of this material. Due to the thermal treatment of the sample, it is amorphous before the last heating. Above the glass transition temperature, the increased molecular motion allows the sample to crystallize. At some higher temperature, the polymer melts. The heats of crystallization, ΔH_c, and the heat of fusion, ΔH_f, determined by the areas under the curves must be equal, of course.

This type of difficultly crystallizable copolymer of poly(ethylene terephthalate), is widely used in today's two-liter soda pop bottles. At room temperature the polymer is in its glassy state, with some order but little actual cystallinity. Besides being transparent and mechanically tough, such copolymers have very low permeability to both oxygen and carbon dioxide; see Table 4.5.

A related technique, thermogravimetric analysis (TGA), must be introduced at this point. In using TGA, the weight of the sample is recorded continuously as the temperature is raised. Volatilization, dehydration, oxidation, and other chemical reactions can easily be recorded, but the simple transitions are missed, as no weight changes occur.

8.3.3 Mechanical Methods

Since the very notion of the glass–rubber transition stems from a softening behavior, the mechanical methods provide the most direct determination of the transition temperature. Two fundamental types of measurement prevail—the static or quasistatic methods, and the dynamic methods.

Results of the static type of measurement have already been shown in Figure 8.2. For amorphous polymers and many types of semicrystalline polymers in which the crystallinity is not too high, stress relaxation, Gehman, and/or Glash–Berg instrumentation provide rapid and inexpensive scans of the temperature behavior of new polymers before going on to more complex methods.

Several instruments are employed to measure the dynamic mechanical spectroscopy (DMS) behavior (see Table 8.5). The Rheovibron (30) requires a sample that is self-supporting and that yields absolute values of the storage modulus and tan δ. The value of E'' is calculated by equation (8.17). Typical data are shown in Figure 8.11 (31). Although the instrument operates at several fixed frequencies, 110 Hz is most often employed. The sample size is about that of a paper match stick. This method provides excellent results with thermoplastics (30) and preformed polymer networks (31).

An increasingly popular method for studying the mechanical spectra of all types of polymers, especially those that are not self-supporting, is torsional braid analysis (TBA). In this case the monomer, prepolymer, polymer solution, or melt is dipped onto a glass braid, which supports the sample. The braid is set into a torsional motion. The sinusoidal decay of the twisting action is recorded as a function of time as the temperature is changed (32–35). Because the braid acts as a support medium, the absolute magnitudes of the transitions are not obtained; only their temperatures and relative intensities are recorded. The TBA method appears to have largely replaced the torsional pendulum (33).

Figure 8.12 shows typical TBA data for cellulose triacetate. Also shown by way of comparison are DTA and TGA results. In Figure 8.12, p represents the

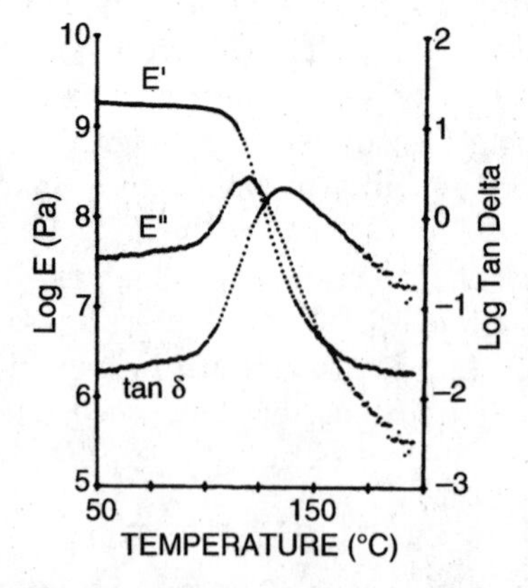

Figure 8.11 Dynamic mechanical spectroscopy on polystyrene cross-linked with 2% divinyl benzene. Data taken with a Rheovibron at 110 Hz. Note that the tan δ peak occurs at a slightly higher temperature than the E'' peak (31). Experiment run by J. J. Fay, Lehigh University.

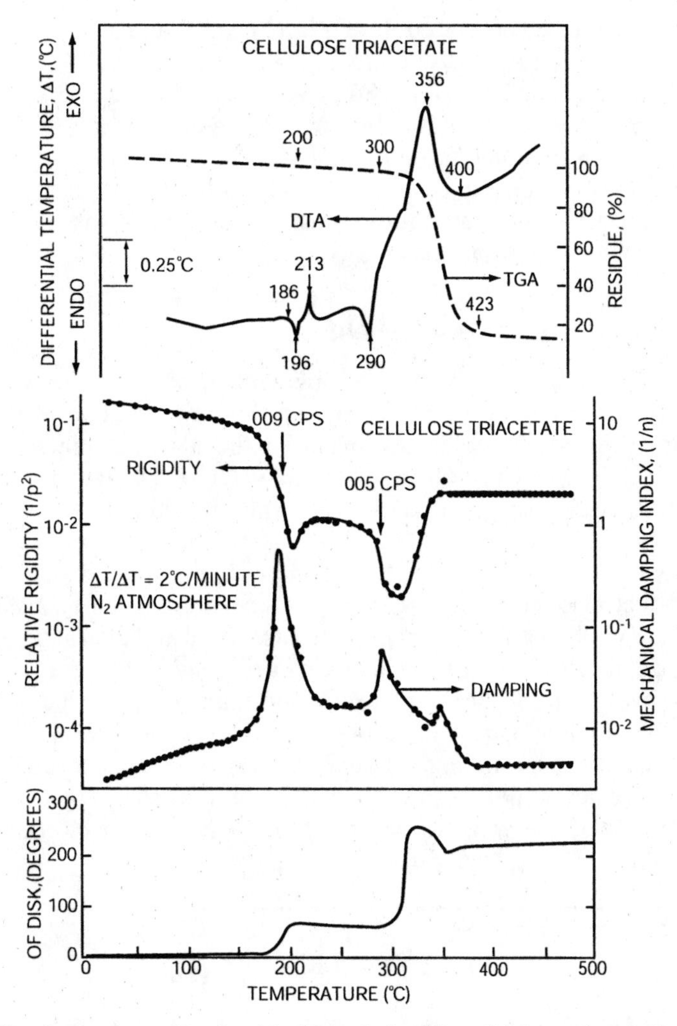

Figure 8.12 Comparison of torsional braid analysis, differential thermal analysis, and thermogravimetric analysis data for cellulose triacetate. The bottom figure shows the twisting of the sample in the absence of oscillations as a result of expansion or contraction of the sample at T_g and T_f (35).

period of oscillation—that is, the inverse of the natural frequency of the sample, braid, and attachments. The quantity $1/p^2$ is a measure of the stiffness of the system, proportional to the modulus. Since the modulus of the glass braid stays nearly constant through the range of measurement, all changes are representative of the polymer. The quantity n represents the number of oscillations required to reduce the angular amplitude, A, by a fixed ratio. In this case, $A_i/A_{i+n} = 20$ (34). By way of comparison, earlier measurements via volume–temperature had placed T_g at 172°C (36,37) and T_f at 307°C (36), in good agreement with the TBA results in Figure 8.12.

The DTA results in Figure 8.12 show T_f at 290°C while the TBA shows first an exothermic, then an endothermic decomposition at 356°C and 400°C, respectively, corresponding to the weight loss shown by the TGA study. Cellulose triacetate is known to have three second-order transitions (37); it is not clear whether the lower temperature transitions associated with the DTA plot represent these or other motions (38).

8.3.4 Dielectric and Magnetic Methods

As stated previously, part of the work performed on a sample will be converted irreversibly into random thermal motion by excitation of the appropriate molecular segments. In Section 8.3.3 the loss maxima so produced through mechanical means were used to characterize the glass transition. The two important electromagnetic methods for the characterization of transitions in polymers are dielectric loss (3) and broad-line nuclear magnetic resonance (NMR) (39–41).

The dielectric loss constant, ε'', or its associated tan δ can be measured by placing the sample between parallel plate capacitors and alternating the electric field. Polar groups on the polymer chain respond to the alternating field. When the average frequency of molecular motion equals the electric field frequency, absorption maxima will occur.

If the dielectric measurements are carried out at the same frequency as the DMS measurements, the transitions will occur at the same temperatures (see Figure 8.13) (41). The glass transition for the polytrifluorochloroethylene at

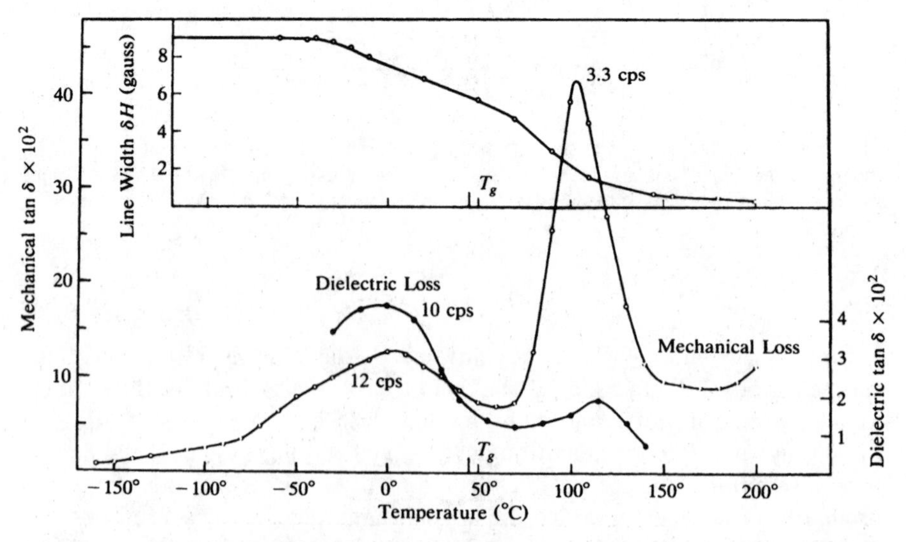

Figure 8.13 Mechanical and dielectric loss tangent tan δ and NMR absorption line width δH (maximum slope, in gauss) of polytrifluorochloroethylene (Kel-F) (41).

52°C shown is due to static measurements. The values at 100°C shown are close to those reported for dynamic measurements (15).

Broad-line NMR measurements depend on the fact that hydrogen nuclei, being simply protons, possess a magnetic moment and therefore precess about an imposed alternating magnetic field, especially at radio frequencies. Stronger interactions exist between the magnetic dipoles of different hydrogen nuclei in polymers below the glass transition temperature, resulting in a broad signal. As the chain segments become more mobile with increasing temperature through T_g, the distribution of proton orientations around a given nucleus becomes increasingly random, and the signal sharpens. The behavior of NMR spectra below and above T_g is illustrated in Figure 8.14 (39). The narrowing of the line width through the glass transition for polytrifluorochloroethylene is shown in the inset of Figure 8.13. A number of other methods of observing T_g are discussed by Boyer (9).

8.3.5 A Comparison of the Methods

All the methods of measuring T_g depend on either a basic property or some derived property. The principal ones have already been discussed.

Basic Property	Derived Property
Volume	Refractive index
Modulus	Penetrometry
Dielectric loss	Resistivity

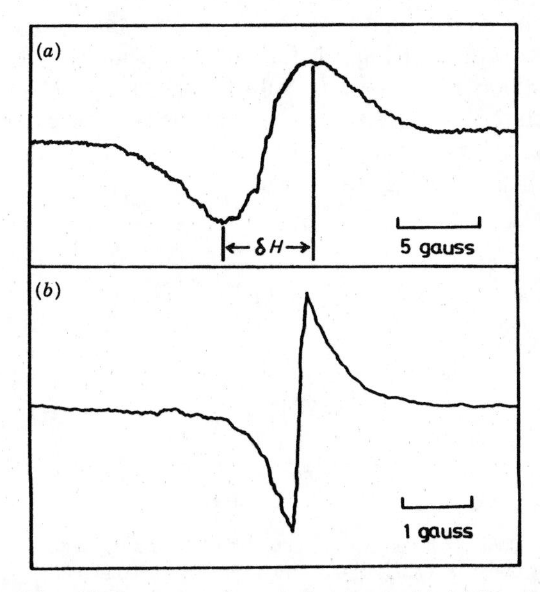

Figure 8.14 Broad-line NMR spectra of a cured epoxy resin. (*a*) Broad line at 291 K; (*b*) motionally narrowed line at 449 K (T_g + 39 K) (39).

From a practical point of view, since the derived property frequently represents the quantity of interest, it is measured rather than the basic property.

The methods most commonly used at the present time (9) include direct-recording DSC units, the Rheometric Scientific ARES System, and the torsional braid. The special value of the mechanical units lies in the fact that loss and storage moduli are frequently of prime engineering value. Thus the instrument supplies basic scientific information about the transitions while giving information about the damping and stiffness characteristics.

On the other hand, DSC supplies thermodynamic information about T_g. Of particular interest is the change in the heat capacity, which reflects fundamental changes in molecular motion. Thus values of C_p are of broad theoretical significance, as described in Section 8.6.

8.3.6 The Cole–Cole Plot

The Cole–Cole plot usually presents a loss term *vs.* a storage term as a function of frequency or time element (42). Figure 8.15 (43) illustrates a plot of Young's loss modulus against Young's storage modulus of an epoxy network as a function of frequency through the glass transition region. The epoxy was based on diglycidyl ether of bisphenol A and diaminodiphenyl methane, the latter serving as the cross-linker. Note the characteristic near-semicircular appearance of this plot.

Although the Cole–Cole plot was first introduced in the context of a dielectric relaxation spectrum, it helped discover that the molecular mechanism underlying both dielectric relaxation and stress relaxation are substantially identical (44). Figure 8.13 provides an illustration, with temperature instead of frequency. Specifically, the same molecular motions that generate a frequency dependence for the dielectric spectrum are also responsible for the relaxation of orientation in polymers above T_g. Subsequently the Cole–Cole type of plot has been applied to the linear viscoelastic mechanical properties of polymers, especially in the vicinity of the glass transition, including the dynamic compliance and dynamic viscosity functions.

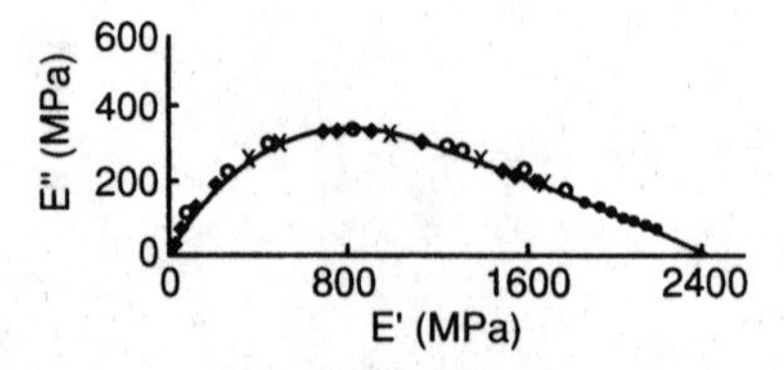

Figure 8.15 Cole–Cole plot for an epoxy network through the glass transition region. Usually, these plots illustrate dynamic mechanical behavior, where the storage quantity is plotted against the loss quantity. Inverted, nearly semicircular curves as shown are common.

8.4 OTHER TRANSITIONS AND RELAXATIONS

As the temperature of a polymer is lowered continuously, the sample may exhibit several second-order transitions. By custom, the glass transition is designated the α transition, and successively lower temperature transitions are called the $\beta, \gamma, \ldots$ transitions. One important second-order transition appears above T_g, designated the T_{ll} (liquid–liquid) transition. Of course, if the polymer is semicrystalline, it will also melt at a temperature above T_g.

8.4.1 The Schatzki Crankshaft Mechanism

8.4.1.1 Main-Chain Motions There appear to be two major mechanisms for transitions in the glassy state (45). For main-chain motions in hydrocarbon-based polymers such as polyethylene, the Schatzki crankshaft mechanism (46), Figure 8.16 (47), is thought to play an important role. Schatzki showed that eight —CH_2— units could be lined up so that the 1–2 bonds and the 7–8 bonds form a collinear axis. Then, given sufficient free volume, the intervening four —CH_2— units rotate more or less independently in the manner of an old-time automobile crankshaft. It is thought that at least four —CH_2— units in succession are required for this motion. The transition of polyethylene occurring near –120°C is thought to involve the Schatzki mechanism.

It is interesting to consider the basic motions possible for small hydrocarbon molecules by way of comparison. At very low temperatures, the CH_3— groups in ethane can only vibrate relative to the other. At about 90 K ethane undergoes a second-order transition as detected by NMR absorption (48), and the two CH_3— units begin to rotate freely, relative to one another. For propane and larger molecules, the number of motions becomes more complex (49), as now three-dimensional rotations come into play. One might imagine that *n*-octane itself might have the motion illustrated in Figure 8.16 as one of its basic energy absorbing modes.

8.4.1.2 Side-Chain Motions The above considers main-chain motions. Many polymers have considerable side-chain "foliage," and these groups can, of course, have their own motions.

A major difference between main-chain and side-chain motions is the toughness imparted to the polymer. Low-temperature main-chain motions act

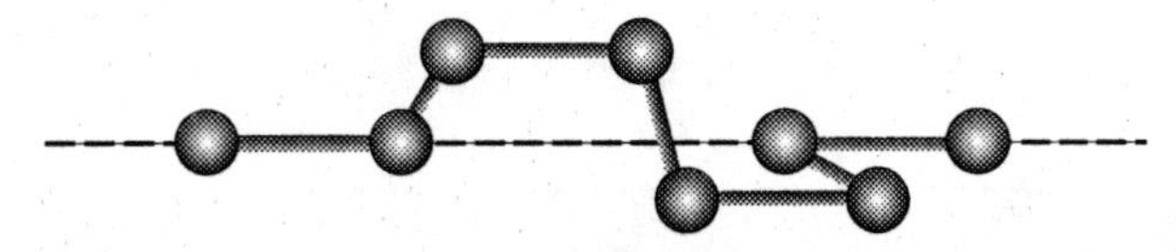

Figure 8.16 Schatzki's crankshaft motion (41) requires at least four —CH_2— groups in succession. As illustrated, for eight —CH_2— groups, bonds 1 and 7 are collinear and intervening —CH_2— units can rotate in the manner of a crankshaft (44).

to absorb energy much better than the equivalent side-chain motions, in the face of impact blows. When the main-chain motions absorb energy under these conditions, they tend to prevent main-chain rupture. (The temperature of the transition actually appears at or below ambient temperature, noting the equivalent "frequency" of the growing crack. The frequency dependence is discussed in Section 8.5.) Toughness and fracture in polymers are discussed in Chapter 11.

8.4.2 The T_{ll} Transition

As illustrated in Figure 8.17 (50), the T_{ll} transition occurs above the glass transition and is thought to represent the onset of the ability of the entire polymer molecule to move as a unit (9,51,52). Above T_{ll}, physical entanglements play a much smaller role, as the molecule becomes able to translate as a whole unit.

Although there is much evidence supporting the existence of a T_{ll} (51–53), it is surrounded by much controversy (54–57). Reasons include the strong dependence of T_{ll} on molecular weight and an analysis of the equivalent

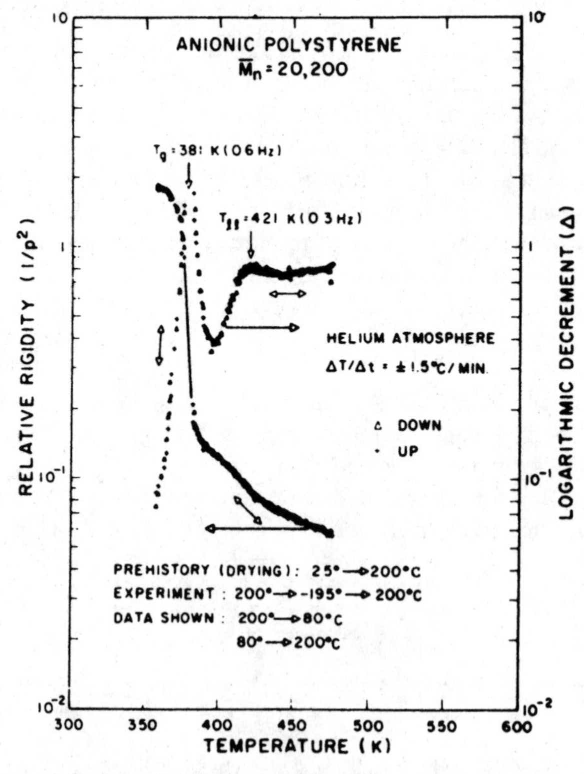

Figure 8.17 Thermomechanical spectra (relative rigidity and logarithmic decrement versus temperature (K) of anionic polystyrene, M_n = 20,200 (50).

Table 8.6 Multiple transitions in polystyrene and other amorphous polymers

Temperature	Transitions	Polystyrene Mechanism	General Mechanism
433 K (160°C)	T_{ll}	Liquid$_1$ to liquid$_2$	Boundary between rubber elasticity and rubbery flow states
373 K (100°C)	T_g	Long-range chain motions, onset of reptation	Cooperative motion of several Kuhn segments, onset of reptation
325 K (50°C)	β	Torsional vibrations of phenyl groups	Single Kuhn segment motion
130 K	γ	Motion due to four carbon backbone moieties	Small-angle torsional vibrations, 2–3 mers
38–48 K	δ	Oscillation or wagging of phenyl groups	Small-angle vibrations, single mer

behavior of spring and dashpot models (see Section 10.1). The critics contend that T_{ll} is an instrumental artifact produced by the composite nature of the specimen in torsional braid analysis (TBA), since TBA instrumentation is the principal method of studying this phenomenon (see Figure 8.17). The T_{ll} transition may be related to reptation.

Many polymers show evidence of several transitions besides T_g. Table 8.6 summarizes the data for polystyrene, including the proposed molecular mechanisms for the several transitions. The General Mechanisms column in Table 8.6 follows the results described by Bershtein and Ergos (58) on a number of amorphous polymers. Clearly, different polymers may have somewhat different mechanistic details for the various transitions, especially the lower temperature ones. However, the participating moieties become smaller in size at lower temperatures. The onset of de Gennes reptation is probably associated with T_g, the motions being experimentally identified at $T_g + 20°C$.

8.5 TIME AND FREQUENCY EFFECTS ON RELAXATION PROCESSES

So far the discussion has implicitly assumed that the time (for static) or frequency (for dynamic) measurements of T_g were constant. In fact the observed glass transition temperature depends very much on the time allotted to the experiment, becoming lower as the experiment is carried out slower.

For static or quasi-static experiments, the effect of time can be judged in two ways: (a) by speeding up the heating or cooling rate, as in dilatometric experiments, or (b) by allowing more time for the actual observation. For example, in measuring the shear modulus by Gehman instrumentation, the sample may be stressed for 100 s rather than 10 s before recording the angle of twist.

In the case of dynamic experiments, especially where the sample is exposed to a sinusoidal motion, the frequency of the experiment can be varied over wide ranges. For dynamic mechanical spectroscopy, the frequency range can be broadened further by changing instrumentation. For example, DMS measurements in the 20,000-Hz range can best be carried out by employing sound waves. In dielectric studies, the frequency of the alternating electric field can be varied.

The inverse of changing the frequency of the experiment, making measurements as a function of time at constant temperature, is called stress relaxation or creep and is discussed in Section 8.5.2 and Chapter 10. In the following paragraphs a few examples of time and frequency effects are given.

8.5.1 Time Dependence in Dilatometric Studies

It was pointed out in Section 8.3.1 that the elbow in volume–temperature studies constitutes a fundamental measure of the glass transition temperature, since the coefficient of expansion increases at T_g. The heating or cooling rate is important in determining exactly where the transition will be observed, however. As illustrated in Figure 8.18 (59), measuring the result twice after

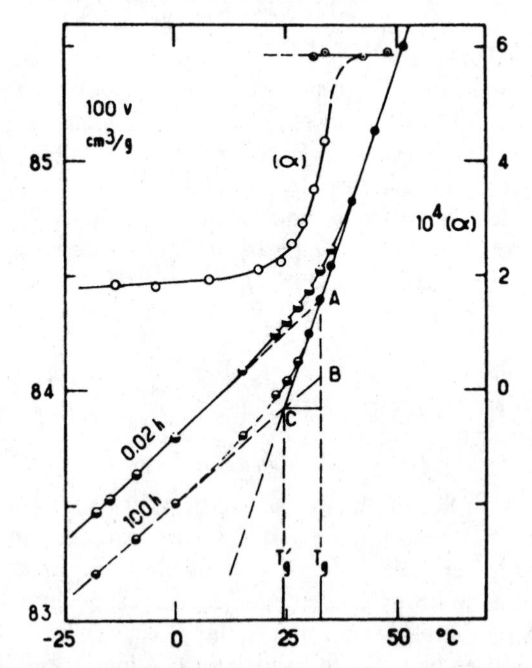

Figure 8.18 Isochronous volumes of poly(vinyl acetate) at two times, 0.02 h and 100 h, after quenching to various isothermal temperatures (59). Also shown is the cubic coefficient of expansion, α, measured at the 0.02 h cooling rate.

differing by a factor of 5000 reduces the glass transition temperature by about 8°C. Similarly T_g varies with heating rate in DSC studies.

8.5.2 Time Dependence in Mechanical Relaxation Studies

If a polymer sample is held at constant strain and measurements of stress are recorded as a function of time, stress relaxation of the type shown in Figure 8.19 (60) will be observed. The shape of the curve shown in Figure 8.19 bears comparison with those in Figure 8.2. In the present case, log time has replaced temperature in the x axis, but the phenomenon is otherwise similar. As time is increased, more molecular motions occur, and the sample softens.

It must be emphasized that the sample softens only after the time allowed for relaxation. For example, on a given curve of the type shown in Figure 8.19, the modulus might be 1×10^7 dynes/cm^2 after 10 years, showing a rubbery behavior. Someone coming up and pressing his thumb into the material after 10 years will report the material as being much harder; however, it must be remembered that pressing one's thumb into a material is a short-time experiment, of the order of a few seconds. (This assumes physical relaxation phenomena only, not true chemical degradation.)

The slope corresponding to the glass transition has been quantitatively treated by Aklonis and co-workers (61–63) for relaxation phenomena. Aklonis defined a steepness index (SI) as the maximum of the negative slope of a stress relaxation curve in the glass transition region. They found that while

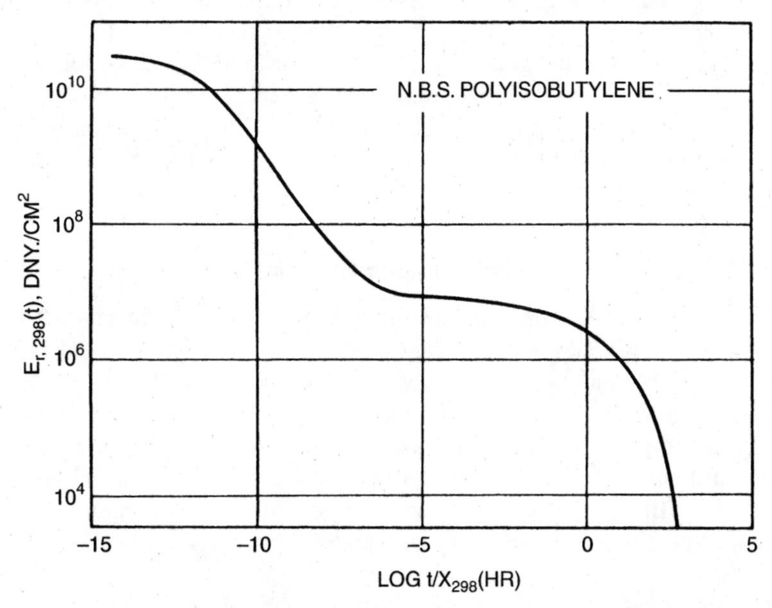

Figure 8.19 Master curve for polyisobutylene (60). The shift factor X, selected here so that $\log X_{298} = 0$, is equivalent to the experimentally determined WLF A_t (see Section 8.6.1.2).

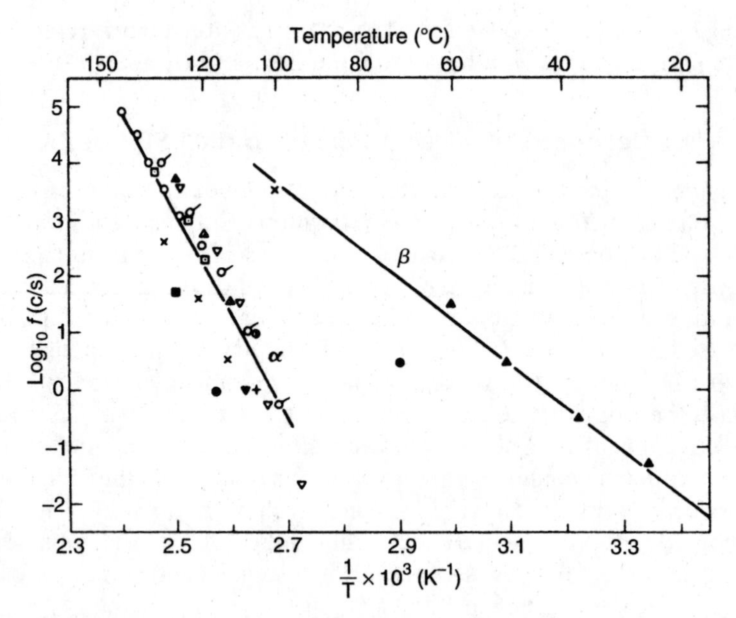

Figure 8.20 The frequency dependence of the α and β transitions of polystyrene (42). Composite of both dynamic mechanical and dielectric studies of several researchers.

polyisobutylene has a SI of about 0.5, poly(methyl methacrylate) has a value of about 1.0, and polystyrene is close to 1.5. Aklonis treated the data theoretically, using the Rouse–Bueche–Zimm bead and spring model (64–66), based in turn on the Debye damped torsional oscillator model (67,68). Aklonis concluded that values of SI equal to 0.5 represented a predominance of intramolecular forces and that a SI of 1.5 represented a predominance of intermolecular forces. A SI of 1.0 was an intermediate case. Stress relaxation is treated in greater detail in Chapter 10.

8.5.3 Frequency Effects in Dynamic Experiments

The loss peaks such as those illustrated in Figure 8.7 and Figure 8.12 can be determined as a function of frequency. The peak frequency can then be plotted against $1/T$ to obtain apparent activation energies.

Figure 8.20 (42) shows the $\alpha^{\dagger}$ transition, T_g, of polystyrene increasing steadily in temperature as the frequency of measurement is increased. Both DMS and dielectric measurements are included. Since T_g is usually reported at 10 s (or 1×10^{-1} Hz), a glass transition temperature of 100°C may be deduced from Figure 8.20, which is, in fact, the usually reported T_g.

†The peaks in the loss spectrum are sometimes labeled $\alpha, \beta, \gamma, \ldots$, with α being the highest temperature peak (T_g).

The straight line for the α relaxation process, as drawn, corresponds to an apparent activation energy of 84 kcal/mol (69). The β relaxation possesses a corresponding apparent energy of activation of 35 kcal/mol. Section 8.6.2.3 discusses methods of calculating these values. As a first approximation for many polymers, $T_\beta \cong 0.75T_g$ (70) at low frequencies. Bershtein et al. (71) point out that T_g and T_β frequently merge at a frequency of 10^6 to 10^8 Hz.

The WLF equation (Section 8.6.1.2) says that T_g will change 6 to 7°C per decade of frequency. Figure 8.19 yields about 5°C change in T_g for a factor of 10 increase in the time scale, and Figure 8.20 yields about 7.5°C per decade. Obviously this depends on the apparent energy of activation of the individual polymer, but many of the common carbon-backbone polymers have similar energies of activation. The effect of frequency on mechanical behavior is discussed further in Section 10.3.

While each of these second-order transitions has a frequency dependence, the corresponding first-order melting transition for semicrystalline polymers does not.

8.6 THEORIES OF THE GLASS TRANSITION

The basic experimental behavior of polymers near their glass transition temperatures was explored in the preceding phenomenological description. In Section 8.5, T_g was shown to decrease steadily as the time allotted to the experiment was increased. One may raise the not so hypothetical question, is there an end to the decrease in T_g as the experiment is slowed? How can the transition be explained on a molecular level? These are the questions to which the theories of the glass transition are addressed.

The following paragraphs describe three main groups of theories of the glass transition (1–4,72): free-volume theory, kinetic theory, and thermodynamic theory. Although these three theories may at first appear to be as different as the proverbial three blind men's description of an elephant, they really examine three aspects of the same phenomenon and can be successfully unified, if only in a qualitative way.

8.6.1 The Free-Volume Theory

As first developed by Eyring (73) and others, molecular motion in the bulk state depends on the presence of holes, or places where there are vacancies or voids (see Figure 8.21). When a molecule moves into a hole, the hole, of course, exchanges places with the molecule, as illustrated by the motion indicated in Figure 8.21. (This model is also exemplified in the children's game involving a square with 15 movable numbers and one empty place; the object of the game is to rearrange the numbers in an orderly fashion.) With real materials, Figure 8.21 must be imagined in three dimensions.

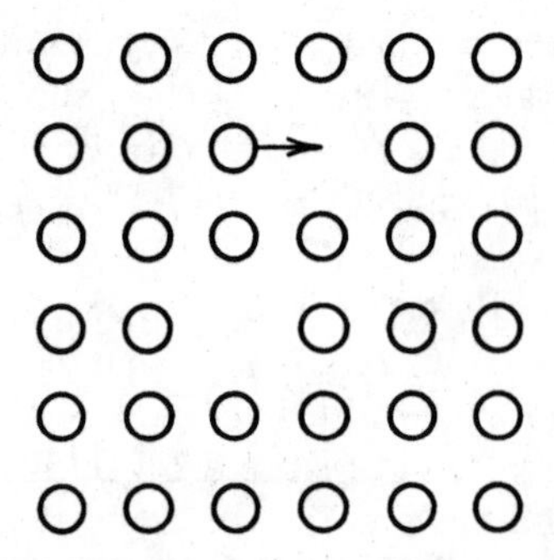

Figure 8.21 A quasi-crystalline lattice exhibiting vacancies, or holes. Circles represent molecules; arrow indicates molecular motion.

Although Figure 8.21 suggests small molecules, a similar model can be constructed for the motion of polymer chains, the main difference being that more than one "hole" may be required to be in the same locality, as cooperative motions are required (see reptation theory, Section 5.4). Thus, for a polymeric segment to move from its present position to an adjacent site, a critical void volume must first exist before the segment can jump.

The important point is that molecular motion cannot take place without the presence of holes. These holes, collectively, are called free volume. One of the most important considerations of the theory discussed below involves the quantitative development of the exact free-volume fraction in a polymeric system.

8.6.1.1 T_g as an Iso–Free-Volume State In 1950 Fox and Flory (74) studied the glass transition and free volume of polystyrene as a function of molecular weight and relaxation time. For infinite molecular weight, they found that the specific free volume, v_f, could be expressed above T_g as

$$v_f = K + (\alpha_R - \alpha_G)T \tag{8.19}$$

where K was related to the free volume at 0 °K, and α_R and α_G represented the cubic (volume) expansion coefficients in the rubbery and glassy states, respectively (see Section 8.3.1). (The linear coefficients of expansion are 1/3 of the volumetric values.) Fox and Flory found that below T_g the same specific volume–temperature relationships held for all the polystyrenes, independent of molecular weight. They concluded that (a) below T_g the local conformational arrangement of the polymer segments was independent of both molecular weight and temperature and (b) the glass transition temperature was an iso–free-volume state. Simha and Boyer (75) then postulated that the free volume at $T = T_g$ should be defined as

$$v - (v_{0,R} + \alpha_G T) = v_f \tag{8.20}$$

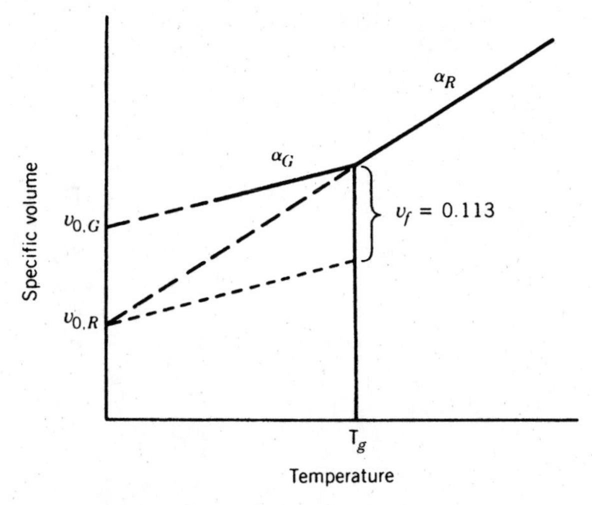

Figure 8.22 A schematic diagram illustrating free volume as calculated by Simha and Boyer.

Figure 8.22 illustrates these quantities.

Substitution of the quantity

$$v = v_{0,R} + \alpha_R T \tag{8.21}$$

leads to the relation

$$(\alpha_R - \alpha_G)T_g = K_1 \tag{8.22}$$

In the expressions above, v is the specific volume, and $v_{0,G}$ and $v_{0,R}$ are the volumes extrapolated to 0 K using α_G and α_R as the coefficients of expansion, respectively. Based on the data in Table 8.7 (75), Simha and Boyer concluded that

$$K_1 = (\alpha_R - \alpha_G)T_g = 0.113 \tag{8.23}$$

Equation (8.23) leads directly to the finding that the free volume at the glass transition temperature is indeed a constant, 11.3%. This is the largest of the theoretical values derived, but the first. (It should be pointed out that many simple organic compounds have a 10% volume increase on melting.) Other early estimates placed the free volume at about 2% (72,74).

The use of α_G in equation (8.20) results from the conclusion that expansion in the glassy state occurs at nearly constant free volume; hence $\alpha_G T$ is proportional to the occupied volume. The use of $\alpha_R T_g$ in Table 8.7 arises from the less exact, but simpler relationship

$$\alpha_R T_g = K_2 = 0.164 \tag{8.24}$$

Table 8.7 Glass transition temperature as an iso–free-volume state (75)

Polymer	T_g, K	$\alpha_G \times 10^4$ K^{-1}	$\alpha_R \times 10^4$ K^{-1}	$\alpha_R T_g$	$(\alpha_R - \alpha_G)$	$(\alpha_R - \alpha_G)\ T_g$
Polyethylene	143	7.1	13.5	0.192	7.4	0.105
Poly(dimethyl siloxane)	150	—	12	0.180	9.3	0.140
Polytetrafluoro-ethylene	160	3.0	8.3	0.133	7.0	0.112
Polybutadiene	188	—	7.8	0.147	5.8	0.109
Polyisobutylene	199.4	—	6.18	0.123	4.70	0.094
Hevea rubber	201	—	6.16	0.124	4.1	0.082
Polyurethane	213	—	8.02	0.171	6.04	0.129
Poly(vinylidene chloride)	256	—	5.7	0.146	4.5	0.115
Poly(methyl acrylate)	282	—	5.6	0.158	2.9	0.082
Poly(vinyl acetate)	302	2.1	5.98	0.18	3.9	0.118
Poly(4-methyl pentene-1)	302	3.4	7.61	0.23	3.78	0.114
Poly(vinyl chloride)	355	2.2	5.2	0.185	3.1	0.110
Polystyrene	373	2.0	5.5	0.205	3.0	0.112
Poly(methyl methacrylate)	378	2.6	5.1	0.182	2.80	0.113

Source: See also J. Brandrup and E. H. Immergut, eds., *Polymer Handbook*, 3rd ed., Wiley, New York, 1989.

Note: The quantity α represents the volume coefficient of expansion. The linear coefficients of expansion are approximately 1/3 of the volume coefficients of expansion. The quantities α_G (determined at 20°C) and α_R represent the glassy and rubbery states, respectively, the former being a typical value for the polymer (semicrystalline for polyethylene and polytetrafluoroethylene) while the latter is calculated for the 100% amorphous polymer. However, the quantities $\alpha_R - \alpha_G$ and $(\alpha_R - \alpha_G)T_g$ are calculated on the basis of 100% amorphous polymer.

The quantities K_1 and K_2 provide a criterion for the glass temperature, especially for new polymers or when the value is in doubt. This latter arises in systems with multiple transitions, for example, and in semicrystalline polymers, where T_g may be lessened or obscured.

The relation expressed in equation (8.23), though approximate, has been a subject of more recent research. Simha and co-worker (76,77) found equation (8.23) still acceptable, while Sharma et al. (78) found $\alpha_R - \alpha_G$ roughly constant at 3.2×10^{-4} deg^{-1} (see below).

8.6.1.2 The WLF Equation Section 8.2.7 illustrates how polymers soften and flow at temperatures near and above T_g. Flow, a form of molecular motion, requires a critical amount of free volume. This section considers the analytical relationships between polymer melt viscosity and free volume, particularly the WLF (Williams–Landel–Ferry) equation. The WLF equation is derived here because the free volume at T_g arises as a fundamental constant. The application of the WLF equation to viscosity and other polymer problems is considered in Chapter 10.

Early work of Doolittle (79) on the viscosity, η, of nonassociated pure liquids such as n-alkanes led to an equation of the form

$$\ln\eta = B\left(\frac{v_0}{v_f}\right) + \ln A \tag{8.25}$$

where A and B are constants and v_0 is the occupied volume, and as before v_f is the specific free volume. The Doolittle equation can be derived by considering the molecular transport of a liquid consisting of hard spheres (80–84).

An important consequence of the Doolittle equation is that it provides a theoretical basis for the WLF equation (85). One derivation of the WLF equation begins with a consideration of the need of free volume to permit rotation of chain segments, and the hindrance to such rotation caused by neighboring molecules.

The quantity P is defined as the probability of the barriers to rotation or cooperative motion per unit time being surmounted (86). An Arrhenius-type relationship is assumed, where ΔE_{act} is the free energy of activation of the process:

$$P = \exp\left(-\frac{\Delta E_{\text{act}}}{kT}\right) \tag{8.26}$$

Of course, P increases with temperature.

Next the time ("time scale") of the experiment is considered. Long times, t, allow for greater probability of the required motion, and P thus also increases. The theory assumes that tP must reach a certain value for the onset of the motion, and for the associated transition to be recorded:

$$\ln tP = \text{constant} = -\frac{\Delta E_{\text{act}}}{kT} + \ln t \tag{8.27}$$

hence

$$\ln t = \text{constant} + \frac{\Delta E_{\text{act}}}{kT} \tag{8.28}$$

Equation (8.28) equates the logarithm of time with an inverse function of the temperature. Taking the differential obtains

$$\Delta \ln t = -\frac{\Delta E_{\text{act}}}{kT^2}\Delta T \tag{8.29}$$

and the relationships become clearer: an increase in the logarithm of time is equivalent to a decrease in the absolute temperature. This must be understood in the context of the time–temperature relationship for the onset of a particular cooperative motion.

The quantity ΔE_{act} is associated with free volume and qualitatively would be expected to decrease as the fractional free volume increases. It is assumed that

$$\frac{\Delta E_{act}}{kT} = \frac{B'}{f} \tag{8.30}$$

where B' is a constant and f is the fractional free volume. Noting the similarity of form between equations (8.30) and (8.25), we take B' as equal to B. Then, instead of the Arrhenius relation, we have

$$P = \exp\left(-\frac{B}{f}\right) \tag{8.31}$$

The quantity tP still remains constant for the particular set of properties to be observed (not necessarily T_g):

$$\ln tP = \text{constant} = -\frac{B}{f} + \ln t \tag{8.32}$$

Taking the differential,

$$\Delta \ln t = B\Delta\left(\frac{1}{f}\right) \tag{8.33}$$

which states that a change in the fractional free volume is equivalent to a change in the logarithm of the time scale of the event to be observed.

In Section 8.6.1.1, it was concluded (74) that the expansion in the glassy state occurs at constant free volume. (Actually, free volume must increase slowly with temperature, even in the glassy state.) As illustrated in Figures 8.9 and 8.22, the coefficient of expansion increases at T_g, allowing for a steady increase in free volume above T_g. Setting α_f equal to the expansion coefficient of the free volume, and f_0 as the fractional free volume at T_g or other point of interest, the dependence of the fractional free volume on temperature may be written

$$f = f_0 + \alpha_f (T - T_0) \tag{8.34}$$

where T_0 is a generalized transition temperature. Equation (8.33) may be differentiated as

$$\Delta \ln t = B\left(\frac{1}{f} - \frac{1}{f_0}\right) \tag{8.35}$$

Substituting equation (8.34) into (8.35),

$$\Delta \ln t = B\left[\frac{1}{f_0 + \alpha_f(T - T_0)} - \frac{1}{f_0}\right] \tag{8.36}$$

Cross-multiplying yields

$$\Delta \ln t = B\left\{\frac{f_0 - [f_0 + \alpha_f(T - T_0)]}{f_0[f_0 + \alpha_f(T - T_0)]}\right\} \tag{8.37}$$

$$\Delta \ln t = -\frac{B\alpha_f(T - T_0)/f_0}{f_0 + \alpha_f(T - T_0)} \tag{8.38}$$

Dividing by α_f yields

$$\Delta \ln t = -\frac{(B/f_0)(T - T_0)}{f_0/\alpha_f + (T - T_0)} \tag{8.39}$$

Consider the meaning of $\Delta \ln t$:

$$\Delta \ln t = \ln t - \ln t_0 = \ln\left(\frac{t}{t_0}\right) = \ln A_T \tag{8.40}$$

where A_T is called the reduced variables shift factor (1–3). The quantity A_T will be shown to relate not only to the time for a transition with another time but also to many other time-dependent quantities at the transition temperature and another temperature. The most important of these quantities is the melt viscosity, described below and in Section 10.4.

The theoretical form of the WLF equation can now be written:

$$\ln A_T = -\frac{(B/f_0)(T - T_0)}{f_0/\alpha_f + (T - T_0)} \tag{8.41}$$

Or in log base 10 form, it is

$$\log A_T = -\frac{B}{2.303 f_0}\left[\frac{T - T_0}{f_0/\alpha_f + (T - T_0)}\right] \tag{8.42}$$

Equations (8.41) and (8.42) show that a shift in the log time scale will produce the same change in molecular motion as will the indicated nonlinear change in temperature.

The derivation leading to equations (8.41) and (8.42) suggests a generalized time dependence. Before proceeding with an interpretation of the constants in these equations, it is useful to consider the derivation originally presented by Williams, Landel, and Ferry (85).

Beginning with the Doolittle equation, equation (8.25), they note that for small v_f,

$$\frac{v_f}{v_0} \simeq \frac{v_f}{v_0 + v_f} = f \tag{8.43}$$

where $v_0 + v_f$ is the specific volume, and equation (8.43) provides a quantitative definition for f. Equation (8.25) may now be written in terms of the melt viscosity,

$$\ln \eta = \ln A + \frac{B}{f} \tag{8.44}$$

Subtracting conditions at T_0 (or T_g),

$$\ln \eta - \ln \eta_0 = \ln A - \ln A + \frac{B}{f} - \frac{B}{f_0} \tag{8.45}$$

$$\ln\left(\frac{\eta}{\eta_0}\right) = B\left(\frac{1}{f} - \frac{1}{f_0}\right) \tag{8.46}$$

The viscosity is a time (shear rate)-dependent quantity,

$$\ln\left(\frac{\eta}{\eta_0}\right) = \ln A_T = \ln\left(\frac{t}{t_0}\right) \tag{8.47}$$

Note that by equation (8.40), this leads directly back to equation (8.35). Thus equations (8.41) and (8.42) follow directly from the original Doolittle equation, although in a somewhat more limited form.

Now the constants in equation (8.42) may be evaluated. Experimentally, for many linear amorphous polymers above T_g, independent of chemical structure,

$$\log\left(\frac{\eta}{\eta_g}\right) = -\frac{17.44(T - T_g)}{51.6 + T - T_g} \tag{8.48}$$

where T_0 has been set as T_g. (For T_0 equal to an arbitrary temperature, T_s, about 50°C above T_g, the constants in the WLF equation read

$$\log\left(\frac{\eta}{\eta_s}\right) = -\frac{8.86(T - T_s)}{101.6 + T - T_s} \tag{8.49}$$

in an alternately phrased mode of expression.) Comparing equation (8.48) with (8.42), we have

$$\frac{B}{2.303f_0} = 17.44 \tag{8.50}$$

$$\frac{f_0}{\alpha_f} = 51.6 \tag{8.51}$$

Here, three unknowns and two equations are shown, which can be solved by assigning the constant B a value of unity (85), consistent with the viscosity data of Doolittle. Then $f_0 = 0.025$, and $\alpha_f = 4.8 \times 10^{-4}$ deg^{-1}.

The value of α_f may be verified in a rough way through equation (8.23). Here, if the free volume is constant in the α_G region, then $\alpha_R - \alpha_G \simeq \alpha_f$. The value of $\alpha_f = 4.8 \times 10^{-4}$ deg^{-1} leads to a temperature of -38°C, a temperature at least in the range of the T_g's observed for many polymers. Sharma et al. (78) found $\alpha_f = 3.2 \times 10^{-4}$ deg^{-1}.

The finding of $f_0 = 0.025$ is more significant. It assigns the value of the free volume at the T_g of any polymer at 2.5%. This approximate value has stood the test of time. Wrasidlo (72) suggested a value of 2.35%, based on thermodynamic data, in relatively good agreement with the WLF value.

For numerical results, it must be emphasized that the WLF equation is good for the range T_g to $T_g + 100$. In equations (8.48) and (8.49), T must be larger than T_g or T_0. Its power lies in its generality: no particular chemical structure is assumed other than a linear amorphous polymer above T_g. For a generation of polymer scientists and rheologists, the WLF equation has provided a mainstay both in utility and theory.

8.6.1.3 An Example of WLF Calculations The WLF equation, equation (8.48), can be used to calculate melt viscosity changes with temperature. Suppose a polymer has a glass transition temperature of 0°C. At 40°C, it has a melt viscosity of 2.5×10^5 poises (P) (2.5×10^4 Pa·s). What will its viscosity be at 50°C?

First calculate η_g:

$$\log\left(\frac{\eta}{\eta_g}\right) = -\frac{17.44(T - T_g)}{51.6 + (T - T_g)}$$

$$\log \eta_g = \log 2.5 \times 10^5 + \frac{17.44(313 - 273)}{51.6 + (313 - 273)}$$

$$\log \eta_g = 13.013$$

Polymers often have melt viscosities near 10^{13} P at their glass transition temperature, 1.03×10^{13} P in this case.

Now calculate the new viscosity:

$$\log \eta = 13.013 - \frac{17.44(323-273)}{51.6+(323-273)}$$

$$\eta = 2.69\times10^{4}\,\text{P or } 2.69\times10^{3}\,\text{Pa}\cdot\text{s}$$

Thus a 10°C increase in temperature has decreased the melt viscosity by approximately a factor of 10 in this range. The WLF equation can be used to calculate shift factors (Section 8.6.1.2), failure envelopes (Section 11.2.5.2), and much more.

8.6.2 The Kinetic Theory of the Glass Transition

The free-volume theory of the glass transition, as developed in Section 8.6.1, is concerned with the introduction of free volume as a requirement for coordinated molecular motion, leading to reptation. The WLF equation also serves to introduce some kinetic aspects. For example, if the time frame of an experiment is decreased by a factor of 10 near T_g, equations (8.47) and (8.48) indicate that the glass transition temperature should be raised by about 3°C:

$$\lim_{T\to T_g}\left(\frac{\log A_T}{T-T_g}\right) = -0.338 \tag{8.52}$$

$$T-T_g = \frac{-1.0}{-0.338} = +3.0 \tag{8.53}$$

For larger changes in the time or frequency frame, values of 6–7°C are obtained from equation (8.48), in agreement with experiment. For example, if $A_T = 1\times10^{-10}$, an average value of 6.9°C per decade change in T_g is obtained. The kinetic theory of the glass transition, to be developed in this section, considers the molecular and macroscopic response within a varying time frame.

8.6.2.1 Estimations of the Free-Volume Hole Size in Polymers Free volume has long been proposed to explain both the molecular motion and physical behavior of polymers. In general, an expression of the free volume, V_f, can be written as the total volume, V_t, minus the occupied volume, V_o. The quantity V_t is usually defined as the specific volume (87). However, V_o has at least three different definitions:

1. Calculated via the van der Walls excluded volume,
2. The crystalline volume at 0 K,
3. The fluctuation volume swept out by the center of gravity of the molecules as the result of thermal motion.

Because of the different ways of defining the free volume, values for it may vary by an order of magnitude!

Free-volume concepts have a long standing in the literature. It goes back to the times of van der Waals and the ideas of molecular mobility in the description of transport phenomena. Next came the Doolittle ideas and the WLF equation (Section 8.6.1). Simha and Carri (88) examined free volume from an equation of state point of view (see Section 4.3.4), arriving at a general equation between V_f and the hole fraction, h,

$$V_f = a + bh(T/T^*) \tag{8.54}$$

where T^* is given by $T/\tilde{T}$, and $\tilde{T}$ is the reduced temperature.

Misra and Mattice (89) utilized atomistic modeling via computer analysis, assuming hard spheres for the atoms. Central to the analysis was the use of mathematical *probes*, which analyzed the properties of the holes. Figure 8.23 (89) shows that the maximum probe radius is around 1.5 Å for polybutadiene at 300 K.

8.6.2.2 Positron Annihilation Lifetime Spectroscopy The principal experiment utilized in examination of the free-volume hole size has been positron annihilation lifetime spectroscopy (PALS), first developed by Kobayashi and co-workers (90,91). Positrons from a ^{22}Na source are allowed to penetrate the polymer, and the lifetime of single positrons is registered. The

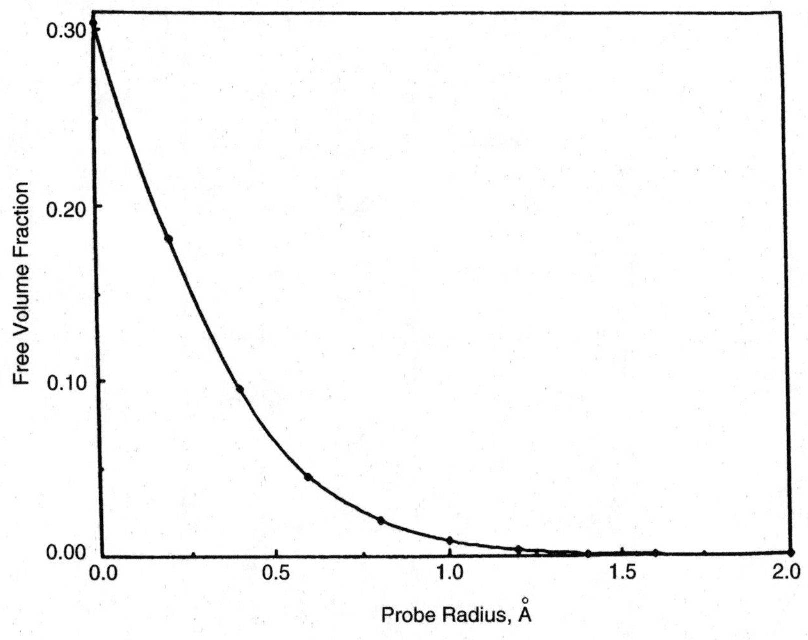

Figure 8.23 Theoretical free volume fraction as a function of the probe sized for polybutadiene.

positron can annihilate as a free positron with an electron or form a metastable state, called positronium, together with an electron. If the spins of the positron and the electron are antiparallel, the species is called *para*-positronium, and if they are parallel, *ortho*-positronium, *o*-PS. Of particular interest is the *o*-PS lifetime, which is sensitive to the free-volume hole sizes in polymer (and other) materials.

Using PALS, Dammert et al. (92) and Yu et al. (93) examined the hole volume of a series of polystyrenes of different tacticity. They arrived at a free-volume hole size distribution maximum of around 110 Å^3 at room temperature; see Figure 8.24 (92). This corresponds to an effective spherical hole radius of approximately 3 Å. While this radius is somewhat larger than the theoretical value of 1.5 Å found above, if the holes are actually irregular in shape the values are seen to agree quite well.

8.6.3 Thermodynamic Theory of T_g

All noncrystalline polymers display what appears to be a second-order transition in the Ehrenfest sense (94): the temperature and pressure derivatives of both volume and entropy are discontinuous when plotted against T or P, although the volume and the entropy themselves remain continuous (see Figure 8.5).

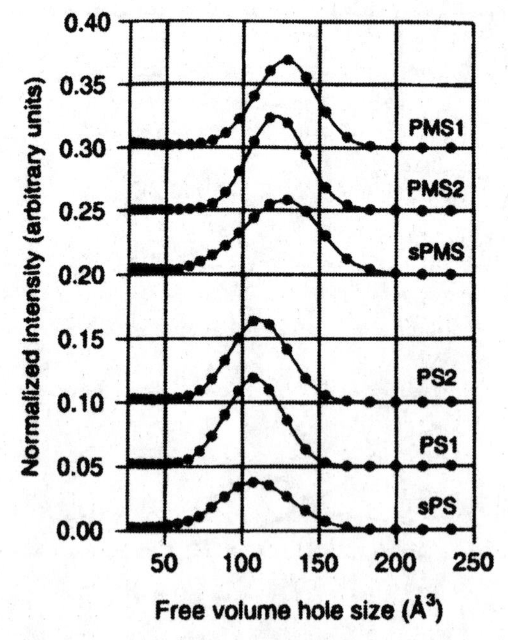

Figure 8.24 PALS study of free volume hole size distribution in polystyrenes and poly(*p*-methylstyrenes) calculated from lifetime distributions of *o*-positronium.

In Section 8.6.2 it was argued that the transition is primarily a kinetic phenomenon because (a) the temperature of the transition can be changed by changing the time scale of experiment, slower measurements resulting in lower T_g's, and (b) the measured relaxation time near the transition approach the time scale of the experiment.

One can ask what equilibrium properties these glass-forming materials have, even if it is necessary to postulate infinite time scale experiments. A thermodynamically based answer was provided by Gibbs and DiMarzio (95–99), based on a lattice model.

8.6.3.1 The Gibbs and DiMarzio Theory Gibbs and DiMarzio (95–99) argued that although the observed glass transitions are indeed a kinetic phenomenon, the underlying true transitions can possess equilibrium properties that may be difficult to realize. At infinitely long times, Gibbs and DiMarzio predict a true second-order transition, when the material finally reaches equilibrium. In infinitely slow experiments, a glassy phase will eventually emerge whose entropy is negligibly higher than that of the crystal. The temperature dependence of the entropy at the approach of T_g is shown in Figure 8.25. This true T_g, designated T_2, as will be shown later, lies some 50°C below the T_g observed at ordinary times.

The central problem of the Gibbs–DiMarzio theory is to find the configurational partition function, Q, from which the expression for the configurational entropy can be calculated. In a manner similar to the kinetic theory

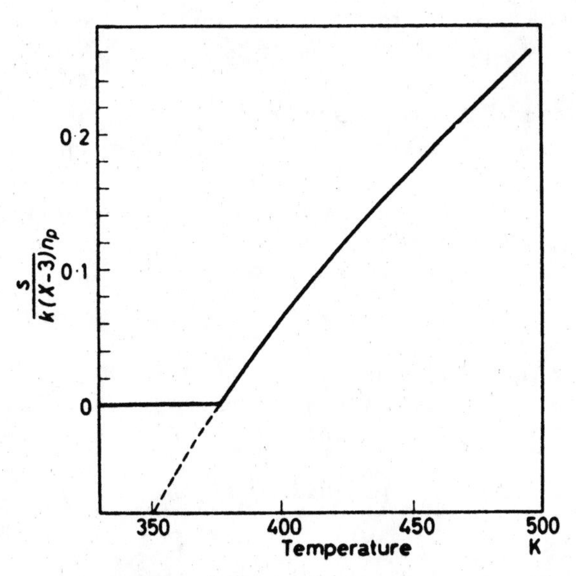

Figure 8.25 Schematic diagram of the conformational entropy of a polymer as a function of temperature according to the Gibbs–DiMarzio theory (97).

described in Section 8.6.2, hindered rotation in the polymer chain is assumed to arise from two energy states: ε_1 is associated with one possible orientation, and ε_2 is associated with all the remaining orientations; $\varepsilon_1 - \varepsilon_2 = \Delta E$, a coordination number, Z, of 4 being assumed. The intermolecular energy is given by the hole energy, α.

The application of the partition function assumes an equilibrium state:

$$Q = \sum_{f_i n_x, n_0} W(f_1 n_x \ldots, f_i n_x \ldots, n_0) \exp\left[-\frac{E(f_1 n_x \ldots, f_i n_x \ldots, n_0)}{kT}\right] \tag{8.55}$$

where $f_i n_x$ is the number of molecules packed in conformation i, and W is the total number of ways that the n_x (x degree of polymerization) molecules can be packed into $xn_x + n_0$ sites on the quasi-lattice, with n_0 being the number of holes. An expression for W was derived earlier by Flory (100) for n_x polymer chains and n_0 solvent molecules, which was used by Gibbs and DiMarzio in their calculations (see Section 3.3).

Once Q is formulated [see equation (8.55)], statistical thermodynamics provides the entropy:

$$S = kT\left(\frac{\partial \ln Q}{\partial T}\right)_{V,n} + k \ln Q \tag{8.56}$$

from which all necessary calculations can be made (96). This theory has been applied to the variation of the glass transition temperature with the molecular weight (89) (see Section 8.7), random copolymer composition (101,102) (see Section 8.8), plasticization (103), extension (92), and cross-linking (92). This last is briefly explored in Section 8.6.3.2.

8.6.3.2 *Effect of Cross-link Density on T_g* The criterion of the second-order transition temperature is that the temperature-dependent conformational entropy, S_c, becomes zero. If S_0 is the conformational entropy for the un-cross-linked system, and ΔS_R is the change in conformational entropy due to adding cross-links (99),

$$S_c = S_0 + \Delta S_R = 0 \tag{8.57}$$

Since cross-linking decreases the conformational entropy, qualitatively it may be concluded that the transition temperature is raised. The final relation may be written

$$\frac{T(\chi') - T(0)}{T(0)} = \frac{KM\chi'/\gamma}{1 - KM\chi'/\gamma} \tag{8.58}$$

where χ' is the number of cross-links per gram, M is the mer molecular weight, and γ is the number of flexible bonds per mer, backbone, and side chain. The quantity K is found by experiment and, interestingly enough, appears independent of the polymer (see Table 8.8).

An alternate relation dates back to Ueberreiter and Kanig (13)

$$\Delta T_{g,c} = Z\chi' \tag{8.59}$$

where the change in the glass temperature with increasing cross-linking, $\Delta T_{g,c}$, is equal to the cross-link density, χ', times a constant, Z. Recently, Glans and Turner (104) compared equations (8.58) and (8.59), using cross-linked polystyrene. The glass transition elevation was observed via DSC analysis (an endothermal peak was reported at T_g; see Section 8.6.2). Plots of straight lines were obtained for $\Delta T_{g,c}$ versus cross-link density, verifying equation (8.59). Some values for Z, with χ' in units of moles per gram and ΔT_c in K, are also shown in Table 8.8. For tetra functional cross-links, $\chi' = 2n$, where n is the number of network chains per unit volume (usually one cm^3).

8.6.3.3 A Summary of the Glass Transition Theories

In the preceding section, three apparently disparate theories of the glass transition were presented. The basic thrust of each is summarized conveniently here and in Table 8.9.

Table 8.8 Constants for cross-link effect on T_g (92,97)

Polymer	γ	M/γ	$K \times 10^{23}$	$Z \times 10^4$
Natural rubber	3	22.7	1.30	3.2
Polystyrene	2	52	1.20	4.6
Poly(methyl methacrylate)	4	25	1.38	1.8

Table 8.9 Glass transition theory box scores

Theory	Advantages	Disadvantages
Free-volume theory	1. Time and temperature of viscoelastic events related to T_g 2. Coefficients of expansion above and below T_g related	1. Actual molecular motions poorly defined
Kinetic theory	1. Shifts in T_g with time frame quantitatively determined 2. Heat capacities determined	1. No T_g predicted at infinite time scales
Thermodynamic theory	1. Variation of T_g with molecular weight, diluent, and cross-link density predicted 2. Predicts true second-order transition temperature	1. Infinite time scale required for measurements 2. True second-order transition temperature poorly defined

1. The free-volume theory introduces free volume in the form of segment-size voids as a requirement for the onset of coordinated molecular motion. This theory provides relationships between coefficients of expansion below and above T_g and yields equations relating viscoelastic motion to the variables of time and temperature.
2. The kinetic theory defines T_g as the temperature at which the relaxation time for the segmental motions in the main polymer chain is of the same order of magnitude as the time scale of the experiment. The kinetic theory is concerned with the rate of approach to equilibrium of the system, taking the respective motions of the holes and molecules into account. The kinetic theory provides quantitative information about the heat capacities below and above the glass transition temperature and explains the 6 to 7°C shift in the glass transition per decade of time scale of the experiment.
3. The thermodynamic theory introduces the notion of equilibrium and the requirements for a true second-order transition, albeit at infinitely long time scales. The theory postulates the existence of a true second-order transition, which the glass transition approaches as a limit when measurements are carried out more and more slowly. It successfully predicts the variation of T_g with molecular weight and cross-link density (see Section 8.7), diluent content, and other variables.

A summary of the free-volume numbers of the various theories can be made:

Theory	Free-Volume Fraction
WLF	0.025
Hirai and Eyring	0.08
Miller†	0.12
Simha-Boyer	0.113

The analytical development of these theories illustrates the power of statistical thermodynamics in providing solutions to important polymer problems. However, much remains to be done. It has been said that less than 5% of all fundamental knowledge has been wrested from nature. This is certainly true in the study of polymer glass transitions. Insofar as research in this area remains highly active, it is highly probable that new insight will provide an integrated theory in the near future. Attempts to do so up until now are summarized in the following section.

8.6.3.4 A Unifying Treatment Adam and Gibbs (103) attempted to unify the theories relating the rate effect of the observed glass transition and the equilibrium behavior of the hypothetical second-order transition. They pro-

† A. A. Miller, *J. Chem. Phys.*, **49**, 1393 (1968); *J. Polym. Sci.*, **A-2** (6), 249, 1161 (1968).

posed the concept of a "cooperatively rearranging region," defined as the smallest region capable of conformational change without a concomitant change outside the region. At T_2 this region becomes equal to the size of the sample, since only one conformation is available.

Adam and Gibbs rederived the WLF equation, putting it in terms of the potential energy hindering the cooperative rearrangement per mer, the molar conformational entropy, and the change in the heat capacity at T_g. By choosing the temperature T in the WLF equation to be T_s [see equation (8.49)] and suitable rearrangements of the WLF formulation to isolate T_2, they found that

$$\frac{T_g}{T_2} = 1.30 \pm 8.4\% \tag{8.60}$$

for a wide range of glass-forming systems, both polymeric and low molecular weight.

For low-temperature elastomers such as the polybutadiene family, $T_g \cong 200$ K. According to equation (8.60), $T_2 \cong 154$ K, or about 50 K below T_g. According to the WLF equation, equation (8.48), the viscosity becomes infinite at $T - T_g = -51.6$°C, which is about the same number. Although this simplified approach yields less quantitative agreement at higher temperatures, the ideas still are interesting.

8.7 EFFECT OF MOLECULAR WEIGHT ON T_g

8.7.1 Linear Polymers

Studies of the increase in T_g with increasing polymer molecular weight date back to the works of Ueberreiter in the 1930s (105). The theoretical analysis of Fox and Flory (74) (see Section 8.6.1.1) indicated that the general relationship between T_g at a molecular weight M was related to the glass temperature at infinite molecular weight, $T_{g\infty}$, by

$$T_g = T_{g\infty} - \frac{K}{(\alpha_R - \alpha_G)M} \tag{8.61}$$

with K being a constant depending on the polymer. Equation (8.61) follows from the decrease in free volume with increasing molecular weight, caused in turn by the increasing number of connected mers in the system, and decreased number of end groups.

The ubiquitous polystyrene seems to have been investigated more than any other polymer (74,105,106). DSC data, first extrapolated to low heating rate, are shown in Figure 8.26 (99). (These data also show an endothermic peak at T_g; see earlier discussions.) The equation for slow heating rates may be expressed

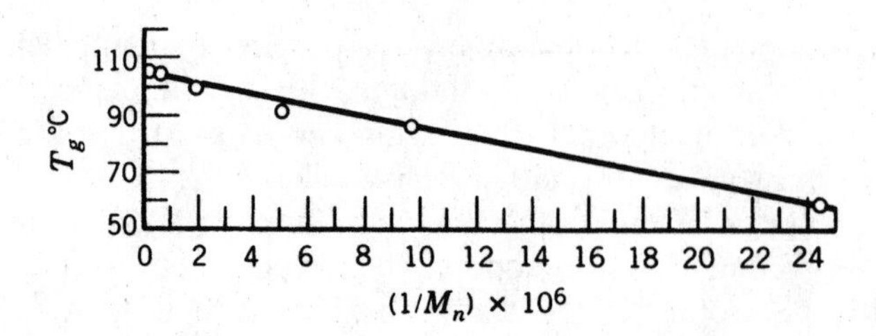

Figure 8.26 The glass transition temperature of polystyrene as a function of $1/M_n$ (106).

$$T_g = 106°\text{C} - \frac{2.1\times10^5}{M_n} \tag{8.62}$$

For heating rates normally encountered (74)

$$T_g = 100°\text{C} - \frac{1.8\times10^5}{M_n} \tag{8.63}$$

The molecular weight in equation (8.63) is for fractionated polystrene. For slow experiments, these equations suggest a 6°C increase in T_g at infinite molecular weight.

8.7.2 Effect of T_g on Polymerization

According to equation (8.61) the glass transition depends on the molecular weight. What happens during an isothermal polymerization? When the polymerization begins, the monomers are always in the liquid state. Sometimes, however, the system may go through T_g and the polymer may vitrify as the reaction proceeds. Since molecular motion is much reduced when the system is below T_g, the reaction substantially stops.

Two conditions can be distinguished. First, during a chain polymerization, the monomer effectively acts like a plasticizer for the nascent polymer. An example relates to the emulsion polymerization of polystyrene, often carried out at about 80°C. The reaction will not proceed quite to 100% conversion, because the system vitrifies.

Second, during stepwise polymerization, the molecular weight is continually increasing. An especially interesting case involves gelation. Taking epoxy polymerization as an example, the resin† is simultaneously polymerizing and cross-linking (see Section 3.7.3).

Gillham (107–112) pointed out the need to postcure the polymer above $T_{g\infty}$, the glass transition temperature of the fully cured‡ system. He developed a time–temperature–transformation (TTT) reaction diagram that may be used

† Resin is an early term for polymer, often used with epoxies.
‡ Cure is an early term for cross-linking, also frequently used with epoxies.

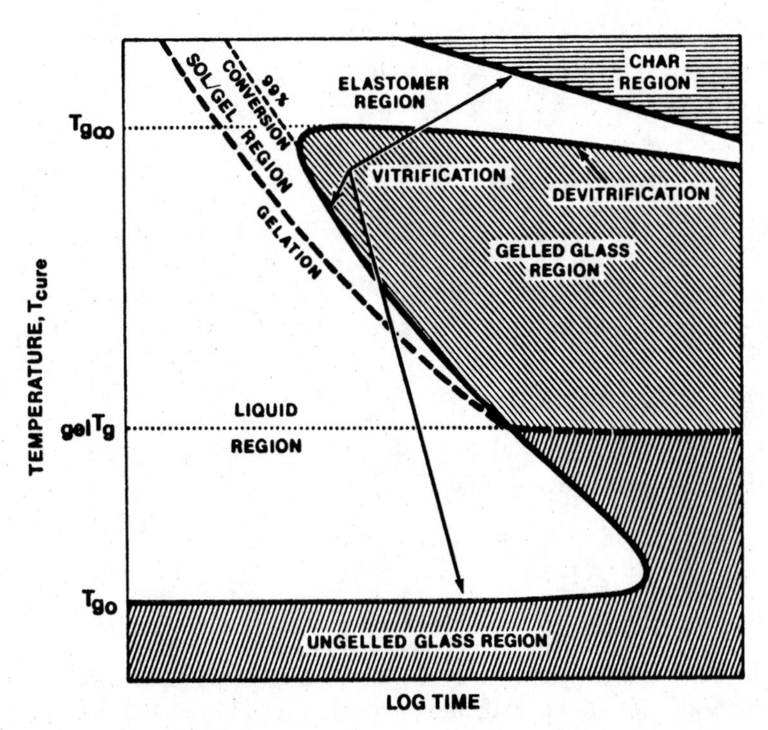

Figure 8.27 The thermosetting process as illustrating by the time–temperature–transformation reaction diagram (112).

to provide an intellectual framework for understanding and comparing the cure and glass transition properties of thermosetting systems. Figure 8.27 illustrates the TTT diagram. Besides $T_{g\infty}$, the diagram also displays $_{gel}T_g$, the temperature at which gelation and vitrification occur simultaneously, and T_{g0}, the glass transition temperature of the reactants. The particular S-shaped curve between T_{g0} and $T_{g\infty}$ results because the reaction rate is increased with increasing temperature. At a temperature intermediate between $_{gel}T_g$ and $T_{g\infty}$, the reacting mass first gels, forming a network. Then it vitrifies, and the reaction stops, incomplete. To the novice, the reaction products may appear complete. This last may result in material failure if the temperature is suddenly raised.

The TTT diagram explains why epoxy and similar reactions are carried out in steps, each at a higher temperature. The last step, the postcure, must be done above $T_{g\infty}$. Other points shown in Figure 8.27 include the devitrification region, caused by degradation, and the char region, at still higher temperatures.

8.8 EFFECT OF COPOLYMERIZATION ON T_g

The discussion above relates to simple homopolymers. Addition of a second component may take the form of copolymerization or polymer blending.

Addition of low-molecular-weight compounds results in plasticization. Experimentally, two general cases may be distinguished: where one phase is retained and where two or more phases result.

8.8.1 One-Phase Systems

Based on the thermodynamic theory of the glass transition, Couchman derived relations to predict the T_g composition dependence of binary mixtures of miscible high polymers (113) and other systems (114–116). The treatment that follows is easily generalized to the case for statistical copolymers (113).

Consider two polymers (or two kinds of mers, or one mer and one plasticizer) having pure-component molar entropies denoted as S_1 and S_2, and their respective mole fractions (moles of mers for the polymers) as X_1 and X_2. The mixed system molar entropy may be written

$$S = X_1 S_1 + X_2 S_2 + \Delta S_m \tag{8.64}$$

where ΔS_m represents the excess entropy of mixing. For later convenience, S_1 and S_2 are referred to their respective pure-component glass transition temperatures of T_{g1} and T_{g2}, when their values are denoted as S_1^0 and S_2^0.

Heat capacities are of fundamental importance in glass transition theories, because the measure of the heat absorbed provides a direct measure of the increase in molecular motion. The use of classical thermodynamics leads to an easy introduction of the pure-component heat capacities at constant pressure, C_{p1} and C_{p2}:

$$S = X_1\left\{S_1^0 + \int_{T_{g1}}^{T} C_{p1}\, d\ln T\right\} + X_2\left\{S_2^0 + \int_{T_{g2}}^{T} C_{p2}\, d\ln T\right\} + \Delta S_m \tag{8.65}$$

The mixed-system glass transition temperature, T_g, is defined by the requirement that S for the glassy state be identical to that for the rubbery state, at T_g. This condition and the use of appropriate superscripts G and R lead to the equation

$$\begin{aligned} &X_1^G\left\{S_1^{0,G} + \int_{T_{g1}}^{T_g} C_{p1}^G\, d\ln T\right\} + X_2^G\left\{S_2^{0,G} + \int_{T_{g2}}^{T_g} C_{p2}^G\, d\ln T\right\} + \Delta S_m^G \\ &\quad = X_1^R\left\{S_1^{0,R} + \int_{T_{g1}}^{T_g} C_{p1}^R\, d\ln T\right\} + X_2^R\left\{S_2^{0,R} + \int_{T_{g2}}^{T_g} C_{p2}^R\, d\ln T\right\} + \Delta S_m^R \end{aligned} \tag{8.66}$$

Since $S_i^{0,G} = S_i^{0,R}$ $(i = 1, 2)$ and $X_i^G = X_i^R = X_i$, equation (8.66) may be simplified:

$$X_1\left\{\int_{T_{g1}}^{T_g} (C_{p1}^G - C_{p1}^R)\, d\ln T\right\} + X_2\left\{\int_{T_{g2}}^{T_g} (C_{p2}^G - C_{p2}^R)\, d\ln T\right\} + \Delta S_m^G - \Delta S_m^R = 0 \tag{8.67}$$

In regular small-molecule mixtures, ΔS_m is proportional to $X \ln X + (1 - X) \ln(1 - X)$, where X denotes X_1 and X_2. Similar relations hold for polymer–solvent (plasticizer) and polymer–polymer combinations. Combined with the continuity relation, $\Delta S_m^G = \Delta S_m^R$. For random copolymers these quantities are also equal. Then

$$X_1 \int_{T_{g1}}^{T_g} \Delta C_{p1} d \ln T + X_2 \int_{T_{g2}}^{T_g} \Delta C_{p2} d \ln T = 0 \tag{8.68}$$

where Δ denotes transition increments. Again, the increase in the heat capacity at T_g reflects the increase in the molecular motion and the increased temperature rate of these motions.

After integration the general relationship emerges,

$$X_1 \, \Delta C_{p1} \ln\left(\frac{T_g}{T_{g1}}\right) + X_2 \, \Delta C_{p2} \ln\left(\frac{T_g}{T_{g2}}\right) = 0 \tag{8.69}$$

For later convenience the X_i are exchanged for mass (weight) fractions, M_i (recall that the ΔC_{pi} are then per unit mass), and equation (8.69) becomes

$$\ln T_g = \frac{M_1 \, \Delta C_{p1} \ln T_{g1} + M_2 \, \Delta C_{p2} \ln T_{g2}}{M_1 \, \Delta C_{p1} + M_2 \, \Delta C_{p2}} \tag{8.70}$$

or equivalently

$$\ln\left(\frac{T_g}{T_{g1}}\right) = \frac{M_2 \, \Delta C_{p2} \ln(T_{g2}/T_{g1})}{M_1 \, \Delta C_{p1} + M_2 \, \Delta C_{p2}} \tag{8.71}$$

Equation (8.71) is shown to fit T_g data of thermodynamically miscible blends (see Figure 8.28). Four particular nontrivial cases of the general mixing relation may be derived.

Making use of the expansions of the form $\ln(1 + x) = x$, for small x, and noting that T_{g1}/T_{g2} usually is not greatly different from unity yield

$$T_g \simeq \frac{M_1 \, \Delta C_{p1} \, T_{g1} + M_2 \, \Delta C_{p2} \, T_{g2}}{M_1 \, \Delta C_{p1} + M_2 \, \Delta C_{p2}} \tag{8.72}$$

which has the same form as the Wood equation (117), originally derived for random copolymers.

If $\Delta C_{pi} \, T_{gi}$ = constant (76–78,118), the familiar Fox equation (119) appears after suitable crossmultiplying:

$$\frac{1}{T_g} = \frac{M_1}{T_{g1}} + \frac{M_2}{T_{g2}} \tag{8.73}$$

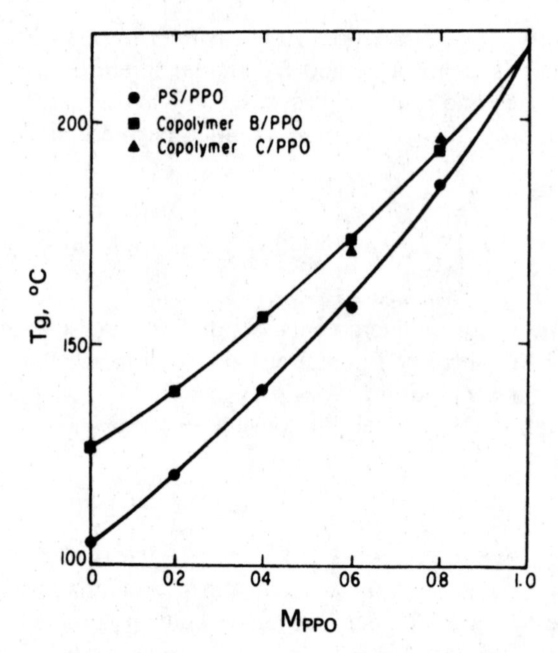

Figure 8.28 Glass-transition temperatures, T_g, of poly(2,6-dimethyl-1,4-phenylene oxide)–*blend*–polystyrene (PPO/PS) blends versus mass fraction of PPO, M_{PPO}. The full curve was calculated from equation (8.71) as circles. ΔC_{p1} = 0.0671 cal $K^{-1} \cdot g^{-1}$; ΔC_{p2} = 0.0528 cal $K^{-1} \cdot g^{-1}$; T_{g1} = 378 K, T_{g2} = 489 K. PPO was designated as component 2 (114,120).

The Fox equation (119) was also originally derived for statistical copolymers (120). This equation predicts the typically convex relationship obtained when T is plotted against M_2 (see Figure 8.28). If $\Delta C_{p1} \simeq \Delta C_{p2}$, the equation of Pochan et al. (121) follows from equation (8.70):

$$\ln T_g = M_1 \ln T_{g1} + M_2 \ln T_{g2} \tag{8.74}$$

Finally, if both pure-component heat capacity increments have the same value and the log functions are expanded,

$$T_g = M_1 T_{g1} + M_2 T_{g2} \tag{8.75}$$

which predicts a linear relation for the T_g of the blend, random copolymer, or plasticized system. This equation usually predicts T_g too high. Equations (8.73) and (8.75) are widely used in the literature. Couchman's work (113–116) shows the relationship between them. Previously they were used on a semiempirical basis.

These equations also apply to plasticizers, a low-molecular-weight compound dissolved in the polymer. In this case the plasticizer behaves as a compound with a low T_g. The effect is to lower the glass transition temperature. A secondary effect is to lower the modulus, softening it through much of the temperature range of interest. An example is the plasticization of poly(vinyl chloride) by dioctyl phthalate to make compositions known as "vinyl."

8.8.2 Two-Phase Systems

Most polymer blends, as well as their related graft and block copolymers and interpenetrating polymer networks, are phase-separated (122) (see Section 4.3). In this case each phase will exhibit its own T_g. Figure 8.29 (123,124) illustrates two glass transitions appearing in a series of triblock copolymers of different overall compositions. The intensity of the transition, especially in the loss spectra (E''), is indicative of the mass fraction of that phase.

The storage modulus in the plateau between the two transitions depends both on the overall composition and on which phase is continuous. Electron microscopy shows that the polystyrene phase is continuous in the present case. As the elastomer component increases (small spheres, then cylinders, then alternating lamellae), the material gradually softens. When the rubbery phase becomes the only continuous-phase, the storage modulus will decrease to about 1×10^8 dynes/cm^2.

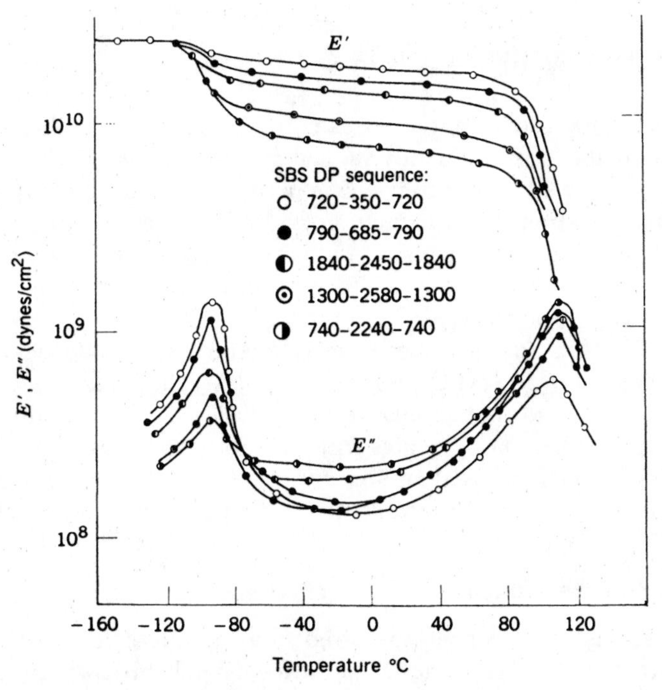

Figure 8.29 Dynamic mechanical behavior of polystyrene-*block*-polybutadiene-*block*-polystyrene, a function of the styrene–butadiene mole ratio (123,124).

Table 8.10 Phase composition of epoxy/acrylic simultaneous interpenetrating networks (125)

Glycidyl Methacrylate[a] (%)	Dispersed Phase Weight Fraction		Matrix Phase Weight Fraction	
	PnBA[b]	Epoxy	PnBA	Epoxy
0	0.97	0.03	0.09	0.91
0.3	0.82	0.18	0.12	0.88
3.0	—	—	0.30	0.70

[a] Grafting mer, increases mixing.
[b] Poly(*n*-butyl acrylate).

If appreciable mixing between the component polymers occurs, the inward shift in the T_g of the two phases can each be expressed by the equations of Section 8.8.1 (125). Using equation (8.73), the extent of mixing within each phase in a simultaneous interpenetrating network of an epoxy resin and poly(*n*-butyl acrylate) was calculated (see Table 8.10). The overall composition was 80/20 epoxy/acrylic, and glycidyl methacrylate is shown to enhance molecular mixing between the chains. Chapter 13 provides additional material on the glass transition behavior of multicomponent materials.

8.9 EFFECT OF CRYSTALLINITY ON T_g

The previous discussion centered on amorphous polymers, with atactic polystyrene being the most frequently studied polymer. Semicrystalline polymers such as polyethylene or polypropylene or of the polyamide and polyester types also exhibit glass transitions, though only in the amorphous portions of these polymers. The T_g is often increased in temperature by the molecular-motion restricting crystallites. Sometimes T_g appears to be masked, especially for highly crystalline polymers.

Boyer (9) points out that many semicrystalline polymers appear to possess two glass temperatures: (a) a lower one, $T_g(L)$, that refers to the completely amorphous state and that should be used in all correlations with chemical structure (this transition correlates with the molecular phenomena discussed in previous sections), and (b) an upper value, $T_g(U)$, that occurs in the semicrystalline material and varies with extent of crystallinity and morphology.

8.9.1 The Glass Transition of Polyethylene

Linear polyethylene, frequently referred to as polymethylene, offers a complete contrast with polystyrene in that it has no side groups and has a high degree of crystallinity, usually in excess of 80%. Because of the high degree of crystallinity, molecular motions associated with T_g are partly masked, leading

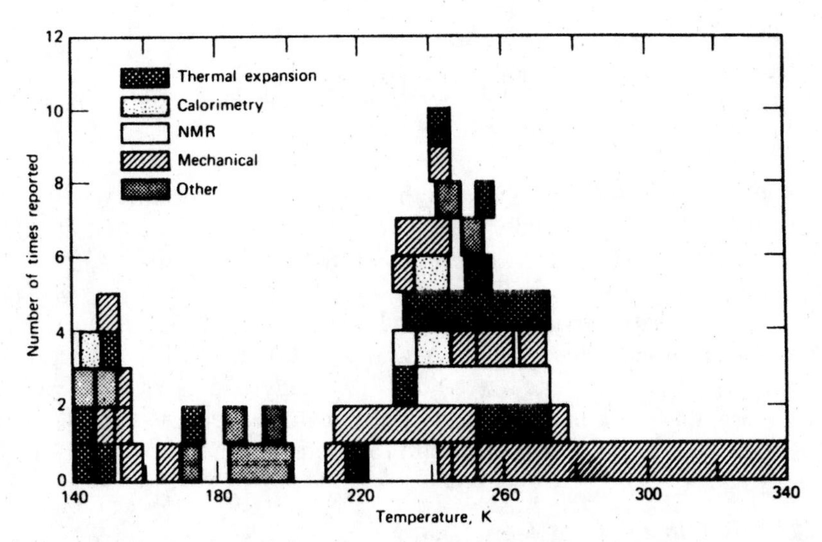

Figure 8.30 Histogram showing number of times a given value of T_g for linear polyethylene has been reported in the literature by various standard methods indicated (126).

to a confusion with other secondary transitions (see Figure 8.30) (126). Thus various investigators consider the T_g of polyethylene to be in three different regions: –30°C, –80°C, or –128°C.

Davis and Eby support the –30°C value on the basis of volume–time measurements; Stehling and Mandelkern (127) favor the –128°C value based on mechanical measurements. Illers (128) and Boyer (9) support the value of –80°C based on extrapolations of completely amorphous ethylene–vinyl acetate copolymer data with copolymer–T_g relationships. Boyer (9) supports the position that –80° is $T_g(L)$ and –30°C represents $T_g(U)$. The transition at –128°C is thought to be related to the Schatzki crankshaft motion (Section 8.4.1), although the situation apparently is more complicated (128). Tobolsky (129) obtained –81°C for amorphous polyethylene based on a Fox plot [see equation (8.73)] of statistical copolymers of ethylene and propylene, *it*-polypropylene having a T_g of –18°C.

8.9.2 The Nylon Family Glass Transition

Two subfamilies of aliphatic nylons (polyamides) exist:

$$\left[-NH\!-\!(CH_2)_x\!-\!NH-\overset{\displaystyle O}{\overset{\|}{C}}\!-\!(CH_2)_y\!-\!\overset{\displaystyle O}{\overset{\|}{C}}- \right]_n \tag{8.76}$$

from diacids and dibases, and

$$\left[\text{—NH} \!\left(\text{CH}_2\right)_x \overset{\text{O}}{\overset{\|}{\text{C}}} \text{—} \right]_n \tag{8.77}$$

originating from ω-amino acids. Both subfamilies are semicrystalline; of course, they form commercially important fibers.

The usually stated T_g range is $T_g \simeq +40°C$ for polyamide 612 to $T_g \simeq 60°C$ for polyamide 6 (9); however, T_g depends on the crystallinity of the particular sample. *N*-methylated polyamides, with a lower hydrogen bonding, have lower T_g's (130). As x and y increase in equations (8.76) and (8.77), the structure becomes more polyethylene-like, and T_g gradually decreases. Interestingly, when $x > 4$, there is a characteristic mechanical loss peak at about –130°C, again suggestive of the Schatzki motion (Section 8.4.1).

8.9.3 Relationships between T_g and T_f

The older literature (131) suggested two relationships between T_g and T_f: $T_g/T_f \simeq \frac{1}{2}$ for symmetrical polymers, and $T_g/T_f \simeq \frac{2}{3}$ for nonsymmetrical polymers. Definitions of symmetry differ, however. One method uses the appearance of atoms down the chain: if a central portion of the chain appears the same when viewed from both ends, it is symmetrical. However, even from the beginning, there were many exceptions to the above. The only rule obeyed in this regard is that T_g is always lower than T_f for homopolymers. This is because (a) the same kinds of molecular motion should occur at T_g and T_f, and (b) short-range order exists at T_g, but long-range order exists at T_f.

Boyer (9) has prepared a cumulative plot of T_g/T_f (see Figure 8.31). Region A (the old $T_g/T_f \simeq \frac{1}{2}$) contains most of the polymers which are free from side groups other than H and F (and hence symmetrical) and contain such polymers as polyethylene, poly(oxymethylene), and poly(vinylidene fluoride). Region B contains most of the common vinyl, vinylidene, and condensation polymers such as the nylons. About 55% of all measured polymers lie in the band $T_g/T_f = 0.667 \pm 0.05$ (9). Region C contains poly(α-olefins) with long alkyl side groups as well as other nontypical polymers such as poly(2,6-dimethylphenylene oxide), which has T_g/T_f approximately equal to 0.93. For an unknown polymer, then, the relationship $T_g/T_f = \frac{2}{3}$ is a good way of providing an estimate of one transition if the temperature of the other is known.

8.9.4 Heat Distortion Temperature

While the glass transition and melting temperatures define the behavior of polymers from a scientific point of view, the engineers frequently depend on more practical tests. These tests work well for plasticized polymers, blends and composites of various types, and thermosets. These tests originated from the old idea of a *softening temperature*, sometimes defined as the temperature in

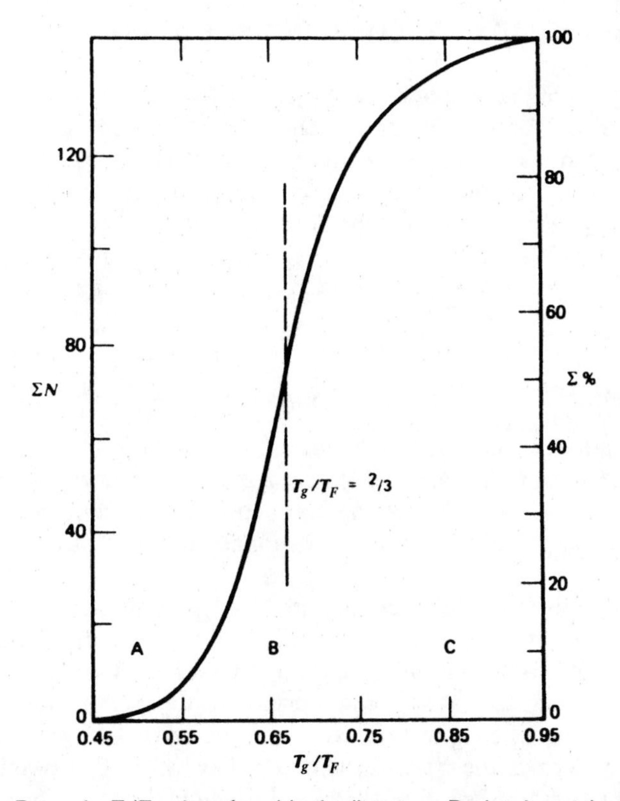

Figure 8.31 Range in T_g/T_f values found in the literature. Region A contains unsubstituted polymers. Region C includes poly(α-olefins) with long side chains. Region B contains the majority of vinyl, vinylidene, and condensation polymers. The left ordinate is cumulative number, N, and the right ordinate is cumulative percentage of all examples reported as having the indicated T_g/T_f values (9).

which a specimen could be easily penetrated with a needle. One such quantitative test is called the Vicat test, where a needle under 1000 g load penetrates the specimen 1 mm (132).

One of the more important of the practical tests is the *heat distortion temperature* (HDT). The HDT is defined as the temperature at which a 100 mm length, 3 mm thick specimen bar at 1.82 MPa in a three-point bending mode deflects 0.25 mm. Young's modulus at the HDT is 0.75 Gpa (133,134). For unfilled polymers, both the Vicat and the HDT tests usually record a temperature just above the glass transition temperature, or for melting conditions, just below the temperature of final disappearance of crystallinity. For polymer blends, both the Vicat and the HDT will tend to reflect the properties of the continuous phase. If the polymer contains filler which raises the modulus, the HDT will be somewhat increased.

8.10 DEPENDENCE OF T_g ON CHEMICAL STRUCTURE

In Section 8.9 some effects of crystallinity and hydrogen bonding on T_g were considered. The effect of molecular weight was discussed in Section 8.7, and the effect of copolymerization was discussed in Section 8.8. This section discusses the effect of chemical structure in homopolymers.

Boyer (135) suggested a number of general factors that affect T_g (see Table 8.11). In general, factors that increase the energy required for the onset of molecular motion increase T_g; those that decrease the energy requirements lower T_g.

8.10.1 Effect of Aliphatic Side Groups on T_g

In monosubstituted vinyl polymers and at least some other classes of polymers, flexible pendant groups reduce the glass transition of the polymer by acting as "internal diluents," lowering the frictional interaction between chains. The total effect is to reduce the rotational energy requirements of the backbone.

The aliphatic esters of poly(acrylic acid) (136), poly(methacrylic acid) (137), and other polymers (138) (see Figure 8.32) (139) show a decline in T_g as the number of —CH_2— units in the side group increases. At still longer aliphatic side groups, T_g increases as side-chain crystallization sets in, impeding chain motion. In this latter composition range the materials feel waxy. In the ultimate case, of course, the polymer would behave like slightly diluted polyethylene. For cellulose triesters (37) the minimum in T_g is observed at the triheptanoate, probably because of the increased basic backbone stiffness.

8.10.2 Effect of Tacticity on T_g

The discussion so far in this chapter has assumed atactic polymers, which with a few exceptions are amorphous. Other stereo isomers include isotactic and syndiotactic polymers (see Section 2.3).

The effect of tacticity on T_g may be significant, as illustrated in Table 8.12 (140,141). Karasz and MacKnight (141) noted that the effect of tacticity on T_g is expected in view of the Gibbs–DiMarzio theory (Section 8.6.3.1). In disubstituted vinyl polymers, the energy difference between the two predominant rotational isomers is greater for the syndiotactic configuration than for the iso-

Table 8.11 Factors affecting T_g (135)

Increase T_g	Decrease T_g
Intermolecular forces	In-chain groups promoting flexibility (double-bonds and ether linkages)
High CED	Flexible side groups
Intrachain steric hindrance	Symmetrical substitution
Bulky, stiff side groups	

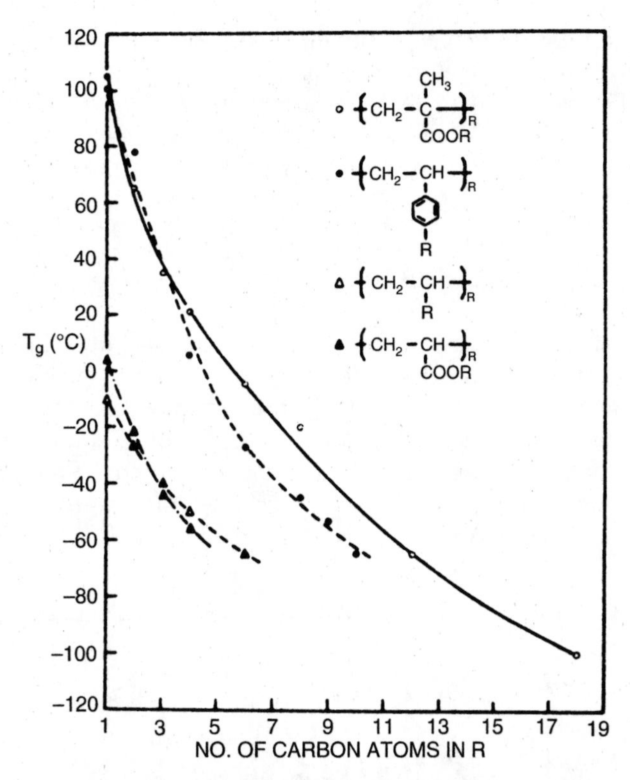

Figure 8.32 Effect of side-chain lengths on the glass transition temperatures of polymethacrylates [○ (S. S. Rogers and L. Mandelkern, *J. Phys. Chem.*, **61**, 985, 1957)]; poly-*p*-alkyl styrenes [• (W. G. Bard, *J. Polym. Sci.*, **37**, 515, 1959)]; poly-α-olefins [△ (M. L. Dannis, *J. Appl. Polym. Sci.*, **1**, 121, 1959; K. R. Dunham, J. Vandenbergh, J. W. H. Farber, and L. E. Contois., *J. Polym. Sci.*, **1A**, 751, 1963)]; and polyacrylates [▲ (J. A. Shetter, *Polym. Lett.*, **1**, 209, 1963)] (139).

Table 8.12 Effect of tacticity on the glass transition temperatures of polyacrylates and polymethacrylates (141)

	T_g (°C)				
	Polyacrylates		Polymethacrylates		
Side Chain	Isotactic	Dominantly Syndiotactic	Isotactic	Dominantly Syndiotactic	100% Syndiotactic
Methyl	10	8	43 (50)*	105 (123)*	160
Ethyl	–25	–24	8	65	120
n-Propyl	—	–44	—	35	—
Iso-Propyl	–11	–6	27	81	139
n-Butyl	—	–49	–24	20	88
Iso-Butyl	—	–24	8	53	120
Sec-Butyl	–23	–22	—	60	—
Cyclo-Hexyl	12	19	51	104	163

* T_g (M = ∞), K. Ute, N. Miyatake, and K. Hatada, Polymer, **36**, 1415 (1995).

tactic configuration. In monosubstituted vinyl polymers, where the other substituent is hydrogen, the energy difference between the rotational states of the two pairs of isomers is the same. Thus the acrylates in Table 8.12 have the same T_g for the two isomers, whereas the methacrylates show distinctly different T_g's, with the isotactic form always having a lower T_g than the syndiotactic form.

8.11 EFFECT OF PRESSURE ON T_g

The discussion above has assumed constant pressure at 1 atm (1 bar). Since an increased pressure causes a decrease in the total volume [see equation (8.8)], an increase in T_g is expected based on the prediction of decreased free volume.

Tamman and Jellinghaus (142) showed that a plot of volume versus pressure at a temperature near the transition shows an elbow reminiscent of the volume–temperature plot (see Figure 8.9). If the temperature is raised at elevated pressures, T_g will in fact show a corresponding increase (see Figure 8.33) (143).

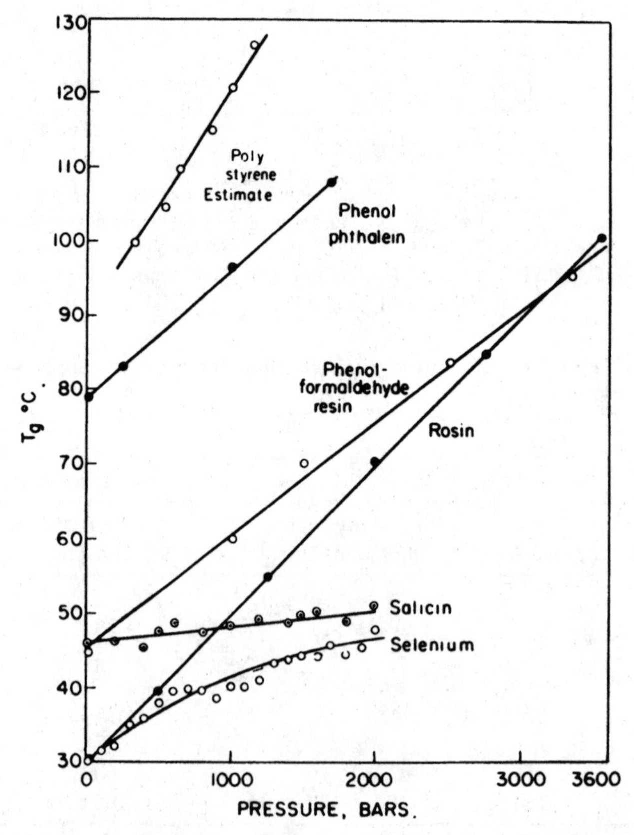

Figure 8.33 Glass transition versus pressure for various substances (143).

The results in Figure 8.33 can easily be interpreted in terms of the free-volume theory of T_g. In developing the WLF equation (Section 8.6.1.2), it was shown that the free-volume fraction at any temperature above T_g could be expressed $f = f_0 + \alpha_f(T - T_g)$. If the free-volume compressibility is β_f, then

$$f_{t,p} = f_0 + \alpha_f[T - T_g(0)] - \beta_f P \tag{8.78}$$

where $T_g(0)$ refers to the glass transition at zero pressure. Under particular glass transition temperature and pressure conditions, $f_{t,p} = f_0$ and equation (8.78) becomes

$$\alpha_f[T_g - T_g(0)] = \beta_f P \tag{8.79}$$

By differentiating with respect to pressure (144–147)

$$\left(\frac{\partial T_g}{\partial P}\right)_f = \frac{\Delta\beta_f}{\Delta\alpha_f} \tag{8.80}$$

a relation strongly reminiscent of Ehrenfest's (90) relation for the change of a second-order transition temperature with pressure,

$$\frac{TV\,\Delta\alpha}{\Delta C_p} = \frac{\Delta\beta}{\Delta\alpha} = \frac{dT_g}{dP} \tag{8.81}$$

where the Δ sign refers to changes from below to above T_g [see also equation (8.69)]. Several representative values of $\partial T_g/\partial P$ are shown in Table 8.13 (139). Since $\Delta\alpha \simeq \alpha_f \simeq 4.8 \times 10^{-4}$ deg^{-1}, $\beta_f \simeq \Delta\beta$ may be estimated. For polystyrene, Table 8.13 predicts a T_g rise of 31°C for a rise in pressure per 1000 atm, in agreement with Figure 8.33.

Table 8.13 Pressure coefficients of the glass transition temperatures for selected materials (139)

Material	T_g (°C)	dT_g/dP (K/atm)
Natural rubber	−72	0.024
Polyisobutylene	−70	0.024
Poly(vinyl acetate)	25	0.022
Rosin	30	0.019
Selenium	30	0.015–0.004[a]
Salicin	46	0.005
Phenolphthalein	78	0.019
Poly(vinyl chloride)	87	0.016
Polystyrene	100	0.031
Poly(methyl methacrylate)	105	0.020–0.023
Boron trioxide	260	0.020

[a] The variation is probably due to the different compressibilities of ring and chain material.

For polyurethanes, Quested et al. (148) found that $\Delta\beta_f/\Delta\alpha_f$ was greater than dT_g/dP, except at pressures close to 1 bar. At high pressures, dT_g/dP reached a limiting value of 10.4°C/kbar. The effect of pressure has been studied for ultrasonic frequencies (149) and fracture stress differences (150).

In the above, it was demonstrated that an increase in pressure can bring about vitrification. This result is important in engineering operations such as molding or extrusion, where operation too close to T_g (1 bar) can result in a stiffening of the material.

Thus we may refer to a glass transition pressure. In a broader sense, the glass transition is multidimensional. We could also refer to the glass transition molecular weight (Section 8.7), the glass transition concentration (for diluted or plasticized species), and so forth.

The solubility parameter can also be estimated with the aid of measurements as a function of hydrostatic pressure. Thus

$$\delta = \left\{T\left(\frac{\alpha}{\beta}\right)\right\}^{1/2} \tag{8.82}$$

where α represents the isobaric volume thermal expansion coefficient and β the isothermal compressibility; see Section 8.1.4. Since β is the most difficult of the three terms in equation (8.82), it is the most likely to be the unknown.

8.12 DAMPING AND DYNAMIC MECHANICAL BEHAVIOR

When the loss modulus or loss tangent is high, as in the glass transition region, the polymers are capable of damping out noise and vibrations, which, after all, are a particular form of dynamic mechanical motion. This section describes some of the aspects of behavior of a polymer under sinusoidal stresses at constant amplitude.

If an applied stress varies with time in a sinusoidal manner, the sinusoidal stress may be written

$$\sigma = \sigma_0 \sin \omega t \tag{8.83}$$

where ω is the angular frequency in radians, equal to $2\pi' \times$ frequency. For Hookian solids, with no energy dissipated, the strain is given by

$$\varepsilon = \varepsilon_0 \sin \omega t \tag{8.84}$$

For real materials, the stress and strain are not in phase, the strain lagging behind the stress by the phase angle δ. The relationships among these parameters are illustrated in Figure 8.34. Of course, the phase angle defines an in-phase and out-of-phase component of the stress, σ' and σ'', as defined in Section 8.1.

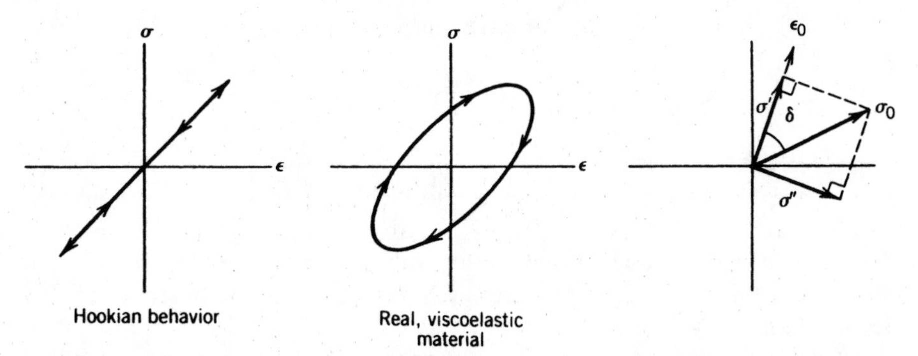

Figure 8.34 Simple dynamic relationships between stress and strain, illustrating the role of the phase angle.

Then the relationships between the in-phase and out-of-phase components and δ are given by

$$\sigma' = \sigma_0 \cos\delta \tag{8.85}$$

$$\sigma'' = \sigma_0 \sin\delta \tag{8.86}$$

The dynamic moduli may now be written

$$E' = \frac{\sigma'}{\varepsilon_0} = E^* \cos\delta \tag{8.87}$$

$$E'' = \frac{\sigma''}{\varepsilon_0} = E^* \sin\delta \tag{8.88}$$

$$E^* = \frac{\sigma_0}{\varepsilon_0} = (E'^2 + E''^2)^{1/2} \tag{8.89}$$

In terms of complex notation,

$$E^* = E' + iE'' \tag{8.90}$$

(see Section 8.1) and

$$E = |E^*| = (E'^2 + E''^2)^{1/2} \tag{8.91}$$

where, of course, $E''/E' = \tan\delta$.

Again, the logarithmic decrement, Δ, may be defined as the natural logarithm of the amplitude ratio between successive vibrations (see Section 8.3.4). This is a measure of damping. The quantity $\tan\delta$ is related to the logarithmic decrement,

$$\Delta \cong \pi' \tan\delta$$

$$\Delta \cong \frac{\pi' E''}{E'} \tag{8.92}$$

where $\pi' = 3.14$. The heat energy per unit volume per cycle is $H = \pi' E'' A_0^2$, where A_0 is the maximum amplitude, and Hooke's law is valid. Further relationships are given in Section 10.2.4, where damping is related to models. Hence both the loss tangent and the logarithmic decrement are proportional to the ratio of energy dissipated per cycle to the maximum potential energy stored during a cycle.

These loss terms are at a maximum near the glass transition or a secondary transition. This phenomenon is widely used in engineering for the construction of objects subject to making noise or vibrations. Properly protected by high damping polymers, car doors close more quietly, motors make less noise, and mechanical damage to bridges through vibrations is reduced.

It must be mentioned that the glass transition can be broadened or shifted through various chemical and physical means. These include the use of plasticizers, fillers, or fibers, or through the formation of interpenetrating polymer networks. In this last case the very small phases are formed with variable composition and molecular chains trapped by cross-links promote juxtaposition of the two molecular species. A very broad glass transition may result, spanning the range of the two homopolymer transitions. Thus with *cross*–poly(ethyl acrylate)–*inter*–*cross*–poly(methyl methacrylate), which is composed of chemically isomeric polymers, glass transition behavior 100°C wide is obtained (31,151,152) (see Figure 8.35) (151). Conversely, objects such as rubber tires must be built of low damping elastomers, lest they overheat in service and blow out; see Section 9.16.2.

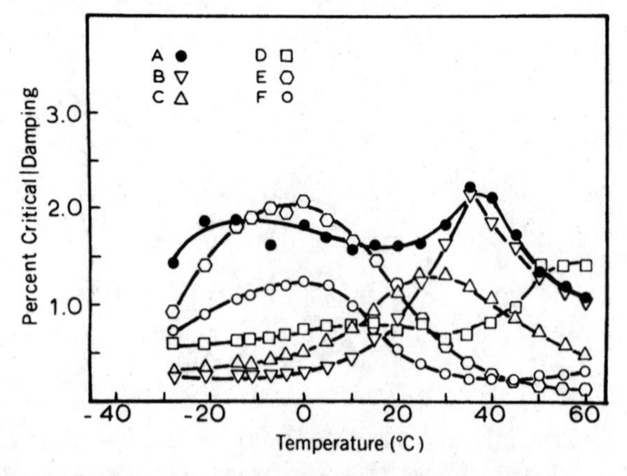

Figure 8.35 Damping as a function of temperature for several polymers. Composition A is an IPN. Note that it damps nearly evenly over a broad temperature range. Compositions B through F are various homopolymers and copolymers that damp over narrow temperature ranges (151).

8.13 DEFINITIONS OF ELASTOMERS, PLASTICS, ADHESIVES, AND FIBERS

This chapter began with an overall approach to the behavior of polymers as a function of temperature. Cast in another way, at ambient temperatures a polymer will be above, below, or at its glass transition temperature, with concomitant properties. Other ways of dividing polymers are according to the presence or absence of crystallinity (Chapter 6) or according to the presence or absence of cross-linking (Chapter 9).

In a certain simplified sense, definitions are given below that will help identify polymers found in the everyday world with their scientific properties.

1. An elastomer is a cross-linked, amorphous polymer above its T_g. An example is the common rubber band.
2. An adhesive is a linear or branched amorphous polymer above its T_g. It must be able to flow on a molecular scale to "grip" surfaces. (This definition is not to be confused with polymerizable adhesive materials, present in monomeric form. These are "tacky" or "sticky" only in the partly polymerized state. Frequently they are cross-linked "thermoset," finally. Contact with the surface to be adhered must be made before gelation, in order to work.) An example is the postage stamp adhesive, composed of linear poly(vinyl alcohol), which is plasticized by water (or saliva) from below its T_g to above its T_g. On migration of the water away from the adhesive surface, it "sticks."
3. A plastic is usually below its T_g if it is amorphous. Crystalline plastics may be either above or below their T_g's.
4. Fibers are always composed of crystalline polymers. Apparel fiber polymers are usually close to T_g at ambient temperatures to allow flexibility in their amorphous portions.
5. House coatings and paints based on either oils or latexes are usually close to T_g at ambient temperatures. A chip of such material is usually flexible but not rubbery, being in the leathery region (see Section 8.2.2).

REFERENCES

1. A. Kaye, R. F. T. Stepto, W. J. Work, J. V. Aleman, and A. Ya. Malkin, *Pure and Applied Chem.*, **70**, 701 (1998).
2. J. D. Ferry, *Viscoelastic Properties of Polymers*, 3rd, ed., Wiley, New York, 1980, Chap. 1.
3. L. E. Nielsen and R. F. Landel, *Mechanical Properties of Polymers and Composites*, 2nd ed., Dekker, New York, 1994.
4. B. Du, J. Liu, Q. Zhang, and T. He, *Polymer*, **42**, 5901 (2001).
5. R. Glazebrook, ed., *Dictionary of Applied Physics*, Vol. 1, Macmillian, London, 1922, pp. 175–176, 244–245.

6. C. K. Schoff, *Encyclopedia of Polymer Science and Engineering*, Vol. 14, J. I. Kroschwitz, ed., Wiley, New York, 1988.
7. A. V. Tobolsky and J. R. McLoughlin, *J. Polym. Sci.*, **8**, 543 (1952).
8. J. J. Aklonis, *J. Chem. Ed.*, **58** (11), 892 (1981).
9. R. F. Boyer, in *Encyclopedia of Polymer Science and Technology*, Suppl. Vol. 2, N. M. Bikales, ed., Interscience, New York, 1977, p. 745; (a) pp. 822–823.
10. T. J. Shimanouchi, M. Asahina, and S. J. Enomoto, *J. Polym. Sci.*, **59**, 93 (1962).
11. D. Katz and I. G. Zervi, *J. Polym. Sci.*, **46C**, 139 (1974).
12. D. Katz and G. Salee, *J. Polym. Sci.*, **A-2** (6), 801 (1968).
13. K. Ueberreiter and G. Kanig, *J. Chem. Phys.*, **18**, 399 (1950).
14. G. M. Martin and L. Mandelkern, *J. Res. Natl. Bur. Stand.*, **62**, 141 (1959).
15. J. Brandruys and E. H. Immergut, eds., *Polymer Handbook*, 2nd ed., Wiley, New York, 1975; III-139; (a) III-150.
16. A. V. Tobolsky and H. Yu, unpublished.
17. H. J. Stern, *Rubber: Natural and Synthetic*, 2nd ed., Palmerton Publishing, New York, 1967, Chap. 1.
18. A. F. Yee and M. T. Takemori, *J. Polym. Sci. Polym. Phys. Ed.*, **20**, 205 (1982).
19. G. O. Jones, *Glass*, Methuen, London, 1956.
20. R. F. Boyer and R. S. Spencer, *J. Appl. Phys.*, **16**, 594 (1945).
21. D. R. Lide, ed., *CRC Handbook of Chemistry and Physics*, CRC Press, Boca Raton, FL, 1999, sec. 12, p. 119.
22. L. H. Sperling, *J. Polym. Sci. Polym. Symp.*, **60**, 175 (1977).
23. M. C. O. Chang, D. A. Thomas, and L. H. Sperling, *J. Polym. Sci. Part B Polym. Phys.*, **26**, 1627 (1988).
24. P. Meares, *Trans. Faraday Soc.*, **53**, 31 (1957).
25. M. Salmerón, C. Torregrosa, A. Vidaurre, J. M. Meseguer Dueñas, M. Monleón Pradas, and J. L. Gómez Ribelles, *Colloid Polym. Sci.*, **277**, 1033 (1999).
26. J. A. Victor, S. D. Kim, A. Klein, and L. H. Sperling, *J. Appl. Polym. Sci.*, **73**, 1763 (1999).
27. C. Kow, M. Morton, and L. J. Fetters, *Rubber Chem. Technol.*, **55** (1), 245 (1982).
28. K. C. Frisch, D. Klempner, S. Migdal, H. L. Frisch, and H. Ghiradella, *Polym. Eng. Sci.*, **14**, 76 (1974).
29. W. Meesiri, J. Menczel, U. Guar, and B. Wunderlich, *J. Polym. Sci. Polym. Phys. Ed.*, **20**, 719 (1982).
30. (a) M. Takayanagi, *Proc. Polym. Phys. (Jpn.)*, 1962–1965. (b) Toya Baldwin Co., Ltd. Rheovibron Instruction Manual, 1969.
31. L. H. Sperling, in *Sound and Vibration Damping with Polymers*, R. D. Corsaro and L. H. Sperling, eds., ACS Books Symp. Ser. 424, American Chemical Society, Washington, DC, 1990.
32. Bordon Award Symposium honoring J. K. Gillham, all the papers in *Polym. Eng. Sci.*, **19** (10) (1979).
33. J. K. Gillham, *Polym. Eng. Sci.*, **19**, 749 (1979).
34. R. A. Venditti and J. K. Gillham, *J. Appl. Polym. Sci.*, **64**, 3 (1997).

35. J. K. Gillham, *AICHE J.*, **20**, 1066 (1974).
36. J. Russell and R. G. Van Kerpel, *J. Polym. Sci.*, **25**, 77 (1957).
37. A. F. Klarman, A. V. Galanti, and L. H. Sperling, *J. Polym. Sci.*, **A-2** (7), 1513 (1969).
38. C. J. Malm, J. W. Mench, D. L. Kendall, and G. D. Hiatt, *Ind. Eng. Chem.*, **43**, 688 (1951).
39. L. Banks and B. Ellis, *J. Polym. Sci. Polym. Phys. Ed.*, **20**, 1055 (1982).
40. H. G. Elias, *Macromolecules: Structure and Properties*, Vol. 1, Plenum, New York, 1977, Chap. 10.
41. N. Saito, K. Okano, S. Iwayanagi, and T. Hideshima, in *Solid State Physics*, Vol. 14, F. Seitz and D. Turnbull, eds., Academic Press, Orlando, 1963, p. 344.
42. K. S. Cole and S. Cole, *J. Chem. Phys.*, **9**, 341 (1941).
43. A. Tcharkhtchi, A. S. Lucas, J. P. Trotignon, and J. Verdu, *Polymer*, **39**, 1233 (1998).
44. C. A. Garcia-Franco and D. W. Mead, *Rheological Acta*, **38**, 34 (1999).
45. N. G. McCrum, B. E. Read, and G. Williams, *Anelastic and Dielectric Effects in Polymeric Solids*, Wiley, New York, 1967.
46. (a) T. F. Schatzki, *J. Polym. Sci.*, **57**, 496 (1962); (b) J. J. Aklonis and W. J. MacKnight, *Introduction of Polymer Viscoelasticity*, 2nd ed., Wiley-Interscience, New York, 1983, p. 81.
47. H. A. Flocke, *Kolloid Z.*, **180** 118 (1962).
48. H. S. Gutawsky, G. B. Kistiakowsky, G. E. Pake, and E. M. Purcell, *J. Chem. Phys.*, **17** (10), 972 (1949); see also J. G. Powles and H. S. Gutowsky, *J. Chem. Phys.*, **21**, 1695 (1953), and W. P. Slichter and E. R. Mandell. *J. Appl. Phys.*, **29**, 1438 (1958).
49. J. V. Koleske and J. A. Faucher, *Polym. Eng. Sci.*, **19** (10), 716 (1979).
50. S. J. Stadnicki, J. K. Gillham, and R. F. Boyer, *J. Appl. Polym. Sci.*, **20**, 1245 (1976).
51. J. K. Gillham, J. A. Benci, and R. F. Boyer, *Polym. Eng. Sci.*, **16**, 357 (1976).
52. R. F. Boyer, *Polym. Eng. Sci.*, **19** (10), 732 (1979).
53. S. Hedvat, *Polymer*, **22**, 774 (1981).
54. G. D. Patterson, H. E. Bair, and A. Tonelli, *J. Polym. Sci. Polym. Symp.*, **54**, 249 (1976).
55. L. E. Nielsen, *Polym. Eng. Sci.*, **17**, 713 (1977).
56. R. M. Neumann, G. A. Senich, and W. J. MacKnight, *Polym. Sci. Eng.*, **18**, 624 (1978).
57. J. Heijboer, *Polym. Eng. Sci.*, **19** (10), 664 (1979).
58. V. A. Bershtein and V. M. Ergos, *Differential Scanning Calorimetry*, Ellis Horwood, Chichester, England, 1994.
59. A. J. Kovacs, *J. Polym. Sci.*, **30**, 131 (1958).
60. E. Catsiff and A. V. Tobolsky, *J. Polym. Sci.*, **19**, 111 (1956).
61. V. B. Rele and J. J. Aklonis, *J. Polym. Sci.*, **46C**, 127 (1974).
62. K. C. Lin and J. J. Aklonis, *Polym. Sci. Eng.*, **21**, 703 (1981).
63. J. J. Aklonis, *IUPAC Proceedings*, University of Massachusetts, Amherst, July 12–16, 1982, p. 834.
64. P. E. Rouse, *J. Chem. Phys.*, **21**, 1272 (1953).
65. F. Bueche, *J. Chem. Phys.*, **22**, 603 (1954).

66. B. H. Zimm, *J. Chem. Phys.*, **24**, 269 (1956).
67. A. V. Tobolsky and D. B. DuPre, *Adv. Polym. Sci.*, **6**, 103 (1969).
68. A. V. Tobolsky and J. J. Aklonis, *J. Phys. Chem.*, **68**, 1970 (1964).
69. R. F. Boyer, in *Encyclopedia of Polymer Science and Technology*, Vol. 13, N. M. Bikales, ed., Interscience, New York, 1970, p. 277.
70. R. Boyer, in *Encyclopedia of Polymer Science Technology* Suppl. Vol. II, p. 765, Wiley, New York, 1977.
71. V. A. Bershtein, V. M. Egorov, L. M. Egorova, and V. A. Ryzhov, *Thermochim. Acta*, **238**, 41 (1994).
72. W. Wrasidlo, *Thermal Analysis of Polymers, Advances in Polymer Science*, Vol. 13, Springer-Verlag, New York, 1974. p. 3.
73. H. Eyring, *J. Chem. Phys.*, **4**, 283 (1936).
74. T. G. Fox and P. J. Flory, *J. Appl. Phys.*, **21**, 581 (1950); T. G. Fox and P. J. Flory, *J. Polym. Sci.*, **14**, 315 (1954).
75. R. Simha and R. F. Boyer, *J. Chem. Phys.*, **37**, 1003 (1962).
76. R. Simha and C. E. Weil, *J Macromol. Sci. Phys.*, **B4**, 215 (1970).
77. R. F. Boyer and R. Simha, *J. Polym. Sci.*, **B11**, 33 (1973).
78. S. C. Sharma, L. Mandelkern, and F. C. Stehling, *J. Polym. Sci.*, **B10**, 345 (1972).
79. A. K. Doolittle, *J. Appl. Phys.*, 1471 (1951).
80. D. Turnbull and M. H. Cohen, *J. Chem. Phys.*, **31**, 1164 (1959).
81. D. Turnbull and M. H. Cohen, *J. Chem. Phys.*, **34**, 120 (1961).
82. F. Bueche, *J. Chem. Phys.*, **21**, 1850 (1953).
83. F. Bueche, *J. Chem. Phys.*, **24**, 418 (1956).
84. F. Bueche, *J. Chem. Phys.*, **30**, 748 (1959).
85. M. L. Williams, R. F. Landel, and J. D. Feery, *J. Am. Chem. Soc.*, **77**, 3701 (1955).
86. E. H. Andrews, *Fracture in Polymers*, American Elsevier, New York, 1968, pp. 9–16.
87. J. Liu, Q. Deng, and Y. C. Jean, *Macromolecules*, **26**, 7149 (1993).
88. R. Simha and G. Carri, *J. Polym. Sci.: Part B: Polym. Phys.*, **32**, 2645 (1994).
89. S. Misra and W. L. Mattice, *Macromolecules*, **26**, 7274 (1993).
90. Y. Kobayashi, K. Haraya, Y. Kamiya, and S. Hattori, *Bull. Chem. Soc. Jpn.*, **65**, 160 (1992).
91. Y. Kobayashi, *J. Chem. Soc. Faraday Trans.*, **87**, 3641 (1991).
92. R. M. Dammert, S. L. Maunu, F. H. J. Maurer, I. M. Neelow, S. Nievela, F. Sundholm, and C. Wastlund, *Macromolecules*, **32**, 1930 (1999).
93. Z. Yu, Y. Yahsi, J. D. McGervey, A. M. Jamieson, and R. Simha, *J. Polym. Sci.: Part B: Polym. Phys.*, **32**, 2637 (1994).
94. P. Ehrenfest, *Leiden Comm. Suppl.*, 756 (1933).
95. J. H. Gibbs, *J. Chem. Phys.*, **25**, 185 (1956).
96. J. H. Gibbs and E. A. DiMarzio, *J. Chem. Phys.*, **28**, 373 (1958).
97. J. H. Gibbs, in *Modern Aspects of the Vitreous State*, J. D. Mackenzie, ed., Butterworth, London, 1960.
98. E. A. DiMarzio and J. H. Gibbs, *J. Polym. Sci.*, **A1**, 1417 (1963).
99. E. A. DiMarzio, *J. Res. Natl. Bur. Studs.*, **68A**, 611 (1964).

100. P. J. Flory, *Proc. R. Soc. (Lond.)*, **A234**, 60 (1956).
101. E. A. DiMarzio and J. H. Gibbs, *J. Polym. Sci.*, **40**, 121 (1959).
102. E. A. DiMarzio and J. H. Gibbs, *J. Polym. Sci.*, **1A**, 1417 (1963).
103. G. Adam and J. H. Gibbs, *J. Chem. Phys.*, **43**, 139 (1965).
104. J. H. Glans and D. T. Turner, *Polymer*, **22**, 1540 (1981).
105. E. Jenckel and K. Ueberreiter, *Z. Phys. Chem.*, **A182**, 361 (1938).
106. L. P. Blanchard, J. Hess, and S. L. Malhorta, *Can. J. Chem.*, **52**, 3170 (1974).
107. J. K. Gillham, *Polym. Eng. Sci.*, **19**, 676 (1979).
108. M. T. DeMuse, J. K. Gillham, and F. Parodi, *J. App. Polym. Sci.*, **64**, 15 (1997).
109. J. K. Gillham, in *The Role of the Polymer Matrix in the Processing and Structural Properties of Composite Materials*, J. C. Seferis and L. Nicolais, eds., Plenum, New York, 1983, pp. 127–145.
110. J. B. Enns and J. K. Gillham, in *Polymer Characterization: Spectroscopic, Chromatographic, and Physical Instrumental Methods*, C. D. Craver, ed., Advances in Chemistry Series No. 203, American Chemical Society, Washington, DC, 1983, pp. 27–63.
111. J. B. Enns and J. K. Gillham, *J. Appl. Polym. Sci.*, **28**, 2567 (1983).
112. J. K. Gillham, *Encyclopedia of Polym. Sci. Tech.*, **4**, 519 (1986).
113. P. R. Couchman, *Macromolecules*, **11**, 1156 (1978).
114. P. R. Couchman, *Polym. Eng. Sci.*, **21**, 377 (1981).
115. P. R. Couchman, *J. Mater. Sci.*, **15**, 1680 (1980).
116. P. R. Couchman and F. E. Karasz, *Macromolecules*, **11**, 117 (1978).
117. J. M. Bardin and D. Patterson, *Polymer*, **10**, 247 (1969); L. A. Wood, *J. Polym. Sci.*, **28**, 319 (1958).
118. R. F. Boyer, *J. Macromol. Sci. Phys.*, **7**, 487 (1973).
119. T. G. Fox, *Bull. Am. Phys. Soc.*, **1**, 123 (1956).
120. J. R. Fried, F. E. Karasz, and W. J. MacKnight, *Macromolecules*, **11**, 150 (1978).
121. J. M. Pochan, C. L. Beatty, and D. F. Hinman, *Macromolecules*, **11**, 1156 (1977).
122. L. H. Sperling, *Polymeric Multicomponent Materials: An Introduction*, Wiley, New York, 1997.
123. M. Matsuo, *Jpn. Plastics*, **2**, 6 (1958).
124. M. Matsuo, T. Ueno, H. Horino, S. Chujya, and H. Asai, *Polymer*, **9**, 425 (1968).
125. P. R. Scarito and L. H. Sperling, *Polym. Eng. Sci.*, **19**, 297 (1979).
126. G. T. Davis and R. K. Eby, *J. Appl. Phys.*, **44**, 4274 (1973).
127. F. C. Stehling and L. Mandelkern, *Macromolecules*, **3**, 242 (1970).
128. K. H. Illers, *Kolloid Z. Z. Polym.*, **190**, 16 (1963); **231**, 622 (1969); **250**, 426 (1972).
129. A. V. Tobolsky, *Properties and Structure of Polymers*, Wiley, New York, 1960, App. K.
130. G. Champetier and J. P. Pied, *Makromol. Chem.*, **44**, 64 (1961).
131. R. F. Boyer, *J. Appl. Phys.*, **25**, 825 (1954).
132. ASTM, 08.01, D 1525, American Society for Testing and Materials, 1998.
133. ASTM, *8.01–8.03*, D 648, American Society for Testing and Materials, 1996.
134. M. T. Takemori, *Polym. Eng. Sci.*, **19**, 1104 (1979).

135. R. F. Boyer, *Rubber Chem. Technol.*, **36**, 1303 (1963).
136. J. A. Shetter, *Polym. Lett.*, **1**, 209 (1963).
137. S. S. Rogers and L. Mandelkern, *J. Phys. Chem.*, **61**, 985 (1957).
138. W. G. Barb, *J. Polym. Sci.*, **37**, 515 (1957).
139. M. C. Shen and A. Eisenberg, *Prog. Solid State Chem.*, **3**, 407 (1966); *Rubber Chem. Technol.*, **43**, 95, 156 (1970).
140. S. Bywater and P. M. Toporawski, *Polymer*, **13**, 94 (1972).
141. F. E. Karasz and W. T. MacKnight, *Macromolecules*, **1**, 537 (1968).
142. G. Tamman and W. Jellinghaus, *Ann. Phys.*, [5] **2**, 264 (1929).
143. A. Eisenberg, *J. Phys. Chem.*, **67**, 1333 (1963).
144. J. D. Ferry and R. A. Stratton, *Kolloid Z.*, **171**, 107 (1960).
145. J. M. O'Reilly, *J. Polym. Sci.*, **57**, 429 (1962).
146. J. E. McKinney, H. V. Belcher, and R. S. Marvin, *Trans. Soc. Rheol.*, **4**, 347 (1960).
147. M. Goldstein, *J. Chem. Phys.*, **39**, 3369 (1963).
148. D. L. Quested, K. P. Pae, B. A. Newman, and J. I. Scheinbaum, *J. Appl. Phys.*, **51** (10), 5100 (1980).
149. D. L. Quested and K. D. Pae, *Ind. Eng. Prod. Res. Dev.*, **22**, 138 (1983).
150. Y. Kaieda and K. D. Pae, *J. Mater. Sci.*, **17**, 369 (1982).
151. L. H. Sperling, T. W. Chiu, R. G. Gramlich, and D. A. Thomas, *J. Paint Technol.*, **46**, 47 (1974).
152. J. A. Grates, D. A. Thomas, E. C. Hickey, and L. H. Sperling, *J. Appl. Polym. Sci.*, **19**, 1731 (1975).

GENERAL READING

V. A. Bershtein and V. M. Egorov, *Differential Scanning Calorimetry of Polymers: Physics, Chemistry, Analysis*, Ellis-Horwood, New York, 1994.

R. D. Corsaro and L. H. Sperling, eds., *Sound and Vibration Damping with Polymers*, ACS Books, Symp. Ser. No. 424, American Chemical Society, Washington, DC, 1990.

J. D. Ferry, *Viscoelastic Properties of Polymers*, 3rd ed., Wiley, New York, 1980.

J. A. Forrest and K. Dalnoki-Veress, *Adv. Coll. Interf. Sci.*, **94**, 167 (2001). Reviews the glass transition behavior of thin films.

L. E. Nielsen and R. F. Landel, *Mechanical Properties of Polymers and Composites*, 2nd ed., Dekker, New York, 1994.

J. P. Silbia, ed., *A Guide to Materials Characterization and Chemical Analysis*, VCH, New York, 1988.

A. K. Sircar, M. L. Galaska, S. Rodrigues, and R. P. Chartoff, *Rubber Chem. Tech.*, **72**, 513 (1999).

STUDY PROBLEMS

1. Name the five regions of viscoelastic behavior, and give an example of a commercial polymer commonly used in each region.

2. Name and give a one-sentence definition of each of the three theories of the glass transition.

3. Polystyrene homopolymer has a T_g = 100°C, and polybutadiene has a T_g = −90°C. Estimate the T_g of a 50/50 w/w statistical copolymer, poly(styrene–*stat*–butadiene).

4. A new linear amorphous polymer has a T_g of +10°C. At +25°C, it has a melt viscosity of 6×10^8 poises. Estimate its viscosity at 40°C.

5. Define the following terms: free volume; loss modulus; tan δ; stress relaxation; plasticizer; Schatzki crankshaft motions; A_t; WLF equation; compressibility; Young's modulus.

6. As the newest employee of Polymeric Industries, Inc., you are attending your first staff meeting. One of the company's most respected chemists is speaking:
   "Yesterday, we completed the preliminary evaluation of the newly synthesized thermoplastic, poly(wantsa cracker). The polymer has a melt viscosity of 1×10^5 Pa·s at 140°C. Our characterization laboratory reported a glass transition temperature of 110°C."
   "You know our extruder works best at 2×10^2 Pa·s," broke in the mechanical engineer, "and you know poly(wantsa cracker) degrades at 160°C. Therefore we won't be able to use poly(wantsa cracker)!"
   As you reach for your trusty calculator to estimate the melt viscosity of poly(wantsa cracker) at 160°C to make a reasoned decision of your own, you realize suddenly that all eyes are on you.

   (a) What is the melt viscosity of poly(wantsa cracker) at 160°C?
   (b) Can Polymeric Industries use the polymer? If not, what can they do to the polymer to increase usability?
   (c) What is the structure of poly(wantsa cracker), anyway?

7. Draw a log E–temperature plot for a linear, amorphous polymer.

   (a) Indicate the position and name the five regions of viscoelastic behavior.
   (b) How is the curve changed if the polymer is semicrystalline?
   (c) How is it changed if the polymer is cross-linked?
   (d) How is it changed if the experiment is run faster—that is, if measurements are made after 1 s rather than 10 s?

   In parts (b), (c), and (d), separate plots are required, each change properly labeled. E stands for Young's modulus.

8. During a coffee break, two chemists, three chemical engineers, and an executive vice president began discussing plastics. "Now everyone knows that such materials as plastics, rubber, fibers, paints, and adhesives have very little in common," began the executive vice president. "For

example, nobody manufactures plastics and paints with the same equipment . . ." Even though you are the most junior member of the group, you interrupt; "That last may be so, but all those materials are closely related because . . ."
Complete the statement in 100 words or less. If you think that the above materials are in fact *not* related, you have 100 words to prove the executive vice president correct.

9. Briefly discuss the salient points in the derivation of the WLF equation.

10. Prepare a "box score" table, laying out the more important advantages and disadvantages of the three theories of the glass transition.

11. A new polymer has been synthesized in your laboratory, and you are proudly discussing the first property studies when your boss walks in. "We need a polymer with a cubic coefficient of thermal expansion of less than $4 \times 10^{-4}\,\text{deg}^{-1}$ at 50°C. Can we consider your new stuff?"
Your technician hands you the sheet of paper with available data:

$$T_g = 100°\text{C}$$
$$\alpha_R = 5.5\times10^{-4}\,\text{deg}^{-1} \text{ at } 150°\text{C}$$

Your boss adds, "By the way, the Board meets in 30 minutes. Any answer by then would surely be valuable." You begin to tear your hair out by its roots, wondering how you can solve this one so fast without going back into the lab, since there really isn't much time.

12. The T_g of poly(vinyl acetate) is listed as 29°C. If 5 mol% of divinyl benzene is copolymerized in the polymer during polymerization, what is the new glass transition temperature?

13. Noting the instruments mentioned in Section 8.3, what instrument would you most like to have in your laboratory if you were testing (a) each of the three theories of T_g, (b) the molecular weight dependence of T_g, (c) the effect of cross-linking on T_g. Defend your choice.

14. A new atactic polymer has a T_g of 0°C. Your boss asks, "If we made the isotactic form, what is its melting temperature likely to be?" Suddenly, you remember that back in college you took physical polymer science. . . .

15. Rephrase the definitions in Section 8.13, and use other examples.

16. Your assistant rushes in with a new polymer. "It softens at 50°C," she says. "Is it a glass transition or a melting temperature?" you ask.
"How would I know?" she answers. "I never took physical polymer science!"
Describe two simple but foolproof experiments to distinguish between the two possibilities.

17. A piece of polystyrene is placed under 100 atm pressure at room temperature. What is the fractional volume decrease?

18. What is the free volume of polystyrene at 100°C? What experimental or theoretical evidence supports your conclusion?

19. A rubber ball is dropped from a height of 1 yard and bounces back 18 in. Assuming a perfectly elastic floor, approximately how much did the ball heat up? The heat capacity, C_p, of SBR rubber is about 1.83 kJ $kg^{-1} \cdot K^{-1}$.

20. Write a 100- to 125-word essay on the importance of free volume in polymer science. This essay is to be accompanied by at least one figure, construction, or equation illustrating your thought train.

21. A new polymer was found to soften at 50°C. Several experiments were performed to determine if the softening was a glass transition or a melting point.

(a) In interpreting the results for each experiment separately, was it a glass transition? a melting transition? cannot be determined for sure? or was there some mistake in the experiment?

(b) What is your reasoning for each decision?

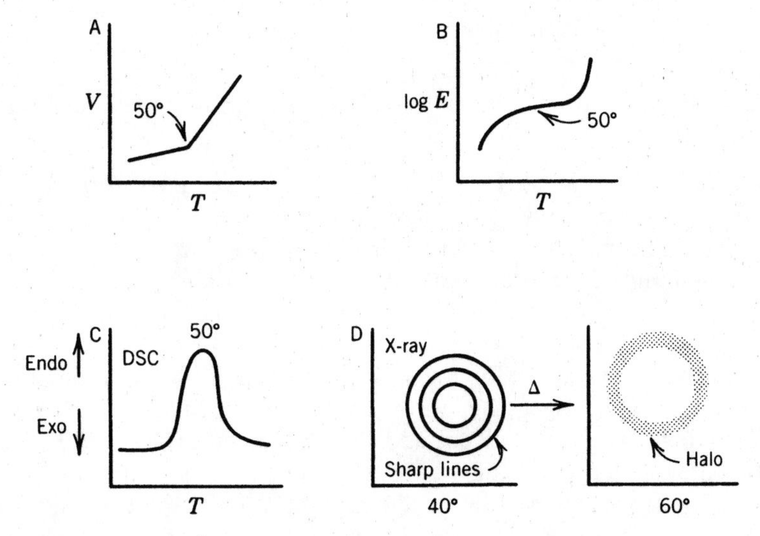

Figure P8.21 Laboratory studies of transitions.

APPENDIX 8.1 MOLECULAR MOTION NEAR THE GLASS TRANSITION†

The following experiment is intended for use as a classroom demonstration and takes about 30 minutes. Place the Superball® in the liquid nitrogen 30

†Reprinted in part from L. H. Sperling. *J. Chem. Ed.*, **59**, 942 (1982).

minutes before class begins. Have a second such ball for comparison of properties.

Experiment

Time: Actual laboratory time about 1 hour
Level: Physical Chemistry
Principles Illustrated:

1. The onset of molecular motion in polymers.
2. The influence of molecular motion on mechanical behavior.

Equipment and Supplies:

One solid rubber ball (a small Superball® is excellent)
One hollow rubber ball (optional)
One Dewar flask of liquid nitrogen, large enough to hold above rubber balls
One ladle or spoon with long handle, to remove frozen balls (alternately, tie balls with long string)
One yardstick (or meter stick)
One clock (a watch is fine)
One hard surface, suitable for ball bouncing (most floors or desks are suitable)

First, place the yardstick vertical to the surface and drop Superball® from the top height; record percent recovery (bounce). Remove Superball® from liquid nitrogen, and record percent bounce immediately and each succeeding minute (or more often) for the first 15 minutes, then after 20 and 25 minutes and each 5 minutes thereafter until recovery equals that first obtained.

For the hollow rubber ball, first observe toughness at room temperature, then cool in liquid nitrogen, then throw hard against the floor or wall. Observe the glassy behavior of the pieces and their behavior as they warm up. The experimenter should wear safety glasses.

For extra credit, obtain three small dinner bells; coat two of them with any latex paint. (Most latex paints have T_g near room temperature.) After drying, ring the bells. Place one coated bell in a freezer, and compare its behavior cold to the room-temperature coated bell. How can you explain the difference observed?

Another related experiment involves dipping adhesive tape into liquid nitrogen. Outdoor gutter drain tapes are excellent because of their size. Compare the stickiness of the tape before and after freezing. (Note again the definition of an adhesive.)

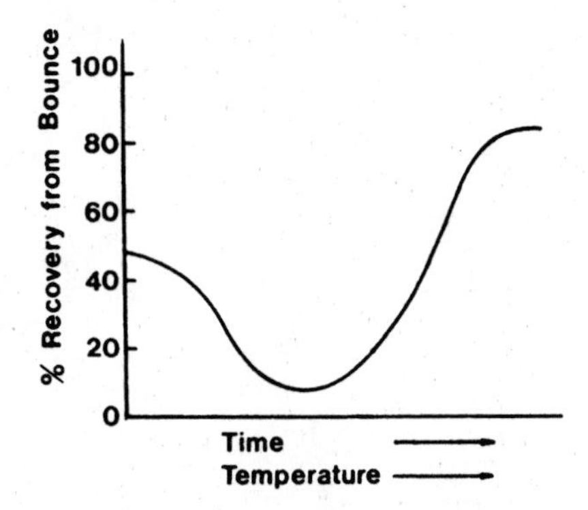

Figure A8.1.1 Ball bounce recovery.

On warming the frozen solid rubber ball, the percent recovery (bounce) versus time will go through a minimum at T_g, as shown in Figure A8.1.1.

Below T_g, the ball is glassy and bounces much like a marble. At T_g, the bounce is at a minimum owing to conversion of kinetic energy to heat. (The ball actually warms up slightly.) Above T_g, normal rubber elasticity and bounce characteristics are observed.

It should be mentioned that the Superball® is based on cross-linked polybutadiene. Thus, its glass transition temperature is close to –85°C. The Superball® gets its extra bounce because it is made under compression (A1). When the pressure is released, the surface of the Superball® attains a certain degree of orientation and stretching on expansion. This results in a phenomenon related to the *trampoline effect*. See Chapter 9 for more detail on rubber elasticity.

REFERENCE

A1. N. A. Stingley, U.S. Pat. 3,241,834 (1966).

For an experiment emphasizing rolling friction near T_g, see G. B. Kauffman, S. W. Mason, and R. B. Seymour, *J. Chem. Ed.*, **67**, 198 (1990).

9

CROSS-LINKED POLYMERS AND RUBBER ELASTICITY

An elastomer is defined as a cross-linked amorphous polymer above its glass transition temperature. Elastomers may be stretched substantially reversibly to several hundred percent. While most of this chapter explores the behavior of elastomers, the study of cross-linking is more general. If the cross-linked polymer is glassy, it is often called a thermoset. Below, the terms elastomer and rubber are often used interchangably.

9.1 CROSS-LINKS AND NETWORKS

During reaction, polymers may be cross-linked to several distinguishable levels. At the lowest level, branched polymers are formed. At this stage the polymers remain soluble, sometimes known as the sol stage. As cross-links are added, clusters form, and cluster size increases. Eventually the structure becomes infinite in size; that is, the composition gels. At this stage a Maxwellian demon could, in principle, traverse the entire macroscopic system stepping on one covalent bond after another. Continued cross-linking produces compositions where, eventually, all the polymer chains are linked to other chains at multiple points, producing, in principle, one giant covalently bonded molecule. This is commonly called a polymer network.

9.1.1 The Sol–Gel Transition

The reaction stage referred to as the sol–gel transition (1–3) is called the gel point. At the gel point the viscosity of the system becomes infinite, and the

Introduction to Physical Polymer Science, by L.H. Sperling
ISBN 0-471-70606-X

equilibrium modulus climbs from zero to finite values. In simple terms the polymer goes from being a liquid to being a solid. There are three different routes for producing cross-linked polymers:

1. Step polymerization reactions, where little molecules such as epoxies (oxiranes) react with amines, or isocyanates react with polyols with functionality greater than two to form short, branched chains, eventually condensing it into epoxies or polyurethanes, respectively. Schematically

$$n\text{A}\sim\text{A} + m\text{B}\overset{\text{B}}{\underset{|}{\sim}}\text{B} \rightarrow \begin{array}{c} \text{BA}\sim\text{AB}\sim \\ | \\ \sim\text{AB}\sim\text{BA}\sim\text{AB}\sim \end{array} \tag{9.1}$$

2. Chain polymerization, with multifunctional molecules present. An example is styrene polymerized with divinyl benzene.
3. Postpolymerization reactions, where a linear (or branched) polymer is cross-linked after synthesis is complete. An example is the vulcanization of rubber with sulfur, which will be considered further below.

The general features of structural evolution during gelation are described by percolation (or connectivity) theory, where one simply connects bonds (or fills sites) on a lattice of arbitrary dimension and coordination number (4–6). Figure 9.1 (6) illustrates a two-dimensional system at the gel point. It must be noted that gels at and just beyond the gel point usually coexist with sol clusters. These can also be seen in Figure 9.1. It is common to speak of the con-

Figure 9.1 A square lattice example of percolation, at the gel point (6). Note structures that span the whole "sample."

version factor, p, which is the fraction of bonds that have been formed between the mers of the system; see Section 3.7. For the two-dimensional schematic illustrated in Figure 9.1, $p = \frac{1}{2}$ yields the gel point.

9.1.2 Micronetworks

A special type of network exists that involves only one or a few polymer chains. Thus microgels may form during specialized reaction conditions where only a few chains are interconnected.

Globular proteins constitute excellent examples of one-molecule micronetworks (7), where a single polymer chain is intramolecularly cross-linked; see Figure 9.2 (8). Here, disulfide bonds help keep the three-dimensional structure required for protein biopolymer activity.

Such proteins may be denatured by any of several mechanisms, especially heat. Thus, when globular proteins are cooked, the intramolecular cross-links become delocalized, forming intermolecular bonds instead. This is the major difference between raw egg white and hard-boiled egg white, for example.

Fully cross-linked on a macroscopic scale, polymer networks fall into different categories. It is convenient to call such polymers, which are used below

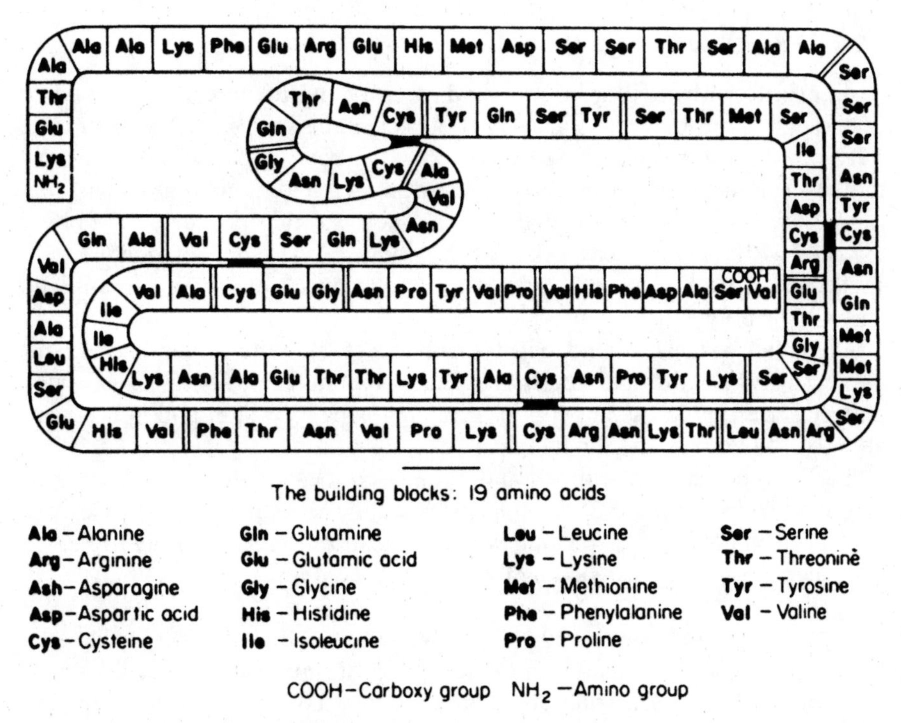

Figure 9.2 The structure of ribonuclease A, a micronetwork of one chain. Note cysteine–cysteine bonds (8).

their glass transition temperature, thermosets, since they are usually plastics incapable of further flow. On the other hand, those networks that are above their glass transition temperature are rubbery unless very heavily cross-linked. While all these kinds of cross-linked polymers are important, the remainder of this chapter is devoted primarily to amorphous, continuous polymer networks above their glass transition temperature, the regime of rubber elasticity.

While the rubber elasticity theory to be described below presumes a randomly cross-linked polymer, it must be noted that each method of network formation described above has distinctive nonuniformities, which can lead to significant deviation of experiment from theory. For example, chain polymerization leads first to microgel formation (9,10), where several chains bonded together remain dissolved in the monomer. On continued polymerization, the microgels grow in number and size, eventually forming a macroscopic gel. However, excluded volume effects, slight differences in reactivity between the monomer and cross-linker, and so on lead to systematic variations in cross-link densities at the 100- to 500-Å level.

9.2 HISTORICAL DEVELOPMENT OF RUBBER

9.2.1 Early Developments

A simple rubber band may be stretched several hundred percent; yet on being released, it snaps back substantially to its original dimensions. By contrast, a steel wire can be stretched reversibly for only about a 1% extension. Above that level, it undergoes an irreversible deformation and then breaks. This long-range, reversible elasticity constitutes the most striking property of rubbery materials. Rubber elasticity takes place in the third region of polymer viscoelasticity (see Section 8.2) and is especially concerned with cross-linked amorphous polymers in that region.

Columbus, on his second trip to America, found the American Indians playing a game with rubber balls (11,12) made of natural rubber. These crude materials were un-cross-linked but of high molecular weight and hence were able to hold their shape for significant periods of time.

The development of rubber and rubber elasticity theory can be traced through several stages. Perhaps the first scientific investigation of rubber was by Gough in 1805 (13). Working with unvulcanized rubber, Gough reached three conclusions of far-reaching thermodynamic impact:

1. A strip of rubber warms on stretching and cools on being allowed to contract. (This experiment can easily be confirmed by a student using a rubber band. The rubber is brought into contact with the lips and stretched rapidly, constituting an adiabatic extension. The warming is easily perceived by the temperature-sensitive lips.)

2. Under conditions of constant load, the stretched length decreases on heating and increases on cooling. Thus it has more retractive strength at higher temperatures. This is the opposite of that observed for most other materials.
3. On stretching a strip of rubber and putting it in cold water, the rubber loses some of its retractile power, and its relative density increases. On warming, however, the rubber regains its original shape. In the light of present-day knowledge, this last set of experiments involved the phenomenon known as strain-induced crystallization, since unvulcanized natural rubber crystallizes easily under these conditions.

In 1844 Goodyear vulcanized rubber by heating it with sulfur (14). In modern terminology, he cross-linked the rubber. (Other terms meaning cross-linking include "tanning" of leather, "drying" of oil-based paints, and "curing" of inks.) Vulcanization introduced dimensional stability, reduced creep and flow, and permitted the manufacture of a wide range of rubber articles, where before only limited uses, such as waterproofing, were available (15). (The "MacIntosh" raincoat of that day consisted of a sandwich of two layers of fabric held together by a layer of unvulcanized natural rubber.)

Using the newly vulcanized materials, Gough's line of research was continued by Kelvin (16). He tested the newly established second law of thermodynamics with rubber and calculated temperature changes for adiabatic stretching. The early history of rubber research has been widely reviewed (17,18).

All of the applications above, of course, were accomplished without an understanding of the molecular structure of polymers or of rubber in particular. Beginning in 1920, Staudinger developed his theory of the long-chain structure of polymers (19,20). [Interestingly Staudinger's view was repeatedly challenged by many investigators tenaciously adhering to ring formulas or colloid structures held together by partial valences (21).] See Appendix 5.1.

9.2.2 Modern Developments

In the early days the only elastomer was natural rubber. Starting around 1914, a polymer of 2,3-dimethylbutadiene known as methyl rubber was made in Germany. This was replaced by a styrene–butadiene copolymer called Buna-S (butadiene–natrium–styrene), where natrium is, of course, sodium. This sodium-catalyzed copolymer, as manufactured in Germany in the period from about 1936 to 1945 had about 32% styrene monomer (22).

In 1939 the U.S. government started a crash program to develop a manufactured elastomer, called the Synthetic Rubber Program (23). The new material was called GR-S (government rubber-styrene). GR-S was made by emulsion polymerization. While the Bunas was catalyzed by sodium, the latter was catalyzed by potassium persulfate. Incidentally, the emulsifier in those days was ordinary soap flakes. Both materials played crucial roles in World War II, a story told many times (23–25).

While Buna-S had about 32% styrene monomer (22), the GR-S material, started a few years later with the benefit of the German recipe, had about 25% styrene (26). This difference in the composition was important for lowering the glass transition temperature. Using the Fox equation, equation (8.73), with a T_g of polystyrene of 373 K, and that of polybutadiene of 188 K, Table 8.7, values of T_g for Buna-S and GR-S are estimated at −47 and −58°C, respectively. Noting that winter temperatures in European Russia reach −40 to −51°C (27), lore has it that use of Buna-S seriously influenced the winter Russian campaigns.

Today, SBR elastomers are widely manufactured with only minor improvements in the GR-S recipe. One such is the use of synthetic surfactants as emulsifiers. These and other improvements allow the production of more uniform latex particles.

9.3 RUBBER NETWORK STRUCTURE

Once the macromolecular hypothesis of Staudinger was accepted, a basic understanding of the molecular structure was possible. Before cross-linking, rubber (natural rubber in those days) consists of linear chains of high molecular weight. With no molecular bonds between the chains, the polymer may flow under stress if it is above T_g.

The original method of cross-linking rubber, via sulfur vulcanization, results in many reactions. One such may be written

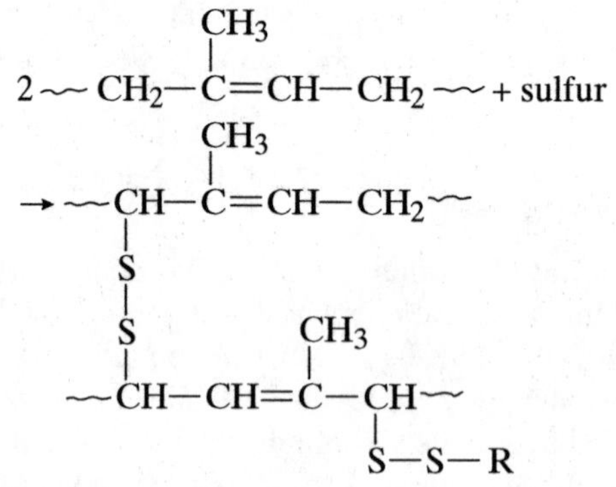

```
                CH3
                |
2 ~~ CH2—C=CH—CH2 ~~ + sulfur
                CH3
                |
→ ~~CH—C=CH—CH2~~
      |
      S                                   (9.1a)
      |
      S          CH3
      |          |
  ~~CH—CH=C—CH~~
                    |
                    S—S—R
```

where R represents other rubber chains.

Two other methods of cross-linking polymers must be mentioned here. One is radiation cross-linking, with an electron beam or gamma irradiation. Using polyethylene as an example,

$$
2(\sim CH_2-CH_2-CH_2-CH_2\sim) \xrightarrow{h\nu} \begin{array}{l} \sim CH_2-CH-CH_2-CH_2\sim \\ \qquad\quad | \\ \sim CH_2-CH-CH_2-CH_2\sim \end{array} + H_2 \tag{9.2}
$$

Another method involves the use of a multifunctional monomer in the simultaneous polymerization and cross-linking of polymers. Taking poly(ethyl acrylate) as an example, with divinyl benzene as cross-linker,

$$
2\sim CH_2-\underset{\underset{O-C_2H_5}{|}}{\underset{C=O}{\underset{|}{\overset{\overset{H}{|}}{C}\cdot}}} + \text{(H)C}{=}CH_2-C_6H_4-CH{=}CH_2 \rightarrow \begin{array}{l} \sim CH_2-C(H)(C(=O)OC_2H_5)-CH(H)-CH\cdot \ (\text{attached to } C_6H_4) \\ \sim CH_2-C(H)(C(=O)OC_2H_5)-CH(H)-CH\cdot \ (\text{attached to } C_6H_4) \end{array} \tag{9.3}
$$

where the upper and lower reactions take place independently in time.

After cross-linking, flow of one molecule past another (viscoelastic behavior) is suppressed. Excluding minor impurities, an object such as a rubber band can be considered as one huge molecule. [It fulfills the two basic requirements of the definition of a molecule: (a) every atom is covalently bonded to every other atom, and (b) it is the smallest unit of matter with the characteristic properties of rubber bands.]

The structure of a cross-linked polymer may be idealized (Figure 9.3). The primary chains are cross-linked at many points along their length. For materials such as rubber bands, tires, and gaskets, the primary chains may have molecular weights of the order of 1×10^5 g/mol and be cross-linked (randomly) every 5 to 10×10^3 g/mol along the chain, producing 10 to 20 cross-links per primary molecule. It is convenient to define the average molecular weight between cross-links as M_c and to call chain portions bound at both ends by cross-link junctions active network chain segments.

In the most general sense, an elastomer may be defined as an amorphous, cross-linked polymer above its glass transition temperature (see Section 8.12). The two terms "rubber" and "elastomer" mean nearly the same thing. The term rubber comes from the "rubbing out" action of an eraser. Originally, of course, rubber was natural rubber, *cis*-polyisoprene. The term elastomer is more general and refers to the elastic-bearing properties of the materials.

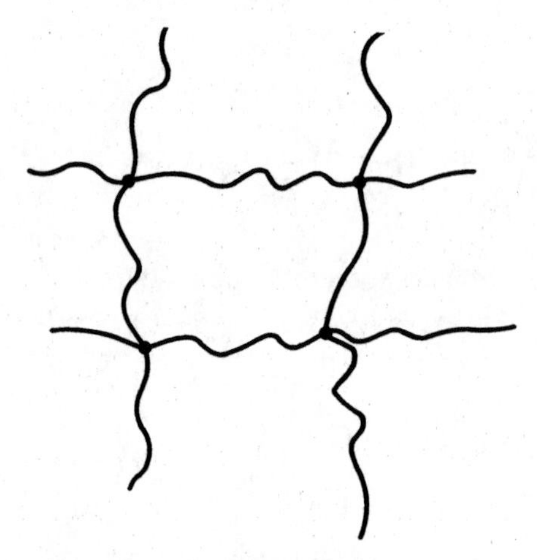

Figure 9.3 Idealized structure of a cross-linked polymer. Wavy lines, polymer chains; circles, cross-links.

9.4 RUBBER ELASTICITY CONCEPTS

The first relationships between macroscopic sample deformation, chain extension, and entropy reduction were expressed by Guth and Mark (28) and by Kuhn (29,30) (see Section 5.3). Mark and Kuhn proposed the model of a random coil polymer chain (Figure 9.4) which forms an active network chain segment in the cross-linked polymer. When the sample was stretched, the chain had extended in proportion, now called an affine deformation. When the sample is relaxed, the chain has an average end-to-end distance, r_0 (Figure 9.4), which increases to r when the sample is stretched. (Obviously, if the sample is compressed or otherwise deformed, different chain dimensional changes will occur.)

Through the research of Guth and James (31–35), Treloar (36), Wall (37), and Flory (38), the quantitative relations between chain extension and entropy reduction were clarified. In brief, the number of conformations that a polymer chain can assume in space were calculated. As the chain is extended, the number of such conformations diminishes. (A fully extended chain, in the shape of a rod, has only one conformation, and its conformational entropy is zero.)

The idea was developed, in accordance with the second law of thermodynamics, that the retractive stress of an elastomer arises through the reduction of entropy rather than through changes in enthalpy. Thus long-chain molecules, capable of reasonably free rotation about their backbone, and joined together in a continuous, monolithic network are required for rubber elasticity.

In brief, the basic equation relating the retractive stress, σ, of an elastomer in simple extension to its extension ratio, α, is given by

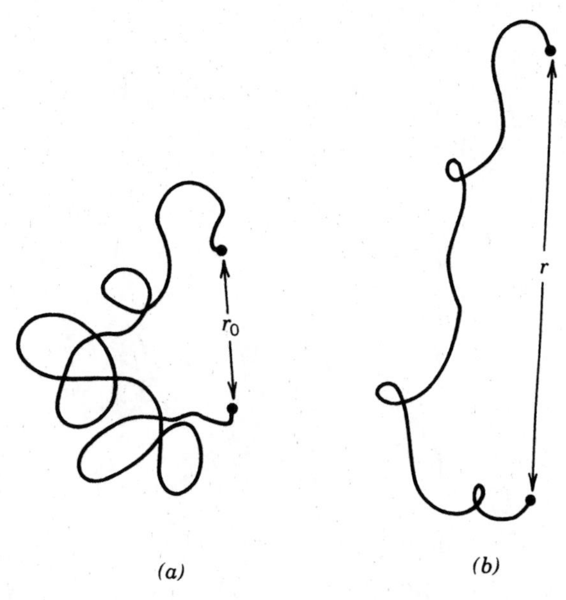

Figure 9.4 A network chain segment: (*a*) relaxed, with a random coil conformation, and (*b*) extended, owing to an external stress.

$$\sigma = nRT\left(\alpha - \frac{1}{\alpha^2}\right) \tag{9.4}$$

where the original length, L_0, is increased to L ($\alpha = L/L_0$), and RT is the gas constant† times the absolute temperature. The quantity n represents the number of active network chain segments per unit volume (39–42). The quantity n equals ρ/M_c, where ρ is the density and M_c is the molecular weight between cross-links.

Several quantities must be distinguished here:

1. The quantity M_c in a polymer network (the present case) is the number-average molecular weight between cross-links.
2. For linear polymers the entanglement molecular weight, M_c', is the value given later in this book by the elbows in Figure 10.16. (Note that some texts list this as M_c, leading to confusion.)
3. The quantity M_e is the molecular weight between the entanglements. Wool (43) defines $M_e = (4/9)M_c'$ for linear polymers but points out that experimentally $M_e \cong (1/2)M_c'$.

† Convenient values of R for calculation purposes are 8.31×10^7 dynes·cm/mol·K, when the stress has units of dynes/cm^2, and 8.31 Pa·m^3/mol·K, when stress has units of Pa.

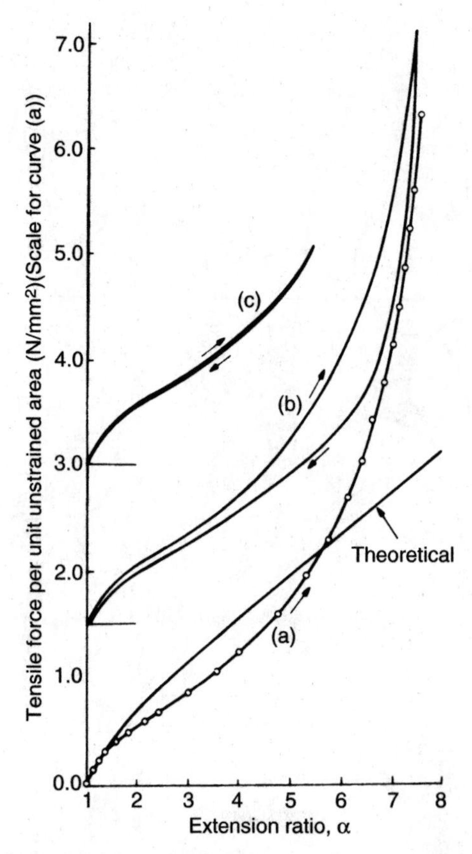

Figure 9.5 Stress–strain behavior of lightly cross-linked natural rubber at 50°C. Curve (*a*), experimental. Theoretical is equation (9.4). Curve (*c*) illustrates the reversible nature of the extension up to $\alpha = 5.5$. At higher elongation, curve (*b*), hysteresis effects become important. The theoretical curve has been fitted to the experimental data in the region of small extensions, with $nRT = 0.39$ N/mm^2 (47,48).

As described in Section 9.10.5, real polymer networks contain both chemical and physical cross-links, the latter being the various kinds of entanglements. It will be observed that equation (9.4) is nonlinear; that is, the Hookean simple proportionality between stress and strain does not hold. Young's modulus is often close to 2×10^6 Pa.

Equation (9.4) is compared to experiment in Figure 9.5 (44,45). The theoretical value of M_c was chosen for the best fit at low extensions. The sharp upturn of the experimental data above $\alpha = 7$ is due to the limited extensibility of the chains themselves,[†] which can be explained in part by more advanced theories (46).

[†] Part of the effect is usually strain-induced crystallinity especially for natural rubber and *cis*-polybutadiene.

In the following sections the principal equations of the theory of rubber elasticity are derived, emphasizing the relationships between molecular chain characteristics, stress, and strain.

9.5 THERMODYNAMIC EQUATION OF STATE

As a first approach to the equation of state for rubber elasticity, we analyze the problem via classical thermodynamics. The Helmholtz free energy, F, is given by

$$F = U - TS \tag{9.5}$$

where U is the internal energy and S is the entropy.

The retractive force, f, exerted by the elastomer depends on the change in free energy with length:

$$f = \left(\frac{\partial F}{\partial L}\right)_{T,V} = \left(\frac{\partial U}{\partial L}\right)_{T,V} - T\left(\frac{\partial S}{\partial L}\right)_{T,V} \tag{9.6}$$

For an elastomer, Poisson's ratio is nearly 0.5, so the extension is nearly isovolume. When the experiment is done isothermally, the analysis becomes significantly simplified; note the subscripts to equation (9.6).

According to the statistical thermodynamic approach to be developed below, each conformation that a network chain segment may take is equally probable. The number of such conformations depends on the end-to-end distance, r, of the chain, reaching a rather sharp maximum at r_0. The retractive force of an elastomer is developed by the thermal motions of the chains, statistically driven toward their most probable end-to-end distance, r_0.

The changes in numbers of chain conformations can be expressed as an entropic effect. Thus, for an ideal elastomer, $(\partial U/\partial L)_{T,V} = 0$.

By contrast, most other materials develop internal energy-driven retractive forces. For example, on extension, in a steel bar the iron atoms are forced farther apart than normal, calling into play energy well effects and concomitant increased atomic attractive forces. Such a model assumes the opposite effect, $(\partial S/\partial L)_{T,V} = 0$.

As derived by Wall (46), there is a perfect differential mathematical relationship between the entropy and the retractive force:

$$-\left(\frac{\partial S}{\partial L}\right)_{T,V} = \left(\frac{\partial f}{\partial T}\right)_{L,V} \tag{9.7}$$

Equation (9.6) can then be expressed as

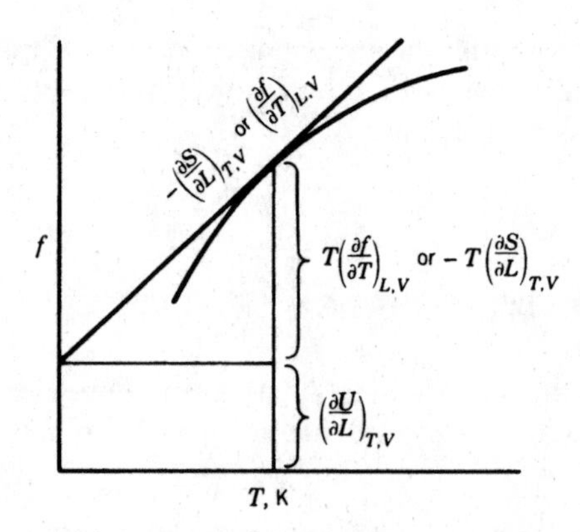

Figure 9.6 An analysis of the thermodynamic equations of state for rubber elasticity (18b).

$$f = \left(\frac{\partial U}{\partial L}\right)_{T,V} + T\left(\frac{\partial f}{\partial T}\right)_{L,V} \tag{9.8}$$

which is sometimes called the thermodynamic equation of state for rubber elasticity.

Equation (9.8) can be analyzed with the aid of a construction due to Flory (18b) (see Figure 9.6), which illustrates a general experimental curve of force versus temperature. The tangent line is extended back to 0 K. For an ideal elastomer, the quantity $(\partial U/\partial L)_{T,V}$ is zero, and the entropic portion of the tangent line goes through the origin. Of course, the experimental line is straight in the ideal case, the slope being proportional to $-(\partial S/\partial L)_{T,V}$ or $(\partial f/\partial T)_{L,V}$.

The first term on the right of equation (9.8) expresses the energetic portion of the retractive force, f_e, and the second term on the right expresses the entropic portion of the force, f_s. Thus

$$f = f_e + f_s \tag{9.9}$$

Equations (9.8) and (9.9) call for stress–temperature (isometric) experiments (47).[†] While a detailed analysis of such experiments will be presented below in Section 9.10, Figure 9.7 (47) shows the results of such an isometric study. The quantity f_s accounts for more than 90% of the stress, whereas f_e hovers near zero. The turndown of f_e above 300% elongation may be due to incipient crystallization.

Incidentally, equation (9.4) fits Figure 9.7 much better than it does Figure 9.5. There are two reasons: (a) Figure 9.7 represents a much closer approach to an equilibrium stress–strain curve, and (b) the much higher level of sulfur

[†] Of course, the term "stress" refers to the force per unit initial cross section (see Section 8.1). Much of the early literature talks about force but measures stress. However, f refers to force.

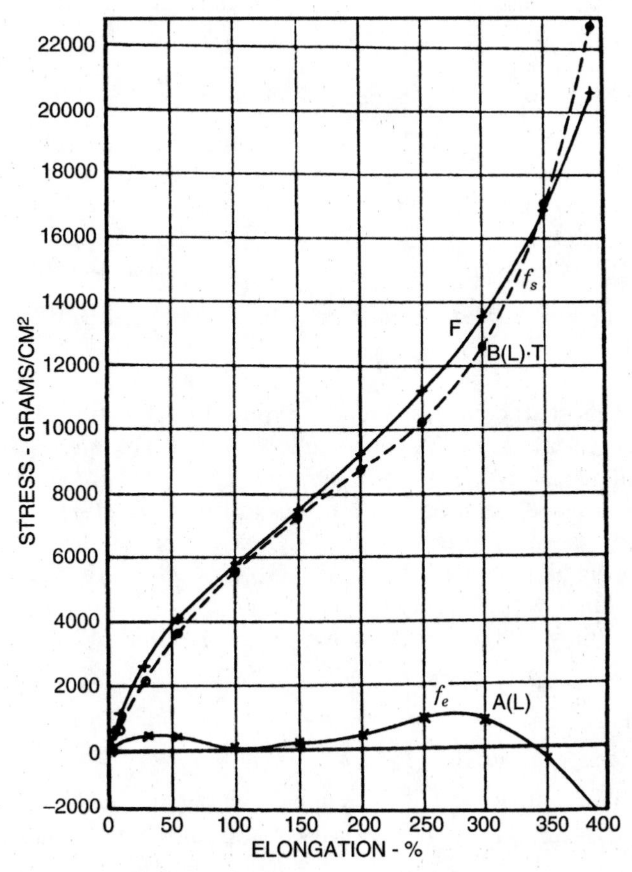

Figure 9.7 The total retractive force and its entropic, f_s, and energetic components, f_e, as a function of elongation. Natural rubber vulcanized with 8% sulfur; values at 20°C (47).

used in the vulcanization (8% vs. 2%) reduces the quantity of crystallization at high elongations.

9.6 EQUATION OF STATE FOR GASES

The equation of state of rubber elasticity will now be calculated via statistical thermodynamics, rather than the classical thermodynamics of Section 9.5. Statistical thermodynamics makes use of the probability of finding an atom, segment, or molecule in any one place as a means of computing the entropy. Thus tremendous insight is obtained into the molecular processes of entropic phenomena, although classical thermodynamics illustrates energetic phenomena adequately.

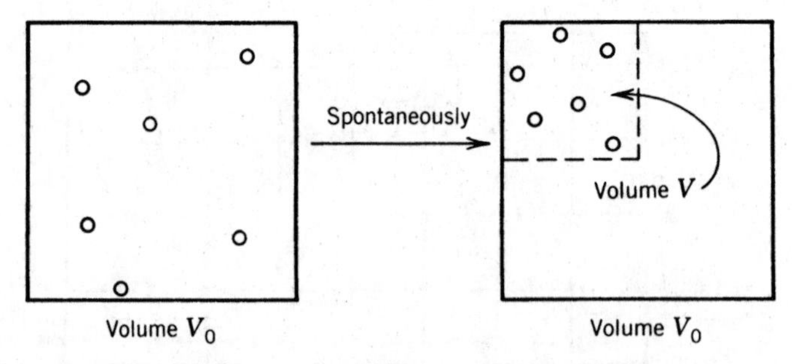

Figure 9.8 Ideal gas molecules spontaneously moving from a volume V_0 to a volume V.

However, most students not broadly exposed to statistical thermodynamics find such calculations difficult to follow at first. For this reason we first derive the ideal gas law via the very same principles that are employed in calculating the stress–elongation relationships in rubber elasticity.

Consider a gas of ν molecules in an original volume of V_0 (18c). Let us calculate the probability Ω that all the gas molecules will move spontaneously to a smaller volume, V (see Figure 9.8).

The probability of finding one molecule in the volume V is given by

$$p_1 = \frac{V}{V_0} \tag{9.10}$$

Neglecting the volume actually occupied by the molecules (an ideal gas assumption), the probability of finding two molecules in the volume V is given by

$$p_2 = \left(\frac{V}{V_0}\right)^2 \tag{9.11}$$

and the probability of finding all the molecules (spontaneously) in the volume V is given by

$$\Omega = \left(\frac{V}{V_0}\right)^\nu \tag{9.12}$$

The change in entropy, ΔS, is given by the Boltzmann relation†

$$\Delta S = +k \ln \Omega \tag{9.13}$$

†The Boltzmann hypothesis *assumes* rather than derives a logarithmic relationship between the probability of a state (or the number of such states, the inverse) and the entropy. Boltzmann's constant, k, is defined as the constant of proportionality.

which yields

$$\Delta S = k\nu \ln\left(\frac{V}{V_0}\right) \tag{9.14}$$

The pressure of the gas is given by

$$P = -\left(\frac{\partial F}{\partial V}\right)_T = -\left(\frac{\partial U}{\partial V}\right)_T + T\left(\frac{\partial S}{\partial V}\right)_T \tag{9.15}$$

where F is the Helmholtz free energy. For a perfect gas, $(\partial U/\partial V)_T = 0$, and

$$P = T\left(\frac{\partial S}{\partial V}\right)_T = \frac{k\nu T}{V} \tag{9.16}$$

If moles instead of molecules are considered, k becomes R and ν becomes n, yielding the familiar

$$PV = nRT \tag{9.17}$$

If the internal energy is not required to be zero, more complex equations arise. For example, van der Waal's law per mole gives

$$P = -\frac{a}{V^2} + \frac{nRT}{V - b} \tag{9.18}$$

where the first term on the right indicates an energetic (attractive force) term and the second term on the right is the entropic term corrected for the molar volume of the gas (47).

Returning to Figure 9.8, if any large number of molecules are involved (e.g., 1 mol), the probability of the gas molecules spontaneously moving to a much smaller volume is very small. Of course, if they are arbitrarily so moved, a pressure P is required to keep them there.

In an analogous strip of rubber, the corresponding situation would be the spontaneous elongation of the strip (a rubber band stretching itself). On a molecular level, such motions are spontaneous, but because of the low probabilities involved, are but momentary. Again, the phenomenon is possible but unlikely. Instead of a pressure P to hold the gas in the volume V, a stress σ (force per unit area) will be required to keep the elastomer stretched from L_0 to L ($\alpha = L/L_0$).

In both problems, to the first approximation, the internal energy component can be assumed to be zero. Table 9.1 compares the concepts of an ideal gas with those of an ideal elastomer, to be developed below.

Table 9.1 Corresponding concepts in ideal gases and ideal elastomers

$PV = nRT$	$G = nRT$
Entropy calculated from probabilities of finding n molecules in a given volume	Entropy calculated from probability of finding end-to-end distance r at r_0
Probability of the gas volume spontaneously decreasing $(\partial U/\partial V)_T =$ 0; internal energy assumed zero	Probability of an elastomer strip spontaneously elongating $(\partial U/\partial V)_{T,V} =$ 0; internal energy assumed constant
Molar volume of gas assumed zero	Elastomer assumed incompressible (molar volume is constant)
Pressure P given by $-(\partial F/\partial V)_T$	Retractive force f given by $-(\partial F/\partial V)_{T,V}$

9.7 STATISTICAL THERMODYNAMICS OF RUBBER ELASTICITY

It is useful to consider again the freely jointed chain (Section 5.3.1). In this case the root-mean-square end-to-end distance is given by $(\overline{r_f^2})^{1/2} = ln^{1/2}$. In a real random coil, with fixed bond angles, the quantity $(\overline{r_f^2})^{1/2}$ is larger but still obeys the $n^{1/2}$ relationship. For a given value of n, however, the root-mean-square end-to-end distance can vary widely, from zero, where the ends touch, to nl, the length of the equivalent rod. The probability of finding particular values of r underlies the following subsections.

9.7.1 The Equation of State for a Single Chain

It is convenient to divide the derivation of equations such as (9.4) into two parts. First, the equation of state for a single chain in space is derived. Then we show how a network of such chains behaves.

It is convenient to start again with the general equation for the Helmholtz free energy, equation (9.5):

$$F = U - TS$$

This can be rewritten in statistical thermodynamic notation:

$$F = \text{constant} - kT \ln \Omega(r, T) \tag{9.19}$$

where the quantity $\Omega(r, T)$ [see equations (9.12) and (9.13)] now refers to the probability that a polymer molecule with end-to-end distance r at temperature T will adopt a given conformation.†

From a quantitative point of view, at each particular end-to-end distance, all possible conformations of the chain need to be counted, holding the ends

†The term "conformation" refers to those arrangements of a molecule that can be attained by rotating about single bonds. Configurations refer to tacticity, steric arrangement, *cis* and *trans*, and so on; see Chapter 2.

fixed in space. Then the sum of all such conformations as the end-to-end distance is varied needs to be calculated. (Later the sum of all conformations of a distribution of molecular weights is considered.)

The retractive force is given by

$$f = \left(\frac{\partial F}{\partial r}\right)_{T,V} = -kT\left(\frac{\partial \ln \Omega(r,T)}{\partial r}\right)_{T,V} \tag{9.20}$$

As before, the quantity U, assumed to be constant (or zero), drops out of the calculation, leaving only the entropic contribution. In this case, for a single chain, the quantity f for force must be used. The cross section of the individual chain, necessary for a determination of the stress, remains undefined.

A particular direction in space is selected first. Then, using vector notation, the probability that r lies between r and $r + dr$ in that direction is given by

$$W(r)\,dr = \frac{\Omega(r,T)\,dr}{\int_0^\infty \Omega(r,T)\,dr} \tag{9.21}$$

where the denominator serves as a normalizing factor. Of course, the integral does not need to extend to infinity; in reality different conformations only go to the fully extended, rodlike chain.

Removing the directional restrictions on r,

$$W(r)\,dr = \frac{\Omega(r,T)\,dr}{\int_0^\infty \Omega(r,T)\,dr} 4\pi r^2 \tag{9.22}$$

A spherical shell between r and $r + dr$ is generated (see Figure 9.9), depicted in Cartesian coordinates.

Rearranging equation (9.22) and taking logarithms,

$$\ln \Omega(r,T) = \ln W(r) + \ln \int_0^\infty \Omega(r,T)\,dr - \ln 4\pi - \ln r^2 \tag{9.23}$$

Differentiating with respect to r,

$$\left(\frac{\partial \ln \Omega(r,T)}{\partial r}\right)_{T,V} = \left(\frac{\partial \ln W(r)}{\partial r}\right)_{T,V} - \frac{2}{r} \tag{9.24}$$

since $\int_0^\infty \Omega(r,T)\,dr$ is independent of r.

The quantity $W(r)$ can be expressed as a Gaussian distribution (18):

$$W(r) = \left(\frac{\beta}{\pi^{1/2}}\right)^3 e^{-\beta^2 r^2} 4\pi r^2 \tag{9.25}$$

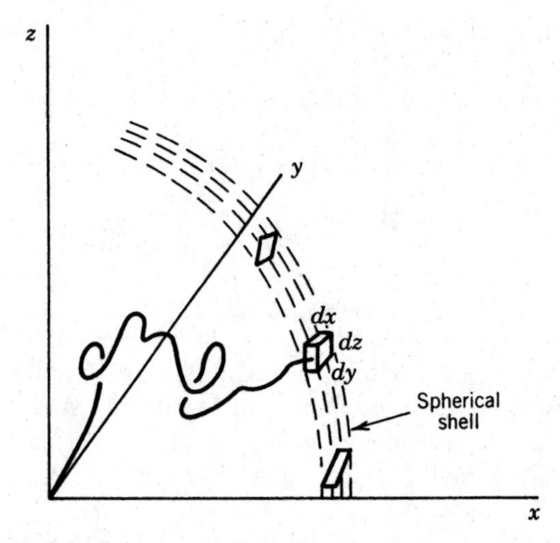

Figure 9.9 A spherical shell at a distance r (inner surface) and $r + dr$ (outer surface), defining all conformations in space having that range of r.

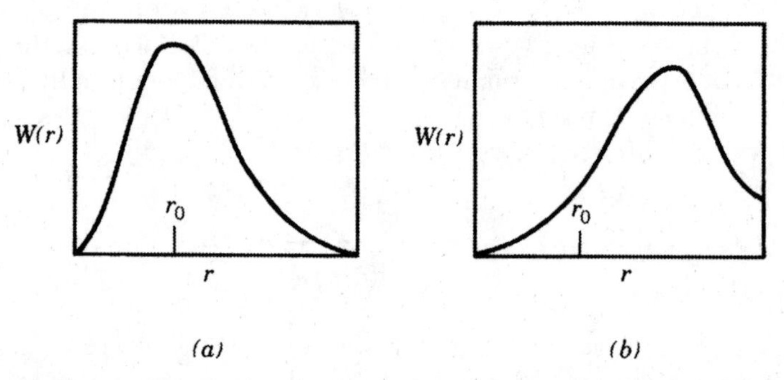

Figure 9.10 The radial distribution function $W(r)$ as a function of r for (a) a relaxed chain, where the most probable end-to-end distance is r_0, and (b) for a chain under an extensive force, f.

In molecular terms, $\beta^2 = 3/(2\overline{r_0^2})$, where $\overline{r_0^2}$ represents the average of the squares of the relaxed end-to-end distances.

The quantity $W(r)$ is shown as a function of r in Figure 9.10. The radial distribution function $W(r)$ is shown for a relaxed chain and a chain extended by a force f.

Equation (9.24) can now be expressed as

$$\left(\frac{\partial \ln \Omega(r,T)}{\partial r}\right)_{T,V} = -2\beta^2 r \tag{9.26}$$

Substituting equation (9.26) into equation (9.20),

$$f = \frac{3kTr}{\overline{r_0^2}} \tag{9.27}$$

and the equation of state for a single chain is obtained. The force appears to be zero at $r = 0$, because of the spherical shell geometry assumed (Figure 9.9). Again, the quantity r_0 is the isotropic end-to-end distance for a free chain in space. This approximates the end-to-end distance expected both in θ-solvents and in the bulk state, for linear chains.

9.7.2 The Equation of State for a Macroscopic Network

Equation (9.27) expresses the retractive force of a single chain on extension. Since the retractive force is proportional to the quantity r, the chain behaves as a Hookean spring on extension. The problem now is to link a large number, n, of these chains together per unit volume to form a macroscopic network.

The assumption of affine deformation is required; that is, the junction points between chains move on deformation as if they were embedded in an elastic continuum (Figure 9.11). The mean square length of a chain in the strained state is given by

$$\overline{r^2} = \tfrac{1}{3}(\alpha_x^2 + \alpha_y^2 + \alpha_z^2)\overline{r_i^2} \tag{9.28}$$

The alphas represent the fractional change in shape in the three directions, equal to $(L/L_0)_i$, where $i = x, y$, or z.

The quantity $\overline{r_i^2}$ represents the isotropic, unstrained end-to-end distance in the network. The two quantities $\overline{r_i^2}$ and $\overline{r_0^2}$ [see equation (9.27)] bear exact comparison. They represent the same chain in the network and un-cross-linked states, respectively. Under many circumstances the quantity $\overline{r_i^2}/\overline{r_0^2}$ approximately equals unity. In fact the simpler derivations of the equation of state for rubber elasticity do not treat this quantity, implicitly assuming it to be unity (42). However, deviations from unity may be caused by swelling, cross-linking while in tension, changes in temperature, and so on, and play an important role in the development of modern theory.

The difference between r_i and r_0 may be illustrated by way of an example. Assume a cube of dimensions $1 \times 1 \times 1$ cm. Its volume, of course, is 1 cm^3. If it is swelled to 10 cm^3 volume, then each linear dimension in the sample is increased by $(10)^{1/3}$, the new length of the sides, in this instance. Assuming an affine deformation, the end-to-end distance of the chains will also increase by $(10)^{1/3}$; that is, $r_i = (10)^{1/3} r_0$. The value of r_0 does not change, because it is the end-to-end distance of the equivalent free chain. The value of r_i is determined by the distances between the cross-link sites binding the chain. Again, neither r_i nor r_0 represents the end-to-end distance of the whole primary chain but rather the end-to-end distance between cross-link junctions.

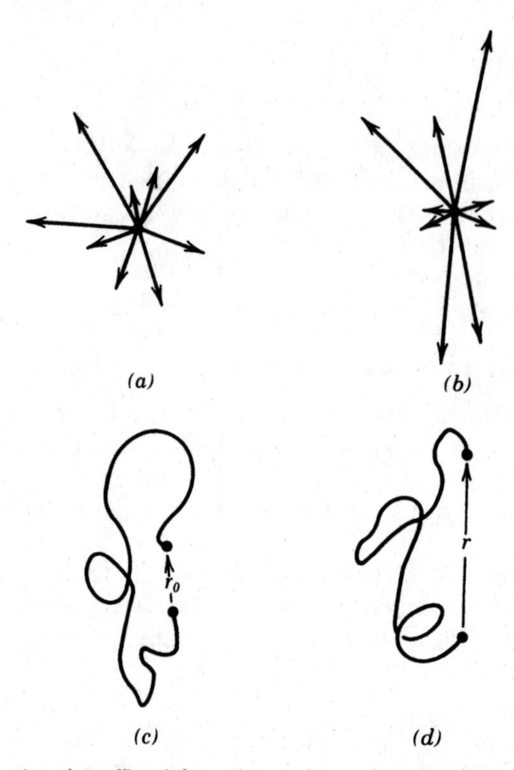

Figure 9.11 Illustration of an affine deformation, end-to-end vectors drawn from a central point. (*a*) End-to-end vectors have spherical symmetry. (*b*) After extending the macroscopic sample, end-to-end vectors have ellipsoidal symmetry. Parts (*c*) and (*d*) illustrate the corresponding effects on a single chain.

For the work done by the n network chains,

$$-W = \Delta F_{\text{el}} \tag{9.29}$$

where ΔF_{el} represents the change in the Helmholtz free energy due to elastic deformation. Returning to equation (9.27), for n chains,

$$\Delta F_{\text{el}} = \frac{3nRT}{\overline{r_0^2}} \int_{\left(\overline{r_i^2}\right)^{1/2}}^{\left(\overline{r^2}\right)^{1/2}} r\, dr \tag{9.30}$$

Integrating and substituting equation (9.28) yields

$$\Delta F_{\text{el}} = \frac{nRT}{2} \frac{\overline{r_i^2}}{\overline{r_0^2}} \left(\alpha_x^2 + \alpha_y^2 + \alpha_z^2 - \alpha_{x0}^2 - \alpha_{y0}^2 - \alpha_{z0}^2\right) \tag{9.31}$$

By definition, $\alpha_{x_0}^2 = \alpha_{y_0}^2 = \alpha_{z_0}^2 = 1$, since these terms deal with the unstrained state.

Since it is assumed that there is no volume change on deformation (Poisson's ratio very nearly equals 0.5, i.e., an incompressible solid),

$$\alpha_x \alpha_y \alpha_z = 1 \tag{9.32}$$

If α_x is taken simply as α, then $\alpha_z = \alpha_y = 1/\alpha^{1/2}$, at constant volume.

After making the above substitutions, the work on elongation equation may be written, per unit volume,

$$-W = \Delta F_{\text{el}} = \frac{nRT}{2} \frac{\overline{r_i^2}}{\overline{r_0^2}} \left(\alpha^2 + \frac{2}{\alpha} - 3 \right) \tag{9.33}$$

The stress is given by

$$\sigma = \left(\frac{\partial F}{\partial \alpha} \right)_{T,V} = nRT \frac{\overline{r_i^2}}{\overline{r_0^2}} \left(\alpha - \frac{1}{\alpha^2} \right) \tag{9.34}$$

which is the equation of state for rubber elasticity. Equation (9.34) bears comparison with equation (9.4). The quantity $\overline{r_i^2}/\overline{r_0^2}$ is known as the "front factor."

The quantity n in the above represents the number of active network chains per unit volume, sometimes called the network or cross-link density. The number of cross-links per unit volume is also of interest. For a tetrafunctional cross-link (see Figure 9.3), the number of cross-links is one-half the number of chain segments (see Section 9.9.2). Equation (9.34) holds for both extension and compression.

Several other relationships may be derived immediately. Young's modulus can be written

$$E = L \left(\frac{\partial \sigma}{\partial L} \right)_{T,V} \tag{9.35}$$

which yields

$$E = nRT \frac{\overline{r_i^2}}{\overline{r_0^2}} \left(2\alpha^2 + \frac{1}{\alpha} \right) \cong 3n \frac{\overline{r_i^2}}{\overline{r_0^2}} RT \tag{9.36}$$

for small strains. This is the engineering modulus, which utilizes the actual cross section at α rather than the relaxed value (i.e., $\alpha = 1$). At low extensions, the two moduli are nearly identical. The shear modulus may be written

$$G = E/2(1+v) \tag{9.37}$$

Poisson's ratio, v, for rubber is approximately 0.5 (incompressibility assumption), so

$$G = n\frac{\overline{r_i^2}}{\overline{r_0^2}}RT \tag{9.38}$$

Then, to a good approximation

$$\sigma = G\left(\alpha - \frac{1}{\alpha^2}\right) \tag{9.39}$$

thus defining the work to stretch, the stress–strain relationships, and the modulus of an ideal elastomer. As equations (9.34) and (9.39) illustrate, the stress–strain relationships are non-Hookean; that is, the strain is not proportional to the stress. Again, these equations yield curves of the type illustrated in Figures 9.5 and 9.7. These same equations can also be used for compression. For compression, of course, $\alpha < 1$ (49).

Another relationship treats biaxial extension (50). If equibiaxial extension is assumed, such as in a spherical rubber balloon, then

$$\sigma = nRT(\alpha^2 - \alpha^{-4}) \tag{9.40}$$

assuming $\bar{r}_i^2/\bar{r}_0^2 \cong 1$, and the volume changes of the elastomer on biaxial extension are nil.

Other moduli are occasionally used to characterize polymeric materials. Appendix 9.1 describes the use of the ball indentation method to characterize the cross-link density of gelatin. In general, this method can be used for sheet rubber and other large sample types.

9.7.3 Some Example Calculations

9.7.3.1 An Example of Rubber Elasticity Calculations Suppose that an elastomer of 0.1 cm × 0.1 cm × 10 cm is stretched to 25 cm length at 35°C, a stress of 2×10^7 dynes/cm² being required. What is the concentration of active network chain segments?

Use equation (9.34), assuming that $\overline{r_i^2} = \overline{r_0^2}$.

$$n = \frac{\sigma}{RT(\alpha - 1/\alpha^2)}$$

$$= \frac{2 \times 10^7 \text{ dynes/cm}^2}{8.31 \times 10^7 \text{ (dynes} \cdot \text{cm/mol} \cdot \text{K)} \times 308 \text{ K} \times (2.5 - 1/2.5^2)}$$

noting that $\alpha = L/L_0 = 25/10 = 2.5$.

$$n = 3.34 \times 10^{-4}\ \text{mol/cm}^3$$

Note that the rubber band in Appendix 9.2 has $n = 1.9 \times 10^{-4}$ mol/cm^3.

9.7.3.2 An Example of Work Done during Stretching The elastomer strip in Section 9.7.3.1 was stretched to 45 cm length at 25°C. How much work was required?

Equation (9.33) provides the basis. The quantity $\bar{r}_i^2/\bar{r}_0^2$ is taken as substantially unity. The quantity $n = 3.34 \times 10^{-4}$ mol/cm^3.

$$\begin{aligned} -W &= 3.34 \times 10^{-4}\ (\text{mol/cm}^3) \times 8.31 \times 10^7\ (\text{dyn} \cdot \text{cm/mol} \cdot \text{K}) \\ &\quad \times 298\ \text{K}(4.5^2 + 2/4.5 - 3) \\ -W &= 1.46 \times 10^8\ \text{erg/cm}^3, \text{noting that dyn} \cdot \text{cm} \equiv \text{erg} \end{aligned}$$

But noting that there are 10 cm^3 of elastomer,

$$-W = 1.46 \times 10^9\ \text{erg or } 1.46 \times 10^2\ \text{J}$$

This calculation ignores the mass of the elastomer itself.

Note that work, as given in equation (9.33), must have units that match the units of the gas constant R.

9.7.3.3 An Example Using M_c In Section 9.4, the quantity n, the number of active network chain segments per unit volume, was shown to be equal to the density over the molecular weight between cross-links, ρ/M_c. Suppose an amorphous polymer of $T_g = -10$°C, and of density $\rho = 1.10$ g/cm^3 was chemically cross-linked, such that a cross-link was placed every 10,000 g/mol of chain. What is Young's modulus at 25°C? (Note that since T_g is well below 25°C and the polymer is cross-linked, the polymer is in the rubbery plateau region.)

In this case, $M_c = 1 \times 10^4$ g/mol.

$$\begin{aligned} n &= \rho/M_c = 1.10\ \text{g/cm}^3/1 \times 10^4\ \text{g/mol} \\ n &= 1.1 \times 10^{-4}\ \text{mol/cm}^3 = 1.1 \times 10^2\ \text{mol/m}^3 \end{aligned}$$

Young's modulus is given by equation (9.36):

$$\begin{aligned} E &= 3nRT = (3 \times 1.1 \times 10^2\ \text{mol/m}^3) \times (8.314\ \text{Pa-m}^3/\text{mol} - \text{K}) \times 298\ \text{K} \\ E &= 8.16 \times 10^5\ \text{Pa} = 8.16 \times 10^6\ \text{dyn/cm}^2 \end{aligned}$$

This result suggests a soft elastomer, but the effects of physical entanglements were ignored, see Section 9.10.5, so the final Young's modulus will be somewhat higher.

9.8 THE "CARNOT CYCLE" FOR ELASTOMERS

In elementary thermodynamics, the Carnot cycle illustrates the production of useful work by a gas in a heat engine. This section outlines the corresponding thermodynamic concepts for an elastomer and illustrates a demonstration experiment.

The conservation of energy for a system may be written

$$dU = V\,dp + T\,dS + \sigma\,dL + \cdots \tag{9.41}$$

where the internal energy, U, is equated to as many variables as exist in the system. For an ideal gas (Section 9.6), P–V–T variables are selected. The corresponding variables for an ideal elastomer are σ–L–T [see equation (9.34)]. Since Poisson's ratio is nearly 0.5 for elastomers, the volume is substantially constant on elongation.

By carrying a gas, elastomer, or any material through the appropriate closed loop with a high- and low-temperature portion, they may be made to perform work proportional to the area enclosed by the loop. A system undergoing such a cycle is called a heat engine.

9.8.1 The Carnot Cycle for a Gas

In the Carnot heat engine, a gas is subjected to two isothermal steps, which alternate with two adiabatic steps, all of which are reversible (see Figure 9.12) (48). Briefly, the gas undergoes a reversible adiabatic compression from state 1 to state 2. The temperature is increased from T_1 to T_2. During this step the surroundings do work $|\omega_{12}|$ on the gas. The absolute signs are used because conventions require that the signs on some of the algebraic quantities herein be negative.

Next the gas undergoes a reversible isothermal expansion from state 2 to state 3. While expanding, the gas does work $|\omega_{23}|$ on the surroundings while absorbing heat $|q_2|$. Then there follows a reversible adiabatic expansion of the gas from state 3 to state 4, the temperature dropping from T_2 to T_1. During this step, the gas does work $|\omega_{34}|$ on the surroundings.

Last, there is an isothermal compression of the gas from state 4 to state 1 at T_1. Work $|\omega_{41}|$ is performed on the gas, and heat $|q_1|$ flows from the gas to the surroundings.

9.8.2 The Carnot Cycle for an Elastomer

For an elastomer, the rubber goes through a series of stress–length steps, two adiabatically and two isothermally, as in the Carnot cycle (see Figure 9.13) (51). Beginning at length L_1 and temperature T^{I}, a stress, σ, is applied stretching the elastomer adiabatically to L_2. The elastomer heats up to T^{II}. The quantity σ is related to the length by the nonlinear equation

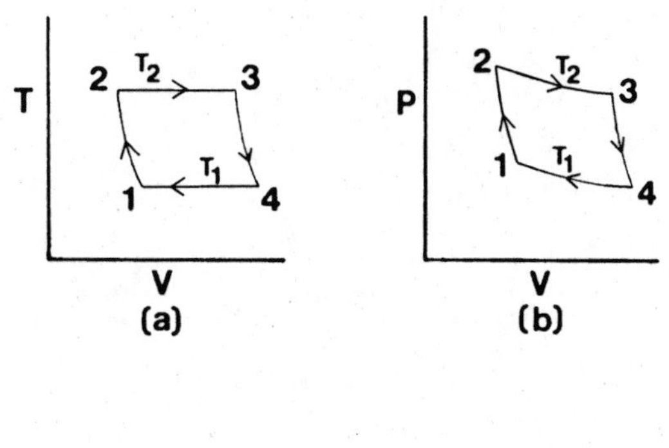

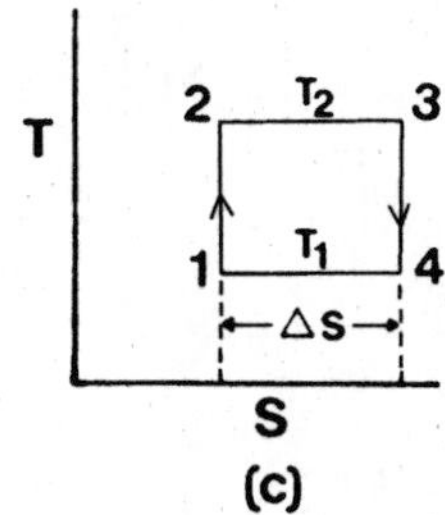

Figure 9.12 Carnot cycle for a gas (48).

$$\sigma = nRT\left[\frac{L}{L_0} - \left(\frac{L_0}{L}\right)^2\right] \tag{9.42}$$

[see equation (9.34)]. In this step work is done on the elastomer.

At T^{II}, the elastomer is allowed to contract isothermally to L_3. It absorbs heat from its surroundings in this step and does work. As the length decreases, its entropy increases by ΔS (see Figure 9.13*c*). The elastomer then is allowed to contract adiabatically to L_4, doing work, and its temperature falls to T^{I} again. The length of the sample is then increased isothermally from L_4 to L_1, work being done on the sample, and heat is given off to its surroundings. This step completes the cycle.

An increase in the volume of the gas, however, corresponds to a decrease in the length of a stretched elastomer. It is important to note that at no time does the elastomer come to its rest length, L_0. Interestingly the corresponding "rest volume" of a gas is infinitely large.

9.8.3 Work and Efficiency

The equations governing the work done during the two cycles may also be compared. For a gas,

$$\omega_g = -\oint P\, dV \tag{9.43}$$

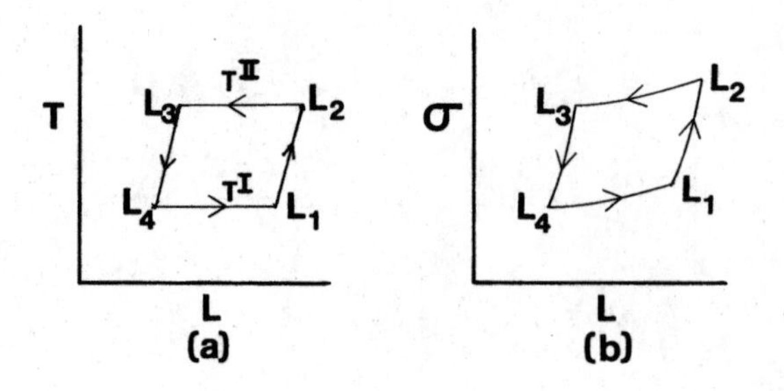

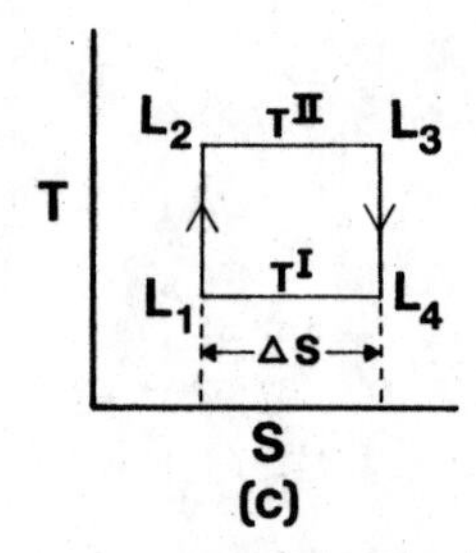

Figure 9.13 Thermal cycle for an elastomer (51).

For an elastomer,

$$\omega_e = -\oint \sigma \, dL \tag{9.44}$$

In both cases the cyclic integral measures the area enclosed by the four steps in Figures 9.12 and 9.13.

The efficiencies, $\overline{\eta}$, of the two systems may also be compared. For a gas,

$$\overline{\eta}_g = \frac{q_1 + q_2}{q_2} \tag{9.45}$$

where q_1 and q_2 are the heat absorbed and released (opposite signs), as above. For the elastomer,

$$\overline{\eta}_e = \frac{\oint \sigma \, dL}{Q_{\mathrm{II}}} = \frac{(T^{\mathrm{II}} - T^{\mathrm{I}})\Delta S}{Q_{\mathrm{II}}} = \frac{Q_{\mathrm{I}} + Q_{\mathrm{II}}}{Q_{\mathrm{II}}} \tag{9.46}$$

or in a different form,

$$\overline{\eta}_e = \frac{T^{\mathrm{II}} - T^{\mathrm{I}}}{T^{\mathrm{II}}} \tag{9.47}$$

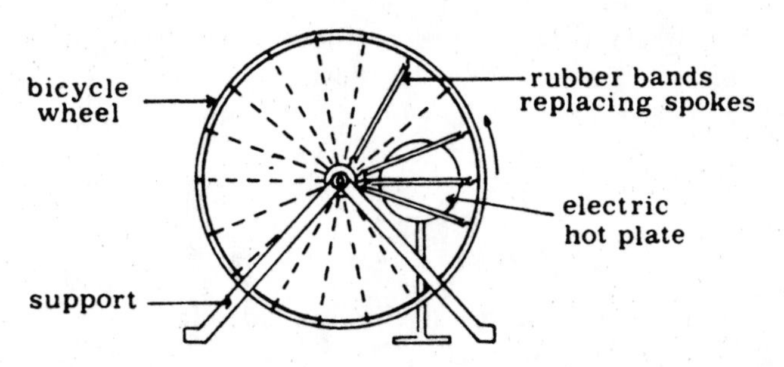

Figure 9.14 A thermally rotated wheel, employing an elastomer as the working substance (51).

where Q_{I} and Q_{II} are the amounts of heat released to the low-temperature reservoir (T^{I}) and absorbed from the high-temperature reservoir (T^{II}), respectively.

While the entropy change is zero for either system during the reversible adiabatic steps (see Figures 9.12*c* and 9.13*c*), it must be emphasized that the entropy change is greater than zero for an irreversible adiabatic process. An example for an elastomer is "letting go" of a stretched rubber band.

9.8.4 An Elastomer Thermal Cycle Demonstration

The elastomer thermal cycle is demonstrated in Figure 9.14 (51). A bicycle wheel is mounted on a stand, with a source of heat on one side only. Stretched rubber bands replace the spokes. On heating, the stress that the stretched rubber bands exert is increased so that the center of gravity of the wheel is displaced toward 9 o'clock in the drawing. The wheel then rotates counterclockwise (52).

Each of the steps in Figure 9.13 may be traced in Figure 9.14, although none of the steps in Figure 9.13 are purely isothermal or adiabatic, and then of course they are not strictly reversible. Steps 1 to 2 in Figure 9.13 occur at 6 o'clock in Figure 9.14, where there is a (near) adiabatic length increase due to gravity. At 3 o'clock, at T^{II}, heat is absorbed (nearly) isothermally, and the length decreases, doing work. At 12 o'clock, corresponding to steps 3 to 4, there is an adiabatic length decrease due to gravity. Last, at 9 o'clock, steps 4 to 1, there is a (nearly) isothermal length increase, and heat is given off to the surroundings at T^{I}, and work is done on the elastomer.

9.9 CONTINUUM THEORIES OF RUBBER ELASTICITY

9.9.1 The Mooney–Rivlin Equation

The statistical theory of rubber elasticity is based on the concepts of random chain motion and the restraining power of cross-links; that is, it is a molecular theory. Amazingly, similar equations can be derived strictly from phenomenological approaches, considering the elastomer as a continuum. The best known such equation is the Mooney–Rivlin equation (53–56),

$$\sigma = 2C_1\left(\alpha - \frac{1}{\alpha^2}\right) + 2C_2\left(1 - \frac{1}{\alpha^3}\right) \tag{9.48}$$

which is sometimes written in the algebraically identical form,

$$\sigma = \left(2C_1 + \frac{2C_2}{\alpha}\right)\left(\alpha - \frac{1}{\alpha^2}\right) \tag{9.49}$$

Equations (9.48) and (9.49) appear to have a correction term for equation (9.34), with an additional term being added. However, they are derived from quite different principles.

According to equation (9.34), the quantity $\sigma/(\alpha - 1/\alpha^2)$ should be a constant. Equation (9.49), on the other hand, predicts that this quantity depends on α:

$$\frac{\sigma}{\alpha - 1/\alpha^2} = 2C_1 + \frac{2C_2}{\alpha} \tag{9.50}$$

Plots of $\sigma/(\alpha - 1/\alpha^2)$ versus $1/\alpha$ are found to be linear, especially at low elongation (see Figure 9.15) (57). The intercept on the $\alpha^{-1} = 0$ axis yields $2C_1$, and the slope yields $2C_2$. The value of $2C_1$ varies from 2 to 6 kg/cm^2, but the value of $2C_2$, interestingly, remains constant near 2 kg/cm^2. Appendix 9.2 describes a demonstration experiment that illustrates both rubber elasticity [see equation (9.34)] and the nonideality expressed by equation (9.50).

On swelling, the value of $2C_2$ drops rapidly (see Figure 9.16) (57), reaching a value of zero near v_2 (volume fraction of polymer) equal to 0.2. This same dependence is observed for the same polymer in different solvents, different levels of cross-linking for the same polymer, or (as shown) different polymers entirely (57).

The interpretation of the constants $2C_1$ and $2C_2$ has absorbed much time; the results are inconclusive (42). It is tempting but generally considered incorrect to equate $2C_1$ and $nRT(\overline{r_i^2}/\overline{r_0^2})$. The original derivation of Mooney (46) shows that $2C_2$ has to be finite, but it does not indicate its value relative to $2C_1$. According to Flory (41), the ratio $2C_2/2C_1$ is related to the looseness with which the cross-links are embedded within the structure. Trifunctional cross-links have larger values of $2C_2/2C_1$ than tetrafunctional cross-links, for example (58).

As indicated above, $2C_2$ decreases with the degree of swelling. Furthermore Gee (59) showed that during stress relaxation, swelling increased the rate of approach to equilibrium. Ciferri and Flory (60) showed that $2C_2$ is markedly

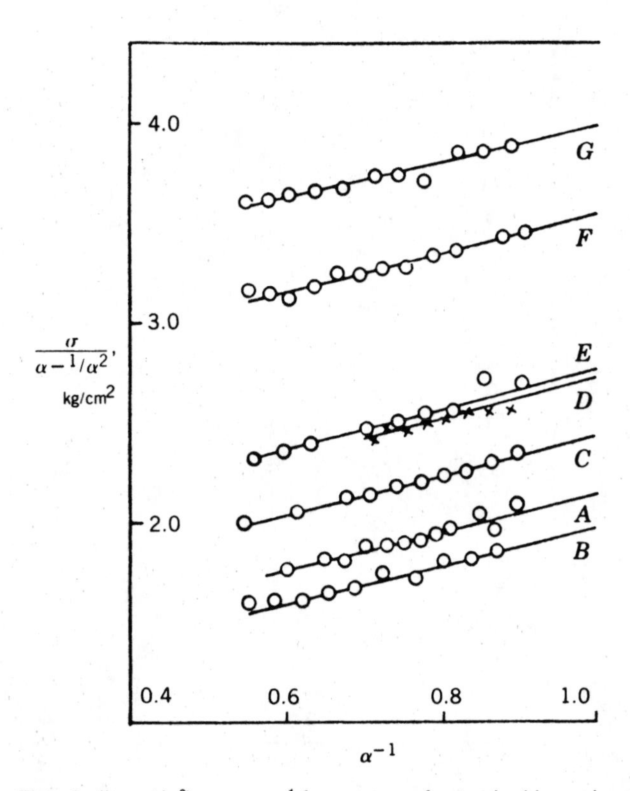

Figure 9.15 Plot of $\sigma/(\alpha - 1/\alpha^2)$ versus α^{-1} for a range of natural rubber vulcanizates. Sulfur content increases from 3 to 4%, with time of vulcanization and other quantities as variables (57).

reduced by swelling and deswelling the sample at each elongation. The samples, actually measured dry, had $2C_2$ values about half as large after the swelling–deswelling operation as those measured before. These results suggest that the magnitude of $2C_2$ is caused by nonequilibrium phenomena. Gumbrell et al. (57) stated it in terms of the reduced numbers of conformations available in the dry state versus the swollen state.

Other possible explanations include non-Gaussian chain or network statistics (see Section 9.10.6) and internal energy effects (42). The latter, bearing on the front factor, will be treated in Section 9.10.

9.9.2 Generalized Strain–Energy Functions

Following the work of Mooney, more generalized theories of the stress–strain relationships in elastomers were sought. The central problem was how to calculate the work, W, stored in the body as strain energy.

Rivlin (56) considered the most general form that such strain–energy functions could assume. As basic assumptions, he took the elastomer to be incom-

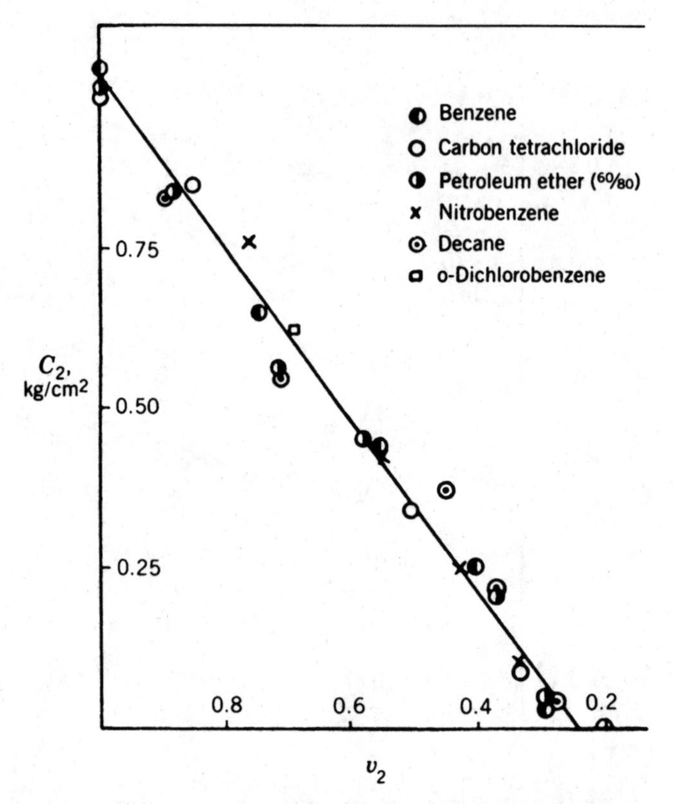

Figure 9.16 Dependence of C_2 on v_2 for synthetic rubber vulcanizates (57). Polymers: ○, butadiene–styrene, (95/5); ◑, butadiene–styrene, (90/10); ◐, butadiene–styrene, (85/15); ●, butadiene–styrene, (75/25); ⊙, butadiene–styrene, (70/30); X, butadiene–acrylonitrile, (75/25).

pressible and isotropic in the unstrained state. Symmetry conditions required that the three principal extension ratios, α_1, α_2, and α_3, depend only on even powers of the α's. In three dimensions (see Figure 9.17), the simplest functions that satisfy these requirements are

$$I_1 = \alpha_1^2 + \alpha_2^2 + \alpha_3^2 \tag{9.51}$$

$$I_2 = \alpha_1^2\alpha_2^2 + \alpha_2^2\alpha_3^2 + \alpha_3^2\alpha_1^2 \tag{9.52}$$

$$I_3 = \alpha_1^2\alpha_2^2\alpha_3^2 \tag{9.53}$$

where I_1, I_2, and I_3 are termed strain invariants.

The third strain invariant is equal to the square of the volume change,

$$I_3 = \left(\frac{V}{V_0}\right)^2 = 1 \tag{9.54}$$

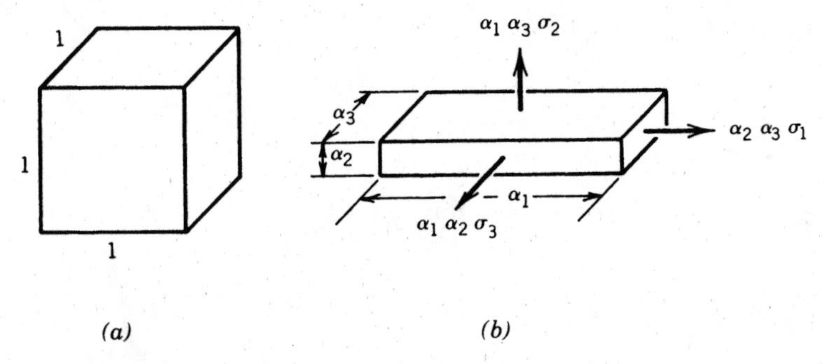

Figure 9.17 An elastomeric cube. (*a*) Undeformed and (*b*) deformed states, showing principal stresses and strains.

which under the assumption of incompressibility equals unity. Alternate formulations have been proposed by Valanis and Landel (61) and by Ogden (62), which have been reviewed by Treloar (42).

Consider the deformation of a cube (Figure 9.17). The work that is stored in the body as strain energy can be written (63).

$$W(\alpha) = \int \sigma_1 \, d\alpha_1 + \int \sigma_2 \, d\alpha_2 + \int \sigma_3 \, d\alpha_3 \tag{9.55}$$

where the σ's are the stresses.

The work, in a more general form, can be expressed as a power series (56):

$$W = \sum_{i,j,k=0}^{\infty} C_{ijk}(I_1 - 3)^i (I_2 - 3)^j (I_3 - 1)^k \tag{9.56}$$

Equation (9.56) is written so that the strain energy term in question vanishes at zero strain.

For the lowest member of the series, $i = 1, j = 0$, and $k = 0$,

$$W = C_{100}(I_1 - 3) \tag{9.57}$$

which is functionally identical to the free energy of deformation expressed in equation (9.31). For the case of uniaxial extension,

$$\alpha_1 = \alpha \tag{9.58}$$

and noting equation (9.32) and equation (9.54),

$$\alpha_2 = \alpha_3 = \left(\frac{1}{\alpha}\right)^{1/2}$$

Equation (9.57) can now be written

$$W = C_{100}\left(\alpha^2 + \frac{2}{\alpha}\right) \tag{9.59}$$

and the stress can be written [see equation (9.54)]

$$\sigma = \frac{\partial W}{\partial \alpha} = 2C_{100}\left(\alpha - \frac{1}{\alpha^2}\right) \tag{9.60}$$

This equation is readily identified with equation (9.34), suggesting (for this case only) that

$$2C_{100} = nRT\frac{\overline{r_i^2}}{r_0^2} \tag{9.61}$$

On retention of an additional term in equation (9.55), with $i = 0$, $j = 1$, and $k = 0$,

$$W = C_{100}(I_1 - 3) + C_{010}(I_2 - 3) \tag{9.62}$$

which leads directly to the Mooney–Rivlin equation, equation (9.49).

Interestingly, if we retain one more term, C_{200}, an equation of the form (63)

$$\sigma = \left(C + \frac{C'}{\alpha} + C''\alpha^2\right)\left(\alpha - \frac{1}{\alpha^2}\right) \tag{9.63}$$

can be written, where

$$C = 2(C_{100} - 6C_{200}) \tag{9.64}$$

$$C' = 2(4C_{200} + C_{010}) \tag{9.65}$$

and

$$C'' = 4C_{200} \tag{9.66}$$

Equation (9.63), with two additional terms over the statistical theory of rubber elasticity, fits the data quite well (see Figure 9.18) (64).

Because no particular molecular model was assumed, theoretical values cannot be assigned to C, C', and C'', nor can any molecular mechanisms be assigned. These phenomenological equations of state, however, accurately express the form of the experimental stress–strain data.

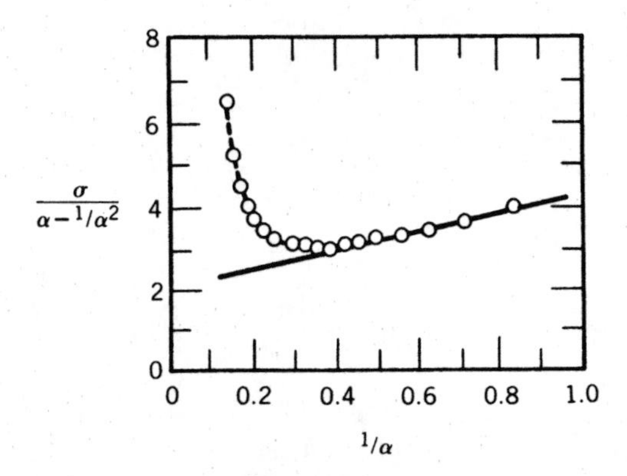

Figure 9.18 Mooney–Rivlin plot for sulfur-vulcanized natural rubber. Solid line, equation (9.48); dotted line, equation (9.63).

9.10 SOME REFINEMENTS TO RUBBER ELASTICITY

The statistical theory of rubber elasticity has undergone significant and continuous refinement, resulting in a series of correction terms. These are sometimes omitted and sometimes included in scientific and engineering research, as the need for them arises. In this section we briefly consider some of these.

9.10.1 The Inverse Langevin Function

The Gaussian statistics leading to equation (9.34) are valid only for relatively small strains—that is, under conditions where the contour length of the chain is much more than its end-to-end distance. In the region of high strains, where the ratio of the two parameters approaches $\frac{1}{3}$ to $\frac{1}{2}$, this limit is exceeded.

Kuhn and Grün (65) derived a distribution function based on the inverse Langevin function. The Langevin function itself can be written

$$L(x) = \coth x - \frac{1}{x} \tag{9.67}$$

and was first applied to magnetic problems. In this case

$$L(\beta') = \frac{r}{n'l} \tag{9.68}$$

where n' is the number of links of length l. (It must be pointed out that the quantity n' in this case need not be identical with the number of mers in the chains.) Thus the quantity $n'l$ represents a measure of the contour length of the chain, and $r/n'l$ is the fractional chain extension. Of course, for the inverse Langevin function of interest here,

$$L^{-1}\left(\frac{r}{n'l}\right)=\beta' \tag{9.69}$$

The stress of an elastomer obeying inverse Langevin statistics can be written (42,63)

$$\sigma = nRT(n')^{1/2}\left[L^{-1}\left(\frac{\alpha}{(n')^{1/2}}\right)-\alpha^{-3/2}L^{-1}\left(\frac{1}{\alpha^{1/2}(n')^{1/2}}\right)\right] \tag{9.70}$$

At intermediate values of α (and hence of $n'l$), equation (9.70) predicts a sharp upturn in the stress at α's greater than 4, as observed in experiments. Because of the complexity of equation (9.70), the Gaussian-based (9.34) is preferred where possible.

9.10.2 Cross-link Functionality

In order to form a network, at least some of the mers need to have a functionality greater than 2; that is, more than two chain portions must emanate from those mers. In the structure depicted in Figure 9.3, the functionality of each cross-link is 4. When divinyl benzene or sulfur is used as a cross-linker, the functionality will indeed be 4.

Suppose, however, that glycerol is used as the cross-linker in the synthesis of a polyester. Then the functionality of the cross-link site will be 3. Use of trimethylol propane trimethacrylate or pentaerythritol tetramethacrylate results in functionalities of 6 and 8, respectively (66).

Duiser and Staverman (67) and Graessley (70) have shown that the front factor depends on the functionality of the network. Representing the network functionality as f^*, equation (9.34) can be written

$$\sigma = \left(\frac{f^*-2}{f^*}\right)nRT\frac{\overline{r_i^2}}{r_0^2}\left(\alpha-\frac{1}{\alpha^2}\right) \tag{9.71}$$

For tetrafunctional cross-links, defined as four chain segments emanating from each cross-link site [the same type as obtained with the use of divinyl benzene (see Figure 9.3)], $f^* = 4$, equation (9.71) predicts one-half the stress that equation (9.34) predicts.

Another way of writing the correction for cross-link functionality is (41,69)

$$\sigma = (n-\mu)RT\frac{\overline{r_i^2}}{r_0^2}\left(\alpha-\frac{1}{\alpha^2}\right) \tag{9.72}$$

where n and μ are the number densities of elastically active strands and junctions. A junction is elastically active if at least three paths leading away from it are independently attached to the network. A strand, meaning a polymer chain segment, is elastically active if it is bound at each end by elastically active junctions (70). Equation (9.72) also predicts a front-factor correction of $\frac{1}{2}$ for a tetrafunctional network, since there are half as many cross-links as there are chain segments.

9.10.3 Network Defects

There are two major types of network defects: (a) the formation of inactive rings or loops, where the two ends of the chain segment are connected to the same cross-link junction, and (b) loose, dangling chain ends, attached to the network by only one end (71–74) (see Figure 9.19).

Both of these defects tend to decrease the retractive stress, because they are not part of the network. The equation in use to correct for dangling ends may be written

$$\sigma = nRT\left(1 - \frac{2M_c}{M}\right)\frac{\overline{r_i^2}}{\overline{r_0^2}}\left(\alpha - \frac{1}{\alpha^2}\right) \tag{9.73}$$

where M_c is the molecular weight between cross-links and M is the primary-chain molecular weight. Where $M \gg M_c$, the correction becomes negligible.

9.10.4 Volume Changes and Swelling

If a polymer network is swollen with a "solvent" (it does not dissolve), equation (9.32) may be rewritten

$$\alpha_x \alpha_y \alpha_z = \frac{1}{v_2} \tag{9.74}$$

where v_2 is the volume fraction of polymer in the swollen material. Of course, v_2 is less than unity, and $\alpha_x\alpha_y\alpha_z$ is larger than unity, as is commonly experienced.

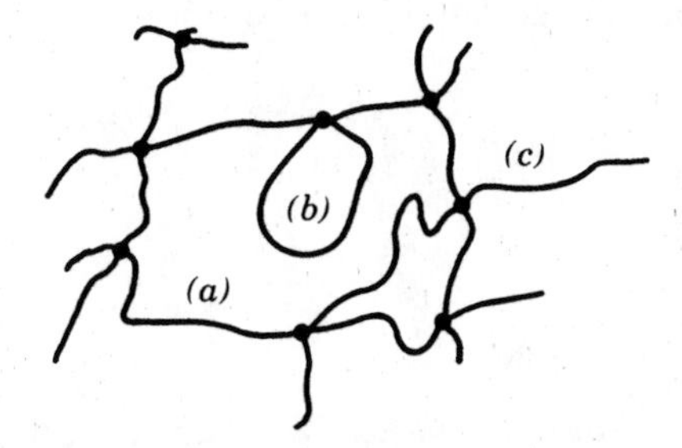

Figure 9.19 Network structure and defects: (*a*) elastically active chain, (*b*) loop, and (*c*) dangling chain end.

The detailed effect has two parts:

1. Effect on the front factor, $\overline{r_i^2}/\overline{r_0^2}$. The quantity $\overline{r_i^2}$ increases with the volume V to the two-thirds power, since r_i itself is a linear quantity. Of course, $\overline{r_0^2}$ remains constant. Thus

$$\left(\frac{\overline{r_i^2}}{\overline{r_0^2}}\right)_s = \left(\frac{V}{V_0}\right)^{2/3}\left(\frac{\overline{r_i^2}}{\overline{r_0^2}}\right) = \frac{1}{v_2^{2/3}}\left(\frac{\overline{r_i^2}}{\overline{r_0^2}}\right) \tag{9.75}$$

where the subscript s refers to the swollen state and where V_0 is the volume of the unswollen polymer.

2. Effect on the number of network chain segments concentration, n. The quantity n decreases with volume:

$$\left(\frac{V_0}{V}\right)n = n_s \tag{9.76}$$

where n_s is the chain segment concentration in the swollen state.

$$\left(\frac{V_0}{V}\right)n = v_2 n \tag{9.77}$$

Incorporating the right-hand sides of equations (9.75) and (9.77) into equation (9.34) leads to an equation of the form (75,76)

$$\sigma = nRTv_2^{1/3}\frac{\overline{r_i^2}}{\overline{r_0^2}}\left(\alpha - \frac{1}{\alpha^2}\right) \tag{9.78}$$

The stress, defined as force per unit actual cross section, is decreased by $v_2^{1/3}$, since the number of chains occupying a given volume has decreased.

When the volume change caused by deformation alone is considered (usually less than 1%), the equation of state can be written (77–80)

$$\sigma = nRT\left(\frac{V_0}{V}\right)^{2/3}\frac{\overline{r_i^2}}{\overline{r_0^2}}\left(\alpha - \frac{1}{\alpha^2}\frac{V}{V_0}\right) \tag{9.79}$$

9.10.5 Physical Cross-links

9.10.5.1 Trapped Entanglements So far the discussion has been restricted to ordinary covalent cross-links. There are, however, several types of physical cross-links that exist as permanent loops or entanglements existing in the network structure. (They may slide, however, yielding a mode of stress relaxation also.)

Three types of trapped entanglements are shown in Figure 9.20 (81–86). They each portray the same phenomenon, but with increasing rigor of definition.

Early works referred to the chemical and physical cross-links in a simple manner,

$$\sigma = (n_c + n_p)RT\frac{\overline{r_i^2}}{\overline{r_0^2}}\left(\alpha - \frac{1}{\alpha^2}\right) \tag{9.80}$$

where n_c and n_p are the concentration of chains bound by chemical and physical cross-links, respectively. It had been established early, for example, that the retractive stress was higher than expected by nearly a constant amount; indeed, for short relaxation times even linear polymers above T_g behaved as if they had some type of cross-linking (87,88) (see also Section 8.2).

Figure 9.21 illustrates the rubbery plateau (see Section 8.2) for a dynamic mechanical study of polystyrene as a function of frequency. The plateau shear modulus, near 3×10^6 dynes/cm^2, corresponds to a number of active network chains of near 1×10^{-4} mol/cm^3, nearly independent of the molecular weight of the polymer.

A more recent approach uses the concept of the potential entanglements that have been trapped by the cross-linking process. Langley (85) defines the quantity T_e as the fraction (or probability) that an entanglement is trapped in this manner.

Two theories, developed by Flory (89) and Scanlan (90), yield the calculation of chemical cross-links (58,59). For Flory's theory, the total number of effective cross-links is

$$n_{\text{tot}} = n_c W_g T_e^{1/2} + n_e T_e \tag{9.81}$$

where n_e is the concentration of potential entanglement strands and W_g is the weight fraction of gel. For $W_g = 1$, equation (9.81) reduces to $n_c + n_p$ [see equation (9.79)].

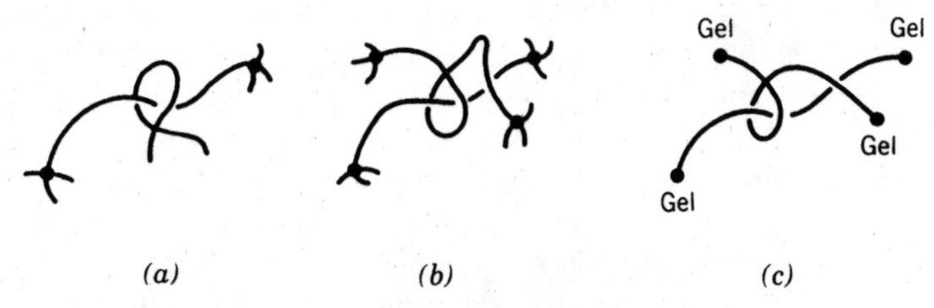

Figure 9.20 Three types of trapped entanglements: (*a*) The Bueche trap (81–86), (*b*) the Ferry trap (84), and (*c*) the Langley trap (77). The black circles are chemical cross-link sites. After Ferry (86).

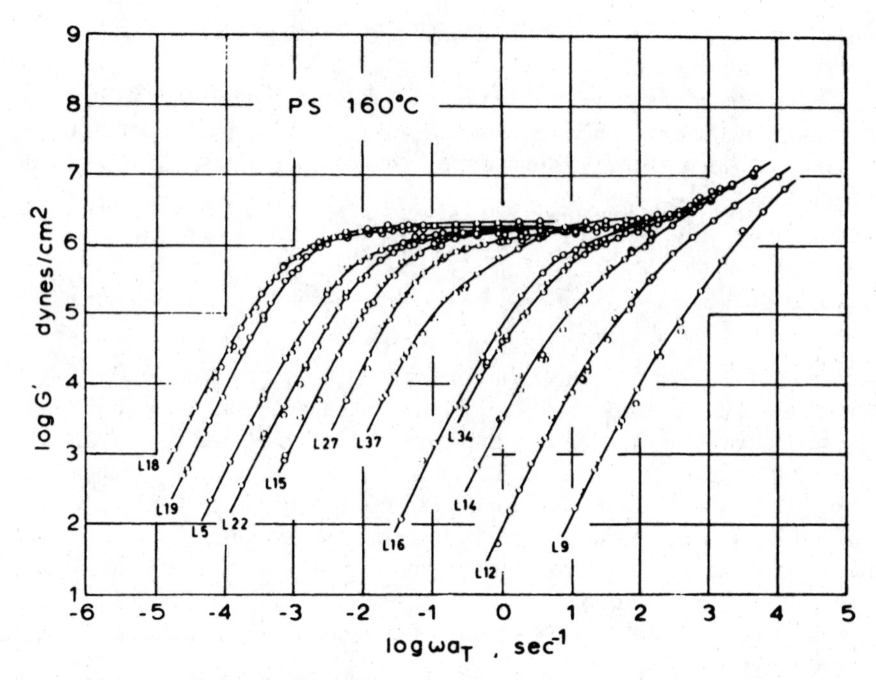

Figure 9.21 Shear storage modulus versus frequency for narrow molecular-weight polystyrenes at 160°C. Molecular weights range from $M_W = 8900$ g/mol (L9) to $M_W = 581{,}000$ g/mol (L18) (88).

With the Scanlan criterion,

$$n_{\text{tot}} = \tfrac{1}{2} n_c T_e^{1/2}(3W_g - T_e^{1/2}) + n_e T_e \tag{9.82}$$

which also reduces to $n_c + n_p$ for $W_g = 1$. The value of these relationships, of course, is that for real networks, $W_g < 1$, and the way is open to evaluate n_c and n_e. Some calculations are shown in Figure 9.22. As they illustrate, the effective (permanent) physical cross-links start out at zero when the system is linear and increase rapidly to a plateau level (91).

9.10.5.2 The Phantom Network It must be remarked that considerable controversy exists over the existence of physical cross-links (40,41,87,92). A theory has been proposed by Flory (40,41) using mathematics of a simplified network, called the "phantom network."

The model consists of a network of Gaussian chains connected in any arbitrary manner. The physical effect of the chains is assumed to be confined exclusively to the forces they exert on the junctions to which they are attached.

For a perfect phantom network of functionality f^*, the front factor contains the term $(f^* - 2)/f^*$, leading to equation (9.71), which considers chemical cross-links of arbitrary functionality, but no physical cross-links.

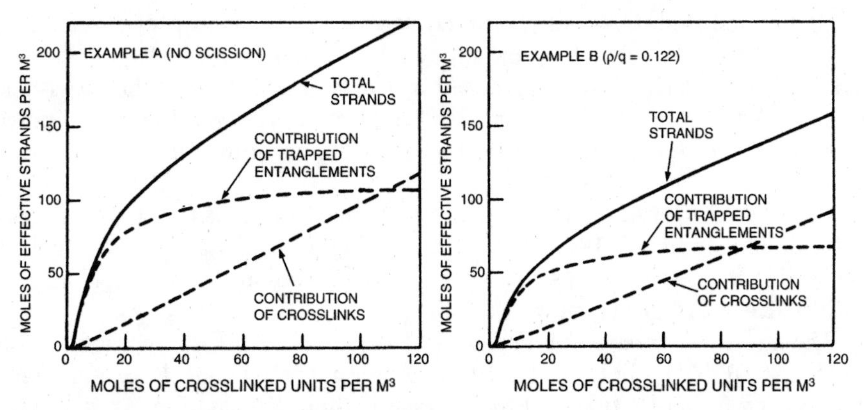

Figure 9.22 Contributions of chemical cross-links and trapped entanglements to the total cross-link level, for polystyrene, $M_n = 5 \times 10^5$ g/mol, $M_w = 1 \times 10^6$ g/mol, based on equation (9.81) (91).

Flory (40) argues that the presence or absence of the term $(f^* - 2)/f^*$ may depend on the magnitude of the strain. At small deformations the displacement of junctions may conform more nearly to the older assumptions (i.e., affine in the macroscopic sense), and hence the term $(f^* - 2)/f^*$ might not appear for real chains that do have entanglements.

9.10.6 Small-Angle Neutron Scattering

In the preceding text the chains were assumed to deform in an affine manner when the networks were swelled or stretched. Until recently there was no way to approach this problem experimentally. With the advent of small-angle neutron scattering (SANS), the conformation of the chains in the bulk state could be investigated (see Section 5.2.2.1).

On swelling, chain dimensions increase, usually isotropically. When SANS studies are done on the stretched elastomers, the scattering pattern yields greater dimensions in one direction than the other, because the chains are anisotropic (see Section 5.2) (94–106).

The question arose whether polymer chains in a network had the same conformation as in the melt before cross-linking. Beltzung et al. (93) prepared well-defined poly(dimethyl siloxane) (PDMS) chains containing Si—H linkages in the α and ω positions. Blends of PDMS(H) and PDMS(D) were prepared, where H and D, of course, represent the protonated and deuterated analogues. These blends were end-linked by tetrafunctional or hexafunctional cross-linkers under stoichiometric conditions.

Neutron scattering was carried out on both the PDMS melts and the corresponding networks. The principal results were that (a) the Gaussian character of the network chains in the undeformed state was confirmed, and (b) the chain dimensions were not changed by the cross-linking process.

Benoit and others (94–98) prepared two types of tagged polystyrene networks: (a) type "A" networks containing labeled (deuterated) cross-link sites. This permitted a characterization of the spatial distribution of the cross-link points. (b) Type "B" networks containing a few percent of perdeuterated polystyrene chains (see Figure 9.23) (94). Cross-linking utilized divinyl benzene (DVB).

Benoit et al. (94) studied these polystyrene networks as is, swollen in several solvents, and stretched. For the latter, stretching was done above T_g, followed by cooling in the stretched state. They found a maximum in the angular scattering curves of type "A" networks.

The mean pair separation distance between chain ends, h, was found to be proportional of $M_c^{0.5}$ both in the dry state and in the swollen state. On extension, $h_{\parallel}$ and $h_{\perp}$ values followed the expected affine deformation.

The "B" network, on the other hand (94), appeared to deviate significantly from the affine. As illustrated in Figure 9.24 (94) the chain radius of gyration

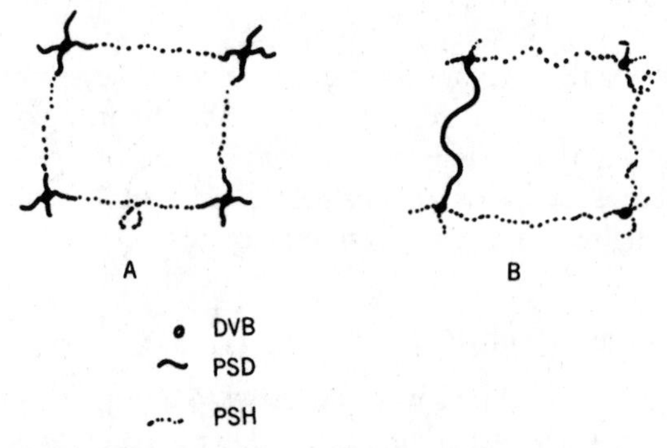

Figure 9.23 Schematic representation of labeled polystyrene networks. (*A*) Cross-linking points labeled. (*B*) Random labeled chains added (96).

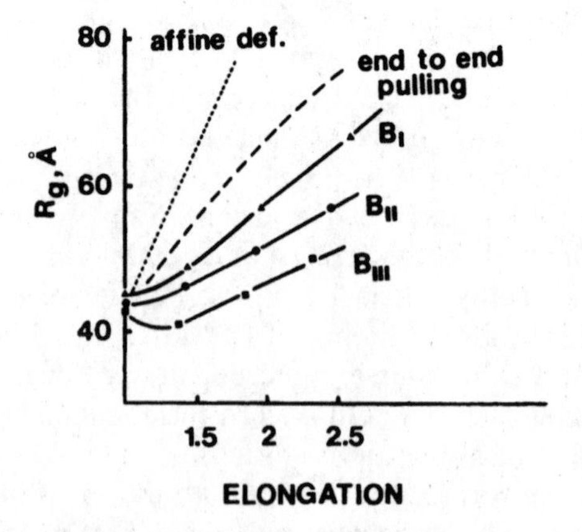

Figure 9.24 Variation of the radius of gyration of type B networks of different functionalities. Dotted line, theoretical behavior for affine deformation; dashed line, theoretical behavior for the end-to-end pulling mechanism (96), for polystyrene.

increased on swelling far less than predicted by the affine deformation mechanism. The values of R_g were also less than predicted by affine deformation, leading to the "end-to-end pulling mechanism" (97), which accentuates the extension of the end portions of the chains rather than the central section. One might imagine that entanglements prevent the motion of the central portions of long chains. Another possible explanation (see below) is that the chain's cross-link junction points rearrange to yield the system with the lowest free energy. This rearrangement minimizes the actual extension of the chain.

More recent SANS experiments on stretched networks were performed by Hinkley et al. (99) and by Clough et al. (100,101). Hinkley et al. (99) prepared blends of polybutadiene and polybutadiene-d_6. Both polymers were made by the "living polymer" technique, end-capped with ethylene oxide, and water-washed to yield the dihydroxy liquid prepolymer. Uniform networks were prepared by reacting the prepolymers with stoichiometric amounts of triphenyl methane triisocyanate. The value of using polybutadiene over polystyrene, of course, is that the networks are elastomeric at ambient temperatures.

These networks (99) were extended up to $\alpha = 1.6$ and characterized by SANS. Owing to large experimental error, no definitive conclusion could be reached, although the data fit the junction affine model better than either the chain affine model or the phantom network model (Section 9.10.5.2).

Random types of cross-linking are of special interest for real systems. Clough et al. (100,101) blended anionically polymerized polystyrene with PS-d_8 and cross-linked the mutual solution with ^{60}Co γ-radiation. Bars of the cross-linked polystyrene were elongated at 145°C and cooled. Specimens were cut in both the longitudinal and transverse directions and characterized by SANS. The quantity

$$\boldsymbol{R} = \frac{R_g(\text{stretched})}{R_g(\text{unstretched})} \tag{9.83}$$

was plotted versus α, as illustrated in Figure 9.25 (100,101). The transverse measurements are divided into "anisotropic" and "end-on." The anisotropic measurements refer to the case where the beam was perpendicular to the direction of orientation (101), and the end-on measurements refer to the case where the beam was parallel to the stretch direction. In neither of these cases was affine chain deformation followed. Both of these experiments, however, yielded identical results within experimental error, confirming important macromolecular hypotheses.

In Figure 9.25 the quantity $\boldsymbol{R} = [(\alpha^2 + 3)/4]^{0.5}$ was obtained from the dependence predicted for a tetrafunctional phantom network (101). The quantity $\boldsymbol{R} = [(\alpha^2 + 1)/2]^{0.5}$ represents the affine junction case.

The constant value of $R_\perp$ up to $\alpha = 2$ suggests that the chains are deforming far less than the junctions. These results follow neither the $\boldsymbol{R} = \alpha$ (parallel) nor the $\boldsymbol{R} = \alpha^{-0.5}$ (perpendicular) prediction but rather support Benoit et al. (94) that affine chain behavior is not followed.

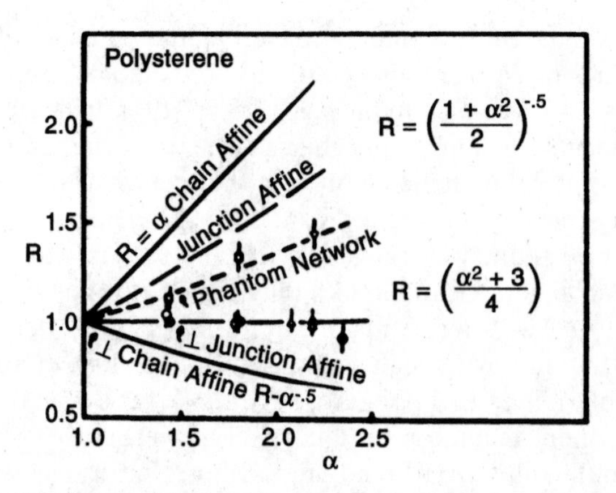

Figure 9.25 The ratio $R = R_g$ (stretched)/R_g (unstretched) as a function of sample elongation (100): □, longitudinal; ∇, longitudinal; ○, transverse, anisotropic; ●, transverse, "end-on"; △, transverse, anisotropic; ▲, transverse, "end-on."

In a series of theoretical papers, Ullman (103–105) reexamined the phantom network theory of rubber elasticity, especially in the light of the new SANS experiments. He developed a semiempirical equation for expressing the lower than expected chain deformation on extension:

$$\lambda^{*2} = \lambda^2(1-\alpha')+\alpha' \tag{9.84}$$

as the basis of a network unfolding model. The quantity λ^* is defined as the ensemble average of the deformation of junction pairs connected by a single submolecule in the network, and λ is the corresponding quantity calculated for the phantom network model. The quantity α' expresses the fractional deviation from ideality. The phantom network corresponds to $\alpha' = 0$. If $\alpha' = 1$, the chain does not deform at all on network stretching or swelling. From the data of Clough et al. (100,101), Ullman (103,104) concluded that α' was in the range of 0.36 to 0.53.

In the above, Hinkley et al. (99), Clough et al. (100,101), and Benoit et al. (94,95) utilized end-linked networks. Ullman (103,104) delineated the differences between the two types of network. He pointed out that randomly cross-linked chains deform to a greater extent than end-linked chains, that sensitivity to network functionality is much greater for end-linked chains, and that for high cross-linking levels, the randomly cross-linked chain approaches the macroscopic deformation of the sample. Ullman (105) recently reviewed these and other SANS experiments on the deformation of polymer networks.

Recently Hadziioannou et al. (106) prepared amorphous polystyrene with extrusion ratios up to 10, using a solid-state coextrusion technique. Their poly-

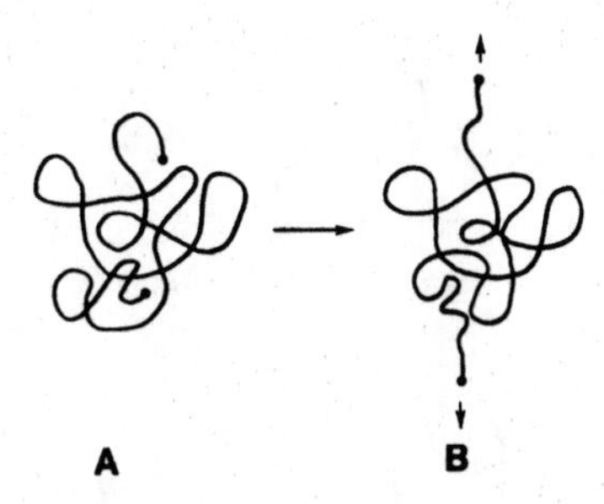

Figure 9.26 Model of the end-pulling mechanism, showing how $R_{\parallel}$ increases, while $R_{\perp}$ remains nearly constant: (*A*) relaxed and (*B*) stretched.

styrene had a molecular weight of 5×10^5 g/mol. The anisotropy of the R_g values agreed with those predicted on the basis of a chain affine model.

While general conclusions appear to be premature, it appears that the cross-link sites rearrange themselves during deformation to achieve their lowest free-energy states; thus the chains deform less than the affine mechanisms predicts. A modified end-pulling mechanism is also possible. A possible molecular mechanism, which results in minimal changes in $R_{\perp}$, is illustrated in Figure 9.26 (107). The debate over the exact molecular mechanism of deformation is sure to continue.

9.11 INTERNAL ENERGY EFFECTS

9.11.1 Thermoelastic Behavior of Rubber

In Section 9.5 some of the basic classical thermodynamic relationships for rubber elasticity were examined. Now the classical and statistical formulations are combined (108,109).

Rearranging equation (9.8),

$$f_e = f - T\left(\frac{\partial f}{\partial T}\right)_{L,V} \tag{9.85}$$

Dividing through by f and rearranging,

$$\frac{f_e}{f} = 1 - \left(\frac{\partial \ln f}{\partial \ln T}\right)_{L,V} \tag{9.86}$$

Rewriting equation (9.79) in terms of force, and substituting equation (9.38), we find that

$$f = GA_0\left(\alpha - \frac{V}{V_0}\frac{1}{\alpha^2}\right) \tag{9.87}$$

where A_0 is the initial cross-sectional area; substituting equation (9.87) into the right-hand side of equation (9.86) and carrying out the partial derivative yields

$$\left(\frac{\partial \ln f}{\partial \ln T}\right)_{L,V} = \frac{d \ln G}{d \ln T} + \frac{\beta T}{3} \tag{9.88}$$

where β is the isobaric coefficient of bulk thermal expansion, $(1/V)(\partial V/\partial T)_{L,P}$.

Substituting equation (9.88) into equation (9.86),

$$\frac{f_e}{f} = 1 - \frac{d \ln G}{d \ln T} - \frac{\beta T}{3} \tag{9.89}$$

Returning to equation (9.38), and differentiating the natural logarithm of the network end-to-end distance with respect to the natural logarithm of the temperature obtains

$$\frac{d \ln \overline{r_0^2}}{d \ln T} = 1 - \frac{d \ln G}{d \ln T} - \frac{\beta T}{3} \tag{9.90}$$

Nothing that the right-hand sides of equations (9.89) and (9.90) are identical,

$$\frac{f_e}{f} = \frac{d \ln \overline{r_0^2}}{d \ln T} = \frac{1}{T}\frac{d \ln \overline{r_0^2}}{dT} \tag{9.91}$$

which expresses the fractional force due to internal energy considerations in terms of the temperature coefficient of the free chains end-to-end distance.

9.11.2 Experimental Values

Values of f_e/f are usually derived by applying the equations above to force–temperature data of the type presented in Figure 9.27 (110). These data, carefully taken after extensive relaxation at elevated temperatures, are reversible within experimental error; that is, the same result is obtained whether the temperature is being lowered (usually first) or raised.

Some values of f_e/f are shown in Table 9.2. For most simple elastomers, f_e/f is a small fraction, near ±0.20 or less. This indicates that some 80% or more of the retractive force is entropic in nature, as illustrated from early data in

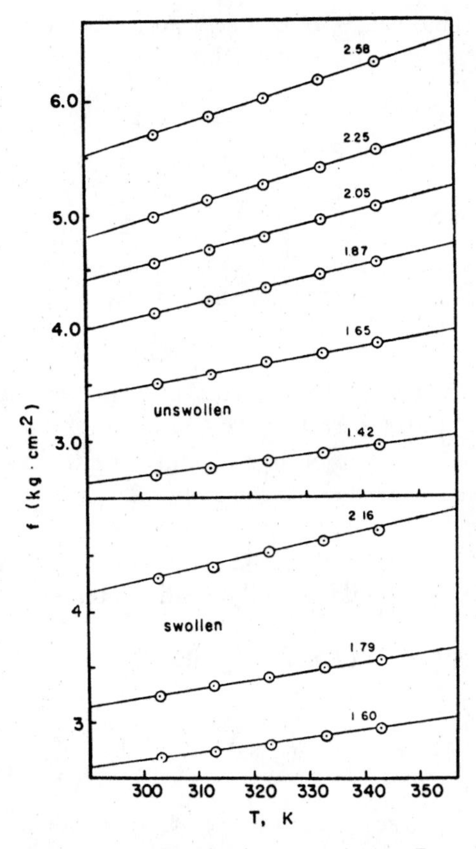

Figure 9.27 Force–temperature relationships for natural rubber. Extension ratios, α, are indicated by the numbers associated with the lines (110).

Table 9.2 Thermoelastic behavior of various polymers

Polymer	f_e/f	Reference
Natural rubber	0.12	(a)
trans-Polyisoprene	0.17	(b)
cis-Polybutadiene	0.10	(c)
Polyethylene	−0.42	(d)
Poly(ethyl acrylate)	−0.16	(e)
Poly(dimethyl siloxane)	0.15	(e)

References: (a) G. Allen, M. J. Kirkham, J. Padget, and C. Price, *Trans. Faraday Soc.*, **67**, 1278 (1971). (b) J. A. Barrie and J. Standen, *Polymer*, **8**, 97 (1967). (c) M. Shen, T. Y. Chem. E. H. Cirlin, and H. M. Gebhard, in *Polymer Networks, Structure, and Mechanical Properties*, A. J. Chompff and S. Newman, eds., Plenum Press, New York, 1978. (d) A. Ciferri, C. A. J. Hoeve, and P. J. Flory, *J. Am. Chem. Soc.*, **83**, 1015 (1961). (e) L. H. Sperling and A. V. Tobolsky, *J. Macromol. Chem.*, **1**, 799 (1966).

Figure 9.7. These same values, of course, lead to temperature coefficients of polymer chain expansion [equation (9.91)].

9.12 THE FLORY–REHNER EQUATION

9.12.1 Causes of Swelling

The equilibrium swelling theory of Flory and Rehner (76) treats simple polymer networks in the presence of small molecules. The theory considers forces arising from three sources:

1. The entropy change caused by mixing polymer and solvent. The entropy change from this source is positive and favors swelling.
2. The entropy change caused by reduction in numbers of possible chain conformations on swelling. The entropy change from this source is negative and opposes swelling.
3. The heat of mixing of polymer and solvent, which may be positive, negative, or zero. Usually it is slightly positive, opposing mixing.

The Flory–Rehner equation may be written

$$-\left[\ln(1-v_2)+v_2+\chi_1 v_2^2\right]=V_1 n\left[v_2^{1/3}-\frac{v_2}{2}\right] \quad (9.92)$$

where v_2 is the volume fraction of polymer in the swollen mass, V_1 is the molar volume of the solvent, and χ_1 is the Flory–Huggins polymer–solvent dimensionless interaction term. Appendix 9.3 describes the application of the Flory–Rehner theory. This theory, of course, is also related to the thermodynamics of solutions (see Section 3.2). As a rubber elasticity phenomenon, it is an extension in three dimensions.

The value of equation (9.92) here lies in its complementary determination of the quantity n [see equation (9.4) for simplicity]. Both equations (9.4) and (9.92) determine the number of elastically active chains per unit volume (containing, implicitly, corrections for front factor changes). By measuring the equilibrium swelling behavior of an elastomer (χ_1 values are known for many polymer–solvent pairs), its modulus may be predicted. Vice versa, by measuring its modulus, the swelling behavior in any solvent may be predicted.

Generally, values from modulus determinations are somewhat higher, because physical cross-links tend to count more in the generally less relaxed mechanical measurements than in the closer-to-equilibrium swelling data. However, agreement is usually within a factor of 2, providing significant interplay between swelling and modulus calculations (111–113).

Simple elastomers may swell a factor of 4 or 5 or so, leading to a quantitative determination of n. However, two factors need to be considered before the final numerical results are accepted:

1. The front-factor, not explicitly stated in the Flory–Rehner equation, may be significantly different from unity (114).
2. While step polymerization methods lead to more or less statistical networks and good agreement with theory, addition polymerization and vulcanization nonuniformities lead to networks that may swell as much as 20% less than theoretically predicted (115,116).

9.12.2 Example Calculation of Young's Modulus from Swelling Data

At equilibrium, a sample of poly(butadiene–*stat*–styrene) swelled 4.8 times its volume in toluene at 25°C. What is Young's modulus at 25°C?

This material is a typical elastomer, widely used for rubber bands, gaskets, and rubber tires. Table 3.4 gives $\chi_1 = 0.39$. The molar volume of toluene can be calculated from its density, 0.8669 g/cm^3, Table 3.1. A molecular weight of 92 g/mol for toluene yields a molar volume of 106 cm^3/mol. The quantity $v_2 = 1/4.8 = 0.208$. Substituting into equation (9.92) yields

$$n = \frac{-[\ln(1-0.208)+0.208+0.39\times 0.208^2]}{(106\ \text{cm}^3/\text{mol})\{0.208^{1/3}-0.208/2\}}$$

$$n = 1.5\times 10^{-4}\ \text{mol/cm}^3$$

Young's modulus is given by equation (9.36),

$$E = 3\times 1.5\times 10^{-4}(\text{mol/cm}^3)\times 8.31\times 10^7\ (\text{dyn}\cdot\text{cm/mol}\cdot\text{K})\times 298(\text{K})$$

$$E = 1.1\times 10^7\ \text{dyn/cm}^2 \quad \text{or} \quad 1.1\ \text{MPa}$$

This is a typical Young's modulus for such elastomers. Of course, the reverse calculation can be performed, starting with the modulus, and estimating the equilibrium swelling volume.

9.13 GELATION PHENOMENA IN POLYMERS

Gelation in polymers may be brought about in several ways: temperature changes, particularly important in protein gelation formation; polymerization with cross-links; phase separation in block copolymers; ionomer formation; or even crystallization. Such materials are usually thermoreversible for physical cross-links, or thermoset through the advent of chemical cross-links. Of course, there must be at least two cross-link sites per chain to induce gelation. A major task in polymer science centers on obtaining unambiguous measures of the gelation point.

9.13.1 Gelation in Solution

Let us consider a polymeric network that contains solvent, usually called a polymeric gel. There are several types of gels. A previously cross-linked polymer subsequently swollen in a solvent follows the Flory–Rehner equation (Section 9.12). If the network was formed in the solvent so that the chains are relaxed, the Flory–Rehner equation will not be followed, but rubber elasticity theory can still be used to count the active network segments.

Gels may be prepared using either chemical or physical cross-links. Physical cross-links in gels may involve dipole–dipole interactions, traces of crystallinity, multiple helices, and so on, and thus vary greatly with the number and strength of the bonds. The number of physical cross-links present in a given system depends on time, pressure, and temperature. Many such gels are thermoreversible; that is, the bonds break at elevated temperature and reform at lower temperatures.

A common thermoreversible gel is gelatin in water. Gelatin is made by hydrolytic degradation, boiling collagen in water. Collagen forms a major constituent of the skin and bones of animals. The gelatin made from it forms the basis for many foods and desserts. (The latter is usually made by dissolving gelatin plus sugar in hot water, then refrigerating to cause gelation.)

Thermoreversible gels may be bonded at single points, called point cross-links; junction zones, where the chains interact over a portion of their length; or in the form of fringe micelles, see Figure 9.28 (115). Gelatin has the junction zone type of cross-links, where the chains form multiple helices. Natural collagen is in the form of a triple helix. In hot aqueous solution, the denatured protein forms random coils. The helix–coil transition is at about 40°C at about 0.5% concentration, as commonly used in foods. Under these conditions gelatin forms either double or triple helices, a subject of current research. The critical nucleation length is about 20 to 50 peptide (mer) units, the most important interactions being among proline, hydroxyproline, and glycine.

Gels can be made from a number of polymers, including poly[acrylamide–*stat*–(acrylic acid)], poly(vinyl acetate), poly(dimethyl siloxane), and polyisocyanurates. The poly[acrylamide–*stat*–(acrylic acid)] hydrogels can be made to swell up to 20,000 times, v/v. Thermoreversible gels can be prepared in organic solvents from polyethylene and *i*-polystyrene, both of which crystallize on cooling but go back in solution on heating.

A type of gel formation involves the globular proteins. Egg white is a typical example. On cooking, it goes from a sol to a gel irreversibly, becoming denatured. Current models (117) show the hydrophobic groups partially flipping out in the range of 75 to 80°C, making contact with similar groups on other globules. Above 85°C, the —S—S— intramolecular cross-links in cysteine become labile, partially interchanging to form intermolecular cross-links. As

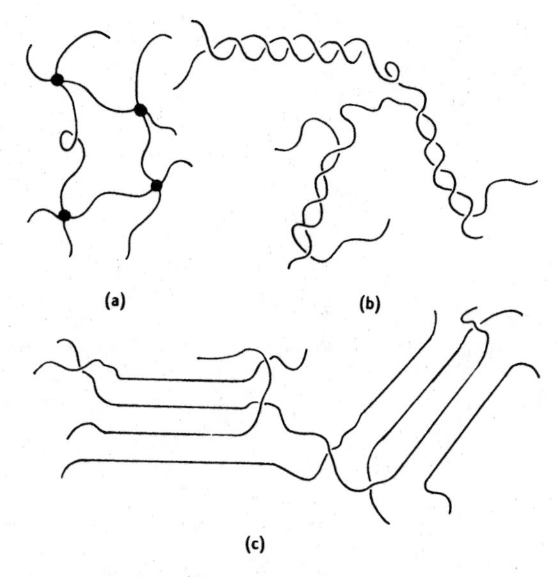

Figure 9.28 Types of thermoreversible cross-links: (*a*) point cross-links, (*b*) junction zones, and (*c*) fringed micelles.

is well known, hard-boiled egg white is highly elastic. The white appearance is caused by the nonuniform distribution of protein and water, causing intense light-scattering.

Gels sometimes undergo the phenomenon known as syneresis, where the solvent is exuded from the gel. Two types of syneresis are distinguishable (118): (a) The χ-type, where the polymer phase separates from the solvent due to poor thermodynamics of mixing. Spinodal decomposition is common in such circumstances. Such gels may be turbid in appearance. (b) The n-type, which exudes solvent because of increasing cross-link density. The polymeric gel still forms one phase with the solvent, but its equilibrium swelling level decreases. Such gels remain clear. In both cases, various amounts of fluid surround the gel.

Fully formed gels may be highly elastic; see Appendix 9.1.

9.13.2 Gelation during Polymerization

If cross-linkers are present during a polymerization, the material may gel at a certain point, its viscosity going to infinity; see Figure 9.29 (119). Before gelation the material is fluid, not having an equilibrium modulus. Thus the definition of a gel point requires an infinite steady-state shear viscosity and an equilibrium modulus of zero. At least one of the molecules of the cross-linking polymer must be very large, having grown to the dimensions of the order of the macroscopic sample.

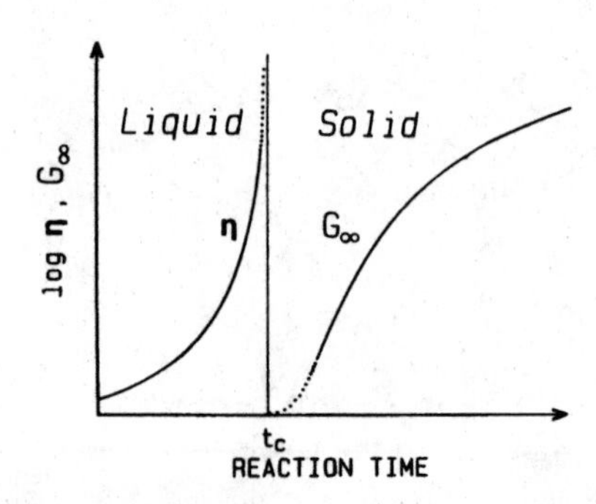

Figure 9.29 Behavior of melt viscosity and shear modulus near the gel point in a cross-linking polymerization as a function of reaction time. The steady shear viscosity goes to infinity, while the equilibrium shear modulus rises from zero, eventually reaching a plateau for the fully reacted material.

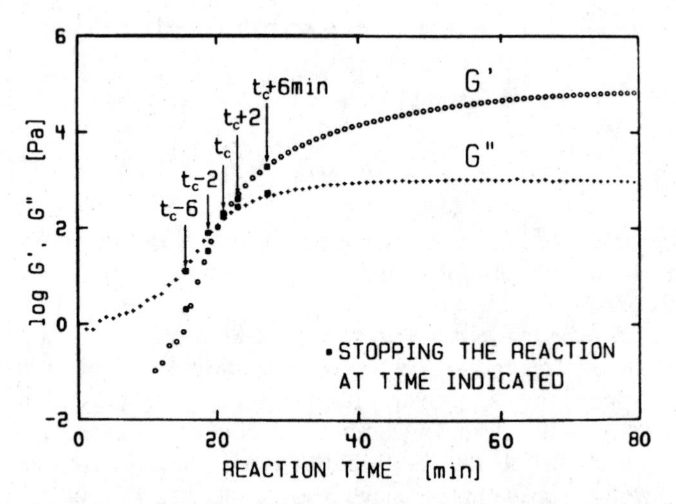

Figure 9.30 Dynamic mechanical experiments of gelling systems such as illustrated in Figure 9.29 often show a crossing of the storage and loss shear modulus. The crossing point is then taken as the gelation point. Illustrated is the behavior of a cross-linking of poly(dimethyl siloxane) in an oscillatory shear experiment at constant frequency, ω.

During a polymerization with cross-links, the storage and loss shear moduli, G' and G'', respectively, may cross. While this phenomenon is not entirely universal, it has been frequently taken as the gelation point, see Figure 9.30 (120). The viscous behavior of the oligomeric material dominates in the initial part of the polymerization. With increasing molecular weight, the loss modulus increases while the storage modulus rises sharply until it intersects, then exceeds the loss modulus. By this definition, the gelation point is the time where the sample exhibits congruent $G' = G''$ behavior, independent of frequency and temperature.

9.13.3 Hydrogels

Hydrogels, or water-containing gels, are polymeric materials characterized by both hydrophilicity and insolubility in water (121). Hydrogels that are capable of absorbing very large amounts of aqueous fluids are sometimes referred to as superwater adsorbents.

Their hydrophilicity arises from the presence of water-solubilizing groups such as —OH, —COOH, —$CONH_2$, —CONH—, —SO_3H, and so on. Alternately, the main chain of the polymer may be water soluble, such as poly(ethylene oxide) derivatives (122), or based on acrylic, acrylamide, *N*-vinyl-2-pyrrolidinone, poly(vinyl alcohol), ionomers or glycopolymers (123). The insolubility arises from the presence of a three-dimensional network. The cross-links may be covalent, electrostatic, hydrophobic, or dipole–dipole in character. The extent of hydration may exceed a factor of 100, based on dry gel weight.

Many ionic hydrogels exhibit a first-order volume-phase transition (124). In such a transition the equilibrium extent of swelling can change dramatically with only a small change in conditions. The transition may be brought about by changing a range of variables, see Figure 9.31 (124). A characteristic manifestation of the transition involves the coexistence of two gel phases, and hence the presence of a first-order transition between them. Owing to the high values of hydrogel swelling, the inverse Langevin function (Section 9.10) is used to describe the chain conformations at near full extension (125). The

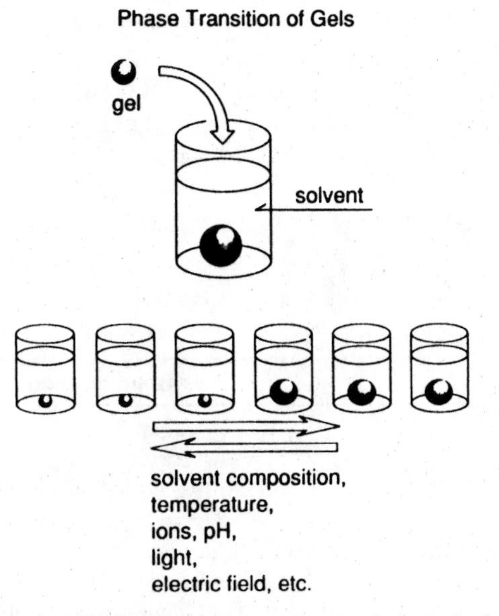

Figure 9.31 Illustration of the first-order phase transition in ionic hydrogels. The extent of swelling may change dramatically with environmental factors, as shown.

Debye-Hückel theory of ionic osmotic pressures is utilized. Of course, thermodynamics ultimately governs the equilibrium swelling (126).

Applications for hydrogels depending on their high-swelling ability include soft contact lenses based on poly(2-hydroxyethyl methacrylate), polyHEMA, and a number of drug delivery systems of various composition (121). While simple entrapment of the medication in the hydrogel works, controlled drug-release systems include eroding reservoir devices and other strategies. Hydrogels based on poly(sodium acrylate) (127,128) form the absorbent portion of disposable diapers (124). Another application of this technology involves switching materials, where *on* and *off* are two extents of swelling, as described above.

9.14 GELS AND GELATION

Most of the above cross-linked polymers were considered in the dry state, although the Flory–Rehner theory (Section 9.12) made use of equilibrium swollen gels in the evaluation of the cross-link density. Generally, a polymeric gel is defined as a system consisting of a polymer network swollen with solvent. It must be understood that the solvent is dissolved in the polymer, not the other way around.

Polymeric gels may be categorized into two major classes (129): thermoreversible gels and permanent gels. The thermoreversible gels undergo a transition from a solid-like form to a liquid-like form at a certain characteristic temperature. The links between the polymeric chains are transient in nature and support a stable polymeric network only below a characteristic "melting" point.

The permanent gels consist of solvent-logged covalently bonded polymer networks. One family of such networks is formed by cross-linking preexisting polymer chains, such as by vulcanization. Another family makes use of simultaneous polymerization and cross-linking. Some of the more important types of gels are delineated in Table 9.3 some specific examples include Jello® (com-

Table 9.3 Types of gels

Class	Bonding Mode
Thermoreversible	Hydrogen bonds
	Microcrystals
	Entwined helical structures
	Specific interactions
Permanent gels	Vulcanization
	Multifunctional monomers
	(a) Chain polymerizable
	(b) Stepwise polymerizable

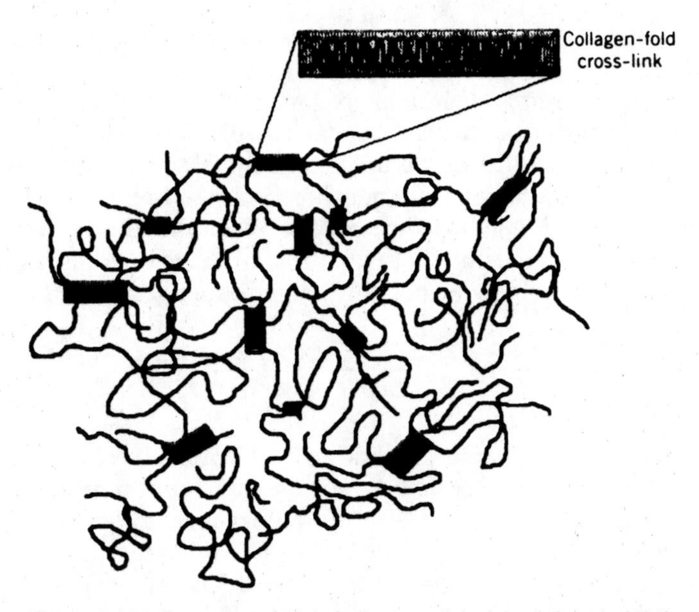

Figure 9.32 Formation of triple-helix cross-links in gelatin gels (131).

prised of approximately 3% of collagen-derived protein gelatin plus colored, flavored, and sweetened water), vitreous humor that fills the interior of the eye, membranes (both natural and synthetic), and soft-contact lenses (130).

Taking gelatin as an example, the thermal naturation and denaturation transition occurs near 40°C in water for native soluble collagen, and near 25°C for partially renatured gelatin. After setting, the initial growth rate of the modulus is nearly third order in gelatin concentration. The mechanism of gelation thus is thought to involve slow association of three rapidly formed single-helix segments (Figure 9.32) (131). A demonstration experiment measuring the modulus of gelatin as a function of concentration is given in Appendix 9.1.

The gelatin-type of triple helix may be considered a type of crystal; indeed, the denaturing and renaturing of these gels is a first-order transition. However, it has been modeled as a nucleated, one-dimensional crystallization, as opposed to the ordinary three-dimensional crystallization of bulk materials.

9.15 EFFECT OF STRAIN ON THE MELTING TEMPERATURE

Most elastomers are amorphous in use. Indeed, significant crystallinity deprives the polymer of its rubbery behavior. However, some elastomers crystallize during strains such as extension. The most important of these are *cis*-polybutadiene, *cis*-polyisoprene, and *cis*-polychloroprene. Crystallization on extension can be responsible for a rapid upturn in the stress–strain curves at high elongation; see Figure 9.5.

Such crystallization can result in significant engineering advantage. Consider, for example, the wear mechanisms in automotive tires. It turns out that abrasion is the most important mode of loss of tread. Tiny shreds of rubber are torn loose, hanging at one end. In contact with the road, these shreds are strained at each revolution of the tire, gradually tearing off more rubber. If the rubber crystallizes during extension, it becomes self-reinforcing when it is needed most, thus slowing the failure process. When the strain is released, the crystallites melt, returning the rubber to its amorphous state reversibly.

The elevation in melting temperature can be expressed (132) as

$$\frac{1}{T_f} = \frac{1}{T_f^0} - \frac{R}{2n\,\Delta H_f}\left(\alpha^2 + \frac{2}{\alpha} - 3\right) \tag{9.93}$$

where T_f and T_f^0 are the thermodynamic melting temperatures for the strained and unstrained polymers, respectively, n is the number of repeating units per network chain, and ΔH_f is the heat of fusion per mer. The student will note the relationship to equations (6.38) to (6.43), albeit that these equations treat melting point depressions, and the present case treats melting point elevation. Also note equation (9.33).

Equation (9.93) was derived from the assumption that the system can be treated as being composed of two independent phases, T_f obtained directly by equating the chemical potentials of the strained amorphous and crystalline phases. The function of the extension ratio, given in equation (9.33) yields the dependence of the change in free energy on elongation.

Figure 9.33 (132) illustrates the application of the equation to natural rubber and polychloroprene (see Section 9.15 for structures). The equation fits the data for natural rubber somewhat better than for polychloroprene, although both polymers are fit for reasonable extensions. It must be pointed out that while the pure polymers both crystallize somewhat above room temperature, ordinary vulcanization and compounding lower the melting temperatures significantly, shifting the data in Figure 9.33 upward. Typical numerical values for equation (9.93) are as follows:

Polymer	ΔH_f (cal/mol)	n
Natural rubber	1000	73
Polychloroprene	2000	37

9.16 ELASTOMERS IN CURRENT USE

The foregoing sections outline the theory of rubber elasticity. This section describes the classes of elastomers in current use. While many of these materials exhibit low modulus, high elongation, and rapid recovery from deformation and obey the theory of rubber elasticity, some materials deviate

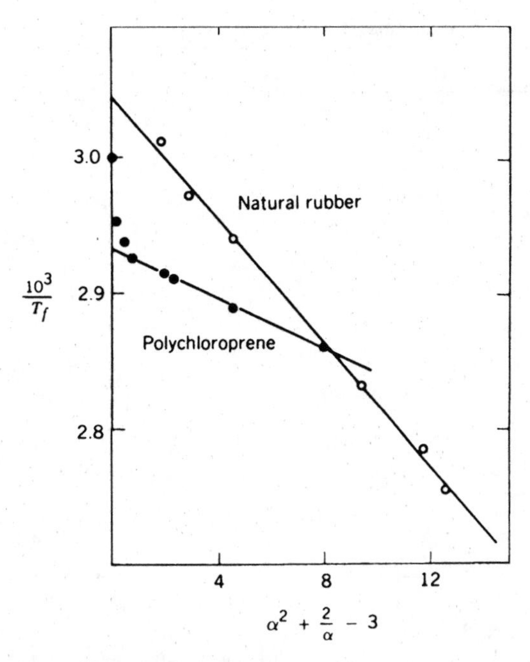

Figure 9.33 Fit of equation (9.93) to melting point elevation (129).

significantly, have limited extensibility or poorly defined rubbery plateaus, but are considered elastomers.

9.16.1 Classes of Elastomers

9.16.1.1 Diene Types The diene elastomers are based on polymers prepared from butadiene, isoprene, their derivatives and copolymers. The oldest elastomer, natural rubber (polyisoprene), is in this class (see Section 9.2). Polybutadiene, polychloroprene, styrene–butadiene rubber (SBR), and acrylonitrile–butadiene rubber (NBR) are also in this class.

The general polymerization scheme may be written

$$n\mathrm{CH_2{=}\underset{\displaystyle |}{\overset{\displaystyle X}{C}}{-}CH{=}CH_2} \rightarrow \left(\mathrm{CH_2{-}\overset{\displaystyle X}{\underset{}{C}}{=}CH{-}CH_2}\right)_n \tag{9.94}$$

where X— may be H—, CH_3—, Cl—, and so on (see Table 9.4).

The diene double bond in equation (9.94) may be either *cis* or *trans*. The *cis* products all have lower glass transition temperatures and/or reduced crystallinity, and they make superior elastomers. A random copolymer of butadiene and styrene is polymerized to form SBR (styrene–butadiene rubber). This copolymer forms the basis for tire rubber (see below). The *trans* materials, such

Table 9.4 Structures of elastomeric materials

Name	Structure
A. Diene elastomers	$-\!\!\left[CH_2-C(X)=CH-CH_2 \right]\!\!-_n$
Polybutadiene	$X- = H-$
Polyisoprene	$X- = CH_3-$
Polychloroprene	$X- = Cl-$
B. Acrylics	$-\!\!\left[CH_2-CH(C(=O)-O-X) \right]\!\!-_n$
Poly(ethyl acrylate)	$X- = CH_3CH_2-$
C. EPDM[a]	$-\!\!\left[CH_2-CH_2 \right]\!\!-_n -\!\!\left[CH_2-CH(CH_3) \right]\!\!-_m$
D. Thermoplastic elastomers	ABA
Poly(styrene–*block*–butadiene–*block*–styrene)	A = polystyrene B = polybutadiene
Segmented polyurethanes	$-\!\!\left[AB \right]\!\!-_n$A = polyether (soft block) B = aromatic urethane (hard block) A = poly(butylene oxide) B = poly(terephthalic acid–ethylene glycol)
E. Inorganic elastomers Silicone rubber	$-\!\!\left[Si(CH_3)_2-O \right]\!\!-_n$
Polyphosphazenes	$-\!\!\left[N=P(R)(R') \right]\!\!-_n$

[a] Ethylene-propylene diene monomer.

as the balata and gutta percha polyisoprenes (12), are highly crystalline and make excellent materials such as golf ball covers.

Natural rubber is widely used in truck and aircraft tires, which require heavy duty. They are self-reinforcing because the rubber crystallizes when stretched.

9.16.1.2 Saturated Elastomers The polyacrylates exemplify these materials:

$$nCH_2=CH(C(=O)-O-X) \rightarrow -\!\!\left[CH_2-CH(C(=O)-O-X) \right]\!\!-_n \qquad (9.95)$$

where *X*— may be CH_3—, CH_3CH_2—, and so on (see Table 9.4). Ethyl and butyl are the two most important derivatives, with glass transition temperatures in the range of −22°C and −50°C, respectively. The main advantage of the saturated elastomers is resistance to oxygen, water, and ultraviolet light, which attack the diene elastomers in outdoor conditions (133).

An important saturated elastomer is based on a random copolymer of ethylene and propylene, EPDM, or ethylene–propylene–diene monomer. The diene is often a bicyclic compound introduced at the 2% level to provide cross-linking sites:

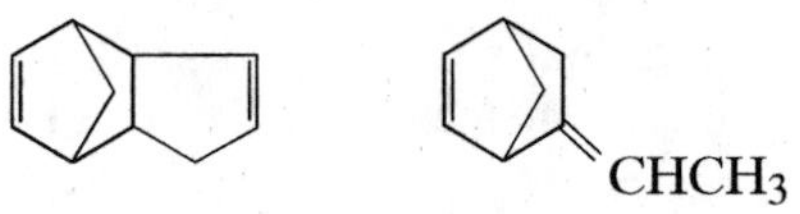

CHCH$_3$ (9.96)

It must be pointed out that although both polyethylene and polypropylene are crystalline polymers, the random polymer at midrange compositions is totally amorphous. These materials, with a glass transition of −50°C, make especially good elastomers for toughening polypropylene plastics (134–136).

9.16.1.3 Thermoplastic Elastomers These new materials contain physical cross-links rather than chemical cross-links. A physical cross-link can be defined as a non-covalent bond that is stable under one condition but not under another. Thermal stability is the most important case. These materials behave like cross-linked elastomers at ambient temperatures but as linear polymers at elevated temperatures, having reversible properties as the temperature is raised or lowered.

The most important method of introducing physical cross-links is through block copolymer formation (137–140). At least three blocks are required. The simplest structure contains two hard blocks (with a T_g or T_f above ambient temperature) and a soft block (with a low T_g) in the middle (see Figure 4.16). The soft block is amorphous and above T_g under application temperatures, and the hard block is glassy or crystalline.

The thermoplastic elastomers depend on phase separation of one block from the other (see Chapter 13), which in turn depends on the very low entropy gained on mixing the blocks. The elastomeric phase must form the continuous phase to produce rubbery properties; thus the center block has a higher molecular weight than the two end blocks combined.

Examples of the thermoplastic elastomers include polystyrene-*block*-polybutadiene-*block*-polystyrene (SBS) or the saturated center block counterpart (SEBS). In the latter, the EB stands for ethylene–butylene, where a combination of 1,2 and 1,4 copolymerization of butadiene on hydrogenation presents the appearance of a random copolymer of ethylene and butylene (see Table 9.4).

When the polymer illustrated in Figure 4.16 is of the SBS or SEBS type, it is sold under the trademark Kraton.®

Important applications of the triblock and its cousins, the starblock copolymer thermoplastic elastomers, include the rubber soles of running shoes and sneakers (140,141) and hot melt adhesives. In the former application, sliding friction generated heat momentarily turns the elastomer into an adhesive (see Section 8.2), reducing slips and falls. When sliding stops, the sole surface cools again, regenerating the rubber.

The so-called segmented polyurethanes form thermoplastic elastomers of the $\text{-}(AB)\text{-}_n$ type, where A is usually a polyether such as poly(ethylene oxide) and B contains aromatic urethane groups (142,143). These polyurethanes make excellent elastic fibers that stretch about 30% and are widely used in undergarments under the trade names Spandex® and Lycra®.

The literature distinguishes various kinds of polyurethanes. There are those that are not thermoplastic elastomers, often densely cross-linked. These need not be elastomeric at all.

The thermoplastic elastomer polyurethanes, TPU, may be of two general types: partly crystalline elastic fibers (see preceding paragraph) or softer elastomers, depending on the relative length of the soft segment. Those polyurethanes based on polyester soft segments tend to be more resistant to hydrocarbons, while the polyether types are more resistant to hydrolysis but tend to swell more in aqueous environments (144). The block copolymer characteristics of polyurethanes are discussed further in Chapter 13.

A newer type of $(AB)_n$ block copolymer, known as the poly(ether–ester) elastomers and sold as Hytrel®, contains alternating blocks of poly(butylene oxide) and butylene terephthalate as the soft and hard blocks, respectively (144–147):

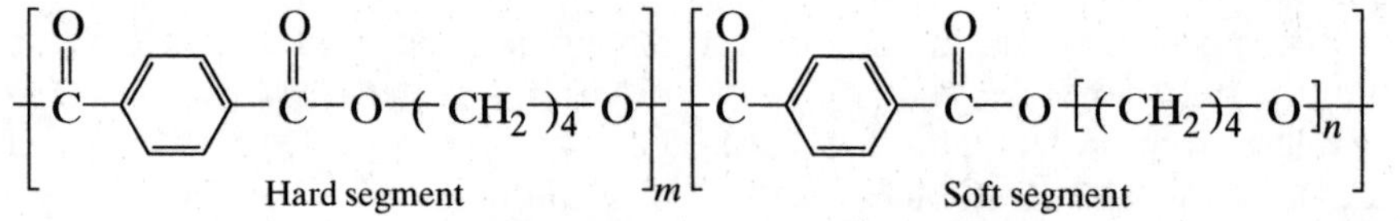

$$\underbrace{\left[\text{-}\overset{O}{\overset{\|}{C}}\text{-}C_6H_4\text{-}\overset{O}{\overset{\|}{C}}\text{-}O\text{-}(CH_2)_4\text{-}O\text{-} \right]_m}_{\text{Hard segment}} \underbrace{\left[\text{-}\overset{O}{\overset{\|}{C}}\text{-}C_6H_4\text{-}\overset{O}{\overset{\|}{C}}\text{-}O\text{-}[(CH_2)_4\text{-}O]_n\text{-} \right]}_{\text{Soft segment}} \tag{9.97}$$

Usually $m = 1$ or 2, and $n = 40–60$; thus the soft segment is much longer than the hard segment (146). The hard blocks crystallize in this case.

9.16.1.4 Inorganic Elastomers The major commercial inorganic elastomer is poly(dimethyl siloxane), known widely as silicone rubber (see Table 9.4). This specialty elastomer has the lowest known glass transition temperature, $T_g = -130°C$ (148); it also serves as a high-temperature elastomer. A common application of this elastomer is as a caulking material. It cross-links on exposure to air.

Another covalently bonded inorganic elastomer class is the polyphosphazenes (148,149),

$$\left[N = \underset{R'}{\overset{R}{\underset{|}{\overset{|}{P}}}} \right]_n \qquad (9.98)$$

In elastomeric compositions R and R′ are mixed substituent fluoroalkoxy groups. The current technological applications depend on the oil resistance and nonflammability of these elastomers; low T_g's are also important. Gaskets, fuel lines, and O-rings are made from this class of elastomer.

Another interesting inorganic elastomer is polymeric sulfur (150). Under equilibrium conditions at room temperature, rhombic sulfur consists mainly of eight-membered rings. On heating, it melts at 113°C to a relatively low viscosity, reddish-yellow liquid. Above approximately 160°C, the viscosity increases suddenly as the eight-membered rings open into long, linear chains of about 1×10^5 degree of polymerization. On further heating, the viscosity declines, as the polymer depolymerizes again. The sudden polymerization on heating is called a *floor temperature*, below which the free energy of polymerization is positive, and above which the free energy of polymerization is negative. (This is the opposite of many organic polymers, which exhibit *ceiling temperatures*, above which they depolymerize back to the monomer.)

If the polymerized form of elemental sulfur is quenched, it becomes highly elastomeric in the vicinity of room temperature. The situation is complicated by the presence of unpolymerized S_8 rings, which behave as plasticizers. In addition the polymer is only metastable, reverting back to S_8 rings in a relatively short time.

The above-mentioned heating and subsequent quenching of sulfur has long been a favorite laboratory demonstration both in high school and college chemistry (151,152). Since the sulfur exhibits both simple liquid and polymeric properties at laboratory temperatures without complex equipment or other chemicals, it illustrates several principles of science easily.

9.16.2 Reinforcing Fillers and Other Additives

Natural rubber has a certain degree of self-reinforcement, since it crystallizes on elongation (153). The thermoplastic elastomers also gain by the presence of hard blocks (140). However, nearly all elastomeric materials have some type of reinforcing filler, usually finely divided carbon black or silicas (137).

These reinforcing fillers, with dimensions of the order of 100 to 200 Å, form a variety of physical and chemical bonds with the polymer chains. Tensile and tear strength are increased, and the modulus is raised (154–158). The reinforcement can be understood through chain slippage mechanisms (see Figure 9.34) (137). The filler permits local chain segment motion but restricts actual flow.

While it is beyond the scope of this book to treat all the other components of commercial elastomers, a few must be mentioned. Frequently, an extender

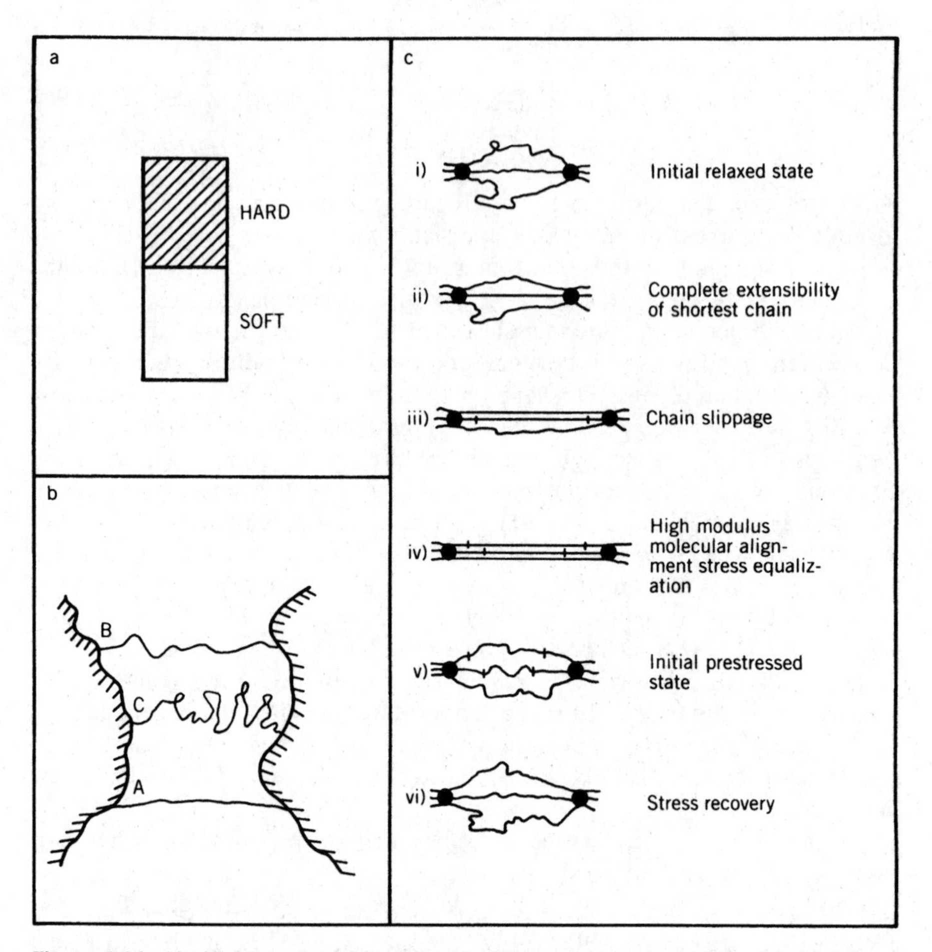

Figure 9.34 Mechanisms of reinforcement of elastomers by carbon black: (*a*) Takayanagi model approach; (*b*) on stretching, some chains (A) will become taut before others (B, C); (*c*) chain slippage on filler surface maintains polymer–filler bonds. From J. A. Harwood, L. Mullins, and A. R. Payne, *J. IRI*, **1**, 17 (1967).

is added. This is a high-molecular-weight oil that reduces the melt viscosity of the polymer while in the linear state, easing processing, and lowers the price of the final product.

Antioxidants or antiozonites are added, especially to diene elastomers, to slow attack on the double bond. Ultraviolet screens are used for outdoor application (133).

The major application of carbon-black-reinforced elastomers is in the manufacture of automotive tires. Table 9.5 illustrates an automotive tire recipe involving synthetic rubber. The synthetic-rubber recipes usually contain two or more elastomers that are blended together with the other ingredients and then covulcanized.

Table 9.5 A modern tire tread recipe

Ingredient	phr[a]	Function
Styrene–butadiene rubber	50	Elastomer
Natural rubber	50	Elastomer
Carbon black	45	Reinforcing filler
Aromatic oil	9	Extender
Paraffin wax	1	Processing aid and finish
Antiozonant	2	Reduce chain scission
Aryl diamine	1	Antioxidant
Stearic acid	3	Accelerator-activator
Zinc oxide	3	Accelerator-activator
Sulfur	1.6	Vulcanizing agent
N-Oxydiethylene benzothiazo-ε-2-sulfanamide	0.8	Accelerator, delayed action type
N,N-Diphenyl guanidine	0.4	A secondary accelerator

Source: The Vanderbilt Rubber Handbook, R. F. Ohm, Ed., R. T. Vanderbilt Co., Norwalk, CT, 14 Ed., 1995 (CD ROM version).

[a] Parts per hundred parts of rubber, by weight.

Current problems in tire usage involve rolling resistance, traction, and skid resistance. High values of tan δ can cause heat generation and fatigue problems. Rolling resistance involves a whole tire frequency of 10 to 100 Hz at 50 to 80°C, the average running temperature of tires.

In the case of skid resistance, stress is generated by friction with the road surface and concomitant motions of the rubber at the surface of the tire tread. The equivalent frequency of this movement is higher, depending on the roughness of the road surface, but is around 10^4 to 10^7 Hz at room temperature. High values of tan δ are desired for gripping the road under wet or icing conditions. By way of the WLF equation (Section 8.6.1.2), a reduced temperature for different tire properties at 1 Hz can be used as the criterion for polymer and filler development for tire compounds; see Figure 9.35 (159).

From a viscoelastic point of view, a high-performance tire should have a low tan δ value at 50 to 80°C to reduce rolling resistance and save energy. The ideal material should also possess high hysteresis from –20 to 0°C, for high skid resistance and wet grip.

While Figure 9.35 reads on the glass transition behavior of the elastomer, the filler also greatly influences the result. Today, carbon black is the principal reinforcing filler added to tire rubber; see item three in Table 9.5. However, newer commercial materials, such as a carbon/silica dual phase filler (160,161), improve the behavior especially with regard to the properties needed in Figure 9.35. The new filler consists of two phases, a carbon black phase, with a finely divided silica phase dispersed therein. The new filler causes stronger filler-polymer interactions in comparison with either conventional carbon black or silica alone having comparable surface areas. This results in a higher *bound*

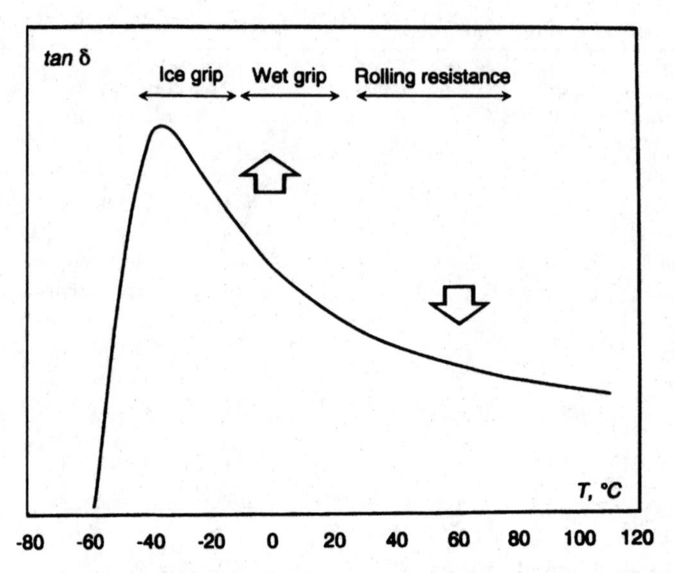

Figure 9.35 Desired values of tan δ as a function of temperature for tire performance. Frequencies reduced to 1 Hz. While a high value of tan δ is needed at low temperatures to reduce skidding on ice and water, low values of tan δ at higher temperatures reduce fuel consumption, tire overheating, and tire wear.

rubber content, which is the fraction of rubber bound to the filler tightly enough that it cannot be easily extracted by solvents. The effect is related to the introduction of greater numbers of surface defects in the graphitic crystal lattice of the carbon phase. The tighter binding of the rubber does not affect the value of the tan δ maximum at low temperatures, but the reduced chain slippage causes a lower value of tan δ at higher temperatures.

9.17 SUMMARY OF RUBBER ELASTICITY BEHAVIOR

When an amorphous cross-linked polymer above T_g is deformed and released, it "snaps back" with rubbery characteristics. The dependence of the stress necessary to deform the elastomer depends on the cross-link density, elongation, and temperature in a way defined by statistical thermodynamics.

The theory of rubber elasticity explains the relationships between stress and deformation in terms of numbers of active network chains and temperature but cannot correctly predict the behavior on extension. The Mooney–Rivlin equation is able to do the latter but not the former. While neither theory covers all aspects of rubber deformation, the theory of rubber elasticity is more satisfying because of its basis in molecular structure.

The theory of rubber elasticity is one of the oldest theories in polymer science and has played a central role in its development. Now one of the key assumptions of rubber elasticity theory, the concept of affine deformation, is under question in the light of small-angle neutron scattering experiments. The SANS experiments suggest a nonaffine chain deformation via an end-pulling mechanism.

If these new results stand the test of time, they may help answer one of the important challenges to the Mark–Flory random coil model (see Section 5.3). In this challenge the considerable discrepancy between rubber elasticity theory and experiment is blamed on the presumed nonrandom coiling of the polymer chain in space. These discrepancies may, however, lie rather in the realm of the mode of chain disentanglement on extension.

An understanding of how an elastomer works, however, has led to many new materials and new types of elastomers. A leading new type of elastomer is based on physical cross-links rather than chemical cross-links. The new materials are known as thermoplastic elastomers.

REFERENCES

1. D. Adolf, J. E. Martin, and J. P. Wilcoxon, *Macromolecules*, **23**, 527 (1990).
2. D. Adolf and J. E. Martin, *Macromolecules*, **23**, 3700 (1990).
3. J. E. Martin, D. Adolf, and J. P. Wilcoxon, *Phys. Rev. A*, **39**, 1325 (1989).
4. P. G. de Gennes, *J. Phys. Lett.*, **37**, 61 (1976).
5. D. J. Stauffer, *J. Chem. Soc. Faraday Trans. 2*, **72**, 1354 (1976).
6. D. Stauffer, A. Coniglio, and M. Adam, *Adv. Poly. Sci.*, **44**, 103 (1982). [*Polymer Networks*, K. Dusek, ed., Springer, New York, 1982.]
7. B. Gutte and R. B. Merrifield, *J. Am Chem. Soc.*, **91**, 501 (1969).
8. F. Rodriguez, *Principles of Polymer Systems*, 3rd ed., Hemisphere, New York, 1988, Chap. 14.
9. K. Dusek, in *Advances in Polymerization. 3*, R. N. Haward, ed., Applied Science Publishers, London, 1982.
10. K. Dusek and W. J. MacKnight, in *Cross-Linked Polymers: Chemistry, Properties, and Applications*, R. A. Dickie, S. S. Labana, and R. S. Bauer, eds., ACS Symposium Series 367, ACS Books, Washington, DC, 1988.
11. H. F. Mark, *Giant Molecules*, Time Inc., New York, 1966, p. 124.
12. H. J. Stern, *Rubber: Natural and Synthetic*, 2nd ed., Palmerton, New York, Chap. 1.
13. J. Gough, *Mem. Lit. Phil. Soc. Manchester*, **1**, 28 (1805).
14. C. Goodyear, U.S. Patent 3, 633 (1844).
15. J. L. White, *Rubber Industry*, August 1974, p. 148.
16. Lord Kelvin (W. Thompson), *J. Pure Appl. Math.*, **1**, 57 (1857).
17. C. Price, *Proc. R. Soc. Lond.*, **A351**, 331 (1976).
18. P. J. Flory, *Principles of Polymer Chemistry*, Cornell University Press, Ithaca, NY, 1953, (a) pp. 434–440, (b) p. 442, (c) p. 464.
19. H. Staudinger, *Ber. Otsch. Chem. Ges.*, **53**, 1073 (1920).

20. H. Staudinger, *Die hochmolekularen organische Verbindungen*, Verlag von Julius Springer, Berlin, 1932.
21. Reference 18, Chap. 1.
22. H. J. Stern, in *Rubber Technology and Manufacture*, C. M. Blow and C. Hepburn, eds., Butterworth Scientific, London, 1982.
23. Anonymous, *Chemical Heritage*, **17**(1), 4 (1999).
24. Y. Furukawa, *Inventing Polymer Science*, University of Pennsylvania Press, Philadelphia, 1998.
25. P. J. T. Morris, in *The American Synthetic Rubber Research Program*, University of Pennsylvania Press, Philadelphia, 1989.
26. W. M. Saltman, in *Rubber Technology*, M. Morton, ed., Van Nostrand Reinhold, New York, 1973.
27. R. A. French, in *Encyclopedia Britannica*, Vol. 22, William Benton, Publisher, Chicago, 1968, p. 506.
28. E. Guth and H. Mark, *Monatsh. Chem.* **65**, 93 (1934).
29. W. Kuhn, *Angew. Chem.*, **51**, 640 (1938).
30. W. Kuhn, *J. Polym. Sci.*, **1**, 380 (1946).
31. E. Guth and H. M. James, *Ind. Eng. Chem.*, **33**, 624 (1941).
32. E. Guth and H. M. James, *J. Chem. Phys.*, **11**, 455 (1943).
33. H. M. James, *J. Chem. Phys.*, **15**, 651 (1947).
34. H. M. James and E. Guth, *J. Chem. Phys.*, **15**, 669 (1947).
35. H. M. James and E. Guth, *J. Polym. Sci.*, **4**, 153 (1949).
36. L. R. G. Treloar, *Trans. Faraday Soc.*, **39**, 36, 241 (1943).
37. F. T. Wall, *J. Chem. Phys.*, **11**, 527 (1943).
38. P. J. Flory, *Trans. Faraday Soc.*, **57**, 829 (1961).
39. G. Allen, *Proc. R. Soc. Lond.*, **A351**, 381 (1976).
40. P. J. Flory, *Polymer*, **20**, 1317 (1979).
41. P. J. Flory, *Proc. R. Soc. Lond.*, **A351**, 351 (1976).
42. L. R. G. Treloar, *The Physics of Rubber Elasticity*, 3rd ed., Clarendon Press, Oxford, 1975.
43. R. P. Wool, *Polymer Interfaces: Structure and Strength*, Hanser, Munich, 1995.
44. L. R. G. Treloar, *Trans. Faraday Soc.*, **40**, 49 (1944).
45. L. R. G. Treloar, *Trans. Faraday Soc.*, **42**, 83 (1946).
46. F. T. Wall, *Chemical Thermodynamics*, 2nd ed., Freeman, San Francisco, 1965, p. 314.
47. R. L. Anthony, R. H. Caston, and E. Guth, *J. Phys. Chem.*, **46**, 826 (1942).
48. K. J. Laidler, *The World of Physical Chemistry*, Oxford University Press, Oxford, 1995.
49. J. E. Mark and B. Erman, *Rubber-like Elasticity: A Molecular Primer*, Wiley-Interscience, NY, 1988, p. 44.
50. J. E. Mark and B. Erman, *Rubberlike Elasticity: A Molecular Primer*, Wiley, New York, 1988.
51. E. Pines, K. L. Wun, and W. Prins, *J. Chem. Ed.*, **50**, 753 (1973).

52. K. J. Mysels, *Introduction to Colloid Chemistry*, Wiley-Interscience, New York, 1965.
53. M. Mooney, *J. Appl. Phys.*, **11**, 582 (1940).
54. M. Mooney, *J. Appl. Phys.*, **19**, 434 (1948).
55. R. S. Rivlin, *Trans. R. Soc. (Lond.)*, **A240**, 459, 491, 509 (1948).
56. R. S. Rivlin, *Trans. R. Soc. (Lond.)*, **A241**, 379 (1948).
57. S. M. Gumbrell, L. Mullins, and R. S. Rivlin, *Trans. Faraday Soc.*, **49**, 1495 (1953).
58. J. E. Mark, R. R. Rahalkar, and J. L. Sullivan, *J. Chem. Phys.*, **70**, 1794 (1979).
59. G. Gee, *Trans. Faraday Soc.*, **42**, 585 (1946).
60. A. Ciferri and P. J. Flory, *J. Appl. Phys.*, **30**, 1498 (1959).
61. K. C. Valanis and R. F. Landel, *J. Appl. Phys.*, **38**, 2997 (1967).
62. R. W. Ogden, *Proc. R. Soc.*, **A326**, 565 (1972).
63. M. Shen, in *Science and Technology of Rubber*, F. R. Eirich, ed., Academic Press, Orlando, 1978.
64. Y. Sato, *Rep. Prog. Polym. Phys. Jpn.*, **9**, 369 (1969).
65. W. Kuhn and F. Grün, *Kolloid Z.*, **101**, 248 (1942).
66. J. K. Yeo, L. H Sperling, and D. A. Thomas, *J. Appl. Polym. Sci.*, **26**, 3977 (1981).
67. J. A. Duiser and J. A. Staverman, in *Physics of Noncrystalline Solids*, J. A. Prins, Ed., North-Holland, Amsterdam, 1965.
68. W. W. Graessley, *Macromolecules*, **8**, 186 (1975).
69. K. J. Smith Jr., and R. J. Gaylord, *J. Polym. Sci. Polym. Phys. Ed.*, **13**, 2069 (1975).
70. D. S. Pearson and W. W. Graessley, *Macromolecules*, **13**, 1001 (1980).
71. P. J. Flory, *Ind. Eng. Chem.*, **138**, 417 (1946).
72. L. Mullins and A. G. Thomas, *J. Polym. Sci.*, **43,** 13 (1960).
73. A. V. Tobolsky, D. J. Metz, and R. B. Mesrobian, *J. Am. Chem. Soc.*, **72**, 1942 (1950).
74. J. Scanlan, *J. Polym. Sci.*, **43**, 501 (1960).
75. H. M. James and E. Guth, *J. Chem. Phys.*, **11**, 455 (1943).
76. P. J. Flory and J. Rehner, *J. Chem. Phys.*, **11**, 521 (1943).
77. H. M. James and E. Guth, *J. Polym. Sci.*, **4**, 153 (1949).
78. P. J. Flory, *Trans. Faraday Soc.*, **57**, 829 (1961).
79. A. V. Tobolsky and M. C. Shen, *J. Appl. Phys.*, **37**, 1952 (1966).
80. L. H. Sperling and A. V. Tobolsky, *J. Macromol. Chem.*, **1**, 799 (1966).
81. A. M. Bueche, *J. Polym. Sci.*, **19**, 297 (1956).
82. L. Mullins, *J. Appl. Polym. Sci.*, **2**, 1 (1959).
83. G. Kraus, *J. Appl. Polym. Sci.*, **7**, 1257 (1963).
84. R. G. Mancke, R. A. Dickie, and J. O. Ferry, *J. Polym. Sci.*, **A-2** (6), 1783 (1968).
85. N. R. Langley, *Macromolecules*, **1**, 348 (1968).
86. J. D. Ferry, *Viscoelastic Properties of Polymers*, 3rd ed., Wiley, New York, 1980, pp. 408–411.
87. O. Kramer, *Polymer*, **20**, 1336 (1979).
88. S. Onogi, T. Masuda, and K. Kitagawa, *Macromolecules*, **3**, 111 (1970).
89. P. J. Flory, *Chem. Rev.*, **35**, 51 (1944).

90. J. Scanlan, *J. Polym. Sci.*, **43**, 501 (1960).
91. N. R. Langley and K. E. Polmanter, *J. Polym. Sci. Polym. Phys. Ed.*, **12**, 1023 (1974).
92. J. D. Ferry, *Polymer*, **20**, 1343 (1979).
93. M. Beltzung, C. Picot, P. Rempp, and J. Hertz, *Macromolecules*, **15**, 1594 (1982).
94. H. Benoit, D. Decker, R. Duplessix, C. Picot, P. Remp, J. P. Cotton, B. Farnoux, G. Jannick, and R. Ober, *J. Polym. Sci. Polym. Phys. Ed.*, **14**, 2119 (1976).
95. H. Benoit, R. Duplessix, R. Ober, M. Daoud, J. P. Cotton, B. Farnoux, and G. Jannick, *Macromolecules*, **8**, 451 (1975).
96. L. H. Sperling, *Polym. Eng. Sci.*, **24**, 1 (1984).
97. L. H. Sperling, A. M. Fernandez, and G. D. Wignall, in *Characterization of Highly Cross-linked Polymers*, S. S. Labana and R. A. Dickie, eds., ACS Symposium Series No. 243, American Chemical Society, Washington, 1984.
98. C. Picot, R. Duplessix, D. Decker, H. Benoit, F. Boue, J. P. Cotton, and P. Pincus, *Macromolecules*, **10**, 436 (1977).
99. J. A. Hinkley, C. C. Han, B. Mozer, and H. Yu, *Macromolecules*, **11**, 837 (1978).
100. S. B. Clough, A. Maconnachie, and G. Allen, *Macromolecules*, **13**, 774 (1980).
101. S. B. Clough, Private communication, December 29, 1982.
102. D. S. Pearson, *Macromolecules*, **10**, 696 (1977).
103. R. Ullman, *Macromolecules*, **15**, 1395 (1982).
104. R. Ullman, *Macromolecules*, **15**, 582 (1982).
105. R. Ullman, in *Elastomers and Rubber Elasticity*, J. E. Mark and J. Lal, eds., ACS Symposium Series No. 193, American Chemical Society, Washington, DC, 1982, Chap. 13.
106. G. Hadziiannou, L. H. Wang, R. S. Stein, and R. S. Porter, *Macromolecules*, **15**, 880 (1982).
107. F. Boue, M. Nierlich, G. Jannick, and R. C. Ball, *J. Physiol. (Paris)*, **43**, 137 (1982).
108. M. Shen and P. J. Blatz, *J. Appl. Physiol.*, **39**, 4937 (1968).
109. M. Shen, *Macromolecules*, **2**, 358 (1969).
110. A. Ciferri, *Macromol. Chem.*, **43**, 152 (1961).
111. K. Dusek, ed., *Polymer Networks*, Springer, New York, 1982.
112. J. E. Mark, B. Erman, and F. R. Eirich, *Science and Technology of Rubber*, 2nd ed., Academic Press, San Diego, 1994.
113. J. E. Mark and J. Lal, eds., *Elastomers and Rubber Elasticity*, American Chemical Society, Washington, DC, 1982.
114. A. V. Galanti and L. H. Sperling, *Polym. Eng. Sci.*, **10**, 177 (1970).
115. B. A. Rozenberg, Presented at *Networks 91*, Moscow, April 1991.
116. B. A. Rozenberg, *Epoxy Resins and Composites II*, Advances in Polymer Science Vol. 75, Springer, Berlin, 1986.
117. S. B. Ross-Murphy, Polymer, **33**, 2622 (1992).
118. L. Z. Rogovina and V. G. Vasil'yev, *Networks 91*, Moscow, April 1991.
119. H. H. Winter and F. Chambon, *J. Rheol.*, **30**, 367 (1986).
120. F. Chambon and H. H. Winter, *Polym. Bull.*, **13**, 499 (1985).

121. V. Kudela, *Encyclopedia of Polymer Science and Engineering*, J. I. Kroschwitz, ed., Wiley, New York, 1987.
122. R. M. Ottenbrite, in *Encyclopedia of Polymer Science and Engineering*, J. I. Kroschwitz, ed., Wiley, New York, 1989.
123. T. Miyata and K. Nakamae, *Trends in Polym. Sci. (TRIP)*, **5**, 198 (1997).
124. M. Shibayama and T. Tanaka, in *Responsive Gels: Volume Transitions I*, K. Dusek, ed., Springer-Verlag, Berlin, 1993.
125. M. Ilavsky, in *Responsive Gels: Volume Transitions I*, K. Dusek, ed., Springer, Berlin, 1993.
126. A. R. Khokhlov, S. G. Starodubtzev, and V. V. Vasilevskaya, in *Responsive Gels: Volume Transitions I*, K. Dusek, ed., Springer, Berlin, 1993.
127. W. C. Buzanowski, S. S. Cutie, R. Howell, R. Papenfuss, and C. G. Smith, *J. Chromatography*, **677**, 355 (1994).
128. D. Benda, J. Snuparek, and V. Cermak, *J. Dispersion Sci. Tech.*, **18**, 115 (1997).
129. S. M. Aharoni and S. F. Edwards, *Macromolecules*, **22**, 3361 (1989).
130. T. Tanaka, in *Encyclopedia of Polymer Science and Engineering*, 2nd ed., Vol., 7, Wiley, New York, 1987, p. 154.
131. P. I. Rose, in *Encyclopedia of Polymer Science and Engineering*, 2nd ed., Vol. 7, Wiley, New York, 1987, p. 488.
132. W. R. Krigbaum, J. V. Dawkins, G. H. Via, and Y. I. Balta, *J. Polym. Sci. A-2*, **4**, 475 (1966).
133. F. Rodriguez, *Principles of Polymer Systems*, 4th ed., Taylor & Francis, Washington, DC, 1996.
134. J. Karger-Kocis, A. Kallo, A. Szafner, G. Bodor, and Z. Senyei, *Polymer*, **20**, 37 (1979).
135. D. W. Bartlett, J. W. Barlow, and D. R. Paul, *J. Appl. Polym. Sci.*, **27**, 2351 (1982).
136. E. Martuscelli, M. Pracela, M. Avella, R. Greco, and G. Ragosta, *Makromol. Chem.*, **181**, 957 (1980).
137. L. H. Sperling, *Polymeric Multicomponent Materials: An Introduction*, Wiley, New York, 1997.
138. J. E. McGrath, *J. Chem. Ed.*, **58** (11), 914 (1981).
139. A. Noshay and J. E. McGrath, *Block Copolymers —Overview and Critical Survey*, Academic Press, Orlando, 1977.
140. G. Holden, *Understanding Thermoplastic Elastomers*, Hanser, Munich, 2000.
141. G. Holden, in *Recent Advances in Polymer Blends, Grafts, and Blocks*, L. H. Sperling, ed., Plenum, New York, 1974.
142. E. Pechhold, G. Pruckmayr, and I. M. Robinson, *Rubber Chem. Technol.*, **53**, 1032 (1980).
143. J. Blackwell and K. H. Gardner, *Polymer, **20***, 13 (1979).
144. R. P. Brentin, *Rubber World*, **208** (1), 22 (1993).
145. L. L. Zhu, G. Wegner, and U. Bandara, *Makromol. Chem.*, **182**, 3639 (1981).
146. A. Lilaonitkul and S. L. Cooper, *Rubber Chem. Technol.*, **50**, 1 (1977).
147. P. C. Mody, G. L. Wilkes, and K. B. Wagener, *J. Appl. Polym Sci.*, **26**, 2853 (1981).

148. J. E. Mark, H. R. Allcock, and R. West, *Inorganic Polymers*, Prentice-Hall, Englewood Cliffs, NJ, 1992.
149. D. P. Tate, *J. Polym. Sci. Polym. Symp.*, **48**, 33 (1974).
150. J. E. Mark, H. R. Allcock, and R. West, *Inorganic Polymers*, Prentice-Hall, Englewood Cliffs, NJ, 1992.
151. H. R. Allcock, *Sci. Am.*, **230** (3), 66 (1974).
152. K. R. Birdwhistell and J. W. Long, *J. Chem. Educ.*, **72**, 56 (1995).
153. J. E. Mark, A. Eisenberg, W. W. Graessley, E. T. Samulski, J. L. Koenig, and G. D. Wignall, *Physical Properties of Polymers*, 2nd ed., ACS Books, Washington, DC, 1993.
154. A. V. Galanti and L. H. Sperling, *Polym. Eng. Sci.*, **10**, 177 (1970).
155. G. Kraus, *Rubber Chem. Technol.*, **51**, 297 (1978).
156. B. B. Boonstra, *Polymer*, **20**, 691 (1979).
157. Z. Rigbi, *Adv. Polym. Sci.*, **36**, 21 (1980).
158. K. E. Polmanteer and C. W. Lentz, *Rubber Chem. Technol.*, **48**, 795 (1975).
159. M.-J. Wang, *Rubber Chem. Technol.*, **71**, 520 (1998).
160. L. J. Murphy, M.-J. Wang, and K. Mahmud, *Rubber Chem. Technol.*, **71**, 998 (1998).
161. L. J. Murphy, E. Khmelnitskaia, M.-J. Wang, and K. Mahmud, *Rubber Chem. Technol.*, **71**, 1015 (1998).

GENERAL READING

W. Burchard and S. B. Ross-Murphy, *Physical Networks: Polymers and Gels*, Elsevier, New York, 1990.

K. Dusek, ed., *Polymer Networks*, Springer, New York, 1982.

B. Erman and J. E. Mark, *Structures and Properties of Rubberlike Networks*, Oxford University Press, Oxford, 1997.

Y. Fukahori, *Rubber Chem. Technol.*, **76(2)**, 548 (2003). (Mechanics and mechanism of carbon black reinforcement.)

W. W. Graessley, *Polymeric Liquids & Networks: Structure and Properties*, Garland Science, New York, 2003.

O. Kramer, ed., *Biological and Synthetic Polymer Networks*, Elsevier, London, 1988.

J. E. Mark, B. Erman, and F. R. Eirich, eds., *Science and Technology of Rubber*, 2nd ed., Academic Press, San Diego, 1994.

J. E. Mark and B. Erman, *Rubber-like Elasticity: A Molecular Primer*, Wiley-Interscience, New York, 1998.

J. E. Mark, *J. Phys. Chem. B*, **107**, 903 (2003). (Some recent theory, experiments and simulations on rubberlike elasticity.)

J. E. Mark, *Macromol. Symp.*, **201**, 77 (2003). (Review of recent developments in rubber elasticity.)

J. E. Mark, *Prog. Polym. Sci.*, **28**, 1205 (2003). (Unusual elastomers and experiments.)

M. Morton, ed., *Rubber Technology*, 3rd ed., Van Nostrand–Reinhold, New York, 1987.

K. te Nijenhuis, *Thermoreversible Networks: Viscoelastic Properties and Structure of Gels*, Advances in Polymer Science Vol. 130, Springer, Berlin, 1997.

STUDY PROBLEMS

1. Why does a rubber band snap back when stretched and released? An explanation including both thermodynamic and molecular aspects is required. (Equations/diagrams/figures and as few words as possible will be appreciated.)

2. A strip of elastomer 1 cm × 1 cm × 10 cm is stretched to 25-cm length at 25°C, a stress of 1.5×10^7 dynes/cm^2 being required.

   (a) Assuming a tetrafunctional cross-linking mode, how many moles of network chains are there per cubic centimeter?

   (b) What stress is required to stretch the sample to only 15 cm, at 25°C?

   (c) What stress is required to stretch the sample to 25-cm length at 100°C?

3. The theory of rubber elasticity and the theory of ideal gas dynamics show that the two equations, $G = nRT$ and $PV = n'RT$, share certain common thermodynamic ideas. What are they?

4. Write the chemical structure for polybutadiene, and show its vulcanization reaction with sulfur.

5. For a swollen elastomer, the equation of state can be written

$$\sigma = nRTv_2^{1/3}\frac{\overline{r_i^2}}{\overline{r_0^2}}\left(\alpha - \frac{1}{\alpha^2}\right)$$

   Explain, qualitatively and very briefly, where the term $v_2^{1/3}$ originates. A derivation is not required.

6. Read any paper, 2004 or more recent, on rubber elasticity and write a 200-word report on it *in your own words*. (Give the reference!) Key figures, tables, or equations may be photocopied. How is the science or engineering of elastomers advanced beyond this book?

7. Show how equations (9.81) and (9.82) reduce to $n = n_c + n_p$ for $W_g = 1$. What do you get when $W_g = 0$?

8. Recent experimental evidence using SANS instrumentation suggests that the ends of a network segment deform affinely, yet the chain itself barely extends in the direction of the stress and contracts in the transverse direction even less. Develop a model to explain the results, and comment on how you think the theory of rubber elasticity ought to be modified to accommodate the new finding.

9. We just had Halloween. Do you believe in “phantom networks”? Why?

10. A sample of vulcanized natural rubber, *cis*-polyisoprene, swells to five times its volume in toluene. What is Young’s modulus of the unswollen elastomer at 25°C? (The interaction parameter, χ_1, is 0.39 for the system *cis*-polyisoprene–toluene.) [*Hint*: See equation (9.92); the molar volume of toluene is 106.3 cm^3/mol.]

11. Young’s modulus for an elastomer at 25°C is 3×10^7 $dynes/cm^2$. What is its shear modulus? What is the retractive stress if a sample 1 cm × 1 cm × 10 cm is stretched to 25 cm length at 100°C?

12. A sample of elastomer, cross-linked at room temperature, is swollen afterward to 10 times its original volume. Then it is stretched. What value of the “front factor” should be used in the calculation of the stress?

13. Two identical 10-cm rubber bands, A and B, are tied together at their ends and stretched to a total of 40 cm length and held in that position. Rubber band A is at 25°C, and rubber band B is at 150°C. How far from the B end is the knot?

14. In the rubber heat engine described in Figure 9.14, the wheel is heated and turns accordingly. Does this experiment have the equivalent of all four steps illustrated in Figure 9.13*a*? [*Hint*: Don’t forget gravity.]

15. A rubber ball is dropped from a height of 1 yard and bounces back 18 in. Assuming a perfectly elastic floor, approximately how much did the ball heat up? The heat capacity, C_p, of SBR rubber is about 1.83 kJ $kg^{-1} \cdot K^{-1}$.

16. Accoding to recent papers published in the *J. Theor. Hypothet. Polym. Sci.*, wheels to round, plastics break, and balls don’t bounce. In one paper of recent vintage, two identical rubber bands, A and B, were dissolved in identical baths A' and B'. The solvent de-cross-linked the rubber bands but otherwise was a simple solvent. Rubber band A was dropped in unstretched. Rubber band B was rapidly stretched from its initial length of 10 cm to 25 cm and instantly placed in the bath while held stretched by a holder of no physical properties during the solution process. Right after the solution process was completed, the poor investigator found he had mixed up the baths. Quick, how would you help him identify the baths? What basic and simple experiment would you perform? Assuming you found the difference, how does this difference change, algebraically, with the extent of stretch of rubber band B? If the two baths are identical, and no difference should exist, write a brief paragraph giving the correct reasons for full credit.

17. A sample of rubber was mixed with sulfur and other curing agents, molded into a sheet, and heated briefly in the relaxed state. A total of n_1 cross-links were introduced. Then the sheet was stretched α times its original length, and the heating continued, introducing n_2 new cross-links. The

sample was then released and cooled. In terms of α, to what extent will the sample remain stretched?

18. You are shipwrecked on a desert island. You find some bushes that have a sticky sap. You also find the island has all kinds of minerals. How do you get off the island?

APPENDIX 9.1 GELATIN AS A PHYSICALLY CROSS-LINKED ELASTOMER†

Introduction

Ordinary gelatin is made from the skins of animals by a partial hydrolysis of their collagen, an important type of protein (A1,A2). At home, a crude type of gelatin can be prepared from the broth of cooked meats and fowl; this material also frequently gels on cooling.

When dissolved in hot water, the gelatin protein has a random coil type of conformation. On cooling, a conformational change takes place to a partial helical arrangement. At the same time, intermolecular hydrogen bonds form, probably involving the N—H linkage. On long standing, such gels may also crystallize locally. The bonds that form in gelatin are known not to be permanent, but rather they relax in the time frame of 10^3 to 10^6 s (A3–A5). The amount of bonding also decreases as the temperature is raised. The purpose of this appendix is to demonstrate the counting of these bonds via modulus measurements.

Theory

By observing the depth of indentation of a sphere into the surface of gelatin, "indentation" modulus is easily determined. The indentation modulus yields its close relative, Young's modulus. The cross-link density and thus the number of hydrogen bonds (simple physical cross-links) are readily determined by treating the gelatin as a hydrogen-bonded elastomer.

Young's modulus may be determined by indentation using the Hertz (A6) equation:

$$E = \frac{3(1-v^2)F}{4h^{3/2}r^{1/2}} \tag{A9.1.1}$$

where F represents the force of sphere against the gelatin surface = mg (dynes), h represents the depth of indentation of sphere (cm), r is the radius of sphere (cm), g represents the gravity constant, and v is Poisson's ratio.

† Reproduced in part from the G. V. Henderson, D. O. Cambell, V. Kuzmicz, and L. H. Sperling, *J. Chem. Ed.*, **62**, 269 (1985).

The ball indentation experiment is the scientific analogue of pressing on an object with one's thumb to determine hardness. The less the indentation, the higher the modulus.

Young's modulus is related to the cross-link density through rubber elasticity theory; see equation (9.36):

$$E = 3nRT \tag{A9.1.2}$$

Assuming a tetrafunctional cross-linking mode (four chain segments emanating from the locus of the hydrogen bond):

$$E = 6\mu RT \tag{A9.1.3}$$

where n represents the number of active chain segments in network and μ is the cross-link density (moles of cross-links per unit volume). For this experiment, the gelatin was at 278.0 K, the temperature of the refrigerator employed.

Experimental

Time: About 30 minutes, the gelatin prepared previously.
Principles Illustrated:

1. Helix formation and physical cross-linking in gelatin.
2. Rubber elasticity in elastomers.
3. Physical behavior of proteins.

Equipment and Supplies:

Five 150 × 75 mm Pyrex® crystallizing dishes or soup dishes
Five 2-cup packets of flavored Jello® brand gelatin (8 g protein per packet)
Eighteen 2-cup packets of unflavored Knox® brand gelatin (6 g protein per packet)
One metric ruler
One steel bearing (1.5-in. diameter and 0.226 kg—or any similar spherical object)
One lab bench
One knife

Five different concentrations (see Table A9.1.1) of gelatin were prepared, each in 600 ml of water, and allowed to set overnight in a refrigerator at 5.0°C. Then indentation measurements were made by placing the steel bearing in the center of the gelatin samples and measuring the depth of indentation, h (see Figure A9.1.1). As it is difficult to see through the gelatin to observe this depth,

Table A9.1.1 Gelatin concentrations

Dish	1	2	3	4	5
Concentration[a]	3.0	2.0	1.0	0.75	0.50
Jello[b]	1	1	1	1	1
Gelatin[c]	8	5	2	1.25	0.5

[a] Concentration = number of times the normal gelatin concentration (each dish contains 600 ml of water).
[b] Jello = number of 2-cup packets of Jello® brand black raspberry flavored gelatin.
[c] Gelatin = number of 2-cup packets of Knox® brand unflavored gelatin.

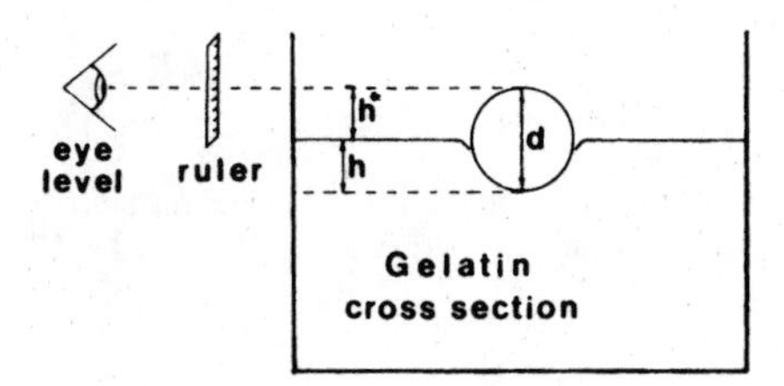

DEPTH OF INDENTATION.
$h(\text{cm}) = d - h^*$

Figure A9.1.1 Schematic of experiment, measuring the indentations of the heavy ball in the gelatin.

it is desirable to measure the height of the bearing from the level surface of the gelatin and subtract this quantity from the diameter of the bearing (Figure A9.1.1).

The measured depth of indentation, the radius of the bearing, and the force due to the bearing are algebraically substituted into equation (A9.1.1). This value of Young's modulus is substituted into equation (A9.1.3) to yield hydrogen bond cross-link density.

Results

A plot of E as a function of gelatin concentration (Figure A9.1.2) demonstrates a linear increase in Young's modulus at low concentrations. The slight upward curvature at high concentrations is caused by the increasing efficiency of the network. However, the line should go through the origin.

Physical cross-link concentrations were determined using equation (A9.1.3), and the results are shown in Table A9.1.2. Assuming a molecular weight of about 65,000 g/mol for the gelatin, there is about 1.2 physical bonds per molecule (see Table A9.1.2).

Using gelatin as a model cross-linked elastomer, its rubber elasticity can also be demonstrated by a simple stretching experiment. Thin slices of the

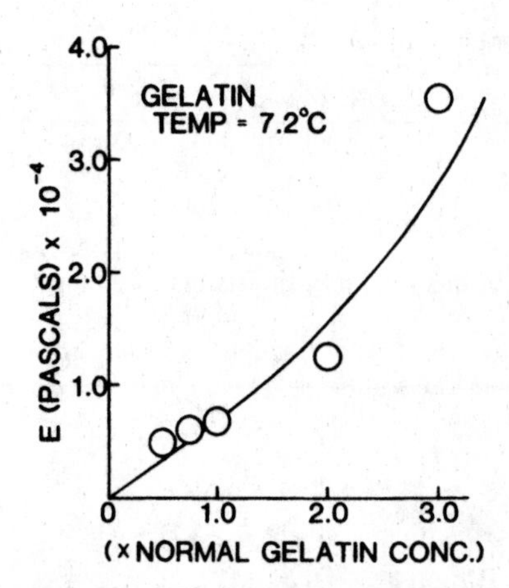

Figure A9.1.2 Modulus of gelatin samples versus concentration. By comparison, a rubber band has a Young's modulus of about 1×10^6 pascals (1 pascal = 10 dynes/cm^2).

Table A9.1.2 Gelatin indentations yield bond numbers

Concentration	h (cm)	μ (mol/cm^3)	N
0.5	1.50	3.5×10^{-7}	1.20
0.75	1.30	4.4×10^{-7}	1.00
1.0	1.20	5.0×10^{-7}	1.00
2.0	0.80	9.1×10^{-7}	0.86
3.0	0.40	2.6×10^{-6}	1.70

Key: h, indentation measured at gelatin temperature of 5°C; C_x, number of hydrogen bonds; N, number of bonds per molecule.

more concentrated gelatin samples were cut and stretched by hand. On release from stretches up to about 50%, the sample snaps back, illustrating the rubberlike elasticity of these materials. At greater elongation the sample breaks, however. The material is weak because the gelatin protein chains are much diluted with water.

Discussion

For rubbery materials, Young's modulus is related to the number of cross-links in the system. In this case the cross-links are of a physical nature, caused by hydrogen bonding. Measurement of the modulus via ball indentation techniques allows a rapid, inexpensive method of counting these bonds. Table A9.1.2 shows that the number of these bonds is of the order of 10^{-7} mol/cm^3. The number of these bonds also was shown to increase linearly with concentration, except at the highest concentrations.

Table A9.1.2 also demonstrates that at each gelatin concentration, the number of bonds per gelatin molecule is relatively constant. This number, of course, is the number of bonds taking part in three-dimensional network formation. Not all the gelatin chains are bound in a true tetrafunctionally cross-linked network. Many dangling chain ends exist at these low concentrations, and the network must be very imperfect.

The gelation molecule is basically composed of short α-helical segments in the form of a triple helix with numerous intramolecular bonds at room temperature; see Section 9.13. The α-helical segments are interrupted by proline and hydroxy proline functional groups. These groups disrupt the helical structure, yielding intervening portions of chain that behave like random coils, and which may be relatively free to develop intermolecular bonds. The subject has been reviewed by Djabourov (A7) and Mel'nichenko et al. (A8).

In this experiment the concentration of sugar was kept constant so as to minimize its effect on the modulus. In concluding, it must be pointed out that if sanitary measures are maintained, the final product may be eaten at the end of the experiment. If gelation five times normal or higher is included in the study, the student should be prepared for his or her jaws springing open after biting down!

REFERENCES

A1. A. Veis, *Macromolecular Chemistry of Gelatin*, Academic Press, Orlando, 1964.

A2. E. M. Marks, in *Encyclopedia of Chemical Technology*, Kirk-Othmer, Interscience, New York, 1966, Vol. 10, p. 499.

A3. J. L. Laurent, P. A. Janmey, and J. D. Ferry, *J. Rheol.*, **24**, 87 (1980).

A4. M. Miller, J. D. Ferry, F. W. Schremp, and J. E. Eldridge, *J. Phys. Colloid Chem.*, **55**, 1387 (1951).

A5. J. D. Ferry, *Viscoelastic Properties of Polymers*, 3rd ed., Wiley, New York, 1980, pp. 529–539.

A6. L. H. Sperling, *Interpenetrating Polymer Networks and Related Materials*, Plenum Press, New York, 1981, p. 177.

A7. M. Djabourov, *Contemp. Phys.*, **29**(3), 273 (1988).

A8. Yu. Mel'nichenko, Yu. P. Gomza, V. V. Shilov, and S. I. Osipov, *Polym. Intern. (Brit. Polym. J.)*, **25**(3), 153 (1991).

APPENDIX 9.2 ELASTIC BEHAVIOR OF A RUBBER BAND†

Stretching a rubber band makes a good demonstration of the stress–strain relationships of cross-linked elastomers. The time required is about 30 minutes.

† Reproduced in part from A. J. Etzel, S. J. Goldstein, H. J. Panabaker, D. G. Fradkin, and L. H. Sperling, *J. Chem. Ed.*, **63**, 731 (1986).

The equipment includes a large rubber band (Star® band size 107, E. Faber, Inc., Wilkes-Barre, PA, is suitable), a set of weights up to 25 kg, and a meter stick. Also required are hooks to attach the weights and a high place from which to hang the rubber band.

First, the rubber band is measured, both in length and cross section, and the hooks are weighed. Increasing weight is hung from the rubber band, its length being recorded at each step. When it nears its breaking length, caution is advised.

A plot of stress (using initial cross-sectional area) as a function of α, Figure A9.2.1, demonstrates the nonlinearity of the stress–strain relationship. Initial values of the slope of the curve yield Young's modulus, E. The sharp upturn of the experimental curve at high elongations is due to the limited extensibility of the chains themselves. The number of active network chains per unit volume can be calculated from equation (9.34) as $1.9 \times 10^2\ \text{mol/m}^3$.

A Mooney–Rivlin plot according to equation (9.50) yields a curve that rapidly increases for values of $1/\alpha$ greater than 0.25; see Figure 9.18. The constants $2C_1$ and $2C_2$ are calculated from the intercept and slope, respectively. Values of 2.3×10^5 Pa and 2.8×10^5 Pa were obtained, respectively.

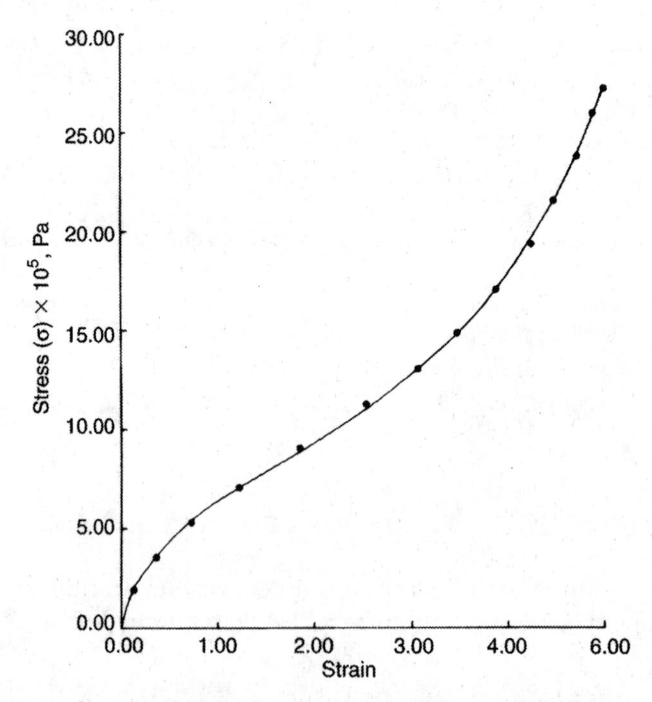

Figure A9.2.1 Simple rubber–elastic behavior of a rubber band under increasing load.

APPENDIX 9.3 DETERMINATION OF THE CROSS-LINK DENSITY OF RUBBER BY SWELLING TO EQUILIBRIUM[†]

The present experiment is based on the rapid swelling of elastomers by organic solvents. Application of the Flory–Rehner equation yields the number of active network chain segments per unit volume, a measure of the extent of vulcanization (C1,C2).

Experiment

Time: About 1 hour
Level: Physical Chemistry
Principles Illustrated:

1. The cross-linked nature of rubber
2. Diffusion of a solvent into a solid

Equipment and Supplies:

One large rubber band
One 600-ml beaker (containing 300 ml toluene)
One ruler or yardstick
One long tweezers to remove swollen rubber band
Paper towels to blot wet swollen rubber band
One clock or watch
One lab bench

First, cut the rubber band in one place to make a long rubber strip. Measure and record its length in the relaxed state. Place in the 600-ml beaker with toluene, making sure the rubber band is completely covered. Remove after 5 to 10 minutes. Blot dry. **Caution: toluene is toxic and can be absorbed through the skin.** Again, measure and record length. Repeat for about 1 hour. Optional: Cover and store overnight. Measure the length of the band the next day.

Expected Results

The rubber band swells to about twice its original length, but then it remains stable. Note that swelling to twice its length means a volume increase of about a factor of 8. Also, *note that the swollen rubber band is much weaker than the dry material and may break if not treated gently*.

[†] Reprinted in part from L. H. Sperling and T. C. Michael, *J. Chem. Ed.*, **59**, 651 (1982).

Chemically most rubber bands and similar materials are composed of a random copolymer of butadiene and styrene, written poly(butadiene–*stat*–styrene), meaning that the placement of the monomer units is statistical along the chain length. Usually this product is made via emulsion polymerization.

Swelling of a Rubber Band with Time

Length, cm	Time, min
16.5	0
24.0	14
26.0	25
27.0	36
28.0	70

Typical results are shown in the table. Over a period of 70 minutes, the length of the rubber band increased from 16.5 to 28.0 cm, for a volume increase of about 4.9. This is sufficiently visible to be seen at the back of an ordinary classroom. The rubber band would continue to swell slowly for some hours, or even days, but for the purposes of demonstrations and classroom calculations, the swelling can be considered nearly complete.

Calculations

For the system poly(butadiene–*stat*–styrene) and toluene, χ_1 is 0.39. Assuming that additivity of volumes v_2 is found from the swelling data to be 0.205. The quantity V_1 is 106.3 cm^3/mol for toluene. Algebraic substitution into equation (9.92) yields n equal to 1.55×10^{-4} mol/cm^3. (Compare result with Appendix 9.2.)

Extra Credit

Two experiments (or demonstrations) can be done easily for extra credit.

1. Obtain some unvulcanized rubber. Most tire and chemical companies can supply this. Put a piece of this material into toluene overnight and observe the results. It should dissolve to form a uniform solution.
2. The quantity n can be used also to predict Young's modulus (the stiffness) of the rubber band. The equation is

$$E = 3nRT \tag{A9.2.1}$$

where E represents Young's modulus, and R in these units is 8.31×10^7 dynes·cm/mol·K. For the present experiment, E is calculated to be 1.1×10^7 dynes/cm^2, typical of such rubbery products.

REFERENCES

C1. P. J. Flory and J. Rehner, *J. Chem. Phys.*, **11**, 521 (1943).

C2. J. E. Mark, A. Eisenberg, W. W. Graessley, L. Mandelkern, E. T. Samnlski, J. E. Koenig, and G. D. Wignall, *Physical Properties of Polymers*, 2nd ed., American Chemical Society, Washington, DC, 1993, Chap. 1.

10

POLYMER VISCOELASTICITY AND RHEOLOGY

The study of polymer viscoelasticity treats the interrelationships among elasticity, flow, and molecular motion. In reality, no liquid exhibits pure Newtonian viscosity, and no solid exhibits pure elastic behavior, although it is convenient to assume so for some simple problems. Rather, all deformation of real bodies includes some elements of both flow and elasticity. Because of the long-chain nature of polymeric materials, their visco-elastic characteristics come to the forefront. This is especially true when the times for molecular relaxation are of the same order of magnitude as an imposed mechanical stress.

Chapters 8 and 9 have introduced the concepts of the glass transition and rubber elasticity. In particular, Section 8.2 outlined the five regions of viscoelasticity, and Section 8.6.1.2 derived the WLF equation. This chapter treats the subjects of stress relaxation and creep, the time–temperature superposition principle, and melt flow. Parts of this topic are commonly called rheology, the science of deformation and flow of matter.

10.1 STRESS RELAXATION AND CREEP

In a stress relaxation experiment the sample is rapidly stretched to the required length, and the stress is recorded as a function of time. The length of the sample remains constant, so there is no macroscopic movement of the body during the experiment. Usually the temperature remains constant also (1).

Introduction to Physical Polymer Science, by L.H. Sperling
ISBN 0-471-70606-X

Creep experiments are conducted in the inverse manner. A constant stress is applied to a sample, and the dimensions are recorded as a function of time. Of course, these experiments can be generalized to include shear motions, compression, and so on.

Section 8.1.6 defined the modulus of a material as a measure of its stiffness, and compliance as a measure of its softness. Under conditions far from a transition, $E \cong 1/J$. Frequently stress relaxation experiments are reported as the time-dependent modulus, $E(t)$, whereas creep experiments are reported as the time-dependent compliance, $J(t)$.

10.1.1 Molecular Bases of Stress Relaxation and Creep

While the exact molecular causes of stress relaxation and creep are varied, they can be grouped into five general categories (2):

1. *Chain scission.* Oxidative degradation and hydrolysis are the primary causes. The reduction in modulus caused by chain scission during stress relaxation can be illustrated by a model where three chains are bearing a load and one is cut:

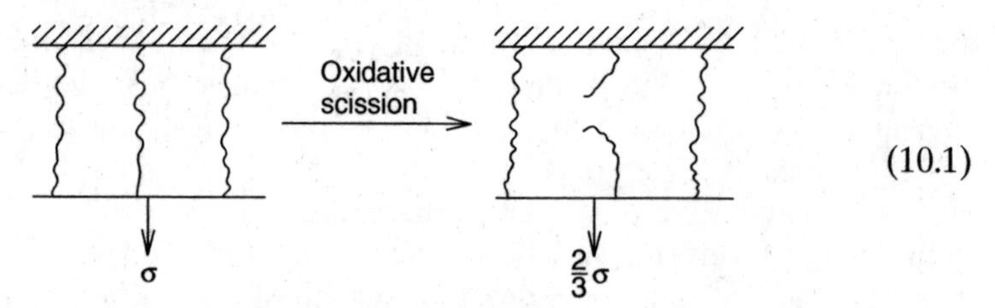

(10.1)

Inversely, this causes an increase in elongation during creep.

2. *Bond interchange.* While this is not a degradation in the sense that the molecular weight is decreased, chain portions changing partners cause a release of stress. Examples of stress relaxation by bond interchange include polyesters and polysiloxanes. Equation (10.2) provides a simple example (3).

   Bond interchange is going on constantly in polysiloxanes, with or without stress. In the presence of a stress, however, the statistical rearrangements tend to reform the chains so that the stress is reduced.

3. *Viscous flow.* Caused by linear chains slipping past one another, this mechanism is responsible for viscous flow in pipes and elongational flow under stress. An example is the pulling out of Silly Putty®.

4. *Thirion relaxation* (4). This is a reversible relaxation of the physical cross-links or trapped entanglements in elastomeric networks. Figure 10.1 illustrates the motions involved. Usually an elastomeric network will relax about 5% by this mechanism, most of it in a few seconds. It must be emphasized that the chains are in constant motion of the reptation type (5) (see Section 5.4).

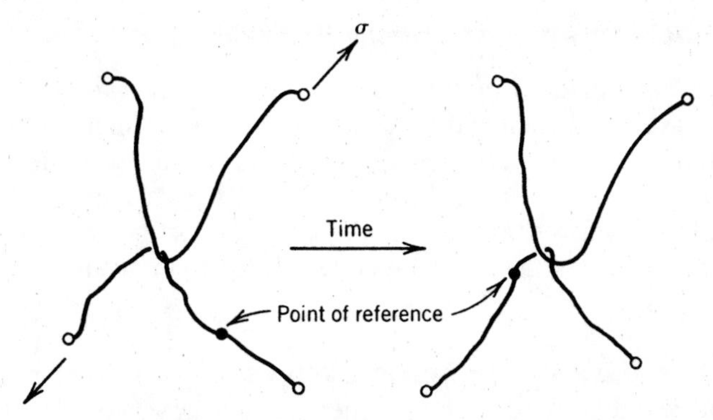

Figure 10.1 An illustration of reversible motion in a trapped entanglement under stress. The marked point of reference moves with time. When the stress is released, entropic forces return the chains to near their original positions.

```
 CH3    CH3  ¦ CH3
  |      |   ¦  |
~Si—O—Si—O┼Si—O~
  |      |   ¦  |
 CH3    CH3  ¦ CH3                → σ = constant
   CH3  H3C  ¦    CH3
    |    |   ¦     |
 ~O—Si—O—Si┼O—Si~
    |    |   ¦     |
   CH3  H3C  ¦    CH3

interchange along dotted line
—————————————————→                                  (10.2)

 CH3    CH3       ¦       CH3
  |      |        ¦        |
~Si—O—Si—CH3 ¦ CH3—Si—O~
  |      |        ¦        |
 CH3     |        ¦        |          → σ = 0
 CH3     O        ¦        O
  |      |        ¦        |
~O—Si—O—Si—CH3 ¦ CH3—Si~
  |      |        ¦        |
 CH3    CH3       ¦       CH3
```

5. *Molecular relaxation, especially near T_g.* This topic was the major subject of Chapter 8, where it was pointed out that near T_g the chains relax at about the same rate as the time frame of the experiment. If the chains are under stress during the experiment, the motions will tend to relieve the stress.

It must be emphasized that more than one of the relaxation modes above may be operative during any real experiment.

10.1.2 Models for Analyzing Stress Relaxation and Creep

To permit a mathematical analysis of the creep and relaxation phenomenon, spring and dashpot elements are frequently used (see Figure 10.2). A spring behaves exactly like a metal spring, stretching instantly under stress and holding that stress indefinitely. A dashpot is full of a purely viscous fluid. Under stress, the plunger moves through the fluid at a rate proportional to the stress. On removing the stress, there is no recovery. Both elements may be deformed indefinitely.

10.1.2.1 The Maxwell and Kelvin Elements The springs and dashpots can be put together to develop mathematically amenable models of viscoelastic behavior. Figure 10.3 illustrates the simplest two such arrangements, the Maxwell and the Kelvin (sometimes called the Voigt) elements. While the spring and the dashpot are in series in the Maxwell element, they are in parallel in the Kelvin element. In such arrangements it is convenient to assign moduli E to the various springs, and viscosities η to the dashpots.

In the Maxwell element both the spring and the dashpot are subjected to the same stress but are permitted independent strains. The inverse is true for the Kelvin element, which is equivalent to saying that the horizontal connecting portions on the right-hand side of Figure 10.3 are constrained to remain parallel.

As examples of the behavior of combinations of springs and dashpots, the Maxwell and Kelvin elements will be subjected to creep experiments. In such an experiment a stress, σ, is applied to the ends of the elements, and the strain, ε, is recorded as a function of time. The results are illustrated in Figure 10.4.

On application of the stress to the Maxwell element, the spring instantly responds, as illustrated by the vertical line in Figure 10.4. The height of the line is given by $\varepsilon = \sigma/E$. The spring term remains extended, as the dashpot gradually pulls out, yielding the slanted upward line. This model illustrates elasticity plus flow.

The spring and the dashpot of the Kelvin element undergo concerted motions, since the top and bottom bars (see Figure 10.3) are constrained to remain parallel. The dashpot responds slowly to the stress, bearing all of it initially and gradually transferring it to the spring as the latter becomes extended.

Figure 10.2 Springs and dashpots are the basic elements in modeling stress relaxation and creep phenomena.

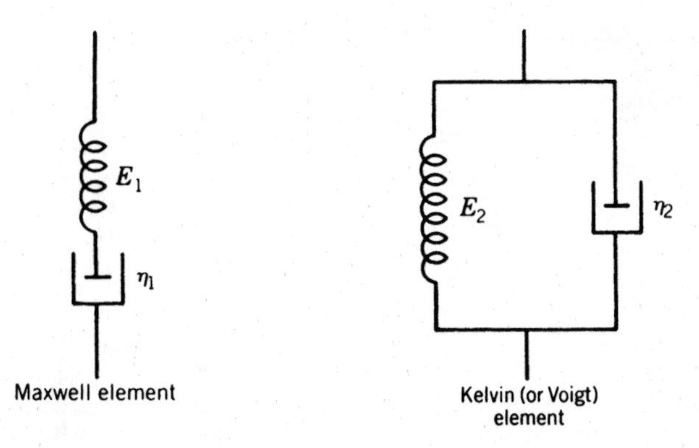

Figure 10.3 The Maxwell and Kelvin (or Voigt) elements, representing simple series and parallel arrays of springs and dashpots.

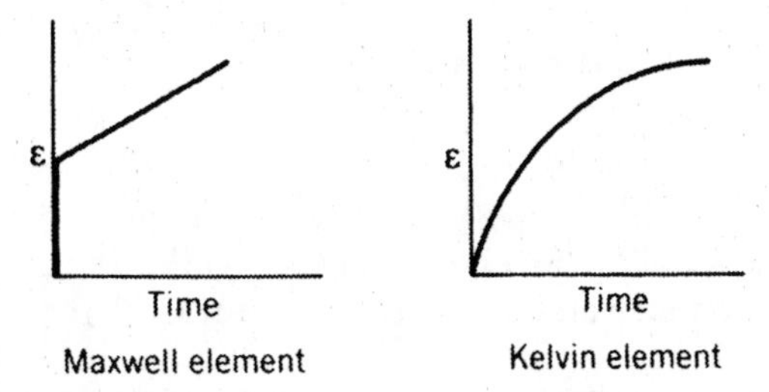

Figure 10.4 Creep behavior of the Maxwell and Kelvin elements. The Maxwell element exhibits viscous flow throughout the time of deformation, whereas the Kelvin element reaches and asymptotic limit to deformation.

The rate of strain of the dashpot is given by

$$\frac{d\varepsilon}{dt} = \frac{\sigma}{\eta} \tag{10.3}$$

When the spring bears all the stress, both the spring and the dashpot stop deforming together, and creep stops. Thus, at long times, the Kelvin element exhibits the asymptotic behavior illustrated in Figure 10.4. In more complex arrangements of springs and dashpots, the Kelvin element contributes a retarded elastic effect.

10.1.2.2 The Four-Element Model While a few problems in viscoelasticity can be solved with the Maxwell or Kelvin elements alone, more often they are used together or in other combinations. Figure 10.5 illustrates the combination of the Maxwell element and the Kelvin element in series, known as the four-element model. It is the simplest model that exhibits all the essential features of viscoelasticity.

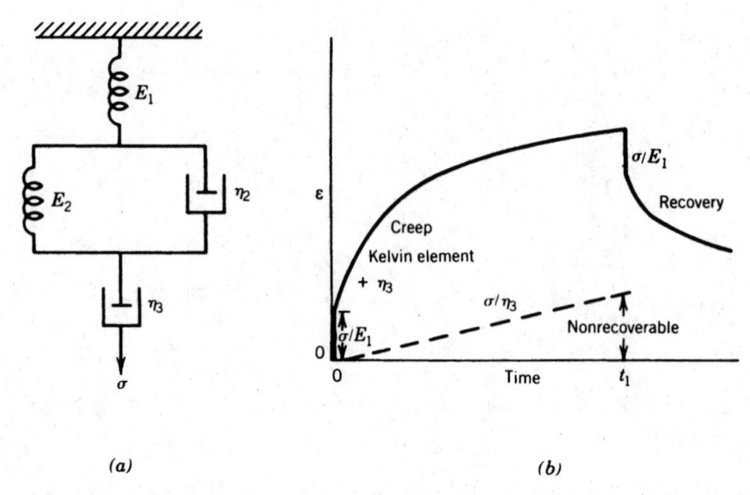

Figure 10.5 (*a*) The four-element model. (*b*) Creep behavior as predicted by this model. At t_1, the stress is relaxed, and the model makes a partial recovery.

On the application of a stress, σ, the model (Figure 10.5*a*) undergoes an elastic deformation, followed by creep (Figure 10.5*b*). The deformation due to η_3, true flow, is nonrecoverable. Thus, on removal of the stress, the model undergoes a partial recovery.

The four-element model exhibits some familiar behavior patterns. Consider the effects of stretching a rubber band around a book. Initially E_1 stretching takes place. As time passes, $E_2 + \eta_2 + \eta_3$ relaxations take place. On removing the rubber band at a later time, the remaining E_1 recovers. Usually the rubber band circle is larger than it was initially. This permanent stretch is due to η_3. Although less obvious, the Kelvin element motions can also be observed by measuring the rubber band dimensions immediately after removal and again at a later time.

The quantities E and η of the models shown above are not, of course, simple values of modulus and viscosity. However, as shown below, they can be used in numerous calculations to provide excellent predictions or understanding of viscoelastic creep and stress relaxation. It must be emphasized that E and η themselves can be governed by theoretical equations. For example, if the polymer is above T_g, the theory of rubber elasticity can be used. Likewise the WLF equation can be used to represent that portion of the deformation due to viscous flow, or for the viscous portion of the Kelvin element.

More complex arrangements of elements are often used, especially if multiple relaxations are involved or if accurate representations of engineering data are required. The Maxwell–Weichert model consists of a very large (or infinite) number of Maxwell elements in parallel (2). The generalized Voigt–Kelvin model places a number of Kelvin elements in series. In each of these models, a spring or a dashpot may be placed alone, indicating elastic or viscous contributions.

10.1.2.3 The Takayanagi Models Most polymer blends, blocks, grafts, and interpenetrating polymer networks are phase-separated. Frequently one phase is elastomeric, and the other is plastic. The mechanical behavior of such a system can be represented by the Takayanagi models (6). Instead of the arrays of springs and dashpots, arrays of rubbery (R) and plastic (P) phases are indicated (see Figure 10.6) (7). The quantities λ and φ or their indicated multiplications indicate volume fractions of the materials.

As with springs and dashpots, the Takayanagi models may also be expressed analytically. For parallel model Figure 10.6*a*, the horizontal bars connecting the two elements must remain parallel and horizontal, yielding an isostrain condition ($\varepsilon_P = \varepsilon_R$). Then

$$\sigma = \sigma_R + \sigma_P \tag{10.4}$$

$$\sigma_i = \varepsilon E_i, \quad i = \text{P, R} \tag{10.5}$$

$$E = (1-\lambda)E_P + \lambda E_R \tag{10.6}$$

Figure 10.6*a* represents an upper bound model, meaning that the modulus predicted is the highest achievable for a two-phased mixture.

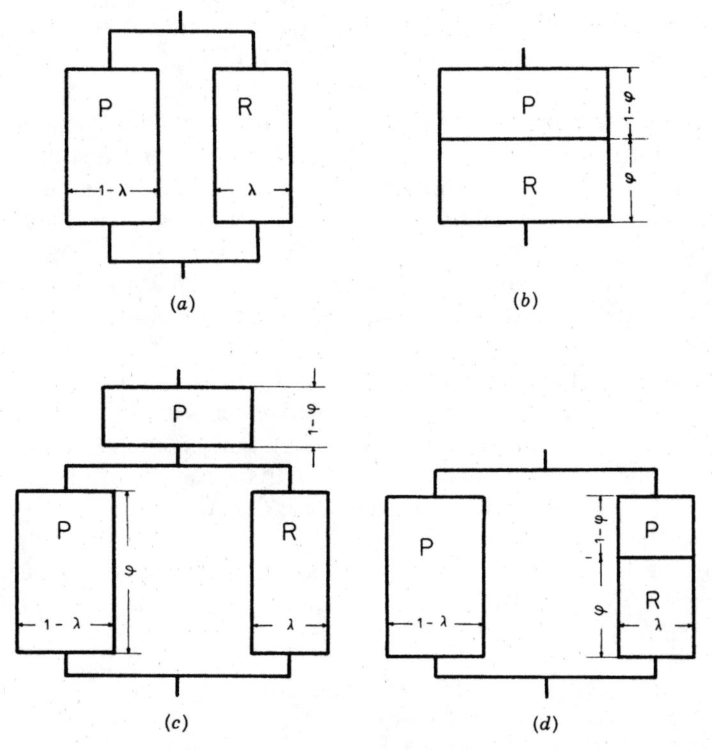

Figure 10.6 The Takayanagi models for two-phase systems: (*a*) an isostrain model; (*b*) isostress model; (*c, d*) combinations. The area of each diagram is proportional to a volume fraction of the phase.

For example, take a 50/50 blend of a plastic and an elastomer. Typical Young's moduli are $E_P = 3 \times 10^9$ Pa, and $E_R = 2 \times 10^6$ Pa. The quantity λ equals 0.5, and equation (10.6) yields 1.5×10^9 Pa. To an excellent approximation, this is half the modulus of the plastic.

For the series model Figure 10.6*b*, an isostress strain condition exists. The strains are additive,

$$\varepsilon = \varepsilon_P + \varepsilon_R \tag{10.7}$$

yielding a Young's modulus of

$$E = \left(\frac{\varphi}{E_R} + \frac{1-\varphi}{E_P} \right)^{-1} \tag{10.8}$$

This is a lower bound modulus, meaning that equation (10.8) yields the lowest modulus any blend material may have. If the example above is used, with $\varphi =$ 0.5, equation (10.8) yields $E = 4 \times 10^6$ Pa, or approximately twice the elastomer value.

Surprisingly, it is possible to make two different 50/50 blends of plastics and elastomers and obtain very nearly both of the results predicted by the above examples. This depends on the morphology, and particularly on phase continuity; see Figure 4.3. When the plastic phase is continuous in space and the elastomer dispersed (Figure 10.6*a*), the material will be stiff, and substantially exhibit upper bound behavior. On the other hand, when the elastomer is continuous and the plastic discontinuous (Figure 10.6*b*), a much softer material results and a lower bound situation prevails. The rubber phase is much more obviously discontinuous in the models represented in Figures 10.6*c* and 10.6*d*.

While the examples above used Young's modulus, many other parameters may be substituted. These include other moduli, rheological functions such as creep, stress relaxation, melt viscosity, and rubber elasticity. Each element may itself be expressed by temperature-, time-, or frequency-dependent quantities. It must also be noted that these models find application in composite problems as well. For example, a composite of continuous fibers in a plastic matrix can be described by Figure 10.6*a* if deformed in the direction of the fibers, and by Figure 10.6*b* if deformed in the transverse direction.

Figures 10.6*c* and 10.6*d* represent a higher level of sophistication, allowing for more precise calculations involving continuous and discontinuous phases. For Figures 10.6*c* and 10.6*d*, respectively, Young's moduli are given by

$$E = \left[\frac{\varphi}{\lambda E_R + (1-\lambda) E_P} + \frac{1-\varphi}{E_P} \right]^{-1} \tag{10.9}$$

$$E = \lambda \left(\frac{\varphi}{E_R} + \frac{1-\varphi}{E_P} \right)^{-1} + (1-\lambda) E_P \tag{10.10}$$

For these latter models, the volume fraction of each phase is given by a function of λ multiplied by a function of φ. The volume fraction of the upper right-hand portion of Figure 10.6*d* is given by $\lambda(1-\varphi)$. Equations (10.9) and (10.10) simplify for numerical analysis if φ is taken equal to λ.

10.2 RELAXATION AND RETARDATION TIMES

The various models were invented explicitly to provide a method of mathematical analysis of polymeric viscoelastic behavior. The Maxwell element expresses a combination of Hooke's and Newton's laws. For the spring,

$$\sigma = E\varepsilon \tag{10.11}$$

the time dependence of the strain may be expressed as

$$\frac{d\varepsilon}{dt} = \frac{1}{E}\frac{d\sigma}{dt} \tag{10.12}$$

The time dependence of the strain on the dashpot is given by

$$\frac{d\varepsilon}{dt} = \frac{\sigma}{\eta} \tag{10.13}$$

Since the Maxwell model has a spring and dashpot in series, the strain on the model is the sum of the strains of its components:

$$\frac{d\varepsilon}{dt} = \frac{1}{E}\frac{d\sigma}{dt} + \frac{\sigma}{\eta} \tag{10.14}$$

10.2.1 The Relaxation Time

For a Maxwell element the relaxation time is defined by

$$\tau_1 = \frac{\eta}{E} \tag{10.15}$$

The viscosity, η, has the units of Pa·s, and the modulus, E, has the units of Pa, so τ_1 has the units of time. Thus τ_1 relates modulus and viscosity.

On a molecular scale the relaxation time of a polymer indicates the order of magnitude of time required for a certain proportion of the polymer chains to relax—that is, to respond to the external stress by thermal motion. It should be noted that the chains are in constant thermal motion whether there is an external stress or not. The stress tends to be relieved, however, when the chains happen to move in the right direction, degrade, and so on.

Alternatively, τ_1 can be a measure of the time required for a chemical reaction to take place. Common reactions that can be measured in this way include bond interchange, degradation, hydrolysis, and oxidation. Combining equations (10.14) and (10.15) leads to

$$\frac{d\varepsilon}{dt} = \frac{1}{E}\frac{d\sigma}{dt} + \frac{\sigma}{\tau_1 E} \tag{10.16}$$

where the first term on the right is important for short-time changes in strain, and the second term on the right controls the longer-time changes.

The stress relaxation experiment requires that $d\varepsilon/dt = 0$; that is, the length does not change, and the strain is constant. Integrating equation (10.16) under these conditions leads to

$$\sigma = \varepsilon_0 E e^{-(t/\tau_1)} \tag{10.17}$$

or

$$\sigma = \sigma_0 e^{-(t/\tau_1)} \tag{10.18}$$

Equations (10.17) and (10.18) predict a straight-line relationship between $\ln \sigma$ or $\log \sigma$ and linear time if a single mechanism controls the relaxation process. If experiments other than simple elongation are done (i.e., relaxation in shear), the appropriate modulus replaces E in equation (10.17).

10.2.2 Applications of Relaxation Times to Chemical Reactions

10.2.2.1 Chemical Stress Relaxation As indicated in Section 10.1.1, stress relaxation can be caused by either physical or chemical phenomena. Examples will be given of each.

Figure 10.7 (8) illustrates the stress relaxation of a poly(dimethyl siloxane) network, silicone rubber, in the presence of dry nitrogen. The reduced stress, $\sigma(t)/\sigma(0)$, is plotted, so that under the initial conditions its value is always unity. Since the theory of rubber elasticity holds (Chapter 9), what is really measured is the fractional decrease in effective network chain segments. The bond interchange reaction of equation (10.2) provides the chemical basis of the process. While the rate of the relaxation increases with temperature, the lines remain straight, suggesting that equation (10.2) can be treated as the sole reaction of importance.

The relaxation times may be estimated from the time necessary for $\sigma(t)/\sigma(0)$ to drop to $1/e = 0.368$. The results are shown in Table 10.1 (8).

Appendix 10.1 derives a relationship between the relaxation time and the energy of activation, ΔE_{act},

$$\tau_1 = \text{constant} \times e^{\Delta E_{act}/RT} \tag{10.19}$$

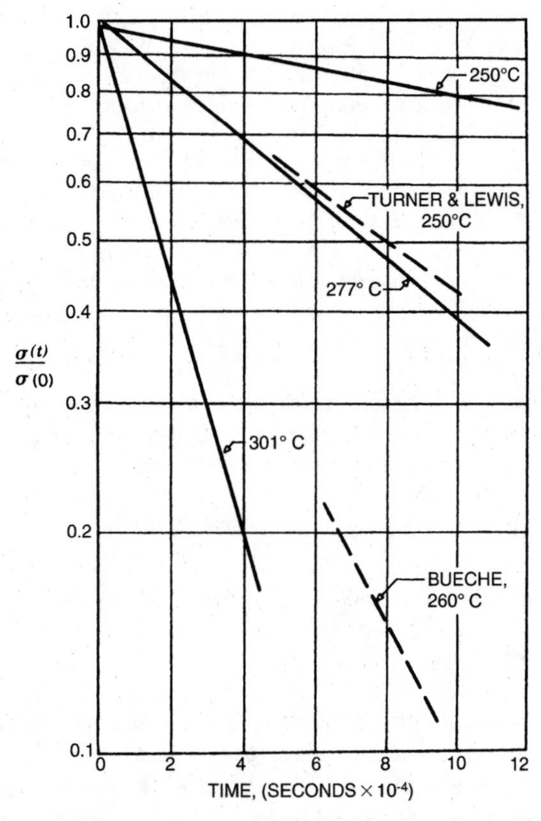

Figure 10.7 Stress relaxation of silicone rubber, poly(dimethyl siloxane). The rate of stress relaxation increases with temperature, but the lines remain straight (8). Equation (10.18) suggests that a logarithmic *y*-axis should produce straight lines.

Table 10.1 Chemical stress relaxation times for silicone rubber (8)

Temperature, °C	$\tau_1 \times 10^{-4}$, s
250	48
277	10.5
301	2.45

A plot of $\ln \tau_1$ versus $1/T$ yields $\Delta E_{act}/R$ for the slope. An apparent energy of activation of 35 kcal/mol was estimated from such a plot. Thus a purely chemical quantity can be deduced from a mechanical experiment.

Stress relaxation experiments were used to determine the mechanism of degradation in synthetic polymers (9) during World War II, when these materials were first being made. Tobolsky later described the results of these famous experiments (2):

It was found that in the temperature range of 100 to 150°C, these vulcanized rubbers showed a fairly rapid decay to zero stress at constant extension. Since in principle a cross-linked rubber network in the rubbery range of behavior should show little stress relaxation, and certainly no decay to zero stress, the phenomenon was attributed to a chemical rupture of the rubber network. This rupture was specifically ascribed to the effect of molecular oxygen since under conditions of *very low* oxygen pressures ($<10^{-4}$ atm) the stress–relaxation rate was markedly diminished. However at moderately low oxygen pressures the rate of chemical stress relaxation was the same as at atmospheric conditions. This result parallels the very long established fact that in the liquid phase the rate of reaction of hydrocarbons with oxygen is independent of oxygen pressure down to fairly low pressures.

10.2.2.2 Procedure X Stress relaxation can also be used to separate and identify two or more reactions causing relaxation, provided the rates are sufficiently different. Consider reactions a and b going on simultaneously:

$$\sigma(t) = \sigma_a(0)e^{-t/\tau_{1a}} + \sigma_b(0)e^{-t/\tau_{1b}} \quad (10.20)$$

If the two relaxation times, chemical or physical, are sufficiently different, two straight lines may be obtained by algebraic analysis. Tobolsky named this method of analysis "procedure X" (2).

10.2.2.3 Continuous and Intermittent Stress Relaxation Another "trick" to separate two reactions involves continuous and intermittent stress relaxation measurements. Figure 10.8 illustrates the separation of degradation and cross-linking in *cis*-1,4-polybutadiene. In an intermittent stress relaxation experiment, the sample is maintained in a relaxed, unstretched condition at a constant temperature. At suitably spaced time intervals the rubber is rapidly stretched to a fixed elongation, and the stress is measured. Then the sample is returned to its unstretched length. Of course, in the continuous stress relaxation experiment, the strain is maintained continuously.

At 130°C, the temperature of the experiment, oxidative scission and cross-linking are both going on all the time. The reactions happen, however, in the condition of the network at the time. This means that the continuous stress relaxation experiment measures only the degradation step, because the new cross-links form in the stretched chains; that is, the second network develops in the extended state, at equilibrium. There is no significant change in conformational entropy on formation of these cross-links, and the change in stress is near zero.

On the other hand, if the oxidative cross-links form when the sample is unstretched, then they can be measured afterward by using the theory of rubber elasticity. The intermittent experiment measures the total number of active chain segments at any given instant of time. As illustrated in Figure 10.8 the cross-linking reaction predominates after 10 hours, and sample actually gets harder. This last is often observed at home, where old rubber materials tend to stiffen up.

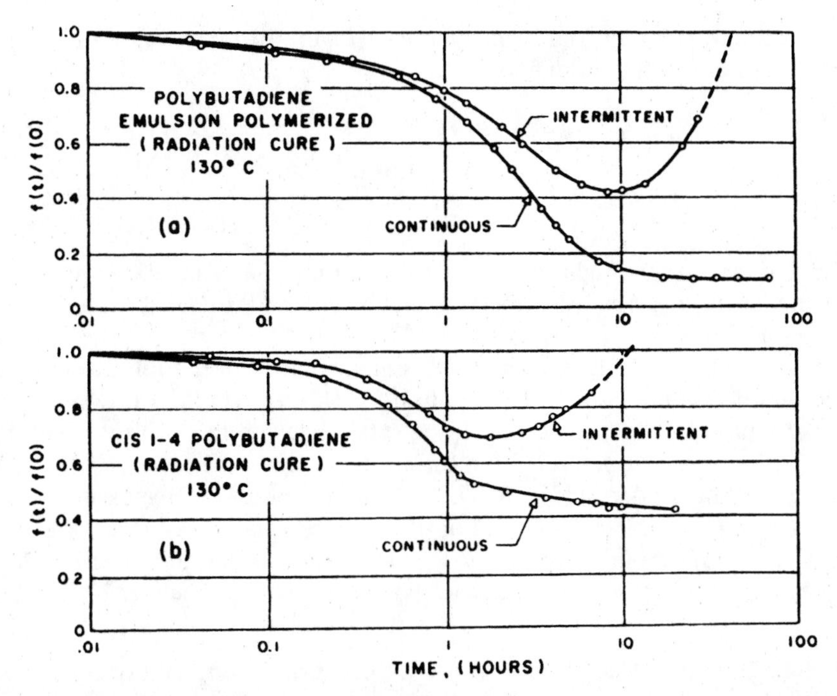

Figure 10.8 Continuous and intermittent stress relaxation at 130°C on radiation cured *cis*-1,4-polybutadiene in the presence of air. The experiment separates oxidative cross-linking and chain scission. The quantity $f(t)$ is the time-dependent retractive force. From A. V. Tobolsky, *Properties and Structure of Polymers*, Wiley, New York, 1960.

10.2.3 The Retardation Time

As the relaxation time, τ_1, is defined for the Maxwell elements, the retardation time, τ_2, is defined for the Kelvin element. The equation for the Kelvin element under stress can be written

$$\sigma = \eta \frac{d\varepsilon}{dt} + E\varepsilon \tag{10.21}$$

Under conditions of constant stress, equation (10.21) can be integrated to

$$\varepsilon = \frac{\sigma}{E}(1 - e^{-(E/\eta)t}) \tag{10.22}$$

which can be rewritten in terms of the retardation time,

$$\varepsilon = \frac{\sigma}{E}(1 - e^{-t/\tau_2}) \tag{10.23}$$

The equation for the four-element model (Figure 10.5*a*) may now be written for the condition of constant stress:

$$\varepsilon = \varepsilon_1 + \varepsilon_2 + \varepsilon_3 \tag{10.24}$$

$$\varepsilon = \frac{\sigma}{E_1} + \frac{\sigma}{E_2}(1 - e^{-t/\tau_2}) + \frac{\sigma}{\eta_3}t \tag{10.25}$$

The first term on the right of equation (10.25) represents an elastic term, the second term expresses the viscoelastic effect, and the third term expresses the viscous effect.

It is of interest to compare the retardation time with the relaxation time. The retardation time is the time required for E_2 and η_2 in the Kelvin element to deform to $1 - 1/e$, or 63.21% of the total expected creep. The relaxation time is the time required for E_1 and η_3 to stress relax to $1/e$ or 0.368 of σ_0, at constant strain. Both τ_1 and τ_2, to a first approximation, yield a measure of the time frame to complete about half of the indicated phenomenon, chemical or physical. A classroom demonstration experiment showing the determination of the constants in the four-element model is shown in Appendix 10.2.

10.2.4 Dynamic Mechanical Behavior of Springs and Dashpots

In addition to stress relaxation and creep, springs and dashpots can model the loss and storage characteristics of polymers undergoing cyclic motions. Since a principal application of such modeling is for noise and vibration damping analysis (see Section 8.12) where the motions are of a shearing nature, the equations for shear are emphasized below. The viscoelastic motions of a Maxwell element in shear may be written for an angular frequency ω (rad/s) (10):

$$J(t) = J + \frac{t}{\eta} \tag{10.26}$$

$$G(t) = Ge^{-t/\tau_1} \tag{10.27}$$

$$G'(\omega) = \frac{G\omega^2\tau_1^2}{1+\omega^2\tau_1^2} \tag{10.28}$$

$$G''(\omega) = \frac{G\omega\tau_1}{1+\omega^2\tau_1^2} \tag{10.29}$$

$$J'(\omega) = J \tag{10.30}$$

$$J''(\omega) = \frac{J}{\omega\tau_1} = \frac{1}{\omega\eta_1} \tag{10.31}$$

$$\tan\delta = \frac{1}{\omega\tau_1} \tag{10.32}$$

The scaling of time in relaxation processes is achieved by means of the "Deborah number," which is defined as

$$D_e = \tau_e / t \tag{10.33}$$

where t is a characteristic time of the deformation process being observed and τ_e is a characteristic time of the polymer. For a Hookian elastic solid, τ_e is infinite, and for a Newtonian viscous liquid, τ_e is zero; see Section 10.2.1. For polymer melts, τ_e is often of the order of a few seconds. Of course, τ_e can be the relaxation time, or the retardation time of the polymer. The Deborah number is widely used in engineering as a single-number approximation to the several molecular phenomena involved in the deformation and flow of many materials, including polymers.

The loss tangent is seen to be a maximum when $\tau_l = 1/\omega$—that is, when the time required for one cycle of the experiment equals the relaxation time. Of course, $G = 1/J$ far from a transition (Chapter 8).

The corresponding quantities for the Kelvin element are as follows (10):

$$J(t) = J(1 - e^{-t/\tau_2}) \tag{10.34}$$

$$G(t) = G \tag{10.35}$$

$$G'(\omega) = G \tag{10.36}$$

$$G''(\omega) = G\omega\tau_2 = \omega\eta_2 \tag{10.37}$$

$$\eta'(\omega) = \eta \tag{10.38}$$

$$J'(\omega) = \frac{J}{1 + \omega^2 \tau_2^2} \tag{10.39}$$

$$J''(\omega) = \frac{J\omega\tau_2}{1 + \omega^2 \tau_2^2} \tag{10.40}$$

$$\tan \delta = \omega\tau_2 \tag{10.41}$$

If springs are equated to capacities and dashpots to resistances, the storage and dissipative units are seen to correspond to time-dependent electrical behavior. However, the topology is backward; that is, series electrical connections correspond to parallel mechanical connections (10).

10.2.5 Molecular Relaxation Processes

The preceding sections describe the mechanical behavior of a polymeric sample in terms of creep and stress relaxation. Both elastomeric and plastic materials can be modeled by combinations of springs and dashpots. However, these are only models. Ultimately stress relaxation and creep derive from molecular origins, and it is in this area that more recent studies have been concentrated (11–15).

The most important method of characterizing the molecular relaxation processes has been through the use of small-angle neutron scattering (see Figure 10.9) (16). Boue et al. (15) investigated linear polystyrene of medium and high molecular weight. In each case, blends of deuterated and protonated polystyrenes were prepared, and the samples stretched up to an α value of 3 at temperatures above T_g in the range of 113 to 134°C. The samples were held for various periods of time at that extension, the stress being recorded (15), and then the samples were quickly quenched to the glassy state at room temperature. Then SANS measurements were made in both the perpendicular and parallel directions.

Figure 10.10 (15) shows the changes in the transverse radius of gyration with time, presented in the form of a master curve (see next section). On an absolute scale, R_g for the isotropic sample was 280 Å, and R_g for the samples

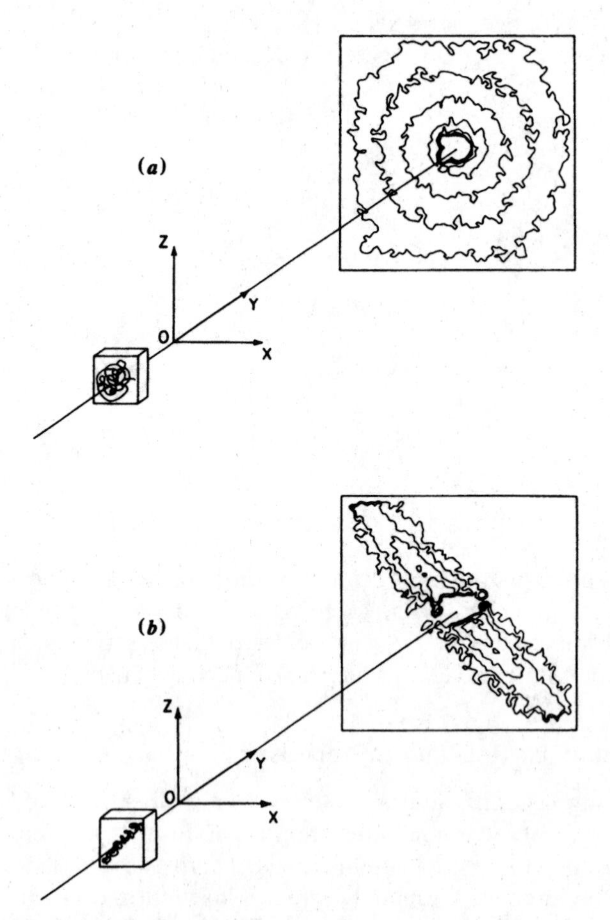

Figure 10.9 Geometry of the SANS experiment and resulting intensity contour plots: (*a*) unoriented amorphous polymer and (*b*) oriented polymer (16).

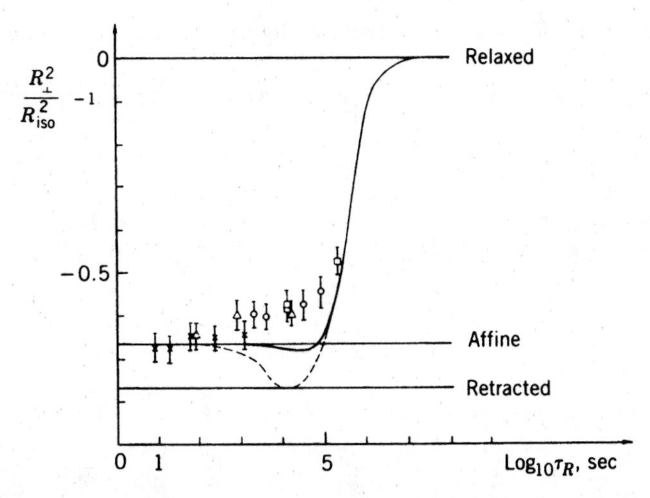

Figure 10.10 Molecular relaxation of polystyrene at different relaxation times and temperatures, as followed by SANS data taken in the transverse direction (15). Data are reduced to 117°C (just above T_g) via the time–temperature superposition principle.

undergoing relaxation ranged from 161 to 210 Å. Boue et al. (15) pointed out that immediately after stretching their sample, the radius of gyration showed affine deformation. This can be interpreted as indicating that the experiment did not fail to capture any major coil relaxation process.

Figure 10.10 (15) also applies the molecular diffusion theories of de Gennes (17), Doi and Edwards (18), and Daoudi (19), which assume that the major mode of molecular relaxation is by reptation. In this way, the chains move back and forth within a hypothetical tube. Relaxation occurs by the chain disengaging itself from the tube, only at the ends, in a backward-and-forward reptation.

Two characteristic times are employed: (a) T_d, defined as the longest of the relaxation times for defect equilibration, proportional to M^2, and (b) T_r, defined as the renewal time for chain conformation, proportional to M^3. This latter time is the time to form a new isotropic tube. For polystyrene of M_w = 650,000 g/mol at 117°C, they found that

$$3 \times 10^3 \text{ s} \leq T_d \leq 4 \times 10^4 \text{ s} \tag{10.42}$$

$$6 \times 10^5 \text{ s} \leq T_r \leq 1 \times 10^7 \text{ s} \tag{10.43}$$

The data in Figure 10.10 are in good agreement with the value of T_r. The authors state that more data are required to determine T_d decisively. The reader is referred to Section 5.4.2 and Appendix 5.2.

By way of contrast, Maconnachie et al. (12), using polystyrene of M_w = 144,000 g/mol, found that most of the relaxation was complete in about 5 minutes at 120°C. However, the chains never quite returned to their initial dimensions, both $R_{\parallel}$ and $R_{\perp}$ remaining slightly higher than the unstretched polymer dimensions even after 1×10^4 s. The deformation appeared to behave as if the affine deformation theory held only for distances separating effective cross-links.

For better comparisons, these data are in the rubbery flow portion of the spectrum. The SANS data in Figure 10.10 correspond to the earlier stress relaxation studies by Tobolsky (20) on almost identical polystyrenes.

More recently Wool (21) described three relaxation times. Referring to Figure 10.11, the relaxation time is given by

$$T_e \sim M_e^2 \tag{10.44}$$

which is constant for a given polymer. The quantity M_e refers to the segment molecular weight involved in movement about the diameter of the equivalent tube. It is also the molecular weight between entanglements. The Rouse relaxation time of the whole chain is given by

$$T_R \sim M^2 \tag{10.45}$$

which expresses the time for one end of the chain to respond to or communicate with the other end. If one end of the chain is stretched, this is the characteristic time for the other end to retract. The reptation time is given by

$$T_r \sim M^3 \tag{10.46}$$

similar to the de Gennes theory, where M is the molecular weight of the whole chain.

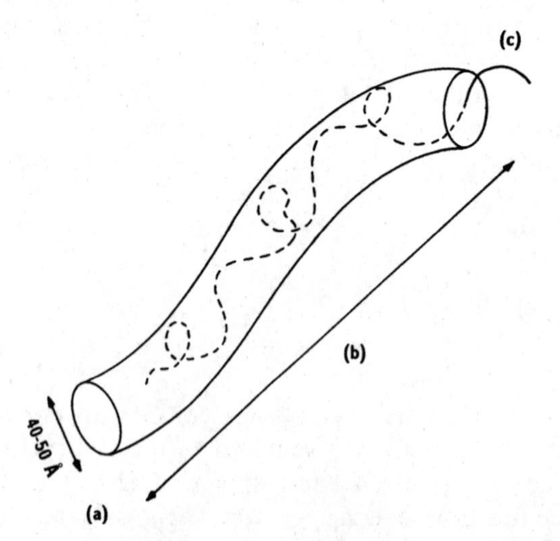

Figure 10.11 Tube model illustrating three relaxation times, each with a specific motion: (*a*) across the tube, (*b*) along the tube, and (*c*) in and out of the tube.

10.2.6 Diffusion at Short Times

Modern theory differentiates two cases for the time dependence of diffusion. The steady state condition first described by de Gennes (Section 5.4.2) holds for times longer than T_r,

$$X(t) \sim t^{1/2} M^{-1} \tag{10.47}$$

where $X(t)$ represents the average mer interpenetration depth into new territory. For times shorter than T_r,

$$X(t) \sim t^{1/4} M^{-1/4} \tag{10.48}$$

Equation (10.48) is particularly useful for studies of crack healing, welding time in molding operations, and film formation from latexes (21). Alternately, equations (10.47) and (10.48) can be cast in the form,

$$X(t) = X_\infty \left(\frac{t}{T_r}\right)^{1/2} \tag{10.49}$$

and

$$X(t) = X_\infty \left(\frac{t}{T_r}\right)^{1/4} \tag{10.50}$$

respectively. Equations (10.48) and (10.50) provide a nonclassical mode of diffusion, called minor chain reptation (21), before the chain completely escapes its tube.

The value of X_∞ is approximately $0.4R_g$ (22). The actual break point between $t^{1/4}$ and $t^{1/2}$ depends on the model (23). If center-of-mass motion from one side of the interface is examined, where the center-of-mass of the chains at the interface are about $0.4R_g$ from the interface, then Wool (21) calculates that the interface is completely healed at about $0.8R_g$ total interdiffusion distance. However, interdiffusion is measured from the surface, rather than from $0.4R_g$ inside, leading to the experimental value of $0.4R_g$ (24). With symmetrical interfaces, interdiffusion from both sides simultaneously leads to a model of $0.4R_g + 0.4R_g = 0.8R_g$ total motion. From a physical point of view, after about $0.8R_g$ total interdiffusion, the chains overlap such that the interface is healed.

Therefore the interdiffusion action described by equation (10.48) or (10.50) is substantially complete when the chains have diffused a distance of approximately $0.8R_g$ (21). At that point, the material has gained substantially all of its mechanical properties, such as tensile strength, impact resistance, and so on, which is to say, the material is healed; see Section 11.5. Further interdiffusion merely results in chains wandering randomly in the material.

10.2.7 Relationships among Molecular Parameters

Fetters et al. (25) summarized several molecular relationships, which broadly express values for random coil polymers,

$$G^0 = 10.52\ \text{MPa}(\text{Å})^3\left(\frac{r^2}{M}\rho N_a\right)^3 \tag{10.51}$$

$$M_e = 2225.8\frac{\text{cm}^3}{(\text{Å})^3\text{mol}}\left(\frac{r^2}{M}\right)^{-3}\rho^{-2}N_a^{-3} \tag{10.52}$$

$$d_t = 19.36\left(\frac{r^2}{M}\right)^{-1}\rho^{-1}N_a^{-1} \tag{10.53}$$

$$\frac{d_t^2}{M_e} = \frac{r^2}{M} \tag{10.54}$$

where G^0 represents the shear plateau modulus, ρ the polymer density, M_e the molecular weight between entanglements, d_t the reptation tube diameter (see Figure 10.11), r the end-to-end distance of the unperturbed (relaxed) chain, M the molecular weight of the polymer chain, and N_a is Avogadro's number.

The ratio r^2/M is a constant, see Table 5.4, where $r^2/M = 6R_g^2/M$. The constants shown in equations (10.51) to (10.54) are universal for random coil polymers of high enough molecular weight. Typical values for M_e and d_t are shown in Table 10.2 (4).

10.2.7.1 *Example Problem for the Plateau Modulus of Polystyrene*

Values for polystyrene are $r^2/M = 0.275^2 \times 6 = 0.434$ Å^2 mol/g (from Table 5.4), and $\rho = 1.06$ g/cm^3 (Table 3.2). Then $G^0 = 2.24 \times 10^5$ Pa, $M_e = 1.13 \times 10^4$ g/mol, and $d_t = 70.0$ Å. The value of G^0 should be compared with the plateau value in Figure 9.21, where G^0 equals about 2×10^5 Pa, providing excellent agreement with experiment. The idea of a plateau modulus implies that the shear

Table 10.2 Molecular characteristics of linear polymers at 413 K

Polymer	M_e, g/mol	d_t, Å
Polyethylene	828	32.8
Poly(ethylene oxide)	1,624	37.5
Poly(methyl methacrylate)	10,013	67.0
Polystyrene	13,309	76.5
Poly(dimethyl siloxane)	12,293	78.6
1,4-Polybutadiene	1,815	44.4
1,4-Polyisoprene	6,147	62.0
a-Polypropylene	4,623	60.7

Source: Values taken from Table 1, L. J. Fetters, D. J. Lohse, D. Richter, T. A. Witten, and A. Zirkel, *Macromolecules*, **27**, 4639 (1994).

rate or frequency is higher than the relaxation rate of the chains, which means that they do not become disentangled during the experiment. This differs from the concept of an equilibrium, infinite time modulus, which is zero for linear amorphous polymers above their glass transition temperature, or cross-linked polymers short of their gel point.

Thus the conformational statistics of chain molecules are now increasingly able to provide a basis for estimating the rheological and viscoelastic behavior of linear amorphous polymers above their glass transition temperatures.

10.2.7.2 Example Problem Relating to Adhesives Many modern adhesives, particularly pressure sensitive adhesives (see Section 12.8.4), deliberately utilize partly cross-linked, partly linear polymers (26). An important factor relating to the strength of the adhesive is whether single chains are able to diffuse into the cross-linked portions, thus forming a linear chain-cross-linked chain entanglement. It is known that strong contact adhesives require $M_c > M_e$, where the first term reads on the cross-linked portion, and the second term reads on the linear polymer (26). (Usually the linear polymer and the cross-linked polymer have similar compositions.)

In a simplified model, M_c assumes monodisperse values in a regular tetrafunctionally cross-linked network. The tube diameter of the linear polymer, d_t, must fit into the net. One side of the net then must be larger than d_t to permit interdiffusion. If as a minimum requirement d_t is taken equal to the equivalent radius of gyration, R_g, then

$$d_t = R_g$$

$$R_g = (KM_c)^{1/2}$$

values of K are shown in Table 5.4; multiply by $6^{1/2}$ for end-to-end distances, and

$$d_t = (M_e K)^{1/2}$$

yielding

$$M_e = M_c$$

for this simplified model.

Taking poly(methyl methacrylate) as an example, $M_e = 10{,}013$ g/mol (Table 10.2), M_c should be equal or greater than 10,013 g/mol. Experiments by Zosel and Ley (27) on cross-linked latex films show that values of $M_c \geq M_e$ are required for good mechanical properties, the oversimplified model above providing the minimum net size. Aspects of adhesion are discussed further in Section 12.8.

Figure 10.12 (22) illustrates stress relaxation in poly(methyl methacrylate) resulting from molecular relaxation processes. Here, the logarithm of the

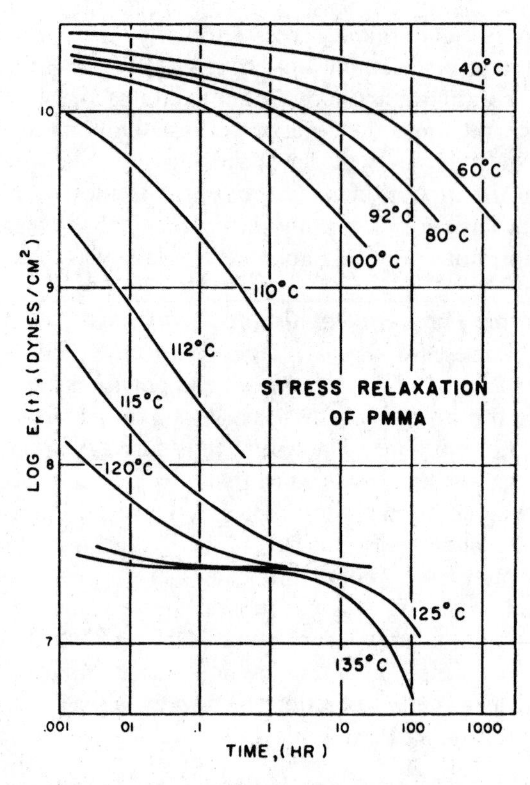

Figure 10.12 Stress–relaxation of poly(methyl methacrylate) with $M_v = 3.6 \times 10^6$ g/mol (20).

relaxation Young's modulus is plotted against the logarithm of time. At low temperatures, the polymer is glassy, and only slow relaxation is observed. As the glass transition temperature is approached (106°C at 10 s), the relaxation rate increases, reaching a maximum just above the classical glass transition temperature range. Then the rate of relaxation decreases as the rubbery plateau is approached. This is exemplified by the data at 125 and 135°C in the range of 0.01 to 1 h. At higher temperatures or longer times, the polymer begins to flow.

The data in Figure 10.12 thus show two distinct relaxation phenomena: first, chain portions corresponding to 10 to 50 carbon atoms are relaxing, which corresponds to the glass–rubber transition; then, at higher temperatures, whole chains are able to slide past one another.

10.2.8 Physical Aging in the Glassy State

While the molecular motion in the rubbery and liquid states involves 10 to 50 carbon atoms, molecular motion in the glassy state is restricted to vibrations, rotations, and motions by relatively short segments of the chains.

The extent of molecular motion depends on the free volume. In the glassy state, the free volume depends on the thermal history of the polymer. When a sample is cooled from the melt to some temperature below T_g and held at constant temperature, its volume will decrease (see Figure 8.18). Because of the lower free volume, the rate of stress relaxation, creep, and related properties will decrease (29–33). This phenomenon is sometimes called physical aging (29,32), although the sample ages in the sense not of degradation or oxidation but rather of an approach to the equilibrium state in the glass.

The effect of physical aging can be illustrated through a programmed series of creep studies. First, the sample is heated to a temperature T_0, about 10 to 15°C above T_g. A period of 10 to 20 minutes suffices to establish thermodynamic equilibrium. Then the sample is quenched to a temperature T_1 below T_g, and kept at this temperature. At a certain elapsed time after the quench, a creep experiment is started. The sample is subjected to a constant stress, σ_0; the resulting strain, ε, is determined as a function of time. The sample is then allowed to undergo creep recovery. Each creep period is short in comparison with the previous aging time as well as the last recovery period preceding it. The student should note the difference between the two time scales: t_e is the aging time, beginning at the time of quenching, whereas t is the creep time, which begins at the moment of loading for each run.

Figure 10.13 (29) illustrates typical creep compliance results for poly(vinyl chloride), which has a glass transition temperature of about 80°C. The most important conclusion obtained from such studies is that the rate of creep slows down as the sample ages. Note that in the first and last curves in Figure 10.13, the polymer reaches a tensile creep compliance of 5×10^{-10} m^2/N after about 10^3 s after being aged for 0.03 days, but after an aging period of 1000 days, 10^7 s are required to reach the same tensile creep compliance level.

The problem of determining the theoretical behavior of polymers in the glassy rates is treated by Curro et al. (34,35). The time dependence of the volume in the glassy state is accounted for by allowing the fraction of unoccupied volume sites to depend on time. This permits the application of the Doolittle equation to predict the shift in viscoelastic relaxation times,

$$\tilde{D} = \tilde{D}_r \exp[-B(f^{-1} - f_r^{-1})] \tag{10.55}$$

where $\tilde{D}$ is the diffusion constant for holes, and the subscript r represents the reference state, taken for convenience at the glass transition temperature, and f is the fractional free volume; see Section 8.6.12.

10.3 THE TIME–TEMPERATURE SUPERPOSITION PRINCIPLE

10.3.1 The Master Curve

As indicated above, relaxation and creep occur by molecular diffusional motions which become more rapid as the temperature is increased. Tempera-

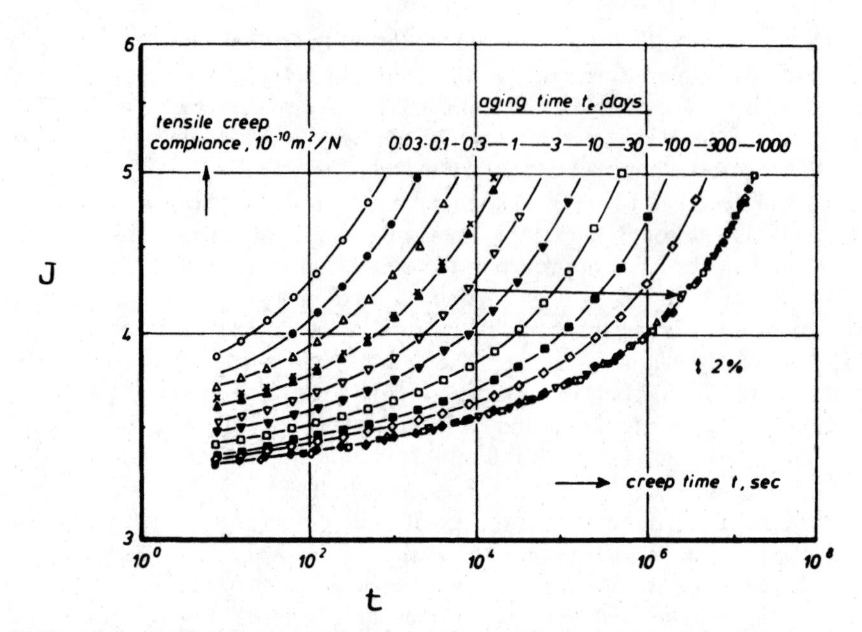

Figure 10.13 Small-strain tensile creep curves for glassy poly(vinyl chloride) quenched from 90°C (about 10°C above T_g) to 40°C and stored at that temperature for four years. The different curves were measured for various times t_e elapsed after the quench. The master curve gives the result of superposition by shifts that were almost horizontal; the shifting direction is shown by the arrow. The crosses refer to another sample quenched in the same way, but only measured for creep at t_e of 1 day (29).

ture is a measure of molecular motion. At higher temperatures, time moves faster for the molecules. The WLF equation, derived in Section 8.6.1.2, expresses a logarithmic relationship between time and temperature. Building on these ideas, the time–temperature superposition principle states that with viscoelastic materials, time and temperature are equivalent to the extent that data at one temperature can be superimposed on data at another temperature by shifting the curves along the log time axis (2).

The importance of these ideas becomes clear when one considers that data can be obtained conveniently over only a narrow time scale, say, from 1 to 10^5 s (see Figure 10.12). The time–temperature superposition principle allows an estimation of the relaxation modulus and other properties over many decades of time.

Figure 10.14 (36–38) illustrates the time–temperature superposition principle using polyisobutylene data. The reference temperature of the master curve is 25°C. The reference temperature is the temperature to which all the data are converted by shifting the curves to overlap the original 25°C curve. Other equivalent curves can be made at other temperatures. The shift factor shown in the inset corresponds to the WLF shift factor. Thus the quantitative shift of the data in the range T_g to $T_g \times 50$°C is governed by the WLF equation, and

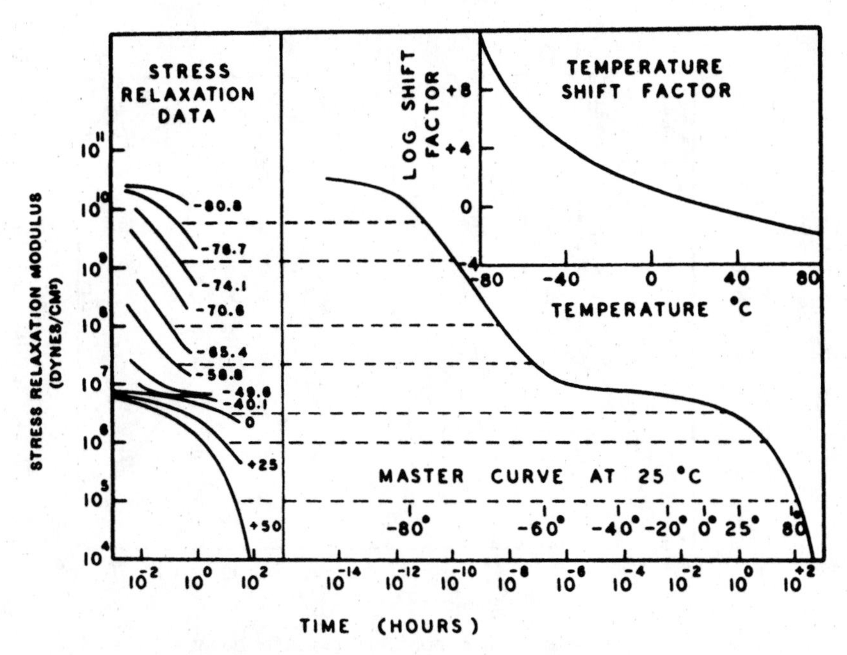

Figure 10.14 The making of a master curve, illustrated with polyisobutylene data. The classical T_g at 10 s of this polymer is −70°C (36–38).

this equation can be used both to check the data and to estimate the shift where data are missing.

Most interestingly, the master curve shown in Figure 10.14 looks much like the modulus temperature curves (see Figures 1.6 and 8.2). This likeness derives, of course, from the equivalence of log time with temperature.

In multicomponent polymeric systems such as polymer blends or blocks, each phase stress relaxes independently (39–41). Thus each phase will show a glass–rubber transition relaxation. While each phase follows the simple superposition rules illustrated above, combining them in a single equation must take into account the continuity of each phase in space. Attempts to do so have been made using the Takayanagi models (41), but the results are not simple.

10.3.2 The Reduced Frequency Nomograph

As illustrated in Figure 10.14, the master curve allows the extrapolation of data over broad temperature and time ranges. Similar master curves can be constructed with frequency as the variable, instead of time. More elegant still is the reduced frequency nomograph, which permits both reduced frequency and temperature simultaneously; see Figure 10.15 (42).

The reduced frequency nomograph (42,43) is constructed as follows. First, the storage and loss modulus (or tan δ) are plotted versus reduced frequency,

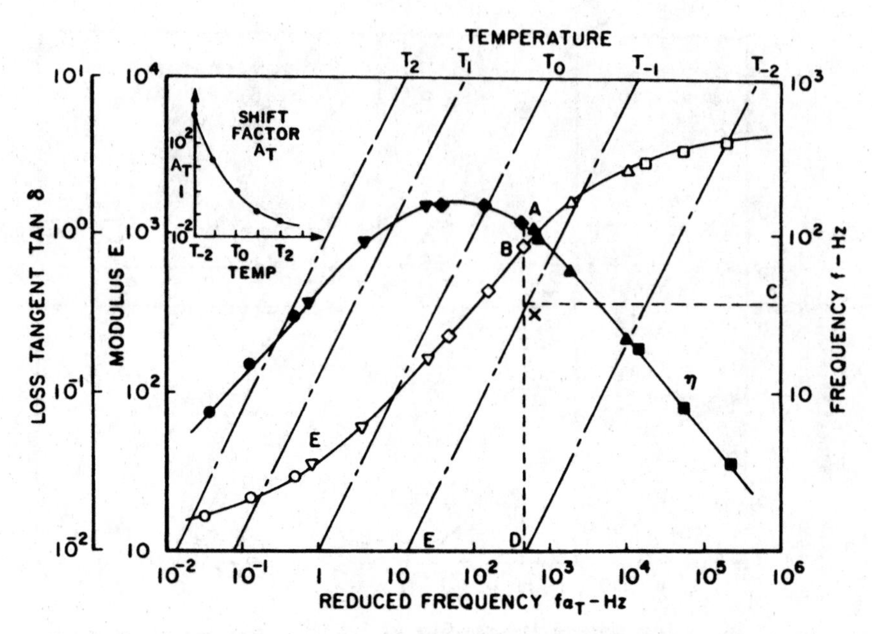

Figure 10.15 The reduced frequency nomograph. Here, the quantity $\eta \equiv \tan \delta$ (42).

making use of the reduced variables shift factor, A_T. The reduced frequency is defined as $f_j A_{T_i}$, where f_j is the frequency and A_{T_i} is the value of A_T at temperature T_i. Then an auxiliary frequency scale is constructed as the ordinate on the right side of the graph; see Figure 10.15. The values of $f_j A_{T_i}$ and f form a set of corresponding oblique lines representing temperature. Weissman and Chartoff (42) point out that computer software has been developed to speed the generation of such nomographs.

Simulated data are used in Figure 10.15 for ease of instruction. Assume that the value of E' and $\tan\delta$ at T_{-1} and some frequency f are needed, illustrated by point C in Figure 10.15. The intersection of the horizontal line CX, with $f_j A_{T_i}$ (point X) defines a value of fA_T at point D of approximately 4×10^2. From this value of fA_T it follows from the plots of E' and $\tan\delta$ that the quantity E' equals 10^{-3} N/m^2, point B, and $\tan\delta$ equals 1.2, point A.

A critical point of general information in Figure 10.15 is that, at high enough frequencies, the storage modulus increases into the glassy range. As the frequency increases beyond the capability of the polymer chains to respond, the polymer glassifies. At low enough frequencies, any linear (or lightly cross-linked polymer) will flow (or be rubbery). Since the reduced frequency is usually plotted with increasing values to the right, this figure appears backward to figures where log time is the x-axis variable. While stress relaxation studies employ time as the variable, dynamic experiments employ frequency.

Applications of the reduced frequency nomograph include sound and vibration damping (see Section 8.12) and earthquake damage control, interdisciplinary fields involving polymer scientists and mechanical engineers, and other disciplines. Frequencies of interest range from about 0.01 Hz (tall buildings) to 10^5 Hz, in the ultrasonic range.

10.4 POLYMER MELT VISCOSITY

10.4.1 The WLF Constants

In regions 4 and 5 of the modulus–temperature curve, linear amorphous polymers are capable of flow if they are subjected to a shear stress. While the melt viscosity, η, of low-molecular-weight substances may be Newtonian [see equation (8.3)], the flow behavior of polymers always contains some elements of viscoelasticity. In Section 8.6.1, the WLF equation was derived:

$$\log\left(\frac{\eta}{\eta_{T_g}}\right) = \frac{-C_1'(T - T_g)}{C_2' + (T - T_g)} \tag{10.56}$$

where C_1' and C_2' are constants, and η_{T_g} is the melt viscosity at the glass transition temperature. If data are not available on the polymer of interest, values of $C_1' = 17.44$ and $C_2' = 51.6$ may be used. However, these constants vary significantly if conditions other than η_{T_g} and T_g are used. Selected values of C_1' and C_2' are tabulated in Table 10.3 (44–46). The universal constants are widely used, as it is believed that the values for the individual polymers differ only by experimental error. Frequently $\eta_{T_g} \cong 1 \times 10^{13}$ poises $= 1 \times 10^{12}$ Pa·s.

10.4.2 The Molecular-Weight Dependence of the Melt Viscosity

10.4.2.1 Critical Entanglement Chain Length Viscoelasticity in polymers ultimately relates back to a few basic molecular characteristics involving the rates of chain molecular motion and chain entanglement. The increasing ability of chains to slip past one another as the temperature is increased governs the temperature dependence of the melt viscosity. One embodiment

Table 10.3 WLF parameters

Polymer	C_1'	C_2'	T_g, K
Polyisobutylene	16.6	104	202
Natural rubber (Hevea)	16.7	53.6	200
Polyurethane elastomer	15.6	32.6	238
Polystyrene	14.5	50.4	373
Poly(ethyl methacrylate)	17.6	65.5	335
“Universal constants”	17.4	51.6	

Source: J. J. Aklonis and W. J. MacKnight, *Introduction to Polymer Viscoelasticity*, Wiley-Interscience, New York, 1983, Table 3-2, p. 48.

of this concept is the WLF equation (10.56). The increased resistance to flow caused by entanglement governs the molecular-weight dependence.

Basic parameters in discussing the molecular weight dependence of the melt viscosity are the degree of polymerization (DP), which represents the number of monomer units linked together, and the number of atoms along the polymer chain's backbone (Z). For styrenics, acrylics and vinyl polymers, $Z = 2$DP, and for diene polymers, $Z = 4$DP. The point is that melt viscosity characteristics depend more on the number of backbone atoms than on the side foliage. It was also found very early that the melt viscosity depends on the weight-average degree of polymerization.

The molecular weight dependence of the melt viscosity exhibits two distinct regions (47), depending on whether the chains are long enough to be significantly entangled (see Figure 10.16). A critical entanglement chain length, $Z_{c,w}$ is defined as the weight-average number of chain atoms in the polymer molecules to cause intermolecular entanglement. Note that Z_c, after transforma-

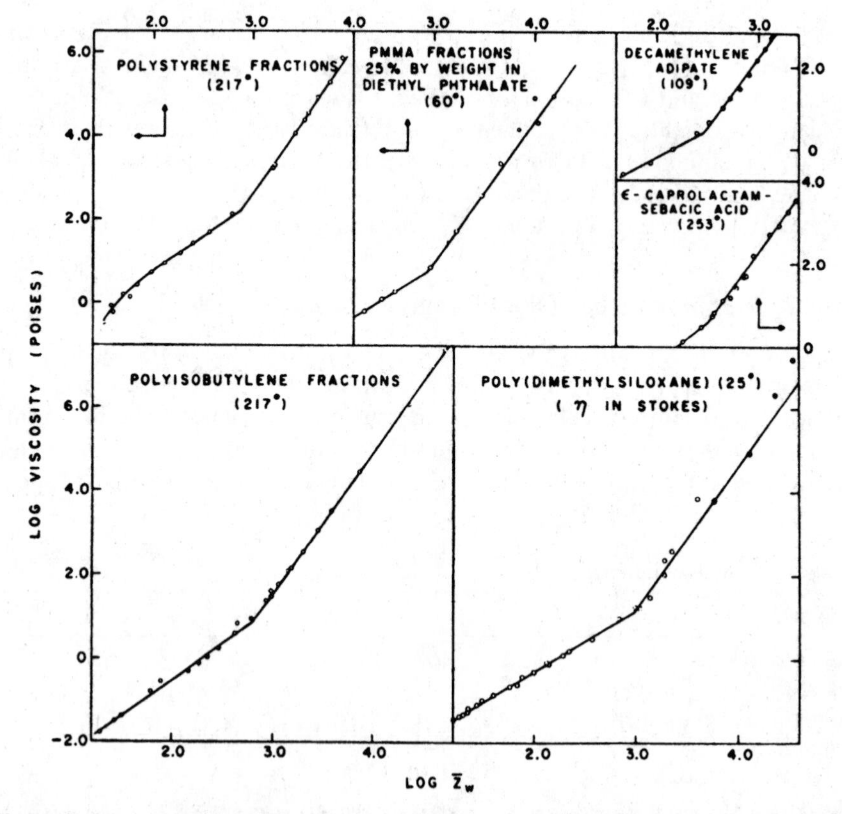

Figure 10.16 A plot of melt viscosity, log η, versus log Z_w for several polymers of importance. Below about $Z_w = 600$, the melt viscosity goes as the first power of Z_w; above about 600, the melt viscosity depends on the 3.4 power of Z_w.

tion to the appropriate molecular weight units, corresponds to the quantity M_c' given in Section 9.4. Below $Z_{c,w}$ the melt viscosity is given by

$$\eta = K_L Z_w^{1.0} \tag{10.57}$$

and above $Z_{c,w}$ the melt viscosity is given by

$$\eta = K_H Z_w^{3.4} \tag{10.58}$$

where K_L and K_H are constants for low and high degrees of polymerization.

The 1.0 power dependence in equation (10.57) represents the simple increase in viscosity as the chains get longer. The dependence of the viscosity on the 3.4 power of the degree of polymerization as shown in equation (10.58) arises from entanglement and diffusion considerations. The algebraic power 3.0 was derived by de Gennes (48), using scaling concepts. According to the reptation model (see Section 5.4), there is a maximum relaxation time of the chains, $\tau_{\max}$, which is the time required for complete renewal of the tube that constrains the chain. That is, $\tau_{\max}$ represents the time necessary for the polymer chain to diffuse out of its original tube and assume a new conformation.

Independently, Wool (49) also derived the exponent to equation (10.58) as 3.0, but by a quite different route. He showed that the exponent of 3.4 should actually be the experimental value obtained for molecular weights of most polymers in the commercial range. However, as the molecular weight of the polymer approaches infinity, the exponent should approach 3.0.

10.4.2.2 An Example Calculation of Melt Viscosity You are the engineer in charge of a polystyrene drinking cup manufacturing unit. Normally you process the material at 160°C, where the viscosity is 1.5×10^3 poises at $Z_w = 800$. Today, your polystyrene has $Z_w = 950$. What changes in processing temperature will bring the viscosity down to 1.5×10^3 again?

T_g for polystyrene is 100°C. The melt viscosity relationships are given by equations (10.58) and (8.48):

$$\eta = K_H Z_w^{3.4}$$

$$\log\left(\frac{\eta}{\eta_{T_g}}\right) = -\frac{17.44(T - T_g)}{51.6 + (T - T_g)}$$

Solving for K_H obtains

$$K_H = \frac{1.5 \times 10^3}{(800)^{3.4}} = 2.02 \times 10^{-7}$$

At 160°C the new viscosity is

$$\eta = 2.02 \times 10^{-7} \times (950)^{3.4} = 2.69 \times 10^3 \text{ P}$$

Solving the WLF equation for η_{T_g} obtains

$$\log\left(\frac{2.69\times10^{3}}{\eta_{T_g}}\right)=-\frac{17.44(433-373)}{51.6+(433-373)}$$

$$\eta_{T_g}=6.39\times10^{12}\ \text{P}$$

Now the new temperature must be calculated, using the WLF equation:

$$\log\left(\frac{1.5\times10^{3}}{6.39\times10^{12}}\right)=-\frac{17.44(T-373)}{51.6+(T-373)}$$

$$T=436.6\ \text{K}\quad\text{or}\quad 163.6^{\circ}\text{C}$$

Thus the temperature should be raised 3.6°C, to return to the same melt viscosity. It is often important in operating polymer melt machinery to work at the same melt viscosity. These two equations, used in tandem, can be used to solve many problems involving simultaneous changes in temperature and molecular weight.

The quantity Z_w may cross the elbow, necessitating a change from equation (10.57) to equation (10.58), and vice versa. The required new constant (K_L or K_H) can be obtained by noting that the viscosity for both equations is the same at the elbow; see Figure 10.16.

Many of the values of $Z_{c,w}$ are in the range of 600 ± 200. In fact, for new polymers or unanalyzed polymers, people still use the value of $Z_{c,w}=600$ as a first estimate of the break in the slope as needed.

After transformation to the appropriate molecular weight units, Z_c corresponds to the quantity M_c' given in Section 9.4. This latter quantity is somewhat easier to utilize in many cases.

10.4.2.3 Analysis of M_c' Most of the numerical values of M_c' are in the range of 4,000 to 31,000 g/mol, see Table 10.4. While this range is fairly broad, as shown above the value of $Z_{c,w}$ remains fairly constant. This suggested an analysis, to find a theoretical relationship for M_c'.

Wool(50) derived the equation,

$$M_c'=30.89(zb/C)^2 jM_0C_\infty \tag{10.59a}$$

where j represents the number of backbone atoms in the mer, z is the number of mers in the c-axis (i.e., equivalent zigzag, 3/1 helix, etc.) of length C, b is the bond length (1.54 Å for C-C bonds), M_0 is the mer molecular weight, and C_∞ is the characteristic ratio defined in Section 5.3.1.1. To a good approximation for vinyl polymers,

$$M_c\approx 30C_\infty M_0 \tag{10.59b}$$

Table 10.4 Experimental values of M_c' for common polymers[a]

Polymer	Experimental Value of M_c', g/mol
Polyethylene	4,000
Polystyrene	31,200
Polypropylene	7,000
Poly(vinyl alcohol)	7,500
Poly(vinyl acetate)	24,500
Poly(vinyl chloride)	11,000
Poly(methyl methacrylate)	18,400
Poly(ethylene oxide)	4,400
Poly(propylene oxide)	5,800
Polycarbonate	4,800
Poly(decamethylene adipate)	5,000
1,4-Polybutadiene	5,900
Polydimethylsiloxane	24,500

[a]Based on data taken from R. P. Wool, *Polymer Interfaces: Structure and Strength*, Hanser Publishers, Munich, 1995, Table 7.2, and J. D. Ferry, *Viscoelastic Properties of Polymers*, 3rd ed., Wiley, New York, 1980, Table 13-II.

10.4.2.4 The Tube Diffusion Coefficient Under a steady force, f, the chain moves with a velocity, v, in the tube. Then the "tube mobility" of the chain, μ_{tube}, is given by (48)

$$\mu_{\text{tube}} = \frac{v}{f} \tag{10.60}$$

If long-range backflow effects are assumed negligible, then the friction force v/μ_{tube} is essentially proportional to the number of atoms in the chain,

$$\mu_{\text{tube}} = \frac{\mu_1}{Z} \tag{10.61}$$

where μ_1 is independent of Z. Similarly the tube diffusion coefficient, D_{tube}, is related to μ_{tube} and Z through an Einstein relationship $D_{\text{tube}} = \mu_1 T/Z$

$$D_{\text{tube}} = \frac{D_1}{Z} \tag{10.62}$$

Since the time necessary to diffuse a certain distance depends on the square of the distance, the time necessary to completely renew the tube depends on the length of the tube, L, squared, Then

$$\tau_{\max} \cong \frac{L^2}{D_{\text{tube}}} \tag{10.63}$$

and

$$\tau_{\max} \cong \frac{ZL^2}{D_1} \tag{10.64}$$

since L is proportional to Z,

$$\tau_{\max} = \tau_1 Z^3 \tag{10.65}$$

The viscosity of the system is given by equation (10.15), $\eta = \tau E$. According to the reptation model, the modulus E depends on the distance between obstacles and does not depend on the chain length. Therefore

$$\eta \propto Z^3 \tag{10.66}$$

Equation (10.66) should be compared with equation (10.58). While the power dependence is not quite correct in this simple derivation, it illustrates the principal molecular-weight dependence of the viscosity (47).

10.5 POLYMER RHEOLOGY

Rheology is the study of the deformation and flow of matter. As such, the reader will recognize that a significant fraction of this book already involves rheological concepts. Some important areas not yet considered include shear rate dependence of flow and the effect of normal stress differences (51,52).

The range of melt viscosities ordinarily encountered in materials is given in Table 10.5 (51). Polymer latexes and suspensions are aqueous dispersions with viscosities dependent on solid content and additives. Polymer solutions may be much more viscous, depending on the concentration, molecular weight, and temperature.

Table 10.5 Viscosities of some common materials (51)

Composition	Viscosity, Pa·s	Consistency
Air	10^{-5}	Gaseous
Water	10^{-3}	Fluid
Polymer latexes	10^{-2}	Fluid
Olive oil	10^{-1}	Liquid
Glycerine	10^{0}	Liquid
Golden syrup	10^{2}	Thick liquid
Polymer melts	10^{2}–10^{6}	Toffee-like
Pitch	10^{9}	Stiff
Plastics	10^{12}	Glassy
Glass	10^{21}	Rigid

The temperature dependence of the viscosity is most easily expressed according to the Arrhenius relationship:

$$\eta = Ae^{-B/T} \tag{10.67}$$

where T represents the absolute temperature and A and B are constants of the liquid. The Arrhenius equation may easily be shown to be an approximation of the WLF equation far above the glass transition temperature.

The rate of energy dissipation per unit volume of the sheared polymer may be expressed as either the product of the shear stress and the shear rate or, equivalently, the product of the viscosity and the square of the shear rate. The rate of heat generated during viscous flow may be significant. Thus the heat generated may actually reduce the viscosity of the material. From an engineering point of view, these quantities provide an important measure of the power necessary to maintain a given flow rate.

All real liquids have both viscous and elastic components, although one or the other may predominate. For example, water behaves as a nearly perfect viscous medium, while a rubber band is a nearly perfect elastomer. A polymer solution of, for example, polyacrylamide in water may exhibit various ranges of viscoelasticity, depending on the concentration and temperature. According to Weissenberg (53), when an elastic liquid is subjected to simple shear flow, there are two forces to be considered:

1. The shear stress, characteristic of ordinary viscosity.
2. A normal force, observed as a pull along the lines of flow.

The several aspects of polymer rheology are developed below.

10.5.1 Shear Dependence of Viscosity

Any liquid showing a deviation from Newtonian behavior is considered non-Newtonian. As soon as reliable viscometers became available, workers found departures from Newtonian behavior for polymer solutions, dispersions, and melts. In the vast majority of cases, the viscosity decreases with increasing shear rate, giving rise to what is often called "shear-thinning."

An example for xanthan gum solutions is given in Figure 10.17 (54). Note especially the wide range of shear rates obtainable. Xanthan gum is a high-molecular-weight polysaccharide (MW = 7.6×10^6 g/mol in Figure 10.17), a biopolymer used in food applications, oil recovery, and textile printing. Less than 1% of xanthan increases water's viscosity by a factor of 10^5 at low shear rates, yet only by a factor of 10 at high shear rates. The viscosity reduction is caused by the reduced number of chain entanglements as the chains orient along the lines of flow. Note that water has 0.01 poise (1×10^{-3} Pa·s), shown as the base line in Figure 10.17.

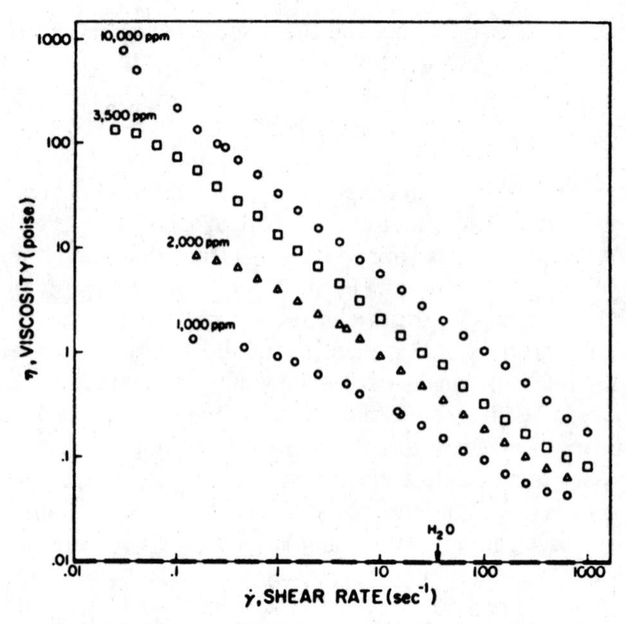

Figure 10.17 Viscosity of polymer melts and solutions usually decreases with increasing shear rate (54). Zanthan gum solutions.

The general case of shear rate dependence includes a limit of constant viscosity at very low shear rates, and a lower constant viscosity at the limit of very high shear rates. While this is not often observed with polymeric materials, a suggestion of both limits can be seen in Figure 10.17. The reason is that, even at the lowest shear rates, the chains begin to orient, and entanglements slip easier. At the limit of very high shear rates, polymers degrade. An extremely useful relationship is the well known "power-law" model,

$$\eta = K_2 \dot{\gamma}^{n-1} \tag{10.68}$$

where $\dot{\gamma}$ is the shear rate, n is the power-law index, and K_2 is called the "consistency." When n equals zero, a form of Newton's law is generated; see Section 8.2. The quantity $\dot{\gamma}$ is identical to ds/dt, equation (8.3), with units of reciprocal seconds. Typical values of n and K_2 for a number of materials are given in Table 10.6 (52). Many practical materials, such as skin creams and inks, contain polymers.

10.5.2 Normal Stress Differences

Consider a small plane surface of area ΔA drawn in a deforming medium; see Figure 10.18. The material flows or deforms through ΔA. For analytic purposes, the vector direction of motion is divided among x, y, and z. The direction of motion is not necessarily normal to the plane of ΔA.

Table 10.6 Typical power-law parameters of a selection of well-known materials for a particular range of shear rates (52)

Material	K_2 (Pa·s^n)	n	Shear rate range (s^{-1})
Ball-point pen ink	10	0.85	10^0–10^3
Fabric conditioner	10	0.6	10^0–10^2
Polymer melt	10,000	0.6	10^2–10^4
Molten chocolate	50	0.5	10^{-1}–10
Synovial fluid	0.5	0.4	10^{-1}–10^2
Toothpaste	300	0.3	10^0–10^3
Skin cream	250	0.1	10^0–10^2
Lubricating grease	1,000	0.1	10^{-1}–10^2

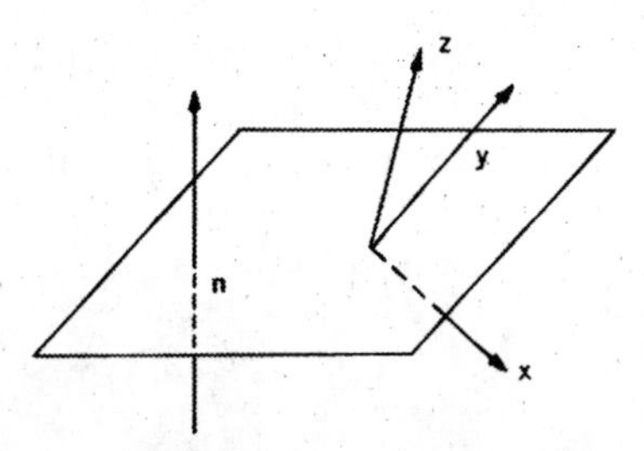

Figure 10.18 Arbitrary infinitesimal plane of area ΔA; n is the direction perpendicular to the plane. The quantities x, y, and z are orthogonal coordinates.

The stress components may be written γ_{nx}, γ_{ny}, and γ_{nz}, the first index referring to the orientation of plane surface, and the second to the direction of the stress. Then the quantity γ_{xx} means that a stress perpendicular to the plane ΔA is under consideration. There are three quantities that are rheologically relevant (55)—the shear stress, γ_{yx}, and the first and second normal stress differences, N_1 and N_2, respectively:

$$N_1 = \gamma_{xx} - \gamma_{yy} \tag{10.69}$$

$$N_2 = \gamma_{yy} - \gamma_{zz} \tag{10.70}$$

Both N_1 and N_2 are shear rate dependent. From a physical point of view the generation of unequal normal stress components arises from the anisotropic structure of a polymer fluid undergoing flow; that is, the polymer becomes oriented. The largest of the three normal stress components is always γ_{xx}, the component in the direction of flow. However, since the molecular structures undergoing flow are anisotropic, the forces are also anisotropic. The deformed chains often have the rough shape of ellipsoids, which have their major axis tilted toward the direction of flow. Thus the restoring force is greater in this direction than in the two orthogonal directions. It is these restoring forces that give rise to the normal forces (52).

The observable consequences of the normal forces are quite dramatic. Perhaps the best known is the so-called Weissenberg effect; see Figure 10.19 (52). It is produced when a rotating rod is placed into a vessel containing a viscoelastic fluid such as a concentrated polymer solution. Whereas a Newtonian liquid would be forced toward the rim of the vessel by inertia, the viscoelastic fluid moves *toward* the rod, climbing it. The chains are extended yet remain partly entangled, which causes a rubber elasticity type of retractive force; see Figure 10.20. This "hoop stress" around the rod forms the basis for the Weissenberg effect.

Another phenomenon, seen in fiber and plastic extruding operations, is "die swell"; see Figure 10.21 (56). Here a viscoelastic fluid is oriented in a tube during flow. When it is extruded, it flows from the exit of the tube and retracts. This results in swelling to a much greater diameter than that of the hole or slit. Die swell becomes much greater when the polymer chains are long enough to contain significant numbers of entanglements, being a direct consequence of the elastic energy stored by the polymer in the tube.

Figure 10.19 The Weissenberg effect illustrated by rotating a rod in a solution of polyisobutylene in polybutene (52).

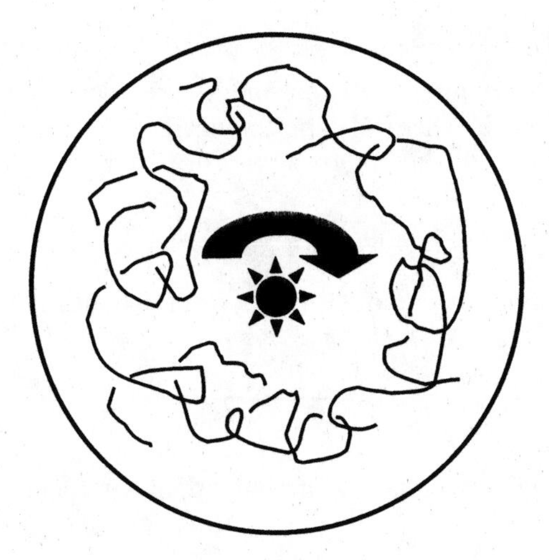

Figure 10.20 Molecular forces involved in the Weissenberg effect. Note polymer orientation and entanglements.

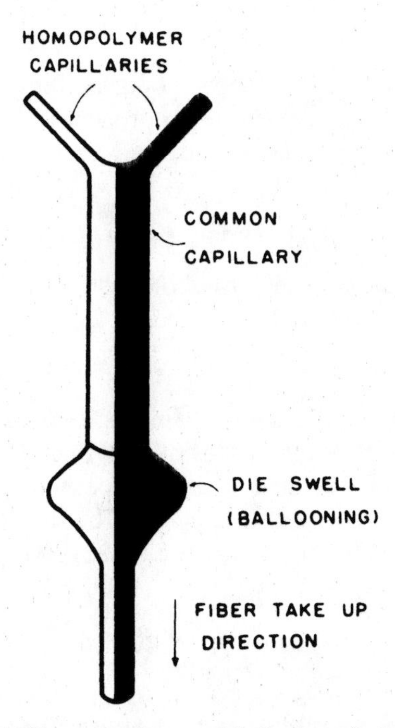

Figure 10.21 Bicomponent fibers are often used in the manufacture of rugs, because a mismatch in the expansion coefficients causes curling on cooling. However, the second normal forces must match, so that the die swell on both sides is the same (56).

10.5.3 Dynamic Viscosity

Periodic or oscillatory viscosity experiments provide a powerful rheological analogue to dynamic mechanical experiments. The theoretical relations are, in fact, closely interlocked. The complex viscosity, η^*, is defined as (10)

$$|\eta^*| = (\eta'^2 + \eta''^2)^{1/2} \tag{10.71}$$

The complex viscosity can be determined from dynamic mechanical data,

$$\eta^* = \frac{\left[(G')^2 + (G'')^2\right]^{1/2}}{\omega} \tag{10.72}$$

where ω is the angular frequency. The individual components are given by

$$\eta' = \frac{G''}{\omega} \tag{10.73}$$

and

$$\eta'' = \frac{G'}{\omega} \tag{10.74}$$

Thus the phase relations are the opposite of those for G^*. The in-phase, or real component, η' for a viscoelastic liquid approaches the steady-state flow viscosity as the frequency approaches zero.

10.5.4 Instruments and Experiments

There are several different types of viscometers, of varying complexity, good for specific purposes and/or ranges of viscosity (57,58). The three major classes are described below.

Falling-Ball Viscometers. This is the simplest type of viscometer, useful for determining relatively low viscosities. In one simple case, a graduated cylinder is filled with the fluid of interest (perhaps a concentrated polymer solution), and a steel ball is dropped in. The time to fall a given distance is recorded. According to Stokes, the terminal velocity is given by

$$\bar{u} = \frac{2}{9}\frac{R^2}{\eta}g(\rho - \rho_0) \tag{10.75}$$

where R is the sphere radius, ρ and ρ_0 are the densities of the sphere and the medium, respectively, and g is the gravitational constant. Within its range of operation, it is inexpensive, easily used, and the viscosity is absolute.

Capillary Viscometers. These instruments are used for intrinsic viscosities and also for more viscous melts, solutions, and dispersions. The viscosity is given by the Hagen–Poiseuille expression,

$$\eta = \frac{\pi' r^4 \Delta pt}{8VL} \tag{10.76}$$

where r is the capillary radius, Δp represents the pressure drop, and V is the volume of liquid that flows through the capillary of length L in time t. For a given viscometer with similar liquids,

$$\eta = Kt \quad \text{or} \quad \bar{v} = \eta/\rho_0 = K't \tag{10.77}$$

where $\bar{v}$ is the kinematic viscosity.

Rotational Viscometers. These relatively complex instruments can be used in the steady state or in an oscillatory, dynamic mode. Some are useful up to the glassy state of the polymer. The working mechanism, in all cases, is one part that moves past another. Designs include concentric cylinders (cup and bob), cone-and-plate, parallel-plate, and disk, paddle, or rotor in a cylinder.

The most important device is the cone-and-plate viscometer; see Figure 10.22 (52). The advantage of the cone-and-plate geometry is that the shear rate is very nearly the same everywhere in the fluid, provided the gap angle, θ_0, is small. The shear rate in the fluid is given by

$$\dot{\gamma} = \frac{\Omega_1}{\theta_0} \tag{10.78}$$

where Ω_1 represents the angular velocity of the rotating platter. The viscosity is given by

$$\eta = \frac{3f\theta_0}{2\pi a^3 \Omega_1} \tag{10.79}$$

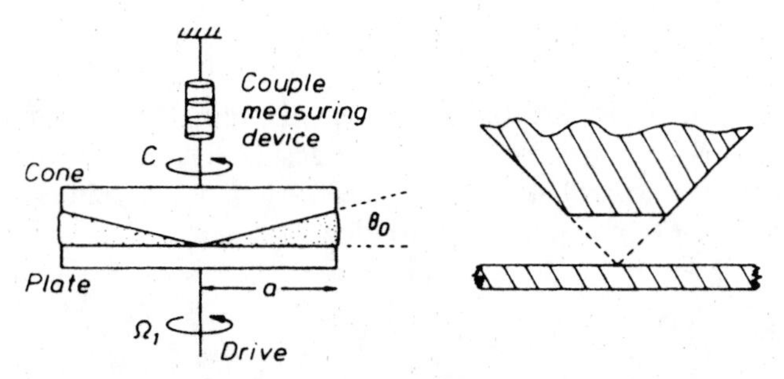

Figure 10.22 The cone-and-plate viscometer, showing the cone on top, a rotating plate, and couple attached to the cone. The inset shows a form of truncation employed in many instruments (52).

Table 10.7 Rheological instrumentation

Instrument	Manufacture	Application
Parr Physica Modular Compact Rheometer, stress controlled Model MCR 500	Parr Physica	Flow characteristics of polymers
ThermoHaake Rheoscope Imaging Rheometer	ThermoHaake	Dynamic oscillatory and steady shear
C-VOR-200 Rheometer	Bohlin	Strain-controlled creep experiments
RFS III Controlled Strain Rheometer	Rheometrics	Strain-controlled oscillatory and steady shear studies
ASTRA Controlled Stress Rheometer with Optical Analysis Module	Rheometrics	Measures stress controlled birefringence
Model No. MCR 500 Modular Compact Rheometer	Paar Physica	Polymer rheological measurements
CaBER 1 Extensional Rheometer	ThermoHaake	Uniaxial extension capillary geometry with laser micrometer

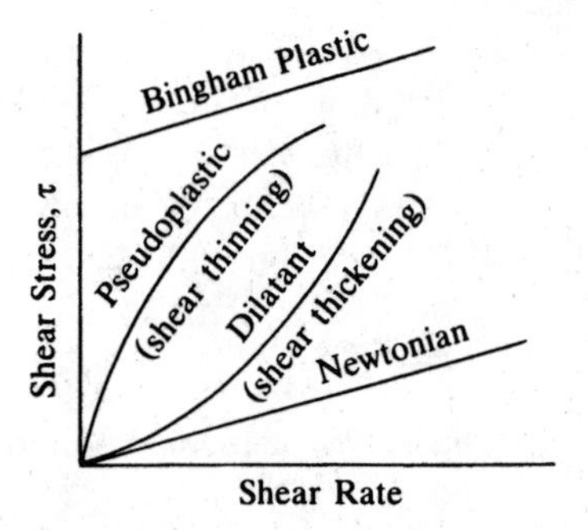

Figure 10.23 Several shear-rate-dependent rheological phenomena.

where the shear stress τ is measured by the couple C on the cone, and a is the radius of the cone-and-plate. An important factor in deciding on a viscometer, of course, is the viscosity of the fluid and the information desired.

Table 10.7 provides some of the current instrumentation for rheological measurements. Note that some of them are designed for flow, while others are designed for dynamic oscillatory measurements, while still others are basically uniaxial extension or creep instrumentation.

10.5.5 General Definitions and Terms

Several shear-rate-dependent phenomena are illustrated in Figure 10.23. A Bingham plastic has a linear relationship between shear stress and shear strain

Table 10.8 Flow models and equations (58)

Model	Equation
Newtonian	$\tau = \eta\dot{\gamma}$
Bingham plastic	$\tau - \tau_0 = \eta\dot{\gamma}$
Power law	$\tau = k\dot{\gamma}^n$
Power law with yield value	$\tau - \tau_0 = k\dot{\gamma}^n$
Casson fluid	$\tau^{1/2} - \tau_0^{1/2} = \eta_\infty^{1/2}\dot{\gamma}^{1/2}$
Williamson	$\eta - \eta_\infty = \dfrac{\eta_0 - \eta_\infty}{1 + \tau/\tau_m}$
Cross	$\eta - \eta_\infty = \dfrac{\eta_0 - \eta_\infty}{1 + \alpha\dot{\gamma}^n}$

but does not go through the origin. It must be realized that this is an idealization. Pseudoplastic materials, which include most polymer solutions and melts, exhibit shear thinning. Dilatant materials shear thicken.

There are several terms that apply to time-dependent behavior. Thixotropic fluids possess a structure that breaks down as a function of time and shear rate. Thus the viscosity is lowered. A famous example of this reversible phenomenon is the ubiquitous catsup bottle, yielding its contents only after sharp blows. The opposite effect, although rarely observed, is called antithixotropic, or rheopectic behavior, where materials "set up" as a function of time and shear rate; that is, the viscosity increases or the material gels.

Viscoelastic fluids exhibit elastic recovery from deformations that occur during flow. The Weissenberg effect and die swell have been discussed previously. Another very simple example is the stirring of concentrated polymer solutions rapidly, then stopping. Elastic effects make the fluid move backward for a time.

Several flow equations are summarized in Table 10.8 (58) for various models. Here, f_0 is the yield stress, η_0 the viscosity at low shear rates and η_∞ at high shear rates, and α and n are constants.

10.6 OVERVIEW OF VISCOELASTICITY AND RHEOLOGY

This chapter has illustrated how stress relaxation, creep, and rheology in polymers depend on the rate of molecular motion of the chains and on the presence of entanglements. It must be remembered that all macroscopic deformations of matter depend ultimately on molecular motion. In the case of high polymers, the chain's radius of gyration is changed during initial deformation or flow. Thermal motions tend to return the polymer to its initial conformation, thus raising its entropy. Clearly, there is a direct relationship between the mechanical or viscous behavior of polymeric materials and their molecular behavior.

REFERENCES

1. J. J. Aklonis, *J. Chem. Ed.*, **58**, 893 (1981).
2. A. V. Tobolsky, *Properties and Structure of Polymers*, Wiley, New York, 1960.
3. L. H. Sperling, S. L. Cooper, and A. V. Tobolsky, *J. Appl. Polym. Sci.*, **10**, 1735 (1966).
4. P. Thirion and R. Chasset, *Proceedings of the 4th Rubber Technology Conference*, London, 1962, p. 338.
5. P. G. de Gennes, *J. Chem. Phys.*, **55**, 572 (1971).
6. M. Takayanagi, *Mem. Fac. Eng. Kyushu Univ.*, **23**, 11 (1963).
7. L. H. Sperling, *Polymeric Multicomponent Materials: An Introduction*, Wiley, New York, 1997.
8. T. C. P. Lee, L. H. Sperling, and A. V. Tobolsky, *J. Appl. Polym. Sci.*, **10**, 1831 (1966).
9. A. V. Tobolsky, I. B. Prettyman, and J. H. Dillon, *J. Appl. Phys.*, **15**, 380 (1944).
10. J. D. Ferry, *Viscoelastic Properties of Polymers*, 3rd ed., Wiley, New York, 1980.
11. C. Picot, R. Duplessix, D. Decker, H. Benoît, F. Boue, J. P. Cotton, M. Daoud, B. Farnoux, G. Jannick, M. Nierloch, A. J. deVries, and P. Pincus, *Macromolecules*, **10**, 957 (1977).
12. A. Maconnachie, G. Allen, and R. W. Richards, *Polymer*, **22**, 1157 (1981).
13. F. Boue and G. Jannink, *J. Phys. Colloq.*, **39**, C2-183 (1978).
14. F. Boue, M. Nierlich, G. Jannink, and R. C. Ball, *J. Phys. Lett.* (*Paris*), **43**, 593 (1982).
15. F. Boue, M. Nierlich, G. Jannink, and R. C. Ball, *J. Phys.* (*Paris*), **43**, 137 (1982).
16. G. Hadziiannou, L. H. Wang, R. S. Stein, and R. S. Porter, *Macromolecules*, **15**, 880 (1982).
17. P. G. de Gennes, J. Chem. *Phys.*, **55**, 572 (1971).
18. M. Doi and J. F. Edwards, J. *Chem. Soc. Faraday Trans.*, **74**, 1789, 1802, 1818 (1978).
19. S. Daoudi, *J. Phys.*, **38**, 731 (1977).
20. A. V. Tobolsky, *J. Polym. Sci. Lett.*, **2**, 103 (1964).
21. R. P. Wool, *Polymer Interfaces: Structure and Strength*, Hanser, Munich, 1995.
22. K. A. Welp, R. P. Wool, G. Agrawal, S. K. Satija, S. Pispas, and J. Mays, *Macromolecules*, **32**, 5127 (1999).
23. S. D. Kim, A. Klein, and L. H. Sperling, *Macromolecules*, **33**, 8334 (2000).
24. K. D. Kim, L. H. Sperling, A. Klein, and B. Hammouda, *Macromolecules*, **27**, 6841 (1994).
25. L. J. Fetters, D. J. Lohse, D. Richter, T. A. Witten, and A. Zirkel, *Macromolecules*, **27**, 4639 (1994).
26. S. Tobing, A. Klein, L. H. Sperling, and B. Petrasko, *J. Appl. Polym. Sci.*, **81**, 2109 (2001).
27. A. Zosel and G. Ley, in *Film Formation in Water Borne Coatings*, T. Provder, M. Winnik and M. Urban, eds., ACS Books, Washington, DC, 1996.
28. J. R. McLaughlin and A. V. Tobolsky, *J. Colloid Sci.*, **7**, 555 (1952).
29. L. C. E. Struik, *Physical Aging in Amorphous Polymers and Other Materials*, Elsevier, New York, 1978.
30. H. C. Booij and J. H. M. Palmen, *Polym. Eng. Sci.*, **18**, 781 (1978).

31. S. Matsuoka, H. E. Bair, S. S. Bearder, H. E. Kern, and J. T. Ryan, *Polym. Eng. Sci.*, **18**, 1073 (1978).
32. F. H. J. Maurer, J. H. M. Palmen, and H. C. Booij, *Polym. Mater. Sci. Eng. Prepr.*, **51**, 614 (1984).
33. F. H. J. Maurer, in *Rheology, Vol 3: Applications*, G. Astarita, G. Marrucci, and L. Nicolais, Eds., Plenum, New York, 1980.
34. J. G. Curro, R. R. Lagasse, and R. Simha, *J. Appl. Phys.*, **52**, 5892 (1981).
35. J. G. Curro, R. R. Lagasse, and R. Simha, *Macromolecules*, **15**, 1621 (1982).
36. E. Castiff and A. V. Tobolsky, *J. Colloid Sci.*, **10**, 375 (1955).
37. E. Castiff and A. V. Tobolsky, *J. Polym. Sci.*, **19**, 111 (1956).
38. L. E. Nielsen, *Mechanical Properties of Polymers*, Reinhold, New York, 1962.
39. T. Horino, Y. Ogawa, T. Soen, and H. Kawai, *J. Appl. Polym. Sci.*, **9**, 2261 (1965).
40. R. E. Cohen and N. W. Tschoegl, *Int. J. Polym. Mater.*, **2**, 205 (1973).
41. D. Kaplan and N. W. Tschoegl, *Polym. Eng. Sci.*, **14**, 43 (1974).
42. P. T. Weissman and R. P. Chartoff, in *Sound and Vibration Damping with Polymers*, R. D. Corsaro and L. H. Sperling, eds., ACS Symposium Series No. 424, American Chemical Society, Washington, DC, 1991.
43. D. I. G. Jones, *Shock Vibration Bull.*, **48**(2), 13 (1978).
44. M. L. Williams, R. F. Landel, and J. D. Ferry, *J. Am. Chem. Soc.*, **77**, 3701 (1955).
45. J. D. Ferry, *Viscoelastic Properties of Polymers*, 3rd ed., Wiley, New York, 1980, Chap. 11.
46. J. J. Aklonis and W. J. MacKnight, *Introduction to Polymer Viscoelasticity*, 2nd ed., Wiley, New York, 1983, Chap. 3.
47. T. G. Fox, S. Gratch, and S. Loshaek, in *Rheology*, F. R. Eirich, ed., Academic Press, Orlando, 1956, Vol. 1, Chap. 12.
48. P. G. de Gennes, *Scaling Concepts in Polymer Physics*, Cornell University Press, Ithaca, NY, 1979, Chap. 8.
49. R. P. Wool, *Macromolecules*, **26**, 1564 (1993).
50. R. P. Wool, *Polymer Interfaces: Structure and Strength*, Hanser Publisher, Munich, 1995.
51. F. N. Cogswell, *Polymer Melt Rheology*, G. Godwin, ed., Wiley, New York, 1981.
52. H. A. Barnes, J. F. Hutton, and K. Walters, *An Introduction to Rheology*, Elsevier, New York, 1989.
53. K. Weissenberg, *Nature*, **159**, 310 (1947).
54. P. J. Whitcomb and C. W. Macosko, *J. Rheol.*, **22**, 493 (1978).
55. J. Meissner, R. W. Garbella, and J. Hostettler, *J. Rheol.*, **33**, 843 (1989).
56. J. A. Manson and L. H. Sperling, *Polymer Blends and Composites*, Plenum, New York, 1976.
57. Perry's *Chemical Engineer's Handbook*, 7th ed., R. H. Perry and D. W. Green, eds., McGraw-Hill, New York, 1997.
58. C. K. Schoff, in *Encyclopedia of Polymer Science and Engineering*, Vol. 14, Wiley, New York, 1988, p. 454.

GENERAL READING

H. A. Barnes, J. F. Hutton, and K. Walters, *An Introduction to Rheology*, Elsevier, Amsterdam, 1989.

C. D. Craver and C. E. Carraher, Jr., eds., *Applied Polymer Science: 21st Century*, Elsevier, Amsterdam, 2000.

J. D. Ferry, *Viscoelastic Properties of Polymers*, 3rd ed., Wiley, New York, 1980.

R. K. Gupta, *Polymer and Composite Rheology*, 2nd ed., Dekker, New York, 2000.

F. A. Morrison, *Understanding Rheology*, Oxford University Press, Oxford, 2001.

L. E. Nielsen and R. F. Landel, *Mechanical Properties of Polymers and Composites*, 2nd ed., Dekker, New York, 1994.

M. Rubinstein and R. H. Colby, *Polymer Physics*, Oxford University Press, Oxford, 2003.

L. C. E. Struik, *Physical Aging in Amorphous Polymers and Other Materials*, Elsevier, Amsterdam, 1978.

STUDY PROBLEMS

1. Draw the creep and creep recovery curves for the three-element model consisting of a Kelvin element and a spring in series.

2. If the modulus of the *cis*-1,4-polybutadiene in Figure 10.8 was 2.5×10^7 dynes/cm^2, plot the total number of remaining active chain segments from the original network as a function of time. Also plot the number of active chain segments formed by the oxidative cross-linking.

3. At 200°C, how long would it take for the silicone rubber in Figure 10.7 to relax 50%?

4. Derive equations to express Young's modulus as a function of rubber and plastic composition using the Takayanagi models (*c*) and (*d*) in Figure 10.6.

5. Draw stress relaxation curves for the four-element model.

6. Derive an equation for the creep recovery of a Kelvin element, beginning after a creep experiment extending it to t_1, a later time.

7. A poly(methyl methacrylate) bridge is to be placed across a river in the tropics, where the average temperature is 40°C. The bridge is a simple platform 100 ft long and 5 ft thick and 10 ft wide. The bridge will fail when creep slopes the sides of the bridge more than a 30° angle, so cars get stuck in the middle. How long will the bridge last? [*Hint:* As a simple approximation, the simple beam, center-loaded bending distance Y is given by $Y = fL^3/4a^3bE$, where a is the thickness, b is the width, f is the force, L is the length, and E is Young's modulus. See Figure 10.12.]

8. Prepare a master curve at 110°C for poly(methyl methacrylate), using the data in Figure 10.12.

9. The melt viscosity of a fraction of natural rubber is 2×10^3 Pa·s at 240 K. What is the melt viscosity of this fraction at 250 K?

10. A new polymer with a mer weight of 211 g/mol and five atoms in the chain was found to have a weight-average molecular weight of 300,000 g/mol. Its melt viscosity is 1500 poises. What is the viscosity of the polymer if its molecular weight is doubled?

11. A polymer with a Z_w of 200 was found to have a melt viscosity of 100 Pa·s. What is the viscosity of this polymer when $Z_w = 800$?

12. A polymer with a T_g of 110°C and a Z_w of 400 was found to have a melt viscosity of 5000 Pa·s at 160°C. What is its melt viscosity at 140°C when $Z_w = 900$? [*Hint:* Combine the WLF equation with the DP dependence.]

13. The three-element springs and dashpot model shown is subject to a creep experiment. Show how the length (or strain) increases with time. At time = t, the stress is removed. Show how the sample recovers.

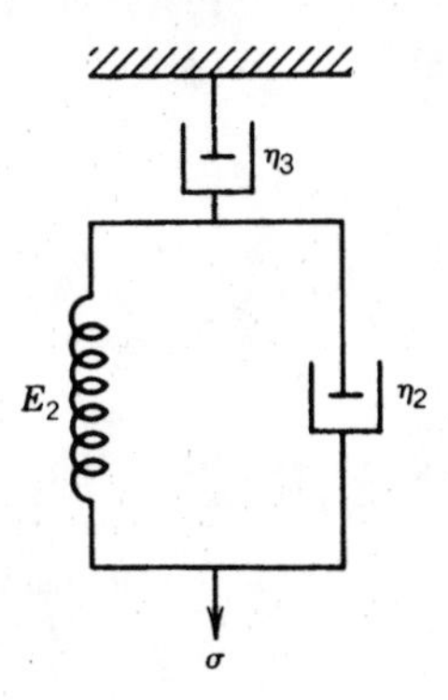

Figure P10.13 The three-element model.

14. A new polymer follows the Kelvin model. The quantity η obeys the WLF eqution, and E obeys rubber elasticity theory. The glass transition temperature of the polymer is 5°C, where it has a viscosity of 1×10^{13} poises. The concentration of active chain segments is 1×10^{-4} mol/cm^3. The temperature of the experiment is 30°C.

(a) How does this polymer creep with time under a stress of 1×10^7 dynes/cm^2? A plot of strain versus time is required.

(b) Briefly discuss two ways to slow down or reduce the rate of creep in part (a).

15. A certain extruder for plastics was found to work best at a melt viscosity of 2×10^4 Pa·s. The polymer of choice had this viscosity at 145°C when its

DP_w was 700. This polymer has a T_g of 75°C. Because of a polymerization kinetics miscalculation by someone who did not take polymer science, today's polymer has a DP_w of 500. At what temperature should the extruder be run so that the viscosity will remain at optimum conditions? (Assume two carbon atoms per mer backbone.)

16. In the spinning of rayon fibers, the viscosity of the viscose solution must be carefully controlled. In the falling ball experiment on a viscose solution, it took a 2 mm lead ball 324 s to fall 10 cm. The density of the viscose was 1.2 g/cm^3. What is its viscosity?

17. A cone-and-plate experiment was performed on a polymer exhibiting first and second normal forces. Will the cone and plate be forced together, pushed apart, pushed sideways, or what? Discuss your reasoning.

18. Based on the concepts of Section 10.2.7, what is the value of G^0 for poly(methyl methacrylate)? How does it compare with the value in Figure 10.12?

APPENDIX 10.1 ENERGY OF ACTIVATION FROM CHEMICAL STRESS RELAXATION TIMES

From first-order chemical kinetics,

$$\frac{-dc_A}{dt} = kC_A \tag{A10.1.1}$$

where C_A is the concentration of species A. Then on integration,

$$-\ln C_A = kt + \text{constant} \tag{A10.1.2}$$

$$C_A = C_{A0}e^{-kt} \tag{A10.1.3}$$

Note that k has the units of inverse time. If $k = 1/\tau_1$, an immediate relationship with equation (10.18) is noted.

From the Arrhenius equation,

$$k = \frac{RT}{Nh}e^{\Delta S/R}e^{-\Delta H/RT} \tag{A10.1.4}$$

For the present purposes,

$$k = se^{-\Delta E_{\text{act}}/RT} \tag{A10.1.5}$$

which may be rewritten

$$\tau_1 = s^{-1}e^{\Delta E_{\text{act}}/RT} \tag{A10.1.6}$$

where ΔE_{act} is the activation energy of the process. Then

$$\ln \tau_1 = \text{constant}^{-1} + \frac{\Delta E_{act}}{RT} \tag{A10.1.7}$$

A plot of $\ln \tau_1$ versus $1/T$ yields $\Delta E_{act}/R$ as the slope.

APPENDIX 10.2 VISCOELASTICITY OF CHEESE†

Cheese is made from milk. While the composition of cheeses varies greatly, ordinary hard cheeses have approximately 31% fat, mostly triglycerides, 25% protein, 1.7% carbohydrates, mostly lactose, about 2.2% ash, mostly sodium salts, and about 40% moisture (B1).

Pasteurized, prepared cheese products such as Velveeta® have about 21% fat, 11% carbohydrates, and about 18% protein. Most of the remaining material is water with some salt. While the morphology of cheese is complex, the water tends to plasticize the protein. The lower protein content is largely responsible for prepared cheese products which are softer (lower modulus) and have greater viscoelasticity, making the material suitable for the present demonstration.

The proteins are largely the biopolymer casein. There are four types of casein present, α_{s1}, β, α_{s2}, and κ. These are present in the molecular ratio of 4:4:1:1, and have 199, 209, 208, and 169 amino acid residues (mers) per chain, respectively (B2). Of course, the structures of all proteins are basically polyamide-2 copolymers,

$$\left(\text{NH}-\underset{\underset{\text{R}}{|}}{\text{CH}}-\underset{\underset{\text{O}}{\|}}{\text{C}} \right)_n$$

where R may take any of 20 structures; see Appendix 2.1. The spatial structure of the protein chain is maintained by hydrogen bonding and internal cross-links (see Figure 9.2), but α_{s1}-casein is known to have a relatively open structure (B3) contributing significantly to the viscoelastic characteristics of the cheese.

The objective of the experiment is to measure the viscoelastic characteristics of cheese, and interpret the data in terms of the four-element model.

The time required is one hour. The level is junior or senior standing. The principles to be investigated are the applicability of the four-element model to polymers in general, and cheese in particular.

†Reprinted in part from Y. S. Chang, J. S. Guo, Y. P. Lee, and L. H. Sperling, *J. Chem. Educat.*, **63**, 1077 (1986).

Equipment and Supplies:

One 1-pound block processed cheese (Velveeta® or similar), at room temperature
Cut the cheese in half, place one-half on top of the other
One clock (stopwatch is fine)
One meter stick
One small, flat plate (insert between cheese and weight)
One flat, hard surface (most desks are suitable)
Weights of about 1 kg (rectangular weights of the same cross-sectional area as the cheese are fine)

The original height of the two layers of cheese is recorded. Then the plate and weight are placed on top of the cheese. Its new height is recorded immediately and each succeeding 5 minutes thereafter. This experiment measures creep in compression.

Replace the weight and deformed cheese with another weight and new cheese, and repeat.

Results

The initial dimensions of the 1-pound block of cheese were 4 cm × 6 cm × 15 cm. Weights of 500 and 700 g were employed. The resulting creep curves are illustrated in Figure A10.2.1. An immediate elastic compression is noted,

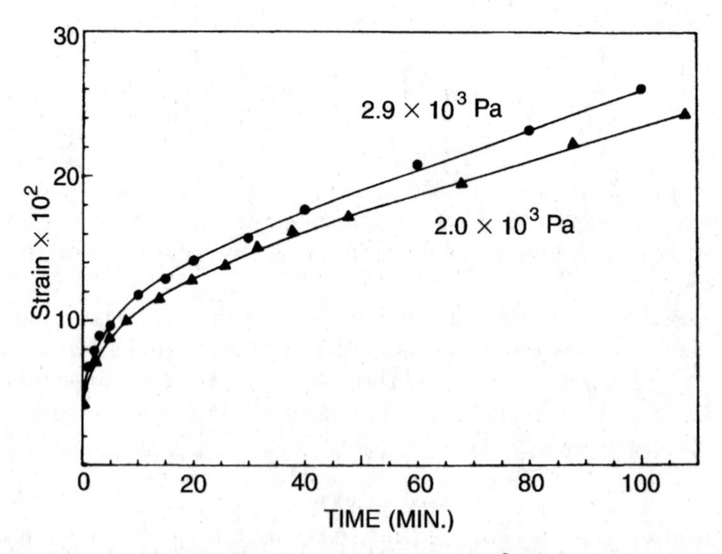

Figure A10.2.1 Experimental creep curves for Velveeta® cheese under two loads.

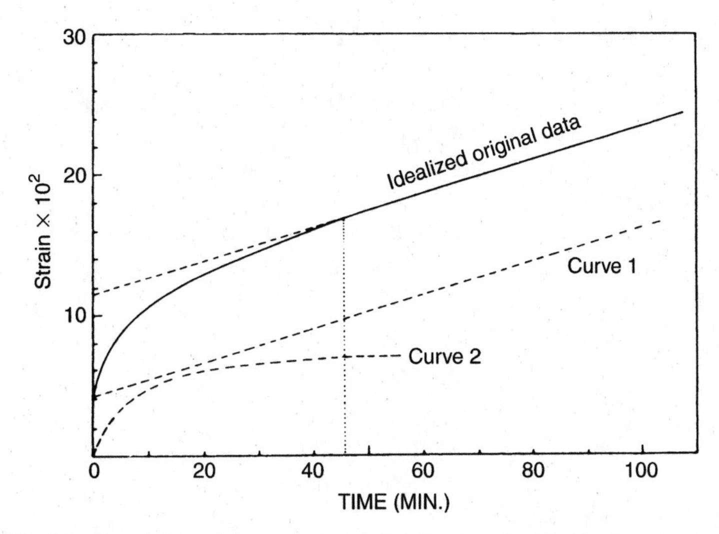

Figure A10.2.2 Resolution of the experimental data according to the four-element model. Curve 1 represents flow according to η_3. Curve 2 illustrates the behavior of the middle element, which is really a Kelvin element. The strain at zero time up to the interception of curve 1 represents the instantaneous deformation according to the spring E_1.

Table A2.10.1 Characteristics of Velveeta cheese at room temperature according to the four-element model

	Applied Weight	
Variable	500 g	700 g
σ (Pa)	2.00×10^3	2.88×10^3
E_1 (Pa)	4.88×10^4	5.18×10^4
E_2 (Pa)	2.82×10^4	4.24×10^4
η_2 (Pa·s)	1.52×10^7	1.78×10^7
η_3 (Pa·s)	1.00×10^8	1.21×10^8
τ_2 (min)	9	7

a measure of the modulus E_1. This is followed by a curved line of strain versus time, and a straight-line portion lasting at least 2 hours; see Figure A10.2.1.

The data were analyzed in Figure A10.2.2 according to the four-element model. First, at zero time, the strain yields E_1. The straight-line portion at long times, separated as curve 1, yields η_3. Then, by subtraction, curve 2 was obtained. By simple curve fittings E_2, η_2, and τ_2 can be determined. The retardation time for Velveeta cheese was found to be about 7 to 9 minutes (see Table A10.2.1).

The value of E_1, near 5×10^4 Pa, places it in the fourth region of viscoelasticity, rubbery flow. The results should be compared to known values of modulus (A4).

REFERENCES

B1. A. R. Hill, *Chemistry of Structure-Function Relationships in Cheese*, E. L. Malin and M. H. Tunick, eds., Plenum, New York, 1995.

B2. W. N. Eigel, J. E. Butler, C. A. Ernstrom, J. H. M. Farrell, V. R. Harwalkar, R. Jenness, and R. M. Whitney, *J. Dairy Sci.*, **67**, 1599 (1984).

B3. E. L. Malin and E. M. Brown, *Int. Dairy J.*, **9**, 207 (1999).

B4. J. H. Prentice, in *Encyclopedia of Food Sci. and Technol.*, Vol. 1, Y. H. Hui, ed., Wiley, 1992, p. 348.

14

MODERN POLYMER TOPICS

The preceding chapters provide an introduction to physical polymer science. However, there are many more topics of great interest not yet covered. While some of the topics to be discussed have been known for some time, a significant part of this chapter describes a series of relatively new developments, often still in their infancy.

14.1 POLYOLEFINS

The polyolefins are those polymers based only on carbon and hydrogen, originating from monomers containing a double bond in the 1-position, sometimes called α-olefins. Principally, these include polyethylene, polypropylene, copolymers of polyethylene containing various comonomers such as 1-butene, 1-hexene, and 1-octene, ethylene–propylene monomer (EPM), and ethylene–propylene–diene–monomer (EPDM). All of these are plastics except EPM and EPDM, which are elastomers.

14.1.1 The Global Picture

Among the synthetic polymers, polyethylene and polypropylene are the largest tonnage synthetic polymers; for comparison, cellulose constitutes the largest tonnage natural polymer; see Table 14.1. These materials are relatively easy to manufacture, have a range of useful properties, and are low priced.

Introduction to Physical Polymer Science, by L.H. Sperling
ISBN 0-471-70606-X

Table 14.1 Production of key global polymers

Polymer	Annual Tonnage, Millions (Late 1990s Quantities)	Reference
Polyethylene	45	(a)
Polypropylene	22	(b)
Cellulose[a]	24	(c)
Poly(vinyl chloride)	25	(b)
Polystyrene	12	(b)

References: (a) *Modern Plastics Encyclopedia '98*, **A-15** (1998). (b) M. S. Reisch, *C&E News*, **75** (21), 14 (1997). (c) G. Stanley, *TAPPI*, **82** (1), 40 (1999).

[a] Quantity for paper pulp only. Rayon, cellophane, and cellulose esters and ethers not included.

Table 14.2 Synthetic methods for polyethylene manufacture[a]

Method	Polyethylene Properties
High pressure (free radical)	Broad molecular weight distribution, both short and long branches along chain, low melting, low density
Ziegler process (coordination catalysts, titanium tetrachloride/triethyl aluminum)	Broad molecular weight distribution, few branches, high density, linear polymers, high melting, comonomers control crystallinity levels
Metallocene catalysis (bis-cyclopentadienyl–metal complexes)	Relatively narrow molecular weight distributions, controlled levels of branching, improved control of comonomer distribution
Metallocene–Ziegler	High comonomer incorporation

[a] Note that anionic syntheses are only used for styrenics, dienes, and some acrylics.

14.1.2 Polyethylene Properties

The melting temperature, extent of crystallinity, modulus, and mechanical behavior depend on the method of manufacture and the addition of comonomers, as well as overall molecular weight. Polyethylene is manufactured by several major processes (1,2): The high-pressure, free-radical polymerization, the Ziegler process, and the newer metallocene-catalyzed polymers and the metallocene–Ziegler processes; see Table 14.2.

While the largest tonnages by far are based on the high-pressure process and the Ziegler process, the newest method, via metallocene catalysis, offers great promise for controlled properties. Note that only the high-pressure process is based on free-radical synthesis. The reader is referred to Brydson (1) and Benedikt and Goodall (2) for catalyst and synthesis details.

The various polymers are named based on their density (Table 14.3). The densities of the polyethylenes decrease with increased side group mole frac-

Table 14.3 Polyethylene properties

Polymer	Designation	Degree of Branching, $CH_3/100C$	Density Range, g/cm^3	Melting Temperature Range, °C
Low-density polyethylene	LDPE	2–7	0.915–0.94	100–129
High-density polyethylene	HDPE	0.1–2	0.94–0.97	108–129
Linear low-density polyethylene	LLDPE	2–6	0.91–0.94	99–108
Metallocene linear low-density polyethylene	*m*-LLDPE	3–7	0.90–0.92	83–102
Ultra-low-density polyethylene	ULDPE	~7	0.86–0.90	~80–85

Source: Data collated from J. Brandrup and E. H. Immergut, eds., *Polymer Handbook*, 3rd ed., Wiley-Interscience, New York, 1989, and G. M. Benedikt and B. L. Goodall, eds., *Metallocene-Catalyzed Polymers: Materials, Properties, Processing, and Markets*, Plastics Design Library, Norwich, NY, 1998.

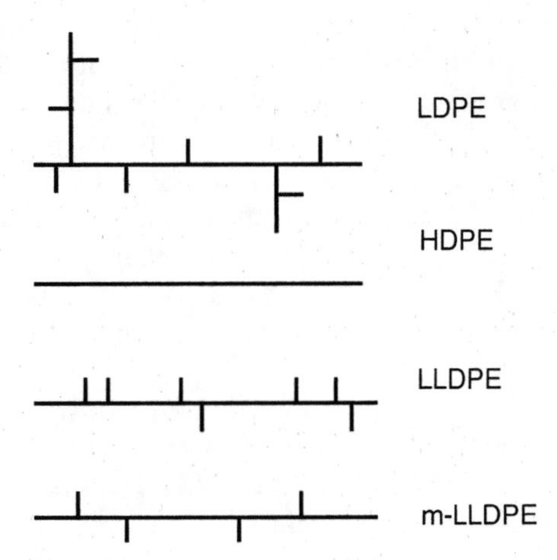

Figure 14.1 Model structures of polyethylenes, illustrating the various types of regularity of side chains.

tion, such as obtained via copolymerization with small amounts of propylene or *n*-butene. In LDPE, short side chains are caused by a *back-biting* phenomenon during polymerization; in addition there are long side chains caused by hydrogen abstraction and subsequent branching. HDPE contains substantially no branches, long or short. In Table 14.3 the LLDPE and *m*-LLDPE are linear polyethylene (no long branches) with controlled quantities of comonomer such as 1-butene added to reduce the crystallinity of the product. The *m* represents metallocene. The presence of short chain branches in LDPE, or their synthetic equivalent added in the form of comonomers in LLDPE disrupt the sequence of ethylene mers; therefore the crystallinity of the ethylene copolymer is reduced. These various compositions are modeled in Figure 14.1.

Below the LLDPEs, there are designations such as ULDPE for still more side groups. The densities in Table 14.3 should be compared to that of 100% crystalline polyethylene, 1.00 g/cm^3; see Table 6.2 and Section 6.3.1. By extrapolation of data from above the melting point, the density of amorphous polyethylene at room temperature is reported as 0.855 g/cm^3 (3), so that the 0.86 g/cm^3 end of the ultra-low-density polyethylene is essentially amorphous.

For some applications, HDPE has too high a modulus; however, the rheological and mechanical properties are improved by having a linear chain. LLDPE and *m*-LLDPE offer the best of both worlds for many applications. Frequently polymers with various extents of comonomers are blended to tailor materials for such properties as improved puncture resistance.

14.1.3 Polypropylene Properties

The major manufacturing process utilizes Ziegler-type catalysts. However, the newer metallocene-based materials offer improved control and specialty properties. The major product is isotactic polypropylene. With the Ziegler-type catalysts, various quantities of atactic polypropylene are included as by-products. The degree of isotacticity of the polypropylene also varies with the exact process. In general, the higher the isotacticity index, the higher the modulus and yield stress, and the lower the elongation to break.

With the advent of metallocene catalysis, a range of tacticities and structures are possible; see Figure 14.2 (4). Thus a range of compositions not easily prepared before, such as syndiotactic, hemi-isotactic, and isoblock copolymers, are now possible.

The melting temperatures of syndiotactic polypropylene were studied as a function of syndiotactic pentad content, *rrrr* (see Section 2.3.4) values determined via ^{13}C NMR spectra; see Figure 14.3 (5). The upper figure utilizes the Hoffman–Weeks extrapolation procedure; see Section 6.8.6. The lower figure shows a second extrapolation of the data to 100% *rrr* content, arriving at a melting temperature of nearly 182°C. By comparison, an isotactic polypropylene with >99% *mmmm* composition has a melting temperature of 170°C (6).

14.1.4 Polyolefin Elastomers

Two elastomers are made based on statistical copolymers of ethylene and propylene: EPDM, in which the diene portion, D, serves as a cross-linking site, and its non-cross-linking counterpart, EPM. Since these materials have few reactive sites, they are relatively impervious to oxidation or hydrolysis. Frequently they are blended with either polyamides or polypropylene to form impact-resistant plastics; see Section 13.8.

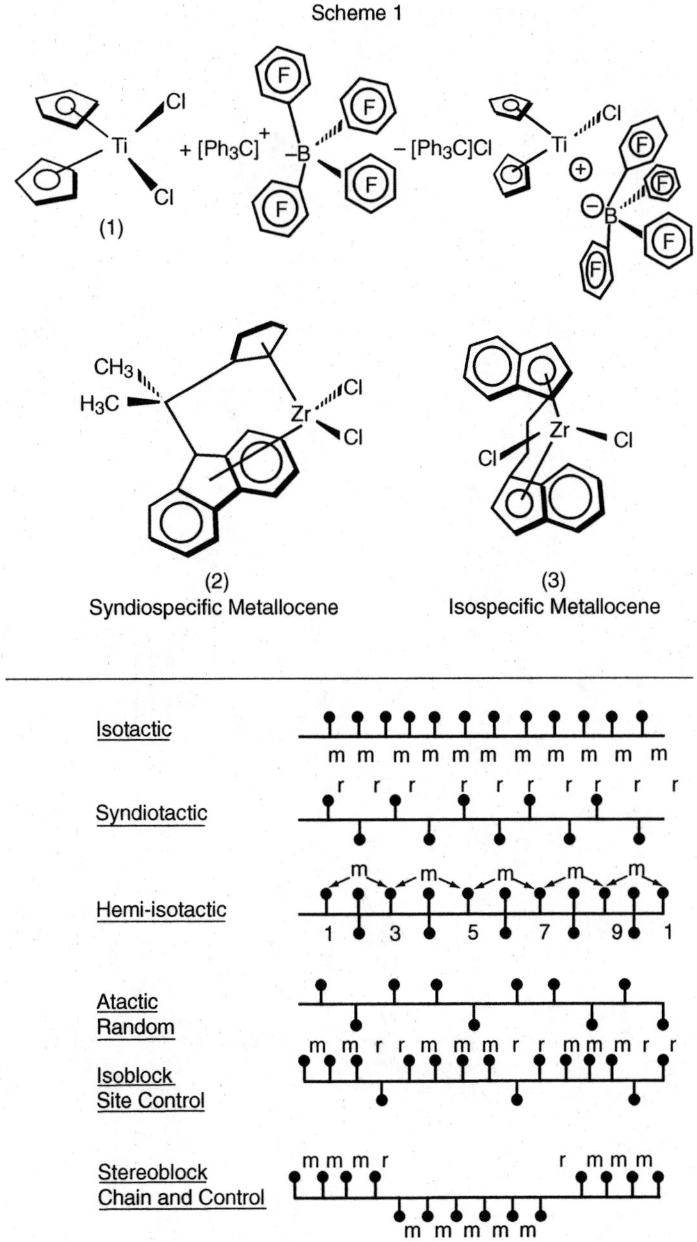

Scheme 1

(1) + $[Ph_3C]^+$ ^-B − $[Ph_3C]Cl$

(2)
Syndiospecific Metallocene

(3)
Isospecific Metallocene

Isotactic

m m m m m m m m m m m

Syndiotactic

r r r r r r r r r r

Hemi-isotactic

m m m m m

1 3 5 7 9 1

Atactic
Random

m m r r m m m r r m m m r r

Isoblock
Site Control

Stereoblock
Chain and Control

m m m r r m m m

m m m m m

Figure 14.2 *Upper*: Illustration of metallocene catalyst structure, which must be specific for each stereo-specific polymer structure. *Lower*: Microstructures of polypropylene; most can be made via metallocene catalysis polymerization.

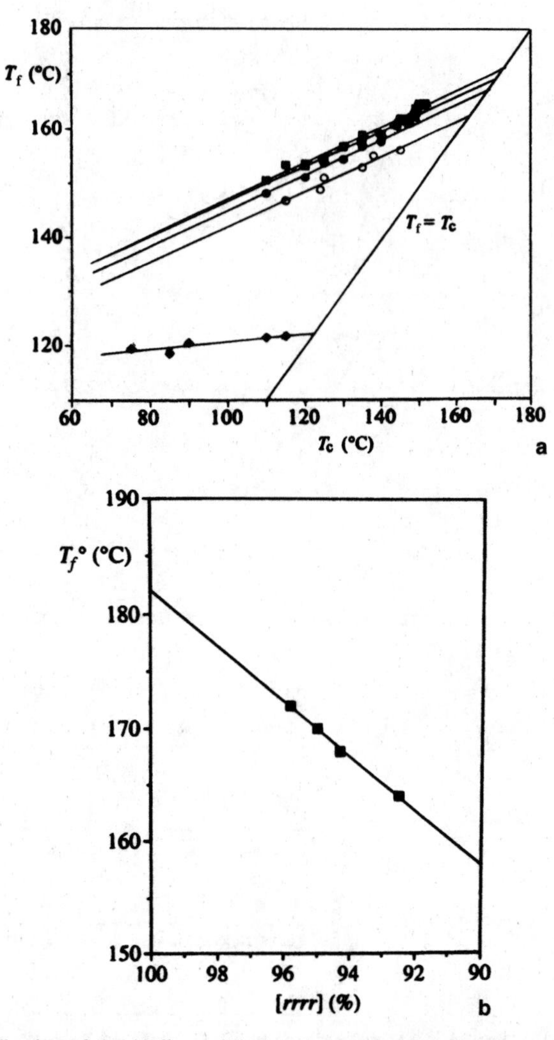

Figure 14.3 Utilization of the Hoffman–Weeks plot to establish the theoretical melting temperature of *rrrr* syndiotactic polypropylene. (*a*) Melting temperatures T_f of melt-crystallized *syn*-polypropylene, taken from DSC scans at 2.5°C/min, as a function of the crystallization temperature, T_c, extrapolated to $T_f = T_c$. (*b*) Extrapolation of the $T_f = T_c$ values in (*a*) to 100% pentad content, *rrrr*.

14.2 THERMOSET POLYMER MATERIALS

A number of plastic, adhesive, and coating materials are utilized in a densely cross-linked form known as thermosets. Some of the more important of these include the phenol–formaldehyde resins, urea-formaldehydes, polyimides, epoxies, amino resins, and the alkyds, among others (8). Because these materials are densely cross-linked, they are amorphous, usually do not exhibit any rubber elasticity behavior, and even their glass transition may be suppressed. This last arises because of insufficient chain mobility. The thermosets are synthesized through step-polymerization kinetics, often through condensation routes.

14.2.1 Phenol–Formaldehyde Resins

These materials are produced by heating phenol and formaldehyde in the presence of either an acid or a base catalyst. With a base catalyst and an excess of formaldehyde, a composition called resole resins are prepared. With an acid catalyst and an excess of phenol, a two-stage resin called novolacs are made. In both cases high-melting or viscous oligomers are made, which react further at elevated temperatures to produce high modulus but brittle materials (9).

The basic structure of these materials in the fully cured (cross-linked) state may be written

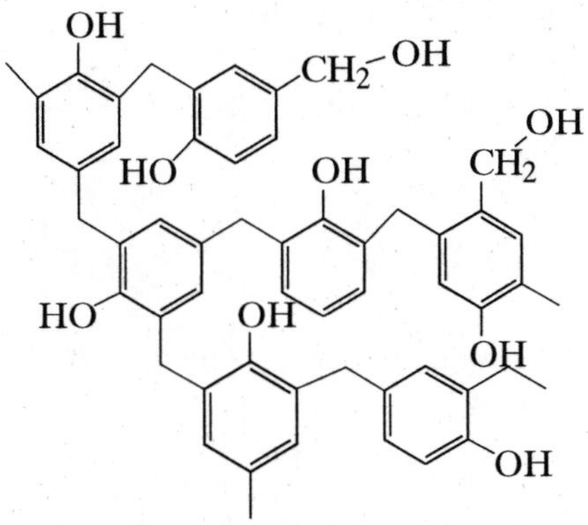

The open bonds indicate points of attachment to other regions of the network. Of course, this structure is only an illustration of the very many possibilities of a very irregular network. Applications include electrical connectors, pot handles, and so on.

14.2.2 Epoxy Resins

One of the most important monomers in epoxy chemistry is the diglycidyl ether of bisphenol A (9a):

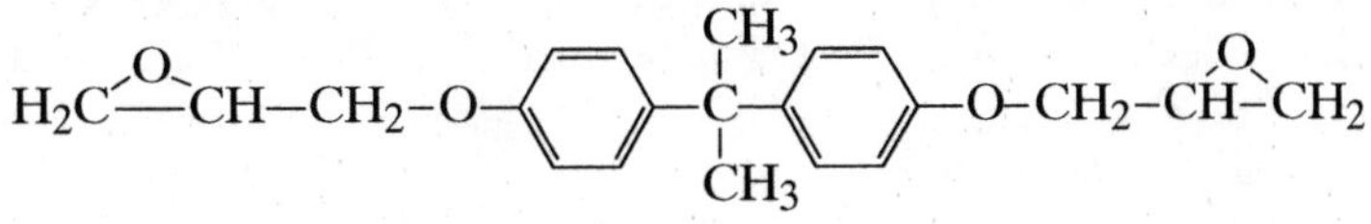

$$H_2C\overset{O}{—}CH-CH_2-O-C_6H_4-C(CH_3)_2-C_6H_4-O-CH_2-CH\overset{O}{—}CH_2$$

A simple way to polymerize these materials involves reacting them with primary or secondary amines. In such reactions the epoxy group is a functionality of one, resulting in ether bonds of the type:

$$R_1-O-CH_2-\underset{\displaystyle |}{\overset{\displaystyle OH}{}}CH-CH_2-\overset{H}{N}-R_2$$

Many epoxides have structures more complex than the diglycidyl ether of bisphenol A, having longer linear structures or more epoxy groups per monomer unit, or both. For some adhesives the amine might be replaced by a low-molecular-weight polyamide. Applications include marine coatings, electronic packaging, and adhesives.

This last is the two-part adhesive sold in hardware stores. One part is based on a solution of a diglycidyl ether derivative, and the other part is based on a solution of a polyamine or polyamide. On mixing, they react with a *pot life* of from 5 to 30 minutes, usually given with the instructions. The pot life is the length of time before gelation, after which it does not flow, and if applied too late may be weak. (See the TTT diagram, Figure 8.27.) The adhesive itself bonds to the two surfaces involved, as well as flowing into the (usually) irregular surface, creating mechanical interlocks.

14.2.3 Polyimides

Polyimides are a class of condensation polymers containing the basic group

$$-\overset{O}{\overset{\|}{C}}-\overset{R}{\overset{|}{N}}-\overset{O}{\overset{\|}{C}}-$$

along the backbone of the main chain. The quantity *R*, as usual, represents any organic group. Typical structures include (7):

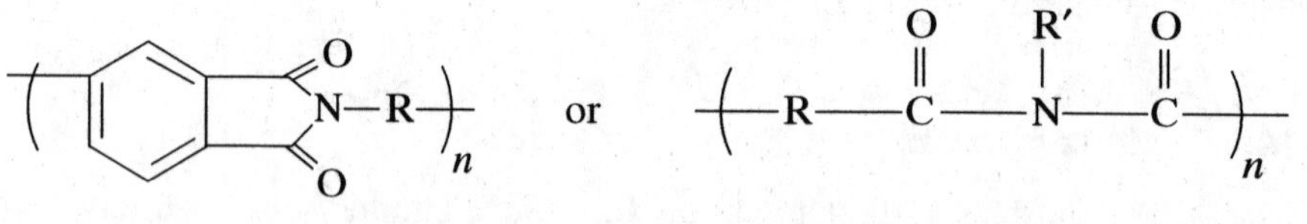

$$\left[\text{phthalimide ring}-N-R\right]_n \quad \text{or} \quad \left[R-\overset{O}{\overset{\|}{C}}-\overset{R'}{\overset{|}{N}}-\overset{O}{\overset{\|}{C}}\right]_n$$

The final product is cross-linked. Polyimides are widely used throughout the electrical, electronics, and aerospace industries as high temperature, tough polymers, see Section 11.6.2. Generally, they exhibit low smoke evolution and low flammability.

Table 14.4 Mechanical behavior of selected thermosets and thermoplastics

Composition	Density (g/cm^3)	Young's Modulus, GPa	Tensile Yield Strength, MPa	Heat Distortion Temperatures, °C
		Thermosetting compositions		
Urea/formaldehyde	1.56	10.0	43	—
Phenol/ formaldehyde	1.36	8.6	50	121
Epoxy resin	1.20	3.6	72	>110
Polyimide	1.40	5.0	72	>243
		Thermoplastic polymers		
Polystyrene	1.05	3.2	46	73
High-density polyethylene	0.96	1.1	32	49
Polyamide 6,6	1.10	2.9	65	75

Source: Based on H. G. Elias, *Macromolecules*, 2nd ed, Plenum, New York, 1984.

14.2.4 Mechanical Behavior

The thermoset resins are important because they have high moduli, and a relatively high range of useful temperatures. Table 14.4 compares the properties of selected thermosets with the corresponding properties of common thermoplastics. The thermosets have higher moduli because of their dense cross-linking. The heat distortion temperature, given as a certain deflection under a load of 1.85 MPa, is somewhat lower than the glass temperature or melting temperature, where they exist. The properties of the thermosets in Table 14.4 are only approximate, as these are representative of classes of the indicated materials.

In commercial use a common application of all of these materials is in the form of glass fiber reinforced polymer composites; see Sections 13.7 and 13.8. The glass fiber provides a significant toughening for these materials. For other applications, a variety of fillers are commonly used.

14.3 POLYMER AND POLYMER BLEND ASPECTS OF BREAD DOUGHS

14.3.1 Polymer Blend Aspects of Bread Making

Bread is made from water plasticized wheat flour, which in turn consists of ground-up wheat seeds. Dry flour consists of approximately 12% of protein, 87% of starch, and approximately 1% of everything else, such as minerals and salts. Since both the protein and the starch are polymeric, bread dough and bread making can be considered from the polymer blend point of view. As with most polymer blends, the starch and protein are immiscible, see Section 4.3 and Chapter 13.

For simplicity, the following discussion will consider bread making from just the recipe of wheat flour, water, and heat; only yeast is added. The yeast, of course, causes bubbles of carbon dioxide to form in the bread, hence the finished product is also a foam. Components such as rye flour, eggs, caraway seeds, and the like, will be considered absent, except where noted.

Wheat starch is composed of two components itself: amylose, a linear, amorphous polymer; and amylopectin, a branched, semicrystalline polymer (10). The protein is known as gluten, also composed of two polymers, gliadin, a low-molecular-weight, soluble polymer; and glutenin, a high-molecular-weight, cross-linked, elastic polymer primarily responsible for the viscoelastic properties of bread doughs.

14.3.2 Morphology Changes during Baking

The morphology of flour consists of a continuous gluten phase with a dispersed phase of starch granules. During the formation of bread dough from flour, water is added—about 34% to 37% water being present in fresh baked breads. During dough formation at room temperature, the water diffuses into the protein phase, the starch remaining essentially unchanged and dispersed. In Figure 14.4 (11) as viewed with an environmental scanning electron microscope (ESEM) the starch granules were large, plump, and round in shape. The hydrated gluten matrix appeared not to constrict the structure significantly. During baking, starch gelatinizes, meaning that it becomes plasticized with the water, between 52 and 66°C.

The gelatinization temperature appears related to the melting temperature of the amylopectin portion of the starch, which drops rapidly with increasing moisture content. Under normal atmospheric conditions most starches contain 10% to 17% moisture. In particular, starch with 11% water content melts at 65°C (12). Thus, on heating, the water is free to swell the starch after it melts. Much of the water then migrates from the minor protein phase into the major starch phase.

Figure 14.5 (11) shows the starch granules to be highly distorted in shape and variable in size, indicating extreme damage due to gelatinization; see Figure 14.5*b* (inset) arrows. The lines between the hydrated, gelatinized starch granules and the surrounding gluten matrix become indistinct, possibly indicating that amylose leached out of the granules and interacted with the gluten matrix. Thus a degree of dual phase continuity is brought about.

The protein portion of the bread also undergoes a transformation on heating. It forms a thermoset network via disulfide cross-linking involving a rearrangement of the cysteine–cysteine amino acids, similar to the chemistry of the hard-boiling of egg whites. This process has sometimes been called a denaturization. (See, e.g., Figure 9.2.)

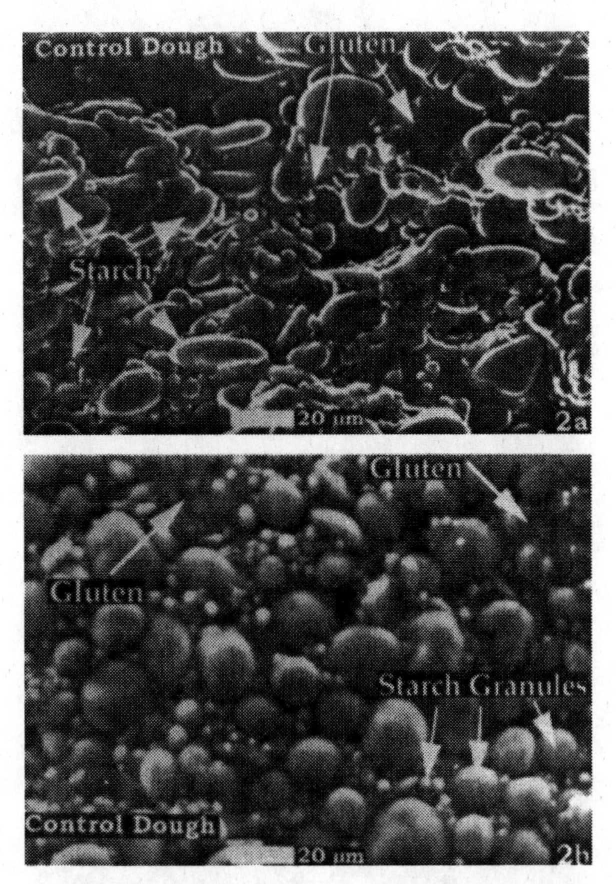

Figure 14.4 Scanning electron microscope studies of bread dough morphology. (*a*) Dried dough showing the starch granules embedded in the gluten matrix. (*b*) Environmental SEM (ESEM) of fresh dough showing hydrated starch granules and the thick gluten matrix that holds the dough together.

On cooling back to room temperature, the whole material is transformed from a viscoelastic dough to a soft solid as the amylopectin begins to recrystallize. Of course, the yeast-caused foam bubbles play a critical role in the final texture.

14.3.3 Mechanical Treatment (Kneading)

Kneading constitutes a form of shearing action. Kneading of the bread dough (13) greatly affects the structure–property relationships by establishing an equilibrium between the co-existing phases and orientation of polypeptide chains of the gluten phase. Thus basic mixing of doughs includes the deformation and breaking down of the liquid and gas-dispersed particles,

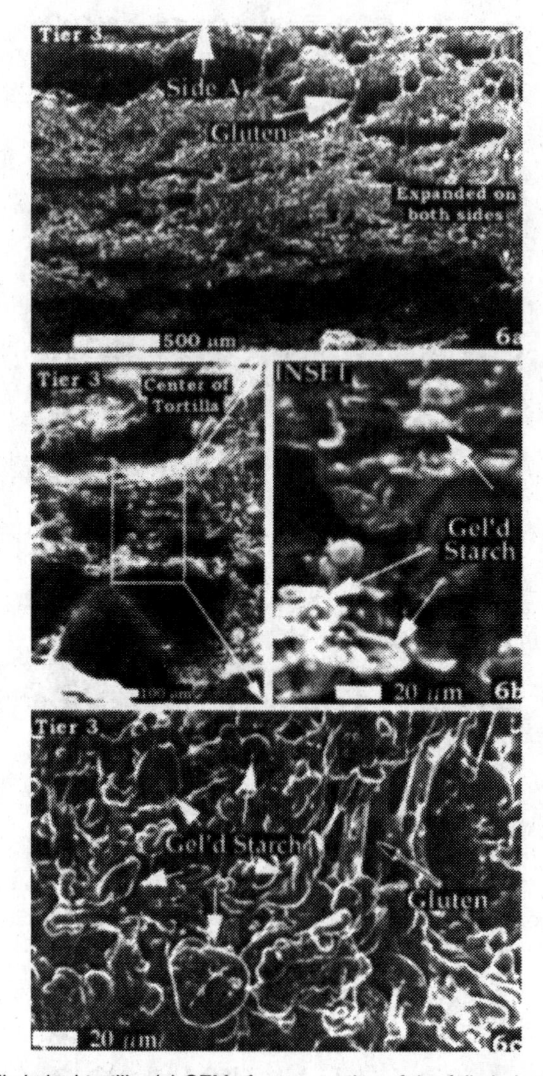

Figure 14.5 Fully baked tortilla. (*a*) SEM of cross section of the fully baked disk showing the considerable expansion of the crumb. (*b*) ESEM illustrating the gelatinization of the starch granules. (*c*) SEM of the crumb of a dried fully cooked tortilla showing the extensive gluten structure and gelatinized starch.

orientation of the gluten, decreasing the size of the liquid and gas-dispersed particles and the thickness of the gluten component, and twisting the starch granules in the shear flow, which provides a higher fluidity due to a "ball-bearing" effect. The starch granules migrate toward the higher-shear regions, providing a decrease in water content of the central layers and formation of substantially starch-empty surface layers of lower stickiness. Kneading also reduces the size of the gas bubbles.

14.3.4 Bread Staling

Increases in amylopectin crystallinity underlies part of the phenomenon known as bread staling (10). However, it has been recently shown that a rise in the glass transition temperature of the gluten, probably due to moisture migration, is more important (14–16).

The glass transition temperature of fresh baked crumb is just below room temperature (17), and the starch phase substantially amorphous, so the bread is soft. Thus, when bread is first baked, the combination of moisture migration and heat destroy much of the crystallinity in the amylopectin. However, on standing at room temperature, the starch undergoes retrogradation; that is, moisture leaves, and crystallinity sets in again. One reheating such breads, as in common experience, the bread may become soft again as the crystallites remelt. This assumes that the staling was not primarily due to moisture evaporation, which can also play a role, and may be substantially irreversible.

Thus large bodies of polymer theory, rheology of dispersed phases, polymer blend and polymer interface interactions, as well as excluded volume effects of polymer solutions are seen to be true for bread doughs, as indeed they must.

14.3.5 Bread and Meatlike Textures

The *texture* or feel of bread-type foods in the mouth arises through having the starch as a continuous phase. If the protein constitutes the continuous phase, the texture becomes meaty to the mouth. While oversimplified, this is the idea behind soy-bean-based texturized products, resembling bacon, chicken, ham, frankfurters, and the like (18).

Soy protein concentrates typically contain 70% to 72% crude protein. For example, bacon strips can be made by a texturization processes involving twin screw extrusion (19). The high-protein content yields the protein as the continuous phase. Note that soy beans, like most seeds, also contain some triglyceride oils, important in cooking the final product.

14.4 NATURAL PRODUCT POLYMERS

Nature produced the first polymers people used: animal skins, wool (see Section 6.10.2), cotton, and wood (see Section 6.10.2), among others. There are many more, some ancient, some the product of modern polymer science, and some still in the laboratory. Below are a few samples.

14.4.1 Silk Fiber Spinning

Nature produces many fine and useful polymeric materials. Important fibers include wool, cotton, and silk. Silk fiber is a natural product protein, produced by the silkworm, *Bombyx mori*, a species of moth, and other insect and arachnid species such as wasps, bees, butterflies, and spiders. The *Bombyx mori* silkworm is a caterpillar that primarily eats mulberry leaves. Silkworms construct the silk cocoons to protect themselves during metamorphosis. The fiber itself is a continuous double monofilament of 1000 to 1600 m in length (20).

Two proteins are synthesized in the silk glands of the silkworm, fibroin and sericin. The double-fibroin monofilaments constitute the main fibrous material. These are covered with sericin, an adhesive. In the solid state, the fibroin may have one of three chain conformations: a random coil, an α-form (silk fibroin I), and an antiparallel-chain pleated-sheet β-form (silk fibroin II). The α-form is a crankshaft pleated structure. The β-form is the important structure observed in nature. The fibroin consists primarily of glycine, alanine, and serine, with important parts of the crystalline regions being the 59-mer sequence Gly–Ala–Gly–Ala–Gly–Ser–Gly–Ala–Ala–Gly–{Ser–Gly–(Ala–Gly)$_2$}$_8$Tyr (21). Of course, while proteins are copolymers, the positioning of each mer is specific and determined genetically.

The silkworm has two glands, one on each side of the larva head. Each gland has three divisions: posterior, middle, and anterior, as well as spinneret region, Figure 14.6 (20). The filament flows from left to right in the figure. The motions

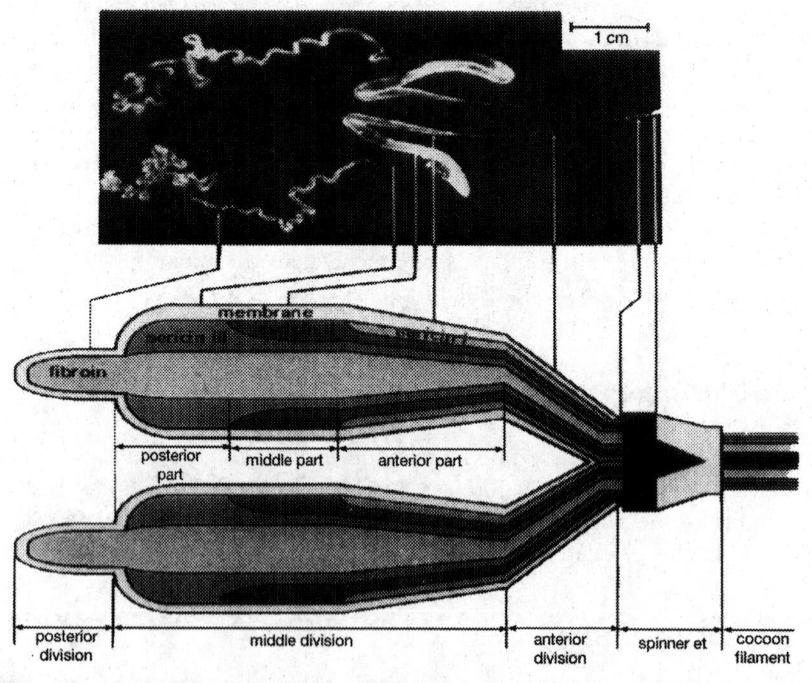

Figure 14.6 Photomicrograph (upper) and schematic diagram (lower) of the silk gland of the larvae of the silkworm, *Bombyx mori*. This is Mother Nature's original solution spinning process.

of the silkworm's head provides a significant drawing action for the orientation of the silk filaments.

The silkworm forms its fibers from the liquid crystalline nematic state (22, 23) (see Chapter 7), which is subsequently transformed into a gel state during spinning. The liquid crystalline state is indicated by the streaming birefringence observed in liquid silk solutions flowing out of the anterior division of the spinning gland. Because of the liquid crystalline organization of the silk solution, its viscosity is thought to be significantly lower than otherwise expected (24).

The fibers are lustrous and strong, with 0.3 to 0.4 GPa tensile strength and about 22% elongation (25); see Figure 14.7 (26). These stress–strain curves are in the same range as many synthetic fibers; see Table 11.2.

Research on silk fiber spinning continues, even intensifying on several bases:

1. Silk fibers remain one of the more important articles of commerce on the planet. The silkworm is being improved genetically, and more efficient production methods are under development,

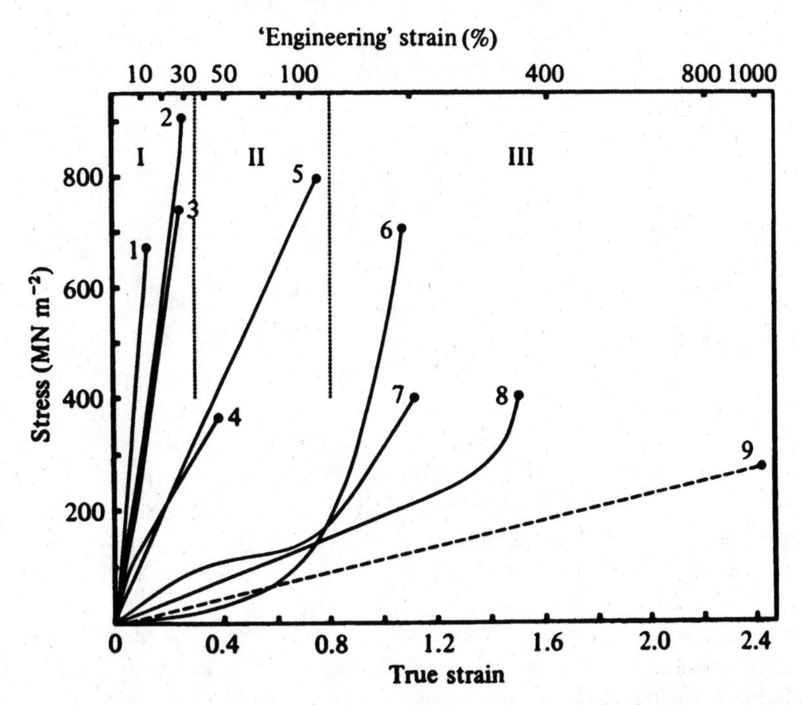

Figure 14.7 Silks from different sources exhibit a wide range of stress–strain properties. The examples shown are divided into three groups on the basis of their extensibility. (1) *Anaphe moloneyi*; (2) *Araneus sericatus* dragline; (3) *Bombyx mori* cocoon; (4) *A. diadematus* cocoon; (5) *Galleria mellonella* cocoon; (6) *A. sericatus* viscid. (7) *Apis mellifora* larval; (8) *Chrysopa carnea* egg stalk; (9) *Meta reticulata* viscid. Data collected from various sources.

2. Other sources of silk, particularly from spiders, continues to draw interest.
3. Modern polymer science is trying to understand and emulate Mother Nature in the liquid crystal manufacture of highly oriented fibers. The natural silk fibers are spun via aqueous solutions (ecologically *green* by today's standards); the resultant materials combine high strength, luster, and light weight.

14.4.2 Bacteria-Produced Polyesters

In 1982, Imperial Chemical Industries, Ltd. (ICI) in England began product development on a new type of thermoplastic polyester. The polymer was to be manufactured by a large-scale fermentation process not unlike the brewing of beer. In this case, it involved the production of the polyester inside of the cells of bacteria grown in high densities and containing as much as 90% of their dry weight as polymer (27).

The bacterium capable of performing this feat was *Wautersia eutropha*, and the commercial polyester was trade named *Biopol*®, now produced by Metabolix, Inc. (28). *Biopol*® is a copolyester containing randomly arranged mers of *R*-3-hydroxybutyrate, HB, and *R*-3-hydroxyvalerate, HV:

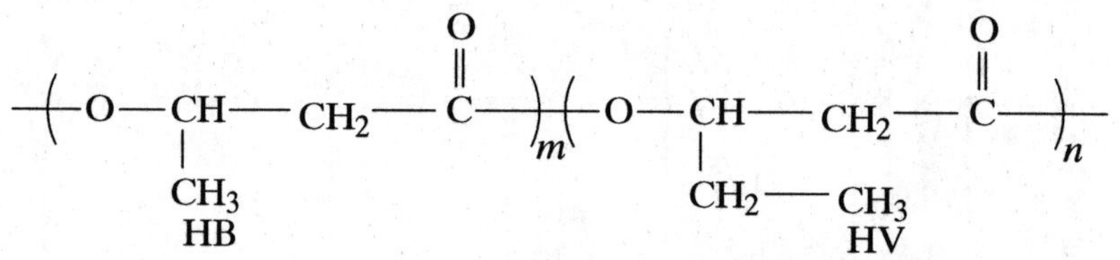

$$\left[O-\underset{\underset{\text{HB}}{CH_3}}{CH}-CH_2-\overset{O}{\overset{\|}{C}} \right]_m \left[O-\underset{\underset{\text{HV}}{CH_2-CH_3}}{CH}-CH_2-\overset{O}{\overset{\|}{C}} \right]_n$$

The natural polymer in many species of bacteria is poly(3-hydroxybutyrate). The polymer serves as a reserve source of food for the bacteria. It has, however, a melting point of 180°C, too high for commercial production because of thermal degradation. However, the above copolymer could be made by feeding the bacteria a mixture of glucose and propionic acid. The composition of the copolymer can be controlled by the diet fed to the bacteria, providing a convenient route to controlling the melting temperature for better processing.

In fact, it is now known that many species of bacteria produce a range of polyesters, some with fairly long side chains (29). These copolymers are all 100% biodegradable, since they are *natural products*.

The HB-HV copolymers can be melt processed into a wide variety of consumer products including plastics, films, and fibers (27). As a special type of fiber, the copolymer provides absorbable sutures for surgery.

14.5 DENDRITIC POLYMERS AND OTHER NOVEL POLYMERIC STRUCTURES

14.5.1 Aspects of Self-Assembly

Classically polymers may be linear, cross-linked, block or graft copolymers, and so on, as delineated throughout this text. However, there are a host of novel structures now synthesized, with some beginning to play important roles in polymer science and engineering. Many fall under the category called *self-assembling* polymers.

Many microorganisms, such as viruses, appear to have important aspects of self-assembly in their structure. Fyfe and Stoddart (30) point out that for the synthetic chemist to be able to build nanosystems, similar to those in the natural world, they must learn the intermolecular, noncovalent bond. This has lead to the field of supramolecular chemistry, as well as a host of self-assembling polymers. These include the dendritic polymers, polycantananes, polyrotaxanes, and others; see Figure 14.8.

14.5.1.1 Polyrotaxanes A polyrotaxane is a polymer chain with rings on it (30). The rings are unable to slip off because the ends are blocked with bulky stopper groups. Since the rings are not chemically attached to the polymer, the rings may be said to be self-assembled onto the polymer chain backbone. A closely related material is the polypseudorotaxane, defined as a molecular thread encircled by one or more macrorings. At least one of the thread's extremities must not possess a bulky stopper group.

There are several methods of synthesizing polyrotaxanes; see Figure 14.9 (31). Synthesis method *a*, by polymerization of a chain in the presence of macrocycles, constitutes one of the more important ways to make a polyrotaxane. For the polypseudorotaxanes, synthesis method *d* leads to an equilibrium between threaded and dethreaded rings. Synthesis method *e* leads to freezing in of the structure. Controlled dethreading may be useful for

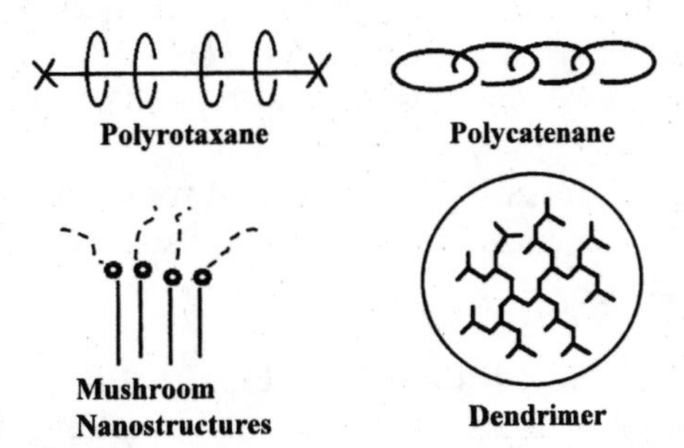

Figure 14.8 Novel polymer structures. These four structures are being examined for applications in various engineering and biomedical areas.

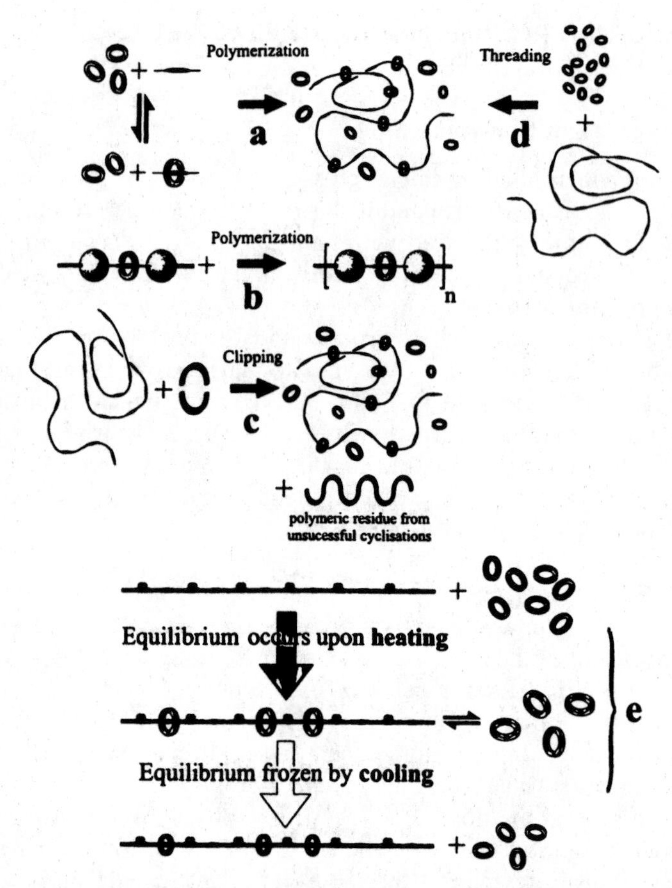

Figure 14.9 Schematic representations of the principal synthetic routes to polyrotaxanes, as illustrated in Figure 14.8. (*a*) Polymerization in the presence of a macrocycle. (*b*) Polymerization of a rotaxane. (*c*) Clipping a macrocycle onto a polymer chain. (*d*) Threading a macrocycle on to a polymer chain. (*e*) Slipping macrocycles onto a polymer by heating then *freezing onto* the polymer chain by cooling.

pharmaceutical drug delivery, for example. As a class, the polyrotaxanes and polypseudorotaxanes are also being considered as molecular switches and sensors.

The catenanes resemble a series of chain links. Each link is bound to its neighbors by physical forces only. Since each one of the links constitutes a separate molecule in the chemical sense, the links also bear some of the features of sell-assembly (32,33).

14.5.1.2 Mushroom Nanostructures Stupp et al. (34) have evolved strategies to create supramolecular units of various sizes and shapes via self-assembly. They found that rod–coil block copolymers can self-assemble into long striplike aggregates measuring 1 μm or more in length and a few nanometers in other dimensions. The mushroom nanostructures in Figure 14.8 constitute yet another example of a self-assembling nanostructure. These are triblock copolymers of polystyrene–*block*–polyisoprene–*block*–diphenylester, the latter exhibiting a rodlike structure. These materials have molecular weights in the range of 200,000 g/mol. They exhibit strong second harmonic generation nonlinear optical activity because of their lack of a center of inversion; see Section 14.8.

14.5.1.3 Dendrimers and Hyperbranched Polymers The term *dendrimer* derives from the Greek words *dendron*, meaning tree, and *meros*, meaning part. Dendritic macromolecules are hyperbranched, fractal-like structures emanating from a central core and containing a large number of terminal groups. There are two synthetic approaches, the divergent and the convergent growth approaches (35). In the divergent approach, the synthesis starts with the central core and works outward. In the convergent approach, slices of the "pie" are synthesized separately, and assembled by reaction with the core later. Both kinds of dendrimers are globular in nature, even to the point of mimicking some globular proteins. Most dendrimer mer structures have the organization AB_2, meaning that each new mer is attached to the dendrimer core at one location, and have two branch points.

During the synthesis of dendrimers, each successive reaction step leads to an additional *generation*, G, of branching (36). Sometimes the generations are numbered $G-1, G-2, G-3, \ldots$. Ideally dendrimers exhibit monodispersity. However, because the synthesis of higher generation materials requires numerous steps, the final products often contain defects. Materials containing up to 10 generations are known (37). Figure 14.10 (35) illustrates the structure of a fourth-generation convergent aromatic polyether dendrimer. With every new generation, the dendrimer doubles the number of terminal groups, and also approximately doubles its molecular weight. The mass of a dendrimer increases as the sum of 2^G, while the volume available for the mers only increases as G^3. Thus the local density increases with each generation beyond the fourth generation, leading de Gennes and Hervet (38) to predict that a maximum generation number exists, beyond which a perfect growth dendrimer cannot be made.

Besides spherical shapes, dendrimers can also be self-assembled into cylindrical shapes, using tapered monodendrons (39). These materials emulate the protein coats surrounding the nucleic acid in viruses.

In contrast to monodendrons, hyperbranched polymers are synthesized in a single-step reaction. As their name suggests, the products are highly branched and have varying degrees of irregularity.

As an example of the broad class of these new materials, the properties of the dendrimers will be briefly explored.

14.5.2 Intrinsic Viscosity of Dendrimers

Figure 14.11 (27) shows the intrinsic viscosity of several generations of the aromatic polyether dendrimers. The maximum in viscosity with generation is a

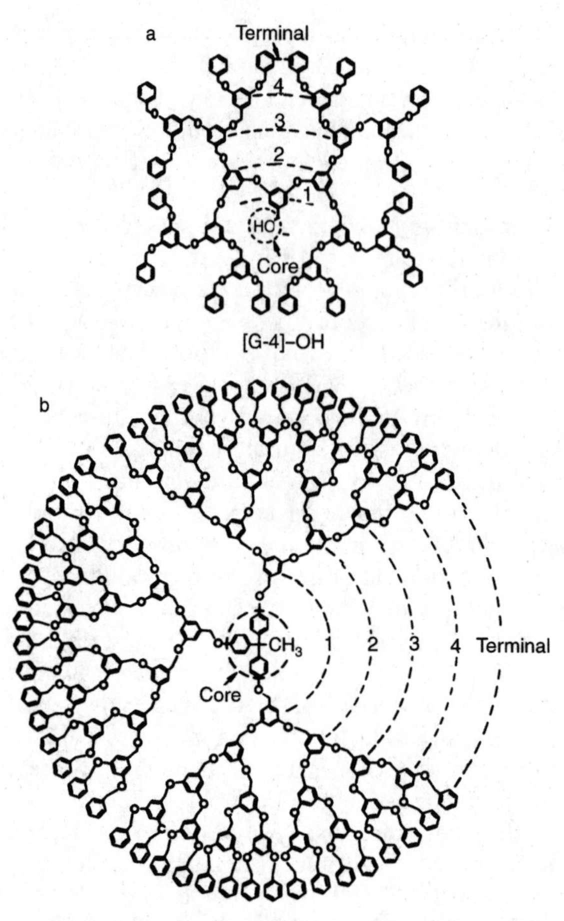

Figure 14.10 Dendrimer structures. (*a*) {*G*–4}–OH monodendron obtained by convergent growth. (*b*) $\{G\text{–}4\}_3$–{C} tridendron obtained by convergent growth.

fairly common observation (40,41) with dendrimers, but except for the liquid crystalline polymers (Chapter 7), rarely observed in the polymer world. However, the quantity of interest is the numerical value of the intrinsic viscosity. Note that a value of $\ln[\eta]$ of −3.7 corresponds to an intrinsic viscosity of 0.025 dL/g, the viscosity of Einstein spheres (see Section 3.8.2). These values are only slightly higher, and they might be expected to level off to the Einstein sphere value at about generation 10. In fact Aharoni et al. (42) found this theoretical value for their materials.

14.5.3 Electron Microscopy Studies of Dendrimers

Jackson et al. (37) investigated the transmission electron microscopy images of a series of generations of dendrimers based on a tetrafunctional core of

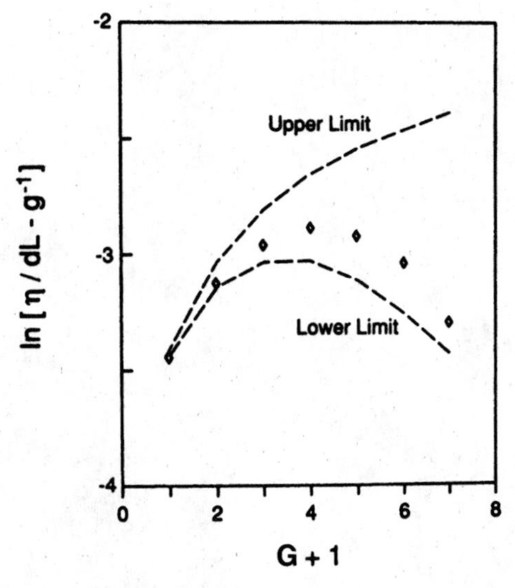

Figure 14.11 Intrinsic viscosity of convergent growth dendrimers. Dashed lines are upper and lower limits calculated from the dimensions of the mers.

ethylenediamine and successive additions of methyl acrylate and ethylenediamine, called PAMAM; see Figure 14.12. The tenth generation dendrimer, shown in Figure 14.12*a*, has a mean diameter of 14.7 nm, with a theoretical molecular weight of 934,721 g/mol.

14.5.4 Applications of Dendrimers

There are several applications, actual and proposed, for the dendrimers (43). These include:

1. Drug delivery vehicles. Characteristics include prolonged circulation in the blood, interactions with cell membranes, targeting with antibodies, encapsulation and solubilization of hydrophobic drugs, and so on. The dendritic polymers have two locations for placing drugs: First, the outer shell can be made with reactive species; these can be used to bond drugs. Second, some dendrimers have a lower than average density inside, with high density on the surface. Then, a dendrimer molecule can serve as a host for guest drug molecules, which slowly diffuse out. Similarly such a material can serve as a catalyst carrier, and so on. Working with novel dendrimers, Estfand and co-workers (44) are developing controlled delivery strategies to improve the aqueous solubility of hydrophobic guest molecules via the formation of inclusion complexes. The objective involves tailor-made dendritic hydrophobic cavities for specific drug compounds, resulting in improved and controlled drug delivery.

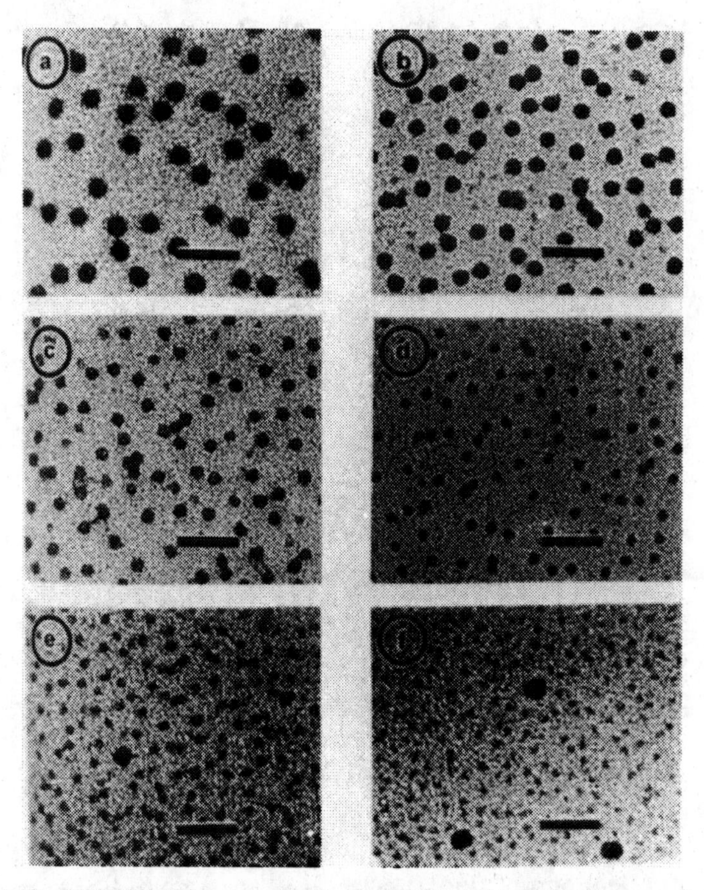

Figure 14.12 PAMAM dendrimers positively stained with a 2% aqueous sodium phosphotungstate imaged by conventional TEM. (*a*) *G*10, (*b*) *G*9, (*c*) *G*8, (*d*) *G*7, (*e*) *G*6, (*f*) *G*5. The scale bars are 50 nm. For *G*6 and *G*5, a small amount of *G*10 was added as a focusing aid.

2. Production of metal and semiconductor nanoparticles in polymer systems. Dendrimers are used for the synthesis of metallic nanoparticles *via* dissolution of a metal precursor in a supercritical fluid followed by reduction.
3. Dendrimer complexes used as ligands for selective binding of toxic metal ions. The system could potentially be used for remediation of contaminated water and soils.
4. Plastics toughened against fracture by increasing the critical energy release rate.
5. Ion-exchange chromatography. Use of specific surface chemistry coupled with the use of different dendrimer generations allows separation selectivity to be modulated.

14.5.5 Hyperbranched Polymer Characterization

Hyperbranched polymers are more simple to produce on a large scale than dendrimers. Generally, a one-pot synthesis is used, yielding fewer regular structures and very broad molecular weight distributions (45).

The average degree of branching, DB, of AB_2 hyperbranched polymers is given by

$$\mathrm{DB} = \frac{D+T}{D+L+T} \tag{14.1}$$

where D, L, and T represent the fractions of dendritic, linear, and terminal mers, respectively.

The maximum possible number of growth directions in AB_2 systems is given by

$$\mathrm{DB} = \frac{2D}{2D+L} \tag{14.2}$$

The DB values range from zero for the linear polymer to unity for the perfect dendrimer structure.

14.6 POLYMERS IN SUPERCRITICAL FLUIDS

14.6.1 General Properties of Supercritical Fluids

Classically pure substances are solid, liquid, or gaseous. A supercritical fluid is a pure substance compressed and heated above its critical point; see Figure 14.13 (46). The major characteristics of a supercritical fluid are liquidlike density, gaslike diffusivity and viscosity, and zero surface tension.

The solvent power of a supercritical fluid can usually be increased by increasing the pressure. The decaffination of coffee provides a simple commercial example (47,48). Here, supercritical carbon dioxide is contacted with ground coffee. The caffeine preferentially dissolves in the carbon dioxide, which is subsequently moved to another location. Then, when the pressure is lowered, the caffeine precipitates out, and the cycle repeated.

Table 14.5 shows some supercritical fluids used to dissolve polymers. Polymers can be synthesized in supercritical fluids in a manner similar to that in ordinary aqueous or organic fluids (49,50). Polymers in supercritical fluids exhibit the same solution properties as discussed in Chapter 3: The molecular weight can be determined by light-scattering or neutron scattering, for example. However, values of χ or A_2 can now be varied by altering the pressure as well as the temperature. For many systems, changing the pressure is far easier than changing the temperature (46).

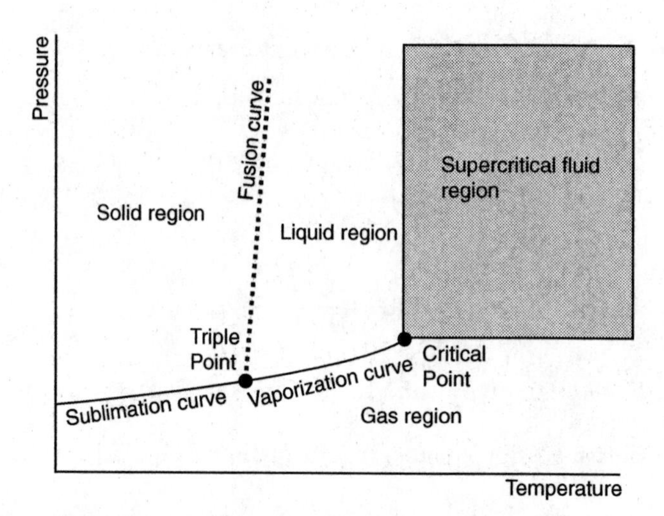

Figure 14.13 The pressure–temperature phase diagram of a pure substance, emphasizing the supercritical fluid region. The critical point is the highest pressure and temperature at which a pure substance can exist in a vapor–liquid equilibrium.

Table 14.5 Supercritical solvents for polymers

Solvent	T_c, C	P_c, Bar	Polymers Dissolved	Reference
Carbon dioxide	31.0	73.8	Poly(dimethyl siloxane), fluoropolymers	(a)
Propane	96.6	42.5	Polypropylene, EPDM	(b)
1,1,1,2-tetrafluoroethylene	101.1	40.6	Polystyrene	(c)

References: (a) E. Buhler, A. V. Dobrynin, J. M. DeSimone, and M. Rubinstein, *Macromolecules*, **31**, 7347 (1998). (b) S. J. Han, D. J. Lohse, M. Radosz, and L. H. Sperling, *Macromolecules*, **31**, 5407 (1998). (c) C. Kwag, C. W. Manke, and E. Gulari, *J. Polym. Sci.: Part B: Polym. Phys.*, **37**, 2771 (1999).

14.6.2 Dispersion Polymerization via Supercritical Fluids

Dispersion polymerization starts as a one-phase, homogeneous system such that both the monomer and the polymerization initiator are soluble in the polymerization medium (the supercritical fluid here), but the resulting polymer is not. The polymerization is initiated homogeneously with the resulting polymer, being insoluble, phase separating into primary particles. Once nucleated, these primary particles become stabilized by added amphipathic polymer molecules that prevent particle flocculation and aggregation. The particles, forming in the colloid size range, are usually stabilized by a *steric* mechanism as opposed to the electrostatic mechanism common in aqueous environments; see Section 4.5. The amphipathic polymer molecules impart steric stabilization because part of the chain becomes adsorbed onto the surface of the dispersed phase, becoming the anchoring segment, and part of the polymer chain projects into the continuous supercritical phase. This projecting moiety prevents flocculation by mutual excluded volume repulsion during Brownian collisions, thereby imparting stability to the colloidal particles.

In a carbon dioxide supercritical fluid, polystyrene can be polymerized by such a dispersion polymerization, using polystyrene–*block*–poly(1,1-dihydroperfluorooctyl acrylate), FOA, as the stabilizer. The perfluorinated moiety, soluble in supercritical carbon dioxide, provides the steric stabilization. Figure 14.14 (51) illustrates the particles of polymer produced in this way. With increasing molecular weight of the stabilizing block copolymer, the particles become smaller and more uniform.

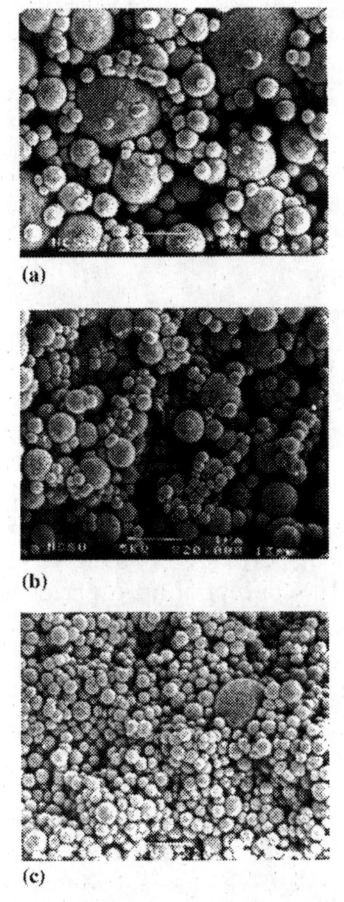

Figure 14.14 SEM of polystyrene particles synthesized via dispersion polymerization utilizing (*a*) 3.7 K/17 K, (*b*) 4.5 K/25 K, and (*c*) 6.6 K/35 K polystyrene–*block*–poly(FOA) stabilizer.

14.6.3 Applications of Polymers in Supercritical Fluids

There are several applications for this new technology:

1. Environmentally green solvents. Supercritical carbon dioxide presents an environmentally benign medium for polymerizations (and other chemical operations), minimizing pollution from organic solvents and facilitating the isolation of the polymeric product.
2. Assisting in blending by lowering the melt viscosity of the polymers (52).
3. Extraction of dyes and other molecules from water into liquid or supercritical carbon dioxide (53) using dendritic surfactants.
4. Serving as a polymerization medium.

14.7 ELECTRICAL BEHAVIOR OF POLYMERS

Most polymers are insulators. In fact, polyethylene and poly(vinyl chloride) are widely used as the insulating materials for electrical wiring, because they are highly insulating and weather resistant. However, to say a polymer is an insulator is a qualitative statement. To what extent does it conduct electricity? Recently, families of conducting polymers were invented, resulting in the 2000 Nobel Prize in Chemistry.

14.7.1 Basic Electrical Relationships

There are several basic relationships governing the electrical behavior of all materials. Ohm's law can be written

$$I = \frac{V}{R} \tag{14.3}$$

where I is the current in amperes (A), V is the voltage in volts, and R is the resistance in ohms (Ω). The concept of the capacitance, C, involves storage of electrical charge,

$$C = \frac{Q}{V} \tag{14.4}$$

where Q is the charge in coulombs (C) (54).

When an alternating voltage is applied to an imperfectly conducting material, a current flows that is displaced in time in such a way that it is out of phase with the voltage by an angle δ. The tangent of this angle plays a role similar to that described in Chapter 8 for mechanical behavior and is referred to as the dissipation factor. The power loss is given by

$$W_{\text{loss}} = 2\pi' f C_p (\tan\delta) V^2 \tag{14.5}$$

where W_{loss} is loss in watts, f is the frequency of the applied voltage, and C_p represents the parallel capacitance of the material. The relative dielectric constant (permittivity) is given by

$$\varepsilon' = \frac{C_p}{C_v} \tag{14.6}$$

where C_v is the capacitance of vacuum. The dielectric loss constant, ε'', sometimes called the loss factor, is given by

$$W_{\text{loss}} \cong \varepsilon' \tan\delta = \varepsilon'' \tag{14.7}$$

These quantities can be used to describe the onset of molecular motion at the glass transition temperature; see Section 8.3. In the present setting, we are interested in determining the electrical behavior of polymers.

Two other quantities of interest are the resistivity, the resistance per unit distance, with units of Ω/cm, and its inverse, conductivity, with units of $\Omega^{-1}/\text{cm}^{-1}$, or siemens per centimeter (S/cm).

14.7.2 Range of Polymer Electrical Behavior

The range of conductivities available for a range of polymers and other materials is given in Figure 14.15 (55). An insulator has conductivities in the range of 10^{-18} to 10^{-5} S/cm. A semiconductor is in the range of 10^{-7} to 10^{-3} S/cm, and a conductor is usually given as 10^{-3} to 10^{6} S/cm. As seen in Figure 14.15, polystyrene, polyethylene, nylon, and a host of other ordinary polymers are insulators.

On the other end of the scale are the conducting and semiconducting polymers. Several types exist. For example, conducting fillers may be added, such as short metallic fibers that touch each other or carbon black. Alternately, an ionic polymer may be employed, or a salt may be added to the polymer. Conductivity in the latter systems depends on the moisture content of the polymer.

14.7.3 Conducting Polymers

In the fall of 2000, Drs. A. J. Heeger, A. G. MacDiarmid, and H. Shirakawa (56–58) were awarded the Nobel Prize in Chemistry for the discovery and development of conducting polymers. These materials, based on doped polyacetylene and other conjugated polymers, are sometimes called *synthetic metals*. These materials are widely used as anti-static agents, shields for computer screens against unwanted electromagnetic radiation, and for "smart" windows that can exclude sunlight, light-emitting diodes, solar cells, and

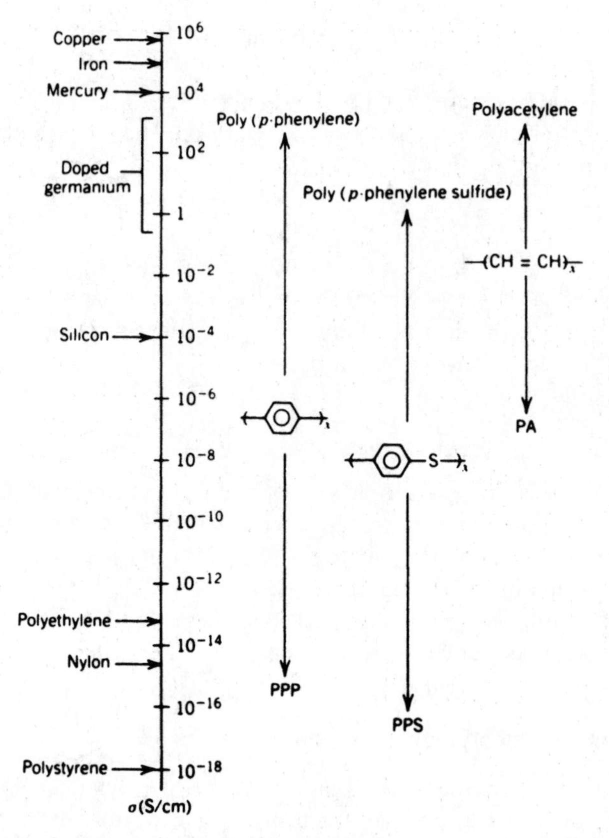

Figure 14.15 Conductivity scale comparing several polymers doped with AsF_5 with conventional materials (55).

electronic displays. Some doped alternating double bond systems are shown in Table 14.6 and Figure 14.15 (59). However, conjugated organic polymers in their pure state are still electrical insulators. Chemical dopants include oxidative materials such as AsF_5 and I_2. Some newer doping materials include SbF_5, $AlCl_3$, Br_2, O_2, and a host of others. The identities of the anionic counterions derived from these dopants show a variety of structures such as $Sb_2F_{11}^-$, AsF_6^-, and I_3^-.

The basic mechanism of electronic conduction is illustrated in Figure 14.16 (55). The importance of π-bonds in electronic conduction must be emphasized, as the overlapping electronic clouds contribute to the conduction. Strong interactions among the π-electrons of the conjugated backbone are indicative of a highly delocalized electronic structure and a large valence bandwidth. For good conductivity, the ionization potential, IP, must be small. The electron affinity, EA, reflects the ease of addition of an electron to the polymer, especially through polymer doping. The bandgap, E_g, correlates with the

Table 14.6 Structures and conductivity of doped conjugated polymers (59)

Polymer	Structure	Typical Methods of Doping	Typical Conductivity, S/cm
Polyacetylene	[structure]	Electrochemical, chemical (AsF_5, Li, K)	500–1.5×10^3
Polyphenylene	[structure]	Chemical (AsF_5, Li, K)	500
Poly(phenylene sulfide)	[structure]	Chemical (AsF_5)	1
Polypyrrole	[structure]	Electrochemical	600
Polythiophene	[structure]	Electrochemical	100
Poly(phenyl quinoline)	[structure]	Electrochemical, chemical (sodium naphthalide)	50

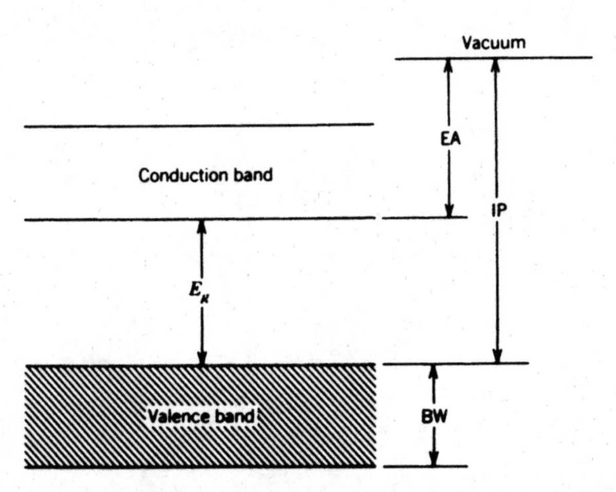

Figure 14.16 Polymer π-electron band structure. E_g represents the optical bandgap; *BW* is the bandwidth of the fully occupied valence band; *EA* is the electron affinity; and *IP* is the ionization potential (55).

optical-absorption threshold. The bandwidth, BW, correlates with carrier mobility; a large bandwidth suggests a high intrachain mobility, which favors high conductivity.

The mechanisms by which these polymers conduct electricity have been a source of controversy ever since conducting polymers were first discovered. At first, doping was assumed to remove electrons from the top of the valence band, a form of oxidation, or to add electrons to the bottom of the conduction band, a form of reduction. This model associates charge carriers with free spins, unpaired electrons. This results in theoretical calculations of conduction that are much too small (59). To account for spinless conductivity, the concept of transport via structural defects in the polymer chain was introduced. From a chemical viewpoint, defects of this nature include a radical cation for oxidation effects, or radical anion for the case of reduction. This is referred to as a polaron. Further oxidation or reduction results in the formation of a bipolaron. This can take place by the reaction of two polarons on the same chain to produce the bipolaron, a reaction calculated to be exothermic; see Figure 14.17 (55). In the bulk doped polymer, both intrachain and intrachain electronic transport are important.

14.8 POLYMERS FOR NONLINEAR OPTICS

The development of nonlinear optic, NLO, materials constitutes one of the newest and most exciting interdisciplinary branches of research, demanding cooperation among polymer scientists, chemists, physicists, and materials scientists. Nonlinear optics is basically concerned with the interaction of optical frequency electromagnetic fields with materials, resulting in the alteration of the phase, frequency, or other propagation characteristics of the incident light. Some of the more interesting embodiments of second-order NLO include

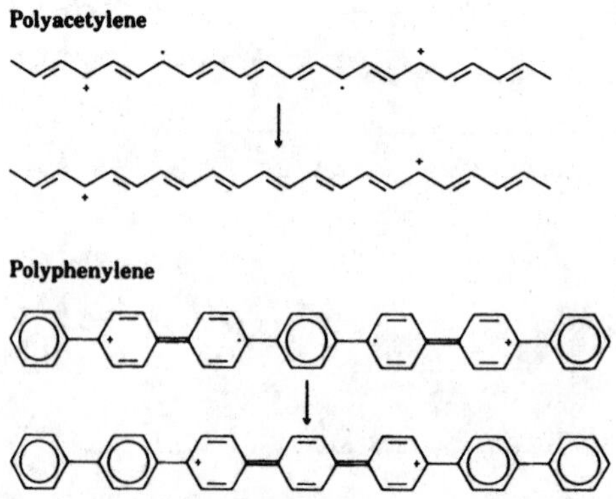

Polyacetylene

Polyphenylene

Figure 14.17 Structures of *trans*-polyacetylene and poly(*p*-phenylene) showing reactions of two polarons (radical cations) to produce a bipolaron (dication) (55).

second harmonic generation, involving the doubling of the frequency of the incoming light; frequency mixing, where the frequency of two beams is either added or subtracted; and electrooptic effects, involving rotation of polarization and frequency and amplitude modulation (60). Third-order NLO involves three photons, for similar effects.

Nonlinear optical effects have been explored in a range of materials—inorganic, organic, and polymeric. The basic structural requirement for second-order NLO activity is a lack of an inversion center, which is derived principally from asymmetric molecular configurations and/or poling. Poling, the application of an electric field of high voltage across the material, orients all or a portion of the molecule, usually in the direction of the field. Third-order NLO behavior does not require asymmetry. Active materials include inorganic crystals such as lithium niobate and gallium arsenide (61,62), organic crystals such as 2-methyl-4-nitroaniline, MNA (63,64), polymers such as polydiacetylenes (65), and a plethora of polymers with liquid crystal side chains (60,66–69). The organics and polymers with mesogenic side chains were found to be many times more active than lithium niobate (64), one of the best inorganic materials.

14.8.1 Theoretical Relationships

The starting point for nonlinear optics involves the constitutive relationship between the polarization induced in a molecule, P, and the electric field components, E, of the applied constant, low frequency, or optical fields (60). Ignoring magnetic dipoles and higher order multipoles, the induced polarization yields the approximation

$$P_i = \alpha_{ij} E_j + \beta_{ijk} E_j E_k + \gamma_{ijkl} E_j E_k E_l + \cdots \tag{14.8}$$

The quantities P and E are vectors, and α, β, and γ are tensors. The subscripts arise through the introduction of Cartesian coordinates.

A similar expression can be written for the polarization induced in an ensemble of molecules, whether in the gas, liquid, or solid state. For this case, the polarization P can be written

$$P_i = \chi_{ij}^{(1)} E_j + \chi_{ijk}^{(2)} E_j E_k + \chi_{ijkl} E_j E_k E_l + \cdots \tag{14.9}$$

where the coefficients $\chi_{ij}^{(1)}, \ldots,$ are tensors describing the polarization induced in the ensemble. The quantity $\chi_{ijk}^{(2)}$ will yield a zero contribution to nonlinear polarization if the system is centrosymmetric. For nonzero values, some type of asymmetry must be induced in the medium. For polymeric materials, asymmetry is generally induced by the generation of a polar axis in the medium through electric field poling.

Several of the effects occurring through $\chi_{ijk}^{(2)}$ are shown in Figure 14.18 (60). The combination of two photons of frequency ω produces a single photon at

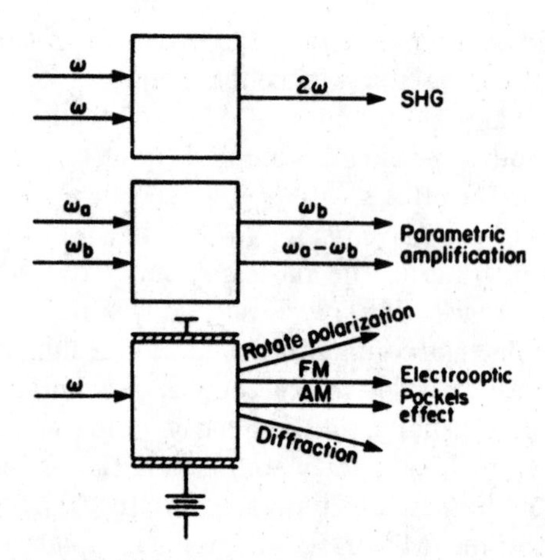

Figure 14.18 Schematic portrayal of some of the more exciting experiments that compose the field of nonlinear optics. Note the frequency doubling in second harmonic generation, *SHG* (60).

2ω in second harmonic generation. Basically, for second harmonic generation, two photons must arrive at nearly the same time at a point in a nonsymmetric sample capable of absorbing them. Then, the combined energy is released as one photon. The other effects in Figure 14.18 depend on other relationships. For example, the linear electrooptic effect follows from the interaction between an optical field and a direct current field in the nonlinear medium. This changes the propagation characteristics of the electromagnetic waves in the medium.

14.8.2 Experimental NLO Studies

When a polymer is subject to an intense sinusoidal electric field such as that due to an intense laser pulse, Fourier analysis of the polarization response can be shown to contain not only terms in the original frequency ω, but also terms in 2ω and 3ω (60). The intensity of the nonlinear response depends on the square of the intensity of the incident beam for 2ω, and the third power for 3ω. For the second-order effects, the system must have some asymmetry, as discussed previously. For poling, this means both high voltage and a chemical organization that will retain the resulting polarization for extended periods of time. Polymeric systems investigated have been of three basic types:

1. Molecules exhibiting large NLO effects are dissolved or dispersed in the polymer, which then merely acts as a carrier (69). An example is 2-methyl-4-nitroaniline dissolved or dispersed in vinylidene fluoride copolymers (70).
2. A segment of the polymer chain backbone is the active portion. This method is clearly the most efficient, taking into account the weight of

the system. If the polymer is poled above T_g, then cooled under poling, the orientation will have temporal longevity (71).

3. A special side-chain group can be attached to the polymer backbone. The advantage of side groups is their relatively easy orientation under electric fields. Some types of side groups can be bonded not only to the main chain but at a second point, providing cross-linking. This latter then holds the active portion of the chain in position for temporal longevity (72).

Typical responses for second-order effects are values in the range of $0.2–20 \times 10^{-9}$ esu. Williams (60) remarked that uniquely useful applications for organic and polymeric materials will require a $\chi^{(2)}$ of 10^{-7} esu or greater. In addition, a variety of additional application-dependent properties and attributes must be present. These include uniform birefringence, minimized scattering losses, transparency, stability, and processibility. The relaxation time of the poled structures must be increased significantly.

NLO materials based on lithium niobate, for example, are now in service for devices for second-harmonic generators, optical switches and routing components. One of the great advantages of electromagnetic waves instead of electrons in communications is the reduction in cross-talk. Effects in the picosecond time range have been investigated by Prasad and co-workers (73,74) and Garito (75). With better polymer NLO materials, perhaps they will compete with the inorganic materials.

14.9 LIGHT-EMITTING POLYMERS AND ELECTROACTIVE MATERIALS

14.9.1 Electroluminescence

Another recent discovery of the interaction of polymers with electricity involves polymers that emit light. Electroluminescence, EL, the generation of light by electrical excitation, is a phenomenon that has been seen in a wide range of semiconductors. It was first reported for anthracene singe crystals in the 1960s. The basic phenomenon requires the injection of electrons from one electrode and holes (i.e., the withdrawal of electrons) from the other, followed by the capture of the now oppositely charged carriers by recombination. This capture produces a radiative decay of the excited electron-hole state produced by this recombination process (76).

A display based on polymer light-emitting diode (LED) technology is illustrated in Figure 14.19 (76). Here, poly(p-phenylene vinylene), PPV, was the first such successful polymer discovered (77); it can be replaced in the diagram now by a significant number of other properly conjugated aromatic polymers. The indium-tin oxide, ITO, layer functions as a transparent electrode, allowing the light generated within the diode to leave the device. The layer of

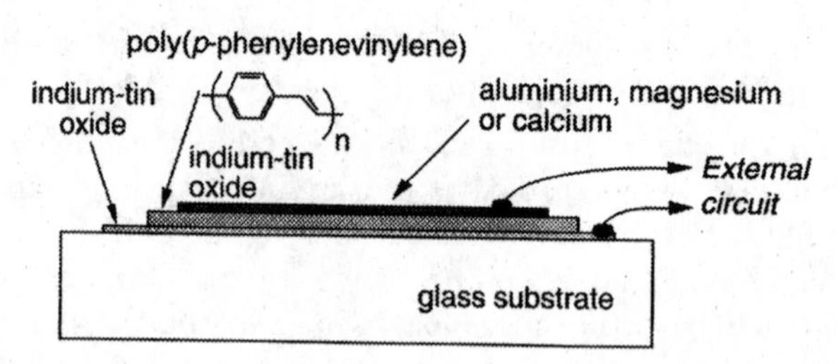

Figure 14.19 The device structure of a single-layer polymer electroluminescent diode.

polymer utilized typically is less than 100 nm thick, generated *via* spin-coating from solution.

Much of the interest in conjugated polymers has been in their properties as conducting materials, usually achieved at high levels of chemical doping; see Section 14.7. However, polyacetylene, the most widely studied of these materials, shows only very weak EL. Conjugated polymers with larger semiconductor gaps are required. PPV has an energy gap between its π and π^* states of about 2.5 eV, producing a yellow-green luminescence. Some recent structures exhibiting EL in the blue region are shown in Figure 14.20 (78).

The EL spectrum of another new polymer is shown in Figure 14.21 (79). Its fluorescence quantum efficiency was measured to be 63%, with peaks at 445 and 473 nm.

14.9.2 Electroactive Polymers

Recently, new polymers have emerged that respond to electrical stimulation with a significant shape or size change. This section will briefly describe several such effects.

14.9.2.1 Piezoelectric Polymers Ferroelectric polymers, because they are easily processed, inexpensive, lightweight, and conform to various shapes and surfaces, are of great interest. One such class of materials exhibiting piezoelectric behavior is poly(vinylidene fluoride), PVDF, and copolymers (80). These materials are semi-crystalline. With proper treatments a ferroelectric phase (the β-phase), which has an all *trans* conformation can be induced in the crystalline region of these polymers. Then a strong dipole arises vertical to the orientation of the chain, where (say) the fluorine groups are all pointing down, and the hydrogens are pointing up.

Typically, PVDF is produced in the form of thin films of approximately 1×10^{-2} mm thickness. A thin layer of nickel, copper, or aluminum is deposited on both surfaces of the film to provide electrical conductivity when an electrical field is applied, or to allow measurements of the charge induced by mechanical deformations (81).

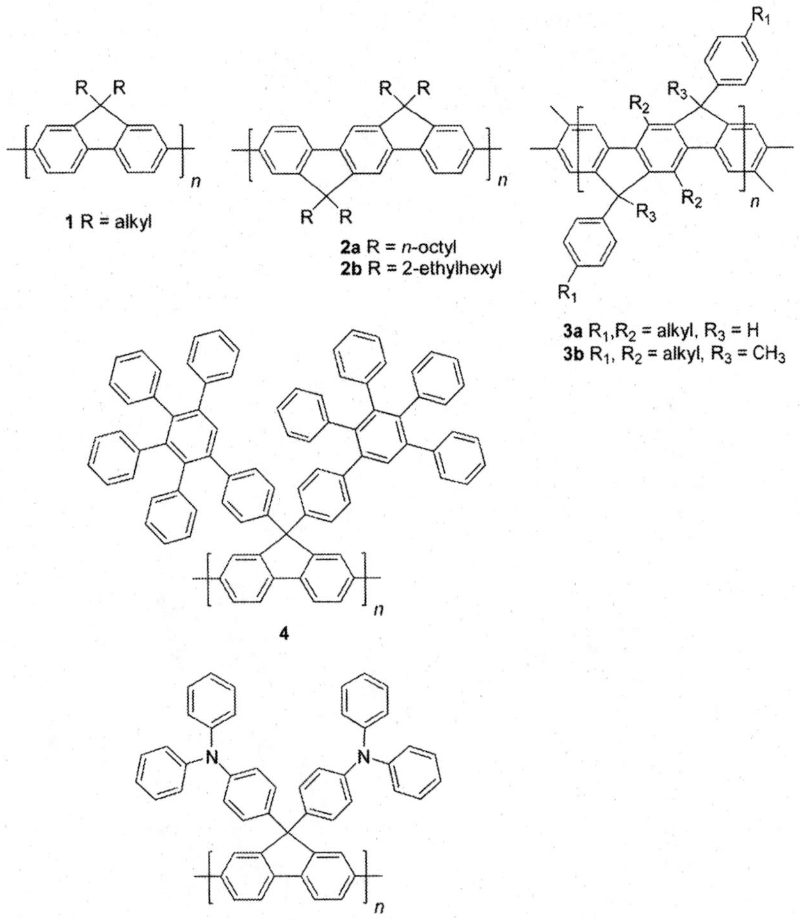

1 R = alkyl

2a R = *n*-octyl
2b R = 2-ethylhexyl

3a R_1,R_2 = alkyl, R_3 = H
3b R_1, R_2 = alkyl, R_3 = CH_3

4

5

Figure 14.20 Structures of typical blue-emitting conjugated polymers.

Figure 14.22 (80) illustrates the strain response of a PVDF copolymer. At room temperature under an electric field of 150 MV/m, the longitudinal strain (in the film direction) reached some 4% with low hysteresis.

Since the discovery of high piezoelectricity more than 30 years ago, there have been many improvements. These include preparation of copolymers with trifluoroethylene (TrFE) and chlorofluoroethylene. Recently, it was demonstrated that, by appropriate high-energy electron irradiation treatment, the piezoelectric P(VDF-TrFE) copolymer can be converted into an electrostrictive polymer with still greater sensitivity (82).

Polymers for electromechanical applications offer many unique and inherent advantages when compared with other materials, because they are lightweight, flexible, and relatively easy to process. Of course, these materials make

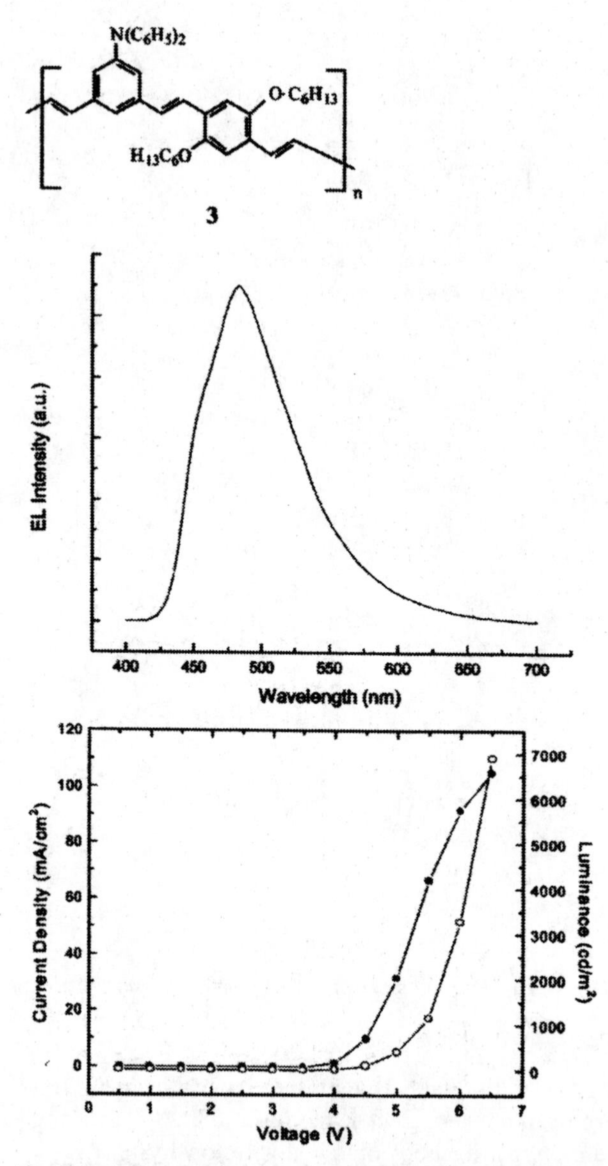

Figure 14.21 Electroluminescence chemistry and spectrum (top) and (bottom) current density (●) voltage-luminance (○) characteristics for indium-tin oxide/polymer 3 (above).

up the action part of the piezoelectric buttons widely used today to run copy machines, and many other modern devices.

14.9.2.2 Low Mass Muscle Actuators Here, the intent is to develop efficient miniature actuators that are light, compact, and driven by low power.

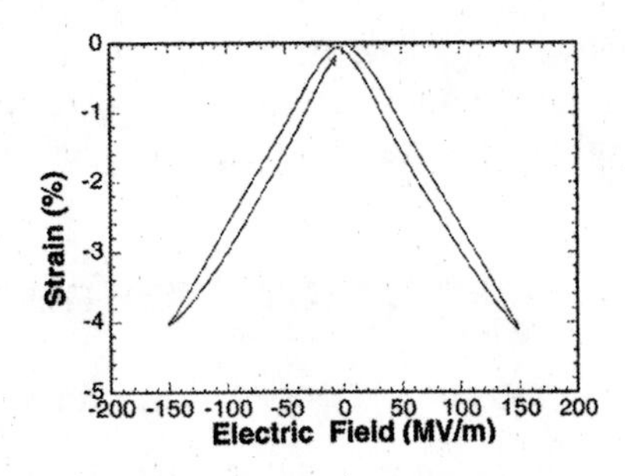

Figure 14.22 The strain-field dependence of poly(vinyl difluoride-*stat*-trifluoroethylene) 50/50 copolymer after irradiation with 4×10^5 Gy at 120°C.

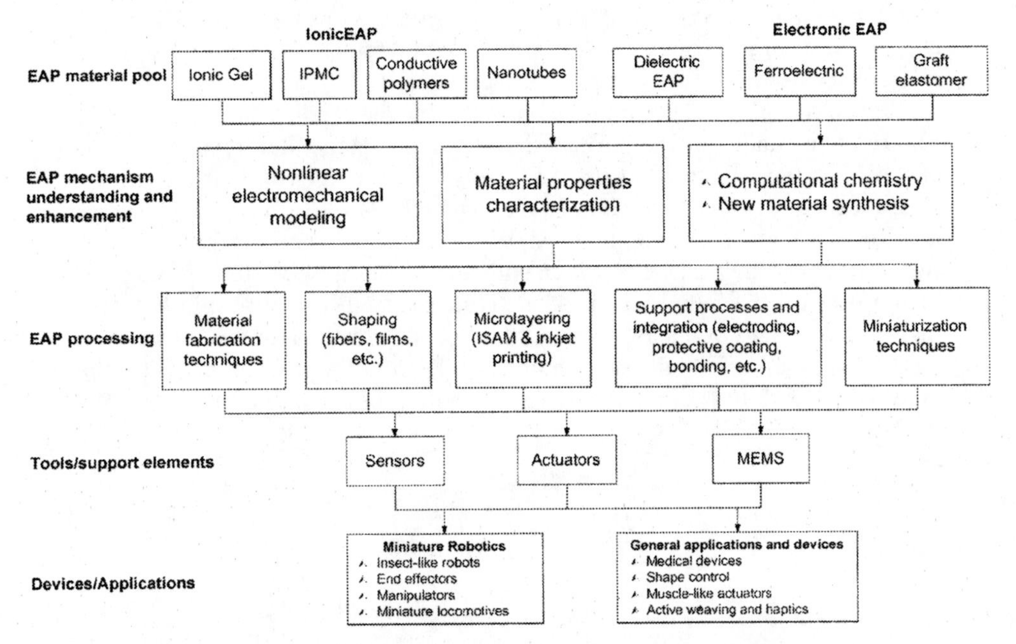

Figure 14.23 Electroactive polymers, EAP, infrastructure, and areas needing further research.

One polymer preparation being investigated is a perfluorinated ion-exchange membrane with chemically deposited platinum electrodes on both its sides (83). When an external voltage of 2 V or higher is applied on such a perfluorinated ion-exchange membrane platinum composite film, it bends toward the anode. When an alternating current is applied the film undergoes movement like a swing, the displacement level depends on the voltage and on the frequency. Applications include space travel and medical prosthetics.

Today, this is a rich area for research and development, encompassing many areas. Figure 14.23 (84) summarizes some of the ongoing work and possibilities. Electroactive polymers, EAP, have a complex positioning, with ionomeric polymer-metal composites, IPMC, and microelectromechanical systems, MEMS, and other materials and methods described.

14.10 OPTICAL TWEEZERS IN BIOPOLYMER RESEARCH

14.10.1 The Physics of Optical Tweezer Experiments

Optical tweezers, otherwise known as optical traps, laser tweezers, magnetic tweezers, or Raman tweezers, constitute something of a strange area for polymer science and engineering. The method was developed by physicists to study motions of organelles in living single cell material. Optical tweezers were first developed by Askin, et al. (85,86), who pointed out that dielectric particles in the size range of 10 μm down to about 25 nm could be stably trapped in aqueous dispersion by properly focused light.

Such particles, typically, are small polystyrene latex, silica, or glass spheres. The refractive index of the particle and the aqueous medium must be different. A laser beam is focused in such a way that it enters the particle with a shallower angle than when it leaves; see Figure 14.24 (86). Due to the changes in direction, with concomitant changes in momentum of the light as it is refracted twice, the particle is driven with a backward net trapping force component toward the initial beam focus. (Note Einstein's relationship between mass and any kind of energy.) This allows for a controlled motion of the particle. Of course, varying the beam intensity (87), and/or position provides the investigator with further control over the motion of the particle.

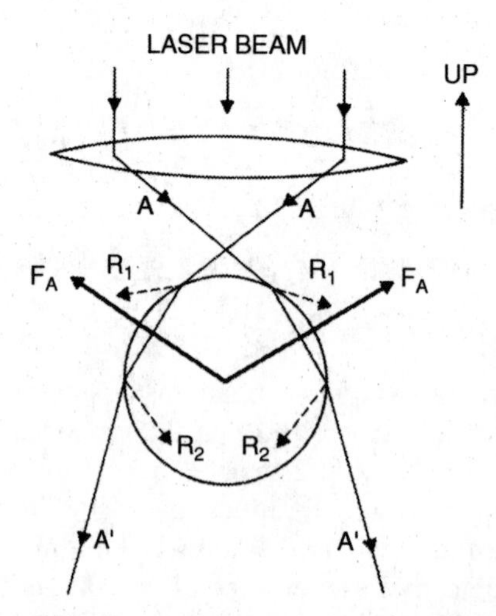

Figure 14.24 Diagram showing the ray optics of a spherical Mie particle trapped in water by a highly convergent light of a single-beam gradient force trap. (A Mie particle is one that is much larger than the wavelength of electromagnetic energy striking it.)

Figure 14.25 Schematic of colloidal microspheres coated with adsorbed polymer, not to scale.

14.10.2 Range of Investigations in Protein and DNA Research

Optical tweezers have proved useful for the study of lateral heterogeneity and order in cell structure and protein motion. For example, Bustamante, et al. (88) studied the single molecule mechanics of DNA uncoiling. The use of optical tweezers has proved especially useful for the study of lateral heterogeneity and order of lipid layers (89), and a host of problems relating to protein motions; see Section 14.11.4.

14.10.3 Studies on Synthetic Polymers: Bound Poly(ethylene oxide)

There has been some research using synthetic polymers, as well. Owen, et al. (90) employed optical tweezers and video microscopy. Here, poly(ethylene oxide), a water soluble polymer, was adsorbed onto 1.1 μm diameter silica microspheres in aqueous media, and the pair interaction potential was examined. This modeled the stabilization of colloidal matter by adsorbed polymer. Figure 14.25 (90) provides a schematic of the system. Four molecular weights of poly(ethylene oxide) were examined, ranging from 4.52×10^5 to 1.58×10^6 g/mol.

Defining a as the particle radius, and δa as the effective thickness of the adsorbed polymer layer, a linear fit of $\delta a = 2.1R_g$ was found, see Figure 14.26 (90). If the bound chains are assumed to be roughly spherical or perhaps mushroom in shape (see Figure 12.21), then δa corresponds to the effective diameter of the bound chains.

14.11 THE 3-D STRUCTURE AND FUNCTION OF BIOPOLYMERS

While most of the biopolymer investigations are properly part of biochemistry, molecular biology, and/or structural biology, molecules such as DNA, proteins, starch, and cellulose are also polymeric in that they are composed of one

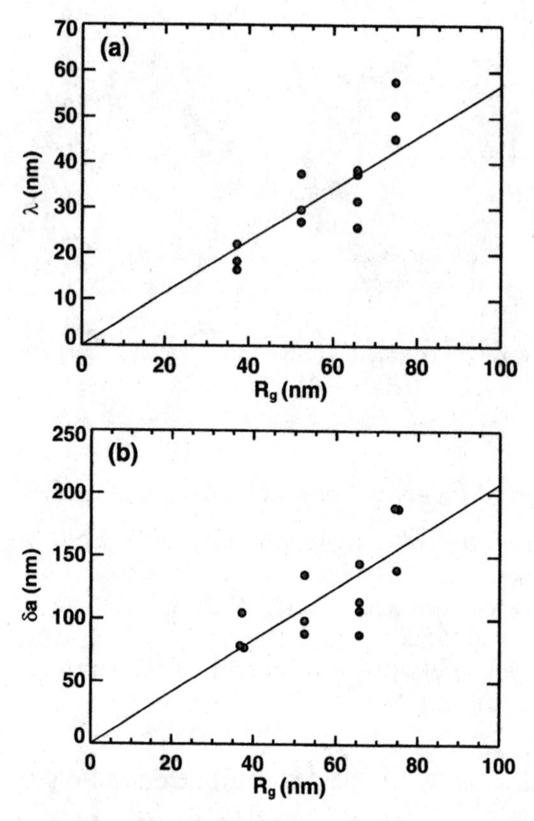

Figure 14.26 (a) Exponential decay length, λ, vs. R_g for four poly(ethylene oxide) samples. (b) The increase in apparent particle radius, δa, as a function of R_g. A linear fit gives $\delta a = 2.1R_g$.

or more repeating units in the form of chains. The subject of biopolymers deals with the polymeric structures of living organisms.

Cellulose and starch are frequently referred to as *renewable resources*, in that they are not derived from petrochemical resources. These two materials are discussed in other parts of this text; see Appendix 2.1, and Sections 6.10, 6.11, 14.3, and 14.4.

DNA and proteins are sometimes referred to as the polymers of life. They are more complex in nature and performance than cellulose and starch, and demand a separate examination. In the living cell, both DNA and proteins take very specific 3-D structures, and do not form either the random coils or the rod-shaped structures of synthetic polymers. The following subsections will describe a few characteristic structures and motions of these natural polymers.

14.11.1 The Polymeric Structure and Organization of DNA

DNA, deoxyribonucleic acid, and its associated molecules of RNA, ribonucleic acid, are molecules that hold the genetic information of each cell. The DNA strands store information, while the RNA molecules take the information from the DNA, transfer it to different locations in the cell for decoding or reading the information. Sections of DNA make up *genes*, each of which contains the information to make a specific protein, and a whole molecule makes up a *chromosome.*

On the molecular level, there are four bases, adenine, guanine, thymine, and cytosine. These four bases are frequently symbolized by the letters *A, G, T, and C.* In the closely related RNA, uracil (*U*) replaces thymine and ribose replaces deoxyribose. The chemical structures are delineated in Figure 14.27 (91).

The DNA bases pair such that *A* links only to *T* and *G* links only to *C.* The basic structure of the helix is modeled in Figure 14.28 (92). Specific groups of three bases, termed *codons*, provides the information for each of the 20 amino acids, see Section 14.2.

Morphologically, DNA takes the form of a double helix, with repeated sugar-phosphate backbone moieties on the outside, and paired bases making up the key steps in the middle, as shown further in Figure 14.28. Thus, DNA consists of a double backbone of phosphate and sugar molecules between which the complementary pairs of bases are connected by weak bonds. The most common conformation is a right-handed double helix of about 2 nm in diameter. One full turn of the helix takes about 3.5 nm, corresponding to about 10 to 10.5 base pairs (92).

The molecular weight of DNA is astronomically high. If it were possible to stretch out a single DNA molecule into one continuous thread, it would be an inch or two in length. It is said that if it were possible to assemble the DNA in a single human cell into one such continuous thread, it would be about a yard long (91)! Of course, the exact molecular weight depends on both the living source as well as the chromosome selected.

In 1962, J. D. Watson, F. H. C. Crick, and M. H. F. Wilkins were awarded the Nobel Prize in Medicine or Physiology for their analysis of the DNA structure and their model of the double helix structure of the paired bases. The double helix structure was deduced by X-ray analysis done by Wilkins and his co-worker, R. Franklin. Unfortunately, Franklin passed away before the Nobel prize was awarded, baring her from receiving it (93). The double helix model itself was deduced by Watson and Crick (94).

At the time of their discovery, Watson and Crick (94) wrote that the novel feature of the DNA structure was the manner in which the two chains were held together *via* purine and pyrimidine bases. They pointed out that they are joined together in pairs, a single base from one chain being hydrogen-bonded to a single base from the other chain.

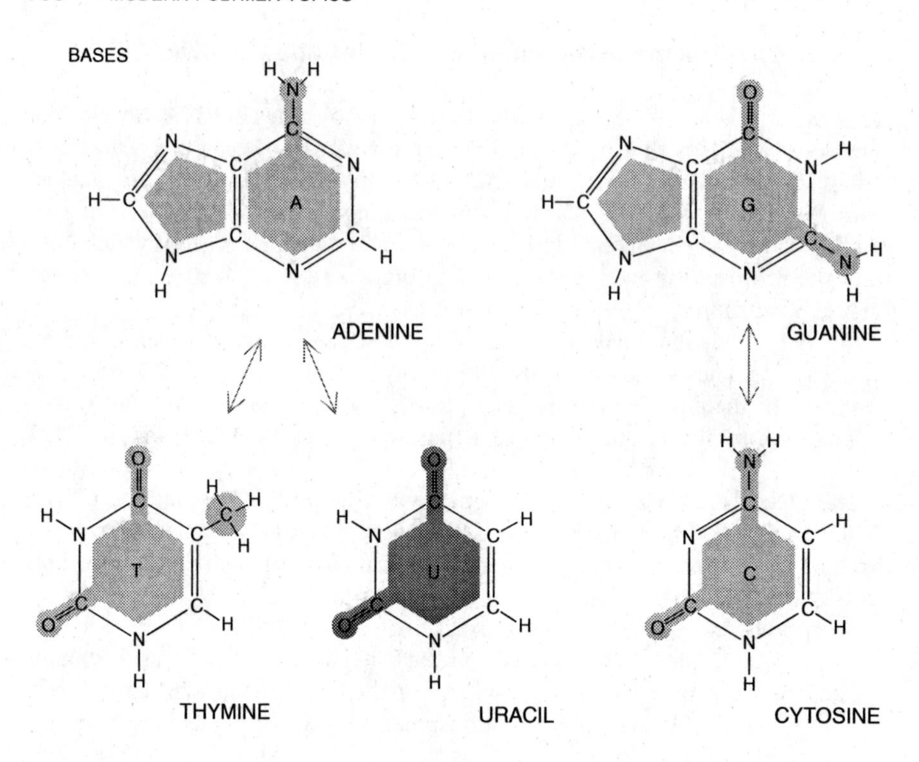

CHAIN COMPONENTS

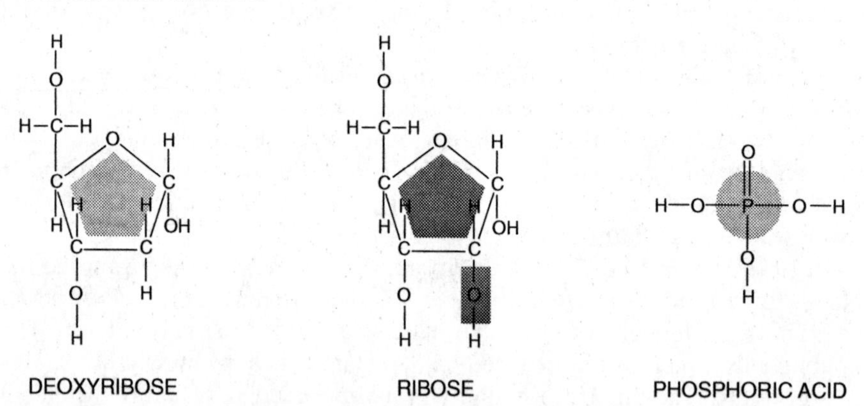

Figure 14.27 The components of DNA are four bases, adenine, guanine, thymine, and cytosine, symbolized by A, G, T, and C. Other components, deoxyribose and phosphoric acid, form chains to which bases attach. In closely related RNA, uracil, U, replaces thymine and ribose replaces deoxyribose.

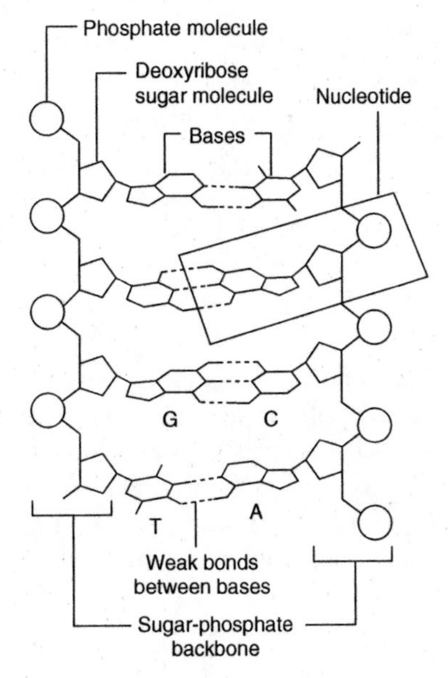

Figure 14.28 A model of the organization of the double helix.

Additionally, they pointed out that only specific pairs of bases bond together, adenine with thymine and guanine with cytosine. They then noted that if the sequence of bases on one chain is determined, then the sequence on the other chain is given. This led to one of the most famous statements in all science (94): "It has not escaped our notice that the specific pairing we have postulated immediately suggests a possible copying mechanism for the genetic material."

While the model itself is one of the more important deductions scientists have ever made, the last sentence quoted is truly remarkable.

While DNA contains the basic genetic code, the information is first transferred to *messenger* RNA, called mRNA. Each three base combination of the mRNA, in turn, codes for a specific *transfer* RNA, tRNA. The tRNA codes specifically for only one amino acid, and links it onto the growing amino acid chain, forming the final protein.

Before returning to the polymeric aspects of DNA and proteins, it is interesting to note that synthetic versions of DNA are being considered for nano-based applications (92). Central to nanotechnology are molecular-scale machines. Nanoscopic tweezers have already been made, as well as nano-computational devices. Rotating shaft devices are being considered.

14.11.2 The Amino Acids

There are 20 amino acids, all of which form proteins that are technically copolymers of polyamide 2. The basic structure of the mer is

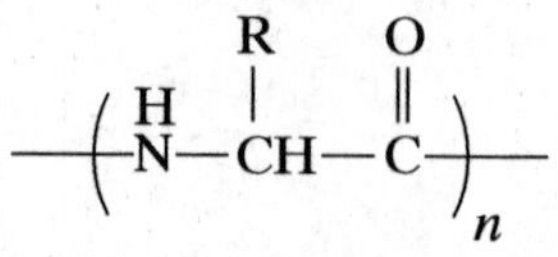

$$\left(\overset{\text{H}}{\text{N}} - \overset{\text{R}}{\underset{}{\text{CH}}} - \overset{\text{O}}{\overset{\|}{\text{C}}} \right)_n$$

The 20 side groups define the different amino acids, as shown in Figure 14.29 (95). Note that each one has both an abbreviation and a letter code attached to it. Both are widely used to describe sequences of amino acids. The letter code is especially space saving if many amino acids in sequence are described. It must be noted that all proteins are composed of all *L isomers* of these mers, except glycine, which has two hydrogens on the central carbon. The *D isomer* does not exist in nature.

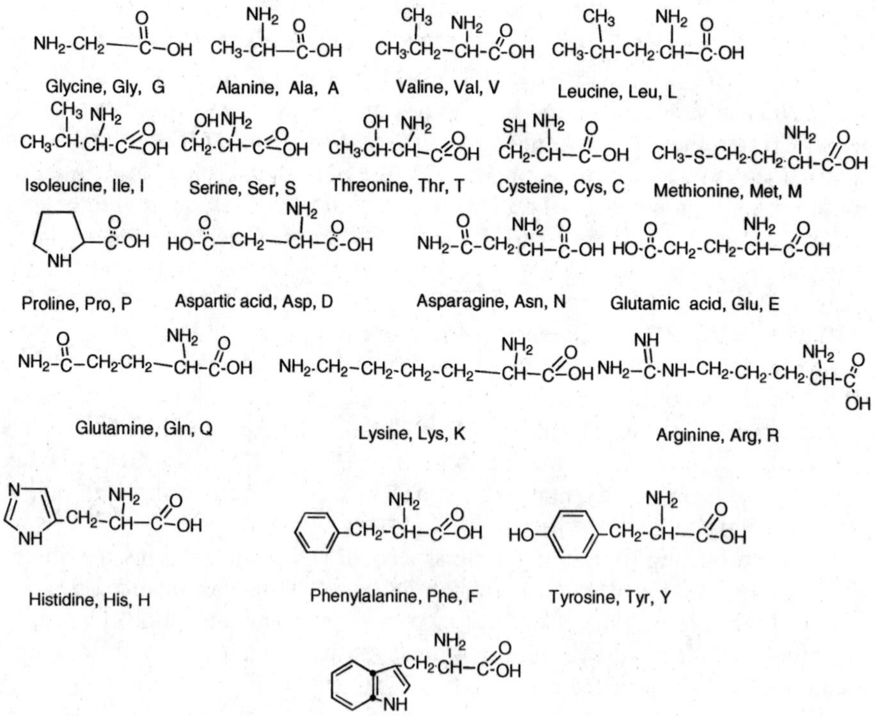

Glycine, Gly, G — NH_2-CH_2-C(=O)-OH

Alanine, Ala, A — CH_3-CH(NH_2)-C(=O)-OH

Valine, Val, V — CH_3CH_2-CH(CH_3)... -CH(NH_2)-C(=O)-OH

Leucine, Leu, L — CH_3-CH(CH_3)-CH_2-CH(NH_2)-C(=O)-OH

Isoleucine, Ile, I — CH_3-CH-CH(CH_3)(NH_2)-C(=O)-OH

Serine, Ser, S — CH_2(OH)-CH(NH_2)-C(=O)-OH

Threonine, Thr, T — CH_3-CH(OH)-CH(NH_2)-C(=O)-OH

Cysteine, Cys, C — CH_2(SH)-CH(NH_2)-C(=O)-OH

Methionine, Met, M — CH_3-S-CH_2-CH_2-CH(NH_2)-C(=O)-OH

Proline, Pro, P — (pyrrolidine ring, NH)-C(=O)-OH

Aspartic acid, Asp, D — HO-C(=O)-CH_2-CH(NH_2)-C(=O)-OH

Asparagine, Asn, N — NH_2-C(=O)-CH_2-CH(NH_2)-C(=O)-OH

Glutamic acid, Glu, E — HO-C(=O)-CH_2-CH_2-CH(NH_2)-C(=O)-OH

Glutamine, Gln, Q — NH_2-C(=O)-CH_2CH_2-CH(NH_2)-C(=O)-OH

Lysine, Lys, K — NH_2-CH_2-CH_2-CH_2-CH_2-CH(NH_2)-C(=O)-OH

Arginine, Arg, R — NH_2-C(=NH)-NH-CH_2-CH_2-CH_2-CH(NH_2)-C(=O)-OH

Histidine, His, H — (imidazole ring, N, NH)-CH_2-CH(NH_2)-C(=O)-OH

Phenylalanine, Phe, F — (benzene ring)-CH_2-CH(NH_2)-C(=O)-OH

Tyrosine, Tyr, Y — HO-(benzene ring)-CH_2-CH(NH_2)-C(=O)-OH

Tryptophan, Trp, W — (indole ring, NH)-CH_2-CH(NH_2)-C(=O)-OH

Figure 14.29 Models of the amino acid structures that make up proteins.

In determining structure and function of the whole protein as it exists in the aqueous body fluids, it is important to note that Ala, Val, Phe, Pro, Met, Ile, and Leu are hydrophobic in nature, Ser, Thr, Tyr, His, Cys, Asn, Gln, and Trp are polar, and hence hydrophilic, and Asp, Glu, Lys, and Arg are usually charged in body fluids, and thus very hydrophilic.

14.11.3 The Relationship of DNA to Protein Sequencing

As described above, DNA determines the structure of mRNA. The mRNA has 64 possible triplet codons. However, there are only 20 amino acids. The code is actually highly degenerate; most amino acid residues are designated by more than one triplet. Only Trp and Met are designated by single codons. Table 14.7 (95) provides the genetic code of protein biosynthesis. Note that three combinations indicate a termination of the translation and the carboxyl end of the protein chain.

In 1968, M. W. Nirenberg, H. G. Khorana, and R. W. Holley were awarded the Nobel Prize in Physiology or Medicine for their work on the genetic code. Working independently, Khorana had mastered the synthesis of nucleic acids, and Holley had discovered the exact chemical structure of tRNA.

Table 14.7 The genetic code of protein biosynthesis, mRNA*

First Position	Second Position				Third Position
	U	C	A	G	
U	Phe	Ser	Tyr	Cys	U
	Phe	Ser	Tyr	Cys	C
	Leu	Ser	Terminate	Terminate	A
	Leu	Ser	Terminate	Trp	G
C	Leu	Pro	His	Arg	U
	Leu	Pro	His	Arg	C
	Leu	Pro	Gln	Arg	A
	Leu	Pro	Gln	Arg	G
A	Ile	Thr	Asn	Ser	U
	Ile	Thr	Asn	Ser	C
	Ile	Thr	Lys	Arg	A
	Met	Thr	Lys	Arg	G
G	Val	Ala	Asp	Gly	U
	Val	Ala	Asp	Gly	C
	Val	Ala	Glu	Gly	A
	Val	Ala	Glu	Gly	G

*T. E. Creighton, *Proteins: Structures and Molecular Properties,* 2nd ed., W. H. Freeman & Co., 1993.

Nirenberg had deciphered the genetic code using synthetic RNA (96). Nirenberg and co-workers prepared a special RNA made up only of uracil, called poly-U, an added it to 20 test tubes containing the sap derived from *E. coli* bacteria and one of the amino acids. The experiment showed that a chain of the repeating bases uracil forced the synthesis of a protein chain composed of one repeating amino acid, phenylalanine. Thus, UUU = phenylalanine, breaking the code.

The biochemistry nomenclature for proteins differs significantly in places from the nomenclature of polymers. Some corresponding terms are given in Table 14.8.

14.11.4 The Mechanics and Thermodynamics of Protein Motions

14.11.4.1 Kinesin Motions While very large portions of protein activities in living systems are highly biochemical in nature such as the action of enzymes, another portion relates to physical activities such as muscle motion. For example, Kawaguchi and Ishiwata (97) used optical tweezers to examine the forces involved in motions of *kinesin*, a protein that transports membrane-bound vesicles and organelles in microtubules of various cells. *Kinesin* takes steps of 8 nm, such steps being associated with cycles of ATP hydrolysis. They found that a force of 14 pN was required via motion of a bead under the optical tweezers experiment. Since the hydrolysis of one ATP molecule generates 18.4 kJ/mol (98), equation (14.9a), this yields a total of 3.6 ATP molecules per completed motion of one molecule of the *kinesin*. (The reader should note that this calculation involves the enthalpy of hydrolysis of ATP to ADP, adenine diphosphate, not the free energy. The entropic portion is considered lost.) The authors point out the motion has multiple parts. While the results do not match a small whole number, they are at least of the same order of magnitude.

Table 14.8 Corresponding nomenclature for proteins and polymers

Protein	Polymer
Amino acid	Monomer
Peptide bond	Amide bond
Polypeptide	Polyamide
Protein	Polyamide, or small number of such chains bonded or connected in some way
Cystine	Cross-link site. (Two cysteine mers with a disulfide bond[a])

[a] See Figure 9.2 for an example.

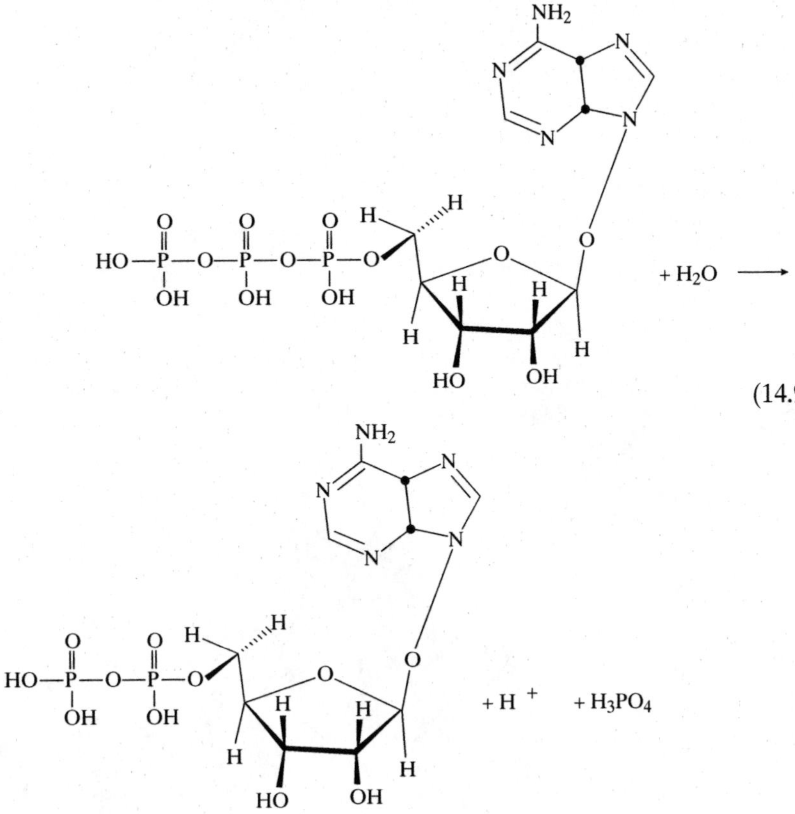

$$+ H_2O \longrightarrow \qquad (14.9a)$$

$$+ H^+ \quad + H_3PO_4$$

Structure of ATP

14.11.4.2 ***Actin and Myosin Motions in Muscle Tissue*** The relationship between the macroscopic structure of muscle tissue and its key proteins actin and myosin is illustrated in Figure 14.30 (99).

The myosin protein is a dimer consisting of two heavy chains and four light chains, forming a head, a neck, and a 140 nm long tail, with an approximate molecular weight of 223,000 g/mol, depending on source. Myosin actually is a super-family of protein molecules, of which there are 15 distinct classes. Myosin II is the most important type for muscle activity, and unless otherwise specified, is the one discussed herein.

The structure of the head and neck portion of myosin is illustrated in Figure 14.31 (100). In this ribbon diagram, the organization of the molecule is presented in 3-D. The head is at the top of the figure, while the long helix in the lower portion is the neck.

Considerable interest has focused on the mechanism of the power stroke of the myosin that propels the actin. Most models suggest that that the neck

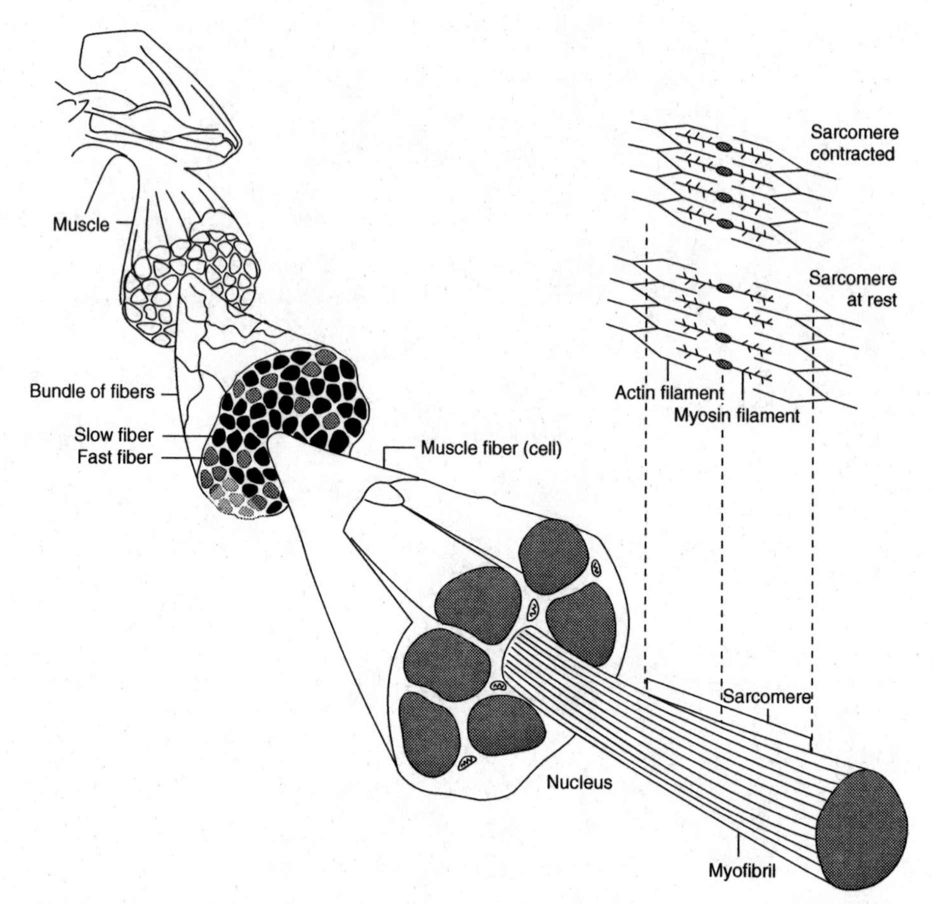

Figure 14.30 The structure of muscle at several levels of magnification, ending with the proteins actin and myosin.

region of the myosin molecule functions as a lever arm to amplify small conformational changes resulting in a working power stroke of about 10 nm in length (100).

Much research has been directed to determine the energetics of the myosin-actin interactions. Polystyrene latex beads were attached to *actin* filaments, and the unbinding force needed to rupture the bond to myosin was determined using optical tweezers in the absence of ATP (101). The average unbinding force was found to be 9.2 pN.

Huxley and Simmons (102) found the isometric force per bond between the head of myosin and actin to be 2.0 pN. While this differs from the kinesin studies (Section 14.11.4.1) somewhat, the proteins were different. There are about 1.5×10^{17} such bridges/m^2. Their calculation yielded 18.4 kJ/mol, corresponding to one ATP molecule split.

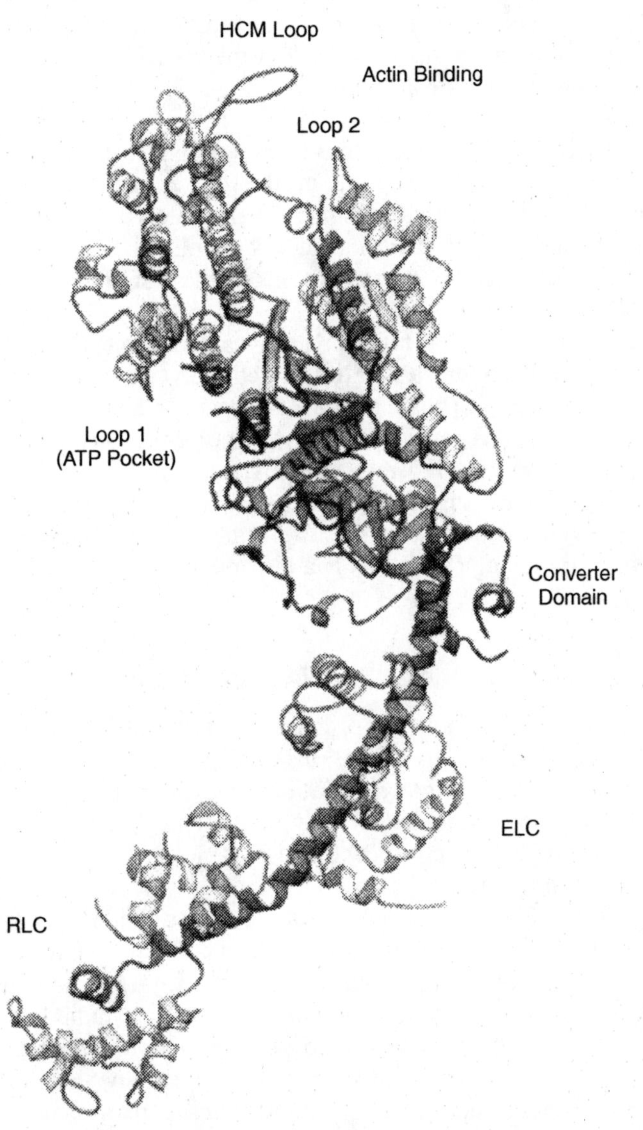

Figure 14.31 A ribbon diagram of the chicken skeletal muscle protein S-1. A ribbon diagram portrays the 3-D α-helix and other twists and turns, yielding insight as to how the protein moves to provide locomotion for the animal.

Kitamura, et al. (103) found single mechanical steps of 5.3 nm with two to five such steps in rapid succession for myosin. Only one ATP is consumed. Most interestingly, their results indicate that the myosin head may be able to store energy from ATP hydrolysis, and release it later in several packets of productive work.

Other myosins have also been investigated. For myosin V, for example, a popular model to explain the stepping motion of this protein is the lever arm model, which predicts that small changes in the catalytic head of the motor are amplified to large, directed displacements, enabling it to span a 36 nm step. Myosin VI is hypothesized to carry proteins to the leading edge of a migrating cell (104).

Currently, it is thought that myosin II moves during muscle contraction by a walking motion, including a rotation. Each step moves one molecule by about 8–10 nm, causing a translocation of an actin filament (105), and each such step requiring the energy of one ATP molecule. One problem is that all of the quantitative research involving myosin and actin has been *in vitro* rather than *in vivo*. Also, different sources of the proteins were used. Optical tweezers were used in many of the experiments to provide a controlled motion of individual molecules. This is a fast moving field.

14.11.5 Laboratory Synthesis of DNA and Polypeptides

Today, methods are available to prepare both synthetic DNA and polypeptides (98). Modern studies utilize the solid phase method introduced by Merrifield (106), including many variations and improvements. With these synthetic methods, it is possible not only to simulate Mother Nature, but to alter the structure of these materials, and also to make DNA and polypeptides that do not exist in nature.

For example, studies on the synthetic polypeptide poly(γ-benzyl-L-glutamate), PBLG, uses an amino acid that does not exist in nature. See equation (7.3) for its structure. This material assumes a rigid helical conformation in a variety of organic solvents. Tadmor, et al. (107) studied the gelation kinetics of isotropic PBLG solutions by SANS, and found evidence for a nucleation and growth mechanism with an Avrami exponent of about 2. This suggests a one-dimensional growth of fibrils. Such a growth may be slightly similar to that observed when toothpicks are placed on a water surface in increasing concentration.

On the biochemistry side, there are many more such studies. In one such, Tian, et al. (108) altered the structure of a protein known as the leucine zipper, which is thought to play an important role in coordinating the assembly of ion channels in cells. The ongoing objective of this and related studies is to discover exactly how these materials work on the molecular level.

14.12 FIRE RETARDANCY IN POLYMERS

14.12.1 Basic Concepts of Polymer Combustion

Plastics are made either from petroleum oil or from natural products. As such, when heated, they may burn. An important objective in modern polymer science relates to methods of reducing flammability and rate of fire spread.

Combustion in polymeric materials involves two steps: First, through a series of mostly thermal decomposition free radical chemical reactions (109), the solid polymer is reduced to monomer, dimer, trimer, and other components. These small molecules vaporize. Heat energy is absorbed in these steps from the surroundings.

Then, in the gaseous state, these small molecules react with oxygen, an exothermic step, producing carbon dioxide, water vapor, and other small molecules depending on the composition of the original polymer. Under fire conditions, part of the heat evolved is returned to the condensed phase polymer to continue the degrading process.

14.12.2 Current Methods of Retarding Fire in Polymers

One of the major objectives in fire retardancy emphasizes a reduction in the peak heat release rate. This, in turn, reduces the fire propagation rate.

The most important method utilizes halogenated chemicals. These compounds are inherently flame retardant because halogen radicals act as free radical scavengers, thus inhibiting combustion (110). Halogen-containing fire retardants may function in either the vapor phase and/or in the condensed phase (111). In halogen-containing organic compounds, the C-X bond will be the first to break, releasing a halogen radical and an organic radical. Reactions in the vapor phase are often the more effective. Halogen free radicals inhibit the radical-chain oxidation reactions that take place in the flame by reacting with the most active chain carriers, hydrogen and hydroxy radicals.

The most important halogen used is bromine. Some chlorine compounds are also used. The bromine and chlorine compounds used decompose in the temperature range of 150 to 350°C, well above ordinary temperatures, but below 500°C, the lowest decomposition temperature of the carbon-hydrogen bonds. Iodine compounds are not sufficiently stable to be generally useful, and fluorine compounds are too stable.

Other flame retardant materials in use include ammonium polyphosphate, organophosphates (112), bromoaromatic phosphates, red phosphorus, aluminum oxide trihydrate, antimony oxide, melamine, and synergistic combinations of two or more of these (110,111). Currently, people are focusing more on phosphorous containing compounds relative to halogens (113–115). The percentage of each of these materials in use today, based on dollar value, is given in Table 14.9.

Table 14.9 Percent of major families of fire retardants in use*

Composition	Percent of Total
Bromine compounds	39
Chlorine compounds	10
Inorganics	22
Melamine	6
Phosphorous compounds	23

*Based on 1997 dollar values. Derived from data in P. Georlette and J. Simons, in *Fire Retardancy of Polymeric Materials*, A. F. Grand and C. A. Wilkie, eds., Marcel Dekker, Inc., New York, 2000.

Another approach involves intumescence. On heating, intumescent systems give rise to a swollen multicellular char capable of protecting the underlying material from the action of the flame (116). A common compound utilized for producing this foamed state is ammonium polyphosphate.

14.12.3 The Limiting Oxygen Index

A classical method for evaluating the burning behavior of different polymers utilizes the limiting oxygen index (110), the minimum fraction of oxygen in an oxygen-nitrogen atmosphere that is just sufficient to maintain combustion (after ignition) of the material. A material must be considered flammable if the limiting oxygen index value is less than 0.26. This test tends to put polymers into three classes. Class I incorporates aliphatic, cycloaliphatic, and partially aliphatic polymers containing some aromatic groups. These burn with little or no char residue. Class II covers high temperature polymers. Structurally, they are characterized by the presence of either wholly aromatic or heterocyclic-aromatic mers. Higher temperatures are required for their decomposition. They also generate high char residues. The formation of char takes place at the expense of flammable products, and insulating effects of char also reduces flammability. Class III consists of halogen-containing polymeric materials, acting as described above.

Very importantly, there is a significant correlation between the char residue (at 850°C) and the oxygen index of the polymers, as illustrated in Figure 14.32 (117). Polymers such as polyethylene and polyisoprene leave little or no char residue. Note that carbon itself leaves a 100% char residue, important for the carbon nanotube discussion below.

14.12.4 The Nanocomposite Approach to Fire Retardancy

Newer methods are based on nanocomposites. Two materials of great interest include exfoliated montmorillonite-type clay and multi-walled carbon

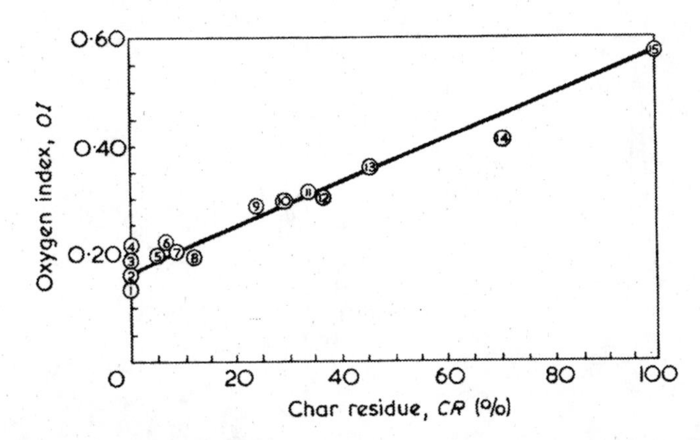

Figure 14.32 The flame resistance of polymeric materials, indicated by the oxygen index. 1, polyformaldehyde; 2, polyethylene, polypropylene; 3, polystyrene, polyisoprene; 4, polyamide; 5, cellulose; 6, poly(vinyl alcohol); 7, poly(ethylene terephthalate); 8, polyacrylonitrile; 9, poly(phenylene oxide); 10, polycarbonate; 11, aromatic nylon; 12, polysulfone; 13, Kynol®; 14, polyimide; 15, carbon. Polymers producing large values of char residue are more fire resistant.

nanotube nanocomposites. (See Section 13.8 for a description of these nanomaterials.) As will be seen below, both are effective as fire retardants, but they act by somewhat different mechanisms.

Two important instruments for evaluating fire retardancy include (118):

1. The cone calorimeter. This instrument measures ignition characteristics, heat release rate, and sample mass loss rate. For the experiments reported below, an external radiant heat flux of $50\,kW/m^2$ was applied.
2. A radiant gasification apparatus, somewhat similar to the cone calorimeter above, measures the mass loss rate and temperatures of the sample exposed to a fire-like heat flux in a nitrogen atmosphere. Thus, there is no burning.

Kashiwagi, et al. (10) used both the cone calorimeter and the radiant gasification apparatus to study the thermal and flammability properties of polypropylene/multi-walled carbon nanotubes, PP/MWNT. The samples themselves were compression molded at 190°C to make 75 mm diameter by 8 mm thick disks.

Figure 14.33 (118) compares the mass loss rate results of these two instruments using pure polypropylene, and polypropylene containing 0.5% and 1.0% MWNT. First of all, the mass loss rate when burning in the cone calorimeter is much reduced. Secondly, the no-flaming mass loss rate in nitrogen exhibits the same general characteristics of the actual burning experiment. This indicates that the observed flame retardant performance of the PP/MWNT

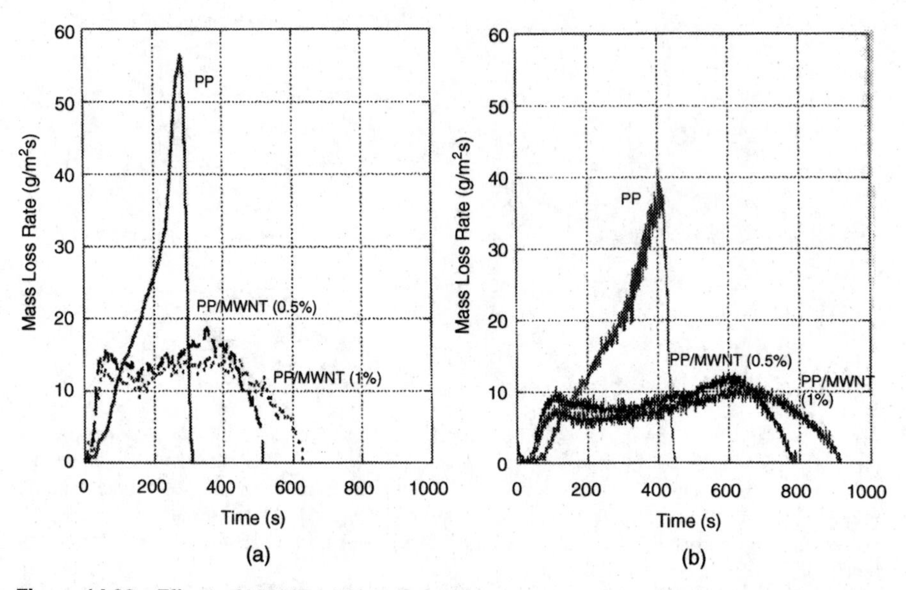

Figure 14.33 Effects of MWNT addition on mass loss rate of polypropylene at an external flux of 50 kW/m^2. (a) Burning in the cone calorimeter, and (b) no-flaming mass loss rate in nitrogen in the gasification device.

nanocomposites is mainly due to chemical and/or physical process in the condensed phase. While they burn slower, they burn nearly completely.

Separate experiments (118) showed that the radiant flux absorptivity at infrared wavelengths is much greater than for pure polypropylene, explaining the faster initial mass loss for the composites over that of pure polypropylene.

At longer times, however, as the nanocomposite sample degrades and its uppermost surface gradually regresses below the original surface position, it forms a MWNT network layer. Close to 50% of the incident heat flux is lost to open space by the emission from the hot nanotube surface layer, and the remainder of the flux is transferred to the nanotube network and material below. Therefore, the role of the nanotube network layer appears to be a radiation emitter from the surface, consequently acting as a radiation shield. Similar effects were found with poly(methyl methacrylate) containing single-walled carbon nanotubes (119).

How does ordinary carbon black compare with the MWNT as a fire retardant material? Separate experiments (118) using two different carbon black fillers showed that while the peak mass loss rate was reduced, the reduction was not nearly as great as that of the MWNT composite materials. This result indicates that the flame retardant effectiveness of the PP/MWNT nanocomposites is mainly due to the extended shape of the MWNTs.

Corresponding experiments using exfoliated montmorillonite clays, also dispersed in polypropylene, were carried out by Bartholmai and Schartel

(120). The polypropylene was grafted with a mass fraction of 0.6 wt-% maleic anhydride to increase bonding to the silica. Silica contents of 2.5, 5, 7.5, and 10 wt-% were used with three different kinds of modified clays.

The results are shown in Figure 14.34 (120). Note that superficially, the results are quite similar to those illustrated in Figure 14.33. However, the percent of clay used is higher, and the results get better as the concentration of clay is increased.

Bartholmai and Schartel (120) concluded that the most dominant mechanisms in reducing the heat release rate were: (1) an accumulation of layered silicate at the surface of the condensed phase working as a barrier and (2) an increase in viscosity and therefore reduced dripping behavior.

While neither the MWNT elongated cylinders nor the exfoliated clay sheets resulted in self-extinguishing behavior, they did slow the rate of fire spread by reducing the rate of heat evolution. Both of these nano materials form a kind of synthetic char, more effectively than larger particulate fillers, and act as heat shields. It may be that combinations of several types of materials will prove most effective in reducing fire spread.

As one would expect, fire resistance in polymers is also the subject of both technical standards (121) and government regulation. Organizations such as ASTM, UL, and others have developed a range of standards depending on the polymer and its application.

14.13 POLYMER SOLUTION-INDUCED DRAG REDUCTION

One of the more interesting discoveries in polymer solution research has been that extremely dilute solutions of very high molecular weight polymers exhib-

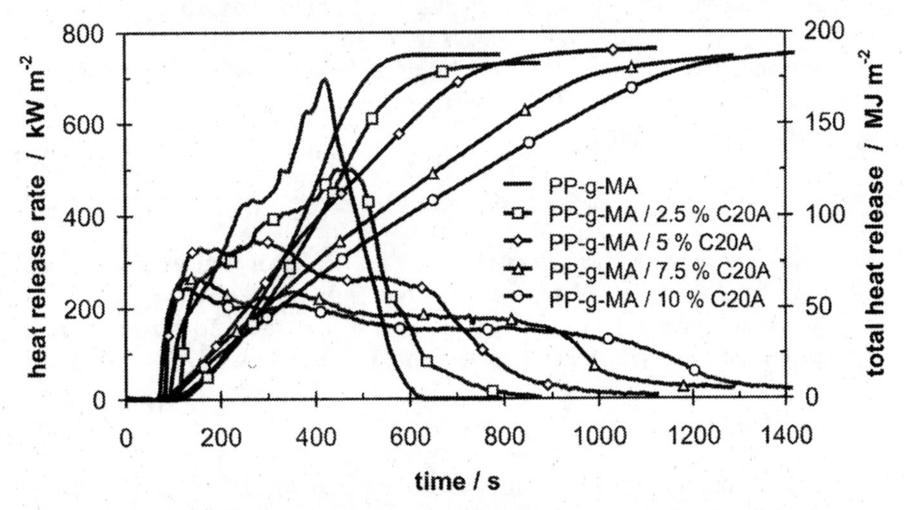

Figure 14.34 Heat release rate and total heat release for maleic anhydride bonded polypropylene with various concentrations of exfoliated montmorillonite added.

ited greatly reduced turbulence and cavitation during flow. This results in much reduced frictional resistance, with concomitant lower pumping power requirements and/or greater flow throughput (122).

14.13.1 The Physics of Drag Reduction

Polymers useful for drag reduction usually have molecular weights in excess of 1×10^6 g/mol. The polymers used are linear, and of course they have to be soluble in the flowing fluid.

While most authors agree that the phenomenon itself is incompletely understood, it seems to have its roots in the dimensions of the polymer chain relative to the size of incipient turbulent bursts. Such polymers in the relaxed state are random coils (see Section 5.3.2), but during flow they become highly extended. The highly extended chains possess significant elastic energy. At the onset of drag reduction, the duration of a turbulent burst is of the order of the terminal relaxation time of the polymer chains in question.

Although the radius of gyration of the chains is believed to be only 1×10^{-3} times that of the turbulent burst length scale, the viscosity of the solution in the immediate vicinity of a polymer chain is very significantly higher than the surroundings. Also, even in very dilute solutions, usually 10–100 ppm as used in drag reduction, there is some chain entanglement. These effects, qualitatively, tend to quell incipient turbulent bursts, preventing laminar flow from becoming turbulent, and reducing the inception and appearance of cavitation. (Note that cavitation itself is the formation of vapor bubbles in the flow, caused by low local pressures, and often produces erosion and noise.) People have suggested that an important part of the interference of the polymer chain with the turbulent bursting process occurs near the surface of the flowing fluid; in the case of pipes, it is near the pipe walls. The total effect is to keep the fluid in laminar flow, preventing the appearance of turbulence.

The basic concept of the energy required to cause flow is expressed by Reynold's number, R_e, the ratio of inertia forces divided by the viscous forces. In terms of flow in a pipe,

$$R_e = VD/\nu \tag{14.10}$$

where V is the average flow velocity, D the pipe diameter, and ν the kinematic viscosity of the fluid. Turbulent flow usually occurs when R_e exceeds 2300.

In terms of universal coordinates, the flow is illustrated for several conditions in Figure 14.35 (123). For conventional laminar flow,

$$u^+ = y^+ \tag{14.11}$$

where the quantity u^+ represents the normalized mean velocity in the x-direction (the flow direction), and y^+ represents the normalized distance in the y-direction (normal to the pipe direction), and

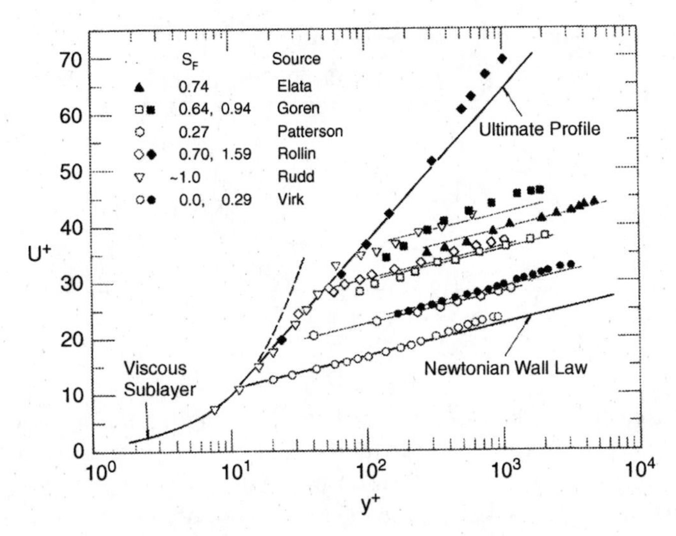

Figure 14.35 Mean velocity profiles during drag reduction. See Table 14.10 for details.

$$u^+ = 2.5 \ln y^+ + 5.5 \tag{14.12}$$

represents the solvent flow condition without polymer added, sometimes called the Newtonian wall law (see Section 10.5.5), and

$$u^+ = 11.7 \ln y^+ - 17.0 \tag{14.13}$$

arises from the maximum drag reduction asymptote, the ultimate profile. Note that the objective is to raise the value of u^+ for the same value of y^+. Turbulent friction can be reduced as much as 70 to 80%.

14.13.2 Polymers in Drag Reduction

Polymers used in drag reduction are of two broad types: Water soluble, and oil soluble. Some water soluble polymers include poly(ethylene oxide), PEO, polyacrylamide, PAM, partly hydrolyzed polyacrylamide, PAMH, and guar gum, GGM. Some oil soluble polymers include polyisobutylene, PIB, poly(methyl methacrylate), PMMA, and polydimethylsiloxane, PDMS. Table 14.10 (123) summarizes the molecular weight and concentration (in terms of weight parts per million, wppm) of the polymers illustrated in Figure 14.35 (123).

14.13.3 Applications of Drag Reducing Polymers

The most spectacular success of polymer drag reducers has been the use of oil-soluble polymers in the trans-Alaska pipeline (122). Flow through the pipeline has been greatly increased by appropriate polymer addition.

Table 14.10 Polymers used for turbulence control*

Polymer (Author)	Molecular Weight, g/mol	WPPM
GGM (Elata)	1.7×10^6	400
PEO (Goren)	5×10^6	2.5
PEO (Virk)	0.7×10^6	1000
PIB (Patterson)	2.3×10^6	2000
PAMH (Rollin)	3×10^6	100
PAMH (Rudd)	3×10^6	100

*Based on data in P. S. Virk, *AIChE J.*, **21**, 625 (1975), Table 5.

Other applications include storm-sewer flow augmentation, where added polymers can temporarily increase the flow capacity of storm sewers in times of heavy rain, etc. Another application has been for fire fighting. Water flow rates can be increased, smaller hose lines used, and nozzle pressures raised by the addition of polymer.

14.13.4 Current Status

As mentioned above, scientists and engineers still have only a qualitative understanding of the phenomenon of polymer-induced drag reduction. However, recent advances in computational power and numerical methods have made possible the simulation of turbulent viscoelastic flows from first principles, utilizing a numerical integration of the governing partial differential equations (124). Great attention has been devoted to improved theory. Besides pipe flow, significant research is being done on the Weissenberg effect (see Section 10.5.2), which is also troubled by turbulence.

14.14 MODERN ENGINEERING PLASTICS

One definition of an engineering plastic is one that has a high glass transition or a high melting temperature, for use at elevated temperatures. Other points of importance mention impact resistance, toughness, tensile strength, heat deflection temperature under load, creep, dimensional stability, and mold shrinkage.

Very few polymers are used commercially in the "pure" state. Some polyethylenes and polystyrenes are sold as homopolymers without any additives. However, several types of materials are added to most polymers to improve their properties:

1. Finely divided rubber is added to brittle plastics to toughen them.
2. Composites with glass, carbon, or boron fibers are made for high modulus and high strength.

3. Carbon black or silicas are added to rubber formulations to improve tear resistance and raise the modulus.
4. Various plasticizers may be added to lower the glass transition or reduce the amount of crystallinity. The main effect is to soften the final product.
5. Properties may be improved by adding silanes or other bonding agents to composites, to improve the bonding between polymer and other solid phases such as glass fibers.
6. Many types of polymers, both glassy and rubbery, are cross-linked to improve elastomer behavior or to control swelling.
7. Other items include fire retardants, colorants, and fillers to reduce price.

In each case, an understanding of how polymers behave and the effect of the various interactions allows the scientist and engineer to approach individual cases scientifically, rather than empirically.

14.15 MAJOR ADVANCES IN POLYMER SCIENCE AND ENGINEERING

While the use of natural polymers dates back to the dawn of civilization, polymer science developed rather recently. Tables 14.11, 14.12, and 14.13 delineate three distinct stages in the discovery, invention, and theory. The vulcanization of natural rubber and the synthesis of phenol-formaldehyde plastics

Table 14.11 Early polymer discoveries, inventions, and advances

Discovery	Reference
Vulcanization of rubber	C. Goodyear, U.S. 3,633 (1844).
Phenol–formaldehyde thermosets (phenolics)	L. H. Baekeland, U.S. 1,019,406 and U.S. 1,019,407 (1912)
Macromolecular Hypothesis	H. Staudinger, *Ber. Dtsch. Chem. Ges.,* **53**, 1074 (1920); summarized by H. Staudinger, *Die Hochmolekularen Organischen Verbindung*, Springer, Berlin, 1932.
X-ray confirmation of the Macromolecular Hypothesis	K. H. Meyer and H. Mark, *Ber. Dtsch. Chem. Ges.,* **61**, 593 (1928).
Synthesis of nylon (polyamide)	W. H. Carothers, U.S. 2,130,947 (1938).
Mark–Houwink–Sakurada equation	H. Mark, *Der Feste Koerper*, Kirzel, Leipzig, 1938.
Glass transition free volume theory	R. F. Boyer and R. S. Spencer, *J. Appl. Phys.*, **15**, 398 (1944).
Rubber elasticity theory	E. Guth and H. Mark, *Monatschefte Chemie*, **65**, 93 (1934); H. M. James and E. Guth, *J. Chem. Phys.,* **15**, 669 (1947).

Table 14.12 Development of classical polymer science and engineering

Discovery	Reference
Random-coil and polymer solution thermodynamics	P. J. Flory, *J. Chem. Phys.,* **10**, 51 (1942); summarized by P. J. Flory, *Principles of Polymer Chemistry*, Cornell University Press, Ithaca, NY, 1953.
Smith–Ewart theory of emulsion polymerization	W. V. Smith and R. W. Ewart, *J. Chem. Phys.*, **16**, 592 (1948).
High-impact plastics	J. L. Amos, J. L. McCurdy, and O. McIntire, U.S. 2,694,692 (1954).
Folded-chain single crystals	A. Keller, *Philos. Mag.*, **2**, 1171 (1957).
WLF equation and viscoelasticity	M. L. Williams, R. F. Landel, and J. D. Ferry, *J. Am. Chem. Soc.,* **77**, 3701 (1955).
Stereospecific polymerization	K. Ziegler, *Angew. Chem.*, **64**, 323 (1952); **67**, 541 (1955). G. Natta and P. Corradini, *J. Polym. Sci.*, **39**, 29 (1959).

Table 14.13 Modern era of polymer science and engineering

Discovery	Reference
Liquid crystalline polymers	S. L. Kwolek, P. W. Morgan, and J. R. Schaefgen, *Encyclopedia of Polymer Science and Engineering,* Vol. 9, Wiley, New York, 1987.
Polymer chain reptation	P. G. de Gennes, *J. Chem. Phys.*, **55**, 572 (1971).
Equation of state theories of polymer blends	P. J. Flory, *J. Am. Chem. Soc.,* **87**, 1833 (1965); I. C. Sanchez, Chap. 3 in *Polymer Blends*, Vol. 1, D. R. Paul and S. Newman, eds., Academic Press, Orlando, FL, 1978.
Polymer interface science	P. G. de Gennes, *Adv. Colloid Interface Sci.,* **27**, 1898 (1987).
Biopolymers	J. D. Watson and F. H. Crick, *Nature*, **171**, 737 (1953); J. R. Sellers, *Myosins*, 2nd ed., Oxford University Press, Oxford, England, 1999.
Nanotechnology	S. Iijima and T. Ichihashi, *Nature*, **363**, 603 (1993); T. D. Fornes and D. R. Paul, *Polymer*, **44**, 4993 (2003).
Conducting, semiconducting, and LED polymers	H. Shirakawa, E. J. Louis, A. G. MacDiarmid, C. K. Chaing, and A. J. Heeger, *J. Chem. Soc. Chem. Comm.*, 579 (1977); L. Edman, M. A. Summers, S. K. Buratto, and A. J. Heeger, *Phys. Rev.*, **B, 70**, Art. No. 115212 (2004).

(Table 14.11) were marked by trial and error, the basic idea of a polymer molecule not yet existing. Staudinger's Macromolecular Hypothesis (see Section 1.6) marked the first theoretical understanding of polymer structure.

In the classical period (Table 14.12) a basic understanding of polymer synthesis and structure was achieved. Concepts like the random coil, folded-chain crystals, and viscoelasticity involve many of the pages of this text. The synthetic advances play equally important roles in modern books on polymer synthesis, as well as general polymer science books. The invention of high-impact plastics (Chapter 13) played a key role in numerous applications of low-priced plastics.

The modern era (Table 14.13) includes the discovery of liquid crystalline polymers; see Chapter 7. The newest major scientific development is polymer interface science. While a number of papers on polymer surfaces and interfaces date back into the classical era, the science itself dates from only 1989, when several key papers were published; see Chapter 12. These described the shape of the polymer chain at interfaces, and their motion and entanglement across the interfaces. Other major advances in modern polymer science and engineering include biopolymers, conducting polymers, and nanotechnology.

Polymer science and engineering continues to develop at a rapid and even accelerating pace. The future of the field, and new directions lie with the imagination of the readers.

REFERENCES

1. J. A. Brydson, *Plastics Materials*, 6th ed., Butterworth-Heinemann, Oxford, England, 1995.
2. G. M. Benedikt and B. L. Goodall, *Metallocene-Catalyzed Polymers: Materials, Properties, Processing & Markets*, Plastics Design Library, Norwich, NY, 1998.
3. A. Turner-Jones and A. J. Cobbold, *J. Polym. Sci.*, **B6**, 539 (1968).
4. E. S. Shamshaum and D. Rouscher, in *Metallocene-Catalyzed Polymers: Materials, Properties, Processing and Markets*, G. M. Denedikt and B. L. Goodall, eds., Plastics Design Library, Norwich, NY, 1998.
5. C. D. Rosa, F. Auriemma, V. Vinti, and M. Galimberti, *Macromolecules*, **31**, 6206 (1998).
6. W. Kaminsky, *Pure Appl. Chem.*, **70**, 1229 (1998).
7. J. W. Verbicky, Jr., *Encyclopedia of Polymer Science and Engineering*, Vol. 12, J. I. Kroschwitz, ed., Wiley, New York, 1988.
8. S. S. Labana, *Encyclopedia of Polymer Science and Engineering*, J. I. Kroschwitz, ed., Wiley, New York, 1986.
9. J. R. Brown and N. A. S. John, *Trends Polym. Sci.* (*TRIP*), **4**, 416 (1996).

9a. B. Ellis, ed., Chemistry and Technology of Epoxy Resins, Blackie Academic and Professional, London, 1993.

10. Y. H. Roos, *Phase Transitions in Foods*, Academic Press, San Diego, 1995.
11. C. M. McDonough, K. Seetharaman, R. D. Waniska, and L. W. Rooney, *J. Food Sci.*, **61**, 995 (1996).
12. S. H. Imam, S. H. Gordon, R. V. Greene, and K. A. Nino, *Polymeric Materials Encyclopedia*, J. C. Salamone, ed., CRC Press, Boca Raton, FL, 1996.
13. V. Tolstoguzov, *Food Colloids*, **11**, 181 (1997).

14. M. LeMeste, V. T. Huang, J. Panama, G. Anderson, and R. Lentz, *Cereal Foods World*, **37**, 264 (1992).
15. L. M. Hallberg and P. Chinachoti, *J. Food Sci.*, **57**, 1201 (1992).
16. J. H. Jagannath, K. S. Jayaraman, and S. S. Arya, *J. Appl. Polym. Sci.*, **71**, 1147 (1999).
17. A. Schiraldi, L. Piazza, O. Brenna, and E. Vittadini, *J. Thermal Analysis*, **47**, 1339 (1996).
18. E. W. Lusas and M. N. Riaz, *J. Nutrition*, **125(3)**, S573 (1995).
19. J. H. Litchfield, *Encyclopedia of Chemical Technology*, J. I. Kroschwitz, ed., Wiley, New York, 1994.
20. J. Magoshi, Y. Magoshi, M. A. Becker, and S. Nakamura, *Polymeric Materials Encyclopedia*, J. C. Salamone, ed., CRC Press, Boca Raton, FL, 1996, p. 669.
21. D. Strydom, T. Haylett, and R. Stead, *Bioch. Biophys. Res. Com.*, **79(3)**, 932 (1977).
22. J. Magoshi, Y. Magoshi, and S. Nakamura, *Repts. Prog. Polym. Phys. Japan*, **23**, 747 (1973).
23. C. Viney, A. E. Huber, D. L. Dunaway, K. Kerkam, and S. T. Case, *Silk Polymers: Materials Science and Biotechnology*, D. Kaplan, W. W. Adams, B. Farmer, and C. Viney, eds., American Chemical Society, Washington, DC, 1994.
24. D. L. Kaplan, C. Mello, S. Fossey, and S. Arcidaicono, *Encyclopedia of Chemical Technology*, 4th ed., J. I. Kroschwitz, ed., Wiley, New York, 1997.
25. M. Tsukada, G. Freddi, and N. Minoura, *J. Appl. Polym. Sci.*, **51**, 823 (1994).
26. M. W. Denny, in *The Mechanical Properties of Biological Materials*, Cambridge University Press, Cambridge, 1980.
27. R. W. Lenz and R. H. Marchessault, *Biomacromolecules*, **6**, 1 (2005).
28. Anonymous, *http://www.metabolix.com* (2005).
29. A. J. Anderson and E. A. Dawes, *Microbiol. Rev.*, **54**, 450 (1990).
30. M. C. T. Fyfe and J. F. Stoddart, *Acc. Chem. Res.*, **30**, 393 (1997).
31. P. E. Mason, W. S. Bryant, and H. W. Gibson, *Macromolecules*, **32**, 1559 (1999).
32. Y. Geerts, D. Muscat, and K. Mullen, *Macromol. Chem. Phys.*, **196**, 3425 (1995).
33. S. Menzer, A. J. P. White, D. J. Williams, M. Belohradsky, C. Hamers, F. M. Raymo, A. N. Shipway, and J. F. Stoddart, *Macromolecules*, **31**, 295 (1998).
34. S. I. Stupp, V. LeBonheur, K. Walker, L. S. Li, K. E. Huggins, M. Keser, and A. Amstutz, *Science*, **276**, 384 (1997).
35. T. H. Mourey, S. R. Turner, M. Rubinstein, J. M. J. Frechet, C. J. Hawker, and K. L. Wooley, *Macromolecules*, **25**, 2401 (1992).
36. M. Freemantle, *C&E News*, **77**, 27 (1999).
37. C. L. Jackson, H. D. Chanzy, F. P. Booy, B. J. Drake, D. A. Tomalia, B. J. Bauer, and E. J. Amis, *Macromolecules*, **31**, 6259 (1998).
38. P. G. de Gennes and H. Hervet, *J. Phys., Lett.*, **44**, L-351 (1983).
39. V. Percec, C.-H. Ahn, G. Ungar, D. J. P. Yeardley, M. Moller, and S. S. Sheiko, *Nature*, **391**, 161 (1998).
40. J. M. J. Frechet, *Science*, **263**, 1710 (1994).
41. L. J. Hobson and W. J. Feast, *J. Chem. Soc. Chem. Commun.*, 2067 (1997).

42. S. M. Aharoni, C. R. Crosby III, and E. K. Walsh, *Macromolecules*, **15**, 1093 (1982).

43. Anonymous, in *Workshop on Properties and Applications of Dendritic Polymers, NIST, Gaithersburg, MD* (July 9–10, 1998).

44. R. Estfand, D. A. Tomalia, E. A. Beezer, J. C. Mitchell, M. Hardy, and C. Orford, *Polym. Prepr.*, **41(2)**, 1324 (2000).

45. E. Žagar and M. Žigon, *Macromolecules*, **35**, 9913 (2002).

46. S. J. Han, *Ph.D. Dissertation, Processing of Polyolefin Blends in Supercritical Propane Solution*, Lehigh University, 1998; S. J. Han, D. J. Lohse, M. Radosz, and L. H. Sperling, *Macromolecules*, **31**, 5407 (1998).

47. H. Graham, in *Encyclopedia of Food Science and Technology*, Y. H. Hui, ed., Wiley, New York, 1992.

48. S. N. Katz, U.S. 4,820,356, 1989.

49. J. M. DeSimone, Z. Guan, and C. S. Elsbernd, *Science*, **257**, 945 (1992).

50. J. M. DeSimone, E. E. Maury, Y. Z. Menceloglu, J. B. McClean, T. J. Romack, and J. R. Combes, *Science*, **265**, 356 (1994).

51. D. A. Canelas, D. E. Betts, and J. M. DeSimone, *Macromolecules*, **29**, 2818 (1996).

52. M. D. Elkovitch, *CAPCE Newsletter (Ohio State University)*, **2**, 2 (1999).

53. A. I. Cooper, J. D. Londono, G. Wignall, J. B. McClain, E. T. Samulski, J. S. Lin, A. Dobrynin, M. Rubinstein, A. L. C. Burke, J. M. J. Frechet, and J. M. DeSimone, *Nature*, **389**, 368 (1997).

54. K. N. Mathes, in *Encyclopedia of Polymer Science and Engineering*, Vol. 5, J. I. Kroschwitz, ed., Wiley, New York, 1986.

55. J. E. Frommer and R. R. Chance, in *Encyclopedia of Polymer Science and Engineering*, Vol. 5, J. I. Kroschwitz, ed., Wiley, New York, 1986.

56. H. Shirakawa, E. J. Louis, A. G. MacDiarmid, C. K. Chiang, and A. J. Heeger, *J. Chem. Soc. Chem. Comm.*, 579 (1977).

57. C. K. Chiang, M. A. Druy, S. C. Gau, A. J. Heeger, E. J. Louis, A. G. MacDiarmid, Y. W. Park, and H. Shirakawa, *J. Am. Chem. Soc.*, **100**, 1013 (1978).

58. R. B. Kaner and A. C. MacDiarmid, *Sci. Am.*, p. 106, Feb., 1988.

59. M. J. Bowden, in *Electronic and Photonic Applications of Polymers*, M. J. Bowden and S. R. Turner, eds., Advances in Chemistry Series No. 218, American Chemical Society, Washington, DC, 1988.

60. D. Williams, in *Electronic and Photonic Applications of Polymers*, M. J. Bowden and S. R. Turner, eds., Advances in Chemistry Series No. 218, American Chemical Society, Washington, DC, 1988.

61. D. J. Williams, ed., *Nonlinear Optical Properties of Organic and Polymeric Materials*, ACS Symposium Series No. 283, American Chemical Society, Washington, DC, 1983.

62. C. Lee, D. Haas, H. T. Man, and V. Mechensky, *Photonics Spectra*, p. 171, April 1979.

63. B. F. Levine, C. G. Bethea, C. D. Thurmond, R. T. Lynch, and J. L. Bernstein, *J. Appl. Phys.*, **50**, 2523 (1979).

64. B. I. Breene, J. Orenstein, and S. Schmitt-Rink, *Science*, **247**, 679 (1990).
65. C. R. Meredith, J. G. VanDusen, and D. J. Williams, *Macromolecules*, **15**, 1385 (1982).
66. R. N. DeMartino, H. N. Yoon, J. R. Stamatoff, and A. Buckley, Eur. Pat. Applic. 0231770 (1987), to Celanese Corporation.
67. D. E. Stuetz, Eur. Pat. Applic. 017212 (1986).
68. N. A. Plate, R. V. Talroze, and V. P. Shibaev, in *Polymer Yearbook 3*, R. A. Pethrick and G. E. Zaikov, eds., Harwood, London, 1986.
69. B. F. Levine, C. G. Bethea, C. D. Thurmond, R. T. Lynch, and J. L. Bernstein, *J. Appl. Phys.*, **50**, 2523 (1979).
70. P. Pantelis and G. J. Davies, U.S. Pat. 4,748,074 (1988), to British Telecommunications.
71. B. I. Breene, J. Orenstein, and S. Schmitt-Rink, *Science*, **247**, 679 (1990).
72. M. Eich, B. Reck, D. Y. Yoon, C. G. Wilson, and G. C. Bjorklund, *J. Appl. Phys.*, **66**, 3241 (1989).
73. D. N. Rao, R. Burzynski, X. Mi, and P. N. Prasad, *Appl. Phys. Lett.*, **48**, 387 (1986).
74. D. N. Rao, J. Swiatkiewicz, P. Chopra, S. K. Chosal, and P. N. Prasad, *Appl. Phys. Lett.*, **48**, 1187 (1986).
75. A. F. Garito, presented at Pacifichem '89, Honolulu, Hawaii, December 1989.
76. R. H. Friend, R. W. Gymer, A. B. Holmes, J. H. Burroughes, R. N. Marks, C. Taliani, D. D. C. Gradley, D. A. D. Santos, J. L. Brédas, M. Lögdlund, and W. R. Salaneck, *Nature*, **397**, 121 (1999).
77. J. H. Burroughes, D. D. C. Bradley, A. R. Brown, R. N. Marks, K. Mackay, R. H. Friend, P. L. Burns, and A. B. Holmes, *Nature*, **347**, 539 (1990).
78. J. Jacob, J. Zhang, A. C. Grimsdale, K. Müllen, M. Gaal, and E. J. W. List, *Macromolecules*, **36**, 8240 (2003).
79. L. Liao, Y. Pang, L. Ding, and F. E. Karasz, *Macromolecules*, **37**, 3970 (2004).
80. Z. M. Zhang, V. Bharti, and X. Zhao, *Science*, **280**, 2101 (1998).
81. A. M. Vinogradov and S. C. Schumacher, *J. Spacecraft and Rockets*, **39**, 839 (2002).
82. R. J. Klein, J. Runt, and Q. M. Zhang, *Macromolecules*, **36**, 7220 (2003).
83. Y. Bar-Cohen, T. Xue, B. Joffe, S.-S. Lih, M. Shahinpoor, J. Simpson, J. Smith, and P. Willis, *SPIE International Conference*, San Diego, 1997.
84. Y. Bar-Cohen, *J. Spacecraft and Rockets*, **39**, 822 (2002).
85. A. Askin, *Phys. Rev. Lett.*, **40**, 729 (1978).
86. A. Askin, J. M. Dziedzic, J. E. Bjorkholm, and S. Chu, *Optics Lett*, **11**, 288 (1986).
87. L. A. Hough and H. D. Ou-Yang, *Phys. Rev. E*, **6502**, 1906 (2002).
88. C. Bustamante, Z. Bryant, and S. B. Smith, *Nature*, **421**, 423 (2003).
89. T. G. D'Onofrio, A. Hatzor, A. E. Counterman, J. H. Heetderks, M. J. Sandel, and P. S. Weiss, *Langmuir*, **19(5)**, 1618 (2003).
90. R. J. Owen, J. C. Crocker, R. Verma, and A. G. Yodh, *Phys. Rev. E*, **64(1)**, 011401 (2001).
91. M. W. Nirenberg, *Sci. Am.*, **208(3)**, 80 (1963).
92. N. A. Seeman, *Sci. Am.*, **290(6)**, 64 (2004).

93. A. Piper, *Trends Biochem. Sci.*, **23**, 151 (1998).
94. J. D. Watson and F. H. Crick, *Nature*, **171**, 737 (1953).
95. T. E. Creighton, *Proteins: Structures and Molecular Properties*, 2nd ed., W. H. Freeman & Co., New York (1993).
96. Anonymous, *http://history.nih.gov/exhibits/nirenberg/HS4_polyU.htm* (2004).
97. I. Kawaguchi and S. Ishiwata, *Science*, **291**, 667 (2001).
98. D. E. Metzler, *Biochemistry: The Chemical Reactions of Living Cells*, 2nd ed., Academic Press, San Diego (2001).
99. H. L. Sweeney, *Sci. Am.*, **291(1)**, 63 (2004).
100. J. R. Sellers, *Myosins*, 2nd ed., Oxford University Press, Oxford, England (1999).
101. T. Nishizaka, H. Miyata, H. Yoshikoawa, and S. Ishiwata, *Nature*, **377**, 251 (1995).
102. A. F. Huxley and R. M. Simmons, *Nature*, **233**, 533 (1971).
103. K. Kitamura, M. Tokunaga, A. H. Iwane, and T. Yanagida, *Nature*, **397**, 129 (1999).
104. D. Altman, H. L. Sweeney, and J. A. Spudich, *Cell*, **116**, 737 (2004).
105. J. J. K. Kinosita, in *Mechanisms of Work Production and Work Absorption in Muscle*, H. Sugi and G. H. Pollack, eds., Plenum Press, New York, 1998.
106. B. Merrifield, *Science*, **232**, 341 (1986).
107. R. Tadmor, R. L. Khalfin, and Y. Cohen, *Langmuir*, **18**, 7146 (2002).
108. J. Tian, L. S. Coghill, S. H.-F. MacDonald, D. L. Armstrong, and M. J. Shipston, *J. Biol. Chem.*, **278**, 8669 (2002).
109. B. N. Jang and C. A. Wilkie, *PMSE Prepr.*, **91**, 88 (2004).
110. Y. P. Khanna and E. M. Pearce, *Applied Polymer Science*, 2nd ed., R. W. Tess and G. W. Poehlein, eds., American Chemical Society, Washington, D.C., 1985.
111. P. Georlette, J. Simons, and L. Costa, *Fire Retardancy of Polymeric Materials*, A. F. Grand and C. A. Wilkie, eds., Marcel Dekker, New York, 2000.
112. J. Green, *Fire Retardancy of Polymeric Materials*, A. F. Grand and C. A. Wilkie, eds., Marcel Dekker, New York, 2000.
113. G. Chigwada and C. A. Wilkie, *Polym. Degrad. Stabil.*, **81**, 551 (2003).
114. B. Schartel, U. Braun, U. Schwarz, and S. Reinemann, *Polymer*, **44**, 6241 (2003).
115. U. Braun and B. Schartel, *J. Fire Sci.*, **23**, 5 (2005).
116. G. Camino and R. Delobel, *Fire Retardancy of Polymeric Materials*, A. F. Grand and C. A. Wilkie, eds., Marcel Dekker, New York, 2000.
117. D. W. van Krevelen, *Polymer*, **16**, 615 (1975).
118. T. Kashiwagi, E. Grulke, J. Hilding, K. Groth, R. Harris, K. Butler, J. Shields, S. Kharchenko, and J. Douglas, *Polymer*, **45**, 4227 (2004).
119. T. Kashiwagi, F. Du, K. I. Winey, K. M. Groth, J. R. Shields, J. R. H. Harris, and J. F. Douglas, *PMSE Prepr.*, **91**, 90 (2004).
120. M. Bartholmai and B. Schartel, *Polym. Adv. Technol.*, **15**, 354 (2004).
121. L. B. Ingram, *ANTEC, Chicago*, Paper 402 (May, 2004).
122. J. W. Hoyt, *Encyclopedia of Polymer Science and Engineering*, 2nd ed., J. I. Kroschwitz, ed., Wiley, New York, 1986.
123. P. S. Virk, *AIChE J*, **21**, 625 (1975).
124. K. D. Housiadas and A. N. Beris, *Phys. Fluids*, **15**, 2369 (2003).

GENERAL READING

R. A. Archer, *Inorganic and Organometallic Polymers*, Wiley, New York, 2001.

D. Avnir, *The Fractal Approach to Heterogeneous Chemistry: Surface, Colloids, Polymers*, Wiley, Chichester, England, 1989.

G. M. Benedickt and B. L. Goodall, eds., *Metallocene-Catalyzed Polymers: Materials, Processing and Markets*, Plastics Design Library, Norwich, NY, 1998.

M. I. Bessonov and V. A. Zubkov, eds., *Polyamic Acids and Polyimides: Synthesis, Transformations, and Structure*, CRC Press, Boca Raton, FL, 1993.

H. B. Bohidar, P. Dubin, and Y. Osada, eds., *Polymer Gels: Fundamentals and Applications*, ACS Symp. Ser. No. 833, American Chemical Society, Washington, D.C., 2003.

M. J. Bowden and S. R. Turner, eds., *Electronic and Photonic Applications of Polymers*, Adv. Chem. Ser. No. 218, American Chemical Society, Washington, DC, 1988.

C. Branden and J. Tooze, *Introduction to Protein Structure*, 2nd ed., Garland Pub., New York, 1999.

J. A. Brydson, *Plastic Matericals*, Butterworths, London, 6th ed., 1995.

E. Chiellini, H. Gil, G. Braunegg, J. Buchert, P. Gatenholm, and M. van der Zee, eds., *Biorelated Polymers*, Kluwer, New York, 2001.

P. Chinachoti and Y. Vodovitz, eds., *Bread Staling*, CRC Press, Boca Raton, 2001.

A. Ciferri, ed., *Supramolecular Polymers*, Dekker, New York, 2000.

T. E. Creighton, *Proteins: Structures and Molecular Properties*, 2nd ed., W. H. Freeman and Co., New York, 1993.

H. J. R. Dutton, *Understanding Optical Communications*, Prentice Hall, Englewood Cliffs, 2000.

J. M. J. Fréchet and D. A. Tomalia, eds., *Dendrimers and Other Dentritic Polymers*, Wiley, Chichester, England, 2001.

A. F. Grand and C. A. Wilkie, *Fire Retardancy of Polymeric Materials*, Marcel Dekker, Inc., New York, 2000.

R. A. Gross and C. Scholz, eds., *Biopolymers from Polysaccharides and Agroproteins*, ACS Symp. Ser. No. 786, American Chemical Society, Washington, DC, 2001.

R. A. Hann and D. Bloor, eds., *Organic Materials for Non-Linear Optics*, Royal Society of Chemistry, London, 1989.

A. J. Heeger, J. Orenstein, and D. R. Ulrich, eds., *Nonlinear Optical Properties of Polymers*, Materials Research Society, Pittsburgh, 1988.

D. Kaplan, W. W. Adams, B. Farmer, and C. Viney, eds., *Silk Polymers: Materials Science and Biotechnology*, American Chemical Society, Washington, DC, 1994.

C. C. Ku and R. Liepins, *Electrical Properties of Polymers*, Hanser, Munich, 1987.

J. H. Lai, ed., *Polymers for Electronic Applications*, CRC Press, Boca Raton, FL, 1989.

R. W. Lenz and R. H. Marchessault, *Biomacromolecules*, **6**, 1 (2005), reviews bacterial polyesters.

A. G. MacDiarmid, *Rev. Mod. Phys.*, **73**, 701 (2001). (Nobel lecture).

C. W. Macosko, *RIM Fundamentals of Reaction Engineering*, Hanser, New York, 1989.

M. K. Mishra and S. Kobayashi, eds., *Star and Hyperbrancehed Polymers*, Dekker, New York, 1999.

H. S. Nalwa and S. Miyata, eds., *Nonlinear Optics of Organic Molecules and Polymers*, CRC Press, Boca Ratton, FL, 1997.

G. R. Newkome, C. N. Moorefield, and F. Vögtle, *Dendrimers and Dendrons*, Wiley-VCH, Weinheim, Germany, 2001.

P. N. Prasad and D. J. Williams, *Nonlinear Optical Effects in Molecules and Polymers*, Wiley, New York, 1991.

T. Radeva, ed., *Physical Chemistry of Polyelectrolytes*, Marcel Dekker, New York, 2001.

S. Roth, *One-Dimensional Metals*, VCH, Weinheim, 1995.

J. Scheirs, *Modern Polyesters*, Wiley, Hoboken, NJ, 2003.

J. Scheirs, *Polymer Recycling*, Wiley, Chichester, England, 1998.

J. Scheirs and W. Kaminsky, *Metallocene-Based Polyolefins: Preparation, Properties, and Technology*, Wiley, New York, 2000.

J. Scheirs and W. Kaminsky, *Metallocene-Based Polymers*, Vols. 1 and 2, Wiley, Chichester, England, 2000.

R. B. Seymour and T. Cheng, eds., *History of Polyolefins*, Reidel, Dordrecht, Holland, 1986.

S. W. Shalaby and K. J. L. Burg, eds., *Absorbable and Biodegradable Polymers*, CRC Press, Boca Raton, FL, 2003.

G. O. Shonaike and S. G. Advani, eds., *Advanced Polymeric Materials: Structure Property Relationships*, CRC Press, Boca Raton, 2003.

R. W. Siegel, E. Ha, and M. C. Roco, eds., *Nanostructure Science and Technology*, Kluwer Academic, Dordrecht, 1999.

G. G. Wallace, G. M. Spinks, L. A. P. Kane-Maguire, and P. R. Teasdale, *Conductive Electroactive Polymers: Intelligent Materials Systems*, 2nd ed., CRC Press, Boca Raton, 2003.

E. S. Wilks, *Polym. Prepr.*, **40**(2), 6 (1999). (References for catenanes, rotaxanes, and dendritic polymers.)

N. Yui, R. J. Mrsny, and K. Park, *Reflexive Polymers and Hydrogels: Understanding and Designing Fast Responsive Polymeric Systems*, CRC Press, Boca Raton, 2004.

STUDY PROBLEMS

1. Read any scientific or engineering paper concerned with the topics of Chapter 14 published since 2004. What is the import of the paper? In what ways does it update the text? Please provide the full reference.

2. Assuming a spherical model, what is the theoretical density of the 4th, 6th, 8th, 10th and 12th generation dendrimers having the structure illustrated in Figure 14.10? Hint: Assume the bulk material has unit density. Why can't one make a 12th generation dendrimer?

3. In order to improve nutrition, you want to add gelatin proteins to bread doughs. How will you do this? What are the morphological and texture consequences of your actions?

4. Why have polyethylene and polypropylene become the largest tonnage synthetic polymers? (Check the references and/or speculate.)

5. What are the advantages and disadvantages of using increased tonnages of cellulose, rather than polyethylene, polypropylene, polystyrene, and poly(vinyl chloride) for modern plastics? (Check recent literature, and speculate.)

6. Could a good thermoset polymer be made out of copolymers of divinyl benzene and styrene? What would be its modulus, glass transition temperature, and impact strength as a function of copolymer composition? (Draw figures illustrating probable behavior.)

7. You have just joined a silk research institute. Your supervisor asks for your research ideas. Write a one-page double spaced summary of a research proposal dealing with any aspect of silk.

8. For recycling purposes, used consumer plastics need to be separated. Based on superfluid technology, how would you separate polystyrene from polypropylene?

9. Why is polyacetylene such an interesting semiconductor? How does it work?

10. A certain application requires blue light with a wavelength of 400 nm. The only monochromatic light source available is an infrared laser with a wavelength of 800 nm. How will you accomplish this task?

11. If you had listened to Staudinger expound on the Macromolecular Hypothesis in the year 1920, what would you have suggested for a research project to prove or disprove his ideas? (Assume you have all the modern instruments available at your disposal, however.)

12. By surprise almost, you discover you are a consultant to the chicken packing industry. Their problem, they tell you, is that they are generating over two billion pounds of feathers per year. They don't know what to do with the stuff. All they could tell you is that chicken feathers are mostly made of keratin, a protein. "Proteins are polymers, aren't they?" they ask. They wanted your services because you have a reputation of thinking *green*, and an expert in polymers.

    How do you propose to process this new material? What do you propose to make out of it?

13. Muscles do work in lifting weights.

    (a) How many moles of ATP must be expended to lift 10 kg to 15 cm height?

(b) In part (a), assume 1 kg of muscle containing 5% myosin of 223,000 g/mol, 5% other proteins, and 90% water and other non-proteinaceous materials do the lifting. What is the efficiency of the muscle? (For simplicity, assume the muscle itself has minimal extensive motions.)